Biotechnology for Fuels and Chemicals
The Twenty-Seventh Symposium

Presented as Volumes 129–132
of *Applied Biochemistry and Biotechnology*

Proceedings of the Twenty-Seventh Symposium
on Biotechnology for Fuels and Chemicals
Held May 1–May 4, 2005, in Denver, Colorado

Sponsored by

US Department of Energy's Office of the Biomass Program
US Department of Agriculture, Agricultural Research Service
National Renewable Energy Laboratory
Oak Ridge National Laboratory
Idaho National Laboratory
Abengoa Bioenergy R&D, Inc.
Archer Daniels Midland Company
Battelle Nanotechnology Innovation Network
BBI International
Cargill
E. I. DuPont de Nemours and Company, Inc.
Genencor International, Inc.
Iogen Corporation
Katzen International, Inc.
Luca Technologies
Natural Resources Canada
Novozymes
Purevision Technology, Inc.
Tate and Lyle PLC
Wynkoop Brewing Company

Editors
James D. McMillan and William S. Adney
National Renewable Energy Laboratory

Jonathan R. Mielenz
Oak Ridge National Laboratory

K. Thomas Klasson
Southern Regional Research Center, USDA–ARS

Applied Biochemistry and Biotechnology
Volumes 129–132, Complete, Spring 2006
Copyright © 2006 Humana Press Inc.
All Rights Reserved.

No part of this publication may be reproduced or transmitted in any form or by any means, electronic or mechanical, including photocopy, recording, or any information storage and retrieval system, without permission in writing from the copyright owner.

Applied Biochemistry and Biotechnology is abstracted or indexed regularly in *Chemical Abstracts, Biological Abstracts, Current Contents, Science Citation Index, Excerpta Medica, Index Medicus,* and appropriate related compendia.

Introduction to the Proceedings of the Twenty-Seventh Symposium on Biotechnology for Fuels and Chemicals

JAMES D. MCMILLAN

National Renewable Energy Laboratory

Golden, CO 80401-3393

The Twenty-Seventh Symposium on Biotechnology for Fuels and Chemicals was held May 1–4, 2005 in Denver, Colorado. Continuing to foster a highly interdisciplinary focus on bioprocessing, this symposium remains the preeminent forum for bringing together active participants and organizations to exchange technical information and update current trends in the development and application of biotechnology for sustainable production of fuels and chemicals. This annual symposium emphasizes advances in biotechnology to produce high-volume, low-price products from renewable resources, as well as to improve the environment. Topical foci include advanced feedstock production and processing, enzymatic and microbial biocatalysis, bioprocess research and development, opportunities in biorefineries, commercialization of biobased products, as well as other special topics.

Advances in commercialization of bioproducts continued apace this year, and the level of interest and excitement in expanding the use of renewable feedstocks continued to grow. Nonetheless, significant techno-economic challenges must be overcome to achieve widespread commercialization of biotechnological fuels and chemicals production, particularly to move the feedstock base beyond primarily sugar crops and cereal grains (starch) to include holocellulose (cellulose and hemicellulose) from fibrous lignocellulosic plant materials.

Participants from academic, industrial, and government venues gathered to discuss the latest research breakthroughs and results in biotechnology to improve the economics of producing fuels and chemicals. The total of 459 attendees represented an all-time conference high; this is almost a 33% increase over the 2004 conference attendance in Chattanooga. Of this total, approximately 45% of attendees were from academia (about half of this, more than 21% of the total attendees, were students), 31% were from industry, and 22% were from government. A total of 71 oral presentations (including Special Topic presentations) and 329 poster presentations were delivered. The high number of poster submissions required splitting the poster session into two evening sessions. (Conference details are posted at http://www.eere.energy.gov/biomass/biotech_symposium/.)

Almost 35% of the attendees were international, showing the strong and building worldwide interest in this area. Nations represented included Australia, Austria, Belgium, Brazil, Canada, Central African Republic, China, Denmark, Finland, France, Gambia, Germany, Hungary, India, Indonesia, Italy, Japan, Mexico, The Netherlands, New Zealand, Portugal, South Africa, South Korea, Spain, Sweden, Thailand, Turkey, United Kingdom, and Venezuela, as well as the United States.

One of the focus areas for bioconversion of renewable resources into fuels is conversion of lignocellulose into sugars and the conversion of sugars into fuels and other products. This focus is continuing to expand toward the more encompassing concept of the integrated multiproduct biorefinery—where the production of multiple fuel, chemical, and energy products occurs at one site using a combination of biochemical and thermochemical conversion technologies. The biorefinery concept continues to grow as a unifying framework and vision, and the biorefinery theme featured prominently in many talks and presentations. However, another emerging theme was the importance of examining and optimizing the entire biorefining process rather than just its bioconversion-related elements.

The conference continued to include two Special Topics sessions devoted to discussing areas of particular interest. This year the two topics were international biofuels developments and the evolving attitudes about biomass as a sustainable feedstock for fuels, chemicals and energy production. The first Special Topic session was entitled "International Energy Agency (IEA) Task #39–Liquid Biofuels." This session focused on recent international progress on production of liquid biofuels and was chaired by Jack Saddler of the University of British Columbia. The second Special Topic session was entitled, "'Outside of a Small Circle of Friends': Changing Attitudes about Biomass as a Sustainable Energy Supply," and was chaired by John Sheehan of NREL. This session focused on the evolving perceptions within the agricultural producer and environmental and energy efficiency advocacy communities that biomass has the potential to be a large volume renewable resource for sustainable production of a variety of fuel, chemical, and energy products.

The Charles D. Scott award for Distinguished Contributions in the field of Biotechnology for Fuels and Chemicals was created to honor Symposium founder Dr. Charles D. Scott who chaired this Symposium for its first ten years. This year, the Charles D. Scott award was presented to Lee R. Lynd. Dr. Lynd is a Professor of Engineering and an Adjunct Professor of Biology at Dartmouth College, as well as a Professor Extraordinary of Microbiology at the University of Stellenbosch, South Africa. He has made many pioneering contributions to bioenergy and biomass conversion. Most impressively, his activities and accomplishments span the science, technol-

Introduction v

ogy and policy domains. Highlights include improving our fundamental understanding of microbial cellulose utilization, advancing the design and evaluation of biomass conversion processes and providing a variety of critical analyses and inputs to policy makers in support of bioenergy. An active consultant and frequently invited presenter on technical and strategic aspects of biomass energy, Dr. Lynd has twice testified before the US Senate. He recently co-led a large multi-institutional project entitled *The Role of Biomass in America's Energy Future*. The author of more than sixty peer-reviewed papers and several comprehensive reviews, and the holder of five patents, the field of biotechnology for fuels and chemicals would not be the same were it not for Dr. Lynd's tireless and inspired efforts.

Session Chairpersons

Session 1A: Feedstock Supply and Logisitics
Chairs: Peter C. Flynn, *University of Alberta, Edmonton, Alberta, Canada*
Foster Agblevor, *Virginia Polytechnic Institute and State University, Blacksburg, VA*

Session 1B: Enzyme Catalysis and Engineering
Chairs: Joel R. Cherry, *Novozymes, Inc., Davis, CA*
Kevin Gray, *Diversa Corporation, San Diego, CA*

Session 2: Today's Biorefineries
Chairs: Robert Benson, *Tembec Chemical Products Group, North Bay, Ontario, Canada*
Paris Tsobanakis, *Cargill, Inc., Wayzata, MN*

Session 3A: Plant Biotechnology and Feedstock Genomics
Chairs: Wilfred Vermerris, *Purdue University, West Lafayette, IN*
Sean Simpson, *LanzaTech, Auckland, New Zealand*

Session 3B: Biomass Pretreatment and Hydrolysis
Chairs: Richard T. Elander, *National Renewable Energy Laboratory, Golden, CO*
Mohammed Moniruzzaman, *Genencor International, Beloit, WI*

Session 4: Industrial Biobased Products
Chairs: Ray Miller, *E. I. DuPont de Nemours and Co., Inc., Wilmington, DE*
Matt Tobin, *Codexis, Redwood City, CA*

Session 5: Microbial Catalysis and Metabolic Engineering
Chairs: Lisbeth Olsson, *BioCentrum–DTU, Technical University of Denmark, Lyngby, Denmark*
Aristos Aristidou, *Natureworks LLC, Minnetonka, MN*

Session 6: Bioprocess Research and Development
Chairs: Michael R. Ladisch, *Purdue University, West Lafayette, IN*
Peter Yu, *Hong Kong Polytechnic University, Hong Kong, P. R. China*

Organizing Committee

Jim McMillan, Conference Chair, *National Renewable Energy Laboratory, Golden, CO*
Brian Davison, Conference Co-Chair, *Oak Ridge National Laboratory, Oak Ridge, TN*
Mark Finkelstein, Conference Co-Chair, *Luca Technologies, Denver, CO*
Doug Cameron, *Cargill, Minneapolis, MN*
Jim Duffield, *National Renewable Energy Laboratory, Golden, CO*
Don Erbach, *USDA Agricultural Research Service, Beltsville, MD*
Tom Jeffries, *USDA Forest Service, Madison, WI*
K. Thomas Klasson, *USDA Agricultural Research Service, Southern Regional Research Laboratory, New Orleans, LA*
Ann Luffman, *Oak Ridge National Laboratory, Oak Ridge, TN*
Lee Lynd, *Dartmouth College, Hanover, NH*
Amy Miranda, *USDOE Office of the Biomass Program, Washington, DC*
Dale Monceaux, *Katzen International, Inc., Cincinnati, OH*
Jack Saddler, *University of British Columbia, Vancouver, British Columbia, Canada*
Sharon Shoemaker, *University of California, Davis, CA*
David Short, *E.I. DuPont de Nemours & Co., Newark, DE*
Dave Thompson, *Idaho National Laboratory, Idaho Falls, ID*
Jeff Tolan, *Iogen Corporation, Ontario, Canada*
Charles Wyman, *Dartmouth College, Hanover, NH*
Guido Zacchi, *Lund University, Lund, Sweden*
Gisella Zanin, *State University of Maringa, Maringa, PR, Brazil*

Acknowledgments

The continued success of the Symposium is due to the many participants, organizers, and sponsors, but is also the result of significant contributions by numerous diligent, creative and talented staff. In particular, Jim Duffield of NREL, conference secretary, provided timely advice and heroic persistence while maintaining an unfailingly upbeat attitude. Connie Neuber, Kay Vernon, and Howard Brown of NREL contributed greatly to the preparation of the program and abstracts book, website design and implementation, and call for papers brochure, respectively. David Glickson and Kim Hutto coordinated and tracked logistical and financial conference aspects, respectively. Other NREL staff assisting with conference set up and execution included Nancy Farmer, Lyn Lumberg, Nick Nagle, Dan Schell, Ivilina Thornton, Bob Wallace, Wendy Zamudio, and Millie Zuccarello.

The National Renewable Energy Laboratory is operated for the US Department of Energy by Midwest Research Institute and Battelle under contract DE-AC36-99GO10337.

Introduction

Oak Ridge National Laboratory is operated for the US Department of Energy by UT-Battelle, LLC under contract DE-AC05-00OR22725.

The submitted Proceedings have been authored by a contractor of the US Government under contract DE-AC36-99GO10337. Accordingly, the US Government retains a nonexclusive, royalty-free license to publish or reproduce the published form of this contribution, or allow others to do so, for US Government purposes.

Other Proceedings in this Series

1. "Proceedings of the First Symposium on Biotechnology in Energy Production and Conservation" (1978), *Biotechnol. Bioeng. Symp.* **8.**
2. "Proceedings of the Second Symposium on Biotechnology in Energy Production and Conservation" (1980), *Biotechnol. Bioeng. Symp.* **10.**
3. "Proceedings of the Third Symposium on Biotechnology in Energy Production and Conservation" (1981), *Biotechnol. Bioeng. Symp.* **11.**
4. "Proceedings of the Fourth Symposium on Biotechnology in Energy Production and Conservation" (1982), *Biotechnol. Bioeng. Symp.* **12.**
5. "Proceedings of the Fifth Symposium on Biotechnology for Fuels and Chemicals" (1983), *Biotechnol. Bioeng. Symp.* **13.**
6. "Proceedings of the Sixth Symposium on Biotechnology for Fuels and Chemicals" (1984), *Biotechnol. Bioeng. Symp.* **14.**
7. "Proceedings of the Seventh Symposium on Biotechnology for Fuels and Chemicals" (1985), *Biotechnol. Bioeng. Symp.* **15.**
8. "Proceedings of the Eigth Symposium on Biotechnology for Fuels and Chemicals" (1986, *Biotechnol. Bioeng. Symp.* **17.**
9. "Proceedings of the Ninth Symposium on Biotechnology for Fuels and Chemicals" (1988), *Appl. Biochem. Biotechnol.* **17,18.**
10. "Proceedings of the Tenth Symposium on Biotechnology for Fuels and Chemicals" (1989), *Appl. Biochem. Biotechnol.* **20,21.**
11. "Proceedings of the Eleventh Symposium on Biotechnology for Fuels and Chemicals" (1990), *Appl. Biochem. Biotechnol.* **24,25.**
12. "Proceedings of the Twelfth Symposium on Biotechnology for Fuels and Chemicals" (1991), *Appl. Biochem. Biotechnol.* **28,29.**
13. "Proceedings of the Thirteenth Symposium on Biotechnology for Fuels and Chemicals" (1992), *Appl. Biochem. Biotechnol.* **34,35.**
14. "Proceedings of the Fourteenth Symposium on Biotechnology for Fuels and Chemicals" (1993), *Appl. Biochem. Biotechnol.* **39,40.**
15. "Proceedings of the Fifteenth Symposium on Biotechnology for Fuels and Chemicals" (1994), *Appl. Biochem. Biotechnol.* **45,46.**
16. "Proceedings of the Sixteenth Symposium on Biotechnology for Fuels and Chemicals" (1995), *Appl. Biochem. Biotechnol.* **51,52.**
17. "Proceedings of the Seventeenth Symposium on Biotechnology for Fuels and Chemicals" (1996), *Appl. Biochem. Biotechnol.* **57,58.**
18. "Proceedings of the Eighteenth Symposium on Biotechnology for Fuels and Chemicals" (1997), *Appl. Biochem. Biotechnol.* **63–65.**
19. "Proceedings of the Nineteenth Symposium on Biotechnology for Fuels and Chemicals" (1998), *Appl. Biochem. Biotechnol.* **70–72.**

20. "Proceedings of the Twentieth Symposium on Biotechnology for Fuels and Chemicals" (1999), *Appl. Biochem. Biotechnol.* **77–79.**
21. "Proceedings of the Twenty-First Symposium on Biotechnology for Fuels and Chemicals" (2000), *Appl. Biochem. Biotechnol.* **84–86.**
22. "Proceedings of the Twenty-Second Symposium on Biotechnology for Fuels and Chemicals" (2001), *Appl. Biochem. Biotechnol.* **91–93.**
23. "Proceedings of the Twenty-Third Symposium on Biotechnology for Fuels and Chemicals" (2002), *Appl. Biochem. Biotechnol.* **98–100.**
24. "Proceedings of the Twenty-Fourth Symposium on Biotechnology for Fuels and Chemicals" (2003), *Appl. Biochem. Biotechnol.* **105–108.**
25. "Proceedings of the Twenty-Fifth Symposium on Biotechnology for Fuels and Chemicals" (2004), *Appl. Biochem. Biotechnol.* **113–116.**
26. "Proceedings of the Twenty-Sixth Symposium on Biotechnology for Fuels and Chemicals" (2005), *Appl. Biochem. Biotechnol.* **121–124.**

This symposium has been held annually since 1978. We are pleased to have the proceedings of the Twenty-Seventh Symposium currently published in this special issue to continue the tradition of providing a record of the contributions made.

The Twenty-Eighth Symposium will be April 30–May 3, 2006 in Nashville, Tennessee. More information on the 27th and 28th Symposia is available at the following websites: [http://www.eere.energy.gov/biomass/biotech_symposium/] and [http://www.simhq.org/html/meetings/]. We encourage comments or discussions relevant to the format or content of the meeting.

Contents

Introduction
James D. McMillan iii

Session 1A: Feedstock Supply and Logistics

Introduction to Session 1A
Peter C. Flynn 1

Agricultural Residue Availability in the United States
Zia Haq and James L. Easterly 3

Canadian Biomass Reserves for Biorefining
Warren E. Mabee,* Evan D. G. Fraser, Paul N. McFarlane,
and John N. Saddler 22

Availability of Crop Residues as Sustainable Feedstock
for Bioethanol Production in North Carolina
Abolghasem Shahbazi and Yebo Li 41

Updates on Softwood-to-Ethanol Process Development
Warren E. Mabee, David J. Gregg, Claudio Arato, Alex Berlin,
Renata Bura, Neil Gilkes, Olga Mirochnik, Xuejun Pan,
E. Kendall Pye, and John N. Saddler 55

Development of a Multicriteria Assessment Model for Ranking
Biomass Feedstock Collection and Transportation Systems
Amit Kumar, Shahab Sokhansanj, and Peter C. Flynn 71

Rail vs Truck Transport of Biomass
Hamed Mahmudi and Peter C. Flynn 88

Corn Stover Fractions and Bioenergy: *Chemical Composition,
Structure, and Response to Enzyme Pretreatment*
Danny E. Akin, W. Herbert Morrison, III, Luanne L. Rigsby,
Franklin E. Barton, II, David S. Himmelsbach,
and Kevin B. Hicks 104

Separate and Simultaneous Enzymatic Hydrolysis and Fermentation
of Wheat Hemicellulose With Recombinant Xylose Utilizing
Saccharomyces cerevisiae
L. Olsson, H. R. Soerensen, B. P. Dam, H. Christensen,
K. M. Krogh, and A. S. Meyer 117

Biofiltration Methods for the Removal of Phenolic Residues
Luiz Carlos Martins Das Neves, Tábata Taemi Miazaki Ohara
Miyamura, Dante Augusto Moraes, Thereza Christina Vessoni
Penna, and Attilio Converti 130

The BTL2 Process of Biomass Utilization Entrained-Flow
  Gasification of Pyrolyzed Biomass Slurries
  *Klaus Raffelt, Edmund Henrich, Andrea Koegel, Ralph Stahl,
  Joachim Steinhardt, and Friedhelm Weirich* 153

Emission Profile of Rapeseed Methyl Ester and Its Blend
  in a Diesel Engine
  Gwi-Taek Jeong, Young-Taig Oh, and Don-Hee Park 165

SESSION 1B: ENZYME CATALYSIS AND ENGINEERING

Introduction to Session 1B
  Joel R. Cherry and Kevin Gray 179

Properties and Performance of Glucoamylases for Fuel Ethanol
  Production
  *Bradley A. Saville, Chunbei Huang, Vince Yacyshyn,
  and Andrew Desbarats* 180

Heterologous Expression of *Trametes versicolor* Laccase
  in *Pichia pastoris* and *Aspergillus niger*
  *Christina Bohlin, Leif J. Jönsson, Robyn Roth,
  and Willem H. van Zyl* 195

Lactose Hydrolysis and Formation of Galactooligosaccharides
  by a Novel Immobilized β-Galactosidase From the Thermophilic
  Fungus *Talaromyces thermophilus*
  Phimchanok Nakkharat and Dietmar Haltrich 215

Evaluation of Cell Recycle on *Thermomyces lanuginosus* Xylanase A
  Production by *Pichia pastoris* GS 115
  *Verônica Ferreira, Patricia C. Nolasco, Aline M. Castro,
  Juliana N. C. Silva, Alexandre S. Santos, Mônica C. T. Damaso,
  and Nei Pereira, Jr.* 226

Evaluation of Solid and Submerged Fermentations for the Production
  of Cyclodextrin Glycosyltransferase by *Paenibacillus campinasensis*
  H69-3 and Characterization of Crude Enzyme
  *Heloiza Ferreira Alves-Prado, Eleni Gomes,
  and Roberto Da Silva* 234

Effect of β-Cyclodextrin in Artificial Chaperones Assisted Foam
  Fractionation of Cellulase
  Vorakan Burapatana, Aleš Prokop, and Robert D. Tanner 247

RSM Analysis of the Effects of the Oxygen Transfer Coefficient
  and Inoculum Size on the Xylitol Production
  by *Candida guilliermondii*
  *Mariana Peñuela Vásquez, Maurício Bezerra De Souza, Jr.,
  and Nei Pereira, Jr.* 256

Introduction ix

Enzymatic Synthesis of Sorbitan Methacrylate According
   to Acyl Donors
   Gwi-Taek Jeong, Hye-Jin Lee, Hae-Sung Kim, and Don-Hee Park 265

Effect of Inhibitors Released During Steam-Explosion Pretreatment
   of Barley Straw on Enzymatic Hydrolysis
   *Mª Prado García-Aparicio, Ignacio Ballesteros, Alberto González,
   José Miguel Oliva, Mercedes Ballesteros, and Mª José Negro* 278

Purification and Characterization of Two Xylanases
   From Alkalophilic and Thermophilic *Bacillus licheniformis* 77-2
   *Valquiria B. Damiano, Richard Ward, Eleni Gomes,
   Heloiza Ferreira Alves-Prado, and Roberto Da Silva* 289

Oxidation Capacity of Laccases and Peroxidases as Reflected
   in Experiments With Methoxy-Substituted Benzyl Alcohols
   Feng Hong, Leif J. Jönsson, Knut Lundquist, and Yijun Wei 303

Obtainment of Chelating Agents Through the Enzymatic Oxidation
   of Lignins by Phenol Oxidase
   Gabriela M. M. Calabria and Adilson R. Gonçalves 320

Reuse of the Xylanase Enzyme in the Biobleaching Process
   of the Sugarcane Bagasse Acetosolv Pulp
   *Luís R. M. Oliveira, Regina Y. Moriya,
   and Adilson R. Gonçalves* 326

Detection of Nisin Expression by *Lactococcus lactis* Using Two
   Susceptible Bacteria to Associate the Effects of Nisin With EDTA
   *Thereza Christina Vessoni Penna, Angela Faustino Jozala,
   Thomas Rodolfo Gentille, Adalberto Pessoa, Jr.,
   and Olivia Cholewa* 334

SESSION 2: TODAY'S BIOREFINERIES

Introduction to Session 2
   Paris Tsobanakis 347

Existing Biorefinery Operations That Benefit From Fractal-Based
   Process Intensification
   Vadim Kochergin and Mike Kearney 349

The Importance of Utility Systems in Today's Biorefineries
   and a Vision for Tomorrow
   Tim Eggeman and Dan Verser 361

Extraction of Hyperoside and Quercitrin From Mimosa
   (*Albizia julibrissin*) Foliage
   *Adam K. Ekenseair, Lijan Duan, Danielle Julie Carrier,
   David I. Bransby, and Edgar C. Clausen* 382

Foam Control in Fermentation Bioprocess:
From Simple Aeration Tests to Bioreactor
  A. Etoc, F. Delvigne, J. P. Lecomte, and P. Thonart 392

Optimization of Biodiesel Production From Castor Oil
  *Nivea de Lima da Silva, Maria Regina Wolf Maciel,
  César Benedito Batistella, and Rubens Maciel Filho* 405

SESSION 3A: PLANT BIOTECHNOLOGY AND FEEDSTOCK GENOMICS

Manipulating the Phenolic Acid Content and Digestibility of Italian
  Ryegrass (*Lolium multiforum*) by Vacuolar-Targeted Expression
  of a Fungal Ferulic Acid Esterase
  *Marcia M. de O. Buanafina, Tim Langdon, Barbara Hauck,
  Sue J. Dalton, and Phil Morris* 416

Variation of S/G Ratio and Lignin Content in a *Populus* Family
  Influences the Release of Xylose by Dilute Acid Hydrolysis
  *Brian H. Davison, Sadie R. Drescher, Gerald A. Tuskan,
  Mark F. Davis, and Nhuan P. Nghiem* 427

Enhanced Secondary Metabolite Biosynthesis by Elicitation
  in Transformed Plant Root System: *Effect of Abiotic Elicitors*
  Gwi-Taek Jeong and Don-Hee Park 436

SESSION 3B: BIOMASS PRETREATMENT AND HYDROLYSIS

Preliminary Results on Optimization of Pilot Scale Pretreatment
  of Wheat Straw Used in Coproduction of Bioethanol
  and Electricity
  *Mette Hedegaard Thomsen, Anders Thygesen,
  Henning Jørgensen, Jan Larsen, Børge Holm Christensen,
  and Anne Belinda Thomsen* 448

The Combined Effects of Acetic Acid, Formic Acid, and Hydroquinone
  on *Debaryomyces hansenii* Physiology
  *Luís C. Duarte, Florbela Carvalheiro, Joana Tadeu,
  and Francisco M. Gírio* 461

Bioethanol From Cellulose With Supercritical Water Treatment
  Followed by Enzymatic Hydrolysis
  Toshiki Nakata, Hisashi Miyafuji, and Shiro Saka 476

Enhancement of the Enzymatic Digestibility of Waste Newspaper
  Using Tween
  Sung Bae Kim, Hyun Joo Kim, and Chang Joon Kim 486

Introduction xiii

Ethanol Production From Steam-Explosion Pretreated Wheat Straw
*Ignacio Ballesteros, Mª José Negro, José Miguel Oliva,
Araceli Cabañas, Paloma Manzanares,
and Mercedes Ballesteros* 496

Catalyst Transport in Corn Stover Internodes: *Elucidating Transport
Mechanisms Using Direct Blue-I*
*Sridhar Viamajala, Michael J. Selig, Todd B. Vinzant,
Melvin P. Tucker, Michael E. Himmel, James D. McMillan,
and Stephen R. Decker* 509

Evaluation of Cellulase Preparations for Hydrolysis of Hardwood
Substrates
*Alex Berlin, Neil Gilkes, Douglas Kilburn, Vera Maximenko,
Renata Bura, Alexander Markov, Anton Skomarovsky,
Alexander Gusakov, Arkady Sinitsyn, Oleg Okunev,
Irina Solovieva, and John N. Saddler* 528

Steam Pretreatment of Acid-Sprayed and Acid-Soaked Barley
Straw for Production of Ethanol
Marie Linde, Mats Galbe, and Guido Zacchi 546

Reaction Kinetics of Stover Liquefaction in Recycled Stover Polyol
*Fei Yu, Roger Ruan, Xiangyang Lin, Yuhuan Liu, Rong Fu,
Yuhong Li, Paul Chen, and Yinyu Gao* 563

Liquefaction of Corn Stover and Preparation of Polyester
From the Liquefied Polyol
*Fei Yu, Yuhuan Liu, Xuejun Pan, Xiangyang Lin, Chengmei Liu,
Paul Chen, and Roger Ruan* 574

Enzymatic Production of Xylooligosaccharides From Corn Stover
and Corn Cobs Treated With Aqueous Ammonia
*Yongming Zhu, Tae Hyun Kim, Y. Y. Lee, Rongfu Chen,
and Richard T. Elander* 586

Optimal Conditions for Alkaline Detoxification of Dilute-Acid
Lignocellulose Hydrolysates
*Björn Alriksson, Anders Sjöde, Nils-Olof Nilvebrant,
and Leif J. Jönsson* 599

Reintroduced Solids Increase Inhibitor Levels in a Pretreated
Corn Stover Hydrolysate
R. Eric Berson, John S. Young, and Thomas R. Hanley 612

Modeling of a Continuous Pretreatment Reactor Using
Computational Fluid Dynamics
R. Eric Berson, Rajesh K. Dasari, and Thomas R. Hanley 621

Ethanol Production From Pretreated Olive Tree Wood
and Sunflower Stalks by an SSF Process
*Encarnación Ruiz, Cristóbal Cara, Mercedes Ballesteros,
Paloma Manzanares, Ignacio Ballesteros, and Eulogio Castro* 631

SESSION 4: INDUSTRIAL BIOBASED PRODUCTS

The Development of Cement and Concrete Additive:
Based on Xylonic Acid Derived Via Bioconversion of Xylose
**Byong-Wa Chun, Benita Dair, Patrick J. Macuch, Debbie Wiebe,
Charlotte Porteneuve, and Ara Jeknavorian** 645

Production of *Bacillus sphaericus* Entomopathogenic Biomass
Using Brewery Residues
**Cristiane Darco Cruz Martins, Paula Fernandes De Aguiar,
and Eliana Flavia Camporese Sérvulo** 659

Batch (One- and Two-Stage) Production of Biodiesel Fuel
From Rapeseed Oil
Gwi-Taek Jeong and Don-Hee Park 668

Optimization of Distilled Monoglycerides Production
**Leonardo Vasconcelos Fregolente, César Benedito Batistella,
Rubens Maciel Filho, and Maria Regina Wolf Maciel** 680

Production of Lactic Acid From Cheese Whey by Batch
and Repeated Batch Cultures of *Lactobacillus* sp. RKY2
**Hyang-Ok Kim, Young-Jung Wee, Jin-Nam Kim, Jong-Sun Yun,
and Hwa-Won Ryu** 694

Production of Bacterial Cellulose by *Gluconacetobacter* sp. RKY5
Isolated From Persimmon Vinegar
**Soo-Yeon Kim, Jin-Nam Kim, Young-Jung Wee, Don-Hee Park,
and Hwa-Won Ryu** 705

Natural Compounds Obtained Through Centrifugal Molecular
Distillation
**Vanessa Mayumi Ito, Patricia Fazzio Martins,
César Benedito Batistella, Rubens Maciel Filho,
and Maria Regina Wolf Maciel** 716

Biosurfactants Production by *Pseudomonas aeruginosa* FR
Using Palm Oil
**Fernando J. S. Oliveira, Leonardo Vazquez,
Norberto P. de Campos, and Francisca P. de França** 727

Novel Approach of Corn Fiber Utilization
G. Kálmán, K. Recseg, M. Gáspár, and K. Réczey 738

Introduction

Stimulation of Nisin Production From Whey by a Mixed Culture
   of *Lactococcus lactis* and *Saccharomyces cerevisiae*
   Chuanbin Liu, Bo Hu, Yan Liu, and Shulin Chen 751

Biochar As a Precursor of Activated Carbon
   R. Azargohar and A. K. Dalai 762

Moisture Sorption, Transport, and Hydrolytic Degradation
   in Polylactide
   **Richard A. Cairncross, Jeffrey G. Becker, Shri Ramaswamy,
   and Ryan O'Connor** 774

SESSION 5: MICROBIAL CATALYSIS AND METABOLIC ENGINEERING

Zymomonas mobilis As Catalyst for the Biotechnological Production
   of Sorbitol and Gluconic Acid
   Gilmar Sidney Erzinger and Michele Vitolo 787

Metabolic Engineering of *Saccharomyces cerevisiae* for Efficient
   Production of Pure L-(+)-Lactic Acid
   **Nobuhiro Ishida, Satoshi Saitoh, Toru Ohnishi, Kenro Tokuhiro,
   Eiji Nagamori, Katsuhiko Kitamoto, and Haruo Takahashi** 795

A Unique Feature of Hydrogen Recovery in Endogenous
   Starch-to-Alcohol Fermentation of the Marine Microalga,
   Chlamydomonas perigranulata
   Koyu Hon-Nami 808

Detailed Analysis of Modifications in Lignin After Treatment
   With Cultures Screened for Lignin Depolymerizing Agents
   **Aarti Gidh, Dinesh Talreja, Todd B. Vinzant,
   Todd Clint Williford, and Alfred Mikell** 829

Optimization of L-(+)-Lactic Acid Production Using Pelletized
   Filamentous *Rhizopus oryzae* NRRL 395
   Yan Liu, Wei Liao, Chuanbin Liu, and Shulin Chen 844

A Simple Method to Generate Chromosomal Mutations
   in *Lactobacillus plantarum* Strain TF103 to Eliminate
   Undesired Fermentation Products
   Siqing Liu 854

Production of Insoluble Exopolysaccharide of *Agrobacterium* sp.
   (ATCC 31749 and IFO 13140)
   **Márcia Portilho, Graciette Matioli, Gisella Maria Zanin,
   Flávio Faria de Moraes, and Adilma Regina Pippa Scamparini** 864

Selective Utilization of Fructose to Glucose by *Candida magnoliae*,
an Erythritol Producer
*Ji-Hee Yu, Dae-Hee Lee, Yong-Joo Oh, Ki-Cheol Han,
Yeon-Woo Ryu, and Jin-Ho Seo* 870

Biosurfactant Production by *Rhodococcus erythropolis* Grown
on Glycerol As Sole Carbon Source
*Elisa M. P. Ciapina, Walber C. Melo, Lidia M. M. Santa Anna,
Alexandre S. Santos, Denise M. G. Freire, and Nei Pereira, Jr.* 880

Methane Production in a 100-L Upflow Bioreactor by Anaerobic
Digestion of Farm Waste
*Abhijeet P. Borole, K. Thomas Klasson, Whitney Ridenour,
Justin Holland, Khursheed Karim, and Muthanna H. Al-Dahhan* 887

Biomodification of Coal to Remove Mercury
*K. Thomas Klasson, Abhijeet P. Borole,
Catherine K. McKeown, and Choo Y. Hamilton* 897

SESSION 6: BIOPROCESS RESEARCH AND DEVELOPMENT

Introduction to Session 6
Michael R. Ladisch 909

Enzymatic Conversion of Waste Cooking Oils Into Alternative
Fuel—Biodiesel
Guanyi Chen, Ming Ying, and Weizhun Li 911

Inulin-Containing Biomass for Ethanol Production:
Carbohydrate Extraction and Ethanol Fermentation
*Mª José Negro, Ignacio Ballesteros, Paloma Manzanares,
José Miguel Oliva, Felicia Sáez, and Mercedes Ballesteros* 922

Production of Medium-Chain-Length Polyhydroxyalkanoates
by *Pseudomonas aeruginosa* With Fatty Acids and Alternative
Carbon Sources
Pui-Ling Chan, Vincent Yu, Lam Wai, and Hoi-Fu Yu 933

Production and Rheological Characterization of Biopolymer
of *Sphingomonas capsulata* ATCC 14666 Using Conventional
and Industrial Media
*Ana Luiza da Silva Berwanger, Natalia Molossi Domingues,
Larissa Tonial Vanzo, Marco Di Luccio,
Helen Treichel, Francine Ferreira Padilha,
and Adilma Regina Pippa Scamparini* 942

Inulinase Production by *Kluyveromyces marxianus* NRRL Y-7571
Using Solid State Fermentation
*João Paulo Bender, Marcio Antônio Mazutti,
Débora de Oliveira, Marco Di Luccio, and Helen Treichel* 951

Introduction xvii

Macroscopic Mass and Energy Balance of a Pilot Plant Anaerobic
    Bioreactor Operated Under Thermophilic Conditions
    *Teodoro Espinosa-Solares, John Bombardiere, Mark Chatfield,
    Max Domaschko, Michael Easter, David A. Stafford,
    Saul Castillo-Angeles, and Nehemias Castellanos-Hernandez* 959

Ethyl Alcohol Production Optimization by Coupling Genetic
    Algorithm and Multilayer Perception Neural Network
    *Elmer Ccopa Rivera, Aline C. da Costa,
    Maria Regina Wolf Maciel, and Rubens Maciel Filho* 969

Lactic Acid Recovery From Cheese Whey Fermentation Broth Using
    Combined Ultrafiltration and Nanofiltration Membranes
    Yebo Li and Abolghasem Shahbazi 985

Fermentation of Rice Straw/Chicken Manure to Carboxylic Acids
    Using a Mixed Culture of Marine Mesophilic Microorganisms
    Frank K. Agbogbo and Mark T. Holtzapple 997

Construction of Recombinant *Bacillus subtilis* for Production
    of Polyhydroxyalkanoates
    *Yujie Wang, Lifang Ruan, Wai-Hung Lo, Hong Chua,
    and Hoi-Fu Yu* 1015

Microorganism Screening for Limonene Bioconversion
    and Correlation With RAPD Markers
    *Geciane Toniazzo, Lindomar Lerin, Débora de Oliveira,
    Cláudio Dariva, Rogério L. Cansian, Francine Ferreira Padilha,
    and Octávio A. C. Antunes* 1023

Use of Different Adsorbents for Sorption and *Bacillus polymyxa*
    Protease Immobilization
    *Irem Kirkkopru, Cenk Alpaslan, Didem Omay,
    and Yüksel Güvenilir* 1034

Simulation and Optimization of a Supercritical Extraction Process
    for Recovering Provitamin A
    *Elenise Bannwart de Moraes, Mario Eusebio Torres Alvarez,
    Maria Regina Wolf Maciel, and Rubens Maciel Filho* 1041

Affinity Foam Fractionation of *Trichoderma* Cellulase
    Qin Zhang, Chi-Ming Lo, and Lu-Kwang Ju 1051

Molecular Distillation: *A Powerful Technology for Obtaining
    Tocopherols From Soya Sludge*
    *Elenise Bannwart de Moraes, Patricia Fazzio Martins,
    César Benedito Batistella, Mario Eusebio Torres Alvarez,
    Rubens Maciel Filho, and Maria Regina Wolf Maciel* 1066

Application of Two-Stage Biofilter System for the Removal
 of Odorous Compounds
*Gwi-Taek Jeong, Don-Hee Park, Gwang-Yeon Lee,
 and Jin-Myeong Cha* **1077**

Author Index **1089**

Subject Index **1093**

Instructions for Authors and Reviewers **1097**

SESSION 1A
Feedstock Supply and Logistics

Introduction to Session 1A

PETER C. FLYNN

Department of Mechanical Engineering, University of Alberta, Edmonton, Canada T6G 2G8

Using biomass as a source of energy and chemicals presents challenges in sourcing, moving and processing the biomass, and in using the products. Session 1A of the *27th Symposium on Biotechnology for Fuels and Chemicals* illustrates the range of research across this broad spectrum.

Three articles from this session address biomass availability: is there enough biomass to make a meaningful impact in a new energy economy? Haq and Easterly look at the availability of agricultural residues across the United States, and Shahbazi and Li focus more narrowly on whether there is a sustainable source of crop residues to support bioethanol in the State of North Carolina. Mabee and his coworkers report on biomass reserves in Canada, including both agricultural and woody biomass. In general, we can extrapolate that availability is not a limiting factor: there are abundant residues that can feed biomass processes.

Transportation of biomass is a critical cost factor because biomass has both a lower energy density (MJ/kg) than fossil fuels and a lower bulk density (kg/cubic meter). One article from this session by Mahmudi and Flynn takes a detailed look at the relative economics of rail vs truck transport, and in particular, focuses on the extra cost incurred when biomass is unloaded from a truck and trans-shipped by rail. A second article by Kumar, Sokhansanj, and Flynn develops a methodology to rank the relative merits of alternatives to collect and transport biomass when multiple criteria form the basis of selection, for example cost and environmental or community impact.

Separation and characterization of fractions of biomass and it processed products is another area of active research. Akin and his coworkers have studied corn stover fractions in detail, characterizing their chemical composition and structure. Phenols are a troublesome byproduct in many biomass processing schemes, and das Neves and his coworkers report on the removal of phenolic residues by biofiltration methods.

Three process-oriented articles came from this session. Olsson and her coworkers report on separate enzymatic hydrolysis and fermentation of wheat hemicellulose and compare it with combining these steps simultaneously. Mabee and his coworkers report on developments in converting softwood biomass, e.g., from pine and spruce, into ethanol, overcoming problems of more refractory composition of the biomass. And, Raffelt and his coworkers report on a two-step processing of biomass in which a pyrolysed slurry is produced in distributed centers and the slurry is then transported to a central large gasifier. These articles illustrate the broad range of processes under active consideration by researchers.

Finally, Jeong and Park look at the emissions profile from using rapeseed methyl esters, i.e., bio-oil, as a diesel fuel.

Together, this session's articles reflect the broad range of technical research issues that arise from sourcing, transporting and processing biomass to energy and chemicals.

Agricultural Residue Availability in the United States

ZIA HAQ*,[1] AND JAMES L. EASTERLY[2]

[1]Energy Information Administration, Mail Stop EI-82, 1000 Independence Ave., SW Washington, DC 20585, E-mail: zia.haq@eia.doe.gov; and [2]Easterly Consulting, Fairfax, VA

Abstract

The National Energy Modeling System (NEMS) is used by the Energy Information Administration (EIA) to forecast US energy production, consumption, and price trends for a 25-yr-time horizon. Biomass is one of the technologies within NEMS, which plays a key role in several scenarios. An endogenously determined biomass supply schedule is used to derive the price–quantity relationship of biomass. There are four components to the NEMS biomass supply schedule including: agricultural residues, energy crops, forestry residues, and urban wood waste/mill residues. The EIA's *Annual Energy Outlook 2005* includes updated estimates of the agricultural residue portion of the biomass supply schedule. The changes from previous agricultural residue supply estimates include: revised assumptions concerning corn stover and wheat straw residue availabilities, inclusion of non-corn and non-wheat agricultural residues (such as barley, rice straw, and sugarcane bagasse), and the implementation of assumptions concerning increases in no-till farming. This article will discuss the impact of these changes on the supply schedule.

Index Entries: Agricultural residues; corn stover; wheat straw; feedstock cost; biomass supply.

Introduction

The Energy Information Administration (EIA) estimates that there is 491 million dry tons (t) (445 million dry metric tons [mt]) of biomass available in the United States on an annual basis. EIA has compiled available biomass resource estimates from Oak Ridge National Laboratory (ORNL) *(1)*, Antares Group, Inc. *(2)*, and the US Department of Agriculture (USDA) *(3)*. This article discusses how these data are used for forecasting purposes by the National Energy Modeling System (NEMS). One of the key determinants for the growth of biomass is the price–quantity relationship of biomass feedstocks. The raw data for the supply curves are available at the state or county level and these are aggregated to form regional supply schedules. Supply data are available for four fuel types: agricultural residues, energy crops, forestry residues, and urban wood waste/mill residues.

*Author to whom all correspondence and reprint requests should be addressed.

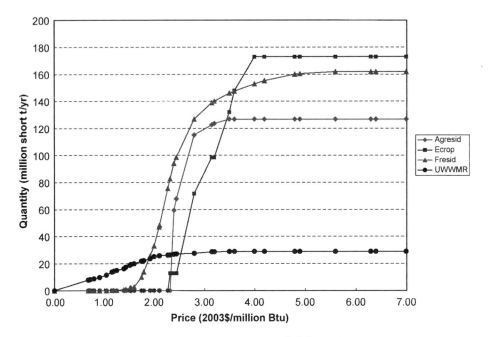

Fig. 1. Biomass resource availability, 2025.

Figure 1 shows the variation in biomass resource as a function of price. A relatively small portion of biomass supply is available at $1.50/million Btu ($1.42/GJ) or less. As a point of comparison, EIA's *Annual Energy Outlook 2005 (AEO2005) (4)* projects coal prices to remain relatively stable (compared with natural gas prices) at $1.28/million Btu ($1.21/GJ) in 2003 to $1.31/million Btu ($1.24/GJ) (in real 2003$) by 2025. Feedstock cost is a major factor that limits biomass growth under AEO2005 reference case assumptions. The available low-cost feedstock (at less than $1.50/million Btu [$1.42/GJ]) is almost exclusively urban wood waste/mill residue. This category of biomass continues to be the only significant resource available at prices up to approx $2/million Btu ($1.90/GJ). At $2/million Btu ($1.90/GJ) and higher, agricultural residues become viable as a second source of biomass. Energy crops and forestry residues begin to make significant contributions at prices around $2.30/million Btu ($2.18/GJ) or higher.

Agricultural Residues

The underlying assumption behind the agricultural residue supply curve is that after each harvesting cycle of agricultural crops, a portion of the stalks can be collected and used for energy production. Agricultural residues cannot be completely extracted, because some of them have to remain on the soil to maintain soil quality (i.e., for erosion control, carbon content, and long-term productivity). The Department of Energy (DOE)

Agricultural Residue Availability

Biomass Program is currently focusing on agricultural residues as the primary (and most likely) source of biomass feedstock supplies for the growing bioenergy industry over the next 10–15 yr. Given the importance of agricultural residues with respect to bioenergy commercialization, EIA decided to update the agricultural residue component of their biomass supply curve in modeling projected energy supplies for AEO2005 and other service requests. Specifically, three aspects of the agricultural residue supply were revised: updated corn stover availability, inclusion of residues other than corn stover and wheat straw, and incorporation of assumptions regarding no-till farming practices in the United States.

Over the last few years a substantial amount of effort has been devoted to developing new county-level estimates of potential corn stover residues, taking into account environmental considerations regarding the amount of corn stover that can be harvested when soil erosion constraints are considered. New estimates have also been made regarding the potential increase in corn stover resources that could be available if no-till cultivation practices were to be more widely adopted (currently 20% of US corn grain is produced using no-till cultivation [5]). No-till farming generally allows for a greater portion of the corn stover to be removed because erosion problems and constraints are substantially reduced. Since 1990, the number of acres of farmland using no-till cultivation has increased by about 1%/yr on average. The Conservation Technology Information Center (CTIC) notes that "50% of cropland acres are suitable for some form of conservation tillage to mitigate soil loss" (6).

Corn Stover Revisions

ORNL recently completed new county-level estimates of available and sustainably removable corn stover for the United States. (7). These estimates include projected costs for the stover at the "farm gate."* These costs include nutrient replacement costs (estimated at $6.50/dry t [$7.17/dry mt] of stover removed), as well as fixed and variable collection costs for producing and delivering round bales of corn stover (stems/leaves/cobs) wrapped with twine and left at the edge of the field. Payments for a farmer premium/profit, as well as transportation costs from the farm-gate to a conversion facility, were treated as separate additional costs. Supply has been constrained by equipment harvest efficiency (75% of gross) and the need to leave residues to limit rain and wind erosion to tolerable losses and to maintain soil moisture in rain-limited regions.

Two sets of new corn stover availability estimates were obtained from ORNL: (1) A base-case assuming corn is produced with the current mix of agricultural tillage and crop rotation practices and; (2) upper-bound case assuming all corn grain would be produced using no-till practices. There

---

*These costs do not include transportation and handling costs for delivering the stover from the farms to a conversion facility.

are various farm-specific soil and crop rotation constraints that limit the maximum percent of overall no-till acres that can be adopted in the United States. The all no-till scenario provides a useful upper-level benchmark in estimating potential future stover supplies. As noted earlier (6), approx 50% of US farms could use conservation tillage practices such as no-till, thus we viewed 50% no-till as the practical upper limit for this cultivation approach. In our analysis we assumed that no-till corn production would reach a level of 30% by the year 2025; this would be a 10% increase in no-till cultivation practices as compared with current practices in which about 20% of corn production is via no-till. A continuation in the trend toward increased no-till farming practices is considered likely owing to soil conservation requirements under US Farm Bill programs, and growth of markets for the production of biofuels and bioproducts from cellulosic feedstocks such as corn stover.

Base-case estimates for corn stover are shown in Table 1. The total amount of corn stover available with current tillage practices is about 64 million dry t/yr (58 million dry mt/yr) (30% of the gross amount, after taking into account the need to leave some of the residue for erosion protection and other soil quality concerns). For the all no-till scenario, Table 1 shows an estimated 111 million dry t/yr (101 million dry mt/yr) of sustainably removable corn stover (51.5% of the gross amount, taking into account the fact that less corn stover would need to be left in the field with no-till practices) this amount of sustainably removable corn stover is shown in the column labeled "total available supply" in Table 1 (note that most, but not all of this amount is estimated to be available at less than $40/dry t [$44/dry mt]; a small fraction of the total is estimated to cost more than $40/dry t [$44/dry mt]).

The base-case numbers from the prior ORNL year 2000 estimate of corn stover availability (1) indicated that a maximum of 119 million dry t/yr (108 million dry mt/yr) of corn stover was available. The new ORNL base-case numbers reflect a significant reduction in anticipated corn stover availability, now that in-depth county-level considerations regarding erosion constraints have been addressed. For the new EIA biomass supply curve, the old maximum of 119 million dry t (108 million dry mt) of stover has been replaced with the new estimate of 62.7 million dry t (56.9 million dry mt) of stover available at less than $40/dry t ($44/dry mt) (farm-gate costs) for the year 2005. Anticipating that no-till practices for corn production will increase over time, the new EIA biomass supply curve assumes that no-till practices will increase from the current level of 20% no-till in year 2005 to 30% no-till in 2025. As a result, the new biomass supply curve has corn stover supplies increasing to 68.4 million dry t/yr (62.0 million dry mt/yr) by 2025 at stover costs of less than $40/dry t ($44/dry mt).* The corn stover supply and cost

---

*An increase of (30%–20%)/(100%–20%) = 1/8th of the potential increase from current practices relative to 100% no-till practices.

Agricultural Residue Availability

values used in the new EIA biomass supply curve for year 2025 are provided in Table 2, based on the assumption of 30% no-till practices.

At $2/million Btu ($1.90/GJ) (equivalent to $31/dry t [$34/dry mt], assuming an energy content of 15.5 million Btu/dry t [18.0 GJ/dry mt]), approx 60 million dry t (54 million dry mt) of corn stover would be available under current tillage practices. This amount of corn stover would be equivalent to 0.93 Quads (0.98 EJ) of energy. For comparison purposes, coal use in 2004 amounted to 22.92 Quads (24.18 EJ) of energy. Therefore, at $2/million Btu ($1.90/GJ), corn stover using current tillage practices could displace 4% of the energy provided by coal in the United States if all corn stover were to be used for electricity generation.

Over the last 30 yr, corn productivity has been increasing by about 1%/yr on average (in terms of the bushels of corn grain produced per acre each year). If this trend continues into the future, it is possible that corn stover quantities will also increase over time, along with corn grain productivity. This potential increase in stover availability has not been included in the newly revised biomass supply curve, pending further input and analysis regarding the likelihood that the trend will continue into the future, and the need for further clarification regarding the anticipated relationship between the amount of stover available per pound of grain produced in the future. More evaluation is needed concerning whether the current ratio of about 1 pound of stover produced per pound of corn grain produced is likely to stay the same or change if corn productivity continues to increase in the future.

There has been a substantial amount of debate regarding the appropriate farmer premium that should be included in determining the total delivered price for corn stover as well as the optimum approach and technology for harvesting and storing stover (8). The bulk of the corn stover supply is anticipated to be available at a cost of $30/dry t ($33/dry mt) at the farm gate. Assuming an average transportation distance of 40 miles (64 km) to deliver round bales from the field edge to a biomass conversion site via flat bed truck, ORNL staff has estimated the transportation cost to be about $7.75/dry t ($8.54/dry mt) of stover (9).

Considering a range of factors, the new EIA biomass supply curve assumes an additional fixed cost of $12/dry t ($13/dry mt) on top of the farm-gate costs in calculating the total delivered price for supplying corn stover to conversion facilities. The $12/dry t ($13/dry mt) fixed cost reflects an adjustment to cover transportation and handling costs, plus farmer premium payments. It is recognized that these costs could be higher than $12/dry t ($13/dry mt). However, this estimation is based on the assumption that cost savings and cost containment will be achieved as integrated harvest and supply operations benefit from experience and operational enhancements (in which custom harvesters are likely to play an important role), in combination with anticipated harvesting technology improvements. With a typical $30/dry t ($33/dry mt) farm-gate cost, plus an additional

Table 1
ORNL Estimates Regarding Corn Stover Availability and Costs (7) (in dry t)

State	Corn acres (000)	With current tillage practices (i.e., approx 20% no-till)						If 100% no-till corn production					
		<$25/t (000)	<$30/t (000)	<$35/t (000)	<$40/t (000)	Total available supply t (000)	Gross stover produced t (000)	$25/t (000)	<$30/t (000)	<$35/t (000)	<$40/t (000)	Total available supply t (000)	Gross stover produced t (000)
Alabama	205	0	0	0	0	2	394	0	0	0	0	6	394
Arkansas	142	23	30	30	31	37	404	137	194	194	197	201	404
Arizona	31	0	0	0	0	0	132	0	0	0	0	0	132
California	220	0	0	0	0	0	855	0	0	0	0	0	855
Colorado	1012	1	93	99	100	101	3256	11	797	981	982	983	3256
Connecticut	0	0	0	0	0	0	0	0	0	0	0	0	0
Delaware	154	228	268	268	268	268	428	280	288	288	288	288	428
Florida	55	1	2	3	3	4	105	1	24	31	32	36	105
Georgia	358	8	48	55	55	65	832	58	208	254	259	270	832
Iowa	11,983	10,474	14,465	14,745	14,928	15,116	39,619	16,580	21,618	21,924	22,059	22,311	39,619
Idaho	47	0	0	0	0	0	167	1	1	1	1	1	167
Illinois	10,667	6391	10,916	11,178	11,293	11,563	34,137	16,132	20,049	20,172	20,399	20,542	34,137
Indiana	5535	3005	5717	5941	6038	6207	16,916	7977	9499	9677	9791	9888	16,916
Kansas	2658	7	370	444	579	658	8786	101	2003	2275	2348	2432	8786
Kentucky	1176	0	33	49	62	70	3187	0	73	100	120	133	3187
Louisiana	387	1	56	63	67	69	1009	17	256	256	264	269	1009
Massachusets	0	0	0	0	0	0	0	0	0	0	0	0	0
Maryland	403	141	285	300	303	311	1107	219	414	421	424	430	1107
Maine	0	0	0	0	0	0	0	0	0	0	0	0	0
Michigan	2074	1702	2946	3062	3096	3139	5639	2463	4067	4067	4067	4067	5639
Minnesota	6582	10,637	12,829	12,917	12,964	13,036	21,419	13,885	15,305	15,305	15,305	15,334	21,419
Missouri	2402	384	549	577	588	633	6767	600	1083	1131	1150	1265	6767
Mississippi	400	0	10	11	11	16	960	2	73	85	85	112	960

State												
Montana	16	0	0	0	0	0	46	0	0	0	0	46
North Caroli	754	25	528	632	651	684	1649	58	976	986	990	995	1649
North Dakota	685	0	2	3	14	25	1664	0	10	14	200	403	1664
Nebraska	8241	1969	5298	5759	5961	6134	25,937	7576	13,653	13,912	14,001	14,080	25,937
New Hampshire	0	0	0	0	0	0	0	0	0	0	0	0	0
New Jersey	82	0	0	0	0	1	197	0	0	0	0	6	197
New Mexico	80	0	9	9	10	10	322	0	65	67	76	83	322
Nevada	0	0	0	0	0	0	0	0	0	0	0	0	0
New York	579	0	64	102	102	138	1444	5	591	707	719	738	1444
Ohio	3214	2061	2737	2812	2828	2885	9938	3414	4935	5046	5077	5126	9938
Oklahoma	207	0	20	20	20	20	684	2	201	205	207	210	684
Oregon	30	6	12	12	12	12	124	25	25	25	25	25	124
Pennsylvania	1005	0	18	39	54	89	2491	0	180	251	300	430	2491
Rhode Island	0	0	0	0	0	0	0	0	0	0	0	0	0
South Caroli	299	0	146	172	177	195	525	0	282	296	300	310	525
South Dakota	3354	38	478	478	478	659	8323	320	3281	3283	3283	4562	8323
Tennessee	594	11	25	25	36	42	1519	27	50	50	62	72	1519
Texas	1810	0	43	46	91	106	5118	0	645	699	765	868	5118
Utah	21	0	0	0	0	0	67	0	0	0	0	0	67
Virginia	269	58	104	114	115	118	681	60	135	148	150	153	681
Vermont	0	0	0	0	0	0	0	0	0	0	0	0	0
Washington	90	22	26	27	27	27	393	126	179	179	179	179	393
Wisconsin	2940	322	1525	1634	1709	1799	8905	2657	4518	4541	4622	4660	8905
West Virgini	29	0	0	0	0	0	69	0	0	0	0	0	69
Wyoming	51	0	0	1	1	1	151	0	4	40	40	44	151
United States	70,840	37,517	59,652	61,630	62,673	64,242	216,366	72,732	105,692	107,621	108,777	111,526	216,366
Quads equiv.		0.58	0.92	0.96	0.97	1.00							

Table 2
Estimated Corn Stover Availability for Year 2025

State	Current stover supplies (20% no-till)				Year 2025 stover supplies (30% no-till)			
	<$25/t (000)	<$30/t (000)	<$35/t (000)	<$40/t (000)	<$25/t (000)	<$30/t (000)	<$35/t (000)	<$40/t (000)
AL	0	0	0	0	0	0	0	0
AR	23	30	30	31	37	51	51	51
AZ	0	0	0	0	0	0	0	0
CA	0	0	0	0	0	0	0	0
CO	1	93	99	100	2	181	209	210
CT	0	0	0	0	0	0	0	0
DE	228	268	268	268	235	271	271	271
FL	1	2	3	3	1	5	6	6
GA	8	48	55	55	15	68	80	81
HI	0	0	0	0	0	0	0	0
IA	10,474	14,465	14,745	14,928	11,237	15,359	15,642	15,820
ID	0	0	0	0	0	0	0	0
IL	6391	10,916	11,178	11,293	7609	12,058	12,303	12,431
IN	3005	5717	5941	6038	3627	6190	6408	6507
KS	7	370	444	579	19	574	673	800
KY	0	33	49	62	0	38	55	70
LA	1	56	63	67	3	81	87	92
MA	0	0	0	0	0	0	0	0
MD	141	285	300	303	151	301	315	318
ME	0	0	0	0	0	0	0	0
MI	1702	2946	3062	3096	1798	3086	3188	3217
MN	10,637	12,829	12,917	12,964	11,043	13,139	13,215	13,256
MO	384	549	577	588	411	616	647	658
MS	0	10	11	11	1	18	20	20
MT	0	0	0	0	0	1	1	2
NC	25	528	632	651	30	584	677	694
ND	0	2	3	14	0	3	4	38
NE	1969	5298	5759	5961	2670	6343	6778	6966
NH	0	0	0	0	0	0	0	0
NJ	0	0	0	0	0	0	0	0
NM	0	9	9	10	0	16	16	18
NV	0	0	0	0	0	0	0	0
NY	0	64	102	102	1	130	177	179
OH	2061	2737	2812	2828	2230	3012	3091	3109
OK	0	20	20	20	0	42	43	43
OR	6	12	12	12	9	13	13	13
PA	0	18	39	54	0	38	65	85
RI	0	0	0	0	0	0	0	0
SC	0	146	172	177	0	163	188	192
SD	38	478	478	478	74	829	829	829
TN	11	25	25	36	13	29	29	39
TX	0	43	46	91	0	118	127	176

(Continued)

Table 2 (Continued)

State	Current stover supplies (20% no-till)				Year 2025 stover supplies (30% no-till)			
	<$25/t (000)	<$30/t (000)	<$35/t (000)	<$40/t (000)	<$25/t (000)	<$30/t (000)	<$35/t (000)	<$40/t (000)
UT	0	0	0	0	0	0	0	0
VA	58	104	114	115	58	108	118	119
VT	0	0	0	0	0	0	0	0
WA	22	26	27	27	35	45	46	46
WI	322	1525	1634	1709	614	1899	1998	2073
WV	0	0	0	0	0	0	0	0
WY	0	0	1	1	0	1	6	6
US	37,517	59,652	61,630	62,673	41,919	65,407	67,379	68,436
Quads	0.58	0.92	0.96	0.97	0.65	1.01	1.04	1.06

$12/dry t ($13/dry mt) in transport and miscellaneous costs, the total delivered price used in EIA's new biomass supply curve is about $42/dry t ($46/dry mt) for the bulk of the stover supplies. In comparison, experience with corn stover harvesting and delivery during 1997–1999 illustrated a range in delivered corn stover prices of between $31.60 and $35.70/dry t ($34.84–$39.36/dry mt) (10).

Non-Corn and Non-Wheat-Based Agricultural Residue Supply Estimates

Although corn stover and wheat straw are anticipated to be the largest potential sources of agricultural residues, there are many other types of crops that could potentially supply biomass residues. Although these other crop residues may tend to represent niche opportunities, on a national aggregate level they offer an expansion in the geographic range and supply for future bioenergy facilities beyond the Corn Belt and Great Plains states. Figure 2 illustrates the limited geographic concentration of corn stover supplies in the United States.

Crop residue supply estimates have been developed for nine crops: sorghum, barley, oats, rye, cotton field trash, cotton gin trash, rice straw, bagasse (the residue from sugar cane processing), and orchard prunings (3,11). Although a large amount of soybeans are produced in the United States, the field residues from this crop are comparatively modest and readily decompose in the field, making collection of soybean plant residues unattractive (at least with the variety of soybean plants currently used by farmers).

In order to reduce the effects of varying yearly crop yields, for each of the "other" crop categories average annual crop production in all US states was calculated over a 3-yr span (1998–2000). The rules-of-thumb used for estimating the dry crop residues produced per pound of crop harvested

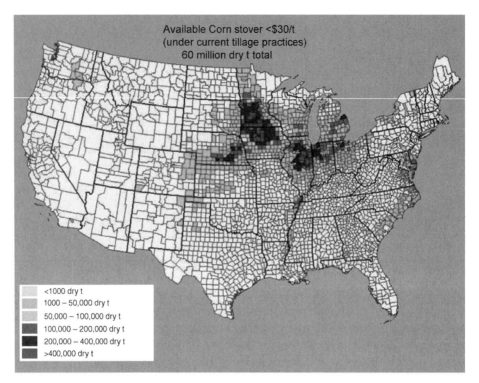

Fig. 2. Geographic distribution of corn stover supplies in the United States (Graham, 2004).

and the estimated percent of sustainably harvestable residues for each crop type were obtained from a variety of sources.*

- Barley: 48 pounds of barley/bushel (0.62 kg/L); 1.67 dry pounds of barley straw/pound of barley; 50% of barley straw harvested (net after erosion requirements and livestock use).
- Rye: 56 pounds of rye/bushel (0.72 kg/L); 1.67 dry pounds of rye straw/pound of rye; 40% of rye straw harvested (net after erosion requirements and livestock use).
- Oats: 32 pounds of oat/bushel (0.41 kg/L); 1.67 dry pounds oat straw/pound of oat; 40% of oat straw harvested (net after erosion requirements and livestock use).
- Sorghum: 56 pounds of sorghum/bushel (0.72 kg/L); 0.74 dry pounds of sorghum stover/pound of sorghum; 30% of sorghum stover harvested (net after erosion requirements and livestock use).

*For barley, rye, oats, sorghum, rice straw, and bagasse, the rules-of-thumb are from the USDA agricultural residue report *(11)*. Percent harvestable factors were derived by averaging state values in the Gallagher report, taking into account limits related to erosion and competing livestock demand for the residues. The factor for orchard prunings is an average for orchard prunings from a California Energy Commission report on biomass residues *(12)*. Cotton gin trash and cotton field residue factors are based in input from staff at the USDA Cotton Ginning Research Laboratory *(13)*.

- Rice: 0.845 dry pounds of rice straw per pound of rice available, with 100% harvested.
- Bagasse: 0.25 dry pounds of bagasse per pound of sugarcane yield.
- Cotton Gin Trash: 0.9 dry pounds of cotton gin trash per pound of cotton harvested using "stripper" type harvesters.*
- Cotton Field Trash: 0.6 dry pounds of cotton field trash remain per pound of cotton harvested for acreage harvested using spindle-type harvesters.†
- Orchard Prunings/Thinnings: 0.7 dry pounds/acre (0.78 dry kg/he) (average for all types of fruit and nut trees).

The USDA report on crop residues (11) used detailed county level data to determine erosion constraints, and detailed livestock data to estimate competing demands for residues. In compiling the new EIA biomass supply curve, the USDA data was used in those states in which the report provided crop residue estimates. For states where the USDA report did not provide crop residue data, estimates were made using USDA-NASS data on average crop production by state, in combination with the rules-of-thumb summarized earlier. The specific states where crop residue estimates from the USDA agricultural residue report were used in the new EIA agricultural residue supply curve are as follows:

- Sorghum–CA, CO, KS, MT, ND, NE, OK, and SD;
- Barley–CA, CO, KS, MT, ND, OK, OR, SD, and WA;
- Oats–CA, CO, KS, MT, ND, NE, OK, OR, SD, and WA;
- Rice Straw–AR, LA, MO, and MS.

For pricing delineation, the "other" (non-corn and non-wheat) agricultural residues were separated into two categories:

1. Lower cost residues including bagasse, cotton gin trash, rice straw, and orchard prunings. These are categorized as "under $25/dry t" ($28/dry mt) at the farm gate, as there are either negligible added costs to harvest these residues or the costs for harvesting them are covered through normal crop management practices. The $12/dry t ($13/dry mt) adder for transportation, handling, and profit was used for these lower cost residues in order to simplify modeling functions. However, adding this cost may overstate the actual price of these residues since they could potentially be converted to energy at the locations where they are produced (especially regarding bagasse residues).

*This as a "blended" number considering that a portion of the cotton harvest is field cleaned with stripper harvesting. About 85% of the cotton in Texas and Oklahoma is harvested with stripper-type harvesters and 0.3 dry pounds of cotton gin trash is produced per pound of cotton harvested using "spindle" type harvesters. Spindle harvesters are used for essentially all of the cotton produced, except for Texas and Oklahoma. The 0.3 dry pounds number is based on the assumption that the difference between the trash produced by the stripper versus the spindle type harvesters (0.9 – 0.6 = 0.3 extra pounds of trash left in the field by spindle-type harvesters) is left in the field in cotton producing states other than Texas and Oklahoma.

†Cotton stalks must currently be ground down to avoid bole weevil pest problems in the following year.

2. Higher cost agricultural residues including sorghum, barley, oats, rye, and cotton field trash in which farmers would have to make an extra effort to collect residues, similar to wheat straw collection. For this category of residues similar cost factors were used as for wheat straw in the existing EIA biomass supply curve. Seventy-five percent of the higher cost residues were assumed to be available at a cost of $30/dry t ($33/dry mt) or less at the farm gate; 88% is assumed to be available at less than $35/dry t ($39/dry mt), and 100% is assumed to be available at $40/dry t ($44/dry mt) at the farm gate. Similar to corn stover residues, an additional $12/dry t ($13/dry mt) were added to the "farm gate" costs to reflect delivered costs, taking into account various factors such as transportation, handling, and farmer profit.

Wheat Straw Estimates and Adjustments

The wheat straw estimates were generally kept unchanged as they had been in the EIA supply curve, pending revised county-level estimates. However, for the state of Oklahoma the amount of available wheat straw available was reduced substantially in the new EIA biomass supply curve. The old EIA supply curve showed only one state, Oklahoma, with wheat straw available at the lower cost category of less than $20/dry t ($22/dry mt) at the farm gate. It showed a substantial (3.2 million dry t/yr [2.9 million dry mt/yr]) of wheat straw available in Oklahoma in this low-price range. The USDA agricultural residue report *(11)* shows only 565,000 dry t (512,000 dry mt) of wheat straw available in Oklahoma, taking into account wind erosion as well as rain erosion constraints, whereas the source document for the existing EIA wheat straw estimates *(1)* only accounted for rain erosion limits, not wind erosion, which is a major consideration in Oklahoma. Based on this observation, the new EIA biomass supply curve includes a revised estimate for wheat straw availability in Oklahoma, now set at 565,000 dry t available/yr (512,000 dry mt/yr) at less than $30/dry t ($33/dry mt) at the farm gate (with none available at less than $20/dry t [$22/dry mt] at the farm gate). Another change regarding wheat straw is that transportation costs (plus some handling and profit costs) have been changed to $12/dry t ($13/dry mt) whereas the old EIA biomass curve had assumed an additional $10/dry t ($11/dry mt) cost adder for transportation in determining delivered costs for wheat straw.

Biomass Bulk Density and Transportation Issues and Costs

Transportation costs represent a significant fraction of the delivered cost of biomass feedstocks. Transportation costs are impacted by the bulk density of biomass and the transportation mode (truck or rail). A substantial amount of analysis is underway by various organizations (such as ORNL) to determine optimized and improved approaches and technologies for harvesting, handling, storing, and transporting biomass such as agricultural residues. As noted earlier, the agricultural residue supply estimates for the new EIA biomass supply curve assumes that round bales are produced and that these bales will be stored at the farm field edge for later transport directly to conversion facilities, with no intermediate storage. The density of round corn stover bales is around 9 dry pounds/ft^3 (144 kg/m^3) *(14)*.

Rather than storing bales at the "farm gate," an alternative is to store them at one or more central storage sites. Round bales can be stored in the open or in covered storage (round bales shed rainfall and can tolerate exposure to the weather), or square bales can be produced that are easier to stack, but are more susceptible to weather damage and thus need to be in covered storage. One alternative approach being investigated is the transport of loose bulk residues to an intermediate wet storage site for later transport to a conversion facility, probably via rail *(15)*. The bulk density of corn stover ejected by a standard corn combine is approx 3 dry pounds/ft^3 (48 dry kg/m^3), which could be increased to about 6 dry pounds/ft^3 (96 dry kg/m^3) with a forage chop approach. The higher density could help to reduce costs for transporting bulk residues to an intermediate storage site *(16)*. The anticipated density of stover removed from a wet storage site is around 12 dry pounds/ft^3 (192 dry kg/m^3). The higher bulk density of the feedstock, as well as other logistics benefits with the dispatch and transport of residues to conversion sites may offset the added costs associated with operating an intermediate storage site.

Densification of agricultural residues to pellets or cubes could increase the bulk density of biomass to as high as 28–40 pounds/ft^3 (448–640 dry kg/m^3) *(17)*. A primary drawback to densification is that it increases the cost for biomass in comparison with conventional approaches such as baling. One recent analysis estimated that densification costs might be in the range of $10 or $11/dry t ($11 or $12/dry mt) *(17)*; however, there are many alternatives for densification (such as the use of various binder additives) that will significantly impact costs. The added benefits of densification in terms of handling, storage, transport, and use may make this an approach attractive in certain applications particularly, whereas competing conventional energy options are expensive.

The approach and equipment used for harvesting, handling, storing, and transporting biomass will also have an impact on the amount of dry matter losses that can occur. For example, one evaluation of bunker storage options for agricultural materials found that storage losses could range from 10 to 16% *(18)*.

Although it is beyond the scope of this article to present and evaluate all of the many options and parameters that could impact biomass supply systems, the observations above help provide a sense of the range of factors that could impact biomass feedstock costs in the future.

Integrated Agricultural Residue Supply Curve

Table 3 provides a state-by-state summary of the agricultural residue supplies that have been included in the new EIA biomass supply curve.

Table 3
Maximum Agricultural Residue Supplies Available in the United States on a Sustainable[a] Basis (1000 dry t/yr)

State	Wheat straw	Corn stover (current till)	Sorghum	Barley	Oats	Rye	Rice straw	Bagasse	Cotton gin trash	Cotton field trash	Orchard prunings/residues	Total
AL	19	2	2	0	6	0	0	0	41	83	20	174
AR	857	37	74	0	6	0	3017	0	97	195	10	4293
AZ	169	0	6	116	0	0	0	0	52	104	44	491
CA	887	0	0	1	0	0	1572	0	150	299	2010	4919
CO	252	101	14	4	0	1	0	0	0	0	4	376
CT	0	0	0	0	0	0	0	0	0	0	2	2
DE	301	268	1	42	0	0	0	0	0	0	1	613
FL	15	4	0	0	0	0	0	2080	7	14	626	2746
GA	343	65	11	0	19	31	0	0	114	229	102	914
HI	239	0	0	0	0	0	0	339	0	0	25	603
IA	1123	15,116	0	0	122	0	0	0	0	0	2	16,363
ID	1456	0	0	1126	19	0	0	0	0	0	5	2606
IL	713	11,563	71	0	43	5	0	0	0	0	6	12,401
IN	0	6207	0	0	18	1	0	0	0	0	4	6230
KS	4456	658	853	0	0	6	0	0	2	3	5	5983
KY	1163	70	6	11	0	0	0	0	0	0	3	1253
LA	80	69	125	0	0	0	775	1696	59	118	11	2933
MA	0	0	0	0	0	0	0	0	0	0	4	4
MD	273	311	6	69	2	3	0	0	0	0	3	667
ME	0	0	0	23	20	0	0	0	0	0	3	46
MI	682	3139	0	17	52	11	0	0	0	0	83	3984
MN	1432	13,036	0	488	74	15	0	0	0	0	3	15,048
MO	1224	633	206	0	11	0	226	0	33	65	15	2414
MS	38	16	39	0	0	0	525	0	117	234	11	980

MT	370	0	0	492	2	0	0	0	0	1	865
NC	475	684	5	27	18	10	0	78	157	9	1463
ND	3715	25	0	3670	423	45	0	0	0	1	7878
NE	327	6134	122	5	2	6	0	0	0	2	6597
NH	0	0	0	0	0	0	0	0	0	0	2
NJ	33	1	0	6	0	3	0	0	0	8	51
NM	214	10	34	0	0	0	0	8	15	30	311
NV	15	0	0	7	0	0	0	0	0	0	22
NY	130	138	0	12	54	10	0	0	0	69	413
OH	1374	2885	0	9	73	3	0	0	0	9	4353
OK	566	20	21	1	0	38	0	28	3	62	739
OR	94	12	0	115	7	0	0	0	0	67	295
PA	1031	89	3	102	88	10	0	0	0	35	1358
RI	0	0	0	0	0	0	0	0	0	0	0
SC	240	195	2	3	18	8	0	24	48	17	555
SD	827	660	2	62	115	28	0	0	0	0	1694
TN	302	42	11	0	0	0	0	44	89	2	490
TX	1529	106	1218	0	57	8	81	824	92	157	4232
UT	154	0	0	129	5	0	0	0	0	6	294
VA	463	118	4	100	0	4	0	11	22	18	740
VT	0	0	0	0	0	0	0	0	0	2	2
WA	955	27	0	262	9	0	0	0	0	218	1471
WI	155	1799	0	67	199	7	0	0	0	7	2234
WV	12	0	0	0	1	0	0	0	0	6	19
WY	84	1	0	149	16	0	0	0	0	0	250
Total	28,787	64,242	2836	7112	1480	253	6195	1691	1771	3731	122,374

[a]Taking into account erosion and soil maintenance requirements.

Table 4
2025 Supply and Cost Estimates for All Agricultural Residues Assuming 30% No-Till Corn Tillage Practices (1000 dry t/yr)

State	Corn stover				Other crop residues				All agricultural residues			
	<$25/t	<$30/t	<$35/t	<$40/t	<$25/t	<$30/t	<$35/t	<$40/t	<$25/t	<$30/t	<$35/t	<$40/t
AL	0	0	0	0	61	61	116	171	61	61	116	171
AR	37	50	51	51	3124	4260	4260	4260	3161	4310	4310	4310
AZ	0	0	0	0	96	491	491	491	96	491	491	491
CA	0	0	0	0	3731	4919	4919	4919	3731	4919	4919	4919
CO	2	181	209	210	4	275	275	275	6	456	484	485
CT	0	0	0	0	2	2	2	2	2	2	2	2
DE	235	271	271	271	0	101	223	344	235	373	494	615
FL	1	5	6	6	2714	2743	2743	2743	2714	2748	2749	2749
GA	15	68	80	81	216	851	851	851	231	919	931	931
HI	0	0	0	0	364	603	603	603	364	603	603	603
IA	11,237	15,359	15,642	15,820	2	1248	1248	1248	11,239	16,607	16,890	17,068
ID	0	0	0	0	5	2606	2606	2606	5	2606	2606	2606
IL	7608	12,058	12,303	12,431	6	838	838	838	7614	12,895	13,140	13,269
IN	3626	6190	6408	6507	4	23	23	23	3630	6213	6431	6530
KS	19	574	673	800	6	5325	5325	5325	25	5899	5998	6125
KY	0	38	55	70	3	482	832	1183	3	519	888	1253
LA	3	81	87	92	2541	2864	2864	2864	2543	2945	2951	2956
MA	0	0	0	0	4	4	4	4	4	4	4	4
MD	151	301	315	318	3	356	356	356	154	657	671	674
ME	0	0	0	0	3	46	46	46	3	46	46	46
MI	1797	3086	3188	3217	83	843	843	845	1880	3929	4031	4062
MN	11,043	13,138	13,215	13,256	3	2011	2011	2011	11,046	15,150	15,227	15,268
MO	411	616	646	658	274	1756	1768	1781	685	2372	2415	2439

MS	1	18	20	20	653	653	809	964	654	671	829	985
MT	0	1	1	2	1	865	865	865	1	866	866	866
NC	30	584	677	694	88	777	777	780	117	1361	1454	1473
ND	0	3	4	38	0	30	3941	7853	0	33	3945	7891
NE	3	6	7	7	1	463	463	463	2671	6805	7241	7429
NH	0	0	0	0	2	2	2	2	2	2	2	2
NJ	0	0	0	0	8	50	50	50	8	50	50	50
NM	0	16	16	18	38	301	301	301	38	317	317	319
NV	0	0	0	0	0	22	22	22	0	22	22	22
NY	1	130	177	179	69	275	275	275	70	405	452	454
OH	2230	3012	3091	3109	9	1468	1468	1468	2240	4480	4559	4577
OK	0	42	43	43	90	718	718	718	90	761	762	762
OR	8	13	13	13	67	283	283	283	76	297	297	297
PA	0	38	65	85	35	272	771	1270	35	310	836	1355
RI	0	0	0	0	0	0	0	0	0	0	0	0
SC	0	162	188	192	41	360	360	360	41	523	548	553
SD	74	829	829	829	0	1035	1035	1035	74	1863	1863	1863
TN	13	28	28	39	47	448	448	448	60	476	476	488
TX	0	118	127	175	1223	4126	4126	4126	1223	4244	4254	4302
UT	0	0	0	0	6	294	294	294	6	294	294	294
VA	58	108	118	119	29	411	517	622	87	519	635	742
VT	0	0	0	0	2	2	2	2	2	2	2	2
WA	35	45	46	46	218	1443	1443	1443	253	1488	1490	1490
WI	614	1899	1998	2073	7	435	435	435	620	2334	2433	2508
WV	0	0	0	0	7	20	20	20	7	20	20	20
WY	0	0	6	6	0	249	249	249	0	249	255	255
US	41,919	65,407	67,379	68,436	15,893	47,710	52,921	58,133	57,812	113,117	120,300	126,568

The corn stover quantities are based on current tillage and crop rotation practices. As discussed earlier, the wheat straw quantities are essentially the same values that were in the old EIA biomass supply curve, with updated Oklahoma values.

The "other" non-wheat/non-corn residues account for about 24% of total potential agricultural residue supplies. Although the other crop residues are dispersed in relatively small amounts, there are a few states in which these resources are concentrated, with the potential to supply larger biomass conversion facilities at these locations. For the most part, however, the "other" agricultural residue supplies will require small modular biomass conversion systems in order to be utilized.* Table 4 provides the new agricultural price–quantity pairs for the EIA biomass supply curve data for year 2025.

Conclusion

Although a significant amount of effort has gone into estimating the available quantities of agricultural residues, the amount of residues that can be sustainably removed is an issue that continues to be evaluated. Further analysis and field experience in the farming community is needed to solidify consensus views regarding the amount of residues that need to remain in the field, and the associated costs for harvesting and supplying these residues for use as energy feedstocks. Given these uncertainties, the current supply curves represent our best understanding of the availability of biomass at this point in time.

Summary

The National Energy Modeling System (NEMS) is used by the Energy Information Administration (EIA) to forecast US energy production, consumption, and price trends. Biomass is one of the technologies within NEMS, which plays a key role in several scenarios. An endogenously determined biomass supply schedule is used to derive the price–quantity relationship of biomass. The EIA's *Annual Energy Outlook 2005* includes updated estimates of the agricultural residue portion of the biomass supply schedule. This article had discussed the impact of these changes on the supply schedule.

References

1. Walsh, M., Perlack, R., Turhollow, A., et al. (2000), *Biomass Feedstock Availability in the United States: 1999 State Level Analysis*, http://bioenergy.ornl.gov/resourcedata/index.html.
2. Antares Group, Inc. (1999), *Biomass Residue Supply Curves for the United States (Update)*, Report for the U.S. Department of Energy and the National Renewable Energy Laboratory.

*For example, a smaller-scale gasifier coupled to an engine-generator (e.g., 100–300 kWe) currently has a typical heat rate of around 20,000 Btu/kW h (21.1 mJ/kW h) *(19)*, and would need roughly 1000 dry t/yr (907 dry mt/yr) of biomass feedstock to fuel a 100 kW system.

3. U.S. Department of Agriculture (2004), *National Agricultural Statistical Service (NASS)*, Agricultural Statistics Data Base, http://www.nass.usda.gov.
4. Energy Information Administration (2005), *Annual Energy Outlook 2005 With Projections to 2025*, http://www.eia.doe.gov/oiaf/aeo/index.html.
5. Conservation Technology Information Center (2004), *Percentage No-Till Acres by Crop 2002*, http://www.ctic.purdue.edu/CTIC/PercentChart.html.
6. Hassell, J. (2001), Conservation Technology Information Center, National Association of Conservation Districts, *Statement to the Senate Committee on Agriculture, Nutrition and Forestry, relative to conservation programs in the Farm Bill*, http://www.ctic.purdue.edu/core4/Testimony.html.
7. Graham, R., Nelson, R., Tharp, L., et al. (2004), *Estimating the U.S. Corn Stover Supply: A Review of Methods & Results*, Oct. 9, 2003 presentation to DOE & USDA, and July 2004 update (e-mail with spreadsheet).
8. U.S. Department of Energy, Biomass Program (2003), *Roadmap for Agricultural Biomass Feedstock Supply in the United States*, http://www.bioproducts-bioenergy.gov/pdf/Ag_Roadmap.pdf.
9. Turhollow, A. and Sokhansanj, S. (2006), *Costs of Harvesting, Storing, and Transporting Corn Stover in Wet Form*, final draft, *Appl. Eng. Agri. J.*, in press.
10. Schechinger, T. and Hettenhaus, J. (2004), *Corn Stover Harvesting: Grower, Custom Operator, and Processor Issues and Answers – Report on Corn Stover Harvest Experiences in Iowa and Wisconsin for the 1997–98 and 1998–99 Crop Years*, Oak Ridge National Laboratory.
11. Gallagher, P., Dikeman, M., Fritz, J., Wailes, E., Gauther, W., and Shapouri, H. (2003), *Biomass From Crop Residues: Cost and Supply Estimates*, published by the USDA Office of Energy Policy and New Uses, Agricultural Economic Report No. 819.
12. California Energy Commission (CEC), Prepared by Von Bernath, H., Jenkins, B., et al. (2004), *An Assessment of Biomass Resources in California*.
13. Holt, G. and Funk, P. (August 8, 2004), Southwestern Cotton Ginning Research Laboratory, USDA Agricultural Research Service, personal communication.
14. Sokhansanj, S., Turhollow, A., and Perlack, R. (2002), *Stochastic Modeling of Costs of Corn Stover Costs Delivered to an Intermediate Storage Facility*, American Society of Agricultural Engineers, Annual International Meeting/CIGR XVth World Congress, Chicago, IL.
15. Atchison, J. and Hettenhaus, J. (2003), *Innovative Methods for Corn Stover Collecting, Handling, Storing and Transporting*, prepared for the National Renewable Energy Laboratory.
16. Turhollow, A. (2004), Personal communication, August 11, 2004.
17. Sokhansanj, S. and Turhollow, A. (July 2004), *Biomass Densification – Cubing Operations and Costs for Corn Stover*, *Appl. Eng. Agri. J.* **20(4)**, 495–499.
18. Miller, D. and Rotz, A. (1995), In: Barnes, Robert, F., Miller, Darrell, A., and Jerry Nelson, C. (eds.), *Harvesting and Storage*. Forages 5th ed., Vol I. *An Introduction to Grassland Science*, Iowa State University Press, Ames, IA, pp. 163–174.
19. Miles, T. (2004), *Gasifier Suitability Analysis: Selection of Gasification Systems for BERC Demonstration/Commercialization*, Final Report, for the Biomass Energy Resource Center.

Canadian Biomass Reserves for Biorefining

Warren E. Mabee,*[,1] Evan D. G. Fraser,[2] Paul N. McFarlane,[1] and John N. Saddler[1]

[1]*Department of Wood Science, University of British Columbia, 2424 Main Mall, Vancouver, BC, V6T 1Z4 Canada, E-mail: warren.mabee@ubc.ca; and [2]School of Earth & Environment, University of Leeds, Leeds LS2 9JT, UK*

Abstract

A lignocellulosic-based biorefining strategy may be supported by biomass reserves, created initially with residues from wood product processing or agriculture. Biomass reserves might be expanded using innovative management techniques that reduce vulnerability of feedstock in the forest products or agricultural supply chain. Forest-harvest residue removal, disturbance isolation, and precommercial thinnings might produce $20-33 \times 10^6$ mt/yr of feedstock for Canadian biorefineries. Energy plantations on marginal Canadian farmland might produce another 9–20 mt. Biomass reserves should be used to support first-generation biorefining installations for bioethanol production, development of which will lead to the creation of future high-value coproducts. Suggestions for Canadian policy reform to support biomass reserves are provided.

Index Entries: Biorefining; lignocellulosic biomass; forestry; agriculture; energy plantations; policy reform.

Introduction

Biomass reserves may be created to supply inexpensive and secure feedstocks to the biorefinery, allowing the substitution of renewable biomass for fossil petroleum in the production of energy, fuels, and chemical or material products. In the literature, the term "biomass reserve" is applied to natural biomass accumulations in both terrestrial and marine ecosystems *(1,2)*, and to human-designated zones used in the management of terrestrial and marine biomass resources *(3)*. The latter type of reserve may be created using policy instruments, and is often associated with resource protection. A good example of a legislated biomass reserve is the farmland under the Conservation Reserve Program in the United States, a voluntary program designed to encourage land-owners for farming land that is prone to erosion or that represents rare or declining habitat.

---

*Author to whom all correspondence and reprint requests should be addressed.

Although biomass reserves for biorefining operations have not been discussed specifically in the literature, the concept of reserves dedicated to supplying energy needs has been referred to by a number of authors. There are two distinct approaches to creating these reserves. Some articles refer to an area-based approach in which dedicated crops or trees are grown on a landbase that is unsuitable for traditional agriculture or forestry (3–8). Other authors (9–11) have considered a volume-based approach in which agricultural (herbaceous) and forest (woody) residues are recovered from existing operations—in practice, an intensification of current activities. These approaches may be considered as complimentary, as each serves a valuable role in supplying biomass and thereby increasing energy independence and national security. The choice of approach will be largely determined by resource availability, moreover, it is also controlled by desired energy-substitution levels, which vary nation-wise, and is a matter of policy.

There is some consensus that initial biomass-for-energy reserves will be volume-based and will take advantage of residue availability (5,10,12), whereas future reserves might use either area- or volume-based approaches, or a combination of the two. In countries that have large populations, large amounts of biomass are required to affect significant levels of energy substitution, and area-based approaches are favored. Studies in the United States and Europe have concentrated on identifying agricultural land that might be dedicated to energy production (4,5). A recent study by the International Institute for Applied Systems Analysis (IIASA) examining 22 nations in Eastern Europe and North-Central Asia has identified 33.8×10^6 ha of land that might be used for energy crops or plantations, supplying approx 12% of total per capita energy demand (8). Conservative estimates for the United States indicate that $(15–17) \times 10^6$ ha of marginal farmland might be used for energy crop purposes (7), potentially replacing a large portion of transportation fuel demand (5). In part, the area-based approach is dictated in these countries by the assumption that a large portion of energy demand might be met with biomass-based energy.

An area-based approach for the United States exploits the fact that growing seasons and agricultural advances favor short-rotation crops in many of the US states, and recognizes that converting existing forest land to energy crop production has a negative environmental impact (13). It also takes advantage of the available marginal land. Significant areas of such land are registered in the Conservation Reserve Program, which controls approx 14.6×10^6 ha for the purpose of watershed and habitat protection. Other federal programs may contribute up to 10×10^6 ha of additional farmland depending on the year (5). Using these lands for energy crops or plantations may have additional benefits for the United States, in addition to increased energy security. It has been predicted that switchgrass crops for energy has the potential to provide higher profits than conventional crops on 16.9×10^6 ha (14). In addition, Murray and Best (6) showed that a switchgrass biomass program in the West America could create new habitat for grassland birds.

Few articles have explored biomass reserve strategies for Canada, where a lower population, lower total energy demands, shorter growing season, and an abundance of natural forest resources dictate a different approach. In Canada, a biomass reserve strategy will likely include elements of the area- and volume-based approaches, and might include short-rotation bioenergy plantations as well as forest residues. A strategy with similar elements is defined by Van Belle et al. as a three-tier system involving small, medium, and large-scale logging operations working within softwood and hardwood forests *(11)*.

It is not proposed that biomass reserves be created to meet Canada's total energy, fuel, or material demand. Rather, this strategy is suggested as a means of overcoming barriers to developing bio-based processes and products. Fischer et al. *(8)* identified two such barriers including the low cost of competing fossil energy sources and the high cost of land suitable for high-yield production of biomass. In order to address these barriers in the long term, the biorefining strategy must focus on creating value-added products in addition to energy or fuels. This article postulates that biomass reserves could be used to support the first generation of biorefineries, allowing for continued research and development at the demonstration and commercial scales into these types of products. Using biomass reserves, biomass-to-energy capacity may be developed in advance of energy demand, with the assumption that fossil-based energy will continue to rise in cost. A biomass reserve strategy will not necessarily going to reduce the cost of biomass under current market value, but it will serve to guarantee long-term access to the resource, a necessary step in securing funding and investor support. Development of a product mix with higher value will eventually ensure greater access to biomass resources in the longer term.

Biorefineries

Biorefineries can offer many environmental, economic, and security-related benefits to global society, and particularly to Canada. For instance, energy, fuels, and chemicals made from renewable biomass are characterized by reduced carbon dioxide emissions when compared with petroleum use, and thus can play a role in meeting the challenges of climate change *(15,16)*. Canada can use biorefinery products in order to fulfill national obligations under the Kyoto Protocol, an agreement scheduled to come into force in February 2005 after Russia formally ratified the pact on November 18, 2004. The processing facilities required to convert biomass into value-added products create direct and indirect jobs, provide regional economic development, and can increase farm and forestry incomes, particularly in rural areas *(17,18)*. In some cases, substituting sustainable biomass for fossil resources may serve to increase the security of energy, fuel, and chemical supplies *(19)*. With this technology, a biomass-rich nation such as Canada might be able to reduce its reliance on foreign-owned oil supplies, which are subject to political uncertainty and conflict epitomized by the current conflict in Iraq. The argument for developing renewable substitutes for petroleum products is particularly compelling as current reserves of fossil oil are being consumed at an increasing rate, whereas the discovery of new reserves is in decline *(20)*.

There are several available technological approaches or platforms that may be applied in a biorefinery in order to separate different products from the biomass feedstock. In this paper, the bioconversion platform is considered. This platform differs from others such as pyrolysis or gasification in that it is designed to recover intermediate products from biomass. Instead of thermal degradation, bioconversion uses acid or enzymatic hydrolysis to release complex chemical compounds, such as sugars, which are contained in many sources of biomass. Existing biorefining operations built on the bioconversion platform utilize agricultural crops and food processing residues that have high starch contents and are relatively homogenous in composition *(21)*. Starch is ideal for bioconversion because it comprises a single sugar (glucose), which simplifies process design and complexity *(22)*. Examples of existing operations include the Archer Daniels Midland facility in Decatur, Illinois, which utilizes corn grain and fiber removed during the processing of corn.

The bioconversion platform may also be applied to inedible biological materials, including the structural components of agricultural (herbaceous) and forest (woody) biomass. These structural components are largely composed of lignocellulosic materials, a matrix that blends α-cellulose, hemicellulose, and lignin *(23,24)*. Implementing a biorefining system utilizing lignocellulosic feedstocks is more challenging than processing starch-based materials, as it requires a separation of carbohydrates from aromatic compounds and the subsequent hydrolysis of a range of carbohydrates, including glucose, galactose, mannose, xylose, and arabinose *(22,24)*. This means that lignocellulosic-based bioconversion is more difficult and therefore more expensive than starch-based operations. The bioconversion process has evolved to include several stages that can separate and recover the various structural and chemical components of lignocellulosic biomass, including fiber, carbohydrates, and aromatic compounds from the lignin and extractives. This creates the potential for additional chemical outputs that can serve as coproducts from the biorefinery. Several new facilities, each utilizing variations on the lignocellulosic-based bioconversion platform, are currently in existence or are under development, indicating that this process is approaching commercialization. These facilities include the Etek Etanolteknik pilot facility in Sweden, the Abengoa demonstration plant in Spain, and the Iogen demonstration plant in Canada.

Current Lignocellulosic Biomass Availability in Canada

Initial lignocellulosic biomass reserves should be created using biomass that is easy to access and relatively centralized. The most likely immediate feedstocks for bioconversion operations in Canada are residues from wood product processing facilities, and from agricultural harvesting operations. Wood product residues have the advantage of being collected at the mill site, whereas agricultural residues are readily accessible through existing permanent road networks. In 2003, approx 86×10^6 m^3 of chips and particles were generated in Canadian primary wood product processing, but 78×10^6 m^3 of this material was subsequently used in the manufacture of secondary wood products, such as pulp and paper *(25–29)*. Only 8×10^6 m^3, or approx $4–5 \times 10^6$ mt, could be considered surplus for biorefinery operations, given average wood densities for Canadian tree species *(30)*. Furthermore, this surplus of material has been shown to be rapidly diminishing. A recent report by McCloy *(31)* indicates that new ventures, such as wood pelletization for energy generation, will have claimed approx 1 mt/yr of this material in British Columbia alone by 2005.

Lignocellulosic residues from agricultural harvesting operations include wheat straw and corn stover. Statistics indicate that approx 37 mt of residues from cereal and oilseed crops were produced in Canada in 2003, much of which is left on the field *(27,32)*. Given that a portion of this material must be set aside for soil conservation and nutrient recovery, estimates of surplus lignocellulosic residues for Canada range from 5.3 to 14.1 mt/yr *(26,33–35)*. To use this material, the biorefining industry must be willing to pay for the retrieval and transport of this material.

The data indicate that 9–19 mt of lignocellulosic material might be easily accessible in Canada. By comparison, the US starch-based ethanol industry processes approx 25 mt/yr of corn. It is clear that for the Canadian biorefining industry to compete in the global marketplace, additional sources of lignocellulosic biomass must be considered, and ways in which costs might be reduced must be considered. A means of creating new sources of inexpensive biomass could be innovative management strategies that harvest lignocellulosic material that is inappropriate for existing forestry or agricultural operations. It is theoretized that these strategies may be applied in order to reduce the impact of disturbances, including insects, disease, fire, and drought, on total biomass supply, which could entail savings (in terms of avoided expenditures) to both sectors of the economy.

Insects, Disease, and Fire

Disturbances including insects, disease, and fire can reduce biomass production in both agricultural and forest operations. These disturbances often occur in a linked progression, with insects infesting a field or stand of wood and introducing disease, leading to mortality in standing timber which then dries and becomes highly flammable. In agricultural systems, diseases carried by insect hosts are estimated to cause up to 20% of the world's crop failures *(36)*.

In North America, a number of preventative measures, including insect control and the development of disease-resistant crops, has a limited the impact of insects and disease on crop production. However, insects and fire have been identified as the primary modes of disturbance within North American forest ecosystems *(37)*. In any given forest stand, the combination of species composition, local climate, and management practices will determine the frequency of individual insect or fire outbreaks. In some ecosystems, such as the interior forests of British Columbia, bark beetle outbreaks may be seen as often as every 5 yr, and may persist from 2 to 4 yr *(38)*. In the western United States, western spruce budworm outbreaks occurred every 20–33 yr and lasted approx 11 yr *(39)*. The longest outbreaks coincide with dryer ecosystems and periods of drought *(38)*. In a sedimentary record in Western Canada, peak fire events were found to occur 1–3 times per century, with extreme fires occurring approximately every 200 yr *(40)*.

The impacts of insects, disease, and fire can be significant, but insects tend to prove the most damaging to timber stocks. In 1952, approx 8.6×10^9 board feet of timber was estimated to be lost per year to insects in the United States, along with 4.3×10^9 board feet lost to disease, and 1.2×10^9 board feet lost to fire *(41)*. In four beetle outbreaks in western Oregon and Washington, occurring between 1950 and 1969, 7.4×10^9 board feet of timber were killed *(38)*. In Canada, the worst insect outbreak in modern times is the ongoing mountain-pine-beetle outbreak centered in British Columbia. This disturbance has been underway since 1994, and has claimed 173.5×10^6 m^3 of timber cumulatively, affecting 65.8×10^6 m^3 (4×10^6 ha) in 2003 alone *(25,42)*. By comparison, approx 482,000 ha of timberland has been damaged or destroyed by fire in British Columbia over the same timeframe *(43)*. The cost of the mountain-pine-beetle epidemic to date is approx $\$18 \times 10^9$ in lost revenue from forest products, whereas the cost of fighting forest fires over the same period is approx $\$1.1 \times 10^9$, as well as the loss of potential products from the damaged or destroyed timber resource *(42,43)*.

Climate change may play a role in increasing the frequency and range of insect outbreaks and fire in the forest. The frequency and severity of extreme weather events is likely to increase owing to global warming *(44,45)*. Climate change is also expected to lengthen the forest-fire season parts of Canada owing to longer growing seasons and increased summer temperatures. The link between climate change and insect outbreaks is less certain. The synchrony between host species and insect pest will likely be altered, particularly in the spring and autumn. The predicted temperature rise should generally, but not always, favor insect growth and reduce winter mortality. Recent outbreaks of budworm in the American west may be related to climate change, although they may also be related to changes in forest structure owing to increased silvicultural management *(39)*. Deforested areas and highly stocked, second-growth forests have been shown to be more flammable than primary growth forests *(44)*. The trends for most of these disturbances seem to be increasing, and it is likely that in the future, more forest biomass will be lost to natural disturbances such as fire, storms, and insect outbreaks *(46)*.

Drought

Biomass production is highly dependent on water availability, and even short droughts have the potential to cause significant interruptions in feedstock supply to the biorefinery. Agricultural operations are particularly sensitive to drought, and can be affected by relatively short disturbances that occur over weeks or months in the growing season. Long-term drought may also increase mortality in forests, as well as increasing the risk of other disturbances including insect outbreaks and fire within forest ecosystems. By studying tree rings, freshwater and marine sediments, and ice-core records, the cyclical nature of drought in North America has been well documented *(40,47,48)*. These environmental records show that drought and other climatic phenomena recur on a regular basis, and it is reasonable to expect that they will affect biomass supply in the future.

The frequency of drought varies with location. The central Plains/Prairie region of the North American continent, including the southern reaches of Alberta, Saskatchewan, and Manitoba, are relatively prone to periods of aridity *(47)*. In two independent studies carried out in New Mexico and in the foothill region of Alberta, respectively, tree ring analysis indicates that less than average rainfall occurred about half the time between 1700 and 2000 *(48,49)*. These analyses further reveal that significant, extended drought periods may recur about once every half-century, and that these disturbances may last for over a decade. Some periods of drought have been especially severe; for instance, the Grissino-Mayer data indicate that a catastrophic, 50-yr drought was experienced in early 1700s, and Cook et al. *(50)* reported a period of aridity that extended from 900 to 1300. In other regions of North America, drought cycles occur at a lower frequency. In the interior forests of British Columbia, lake sediment analyses indicate that two significant droughts have occurred since 1700, each lasting for extended periods of 10–20 yr *(40)*. Data show that a new drought cycle has begun in the western regions of Canada and the United States *(40,50)*.

In agricultural systems, even a short drought may tremendously reduce the productivity of crops. During one extended drought event in 1950s, it is estimated that total production of corn in South Dakota was diminished by up to 60%/yr *(47)*. In 2003, severe droughts in Europe reduced wheat production by 10% and corn production by 20% under 2002 figures *(51)*. In a drought year, the agricultural economy of the Canadian

Prairies may lose millions or billions of dollars *(52)*. In 2002 alone, ongoing drought conditions in the Prairies was estimated to have cost the regional economy 2.774×10^9, on approx 23×10^6 ha of land *(53)*.

Tree species tend to be more resilient to drought than herbaceous crops, although droughts can reduce forest productivity, and over longer periods increase the risk of other disturbances. In British Columbia, interior forests experienced exceptionally hot and dry summers between 2001 and 2004 *(54)*. As a result, lake levels in this region were the lowest since records were set in 1920s *(55)*. Elevated temperatures during this dry period increased evaporation from lakes and transpiration from forests and soils, which further reduced the water table. The extreme aridity across the province created forest-fire threats, raised concerns about drinking water quality, and adversely affected fisheries *(56)*.

Although historical drought is characterized by cyclical, repeating events, future patterns may not be so easy to predict. A rise in greenhouse gases within the atmosphere owing to industrial activities will likely exacerbate the severity and duration of drought events. In the Great Plains region of the United States, for instance, models suggest that doubling the concentration of CO_2 in the atmosphere will result in a marked increase in the occurrence of extreme droughts *(57)*. Thus, prudent policy for biomass generation should promote systems that are highly resistant to this phenomenon.

Innovative Management Strategies for Forest Ecosystems

Some management strategies proposed to reduce forest ecosystem vulnerability to insects, disease, and fire disturbances involve biomass removal *(58,59)*. These strategies are volume-based, and involve reducing and controlling stand densities, effectively reducing their vulnerability to disturbance by limiting habitat for insects or fuel loading for fire.

One strategy is to remove biomass from the growing stand via commercial or precommercial thinning activities. Precommercial thinnings are ideal for the biorefining strategy, as this removes stems that are of little value to the forest product industry. High-intensity thinning treatments, leaving relatively few stems per hectare, have been shown to be most effective in reducing the vulnerability of stands to mountain-pine-beetle infestations and fire *(58,60,61)*, and provide a range of ancillary benefits to the mature stand. In a British Columbia study examining paper birch (*Betula papyrifera* Marsh.), high-intensity precommercial thinnings in 5-yr-old stands were shown to reduce mortality in young stands and to significantly increase the size of remaining trees on the site as they matured *(62)*. Even low-intensity thinning treatments may lead to reduced fire risk, and provide maximum stand volumes in mature stands when compared with high-intensity treatments.

One disadvantage to the precommercial thinning strategy is that it entails additional harvest and transportation costs. Precommercial thinnings at any intensity yield relatively little material per hectare, which means that these costs are relatively high. In the study by Simard et al. *(62)*, yields from precommercial thinnings varied from 8.5 to 13.3 m^3/ha (approx 2.7–4.3 t/ha, given average wood specific gravities of approx 0.325). Therefore, if 100,000 ha were thinned in Canada on an yearly basis, approx 0.3–0.4 mt of lignocellulosic biomass could be realized. The relatively low returns on a per-hectare basis means that this option may only be of interest in areas where disturbances may cause additional economic hardship, such as in forest-urban interface zones where residential housing exists. Dellasala *(58)* provided a conceptual framework that might be utilized to prioritize precommercial thinning treatments in the forest-urban interface zone, rather than in wild areas.

A second approach to reducing vulnerability in forest ecosystems is the removal of logging slash. Models indicate that slash reduction reduces the risk of fire and insect outbreaks *(60)*. An advantage to utilizing slash is that the equipment is already present and the material is already harvested, meaning that the only additional costs are transportation to the processing facility. In 2003, approx 186×10^6 m^3 of forest biomass, in the form of industrial roundwood (approx 130×10^6 mt), was harvested in Canada *(28,29)*. Forest-harvesting operations generally leave behind variable amounts of lignocellulosic residue, depending on forest type, stand composition, and site characteristics *(63)*. Residue-generation rates range from 15% to 25% of the total above-ground forest harvest *(10,64)*. Based on the typical residue-generation rates, it can be assumed that Canada generates 19.5–32.5 mmt of lignocellulosic residues on an yearly basis.

Total removal of forest-harvest residues may trigger erosion and nutrient deficiencies, which will increase the long-term cost of harvests and the difficulty associated with forest regeneration *(65–67)*. The amount of forest residue that may be safely recovered for the biorefinery will thus be less than the total biomass available. Few studies are available that describe the impacts of forest-residue recovery on site characteristics, but two studies carried out within the Canadian boreal forest indicate that high levels of removal may be possible, within the boreal setting. One study by Brais et al. *(68)* suggested that slash removal alone has little impact on soil nitrogen, although the authors speculate that long-term use of harvesting techniques which increase slash removal may adversely reduce nitrogen levels. Similarly, decreased slash removal associated with whole-tree harvests was found to have little impact on the regeneration of black spruce (*Picea mariana* [Mill.] BSP) *(69)*. On sites prone to erosion, safe residue recovery may be reduced or curtailed.

These strategies would have little impact on existing biomass supplies to forest operations, but have the potential to reduce forestry losses during periods of high insect and fire risk. The decade-long mountain-pine-beetle disturbance in British Columbia had cost approx $\$18 \times 10^9$ in lost revenue from forest products, and fire management had cost an additional $\$1.1 \times 10^9$ over the same period. This means that, if innovative management techniques had reduced the occurrence of these disturbances by 10%, up to 19×10^7/yr could have been saved in British Columbia alone.

Innovative Management Strategies for Agricultural Systems

In Canada, farming operations have become increasingly concentrated on the most productive soil types over the past 50 yr, with a corresponding decline in use of less productive areas. This has led to a decline in the total amount of farmland in use in Canada from approx 70 $\times$ 10^6 ha in 1951 to 67 $\times$ 10^6 ha in 1991 *(70)*, with the greatest areas of marginal land being freed in Ontario, Quebec, New Brunswick, and Nova Scotia *(71)*. Of the remaining arable farmland, approx 25 $\times$ 10^6 ha is considered to be only marginally viable with severe restrictions on its use, under the Canadian Land Inventory *(70,71)*. The location of a significant proportion of this marginal farmland corresponds to the "dry belt" region of the Palliser triangle—a part of Alberta and Saskatchewan, known to be subject to relatively frequent and extended drought activity *(72)*. If 10% of the most vulnerable farmland in the Prairies, or approx 2.3 $\times$ 10^6 ha, was removed from agricultural service and turned over to energy plantations, it may be estimated that at least 277×10^6 in losses might be avoided during a drought year *(53)*. Innovative management strategies might provide incentives to replace marginal farmland in drought-prone areas with forest cover that is more drought resistant.

On marginal farmland, alternatives to herbaceous crops might be considered. With much longer periods of growth, trees may weather short droughts with less resulting impact on total biomass accumulation. Some tree species are naturally more drought tolerant than others, including species of pine as well as rocky mountain or interior Douglas-fir (*Pseudotsuga menziesii* var. *glauca* [Beissn.] Franco) *(73,74)*. Other species may be modified to be more drought resistant, including loblolly pine (*Pinus taeda* L.) and specific genotypes of hybrid poplar (*Populus spp.*) *(75)*. Simulation models have shown that adaptive forest management strategies that utilize drought-tolerant tree species may mitigate the impact of drought on biomass production *(76)*.

The potential of agroforestry plantations on marginal farmland to contribute biomass to the biorefinery strategy is difficult to estimate, as growth rates will change depending on year-to-year conditions, species considered, and initial soil conditions. Relatively conservative yield estimates for fast-growing hybrid poplar in Canada range from 10 to 12 m^3/ha/yr, given 20-yr rotations. More ambitious models suggest that yields of up to 20 m^3/ha/yr might be possible in the same rotation *(77)*. Given average wood densities for poplar, this translates approximately to yields ranging from 3 to 6.5 mt/ha/yr *(30)*. If the 3 $\times$ 10^6 ha already removed from active farming in eastern Canada were combined with 10% of vulnerable Prairie farmland,

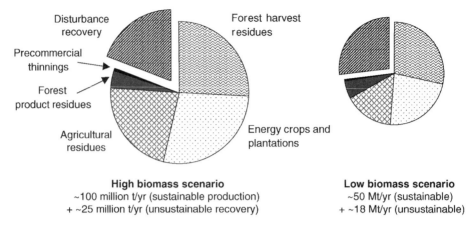

Fig. 1. Two scenarios of lignocellulosic biomass residue availability in Canada.

5.3×10^6 ha could be completely given over to biomass production through energy plantations. This could generate 9–19.5 mmt/yr of feedstock for biorefining operations.

Potential Bioethanol Production

The current levels of availability of six sources of lignocellulosic residues in Canada have been discussed, including sustainable contributions from precommercial thinnings, wood processing residues, agricultural residues, energy crops and plantations, forest-harvest residues, and recovery from natural disturbances. Using the high and low potentials discussed earlier, the contribution of each of these sources to total lignocellulosic biomass availability is shown in Fig. 1.

A calculation of potential bioethanol production from lignocellulosic sources was created based on previous work in the literature. According to Gregg et al. *(78)*, hydrolysis of C_6 sugars can result in yields of 0.5–0.75 g/g cellulose, for softwoods and hardwoods, respectively; fermentation of these sugars can result in ethanol yields of 95%, or 0.48–0.72 g/g cellulose *(78)*. In another study, it was demonstrated that variations in subprocess design have a large impact on total yields. The bioconversion of C_6 sugars through separate hydrolysis and fermentation, and simultaneous saccharification and fermentation yielded 0.61 and 0.86 g/g cellulose, respectively *(79)*. Finally, it has been shown that process ethanol yields from hydrolyzed feedstock containing both C_6 and C_5 sugars (i.e., both cellulose and hemicellulose) can range from 0.32–0.47 g/g for agricultural residues to 0.48–0.50 g/g for hardwood residues *(80,81)*. These yields are based on the fermentation of a hydrolyzed feedstock using recombinant *Zymomonas* yeast, which is capable of hydrolyzing both glucose and xylose effectively. When the chemical characteristics of lignocellulosic residues are considered, it may be estimated that ethanol yields will range from 0.12 to 0.32 L/kg of undried feedstock. In Table 1, two

Table 1
Scenarios of Lignocellulosic Biomass Availability and Potential Ethanol Production

Feedstock availability by scenario (10^6 mt, undried)			Yield scenarios (10^9 L ethanol/yr)			
			Feedstock low (0.12 L/kg EtOH)	Feedstock low (0.32 L/kg EtOH)	Feedstock high (0.12 L/kg EtOH)	Feedstock high (0.32 L/kg EtOH)
Forest residues	Low (mt)	19.5				
	High (mt)	32.5	2.34	6.24	3.90	10.40
Energy plantations	Low (mt)	15.9				
	High (mt)	34.5	1.91	5.09	4.14	11.04
Agricultural residues	Low (mt)	5.3				
	High (mt)	14.1	1.27	3.39	3.38	9.02
Wood product residues	Low (mt)	4				
	High (mt)	5	0.48	1.28	0.60	1.60
Precommercial thinnings	Low (mt)	0.3				
	High (mt)	0.4	0.03	0.09	0.05	0.14
Disturbances	Low (mt)	18.5				
	High (mt)	24.5	2.22	5.92	2.94	7.84

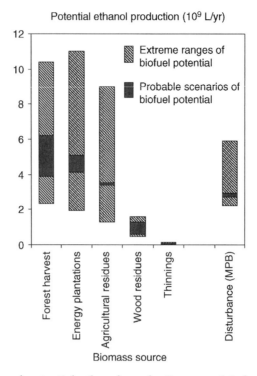

Fig. 2. Scenarios of potential ethanol production associated with lignocellulosic biomass sources.

scenarios of biomass availability are combined with the high and low ethanol yield limits in order to create four scenarios of lignocellulosic-based ethanol availability in Canada.

Figure 2 illustrates the wide variation in potential fuel production from lignocellulosics, ranging from low bioethanol yield and low feedstock availability to high yield and high availability for each category of material. The data indicate that, even at low yields and availability, sustainable biomass reserves in Canada could be significant. By harvesting agricultural residues, $(1.3–9) \times 10^9$ L of bioethanol per year might be provided. Energy plantations on marginal farmland could generate an additional $1.9–11 \times 10^9$ L/yr of bioethanol. Given current levels of residue generation from the wood processing industry, the amount of bioethanol that could be produced on an yearly basis ranges from 480 to 1.6×10^9 L of fuel. Forest residues could be converted to $(2.3–10.4) \times 10^9$ L of bioethanol every year. In addition to sustainable biomass reserves, unsustainable, opportunistic supplies might also be accessed. Disturbances including fire and insect-kill would provide a highly variable amount of wood annually. The ongoing mountain-pine-beetle outbreak in British Columbia and Alberta, for example, could provide $(2.9–7.8) \times 10^9$ L/yr of bioethanol at the anticipated height of the outbreak, which is predicted to occur in 2012.

Canadian Biomass Reserves for Biorefining

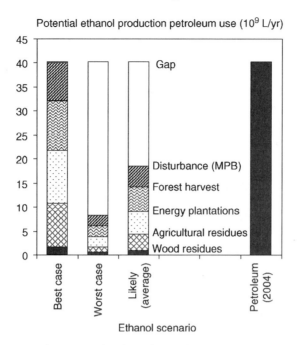

Fig. 3. Scenarios of potential ethanol production compared with Canadian petroleum use.

In Fig. 3, the cumulative potential biofuel production from each scenario is shown and compared with current levels of gasoline consumption. It is clear from this figure that no single biomass option can source Canadian demand for biofuels, but that available biomass sources a significant amount of fuel requirements. At the extreme positive range of these scenarios, almost all fuel consumption in Canada could be substituted with lignocellulosic-based ethanol without impacting agriculture or forestry operations significantly. A biomass reserve might then be an effective strategy for increasing environmental performance and energy security, in addition to providing opportunities for economic growth.

Conclusions and Suggestions for Policy Reform

Canada has an excess of 400×10^6 ha of forested land, approx 93% of which is publicly owned, and 77% of which is controlled by various Provincial Governments (25). Rights to harvest or manage these public resources are allocated by the Provincial Government to private parties in the form of "timber tenures" also referred to as the tenure system (82). A portion of the biomass reserve might access volumes of forest residues, which would require that some revisions to the tenure system must be contemplated. A major challenge is that, although most forests are controlled by Provincial Governments, policy related to the Kyoto Protocol is controlled by the Federal Government. Thus, a national biorefining strategy must seek consensus from the provinces, which is never easy to achieve.

In British Columbia, forest harvests are controlled licenses that regulate either area harvested or volume removed. The area-based tree farm licence and the volume-based forest licence account for over 80% of the timber volume harvested from British Columbia's public lands *(82)*. However, the even flow of wood generated by the tenure system has proven to be insufficient at insulating resource-based communities from boom and bust cycles. From 1979 to 1997, three complete cycles of boom and bust were recorded owing to fluctuations in the international demand for forest products, resulting in temporary and permanent job loss and mill closures *(83)*. A biorefining strategy may be useful in reducing the effect of boom and bust cycles in resource-dependent communities.

One possible reform that would encourage innovative management techniques and create supply for the biorefinery would be to exempt a predetermined volume of biomass used as feedstock for the biorefinery from annual allowable cut quotas. This would support Canada's commitment to the Kyoto Protocol, and would fall within the mandate of Provincial Governments, which is not only to manage the timber supply to generate revenue and jobs, but also to develop policies that address the environmental concerns and socioeconomic values of multiple forest stakeholders *(84)*. Determining the volumes to be exempt would be a challenge; land tenure reform is a hotly contested issue at the time of writing, with advocates spanning all facets of the forestry industry. Supporters of changes to the forest tenure system include royal commissions, existing tenure holders, and smaller timber companies that are out-competed by larger, established firms *(84)*. Change to the existing land tenure systems should include a consideration of a wider range of nontimber uses for the forest. Some have suggested that new tenures should contain incentives for the production of value-added and higher-value products *(84,85)*, an approach that is synergistic with a biorefining strategy.

Agricultural residue availability in Canada is relatively limited, particularly compared with the United States, where large amounts of corn stover and wheat straw may be accessed. Additional biomass production could be achieved if areas of marginal farmland in the Prairie provinces could be set aside and utilized for biomass production. This type of area-based approach is already seen in existing programs that promote plantations on marginal farmland, such as the Saskatchewan Forest Carbon Sequestration Project, which are beginning to address needs beyond shelterbelt or windbreak creation toward the provision of environmental services *(86)*. Perhaps the most significant of these projects is Agriculture and Agri-Food Canada's Prairie Farm Rehabilitation Administration (PFRA). This 60-yr-old project works to ensure sustainable use of the Prairie's soil and water resources, and promotes environmentally responsible use of the land. The Shelterbelt Centre of the PFRA provides seedlings at no cost (save shipping) to eligible farmers, which can be used in afforestation activities on marginal land. As part of Canada's Action Plan 2000 on

Climate Change, the PFRA has launched the Shelterbelt Enhancement Program, which is designed to increase afforestation activities in order to increase greenhouse gas sequestration in the Prairie provinces (87). This program could easily be expanded and linked to biorefinery operations, which could provide farmers with revenue from plantation operations in addition to further contributing to Canada's Kyoto obligations. This type of program would also be useful in parts of the United States where no suitable crop residues are available.

The policy reforms suggested have the potential to be interpreted as a negative force by forestry companies and farmers. However, each has the potential of reducing the vulnerability of ecosystems to disturbance, which reduces costs to the tenure holder or land owner in the long term. As shown in this article, the costs of disturbance can be extremely significant, and avoiding even 1% of these costs would mean tremendous savings for operators. In pursuing this strategy, it is essential that a mechanism should be created that transfers some of the benefit of avoided costs to the companies and individuals who hold stewardship over the land, in order to win public support for the biomass reserve strategy. Potentially, a biomass reserve strategy could also utilize some of these savings as a tool to reduce the cost of biomass to the operators and promote the biorefining sector.

It is clear from the data that no single biomass option can source Canadian demand for biofuels, and that different feedstocks will require significantly different policy approaches. A volume-based policy reform that would create supply for the biorefinery would be to exempt biomass used as feedstock for the biorefinery from annual allowable cut quotas. An area-based approach would identify marginal farmland for energy plantations. Together, these approaches could provide a secure and sustainable biomass source to support first-generation biorefining operations, allowing technological advancement and the development of future value-added products. This approach would contribute to Canada's biofuel requirements in the short term, and support economic growth and environmental sustainability over a long period.

Acknowledgments

We would like to acknowledge the Natural Resources Canada (NRCan) and the Natural Sciences and Engineering Research Council (NSERC) for providing funds that have supported this research and associated projects. We also acknowledge the International Energy Agency (IEA) Bioenergy Program, and particularly the members of Task 39, for financial and intellectual contributions.

References

1. Alcoverro, T., Manzanera, M., and Romano, J. (2001), *Mar. Ecol. Prog. Ser.* **211**, 105–116.
2. Whitehead, S. J. and Digby, J. (1997), *Ann. Appl. Biol.* **131(1)**, 117–131.

3. Guenette, S., Pitcher, T. J., and Walters, C. J. (2000), *B. Mar. Sci.* **66(3),** 831–852.
4. Phillips, V. D., Singh, D., Ra, M., and Ma, K. (1993), *Agr. Syst.* **43(1),** 1–17.
5. Lynd, L. (1996), *Annu. Rev. Energ. Env.* **21**, 403–465.
6. Murray, L. D. and Best, L. B. (2003), *J. Wildlife Manage.* **67(3),** 611–621.
7. Walsh, M. E., Ugarte, D. G. D., Shapouri, H., and Slinsky, S. P. (2003), *Envir. Resour. Econ.* **24(4),** 313–333.
8. Fischer, G., Prieler, S., and van Velthuizen, H. (2005), *Biomass Bioenerg.* **28(2),** 119–132.
9. Kim, S. and Dale, B. E. (2004), *Biomass Bioenerg.* **26(4),** 361–375.
10. Kumar, A., Cameron, J. B., and Flynn, P. C. (2003), *Biomass Bioenerg.* **24(6),** 445–464.
11. Van Belle, J. F., Temmerman, M., and Schenkel, Y. (2003), *Biomass Bioenerg.* **24(4,5),** 401–409.
12. Wyman, C. E. and Goodman, B. J. (1993), In: *Opportunities for Innovation, Biotechnology.* Busche, R. M. (ed.), National Institute for Standards and Technologies, Gaithersburg, pp. 151–190.
13. Christian, D. P., Niemi, G. J., Hanowski, J. M., and Collins, P. (1994), *Biomass Bioenerg.* **6(1,2),** 31–39.
14. McLaughlin, S. B., Ugarte, D. G. D. L., Garten, C. T., et al. (2002), *Environ. Sci. Technol.* **36(1),** 2122–2129.
15. Quirin, M., Gärtner, S. O., Pehnt, M., and Reinhardt, G. A. (2004), *CO_2 Mitigation Through Biofuels in the Transport Sector.* Institute for Energy and Environmental Research, Heidelberg, http://www.ifeu.de.
16. Braune, I. (1998), *Ber. Landwirtsch.* **76(4),** 580–597.
17. Morris, D. (2000), *Carbohydrate Economy Newsletter, Fall 2000 Issue.* Institute for Local Self Reliance, Washington, DC.
18. Evans, M. K. (1997), *Midwestern Governors' Conference*, Lombard, IL.
19. Jolly, L. and Woods, J. (2004), *Int. Sugar J.* **106(1263),** 118.
20. Iledare, O. O. and Pulsipher, A. G. (1999), *J. Petrol. Technol.* **51(11),** 44–48.
21. Pandey, A., Soccol, C. R., Nigam, P., Soccol, V. T., Vandenberge, L. P. S., and Mohan, R. (2000), *Bioresour. Technol.* **74(1),** 81–87.
22. Wyman, C. E. (2003), *Biotechnol. Prog.* **19**, 254–262.
23. Atchison, J. E. (1993), In: *Pulp and Paper Manufacture.* Hamilton, F., Leopold, B., and Kocurek, M. J. (eds.), Secondary Fibers and Non-Wood Pulping, Vol. 3, TAPPI, Atlanta, pp. 4–16.
24. Sjöström, E. (1993), *Wood Chemistry, Fundamentals and Application,* 2nd ed., Academic Press, New York.
25. NRCan (2003), *The State of Canada's Forests 2002–2003.* Natural Resources Canada, Ottawa, ON.
26. Mabee, W. E., Gregg, D. J., and Saddler, J. N. (2003), *Ethanol from Lignocellulosics, Views to Implementation.* IEA, Vancouver, BC.
27. Statistics Canada (2004), *CANSIM II Table 051-0001.* Statistics Canada, Ottawa, ON.
28. CCFM (2004), *Compendium of Canadian Forestry Statistics.* Canadian Council of Forest Ministers, Ottawa, ON.
29. FAO (2005), *FAOStat Forestry Data,* FAO, Rome, Italy, http://www.fao.org/waicent/portal/statistics_en.asp.
30. Panshin, A. J. and de Zeeuw, C. (1980), *Textbook of Wood Technology,* 4th ed., McGraw-Hill, Toronto.
31. McCloy, B. W. (2003), *Estimated Production, Consumption and Surplus Mill Residues in British Columbia—A 2003 Update.* BW McCloy & Associates, Vancouver, BC.
32. Statistics Canada (2003), *Census of Agriculture.* Statistics Canada, Ottawa, ON.
33. Mabee, W. E., Gregg, D. J., and Saddler, J. N. (2005), *Appl. Biochem. Biotechnol.* **121–124**, 765–778.
34. Bowyer, J. L. and Stockmann, V. E. (2001), *For. Prod. J.* **51(1),** 10–21.
35. FAO (2005), *FAOStat Agriculture Data,* FAO, Rome, Italy, http://www.fao.org/waicent/portal/statistics_en.asp.
36. Murry, L. E. (1995), *ACS Sym. Ser.* **605**, 113–123.
37. Howe, E. and Baker, W. L. (2003), *Ann. Assoc. Am. Geogr.* **93(4),** 797–813.

38. Schmitz, R. E. and Gibson, K. E. (1996), *USDA Forest Services Forest Insect & Disease Leaflet 5*, http://www.barkbeetles.org/douglasfir/dfir.html.
39. Swetnam, T. W. and Lynch, A. M. (1993), *Ecol. Monogr.* **63(4)**, 399–424.
40. Hallett, D. J., Mathewes, R. W., and Walker, R. C. (2003), *Holocene* **13(5)**, 751–761.
41. Graham, S. A. and Knight, F. B. (1965), *Principles of Forest Entomology*. 4th ed., McGraw-Hill, Toronto.
42. COFI (2004), *Council of Forest Industries (COFI) Mountain Pine Beetle Task Force*, http://www.mountainpinebeetle.com.
43. BC Ministry of Forests (2003), *Estimates of Total Cost of Fire Fighting*, Ministry of Forests and BC Ministry of Finance, Victoria, BC.
44. Laurance, W. F. and Williamson, G. B. (2001), *Conserv. Biol.* **15(6)**, 1529–1535.
45. Volney, W. J. A. and Fleming, R. A. (2000), *Agr. Ecosyst. Environ.* **82(1–3)**, 283–294.
46. Schelhaas, M. J., Nabuurs, G. J., and Schuck, A. (2003), *Global Change Biol.* **9(11)**, 1620–1633.
47. Waltman, W. J. and Peake, J. S. (2003), In: *Proceedings of Drought Risks and Mitigation on the Agricultural Landscape*. University of Nebraska-Kearney, Kearney, March 7, 2003.
48. Grissino-Mayer, H. D. (1996), In: *Tree Rings, Environment and Humanity, Processes and Relationships*. Dean, J. S., Meko, D. M., and Swetnam, T. W. (eds.), Radiocarbon, University of Arizona, Tucson, pp. 191–204.
49. Watson, E. and Luckman, B. H. (2001), *Holocene* **11**, 203–213.
50. Cook, E. R., Woodhouse, C., Eakin, C. M., Meko, D. M., and Stahle, D. W. (2004), *Science* **306**, 1015–1018.
51. USDA (2003), *World Agricultural Production*, Foreign Agricultural Service Circular Series WAP 09-03, US Department of Agriculture, Washington, September, 2003.
52. Maybank, J., Bonsal, B., Jones, K., et al. (1995), *Atmos. Ocean* **33(2)**, 195–222.
53. CWB (2004), *Canadian Wheat Board (CWB) Report on Drought*, http://www.cwb.ca/en/growing/weather/2002_drought.jsp.
54. Halter, R. (2004), *Worst B.C. Drought in 4 Centuries, Says Expert*, CBC, Vancouver, http://vancouver.cbc.ca/regional/servlet/View?filename=bc_fires20040707.
55. Gooding, D. (2004), *British Columbia said to be Worst Since the Depression*, U.S. Water News Online, June 2004, http://www.uswaternews.com/archives/arcsupply/4drouinxx6.html.
56. Land and Water British Columbia (2004), http://www.lwbc.bc.ca/03water/overview/drought/.
57. Woodhouse, C. A. and Overpeck, J. T. (1998), *B. Am. Meteorol. Soc.* **79(12)**, 2693–2714.
58. Dellasala, D. A., Williams, J. E., Williams, C. D., and Franklin, J. E. (2004), *Conserv. Biol.* **18(4)**, 976–986.
59. Whitehead, R. J., Richardson, G., Begin, E., and DeLong, D. (1999), *Case Study in Adaptive Management, Beetle Proofing Lodgepole Pine in Southeastern British Columbia*, Extension Note EN-039, BC Ministry of Forests, Victoria.
60. Perry, D. A., Hing, H., Youngblood, A., and Oetter, D. R. (2004), *Conserv. Biol.* **18(4)**, 913–926.
61. Johnstone, W. (2002), *Thinning Lodgepole Pine in Southeastern British Columbia: 46-year Results*, Working Paper 63, BC Ministry of Forests, Victoria.
62. Simard, S. W., Blenner-Hassett, T., and Cameron, I. R. (2004), *Forest Ecol. Manag.* **190(2,3)**, 163–178.
63. Nalder, I. A., Wein, R. W., Alexander, M. E., and de Groot, W. J. (1999), *Int. J. Wildland Fire* **9(2)**, 85–99.
64. Matin, A., Collas, P., Blain, D., et al. (2004), *Canada's Greenhouse Gas Inventory, 1990–2002*, Environment Canada Greenhouse Gas Division, Ottawa.
65. Lindstrom, M., Skidmore, E., Gupta, S., and Onstad, C. (1979), *J. Environ. Qual.* **8(4)**, 533–537.
66. Shanahan, J., Smith, D., Stanton, T., and Horn, B. (1999), *Crop residues for livestock feed*, Colorado State University Cooperative Extension, http://www.ext.colostate.edu/pubs/livestk/00551.html.
67. Skog, K. E. and Rosen, H. N. (1997), *Forest Prod. J.* **47(2)**, 63–69.

68. Brais, S., Pare, D., Camire, C., Rochon, P., and Vasseur, C. (2002), *For. Ecol. Manag.* **157(1–3),** 119–130.
69. Waters, I., Kembel, S. W., Gingras, J. F., and Shay, J. M. (2004), *Can. J. Forest Res.* **34(9),** 1938–1945.
70. Parson, H. E. (1999), *Can. J. Regional Sci.* **22(3),** 343–356.
71. Statistics Canada (2000), *Human Activity and the Environment 2000,* Catalog No. 11-509-XPE, Statistics Canada, Ottawa, pp. 128–130.
72. Nemanishen, W. (1998), *Drought in the Palliser Triangle,* Prairie Farm Rehabilitation Administration, http://www.agr.gc.ca/pfra/pub/drprimer.pdf.
73. Moore, J. A., Mika, P. G., and Shaw, T. M. (2000), *Forest Sci.* **46(4),** 531–536.
74. Cienciala, E., Kucera, J., and Lindroth, A. (1999), *Agr. Forest Meteorol.* **98,99,** 547–554.
75. Rahman, M. S., Messina, M. G., and Newton, R. J. (2003), *Forest Ecol. Manag.* **178(3),** 257–270.
76. Lidner, M. (1999), *Forstwiss. Centralbl.* **118(1),** 1–13.
77. McKenney, D. W., Yemshanov, D., Fox, G., and Ramlal, E. (2004), *Forest Policy Econ.* **6(3,4),** 345–358.
78. Gregg, D. J., Boussaid, A., and Saddler, J. N. (1998), *Bioresource Technol.* **63,** 7–12.
79. Wingren, A., Galbe, M., and Zacchi, G. (2003), *Biotechnol. Prog.* **19,** 1109–1117.
80. Lawford, H. G., Rousseau, J. D., and Tolan, J. S. (2001), *Appl. Biochem. Biotechnol.* **91–93,** 133–146.
81. Lawford, H. G., Rousseau, J. D., Mohagheghi, A., and McMillan, J. D. (1999), *Appl. Biochem. Biotechnol.* **77–79,** 191–204.
82. Clogg, J. (2001), *BC Forest Tenure Background Paper,* Kooteney Centre for Forest Alternatives, Kamloops, http://www.kcfa.bc.ca/tenure2.html.
83. Markey, S. P. and Pierce, J. T. (2000), *Forest-Based Communities and Community Stability in British Columbia,* Community Economic Development Centre for Forest-based Communities, Vancouver, http://www.sfu.ca/cscd/forestcomm/fcbackfile/pprworking/comstab.htm.
84. Hadley, M. J. (1998), *Tenure Reform, moving the process forward: A Framework for Tenure Reform in British Columbia,* FORUM, November 2, 1998, http://www.cortex.org/tenrefor.pdf.
85. Drushka, K. (1993), In: *Touch Wood: BC Forests at the Crossroads,* Drushka, K., Nixon, B., and Travers, R. (eds.), Madeira Park, Harbour.
86. Szwaluk, K. (2003), *Tree Planting and Conservation Delivery Organizations, Programs and Projects Across the Prairie Provinces.* Manitoba Forestry Association, Winnipeg.
87. PFRA (2004), Prairie Farm Rehabilitation Administration, Calgary, http://www.agr.gc.ca/pfra.

Availability of Crop Residues as Sustainable Feedstock for Bioethanol Production in North Carolina

ABOLGHASEM SHAHBAZI* AND YEBO LI

Bioenvironmental Engineering Program, Department of Natural Resources and Environmental Design, North Carolina A&T State University, 1601 East Market Street, Greensboro, NC 27411, E-mail: ash@ncat.edu

Abstract

The amount of corn stover and wheat straw that can be sustainably collected in North Carolina was estimated to be 0.64 and 0.16 million dry t/yr, respectively. More than 80% of these crop residues are located in the coastal area. The bioethanol potential from corn stover and wheat straw was estimated to be about 238 million L (63 million gal/yr) in North Carolina. The future location of ethanol plant in North Carolina was estimated based on feedstock demand and collection radius. It is possible to have four ethanol plants with feedstock demand of 400, 450, 500, and 640 dry t/d. The collection radii for these four ethanol plants are 46, 60, 42, and 67 km (28, 37, 26, and 42 miles), respectively. The best location for a bioethanol plant includes four counties (Beaufort, Hyde, Tyrrell, and Washington) with feedstock demand of 500 t/d and collection radius about 26 mile.

Index Entries: Crop residues; ethanol; corn stover; wheat straw.

Introduction

The conversion technology for ethanol production from crop residues has been extensively studied *(1–4)*. The conversion of biomass to ethanol includes the following processes, sometimes in combination:

1. Pretreatment that breaks the long-chain hemicellulose down into five- and six-carbon sugars and makes the cellulosic components of the biomass more accessible to enzymatic attack;
2. Hydrolysis by acids, or by enzymes that break down the long-chain cellulose into six-carbon sugars;
3. Fermentation by microbes, converting the five- and six-carbon sugars to ethanol and other oxygenated chemicals;
4. Distillation and dehydration of the fermentation broth to produce fuel ethanol.

The hydrolysis and fermentation steps can be combined in a process called simultaneous saccharification and fermentation. The solid residue

*Author to whom all correspondence and reprint requests should be addressed.

after fermentation mainly consists of lignin, which can be used as a fuel. Among these processes, a critical step in production of ethanol from crop residue is the conversion of cellulose and hemicellulose to simple sugars usable by fermentation yeast.

Agricultural residue is a low-cost feedstock for near-term bioethanol production. Corn stover and wheat straw are the most plentiful sources of the agricultural residues for bioethanol production. Corn stover contains approx 37.3% cellulose, 20.6% hemicellulose, and 17.5% lignin and wheat straw contains approx 32.6% cellulose, 20.5% hemicellulose, and 16.9% lignin *(5)*. The availability of corn stover as a sustainable feedstock for bioethanol production in the US has been studied by Kadam and McMillan *(6)* on a state by state basis. The amount of corn stover that can be sustainably collected is estimated to be 80–100 million dry t/yr in United States. Perlack and Turhollow *(7)* indicated that corn stover can be collected, stored, and transported for about $43 to $60/dry t using conventional baling equipments for ethanol facilities consuming from 500 to 4000 dry t/d. Transportation, collection, baling, and farmer payments account for over 90% of the total delivered costs *(7)*.

In this article, we have investigated the availability of major crop residues (corn stover and wheat straw) in North Carolina for bioethanol production at the county level. The sustainable amount of residue in each county was calculated based on the residue yield and collection factor. Finally, the counties were clustered for the purposes of identifying potential feedstock supply area for future bioethanol plants.

Methodology

Sustainable Crop Residue

Corn and wheat production data for North Carolina were obtained from the North Carolina Department of Agriculture and Consumer Services *(8)*. Average values of corn and wheat production in North Carolina for 2002–2004 were extracted from agricultural statistics *(8)* and were used in this study. The total production of major crops in millions of bushels in North Carolina for the years of 2002–2004 was as follows: corn (71.70), soybean (41.32), wheat (18.61), barley (0.92), and sorghum (0.64). As these data show, corn, wheat, and soybean are the three major crops in North Carolina. Because soybeans generate a relatively small amount of residue *(9)*, and it rapidly degrades in the field *(10)*, its utility as a feedstock for ethanol production is limited. Therefore, soybean residue was not considered as a potential feedstock for ethanol production.

The amount of crop residue that can be removed sustainably depends on the quantity of residue generated, the portion of the residue that can be removed without damaging soil quality, and site accessibility. The portion of residue that can be removed without damaging soil quality depends on a variety of factors such as: weather (wind and rainfall), crop rotation

Availability of Crop Residues

practices, soil fertility, slope of the land, and tillage practices. We have conducted a survey in the Piedmont and Coastal areas of North Carolina about yields and competitive uses of residue in 2003. Our survey indicates that the average residue yields for corn stover and wheat straw were 0.81 and 0.65 t/ha (2 and 1.6 t/acre), respectively. These survey results are in accord with findings reported by Kadam and McMillan (6) and Perlack and Turhollow (7). Based on these data, grain to stover ratios of 1:1 for corn and 1:0.8 for wheat were used in this study. About 35% of the crop residue can be sustainably collected considering factors such as the condition of the land, accessibility, and competitive uses.

An index of biomass availability was defined as the amount of sustainable crop residue per square kilometer (mile). It is the sustainable amount of crop residue that can be collected per square kilometer (mile). The land area of each county was obtained from the state agricultural report sources (11). Biomass availability is an important index to evaluate the collection cost of biomass for ethanol production.

Bioethanol Production Potential

There is considerable variation in the literature regarding the yield of ethanol from a given substrate. It is affected by the composition of the crop residues and ethanol production technology. The theoretical ethanol yield is about 471 L/dry t (124.4 gal/dry t) using an average corn stover composition and assuming that both hexose and pentose sugars are fermented (12). Estimated ethanol yield of 300 L/dry t (79.3 gal/dry t) were used for corn stover feedstock sources, which is about 64% of theoretical ethanol yield. Ethanol yield of 292 L/dry t (77.1 gal/dry t) has been used by Kadam et al. (6) to estimate the national ethanol potential in the United States from wheat straw. Ethanol yield of 292 L/dry t (77.1 gal/dry t) was used in this study to predict the ethanol potential of wheat straw. The hydrolysis technologies assumed are either dilute acid hydrolysis or enzymatic hydrolysis. The fermentation process should be able to convert hexose and some of the pentose.

Results and Discussion

Biomass Availability

The total sustainable amount of corn stover and wheat straw in North Carolina was estimated to be 638,782 and 159,848 dry t/yr, respectively (Tables 1 and 2). The sustainable amounts of corn stover produced in mountain, piedmont, and coastal area were estimated to be 20,372, 67,783, and 550,626 dry t/yr, respectively (Table 1). About 86.2% of the sustainable corn stover was determined to be available from the coastal area (Fig. 1). The sustainable amount of wheat straw produced each year in mountain, piedmont, and coastal area were estimated to be 2649, 42,362, and 114,838 dry t/yr, respectively (Table 2). About 71.8 and 26.5% of the wheat straw was

Table 1
Sustainable Amount of Corn Stover and Ethanol Potential
in North Carolina by Region

Region	Plant area ha (acres)	Grain production t/yr (bushel/yr)	Residue production dry t/yr	Sustainable residue amount dry t/yr	Ethanol potential l/yr (gal/yr)
Mountain	8195 (20,233)	58,206 (2,286,667)	58,206	20,372	6,111,636 (1,614,524)
Piedmont	34,061 (84,100)	193,667 (7,608,333)	193,667	67,783	20,335,000 (5,371,940)
Coastal	241,245 (595,667)	1,573,218 (61,805,000)	1,573,218	550,626	165,187,909 (43,638,037)
State total	283,500 (700,000)	1,825,091 (71,700,000)	1,825,091	638,782	191,634,545 (50,624,500)

Table 2
Sustainable Amount of Wheat Straw and Ethanol Potential
in North Carolina by Region

Region	Plant area ha (acres)	Grain production t/yr (bushel/yr)	Residue production (dry t/yr)	Sustainable residue amount (dry t/yr)	Ethanol potential l/yr (gal/yr)
Mountain	3510 (8667)	8409 (308,333)	7568	2649	773,468 (204,329)
Piedmont	51,570 (127,333)	134,482 (4,931,000)	121,034	42,362	12,369,638 (3,267,713)
Coastal	139,320 (344,000)	364,564 (13,367,333)	328,107	114,838	33,532,563 (8,858,368)
State Total	194,400 (480,000)	507,455 (18,606,667)	456,709	159,848	46,675,669 (12,330,410)

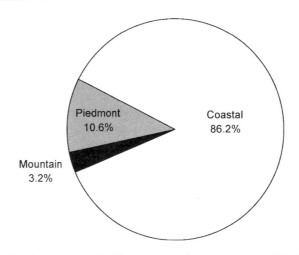

Fig. 1. Distribution of sustainable amount of corn stover in North Carolina.

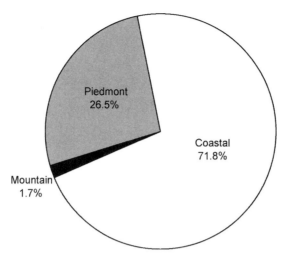

Fig. 2. Distribution of sustainable amount of wheat straw in North Carolina.

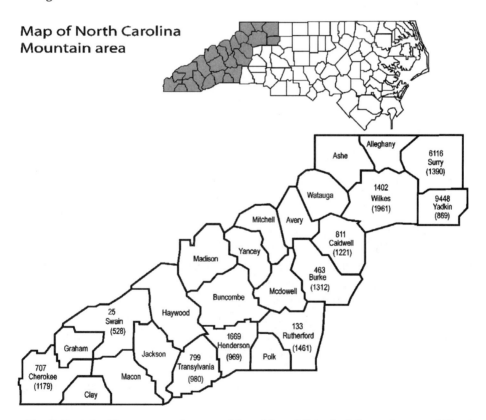

Fig. 3. Sustainable amount of crop residues (dry t/d) in the Mountain area of North Carolina (values in the parenthesis show the land area in km^2).

located in the coastal and piedmont areas (Fig. 2). The sustainable crop residue produced in the Mountain area was negligible.

The amounts of sustainable crop residues (corn stover and wheat straw) for each county of North Carolina are shown in Figs. 3–5. The data presented

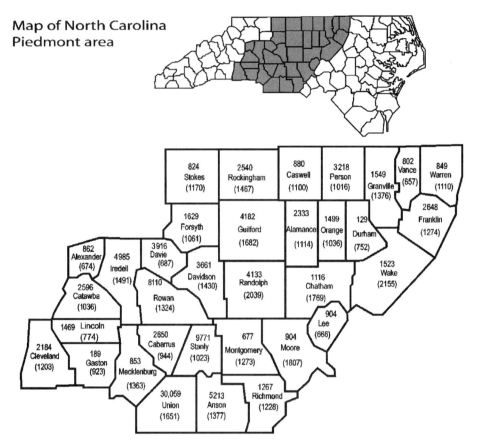

Fig. 4. Sustainable amount of crop residues (dry t/d) in the Piedmont area of North Carolina (values in the parenthesis show the land area in km^2).

in the parentheses is the land area of the county in square kilometer. The biomass availability of each county was distinguished with different color based on the sustainable amount of crop residue per square kilometer in Fig. 6. Pasquotank has the highest biomass availability 52.7 dry t/km^2 (136.4 dry t/square mile). The biomass availability of four counties (Tyrrell, Perquimans, Washington, and Camden) falls between 30 and 40 dry t/km^2 (80–100 dry t/square mile). The biomass availability of five counties (Lenoir, Wayne, Currituck, Duplin, and Beaufort) falls between 20 and 30 dry t/km^2 (50–80 dry t/square mile). The biomass availability of nine counties (Bladen, Columbus, Sampson, Pamlico, Carteret, Greene, Union, Robeson, and Hyde) falls between 10 and 20 dry t/km^2 (25–50 dry t/square mile).

Bioethanol Production Potential

The bioethanol potential in North Carolina was estimated to be about 238 million L (63 million gal)/yr. The bioethanol potential in the coastal area was estimated to be about 200 million L (53 million gal)/yr, which would account for about 83% of the total ethanol potential in North

Availability of Crop Residues

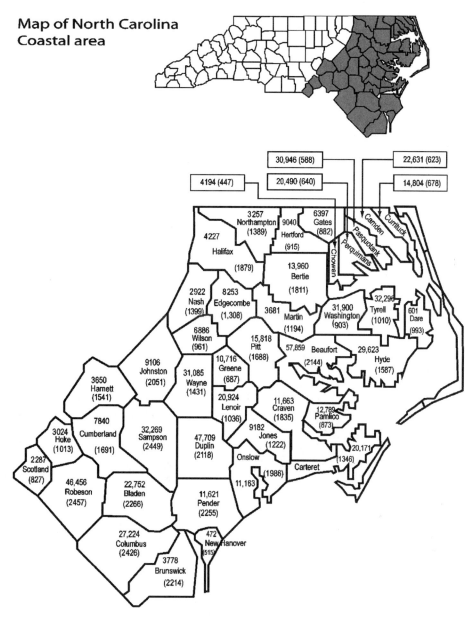

Fig. 5. Sustainable amount (dry t/d) of crop residues in the Coastal area of North Carolina (values in the parenthesis show the land area, km^2).

Carolina. The amounts of ethanol potential for each county in North Carolina are shown in Figs. 7–9. The data presented in the parentheses is the land area of the county in square kilometer.

The delivered cost of biomass is the most important factor for the bioethanol producer. Nearly one-third of the biomass ethanol cost can be attributed to the cost of feedstock *(7)*. The total feedstock cost includes collection, storage, transportation, and fertilizer replacement

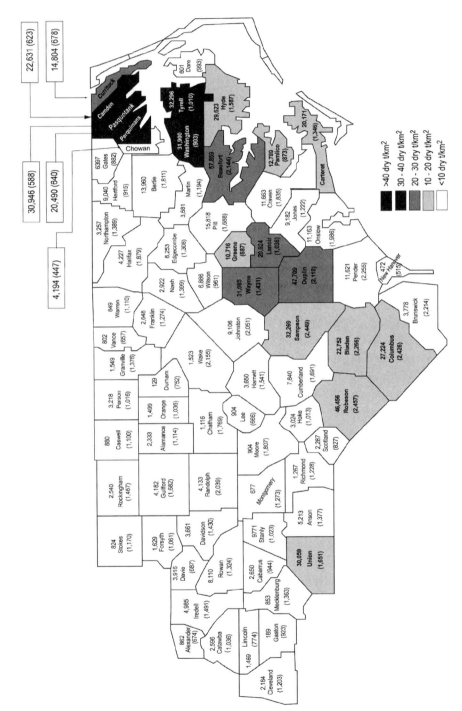

Fig. 6. Biomass availability in the Piedmont and Coastal area of North Carolina.

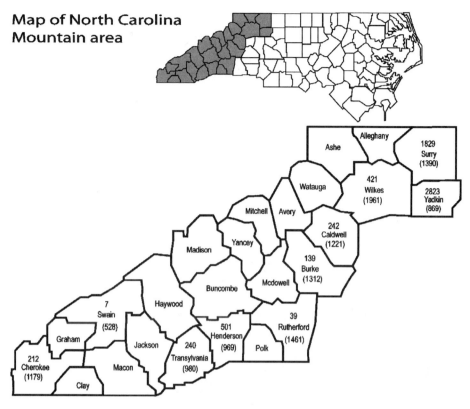

Fig. 7. Ethanol potential (kL) from corn stover and wheat straw in the Mountain area of North Carolina (values in the parenthesis show the land area in km²).

costs, as well as any payment to the landowner. The cost of delivery of feedstock from the farm to an ethanol plant is mainly a function of the transportation distance. Economies of scale are relatively significant for bioethanol conversion facilities. Generally, capital costs for ethanol plants increase by 60–70% for each doubling of output capacity (7). As plants increase in capacity, the cost saving from economies of scale can be offset somewhat by increased transportation costs associated with hauling feedstock to greater distances. Consequently, the ideal location of an ethanol plant involves striking a balance between larger handling capacities and higher feedstock costs.

Selection of Potential Ethanol Plant Sites

Based on considerations of economies of scale and biomass availability, we studied the optimal locations of potential ethanol plants in North Carolina. Feedstock transportation cost was assumed to have a fixed component of $5.50/dry t and a variable component of $0.88/mile ($0.54/km) (13).

Based on the amounts of crop residue and land area of each county, counties with high biomass availability could be clustered to provide four

Fig. 8. Ethanol potential (kL) from corn stover and wheat straw in the Piedmont area of North Carolina (values in the parenthesis show the land area in km^2).

ethanol plants with feedstock demands of 400, 450, 500, and 640 dry t/d, respectively (Fig. 10).

Cluster 1 consists of eight counties including: Bertie, Chowan, Gates, Hertford, Currituck, Camden, Pasquotank, and Perquimans (Fig. 10). These eight counties can sustainably produce about 400 dry t of residues/d, which can supply an ethanol plant of about 36.6 million L (9.7 million gal) ethanol/yr. The equivalent collection radius for this ethanol plant was estimated to be about 46 km (28 mile). The equivalent collection radius was calculated based on the total area of the counties in the cluster.

Cluster 2 consists of four counties including: Tyrell, Washington, Beaufort, and Hyde (Fig. 10). These four counties can sustainably produce about 500 dry t of residues/d, which can supply an ethanol plant of about 45.3 million L (12.0 million gal) ethanol/yr. The equivalent collection radius for this cluster was estimated to be about 42 km (26 miles).

Cluster 3 consists of 10 counties including: Craven, Pitt, Onslow, Jones, Greene, Lenoir, Pamlico, Carteret, Wayne, and Duplin (Fig. 10). These 10 counties can sustainably produce about 640 dry t of residues/d, which

Availability of Crop Residues

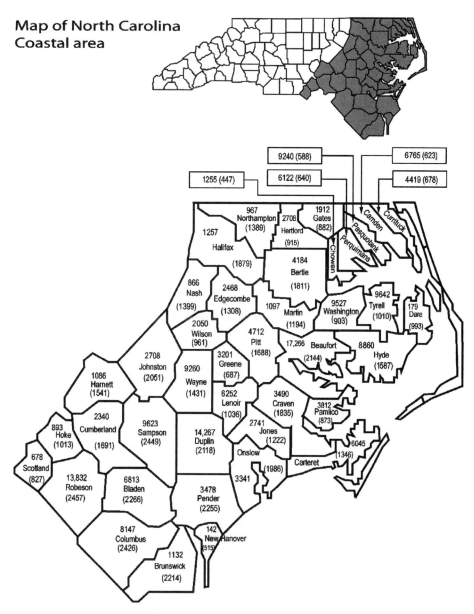

Fig. 9. Ethanol potential (kL) from corn stover and wheat straw in the Coastal area of North Carolina (values in the parenthesis show the land area in km²).

can supply an ethanol plant of about 57.1 million L (15.1 million gal) ethanol/yr. The equivalent collection radius for this cluster was estimated to be about 67 km (42 mile).

Cluster 4 consists of five counties including: Bladen, Columbus, Cumberland, Robeson, and Sampson (Fig. 10). These five counties can sustainably produce about 450 dry t of residues/d, which can supply an ethanol plant of about 40.8 million L (10.8 million gal) ethanol/yr. The

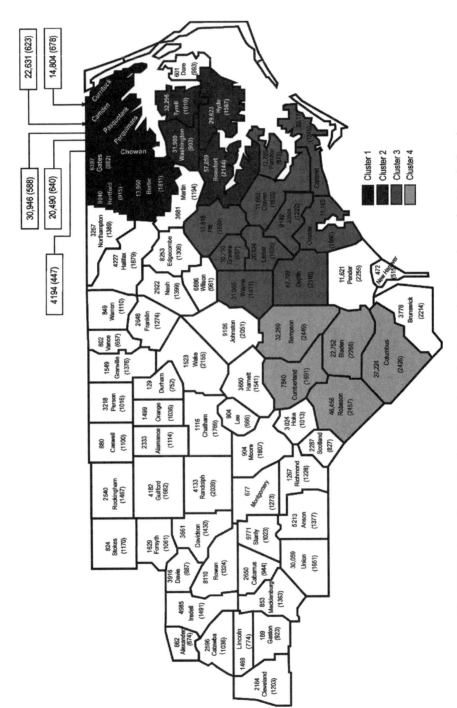

Fig. 10. Clusters of counties for ethanol production in the Coastal area for North Carolina.

Table 3
Estimated Collection Radius, Sustainable Feedstock Amount, and Ethanol Potential for the Proposed Clusters of Counties

		Cluster 1	Cluster 2	Cluster 3	Cluster 4
Equivalent radius	km	45.8	42.4	67.3	59.9
	mile	28.4	26.3	41.8	37.2
Crop residue amount	dry t/d	408.2	505.6	637.4	455.1
Ethanol potential	million L/yr	36.6	45.3	57.1	40.8
	million gal/yr	9.7	12.0	15.1	10.8

equivalent collection radius for this cluster was estimated to be about 60 km (37 miles).

The summary of the estimated collection radius, feedstock amount, and ethanol potential of these proposed clusters of counties are shown in Table 3. The best cluster for ethanol production is cluster 2, which has a minimum collection radius of 26 miles and sustainable feedstock demand about 500 dry t/d.

These clusters of counties are just selected examples of how to combine the counties to provide feedstock for an ethanol plant within a minimum collection radius. The location of any future ethanol plant would be determined using the biomass availability of each county.

Conclusion

1. More than 80% of the crop residues in North Carolina are located in the coastal area. Crop residues of corn stover and wheat straw were considered in this study.
2. Corn stover and wheat straw in the coastal areas of North Carolina can be utilized to produce about 200 million L (53 million gal) of ethanol/yr.
3. The best location for bioethanol production in North Carolina is in the coastal area. A cluster of four counties can provide 500 dry t/d of crop residues for a bioethanol plant of about 45.3 million L (12.0 million gal)/yr. The equivalent collection radius of this cluster was estimated to be 42 km (26 mile).
4. It is possible to have four ethanol plants in North Carolina with feedstock demands of 400, 450, 500, and 640 dry t/d, respectively. The equivalent collection radii of these four plants were estimated to be 46, 60, 42, and 67 km (28, 37, 26, and 37 miles).

Acknowledgment

We gratefully acknowledge the financial support from State Energy Office of North Carolina.

References

1. Li, Y., Ruan, R. R., Chen, P. L., Liu, Z., and Pan, X. (2004), *Trans. ASAE* **47(3),** 821–825.
2. Shahbazi, A., Li, Y., and Mims, M. R. (2005), *Appl. Biochem. Biotechnol.* **121–124,** 973–988.
3. Kim, T. H., Kim, J. S., Sunwoo, S., and Lee, Y. Y. (2003), *Bioresour. Technol.* **90(1),** 39–47.
4. William, E. K. and Holtzapple, M. T. (2000), *Biomass Bioener.* **18,** 189–199.
5. NREL website. http://www.nrel.gov. Accessed on 02/02/2005.
6. Kadam, K. L. and McMillan, J. D. (2003), *Bioresour. Technol.* **88,** 17–25.
7. Perlack, R. D. and Turhollow, A. F. (2003), *Energy* **28,** 1395–1403.
8. North Carolina Department of Agriculture and Consumer Services (2005), http://www.ncagr.com/stats/cnty_est/cnty_est.htm. Accessed on 03/09/2005.
9. Arkansas Agriculture. http://www.aragriculture.org/agengineering/harvesting/soybeans/default.asp. Accessed on 04/09/2005.
10. Walsha, M. E., Perlacka, R. L., Ugarteb, D. T., et al. http://bioenergy.ornl.gov/resourcedata/. Accessed on 03/09/2005.
11. State Library of North Carolina, North Carolina County Development. http://statelibrary.dcr.state.nc.us/iss/gr/counties.htm. Accessed on 03/09/2005.
12. NREL website, Theoretical Ethanol Yield Calculator, http://www.eere.energy.gov/biomass/ethanol_yield_calculator.html. Accessed on 09/01/2005.
13. Kerstetter, J. D. and Lyons, J. K. (2001), http://www.newuses.org/pdf/WSUCEEP 2001084.pdf. Accessed on 01/02/2005.

Updates on Softwood-to-Ethanol Process Development

WARREN E. MABEE,*,[1] DAVID J. GREGG,[1] CLAUDIO ARATO,[2]
ALEX BERLIN,[1] RENATA BURA,[1] NEIL GILKES,[1]
OLGA MIROCHNIK,[1] XUEJUN PAN,[1] E. KENDALL PYE,[2]
AND JOHN N. SADDLER[1]

[1]*Department of Wood Science, University of British Columbia,
2424 Main Mall, Vancouver, British Columbia, Canada, V6T 1Z4,
E-mail: warren.mabee@ubc.ca; and* [2]*Lignol Innovations Corp.,
3650 Wesbrook Mall, Vancouver, British Columbia, Canada, V6S 2L2*

Abstract

Softwoods are generally considered to be one of the most difficult lignocellulosic feedstocks to hydrolyze to sugars for fermentation, primarily owing to the nature and amount of lignin. If the inhibitory effect of lignin can be significantly reduced, softwoods may become a more useful feedstock for the bioconversion processes. Moreover, strategies developed to reduce problems with softwood lignin may also provide a means to enhance the processing of other lignocellulosic substrates. The Forest Products Biotechnology Group at the University of British Columbia has been developing softwood-to-ethanol processes with SO_2-catalyzed steam explosion and ethanol organosolv pretreatments. Lignin from the steam explosion process has relatively low reactivity and, consequently, low product value, compared with the high-value coproduct that can be obtained through organosolv. The technical and economic challenges of both processes are presented, together with suggestions for future process development.

Index Entries: SO_2 steam explosion; softwoods; ethanol; lignin; organosolv.

Introduction

Lignocellulosic biomass comprises the bulk of agricultural and forest residues potentially available for bioconversion processes. In Canada, for example, it has been estimated that the amount of lignocellulosic biomass in these residue streams totals approx 9.4×10^6 mt/yr *(1–9)*. Almost half of these residues (4.1×10^6 mt) are estimated to be generated from forest harvest and wood processing operations, of which nearly 40% (1.6×10^6 mt) are generated in British Columbia. The majority of Canadian forests is softwood (67%) or mixed wood (18%); therefore, softwood biomass from the

*Author to whom all correspondence are reprint requests should be addressed.

forest products industry is the most likely feedstock in these locations (6). The softwood feedstock chosen for this study was Douglas-Fir (*Pseudotsuga menziesii*).

All lignocellulosics, including herbaceous biomass from agricultural plants, and woody biomass from hardwood and softwood tree species, comprise a combination of cellulose, hemicellulose, and lignin. Softwoods, including species of pine, spruce, hemlock, and fir, have a unique chemical composition that differs from agricultural residues or hardwoods. These differences create particular challenges and potential opportunities for bioconversion. There is more hemicellulose in softwoods, with a lower xylose content, and higher mannose content, relative to hardwood species. Softwoods too are characterized by having only two principal phenylpropane units (coumaryl and guaiacyl) that form the basic building blocks of lignin, whereas hardwoods and herbaceous plants have additional syringyl units (10). Interestingly, this simplification of lignin chemistry increases the difficulty of delignification, owing to the enhanced stability of the lignin in condensed form when exposed to acidic conditions (11).

This unique chemistry makes softwood lignocellulosic material an extremely challenging material for bioconversion. This article is focused on pretreatment technologies to overcome problems associated with delignification of Douglas-Fir in an attempt to make the entire softwood-to-ethanol process more commercially viable and feedstock-robust. Ultimately, the results of these investigations may also be applicable to the processing of agricultural residues or hardwoods and the design of bioconversion process suited to a range of substrates. For this reason, corn fiber is included as a reference substrate in the analysis.

The bioconversion of softwood to ethanol can be divided into four individual process elements: pretreatment, hydrolysis to sugars, fermentation of sugars to ethanol, and coproduct recovery. The pretreatment stage promotes the physical disruption of the lignocellulosic matrix in order to facilitate acid- or enzyme-catalyzed hydrolysis. Pretreatments can have significant implications on the configuration and efficiency of the rest of the process and, ultimately, also the economics. This study reviews the effect of changing a feedstock (Douglas-Fir and corn fiber) on the process design of a steam-explosion process and using a different pretreatment (ethanol organosolv) with the same feedstock (Douglas-Fir).

Steam-Explosion Pretreatment

Steam-explosion technology has been investigated over the course of the past 100 yr as a possible alternative to existing mechanical or chemical pulping techniques (12). In this process, high-pressure, high-temperature steam is introduced into a sealed chamber containing woody lignocellulosic material in the form of chips (Fig. 1) or agricultural residues (Fig. 2). After 1–5 min, the pressure is released, causing the steam to expand within

Update on Softwood-to-Ethanol Process Development

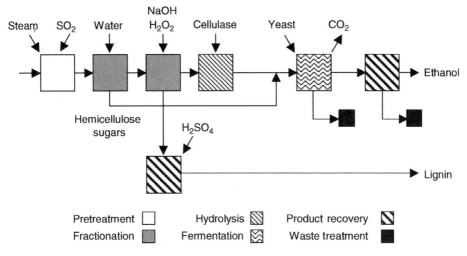

Fig. 1. Douglas-Fir steam explosion process flow diagram.

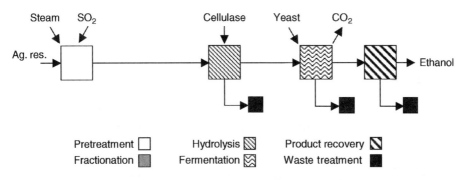

Fig. 2. Corn fiber steam explosion process flow diagram.

the lignocellulosic matrix, separating individual fibers with minimal loss of material. After cooling, the material may be further processed to ethanol as shown in the figures. This process has been shown to work fairly well with agricultural residues, and promising results have been shown with hardwood chips *(13–17)*. We have found steam explosion to be less effective for softwoods, perhaps owing to the impact of lignin condensation, in which dissolved lignin recombines into macromolecules and becomes attached to the surface of cellulose fibers. We have also found it necessary to add a catalyst to aid hemicellulose hydrolysis and removal *(18,19)*.

From our previous work we have determined the optimum conditions, material balances, and process design for SO_2-catalyzed steam explosion of Douglas-Fir *(13,20)* and corn fiber *(21)*. Owing primarily to lignin (amount and type), the processes are very different (Figs. 1 and 2). The Douglas-Fir process requires delignification before enzymatic hydrolysis whereas the corn fiber does not. This study will discuss primarily the economic implications of these differences.

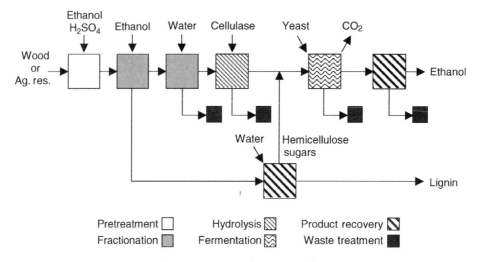

Fig. 3. Douglas-Fir organosolv process flow diagram.

Organosolv Pretreatment

The delignification of wood in a nonaqueous polar solvent is generally referred to as organosolv pulping and was originally developed for pulp and paper applications. Variations of this method have also been used for preparation of lignin in the laboratory (10). The organosolv process has demonstrated certain advantages over traditional chemical pulping options, including the ability to operate at smaller scales than Kraft mills, owing to simplified requirements for recovery of pollutants. Pulp produced using this method tends to be of high quality, and has proven to be competitive with conventional pulps. Important advantages of organosolv processes are the ability to isolate and recover lignin in a relatively unadulterated form and good recovery of hemicellulosic sugars (22).

The ethanol organosolv process (Fig. 3), and related processes using organic solvents, involves chemical breakdown of lignin and solubilization of lignin fragments. Ethanol is added to lignocellulosic material in the form of chips in a reactor vessel, which is then heated to process conditions and agitated to ensure good mixing. The reactor is then cooled and the fibrous lignocellulosic material is recovered for further processing to ethanol. Ultimately, because the product and solvent are the same (ethanol), only one chemical stream needs to be recovered, which simplifies the process design by reducing the number of steps required. Depolymerization of the lignin macromolecule occurs primarily through cleavage of α-aryl ether, β-aryl ether, and β-O-4 linkages, the latter of which influence up to 90% of delignification of softwoods (23,24). It is not yet clear what the relation is between cleavage mechanism and the subsequent differences between pretreatments. Lignin condensation, a major problem in steam-explosion pulping, occurs relatively slowly owing in part to the countering effect of organic solvents that retain the lignin components in solution and slow recombination of macromolecules (25).

Interest in the organosolv process from the pulp industry has been variable. In late 1980s, Repap Enterprises announced a CDN$ 45,000,000 investment in an Alcell demonstration pulp mill in New Brunswick, Canada, using hardwood chips and an aqueous ethanol liquor to produce 33 t of pulp per day (26,27). It was anticipated that this plant would eventually expand into a full-scale operation (28,29). Repap had initiated the construction of a 450-mtpd commercial mill in Atholville, New Brunswick. However, declining pulp prices and a glut in production capacity brought an end to this enterprise (30). More recently, the Alcell pilot facility was bought by Lignol Innovations Corp., a company dedicated to producing high-value lignin products, xylose coproducts, and bioethanol. The pilot plant has been relocated to Vancouver, British Columbia, where it forms the basis of Lignol's research and development facility. Figure 3 shows the Lignol process diagram for Douglas-Fir. It is difficult to generate a full material balance from the pilot plant owing to a lack of sufficient sampling ports. For this study the material balance was derived from the work of Mirochnik (31) using lab-scale equipment described further in the following section.

Methods

Steam Explosion

In this study, SO_2 was chosen as the catalyst for steam explosion in part because it is less corrosive than other options, such as H_2SO_4 (32). Samples of 100 g (dry) Douglas-Fir chips were impregnated overnight with anhydrous SO_2 (4.5% w/w). These samples were then steam-exploded at 195°C with a retention time of 4.5 min. These pretreatment conditions were based on past work that optimized the recovery of carbohydrate (hemicellulose and cellulose) in a hydrolyzable and fermentable form (13,33). In our past work we have shown that without a further lignin removal posttreatment the cellulose hydrolysis yields are very low, i.e., 20–30% (13,20). A variety of these posttreatments have been tried by a number of researchers, although for the purposes of this article a posttreatment of hot alkaline peroxide was used. The conditions for the posttreatment were 1% H_2O_2 on a w/v basis to the steam-exploded biomass, a reaction temperature of 80°C, a pH of 11.5 (adjusted through additions of NaOH), and a retention time of 45 min (31).

The pretreatment for corn fiber was based on more recent work (21) with the same criteria and methods used for Douglas-Fir to produce the optimized conditions (190°C, 5 min, and 3% SO_2 w/w). Posttreatment was shown to be unnecessary with corn fiber. There was no attempt to assess the effect of the processing conditions for Douglas-Fir or corn fiber on the properties of lignin to produce high-value coproducts.

Organosolv Pretreatment

The pretreatment was done at 7 : 1 liquor-to-wood ratio by weight, based on oven-dry wood weight. The concentration was chosen to facilitate good coverage of Douglas-Fir wood chips by liquor. The pretreatment liquor was prepared to be 50% ethanol concentration and the H_2SO_4 was added to adjust pH to 2.4 at room temperature (20°C). The Douglas-Fir wood chips were placed in the reactor, covered with pretreatment liquor, and thoroughly mixed to ensure even liquor distribution. The reactor was sealed and placed in the heating element, and the contents of the reactor were allowed to reach the reaction temperature. The average time to temperature was approx 1 h. Temperature increases following a logarithmic curve with large increases in the early portion of heating, followed by lower increases as the desired temperature was approached. For organosolv pretreatment, experimental conditions were not predetermined and thus a partial factorial design was applied to determine a combination of temperatures (181–202°C) and reaction times (15–90 min). Based on the yield of sugars from the washed pretreated solids, optimum enzymatic hydrolysis conditions were determined to be 188°C for 15 min.

On completion of each reaction, the reactor was rapidly cooled in a bucket of ice and water. Once the temperature was below 80°C, the reactor's outlet valve was fully opened to accelerate depressurizing of the chamber. The lid of the reactor was removed when pressure in the reactor fell to zero and the strong black liquor was decanted from the pretreated substrate. The pretreated substrate was then thoroughly washed with 2 L of 70% ethanol, and weak black liquor was decanted from the substrate. Finally, the pretreated substrate was washed in water, in order to remove residual water-soluble compounds and ethanol. Warm distilled water was added at a volume of 20 times the dry weight of the substrate and the substrate was filtered through a Buchner funnel lined with FisherBrand coarse filter paper. The washed substrate was then stored in an airtight plastic bag in a 4°C refrigerator. The washed pretreated solids were sampled in order to determine moisture content, Klason lignin, acid-soluble lignin, and sugar (carbohydrate) contents.

Enzymatic Hydrolysis

Enzymatic hydrolysis on washed pretreated solids was carried out at 2% consistency in 50 mM acetate buffer solution (pH 4.8) in 100 mL septa vials. Two antibiotics were added to each sample in order prevent bacterial growth, 40 µg/mL of tetracycline and 30 µg/mL of cyclohexamide (Sigma Aldrich, USA). Enzymes were added based on the predetermined sugar content of the samples (IU per cellulose). Two enzyme complexes were added: celluclast (Novozymes, North America [NA]), a commercial cellulase mixture with activity of 43 filter paper units (FPU)/mL, and

Table 1
Feedstock Composition

	Douglas-Fir	Corn fiber
Moisture content	0.5	0.6
Cellulose	0.43	0.46
Hemicellulose	0.23	0.36
Galactose	0.02	0.02
Mannose	0.13	0.07
Xylose	0.03	0.17
Arabinose	0.01	0.10
Lignin	0.28	0.08
Total	0.94	0.9

Novozyme188 (Novozymes, NA), a β-glucosidase solution with activity of 346 CBU/mL. The enzymes were loaded at 7, 10, and 14 FPU/g of cellulose and at a FPU-CBU ratio of 2:1.

Enzymatic hydrolysis reactions were conducted at 45°C and mixing speed of 150 rpm for 48 h. Samples volumes of 0.5 mL were withdrawn at 1, 2, 24, and 48 h. The samples were boiled for 5 min and then centrifuged at 16,500g for 5 min using a microcentrifuge. The supernatant was then removed and stored in the freezer for further sugar analysis.

Measurement of Process Metrics

Samples of Douglas-Fir and corn fiber material were collected before and after pretreatment and characterized using standard methods (Table 1). Sugar concentrations were determined using a Dionex high-performance liquid chromatography system fitted with an anion exchange column (Dionex CarboPac™ PA1) and an ED40 electrochemical detector. Deionized water was used as an eluent at a flow rate of 1 mL/min; 1 M NaOH was used to equilibrate the column after elution of sugars. To optimize baseline stability and detector sensitivity, 0.2 M NaOH was added postcolumn. Klason lignin content was determined according to the TAPPI standard method T-222. The hydrolysate from Klason lignin determination was collected and analyzed for sugars and acid-soluble lignin. This sugar was determined as described earlier, except that sugar standards were autoclaved at 120°C for 1 h to compensate for changes during hydrolysis. Acid-soluble lignin was determined in the hydrolysate according to TAPPI method UM-250.

After pretreatment, cellulose factions were subjected to enzymatic hydrolysis using a mixture of cellulase (Celluclast) and β-glucosidase (Novozym 188) in a ratio of 1:2 (FPU:CBU). Batch hydrolysis experiments were conducted in duplicate at 2% consistency of glucose in 50 mM acetate buffer, pH 4.8, with 4-mg tetracycline per 100 mL buffer as antibiotic. Hydrolysis was performed at 45°C on a rotary shaker at 150 rpm with periodic sampling for sugar analysis. Fermentation was carried out using a spent sulfite liquor-adapted strain of *Saccharomyces cerevisiae*. The yeast was maintained at 4°C on YPD medium (2% glucose, 1% yeast extract, 2% peptone, and 1.8% agar). For each batch of fermentation, the yeast was pregrown in 50 mL of YP medium (1% glucose, 1% yeast extract, and 1% peptone) at 30°C, 150 rpm shaker for 24 h, transferred into 200 mL of fresh medium, and cultivated for a further 24 h. The culture was then centrifuged at 5000 rpm for 10 min. Cells were collected and washed twice with water, and adjusted to a density of 125 g/L for fermentation; fermentation took place in a 1-L vessel. Ethanol production following fermentation was determined by gas chromatography using a Hewlett Packard 5890 gas-chromatography system, as previously described *(33)*. Ethanol yield is expressed as percentage theoretical maximum based on initial sugar content in the washed pretreated solids, based on the assumption that 0.51 g ethanol may be produced per gram of glucose.

Technoeconomic Assessment and Monitoring Model

This study focused on evaluating the effectiveness of the two pretreatment methods described in Methods section in terms of economic performance, within the context of the entire wood-to-ethanol bioconversion process. In order to do so, a model of the bioconversion process was applied. Within the framework of existing lignocellulose-to-ethanol models, our group has developed the technoeconomic assessment and monitoring (TEAM) model, which has appeared in the past with the acronym "STEAM" *(34)*. The TEAM model is built on a Microsoft Excel platform and uses a flow sheet simulator structure to describe the subprocesses in the bioconversion of lignocellulosics to ethanol. By altering the model parameters associated with each subprocess, different scenarios can be assessed for their economic feasibility. The model is essentially a series of individual modules that represent each of the various subprocesses and the calculations associated with each of these elements. Each unit operation or piece of equipment within a subprocess has associated properties, including calculation routines. For example, details for the steam reactor contained within the steam-explosion subprocess can be accessed through the model, and the parameters of temperature, pressure, time, pH, and sample size can be controlled.

The TEAM model was used to simulate the bioconversion process for model facilities using different pretreatments and feedstocks. The model facility was assumed to be a first-generation plant able to process 35,500 metric t of dry raw material yearly, with on-line time being 355 d/yr, and producing ethanol at 94% purity. Material balances using equipment models and laboratory data were used to determine the flow rates and composition of all streams in the process. Energy balances derived from equipment models and laboratory data were used in the steam-explosion pretreatment, enzymatic hydrolysis, and fermentation sections to determine partial energy flows primarily for costing purposes. Sizing of this

Table 2
Costs Used in Evaluations

	Cost/unit (CDN$)
Chemicals	
SO_2	0.25/L
Ethanol	0/L
Defoamer	0.5/kg
NaOH (50%)	0.46/kg
H_2O_2	0.6/kg
Cellulase enzymes	3.3×10^6 FPU
H_2SO_4 (93%)	0.11/kg
Acetic acid	0.72/kg
$(NH_4)_2HPO_4$	1.5/kg
Sodium phosphate	1.6/kg
Sodium acetate	1.6/kg
Corn steep liquor	0.36/kg
Coproducts	
Solid fuel	0.03/kg
CO_2	0/kg
Utilities	
Electricity	36.6/MW h
Cooling water	0.02/L
Process water	0.24/L
Process steam	11.7/t
Other costs	
Property tax and insurance	1.5% of fixed capital
Working capital	8% of fixed capital
Contingency and fee	18% of fixed capital

equipment for the model was carried out through equipment vendor information and rules of thumb (35,36).

All of the costs utilized in the TEAM model are summarized in Table 2, with the exception of labor and maintenance costs, which were both determined on an individual equipment or subprocess basis and adjusted for inflation through the chemical engineering index for that particular type of equipment. Costs for steam-explosion pretreatment, enzymatic hydrolysis, and fermentation were calculated based on laboratory observations as previously reported (34). Capital and operating costs for organosolv pretreatment, lignin recovery, and ethanol recovery stage were quoted by Lignol Innovations. Equipment cost was estimated from various reports, vendor quotations, and adjusted to other capacities using vendor information or six-tenth rule. Fixed capital investment and working capital were estimated as suggested by Ulrich (36). The annual cost of the fixed capital was obtained by multiplying the fixed capital investment by an annualization factor of 0.103, corresponding to an annual interest rate of 6% and a 15-yr project life. It should be noted that the interest rate used in the model is low, and that higher value on capital will skew these results. No salvage value was assumed for the equipment at the end of the pay-back period.

For the purposes of this study the raw material cost was assumed to be zero, in order to determine the influence of process design elements on production cost. Feedstock costs vary widely and are often heavily influenced by factors outside the control of a process engineer, such as transportation distance, weather, and markets. Chemical costs are based on estimates of delivery to Tembec's temiscaming pulping facility. Electricity, water, and steam costs as used in the model (Table 2) are based on British Columbia rates in late 1990s, which are lower than current costs across North America. However, all of the cost comparisons are provided on a basis relative to our current, best lab results for ethanol production from SO_2 steam explosion of Douglas-Fir, which has been assigned a normalized value of 100%. Thus, each of the costs provided in the following section should be interpreted as relative costs.

Results and Discussion

Technical Performance of Pretreatments

In Table 3, the technical metrics of steam-exploded Douglas-Fir, steam-exploded corn fiber, and ethanol organosolv-treated Douglas-Fir are compared. Several significant details should be highlighted. It is clear that the steam-explosion pretreatment is not ideal for Douglas-Fir. The recovery of glucose, hemicellulose sugars, and lignin after pretreatment is significantly lower than found with steam-exploded corn fiber, or with organosolv-treated Douglas-Fir. The slightly lower lignin recovery may serve as further evidence of the difficulty that softwood lignin poses, and the tendency found in the steam-explosion pretreatment process for lignin redeposition and condensation on the surface of fibers. Avoiding the problem of lignin condensation and redeposition is one of the primary drivers for choosing a different pretreatment method for softwoods.

One of our research goals has been to optimize the steam-explosion pretreatment in order to preserve lignin in the biomass, whereas reducing the formation of inhibitors to the hydrolysis process. In a work by Pan et al. (37), it has been shown that 35.5–43.2% of lignin may be removed from steam-exploded Douglas-Fir through the application of 1% NaOH (w/w) at room temperature. This posttreatment was effective in enhancing enzymatic hydrolysis yield from 50% to 85%. Among the conclusions that may be drawn from this work is that an "active" portion of lignin impedes hydrolysis, and that lignin inhibition might be significantly reduced by selective removal or modification of this active fraction.

The inhibitory nature of lignin for enzymatic hydrolysis has being explored by Berlin et al. (38), who have shown that lignin interferes with enzymatic hydrolysis by binding to the enzyme, forming a triple complex

Table 3
Technical Metrics

	Steam explosion		Organosolv
	Douglas-Fir	Corn fiber	Douglas-Fir
Glucose recovery (%)[a]	85	95	90
Hemicellulose recovery (%)[a]	74	95	90[a]
Lignin recovery (%)[a]	94	99	99
Hydrolysis yield (%)	>95	>95	>90
Substrate loading (%)[b]	2	2	2
Enzyme loading (FPU/g cellulose)	20	10	14
Hydrolysis time (h)	48	24	24
Fermentation yield (%)[c]	>90	>90	>90
Fermentation time (h)	48	6	8
Lignin opportunities	Energy only	Energy only	High value

[a]Percentage of original.
[b]Percentage on w/w basis.
[c]Percentage of theoretical.

of enzyme–substrate inhibitor. On a mass basis, lignin inhibition has been found comparable to inhibition by glucose, but not competitive in nature. Engineering of enzymes with lower lignin affinity could provide a useful strategy for improvement of cellulase activity on lignocellulosic substrates *(39)*.

The lower glucose and hemicellulose sugar recovery found with steam-explosion of Douglas-Fir is another problem that is particularly important to resolve. A significant amount of the sugars that could be utilized in ethanol production, as well as in other products, is lost during steam explosion. This may in part have to do with the physical architecture of the steam-explosion system, and the fact that soluble hemicelluloses escape in solution in the wash from this system. Recent work in our laboratory has shown that diverting this hemicellulose-soluble stream, which contains both hexoses and pentoses, to the fermentation stage and combining it with a high-consistency hydrolysate (10% w/v) has the result of doubling the initial sugar concentration to 54.3 g/L, increasing hexose sugar concentration by greater than 50%, and achieving a high-yield ethanol production of 0.46 g/g, or 90% of theoretical *(33)*. Unfortunately, this adds a level of process complexity to the system, which in turn increases costs.

Obviously, one of the goals of pretreatment should be to introduce the least amount of process complexity in order to add minimal cost to the total process. Although effective steam-explosion of Douglas-Fir may not be possible under this criterion, the method remains suitable for other substrates. Table 3 highlights the high recovery rates and relative ease of hydrolysis associated with steam-exploded agricultural residues. Corn fiber can be effectively hydrolyzed in half the time, with half the enzyme loading, when compared with Douglas-Fir. This indicates that steam-explosion pretreatment remains an effective tool for the processing of these types of agricultural wastes. Other softwoods, such as Norway spruce (*Picea abies*), may respond to steam explosion in a manner closer to corn fiber than Douglas-Fir thus eliminating the costly delignification stage *(40)*. Currently, joint efforts between UBC and the University of Lund are underway to confirm this difference in pretreatment response. The relation between pretreatment selection and feedstock characteristics is one of particular interest as the processing of lignocellulosic materials begins to enter the commercial sphere. This work indicates that other options may have to be explored for specific substrates.

Another way in which pretreatment costs may be reduced is by lowering chemical usage or the costs of delignification. Ethanol organosolv pretreatment was first undertaken for the processing of Douglas-Fir because of its ability to simultaneously achieve both of these goals. As shown in Table 3, the organosolv pretreatment with Douglas-Fir substrate compares closely with the steam-exploded corn fiber substrate. The use of ethanol as a pulping liquor means that liquor recovery and product recovery may be combined, reducing process costs and lowering chemical usage; however, it may also act as an inhibitor in the fermentation process. Finally, the lignin recovered through the organosolv process has been shown to have reactive properties that are significantly improved over steam-exploded lignin. This creates opportunities for value-added coproducts that may reduce the overall costs of delignification for this process.

Economic Performance of Pretreatments

In Fig. 4, the process economics for steam-exploded Douglas-Fir, steam-exploded corn fiber, and ethanol organosolv-treated Douglas-Fir are compared. As stated in the methodology, all of the cost comparisons are provided on a relative basis to the current best lab results for the SO_2 steam explosion of Douglas-Fir, which has been made equal to 100%. Thus, the data in Fig. 4 should be interpreted as relative costs to the steam-exploded Douglas-Fir case. In addition, no additional coproduct values, including that of lignin, are included in calculating the overall prices.

The economic performance of steam-exploded Douglas-Fir is significantly worse than corn fiber in large part owing to the fractionation (or delignification) cost associated with the process. As stated before, the unique nature of softwood lignin, and the fact that there is relatively more of this material when compared with hardwoods and agricultural residues, is in large part to blame for this issue. As stated in the preceding section, a major goal of ethanol organosolv pretreatment is to reduce process costs over steam-explosion by reducing some of the delignification costs. As shown in Fig. 4, the organosolv method is highly effective in doing so. Fractionation or delignification costs are almost completely removed in the organosolv process, which to a large degree accounts for

Update on Softwood-to-Ethanol Process Development

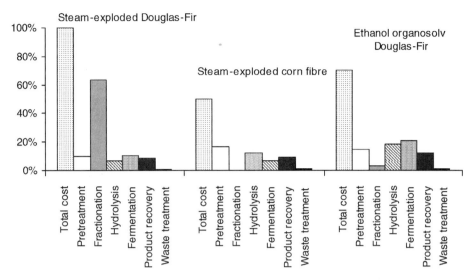

Fig. 4. Process economics for three bioconversion processes.

the reduction in relative price by almost 25% over steam-exploded Douglas-Fir.

The effectiveness of steam explosion for agricultural residues is also shown in Fig. 4. Overall process costs are relatively low for steam-exploded corn fiber, and whereas pretreatment is a significant portion of these costs, it is fairly balanced by the other subprocesses. One important message that can be taken from Fig. 4, and which holds true with each pretreatment and substrate, is the relatively low impact that the enzymatic hydrolysis subprocess has on overall costs. Previous work (41) identified the cost of enzymes as a critical cost issue, but lower enzyme costs have changed this scenario. Our analysis indicates that the overall cost of delignification associated with steam-explosion pretreatment remains a large hurdle to overcome for softwood lignocellulosic substrates. Alternative pretreatment technologies, such as organosolv, may prove to be a more beneficial method for handling softwood substrates.

It is worth restating that no additional coproduct values, including that of lignin, are included in calculating the overall costs shown in Fig. 4. The value of lignin coproducts has been shown to be a crucial factor in making the softwood-to-ethanol process economically viable (13,42). Further economic analysis was carried out, using the TEAM model, to test the sensitivity of the bioconversion of Douglas-Fir using the organosolv pretreatment to variations in the lignin value. We found that ethanol production costs decrease significantly as the value of the lignin rises. Sufficiently high lignin values of approx 1.5 times the normalized ethanol production cost from steam-exploded Douglas-Fir render ethanol production from the bioconversion process cost neutral, as the coproduct value can cover the costs of ethanol generation. These costs include unit operations for pretreatment, fractionation and delignification, hydrolysis, fermentation, product recovery, and waste treatment. However, they do not include the cost of biomass, which could easily double the cost of the process, given current market conditions. The results indicate that high-value lignin coproducts can economically justify ethanol production, and suggest that further research should focus on developing these products.

Conclusions

In our laboratory, we have found that we can reduce lignin condensation, chemical usage, and cost for delignification, and partially reduce lignin inhibition by simply optimizing pretreatments, or through the addition of effective posttreatments. We have found that steam explosion technically works with agricultural residues, hardwoods, and softwoods, to produce ethanol and a burnable form of lignin. The choice of feedstock determines whether it can currently be done economically and the complexity of the process.

When considering softwood biomass as a substrate, lignin separation and utilization becomes a critical issue in addition to the enzymatic hydrolysis step. With steam-explosion pretreatment, a delignification stage is currently required, which is costly, producing a low-value lignin. Our laboratory is working on methods to reduce the cost of delignification with steam-explosion technology. Recent results indicate that ethanol organosolv pretreatment may be useful in entirely avoiding this stage.

Ethanol organosolv of softwoods produces a lignin with more reactive properties, as well as a cellulose fraction with good hydrolysis rates and yields. This is owing to the nature of the substrate after organosolv pretreatment, which is more amenable to hydrolysis, even with relatively high original lignin content. With ethanol organosolv pretreatment, enzyme loadings and recoveries of the solvent and hemicellulose sugars remain issues to be resolved. The ethanol organosolv process has a potential to produce both high-value lignin coproducts and low-cost ethanol, making this process a promising avenue for further investigation.

Acknowledgments

We would like to acknowledge the Natural Resources Canada (NRCan) and the Natural Sciences and Engineering Research Council (NSERC) for operating funds that have supported our research program. Capital funds for the Process Development Unit and analytical support were provided by the Canada Foundation for Innovation (CFI), the British Columbia Knowledge Development Fund (BCKDF), and the University of British Columbia. Finally, we thank the corporations that haves supplied us with materials for research, including Tembec, who provided the SSL-adapted yeast used in the fermentation studies, and Novozymes, who provided the cellulase and β-glucosidase enzymes used in the hydrolysis studies.

References

1. Bowyer, J. L. and Stockmann, V. E. (2001), *Forest Prod. J.* **51(1),** 10–21.
2. CCFM (2003), *Table 5.3—Net Merchantable Volume of Roundwood Harvested by Ownership, Category, Species Group and Province/Territory, 1990–2002, Compendium of Canadian Forestry Statistics,* Canadian Council of Forest Ministers, Ottawa, Ontario, http://nfdp.ccfm.org/default.htm.
3. FAO (2002), *FAOStat Agriculture Data,* Food and Agriculture Organization of the United Nations, Rome, Italy, http://apps.fao.org/page/collections?subset=agriculture.
4. FAO (2002), *FAOStat Forestry Data,* Food and Agriculture Organization of the United Nations, Rome, Italy, http://apps.fao.org/page/collections?subset=forestry.
5. Mabee, W. E., Gregg, D. J., and Saddler, J. N. (2003), *Ethanol from Lignocellulosics: Views to Implementation,* Report to IEA Bioenergy Task 39, Vancouver, Canada.
6. NRCan (2003), *The State of Canada's Forests 2002–2003,* Natural Resources Canada, Ottawa, Ontario, p. 18.
7. Panshin, A. J. and de Zeeuw, C. (1980), *Textbook of Wood Technology,* 4th ed., McGraw-Hill, Toronto, Ontario, p. 213.
8. Statistics Canada (2003), *CANSIM II, Table 051-0001: Estimates of Population, by Age Group and Sex, Canada, Provinces and Territories, Annual,* Statistics Canada, Ottawa, Ontario, http://cansim2.statcan.ca.
9. Statistics Canada (2003), *Census of Agriculture,* Statistics Canada, Ottawa, Ontario, http://www.statcan.ca/english/Pgdb/census.htm.
10. Sjöström, E. (1993), *Wood Chemistry,* Academic Press, San Diego.
11. Shimada, K., Hosoya, S., and Ikeda, T. (1997), *J. Wood Chem. Technol.* **17,** 57–72.
12. Avellar, B. K. and Glasser, W. G. (1998), *Biomass Bioenerg.* **14,** 205–218.
13. Boussaid, A. L., Esteghlalian, A. R., Gregg, D. J., Lee, K. H., and Saddler, J. N. (2000), *Appl. Biochem. Biotechnol.* **84–86,** 693–705.
14. Brownell, H. H., Yu, E. K. C., and Saddler, J. N. (1986), *Biotechnol. Bioeng.* **28,** 792–801.
15. Carrasco, J. E., Saiz, M. C., Navarro, A., Soriano, P., Saez, F., and Martinez, J. M. (1994), *Appl. Biochem. Biotechnol.* **45–46,** 23–34.
16. Nguyen, Q. A., Tucker, M. P., Boynton, B. L., Keller, F. A., and Schell, D. J. (1998), *Appl. Biochem. Biotechnol.* **70–72,** 77–87.
17. Nguyen, Q. A., Tucker, M. P., Keller, F. A., Beaty, D. A., Connors, K. M., and Eddy, F. P. (1999), *Appl. Biochem. Biotechnol.* **77,** 133–142.
18. Boussaid, A., Jarvis, J., Gregg, D. J., and Saddler, J. N. (1997), In: *The Third Biomass Conference of the Americas,* Overend, R. P. and Chornet, E. (eds.), Montreal, Canada, pp. 873–880.
19. Donaldson, L. A., Wong, K. K. Y., and Mackie, K. L. (1988), *Wood Sci. Technol.* **22,** 103–114.
20. Yang, B., Boussaid, A., Mansfield, S. D., Gregg, D. J., and Saddler, J. N. (2002), *Biotechnol. Bioeng.* **77,** 678–684.
21. Bura, R. (2004), *PhD Thesis,* University of British Columbia, Vancouver, Canada.
22. Pye, E. K., Klein, W. R., Lora, J. H., and Cronlund, M. (1987), *The Alcell™ Process. Solvent Pulping—Promises & Problems Conference,* Appleton, WI, pp. 55–67.
23. Bose, S. K. and Francis, R. C. (1999), *J. Pulp Paper Sci.* **25,** 425–430.
24. Meshgini, M. and Sarkanen, K. V. (1989), *Holzforschung* **43,** 239.
25. Evtiguin, D. V., Neto, C. P., and Silvestre, A. J. D. (1997), *J. Wood Chem. Technol.* **17,** 41–55.
26. Lora, J. H., Wu, C. F., Pye, E. K., and Balatinecz, J. J. (1989), In: *Lignin: Properties and Materials.* American Chemical Society Symposium Series, **397,** 312–323.
27. Mullinder, J. (1989), *Pulp Paper J.* **42(7),** 31.
28. Fales, G. (1988), *Paper Age* **104(10),** 40
29. Karl, W. (1988), *Pulp Paper J.* **41(7),** 17, 21.

30. Anonymous (1998), *PIMAs North American Papermaker* **6,** 18.
31. Mirochnik, O. (2004), *MASc Thesis*, University of British Columbia, Vancouver, Canada.
32. Galbe, M. and Zacchi, G. (2002), *Appl. Microbiol. Biotechnol.* **59,** 618–628.
33. Robinson, J., Keating, J. D., Mansfield, S. D., and Saddler, J. N. (2003), *Enzyme Microb. Technol.* **33,** 757–765.
34. Gregg, D. J. and Saddler, J. N. (1997), *Appl. Biochem. Biotechnol.* **63–65,** 609–623.
35. Perry, P. H. and Green, D. (1984), *Perry's Chemical Engineers' Handbook*, 6th ed., McGraw-Hill, New York.
36. Ulrich, G. D. (1984), *A Guide to Chemical Engineering Process Design and Economics*, Wiley and Sons, New York.
37. Pan, X., Arato, C., Gilkes, N., et al. (2005), *Biotechnol. Bioeng.* **90(4),** 473–481.
38. Kurabi, A., Berlin, A., Gilkes, N., et al. (2004), *Appl. Biochem. Biotechnol.* **121–124,** 219–230.
39. Berlin, A., Gilkes, N., Kurabi, A., et al. (2004), *Appl. Biochem. Biotechnol.* **121–124,** 163–170.
40. Wingren, A., Soderstrom, J., Galbe, M., and Zacchi, G. (2004), *Biotechnol. Prog.* **20,** 1421–1429.
41. Aden, A., Ruth, M., Ibsen, K., et al. (2002), *Lignocellulosic Biomass to Ethanol Process Design and Economics Utilizing Co-Current Dilute Acid Prehydrolysis and Enzymatic Hydrolysis for Corn Stover*, National Renewable Energy Lab, Golden, CO, NREL Report No. TP-510-32438.
42. Gregg, D. J., Boussaid, A., and Saddler, J. N. (1998), *Bioresour. Technol.* **63,** 7–12.

Development of a Multicriteria Assessment Model for Ranking Biomass Feedstock Collection and Transportation Systems

AMIT KUMAR,[*,1] SHAHAB SOKHANSANJ,[1] AND PETER C. FLYNN[2]

[1]Department of Chemical and Biological Engineering, University of British Columbia, 2216 Main Mall, Vancouver, Canada V6T 1Z4, E-mail: Amit.Kumar@ualberta.ca; and [2]Department of Mechanical Engineering, University of Alberta, Edmonton, Canada T6G 2G8

Abstract

This study details multicriteria assessment methodology that integrates economic, social, environmental, and technical factors in order to rank alternatives for biomass collection and transportation systems. Ranking of biomass collection systems is based on cost of delivered biomass, quality of biomass supplied, emissions during collection, energy input to the chain operations, and maturity of supply system technologies. The assessment methodology is used to evaluate alternatives for collecting 1.8×10^6 dry t/yr based on assumptions made on performance of various assemblies of biomass collection systems. A proposed collection option using loafer/stacker was shown to be the best option followed by ensiling and baling. Ranking of biomass transport systems is based on cost of biomass transport, emissions during transport, traffic congestion, and maturity of different technologies. At a capacity of 4×10^6 dry t/yr, rail transport was shown to be the best option, followed by truck transport and pipeline transport, respectively. These rankings depend highly on assumed maturity of technologies and scale of utilization. These may change if technologies such as loafing or ensiling (wet storage) methods are proved to be infeasible for large-scale collection systems.

Index Entries: Multicriteria assessment; biomass collection systems; biomass transportation systems; PROMETHEE; ranking; bioenergy; IBSAL model.

Introduction

To select a system among several alternatives, a number of criteria should be taken into account: costs (investment, operating, and manpower), technology (efficiency and energy consumption), social factors

---

*Author to whom all correspondence and reprint requests should be a addressed.

(job creation and quality of life), and environmental factors (emission and discharge). A decision-maker faces the problem of selecting the optimal solution, and cannot focus on one criterion at the expense of others. A multidimensional approach helps in solving this type of complex problem. Assessment of multicriteria for different alternatives is required in order to arrive at a credible solution.

Interest in biomass utilization for energy and chemicals has increased recently, in order to reduce the dependence on fossil fuels and also to reduce greenhouse gas (GHG) emissions. Biomass feedstock collection and transportation is an integral aspect of biomass utilization. Feedstock delivery cost constitutes about 35–50% of the total production cost of ethanol or power *(1–3)*. The actual percentage depends largely on geographical factors such as location, climate, local economy, and the type of systems used for harvesting, collection, processing, and transportation. Many studies have evaluated different biomass harvesting, collection, and transportation systems *(4–8)* but most of these studies were designed to optimize economic factors only, producing options of lower cost. There is a scarcity of work on integration of economical, environmental, and social factors for selection of optimum biomass systems.

Investment in biorefineries must consider economic, social, and environmental factors, particularly in which the public interest is involved. The main objective of this paper is to develop a methodology to rank biomass collection and transportation alternatives using a multicriteria approach. The approach is applied to two cases: (1) biomass collection and (2) biomass transportation. For this analysis we consider financial, environmental, and social factors as criteria. Biomass collection alternatives considered in this study are for agricultural residues (corn stover) and include baling, loafing, and ensiling. Biomass transport options include truck, rail, and pipeline.

Methodology

We used a well-known method called preference ranking organization method for enrichment and evaluations (PROMETHEE). The method is described by Brans and Vincke *(9)*, Brans et al. *(10)*, and Brans and Mareschal *(11)*. PROMETHEE integrates quantitative and qualitative criteria to conduct a paired comparison of alternatives, as described in the following section.

Comparing Two Alternatives Based on One Criterion

Suppose a and b are two alternatives and these alternatives are compared over criterion f. We determine the value of each of these two alternatives with respect to the criterion f and designate them as $f(a)$ and $f(b)$. $f(a)$ and $f(b)$ must be maximized in order to delineate between the two alternatives. In a PROMETHEE method a preference function is applied to the criterion f to map the difference between $f(a)$ and $f(b)$ to a degree of preference of one alternative over another. If the difference is significant, the degree of preference is higher and there is more preference of one alternative over another. The associated degree of preference function $P(a,b)$ is given as

$$P(a,b) = \begin{cases} 0 & \text{if } f(a) \leq f(b) \\ p[f(a), f(b)] & \text{if } f(a) > f(b) \end{cases} \quad (1)$$

where $P(a,b)$ is degree of preference, $f(a)$ and $f(b)$ are the values of alternatives a and b over the criteria f, and $p[f(a), f(b)]$ is the preference function. For practical cases, it is reasonable to assume

$$p[f(a), f(b)] = p[f(a) - f(b)] \quad (2)$$

The degree of preference is a numerical value $P(a, b)$. $P(a, b)$ is estimated using Eq. 1 and is dependent on the value of $p[f(a)-f(b)]$. Let the difference between the values of two criteria be

$$d = f(a) - f(b) \quad (3)$$

$p(d)$, the preference function, is defined by the decision-maker and the value of this function is always

$$0 \leq p(d) \leq 1 \quad (4)$$

Hence Eq. 1 becomes (using Eq. 3 in Eq. 1)

$$P(a,b) = \begin{cases} 0 & \text{if } f(a) \leq f(b) \\ p(d) & \text{if } f(a) > f(b) \end{cases} \quad (5)$$

We consider the preference function such that

$$0 \leq P(a,b) \leq 1 \quad (6)$$

There are four cases that arise when comparing two alternatives a and b. These are

If $P(a,b) = 0$, then there is indifference between a and b (i.e., none is preferred).
If $P(a,b) \sim 0$, then there is weak preference of a over b.
If $P(a,b) \sim 1$, then there is strong preference of a over b.
If $P(a,b) = 1$, then there is strict preference of a over b.

Suppose cars A and B are two alternatives and are compared on distance each travel using 1 L of gasoline. The car with higher traveled distance per unit of gasoline used is selected (maximize the criterion). Alternative a is represented by car A and alternative b is represented by car B.

The criterion is km/L of gasoline. Let distance traveled by cars A and B be 25 and 15 km/L, respectively. Applying the methodology described earlier to this example, $f(a) = 25$, $f(b) = 15$, and $d = f(a) - f(b) = 10$ km/L.

Let us assume a preference function $p(d)$, such that

$$p(d) = 1 \text{ if } d > 0$$
$$p(d) = 0 \text{ if } d < 0$$

Hence in this case, degree of preference, $P(a,b)$, is unity, using Eq. 5. According to the cases defined earlier, there is strict preference of car A over B.

General Case—Comparison of Multiple Alternatives Based on Multiple Criteria

In case of a multicriteria problem, a preference function is defined for each of the criteria. The preference functions translate the difference between the values of two alternatives over a criterion in terms of a degree of preference. The higher the difference between the values of alternatives over a criterion, the larger is the degree of preference. The degree of preference is calculated by comparing pairs of alternatives, using the selected preference function for a criterion. These degrees of preference are used to estimate multicriteria preference index of an alternative over other alternative, considering all the criteria together. This methodology is described later. A preference function also helps in bringing the scale of different criteria to a single uniform scale. For example, if two different criteria, cost ($/t) and emissions (kg C/t), have to be integrated in order to rank two alternatives, the comparison of these criteria need to be converted to the same scale; this is done with the help of a preference function.

PROMETHEE is useful in analyzing problems in which there are multiple criteria and multiple alternatives for solutions. The generalized methodology may be expressed as a ranking of a set of N alternatives based on k criteria (f_i, where $i = 1, \ldots, k$). In this method, a preference index comparing alternatives a and b using multiple $(1, \ldots, k)$ criteria is the weighted average of the degree of preference $P_i(a,b)$ for each individual criterion, and is expressed as $\gamma(a,b)$. The weighting assigned to each criterion (w_i) is based on the relative importance of the criterion.

$$\gamma(a,b) = \frac{\sum_{i=1}^{k} w_i P_i(a,b)}{\sum_{i=1}^{k} w_i} \qquad (7)$$

where $a, b \in N$

N is the set of alternatives and k is the number of criteria.

This study uses a decision-support software, namely, Decision Lab 2000—Executive Edition *(12)*. Six different preference functions are defined in the Decision Lab 2000. In this study we use two preference functions: linear preference, in which the preference of a over b increases linearly, and level preference, in which a threshold denotes the change point between indifference and preference.

The sum of the multicriteria preference indexes $\gamma(a,b)$, is used to calculate a leaving flow $\varphi^+(a)$, an entering flow $\varphi^-(a)$, and a net flow $\varphi(a)$ for the entire system. Leaving flow denotes the dominance of an alternative over other alternatives and is a measure of outranking character. Entering flow is the measure of outranked character. Net flow is the difference between the two flows.

$$\varphi^+(a) = \sum_{x \in N} \gamma(a,x) \tag{8}$$

$$\varphi^-(a) = \sum_{x \in N} \gamma(x,a) \tag{9}$$

$$\varphi(a) = \varphi^+(a) - \varphi^-(b) \tag{10}$$

PROMETHEE provides two ranking methods: PROMETHEE I—the preferences are confirmed by both leaving and entering flow and ranking is based on leaving flow (if the information provided by two flows for two alternatives are contradicting, this method cannot decide the ranking); PROMETHEE II—ranking is based on net flow. In both the methods, greater the value of flow, higher is the ranking. One of the powerful features of PROMETHEE is its ability to integrate both quantitative and qualitative criteria. Qualitative criteria are defined on a scale. A qualitative scale is a series of ordered semantic values. Each semantic value is associated with a numerical value that is used in calculations. We used a five-point scale to assign values to qualitative criteria (very high = 5, very low = 1). Figure 1 shows the flowchart for the ranking methodology used in this study.

Case 1: Biomass Collection Systems

Biomass feedstock supply is characterized by a large collection area, variation in crop maturity with time and weather, a short window of collection, and competition from concurrent harvest operations. It is important to have an optimized supply system for long-term success of a biomass processing facility such as a biorefinery. Biomass supply logistics consists of harvesting, collection, storage, preprocessing, and transportation (13). In this study collection options for agricultural residues (in this study, corn stover) include the following.

1. Baling (conventional round baling)—harvest grain, shred or rake crop residue, bale biomass, transport bales to the field edge and stack, load bales on a truck and transport to a biorefinery, unload, and grind the biomass.

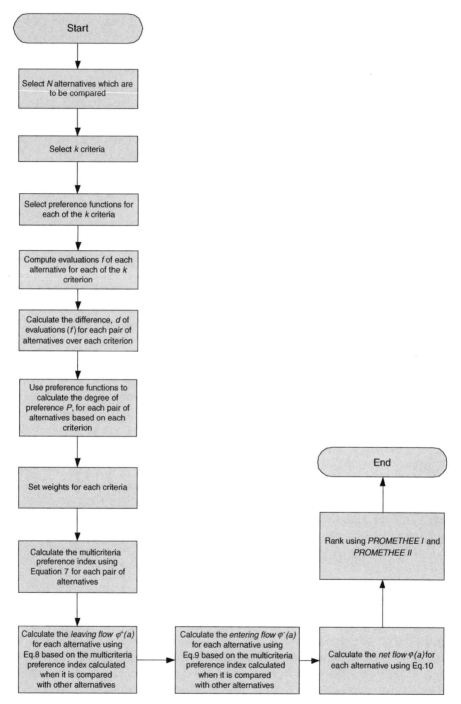

Fig. 1. Flowchart for implementing PROMETHEE.

2. Loafing—harvest grain, shred or rake crop residue, load a stacker (loafer) with biomass, transport the stacker (filled with biomass) to the field edge, unload the stacker, grind the biomass, load the grind into a truck, and transport the filled truck to a biorefinery.

3. Chopping and ensiling—harvest grain, chop stover using a forage harvester, load wagon, transport wagon to the silage, ensile, load, a silage truck, and transport to a biorefinery.

Integrated Biomass Supply Analysis and Logistics

The integrated biomass supply analysis and logistics (IBSAL) model developed by Sokhansanj at the Oak Ridge National Laboratory (ORNL) *(14,15)* is used to estimate the costs, energy use, and emissions for each of the collection options. IBSAL is a dynamic simulation model which simulates flow of biomass from field to a biorefinery. It consists of different submodules detailing discrete process steps, including harvest, baling, grinding, storage, and transportation. The current version of IBSAL model has been developed for straw and corn stover harvesting, collection, storage, preprocessing, and transportation. The model input data include local weather data (average daily temperature, humidity, and precipitations), average yield of biomass, proportion of land that is cultivated to the crop that yields the residue, crop harvest progress data, capacity of the biorefinery, dry matter loss with time in the storage, operating parameters of different agricultural machinery, and capital and operating costs of different agricultural machinery. This model has been built on EXTEND™ platform, available from Imagine That, Inc. *(16)*. Main outputs of the model include delivered cost of biomass to a biorefinery ($/t of biomass delivered), GHG emission (kg C/t of biomass delivered), and energy consumption (GJ/t of biomass delivered). Cost, energy, and emission parameters can be obtained for individual processing steps. The model can also be used to estimate the monthly equipment requirement and the time required to finish each operation. Details of the model can be found in the work of Sokhansanj and Turhollow *(14,15)*. This model runs on the generic data.

Input Data and Assumptions

Five different criteria are considered for comparing and ranking three biomass collection systems. Table 1 shows input data for each criterion considered for corn stover collection systems. Delivered cost of material to a biorefinery represents the economic factor in the analysis. Emissions represent the environmental factor in the analysis. Data for cost, emissions, and energy consumption for different collection systems are calculated using the IBSAL model, for a biorefinery capacity of approx 1.8×10^6 dry t/yr. The values of two qualitative criteria, i.e., quality of material and maturity of the technology, are based on the experience of the authors, in discussion with the industry and in informal consultation with the experts in the area. Note that all cost figures in this study are in 2004 USD.

Table 2 lists the weights (w_i) for each criterion, the preference threshold for criteria differences (α), and the indifference threshold for criteria differences (β) for corn stover collection systems.

Table 1
Input Data for Biomass Collection Systems

Options/criteria	Cost ($/t)	Quality of material	Emissions (kg C/t)	Energy consumption (GJ/t)	Maturity of technology
Corn stover					
Baling	49.77	Very high	29.6	1.35	Very high
Ensiling	49.99	Average	20.4	0.93	Very low
Loafing	27.36	Low	20.7	0.88	Average

Table 2
Assumptions for Biomass Collection Systems

Items	Cost	Quality of material	Energy	Emission	Maturity of technology
Min./Max.	Min.	Max.	Min.	Min.	Max.
Weights[c] (w_i)	0.30	0.25	0.15	0.15	0.15
Preference threshold[a] (α)	5.00	2.50	0.30	5.00	2.50
Indifference threshold[b] (β)	–	0.50	–	–	0.50

[a]Units for values of different quantitative criteria are cost ($/t), energy (GJ/t), and emission (kg C/t). Values for qualitative criteria are in absolute numbers.
[b]Values for qualitative criteria are in absolute numbers.
[c]Sum of all the weights is 1.

Comments on Selected Criteria and Input Parameters

Delivered cost of biomass—delivered cost is the sum of costs of each unit operation in a particular collection system. Many earlier studies report the delivered cost of agricultural residues to a biorefinery (see, e.g., refs. 2,3,6,7,13,17,18). These studies do not consider costs based on time-dependent biomass supply modeling. The IBSAL model is time dependent and takes into account all the aspects of biomass supply from field to biorefinery including the seasonal nature of the biomass.

Emissions—in this study the carbon emissions during the direct operation of agricultural machinery and transportation of biomass by wagons or trucks represent the environmental impact of different collection systems. The values in Table 1 do not include emissions during indirect use of fossil fuels for fertilizer or pesticides' production. It is assumed that all the agricultural machinery use diesel fuel. Earlier studies have estimated emissions during harvesting and transportation of different types of biomass, but most of these have not considered detailed machinery use in each farm operation (see, e.g., refs. 19–23). The IBSAL model simulates the use of all the equipment and farm machineries required on a farm and hence gives more accurate value.

Energy consumption—Table 1 shows the values of energy consumption per tonne of biomass material supplied. Earlier studies have calculated the energy consumption for different biomass supply systems (see, e.g., refs. 19–23). In this study energy consumption represents the fossil fuel consumed in direct process use and does not take into account the energy consumed in indirect process use such as manufacturing of equipment, fertilizer, and so on. The IBSAL model outputs are specific to the collection systems considered.

Quality of material—this is an important criterion for deciding the best collection systems. Quality of biomass affects the efficiency and output of the biomass conversion process. Supply of material in the form of bales helps in minimizing the dirt and soil in the biomass material. In loafing process dirt may be picked up from the ground and this has a negative impact on fuel or product yield from biomass during conversion. It is difficult to assign a quantitative value to each collection system for this criterion, hence a qualitative scale is chosen for comparison.

Maturity of technology—currently, the most common method of collecting biomass in North America is through baling (6,7,13,17,18). This is a mature technology in comparison with loafing and ensiling. Currently, ensiling is not used for biomass collection but research is in progress at the ORNL, Idaho National Laboratory, and elsewhere. Loafing has been studied earlier as collection and storage system but the loafer needs improvement.

Weights of different criteria—weights are critical parameters in a multicriteria assessment. Weighting values assigned within the model depend on the decision-maker and can vary widely. In this study we have assumed weights based on our experience and in consultation with other experts in this area; these are shown in Table 2. A sensitivity analysis, discussed later, illustrates the dependence of the ranking on the relative weighting of criteria.

Preference and indifference threshold—these thresholds represent the limits, above or below which there is a strict preference or indifference. These values are based on experience and in consultation with the experts. A sensitivity analysis for these values is done. The values are shown in Table 2.

Results and Discussion for Case 1

The input and assumption data given in Tables 1 and 2 are used in Decision Lab 2000 to estimate the leaving, entering, and net flows. Table 3 gives the leaving flow $\varphi^+(a)$, entering flow $\varphi^-(a)$, and net flow $\varphi(a)$ for different alternatives.

Based on the PROMETHEE I ranking method, loafing is the best alternative, and we cannot make a judgment between baling and ensiling. According to the results shown in Table 3, φ^+(baling) is greater than φ^+(ensiling) and also φ^-(baling) is greater than φ^-(ensiling). According to

Table 3
Results for Biomass Collection Systems

Options	Leaving flow $\varphi^+(a)$	Entering flow $\varphi^-(a)$	Net flow $\varphi(a)$	Ranking PROMETHEE I	Ranking PROMETHEE II
Corn stover					
Baling	0.31	0.45	−0.14	2	3
Ensiling	0.22	0.34	−0.12	2	2
Loafing	0.50	0.23	0.27	1	1

Table 4
Stability Intervals for Corn Stover Collection Systems

Criteria	Corn stover Unit: percentage values (%)		
	Weight	Min.	Max.
Cost	30	5	49
Emission	15	14	89
Energy consumption	15	14	100
Maturity of technology	15	0	16
Quality of material	25	0	27

PROMETHEE I method, it is difficult to make a decision between baling and ensiling systems. In other words, based on PROMETHEE I, baling and ensiling cases cannot be distinguished. As discussed earlier, the PROMETHEE II ranking is based on net flows. It selects ensiling over baling. Note that the two ranking methods use different calculations to rank the alternatives. A decision-maker would have to look at the actual role of each variable before making a decision. In the case of stover collection systems, loafing is the best system based on both PROMETHEE I and II rankings, which is largely owing to a combination of low costs, emissions, and specific energy consumption. The advantages of these three criteria offset the penalty owing to low quality of material.

A sensitivity analysis on weighting factors is done by establishing stability intervals, which represent the range over which weighting values for different criteria do not influence the PROMETHEE II outcome. Table 4 lists the range of stability intervals for each variable. For example, if we change the weight of cost criteria to any value between 5% and 49% (as shown in Table 4), keeping the values of weights of other criteria the same, the PROMETHEE II ranking of collection systems will not change. This represents a wide range and reflects the low sensitivity of ranking to cost. Similar information is conveyed by stability intervals for emission and energy consumption regarding their impact on ranking. These criteria have low sensitivity on ranking. However, ranking is sensitive to the weights of the criteria—maturity of technology and quality of material.

This is evident by the narrow range of the stability interval for these criteria. The stability interval gives the decision-makers an option of analyzing the impact of changing the weights on ranking of different alternatives.

Similar analysis of biomass collection systems at a lower or higher capacity per year does not change the PROMETHEE I and II rankings. The reason is that collection costs ($/t), emissions (kg C/t) and energy consumption (GJ/t) for collection do not change with capacity.

Case 2: Biomass Transportation Systems

Cost of biomass transportation is a significant component of biomass-delivered cost. Several studies have shown that truck transport cost of agricultural residues' (corn stover) biomass ranges from 20% to greater than 40% of total delivered cost, depending on distance traveled and mode of transportation (2,7,8,17,24). A small-scale biorefinery is not economical in comparison with an oil refinery. At a large scale, biomass transportation by truck may not be physically possible owing to traffic congestion and resulting community opposition. Rail transport of biomass reduces the frequency of loads. Pipeline transport would deliver biomass with minimum ongoing community impact. However, selection of a transportation mode cannot be made based on only one issue. Economical, environmental, social, and technical parameters should be integrated to select the best system.

Input Data and Assumptions

We compare three transportation options: truck transport, rail transport, and pipeline transport. The four criteria include cost of biomass transport, carbon emission during biomass transport, traffic congestion, and maturity of technology. Table 5 lists the input parameters for 4×10^6 dry t/yr transport capacity. The following are comments on input parameters and assumptions for transport analysis:

Cost of biomass transport—Biomass transport cost has two components (fixed cost with respect to distance and distance variable cost). An example of fixed cost, independent of distance of transport, is the cost of loading and unloading a truck, whereas distance variable cost is the "per km" cost of transport, covering fuel, depreciation, maintenance, and labor. The yield is the amount of corn stover that can be removed in a sustainable manner (this takes into account percentage of farmers willing to sell, inaccessible fields, storage, and handling losses) and is derived from earlier study by Perlack and Turhollow (17) of ORNL, US for corn stover. Details of the assumptions regarding the yield estimation are given in the work of Perlack and Turhollow (17).

Transportation cost by truck has been studied in detail by many authors. In the literature, the fixed cost varies from $3.31 to $6.76/dry t and

Table 5
Input Data for Biomass Transportation Systems

Options/criteria	Cost ($/t)	Emissions (kg C/t)	Traffic congestion	Maturity of technology
Truck transport	25.62	2.68	Very high	Very high
Rail transport	64.65	1.40	Average	High
Pipeline transport	73.20	8.22	Very low	Average

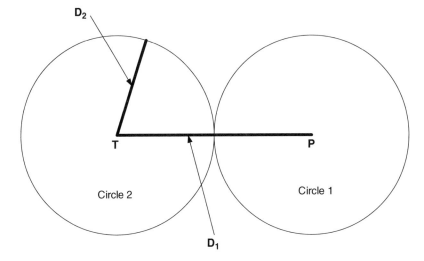

Fig. 2. Configuration of area supplying 4×10^6 dry t of corn stover to a biomass processing facility using rail and pipeline transport.

the variable cost ranges from $0.05 to $0.19/dry t/km. In this study we have taken a blended value of costs reported in earlier studies *(2–4,8)*. A fixed cost of $5.70/dry t and a variable cost of $0.14/dry t/km are used in this analysis. For supplying a biomass processing facility of 4×10^6 dry t/yr capacity by truck only, biomass is required to be collected from a circular area of radius 206 km (based on corn stover yield of 30 dry t/gross km^2).

In this study we have assumed a simple configuration (shown in Fig. 2) of the area for supplying biomass to a 4×10^6 dry t processing facility using rail and pipeline transport. It is important to note that all biomass start their journey on a truck irrespective of its mode of transport afterwards (rail or pipeline).

In Fig. 2 each circular area (circles 1 and 2) has a capacity of supplying 2×10^6 dry t of biomass. In other words, a total of 4×10^6 dry t of corn stover can be collected from the two circular areas. The diameter is a function of biomass yield. In this case, based on a yield of 30 dry t/gross km^2 for corn stover, the radius of each circle is 146 km. A biomass processing facility is located at the center of circle 1 (P) and the rail or pipeline collection terminal is located at the center of circle 2 (T). This is discussed further.

Rail transport of biomass has been studied earlier *(6,25)*. In this study we have used costs reported by a recent study done by Mahmudi and Flynn (published in this volume) for rail transport of straw in North America. The fixed cost of rail transport is $17.01/dry t and the variable cost of $0.03/dry t/km. The fixed cost in case of rail transport includes the capital cost of the rail siding, rail cars, and equipment for loading and unloading biomass. The variable cost includes the charges of the rail company, which include capital recovery and maintenance for track and engines and fuel and operating costs. In this study we estimated the delivered cost of rail transport for the configuration shown in Fig. 2. Biomass is collected from each of the circular areas and transported by trucks to the center of each circle. For biomass collected from circle 1, it is transported to the processing facility directly, whereas for circle 2, biomass is collected and transported by trucks to the rail loading terminal T and from there it is transported by rail to the processing facility P. Hence the cost of biomass transportation in case of rail transport has three components: truck transportation cost to the processing facility P from area of circle 1, truck transportation cost to the rail loading terminal T from the area of circle 2, and rail transportation cost from the loading terminal T to the processing facility P. D_1 (292 km) and D_2 (146 km) represent the distances of transport by rail and truck, respectively. Truck transportation cost is estimated based on average distance of transport. This configuration can be generalized for transportation of biomass to any size of biomass processing facility. Rail transport is discussed in detail in Mahmudi and Flynn *(26)*.

Previous studies have evaluated biomass transport by pipeline *(8,24,26,27)*. Biomass is transported by pipeline in the form of a slurry mixture; the carrier fluid is water. Note that the impact of carrier fluid on the end-use of biomass is discussed in an earlier study *(27)*; water transport to combustion-based processes is precluded by high water uptake by the biomass. In this analysis we have derived the cost of pipeline transport of corn stover at a solids' concentration of 20% (wet basis) using data from previous studies and assumed that the end-use of biomass will be in a biorefinery. Corn stover absorbs water quickly and achieves a moisture level of 80%; so 20% slurry of wet corn stover would be 4% dry matter and 96% water *(26)*. The fixed cost of pipeline transport is $1.82/dry t and variable cost is $0.11/dry t/km, at a pipeline capacity of 2×10^6 dry t/yr. The distance-fixed cost for pipeline transport includes the capital cost of pipeline inlet and outlet equipment. The distance-variable cost includes the capital cost of pipeline and booster stations and the operating and maintenance cost. The delivered cost of biomass by pipeline is calculated similar to the analysis of rail transport. The pipeline inlet terminal is located at the center of circle 2 (T) and biomass is transported by a pipeline to the processing facility (P). Trucks are used to transport biomass from each of the circular area to the center of the circle; dimensions are the same as for the truck plus rail option.

Table 6
Assumptions for Biomass Transport Systems

Items	Cost	Traffic congestion	Emission	Maturity of Technology
Min./Max.	Min.	Min.	Min.	Max.
Weights[a] (w_i)	0.30	0.30	0.25	0.15
Preference threshold (α)[b]	2.00	2.50	1.00	2.50
Indifference threshold (β)[c]	–	0.50	–	0.50

[a]Sum of all the weights is 1.
[b]Units for values of different quantitative criteria are cost ($/t) and emission (kg C/t). Values for qualitative criteria are in absolute numbers.
[c]Values for qualitative criteria are in absolute numbers.

Emissions: in this study only direct carbon emissions are considered. The carbon emissions from truck transport is based on an energy input of 1.3 MJ/t/km (20) and a release of 20 g C/MJ (28). The figure for energy input of rail transport is 0.68 MJ/t/km (20). It is assumed that diesel fuel is used for both truck and rail. The carbon emission from pipeline use is based on the electricity consumption by pumps to transport the biomass slurry. The electrical power is assumed to be produced from a coal power plant; we have used an emission factor of 984.6 g CO_2/kW h (3). The emissions from transport of biomass by each mode are given in Table 5.

Traffic congestion: this is a critical issue when the capacity of biomass processing facility is large. At a capacity of 2×10^6 dry t, a truck would be required every 4 min throughout the year (this calculation is based on a truck capacity of 20-wet t/load, a biomass moisture content of 15%, and plant-operating factor of 0.85, which represents the fraction of time plant runs in 1 yr). This frequency is likely to face public resistance if the plant is close to a community. At the same capacity about 200 car unit trains would be required per day (based on a rail car capacity of 100 wet t). For both the train and pipeline transport options, a significant amount of biomass is still delivered to the biorefinery by truck, but the impact is reduced. In this study we have used traffic congestion as a qualitative criterion. The qualitative assessment for each mode is shown in Table 5.

Maturity of technology: most of the biomass transportation today is by trucks and it is the most common mode of transport. Rail transport is used for transportation of grains for longer distances and a significant quantity of lumber is also shipped by rail. Pipeline transport of biomass is a developing technology. Currently, pipeline is used for transport of pulp in pulp mills on a smaller scale. In this study, we have used maturity of technology as a qualitative criterion. The qualitative assessment for each mode is shown in Table 5. Assumption for each criterion is shown in Table 6.

Table 7
Results for Biomass Transportation Systems

Options	Leaving flow $\varphi^+(a)$	Entering flow $\varphi^-(a)$	Net flow $\varphi(a)$	Ranking PROMETHEE I	Ranking PROMETHEE II
Truck	0.50	0.35	0.15	2	2
Rail	0.51	0.26	0.25	1	1
Pipeline	0.22	0.62	−0.40	3	3

Table 8
Stability Intervals for Biomass Transportation Systems

Criteria	Corn stover Unit: percentage values (%)		
	Weight	Min.	Max.
Cost	30	3	36
Emission	25	17	100
Traffic congestion	30	19	49
Maturity of technology	15	0	29

Results and Discussion for Case 2

Table 7 gives the leaving flow $\varphi^+(a)$, entering flow $\varphi^-(a)$, and net flow $\varphi(a)$ for different transportation alternatives at a capacity of 4×10^6 dry t/yr.

Based on the PROMETHEE I ranking method, rail transport is the best alternative, followed by truck and pipeline transports. PROMETHEE II ranking based on net flows, as discussed earlier, again selects rail over pipeline and truck. Notice that the two ranking systems use different calculations to rank the options. A decision-maker needs to look at both rankings before making a decision. In the case of stover transportation systems, rail transport is the best system based on both PROMETHEE I and II rankings.

Table 8 shows the stability interval for weights of different criteria for biomass transportation systems. In case of transportation systems, stability interval for emission is 17–100%. This represents a wide range and has a small impact of this on the rankings. Ranking is sensitive to weights of cost and maturity of technology within a narrow range. The range of stability interval gives the option to the decision-maker to analyze different situations in which the weights vary. At 4×10^6 dry t, truck traffic congestion will be very high and based on the decision-makers if the weight for traffic congestion is increased more than 49% or decreased lesser than 19%, the ranking will change. Another important factor to note is that in this study we have assumed that electricity for pipeline transport of biomass is generated using coal, which gives a high emission of carbon per tonne of biomass processed. In many regions in North America hydroelectricity and nuclear power are generated. This consideration will greatly influence the emissions, and hence the ranking can change significantly.

Conclusions

The study illustrates a systematic method to rank biomass collection and transportation systems based on a multicriteria assessment. It shows that selection of an option is not only based on economic factors but also on environmental and social factors. This study also illustrates a method to integrate quantitative and qualitative factors in decision-making. PROMETHEE method could be used in various areas for comparison of different systems. In this study focus is on the biomass feedstock systems, but this methodology could be extended to biomass energy conversion systems.

For collection systems, loafing (net flow of 0.27) is the better option than ensiling (net flow of –0.12) and baling (net flow of –0.14). Loafing is ranked at the top using PROMETHEE II methodology. The same methodology for the transportation system at a capacity of 4×10^6 dry t/yr ranks rail transport (net flow of 0.25) at the top followed by truck (net flow of 0.15) and pipeline (net flow of –0.40). This study has used input data specific to the location; the methodology can be used at different locations by inputting regional data.

Acknowledgments

The authors are grateful to the ORNL, and the Natural Sciences and Engineering Research Council of Canada for financial support to carry out this study. All of the conclusions and opinions are solely drawn by the authors, and have not been reviewed or endorsed by any other party.

References

1. Noon, C. E., Zhan, F. B., and Graham, R. L. (2002), *Netw. Spatial Eco.* **2(1),** 79–93.
2. Aden, A., Ruth, M., Ibsen, K., et al. (2002), Report No. NREL/TP-510-32438, http://www.nrel.gov/docs/fy02osti/32438.pdf.
3. Kumar, A., Cameron, J. B., and Flynn, P. C. (2003), *Biomass Bioener.* **24(6),** 445–464.
4. Jenkins, B. M., Dhaliwal, R. B., Summers, M. D., et al. (2000), Presented at the American Society of Agricultural Engineers (ASAE), July 9–12, Paper No. 006035, ASAE, 2650 Niles Road, St. Joseph, MI 49085-8659.
5. Glassner, D., Hettenhaus, J., and Schechinger, T. (1998), In: *Bioenergy'98—Expanding Bioenergy Partnerships: Proceedings,* Vol. 2, Madison, WI, pp. 1100–1110, http://www.ceassist.com/bio98paper.pdf.
6. Atchison, J. E. and Hettenhaus, J. R. (2003), Prepared for National Renewable Energy Laboratory, Colorado, Report No. ACO-1-31042-01, http://www.afdc.doe.gov/pdfs/7241.pdf.
7. Sokhansanj, S. and Turhollow, A. F. (2002), *Appl. Eng. Agri.* **18(5),** 525–530.
8. Kumar, A., Cameron, J., and Flynn, P. (2005), *Bioresour. Technol.* **96(7),** 819–829.
9. Brans, J. P. and Vincke, Ph. (1985), *Manage. Sci.* **31(6),** 647–656.
10. Brans, J. P., Vincke, Ph., and Mareschal, B. (1986), *Eur. J. Oper. Res.* **24,** 228–238.
11. Brans, J. P. and Mareschal, B. (2004), http://www.visualdecision.com.
12. Visual Decision Inc. (2004), http://www.visualdecision.com.
13. Sokhansanj, S., Turhollow, A. F., Cushman, J., and Cundiff, J. (2002), *Biomass Bioener.* **23(5),** 347–355.

14. Sokhansanj, S. and Turhollow, A. F. (2005), *Simulation of Biomass Collection and Supply Systems—Model Development*, internal report, Oak Ridge National Laboratory (unpublished).
15. Sokhansanj, S. and Turhollow, A. F. (2005), *Identifying Competitive Technologies for Stover and Straw Supply Systems*, internal report, Oak Ridge National Laboratory (unpublished).
16. Imagine That, Inc., San Jose, CA, www.imaginethatinc.com.
17. Perlack, R. D. and Turhollow, A. F. (2002), Report No. ORNL/TM-2002/44, //bioenergy.ornl.gov/pdfs/ornltm-200244.pdf.
18. Perlack, R. D. and Turhollow, A. F. (2003), *Energy* **28(14),** 1395–1403.
19. Novem (1996), BTG Biomass Technology Group BV, Enschede, Report No. 9525.
20. Borjesson, P. I. I. (1996), *Energy Convers. Manage.*, **37(6–8),** 1235–1240.
21. Boman, U. R. and Turnbull, J. H. (1997), *Biomass Bioener.* **13(6),** 333–343.
22. Ney, R. A. and Schnoor, J. L. (2002), *Biomass Bioener.* **22(4),** 257–269.
23. Turhollow, A. F. and Perlack, R. D. (1991), *Biomass Bioener.* **1(3),** 129–135.
24. Kumar, A., Cameron, J. B., and Flynn, P. C. (2005), *Appl. Biochem. Biotechnol.* **121(1–3),** 0047–0058.
25. Suurs, R. (2002), University of Utrecht, Copernicus, Department of Science, Technology and Society, The Netherlands, Report No. NWS–E–2002–01, ISBN 90-73958-83-0.
26. Kumar, A., Cameron, J. B., and Flynn, P. C. (2004), *Appl. Biochem. Biotechnol.* **113(1–3),** 27–40.
27. Kumar, A. and Flynn, P. C. (2005), *Fuel Process. Technol.* (in press).
28. Environmental Manual database (1995), developed by German Technical Cooperation (GTZ) with scientific support from Institute for Applied Ecology (Oko—Institute).

Rail vs Truck Transport of Biomass

HAMED MAHMUDI AND PETER C. FLYNN*

Department of Mechanical Engineering, University of Alberta, Edmonton, Alberta, T6G 2G8, Canada, E-mail: peter.flynn@ualberta.ca

Abstract

This study analyzes the economics of transshipping biomass from truck to train in a North American setting. Transshipment will only be economic when the cost per unit distance of a second transportation mode is less than the original mode. There is an optimum number of transshipment terminals which is related to biomass yield. Transshipment incurs incremental fixed costs, and hence there is a minimum shipping distance for rail transport above which lower costs/km offset the incremental fixed costs. For transport by dedicated unit train with an optimum number of terminals, the minimum economic rail shipping distance for straw is 170 km, and for boreal forest harvest residue wood chips is 145 km. The minimum economic shipping distance for straw exceeds the biomass draw distance for economically sized centrally located power plants, and hence the prospects for rail transport are limited to cases in which traffic congestion from truck transport would otherwise preclude project development. Ideally, wood chip transport costs would be lowered by rail transshipment for an economically sized centrally located power plant, but in a specific case in Alberta, Canada, the layout of existing rail lines precludes a centrally located plant supplied by rail, whereas a more versatile road system enables it by truck. Hence for wood chips as well as straw the economic incentive for rail transport to centrally located processing plants is limited. Rail transshipment may still be preferred in cases in which road congestion precludes truck delivery, for example as result of community objections.

Index Entries: Biomass transportation; transportation economics; rail transport; truck transport; straw.

Introduction

In comparison with solid and liquid fossil fuels, biomass is lower in energy density and physical density. Because field harvested biomass has a low energy yield/unit area in comparision with solid fossil fuel sources such as a coal, its initial transport is typically in a transport truck with a 20–40 t capacity. Each of these factors contributes to biomass having a significantly higher cost of transportation per unit of available energy than fossil fuels.

When biomass is transported all the way to its final destination by a transport truck, a further problem with road congestion may arise. Many studies have shown that the optimum size of biomass projects is large

*Author to whom all correspondence and reprint requests should be addressed.

when abundant biomass is available *(1–3)*. A detailed study of three field biomass sources in western Canada showed that optimum power plant size was 900 MW for biomass drawn from harvesting the whole boreal forest, 450 MW for straw, and 130 MW for forest harvest residues (FHR, the branches, tops, and possibly stumps of trees harvested for pulp or lumber); 450 MW is the largest assumed single unit size for boiler and steam turbo-generator in this study *(1)*. At 450 MW, biomass requirements are 2.1 M dry t of biomass per year, equivalent to one truck delivery of straw every 4 min if truck capacity is 17 t of straw per load (typical straw trucks have a nominal capacity of 20 t but are constrained by volume to carry about 17 t/load). This intensity of truck traffic could lead to community resistance in site selection for biomass processing plants.

One alternative to try to reduce transportation costs of field harvested biomass and alleviate truck congestion is to offload biomass from trucks to an alternative transportation mode before delivery to the processing plant. Previous studies have evaluated pipeline transport of biomass in detail *(4,5)*, slurry pipelining of biomass in water has a feasible cost structure for aqueous based processing, such as fermentation or supercritical gasification. Biomass is not amenable to pipeline transport if the end usage is combustion, because uptake of carrier fluid by the biomass is too high.

In this work, we evaluate offloading field harvested biomass onto dedicated unit trains for delivery to large scale power plants. The cost of delivering straw and wood chips from FHR by trucks to rail terminals for further transport by unit train is in comparision with previous studies of power plants supplied by truck alone. We develop an idealized case to explore the critical factors in transshipment of biomass, and then analyze two specific cases in the Province of Alberta, Canada, using existing rail lines. A previous European study looked at transportation costs in Europe by truck, rail, and ship *(6)*, and concluded that higher distances favored first rail, then ship transport. This study did not distinguish fixed and variable costs, a focus of this study, because recovery of incremental fixed costs of transshipment determine the distance at which it is economic.

Shipment of Biomass by a Single Transportation Mode

Many modes of transportation have a similarly shaped profile of cost vs distance shipped, as shown in Fig. 1. The intercept of the line at zero distance, "a," is the fixed cost of shipping a ton biomass regardless of distance; we call this the distance fixed cost (DFC). For example, for trucking in North America a typical cost of loading and unloading a straw or wood chip truck is approx $5/t *(7)*. The slope of the line in Fig. 1, "b," is the distance variable cost (DVC). Most transportation modes have a linear DVC because the distance variable cost components, for example wages, fuel, and capital recovery for the transportation equipment, are directly proportional to the distance traveled.

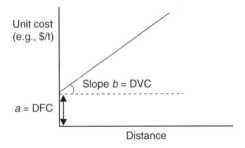

Fig. 1. General plot of unit transportation cost vs distance showing distance fixed and distance variable cost.

Table 1
Values of DFC and DVC Used in this Study

	Truck		Rail		
	DFC ($/dry t)	DVC ($/dry t-km)	DFC ($/dry t)		DVC ($/dry t-km)
			Shipper components	Carrier components	
Straw	4.76	0.1309	6.74	10.27	0.0277
Wood chips	4.98	0.1114	6.35	3.62	0.0306

Truck transport of biomass often requires little or no investment by the shipper, because trucks are owned by the carrier, not the shipper. Straw bales located at the roadside can be loaded on a straw transport truck by equipment located on the truck, and conveying of wood chips into chip trucks has a low fixed cost per ton of wood chips. The situation is different for rail transport in North America, the rail carrier typically owns the main tracks but the shipper owns the siding and all equipment located there, i.e., the shipper is responsible for loading the railcars. In addition, for any long term project such as a power plant supplied by dedicated unit trains the shipper typically owns the railcars. Thus for rail transport of straw or wood chips, DFC has two components to it, the fixed cost charged by the rail carrier and the costs incurred by the shipper for loading the rail cars, including the rail siding and the railcars themselves.

Table 1 shows the values of DFC and DVC used in this study (all costs in this study are reported in 2004 in USD). Truck costs are midrange values drawn from a detailed analysis of previous studies of trucking costs [7]. The estimate of rail carrier costs were drawn from an analysis of estimates for moving straw and wood chips in western Canada provided by a carrier active throughout North America [8]. Wood chips have an assumed moisture level of 45%, and straw an assumed moisture level of 16% [1].

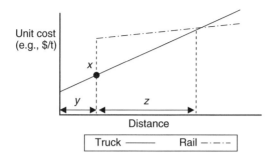

Fig. 2. Unit transportation cost vs distance for truck only and truck plus rail.

Truck and rail carrier DFC is independent of scale, but the shipper component of DFC in Table 1 is calculated based on the "specific case" sizes in this study. The specific wood chip case is a 130 MW power plant burning chipped biomass from FHR; the size of this plant was determined to be optimal from a previous study (1). The specific straw case is a 250 MW power plant burning straw chopped from delivered large round or square bales with a weight of approx 0.6–1 t/bale. The straw power plant size is less than optimal, but the power cost vs size profile is relatively flat between 250 and 450 MW and hence the impact of the smaller plant size is small (1). The wood chip power plant requires 3.8 unit trains/wk, and the comparable figures for straw are 10.1 unit trains/wk. For each of these cases a detailed scope of equipment required by the shipper is developed in order to estimate the shipper's component of DFC. In addition, we study a range of idealized straw and wood chip power plant sizes to estimate the size of power plant at which train transport is justified.

By inspection, rail shipment of biomass has a lower variable cost but a higher fixed cost, this is why most short haul of bulk goods is by truck and long haul by rail. However, the dispersed nature of biomass requires that it start its transportation to a processing plant on a truck. The critical issue for biomass therefore is under what circumstances it is economic to offload truck transported biomass to a train.

Transshipment: Using Two Transportation Modes

Because field sourced biomass must be hauled to a transshipment depot by truck, the total cost of shipment by truck and train is illustrated by Fig. 2. At the point at which biomass is unloaded from the truck, point x in Fig. 2, incremental DFC is incurred. This incremental fixed cost can only be justified if the DVC of the second transportation mode, in this case rail, is lower than that of the first mode, truck. The critical question for transshipment then becomes at what distance, z in Fig. 2, does the lower DVC of the second transportation mode offset the incremental DFC? If the distance for rail shipment of biomass is shorter than z, trans-shipment is not economic. Put another way, is the rail shipment distance far enough so that one can afford to offload biomass to a second transportation mode? In this study we refer to the distance z as the crossover distance, i.e., the minimum rail distance for which transshipment is economic.

Note that the distance y in Fig. 2 is the average truck haul distance to the rail site. This is influenced by the number of transshipment points (in this study, rail sidings) that can offload the biomass. This is discussed later.

Scope of Equipment

Specific Straw Case

For truck only movement of straw, this study assumes that farmers place round or large square bales at roadside and cover them with tarps. The power company contracts with trucking firms to bring straw to the power plant, removing the tarps and leaving them at the roadside for reuse by the farmer. Trucks have self-loading equipment and are contracted year round, so that the annual harvest of straw is primarily stored on public road allowances at the sides of farmer's fields; in western North America road allowances are large and could store all of the straw harvest from adjacent fields. The power plant has at least 1–2 wk of straw storage, and more if seasonal road access is an issue. Trucks are weighed on entry and exit from the plant, and straw moisture is measured; payments to the farmer are calculated on this basis. Straw is removed from trucks by fork lifts; a fleet of 18 is required (including one spare) for the specific case described earlier.

For transshipment of straw to trains, trucks arrive at existing designated grain elevator terminals and are weighed on entry and exit; straw moisture is measured at this time. Rental fees for land usage at grain terminals are based on discussions with industry (9). Straw is stored until 2650 t, an amount sufficient for a 100 rail car unit train, is amassed. Unit trains are dedicated to a single use, and not used for backhaul. Note that 100 car unit trains are standard for carrying grain and supplying coal to some power plants in North America, and that many (but not all) grain elevators have the capability to load 100 car unit trains; rates charged by rail carriers are lowest for 100 car unit trains with short turnaround times, i.e., loading times less than 9 h. Rail cars would be owned by the power plant through purchase or committed long term lease. Straw is loaded on the railcars in 9 h or less by a fleet of 18 forklifts located at each transshipment terminal. The operating crew for the forklifts rotates between the transshipment sites at grain elevators. When unit trains arrive at the power plant they are unloaded by forklifts; there is minimal difference in the requirement for forklifts and other equipment at the power plant between truck and truck plus train delivery of straw.

Specific Wood Chip Case

In western Canada, most trees that are harvested for pulp or lumber are skidded to the side of a logging road whole, delimbed and topped at the roadside. Hence, FHR accumulates at roadsides as long windrows of material. A modified forwarder equipped with a pushing blade and a grapple would consolidate the residues and load the chipper. For truck only delivery of chips, trucks would normally be directly loaded from the chipper, but could also self load from wood chip piles. (Note that an alternative scheme would see residues rolled and bound and transported as "logs," a system developed in Finland for coniferous trees and applied to a variety of species in subsequent trials [10]. This system would require testing to determine its suitability for mixed hardwood and softwood stands found in western Canada.) Trucks would dump chips in the vicinity of dump pockets linked to a conveyor belts at the power plant. A bulldozer and front end loader would consolidate material at the power plant, and would move chips from long term plant site storage to the dump pockets if needed.

For transshipment of chips to trains, specialized sidings on existing rail lines in northern Alberta would be built, equipped with dump pockets and conveyor belts. Trucks would dump chips over the dump pockets; each siding would require a bulldozer and front end loader to consolidate chips. Chips would be accumulated until a full unit train could be loaded from multiple conveyors in 9 h or less. At the power plant rail cars would be rotated over dump pockets, a process currently in use with unit coal trains using gondola cars. There is minimal difference in the requirement for equipment at the power plant between truck and truck plus train delivery of wood chips.

Full equipment and staffing requirements were developed for each case and, capital and operating cost estimates were then calculated. Critical values for each "specific" case are shown in Table 2. Biomass yields are per gross hectare, which allows for uses of land in an area for purposes other than that associated with the biomass, such as communities, roads, and alternate crops. (Details are in ref. [1]). For straw we assume the power plant is able to purchase 80–85% of the available straw in the area in a poor harvest year (lowest quartile) and 60–65% in an average harvest year; note that a study has shown that recovery of straw does not reduce soil carbon in Canadian prairie black soils (11).

Idealized Straw and Wood Chip Cases

In evaluating truck only vs truck plus rail shipment of biomass, we start with an idealized "best" case for straw and wood chips, in which rail sidings are assumed to be located exactly as needed in the center of contiguous sources of biomass, and rail lines direct from the sidings to the power plant are available. An idealized case defines a value of the crossover distance below which transshipment from truck to rail is not economic. If the number of transshipment terminals is optimized to give lowest overall

Table 2
Cost Factors for Biomass Transportation[a]

Fuel type	Straw	Wood chips
Power plant size (MW)	250	130
Biomass yield (dry t/gross ha)[b]	0.416	0.247
Biomass demand (M dry t/yr)	1180	635
Hectares required/yr	2830	77,100
Average driving distance (km)	67.2	350.3
Capital cost at the power plant ($ 000)		
Train cars (12,13)	28,500	16,000
Forklifts	418	–
Trailer buildings	45	15
Front end loader	–	30
Bulldozer	–	100
Building tracks at the plant	–	1200
Operating cost for the power plant ($ 000)		
Salaries	1400	500
Maintenance	318	170
Capital cost per rail transshipment terminal ($ 000)		
Forklifts	418	–
Trailer buildings	45	15
Front end loader	–	30
Bulldozer	–	100
Land for storage	–	23
Building tracks at the terminals	–	1200
Mechanisms (Including conveyor belt, dumping system, and dump pocket)	–	450
Operating cost per rail transshipment terminal ($ 000)		
Salaries	1250	400
Maintenance	318	170
Rent for usage of facilities at terminals (Including land for storage)	150	–
Total capital cost ($ 000)[c]	29,700	22,700
Annual return on capital at 10%	3210	2410
Total operating cost ($ 000)[c]	4730	1410
Shipper component for rail DFC ($/dry t)	6.74	6.35

[a]All costs in 2004 in $ US.
[b]Gross hectares includes all land, including land used for other crops/species, and for non-agricultural or forestry purposes such as roads, communities, and industry.
[c]The capital and operating cost calculations are based on three rail transshipment terminals.

biomass transportation cost, we call this crossover distance the minimum economic rail shipping distance (MERSD).

The first task is to determine the optimum number of transshipment points, i.e., railroad sidings that transfer biomass from truck to storage to train. There is an optimum, because a decrease in transshipment points

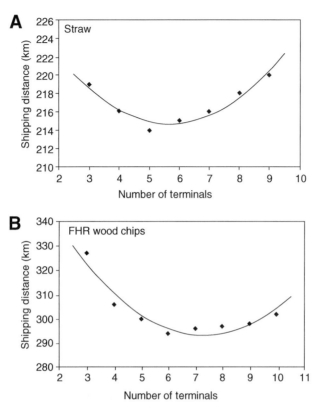

Fig. 3. (A) Total shipping distance vs number of rail transshipment terminals for a 250 MW straw power plant **(B)** and for a 130 MW wood chip power plant.

increases the distance over which biomass is carried at the higher "per km" rate (DVC), i.e., it increases the distance y in Fig. 2, whereas an increase in transshipment points increases the total DFC, because each siding requires a land payment and investment in loading and unloading equipment. The optimum number of terminal is that which gives a minimum total shipping distance, $y + z$ in Fig. 2, which corresponds to the minimum shipping cost. Figure 3 shows the calculated total shipping distance as a function of the number of terminals delivering straw and wood chips. From this we concluded that the optimum biomass shipment per rail terminal was approx 255,000 dry t/yr. We tested this assumption with a larger straw fired power plant, 450 MW, and found a comparable result, shipping 225,000 dry t of straw per terminal is the optimum tradeoff between truck DVC and rail DFC. A value of 255,000 dry t of straw per year per terminal was used in the analysis of the idealized case.

A similar analysis was done for an idealized wood chip case, and the minimum crossover distance corresponds to 100,000 t of wood chips per terminal. This value was used in the subsequent development of the idealized case. Note that the optimum ton per year per terminal is lower for wood chips than for straw; this arises because the gross yield, i.e., the tons

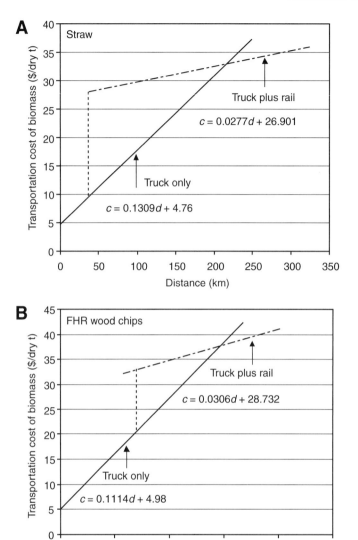

Fig. 4. The cost of delivered biomass for truck only and truck plus rail shipment as a function of distance.

of biomass per gross hectare, is lower for FHR than for straw. Boreal forests have a long rotation cycle, typically 100 yr in Alberta, Canada, and it is this low cutting frequency that gives a low net yield of FHR per gross hectare of forest. Hence to aggregate the same amount of biomass from chipped FHR a longer driving distance is required than for straw, and the optimum configuration of a two mode transportation system is less tonnage per rail terminal for a biomass source with lower gross yield.

Figure 4 shows the cost of delivered biomass in $/t for truck only and truck plus rail shipment for the two cases. Note that the total shipping distance is 215 km for straw and the MERSD distance is 170 km for straw, and yet if centrally located a 250 MW straw power plant would draw from an area of radius of less than 100 km (the biomass draw distance) with an average driving distance of 70 km. The shipping and MERSD distances for straw are so much greater than the biomass draw distance that transshipment to rail is not economic at 250 MW for a centrally located straw power plant; there is not enough haul distance on rail to recover the incremental fixed costs of transshipment. Note that the calculated MERSD distance is consistent with current shipping practice in the grain industry: in discussion an Alberta based grain terminal manager noted that for single rail car quantities of grain (not a unit train) trucks are used to haul grain for distances up to 300 km, even if the truck route parallels existing train tracks (14). Shipment of biomass to a plant that is not centrally located to the draw area is discussed later.

For wood chips from boreal FHR the total shipping distance is 295 km and the MERSD distance is 145 km, whereas the biomass draw distance is 480 km, which gives an average driving distance of 340 km. Hence in an idealized case in which abundant rail lines are available it is more economic to transport boreal FHR wood chips by a combination of truck plus rail; the impact of the lower gross yield of biomass from FHR is to shift the optimum transportation mode to truck plus rail.

DVC for rail transport of straw is slightly less than DVC for rail transport of wood chips, because rail lines for forested areas of Alberta are more remote and presumably have higher maintenance cost. Despite this, the MERSD distance for hauling straw by rail is higher than for hauling wood chips, because the DFC for straw is higher. There is a larger cost in loading straw onto rail cars than wood chips, and a longer distance is required to recover this fixed cost. In general, determination of the minimum economic distance for transshipment requires a specific determination of DVC for both modes of transportation and DFC for switching from one mode to the other.

The size of the power plant determines the biomass draw distance, we analyze a range of plant sizes to determine the point at which, in an ideal case, transshipment to rail is more economic than truck only transport. Figure 5A shows the total delivered cost of straw by truck only and by truck plus rail as a function of power plant size, assuming the plant is centrally located. Not until total plant size reaches 2700 MW does transshipment of straw to rail result in a lower delivered cost of biomass. Previous studies (1) have shown that the optimum size of a straw power plant is in the range of 250–750 MW, and hence the practical potential for transshipment to give a lower biomass cost to centrally located power plants appears to be negligible. Even if straw yield is reduced by a factor of three, say because farmers are willing to sell less than 33% of recoverable straw to a power plant, transport by truck to a 250 MW centrally located power plant is more economic than truck plus rail transport. Figure 5B shows the comparable data for wood chips from boreal FHR. In

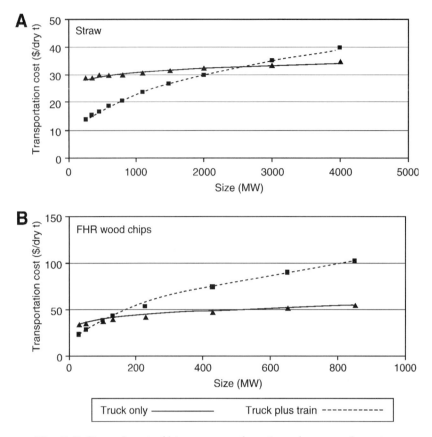

Fig. 5. Delivered cost of biomass as a function of power plant size.

an idealized case, transshipment to train gives a lower delivered cost of biomass to a centrally located plant above 100 MW.

Two Actual Cases in Alberta

The idealized cases assumed that rail lines and sidings were available exactly where needed, and hence from an economic perspective are "best case" analyses. In reality rail lines are well established, and impose their own geographical limitations on plant location. We explore this impact by looking at two specific cases in the Province of Alberta, Canada. Figure 6 shows a map of Alberta, showing existing rail lines. The circles drawn around the straw (S) and wood chip (WC) terminals show the area from which we assume that biomass is drawn. The radius is calculated based on the biomass yield and the size of the plant. For the 250 MW straw plant with three terminals the radius of each draw area is 55 km; for the 130 MW wood chip plant with three terminals it is 270 km.

For the straw case, rail lines in the area of grain growing use Edmonton as a hub. Straw terminals assumed in the specific case are labeled S1–S3 in Fig. 6, and the power plant location at Camrose is labeled

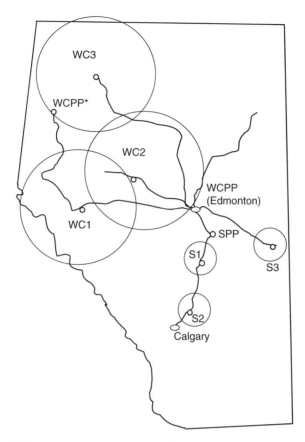

Fig. 6. Map of Alberta showing existing rail lines related to two specific prospective biomass power plants.

SPP. Rail haul distances range from 95 to 215 km. Truck haul distances are far higher than in the ideal case because the straw source is not adjacent to the power plant.

For the wood chip case a large draw area uses three rail lines that also converge on Edmonton. In this case, the only practical location for a wood chip power plant supplied by rail is adjacent to the city of Edmonton. Rail distances are high, and range from 160 to 410 km. The wood chip terminal locations and power plant location are labeled WC and WCPP in Fig. 6. Because roads are more prevalent than rail lines in northern Alberta, a wood chip power plant supplied by truck could have a more central location; WCPP* in Fig. 6 shows the alternate location of a truck supplied wood chip power plant, in Grande Prairie. This is a critical difference between the two transportation alternatives, a more extensive road network allows a more centrally located power plant in comparision with the restrictions imposed by the layout of the rail system.

Table 3 shows the delivered cost of biomass by truck only and truck plus train for the straw and wood chip power plants. Truck only delivery

Table 3
Cost of Biomass Transport by Truck Only and Truck Plus Train for the Straw and Wood Chip Power Plant ($/t)

	Truck only	Truck plus train
Straw plant in Camrose	25.6	33.7
Wood chip plant in Edmonton (rail) or Grande Prairie (truck)	43	44

is less expensive than truck plus train for the straw power plant, even though the straw is being drawn from further away than in the ideal case. Truck only delivery is also less expensive than truck plus train for wood chips, because truck transport enables a more centrally located plant. Thus, although in an ideal case transshipment of boreal FHR wood chips to train gives a lower cost, the geographic constraints of rail line layout shift the balance in favor of truck transport.

Discussion

Field sourced biomass, in comparision with other energy forms, has a low physical and energy density and starts its journey to a processing plant on a truck. For these reasons transportation of biomass is a significant cost, and as biomass processing grows, project developers will place an emphasis on reducing these costs.

Transshipment from truck to any other mode of transportation only makes sense if the second mode has a lower cost per km (DVC) than the originating mode. Train transport has a DVC significantly lower than truck transport. However there is a minimum shipping distance required for transshipment to be economic, because transshipment has incremental fixed costs independent of distance shipped (DFC). Only when the savings in DVC are large enough to offset the incremental DFC is transshipment economic.

DVC and DFC are case specific. DVC depends on the transportation mode and the specific location, and DFC depends on the specific biomass being transported. The values for truck transport cited in this study are representative of North America (7), but would not necessarily apply to Europe, for example, DFC reflects the specific equipment and contractual arrangements involved. For example, truck transport in North America is typically through third party carriers who charge for loading and unloading time, whereas for rail transport in North America it is the shipper, not the carrier, that leases or owns the rail cars and constructs the rail siding and the loading equipment. Thus any analysis of transshipment would have to factor in specific values to determine the minimum economic shipping distance.

There is an optimum number of transshipment terminals for any two mode transportation scheme. A higher number of terminals increases the fixed costs of transshipment, for example, the investment in land and equipment to move biomass from truck to train, but reduces the truck transport distance and thus reduces the overall DVC incurred. In the ideal analysis we assume that the optimum number of terminals is in place, and calculate MERSD based on that number of terminals. The ideal number of terminals and the biomass moved per terminal depends on the biomass gross yield, i.e., the amount of biomass per total hectares in the draw area. A lower biomass gross yield reduces the value of the optimum amount of biomass moved through each terminal, because truck haul distances increase as biomass yield decreases.

For straw or corn stover in North America we estimate that the MERSD to recover fixed costs of loading dedicated unit trains is 170 km. An economically sized centrally located power plant would have a biomass draw area significantly less than the shipping distances associated with rail transshipment, hence using rail would increase, not decrease the overall power cost. For a more diffuse biomass source such as boreal FHR wood chips, we estimate the MERSD to recover fixed costs of loading dedicated unit trains is 145 km. In theory, if rail lines were conveniently located, it would make sense to transship wood chips to rail for transport to an economically sized centrally located power plant.

As this study has shown for one location, the Province of Alberta, Canada, rail lines are usually not ideally located for biomass processing as a fuel or feedstock. Road networks tend to be far more versatile than rail networks for aggregating biomass for processing near the point of origin. Road networks reach more locations than rail lines, reflecting that fact that there are far more motor vehicles than trains in operation at any point in time, and most use of roads is for the transportation of smaller quantities of people or goods than would be appropriate for rail. One consequence of this is that roads allow a more diverse pathway for the movement of goods, and hence give greater flexibility in the location of a power plant supplied by truck as in comparision with one supplied by rail. In this study, the wood chip power plant requires trains arriving from three different rail lines. The existing layout of rail lines in Alberta would dictate an Edmonton location for a rail-based wood chip power plant, because the three rail lines converge there. Locating the wood chip power plant in Northwestern Alberta and supplying it by rail would require much longer overall rail transport distances. However, a truck-based power plant could be located more centrally to the wood chip supply in northwestern Alberta, and the wide availability of roads means that longer transportation distances are not required for this location. This difference in location between a rail supplied and truck supplied plant is enough to shift the economics in favor of truck transport. In general, we conclude from this study that the economic incentive for rail transshipment of biomass to centrally located processing plants is limited at best.

Unit trains provide the least expensive form of rail transport for bulk commodities because the processing of rail cars by the carrier is minimal.

If biomass is being shipped in smaller quantities, and especially if it uses rail cars provided by the carrier, charges will be higher, which would increase the MERSD distance.

Long distance shipment of biomass to a noncentrally located processing plant would justify rail transport at distances above the MERSD distance identified in this study. However, economics will likely favor processing close to the biomass source unless the cost of transporting the products of processing biomass, for example, power or ethanol, are higher than the cost of transporting the biomass itself. Hence, biomass transshipment is theoretically economic but in practical terms we expect it to be cost effective only in limited cases in which long distance transport is required.

The analysis in this study has focused on the cost of truck vs rail delivery of biomass. However, social issues may create a situation in which a higher cost alternative, rail transport, is selected to provide a social good. One potential reason to use rail shipment of biomass even when not lower cost than truck delivery is to avoid a traffic congestion issue that would otherwise preclude the development of a biomass processing plant. Rail shipment by unit train has less impact on communities because rail lines are well established and the additional usage for one or two unit trains/d has a lower impact on people near the transportation corridor than a steady stream of trucks. Truck traffic for economically sized biomass power plants could exceed community tolerance. In this case, a small difference in cost would be offset by the potential failure to receive approval for a truck-supplied project.

Conclusions

The key conclusions from this study are:

- Transshipment of biomass from truck to a second mode of transportation will only be economic if the cost/distance traveled is lower for the second transportation mode. It also requires additional fixed costs independent of the distance (DFC), the investment in land and facilities to transship the biomass. Hence there will always be a minimum economic shipping distance for the second transportation mode, because the savings in DVC must offset the incremental DFC.
- For any two mode transportation scheme there is an optimum number of transshipment terminals that minimizes overall shipping costs. There is a tradeoff between higher DFC and lower DVC as the number of terminals increases. In this study, 255,000 dry t of straw per year and 100,000 dry t of boreal FHR wood chips are the optimal rates of biomass per terminal.
- Alberta, Canada, rail and truck rates are typical of North America. If dedicated unit trains are used for rail transport and the number of terminals is optimized, the MERSD for straw is 170 km, and for boreal FHR wood chips is 145 km.

- A centrally located straw power plant of economic size (250 MW) has a biomass draw area lower than the minimum economic rail distance, and hence transshipment to rail will not be economic for such a plant. It might be warranted if community resistance to truck traffic is a major factor in plant sizing.
- A centrally located boreal FHR wood chip power plant of economic size (130 MW) has a biomass draw area larger than the minimum economic rail distance and associated truck travel, and hence transshipment to rail would be economic if rail lines existed that went to a central location.
- The actual layout of rail lines frequently precludes central location of an economically sized biomass processing plant supplied by rail. Road networks frequently allow more flexible location of processing plants than rail lines. In a specific case analyzed in Alberta, the difference in location for a boreal FHR wood chip power plant tips the balance in favor of truck transport.

Acknowledgments

The authors gratefully acknowledge helpful discussions with Colin Johnson and Roger Stenvold of CN, a North American rail company, and Helgie Eymundson and Ron Simmons of Agricore United, and Marty O'Brian of Cargill Limited, two grain companies operating in Alberta. Amit Kumar of the University of British Columbia provided helpful information on biomass-based power plants. However, all conclusions and opinions are the authors, and have not been reviewed by others.

References

1. Kumar, A., Cameron, J., and Flynn, P. (2003), *Biomass Bioener.* **24,** 445–464.
2. Jenkins, B. (1997), *Biomass Bioener.* **13**(1/2), 1–9.
3. Dornburg, V. and Faaij, A. (2001), *Biomass Bioener.* **21,** 91–108.
4. Kumar, A., Cameron, J., and Flynn, P. (2004), *App. Biochem. Biotechnol* **113,** 27–39.
5. Kumar, A. and Flynn, P. *Fuel Processing Technol.* in press.
6. Borjesson, P. and Gustavsson, L. (1996), *Energy* **21,** 747–764.
7. Kumar, A., Cameron, J., and Flynn, P. (2005), *Bioresource Technol.* **96,** 819–829.
8. Johnson, C. (2004), Personal communication, Canadian National Railway, Edmonton, Canada.
9. Simmons, R. (2004), Personal communication, Agricore United, Winnipeg, Canada.
10. Cuchet, E., Roux, P., and Spinelli, R. (2004), *Biomass Bioener.* **27,** 31–39.
11. Hartman, M. (1999), Agdex, 519–525, Alberta Agriculture, Canada.
12. Laver, C. (2004), Personal communication, Trinity Rail, Huntsville, Canada.
13. Nicholson, H. (2004), Personal communication, National Steel Car Limited, Canada.
14. O'Brian, M. (2004), Personal communication, Cargill Limited, Edmonton, Canada.

Corn Stover Fractions and Bioenergy

Chemical Composition, Structure, and Response to Enzyme Pretreatment

Danny E. Akin,*,[1] W. Herbert Morrison, III,[1] Luanne L. Rigsby,[1] Franklin E. Barton, II,[1] David S. Himmelsbach,[1] and Kevin B. Hicks[2]

[1]Russell Research Center, PO Box 5677, USDA-ARS, Athens, GA 30604, E-mail: deakin@qaru.ars.usda.gov; and [2]Eastern Regional Research Center, 600 E. Mermaid Lane, USDA-ARS Wyndmoor, PA 19038

Abstract

Information is presented on structure, composition, and response to enzymes of corn stover related to barriers for bioconversion to ethanol. Aromatic compounds occurred in most tissue cell walls. Ferulic acid esterase treatment before cellulase treatment significantly improved dry weight loss and release of phenolic acids and sugars in most fractions over cellulase alone. Leaf fractions were considerably higher in dry weight loss and released sugars with esterase treatment, but stem pith cells gave up the most phenolic acids. Results help identify plant fractions more appropriate for coproducts and bioconversion and those more suitable as residues for soil erosion control.

Index Entries: *Zea mays* L.; lignin; phenolic acids; esterase; cellulase.

Introduction

The US ethanol industry produced more than 3.4 billion gal in 2004, up from 2.8 billion gal in 2003 and 2.1 billion gal in 2002. (from the Renewable Fuels Assn. in http://www.eere.doe.gov/biomass/ethanol.html). New ethanol plants with increased production capacity signify further increases in future years. Fuel ethanol, which is generally added at about a 1:9 ratio to gasoline, has potential to be used at considerably higher ratios, such as a blend of 85:15 ethanol to gasoline *(1,2)*. Approximately 95% of this ethanol is produced from corn starch, with the remaining from biomass such as wheat, barley, sugarcane, and wastes from a variety of crop/fiber industries *(1)*. A substantial increase in ethanol production based on corn grain may conflict with use of corn for food and feed and therefore be impractical. Further, the cost of corn is a major factor in the economics of bioethanol.

*Author to whom all correspondence and reprint requests should be addressed.

Lignocellulosic materials potentially offer a low-cost resource to expand ethanol production. Considerable research has been conducted to develop a conversion process and recently the first commercial production of ethanol for biofuel from a lignocellulosic resource was reported by the Iogen Corporation. Corn (*Zea mays* L.) stover, the above ground plant material left after grain harvest, has been identified by the US Department of Energy as a near- to mid-term agricultural residue for lignocellulose-to-ethanol production, and considerable interest and study have been devoted to this purpose *(3)*. Annual production of corn stover in the United States is reported to be well over 200 millon t *(4)*, but use of this large resource requires investigation. Wilhelm et al. *(5)* assessed the complex agronomic issues of corn stover removal for bioenergy as related to soil conditions and sustainable crop production; they concluded that a portion of the corn stover could be harvested within limits for ethanol production. Removal of approx 30% of the stover from the field is often mentioned as an acceptable practice for both biomass conversion and for soil erosion protection, but the amount that should be removed is dependent on several factors *(5)*. Selective use of portions of the corn stover for bioenergy, while leaving enough material for sustainable soil productivity, is a potential strategy, but information on the most advantageous fractions for bioconversion and for soil conservation is required.

Corn stover lignocellulose is recalcitrant to biodegradation, and the sugars comprising plant tissues are not readily available for conversion. Lignin and other aromatics associated with plant cell walls are known generally to be major detriments to biodegradation in plants *(6)*. Pretreatment of the lignocellulose is required before corn stover can be efficiently converted to ethanol *(7)*. Galbe and Zacchi *(8)* reported that the most investigated pretreatment for lignocellulosic conversion is steam explosion, with or without acid. Numerous studies have been undertaken to optimize use of dilute acid pretreatment to improve the enzymatic saccharification of corn stover *(9,10)*. Biological methods, including the use of the delignifying white rot fungi *Cyathus stercoreus* and *Phanerochaete chrysosporium*, have been used to improve enzymatic degradation of corn stover *(11)*.

It is clear from general knowledge of plants and biodegradation *(6)*, and from studies of corn stover particularly *(7)*, that fundamental knowledge on the cell-wall structure and location of particular aromatics is needed to optimize methods to improve fermentation of corn stover. The objective of work presented herein is to identify the location and types of aromatics within particular corn stover fractions and to rank their degradation by enzymes as a basis to assess potential and develop strategies for improved bioconversion. Particularly, in light of the need to leave some of the corn stover to benefit the soil, results should help identify the plant fractions most appropriate for bioconversion and the fractions most suited to leave as residues.

Materials and Methods

Samples

Mature, standing plants of Pioneer 3085 were cut approx 5–10 cm from the soil after grain harvest. The seeds were planted on 28 March 2002, and the corn harvested on 17 July 2002, from an experimental plot at the Pee Dee Research and Education Center, Florence, SC. Five intact, standing plants were harvested in October, 2002, and the various fractions separated and used are described for individual studies.

For baled corn stover, a commercial product was purchased. The corn stover (unknown cultivar) was harvested and shredded after grain harvest, wrapped in plastic, and boxed. Random samples of mixed fractions and pure fractions were selected from one bale.

Separation of Plant Fractions

For fractions of standing plants, the five intact plants of Pioneer 3085 were partitioned into the leaf blades (laminae and midveins), sheaths (laminae and midveins), and stems. The husk and cobs were not considered in this sample. For the shredded and baled corn stover, three representative lots of about 500 g were separated into the three parts listed above and also into husks and cobs. The percentage of each fraction was determined based on the total weight of the sample as is (without further drying). For the baled sample, the percentage of all five fractions was determined; the percentage of the leaf blades, sheaths, and stems was also calculated based on the sum of only these fractions.

Histochemistry

The following fractions were collected from both the third and eight internodes regions from the flower for each of three standing plants: leaf blade lamina, blade midvein, leaf sheath, sheath midvein, and intact stem. The rigid midveins were excised using a razor blade from the lamina of blades and sheaths. Fractions were soaked in water for 1 h under vacuum, and free-hand sections (approx 100-µm thick) were cut with a razor blade. Sections were then treated with a series of histochemical stains emphasizing particular aromatic constituents as follows: coniferyl lignin by acid phloroglucinol *(12)*, syringyl lignin by chlorine-sulfite *(12)*, and phenolic constituents by diazotized sulfanilic acid *(13)*. No additional preparation was done to minimize staining of aromatic amino acids. Stained sections were observed through an Olympus BH-2 light microscope under bright field illumination and scored for intensity as follows: 3 = intense, 2 = definite, 1 = slight, and 0 = none.

Chemical Analyses

Samples of untreated fractions as well as pretreated residues were freeze-dried and ground in a SPEX 5100 Mixer Mill (SPEX Industries, Inc.,

Metuchen, NJ) before chemical analysis. Fractions of the standing plants were treated with 4 M NaOH at 170°C for 2 h to release a variety of mono- or dimethoxylated aromatic compounds. The aromatic compounds, including phenolic acids, were measured as their silyl ethers using N,O-bis(trimethylsilys)trifloroacetamide (BSTFA) using gas-liquid chromatography as previously described *(14)*. The amounts of compounds of similar structure were summed to provide estimates of coniferyl lignin and syringyl lignin, and these compounds were summed with and total *p*-coumaric and ferulic acids to provide concentrations of aromatics with plant fractions. Ester- and ether-linked phenolic acids were determined by treatment of freeze-dried ground material with (a) 2 M NaOH at room temperature for 24 h or (b) 4 M NaOH at 170°C for 2 h, respectively. A follow-up study using sequential treatments of 2 M NaOH and 4 M NaOH under conditions described earlier was carried out to confirm the high amount of ester-linked phenolic acids.

Cellulase Degradation of Corn Fractions

Cellulase (EC 3.2.1.4, catalog No. C-8546, Sigma Chemical Co, St. Louis, MO), which had shown high activity against cellulosic materials in our laboratory, was used to evaluate and rank the biodegradation of various nonpretreated corn fractions. The general potency of the enzyme was verified with 100 × 12 mm filter-paper strips (Whatman No. 541), which broke in 5 h when incubated in 50 mM sodium acetate buffer at pH 5.0. This enzyme was then tested against plant fractions from both the standing plants and the baled corn stover. The cellulase was added to 50 mM sodium acetate buffer, pH 5.0, to give an activity of 20 U/mL (192.3 g/100 mL buffer). Twenty milliliters of enzyme solution were added to 1 g of freeze-dried corn stover material, which had been ground in a Wiley Mill to pass through a 10 mesh and then a 20 mesh screen. Samples were incubated at 37°C for 72 h unless otherwise indicated. Afterward, the tubes were centrifuged, the liquid decanted, and the residue freeze-dried and weighed. Dry weight loss was calculated as follows: 1 − (residual weight/ starting weight) × 100.

Pretreatment of Corn Stover Fractions With Esterases Before Cellulase Hydrolysis

Separated fractions of the corn stover were ground as described and freeze-dried, and 0.5 g samples were used in triplicate for each test. For this study, the rind was manually separated from pith tissue for stems, and these stem parts were tested individually. Samples were treated with the commercial products Depol 740 L, which is one of a range of ferulic acid esterase-containing enzymes, or with TP 692 L from Biocatalysts Ltd. (Pontypridd, UK). Depol 740 L reportedly removes free phenolic acids from plant material and increases amounts of fermentable sugars. Depol 692 L is also used as a macerating agent for plant material and consists of a more complex mixture of ferulic acid esterases, cellulases, and hemicellulases. Typical ferulic acid esterase activities for Depol 740 L and Depol 692 L are 36 U/g and 800 U/g, respectively (product information from Biocatalysts Ltd., UK). These commercial enzymes were used as supplied and in high amounts to ensure maximal loss of ester-linked aromatic acids. Previous work on other lignocellulosic material had indicated that levels of Depol 740 L at 1 g/1 g sample was as sufficient as higher levels in releasing ferulic acid; therefore, 1 g was used. Each sample was mixed with 20 mL of 50 mM sodium acetate buffer, pH 5.0, containing 1.0 g of Depol 740 L or 1.0 g TP 692L, mixed for 5 s, and incubated at 37°C for 24 h in a reciprocal water bath at about 100 back and forth strokes/min. After incubation with esterases, tubes were centrifuged at $730 \times g$ for 2 min and the liquid decanted. The samples were washed by adding the pellet to 20 mL distilled water, mixing for 5 s, centrifuging, and decanting the liquid. Decanted liquids, enzyme mixture and washing, were analyzed for phenolic acids and for sugars. The esterase-treated, washed residues were subsequently incubated as above with cellulase (as described earlier) or freeze-dried and weight loss calculated before subsequent incubation with cellulase. Dry weight loss was calculated as described earlier.

Ferulic and *p*-coumaric acids and sugars in the decanted liquids from the pretreatments were analyzed as their silyl ethers and measured by gas–liquid chromatography as described *(15)*. Data are presented as mg/g amounts of the starting weights of freeze-dried, ground fractions.

Results and Discussion

Plant Fractions

Proportions of the various fractions of standing plants of Pioneer 3085 and of commercially baled corn stover are shown in Table 1. Husk and cob fractions contributed 15% and less than 1%, respectively, of the total weight in the baled stover (not shown). Considering only the leaf blade, sheath, and stem portions, the standing plants had roughly similar amounts of leaf blade and stems but lesser amounts of leaf sheath. In the baled stover, amounts of stem fractions were considerably higher than either of the leaf fractions, which were lower than in the standing plants. Other work has indicated a stem content of near 50%, when all fractions were considered in mature corn *(4,16)*. The shattering of the dry and brittle leaf blade and sheath during harvesting and baling likely increased stem proportions in stored corn stover.

Lignin and Aromatic Content

Lignins in higher plants vary in type based on the functional groups of the aromatic ring *(12)*. We tested for mono and dimethoxylated lignins as well as phenolic acids. Total aromatics of a single sample of mature Pioneer 3085 as determined for coniferyl and syringyl lignin groups plus

Table 1
Proportion of Plant Fractions in Corn

Plant part[a]	Weight[a] (%)	
	Standing plant[b]	Baled stover[c]
Blade	36.1 ± 3.6	30.9 ± 6.7
Sheath	24.5 ± 1.9	16.9 ± 1.23
Stem	39.4 ± 2.8	52.2 ± 8.0

[a]Average and standard deviation of weight as is.
[b]From five plants of Pioneer 3085 planted 28 March 2002, and harvested 17 July 2002.
[c]From a commercial source of harvested, shredded, and baled stover (unknown cultivar) delivered to the laboratory. Data derived from three replicates (approx 500 g) from one bale.

p-coumaric and ferulic acids gave the following amounts (mg/g dry weight): leaf blade, 7.5; leaf sheath 16; stem pith, 21.7; and stem rind, 45.4. The plant fractions varied in aromatic types and linkages, with the rind containing higher levels of all components. The ratio of coniferyl lignin to syringyl lignin varied for the different plant components, and p-coumaric acid was higher than ferulic acid in all components except the leaf blade.

Warm-season grasses like corn often are particularly high in phenolic acids, with high proportions of p-coumaric acid (17). Ester-linked p-coumaric and ferulic acids, as determined by susceptibility to 2 M NaOH, and total phenolic acids (i.e., both ester- and ether-linked acids), as determined by a more rigorous extraction with 4 M NaOH under high temperature and pressure, indicated variable amounts of phenolic acids in the fractions (Table 2). The amounts of p-coumaric and ferulic acids were similar in blade laminae, whereas the stem tissues had considerably more p-coumaric acid. Although the milder 2 M NaOH extraction removes more ester-linked phenolic acids and the more rigorous 4 M treatment releases both ester-and ether-linked acids, little difference occurred between the two treatments. In fact, levels after 2 M treatment tended to be greater, although not significant ($p < 0.05$), except for leaf-sheath midveins. A separate evaluation using sequential treatments of 2 M (room temperature, 24 h) followed by 4 M NaOH (170°C, 2 h) confirmed that most of these phenolics acids were ester-linked; the more rigorous 4 M NaOH treatment destroyed some of the acids based on tests with known amounts of standards. These data provide further evidence of high levels of ester-linked p-coumaric and ferulic acids in the various fractions of corn lignocellulose.

Histochemical Staining and Aromatic Content of Specific Cell Walls

The anatomy of leaf blade and stem is well known (18,19). Leaf laminae have a high amount of thin-walled, living cells that do not stain for aromatics; most of the thickened cell walls are associated with the

Table 2
Ester- and Ether-Linked Phenolic Acids in Corn Fractions[a]

	p-Coumaric acid (mg/g)				Ferulic acid (mg/g)			
	2 M NaOH		4 M NaOH		2 M NaOH		4 M NaOH	
Leaf-blade lamina	1.5	0.5	1.7	0.2	2.5	0.8	2.8	0.5
Leaf-blade midvein	9.3	1.6	6.4	1.1	4.4	1.5	5.0	1.1
Leaf sheath	8.5	2.4	5.4	0.6	8.4	5.0	6.1	1.2
Leaf-sheath midvein	12.3	2.2	7.2	0.6	8.6	1.6	7.2	0.7
Stem pith	28.1	7.8	18.8	2.9	5.7	1.5	7.4	1.4
Stem rind	32.1	11.1	15.4	1.4	6.6	3.2	6.6	0.8

[a]Average ± standard deviation of three plants from Pioneer 3085. Extractions at 2 M were at room temperature for 24 h and extractions at 4 M were at 170°C for 2 h, with samples analyzed as in Table 3. For 2 M extractions, each plant part was run a second time, and the mean value of the two runs was taken as the average for that plant. No differences occurred ($p > 0.05$) between 2 M and 4 M extractions for any part for either acid, with the exception of pCA for leaf-sheath midvein.

lignified vascular tissue and the epidermis (6). Sheaths show a considerable amount of thin-walled cells but a higher proportion of thick-walled cells than leaf blades. Sheaths contain more lignin overall than blades. In stems, an outermost rind consists of thick-walled, subepidermal sclerenchyma and parenchyma cells with a concentration of lignified vascular bundles (18). Pith consists of thin-walled parenchyma cells within which are embedded lignified vascular bundles. The rind, which consists of thickened cell walls in epidermal and subepidermal layers, is considerably thicker in lower internodes; pith cell walls are relatively thin in both upper and lower internodes.

Just as the various fractions of the corn plant differed in aromatics, individual tissues within these fractions varied with the presence of aromatics as determined by histochemical stains. Sarkanen and Ludwig (12) reported that acid phloroglucinol "had universal application to all lignins, although the reaction may be weak or even absent in lignins containing high amounts of syringyl propane units." The chlorine-sulfite stain, however, gives a red color with syringyl units of lignin (12). The application of these two stains, therefore, can identify the location of lignin types within tissues. In addition to stains for coniferyl (acid phloroglucinol) and syringyl (chlorine-sulfite) lignin, diazonium salts react with phenolic compounds in general (13) and are useful to show phenolic acids (presumably ester-linked) in nonlignified, "living" tissues (20). Although these methods stain for aromatic compounds in cell walls, a role for aromatic amino acids was not addressed. The occurrence of aromatic compounds in unlignified cell walls is a major feature of grasses (21). The linkages and compounds of aromatics in unlignified and lignified cell walls of bermudagrass (Cynodon dactylon), another C-4 grass, have been verified using ultraviolet absorption microspectroscopy (22,23). Because maturity and plant position have been shown to affect lignification in plant parts (24), fractions from the third and eighth internodes from the top of the plant were analyzed.

The salient histochemical results are summarized from three replicates of sections from each fraction of standing plants of the third and eighth internodes. Some differences in histochemistry between the two lignin stains occurred among tissues, both in positive reactions and in intensities. In sheath midveins and stem tissues, eighth internodes tissues often gave more intense reactions, especially with chlorine-sulfite staining. Diazotized sulfanilic acid indicated that most of the cell walls contained aromatic compounds. Mesophyll cell walls of leaf-blade laminae, however, lacked a positive reaction for aromatics. In stems, even the thin-walled parenchyma cells gave a strong reaction (score of 2.7 out of a possible 3.0). Eighth internode fractions did not have consistently more intense reactions than the third internodes. These data complement that for total aromatics in indicating that considerable lignification exists in most cell walls of corn stover, but variations are evident in different fractions. Phenolic acids, presumably ester-linked to carbohydrates, are prevalent in most of the tissues.

Incubation With Esterase and Cellulase

A preliminary test was carried out to assess dry weight loss after cellulase treatment of fractions from standing plants and corn stover. Material lost in buffer solutions (without cellulase) was greater in standing plants at approx 30% in all fractions in comparison with 7–16% loss in baled stover. This difference suggested more drying or loss of soluble components from corn stover that was baled and stored. After incubation with cellulase, dry weight loss was low in all samples. Dry weight loss for leaves (blades and sheaths) was 7–16% over that by buffer alone, whereas that for stems was approx 5%. This low biodegradabililty with cellulase was consistent with the recalcitrance owing to high levels of aromatics in cell walls, particularly for stems.

Pretreatment with Depol 740 L was equal to or better than TP 692 L in increasing dry weight loss (not shown). Pretreatment with esterase resulted in significantly greater dry weight loss with cellulase than without esterase pretreatment for all fractions (Table 3). Improvement was greater in leaf blades and sheaths (near 62% dry weight loss) compared with that in stem rind or pith (21–29%). Incubation filtrates (plus washings) from stem fractions contained considerably more phenolic acids than those from leaf fractions (Table 3). The ferulic acid esterase was significantly more effective in releasing both p-coumaric and ferulic acids into the liquid than treatment with cellulase alone. Levels of ferulic acid released were about twofold higher than those of p-coumaric acid in leaf fractions but similar in amounts in stem fractions. These data suggest that some p-coumaryl esterase exists in Depol 740 or the enzyme has crossreactivity with both ester-linked phenolic acids. On subsequent treatment

Table 3
Dry Weight Loss and Phenolic Acids in Filtrate and Residue of Corn Fractions Treated With Enzymes

Fraction	Treatment[a]	Dry weight loss (%)	Phenolic acids (mg/g)				
			Filtrate			Residue	
			p-Coumaric	Ferulic		p-Coumaric	Ferulic
Leaf blade	Buffer 72 h	32.0 ± 0.7^b	$0.12 \pm 0^{b,c}$	0.32 ± 0.10^b		0.92 ± 0.26^b	0.63 ± 0.22^b
	Esterase 24 h		0.24 ± 0.02^d	0.50 ± 0.06^c			
	then cellulase 72h	61.8 ± 1.0^d	0.08 ± 0.06^c	0.08 ± 0.06^d		2.86 ± 0.26^c	1.20 ± 0.10^c
	Cellulase 72 h	40.9 ± 0.4^c	0.14 ± 0.02^b	0.33 ± 0.07^b		1.17 ± 0.48^b	1.22 ± 0.36^c
Leaf sheath	Buffer 72 h	30.8 ± 1.5^b	0.07 ± 0.02^b	0.13 ± 0.06^b		1.94 ± 0.51^b	1.23 ± 0.36^b
	Esterase 24 h		0.36 ± 0^c	0.69 ± 0.03^c			
	then cellulase 72h	62.7 ± 1.8^d	0.16 ± 0.01^a	0.13 ± 0.01^b		6.16 ± 0.47^c	1.82 ± 0.10^c
	Cellulase 72 h	47.1 ± 1.3^c	0.09 ± 0.01^b	0.35 ± 0.02^d		7.44 ± 0.10^d	3.52 ± 0.16^d
Rind	Buffer 72 h	12.8 ± 0.9^b	0.38 ± 0.05	0.14 ± 0.02^b		6.81 ± 2.12^b	0.99 ± 0.29^b
	Esterase 24 h		0.44 ± 0.02	0.25 ± 0.01^c			
	then cellulase 72h	20.5 ± 1.1^d	0.31 ± 0.03	0.14 ± 0.02^b		12.12 ± 32.78^c	1.58 ± 0.40^c
	Cellulase 72 h	16.4 ± 1.0^c	0.23 ± 0.14	0.10 ± 0.05^b		28.07 ± 0.75^d	4.56 ± 0.15^d
Pith	Buffer 72 h	14.1 ± 2.6^b	0.44 ± 0.04	0.37 ± 0.04		25.55 ± 1.91^b	5.91 ± 0.38^b
	Esterase 24 h		0.48 ± 0.01	0.57 ± 0.29			
	then cellulase 72h	28.8 ± 0.7^c	0.46 ± 0.03	0.30 ± 0.03		8.61 ± 2.06^c	1.43 ± 0.34^c
	Cellulase 72 h	16.7 ± 0.7^b	0.26 ± 0.15	0.22 ± 0.11		21.67 ± 0.85^d	4.71 ± 0.15^d

[a]Cellulase, Cat. No. C8546, Sigma-Aldrich; Esterase, ferulic acid esterase (Depol 740L, Biocatalysts Ltd.).
[b-d]Values within columns with different values differ at $p \leq 0.05$.

with cellulase, the esterase-treated residues often gave up more phenolic acids than cellulase treatment alone; this difference was reflected in the higher dry weight losses (Table 3). With the exception of leaf blades, the levels of phenolic acids remaining in the treated residues were lower after esterase plus cellulase compared with cellulase only, based on dry weight of starting material (Table 3). Although the results indicate preferential loss of phenolics with esterase pretreatment, not all phenolic acids were released. Additional physical treatments, such as homogenization or grinding (Akin, unpublished data; 25) may improve the interaction of enzyme and substrate and increase the amount of phenolic acids released.

The release of soluble sugars by esterase and cellulase treatments is shown in Table 4. Release of soluble sugars by esterase pretreatment, either directly or in subsequent cellulase treatments, was particularly effective in removing glucose and xylose from leaves. The additional available glucose from esterase and subsequent cellulase treatments included an increase of 2.53 and 1.44 times in blades and sheaths, respectively. Sugar release from stem fractions was not, or was less, improved with esterase pretreatment. Stem rinds were particularly recalcitrant to esterase pretreatment.

Cell walls of grasses, such as corn, vary from other plant types, for example, woody tissue, dicotyledonous plants, in having a high level of phenolic acids, particularly *p*-coumaric and ferulic acids *(21,26)*. These phenolic acids are often esterified to arabinose and then linked to xylose moieties *(27,28)*; dimers of these compounds commonly occur as recalcitrant bridges in grasses also. Ester-linked phenolic acids appear to be the main aromatic compound in certain unlignified "living" cell walls of grasses *(20,22)*. The present data indicates that a high proportion of the aromatics in all corn stover fractions are ester-linked phenolic acids. These esters, which can inhibit utilization of linked sugars *(29)*, are more readily susceptible to pretreatment than highly lignified (e.g., ether-linked) tissues and can be disrupted by milder treatments that break ester bonds and release phenolic acids. Other recent research has similarly shown that ferulic acid esterases, particularly with xylanases, release phenolic acids from plant cell walls *(30)*. The present research indicated that Depol 740 L was as effective as products with esterase plus other cell wall degrading enzymes (based on information from the supplier); the exact enzyme composition of these products is not available. Further research on specific combinations of esterases and xylanases for particular plant fractions could improve removal of phenolic acids.

The release of phenolic acids from corn stover may provide a high-value coproduct and improve the cost efficiency of bioconversion. Graf *(31)* reviewed several actual or potential uses of ferulic acid:

1. Substrate for conversion to vanillin used as a flavor constituent;
2. UV protection in suntan lotions and cosmetic preparations;
3. allelopathic functions and regulation of plants, and;
4. protection against pests.

Table 4
Carbohydrates in Filtrates of Enzyme-Treated Corn Fractions

Sample	Enzyme treatment	Arabinose (mg/g)	Xylose (mg/g)	Mannose (mg/g)	Galactose (mg/g)	Glucose (mg/g)
Leaf blade	Buffer 72 h	0.8 ± 0.1^a	1 ± 0.6^a	7.6 ± 1.7^a	1.4 ± 0.4^a	4.4 ± 1.4^a
	Esterase 24 h	2.2 ± 0.5^a	20.6 ± 3.9^b	0.4 ± 0.6^b	0.9 ± 0.2^a	35 ± 9.9^b
	Esterase + Cellulase 72 h	5.6 ± 3.1^b	16.1 ± 0.6^c	7.3 ± 0.6^a	7.3 ± 0.6^b	90 ± 3.6^c
	Cellulase 72 h	$3.3 \pm 0.7^{a,b}$	8 ± 0.7^d	5.6 ± 2^a	2.8 ± 0.1^c	53 ± 20.1^b
Leaf sheath	Buffer 72 h	0.8 ± 1.3^a	0^a	27.2 ± 14.5^a	0.2 ± 0.4^a	54.4 ± 31.7^a
	Esterase 24 h	2.5 ± 0.3^a	13.3 ± 2.1^b	0.4 ± 0.4^b	0.4 ± 0.4^a	30.9 ± 3.3^a
	Esterase then Cellulase 72 h	7.8 ± 1.7^b	25.7 ± 3.8^c	$14.1 \pm 3.1^{a,b}$	2.3 ± 0.2^b	150 ± 8.9^a
	Cellulase 72 h	6 ± 0.3^b	11.1 ± 0.2^b	9.4 ± 1.5^b	1.4 ± 0.2^c	125.3 ± 99.1^a
Stem rind	Buffer 72 h	1 ± 0.9^a	3.5 ± 3.2^a	0.7 ± 0.6^a	0.5 ± 0.4^a	8.6 ± 8.1^a
	Esterase 24 h	2.1 ± 1.9^a	7.5 ± 1.0^b	0^b	0^b	11.9 ± 1.6^a
	Esterase then Cellulase 72 h	1.7 ± 1.4^a	7.3 ± 1.8^b	1.5 ± 0.1^c	0.8 ± 0.1^a	22.7 ± 3.1^b
	Cellulase 72 h	2.7 ± 0.3^a	10.4 ± 0.6^b	$1 \pm 0.3^{a,c}$	1.4 ± 0.1^c	39.2 ± 2.8^c
Stem pith	Buffer 72 h	5 ± 0.7^a	17.8 ± 1.8^a	0.6 ± 0.1^a	1.6 ± 0.2^a	27.3 ± 2.5^a
	Esterase 24 h	3.2 ± 0.6^b	16.2 ± 2.3^a	0.3 ± 0.2^a	3.9 ± 4.4^a	27.5 ± 11.1^a
	Esterase then Cellulase 72 h	$4.3 \pm 0.6^{a,b}$	17.8 ± 1.4^a	0.5 ± 0.5^a	1.5 ± 0.1^a	56.3 ± 4.5^b
	Cellulase 72 h	6.4 ± 0.4^c	23 ± 0.7^b	1.1 ± 0.1^b	2.4 ± 0.3^a	68.6 ± 1.5^c

[a-d]Values within columns with different values differ at $p \leq 0.05$.

Much of the usefulness of ferulic acid and other plant phenolics is owing to their antioxidant and antimicrobial properties *(32)*. These phenolic monomers, when released from cell walls, can be toxic to biological systems *(33)* and become a detriment to bioconversion unless they are removed.

Leaves of corn stover showed the greatest improvement in dry weight loss with esterase pretreatment. Stem rind cells are the most heavily lignified and recalcitrant and may be of little value without harsh, chemical pretreatment. A physical processing method (e.g., shredding and rubbing) on stems could separate rind from pith cells, which differ in lignification and release of ferulic acid and are more amenable to pretreatments. The more recalcitrant fractions could be left in the field for soil erosion control and soil organic carbon *(5)*, thereby reducing transport weight and concentrating the most useful resource for bioconversion.

Acknowledgments

We thank Roy Dodd, Clemson University, for samples of intact plants of Pioneer 3085. We also thank Biocatalysts Ltd., Pontypridd, Wales, UK, for their generous gift of ferulic acid esterases.

Mention of trade names does not constitute an endorsement of one commercial product over another but is used only for identification purposes.

References

1. Agricultural Research. (April, 2002). vol 50(4). US Department of Agriculture, Washington, DC, 23p.
2. Sun, Y. and Cheng, J. (2002), *Bioresour Technol.* **83**, 1–11.
3. McAloon, A., Taylor, F., Yee, W., Ibsen, K., and Wooley, R. (2000), NREL Technical Report NREL/TP-580-28893. National Renewable Energy Laboratory, Golden, CO.
4. Sokhansanj, S., Turhollow, A., Cushman, J., and Cundiff, J. (2002), *Biomass Bioenergy* **23**, 347–355.
5. Wilhelm, W. W., Johnson, J. M. F., Hatfield, J. L., Voorhees, W. B., and Linden, D. R. (2004), *Agron. J.* **96**, 1–17.
6. Akin, D. E. (1989), *Agron. J.* **81**, 17–25.
7. McMillan, J. D. (1994), In: *Enzymatic Conversion of Biomass for Fuels Production*, Himmel, M. E., Baker, J. O., and Overend, R. P. (eds.), American Chemical Society, Washington, DC, pp. 292–324.
8. Galbe, M. and Zacchi, G. (2002), *Appl. Microbiol. Biotechnol.* **59**, 618–628.
9. Schell, D. J., Farmer, J., Newman, M., and McMillan, J. D. (2003), *Appl. Biochem. Biotechnol.* **105–108**, 69–85.
10. Tucker, M. P., Kim, K. H., Newman, M. M., and Nguyen, Q. A. (2003), *Appl. Biochem. Biotechnol.* **105–108**, 165–177.
11. Keller, F. A., Hamilton, J. E., and Nguyen, Q. A. (2003), *Appl. Biochem. Biotechnol.* **105–108**, 27–41.
12. Sarkanen, K. V. and Ludwig, C. H. (1971), *Lignins: Occurrence, Formation, Structure, and Reactions*. Wiley-Interscience, New York, pp. 1–18.
13. Harris, P. J., Hartley, R. D., and Barton, G. E. (1982), *J. Sci. Food Agric.* **33**, 516–520.
14. Morrison, W. H. III, Akin, D. E., Ramaswamy, G., and Baldwin, D. (1996), *Textile Res. J.* **66**, 651–656.

15. Anderson, W. F., Peterson, J., Akin, D. E., and Morrison, W. H. III. (2005), *Appl. Biochem. Biotechnol.* **121–124,** 303–310.
16. Tolera, A. and Sundstol, F. (1999), *Anim. Feed Sci.Technol.* **81,** 1–16.
17. Akin, D. E. and Chesson, A. (1989), Proc. Int. Grassl. Congr. **16,** 1753–1760.
18. Esau, K. (1977), *Anatomy of Seed Plants*, 2nd ed. John Wiley & Sons, New York.
19. Stern, K. R., Jansky, S., and Bidlack, J. E. (2003), *Introductory Plant Biology*. McGraw-Hill, New York.
20. Akin, D. E., Hartley, R. D., Morrison, W. H. III., and Himmelsbach, D. S. (1990), *Crop Sci.* **30,** 985–989.
21. Carpita, N. C. (1996), *Ann. Rev. Plant Physiol. Plant Mol. Biol.* **47,** 445–476.
22. Akin, D. E., Ames-Gottfred, N., Hartley, R. D., Fulcher, R. G., and Rigsby, L. L. (1990), *Crop Sci.* **30,** 396–411.
23. Hartley, R. D., Akin, D. E., Himmelsbach, D. S., and Beach, D. C. (1990), *J. Sci. Food Agric.* **50,** 179–189.
24. Akin, D. E. and Hartley, R. D. (1992), *J. Sci. Food Agric.* **59,** 437–447.
25. Li, Y., Ruan, R., Chen, P. L., et al. (2004), *Trans. ASAE* **47,** 821–825.
26. Hartley, R. D. and Ford, C. W. (1989), In: *Plant Cell Wall Polymers: Biogenesis and Biodegradation.* Lewis, N. G. and Paice, M. G. (eds.), American Chemical Society, Washington, DC, pp. 137–145.
27. Borneman, W. S., Hartley, R. D., Himmelsbach, D. S., and Ljungdahl, L. G. (1990), *Anal. Biochem.* **190,** 129–133.
28. Grabber, J. H., Ralph, J., and Hatfield, R. D. (2002), *J. Agric. Food Chem.* 50, 6008–6016.
29. Akin, D. E., Borneman, W. S., Rigsby, L. L., and Martin, S. A. (1993), *Appl. Environ. Microbiol.* **59,** 644–647.
30. Faulds, C. B., Zanichelli, D., Crepin, V. F., et al. (2003), *J. Cer. Sci.* **38,** 281–288.
31. Graf, E. (1992), *Free Radical Biol. Med.* **13,** 435–448.
32. Garrote, G., Cruz, J. M., Moure, A., Dominguez, H., and Parajo, J. C. (2004), *Food Sci. Technol.* **15,** 191–200.
33. Borneman, W. S., Akin, D. E., and VanEseltine, W. P. (1986), *Appl. Environ. Microbiol.* **52,** 1331–1339.

Separate and Simultaneous Enzymatic Hydrolysis and Fermentation of Wheat Hemicellulose With Recombinant Xylose Utilizing *Saccharomyces cerevisiae*

L. Olsson,*,[1] H. R. Soerensen,[2] B. P. Dam,[1] H. Christensen,[1] K. M. Krogh,[1] and A. S. Meyer[1]

[1]BioCentrum-DTU, Technical University of Denmark, DK-2800 Lyngby, Denmark, E-mail: lo@biocentrum.dtu.dk; and [2]Novozymes A/S, Krogshoejvej 36, DK-2880 Bagsvaerd, Denmark

Abstract

Fermentations with three different xylose-utilizing recombinant *Saccharomyces cerevisiae* strains (F12, CR4, and CB4) were performed using two different wheat hemicellulose substrates, unfermented starch free fibers, and an industrial ethanol fermentation residue, vinasse. With CR4 and F12, the maximum ethanol concentrations obtained were 4.3 and 4 g/L, respectively, but F12 converted xylose 15% faster than CR4 during the first 24 h. The comparison of separate hydrolysis and fermentation (SHF) and simultaneous saccharification and fermentation (SSF) with F12 showed that the highest, maximum ethanol concentrations were obtained with SSF. In general, the volumetric ethanol productivity was initially, highest in the SHF, but the overall volumetric ethanol productivity ended up being maximal in the SSF, at 0.013 and 0.010 g/Lh, with starch free fibers and vinasse, respectively.

Index Entries: Xylose conversion; ethanol; starch free fibers; vinasse.

Introduction

In Europe, wheat is one of the major feedstocks employed in the industrial production of ethanol for potable spirits, technical alcohol, and fuel ethanol. This production of ethanol is based on enzyme-catalyzed conversion of wheat endosperm starch to glucose with subsequent fermentation of glucose to ethanol by the yeast *Saccharomyces cerevisiae* (1). The wheat endosperm cell wall material is currently left behind as an unhydrolyzed, unfermented residue.

Wheat endosperm cell walls comprise various nonstarch polysaccharides notably (1→3)(1→4)-β-D-glucans, glucomannans, cellulose, and arabinoxylans as well as some protein. Arabinoxylans make up approx 70% by

*Author to whom all correspondence and reprint requests should be addressed.

weight of the wheat endosperm polysaccharides (2). Arabinoxylans are pentose polymers that consist of a backbone of (1→4)-linked β-D-xylopyranosyl residues with single α-arabinofuranosyl substituents attached to the C(O)-2, C(O)-3 or to both C(O)-2,3 of the xylose residues with the A : X ratio typically being approx 0.5 : 1 (3). The xylan backbone may additionally be substituted with α-D-glucopyranosyl uronic acid, its 4-O-methyl derivative, and/or acetyl groups (4). The arabinofuranosyl units may be esterified with ferulic acid and/or p-coumaric acid (5).

Utilization of the wheat endosperm cell wall polysaccharides in ethanol production will require both, (1) generation of a fermentable hydrolysate from the cell wall polysaccharides, notably from the arabinoxylan because of its dominance in these cell walls, (2) a microorganism able to utilize the resulting pentoses for ethanol production. Arabinoxylans are readily hydrolyzed to monosaccharides by acid treatment. However, acid hydrolysis unavoidably generates byproducts that inhibit the subsequent microbial fermentation (6). For this reason enzymatic hydrolysis is preferable. Furthermore, this method is considered to be a more economically viable in future bioethanol processes (7,8). Hence, production of ethanol from wheat endosperm cell walls using biological conversion includes both degradation of the polysaccharides by enzymes and microbial conversion of the monomeric sugars (hexoses and pentoses) to ethanol.

As a result of the complexity and heterogeneity of the arabinoxylan structure, complete enzymatic degradation into monosaccharides requires both side-group cleaving and depolymerizing enzyme activities (9). A synergistic interaction during degradation of soluble arabinoxylan using enzyme side-activities in a cellulase preparation from *Trichoderma reesei* (Celluclast 1.5 L) and a hemicellulosic enzyme preparation produced by *Humicola insolens* (Ultraflo L) has recently been identified (10,11). Depending on the degree of entanglement among the polysaccharide structures in the substrate, endoglucanases (EC 3.2.1.4) and β-glucanases (EC 3.2.1.6) may also be required in enzymatic degradation of wheat endosperm cell walls.

Traditionally, yeast, *S. cerevisiae* is widely used for ethanol production, however, it cannot ferment pentoses naturally. A number of natural pentose fermenting microorganisms have been identified and many efforts have been made to genetically engineer different microorganisms, for example, *S. cerevisiae*, *Zymomonas mobilis* and *Escherichia coli* to efficiently produce ethanol from both hexoses and pentoses (for recent reviews *see* refs. [12–15]). Because xylose is the dominant pentose monosaccharide in most lignocellulosic and hemicellulosic hydrolysates, identification and development of efficient xylose fermenting microorganisms have received most attention. Several recombinant, mutated, and evolved xylose-fermenting *S. cerevisiae* strains have thus been developed during the past 10–15 yr. In a recent report, *S. cerevisiae* F12 was identified as being attractive for further investigation as a result of its high robustness to inhibitors in lignocellulosic hydrolysates *(16)*. S. cerevisiae F12 is an industrial strain engineered to use xylose for ethanol production by integration of the xylose reductase (XR) and xylitol dehydrogenase (XDH) encoding genes from *Pichia stipitis*, and by over expression of the endogenous xylulokinase (XKS) gene *(16)*. A major problem of recombinant, xylose fermenting S. cerevisiae is that the ethanol yield on xylose in many constructs is suboptimal. Among many attempts to improve the xylose utilization in recombinant *S. cerevisiae*, two redox manipulated strains, *S. cerevisiae* CR4 (redox manipulated) *(17)* and CB4 (redox and ATP manipulated) *(18)* have been developed. These two strains were selected in addition to *S. cerevisiae* F12 for evaluation in the present study.

Two different process strategies can be taken in order to accomplish the enzymatic degradation of polysaccharides and the microbial conversion of the resulting monosaccharides to ethanol, either the process may be performed as a separate hydrolysis and fermentation (SHF) process or it may be performed as a simultaneous saccharification and fermentation (SSF) process. The main advantage of the SHF process is that both process steps can be carried out at their individual, optimal process conditions, whereas in the SSF process a compromise with respect to reaction conditions must be made. However, a main advantage of the SSF process is that the end products from the enzymatic hydrolysis are not building up as they are further converted by the microorganism. This conversion diminishes the effect of any eventual end-product inhibition in the enzymatic catalysis. Additionally, the SSF process can be accomplished in one process step.

The aim of the present study was to determine the efficiency of ethanol production from xylose in the two different processing strategies, SHF and SSF, on (1) starch free fibers in the stream before industrial ethanol fermentation and on (2) the wheat endosperm residue left behind after the industrial fermentation. A main objective was to assess these two process strategies and substrates, and to unravel any potential differences in the ethanol productivity and yield. A second objective was to compare the xylose conversion efficiency of three genetically engineered *S. cerevisiae* strains: CR4, CB4, and F12.

Materials and Methods

All chemicals were standard, analytical grade chemicals unless otherwise stated.

Strains

The strains *S. cerevisiae* CR4 (*MATαSUC2 MAL2-8c pADH-XYL1 pPGK-XYL2 pPGK-XKS1 gdh1ΔpPGK-GDH2*) *(17)* and *S. cerevisiae* CB4 (*MATa SUC2 MAL2-8 pADH-XYL1 pPGK-XYL2 pPGK-XKS1 pTPIp-gapN*) *(18)* has been constructed in a laboratory strain background (*S. cerevisiae*

CEN.PK), whereas *S. cerevisiae* F12 (*HIS3::YIploxZEO overexpressing XR, XDH, and XKS*) was of an industrial strain background *(16)*.

Preparation of the Precultures

The *S. cerevisiae* strains were stored at –80°C in yeast peptone dextrose (YPD)-glycerol solution. The strain from the frozen stock was streaked on to YPD-agar plates and grown at 30°C. The agar plates were kept at 4°C.

A defined medium containing trace metal elements and vitamins was used for the precultures prepared according to *(19)*. The trace element solution and vitamin solution had the following compositions: Trace element solution: 15 g/L ethylenediaminetetraacetic acid (EDTA); 4.5 g/L $CaCl_2 \cdot 2H_2O$; 4.5 g/L $ZnSO_4 \cdot 7H_2O$; 3 g/L $FeSO_4 \cdot 7H_2O$; 1 g/L H_3BO_3; 0.84 g/L $MnCl_2 \cdot 2H_2O$; 0.3 g/L $CoCl_2 \cdot 6H_2O$; 0.3 g/L $CuSO_4 \cdot 5H_2O$; 0.4 g/L $Na_2MoO_4 \cdot 2H_2O$ and 0.1 g/L KI. The pH of the trace element solution was adjusted to 4.0 with NaOH and autoclaved. Vitamin solution: 50 mg/L d-biotin; 200 mg/L *para*-amino benzoic acid; 1 g/L nicotinic acid; 1 g/L Ca-pentothenate; 1 g/L pyridoxine HCl; 1 g/L thiamine HCl and 25 g/L m-inositol. The pH was adjusted to 6.5 and the solution was stored at 4°C after sterile filtration. The medium used for the precultures had the following composition: 7.5 g/L $(NH_4)_2SO_4$; 14.4 g/L KH_2PO_4; 0.5 g/L $MgSO_4 \cdot 7H_2O$; 2 mL/L trace element solution; 1 mL/L vitamins solution and 0.05 mL/L antifoam 289 (Sigma A-8436, St.Louis, MO). For cultivations on glucose, the concentration was 10 g/L. A colony taken from an YPD-agar plate was inoculated into a 500 mL baffled Erlenmeyer flask containing 100 mL medium. The preculture was placed at 30°C and grown on a rotary shaker at 150 rpm for 24–36 h. The cells were harvested in the late exponential growth phase. Two shake flasks supplied enough cell mass for inoculation of one fermentor.

At harvest, the preculture was transferred to sterile centrifuge glasses and centrifuged at 18,000g for 5 min at 4°C and the supernatant was discarded. The cells were washed with sterile 0.9% sodium chloride solution and centrifuged again. The supernatant was removed and the cells were diluted into 50 mL media used for the fermentation (vinasse or starch free fiber hydrolysate).

Raw Materials

Both substrates, i.e., the starch free fibers and the still bottoms fermentation residue (vinasse), were sampled from an industrial starch processing plant (Tate & Lyle, Amylum, UK). The starch free fiber fraction was extracted from a sample of liquified wheat endosperm stream (WES) entering the conventional, industrial ethanol fermentation. During the starch process at the factory, the WES had undergone a brief hemicellulytic enzyme treatment (to lower the viscosity and ease the early starch separation), and an enzymatic liquifaction by treatment with α-amylase. Before

Table 1
Monosaccharide Composition of the Starch Free Fiber Hydrolysate and the Still Bottoms Fermentation Residue "Vinasse" Employed As Substrates in This Study[a]

Monosaccharide[a]	Starch-free-fibers g/kg DM[b]	Vinasse g/kg DM[b]
Arabinose	147 ± 2	91.4 ± 4.5
Galactose	17.6 ± 0.3	25.4 ± 1.5
Glucose	31.4 ± 0.4	76.9 ± 1.3
Xylose	213 ± 2	168 ± 6

Data are shown as the average of three determinations ± the standard deviation.
[a]The monosaccharide compostions were determined by HPAEC analysis after standardized acid hydrolysis of the substrates (at 0.4 M HCl for 2 h at 100°C). Data are shown as the average of three determinations ± the standard deviation.
[b]DM is dry matter.

use in our study, the WES was subjected to enzymatic saccharification in the form of a treatment with amyloglucosidase for 48 h at 50°C in order to hydrolyze the remaining starch and dextrins to glucose. The starch free fiber fraction was collected as the solid residue after separation of the saccharified material by filtration through a Munktell filter paper (quality 5). Immediately after this filtration, the solid starch free fiber residue was washed with four times its own volume of deionized water in order to remove remaining glucose. The vinasse constituted a sample of the ethanol fermentation residue left behind after the industrial fermentation of the WES. The dry matter content of the starch free fibers and vinasse were 18.8 and 9% by weight, respectively. Monosaccharide compositions are given in Table 1.

Separate Hydrolysis and Fermentation

In SHF, a complete enzymatic hydrolysis was performed before fermentation. In order to achieve a dry matter content of 5% (w/w), the raw materials were diluted with 0.025 M KH_2PO_4 buffer and pH was adjusted to 5.0 with NaOH. Each of the two substrates were then enzymatically hydrolysed for 24 h at 50°C, pH 5.0 with a 50:50 mixture of Ultraflo L (*Humicola insolens*, Novozymes A/S) and Celluclast 1.5 L (*Trichoderma reesei*, Novozymes A/S) at a dosage of 15% w/w enzyme/substrate dry matter (E/S). The enzymatic hydrolysis was stopped by boiling for 10 min. After this, the hydrolysates were cooled on ice until the fermentation was started. These hydrolysates were used for the experiments in which the SHF process was compared with the SSF process using *S. cerevisiae* F12. In the SHF experiments in which the performance of the three *S. cerevisiae* strains were compared using vinasse, a higher dry matter content 9.0% (w/w) and a lower enzyme loading 10% (w/w) were used for the enzymatic hydrolysis step.

Anaerobic batch cultivations were carried out in in-house manufactured 1 L fermenters. The fermentation substrate was added into a sterile fermenter to a volume of 800 mL and to prevent contamination 80 mg Penicillium V (Novo Nordisk) and 80 mg streptomycin sulfate (Sigma) were added to the substrates. The fermentations comparing the different strains were inoculated with 5 g/L dry weight and they were run for 45 h. For the experiments comparing *S. cerevisiae* F12 during SSF and SHF, the fermenter was inoculated to a cell mass concentration of 0.5–1 g biomass/L. These fermentations were run for 185 h.

The fermentations were run at 30°C, at a stirring speed of 300 rpm and with pH maintained at pH 5.0 with 2 M NaOH. The fermenter was continuously sparged with nitrogen at 0.02 L/min to maintain an anaerobic environment. The exhaust gas was led through a condenser.

Simultaneous Saccharification and Fermentation

Before the SSF, the substrates were heated to 95°C for 10 min. Because the enzymes were most efficient at 50°C, and the yeast is most efficient at 30°C, the process was started with a 2 h enzymatic hydrolysis at 50°C with the purpose of ensuring a rapid initial release of fermentable sugars. After 2 h of prehydrolysis, the temperature was lowered to 30°C, and the hydrolysate (including the enzymes) was transferred to the fermenter, whereafter the cell mass was added as described earlier. The SSF was performed at 30°C, pH 5.0 and with an enzyme dosage of (50:50 mixture of Ultraflo L and Celluclast 1.5 L) 15% w/w E/S dry matter.

Sampling and Analysis

Samples were taken regularly from the fermenter. SHF samples were centrifuged at 10,000g for 1 min before filtration through a 0.45-μm filter. SSF samples were treated in two different manners, the samples, which were to be analysed by high performance anonic exchange chromatography (HPAEC) (*see* later) were centrifuged at 10,000g for 1 min, filtered, placed in boiling water for 10 min to inactivate the enzymes in the samples. Samples for high performance liquid chromatography (HPLC) analysis (described later) were centrifuged at 10,000g for 1 min, filtered, and diluted 1:1 with a 10 mM H_2SO_4 solution. All samples were stored at –20°C until analysis.

The monosaccharide composition and concentration were determined by HPAEC as described previously (*see* ref. *10*). The content of xylitol, lactate, acetate, glycerol, and ethanol were determined by HPLC on a Waters (Milford, MA) HPLC system that encompassed a Waters 717 Plus autosampler, a Waters 515 pump, a Waters 2410 RI detector, and was equipped with a precolumn, cation-H 30 × 4.6 mm from Bio-Rad (Hercules, CA), and two serially placed Bio-Rad Aminex HPX 87-H columns (held at 60°C). The eluent used was 5 mM H_2SO_4 at 0.40 mL/min.

Results

Comparison of the Performance of Three Recombinant Xylose Fermenting S. cerevisiae Strains

The performance of the three, metabolically engineered, xylose fermenting *S. cerevisiae* strains, CB4, CR4, and F12 were followed during fermentation of enzymatically prehydrolyzed vinasse. All three strains converted all the present glucose within the first few hours of fermentation, and as a result, the main ethanol production took place during this period (Fig. 1A–C). All three strains were able to convert some of the xylose, but at varying xylose consumption rates and with different ethanol productivities: the CB4 strain converted the xylose very slowly, and the conversion of xylose ceased after approx 24 h of fermentation. As a result, only about one third of the xylose was utilized, during the 45 h of fermentation. Only limited amounts of xylitol were produced by this strain, and the maximum ethanol concentration achieved was 3.1 g/L (Fig. 1A). In contrast, both the CR4 and the F12 strain converted almost 90% of the xylose during the 45 h of fermentation, and maximum ethanol concentrations of 4.3 g/L and 4.0 g/L, respectively, were achieved. The volumetric ethanol productivities achieved during the first 24 h of the fermentations were 0.17 and 0.16 g/Lh for CR4 and F12, respectively. The F12 strain produced higher xylitol levels than the CR4 (final concentration of 1.1 g/L vs 0.7 g/L, respectively Fig. 1B and C). However, a comparison of the rate of xylose consumption revealed that the F12 converted the xylose about 15% faster than the CR4. Thus, assuming a linear conversion rate during the first 24 h of the fermentations, the rates of xylose consumption, (from linear regression of the xylose data) gave 0.17 g/Lh for F12 and 0.15 g/Lh for CR4. This slightly better performance in xylose conversion by F12, coupled with its industrial strain background and previously demonstrated high robustness, led us to choose this strain for further study in the different SHF and SSF fermentations.

Comparison of SHF and SSF in Vinasse and Starch Free Fibers Using S. cerevisiae F12

Initial Glucose and Xylose levels

The initial glucose levels in the fermentation substrates ranged from 1.1 to 4 g/L, whereas the initial xylose levels were higher, and ranged from 2.7 to 6.2 g/L in the different fermentation substrates (Table 2). Irrespective of enzymatic prehydrolysis, the initial glucose concentrations were generally twofold higher in the starch free fibers than in the vinasse (Table 2). The higher glucose level in the starch free fibers could be a result of incomplete glucose removal by filtration and washing after the saccharification treatment during their preparation (*see* Materials and Methods). In accordance with the expectation that both some β-glucan and cellulose

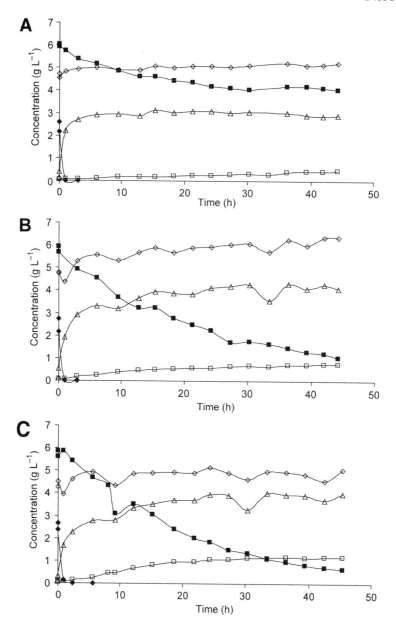

Fig. 1. Time courses for SHF fermentations on vinasse with different recombinant, xylose fermenting *S. cerevisiae* strains. **(A)** *S. cerevisiae* CB4 **(B)** *S. cerevisiae* CR4 **(C)** *S. cerevisiae* F12. Symbols used: ♦ glucose; ■ xylose; △ ethanol; □ xylitol; ◇ glycerol. Before fermentation the vinasse substrate had been enzymatically treated with a 50 : 50 mixture of Ultraflo L and Celluclast added at a dosage of 10% w/w E/S dry matter.

would be hydrolysed during the enzymatic treatment in the separately hydrolyzed samples, the initial glucose levels were higher in both types of the SHF samples than in the SSF substrates at the start of the fermentation (Table 2). The xylose levels in the SHF substrates were similar, but twofold

Table 2
Summary of Fermentation Performance Results Using S. cerevisiae F12

		Vinasse SHF	Vinasse SSF	Starch free fibers SHF	Starch free fibers SSF
Glucose concentration	at start[a]	2.3	1.1	4.0	2.0
Xylose concentration	at start	5.6	2.7	6.2	3.3
	at end	0.9	1.4	1.9	2.2
Ethanol concentration	at start[b]	0	0	0.14	0
	at max.	2.0 (after 88 h)	2.1 (after 88 h)	2.0 (after 40 h)	2.8 (after 185 h)
Xylitol	at start	0	0	0	0
	at end	0.5	1.0	1.2	0.8
Glycerol	at start	3.8	3.7	0.7	0.6
	at end	4.1	4.4	1.2	0.9
Acetate	at start	1.2	1.2	0.9	0.9
	at end	1.3	1.2	1.0	0.9
Lactate	at start	6.0	6.0	0.4	0.3
	at end	5.5	5.7	0.1	0.1
Volumetric ethanol production g/Lh	Phase 1 (0–3 h)	0.40	0.28	0.63	0.35
	Phase 2 (3–88 h)	0.009	0.015	0.003	0.01
Specific ethanol productivity[c] g/gh	Phase 1 (0–3 h)	0.24	0.18	0.74	0.44
	Phase 2 (3–88 h)	0.005	0.010	0.003	0.013

Concentrations are given as g/L and the fermentation were run for 185 h.
[a]The final glucose concentration at the end of the fermentation was 0.0 g/L in all cases; glucose was used up within the first 3 h of the fermentation.
[b]The first ethanol measurement was made 5 min. after inoculation.
[c]Productivity: calculations for the specific ethanol productivities were based on the amount of final cell mass concentration.

higher than the levels of those in the SSF substrates (Table 2). During the fermentations, the glucose consumption rates were high and no glucose was measurable after 3 h of fermentation. In contrast, the xylose was consumed more slowly in all the four fermentations, and xylose was not fully consumed at the end of the fermentations in either of the cases (Table 2).

Ethanol Production Levels and Total Specific Ethanol Productivities

The maximal ethanol level obtained was 2–2.1 g/L except in the simultaneously saccharified and fermented starch free fibers, in which the maximal ethanol level reached 2.8 g/L (after 185 h) (Table 2). As expected, the total volumetric ethanol productivity was in all four cases highest in the first few hours of the fermentation, when the glucose was consumed. The volumetric ethanol productivities during the first few hours of fermentation ranged from 0.28 to 0.63 g/L·h and were highest in the separately hydrolyzed starch free fibers fermentation (starch free fibers SHF, Table 2). These rates corresponded to specific ethanol productivity rates of 0.18–0.74 g/g·h (when using the final level of cell mass for the calculations). The specific ethanol productivity rates were consistently higher in the starch free fibers fermentations than in the vinasse fermentations, with the rate in the starch free fibers SHF being the highest (Table 2). In all cases, the initial glucose was consumed during the first 3 h of the fermentation, and after this period, the volumetric ethanol productivities fell drastically to 0.003–0.015 g/L·h in all four types of fermentation (Table 2). The rates in phase 2 allowed a comparison of the ethanol production from xylose, and because the volumetric ethanol productivity was highest in phase 2 in both vinasse fermentations, the data indicated that xylose was better fermented to ethanol in the vinasse than in the starch free fibers substrate.

Production of Byproducts

Not surprisingly, the initial glycerol levels, relating to microbial biomass production, were much higher in the vinasse fermentations than in the starch free fibers fermentations, as vinasse is a post fermentation waste stream. In all four fermentations, the glycerol levels increased marginally from 0.3 to 0.7 g/L, from start to end of the fermentation. As discussed earlier, xylitol was also produced as a byproduct during the fermentation, and the xylitol concentrations reached levels of 0.5–1.2 g/L with no clear dependence of the substrate or process type (Table 2). The initial lactate levels were relatively high in the vinasse samples (6 g/L), but modest in the starch free fibers (approx 0.3 g/L). The higher levels of lactate found in the vinasse showed that lactate production had taken place before recovering the vinasse residue. This lactate production was probably as a result of growth of lactic acid bacteria as contaminants immediately before the industrial ethanol fermentation or during the industrial fermentation process itself, which is common in industrial ethanol fermentations *(20)*. Small decreases in the lactate concentration were observed (Table 2). The initial acetate levels were also slightly higher in the vinasse than in the starch free fibers substrate samples. However, the acetate levels remained constant throughout all fermentations.

Discussion

Components other than fermentable monosaccharides are often present during fermentation of industrial substrates, originating as residual products from food and agroindustrial processes, such as in vinasse. The presence of these components may negatively influence the fermentation yield and productivity. The three *S. cerevisiae* strains F12, CB4, and CR4, which all perform well in a laboratory medium *(16–18)*, could ferment in vinasse. However, the performance of the strain *S. cerevisiae* CB4 during xylose utilization was poor and the utilization of xylose completely stopped before the fermentation was terminated. As also indicated in an earlier study *(17)*, recombinant strains with decreased ATP yield, such as the *S. cerevisiae* CB4 strain, in which the adenosine triphosphate (ATP) generation has been weakened with a nonATP generating step, encounter problems during xylose fermenting. These problems can get severe, as illustrated in the present study, when fermenting industrial substrates containing components other than monosaccharides, which put further demands on the cellular metabolism. Vinasse contains components like acetate, glycerol, and lactate (Table 2) in addition to monosaccharides. Acetic acid, and to a lesser extent lactic acid, inhibits *S. cerevisiae* because they can diffuse into the cells in their undissociated form, thereby lowering the intracellular pH, which in turn leads to higher ATP requirement for maintenance *(21,22)*. The occurrence of this series of events can explain the bad performance of *S. cerevisiae* CB4.

Comparing the fermentation performance of the *S. cerevisiae* strain F12 with CR4, showed that the redox modulated strain had the lowest production of xylitol resulting in a slightly higher maximum ethanol concentration. Several other studies have also shown that redox modulation of strains can improve the ethanol production from xylose, when the XR-XDH-XKS system is used for introducing xylose utilization in *S. cerevisiae* *(23–25)*.

During the fermentation process glucose was taken up preferentially before xylose, and furthermore, the glucose uptake was a magnitude higher than the xylose uptake both during fermentation of vinasse (Fig. 1), and during fermentation of laboratory medium *(26)*. It has been found that during fermentation of recombinant xylose utilizing *S. cerevisiae* that the specific xylose consumption rate is highest during coconsumption of glucose, after which the rate declines by time *(26)*. Based on this earlier observation, one can make the hypothesis that an SSF process might be advantageous as the monosaccharides are released throughout the process, which might support a coconsumption of glucose and xylose, which based on earlier experience would allow faster xylose consumption. From the present study, it can be concluded that using SSF as process configuration is advantageous as higher final ethanol concentrations were reached on both substrates (Table 2). However, more residual xylose was found in SSF fermentations in comparison with SHF fermentations at the time the fermentation was terminated, indicating that the xylose consumption was not improved in SSF in comparison with SHF.

The most important difference between the two substrates used, vinasse and starch free fibers, was that vinasse contained significantly higher levels of lactate and glycerol than starch free fibers. Organic acids are known to be inhibitory to *S. cerevisiae* as discussed earlier. In the present study, two to threefold higher specific ethanol productivites in phase 1 were found on the starch free fibers substrate as compared with the vinasse both for the SSF and the SHF process, respectively, confirming that the starch free fibers substrate was less inhibitory to *S. cerevisiae* than vinasse was.

In conclusion, the three investigated strains could all ferment xylose from the two evaluated substrates, however, the ethanol productivity on xylose was slow in comparison with utilization of glucose both using SHF or SSF. There are recent reports, were xylose consumption rates have been improved in recombinant xylose fermenting *S. cerevisiae* strains, by using metabolic engineering and/or random mutagenesis *(27,28)*, these strains will probably lead to better performance under SSF conditions. In this work, the SSF strategy on starch free fibers appeared to be the best for obtaining maximum ethanol levels and highest ethanol productivity with the F12 strain. Higher specific ethanol productivities in phase 1 were reached using starch free fibers than vinasse, and the difference is most likely as a result of higher amounts of inhibitory compounds in the latter. To construct an efficient xylose fermenting strain, both the strain background and the metabolic engineering strategy have to be taken into account in relation to the substrate it will be fermenting, i.e., influence on metabolism by potential inhibitory components in the applied substrate.

Acknowledgment

We wish to thank Beatriz Palmarola-Adrados, Lund University, Sweden, for assistance in preparing the starch free fibers.

References

1. Lewis, S. M. (1996), Fermentation alcohol. In: *Industrial Enzymology*. 2nd ed., Godfrey, T., West, S., (eds.), MacMillan Press Ltd, London UK, pp. 11–48.
2. Bacic, A. and Stone, B. A. (1980), *Carbohydr. Res.* **82,** 372–377.
3. Izydorczyk, M. S. and Biliaderis, C. G. (1993), *Cereal Chem.* **70,** 641–646.
4. Smith, M. M. and Hartley, R. D. (1983), *Carbohydr. Res.* **118,** 65–80.
5. Ishii, T. (1997), *Plant Sci.* **127,** 111–127.
6. Saha, B. C. (2000), *Biotechnol. Adv.* **18,** 403–423.
7. Sheehan, J. and Himmel, M. E. (1999), *Biotechnol. Prog.* **15,** 817–827.
8. Wooley, R., Ruth, M., Glassner, D., and Sheehan, J. (1999), *Biotechnol. Prog.* **15,** 794–803.
9. de Vries, R. P., Kester, H. C. M., Poulsen, C. H., Benen, J. A. E., and Visser, J. (2000), *Carbohydr. Res.* **327,** 401–410.
10. Sørensen, H. R., Meyer, A. S., and Pedersen, S. (2003), *Biotech. Bioeng.* **81,** 726–731.
11. Sørensen, H. R., Pedersen, S., Viksø-Nielsen, A., and Meyer, A. S. (2005), *Enz. Microb. Technol.* **36,** 773–784.
12. Ingram, L. O., Gomez, P. F., Lai, X., et al. (1998), *Biotech. Bioeng.* **58,** 204–214.
13. Aristidou, A. and Penttilä, M. (2000), *Curr. Opin. Biotechnol.* **11,** 187–198.
14. Hahn-Hägerdal, B., Wahlbom, C. F., Gardonyi, M., van Zyl, W., Cordero-Otero, R., and Jönsson, L. F. (2001), *Adv. Biochem. Eng. Biotechnol.* **73,** 53–84.
15. Zaldivar, J., Nielsen, J., and Olsson, L. (2001), *Appl. Microbiol. Biotechnol.* **56,** 17–34.
16. Sonderegger, M., Jeppsson, M., Larsson, C., et al. (2004), *Biotech. Bioeng.* **87,** 90–98.
17. Roca, C., Nielsen, J., and Olsson, L. (2003), *Appl. Environ. Microbiol.* **69,** 4732–4736.

18. Bro, C., Regenberg, B., Forster, J., and Nielsen, J. In silico aided metabolic engineering of *Saccharomyces cerevisiae, Metab. Eng.,* in press.
19. Verduyn, C., Postma, E., Scheffers, W. A., and van Dijken, J. P. (1990), *J. Gen. Microbiol.* **136**, 395–403.
20. Stenberg, K., Galbe, M., and Zacchi, G. (2000), *Enzyme Microb. Technol.* **26**, 71–79.
21. Maiorella, B., Blanch, H. W., and Wilke, C. R. (1983), *Biotechnol. Bioeng.* **23**, 103–121.
22. Pampulha, M. E. and Loureiro-Dias, M. C. (1989), *Appl. Microbiol. Biotechnol.* **31**, 547–550.
23. Jeppsson, M., Johansson, B., Hahn-Hägerdal, B., and Gorwa-Grauslund, M. F. (2002), *Appl. Environ. Microbiol.* **68**, 1604–1609.
24. Jeppsson, M., Träff, K., Johansson, B., Hahn-Hägerdal, B., and Gorwa-Grauslund, M. F. (2003), *FEMS Yeast Res.* **3**, 167–175.
25. Verho, R., Londesborough, J., Penttilä, M., and Richard, P. (2003), *Appl. Environ. Microbiol.* **69**, 5892–5897.
26. Roca, C., Haack, M. B., and Olsson, L. (2004), *Appl. Microbiol. Biotechnol.* **63**, 578–583.
27. Wahlbom, C. F., van Zyl, W. H., Jönsson, L. J., Hahn-Hägerdal, B., and Cordero Otero, R. R. (2003), *FEMS Yeast Res.* **3**, 319–326.
28. Kuyper, M., Hartog, M. M. P., Toirkens, M. J., et al. (2005), *FEMS Yeast Res.* **5**, 399–409.

Copyright © 2006 by Humana Press Inc.
All rights of any nature whatsoever reserved.
0273-2289/06/129–132/130–152/$30.00

Biofiltration Methods for the Removal of Phenolic Residues

LUIZ CARLOS MARTINS DAS NEVES,[1] TÁBATA TAEMI MIAZAKI OHARA MIYAMURA,[1] DANTE AUGUSTO MORAES,[1] THEREZA CHRISTINA VESSONI PENNA,[1] AND ATTILIO CONVERTI*,[2]

[1]*Departamento de Tecnologia Bioquímico-Farmacêutica, Faculdade de Ciências Farmacêuticas, Universidade de São Paulo, SP, Brazil; and [2]Università degli Studi di Genova, Dipartimento di Ingegneria Chimica e di Processo, Genoa, Italy, E-mail: converti@unige.it*

Abstract

Industrial effluents from the pharmaceutical industry often contain high concentrations of phenolic compounds. The presence of "anthropogenic" organic compounds in the environment is a serious problem for human health; therefore, it merits special attention by the competent public agencies. Different methods have been proposed in the last two decades for the treatment of this kind of industrial residues, the most important of which are those utilizing absorption columns, vaporization and extraction, and biotechnological methods. Biofiltration is a method for the removal of contaminants present in liquid or gaseous effluents by the use of aerobic microorganisms, which are immobilized on solid or porous supports. Although several bacteria can utilize aromatic compounds as carbon and energy source, only a few of them are able to make this biodegradation effectively and with satisfactory rate. For this reason, more investigation is needed to ensure an efficient control of process parameters as well as to select the suited reactor configuration. The aim of this work is to provide an overview on the main aspects of biofiltration for the treatment of different industrial effluents, with particular concern to those coming from pharmaceutical industry and laboratories for the production of galenicals.

Index Entries: Biofiltration; phenolic residues; bioremediation.

Introduction

Environmental biotechnology is the science that uses biological systems for the destruction of pollutants from industrial effluents or present in the environment. Such technologies show a certain number

*Author to whom all correspondence and reprint requests should be addressed.

of advantages if compared with the traditional chemical and physical methods because of their low cost and impact on the environment *(1,2)*. Through these biotreatments, the organic pollutants are biodegraded to inorganic compounds such as carbon dioxide and water, whereas the traditional methods (vaporization, adsorption, extraction, and so on) imply their simple transfer to other compartments of the environment. The treatment by activated sludge has been a common practice to remove pollutants from industrial liquid effluents for more than 80 yr *(3)*, whereas soil, river, and sea bioremediation goes back only to the 1990s *(4–6)*. The most recent studies on these technologies, dealing with bacteria isolation, their classification and physiological characterization, and molecular analysis of the enzymes responsible for pollutant degradation led to the construction of so-called "super-microorganisms" able to degrade various types of pollutants. They are usually employed as consortia of microorganisms directly isolated from the polluted environment, without any characterization or physiological investigation.

There are two different ways to develop environmental biotechnology studies. According to the former, laboratory-scale studies are focused to the isolation and investigation of the bacterium responsible for the pollutant degradation; although the other has the practical aim of making the treatment method effective and optimizing the related process parameters. A combination of these two approaches is needed to make environmental biotechnology able to overcome the present difficulties associated to the use of these technologies, i.e., the insufficient number of bench-scale studies needed to interpret the phenomena occurring during the treatment and the complexity of systems using more than one microorganism for the effluent treatment.

To obtain satisfactory results with these technologies, it is necessary to get a thorough knowledge of the biodegradation process, to control the process conditions, and to optimize the construction of both the biofilter and the support material *(7,8)*. Cell immobilization is a fundamental step for the success of the process; therefore, the correct choice of the support material is necessary mainly to avoid possible phenomena responsible for high head losses in the bed as well as to increase microbial activity and operative steady-state stability.

The present advances in the isolation, selection, and construction of strains or consortia of various microorganisms, mainly bacteria, have extended the biodegradation processes to the treatment either of anthropogenic or xenobiotic compounds. The basis for the development of an efficient biofiltration process is the capability of microorganisms to adapt themselves to new substrates and the ability of some of them to use xenobiotics.

The isolation of microorganisms able to degrade pollutants is usually done from natural microbial populations by batch cultivations using enriched media containing high pollutant levels. Alternatively, bacteria with high affinity for the polluting substrate can be isolated by continuous cultivation.

Another interesting technique to isolate microorganisms is the dilution of the contaminated effluent with uncontaminated effluent. Marine oligotrophic bacteria have been successfully isolated by this technique consisting in the dilution of seawater with sterile seawater (containing low concentrations of organic nutrients), and pure cultures might be obtained *(9)*. These isolation protocols promoted the discovery and utilization of new microorganisms exhibiting different genotypes and phenotypes with respect to those obtained by the traditional batch cultivation *(10)*.

When compared with traditional techniques, biological means are generally friendlier for the treatment of such residues, because they are usually cheaper and release fewer byproducts. Among the biological methods, biofiltration has been increasingly proving to be particularly effective for the treatment of organic compounds at low concentrations. It is a technology for the biological removal of contaminants using aerobic microorganisms immobilized in solid and porous supports, which can be used for the destruction either of gaseous or liquid pollutants. Table 1 shows the different types of pollutants removed by biofiltration from effluents of different industrial sectors, and Table 2 indicates the degradability of the main polluting compounds.

The main contaminants which can be removed by biofiltration with excellent yields are butadiene, cresols, ethyl benzene, xylene, trimethylamine, alcohols (butanol, ethanol, methanol, and so on), esters, ethers, ketones, organic acids, methyl mercaptan, and some inorganic acids (hydrogen sulfide, hydrochloric acid, and hydrofluoric acid), among others *(11)*.

Studies have been developing on applications of biofilters for the removal of phenols and chlorophenols *(12)*, toluene *(13,14)*, styrene *(15)*, and benzene *(16)*. In all studies, the effluents under investigation were present either in liquid or in gaseous phase, and the reported results seem to be satisfactory for large-scale application of such a technology for the treatment of these effluents.

Several studies dealt with continuous *(17)* and batch processes *(18,19)*, performed either using fixed- *(15,16)* or fluidized-bed reactors *(18,20)*, hence suggesting that the choice of the process and reactor configuration can be influenced by the residue to be treated and the microorganism employed. The most commonly used microorganisms, belonging to the genera *Pseudomonas* sp., *Acinetobacter* sp., and *Bacillus* sp., are immobilized onto different types of inert supports (spheres of glass, calcium alginate, and resins) *(11)*. New immobilization procedures, employing cheaper materials, such as sugarcane bagasse and other solid residues, could be an interesting alternative to further reduce the operating costs of biofiltration *(16)*.

Table 1
Different Types of Pollutants Removed by Biofiltration From the Effluents of Different Industrial Sectors

Industries	Aliphatics	Aromatics	Oxygenated	Sulfuric	Nitrogenous	Halogenated	Inorganics
Foundries	X	X	X				X
Plastics	X	X	X				
Lacquer	X	X	X		X		
Adhesive		X	X				
Oil and fat	X	X		X		X	
Polyester				X	X		
Friction-linings			X	X	X		
Glue		X	X				X
Composting and waste processing	X	X	X	X	X	X	X
Sewage treatment and sludge drying	X	X	X	X	X	X	X

Table 2
Classification of the Main Polluting Compounds According
to Their Biodegradability Using Biofiltration

High	Good	Low
Acetaldehyde, acetone	Acetonitrile, isonitriles	Dichloroethane
Butyric acid	Amides, benzene	Dichloromethane
Hydrochloric acid	Chlorophenols	Dioxane
Hydrofluoric acid	DMS	Carbon disulfide
Hydrogen sulfide	Styrene, hexane	Methane
Ammonia, sulfates	Phenols, toluene	Pentachlorophenol
Butadiene, butanol	Methylisobutylketone	Pentane
Ethanol, methanol	Pyridine	Perchloroethylene
Formaldehyde, cresols	Thiocyanates	Carbon tetrachloride
Ethyl acetate	Thioethers, thiophene	Trichloroethane
Ethyl benzene		Trichloroethylene
Nitrocompounds	Uncertain	
Tetrahydrofurane	Acetylene	Very low
Trimethylamine	Isocyanates	1,1,1-Trichloroethane
Xylenes	Methylmethacrylate	

Biofiltration History

The first reports about the application of biofiltration, going back to 1950s in Germany and 1960s in United States, respectively, demonstrated its ability to remove odors (sulfides, ammonia, and so on). Such biofilters were applied to the control of odors from industrial effluents, plants for waste thermal treatment, activated sludge systems, and so on (21–23). In 1991, the biofilters operating in United States and Canada were less than 50, whereas more than 500 biofilters were present in Germany and the Netherlands. In the same period, this technology had a certain success in Japan where the number of biofilters grew from 50 to 90. Although the biofilters were initially designed for the deodorization of industrial residues, they are growingly employed since the 1980s for the control of pollutant emissions.

Advantages and Drawbacks of Biofiltration

Most of the research-work was carried out to demonstrate the economic feasibility of biofiltration methods, particularly that of Jäger and Jager (24). The biofiltration effectiveness is related to the bed and the distribution of the effluent inside the reactor (25). Thistlethwayte et al. (26) described a biofiltration method for the purification of air contaminated by industrial residues containing sulfur, ammonia, and alcohols, able to ensure removals from 40 to 100%. Helmer (27) utilized a biofilter with humidified earth as a support to extract anaerobic fermentation products

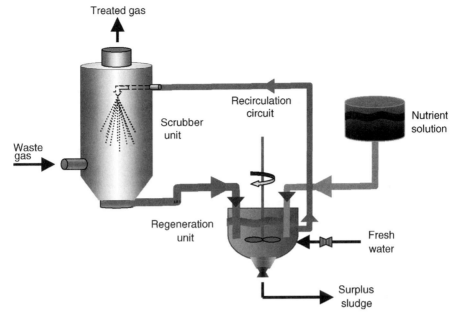

Fig. 1. Schematic setup of a bioscrubber for the removal of gaseous contaminants.

(ammonia, sulfides, alcohols, and aldehydes) from organic materials. Among the main advantages of the use of biofilters, there are the low energy needs because of moderate pressure and temperature, easy control and management, low operating costs, and possibility of performing the process in continuous mode for long time with a very few maintenance and cleaning requirements.

The main advantage of biofiltration is that it does not imply any transfer of the pollutant from one environment compartment to the other, as it occurs with most of the traditional technologies. Through biofiltration, the pollutants are in fact transformed to products belonging to natural cycles without imparting any secondary pollution. On the other hand, the biodegradation of organic compounds by pure microbial cultures can release several toxic intermediates; this problem can become even more serious when using complex microbial consortia as a result of the formation of a wide spectrum of metabolites *(19)*.

Biofiltration and Bioscrubbing

Bioscrubbers (Fig. 1) and biofilters (Figs. 2 and 3) are the main technologies developed for the biological treatment of pollutants.

Bioscrubbing can be defined as a methodology by which microorganisms utilized for the degradation are freely dispersed in the liquid phase (cultivation medium) of the treatment system, whereas they are immobilized in semi-inert supports in biofilters. In conventional bioscrubbers, the scrubber unit consists, in the case of gaseous pollutants, of an absorption

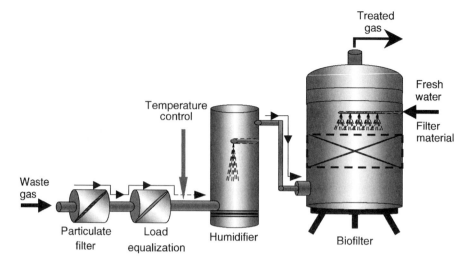

Fig. 2. Schematic setup of a biofilter for the removal of liquid contaminants.

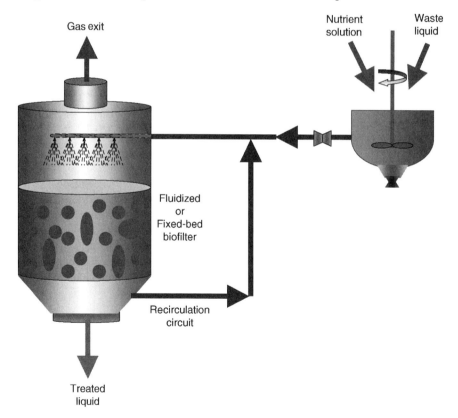

Fig. 3. Schematic setup of a biofilter (biotrickling filter) for the removal of gaseous contaminants.

column in which they are absorbed and partly oxidized in the liquid phase. After passing through the washing column, the effluent is transferred to a regeneration unit made up of a mechanically mixed reactor, in which the pollutants are completely degraded. To increase the effectiveness of the treatment, a portion of the effluent is often recycled to the scrubber unit. Biofiltration systems can be subdivided into two distinct processes, namely biofiltration *sensu latu* (Fig. 2) and biotrickling filtration (Fig. 3), both methodologies consisting, in the case of treatment of gaseous effluents, in the prior absorption of the polluting compounds, and their subsequent biological degradation.

Biofilters

Biofiltration systems are more compact than bioscrubbers and are made up of only one treatment unit (column), in which the polluted effluent passes through a biologically active bed containing microorganisms immobilized in the form of a biofilm. Both organic and inorganic pollutants are degraded by the microorganisms that utilize them as carbon and energy sources.

Biotrickling filtration makes use of a bed of inert materials (porous glass, ceramic, or plastic materials). The effluent passes through a column counter-currently with respect to the cultivation medium containing inorganic nutrients. Either the bioscrubbers or the biotrickling filters are mainly used for effluents containing pollutants which present high or medium solubility in water. On the contrary, because of better mass transfer, the biofilters are more suited in the presence of scarcely hydro-soluble pollutants at concentrations lower than 1000 ppm. The main drawback of the use of biofilters is the difficult control of the operating parameters influencing the degradation efficiency, like pH, temperature, and nutrient concentration.

Reactors for the Biofiltration of Liquid Effluents

Biofilters are usually operated continuously and their configuration strongly depends on the characteristics of the substance to be removed from the gaseous or liquid effluent. Bioreactors utilized for this purpose employ the biocatalyst immobilized inside the system and can be classified according to the type of catalyst (cells or enzymes) or the type of mixing system. In general, studies on the biofiltration of pollutants present in industrial effluents employ immobilized cells. The difficult identification, purification, and characterization of the enzymes (or enzyme mixtures) to be used in the treatment make the employment of immobilized enzymes difficult. Effective biofilter utilization in the treatment of effluents requires that it has the ability of treating the largest amount of residues as possible. For this reason, biofilters containing more than one microorganism are often employed whose degradation metabolism implies a large variety of enzymes.

Bioreactors used for biofiltration either of liquid or gaseous effluents must ensure cell immobilization in the support. The bioreactor configurations allowing for such a condition are the so-called fixed-bed *(15,16)* and fluidized-bed reactors *(18,20)*. The reactor aeration, which is crucial requisite depending on both the residue to be eliminated and the microorganism employed, can be obtained by plant configurations in which the contact between immobilized cells and effluents *(28)* is ensured either by the pressure difference or by the presence of an air flow ("airlift" and bubble column) *(29)*. In addition, the cells can be entrapped between semi-permeable membranes, as it occurs either in plane-membrane bioreactor *(30,31)* or in hollow-fiber bioreactor *(32)*.

The main feature of immobilized-cell systems is the use of some physical structure that compels the cells to keep confined within a specific region of the bioreactor. Among the several advantages of these systems, it should be mentioned the possibility of increasing cell concentration so as to raise the process rate and often the conversion efficiency of the biofilter.

Fixed-Bed Biofilter

Among the existing fixed-bed configurations, the most common is the vertical fixed bed, although there are reports also on the use of horizontal fixed bed and parallel flux bioreactors *(33,34)*. The immobilization takes place usually in this bioreactor configuration by entrapment (hydrophilic gel) or by adsorption (resins) *(34)*. The major drawback of the use of such a bioreactor in the treatment of effluents is the difficult long-term continuous operation, owing to the possible occurrence of excess cell growth in the bed and consequent clogging of the system and formation of preferential channels. These problems, which are responsible for a decrease in system effectiveness with time, can be overcome by the removal of some nutrients that are necessary for cell growth so as to reduce the growth rate as much as possible. In those treatments releasing gases, such as CO_2, the use of the horizontal bed is recommended in order to avoid the increase in pressure drop and the formation of preferential channels in the biofilter.

Some process parameters demonstrated to remarkably influence the performance of microorganisms in the treatment of effluents. Onysko et al. *(35)* investigated the effect of temperature on a mixed microbial culture in fixed bed either in batch or continuous process. Because this parameter was shown to have a strong effect on both cell growth and biodegradation of phenolic compounds, it needs to be optimized so as to ensure, at the same time, a low growth rate in the immobilized-cell reactor as well as a high removal yield of the industrial residue.

Fluidized-Bed Biofilter With Mechanical Mixing

In the biofiltration methods employing fluidized-bed reactors the cells are immobilized onto an expanded bed. These bioreactors consist of vertical columns with circular section in which the particles containing immobilized cells are filled up to about 70% of the working volume and then expanded by introduction of air or some other inert gas from the bottom, or by partial recycle of the effluent from the column or by mechanical mixing. Such a bioreactor configuration is widely used because it allows overcoming the aforementioned problems often encountered with the fixed bed.

Several studies performed using different bioreactor configurations with immobilized cells demonstrated that the fluidized-bed reactor is a promising biotechnology for the treatment of phenolic compounds; however, satisfactory performance of the process can only be ensured by optimization of the concentrations of phenol, biomass, and substrate in the biofilm and of biofilm thickness (18,20,36).

One of the limiting factors in the use of the fluidized bed is the transient phase during the first hours of the process, which can affect the efficiency of a continuous process utilizing this kind of bioreactor. During this phase, the cell growth or substrate consumption can in fact create a kind of inhibition to the pollutants degradation. It has been suggested that adequate removal and growth kinetics (to be established through studies performed with biomass not acclimated to the process conditions) can minimize these problems and increase the final yield of phenolic compounds degradation. For example, Onysko et al. (36), utilizing a fluidized-bed biofilter containing *Pseudomonas putida*, obtained satisfactory removal of phenol at 10°C.

Air-Lift Fluidized-Bed Biofilter

Koch et al. (37) were able to aerobically degrade a mixture of 22 phenolic, heterocyclic and aromatic compounds with yields in the range 59–69% by a mixed culture previously isolated from the soil and acclimated to selected conditions in a continuous fluidized-bed reactor using sand and charcoal as supports. Dluhy et al. (38), comparing conventional and air-lift fluidized-bed bioreactors, concluded that both reactor configurations can be successfully used for the treatment of phenolic residues and developed mathematical models to optimize the process.

Reactors for the Biofiltration of Gaseous Effluents

A large number of biofilter configurations are described in the literature for different applications (39). Open biofilters, consisting of simple compost or porous earth beds having depth of about 1 m, are widely utilized in the removal of odors or volatile organic compounds because of their very low cost and simplicity. However, as they are in direct contact with the atmosphere, their effectiveness is affected by the climate, additional disadvantages being the difficulty of monitoring the process and the surface needed for their installation.

Most biofilters for the treatment of gaseous effluents are covered, contain mixtures of organic materials as supports, and allow controlling and monitoring some process parameters; nevertheless, they exhibit low operating flexibility. The "Multiple-layer" configuration, which consists of beds disposed in series, each one containing a different type of microorganism under optimized growth conditions, allows for a more accurate control of the process inside the bed.

Reports on styrene biodegradation are few and very difficult to compare, owing to the different conditions employed *(15,40–42)*. Although the trickling filter is considered to be the most effective configuration *(43,44)*, it can only be used in short operation owing to excess biomass growth *(45)*. "Perlite-packed biofilters" inoculated with fungi able to degrade styrene were shown to be an interesting alternative to the trickling filters because of better ability to face drying out and bed acidification *(40,46)*.

Microorganisms

Main Microorganisms Utilized for the Biofiltration of Liquid Effluents

Table 3 lists the main microorganisms utilized for the removal of the main pollutants. As far as the biofiltration of liquid effluents is concerned, Onysko et al. *(36)* tried to optimize the treatment at 10°C of phenolic residues by *P. putida* Q5 immobilized in fluidized-bed reactor. Sánchez et al. *(47)* proposed effective mathematical models for the biodegradation at 10°C and 25°C of phenolic compounds utilizing the same strain in the presence of the inhibiting effects exerted by the intermediate metabolites secreted in the medium.

P. putida can be used in biological oxidation of chlorophenols, but these compounds exert toxic effects on microorganisms. Kargi and Eker *(48)* developed a new rotating perforated tubes biofilm reactor that was used in continuous mode for removal of 2,4-dichlorophenol (DCP) from synthetic wastewater and toxicity reduction. A special culture of *P. putida* capable of degrading DCP supplemented with activated sludge was used for this purpose resulting in removal efficiencies up to 96%.

Bacillus subtilis has widely been employed in biofilters for the treatment of contaminants in liquid effluents, with particular concern to the degradation of aromatic hydrocarbons and the production of biosurfactants. Christova et al. *(49)* focused on the simultaneous degradation of *n*-hexadecane and naphthalene, whereas Moran et al. *(50)* discussed on the application of the surfactin produced by the strain *B. subtilis* O9 grown on a sucrose-based medium for the biodegradation of industrial residues. The presence of surfactin in the medium was shown to increase the degradation yield of aliphatic and aromatic hydrocarbons from 20.9% to 35.5% and 41%, respectively, its positive effect on the biodegradation being higher when using residues with long lateral aliphatic chains. The quick production of surfactin by *B. subtilis* strains makes this microorganism of particular interest for the treatment of effluents and bioremediation applications.

Feitkenhauer et al. *(51)* investigated the potential of different *B. thermoleovorans* strains to degrade phenol. Growth rates were about four times higher than those of mesophilic microorganisms such as *P. putida*,

Table 3
Main Xenobiotic Compounds and Microorganisms Utilized
for Their Degradation

Pollutant	Microorganism
Chlorinated hydrocarbons	
Dichloromethane	*Hyphomicrobium* sp.
Methyl chloride	*Pseudomonas* DM1
	Methylobacter DM111
	Hyphomicrobium sp.
1,2-Dichloroethane	*Xanthobacter* GJ10
Vinyl chloride	*Mycobacterium* L1
Epichlorohydrine	*Pseudomonas* AD1
Chlorobenzene	*Pseudomonas* WR1306
1,2-Dichlorobenzene	*Pseudomonas* GJ 60
1,3-Dichlorobenzene	*Pseudomonas* sp.
1,4-Dichlorobenzene	*Alcaligenes* A175
Aromatics	
Benzene	*Pseudomonas* sp.
	Acinetobacter calcoaceticus RJE74
Toluene	*Pseudomonas* sp.
	P. putida
	Bacillus sp.
Monoalkylbenzenes	*Pseudomonas* sp. NCIB 10643
	P. fluorescens
	P. putida RE204
	Acinetobacter lwoffi
m-, *p*-Xylene	*Pseudomonas* sp.
	Nocardia sp.
o-Xylene	*P. stutzeri*
	Corynebacterium sp. C125
Styrene	*Xanthobacter* sp. 124X
	P. putida
	Pseudomonas sp.
Biphenyls	
Biphenyl	*Pseudomonas* sp. NCIB 10643
	Nocardia sp. NCIB 10503
	P. cruciviae
	P. putida
	P. pseudoalcaligenes
	P. aeruginosa
	Escherichia coli
Others	
Naphthalene	*Pseudomonas* sp.
	P. fluorescens
	P. putida

and the high values (about 2.8/h) were detected at phenol concentrations of 15 mg/L. The thermophilic strain *B. thermoleovorans* sp. A2 was found to be insensitive to hydrodynamic shear stress in stirred bioreactor experiments (despite of possible membrane damage caused by phenol) and flourished at an ionic strength of the medium of 15–60 g of NaCl/L. These exceptional properties make this strain an excellent candidate for technical applications.

Main Microorganisms Utilized for the Biofiltration of Gaseous Effluents

The composition of the microflora present in a biofilter depends on that of the gaseous effluent as well as the ability of some contaminants to be catabolized as nutrients.

Before the continuous operation, the microorganisms present in the biofilter are subject to progressive adaptation to the organic pollutants present in the gaseous stream, which usually lasts a period of 10 d. The microorganisms to be used can be furnished either as isolated pure cultures or as an industrial sludge *(15)*. The degradation of aromatic organic compounds from the pharmaceutical industry, like phenolic residues, needs heterogeneous populations to make the parallel degradation of other pollutants possible and then to accelerate the overall process *(52–54)*. Since heterogeneous populations used in biofilters are natural consortia, they are particularly resistant to the pollutants, thus increasing the overall time of biofilter operation. Regarding the specific degradation of styrene, biofilters inoculated with heterogeneous consortia, although characterized by low conversion yields, are usually preferred to mono-culture biofilters because they are more easily operated and managed *(8,42,44,55–59)*.

Pseudomonas sp. is largely employed in styrene degradation that starts with the oxidation of the vinyl chain. After preliminary formation of styrene oxide, such an intermediate is transformed to phenyl-acetaldehyde and phenyl-acetate, which are subsequently converted to intermediates of the tricarboxylic acid (TCA) cycle through a complex series of reactions and then metabolized. Notwithstanding the incomplete characterization of *Xanthobacter* 124X *(60)* and *Pseudomonas* MST *(61)* with respect to their capability to metabolize aromatic compounds, these microorganisms seem to be good alternative styrene degraders. Jang et al. *(62)* isolated *Pseudomonas* sp. SR-5 as a styrene-degrading bacterium immobilized in organic (peat) and inorganic (ceramic) packing material and evaluated its ability to degrade styrene at different concentrations. The effectiveness of styrene degradation varied according to the packing material utilized; a maximum degradation capacity of 236 g/(m^3·h) was attained with peat, corresponding to 90.4% removal.

P. putida has been used to remove toluene and ethanol present in waste air. Lim et al. *(63)* studied the transient behavior of a hybrid system made up of biofilter and photo-catalytic reactor. The system was inoculated with a pure culture of *Burkholderia cepacia* G4 and *P. putida*, whereas a photo-catalytic reactor was made up of 15W ultraviolet (UV)-A lamps and annular tubes packed with glass beads coated with TiO_2 solution before calcination. The maximum elimination capacities of toluene and ethanol turned out to be 130 g/(m^3·h) and 230 g/(m^3·h), although in experiments using only a biofilter, they decreased to 40 g/(m^3·h) and 130 g/(m^3·h), respectively. Marek et al. (64) evaluated xylene and toluene degradation in waste air using a laboratory-scale biofilter containing immobilized *P. putida*. A decrease in bed pH improved the efficiency of toluene degradation, but the simultaneous degradation of both pollutants required higher pH.

The biological treatment of odorous sulfur-containing compounds was reported by Geng et al. (65), who successfully isolated a dimethylsulfide (DMS)-degrading bacterium from activated sludge, using the enrichment isolation technique. The isolate was able to metabolize DMS as well as hydrogen sulfide. Batch tests demonstrated that over half of DMS could be removed by the isolate in 3 h when the initial DMS amount was approx 0.6–1.5 g. Nearly complete removal of H_2S by the isolate was obtained by a continuous test in a 2-L gas-bubbling bottle. Nevertheless, Oyarzun et al. (66) evaluated the use of *Thiobacillus thioparus* for treating a gaseous stream containing high concentrations of H_2S. The biofilter reached an efficiency of almost 100% when fed with 0.5 g/m^3 of the pollutant or less.

Finally, although *Rhodococcus rhodochrous* NCIMB 13259 is the best known microorganism able to degrade styrene (67–69), only a few reports described its application to biofiltration of gaseous effluents (15).

Biofiltration for the Removal of Phenolic Residues From the Pharmaceutical Industry

Phenolic compounds find application in different sectors of chemical, pharmaceutical, and petrochemical industry, being employed as antimicrobial agents (disinfectants) as well as to preserve paints, leather, and some textiles. Because of their widespread use, phenolic residues are often present in industrial effluents in toxic levels, which can give rise to serious environment problems; when contained in industrial effluents without previous treatment or in the absence of adequate treatment, they can in fact contaminate soil and subsoil, being even able to reach groundwater tables and river heads.

The control of emissions of toxic organic compounds to the atmosphere has become a critical and expensive challenge for industry, being necessary to face the new requirements in terms of environmental quality and policy. With special regard to phenolic compounds, which are abundantly released by the pharmaceutical industry, the present legislation of many countries imposes maximum concentration thresholds for these pollutants, owing to their potential hazardous effects on the humans and the environment. For liquid effluents, the maximum threshold values depend, in Brazil, on the water class of the receiving body. Class 1 does refer to water utilized for domestic use without prior treatment, class 2 to water to be conventionally treated before domestic use, irrigation and recreation, class 3 to water addressed to domestic use or in places where there is need of fauna and flora preservation, and class 4 to water that can be utilized for domestic use after significant treatment, shipping, industrial use, irrigation, and other uses requiring less quality standards. Although water belonging to class 1 has to be absolutely free of phenolic residues, for the others the concentration thresholds vary from 0.001 mg/L (classes 2 and 3) to 1 mg/L (class 4) *(70)*. To solve the environmental and human health problems potentially associated to the widespread presence of highly toxic residues in the environment, the United Nations published, through the *World Health Organization for the International Programme* and the *Environmental Health Criteria*, the maximum limits for the exposition to phenolic residues in the air recommended either in the working place or in the environment. For example, a maximum level of 19 mg/m^3 is allowed in Germany, USA, Japan, and Australia in the working place, whereas Russia and Czech Republic established a maximum daily threshold in the environment of 0.001 mg/m^3 *(71)*.

Removal of Phenolic Compounds From Liquid Effluents

The phenolic compounds are pollutants commonly present in concentrations from 5 to 500 mg/L in effluents from refineries, chemical plants for the production of explosives, resins, pesticides, textiles, and pharmaceuticals *(71)*. To describe phenol biodegradation, Haldane *(72)* related the microbial growth to substrate concentration taking into account the possible inhibition exerted by high substrate levels:

$$\mu = \frac{\mu_m \times S}{K_s + S + \left(\dfrac{S^2}{K_i}\right)}$$

being μ_m the maximum specific growth rate, K_S the saturation constant of Monod equation and K_i the inhibition constant. Several authors demonstrated that cell growth on phenol depends on phenol concentration, temperature, pH, and the way of inoculum preparation *(73,74)*. As a general rule, the cultivations exhibited a prolonged lag phase *(75)*.

Using this equation, Monteiro et al. *(19)* estimated the kinetic parameters of batch degradation of phenol at 26°C and pH 6.8 by *P. putida* DSM 548, using phenol as the only carbon source at concentrations in the range 1–100 mg/L (μ_m = 0.463/h, K_S = 6.19 mg/L and K_i = 54.1 mg/L). The maximum specific growth rate was higher for pure rather than for mixed culture, hence evidencing the likely influence of secondary metabolites on microbial growth and phenol degradation.

Most research-work has been performed on the biodegradation of phenolic compounds using different microorganisms either in pure or mixed cultures, among which are *Pseudomonas* sp. *(76–79)*, *Alcaligenes* sp.

(80,81), *Rhodococcus* sp. (82,83), and *Cryptococcus* sp. (84). In addition, aerobic cultivations were shown to be more effective than the anaerobic ones (85).

Recent works have been developed using the microorganisms in a cometabolism system, in which phenol biodegradation is realized simultaneously with the main metabolic route for substrate uptake. Dupasquier et al. (86) demonstrated the viability of the biodegradation of methyl tertiary butyl ether (MTBE) vapors by cometabolism with pentane using a pentane-oxidizing strain of *P. aeruginosa*. As a cosubstrate, MTBE was degraded during the pentane uptake by the cells. The experimental data of pentane and MTBE removal efficiencies compared satisfactorily with the theoretical predictions of the model under steady-state conditions.

The biodegradation of phenolic compounds can also be carried out using mixed cultures. An example of this application is provided by activated sludge, a heterogeneous mixture of unidentified aerobic microorganisms able to oxidize the pollutants to carbon dioxide. Otherwise, mixed cultures of identified microorganisms can be used. For example, Oh et al. (87) utilized with success a mixed culture of *P. putida*, *Flavobacterium* sp., and *Acinetobacter* sp., whereas Wiesel et al. (88) employed a mixture of five different microorganisms for the treatment of polycyclic aromatic hydrocarbons.

By biochemical tests and molecular biological analysis using 16S ribotyping, Andretta et al. (89) identified a 4,5,6-trichloroguaiacol (TCG)-degrading strain of *B. subtilis*. Biodegradation occurred in a mineral salts' medium only when the inoculum was made up of cells in the stationary phase of growth and was accelerated by an additional carbon source, such as glucose, sucrose, glycerol, or molasses. An additional nitrogen source (like ammonium sulfate) did not affect the rate of 4,5,6-TCG removal. No plasmids were detected in the bacterial cells. It was demonstrated that 4,5,6-TCG is not degraded by cometabolism and that the gene encoding this characteristic is probably located on the chromosome. The lack of requirement for additional nitrogen source, the ability to enhance biodegradation by addition of cheap carbon sources such as molasses, and the fact that the trait is likely to be stable as it is encoded on the cell chromosome, are all characteristics that make this organism attractive for treatment of wastes and environments polluted with organochlorinated compounds.

González et al. (90), comparing the biodegradation of phenolic compounds from industrial effluents by a combined bioscrubbing (with mechanical mixing) and biofiltration (fixed-bed reactor [FBR]) process, demonstrated the significance of the reactor configuration either in reducing the treatment costs or improving its yield. Although the use of mechanical mixing allows for easy control and residence time adjustment, it can stress the culture, thus affecting the process and requiring a long time for culture recovery. By the use of immobilized and acclimated cells under steady-state conditions, biofilters minimize this deleterious effect by promoting a quicker recovery of the system. According to Holladay et al. *(91)*, the main drawback of the use of fixed-bed biofilters, with cells immobilized on inert supports, would be the natural growth of biomass, which reduces the reactor operating time. Beg and Hassan *(92)* overcame this disadvantage increasing the energy requirements of endogenous respiration.

González et al. *(90)* confirmed that phenol and 4-chlorophenol are degraded with no appreciable difference by *P. testosteroni* CPW301 through the same metabolic pathway and that the degradation rate is affected by the concentration of the latter substrate in the cultivation medium. Appreciable loss of microbial activity was observed only after several operations of immobilized-biomass recycling, and the rate of continuous degradation and dilution rates applied were higher in FBR rather than in mechanically mixed bioscrubber (regarded as a continuous stirred tank reactor [CSTR]). The use of a coculture of *P. solanacearum* TCP114 and *P. testosteroni* CPW301 made the simultaneous treatment of 2,4,6-trichlorophenol, phenol, and 4-chlorophenol possible. Although the microorganisms were not affected by the presence of phenol in the medium, the rate of 2,4,6-trichlorophenol degradation was increased by 4-clorophenol, but the rate decreased and inhibition took place when phenol concentration exceeded the level of toxicity to the cultures. In the absence of *P. solanacearum* TCP114, phenol and 4-chlorophenol degradations were affected even in the presence of 2,4,6-trichlorophenol. These results demonstrate that the simultaneous treatment of different compounds requires a better knowledge of the toxic compounds and a deep investigation on their reciprocal interference on cell metabolism. Also the type of reactor and biotreatment can strongly affect the degradation of industrial residues, as it was demonstrated by the higher effectiveness of CSTR with respect to FBR.

Promising results were also obtained by the use of fluidized-bed biofilters for the treatment of phenolic compounds from pharmaceutical industry effluents. González et al. *(90,93)* investigated the biodegradation of phenolic compounds either by free or immobilized cells in continuous or batch operation, the best results having been obtained using the fluidized-bed reactor with immobilized-cells (biofiltration).

A way to remarkably increase the effectiveness of industrial effluents biodegradation is the previous adaptation of the microorganism to the toxic compound *(94)*.

Removal of Phenolic Compounds From Gaseous Effluents

As is well known, biofiltration of pollutants in gaseous phase implies the simultaneous transfer of the pollutant from the gaseous phase to the liquid phase by absorption and its biodegradation in the liquid phase thanks to the contact with the immobilized cells.

Biofiltration is increasingly applied worldwide in the treatment of these gaseous effluents *(8,11,56,95)* because of its low operating costs and high removal efficiency. Zilli et al. *(15)*, who investigated the continuous removal of toluene and styrene in synthetic air streams by means of biofilters, obtained for the former pollutant a maximum elimination capacity of 275 mg/(m^3·h), i.e., a value about 10% higher than those previously reported *(96,97)*. The increased capacity was likely owing to the heterogeneous microflora developed in the biofilter, which seems to confirm the better performances of these systems with respect to those using pure cultures. At low air flowrates, the pollutant was almost completely removed from the air stream, and there was a linear relationship between the rates of pollutant removal and gaseous effluent feeding.

Biofiltration and Simultaneous Production of Surfactants

Biosurfactants are, generally, glycolipids, lipopeptides, protein-polysaccharides complex, phospholipids, fatty acids, and some neutral lipids *(98)*. Because of their wide market and large number of applications *(99)*, the surfactants constitute an important class of chemicals.

They are amphiphilic molecules composed by a hydrophilic polar portion and a hydrophobic apolar portion. This property confers them the ability of reducing the superficial and the interfacial tensions, thus forming emulsions so that the hydrocarbons can be dispersed in water or vice versa *(100)*. As most of the common surfactants are produced chemically from oil derivatives, the production of biosurfactants for cleaning and removal of oily residues is gaining increasing interest in the field of environmental biotechnology.

The biosurfactants are a class of surfacing molecules obtained by microbial cultivations able to reduce superficial and interfacial tensions either in water solutions or in mixtures of hydrocarbons. Several interesting properties of these substances, which spontaneously form during cell growth *(101)*, were extensively investigated *(102–104)*. The world consumption of surfactants has been continuously increasing in the last decades *(99)*, and the petrochemical industry accounts for about one half of its whole utilization.

The biosurfactant surfactin has the potential to aid in the recovery of subsurface organic contaminants (environmental remediation) or crude oils (oil recovery). However, high medium and purification costs limit its use in these high-volume applications *(105)*.

They attracted attention as hydrocarbon dissolution agents for the first time in the late 1960s, and their applications have been greatly extended in the past five decades as an improved alternative to chemical surfactants as they are biodegradable *(106)*.

The numerous advantages of biosurfactants, such as mild production conditions, low toxicity, high biodegradability, and environmental compatibility have prompted applications not only in the food, cosmetic, and pharmaceutical industries but also in environmental protection and energy-saving technology *(107,108)*.

It seems that their action is related to the consumption of hydrocarbons *(101)*, so they are mainly produced by microorganisms able to degrade hydrocarbons. The production of biosurfactants by *B. subtilis* ATCC 6633 was investigated using commercial sugar, sugarcane juice and cane molasses, glycerol, mannitol, soybean oil, among others *(109)*. The results showed that the best carbon source was commercial sugar because of minimum surface tension.

Additional advantages with respect to chemical surfactants are higher formation of foam, selectivity and specific activity at high temperature, pH, and salinity *(108)*.

Several lipopeptide surfactants have the most potent antibiotic activity and have been a subject of several studies on the discovery of new antibiotics, including surfactin of *B. subtilis* *(110)*.

Among the various studies on the production of biosurfactants, those dealing with the production of surfactin and iturin A by *B. subtilis* are of particular interest, as this microorganism can also be used for the biodegradation of different phenolic compounds *(98)*.

Concluding Remarks

Besides its traditional application for the control of odors, biofiltration has become a successful technology for the treatment of pollutants contaminating either gaseous or liquid effluents. During the last two decades, thanks to the remarkable progress in the fields of microbiology and process technology, biofiltration has been gaining the interest of various industrial sectors and is increasingly applied. The feasibility of this technique for the treatment of different types of pollutants contained in industrial effluents has been demonstrated, with particular concern to the biodegradation of phenolic compounds. The removal yield is often higher than 90%, mainly in the presence of effluents contaminated by alcohols, ketones, ethers, aldehydes, and volatile aromatic compounds. As far as the phenolic residues are concerned, there are reports on the reduction of their concentration in the effluent from 50 to 1 mg/L, which is a level suitable for the main environmental legislations worldwide. The main advantages of the use of this technology are the low costs, the low energy and space requirements, moderate temperature, pH and pressure conditions, and high compactness.

The applicability of this technology depends on the availability of adequate strains or complex of microorganisms able to perform the biodegradation of one or more pollutants, so as to make it feasible and economically competitive with the traditional physical-chemical methods. More research-work is needed to ensure a more effective control of the operating parameters as well as the selection of the suited reactor configuration, so as to promote the use of biofilters for the treatment of effluents from the pharmaceutical industry contaminated by phenolic compounds and then to reduce their concentrations below the emission limits imposed in Brazil.

References

1. Edgington, S. M. (1994), *Biotechnology* **12**, 1338–1342.
2. Head, I. M. (1998), *Microbiology* **144**, 599–608.
3. Eckenfelder, W. W. and Musterman, J. L. (1994), *Water Sci. Technol.* **29(9)**, 79–88.
4. Liu, S. and Suflita, J. M. (1993), *Trends Biotechnol.* **11**, 344–352.
5. Bragg, J. R., Prince, R. C., Harner, K. J., and Altas, R. M. (1994), *Nature* **368**, 413–418.
6. Swannell, R. P. J., Lee, K., and McDonagh, M. (1996), *Microbiol. Rev.* **60**, 342–365.
7. Ottengraf, S. P. P., van den Oever, A. H. C., and Kempenaars, F. J. C. M. (1984), In: *Innovations in Biotechnology*. Houwink, E. H. and van der Meer R. R., eds., Elsevier, Amsterdam, pp. 157–167.
8. Ottengraf, S. P. P., Meesters, J. J. P., van den Oever, A. H. C., and Rozema, H. R. (1986), *Bioproc. Eng.* **1**, 61–69.
9. Button, D. K., Schut, F., Quang, P., Martin, R. M., and Robertson, B. (1993), *Environ. Microbiol.* **59**, 881–891.
10. Watanabe, K. and Baker, P. W. (2000), *J. Biosci. Bioeng.* **89**, 1–11.
11. Zilli, M. and Converti, A. (1999), In: *The Encyclopedia of Bioprocess Technology: Fermentation, Biocatalysis and Bioseparation*. Flickinger, M. C. and Drew, S. W., eds., Wiley, New York, pp. 305–319.
12. Kim, J. H., Oh, K. K., Lee, S. T., and Kim, S. W. (2002), *Proc. Biochem.* **37**, 1367–1373.
13. Zilli, M., Del Borghi, A., and Converti, A. (2000), *Appl. Microbiol. Biotechnol.* **54**, 248–254.
14. Abu Hamed, T., Bayraktar, E., Mehmetoğlu, Ü., and Mehmetoğlu, T. (2004), *Biochem. Eng. J.* **19**, 137–146.
15. Zilli, M., Palazzi, E., Sene, L., Converti, A., and Del Borghi, M. (2001), *Proc. Biochem.* **37**, 423–429.
16. Sene, L., Converti, A., Felipe, M. G. A., and Zilli, M. (2002), *Biores. Technol.* **83**, 153–157.
17. Moharikar, A. and Purohit, H. (2003), *Int. Biodeter. Biodegr.* **52**, 255–260.
18. Vinod, A. V. and Reddy, G. V. (2005), *Biochem. Eng. J.* **24**, 1–10.
19. Monteiro, Á. A. M. G., Boaventura, R. A. R., and Rodrigues, A. E. (2000), *Biochem. Eng. J.* **6**, 45–49.
20. Clarke, K. L., Pugsley, T., and Hill, T. A. (2005), *Chem. Eng. Sci.* **60**, 6909–6918.
21. Prokop, W. H. and Bohn, H. L. (1985), *J. Air Pollut. Control Assoc.* **35**, 1332–1338.
22. Kampbell, D. H., Wilson, J. T., Read, H. W., and Stocksdale, J. (1987), *J. Air Pollut. Control Assoc.* **37**, 1236–1240.
23. Alfani, F., Cantarella, L., Gallifuoco, A., and Cantarella, M. (1990), *Acqua-Aria* **10**, 877–884.
24. Jäger, B. and Jager, J. (1978), *Müll und Abfall* **2**, 48–54.
25. Hartmann, H. (1977), *Stuttg. Ver. Siedlungswasserwirstsch* **59**, 3–19.
26. Thistlethwayte, B., Hardwick, B., and Goleb, E. E. (1973), *Chimie Ind.* **106**, 795–801.
27. Helmer, R. (1974), *Ges. Ing.* **94**, 21–30.
28. Chen, K. C., Lin, W. H., and Liu, Y. C. (2002), *Enzyme Microbial Technol.* **31**, 490–497.
29. Chiangchun, Q., Hanchang, S., Yongming, Z., and Yi, Q. (2003), *Proc. Biochem.* **38**, 1545–1551.
30. Tsai, H. H., Ravindran, V., and Pirbazari, M. (2005), *Chem. Eng. Sci.* **60**, 5620–5636.
31. Luke, A. K. and Burton, S. G. (2001), *Enzyme Microbial Technol.* **29**, 348–356.
32. Kim, D. J. and Kim, H. (2005), *Proc. Biochem.* **40**, 2015–2020.
33. Zaiat, M., Cabral, A. K. A., and Foresti, E. (1996), *Wat. Res.* **30**, 2435–2439.

34. Schmidell, W. and Facciotti, M. C. R. (2001), In: *Biotecnologia Industrial*. Schmidell, W., Lima, U. A. L., Aquarone, E., and Borzani, W., eds., Edgard Blücher, São Paulo, pp. 179–192.
35. Onysko, K. A., Budman, H. M., and Robinson, C. W. (2000), *Biotechnol. Bioeng.* **70**, 291–299.
36. Onysko, K. A., Robinson, C. W., and Budman, H. M. (2002). *Can. J. Chem. Eng.* **80**, 239–252.
37. Koch, B., Ostermann, M., Hoke, H., and Hempel, D. C. (1991), *Wat. Res.* **25**, 1–8.
38. Dluhy, M., Sefcik, J., and Bales, V. (1994), *Comput. Chem. Eng.* **18**, S725–S729.
39. Swanson, W. J. and Loeher, R. C. (1997), *J. Environ. Eng.* **123**, 538–546.
40. Cox, H. H. J., Houtman, J. H. M., Doddema, H. J., and Harder, W. (1993), *Biotechnol. Lett.* **15**, 737–742.
41. Cox, H. H. J., Houtman, J. H. M., Doddema, H. J., and Harder, W. (1993), *Appl. Microbiol. Biotechnol.* **39**, 372–376.
42. Arnold, M., Reittu, A., von Wright, A., Martikainen, P. J., and Suihko, M. L. (1997), *Appl. Microbiol. Biotechnol.* **48**, 738–744.
43. Sorial, G. A., Smith, F. L., Suidan, M. T., Pandit, A., Biswas, P., and Brenner, R. C. (1998), *Wat. Res.* **32**, 1593–1603.
44. Kim, D., Cai, Z. L., and Sorial, G. A. (2005), *J. Air Waste Manag. Assoc.* **55**, 200–209.
45. Togna, A. P. and Frisch, S. (1993), 86th Meeting of the Air and Waste Management Association, Denver, CO, 14–18 June 1993.
46. Cox, H. H. J., Moerman, R. E., van Baalen, S., and van Gheiningen, W. N. M. (1997), *Biotechnol. Bioeng.* **53**, 259–266.
47. Sánchez, J. L. G., Kamp, B., Onysko, K. A., Budman, H., and Robinson, C. W. (1998), *Biotechnol. Bioeng.* **60**, 560–567.
48. Kargi, F. and Eker, S. (2005), *Proc. Biochem.* **40**, 2105–2111.
49. Christova, N., Tuleva, B., and Nikolova-Damyanova, B. (2004), *J. Biosci.* **59**, 205–208.
50. Moran, A. C., Olivera, N., Commendatore, M., Esteves, J. L., and Sineriz, F. (2000), *Biodegradation* **11**, 65–71.
51. Feitkenhauer, H., Schnicke, S., Muller, R., and Markl, H. (2001), *Appl. Microbiol. Biotechnol.* **57**, 744–750.
52. Oldenhuis, R., Vink, R. L. J. M., Janssen, D. B., and Witholt, B. (1989), *Appl. Environ. Microbiol.* **55**, 2819–2826.
53. Oldenhuis, R., Kuijk, L., Lammers, A., Jannsen, D. B., and Witholt, B. (1989), *Appl. Environ. Microbiol.* **30**, 211–217.
54. Ergas, S. J., Kinney, K., Fuller, M. E., and Scow, K. M. (1994), *Biotechnol. Bioeng.* **44**, 1048–1054.
55. Janssen, D. B., Grobben, G., Hoekstra, R., Oldenhuis, R., and Whitolt, B. (1988), *Appl. Microbiol. Biotechnol.* **29**, 392–399.
56. Ottengraf, S. P. P. (1986), In: *Biotechnology*. Rehm, H. J. and Reed, G., eds., VCH, Weinheim, pp. 425–452.
57. van der Werf, M. J., Swarts, H. J., and de Bont, J. A. M. (1999), *Appl. Environ. Microbiol.* **65**, 2092–2102.
58. Arand, M., Hallberg, B. M., Zou, J. Y., et al. (2003), *EMBO J.* **22**, 2583–2592.
59. Sabo, F., Motz, U., and Fischer, K. (1993), 86th Meeting of the Air and Waste Management Association, Denver, CO, 14–18 June 1993.
60. Hartmans, S., Smits, J. P., van der Werf, M. J., Volkering, F., and De Bont, J. A. M. (1989), *Appl. Environ. Microbiol.* **55**, 2850–2855.
61. Dijk, J. A., Stams, A. J. M., Schraa, G., Ballerstedt, H., de Bont, J. A. M., and Gerritse, J. (2003), *Appl. Microbiol. Biotech.* **63**, 68–74.
62. Jang, J. H., Hirai, M., and Shoda, M. (2004), *Appl. Microb. Biotechnol.* **65**, 349–355.
63. Lim, K. H., Park, S. W., Lee, E. J., and Hong, S. H. (2005), *Korean J. Chem. Eng.* **22**, 70–79.
64. Marek, J., Paca, J., Halecky, M., Koutsky, B., Sobotka, M., and Keshavarz T. (2001), *Folia Microb.* **46**, 205–209.
65. Geng, A. L., Chen, X. G., Gould, W. D., et al. (2004), *Wat. Sci. Technol.* **50(4)**, 291–297.

66. Oyarzun, P., Arancibia, F., Canales, C., and Aroca, G. E. (2003), *Proc. Biochem.* **39**, 165–170.
67. Warhust, A. M. and Fewson, C. A. (1994), *J. Appl. Bacteriol.* **77**, 597–606.
68. Warhust, A. M. and Fewson, C. A. (1994), *Crit. Rev. Biotechnol.* **14**, 29–73.
69. Warhust, A. M., Clarke, K. F., Hill, R. A., Holt, R. A., and Fewson, C. A. (1994), *Appl. Environ. Microbiol.* **60**, 1137–1145.
70. ANVISA—Agência Nacional de Vigilância Sanitária, RDC n210, DE 04/08/2003, Regulamento Técnico das Boas Prácticas de Fabricação de Medicamentos, D.O.U.—Diário Oficial da União; Poder Executivo, Brazilia, 14 August 2003.
71. WHO, Phenol Health and Safety Guide—Environmental Health Criteria 161: Phenol, Published by the World Health Organization for the International Programme on Chemical Safety, UNEP, ILO, WHO. http://www.inchem.org/documents/hsg/hsg/hsg88_e.htm.
72. Haldane, J. B. S (1965), In: *Enzymes*. MIT Press, Cambridge, MA, p. 84.
73. D'Adamo, P. D., Rozich, A. F., and Gaudy, A. F. (1984), *Biotechnol. Bioeng.* **26**, 397–402.
74. Hill, A. and Robinson, C. W. (1975), *Biotechnol. Bioeng.* **17**, 1599–1615.
75. Yang, R. D. and Humphrey, A. E. (1975), *Biotechnol. Bioeng.* **17**, 1211–1235.
76. Dapaah, S. Y. and Hill, G. A. (1992), *Biotechnol. Bioeng.* **40**, 1353–1358.
77. Hinteregger, C., Leitner, R., Loidl, M., Ferschi, A., and Streichsbier, F. (1992), *Appl. Microbiol. Biotechnol.* **37**, 252–259.
78. Chitra, S., Sekaran, G., Padmavathi, S., and Chandrakasan, G. J. (1995), *Gen. Appl. Microbiol.* **41**, 229–237.
79. Spigno, G., Zilli, M., and Nicolella, C. (2004), *Biochem. Eng. J.* **19**, 267–275.
80. Hill, G. A., Milne, B. J., and Nawrocki, P. A. (1996), *Appl. Microbiol. Biotechnol.* **46**, 163–168.
81. Valenzuela, J., Bumann, U., Céspedes, R., Padilla, L., and González, B. (1997), *Appl. Environ. Microbiol.* **63**, 227–232.
82. Apajalahti, J. H. A. and Salkinoja-Salomen, M. S. (1986), *Appl. Microbiol. Biotechnol.* **25**, 62–67.
83. Oh, J. S. and Han, Y. H. J. (1997), *Kor. J. Appl. Microbiol. Biotechnol.* **25**, 459–463.
84. Morsen, A. and Rehm, H. J. (1987), *Appl. Microbiol. Biotechnol.* **26**, 283–288.
85. Kim, J. H., Oh, K. K., Lee, S. T., and Kim, S. W. (2002), *Proc. Biochem.* **37**, 1367–1373.
86. Dupasquier, D., Revaii, S., and Auria, R. (2002), *Environ. Sci. Technol.* **36**, 247–253.
87. Oh, H. M., Ku, Y. H., Ahn, K. H., Jang, K. Y., Kho, Y. H., and Kwon, G. S. (1995), *Korean J. Appl. Microbiol. Technol.* **23**, 755–762.
88. Wiesel, I., Wubker, S. M., and Rehm, H. J. (1993), *Appl. Microbiol. Biotechnol.* **39**, 110–116.
89. Andretta, C. W. S., Rosa, R. M., Tondo, E. C., Gaylarde, C. C., and Henriques, J. A. P. (2004), *Chemosphere* **55**, 631–639.
90. González, G., Herrera, G., García, M. T., and Peña, M. (2001), *Biores. Technol.* **76**, 245–251.
91. Holladay, D. W., Hancher, C. W., Scott, C. D., and Chilcote, D. D. (1978), *J. Wat. Pollut. Control Fed.* **50**, 2573–2588.
92. Beg, S. A. and Hassan, M. M. (1985), *Chem. Eng. J.* **30**, 1–8.
93. González, G. and Herrera, G. (1995), *Acta Microbiol. Polonica* **44**, 285–296.
94. Zilli, M., Converti, A., Lodi, A., Del Borghi, M., and Ferraiolo, G. (1993), *Biotechnol. Bioeng.* **41**, 693–699.
95. Mpanias, C. J. and Baltzis, B. C. (1998), *Biotechnol. Bioeng.* **59**, 328–343.
96. Morales, M., Revah, S., and Auria, R. (1998), *Biotechnol. Bioeng.* **60**, 483–491.
97. Acuña, M. E., Pérez, F., Auria, R., and Revah, S. (1999), *Biotechnol. Bioeng.* **63**, 175–184.
98. Ahimou, F., Jacques, P., and Deleu, M. (2000), *Enzyme Microbial Technol.* **27**, 749–754.
99. Deleu, M. and Paquot, M. (2004), *C. R. Chimie* **7**, 641–646.
100. Desai, J. D. and Banat, I. M. (1997), *Microbiol. Mol. Biol. Rev.* **61**, 47–64.
101. Banat, I. M. (1994), *Biores. Technol.* **51**, 1–12.
102. Cooper, D. G. (1986), *Microbiol. Sci.* **3**, 145–149.

103. Rosenberg, E. (1986), *Crit. Rev. Biotechnol.* **3,** 109–132.
104. Haferburg, D., Hommel, R., Claus, R., and Kleber, H. P. (1986), *Adv. Biochem. Eng. Biotechnol.* **33,** 53–93.
105. Noah, K. S., Fox, S. L., Bruhn, D. F., Thompson, D. N., and Bala, G. A. (2002), *Appl. Biochem. Biotechnol.* **98,** 803–813.
106. Cameotra, S. S., Hommel, R., Claus, R., and Kleber, H. P. (2004), *Curr. Opin. Microbiol.* **7,** 262–266.
107. Banat, I. M., Makkar, R. S., and Cameotra, S. S. (2000), *Appl. Microbiol. Biotechnol.* **53,** 495–508.
108. Cameotra, S. S. and Makkar, R. S. (1998), *Appl. Microbiol. Biotechnol.* **50,** 520–529.
109. Reis, F. A. S. L., Servulo, F. C., and de França, F. P. (2004), *Appl. Biochem. Biotechnol.* **113–116,** 899–912.
110. Peypoux, F., Bonmatin, J. M., and Wallach, J. (1999). *Appl. Microbiol. Biotechnol.* **51,** 553–563.

The BTL2 Process of Biomass Utilization Entrained-Flow Gasification of Pyrolyzed Biomass Slurries

Klaus Raffelt,* Edmund Henrich, Andrea Koegel, Ralph Stahl, Joachim Steinhardt, and Friedhelm Weirich

Forschungszentrum Karlsruhe, Institut für Technische Chemie (ITC-CPV), Hermann-von-Helmholtz-Platz 1, D-76344 Eggenstein-Leopoldshafen, Germany, E-mail: klaus.raffelt@itc-cpv.fzk.de

Abstract

Forschungszentrum Karlsruhe has developed a concept for the utilization of cereal straw and other thin-walled biomass with high ash content. The concept consists of a regional step (drying, chopping, flash-pyrolysis, and mixing) and a central one (pressurized entrained-flow gasification, gas cleaning, synthesis of fuel, and production of byproducts). The purpose of the regional plant is to prepare the biomass by minimizing its volume and producing a stable and safe storage and transport form. In the central gasifier, the pyrolysis products are converted into syngas. The syngas is tar-free and can be used for Fischer-Tropsch synthesis after gas cleaning.

Index Entries: Straw; flash-pyrolysis; slurry; entrained-flow gasification; synthesis gas.

Preliminary Considerations Regarding the Potential

The use of biomass is one important element toward a sustainable energy production in the future. In a former study, it was estimated that biomass will be able to contribute about 20% of the global primary energy in the next century *(1)*. Biomass is unique among the sustainable energy sources, because it is the only renewable carbon source, whereas energy from wind, sea, geothermal sources, and sun can only supply heat and electricity. Carbon-based liquids will probably remain the most important fuels in the mobility sector in this century. Biomass conversion into liquids therefore makes economic sense, even if combustion or cocombustion is cheaper in the short term. Regarding the German energy biomass market, residual wood from forestry and waste wood from industry make up 41% of all organic residues (ash- and water-free mass) *(2)*, but the market of wood residues is tight at the same time and a considerable increase in woody waste is not expected. Nineteen percent of organic residues originate

---

*Author to whom all correspondence and reprint requests should be addressed.

from household and garden wastes, 19% from sewage sludge and liquid manure (both ash- and water-free). These two classes of wastes can be used in garbage incineration, composting or biogas production, for example. Twenty-one percent of organic residues are thin-walled lignocellulose substances like straw and hay. When calculating this percentage, it is taken into account that only half of the straw may be considered waste, whereas other half is used for keeping animals and improving soil quality. The so far not utilized part of straw and hay in Germany corresponds to 6 Mtoe (250 PJ or 70 TWh) and is worth being considered an energy source for the future. Unfortunately, straw is difficult to manage: Its high content of alkalies, chlorine, and other inorganic compounds leads to problems with corrosion and sticking ash *(4)*. Although combustion plants for straw have already been established in Denmark *(3)*, it is not trivial to satisfy the strict emission guidelines in the European Unit. Research on direct gasification of straw is also known *(4,5)*. Gasification and syngas use is a sophisticated technology which is only economically efficient when used on a large scale. The gasifier should reach a power in the order of >1 GW(th). Considering a 3 GW(th) gasifier with 8000 h of annual operation, it should be fed with biomass with an annual energy of 86.4 PJ (24 TWh or 2.06 Mtoe), this would be 6,000,000 t of air-dry straw. In Germany, an average of 720 t of wheat was harvested per km^2 in the year 2000 *(6)*. The grain : straw ratio is variable and considered to be 1:1. Half of the straw is needed for animal keeping and soil improvement, the rest is available for conversion into energy. A 3 GW(th) gasifier can therefore be supplied with straw from an area of 17000 km^2 under cultivation of cereals. Of the total German area, 19.7% is used for cultivation of wheat, rye, barley, and oats all over the country *(7)*. Areas with cereal fields alternate with forests, areas covered by buildings (towns and villages), areas of other agricultural use, and so on. This means that a gasifier may be supplied with straw from an area of 85,000 km^2 on an average and that the distance of transportation is up to 170 km, provided that the gasifier is situated in the center of this area. Leible et al. *(8)* estimated transport costs for straw in Germany. For a distance of 170 km, costs of 63 € (train) and 76 € (truck) per t straw result, this is 4.4 € (train) and 5.3 € (truck) per GJ heat content (lower heating value [LHV]). Because of the high oxygen content of biomass, only 41% of the heat content in the biomass can be converted into oxygen-free diesel or gasoline, the remaining heat content is lost by various reactions like the formation of CO_2 by various equilibrium reactions (e.g., water–gas shift reaction) and partial combustion. The transport costs of straw feed for 1 GJ of Fischer-Tropsch (FT)-diesel from biomass would therefore be 10.7 € (train) and 12.9 € (truck). The actual price of fossil fuels is about the same (tax-free approx 10 €/GJ in August 2005) *(9)*. If the transport costs of straw are as high as the all-inclusive production costs for fossil fuels, FT-diesel from direct gasification of straw cannot compete with fossil fuel, regardless of conversion technology details.

Table 1
Properties of Straw (15% Water Content) and Its Pyrolysis Products

	LHV GJ/t	Density kg/m^3	Energy density GJ/m^3	Costs for transport 250 km by train
Straw	14.4	50–200	0.7–2.6	100 €/t
Pyrolysis liquids	7–22	1100	8–24	
Char	31	200–700	6–22	
Slurry with 30% solids	14–25	1300	17–33	14 €/t

Table 2
Product Yields of Flash Pyrolysis in a Twin-Screw Reactor at 500°C and Atmospheric Pressure

Mass (%)	Ash	Moisture	Char	Gas	Condensates
Wood sawdust	1	5–10	14–18	~15	~70
Wheat straw chops	5	5–10	25–30	~20	50–55
Rice straw chops	~16	5–10	25–30	~20	50–55

Sand/Biomass Mixing Ratio = 6:1 to 20:1.

Why BTL2?

The abbreviation means "Biomass to Liquid in Two Steps," a regional flash-pyrolysis and a central gasification and synthesis. The concept was developed to solve the transport problem explained earlier. The products of flash-pyrolysis are pyrolysis liquids, char, and gas. Tar and char can be mixed to pumpable slurry with a much higher energy density than a straw bale, which results in lower transportation costs per tonne. Although the distance between the pyrolysis plant and the central gasifier may be >170 km, the energy consumption by the transportation of slurry is negligible (<1 %). The data of the energy content, energy density, and transport costs are given in Table 1.

Flash-Pyrolysis

The input material has to be air-dry (15% water content or less) and thin-walled. In an inert gas atmosphere (1 bar or low pressure), the small chopped biomass particles are heated quickly up to 500°C in the reactor with a residence time of the gaseous molecules of up to 1 s and of the solids of 10 s or more. These reaction conditions are chosen to maximize the yield of liquid (see Table 2). Typically, the biomass is converted into 53–78% condensed liquid (tar or so-called "bio-oil"), 12–34% char (with high-carbon content), and 8–20% gas (mainly H_2, CO_2, CO, CH_4, and little C_2H_x).

For flash-pyrolysis, a process development unit (see Fig. 1) was built for a throughput of up to 15 kg/h with the twin-screw mixing reactor

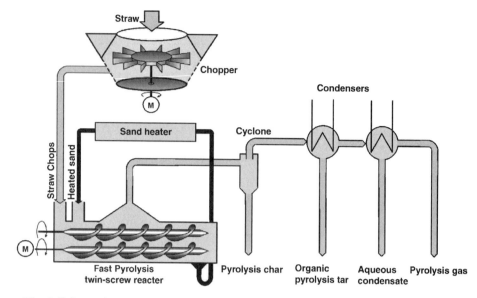

Fig. 1. Schematic representation of the process demonstration unit for flash pyrolysis at Forschungszentrum Karlsruhe.

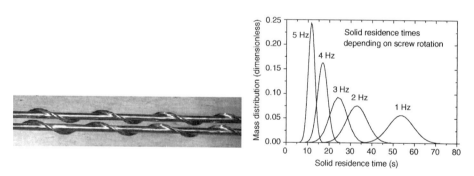

Fig. 2. Twin screws from the reactor, 1.5 m long (top). Solid residence times in the reactor depending on screw rotation (right).

developed by the LURGI company, Frankfurt (Germany) some decades ago. Although many different reactor types are conceivable, mainly two advantages lead to the decision to use this LURGI technology: (1) Feed and heat carrier are fluidized mechanically. Inert gas which dilutes the generated vapor is not needed. (2) Experience from the use of the twin-screw reactor in commercial-sized plants of oil refinery products was gathered over many decades *(10,11)*, and can now be transferred easily to a similar reactor for biomass-pyrolysis. The screws of the process demonstration unit in Karlsruhe have a length of 1.5 m and an inner and outer diameter of 20 and 40 mm, respectively (*see* Fig. 2). They are intermeshing and self-cleaning. Hot sand (500°C) falls onto the straw and both are fluidized mechanically with an excellent heat transfer. The mass ratio of sand and straw was 20:1 in the first experiments and has been lowered to 6:1 so far.

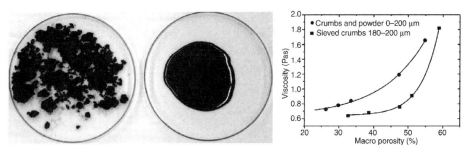

Fig. 3. Mixture of 22% straw char from flash pyrolysis and 78% pyrolysis tar. Left glass: Particle size 1–100 my, directly from the cyclones of the reactor. Right glass: Same mixture after application of a mortar. Diagram: Viscosity of slurries with 20% char coal and different porosities, initial particle size up to 200 µm, two lines for two different char fractions. Porosity reduction via pressure.

Literature data regarding the energy of heat for flash-pyrolysis vary between 0.8 and 3.5 MJ/kg *(12)*. The minimum mass of sand required for a sufficient heat supply has only been estimated roughly so far and needs to be investigated in future experiments. While the straw is pyrolyzed, the gaseous phase (mostly tar vapor) is sucked off by low pressure the coarse char is transported to the outlet at the end of the reactor and simultaneously premilled by abrasion. The fine-sized char particles leave the reactor with the vapors and are separated from the gaseous phase by two cyclones. The vapors are cooled in two condensers in which a highly organic tar is collected (about 10% water) as well as an aqueous solution of water-soluble molecules (about 70% water). The residual gas has a low-heating value only and consists of 45–65% CO_2, 27–45% CO, 5–9% hydrocarbons, and <0.3% H_2. In a commercial plant, this mixture of gases will be combusted for improving the energy balance. Its flammability is low. Consequently, it must be preheated first using low-temperature waste heat from the process. Downstream of the twin screws, the heat carrier is transported upwards by a bucket elevator and falls down again onto fresh straw passing a heat exchanger. A certain part of coarse char in the heat carrier can be accepted, because it will be milled during the following runs in the twin screws. No char accumulation is observed inside the heat carrier.

Slurry Preparation

Dry char is highly reactive and its handling is dangerous as far as powder explosions are concerned. For storage and transportation, char and tar are mixed to a highly viscous suspension (slurry) which has a much higher density than the biomass itself. A mixture of about 20% of straw char and 80% pyrolysis liquid at first represents a crumbly, thick, and wet bulk material (*see* Fig. 3). At surprisingly low-solid concentrations, flowability is lost. Because of its high porosity and wetting ability, the char absorbs the pyrolysis liquid like a sponge. This crumbly material has favorable safety properties and is suitable for long-term storage, but can neither be pumped through the

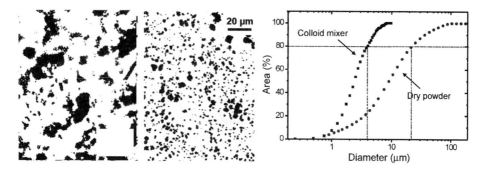

Fig. 4. Agglomerates of dry char particles (left) and single particles in the mixed slurry (center).

pipes of the gasifier nor can it be atomized which is necessary for complete carbon conversion at a low-reaction time. The crumbly substance becomes liquid by means of a colloid mixer *(13)*. The porosity of char is very high because of its tendency of forming stable mechanical agglomerates (Fig. 4) which are destroyed by the high shear forces of colloid mixing. After forming single particles, the liquid absorbed by the agglomerates can contribute to the lubrication of the solid particles inside of the slurry, a highly viscous, pumpable slurry is formed (density approx 1300 kg/m^3). Particle size reduction caused by the milling processes has an additional large effect (Fig. 3), because not only the agglomerates, but also the single particles have a high porosity. Especially the macropores are destroyed by milling, because they act like breaking points. So far, 30 metric t of slurries have been mixed for the gasification experiments, which will be described in this next section.

Entrained-Flow Gasification

The gasification experiments are carried out in cooperation with Future Energy in Freiberg, Germany *(14)*. The pilot-scale entrained flow facility of 3–5 MW(th) has a throughput of 350–500 kg/h and works under pressure (26 bar) (Fig. 5, center). It was designed as part of the Noell waste conversion process *(15)*. An equivalent, but upscaled 130 MW(th) gasifier has been run successfully with various feedstocks at the Sekundaerverwertungszentrum (center of waste recycling) Schwarze Pumpe, Germany, for more than 15 yr (Fig. 5, right). The feed is atomized with technical oxygen and gasified with variable λ-values in order to achieve the desired temperature. High-ash content is needed, because the molten slag settles onto the cooling screen inside. A thin layer of slag near the cooling screen turns into a glass-like solid and protects the reactor from corrosion, whereas the main part of the slag flows down and is discharged via a lock. The sintering point of straw ash was measured to be 770°C, the flow point was 1320°C. In practice, flowability of slag is sufficient at 1250°C. Carbon conversion also depends on temperature and additionally on parameters like residence

The BTL2 Process of Biomass Utilization

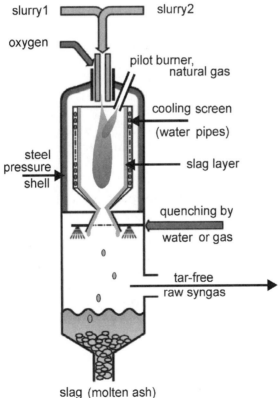

Fig. 5. Pressurized entained flow gasifier: Diagram (bottom) of the reactor and photo (top) of the reactor head at FUTURE-ENERGY in Freiburg, Germany.

time (proportional to throughput^{-1}), atomization quality, energy content of the feed, and elemental composition. Carbon conversion rates of approx 99% (1250°C) and >99.5% (1400°C) were estimated roughly (residual carbon in the quench water). Differences caused by the char particle size were not observed. The x_{80}-value of particle size describes that all particles with an equivalent circle diameter smaller than x_{80} contribute 80% of the total volume. In these experiments particle sizes of x_{80} = 6, 20, and 50 µm, respectively were used. The hot syngas is practically tar-free because of the high-gasification temperature. So far, we have not carried out our own experiments with gas cleaning (alkali metals, chlorine, and sulfur) and FT synthesis, but these processes are commercially available and state of the art (16–18).

Experience Gained From Gasification

Feedstocks used in the test runs in the 3–5 MW(th) gasifier were slurries of commercial pyrolysis products from Chemviron, Bodenfelde, Germany (19). Four different qualities of pyrolysis liquids were used and three chars with identical chemical compositions, but different particle size distributions. Viscosities were up to 2 Pa·s with solid contents of up to 39%. In the mixing unit, the slurry was heated up to 50°C and an even higher temperature (up to 80°C) before atomization with pressurized technical oxygen. By this, the viscosity of the slurry was reduced by the factor of 20–30 as compared with the initial viscosity at 20°C. There were no problems observed like the plugging of nozzles or sedimentation in pipes, which might be caused by slurry properties. So far, three experimental gasification campaigns have been performed, during which 20 separate stationary operating conditions were studied at various temperatures, solid concentrations, particle sizes of solid, slurry liquids, and residence times in the reactor. During the campaigns, about 30 t of different slurries were gasified. In the first experiments, a 0.6 MW(th) pilot burner (50 Nm^3/h) with natural gas was kept for safety reasons. Later, the pilot flame was used during start-up and shut-down procedures only. Atomization was considered to be good when stationary states (gas composition, temperature) were achieved, which turned out to be impossible once only: After reducing the oxygen supply to measure a stationary state at a very low temperature, the O_2 pressure was too low for atomization of a highly viscous slurry. The composition of the dry syngas varied in the ranges of 13–31% CO_2, 21–32% H_2, 27–47% CO, 9–17% N_2, and 0–0.5% CH_4. Gasification efficiencies ranged between 50 and 71%. This will be improved drastically in a large >1 GW gasifier, because the heat loss via the radiation screen will be decreased as will the O_2 consumption. Furthermore, the inert gas flows (now 9–17% N_2) can also be reduced to a few percents. Tables 3–5 show input and output data of five stationary states, each of them maintained for 3 h. Because of the stoichiometry of CO and H_2 formation, the theoretically lowest λ-value for complete gasification of lignocellulose

Table 3
Elemental Composition of the Gasification Feedstock

%	H$_2$O	Ash	C	H	N	O
Pyrolysis liquid 1	6.4	–	58.6	6.2	0.3	28.4
Pyrolysis liquid 2	13.6	–	51.8	6.3	0.4	27.8
Pyrolysis liquid 3	49.7	–	24.3	3.8	0.3	21.9
Pyrolysis liquid 4	67.4	–	10	2.9	0.1	19.6
Char[a]	4.2	5.5[b]	81.1	2.5	0.4	6.2

[a]Char from commercial beech wood pyrolysis (straw char not available on a large scale).
[b]Ash from beech wood char + additional straw ash.

Table 4
Stationary Reaction Conditions in the 3–5 MW(th) Entrained Flow Gasifier

No.	Char size x_{80} in μm	Solid [%] in slurry	Pyrolysis liquid	Oxygen λ	Feed in t/h	Temp. °C
1	94	25	1	0.48	0.50	~1350
2	56	23	2	0.50	0.50	~1250
3	10	23	2	0.46	0.35	~1350
4	10	39	3	0.52	0.50	*
5	10	29	4	0.64	0.35	~1350

*=Temperature not measured.

Table 5
Gas Composition of Raw Syngas Produced in the 3–5 MW(th) Entrained Flow Gasifier at the Stationary States Characterized in Table 4

No.	CO	H$_2$	CO$_2$	N$_2$[a]	CH$_4$
1	46.7	23.5	14.5	15.1	0.15
2	47.1	21.6	16.0	15.3	<0.1
3	47.0	30.2	13.1	9.6	<0.1
4	42.7	32.0	15.4	9.9	0.12
5	26.8	24.6	30.7	17.4	0.5

[a]Purge gas for not used pipes.

biomass is 0.16 (Fig. 6). (λ is defined as the ratio between actual oxygen consumption and theoretical oxygen consumption for stochiometric combustion). In reality, more oxygen than $\lambda = 0.16$ is needed, because the reaction temperature is achieved by partial combustion. The heat loss in a >3 GW gasifier is much lower than in the pilot-scale gasifier of 3 MW. Therefore, the λ-values measured on the pilot scale will be significantly higher than on the large scale. Furthermore, the water–gas shift reaction lowers CO and increases H$_2$ by forming CO$_2$. All in all, $\lambda = 0.32$ can be

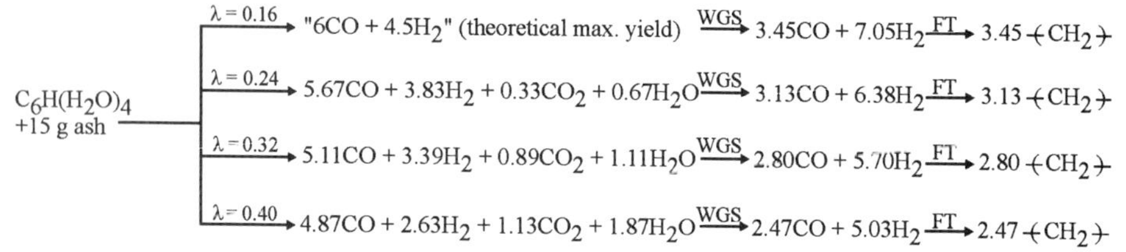

$$C_6H(H_2O)_4 + 15\text{ g ash}$$

- $\lambda = 0.16$: "6CO + 4.5H$_2$" (theoretical max. yield) $\xrightarrow{WGS}$ 3.45CO + 7.05H$_2$ $\xrightarrow{FT}$ 3.45 –(CH$_2$)–
- $\lambda = 0.24$: 5.67CO + 3.83H$_2$ + 0.33CO$_2$ + 0.67H$_2$O $\xrightarrow{WGS}$ 3.13CO + 6.38H$_2$ $\xrightarrow{FT}$ 3.13 –(CH$_2$)–
- $\lambda = 0.32$: 5.11CO + 3.39H$_2$ + 0.89CO$_2$ + 1.11H$_2$O $\xrightarrow{WGS}$ 2.80CO + 5.70H$_2$ $\xrightarrow{FT}$ 2.80 –(CH$_2$)–
- $\lambda = 0.40$: 4.87CO + 2.63H$_2$ + 1.13CO$_2$ + 1.87H$_2$O $\xrightarrow{WGS}$ 2.47CO + 5.03H$_2$ $\xrightarrow{FT}$ 2.47 –(CH$_2$)–

Fig. 6. The elemental composition of lignocellulose biomass can be described approx by C$_6$H$_9$O$_4$ + 15 g inorganics. When λ is decreased, the theoretical yield of the Fischer-Tropsch product increases. At a practically achievable λ-value of 0.32, the mass ratio between biomass input and synfuel product is 4.1:1. *Abbreviations:* WGS, water–gas shift reaction; FT, Fischer-Tropsch synthesis.

expected for a large-scale entrained-flow gasification of biomass. Raw syngas needs a thorough gas cleaning and little additional water which is still present after quenching the reaction heat by atomized water. Before FT synthesis, a H$_2$/CO ratio of approx 2.04 must be adjusted by means of the catalyzed water–gas-shift reaction. The FT product needs further refining like hydrocracking before it may be used as diesel fuel.

Efficiency and Economy of BTL2

The fuel yield of the total process can be calculated on the basis of Fig. 6. The elemental composition of biomass can be described empirically as "C$_6$H$_9$O$_4$ + 15 g inorganics + 18 g H$_2$O" with a mass of 178 g which could theoretically be converted into 2.8 units CH$_2$ and 39.2 g FT fuel (λ = 0.32). Slurry production from biomass results in 15% mass loss (gas yield of flash-pyrolysis), carbon conversion of gasification is assumed to be 98%. FT yield of C$_{5+}$ (hydrocarbons with five or more carbon atoms) in the once-through-then-out mode is 72% *(20)*. If these realistic efficiencies are taken into account, 23.5 g fuel can be produced from the input of Fig. 6. In other words, for 1 t of synfuel, 7.5 t of biomass are needed. During entrained-flow gasification, about 15% of the energy content become high-temperature heat with >1000°C in the hot raw gas. By burning nonconverted residual syngas and nonmarketable secondary products, electricity can be obtained by gas turbines. In the case of 33% efficiency, 14.1 GJ(el) result/1 t fuel. On the other hand, electricity is consumed for generation and compression of O$_2$, for the chopping, milling, and mixing of the feed and for numerous other activities. Therefore, only 4.2 GJ of electricity from the 14.1 GJ is left for sale. The principal dependent costs were simplified by assuming 25% of the unit original value per year. Costs of new installations were taken from literature *(21,22)*, estimated for the desired unit size using a gradual decrease exponent of 0.7, and calculated with customary market rates and payout times. In Table 6, the substantial single contributions are given as production costs for 1 t FT product. Contributions for straw therefore refer to 7.5 t, for slurries to 6.4 t. So-called work overhead costs were

Table 6
Costs for the Production of 1 t Fischer-Tropsch-Diesel
Regarding the German Price Level 2005

Regional flash pyrolysis for 10 t/h slurry production		
Straw bale on field, air-dry (8% H_2O)	7.5 t × 45 €/t →	338 €
Straw transport with tractor	7.5 t × 25 €/t →	187 €
Slurry production	6.4 t × 47€/t →	302 €
Personnel (25/plant, 60 k€/ yr each person)	→	96 €
Central gasification and synthesis, 1 Mt Fischer-Tropsch product per year		
Slurry transport, 200 km by train	6.4 t × 12.2 €/t →	78 €
O_2- (1200 Nm^3 with a price of 6.25 c€/Nm^3, without current)	→	75 €
Production of fuel	→	125 €
Personnel (300/plant, 60 k €/yr each person)	→	18 €
Sum of all costs		Σ 1219 €
Subtraction of credit for electricity (70 €)		1149 €

considered by 2% of fresh capital expenditures and 10% higher personnel expenditure. The production costs for 1 t fuel with the BTL2 process are 1149 €. This is 97 c€/L or 26.9 €/GJ, which is much too high to be competitive with production costs of fossil fuel (10 €/GJ). On the other hand, this cost estimation is calculated for production in Germany, and many different locations can be chosen. Hoogzaad, for example, investigated thoroughly the Ukrainian market situation and production costs *(23)*. Comparing the cost estimation above with his results, it is appropriate to estimate a massive cost reduction for feed, transport, and personnel (90%) and for technical facilities (30%). Then, it would be possible to produce 1 t of diesel at 377 €, which is 32 c€/L or 8.8 €/GJ. This is a competitive price even in a country where biomass-based fuel is not promoted by tax reduction. BTL2 is an interesting option for the large-scale production of bio-synfuels. The last example shows that even now it can be an economic process in countries at the stage of economic take-off, if long-term contracts guarantee the large investments to be safe.

Acknowledgment

We thank the Ministerium Für Ernährung und Ländlichen Raum (MELR), Baden Wuerttemberg, Germany, for financial support.

References

1. Henrich, E., Raffelt, K., Stahl, R., and Weirich, F. (2002), *Pyrolysis 2002*, Conference in Leoben, Austria, 17–20 September 2002.

2. Leible, L., Kaelber, S., Kappler, G., et al. (2004), Second World Conference on Biomass for Energy, Industry and Climate Protection in Rome, 2113–2116.
3. Jensen, J. P., Fenger, L. D., and Huebbe, C. (2002), 12th European Conference on Biomass for Energy, Industry, and Climate Protection, 17–21 June 2002 in Amsterdam, the Netherlands, 45–48.
4. Risnes, H., Fjellerup, J., Henriksen, U., et al. (2003), *Fuel* **82/6**, 641–651.
5. Asikainen, A. H., Kuusisto, M. P., Hiltunen, M. A., and Ruuskanen, J. (2002), *Environ. Sci. Technol.* **36**, 2193–2197.
6. Schmitz, N. ed. (2003), *Bioethanol in Zahlen, Schriftenreihe "Nachwachsende Rohstoffe Bd.* Vol. 21, Landwirtschaftsverlag GmbH, Muenster, Germany, p. 44.
7. Goessler, R. ed. (2003), *Agrarmaerkte in Zahlen Europaeische Union 2003*, ZMP Zentrale Markt- und Preisberichtsstelle GmbH, Bonn, Germany, pp. 77–88.
8. Leible, L., Arlt, A., Fuerniss, B., et al. (2003), Energie aus biogenen Rest- und Abfallstoffen, Wissenschaftliche Berichte, FZKA 6882, Forschungszentrum Karlsruhe, p. 92.
9. Homepage of Aral Aktiengesellschaft, Bochum, Germany, http://www.aral.de. Date accessed: May 2005.
10. Weiss, H., Pagel, J. F., and Jacobson, M. (2000), 16th World Petroleum Congress in Calgary, Canada.
11. Weiss, H. and Schmalfeld, J. (1989), *Sci. Technol. (Wissenschaft & Technik)*, **42(6)**, 235–237.
12. Bridgwater, A. V. et al. (1999), *Fast Pyrolysis of Biomass, A Handbook,* CPL-Press, UK.
13. MAT-Mischanlagentechnik, Illerstraβe 6, 87509 Immenstadt, Germany, www.mat-oa.de. Date accessed: May 2005.
14. Future Energy GmbH, Halsbrücker Straβe 34, 09599 Freiberg, Germany, www.future-energy.de. Date accessed: May 2005.
15. Carl, J. and Fritz, P. (1994), *Noell-Konversionsverfahren zur Verwertung und Entsorgung von Abfällen*, EF-Verlag fuer Energie und Umwelttechnik GmbH, Berlin, Germany.
16. Dry, M. E. (2001), *J Chem. Technol. Biotechnol.* **77**, 43–50.
17. Chang, T. (2000), *Oil Gas J.* **10**, 42–45.
18. Wender, I. (1996), *Fuel Proc. Technol.* **48**, 189–297.
19. Chemviron Carbon GmbH, Uslarer Str. 30, 37194 Bodenfelde, Germany, www.holzkohle.de. Date accessed: May 2005.
20. Tijmensen, M. J. A., Faaij, A. P. C., Hamelinck, C. N., and van Hardeveld, M. R. M. (2002), *Biomass and Bioenergy*, **23**, 129–152.
21. Peacocke, G. V. C. and Bridgwater, A. V. (to be published in 2005), *Science in Thermal and Chemical Biomass Conversion (STCBC)*, 30.8.2004–2.9.2004 in Victoria, BC, Canada.
22. Hamelinck, C. N., Faaij, A. P. C., den Uil, H., and Boerrigter, H. (2004), *Energy* **29**, 1743–1771.
23. Hoogzaad, J. (2004), *PhD. Thesis*, University Utrecht, the Netherlands.

Emission Profile of Rapeseed Methyl Ester and Its Blend in a Diesel Engine

Gwi-Taek Jeong,[1,2] Young-Taig Oh,[7] and Don-Hee Park*,[2-6]

[1]Engineering Research Institute; [2]School of Biological Sciences and Technology; [3]Faculty of Applied Chemical Engineering; [4]Research Institute for Catalysis; [5]Biotechnology Research Institute; [6]Institute of Bioindustrial Technology, Chonnam National University, Gwangju 500-757, Korea, E-mail: dhpark@chonnam.ac.kr; and [7]Department of Mechanical Engineering, Chonbuk National University, Jeonbuk 561-756, Korea

Abstract

Fatty acid methyl esters, also known as biodiesel, have been shown to have a great deal of potential as petro-diesel substitutes. Biodiesel comprise a renewable alternative energy source, the development of which would clearly reduce global dependence on petroleum, and would also help to reduce air pollution. This paper analyzes the fuel properties of rapeseed biodiesel and its blend with petro-diesel, as well as the emission profiles of a diesel engine on these fuels. Fuels performance studies were conducted in order to acquire comparative data regarding specific fuel consumption and exhaust emissions, including levels of carbon monoxide (CO), carbon dioxide (CO_2), smoke density, and NO_X, in an effort to assess the performance of these biodiesel and blend. The fuel consumption amount of oil operations at high loads was similar or greater than that observed during petro-diesel operation. The use of biodiesel is associated with lower smoke density than would be seen with petro-diesel. However, biodiesel and its blend increased the emission of CO, CO_2, and nitrogen oxides, to a greater degree than was seen with petro-diesel. The above results indicate that rapeseed biodiesel can be partially substituted for petro-diesel under most operating conditions, regarding both performance parameters and exhaust, without any modifications having to be made to the engine.

Index Entries: Biodiesel; engine performance; rapeseed oil; exhaust emission.

Introduction

Biodiesel (fatty acid methyl esters [FAMEs]) is an alternative and renewable energy source, the development of which is hoped to reduce

---

*Author to whom all correspondence and reprint requests should be addressed.

global dependence on petroleum, as well as air pollution. Biodiesel derived from diverse vegetable oils and animal fats have been found to exhibit low viscosities, similar to those associated with petro-diesel. Additionally, many of the salient characteristics of biodiesels, most notably volumetric heating value, cetan number, and flash point, are also comparable with those of petro-diesel (1,2). FAMEs, which are derived from vegetable oils and animal fats via alcohol transesterification, are also thought to have great potential as diesel substitutes, and these blends are called biodiesel (3,4). Several processes for the production of biodiesel via acid-, alkali-, and enzyme-catalyzed transesterification reactions have been previously developed (5,6). Transesterification, also called alcoholysis, refers to the displacement of an alcohol from an ester by another alcohol, via a process which is generally similar to hydrolysis. Transesterification occurs through a number of consecutive and reversible reactions. The reaction steps involve the conversion of triglycerides to diglycerides, followed by the conversion of diglycerides to monoglycerides, and then of monoglycerides to glycerol (7).

In general, the applied ratio of FAMEs in mixed biodiesel with petro-diesel ranges between 5 and 30 wt%. Biodiesel has been shown to be compatible with petro-diesel in compression–ignition engines, and mixtures of biodiesel and petro-diesel have been successfully used in engines without any modification. The use of biodiesel would have several advantages over that of petro-diesel, as biodiesel represents both an alternative renewable energy source and a biodegradable nontoxic fuel (8). Biodiesel also carries the advantages of a superior engine combustion emission profile, including low-emission rates. The levels of CO, particulate material, and unburned hydrocarbons (HC), all of which have been proven to exert carcinogenic and mutagenic effects, has been shown to be significantly lower in biodiesel exhaust than in petro-diesel exhaust. Biodiesels can also attenuate the rate of CO_2 recycling in the short-term (6).

Several researchers have observed that the use of biodiesel is associated with a reduction in exhaust emissions. The use of biodiesel in diesel engine has been generally shown to increase nitrogen oxide (NO_x) emissions, and also has been shown to effect a reduction in the HC, CO, and particulate emissions, as compared with petro-diesel. The magnitude of these differences in emission profiles appear to be somewhat dependent on the engine used in the testing (9). The most relevant compositional difference between biodiesel and regular petro-diesel is the oxygen content. Biodiesel contains 10–12% oxygen by weight. However, this high-oxygen content also translates to a lower energy content, which results in reductions in both engine torque and power. However, because biodiesel contains only small amounts of sulphur compounds as in comparison with petro-diesel, biodiesel has also been associated with significant reduced SO_2 emissions (10,11).

In this article, we have attempted to characterize the fuel properties, engine performance, and exhaust emission profiles of rapeseed biodiesel and its blend, using a single cylinder, four-stroke, direct injection (DI), water-cooled diesel engine.

Table 1
Fatty Acid Composition and Characteristics of Rapeseed Oil

Characteristics	Content
Specific gravity	0.917
Moisture content	0.01%
Free fatty acid	0.018%
Unsaponifiable matter	0.39%
Fatty acid % (w/w)	
Palmitic acid ($C_{16:0}$)	5.7%
Stearic acid ($C_{18:0}$)	2.2%
Oleic acid ($C_{18:1}$)	58.5%
Linoleic acid ($C_{18:2}$)	24.5%
Linolenic acid ($C_{18:3}$)	9.1 %

Materials and Methods

Materials

Refined and bleached rapeseed oil was acquired from Onbio Co. Ltd. (Pucheon, Korea). Table 1 shows the fatty acid composition and general characteristics of this rapeseed oil sample. Reference standards of FAMEs, including palmitic, stearic, linolenic, linoleic, and oleic methyl ester (all of >99% purity) were acquired from Sigma-Aldrich Co. Ltd. (St. Louis, MO). The methanol, potassium hydroxide, and other reagents used were of analytical grade.

Transesterification of Rapeseed Oil

The rapeseed oil was transesterified via an alkali process in our laboratory using our developed esterification reaction system as describe below: the heating of the oil, the addition of potassium hydroxide and methyl alcohol, mixture-agitation, glycerol-separation, washing with distilled water, and an additional heating in order to remove remaining water. We used a 30 L reactor system, which was equipped with a mechanical stirrer, a temperature control, electric steam generator, and a condenser. Eighteen liters of rapeseed oil and methyl alcohol were added to the reactor and heated to 50°C with agitation. After a temperature of 50°C was achieved, the prepared potassium hydroxide was dissolved in methanol then rapidly added with stirring, and the reaction continued for an additional 30 min at the same temperature. The reaction was then discontinued, and settled for later separation. Two layers were clearly observed after cooling. The top layer was identified as biodiesel, and the bottom dark denser layer was made up of glycerin. The top layer was then neutralized via the addition of diluted phosphoric acid, distilled with nonreacted methyl alcohol, and finally washed with distilled water.

In our previous report (6), we optimized the production of rapeseed biodiesel via alkali-catalyzed transesterification, using both anhydrous

Table 2
Specification of Test Engine

Item	Specification
Engine model	ND130DIE
Bore × stroke	95 × 95 (mm)
Number of cylinder	1
Displacement	673 (cm^3)
Compression ratio	18
Combustion chamber type	Toroidal
Injection timing	BTDC 23°CA
Injection type	Direct injection
Rated power	13 PS/2400 rpm

methanol and potassium hydroxide. The optimized conditions for alkali-catalyzed transesterification using KOH were found to be as follows: oil to methanol molar ratio, 1:8 to 1:10; catalyst, KOH 1.0% (w/w) by oil weight; reaction temperature, 60°C; and reaction time, 30 min. Under the given conditions, the conversion yield was determined to be approx 98%. From the refined product (rapeseed FAME, and biodiesel), the purity of the product was greater than 99% posttreatment, including washing and centrifugation steps.

Fuel Properties

The fuel properties of biodiesel were assessed in accordance with the test code developed by the Korean Petroleum Quality Inspection Institute (Gwangju, Korea). We assessed the properties of the pure biodiesel (BDF 100), as well as a blend of petro-diesel with 20% biodiesel by volume (BDF 20).

Emissions and Engine Performance

Both the rapeseed biodiesel (BDF 100) and its blend (BDF 20) were tested in a single cylinder, four-stroke, DI, water-cooled diesel engine, with a rated output of 13 PS at 2400 rpm and a compression ratio of 18:1. The details are provided in Table 2. A schematic diagram of the experimental apparatus used to measure the performance and emissions from the engine are shown in Fig. 1. Engine performance and emissions were determined at different engine loads (0, 25, 50, 75, 90, and 100% of the load corresponding to the load at maximum power) at engine speeds ranging from 1000 to 2000 rpm (Table 3). After the engine reached a stabilized condition, emissions including smoke, CO, CO_2, and NO_x, were measured with a smoke meter (Hesbon, HBN-1500, Korea), and an online exhaust gas analyzer (Motor branch, Mod. 588), and were then recorded

Emission Profile of Rapeseed Methyl Ester

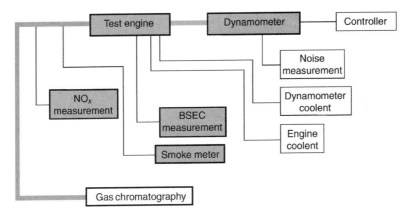

Fig. 1. Schematic diagram of experimental apparatus.

Table 3
Engine Test Condition of Petro-Diesel, Biodiesel, and Its Blend

Fuel	Engine speed (rpm)	Engine load (%)
Petro-diesel	1000	0
BDF 20	1300	25
BDF 100	1500	50
	1700	75
	2000	90
		100

at different engine loads and speeds. Each of these readings was replicated twice in order to acquire reasonable values. Parameters including engine speed, torque, and fuel consumption were also assessed, and from these values, brake power and brake-specific fuel consumption (BSFC) were computed.

Results and Discussion

Fuel Preparation and Its Characteristics

In order to conduct the alkali-catalyzed transesterification process with the rapeseed oil, we applied several different reaction systems. In the alkali-catalyzed transesterification, the amounts of free fatty acid were proposed to be below 0.5% on the basis of oil weight, to ensure a high conversion yield *(12)*.

Figure 2 depicts the time-course of rapeseed oil transesterification at a 1% catalyst concentration (w/w), a molar ratio of 1:6, and a temperature of 60°C. Within the first 5 min, the reaction proceeded quite rapidly. Rapeseed oil was converted to greater than 86% within the first 5 min, and

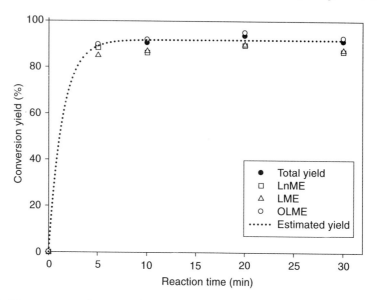

Fig. 2. Time-course of rapeseed oil transesterification on 60°C, 1:6, 1% potassium hydroxide and 600 rpm in 30 L reactor.

conversion reached an equilibrium state after approx 10 min. At a reaction time of 30 min, linolenic and linoleic acid methyl ester were generated at low-conversion rates, whereas oleic methyl ester was produced fairly rapidly. After transesterification under optimal operation conditions, the rapeseed methyl ester changed in color from yellow to brownish-yellow, and, on average, an identical amount of biodiesel was acquired from 18 L of rapeseed oil. The properties of biodiesel and its blend are shown in Table 4. The kinematic viscosity of rapeseed oil was determined to be higher than that of petro-diesel at a temperature of 40°C. After reaction, the kinematic viscosity dropped, becoming lower than that of the prepared rapeseed oil. It was further reduced with an increase in the amount of petro-diesel in the blend. We also observed a similar reduction in specific gravity. The flash points of both the biodiesel and the blend were higher than 100°C, which has been determined to be safe for both storage and handling. Ash, sulfur, cloud point, acid value, water content, and cetan number were adjusted to quality standard.

Engine Performance

The term "brake-specific" refers to quantities which have been normalized by dividing by the engine's power. Therefore, the BSFC is equal to the fuel-flow rate, divided by the power of the engine (13). It has been reported that the increase in the BSFC can be attributed principally to the lower energy content of the blended fuel (14).

The BSFC values of each fuel, according to changes in the speed of the engine, are shown in Fig. 3. At the tested engine speeds, the increases in

Table 4
Fuel Properties of Rapeseed Biodiesel and Its Blend

Property	Standard	BDF 100	Standard	BDF 20	Test method
Flash point (PM, °C)	min. 100	182–186	min. 40	59	KS M2010-99
90% Distillation temp. (°C)	max. 360	–	max. 360	347	ASTM D86
10% Ramsbottom carbon residue (wt%)	max. 0.5	–	max. 0.15	0.05	ASTM D4530
Ash (wt%)	max. 0.01	max. 0.01	max. 0.02	max. 0.01	KS M2044-00
Sulfur (wt%)	max. 0.02	0.01–0.004	max. 0.043	0.019	KS M2027-98
Kinematic viscosity (40°C, cSt)	min. 1.9	4.519–4.570	min. 1.9	3.316	KS M2014-99
	max. 6.0		max. 5.5		
Copper strip corrosion (100°C, 3 h)	max. 1	1	max. 1	1	KS M2018-97
Cetan number	min. 49	–	min. 45	51	KS M2610-01
Water and sediment (vol%)	max. 0.05	max. 0.01	–	–	KS M2115-96
Density (15°C, kg/m^3)	–	–	min. 815	852.4	KS M2002-01
			max. 855		
Cloud point (°C)	–	–	min. 0.0	-17.5	KS M2016-96
Acid value (mg KOH/g)	max. 0.80	0.20	–	–	KS M2004-00

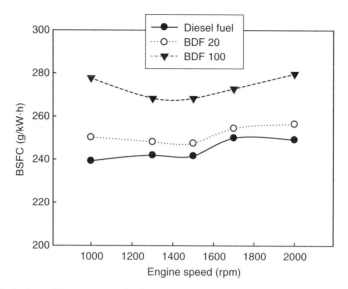

Fig. 3. Variation of brake-specific fuel consumption with engine speed for different fuels tested.

the amount of BDF in the fuel blend evidenced a consistently increasing trend in the BSFC. Variations in the BSFC values according to differences in the biodiesel content of the fuels were found to exhibit a trend similar to that associated with engine speed. In the case of petro-diesel, increases in the speed of the engine were associated with increases in the rate of fuel consumption. BDF 20, which contained 80% petro-diesel, also evidenced a trend similar to that of petro-diesel. In the case of BDF 100, we noted low-BSFC values at 1300–1500 rpm, in comparison with those of the other tested engine speeds. At an average speed of 1300 rpm, the BSFC values for BDF 20 and BDF 100 were determined to be 2.6–10.9% higher than those of petro-diesel. This trend was observed owing to the fact that ester has a lower heating value (lower calorific value) than does petro-diesel, and thus more FAME-based fuel is required for the maintenance of a constant power out put *(13)*.

Emission Studies

Figure 4 shows variations in the density of smoke generated during the tests of the different fuels (petro-diesel, BDF 100, and BDF 20) and engine speeds (1000–2000 rpm). Under engine load <75% conditions, the emitted smoke percentage was low, without relation to the tested fuels or engine speed parameters. At an engine load of >75%, increases in BDF content resulted in markedly lower smoke emission. Under high speed and engine load conditions, smoke emission from the BDF 100 and BDF 20 fuels were lower than those observed with the regular petro-diesel. Smoke was determined to decrease consistently for all of the tested engine conditions,

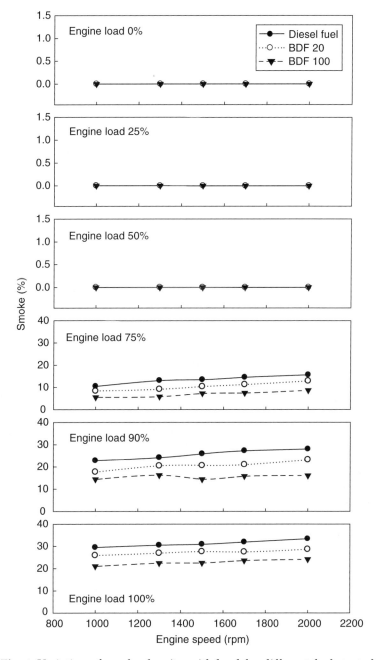

Fig. 4. Variation of smoke density with load for different fuels tested.

along with increases in the amount of biodiesel in the fuel blend. In particular, BDF 100 was associated with a 26.05–28.73% reduction in smoke production in comparison with that of the petro-diesel. Because of the heterogeneous nature of petro-diesel combustion, fuel/air ratios, which affect smoke formation, tend to vary within the cylinder of a diesel engine.

Smoke formation occurs primarily in the fuel-rich zone of the cylinder, at high temperatures and pressures. If the applied fuel is partially oxygenated, locally over-rich regions can be reduced and primary smoke formation can be limited (13).

Nitric oxide (NO) and nitrogen dioxide (NO_2) are usually combined together as NO_x (9). Variations in NO_x emissions according to engine load and speed with the different tested fuels (petro-diesel, BDF 100, and BDF 20) are shown in Fig. 5. Under nonengine load conditions, NO_x emissions were measured to be approx 200 ppm with the BDF 20 and BDF 100, but the petro-diesel did not emit NO_x at any of the tested engine speed ranges. NO_x emissions trended higher directly with increases in the engine load for all of the tested fuels, and were also observed to increase directly with biodiesel content. However, the tested engine speeds did not affect NO_x emission significantly. Regarding biodiesel blends, we noted no significant differences across the range of engine load tested in our study. The NO_x emissions associated with the blends were found to be slightly higher than those of the regular petro-diesel, under both full and partial loads.

The reasons behind these observed increases in NO_x emission remain poorly understood. NO_x emissions have been reported by several researchers to be increased along with the use of biodiesel. Three principal factors affect the emission of NO_x as the result of combustion: oxygen concentration, combustion temperature, and time. The observed NO_x emission increases appeared to have been induced as the result of increases in the temperature of the combustion chamber, which were apparently the result of the 10% oxygen content of the biodiesel (13,14). However, some other studies have reported that NO_x emissions also exhibit a decreasing trend. When operated with pure coconut oil, NO_x emissions were reduced by approx 40% (14).

The variances in CO emissions that resulted from the changes in injection timing, fuel type, and load were found to be significant (9). The emission profiles of CO when running the diesel engine on blends—from BDF 20 to BDF 100—were compared with those associated with petro-diesel, as shown in Fig. 6. CO emissions were determined to be relatively higher in the case of BDF 20 and BDF 100 than with petro-diesel. Under nonload conditions, BDF 100 was determined to emit a greater quantity of CO than did petro-diesel, across the engine speed continuum. However, under full-load conditions, CO emissions increased directly with biodiesel contents at low engine speeds. Also, increases in the engine speed occurred concomitantly with diminutions in the degree to which CO was emitted. Increases in the engine load resulted in greater CO emission. It is generally accepted that the presence of oxygen in the fuel, which facilitates the combustion processes, also results in a reduction of exhaust CO emission, as compared with petro-diesel (13).

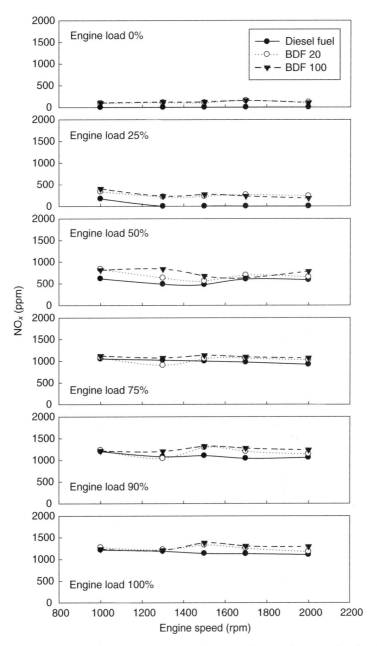

Fig. 5. Nitrogen oxides emission vs engine speed at various engine loads.

Figure 7 profiles CO_2 emissions detected as the result of the operation of a diesel engine running on BDF 20 to BDF 100 and petro-diesel. CO_2 emissions were determined to be relatively high in the case of BDF 20 and BDF 100. Under nonload conditions, BDF 100 emitted a greater quantity of CO_2 than did petro-diesel at all engine speeds, in contrast to the low smoke

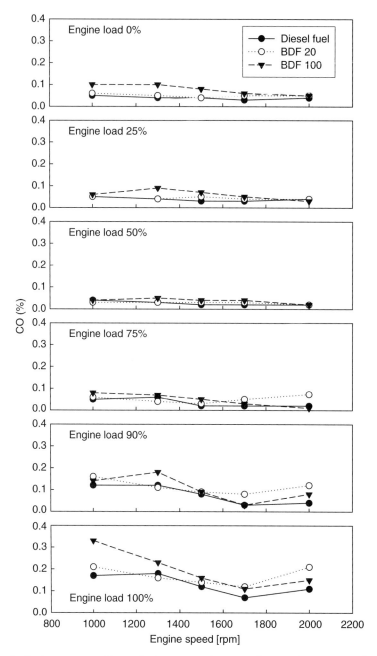

Fig. 6. Carbon monoxide emission vs engine speed at various engine loads.

density detected under such conditions. Under full-load conditions, CO_2 emissions did not vary with biodiesel content at low engine speeds. Also, increases in the engine speed were associated with increases in the emission of CO_2. In general, increases in the engine load were associated with a greater degree of CO_2 emission, whereas engine speed did not affect emission rates.

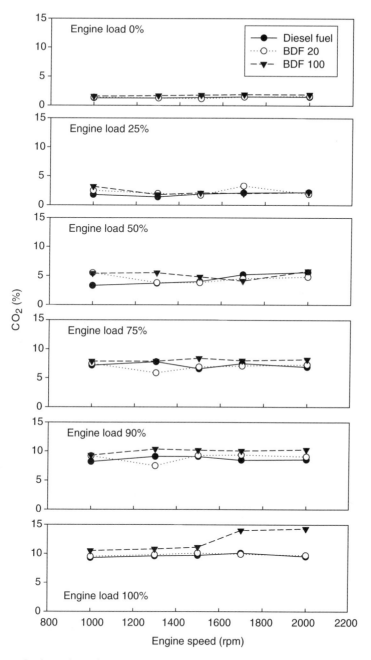

Fig. 7. Carbon dioxide emission vs engine speed at various engine loads.

Conclusions

Biodiesel is a fatty acid alkyl ester, and can be derived from any vegetable oil via transesterification. Biodiesel is a renewable, biodegradable, and nontoxic fuel. Here, we transesterified rapeseed oil with methanol in the presence of alkali, and the biodiesel which was obtained was then investigated regarding its fuel properties and exhaust emission profile.

All of the results which were obtained for the rapeseed biodiesel have been compared with those of petro-diesel. The following conclusions may be drawn from this study:

1. The physiochemical properties and engine performance of rapeseed biodiesel are comparable to those of petro-diesel.
2. The fuel consumption of biodiesel and its blends at high loads was similar or greater than that observed during petro-diesel operation.
3. The use of biodiesel is associated with lower smoke density than would be seen with petro-diesel. However, biodiesel and its blend slightly increased the emission of CO, CO_2, and nitrogen oxides, to a greater degree than was seen with petro-diesel.

The findings in this article, when taken together, indicate that rapeseed biodiesel can be partially substituted for petro-diesel under most operating conditions, regarding both performance parameters and exhaust, without any modifications having to be made to the engine. However, it is suggested that subsequent studies be conducted to elucidate the improvement of diesel engine for the useful application of biodiesel and its blends.

Acknowledgments

The authors would like to thank the Korea Energy Management Corporation for their financial support for this study.

References

1. Lang, X., Dalai, A. K., Bakhshi, N. N., Reaney, M. J., and Hertz, P. B. (2001), *Bioresource Technol.* **80**, 53–62.
2. Cvengros, J. and Povazanec, F. (1996), *Bioresource Technol.* **55**, 145–152.
3. Kim, H. R. (2002), *Prospectives Ind Chem.* **5(1)**, 27–34.
4. Fukuda, H., Kondo, A., and Noda, H. (2001), *J. Biosci. Bioeng.* **92(5)**, 405–416.
5. Freedman, B., Pryde, E. H., and Mounts, T. L. (1984), *J. Am. Oil Chem. Soc.* **61(10)**, 1638–1643.
6. Jeong, G. T., Park, D. H., Kwang, C. H., et al. (2004), *Appl. Biochem. Biotechnol.* **114**, 747–758.
7. Darnoko, D. and Cheryan, M. (2000), *JAOCS* **77(12)**, 1263–1267.
8. Graboski, M. S. and McCormick, R. L. (1998), *Prog. Energ. Combust. Sci.* **24**, 125–164.
9. Monyem, A. and Gerpen, J. H. V. (2001), *Biomass Bioenerg.* **20**, 317–325.
10. Lin, C. Y. and Huang, J. C. (2003), *Ocean Eng.* **30**, 1699–1715.
11. Usta, N. (2005), *Biomass Bioenerg.* **28(1)**, 77–86.
12. Krawczyk, T. (1996), *INFORM* **7**, 801–829.
13. Puhan, S., Vedaraman, N., Ram, B. V. B., Sankaranarayanan, K., and Jeychandran, K. (2005), *Biomass Bioenerg.* **28(1)**, 87–93.
14. Machacon, H. T. C., Shiga, S., Karasawa, T., and Nakamura, H. (2001), *Biomass Bioenerg.* **20(1)**, 63–69.

SESSION 1B
Enzyme Catalysis and Engineering

Introduction to Session 1B

JOEL R. CHERRY[1] AND KEVIN GRAY[2]

[1]Novozymes, Inc., Davis, CA; and [2]Diversa Corp., San Diego, CA

Enzymes are clearly recognized as a keystone technology for the production of fuels and chemicals from renewable feedstocks. Their specificity, performance under mild reaction conditions, and biodegradability make them ideally suited to widespread use in biorefineries around the world, and as the world puts greater and greater value on sustainable processes and environmentally friendly production methods, the further the development of enzyme technology grows in importance. This session focuses on the discovery, production, modification, and use of enzymes by bringing together 6 oral and 64 poster presentations describing the state of the art in enzyme technology.

The plenary session was designed to build from recent technology improvements to the existing biorefineries (wet or dry mills processing corn starch to a variety of products) to improvements that may assist in the development and commercialization of lignocellulose-based biorefineries. Beginning with a presentation on the effect of cellulase addition to a dry mill starch process, progressing to recent improvements in enzymes for the hydrolysis of non-gelatinized starch and lignocellulose, and ending with a presentation on the effect of poly(ethylene glycol) in lignocellulose hydrolysis, existing and future concepts for the use of enzymes in the biorefinery were addressed.

The articles presented in this volume cover a broad range of additional topics related to enzyme production and use. Improved fermentation methods for enzyme production, alternative recovery of active enzymes after fermentation, and formulation to increase enzyme stability and effectiveness are covered in many of the articles. In addition, a number of the articles present enzymatic methods for the production of novel chemicals or materials, which may, one day, be products of a future biorefinery.

Properties and Performance of Glucoamylases for Fuel Ethanol Production

Bradley A. Saville,* Chunbei Huang, Vince Yacyshyn, and Andrew Desbarats

Department of Chemical Engineering and Applied Chemistry, University of Toronto, 200 College Street, Toronto, Ontario, Canada M5S 3E5; E-mail: saville@chem-eng.utoronto.ca

Abstract

Studies were conducted on maltodextrin saccharification and on simultaneous saccharification and fermentation (SSF) with various commercial glucoamylases. In kinetics studies, none of the glucoamylases were able to completely convert maltodextrin into glucose. Typically, about 85% conversion was obtained, and glucose yields were about 75%. Typically, the kinetics were biphasic, with 1 h of rapid conversion, then a significant reduction in rate. Data were consistent with strong product inhibition and/or enzyme inactivation. Some glucoamylases followed first-order kinetics, initially slower at dextrin conversion, but eventually achieving comparable conversion and glucose concentrations. Most of the glucoamylases were more active at 55°C than at 35°C, but pH had little effect on activity. Screening studies in an SSF system demonstrated little difference between the glucoamylases, with a few exceptions. Subsequent targeted studies showed clear differences in performance, depending on the fermentation temperature and yeast used, suggesting that these are key parameters that would guide the selection of a glucoamylase.

Index Entries: Glucoamylase; kinetics; fermentation; saccharification.

Introduction

Glucoamylase is an exoglycosidase responsible for hydrolyzing the terminal α-1,4 glucosidic bonds of dextrins and related oligo- and polysaccharides. The reaction involves a proton transfer by acid catalysis, followed by formation of a transition state analogous to an oxocarbonium ion, and finally, a base-catalyzed nucleophilic attack of water *(1)*. Glutamic acid present in different regions of the enzyme-active site is thought to act as the acid and base catalysts required for the reaction. A typical three step reaction scheme is as follows:

$$E + S \xrightleftharpoons{K_1} ES_1 \xrightleftharpoons[k_{-2}]{k_2} ES_2 \xrightarrow{k_{cat}} E + P \quad (1)$$

*Author to whom all correspondence and reprint requests should be addressed.

where E represents the enzyme, S represents the substrate, P represents the product, ES_1 and ES_2 are enzyme–substrate complexes, k_{cat} is the reaction rate constant for the species in question, and K_1 is its equilibrium binding constant.

Two of the three reaction steps are pH-dependent, involving the protonation of the enzyme before catalysis. The kinetics schemes used to describe maltodextrin saccharification have ranged from simple Michaelis-Menten expressions *(2)* to complex schemes that account for enzyme inhibition *(3)* and glucose condensation reactions into maltose and isomaltose. Additional complexity arises owing to the fact that each reaction product from maltodextrin hydrolysis acts as a substrate for subsequent conversion. For example, assuming a maltodextrin with a degree of polymerization (DP) equal to six, the following hydrolysis reactions may occur:

$$DP6 + H_2O \rightarrow DP5 + DP1$$

$$DP5 + H_2O \rightarrow DP4 + DP1$$

$$DP4 + H_2O \rightarrow DP3 + DP1$$

$$DP3 + H_2O \rightarrow DP2 + DP1$$

$$DP2 + H_2O \rightarrow 2\,DP1$$

where, DP1 = glucose; DP2 = maltose; DP3 = maltotriose; DP4 = maltotetraose; DP5 = maltopentaose; and DP6 = maltohexaose. Each species may possess different binding (K_1) and reaction (k_{cat}) rate constants, in accordance with scheme *(1)*.

Additional products may form owing to condensation reactions between various components in the system. In a typical mash following liquefaction, maltodextrins of size DP15–DP20 may be present.

Peeva and Yankov *(4)* proposed a mathematical model to describe the key steps in the conversion of maltodextrins into ethanol. Reactions to form dextrins DP5 and larger were simulated by a lumped first order model; subsequent reactions to form DP4, DP3, DP2, and DP1 were described using a standard Michaelis-Menten model, with different reaction velocities (V_{max}) and Michaelis constants (K_m) for each species. Finally, the conversion of glucose into cells and into ethanol was described using Monod kinetics, wherein the expression for the specific growth rate of yeast cells was modified to account for inhibition by glucose.

Maltodextrin hydrolysis can be conducted in batch saccharification tanks at about 55°C for about 3–4 h, followed by fermentation, or a simultaneous saccharification and fermentation (SSF) step may be adopted, in which the mash, glucoamylase, yeast and nutrients are all added to the fermenters and incubated at about 33°C for up to 70 h. SSF processing is most common, since it reduces the risk of infection, and also reduces the impact of product inhibition on glucoamylase activity, because glucose/maltose are consumed by yeast, and their concentrations do not build up appreciably. In this latter case, the glucose and maltose (DP1 and DP2) produced via scheme one, above, are converted into ethanol, carbon dioxide and other products by the yeast.

The objective of this study was to compare the activities of various glucoamylases under standard saccharifying conditions, and identify characteristics of importance for ethanol production. Subsequent "mini-fermenter" trials were conducted to assess the performance of these enzymes for ethanol production under SSF conditions.

Materials and Methods

Glucoamylase Kinetics

Dextrin (5 or 10 g), citrate buffer 25 mL and pH 4.8 (0.025 M) and glucoamylase were combined in an 80-mL jacketed glass reactor. The temperature was controlled at 55°C by water flowing through the jacket, and a magnetic stirrer was used to mix the contents of the vessel. The duration of each experiment ranged from 4 to 48 h, depending on whether initial rate data or a full set of kinetics data were desired.

One mililiter samples were collected during the trial at designated time intervals. The sample was immediately inactivated with 0.10 mL HCl solution (pH = 1.05, 25°C). From the inactivated sample, 0.2 mL was added to 1.8 mL of filtered deionized water and 1 mL of 0.075 M mannose (internal standard). This preparation was subsequently centrifuged to eliminate solid impurities. The supernatant was collected and transferred to vials for subsequent high-performance liquid chromatography (HPLC) assay. The HPLC system was equipped with a Bio-Rad HPX 87C carbohydrate analysis column and a Shodex RI-71 refractive index detector. Deionized, filtered (0.2 nm) water was used as the mobile phase, with a flow rate of 0.6 mL/min and temperature of 85°C. The run time for each sample was 14.1 min. For fermentation trials, an HPX 87H column was used, with a run time of 22.5 min. Samples were injected using an autoinjector. Glucoamylases studied in these experiments were obtained from a variety of commercial suppliers. For the purpose of comparison, glucoamylases are labeled Glu-1 through Glu-11.

Effects of pH

The baseline kinetics assay was conducted, except that the buffer pH was set to 3.6, 4.0, 4.4, or 5.2, to complement the data originally collected at pH 4.8.

Effects of Temperature

The baseline kinetics assay was conducted, except that the reaction temperature was set at 35°C, to more closely replicate the conditions in an SSF system.

Mini-Fermentation Trials

These trials were based on the procedure proposed by Allain et al. (5), wherein small-scale fermentations are tracked based on loss of mass from the system. The premise is that, as glucose is fermented to ethanol, carbon dioxide is also produced and emitted from the system, according to the reaction:

$$C_6H_{12}O_6 \rightarrow 2\ CO_2 + 2C_2H_5OH$$

On a mass basis, each 100 g of glucose should produce 48.9 g of carbon dioxide and 51.1 g of ethanol, assuming no side reactions. Thus, a mini-fermenter should experience a decrease in mass equal to about one-half of the original substrate mass added to the vessel. Note that this does not account for the mass gain owing to dextrin hydrolysis. For example, 100 g of 100% maltohexaose (DP6) would produce about 109 g of glucose.

This is a relatively crude method, in that it does not account for mass losses owing to evaporation of volatile components in the vessels, such as ethanol or acetaldehyde. Nonetheless, data collected from these "mass-loss" experiments correlated well with HPLC measurements from parallel experiments wherein samples were collected. Thus, these mass-loss experiments provide a reasonable representation of the SSF process.

In these trials, different glucoamylases were compared under equivalent conditions to assess their efficacy for fermentation. Each trial used a control which received all necessary reaction components except yeast and glucoamylase. In some trials, maltodextrin was used as the substrate, using citrate buffer as the reaction medium. In other trials, corn mash was prepared by 3 h of slurry/liquefaction using α-amylase, and then transferred to the mini-fermenters for a subsequent SSF trial. Before adding yeast or glucoamylase, the mash was cooled to about 35°C, and the pH was adjusted to about 4.5 by addition of HCl.

Initial screening tests were performed to establish the dose of yeast and glucoamylase. Initial experiments were conducted with baker's yeast; subsequent trials used brewer's yeast (Red Star, LeSaffre, Headland, AL).

Reactions were conducted in a 125 mL (unjacketed) vessel. Thus, the temperature followed ambient conditions, which ranged between 23 and 27.5°C. For each 10 mL of liquefied corn mash added to the vessel, five drops of 5% (w/v) HCl and 0.09 g of yeast were added. The glucoamylase dose was varied from 4 to 50 µL. The mass of the empty vessel was recorded, and the mass was also recorded after each item was added. The final mass after adding parafilm and an elastic seal was also recorded. A small (approx 2 mm) hole was punched in the parafilm to allow CO_2 to escape. Typically, several vessels were run simultaneously in a single trial, along with a control vessel (to account for evaporative losses). The mass of each vessel was recorded at several points in time, over 48–120 h. The mass loss was corrected to account for the mass loss in the control vessel (which did not contain enzyme or yeast), and plotted to compare the glucoamylases used in each trial.

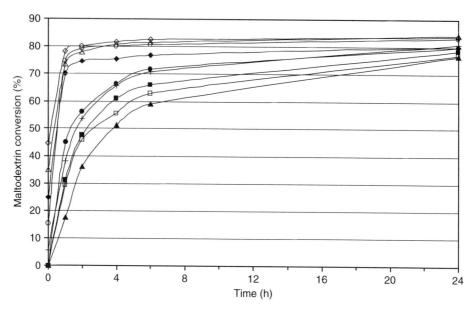

Fig. 1. Dextrin conversion with various glucoamylases at 55°C, 10 g maltodextrin, and 125 µL of glucoamylase at pH 4.8. ♦, Glu-1; ■, Glu-2; ○, Glu-3; ●, Glu-4; □, Glu-5; ▲, Glu-6; +, Glu-7; Δ, Glu-8; ◊, Glu-9.

Owing to slight differences in substrate, mixing conditions and temperature histories encountered in the different trials, comparisons among trials are not justified. However, within each trial, comparisons can be made because all parameters within a trial were kept constant, and each vessel would experience the same temperature history.

Results and Discussion

Glucoamylase Kinetics

Experiments were conducted to compare the different glucoamylases, the effect of enzyme dilution on conversion, and the effect of enzyme dose. Subsequently, the kinetics of each reaction was compared using simple models.

Figure 1 shows the maltodextrin conversion profiles arising from studies at pH 4.8 and 55°C for nine of the glucoamylases. There was a dramatic difference in conversion rate. For example, Glu-1 reached a plateau within 4–5 h, corresponding to about 80% conversion of maltodextrin. Conversely, Glu-4, -5, and -6 converted the dextrin more slowly, but sustained this conversion over 24 h. Ultimately, there was very little difference in the dextrin fractional conversion after a 24-h reaction; all enzymes had converted about 80% of the available dextrin. The observed plateau in substrate conversion indicates either strong end product inhibition or enzyme inactivation. A similar lack of complete conversion was observed by Peeva and Yankov (4), who studied the SSF of maltodextrins from starch using Diazyme L-200.

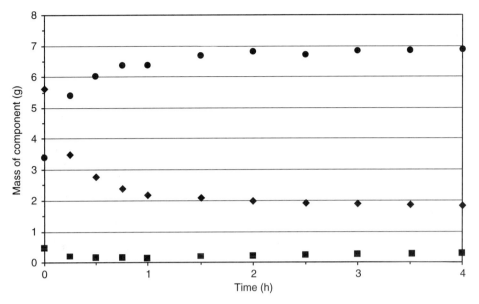

Fig. 2. Maltodextrin conversion by Glu-9, 10 g maltodextrin, and 125 µL of glucoamylase at pH 4.8. ♦, maltodextrin; ■, maltose; ●, glucose.

The glucose production rates arising from the kinetics trials were generally consistent with the dextrin conversion profiles. Some enzymes (e.g., Glu-1, -3, -8, and -9), exhibited very rapid glucose production, with yields approaching 70–75% of the originally available dextrin within as little as 15 min. However, the glucose levels then reached a plateau, with little additional glucose produced over the remaining 24 h of reaction. Such a rapid increase in glucose concentration may not be suitable for fermentation; high glucose levels can inhibit yeast (6), and gradual feeding of glucose to yeast is generally preferred. Conversely, glucose profiles arising from Glu-4, -5, and -6 did not plateau, but continued to increase over the duration of a 24-h reaction. Although the initial glucose production rate was slower with Glu-4, -5, and -6, there was sustained glucose production throughout the reaction, which is thought to be beneficial for fermentation.

Generally, maltodextrin profiles of Glu-1 and Glu-3 were comparable, regardless of enzyme dose or reaction conditions. However, glucose levels were about 5–10% lower when Glu-3 was used and maltose levels were correspondingly higher. Thus, whereas the rate constants (k_{cat} values) for larger dextrins were comparable, the rate constant for conversion of maltose to glucose was clearly greater for Glu-1. Glu-9 showed much more rapid maltodextrin conversion—almost 65% in the first 15 min of reaction, with a corresponding rapid increase in glucose (Fig. 2). However, there was less overall conversion to glucose, with only about 70% of the maltodextrin converted to glucose.

Glu-4, -5, and -6 are different strength formulations of the same enzyme. The dextrin conversion (Fig. 1) and glucose production profiles

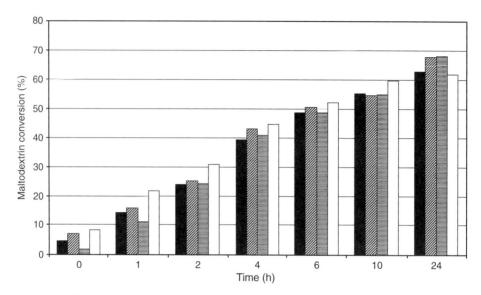

Fig. 3. Effect of pH on dextrin conversion by Glu-5, 10 g maltodextrin, and 125 µL of glucoamylase at 35°C. ■, pH 3.6; ▨, pH 4.4; ▧, pH 4.8; □, pH 5.2.

with Glu-4, -5, and -6 show different initial dextrin conversion rates, but the profiles converge after about 24 h. Similarly, Glu-4 produces more glucose during the early stages of the reaction, but by 24 h, the total glucose produced is essentially equivalent. Glu-6, the weakest formulation of the three, shows the lowest rates of maltodextrin conversion and glucose production, reaching about 80% conversion after about 32 h of reaction. Nonetheless, the different formulations demonstrate the opportunity to control the rate of glucose delivery to the fermenters. There is also a likely opportunity to tailor the formulation to the particular needs of a plant or process, to match the glucose production rate with the glucose requirements demanded by the yeast.

Effects of pH

The effect of pH on the kinetics of maltodextrin conversion at 35°C is shown in Fig. 3, using Glu-5. There is little effect of pH during the first 12 h of reaction, but, at 24 h, the conversion is slightly better at pH 4.4 and 4.8 (approx 68%) than at pH 5.2 and 3.8 (approx 60%). Interestingly, the greater dextrin conversion at pH 4.4 and 4.8 does not correspond to higher glucose concentrations; instead, at about 24 h, glucose concentrations are about 10% higher if the pH is 3.8 or 5.2 (Fig. 4). Collectively, these results imply greater conversion of the maltodextrin substrate at pH 4.4 and 4.8, but reduced conversion of reaction intermediates (maltotriose, maltose, and so on) into glucose. Consequently, k_{cat} values for large polysaccharides exhibit a maximum at about 4.6 as the pH is increased from 3.6 to 5.2, whereas k_{cat} values for smaller oligosaccharides exhibit a minimum over the same pH range. Nonetheless, these differences in dextrin conversion

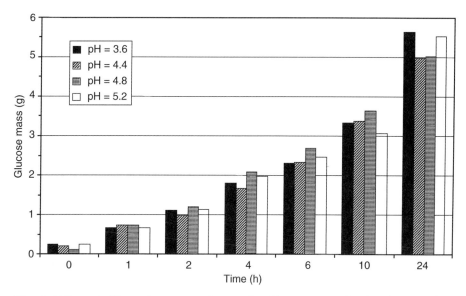

Fig. 4. Effect of pH on glucose production by Glu-5, 10 g maltodextrin, and 125 µL of glucoamylase at 35°C. ■, pH 3.6; ▨, pH 4.4; ▦, pH 4.8; □, pH 5.2.

and glucose production are quite small. Considering that the normal pH during fermentation is between 4.2 and 4.8, it is apparent that the pH-activity profile for Glu-5 is suitable for SSF systems. Although other glucoamylases showed a slightly greater dependence on pH, the overall effect of pH was still fairly mild, affecting dextrin conversion and/or glucose production by up to 20%.

Effects of Temperature

The effects of temperature on maltodextrin conversion and on glucose production by Glu-5, -7, and -8 are shown in Figs. 5 and 6, respectively.

Increasing the temperature from 35°C to 55°C caused the dextrin conversion by Glu-5 to increase from about 68–80% over 24 h of reaction. Glucose concentrations increased by about 40% owing to the increase in temperature, suggesting that temperature had a large effect on k_{cat} values for small oligosaccharides, but a comparatively smaller effect on k_{cat} values for the larger polysaccharides. Glu-8, regardless of temperature, produced glucose much more quickly than Glu-7, and Glu-7 produced nearly twice the amount of glucose at 55°C than at 35°C during the first 5 h of reaction. However, at longer times, the effect of temperature on glucose profiles from Glu-7 was less. Furthermore, Glu-7 eventually caught up with Glu-8; at 55°C, glucose levels are nearly equivalent after about 24 h of reaction. Surprisingly, temperature had little effect on maltodextrin conversion by Glu-8; profiles were nearly equivalent at both temperatures, with a very rapid initial rise, then a plateau (Fig. 5). For Glu-7, the effect of temperature was much more pronounced. The initial maltodextrin conversion rate with Glu-7 was actually faster at 35°C, but beyond the 1 h of

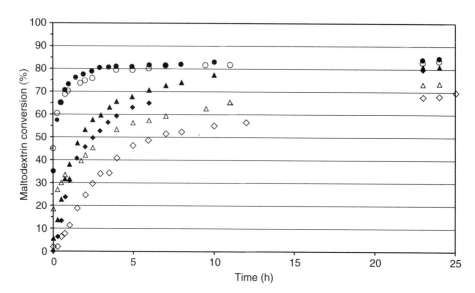

Fig. 5. Effect of temperature on maltodextrin conversion, 10 g maltodextrin, and 125 μL of glucoamylase at pH 4.8. ♦, Glu-5, 55°C; ▲, Glu-7, 55°C; ●, Glu-8, 55°C; ◊, Glu-5, 35°C; △, Glu-7, 35°C; ○, Glu-8, 35°C.

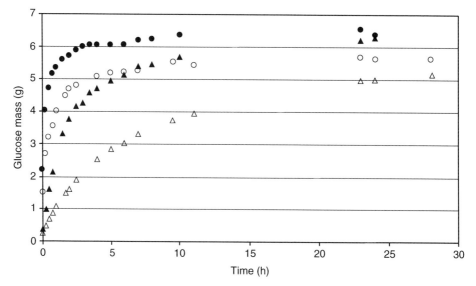

Fig. 6. Effect of temperature on glucose production, 10 g maltodextrin, and 125 μL of glucoamylase at pH 4.8. ▲, Glu-7, 55°C; ●, Glu-8, 55°C; △, Glu-7, 35°C; ○, Glu-8, 35°C.

reaction, the conversion (and rate) was greater at 55°C. The glucose profiles, conversely, consistently showed a greater rate of formation at 55°C vs 35°C. Thus, the higher degree of maltodextrin conversion during the first hour of reaction at 35°C did not translate into additional glucose, but rather, an accumulation of intermediate products.

The kinetics profile for Glu-8 suggests that temperature has little effect on k_{cat} for the larger polysaccharides, but has a significant effect on k_{cat} for

Table 1
Fermentation Efficiency of Various Glucoamylases

Enzyme	Average mass loss rate over 99 h, mg/h
Glu-3	48
Glu-4	45
Glu-7	28
Glu-8	46
Glu-9	46
Glu-11	9
Glu-10	46

the smaller species, such has maltotetraose and maltotriose. Experimentally, there was little difference in maltose concentrations at the two temperatures, but the material balance was much worse at 35°C. Thus, more of the original maltodextrin was unaccounted for at 35°C, which implies greater quantities of maltopentaose and maltotetraose, compounds that are not easily detected by the current HPLC assay. The kinetics profiles for Glu-7 indicate an enzyme designed for greater activity at lower temperatures (as evidenced by the higher initial dextrin conversion rate at 35°C), but also subject to end product inhibition, since the reaction rate slows dramatically as the concentration of glucose increases. Such end product inhibition is more severe at lower temperatures, based on the data obtained.

Mini-Fermentation Trials

Preliminary mini-fermentation (SSF) trials were conducted to compare the performance of various glucoamylases. In these experiments, 50 mL of liquefied corn mash was treated with 20 µL of glucoamylase and 0.45 g of baker's yeast. The mass loss was determined at regular intervals over 99 h, and corrected by subtracting the mass loss in a control vessel (i.e., without enzyme or yeast). The average mass loss produced using each enzyme is shown in Table 1. For the most part, overall mass loss rates were comparable, except for Glu-7 and Glu-11, which were substantially poorer than Glu-3, -4, -8, -9, and -10. Glu-11 consistently lagged all other glucoamylases throughout the experiment, whereas Glu-7 performed comparably with the other glucoamylases over the first 24 h, but lagged thereafter.

Figure 7 demonstrates the effect of enzyme dose on fermentations conducted with Glu-4 and Glu-7. The baseline dose, 20 µL, is close to that used in dry-mill plants, on a volume per substrate mass basis. The reproducibility of the procedure is evident, with a 2–5% variation in mass loss from replicate trials. As expected, the mass loss rate increased with dose; at low doses of Glu-7, the mass loss rate dropped off significantly after the initial 24 h of fermentation. In subsequent studies to examine the effect of dose on the performance of Glu-4 (data not shown), it was observed that each 5 µL increase in enzyme dose increased the average mass loss rate by

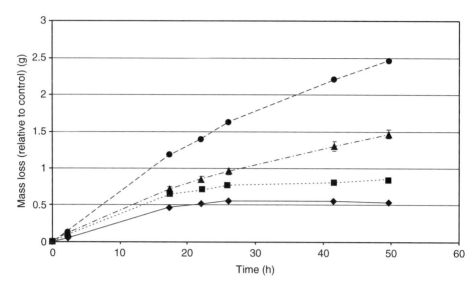

Fig. 7. Fermentation of corn mash with glucoamylase. ■, Glu-7 (8 µL); ♦, Glu-7 (4 µL); ▲, Glu-7 (20 µL); ●, Glu-4 (20 µL). Error bars represent ± SD. Error bars for Glu-4 are contained within the size of the symbol.

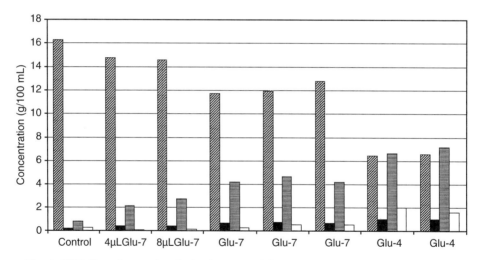

Fig. 8. HPLC analyses of carbohydrates and fermentation products from SSF studies. Liquefied corn mash treated with 0.45 g baker's yeast and 20 µL glucoamylase (unless otherwise specified); Control has no yeast or enzyme. ▨, Maltodextrin; ▫, Glycerol; ▤, Ethanol; ■, Glucose.

about 7 mg/h. Furthermore, at each enzyme dose, Glu-4 outperformed Glu-7, in spite of the observation that, during kinetics studies, Glu-7 produces glucose more rapidly.

Figure 8 shows that the carbohydrate and ethanol profiles (as measured by HPLC) are consistent with the mass loss data shown in Fig. 7. The HPLC results further support the observation that Glu-4 is superior to

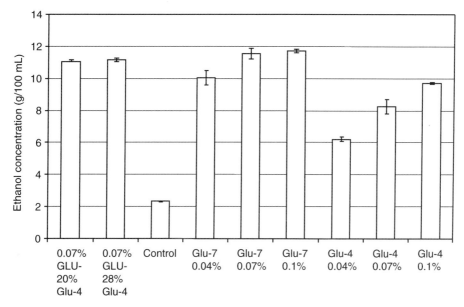

Fig. 9. Effect of glucoamylase, dose, and blends on ethanol production in SSF. Liquefied corn mash (pH 4.5) treated with 0.35 g brewer's yeast; control has no enzyme. Data shown as mean ± SD; For single enzymes, percentages refer to dose in wt%; For blends, the first percentage refers to the total glucoamylase dose (GLU) in wt%, and the second percentage refers to the percentage of Glu-4 in the Glu-4/Glu-7 mixture.

Glu-7, based on these trials with baker's yeast. Interestingly, the HPLC results also show that Glu-4 provided superior dextrin conversion than Glu-7, contrary to prior observations from kinetics studies. However, Fig. 5 also shows that Glu-7 is sensitive to temperature, and thus, the reduced performance of Glu-7 might be owing to the lower fermentation temperatures used in these studies.

Subsequent mini-fermentation studies were performed using brewer's yeast, with the temperature controlled at 33°C. In these studies, the performance of Glu-4 and Glu-7 was compared, and the performance of several blends of these two glucoamylases was also studied, to replicate conditions in some dry-mill plants that use a blend of commercial glucoamylases. The effect of enzyme dose was also examined, with doses ranging from 0.04 to 0.10 wt%. The results from these studies are shown in Figs. 9 and 10. When blends were used, the total dose was maintained at 0.07 wt%, and the percentage of Glu-4 in the blend was varied. Trials using blends are designated with "GLU" in Figs. 9 and 10, with a total glucoamylase (GLU) dose of 0.07 wt%.

It is apparent that, under these conditions, Glu-7 outperforms Glu-4, regardless of dose. Glu-7 produced higher ethanol levels, lower dextrin levels, and lower glucose levels than Glu-4. However, blends of Glu-4 and Glu-7 produced ethanol levels comparable to those obtained when Glu-7 was used alone, in spite of higher dextrin and glucose levels at the end of

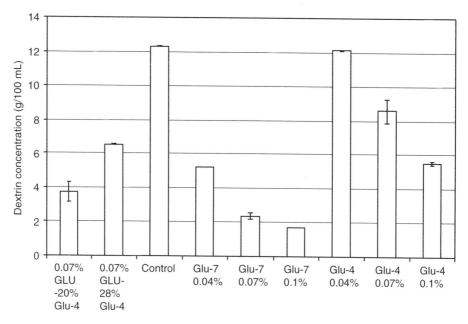

Fig. 10. Effect of glucoamylase, dose, and blends on dextrin conversion in SSF. Liquefied corn mash (pH 4.5) treated with 0.35 g brewer's yeast; control has no enzyme. Data shown as mean ± SD. For single enzymes, percentages refer to dose in wt%; For blends, the first percentage refers to the total glucoamylase dose (GLU) in wt%, and the second percentage refers to the percentage of Glu-4 in the Glu-4/Glu-7 mixture.

fermentation. This result suggests that the yeast is making more effective use of sugars produced using the glucoamylase blend, but the comparably higher dextrin and glucose levels also indicate the need to optimize the fermentation to use all of the available sugars.

The difference in relative performance with Glu-4 and Glu-7 in these mini-fermentation studies underscores the important relationship between glucoamylase function, temperature, and yeast selection. Although Glu-4 proved superior at low temperatures with baker's yeast, Glu-7 was superior in trials at 33°C using brewer's yeast. Temperature has a substantial effect on glucose production rate, and the rate of glucose production will also affect yeast performance. It is expected that some yeasts will be better able to tolerate high levels of glucose than others, and thus, selection of a glucoamylase (or blend) for use in an ethanol plant must also consider the nature of the yeast used during fermentation. These studies show dramatic variations in glucoamylase performance, with some producing glucose very rapidly but unable to sustain the rapid initial production rate, whereas others generate glucose gradually, and maintain that production over an extended period. Ultimately, a combination of glucoamylases may be desirable, aiming to produce sufficient initial glucose to launch the yeast, and then sustaining glucose production to ensure yeast viability. Glucoamylases also differed substantially in their dependence on temperature, confirming that this is also an important parameter to consider when selecting a glucoamylase. Changing the pH may lead to subtle changes in glucoamylase performance; evidence suggests a greater impact on glucose production than on overall dextrin conversion.

Conclusions

During kinetics trials, dextrin conversion did not exceed 85%, with a plateau in the quantity of glucose produced, suggesting product inhibition and/or deactivation. The kinetics of Glu-1 and Glu-3 were virtually equivalent under all reaction conditions and enzyme doses. Conversely, Glu-9 produced glucose much more rapidly than other glucoamylases, but also exhibited lower overall dextrin conversion (approx 70%) and a more rapid onset of the plateaus in glucose and dextrin concentrations. This indicates that Glu-9 is more sensitive to product inhibition, or is less stable than other glucoamylases. Glu-7 and Glu-8 also produced glucose fairly quickly.

Dextrin conversion rates with Glu-4, -5, and -6 were slower than those with other glucoamylases over about the first 4 h, but equivalent conversion was obtained within 12–24 h, depending on the formulation used. The persistent production of glucose suggests either a reduced sensitivity to inhibition or a more stable enzyme formulation, and enables "spoon-feeding" of the yeast if desired. pH variations between 3.6 and 5.2 led to small (approx 10%) changes in dextrin conversion and glucose production. Dextrin conversion was maximized at pH 4.0 to 4.8, whereas more glucose was produced at pH 3.6 and 5.2.

Generally, the glucoamylases were significantly more active at 55°C than at 35°C, with the exception of Glu-8, which had nearly equivalent activity at both temperatures. Temperature had a greater effect on glucose production than on maltodextrin conversion.

The mini-SSF trials demonstrated the key relationships between important fermentation parameters: temperature, glucoamylase, and yeast. Mass-loss measurements were consistent with independent measurements using HPLC. Furthermore, the mass loss consistently increased in response to an increase in glucoamylase dose, indicating that the systems were enzyme-limited.

Results from ambient temperature trials with baker's yeast suggested that Glu-4 was superior to Glu-7, but the opposite conclusion was reached from trials at 33°C with brewer's yeast. The strong effect of temperature on glucose production rate and the susceptibility of some yeast to inhibition by glucose may account for these differences, underscoring the importance of these parameters when contemplating selection of a new yeast and/or glucoamylase. There is also some evidence that blends of certain glucoamylases may be beneficial, particularly in case the blend can produce glucose at a controlled rate over an extended time, minimizing inhibition while maintaining yeast viability.

Acknowledgments

The authors would like to thank NSERC for financial support, and various enzyme manufacturers and industry representatives for fruitful discussions, and for provision of samples.

References

1. Christenson, U. (2000), *Biochem. J.* **349,** 623–628.
2. Nagy, E., Belafi-Bako, K., and Szabo, L. (1992), *Starch/Staerke* **44,** 145–149.
3. Cepeda, E., Hermosa, M., and Ballesteros, A. (2001), *Biotechnol. Bioeng.* **76,** 70–76.
4. Peeva, L. and Yankov, D. (2000), *Bioprocess Eng.* **22,** 397–401.
5. Allain, E. (2004), Lab-Scale Modelling of Fuel Alcohol Fermentations, presented at the 20th Fuel Ethanol Workshop, Madison, WI.
6. Ingledew, M. (1999), In: *The Alcohol Textbook, 3rd edition,* Jacques, K. A., Lyons, T. P., and Kelsall, D.R. (eds.), Alltech, Inc.

Heterologous Expression of *Trametes versicolor* Laccase in *Pichia pastoris* and *Aspergillus niger*

Christina Bohlin,[1] Leif J. Jönsson,*,[1] Robyn Roth,[2,3] and Willem H. van Zyl[3]

[1]*Biochemistry, Division for Chemistry, Karlstad University, SE-651 88 Karlstad, Sweden, E-mail: Leif.Jonsson@kau.se;* [2]*Bio/Chemtek, CSIR, Private Bag X2, Modderfontein 1645, South Africa; and* [3]*Department of Microbiology, University of Stellenbosch, Private Bag X1, Matieland 7602, South Africa*

Abstract

Convenient expression systems for efficient heterologous production of different laccases are needed for their characterization and application. The laccase cDNAs *lcc1* and *lcc2* from *Trametes versicolor* were expressed in *Pichia pastoris* and *Aspergillus niger* under control of their respective glyceraldehyde-3-phosphate dehydrogenase promoters and with the native secretion signal directing catalytically active laccase to the medium. *P. pastoris* batch cultures in shake-flasks gave higher volumetric activity (1.3 U/L) and a better activity to biomass ratio with glucose than with glycerol or maltose as carbon source. Preliminary experiments with fed-batch cultures of *P. pastoris* in bioreactors yielded higher activity (2.8 U/L) than the shake-flask experiments, although the levels remained moderate and useful primarily for screening purposes. With *A. niger*, high levels of laccase (2700 U/L) were produced using a minimal medium containing sucrose and yeast extract. Recombinant laccase from *A. niger* harboring the *lcc2* cDNA was purified to homogeneity and it was found to be a 70-kDa homogeneous enzyme with biochemical and catalytic properties similar to those of native *T. versicolor* laccase A.

Index Entries: Laccase; heterologous expression; *Pichia pastoris*; *Aspergillus niger*.

Introduction

Laccases are phenol-oxidizing enzymes that are of interest in several different applications (reviewed in ref. [1]). Possible applications include textile processing, detoxification of industrial effluents and pollutants, detoxification of lignocellulose hydrolysates in fuel ethanol production, utilization as an environmentally benign oxidant in the production of chemicals, delignification of pulp for paper manufacture, catalysis of grafting

*Author to whom all correspondence and reprint requests should be addressed.

processes in the development of novel polymers, production of fiberboards, use in fuel cells, and utilization in biosensors for monitoring phenolic pollutants and drugs. The properties of different laccases show a great deal of divergence. Laccases with high-redox potential, such as the laccase from the white-rot fungus *Trametes (Polyporus, Coriolus) versicolor* (2), are required for oxidation of recalcitrant substrates. Considering the wide range of applications for laccases (1), there is a need for heterologous expression systems to screen mutated laccases for novel properties or for large-scale production of selected laccases.

Previous studies suggest that yeasts such as *Pichia pastoris* and *Saccharomyces cerevisiae* are convenient systems for rapid expression of laccase genes. However, production levels in yeast have been quite low (up to approx 5 mg/L), whereas filamentous fungi in general have given 2–30 times higher levels (10–135 mg/L) (3). *P. pastoris* is easy to manipulate genetically, easy to use in conventional fed-batch fermentations, secretes low levels of native proteins, and capable of adding both O- and N-linked glycans to secreted proteins (4). Filamentous fungi, such as *A. niger*, have the ability to produce and secrete exceptionally large amounts of properly folded proteins with the correct cofactors incorporated and can produce proteins that contain O- and N-linked glycans without extensive hyperglycosylation (5–7).

It has been shown previously that production of laccase in *P. pastoris* using the *AOX1* promoter system is negatively affected by increasing methanol concentration (8). The *AOX1* promoter requires methanol for induction, which makes it difficult to use lower concentrations because that would negatively affect the mRNA expression level. Therefore, it should be of interest to consider the glyceraldehyde-3-phosphate dehydrogenase promoter system as an option for expression in *P. pastoris,* as well as in *A. niger*.

In this study, we have explored the potential in using *P. pastoris* for screening purposes and *A. niger* for production of selected laccases. The cDNAs *lcc1* and *lcc2* from *T. versicolor* were expressed in *P. pastoris* and *A. niger*, and the effects of different media and cultivation conditions on the laccase production levels were investigated. The recombinant laccase expressed in *A. niger* was purified to homogeneity and its biochemical and catalytic properties were compared to the well-characterized native *T. versicolor* laccases A (LccA) and B (LccB) (2).

Methods

Microbial Strains and Recombinant DNA

The *lcc1* and *lcc2* cDNA genes from *T. (Coriolus, Polyporus) versicolor* (9–11) were used in the construction of plasmids for expression of laccases in *P. pastoris* and *A. niger*.

For the expression in *P. pastoris*, the *lcc1* and *lcc2* cDNA genes were inserted into the vectors pGAPZ A and pGAPZ B, respectively. The expression cassette of the pGAPZ vectors includes the glyceraldehyde-3-phosphate

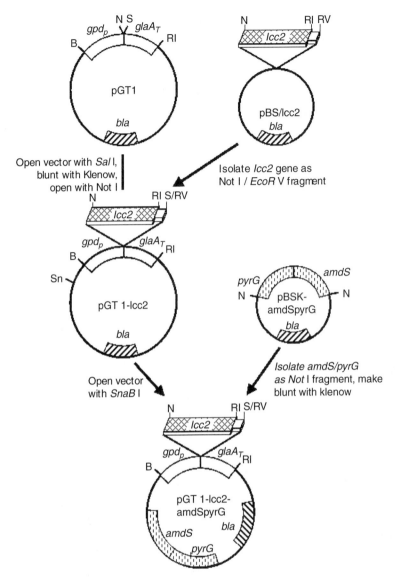

Fig. 1. Construction of plasmids for expression of laccase in *A. niger*. B, *BamH* I; N, *Not* I; RI, *EcoR* I; RV, *EcoR* V; S, *Sal* I; Sn, *SnaB* I. pGT1-*lcc1-amdSpyrG* was made similarly.

dehydrogenase gene (*GAP*) promoter region and the transcriptional terminator of the alcohol oxidase 1 gene (*AOX1*). The pGAPZ vectors are designed for constitutive expression in *P. pastoris* (Invitrogen, Carlsbad, CA). *P. pastoris* SMD1168, a *his4 pep4* strain, was transformed by electroporation.

Figure 1 summarizes the construction of plasmids for the expression in *A. niger* D15 (obtained from TNO, Zeist, The Netherlands). Insertion of the *lcc1* and *lcc2* cDNA genes into the plasmid pGT1, which harbors the *Aspergillus* glyceraldehyde-3-phosphate dehydrogenase (*gpd*) promoter and glucoamylase (*gla*) terminator from pSPORT *(12)*, generated pGT1-*lcc1* and pGT1-*lcc2*, respectively. Insertion of an *amdSpyrG* fragment taken from pBSKII(+)-*amdSpyrG* generated pGT1-*lcc1*-*amdSpyrG* and pGT1-*lcc2*-*amdSpyrG*, respectively (Fig. 1). *A. niger* spheroplasts were formed using Novozyme 234 (Sigma, St. Louis, MO) and transformed in accordance with the method of Punt and van den Hondel *(13)*.

Selection of Laccase-Expressing Transformants

P. pastoris transformants that survived a concentration of 0.1 mg/mL zeocin were spread on BMG agar plates (Invitrogen) containing glucose instead of glycerol, 0.1 mM CuSO$_4$, and 0.2 mM ABTS [2,2'-azino-*bis*(3-ethylbenzothiazoline-6-sulfonic acid)]. The color development was followed at room temperature (approx 22°C). Laccase-expressing transformants were selected and incubated in 50 mL buffered minimal glycerol (BMG) medium with 0.1 mM CuSO$_4$. An addition of 0.3 mL of a 0.6 M solution of potassium phosphate (pH 6.0) and 0.2 mL of a 0.4 M solution of NaOH was made daily. For selection of top laccase producers, samples were taken daily to determine the laccase activity (*see* Analyses of Samples).

A. niger transformants containing the *lcc1* and *lcc2* cDNA constructs as well as a negative control strain obtained from a transformation of *A. niger* D15 with the pBSKII(+)-*amdSpyrG* plasmid (Fig. 1) were plated onto agar plates containing 5% glucose, 0.2% amino acid pool without tyrosine or phenylalanine (200 mg each of adenine, uracil, tryptophan, arginine, methionine, and histidine; 300 mg of lysine; 600 mg of leucine; 2000 mg of threonine; and 65 mg of inositol), 0.5% yeast extract, 0.1 mM CuSO$_4$, 2 mM MgSO$_4$, 0.1% ABTS, 0.1% 1000× trace elements [the 1000X trace element solution contained (per 100 mL deionized water) 2.2 g ZnSO$_4$·7H$_2$O, 1.1 g H$_3$BO$_3$, 0.5 g MnCl$_2$·4H$_2$O, 0.17 g CoCl$_2$·6H$_2$O, 0.16 g CuSO$_4$·5H$_2$O, 0.15 g Na$_2$MoO$_4$·2H$_2$O, 5 g ethylene diamine tetraacetic acid (EDTA), and the pH was 6.5], and 1× AspA with nitrate [50X AspA with nitrate contained (per 500 mL deionized water): 150 g NaNO$_3$, 13 g KCl, 38 g KH$_2$PO$_4$]. The plates were incubated at 30°C for 48 h and the color development was followed. Laccase-expressing *A. niger* transformants were transferred to 50 mL medium containing 0.5% yeast extract, 1% glucose, 0.1% casamino acids (Difco, BD, Franklin Lakes, NJ), 2% MgSO$_4$·7H$_2$O, 1X trace element solution, and 1X AspA with nitrate by inoculation with spores to a concentration of 1×10^6 spores/mL (experimental series 1). The cultures were then grown for 48, 72, or 144 h. The mycelia were harvested by filtering through Miracloth (Calbiochem, San Diego, CA) and the culture supernatants were collected to determine the laccase activity.

Shake-Flask Cultivations of P. pastoris

A selected laccase-producing transformant (designated *P. pastoris/lcc2*) was incubated in BMG medium and also in the same medium but with glucose or maltose instead of glycerol. The initial concentration of carbon source was 0.5–3%. The concentration of YNB (Yeast Nitrogen Base w/o amino acids, a component of BMG medium), was varied in the range 1–3%. In addition, the medium contained 0.4% histidine, 1 mM CuSO$_4$, and

0.8% alanine. The cultivation was allowed to continue for 7 d at 20°C and samples were taken daily to determine the laccase activity.

Parallel Fed-Batch Cultivations of P. pastoris in Multibioreactor

Preparation of Inoculum

A colony of *P. pastoris/lcc2* containing pGAPZ/*lcc2* was transferred from a yeast peptone dextrose (YPD) agar plate into a baffled flask containing 200 mL of yeast peptone dextrose (YPD) (10 g/L yeast extract, 20 g/L peptone, and 20 g/L glucose) and incubated at 30°C with shaking (200 rpm in a G25 orbital shaker, New Brunswick Scientific, Edison, NJ) for 20 h until an OD_{600} (optical density at 600 nm) of greater than 3 was reached.

Batch Phase

A Sixfors multibioreactor system (Infors, Bottmingen-Basel, Switzerland) equipped with three parallel temperature-controlled 500 mL bioreactors with sensors for pH and pO_2 was used. Each bioreactor contained 180-mL basal salt medium (Invitrogen) not including glycerol but supplemented with 4% glucose, 2 g/L histidine, 4.35 mL/L PTM trace salt solution (Invitrogen), and 0.2 mM $CuSO_4$. The pH of the medium was adjusted to 5.0 with 28% ammonium hydroxide before the inoculum (3 mL) was added to a final OD_{600} of 1. The bioreactors were maintained at a temperature of 20°C and at an agitation rate of 1000 rpm. Air was introduced into the bioreactor at a rate of 30 L/h and the pH was automatically maintained at 5.0 by the addition of a 2.8% ammonium hydroxide solution. The batch phase was allowed to continue until the glucose was depleted, which occurred after approx 40 h. Thereafter, the fed-batch phase was initiated.

Fed-Batch Phase

The medium that was fed into the bioreactor was the same as in the batch phase, except that different concentrations of glucose were used. Bioreactors A–C were fed with glucose at a rate of 0.06, 0.17, and 0.43 g/h, respectively. Samples were taken daily and centrifuged in a Mini Spin microcentrifuge (Eppendorf, Hamburg, Germany) at maximum speed for 15 min. The supernatants were transferred to fresh tubes, which were snap-frozen in liquid nitrogen. Samples were analyzed with respect to laccase activity, protein concentration, protease activity, OD_{600}, wet cell weight (WCW), and glucose concentration (*see* Analyses of Samples).

Media Optimization for A. niger Transformants Using Shake-Flask Cultures

Experimental Series 2 and 3

To determine optimal medium composition for laccase production, spores of the *A. niger* transformant expressing the highest level of laccase (designated *A. niger/lcc2*) were used to inoculate different media (Table 1)

Table 1
Media Used for Optimization of Laccase Production by A. niger

Ingredient (g/L)	A	B	C	D	E	F	G
$(NH_4)_2SO_4$	45						3^a
KH_2PO_4	23	1.5^b	1.5^b		1.5^b		
$FeSO_4 \cdot H_2O$	0.1						
$MgSO_4 \cdot 7H_2O$	7	0.49^c	0.49^c	0.49^c	0.49^c	0.49^c	0.98^d
Sucrose	50		50				100
Urea	11						
Yeast extract	5		5		5	5	
Glucose		10		100	10		
1000X trace elements		1 mL	1 mL	1 mL	1 mL	1 mL	2 mL
$NaNO_3$		6^b	6^b	5.95	6^b	6^b	12^a
KCl		0.52^b	0.52^b	0.52	0.52^b	0.52^b	1.04^a
NaH_2PO_4				35.60			
Initial pH	4, 5, 6	4, 5, 6	4, 5, 6	5	5	5	5

^{a}As 40 mL 50X AspA+N salts.
^{b}As 20 mL 50X AspA+N salts.
^{c}As 2 mL of 1 M.
^{d}As 4 mL of 1 M.

to a final concentration of 1×10^6 spores/mL. The inoculum was generated by resuspending spores in a sterile 0.9% NaCl solution. The spore suspension was stored at 4°C until use. The cultures were incubated in a volume of 100 mL at 30°C with shaking (200 rpm). The performance of the transformant in media A–D (experimental series 2) (Table 1) was studied for 4 d. In the experiments with media E–G (experimental series 3) (Table 1), the cultivation was allowed to continue for 10 d.

Experimental Series 4

The media optimization was continued by investigating the effect of the addition of 0.5% yeast extract to the medium, as well as the use of a preculture. The 20-mL precultures were all inoculated to a final concentration of 5×10^6 spores/mL using the A. niger/lcc2 transformant. The precultures were incubated at 30°C for 36 h and then transferred to 80-mL fresh media.

Experimental Series 5

Additional experiments were carried out at pH 4.0–6.0 and at 25°C and 30°C. The laccase activity was measured daily. The results of these and earlier experiments were used to plan the following experiment.

Experimental Series 6

The production of laccase by the transformant A. niger/lcc2 was followed over a time period of 10 d in cultures with three different media in a volume of 250 mL. The first medium consisted of a minimal salt medium

(0.05% $MgSO_4 \cdot 7H_2O$, 0.6% $NaNO_3$, 0.05% KCl, 0.15% KH_2PO_4, 1X trace elements) supplemented with 1% glucose and 0.5% yeast extract. The second medium was a double-strength minimal salt medium (0.1% $MgSO_4 \cdot 7H_2O$, 1.2% $NaNO_3$, 0.1% KCl, 0.3% KH_2PO_4, 2X trace elements) with 10% sucrose. The third medium was the same as the second but supplemented with 0.5% yeast extract. The pH of the first two media at the time of inoculation was 5.0, whereas the pH of the third was 6.0.

Purification of Heterologously Expressed Laccase From A. niger

The transformant *A. niger/lcc2* was cultivated in 250 mL of previously optimized medium (0.1% $MgSO_4 \cdot 7 H_2O$, 1.2% $NaNO_3$, 0.1% KCl, 0.3% KH_2PO_4, 2X trace elements) with 10% sucrose and 0.5% yeast extract, and with the pH adjusted to 6.0. The culture was inoculated to 1×10^6 spores/mL and incubated at 30°C and 200 rpm for 11 d. The extracellular fraction was harvested by filtration. Ammonium sulfate precipitation was carried out at 70% saturation (at pH 4.2). The harvested protein precipitate was resuspended in 20 mM imidazole buffer, pH 7.0, and then snap-frozen in liquid N_2 in aliquots. The aliquots were freeze-dried and stored at –70°C. They were then resuspended in a 20 mM solution of potassium phosphate (pH 6.0) and dialyzed against the same buffer before purification.

The first purification step was done by loading the dialyzed sample onto a DEAE Sepharose Fast Flow column (Amersham Biosciences, Uppsala, Sweden), pre-equilibrated with 100 mM phosphate buffer (pH 6.0). A NaCl gradient from 0 to 500 mM was applied to the column, and the laccase started eluting at 40 mM NaCl. Fractions collected between 40 and 125 mM NaCl were pooled and dialyzed against 20 mM 3-(N-morpholino propanesulfonic acid) MOPS buffer, pH 7.2, overnight.

The second purification step was done by loading the dialyzed DEAE fractions onto a HiTrap Q FF column (Amersham Biosciences) equilibrated with 20 mM MOPS, pH 7.2. The protein was eluted with a 0–500 mM NaCl gradient. The laccase eluted from 150 mM NaCl. Fractions collected between 150 and 250 mM NaCl were pooled and dialyzed overnight against 10 mM MOPS, pH 6.5.

The dialyzed protein solution was concentrated 15 times, using 5-kDa cutoff spin columns (Millipore, Bedford, MA). This preparation was used for kinetic analysis and comparison with two different forms of the native protein, laccase A (LccA) and laccase B (LccB), isolated from *T. versicolor* (2,14).

A further purification step was carried out by loading the HiTrap Q-purified protein onto a MonoP HR 5/20 column (Amersham Biosciences). The column was equilibrated with 20 mM MOPS buffer, pH 7.2, and a pH gradient from 7.2 to 3.0 was applied using 10% PolyBuffer 74 (Amersham Biosciences), pH 3.0, as eluent. A peak was visible for fractions eluting at pH 3.0–4.0 and the relevant fractions were neutralized by adding 100 µL of 1 M Tris-HCl per 2 mL fraction and pooled. The protein solution was dialyzed against 10 m*M* phosphate buffer, pH 6.0, and concentrated using the 5-kDa spin columns.

Analyses of Samples

Laccase Activity

Assays were performed as described previously *(9)*. One unit was defined as the amount of laccase that forms 1 µmol of ABTS radical cation ($\varepsilon = 3.6 \times 10^4/M/cm$ at 414 nm *[15]*). To avoid assay interference by medium components, the laccase activity in samples from *P. pastoris* cultures was measured after purification of the samples using Microspin G-25 columns (Amersham Biosciences, Sweden) equilibrated with 10 m*M* phosphate buffer (pH 6.0).

Protein Concentration

The protein concentrations in samples from *P. pastoris* cultures were determined using Coomassie protein assay reagent (Pierce, Rockford, IL) with bovine-serum albumin as the standard. The protein concentrations in samples from *A. niger* cultures were estimated using the Lowry method *(16)*. During the purification of laccase from *A. niger*, the protein concentration was determined using the BCA protein assay reagent kit (Pierce, IL).

Optical Density

The OD of the samples was measured against distilled water at 600 nm with a ultraviolet-1601PC spectrophotometer (Shimadzu, Kyoto, Japan) after appropriate dilution.

Wet Cell Weight

The WCW was determined by centrifugation of the samples at maximum speed with a Mini Spin centrifuge (Eppendorf, Germany) for 15 min in preweighed test tubes and subsequent removal of the supernatant.

Glucose Concentration

The glucose concentration was determined using a Glucometer Elite XL (Bayer, Leverkusen, Germany) after appropriate dilution in artificial liquor containing (per liter of deionized water) 8.65 mg NaCl, 176.4 mg $CaCl_2 \cdot 2H_2O$, 182.9 mg $MgCl_2 \cdot 6H_2O$, 201.3 mg KCl, and NaOH to a pH of 7.4.

Electrophoresis

Estimation of the purity, size, and pI of the enzyme was obtained by using the PhastGel system (Amersham Biosciences), with a 4–15% gradient gel for sodium dodecyl sulfate-polyacrylamide gel electrophoresis (SDS-PAGE) and an isoelectric focusing (IEF) 3–9 gel for IEF. Two identical IEF gels were run at the same time and one was stained with Coomassie brilliant blue. The other was used for zymogram analysis and was immersed for 5 min in a solution of 0.4 m*M* ABTS in 50 m*M* NaAc, pH 5.0.

Glycosylation

To determine the level of glycosylation of the recombinant laccase and the native LccA sample, deglycosylation was carried out using N-glycosidase F (Roche, Mannheim, Germany). First, 8 μg of laccase (in 5 μL of a solution of 1% SDS, 1% β-mercaptoethanol, and 20 mM sodium phosphate [pH 8.6]) was denatured by boiling for 3 min. The denatured sample was subsequently deglycosylated in a final volume of 20 μL containing 1% Nonidet P-40, 0.5% EDTA, 1% β-mercaptoethanol, 25 mM sodium phosphate buffer (pH 7.2), 0.25% SDS, and 5 U of the deglycosylation enzyme. The reactions were incubated at 37°C for 2 h, and 4 μL was loaded onto a 12% homogeneous SDS-PAGE gel using the PhastGel system. Control reactions without N-glycosidase F addition were done in parallel.

Kinetic Analysis

The K_M values of LccA, LccB, and the recombinant laccase purified from *A. niger* were determined using the substrate 2,4,5-trimethoxylbenzyl alcohol and were performed as described elsewhere (32).

Protease Activity

The protease activity was determined using the QuantiCleave fluorescent protease assay kit (Pierce, IL), which is based on proteolytic digestion of a fluorescein thiocarbamoyl-casein conjugate and measurement of the fluorescence at 538 nm. The proteolytic activity at pH 5.0 was assayed using an LS 55 luminescence spectrometer (PerkinElmer, Wellesley, MA).

Results and Discussion

The expression of two isoenzymes of laccase from *T. versicolor* in *P. pastoris* and *A. niger* was investigated. In both cases, the expression was under the control of the glyceraldehyde-3-phosphate dehydrogenase promoter, the *GAP* promoter of *P. pastoris* (17) and the *gpd* promoter of *Aspergillus* (18). The *GAP* promoter gives constitutive expression, although its strength varies depending on the carbon source (17). This is the first report of a laccase expressed in *P. pastoris* with the *GAP* promoter, as previous attempts have been made using the *AOX1* promoter.

Analysis of shake-flask cultures revealed that *P. pastoris* transformants with the *lcc2* gene gave approx three to four times higher activity than transformants with the *lcc1* gene (Fig. 2A). The transformant giving the highest activity (*P. pastoris*/*lcc2*) (approx 0.6 U/L) (Fig. 2A) was chosen for further studies. The choice of laccase gene for expression appears to have a great impact on the levels of activity reached.

Shake-flask cultures of *P. pastoris*/*lcc2*, in which the type and concentration of carbon source were varied, gave higher laccase activity (1.3 U/L) with glucose than with glycerol (0.7 U/L) or maltose (0.6 U/L) (Fig. 2B). Cells grown on glycerol showed higher growth rate than cells grown on glucose, which in turn showed higher growth rate than cells grown on

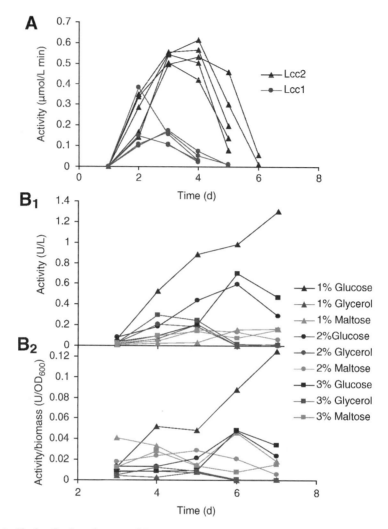

Fig. 2. Shake-flask cultures of *P. pastoris*. **(A)** Comparison of *lcc1* and *lcc2* transformants, **(B₁)** volumetric activity with *P. pastoris/lcc2*, **(B₂)** ratio of activity to biomass with *P. pastoris/lcc2*.

maltose. The cultures reached an OD_{600} of 27 (glycerol), 20 (glucose), and 10 (maltose). Previous results suggest that a slow growth rate is usually better for reaching high-laccase activity *(8)*, but the maltose cultures did not follow that trend. The ratio of activity to biomass was higher with glucose than with maltose or glycerol (Fig. 2B). This result shows that the laccase activity depends on more than just the growth rate. Growth on different carbon sources results in distinct patterns of intracellular proteins *(19)*, which in turn may affect the production of heterologous proteins. For laccase production, a glucose concentration of 1% was better than 2–3% (Fig. 2B). When even lower glucose concentrations were studied, 0.5% glucose gave about the same activity to biomass ratio as 1% glucose (not shown). The activities reached with the *GAP* promoter can be compared

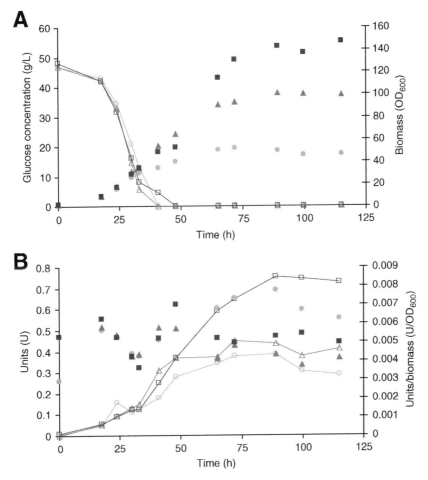

Fig. 3. Bioreactor cultures of *P. pastoris*/*lcc2*. **(A)** Glucose concentration and biomass, **(B)** units (μmol/min) and units/biomass. Filled symbols show biomass and units/biomass [fermentor A (●), fermentor B (▲), and fermentor C (■)]. Open symbols show glucose concentration and units.

with the expression of the same gene but with the *AOX1* promoter, which gave an activity of 0.35 U/L in the X-33 strain.

The influence of different feeding rates of glucose on the selected *P. pastoris*/*lcc2* transformant was investigated in parallel fed-batch cultures using a multibioreactor system (Fig. 3). The batch phase lasted for about 40 h (Fig. 3A) and resulted in a biomass concentration of 40–60 g/L. In the fed-batch phase, glucose was added at different rates for 82 h and the biomass concentrations in the bioreactors started to vary (Fig. 3A). The biomass concentration in fermentor A did not increase that much in the fed-batch phase, whereas fermentors B and C reached biomass concentrations of 80–90 g/L and 130–140 g/L, respectively. Fermentor C reached the highest volumetric activity (2.8 U/L), whereas fermentors B and A reached 1.8 and 1.3 U/L, respectively. The activity increased until approx 90 h, and thereafter the level was constant or decreased slightly (Fig. 3B). The ratio of activity to biomass was slightly higher in fermentor A (Fig. 3B), but the relatively low values compared to the shake-flask cultures (Fig. 2B) indicate that further improvements should be possible. The specific activity (not shown) increased for about 50 h and subsequently leveled off.

Compared to a number of reports documenting high-level expression of foreign genes in *P. pastoris (4,20)*, the concentration of secreted protein in this study (0.1 g/L) was satisfactory. When laccases from *Pleurotus sajor-caju (21)* and *Pycnoporus cinnabarinus (22)* were expressed in *P. pastoris* using the *AOX1* system, the protein concentrations were 0.11 and 0.008 g/L, respectively. Although the total protein concentration in our study was at the same level or higher, the specific activity was much lower than in those studies *(21,22)*. The use of controlled conditions in a bioreactor instead of using shake-flasks generally improves heterologous protein production *(20)*. The laccase production reported in the current study could probably be enhanced further by optimization of medium and cultivation conditions.

When *P. pastoris* was used to express and determine the size of laccases from *T. versicolor (8)*, *P. cinnabarinus (22)*, *Fome lignosus (23)*, and *P. sajor-caju (21)*, the recombinant laccases were found to be between 5% and 36% larger then the native enzymes, strongly indicating hyperglycosylation. However, no reports show that hyperglycosylation affects the activity of the enzyme produced.

The use of a proteinase A mutant (*pep4*) strain (SMD1168) has been reported to be beneficial for the production of secreted recombinant proteins *(24)*, and that has also been observed for laccase *(9)*. In this study, the protease level in the fermentation medium was around 0.6 µg/mL (with N-tosyl-L-phenylalanine chloromethyl ketone (TPCK) trypsin as the standard) after 17 h and was fairly constant throughout the whole fermentation, never reaching a level greater than 1 µg/mL. It has been shown that cells grown on glycerol have lower levels of total protease activity compared with cells grown on methanol as the sole carbon source *(25)*. This could be a problem when laccase is expressed with the *AOX1* promoter and is one of the reasons why alternative carbon sources are of interest to study.

Very few transformants were obtained for *A. niger*. The transformation with pGT1-*lcc1*-*amdSpyrG* and pGT1-*lcc2*-*amdSpyrG* gave rise to nine and seven transformants, respectively. Of these 16 transformants, 11 showed color development on ABTS plates. No development of color was observed with the control transformed with pBSKII(+)-*amdSpyrG*.

To select a top laccase producer, transformants were grown in shake-flasks (experimental series 1). The results obtained showed that most of the activity was lost by 144 h, with the relative activities at 48 and 72 h being very similar (data not shown). Again, the control transformant showed no activity. Specific activities of laccase in the supernatants were calculated indicating that *lcc2* gave higher specific activity than *lcc1*, although the

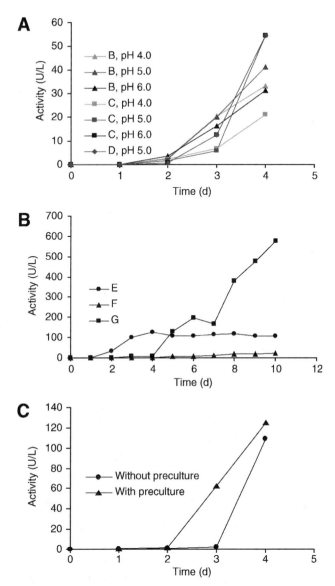

Fig. 4. Media optimization for *A. niger*/*lcc2* in **(A)** experimental series 2 with media A–D, **(B)** experimental series 3 with media E–G, and **(C)** experimental series 4 with and without preculture.

activity also varied between *lcc2* transformants (data not shown). The average volumetric and specific activity was almost 10 times higher for the *lcc2* transformants in comparison to the *lcc1* transformants. An *lcc2* transformant (*A. niger*/*lcc2*) was selected to be used in shake-flask analysis for media optimization.

Laccase production in media A–D was studied during 4 d of cultivation (Fig. 4A). Medium A (high salts) was used by Balasubramaniem et al. (26) for the production of β-fructofuranosidase in *A. niger* NRRL 330.

Growth in medium A gave no extracellular laccase activity, and very low biomass concentration (4–5 g/L after 4 d) was obtained. Medium B gave extracellular laccase activity levels of 31–41 U/L depending on the pH. The biomass production was good (19 g/L at pH 6.0). At low pH, the biomass concentration was lower (only approx 10 g/L at pH 4.0). Medium C also gave extracellular laccase activity (21–55 U/L), with the activity levels increasing as the pH increased. The biomass concentration was 14–15 g/L regardless of pH. Medium D was used by Record et al. (27) for expression of the *P. cinnabarinus* laccase in *A. niger* D15 No. 26, similar to the host strain being used in our experiments. A maximum activity of approx 50 U/L was obtained (Fig. 4A), which is in agreement with the reported levels of *P. cinnabarinus* laccase activity of 50 U/L using the native secretion signal. The biomass production was low, only approx 6 g/L. As the laccase activity was still increasing at the end of the experiment, a longer time period was used in further studies (experimental series 3–6).

Laccase production in media E–G was studied over 10 d (Fig. 4B). Medium E showed excellent laccase activity. By d 4 of the experiment (to compare with media A–D), the activity levels were 128 U/L, which was the maximum level obtained in this medium. The glucose was spent by 48 h, which was the point at which the biomass peaked (6.1 g/L). The cells may have lyzed during the next 2 d, as there was a small increase in extracellular laccase activity until 96 h (128 U/L), after which the activity dipped slightly and remained constant at approx 110–120 U/L until d 10. This may indicate that there is very little protease activity in the medium, as the laccase is not disappearing. *A. niger* D15 (the host organism) is a pH mutant which does not naturally acidify the medium, unlike like wild-type *Aspergillus*, and the acid proteases are not activated (28). Medium F showed very little activity, as well as very little biomass, which was to be expected as the only carbon source added was yeast extract. Medium G combined sucrose as a carbon source with double strength minimal medium. This medium gave by far the best laccase activity, which reached 576 U/L after 10 d (Fig. 4B). The results show that yeast extract has a positive effect on laccase activity when added to the minimal medium (medium B vs medium E). The lag phase before activity was seen could be reduced from 3–4 to 2 d. This phenomenon was investigated further in experimental series 4 using 2× minimal medium with 10% sucrose (medium G) with and without yeast extract, as well as using glucose in place of sucrose in this medium. The presence of yeast extract resulted in laccase production, with low levels seen after 48 h (data not shown). An attempt to reduce the lag phase of laccase activity even further was done by using precultures for inoculation of the shake-flasks. The lag phase was reduced from 3 to 2 d (Fig. 4C).

In experimental series 5, the effects of pH and temperature were determined through the cultivation of *A. niger*/*lcc2* in sucrose 2× minimal medium supplemented with yeast extract. Spores were inoculated into 50-mL medium with the pH adjusted to 4.0, 5.0, or 6.0 and the cultures were

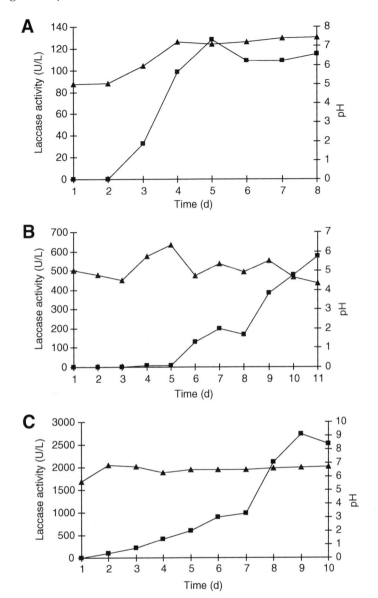

Fig. 5. The production of laccase by the transformant *A. niger/lcc2* in (A) glucose minimal medium supplemented with yeast extract, (B) double-strength sucrose minimal medium, and (C) double-strength sucrose minimal medium supplemented with yeast extract. Activity (■) and pH (▲) are indicated.

grown at 25°C or 30°C. The best results were obtained with the pH 6 medium (data not shown).

Media composition and growth conditions for further studies were selected on basis of results from experimental series 1–5. In experimental series 6, the cultivation of *A. niger/lcc2* was carried out over a period of 10 d in glucose minimal medium supplemented with yeast extract. A rise in pH from 5.0 to 7.0 was seen during the cultivation (Fig. 5A). This should not

Table 2
Purification of Recombinant Laccase From *A. niger/lcc2*

Purification step	Total volume (mL)	Total activity (U)	Total protein (mg)	Specific activity (U/mg)	Yield (%)	Purification (fold)
Resuspended freeze-dried sample	5	56.8	15.55	3.65	100	1
DEAE fractions	2.5	28.4	1.53	18.6	50	5.1
HiTrap fractions	1.5	10.7			19	
Mono P fractions	0.125	1	0.04	25	2	6.9

be a problem as it has been previously shown that laccase from *T. versicolor* has best stability between pH 6.0 and 7.0 (29). The activity increased rapidly between the second and the third day after inoculation to approx 100 U/L and then stayed at that level (Fig. 5A). A higher activity was reached with sucrose 2× minimal medium (Fig. 5B). In that case the activity appeared much later (d 6) and the highest activity was observed after 11 d. The pH dropped to 4.0, which is suboptimal for native laccase, with 70% of the activity remaining in comparison to pH 6.0–7.0 (29). In sucrose 2× minimal medium with yeast extract, the pH was stable at pH 6.0–7.0 during the whole period. The highest activity was reached after 9 d and reached approx 2700 U/L (Fig. 5C). In comparison to sucrose 2× minimal medium, the activity appeared sooner (d 2–3) in sucrose 2× minimal medium with yeast extract. The activity was similar to that determined in the work of Record et al. (27), who expressed a *P. cinnabarinus* laccase in *A. niger* using a similar strain but replaced the native laccase signal peptide with the 24-amino-acid-residue glucoamylase (*gla*) preprosequence from *A. niger*, as the production of laccase with the native signal peptide was quite low. In this study, we demonstrate that it is possible to express high levels of laccase with the native secretion signal. Recombinant Lcc2 protein was purified from the culture medium of *A. niger/lcc2* in three steps (Table 2). The purification was not optimized further regarding the yield, as enough enzyme was obtained to characterize the recombinant protein.

Recombinant Lcc2 from different stages of purification were run on an SDS-PAGE gel, with LccA and LccB preparations from *T. versicolor* as controls. Molecular weight determination indicated that the recombinant Lcc2 was similar to LccA rather than to LccB (Fig. 6A). This was to be expected, as the *lcc2* gene has been associated with the LccA form of the protein in the native host (10). The calibration indicated that the recombinant Lcc2 is 74 kDa vs 68 kDa of LccA. IEF was also carried out, using LccA and LccB as controls (Fig. 6B). The pI of the recombinant Lcc2 protein and LccA appeared identical and very acidic, at about 3.5. LccB showed a different pattern indicating a higher pI and the presence of several forms in the preparation (Fig. 6B). The expected pI ranges of preparations of

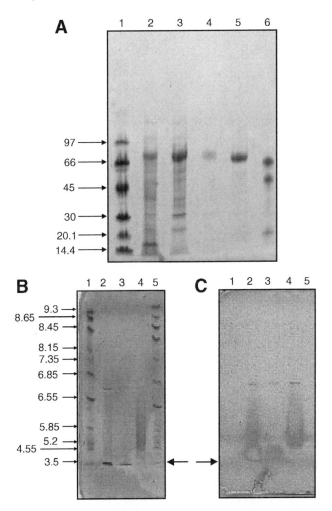

Fig. 6. (A) SDS-PAGE analysis: lane 1, marker proteins (sizes in kDa indicated to the left); lanes 2–4, recombinant Lcc2 protein from *A. niger/lcc2* after $(NH_4)_2SO_4$ precipitatation (lane 2), purification on HiTrap Q (lane 3), and purification on MonoP (lane 4); lanes 5 and 6, native LccA (lane 5) and LccB (lane 6). (B) IEF analysis: lanes 1 and 5, marker proteins (pI indicated to the left); lane 2, recombinant Lcc2 protein from *A. niger/lcc2*; lane 3, native LccA; lane 4, native LccB. (C) Zymogram analysis: lanes 1 and 5, marker proteins; lane 2, native LccA; lane 3, recombinant Lcc2 protein from *A. niger/lcc2* (position indicated by arrow); lane 4, native LccB.

LccA and LccB are 3.07–3.27 and 4.64–6.76, respectively (2). LccB may contain at least ten components. Zymogram analysis confirmed that the recombinant laccase was similar to LccA, whereas most of the LccB activity was found around pH 5.0 (Fig. 6C).

As can be seen in Fig. 7, both recombinant Lcc2 protein and native LccA have similar carbohydrate contents. The calculated carbohydrate content of recombinant Lcc2 protein is 16%, whereas LccA shows a carbohydrate content of 11%. Previous estimates of the carbohydrate content

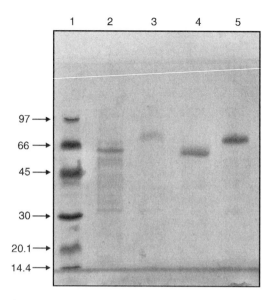

Fig. 7. Deglycosylation of recombinant Lcc2 protein from *A. niger/lcc2* and native LccA. Lane 1, marker proteins (sizes in kDa indicated on left); lane 2, deglycosylated recombinant Lcc2; lane 3, untreated recombinant Lcc2; lane 4, deglycosylated LccA; lane 5, untreated LccA.

of LccA range from 10% to 14% *(2)*. The molecular weight of the unglycosylated laccase as deduced from the amino-acid sequence is 53 kDa *(10)*. The recombinant laccase appeared to be homogeneous. Laccase from *P. cinnabarinus* expressed in *A. niger* gave rise to a 70-kDa enzyme, which was the same as for the native protein *(27)*. Laccase from *Trametes villosa* expressed in *Aspergillus oryzae* showed 0.5% and 10% glycosylation for the native and the recombinant enzyme, respectively *(30)*. Recombinant laccase from *Myceliophthora thermophila* expressed in *A. oryzae* also showed a higher degree of glycosylation *(31)*.

The K_M values were determined using 2,4,5-trimethoxybenzyl alcohol (Hong et al., in this volume) as the substrate. For LccA and the recombinant Lcc2 protein, substrate concentrations of 1–60 mM were tested, whereas for LccB the maximum concentration was increased to 90 mM, as the rate continued to increase at 60 mM. The results suggest that the K_M of LccA and recombinant laccase are similar, whereas the K_M of LccB is considerably higher.

Conclusions

Production of catalytically active laccase from the *T. versicolor* cDNAs *lcc1* and *lcc2* under control of glyceraldehyde-3-phosphate dehydrogenase promoters was achieved in both *P. pastoris* and *A. niger*. Expression of laccase in *P. pastoris* using the *GAP* system gave better results with glucose than with glycerol or maltose as carbon source. The activity obtained with *P. pastoris* was considerably lower than for *A. niger*, but the *P. pastoris* system may still be of interest for screening studies owing to convenience and speed. With *A. niger*, high laccase activity levels (2700 U/L using ABTS as reducing substrate) were obtained with the native secretion signal by using a medium containing sucrose and yeast extract.

Recombinant laccase produced by *A. niger* D15 transformed with the *lcc2* cDNA was purified to homogeneity, and its biochemical and catalytical properties were found to be similar to those of native laccase A from *T. versicolor*. This is an important finding for the applicability of *A. niger* for heterologous production of enzymes, as it is an indication that the activity and characteristics of enzymes produced in this manner can be correlated to the native enzymes.

Acknowledgments

We would like to thank Professor C. A. M. J. J. van den Hondel (TNO, Zeist, The Netherlands) for providing plasmid p3SR2 and *A. niger* D15 as fungal host for this study. This work was supported by the South African-Swedish Research Partnership Programme (NRF/SIDA) and the Swedish Research Council.

References

1. Mayer, A. M. and Staples, R. C. (2002), *Phytochemistry* **60,** 551–565.
2. Reinhammar, B. (1984), In: *Copper Proteins and Copper Enzymes*, Lontie, R. (ed.), CRC Press, Boca Raton, FL, vol. 3, pp. 1–35.
3. Yaver, D. S., Overjero, M. D., Xu, F., et al. (1999), *Appl. Environ. Microbiol.* **65,** 4943–4948.
4. Cereghino, J. L. and Cregg, J. M. (2000), *FEMS Microbiol. Rev.* **24,** 45–66.
5. Punt, P. J., van Biezen, N., Conesa, A., Albers, A., Mangnus, J., and van den Hondel, C. (2002), *Trends Biotechnol.* **20,** 200–206.
6. Maras, M., van Die, I., Contreras, R., and van den Hondel, C. A. M. J. J. (1999), *Glycoconj. J.* **16,** 99–107.
7. Hombergh, J. P. T. W., Vondervoort, P. J. I., Fraissinet-Tachet, L., and Visser, J. (1997), *Tibtech* **15,** 256–263.
8. Hong, F., Meinander, N. Q., and Jönsson, L. J. (2002), *Biotechnol. Bioeng.* **79,** 438–449.
9. Jönsson, L. J., Saloheimo, M., and Penttilä, M. (1997), *Curr. Genet.* **32,** 425–430.
10. Cassland, P. and Jönsson, L. J. (1999), *Appl. Microbiol. Biotechnol.* **52,** 393–400.
11. Jönsson, L., Sjöström, K., Häggström, I., and Nyman, P. O. (1995), *Biochim. Biophys. Acta* **1251,** 210–215.
12. Rose, S. H. and van Zyl, W. H. (2002), *Appl. Microbiol. Biotechnol.* **58,** 461–468.
13. Punt, P. J. and van den Hondel, C. A. M. J. J. (1992), *Methods Enzymol.* **216,** 447–457.
14. Fåhraeus, G. and Reinhammar, B. (1967), *Acta Chem. Scand.* **21,** 2367–2378.
15. Childs, R. E. and Bardsley, W. G. (1975), *Biochem. J.* **145,** 93–103.
16. Lowry, O. H., Rosebrough, N. J., Farr, A. L., and Randall, R. J. (1951), *J. Biol. Chem.* **193,** 265–275.
17. Waterham, H. R., Digan, M. E., Koutz, P. J., Lair, S. V., and Cregg, J. M. (1997), *Gene* **186,** 37–44.
18. Punt, P. J., Dingemanse, M. A., Kuyvenhoven, A., et al. (1990), *Gene* **93,** 101–109.
19. Gellissen, G. (2000), *Appl. Microbiol. Biotechnol.* **54,** 741–750.
20. Cregg, J. M., Vedvick, T. S., and Raschke, W. C. (1993), *Biotechnology (NY)* **11,** 905–910.
21. Soden, D. M., O'Callaghan, J., and Dobson, A. D. W. (2002), *Microbiology* **148,** 4003–4014.

22. Otterbein, L., Record, E., Longhi, S., Asther, M., and Moukha, S. (2000), *Eur. J. Biochem.* **267,** 1619–1625.
23. Liu, W., Chao, Y., Liu, S., Bao, H., and Qian, S. (2003), *Appl. Microbiol. Biotechnol.* **63,** 174–181.
24. Brankamp, R. G., Sreekrishna, K., Smith, P. L., Blankenship, D. T., and Cardin, A. D. (1995), *Protein Express. Purif.* **6,** 813–820.
25. Sinha, J., Plantz, B. A., Inan, M., and Meagher, M. M. (2004), *Biotechnol. Bioeng.* **89,** 102–112.
26. Balasubramaniem, A. K., Nagarajan, K. V., and Paramasamy, G. (2001), *Proc. Biochem.* **36,** 1241–1247.
27. Record, E., Punt, P. J., Chamkha, M., Labat, M., van den Hondel, C. A. M. J. J., and Asther, M. (2002), *Eur. J. Biochem.* **269,** 602–609.
28. Wiebe, M. G., Karandikar, A., Robson, G. D., et al. (2001), *Biotechnol. Bioeng.* **76,** 164–174.
29. Larsson, S., Cassland, P., and Jönsson, L.J. (2001), *Appl. Environ. Microbiol.* **67,** 1163–1170.
30. Yaver, D. S., Xu, F., Golightly, E. J., et al. (1996), *Appl. Environ. Microbiol.* **62,** 834–841.
31. Berka, R. M., Schneider, P., Golightly, E. J., et al. (1997), *Appl. Environ. Microbiol.* **63,** 3151–3157.

Lactose Hydrolysis and Formation of Galactooligosaccharides by a Novel Immobilized β-Galactosidase From the Thermophilic Fungus *Talaromyces thermophilus*

PHIMCHANOK NAKKHARAT[1,2] AND DIETMAR HALTRICH*,[1]

[1]*Division of Food Biotechnology, Department of Food Sciences and Technology, BOKU-University of Natural Resources and Applied Life Sciences, Vienna Muthgasse 18, A-1190, Vienna, Austria, E-mail: dietmar. haltrich@boku.ac.at; and [2]Department of Biotechnology, Faculty of Engineering and Industrial Technology, Silpakorn University, Nakorn Pathom, 73000, Thailand*

Abstract

β-Galactosidase from the fungus *Talaromyces thermophilus* CBS 236.58 was immobilized by covalent attachment onto the insoluble carrier Eupergit C with a high binding efficiency of 95%. Immobilization increased both activity and stability at higher pH values and temperature when compared with the free enzyme. Especially the effect of immobilization on thermostability is notable. This is expressed by the half-lifetime of the activity at 50°C, which was determined to be 8 and 27 h for the free and immobilized enzymes, respectively. Although immobilization did not significantly change kinetic parameters for the substrate lactose, a considerable decrease in the maximum reaction velocity V_{max} was observed for the artificial substrate o-nitrophenyl-β-D-galactopyranoside (oNPG). The hydrolysis of both oNPG and lactose is competitively inhibited by the end products glucose and galactose. However, this inhibition is only very moderate as judged from kinetic analysis with glucose exerting a more pronounced inhibitory effect. It was evident from bioconversion experiments with 20% lactose as substrate, that the immobilized enzyme showed a strong transgalactosylation reaction, resulting in the formation of galactooligosaccharides (GalOS). The maximum yield of GalOS of 34% was obtained when the degree of lactose conversion was roughly 80%. Hence, this immobilized enzyme can be useful both for the cleavage of lactose at elevated temperatures, and the formation of GalOS, prebiotic sugars that have a number of interesting properties for food applications.

Index Entries: Immobilization; Eupergit; β-glycosidase; lactase; transgalactosylation.

---

*Author to whom all correspondence and reprint requests should be addressed.

Introduction

The enzyme β-D-galactoside galactohydrolase (EC. 2.2.1.23), commonly known as β-galactosidase or lactase, is a commercially important enzyme with a well-established use in the dairy and food industries. Its occurrence in nature is quite diverse and it has been found in plants, animals, and microorganisms (1). β-Galactosidase hydrolyzes β-D-galactopyranosides such as lactose, but it also catalyzes transgalactosylation reactions: for example, lactose can serve both as galactosyl donor and as acceptor to yield di-, tri-, or higher oligosaccharides (2). Lactose hydrolysis products, lactose-reduced or lactose-free dairy products, have the advantage of improved quality as they are more easily digested by customers who are lactose intolerant. In whey, lactose hydrolysis offers additional product utilization of this abundant renewable resource which is still considered a waste and discarded frequently. The resulting sweeter product (lactose syrup) has numerous applications in the food and dairy industry. In addition, the undesirable crystallization of lactose in certain products such as ice cream is avoided by its hydrolysis (1,3,4).

Oligosaccharides produced by the reaction of transferase activity of β-galactosidases during lactose hydrolysis, the so-called galactooligosaccharides (GalOS), were already reported in the early 1950s when they were considered undesired byproducts (5). These GalOS typically contain a galactosyl-galactose chain with a terminal glucose residue. They are currently used as a low caloric sweetener in food applications, or as pharmaceutical compounds. More recently, GalOS, also known as "Bifidus growth factor," have been used as a food ingredient beneficial to human health (3,6).

The application of β-galactosidases was described both for soluble enzymes, which are normally used in batch processes, or for immobilized preparations, which predominate in continuous processes. Many different solid supports and immobilization strategies have been developed, including entrapment, crosslinking, adsorption, and covalent attachment. The major industrial application of immobilized β-galactosidase has been for the hydrolysis of lactose or for the production of GalOS, thereby improving the functional properties of dairy products. Immobilization is an attractive tool for obtaining an enzyme preparation that can be easily recovered and used in continuous operations, thus making possible the efficient use of an enzyme in an industrial application (1,7).

This work describes the characterization of the moderately thermophilic β-galactosidase from *Talaromyces thermophilus* immobilized onto Eupergit C and its comparison with the free enzyme. We also investigated the performance of this enzyme preparation for lactose hydrolysis and GalOS formation.

Methods

Materials

T. thermophilus CBS 236.58 (Centraalbureau voor Schimmelcultures, Utrecht, the Netherlands) was used as the source of β-galactosidase. β-Galactosidase solutions were produced by the moderately thermophilic fungus and subsequently purified according to Nakkharat (8). Eupergit C was obtained from Röhm (Darmstadt, Germany). o-Nitrophenyl-β-D-galactopyranoside (oNPG) and lactose were purchased from Sigma (Deisenhofen, Germany).

Immobilization of β-Galactosidase

The immobilization of β-galactosidase was carried out at room temperature (approx 25°C) for 96 h. The soluble enzyme (70 U), which previously had been purified to apparent homogeneity of a specific activity of 68 U/mg, was incubated with 580 mg of Eupergit C in a total volume of 3 mL. After the selected time, both the supernatant and the support were analyzed for enzyme activity. The immobilized enzyme preparation was used without further treatment and stored at 4°C. The amount of enzyme that binds to the matrix (U_{bound}) was calculated as the difference of the total activity (U_{tot}) used for immobilization and the remaining activity measured in the supernatant (U_{sup}) after immobilization. The recovery of activity is calculated as the ratio of $(U_{tot}-U_{sup})/U_{tot}$. The binding efficiency, η, relates the measured value of the immobilized preparation (U_{imm}) to the value expected from the difference in the activity of the free enzyme before and after the immobilization (U_{imm}/U_{bound}).

Enzyme Assay

The activity of β-galactosidase was determined by using the artificial substrate oNPG. An appropriate amount of soluble or immobilized enzyme (approx 1 mg) was mixed with 22 mM oNPG dissolved in 500 µL of 50 mM Na-phosphate buffer, pH 6.5. After incubation at 40°C for 15 min, the reaction was stopped by adding 750 µL of 0.4 M Na$_2$CO$_3$. The absorbance was measured at 420 nm and the concentration of o-nitrophenol (oNP) released was calculated using a standard curve. One unit of β-galactosidase activity is defined as the amount of enzyme that releases 1 µmol of oNP from oNPG/min under the experimental conditions described above. When using lactose as a substrate, 600 mM lactose in 50 mM Na-phosphate buffer, pH 6.5 was incubated under the same condition as above, and the reaction was stopped by heating at 95°C for 10 min. Lactose hydrolysis was monitored by measuring the amount of glucose released using the glucose oxidase assay (9). One unit of β-galactosidase activity refers to 1 µmol of glucose released per minute under the reaction conditions selected.

Determination of Optimum Temperature, pH, and Kinetic Parameters

Optimum temperature and pH were determined by changing individually the condition of the standard β-galactosidase assay (varying the temperature from 35°C to 60°C, and the pH from 4.5 to 8.0). The same amount of free and immobilized enzyme (in terms of protein) was used to determine the kinetic parameters by measuring the enzyme activity with varying substrate concentrations. Kinetic constants were calculated using Lineweaver-Burk plots and nonlinear regression (SigmaPlot 2000; SPSS, Chicago, IL).

Stability

The immobilized enzyme was incubated at various temperatures ranging from 35°C to 60°C. The half-lifetime of the activity was estimated using plots of $\ln(a/a_o)$ as functions of time at the respective temperature, where, a and a_o are the activity at time t and initial activity of the enzyme, respectively.

Lactose Hydrolysis

An immobilized enzyme preparation (2 $U_{lactose}$) was incubated with 20% lactose in Na-phosphate buffer pH 6.5 at 40°C and 1200 rpm using an Eppendorf Thermomixer (Munich, Germany). Aliquots were taken at various times and the enzymatic reaction stopped by a 10-min incubation at 95°C. The sugars were analyzed by capillary electrophoresis (CE) as previously described *(9)*.

Results

Immobilization of β-Galactosidase

Eupergit C consists of macromolecular beads of an acrylic polymer and is a suitable carrier for the immobilization of industrial enzymes by covalent attachment. Because of its mechanical stability, which was described as "highly reactor-compatible," it can be employed in most commonly used reactor types, including stirred tank or fixed bed reactors *(10)*. β-Galactosidase from *T. thermophilus* was covalently attached to Eupergit C with a high binding efficiency η of 95% and 75% recovery of the enzyme activity. In long-term stability tests of this immobilized enzyme preparation, enzyme activity was completely retained when stored for 16 d at 4°C.

pH and Temperature Dependence of Enzyme Activity

The activity-pH profiles of the free and the immobilized enzyme were determined for both substrates, oNPG and lactose (Fig. 1). The optimum pH of the free and the immobilized enzyme were found to be 5.5 and 6.0, respectively, when using lactose as a substrate, and 6.0 for oNPG hydrolysis. Figure 2 shows the temperature optimum determined for both preparations

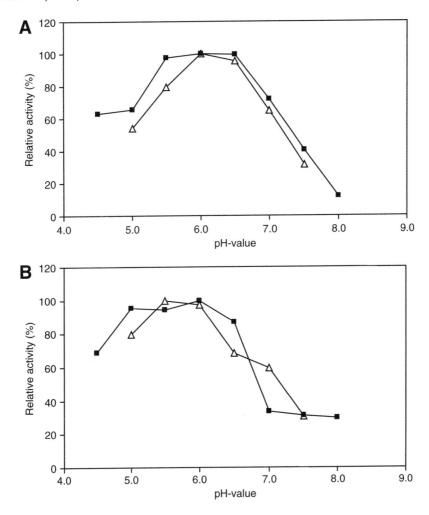

Fig. 1. pH optimum of free (Δ) and immobilized (■) β-galactosidase from *T. thermophilus*. **(A)** oNPG hydrolysis; **(B)** lactose hydrolysis. Activity tests were carried out in 50 mM Na phosphate buffer using the following reaction condition: reaction time, 15 min; *T*, 40°C.

for the 15-min assay. Immobilizing the enzyme increases its temperature optimum from 50°C to 60°C and from 45°C to 50°C for lactose and oNPG as the substrate, respectively. The thermostability of β-galactosidase for both the free and the immobilized preparation was evaluated by determining the half-lifetime of the activity at various temperatures. The plot of half-lifetime as a function of temperature is shown in Fig. 3. The half-lifetimes of activity for the free and immobilized enzyme at 50°C were estimated to be approx 8 and 27 h, respectively.

Kinetic Parameters

The kinetic parameters K_m and V_{max} of the free and immobilized β-galactosidase were determined for the two substrates, oNPG and lactose.

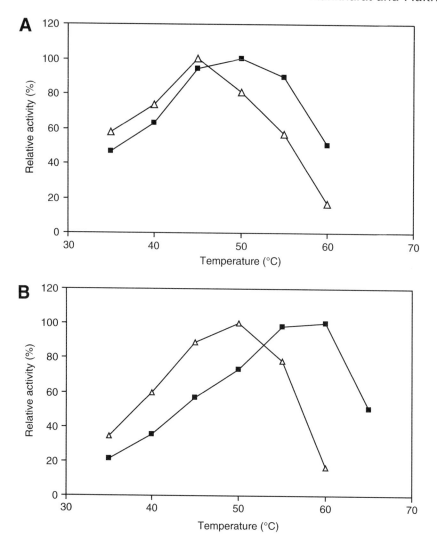

Fig. 2. Temperature optimum of free (△) and immobilized (■) β-galactosidase from *T. thermophilus*. **(A)** oNPG hydrolysis; **(B)** lactose hydrolysis. Activity tests were carried out in 50 mM Na phosphate buffer using the following reaction condition: reaction time, 15 min; pH, 6.5.

In addition, different concentrations of the end products, glucose and galactose, were tested for a possible inhibiting effect. Results are shown in Table 1. When measuring kinetic constants for β-galactosidase immobilized onto Eupergit C, it was found that V_{max} was lowered to approx 150 µmol/min/mg for oNPG as a substrate whereas the K_m-value was decreased approximately threefold to 10 mM. In contrast, the K_m for lactose remained almost unaltered whereas V_{max} even increased to some extent on immobilization. As was evident from this kinetic analysis, the hydrolysis of both oNPG and lactose is competitively inhibited by glucose and galactose for both enzyme preparations. Glucose was found to be a stronger competitive

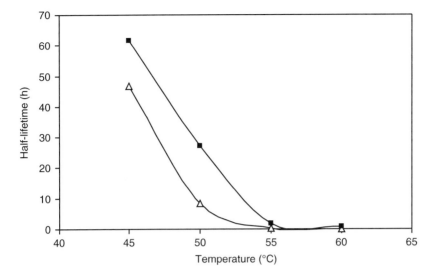

Fig. 3. Half-lifetime of activity of the free (△) and immobilized (■) β-galactosidase from *T. thermophilus*. Activity tests were carried out in 50 m*M* Na phosphate buffer using the following reaction condition: reaction time, 15 min; *T*, 40°C; pH, 6.5.

Table 1
Kinetic Parameters of Free and Immobilized β-Galactosidase

	Substrate	Inhibitor	V_{max} (µmol/min/mg)	K_m (m*M*)	K_i (m*M*)
Free	oNPG	D-glucose	450	35	66
		D-galactose	400	30	370
	lactose	D-galactose	95	19	420
Immobilized	oNPG	D-glucose	150	11	58
		D-galactose	146	10	251
	lactose	D-galactose	145	21	355

Activity tests were carried out in 50 m*M* Na phosphate buffer using the following reaction condition: reaction time, 15 min; *T*, 40°C; pH, 6.5.

inhibitor of oNPG and lactose hydrolysis for both enzyme preparations than galactose, yet based on the kinetic analysis and the inhibition constants K_i calculated, β-galactosidase is only moderately inhibited by both reaction end products.

Lactose Hydrolysis and GalOS Formation

Immobilized β-galactosidase from *T. thermophilus* was able to catalyze transgalactosylation reactions with lactose as substrate. As was analyzed by CE, not only the main hydrolysis products glucose and galactose but

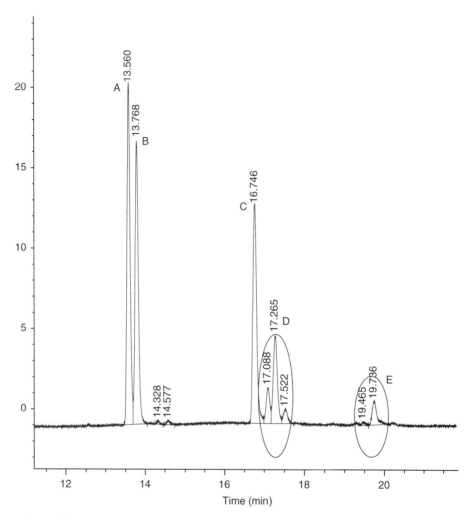

Fig. 4. CE chromatogram of lactose converted by immobilized β-galactosidase from *T. thermophilus* on Eupergit C: glucose (**A**), galactose (**B**), lactose (**C**), disaccharides (**D**), and trisaccharides (**E**). Reaction condition for the transformation: 20% initial lactose condition, immobilized β-galactosidase activity 2 $U_{lactose}$, 70% lactose conversion.

also GalOS are formed during incubation of the enzyme with increased concentrations of lactose. Analysis of product mixtures obtained at a lactose conversion of 70–80% revealed the formation of several GalOS with different nonlactose disaccharides and trisaccharides being the main reaction products (Fig. 4). Figure 5 shows the time-course of lactose hydrolysis and GalOS formation by immobilized enzyme when using 20% lactose as the initial substrate concentration. The maximum amount of GalOS obtained after 24 h incubation was approx 60 mg/mL, and this amount was maintained until the end of the incubation period. Approximately 80% of the lactose was hydrolyzed when the GalOS yield reached its maximum.

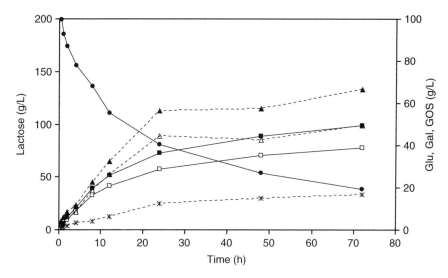

Fig. 5. Time-course of lactose hydrolysis as analyzed by CE: lactose (●), glucose (Glu) (■), galactose (Gal) (□), and total GalOS (▲) including disaccharides (△) and trisaccharides (*). Reaction condition for the transformation: 20% initial lactose condition, reaction time 72 h, immobilized β-galactosidase activity 2 $U_{lactose}$.

Discussion

The aim of the present study was to determine properties of a novel β-galactosidase from the moderately thermophilic fungus *T. thermophilus* when covalently immobilized onto Eupergit C, and to compare these properties to those of the free enzyme, for example, in terms of pH and temperature optimum, thermal stability and kinetic parameters. Recently, soluble β-galactosidase from *T. thermophilus* was characterized and found to exhibit some properties that make this novel enzyme attractive for applications in food technology *(8)*. The immobilized enzyme exhibited a shift of the optimal pH by about 0.5 units toward the alkaline side when compared with soluble β-galactosidase. Similar results were also reported for the immobilization of cyclodextrin glucosyltransferase on Eupergit C *(11)*. The immobilized β-galactosidase also showed a slightly broader pH optimum profile for both substrates as compared to the free enzyme. In addition, the covalent immobilization of β-galactosidase on Eupergit C significantly improves both the temperature optimum and stability as is expressed by the increase in the half-lifetime of activity at 50°C from 8 to 27 h. This improved stability is important for long-term applications of the enzyme, for example, when using it in continuous reactor configurations.

The comparison of the kinetic parameters for oNPG and lactose hydrolysis catalyzed by immobilized and soluble enzyme preparations revealed different effects on substrate turnover and binding as expressed by the values V_{max} and K_m. When using oNPG as the substrate, V_{max} was lowered for the immobilized enzyme. Furthermore, K_m also decreased significantly when oNPG was the substrate indicating improved binding in the active site. In contrast, the K_m values for lactose hydrolysis by free and immobilized β-galactosidase are identical, whereas V_{max} was slightly increased. These results indicate that accessibility to the active site of the enzyme is probably not limited by the covalent attachment to the carrier, for example, by increased diffusional resistance. A possible explanation for these altered kinetic constants could be that immobilization results in slight changes of the conformation of the enzyme and the active site, which causes these differences in binding and/or turnover of the two different β-galactosidase substrates. It is well known that the end products of lactose hydrolysis, glucose and galactose, can have a severe inhibitory effect on β-galactosidases from various sources. Kinetic analysis of oNPG and lactose hydrolysis showed that *T. thermophilus* β-galactosidase is inhibited by both of these products, albeit this inhibition is only moderate as is obvious from the high values of K_i/K_m calculated for the immobilized enzyme preparations. These values were found to be 5.3 and 25 for oNPG hydrolysis, and glucose and galactose, respectively. These values indicate that lactose hydrolysis by β-galactosidase from *T. thermophilus* is only negligibly inhibited by its reaction products under operational conditions.

Eupergit C-immobilized β-galactosidase could be conveniently used for lactose hydrolysis as well as for GalOS formation. The maximum yield of GalOS of approx 34% was obtained when the degree of lactose conversion was approx 80%. From the time-course of hydrolysis (Fig. 5) it is obvious that the galactose concentration is constantly lower than that of glucose during conversion experiments. The glucose-to-galactose ratio during lactose hydrolysis can be used as a good measurement for GalOS formation as galactose moieties are transferred to a suitable acceptor such as another sugar molecule including lactose or even the reaction product galactose instead of being released as the free monosaccharide *(9)*. Maximum values for the glucose-to-galactose ratio of 1.26 were found over a broad range of substrate conversion (50–80%). The main oligosaccharide products that were qualitatively identified by CE are disaccharides and trisaccharides (Fig. 4). This together with the high glucose-to-galactose ratio indicates that galactosyl moieties are not only transferred to lactose, resulting in the formation of galactosyl-lactose, but also to other galactose molecules, and that the main transgalactosylation products are various positional isomers of galactosyl-galactose.

In conclusion, immobilization of *T. thermophilus* β-galactosidase onto Eupergit C is characterized by a high binding efficiency and a significant improvement of the properties of this enzyme with the most notable effect being on thermostability. Because of this increased thermostability the immobilized enzyme preparation can be used more efficiently than the soluble enzyme, for example, when using different continuous reactor configurations. The immobilized enzyme cannot only be used for lactose hydrolysis but also for the formation of prebiotic GalOS, compounds that are of great interest for food and feed applications because of proven and presumed beneficial effects on health and well-being (3).

Acknowledgment

We want to thank the Austrian Academic Exchange Service ÖAD for a Technology Grant to PN and the European Commission for financing of this research (AUNP Programme, project no. 13).

References

1. Gekas, V. and López-Leiva, M. (1985), *Process Biochem.* **20**, 2–12.
2. Prenosil, J. E., Stuker, E. and Bourne, J. R. (1987), *Biotechnol. Bioeng.* **30**, 1026–1031.
3. Nakayama, T. and Amachi, T. (1999), In: *Encylopedia of Bioprocess Technology: Fermentation, Biocatalysis and Bioseparationm,* Flickinger, M. C., Drew S. W. (eds.), Wiley, New York, Vol 3, pp. 1291–1305.
4. Greenberg, N. A. and Mahoney, R. R. (1981), *Process Biochem.* **16**, 2–8.
5. Aronson, M. (1952), *Arch. Biochem. Biophys.* **39**, 370–378.
6. Mahoney, R. R. (1998), *Food Chem.* **63**, 147–154.
7. Mahoney, R. R. (1985), In: *Developments in Dairy Chemistry,* Fox, P. F. (ed.), Elsevier Applied Science, Amsterdam, The Netherlands, Vol 3, pp. 69–108.
8. Nakkharat, P. and Haltrich, D. (2005), *J. Biotechnol.,* in press.
9. Petzelbauer, I., Nidetzky, B., Haltrich, D. and Kulbe, K. D. (1999), *Biotechnol. Bioeng.* **65**, 322–332.
10. Katchalski-Katzir, E. and Krämer, D. M. (2000), *J. Mol. Catal. B Enzym.* **10**, 157–176.
11. Martín, M. T., Plou, F. J., Alcalde, M., and Ballesteros, A. (2003), *J. Mol. Catal. B Enzym.* **21**, 299–308.

Evaluation of Cell Recycle on *Thermomyces lanuginosus* Xylanase A Production by *Pichia pastoris* GS 115

VERÔNICA FERREIRA,[1] PATRICIA C. NOLASCO,[1] ALINE M. CASTRO,[1] JULIANA N. C. SILVA,[1] ALEXANDRE S. SANTOS,[1] MÔNICA C. T. DAMASO,[2] AND NEI PEREIRA, JR.*,[1]

[1]*Escola de Química—Universidade Federal do Rio de Janeiro, Caixa Postal 68542, CEP 21945-970, Rio de Janeiro, RJ, Brazil, E-mail: nei@eq.ufrj.br;* and [2]*Embrapa Agroindústria de Alimentos—Laboratório de Processos Fermentativos, Rio de Janeiro, RJ, Brazil*

Abstract

This work aims to evaluate cell recycle of a recombinant strain of *Pichia pastoris* GS115 on the Xylanase A (XynA) production of *Thermomyces lanuginosus* IOC-4145 in submerged fermentation. Fed-batch processes were carried out with methanol feeding at each 12 h and recycling cell at 24, 48, and 72 h. Additionally, the influence of the initial cell concentration was investigated. XynA production was not decreased with the recycling time, during four cell recycles, using an initial cell concentration of 2.5 g/L. The maximum activity was 14,050 U/L obtained in 24 h of expression. However, when the initial cell concentration of 0.25 g/L was investigated, the enzymatic activity was reduced by 30 and 75% after the third and fourth cycles, respectively. Finally, it could be concluded that the initial cell concentration influenced the process performance and the interval of cell recycle affected enzymatic production.

Index Entries: Xylanase A; *Pichia pastoris*; cell reutilization; heterologous expression.

Introduction

Xylanases are becoming one of the major groups of industrial enzymes, finding significant application in the paper and pulp industry, food and animal feed industries, and textile industry *(1)*. In the pulp and paper industry, the commitment to remove chlorine from the production of the pulps and subsequently to completely eliminate chlorine compounds, producing totally chlorine-free (TCF) pulps, require the study and optimization of new bleaching sequences. The use of xylanases as bleaching

*Author to whom all correspondence and reprint requests should be addressed.

boosters is a recent application of biotechnology to the paper industry (2). The positive results using these enzymes, which significantly increase fiber bleachability, seem to be a consequence of the enzymatic removal of xylan from the fiber surface, which facilitates chemical bleaching of lignin-derived substances. In the food industry, xylanases are used to accelerate the baking of cookies, cakes, crackers, and other foods by helping to break down polysaccharides in the dough. In the animal feed industry, xylanases improve the digestibility of wheat by poultry and swine, by decreasing the feed viscosity (2,3).

Xylan is the second most abundant biopolymer after cellulose and the major hemicellulosic polysaccharide found in plant cell walls. Endo-β-1, 4-D-xylan xylanohydrolase, generally called xylanases, are the key enzymes, because they depolymerize the backbone by cleaving the β-(1,4) glycosidic bonds between the D-xylose residues in the main chain to produce short xylooligosaccharides (4).

Xylanases are produced by several microorganisms mainly by fungi. Damaso (5) demonstrated that *Thermomyces lanuginosus* IOC 4145 secreted cellulase-free xylanase in submerged and solid-state fermentations using corncob as substrate, obtaining the best results with the latter mode of operation (6,7). However, the scale-up of solid-state fermentation is not completely efficient, as a result of the process problems, such as mass and heat transfer limitations (7). Additionally, the enhanced enzyme production and the industrial use of thermostable xylanases and cellulase-free preparations could be facilitated using a heterologous expression system that produces large amounts of secreted protein with an organism that can be grown in an industrial scale fermentor (7).

The methylotrophic yeast *Pichia pastoris* has been developed to be an outstanding host for the production of foreign protein because its alcohol oxidase promoter was isolated and cloned (7). In comparison with other eukaryotic expression systems, it offers many advantages. *P. pastoris* can utilize methanol as a carbon source in the absence of glucose. Its expression system uses the methanol-induced alcohol oxidase (AOX1) promoter, which controls the gene that codes the expression of alcohol oxidase, the enzyme which catalyses the first step in the metabolism of methanol. The most important features of the system are that proteins produced in *P. pastoris* are typically folded correctly and secreted into the medium (8).

Our group has first cloned and expressed the gene of *T. lanuginosus* IOC-4145 in the expression system of *P. pastoris* GS 115, with the aim to produce cellulase-free XynA by submerged fermentation (5). This strategy was adopted to facilitate the bioprocess monitoring when compared with solid state fermentation, as well as its downstream separation, because the recombinant yeast produces fundamentally the enzyme of interest (8).

This work aims to evaluate cell recycle of a strain of *P. pastoris* GS115 bearing the *xilanase A* gene of *T. lanuginosus* IOC-4145, by submerged fermentation. Additionally, the influence of the initial cell concentration and the feeding interval were investigated on the enzymatic production, to evaluate the strain stability to repeated batch operations.

Materials and Methods

Chemicals

All chemicals utilized were of analytical grade. Birchwood xylan and biotin were obtained from Sigma-Aldrich. Methanol and yeast nitrogen base (YNB) were purchased from Merck and Difco, respectively.

Organism and Growth Conditions

P. pastoris GS115 containing xylanase gene was obtained as reported by Damaso *(6)*. *P. pastoris* was maintained in minimal dextrose (MD) medium (in w/v: 1% glucose, 4×10^{-5}% biotin, 1.34% YNB, and 1.5% agar) for 4 d at 30°C and stocked at 4°C. For cell growth, cells were cultured in 1000-mL conical flasks containing 200 mL of buffered minimal glycerol (BMG) medium (100 mM potassium phosphate buffer pH 6.0 in w/v: 1.34% YNB and 4×10^{-5}% biotin in v/v: 1% glycerol), at 30°C and 250 rpm, until a cell concentration of about 9 g/L was reached. Cells were then harvested by centrifugation at 7800g for 18 min and ressuspended in expression medium.

Expression Medium

After centrifugation, cells were ressuspended in 20 mL of buffered minimal methanol (BMM) medium (100 mM potassium phosphate buffer pH 6.0 in w/v: 1.34% YNB and 4×10^{-5} % biotin, in v/v: 0.5% methanol) with an initial cell concentration of 0.25 or 2.5 g/L and incubated in 125-mL conical flasks at 30°C and 250 rpm (throw = 5 cm). At each 12 h, the pH was adjusted to 6.0 and methanol was added to a final concentration of 0.5%. The conditions of all experiments are shown in Table 1. The experiments were carried out in duplicate.

Enzyme Assays

Xylanase activity was assayed with Birchwood xylan as substrate. The solution of xylan (1% w/v) pH 6.0 and the enzyme at appropriated dilution were incubated at 75°C for 3 min, and the reducing sugars were determined by the Somogyi-Nelson procedure *(9)* by measuring absorbance at 540 nm. One unit of enzyme activity was defined as the amount of enzyme capable of releasing 1 µmol of reducing sugars (expressed as xylose) per minute under the assay conditions.

Protein Assays

Protein concentration was measured by the method of Lowry with bovine serum albumin as standard *(10,11)*.

Table 1
Experimental Conditions Carried Out in 125-mL Conic Flasks for Recombinant
XynA Production by *P. pastoris* GS115

Parameters	LCC 24 h	HCC 24 h	HCC 48 h	HCC 72 h
Interval of cell recycle (h)	24	24	48	72
Initial cell concentration (g/L)	0.25	2.50	2.50	2.50
Number of cycles evaluated	5	4	4	3

LCC, lower cell concentration; HCC, higher cell concentration.

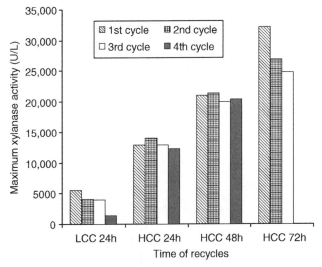

(LCC, lower cell concentration; HCC, higher cell concentration)

Fig. 1. Maximum XynA activity at the end of each cycle.

Results and Discussion

Aiming at investigating the effect of the inoculum size on the production of XynA by *P. pastoris* GS115, two initial cell concentrations were evaluated as 0.25 and 2.5 g/L. In the experiment carried out with X_o of 0.25 g/L and an interval cell recycle of 24 h, the cell growth was maintained for 96 h (four cycles). After that, *P. pastoris* stopped growing, and the maximum xylanase activity reached was 5550 U/L, in the first cycle. However, XynA production was not constant. There was a reduction of approx 30 and 75% in the maximum xylanase activity in the third and fourth cycles, respectively, in relation to the first one (Fig. 1), which corresponded to a decrease of approx 27 and 34% in the yeast productive capacity (Fig. 2).

When X_o of 2.5 g/L was evaluated on the production of XynA, the cell growth was maintained throughout all cycles with a 24 h interval cell recycles (Fig. 3A). The maximum xylanase activity (14,050 U/L) was achieved in the second cycle, decreasing only 10% in the following cycle (Fig. 4A). Comparing these results with those obtained with X_o of 0.25 g/L it was

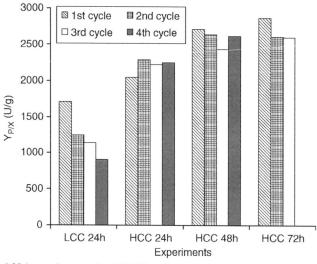

Fig. 2. Productive capacity of *P. pastoris* GS 115 throughout the cycles.

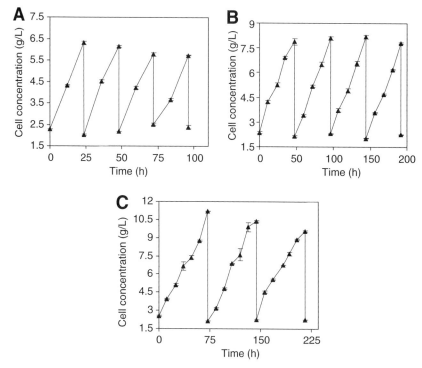

Fig. 3. Growth of *P. pastoris* GS115 in BMM medium throughout the cycles. Cell recycles with intervals of: **(A)** 24 h, **(B)** 48 h, and **(C)** 72 h.

possible to conclude that the initial cell concentration influenced the bioprocess performance. Therefore, $X_o = 2.5$ g/L was chosen for the next experiments. In addition, all experiments performed with $X_o = 2.5$ g/L did

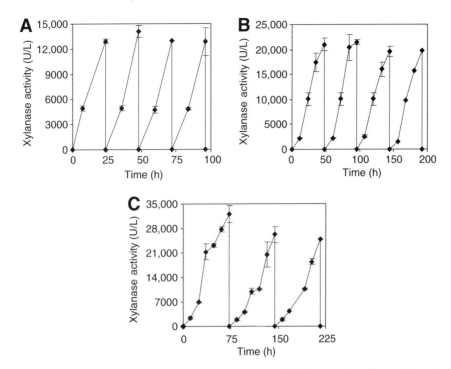

Fig. 4. XynA production by recombinant *P. pastoris* GS115 with cell recycling intervals of **(A)** 24 h, **(B)** 48 h, and **(C)** 72 h.

not have significant decrease in the productive capacity, as was observed when $X_o = 0.25$ g/L was evaluated (Fig. 2).

After selecting the most appropriate initial cell concentration, the influence of the interval of cell recycle on the XynA production was assayed. Additionally, the maximum number of recycles that maintain *P. pastoris* stability was determined. In the experiments with cell recycle at each 48 h, the maximum xylanase activity was obtained at the end of the second cycle was 21,388 U/L (Fig. 4B). Thus, it was concluded that four recycles are the limit when 48-h interval was adopted.

In the experiments carried out with a 72 h interval cell recycle, the xylanase activity did not remain constant, with a maximum value of 32,140 U at the end of the first cycle, and a reduction of 19 and 23% in the following cycles (Fig. 4C). The cell growth showed decreases of 8 and 15% in the second and third cycles, respectively (Fig. 3C). However, the productive capacity ($Y_{P/X}$) had a reduction of only 10%.

The loss of the capacity of growth and consequently of production, observed after three or four cell recycles, may be associated with the phenomenon of senescence, which is a consequence of termination of replication and is therefore intimately linked to cell division and hence the cell cycle, as pointed out by Powell (*12*).

The response parameters of each experiment are displayed in Table 2. It can be observed that volumetric productivity (Q_P) and yield of product

Table 2
Comparison of the Process Parameters Between the Experiments
With Different Initial Cell Concentration and Cell Recycle Times
on the XynA Production of *T. lanuginosus* IOC-4145 by *P. pastoris* GS115

Parameters	LCC 24 h	HCC 24 h	HCC 48 h	HCC 72 h
Maximum enzymatic activity (U/L)	5550 ± 10	14,050 ± 700	21,020 ± 1290	32,140 ± 2370
Q_P (U/Lh)^a	151.6 ± 88	549.6 ± 24	425.2 ± 19.2	385.1 ± 54
$Y_{X/S}$ (g/g)^a	0.429 ± 0.04	0.473 ± 0.05	0.371 ± 0.02	0.339 ± 0.03
$Y_{P/X}$ (U/g)^a	1129 ± 594	3551 ± 329	3563 ± 339	3437 ± 262
$Y_{P/S}$ (U/g)^a	459.8 ± 267	1667 ± 72.6	1290 ± 58.5	1167 ± 165.2
Total time (h)	96	96	196	216

^{a}Average values of the parameters for all cycles and their standard deviation.
LCC, lower cell concentration; HCC, higher cell concentration.

(xylanase activity) in relation to substrate consumed ($Y_{P/S}$) showed considerable differences between experiments with initial cell concentrations of 0.25 and 2.50 g/L, independent of the recycling.

The average values of the volumetric productivity (Q_P) at the end of all cycles for each experiment were 549.6 , 425.2 , and 385.1 ULh for cycles of 24, 48, and 72 h, respectively (Table 2). Comparing the results of volumetric productivity, yield of cell growth in relation to methanol consumed, yield of product (xylanase activity) in relation to cell growth and yield of product in relation to methanol consumed (Table 2), it is possible to conclude that a 24 h interval cell recycle, using X_o of 2.5 g/L, seems to be the best choice for XynA production. These results suggest that the continuous fed batch operation should be the next step to be adopted for XynA production.

Acknowledgment

This work was supported by CNPq and CAPES (Brazilian Councils for Research and Post-graduation Studies).

References

1. Subramaniyan, S. and Prema, P. (2002), *Crit. Rev. Biotechnol.* **22(1),** 33–64.
2. Li, K., Azadi, P., Collins, R., Tolan, J., Kim, J. S., and Eriksson, K. L. (2000), *Enzyme Microb. Technol.* **27,** 89–94.
3. Roncero, M. B., Torres, A. L., Colom, J. F., and Vidal, T. (2000), *Process Biochem.* **36,** 45–50.
4. Berrin, J. G., Williamson, G., Puigserver, A., Chaix, J. C., McLauchlan, W. R., and Juge, N. (2000), *Protein Expr. Purif.* **19,** 179–187.
5. Damaso, M. C. T., Almeida, M. S., Kurtenbach, E., et al. (2003), *Appl. Environ. Microbiol.* **69(10),** 6064–6072.
6. Mitchell, D. A., Krieger, N., Stuart, D. M., and Pandey, A. (2000), *Process Biochem.* **35,** 1211–1225.

7. Chen, Y., Cino, J., Hart, G., Freedman, D., White, C., and Komives, E. A. (1997), *Process Biochem.* **32(2),** 107–111.
8. Cregg, J. M., Vedvick, T. S., and Raschke, W. C. (1993), *Bioresour. Technol.* **11,** 905–910.
9. Somogyi, M. (1952), *J. Biol. Chem.* **195,** 19–23.
10. Lowry, O. H., Rosebrough, N. J., Farr, A. L., and Randall, R. J. (1951), *J. Biol. Chem.* **193,** 256–75.
11. Damaso, M. C. T., Castro, A. M., Castro, R. M., Andrade, C. M. M. C., and Pereira, Jr. (2004), *Appl. Biochem. Biotechnol.* **113–116,** 1003–1012.
12. Powell, C. D., Zandycke, S. M. V., Quain, D. E., and Smart, K. A. (2000), *Microbiology* **146,** 1023–1034.

Evaluation of Solid and Submerged Fermentations for the Production of Cyclodextrin Glycosyltransferase by *Paenibacillus campinasensis* H69-3 and Characterization of Crude Enzyme

HELOIZA FERREIRA ALVES-PRADO,[1,2] ELENI GOMES,[1] AND ROBERTO DA SILVA*,[1]

[1]UNESP-State University of São Paulo, Biochemistry and Applied Microbiology Laboratory, Rua Cristóvão Colombo no. 2265, 15054-000, São José do Rio Preto, SP, Brazil, E-mail: dasilva@ibilce.unesp.br; and [2]UNESP-State University of São Paulo, Biology Institute, Rio Claro, SP, Brazil

Abstract

Cyclodextrin glycosyltransferase (CGTase) is an enzyme that produces cyclodextrins from starch by an intramolecular transglycosylation reaction. Cyclodextrins have been shown to have a number of applications in the food, cosmetic, pharmaceutical, and chemical industries. In the current study, the production of CGTase by *Paenibacillus campinasensis* strain H69-3 was examined in submerged and solid-state fermentations. *P. campinasensis* strain H69-3 was isolated from the soil, which grows at 45°C, and is a Gram-variable bacterium. Different substrate sources such as wheat bran, soybean bran, soybean extract, cassava solid residue, cassava starch, corn starch, and other combinations were used in the enzyme production. CGTase activity was highest in submerged fermentations with the greatest production observed at 48–72 h. The physical and chemical properties of CGTase were determined from the crude enzyme produced from submerged fermentations. The optimum temperature was found to be 70–75°C, and the activity was stable at 55°C for 1 h. The enzyme displayed two optimum pH values, 5.5 and 9.0 and was found to be stable between a pH of 4.5 and 11.0.

Index Entries: Cyclodextrin glycosyltransferase; *Paenibacillus campinasensis*; submerged fermentation; solid-state fermentation.

Introduction

Cyclodextrin glycosyltransferase (CGTase; EC 2.4.1.19) is an enzyme used for cyclodextrin production from starch by an intramolecular

---

*Author to whom all correspondence and reprint requests should be addressed.

transglycosylation reaction. Cyclodextrins (CDs) are cyclic malto-oligosaccharides made up of D-glucose residues linked by α-1,4 bonds. The most common CDs are α-CD, β-CD, and γ-CD types, containing six, seven and eight D-glucose residues, respectively *(1–3)*. CD cavities have hydrophobic interiors and hydrophilic exteriors, which allows hydrophobic molecules or the hydrophobic portion of a molecule to become included within the inner cavity and form an inclusion complex. Because of their ability to form inclusion complexes with organic molecules, CDs and their derivatives have become increasingly useful in pharmaceutical, analytical chemistry, agricultural, cosmetics, food, and biotechnology applications. CDs can be used to capture flavors or odors, to stabilize volatile compounds, to improve the solubility of hydrophobic substances and to protect substances against undesirable modifications *(1–6)*.

A number of microorganisms produce extracellular CGTases. This production occurs mainly through members of the genus *Bacillus*, especially aerobic alkalophilic types *(7–19)*. Other reported mesophilic and thermophilic CGTase producers include: *Klebsiella* sp. *(20,21)*, *Brevibacterium* sp. *(22)*, and *Paenibacillus* sp. *(23)*, *Thermoanaerbacter* sp. *(24,25)*, *Thermoanaerobacterium* sp. *(26)* and hyperthermophilic archae-bacteria *Thermococcus* sp. *(27)*.

The CGTase production is typically conducted by submerged fermentation (SmF) utilizing starch as carbon source and other nutrients. However, solid-state fermentation (SSF) has been exploited extensively on the industrial scale in recent years. The growing conditions in SSF approximate the natural habitat of some microorganisms more closely than those in liquid culture, so these microorganisms are able to grow and excrete large quantities of enzyme *(28,29)*. Both types of fermentation present advantages and disadvantages, the choice will depend on operational limitations, microbial performance and its enzyme production. In most literature reports, the CGTase productions were conducted in SmF with mesophilic microorganisms. Horikoshi *(30)* and Starnes *(24)* reported that industrial CGTase production became possible because of the high level of the enzyme produced by a mesophilic alkalophilic *Bacillus* sp. Mesophilic microorganisms usually produce enzymes with low-thermal stability. In the CD production process thermally stable α-amylase is added during the liquefaction step which is carried out at temperatures between 95°C and 105°C. Thermally stable CGTases would make it possible to increase the temperature for CGTase action, and decrease the cost of CD production. The high production costs of CGTase and CDs are considered a limiting factor for CD application on an industrial scale. Research into the reduction of CGTase production costs is important to enable the economic commercial scale use of CDs, and finding a thermophilic CGTase producing microorganism with high-thermal stability is of commercial interest.

This study reports the isolation of a novel CGTase from *Paenibacillus campinasensis* strain H69-3 isolated from a soil sample in São José do Rio Preto, SP, Brazil. The production of CGTase using submerged and SSF was studied, and the properties of the crude CGTase determined.

Material and Methods

Medium

The selective medium used as reported by Park et al. *(31)* was made up of soluble starch 10 g/L, peptone 5 g/L, yeast extract 5 g/L, K_2HPO_4 1 g/L, $MgSO_4 \cdot 7H_2O$ 0.2 g/L, Na_2CO_3 10 g/L (separately sterilized), phenolphthalein 0.3 g/L, orange methyl 0.1 g/L, and agar 15 g/L, pH 10.0. The colonies surrounded by a yellowish halo were selected for examination on a medium made up of the same ingredients without phenolphthalein, orange methyl and agar *(7)*. Stock cultures were maintained in this culture medium with the addiction of agar 15 g/L at 4°C in a slant tube (maintenance medium) *(8)*.

Microorganism Isolation

Approximately 1 g of shaded soils from market-garden and crop soils were added to flasks with 5-mL culture medium. The samples were incubated at 45°C for 24 h and then inoculated on Petri dishes containing selective medium for CGTase detection. The plates were incubated at 45°C for a period of 144 h. Positive colonies for CGTase enzyme were transferred to tubes containing maintenance medium.

Submerged Fermentation

Two methods based on the culture medium proposed by Nakamura and Horikoshi *(7)*, were used to analyze the SmFs. The culture medium was made up of soluble starch 10 g/L, peptone 5 g/L, yeast extract 5 g/L, K_2HPO_4 1 g/L, $MgSO_4 \cdot 7H_2O$ 0.2 g/L, and Na_2CO_3 5 g/L (separately sterilized), pH 9.6. For the first method, peptone and yeast extract were not included in the culture medium, exchanging the soluble starch for other carbon sources at 10 g/L concentration. These sources were: cassava bran, corn starch, cassava starch, corn flour, wheat bran, soy bran, soy extract, soy, and soy flour. The second method used the same culture medium concentration was only soluble starch was substituted for sweet potato starch, corn starch, cassava starch, cassava bran, and corn flour. The culture medium proposed by Nakamura and Horikoshi *(7)* was used as the enzyme production control.

The preinoculum was prepared using 125-mL Erlenmeyer flasks containing 20 mL of the culture medium and a microbial mass produced in an agar slant tube. Microbial growth was carried out using a rotary shaker at 45°C and 150 cycles/min for 24 h. SmFs were carried out in 125-mL

Erlenmeyer flasks containing 20 mL of culture medium for CGTase production analysis and cellular growth. Shake flasks were inoculated with 0.5-mL cellular suspension obtained from a 24-h preinoculum. The Erlenmeyer flasks were incubated in a rotary shaker at 45°C and 150 cycles/min, for 96 h. After fermentation, the volume of all flasks was centrifuged at 10,000g at 5°C for 15 min. The supernatant liquid, free of cells, was utilized to determine enzymatic activities. The centrifuged cell mass was washed with saline solution and the microbial biomass was quantified by absorbance at 640 nm. All experiments were performed once.

Solid-State Fermentation

Various substrates, such as wheat bran, soy bran, soy flour, corn flour, cassava bran, cassava starch, corn starch, and soluble starch, were used to measure CGTase production in SSFs. In the previous article (32), describing CGTase production from *Bacillus* sp., it was shown that the use of a mixture, made up of a substrate and wheat bran, presented better results than the use of only one substrate each time. Thus, a mixture was prepared, with a 1:1 proportion, with each substrate and wheat bran. These media were prepared using 250-mL Erlenmeyer flasks containing a total of 5 g of solid material (2.5 g of wheat bran and 2.5 g of another substrate). After sterilization, this solid material was incubated with 2.5 mL of preinoculum and 2.5 mL of Na_2CO_3 solution 0.5%. The initial humidity was 70%. The flasks were incubated in an oven at 45°C and in accordance with the literature; SSF lasted 72 h (29,32,33). Growth was detected through visualization and the typical smell. The enzyme was extracted by adding 20 mL of distilled water to the solid media. These solid media were gently cleaved using a glass rod, and then shaken at 100 cycles/min for 1 h at room temperature. The suspension was filtered under vacuum using Whatman filter paper followed by centrifugation at 10,000g at 5°C for 15 min. The enzymatic activities were determined using the supernatant liquid.

Enzymatic Assay Determination

Two methods were used for CGTase determination because CD formation involves both dextrinization and cyclization. The iodine method was applied to determine dextrinization whereas the phenolphthalein method was applied to determine CD formation. Dextrinizing was determined in accordance with Fuwa (34) and Pongasawasdi and Yagisawa (35) with slight modifications. A sample of 0.1 mL of appropriately diluted enzyme was added to 0.3 mL of 0.5% soluble starch prepared in 0.1 M acetate buffer, pH 5.5 and incubated at 55°C for 10 min. The enzyme reaction was stopped by the addition of 4 mL of 0.2 M HCl solution. Then, 0.5 mL of iodine solution prepared with 0.03% I and 0.3 KI was added to the reaction mixture. The absorbance was measured at 700 nm and a decrease in absorbance was verified, when compared with a control tube with heat inactivated enzyme.

One unit of enzyme activity was defined as the quantity of enzyme that reduces the blue color of starch–iodine complex by 10%/min.

The method of Mäkelä et al. *(36)* was slightly modified to determine the cyclization reaction. A sample of 0.1 mL of appropriately diluted enzyme was added to 0.8 mL of 1% soluble starch prepared in 0.1 M acetate buffer, pH 5.5 and incubated at 55°C for 10 min. The enzyme reaction was stopped by the addition of 4 mL of 0.25 M Na_2CO_3 solution and 0.1 mL of 1 mM phenolphthalein solution was added to the reaction mixture. The absorbance was measured at 550 nm and a decrease in absorbance was compared with a control tube with inactive enzyme at 100°C for 30 min. One unit of enzyme activity was defined as the amount of enzyme that produced 1 µmol of β-CD/min.

Protease Activity

For the protease activity determination, a mixture made up of 0.2 mL of enzyme solution and 0.8 mL of the 1% (p/v) casein solution was prepared in acetate buffer 50 mM, pH 5.6. After incubation at 37°C for 60 min, the reaction was stopped by the addition of 1 mL of 10% trichloroacetic acid (TCA). The mixture was centrifuged at 6000g for 5 min and the absorbance of the supernatant was determined at 280 nm. A control tube was prepared by adding the enzymatic solution to the reaction mixture after the addition of TCA solution. One protease activity unit was defined as the quantity of enzyme needed to produce 1 µmol of tyrosine/min *(37)*.

Protein Determination

Protein concentration was estimated according to the Hartree-Lowry method, using bovine-serum albumin as standard *(38)*.

Results and Discussion

Microorganism Isolation

Eighty strains of CGTase producers were isolated. Only three strains presented high-CGTase activity and grew at 45°C. These strains were isolated from the soil in shady sites in cassava crop fields, in São José do Rio Preto. The strains were designated H69-2, H69-3, and H69-5 *(32)*. Strain H69-3 presented the highest CGTase activity (Table 1) and was selected for additional characterization.

The strain designated H69-3 was identified by the *Centro Pluridisciplinar de Pesquisas Químicas, Biológicas e Agrícolas* (CPQBA, Unicamp, Campinas, SP, Brazil). The partial sequence of 16S rRNA revealed 99% homology with the 16S rRNA sequence of *P. campinasensis* strain 324 *(39)*. Based on these results the strain was classified as *P. campinasensis* strain H69-3. The 16S rRNA sequence has been deposited in the Gene Bank database under the accession number DQ153080.

Table 1
CGTase Production After 60 h of Fermentation and Origin of Isolated Strains

Strains	H69-2	H69-3	H69-5
Soil crop	Cassava	Cassava	Cassava
Dextrinizing activity (U/mL)	42.3	85.75	73.15
Phenolphthalein activity (U/mL)	0.81	1.34	0.88

Submerged Fermentations

SmFs were evaluated using two different media compositions. The first consisted of medium without peptone and yeast extract using different substrates. The culture medium of Nakamura and Horikoshi (7) containing yeast extract and peptone was used as a control. It was observed that the microorganism presented a protein content and CGTase activity lower than the control medium in which, after the 48-h fermentation, there was significant protein content and CGTase production. Because of the lack of peptone and yeast extract in the culture medium, there was no significant biomass production, indirectly observed by the protein content, for all media used. Because of this fact, fermentation was extended to 96 h, with no change in the results observed. For each assay, the mean values of enzymatic activities are shown in Fig. 1.

According to Fig. 1, CGTase production was lower compared with the control medium. The medium with wheat bran showed the most significant CGTase production, followed by the medium with soy bran. Surprisingly, CGTase activity was not detected by the phenolphthalein method probably owing to low-enzyme activity. Based on these results, it was inferred that peptone and yeast extract are important nutrients for microbial growth and CGTase production. It is likely that the supply of nitrogen and/or amino acids by the substrates was insufficient for the complete development of the *P. campinasensis* strain H69-3.

For the second experiment, only the starch source was changed, keeping the usual concentrations of peptone and yeast extract. With this composition, growth was complete, showing good CGTase production (Fig. 2). In this case, it was possible to determine significant CGTase activity using both methods for activity determination. The most significant CGTase production was obtained for the medium containing cassava bran as a substrate. The medium with cassava starch also presented significant activity, higher than the control medium. These two substrates have the same vegetal origin, the cassava plant. As the microorganism was isolated from soil of the cassava crop, we can infer that the natural habitat of the microorganism may be a good indicator of the kind of enzyme that it is able to produce. This result also suggests the nutritional importance of yeast extract and peptone for CGTase production of the *P. campinasensis* strain H69-3 studied under SmF.

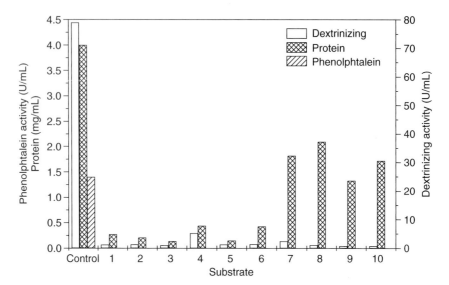

Fig. 1. Cyclodextrin glycosyltransferase production from *Paenibacillus campinasensis* strain H69-3 in submerged fermentation after 96 h. The culture media was without peptone and yeast extract. The control medium was based on Nakamura and Horikoshi (7). Substrates: soluble starch (1), cassava bran (2), cassava starch (3), wheat bran (4), corn starch (5), corn flour (6), soy bran (7), soy extract (8), soy (9), and soy flour (10).

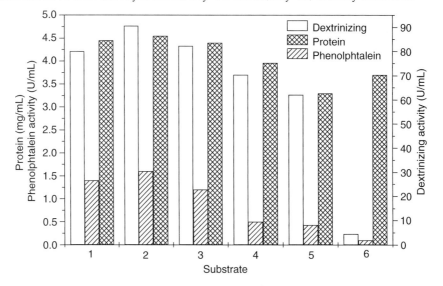

Fig. 2. Cyclodextrin glycosyltransferase production obtained from *Paenibacillus campinasensis* strain H69-3 using the submerged fermentation after 48-h cultivation. Substrate: soluble starch (1), cassava bran (2), cassava starch (3), sweet potato starch (4), corn starch (5), and corn flour (6).

Solid-State Fermentation

For the composition of culture media, the use of traditional commercial sources of carbon (C) and nitrogen (N) leads to high costs for industrial enzyme production *(32,33,40,41)*. So research that investigates enzyme

Characterization of Crude Enzyme

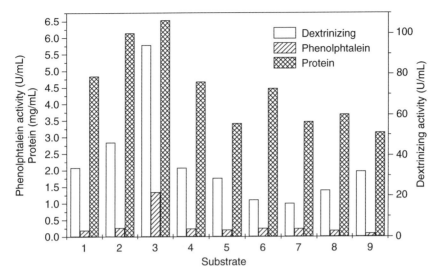

Fig. 3. Cyclodextrin glycosyltransferase production from *Paenibacillus campinasensis* strain H69-3 using solid-state fermentation after 72-h cultivation. Substrate: wheat bran (1), soy bran (2), wheat bran + soy bran (3), wheat bran + soy flour (4), wheat bran + corn flour (5), wheat bran + cassava bran (6), wheat bran + cassava starch (7), wheat bran + corn starch (8), and wheat bran + soluble starch (9).

production using low-cost agroindustrial residues is important to make the use of commercial enzymes economically feasible. In this case, commercial sources of carbon and nitrogen were replaced by low-cost agroindustrial residues (3,7,28,33). Nine different media compositions with solid substrates were analyzed and all demonstrated microbial growth, measured as protein content. Only the medium made up of wheat bran and soy bran showed CGTase production similar to the levels obtained for SmF. The other media presented a CGTase production lower than the medium mentioned. Figure 3 shows that if the soy bran or wheat bran substrates are used separately, the resulting enzymatic activity is lower than the activity obtained with a mixture of these substrates. In this case, the enzymatic activity is increased about two- or threefold. Thus, it can be inferred that some nutrients of wheat bran or soy bran would not be available to the microorganisms if both bran types were not mixed. Few studies have reported CGTase production obtained from solid substrates. This is probably related to the fact that CGTase producing microorganisms belong to the bacilli gender and SSF is a more common characteristic of filamentous fungi. SSF is generally defined as the growth of microorganisms on solid material in the absence or near-absence of free water, which is physiologically more favorable to the cultivation of filamentous fungi (42). Even so, Ramakrishna (29) and Alves-Prado (32) obtained CGTase activity, using SSF, similar to activities using SmF. These researchers used mesophilic species: *Bacillus cereus* and *Bacillus* sp. subgroup *alkalophilus*, respectively.

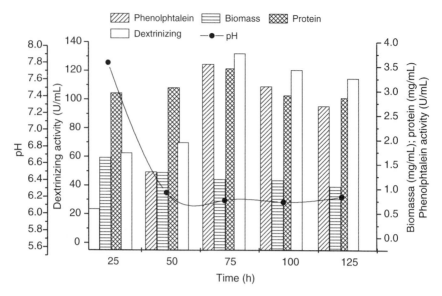

Fig. 4. Cyclodextrin glycosyltransferase production from *Paenibacillus campinasensis* strain H69-3 using a control medium based on Nakamura and Horikoshi *(7)*.

Protease activity was analyzed using all media compositions shown in this article and detected in almost all solid-state media (data not shown). A possible explanation for this detection result is the high level of protein available in these media. There was also protease activity for the second composition of SmF and the highest production level was obtained by using cassava bran which, coincidentally, also presented the largest CGTase activity level. Because of protease presence during CGTase production, the time of these fermentations must not be extended because protease production can interfere with CGTase production. Ramakrishna *(29)* noticed a reduction in CGTase activity that was dependent on the culture time of the media analyzed in their article, suggesting that this reduction was an effect of protease.

Time-Course of Enzyme Production in SmF

Figure 4 shows the CGTase activity, protein content, biomass, and pH variation produced by *P. campinasensis* strain H69-3 as a function of fermentation time, using the control medium proposed by Nakamura and Horikoshi *(7)*. In earlier studies (data not shown), it was indicated that *P. campinasensis* strain H69-3 grew best in 0.5% Na_2CO_3 and at an initial pH of 9.6. During fermentation, the pH of the medium was lowered and stabilized at pH 6.0. The CGTase producers reported in the literature usually show maximum production at approx 40–60 h of fermentation *(3)* and the results shown here are in agreement with these studies. The CGTase activity determined by the dextrinizing and phenolphthalein methods showed an increase during growth in the exponential phase, which occurred in the first 24 h. CGTase activity peaks were (132 U/mL) and (3.5 U/mL) for

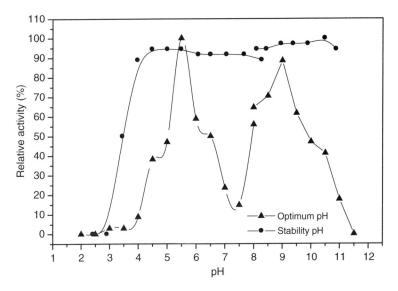

Fig. 5. Effects of pH on CGTase activity (—▲—) and CGTase stability (—●—) from *Paenibacillus campinasensis* strain H69-3. The buffers used were: MacIlvaine (pH 2.5–8.0) and glycine-NaOH (pH 8.0–11.5).

dextrinizing and phenolphthalein methods, respectively, during 72-h fermentation. After these peaks, CGTase activity slowed down and stabilized after approx 80 h of fermentation, coinciding with the growth in the stationary phase. Other authors show that *Bacillus cereus*, *Bacillus stearothermophilus* ET1 and other alkalophilic *Bacillus* sp. CGTase producers reveal a CGTase production peak during the growth stationary phase, but for *B. cereus* this peak occurs after 16–20 h of fermentation. This was probably owing to a high concentration of the initial inoculum at 5% (3,7,18,43).

Effects of pH on Activity and Stability of Enzyme

For characterization of the CGTase from *P. campinasensis* strain H69-3, crude CGTase was produced under SmF for 50 h using the methods of Nakamura and Horikoshi (7). The crude CGTase was characterized through the dextrinizing method under standard activity conditions. The enzyme activity was measured at varying pH values ranging from 2.5 to 11.5 at 70°C. Optimal activity was seen at two peaks of optimum pH 5.5 and 9.0 (Fig. 5), a characteristic also shown by other CGTases. This may suggest that there are two proteins with distinct catalytic activity or that the same enzyme is capable of acting at different pH values. This has also been reported for *Bacillus circulans* no. 79 (44), *Bacillus amiloquefaciens* AL35 (9), *Bacillus firmus* (14–16), alkalophilic *Bacillus* sp. (7,8), and *Thermoanaerobacterium thermosufurigenes* EM1 (26). The pH stability was determined by incubating the crude CGTase at different pH values, ranging from 2.5 to 12.0 for 24 h at 25°C. Then, the residual activity was measured as a standard activity condition. As shown in Fig. 5, the crude CGTase presented a wide range of pH stability from 4.5 to 11.0.

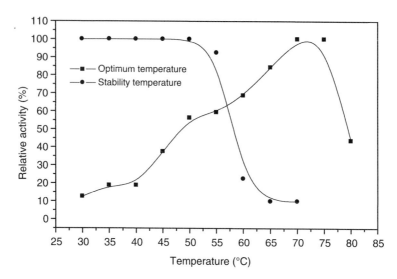

Fig. 6. Temperature effects on Cyclodextrin glycosyltransferase activity and stability from *Paenibacillus campinasensis* strain H69-3. (—■—) optimum temperature; (—●—) thermal stability.

Effect of Temperature on Enzyme Activity and Stability

The activity of crude CGTase was measured at different temperatures, 30–80°C at pH 5.5. Figure 6 shows that CGTase was optimally active at 70–75°C. The effect of temperature on the stability of crude CGTase was also investigated. The enzyme was incubated for 1 h at various temperatures (30–70°C) followed by the measurement of residual activity under standard assay conditions. CGTase activity was maintained up to 55°C, indicating a good thermal stability (Fig. 6).

CGTase from *P. campinasensis* strain H69-3 presented high-optimum temperature and thermal stability. These values are lower than some results obtained with thermophilic microorganisms, for example: *B. stearothermophilus* ET1 *(18)* which was reported to have an optimum temperature of 80°C and to be thermally stable at 60°C, *B. stearothemophilus* R2 *(19)*, with was thermally stable at 70°C after 30-min heat treatment; *Thermoanaerobacter* sp. and *Thermoanaerobacterium thermosufurigenes* EM1, with optimum temperatures of 80°C and 95°C, respectively, and a thermal stability up to 70°C for both enzymes *(25,26)*. However, *P. campinasensis* strain H69-3 showed higher CGTase activity than many other microorganisms reported, indicating a potential industrial application of this CGTase in processes in which thermal stability is required. This enzyme could be used after starch gelatinization without cooling the solution to temperatures lower than 60°C.

Conclusions

The importance of adding yeast extract and peptone to *P. campinasensis* strain H69-3 SmFs for the production of CGTase has been demonstrated. The substrate sources with high-starch contents showed higher CGTase production. *P. campinasensis* strain H69-3 showed good growth using SSF when combined with wheat bran and soy bran. In this case, the CGTase production levels were similar to that obtained for SmF. Crude CGTase activity was stable up to 50°C, and the optimum temperature was in the range of 70–75°C. The optimum pH values were found to be 5.5 and 9.0, and CGTase activity was stable between 6.5 and 10.5. These results suggest that CGTase from the *P. campinasensis* strain H69-3 presents potential for industrial applications.

Acknowledgment

The authors are grateful to the Fundação de Amparo à Pesquisa do Estado de São Paulo (FAPESP) for its financial support.

References

1. Bender, H. (1986), *Adv. Biotechnol. Process* **6**, 31–71.
2. Szejtli, J. (1997), *J. Mater. Chem.* **7**, 575–587.
3. Tonkova, A. (1998), *Enzyme Microb. Technol.* **22**, 678–686.
4. Szejtli, J. (1982), *Starch/Stärke* **34**, 379–385.
5. Pszczola, D. E. (1988), *Food Technol.* January, 96–100.
6. Allegre, M. and Deratani, A. (1994), *Agroo Food Ind. Hi Technol.* January/February, 9–17.
7. Nakamura, N. and Horikoshi, K. (1976), *Agric. Biol. Chem.* **40**, 1785–1791.
8. Nakamura, N. and Horikoshi, K. (1976), *Agric. Biol. Chem.* **40**, 935–941.
9. Yu, E. K. C., Aoki, H., and Misawa, M. (1988), *Appl. Microbiol. Biotechnol.* **28**, 377–379.
10. Tomita, K., Kaneda, M., Kawamura, K., and Nakanishi, K. (1993), *J. Fermen. Bioeng.* **75**, 89–92.
11. Bovetto, L. J., Backer, D. P., Villette, J. R., Sicard, P. J., and Bouquelet, S. J. -L. (1992), *Biotechnol. Appl. Biochem.* **15**, 48–58.
12. Sabioni, J. G. and Park, Y. K. (1992), *Starch/Stärke* **44**, 225–229.
13. Yim, D. G., Sato, H. H., Park, Y. H. E., and Park, Y. K. (1997), *J. Ind. Microbiol. Biotechnol.* **18**, 402–405.
14. Matioli, G., Zanin, G. M., Guimarães, M. F., and Moraes, F. F. (1998), *Appl. Biochem. Biotechnol.* **70–72**, 267–275.
15. Gawande, B. N., Singh, R. K., Chauhan, A. K., Goel, A., and Patkar, A. (1998), *Enzyme Microb. Technol.* **22**, 288–291.
16. Martins, R. F. and Hatti-Kaul, R. (2002), *Enzyme Microb. Technol.* **30**, 116–124.
17. Alves-Prado, H. F., Gomes, E., and DaSilva, R. (2002), *Bol. SBCTA.* **36**, 43–54.
18. Chung, H. J., Yoon, S. H., Lee, M. J., et al. (1998), *J. Agric. Food Chem.* **46**, 952–959.
19. Kabavainova, L., Dobreva, E., and Miteva, V. (1999), *J. Appl. Microbiol.* **86**, 1017–1023.
20. Bender, H. (1977), *Arch. Microbiol.* **111**, 271–281.
21. Gawande, B. N. and Patkar, A. Y. (2001), *Enzyme Microb. Technol.* **28**, 9, 10.
22. Mori, S., Hirose, S., Oya, T., and Kitahata, S. (1994), *Biosc. Biotechnol. Biochem.* **58**, 1968–1972.
23. Larsen, K. L., Duedhal-Olisen, L., Christensen, H. J. S., Mathiesen, F., Pedersen, L. H., and Zimmermann, W. (1998), *Carbohyd. Res.* **310**, 211–219.
24. Starnes, R. L. (1990), *Cereal Foods World* **35**, 1094–1099.
25. Zamost, B. L., Nilsen, H. K., and Starnes, R. L. (1991), *J. Ind. Microbiol.* **8**, 71–82.
26. Wind, R. D., Libl, W., Buitelaar, R. M., et al. (1995), *Appl. Environ. Microbiol.* **61**, 1257–1265.
27. Tachibana, Y., Kuramura, A., Shirasaka, N., et al. (1997), *Appl. Environ. Microbiol.* **65**, 1991–1997.
28. Raimbault, M. (1998), *Eletr. J. Biotechnol.* **1**, 174–188.

29. Ramakrishna, S. V., Saswathi, N., Sheela, R., and Jamuna, R. (1994), *Enzyme Microb. Technol.* **16,** 441–444.
30. Horikoshi, K. (1996), *FEMS Microbiol. Rev.* **18,** 259–270.
31. Park, C. S., Park, K. H., and Kim, S. H. (1989), *Agric. Biol. Chem.* **53,** 1167–1169.
32. Alves-Prado, H. F., Gomes, E., and DaSilva, R. (2002), *Brazilian J. Food Technol.* **5,** 189–196.
33. Dias, A. A. M., Andrade, C. M. M. C., and Linardi, V. R. (1992), *Rev. Microbiol.* **23,** 189–193.
34. Fuwa, H. (1954), *J. Biochem.* **41,** 583–603.
35. Pongsawasdi, P. and Yagisawa, M. (1987), *J. Fermen. Technol.* **65,** 463–467.
36. Mäkelä, M. J., Korpela, T. K., Puisto, J., and Laakso, S. V. (1988), *Agric. Food Chem.* **36,** 83–88.
37. Sarath, G. (1996), In: *Proteolytic Enzymes a Practical Approach.* Beynon, R. J., Bond, J. S., (ed.), Oxford University Press, New York, pp. 25–55.
38. Hartree, E. F. (1972), *Anal. Biochem.* **48,** 422–427.
39. Yoon, J. -H., Yim, D. K., Lee, J. -S., et al. (1998), *Internat. J. System. Bacteriol.* **48,** 833–837.
40. Bailey, J. E. and Ollis, D. F. (1986), *Biochemical Engineering Fundamentals.* 2nd edition, McGraw-Hill International Editions, New York.
41. Damaso, M. C. T., Andrade, C. M. M. C., Pereira, N. Jr, (2000), *Appl. Biochem. Biotechnol.* **84,** 821–834.
42. Pandey, A. (1992), *Process Biochem.* **27,** 109–117.
43. Jamuna, R., Saswathi, N., Sheela, R., and Ramakrishna, S. V. (1993), *Appl. Biochem. Biotechnol.* **43,** 163–176.
44. Salva, T. J. G., Lima, V. B., and Pagan, A. P. (1997), *Rev. Microbiol.* **28,** 157–164.

Effect of β-Cyclodextrin in Artificial Chaperones Assisted Foam Fractionation of Cellulase

Vorakan Burapatana, Aleš Prokop, and Robert D. Tanner*

Chemical Engineering Department, Vanderbilt University, Nashville, TN 37235, E-mail: robert.d.tanner@vanderbilt.edu

Abstract

Foam fractionation has the potential to be a low-cost protein separation process; however, it may cause protein denaturation during the foaming process. In previous work with cellulase, artificial chaperones were integrated into the foam fractionation process in order to reduce the loss of enzymatic activity. In this study, other factors were introduced to further reduce the loss of cellulase activity: type of cyclodextrin, cyclodextrin concentration, dilution ratio cyclodextrin to the foamate and holding time. α-Cyclodextrin was almost as effective as β-cyclodextrin in refolding the foamed cellulase-Cetyltrimethylammonium bromide mixture. β-Cyclodextrin (6.5 mM) was almost as effective as 13 mM β-cyclodextrin in refolding. The dilution ratio, seven parts foamate and three parts β-cyclodextrin solution, was found to be most effective among the three ratios tested (7:3, 1:1, and 3:7). The activity after refolding at this dilution ratio is around 0.14 unit/mL The refolding time study showed that the refolding process was found to be most effective for the short refolding times (within 1 h).

Index Entries: β-cyclodextrin; artificial chaperones; foam fractionation; cellulase; protein denaturation; protein refolding.

Introduction

Foam fractionation is an adsorptive bubble separation process which can be used to concentrate or purify surface active chemicals from a dilute solution. This foam separation technique has been used in wastewater treatment and ore flotation (1). Many studies also show that foam fractionation can be effective in recovering proteins from dilute solution (2–5). Foam fractionation can also recover biosurfactants from a culture broth (6) or even be integrated into a fermentation process to recover the biosurfactant, surfactin, directly from *Bacillus subtilis* broth (7). The low operating cost of foam fractionation makes it an attractive alternative for current protein concentration methods (8). Because foam fractionation works well

*Author to whom all correspondence and reprint requests should be addressed.

with dilute solutions, it is most useful when proteins are in low concentration as in the early stages of a downstream purification process or when coupled with a fermentation process. Foam fractionation usually does not, however, work with proteins that do not form a foam layer when aerated.

Protein denaturation at a gas–liquid interface is a concern when operating a protein foam fractionation process. The hydrophobic and hydrophilic parts of a protein reorient themselves during adsorption onto the surface of a bubble *(9,10)* in such a process, which causes the protein to shift away from its native state. This bubble surface denaturation can result in a loss of enzymatic activity for those proteins which are enzymes. In our previous work, an artificial chaperones system was integrated into the foam fractionation process. Because the system contained a surfactant, along with a detergent-stripping agent, it successfully reduced the amount of denatured cellulase at the end of the process, whereas simultaneously allowing foam to form using cellulase. Reducing the amount of denaturation makes it feasible to concentrate both foaming and nonfoaming enzymes such as cellulases by foam fractionation.

During foam fractionation, surface denaturation occurs because of protein adsorption at gas–liquid interfaces *(11–13)*. Adsorption causes the tertiary structure of proteins to change, and in some cases (e.g., pepsin) the secondary structure can change as well *(14–16)*. Previously, protein denaturation occurring during foam fractionation was minimized by setting the operating conditions to reduce protein adsorption at gas–liquid interfaces *(5,17)*. However, these conditions may not be the best settings strategy for concentrating enzymes since adequate bubble adsorption is a key to high enrichment and mass recovery of proteins. In this article, the effect of several operating variables such as cyclodextrin type, β-cyclodextrin concentration, dilution ratio of β-cyclodextrin to recovered protein and the effect of time on refolding will be evaluated in order to determine an effective combination to achieve the highest cellulase activity following the foam fractionation process.

Materials and Methods

Cellulase from *Trichoderma reesei* (cat. no. C-8546), 3,5-dinitrosalicylic acid (DNS), pluronic F-68, and sodium dodecyl sulfate (SDS) were purchased from Sigma Co. (St. Louis, MO). Bicinchoninic acid (BCA) protein assay reagent, Whatman No. 1 filter paper, Fisher brand 96 well-microplate and β-cyclodextrin were purchased from Fisher Scientific Co. (Pittsburgh, PA). Cetyltrimethylammonium bromide (CTAB) was purchased from Fluka Co. (Buchs, Switzerland).

Foam Fractionation

Batch foam fractionation experiments were carried out in a glass column a small column was used here to minimize the cost of cellulase used (which was $65/g in relatively pure form). The column inside diameter was 2 cm, and the column height was 10 cm as previously described in Burapatana et al. *(18)*. A mixture of cellulase and detergent (pH 5.0; 10 mM phosphate buffer was used as a solvent) was placed in the column, and air from a compressed gas cylinder was introduced continuously through a fritted disk sparger (pore size 40–60 μm) at the bottom of the column. Water loss in the effluent air stream was minimized by humidifying the air before it entered the column. Air continued to flow into the column until no more foam was generated. The produced foam was allowed to collapse into a liquid product (foamate) in the foam collector.

Renaturation of Cellulase After Foam Fractionation

After foam fractionation of a cellulase and CTAB mixture, 350 μL of collected foamate was diluted with 150 μL of 13 mM β-cyclodextrin solution in water. The resulting solution was stored overnight before measuring the cellulase activity.

BCA Assay for Protein Concentration

Foamate samples of 20 μL were placed in a 96 well-microplate in triplicate. Then, 180 μL of BCA reagent *(19)* was added to each well-plate. The microplate was scanned after 30 min at 562 nm for absorbance determination.

Cellulase Activity

The filter paper assay *(20)* was used to determine the cellulase activity. The DNS reducing sugar assay *(21)* was used to measure the amount of sugar produced.

To determine the refolding effectiveness of α-cyclodextrin and γ-cyclodextrin as compared β-cyclodextrin, foamate of a cellulase-CTAB mixture collected from foaming at an air-flow rate of 8 mL/min was diluted with the three cyclodextrins. The concentration of β-cyclodextrin solution initially used was 13 mM. In additional experiments, refolding of the cellulase-CTAB mixture was conducted after dilution with 3.25 and 6.5 mM β-cyclodextrin solutions in order to see if lower concentrations of β-cyclodextrin could be used effectively. The dilution ratio of foamate to β-cyclodextrin was 7:3 initially, but 1:1 and 3:7 were also tested to see if higher refolding could be achieved. To investigate the process dynamics, the dilution ratio of 7:3 was divided into three equal size increments instead of a single increment, with the change in activity after each increment being tracked. Usually, the solution was left overnight after adding cyclodextrin. Because cellulase in solution can degrade over a period of time, the foamate of each cellulase-CTAB mixture (12 mL initial volume, 8 mL/min air-flow rate) was monitored over a period of 12 h following the addition of β-cyclodextrin. This time test was used to determine how long the postcyclodextrin waiting time should be.

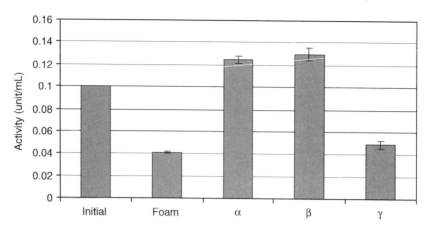

Fig. 1. Cellulase activity before and after foaming and after addition of three types of cyclodextrin to the foamate collected at an air-flow rate of 12 mL/min from a mixture of 200 mg/L cellulase and 100 mg/L CTAB.

Results and Discussion

Effect of Different Cyclodextrins

The artificial chaperones system consists of both a surfactant and a stripping agent, so the effect of different stripping agents on active protein recovery, specifically the type of cyclodextrin, was also investigated in this foam fractionation study. Twelve mililiters of a 200 mg/L cellulase and 100 mg/L CTAB mixture was aerated at an air-flow rate of 12 mL/min. Seven hundred microliters of the resulting foamate was then placed in a 1.5 mL microcentrifuge tube and diluted with 300 µL of a different type of cyclodextrin (α-cyclodextrin, β-cyclodextrin, or γ-cyclodextrin all at 13 mM concentration in water). The solutions were left overnight after adding β-cyclodextrin and the cellulase activities were measured the following day. The activity response results are shown in Fig. 1.

γ-Cyclodextrin was inefficient at renaturing cellulase, as the activity did not improve significantly as compared with the sample after foaming. However, α-cyclodextrin and β-cyclodextrin exhibited similar abilities to renature cellulase and improved the activity by a factor of three as compared with the foamate and even resulting in a higher activity than the initial mixture. This is consistent with another study, which has shown that α-cyclodextrin and β-cyclodextrin renatured chemically denatured pepsin similarly *(22)*.

Effect of Cyclodextrin Dilution Ratio

The dilution ratio of β-cyclodextrin recommended by Rozema *(23)* for artificial chaperones ratio was seven to three (denatured protein-detergent complex to β-cyclodextrin), and this ratio was tested to determine whether this is the optimal dilution ratio for inclusion in a cellulase foam fractionation/refolding system. Twelve mililiters of a 200 mg/L cellulase and 100 mg/L

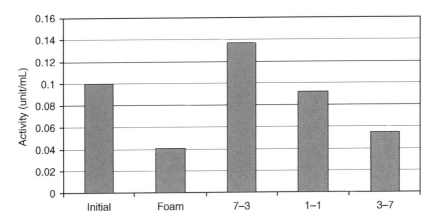

Fig. 2. Effect of the dilution ratio of β-cyclodextrin on cellulase activity before and after foaming and at different dilution ratios after foaming. The foamate was collected at an air-flow rate of 8 mL/min from a mixture of 200 mg/L cellulase and 100 mg/L CTAB.

CTAB mixture was aerated at an air-flow rate of 8 mL/min. For the 7:3 ratio, 700 µL of the foamate was placed in a 1.5 mL microcentrifuge tube and diluted with 300 µL of 13 mM β-cyclodextrin solution. For the 1:1 ratio, 500 µL of the foamate was placed in a 1.5 mL microcentrifuge tube and diluted with 500 µL of β-cyclodextrin solution. To achieve 3:7 ratio, 300 µL of the foamate was placed in a 1.5 mL microcentrifuge tube and diluted with 700 µL of β-cyclodextrin solution. The solutions were left overnight after adding β-cyclodextrin, and the cellulase activity for each was measured the following day. The 7:3 dilution ratio recommended by Rozema (23) for renaturing chemically denatured proteins was the best ratio of the three tested here for renaturing the foamate. Regardless of the dilution ratio, the specific activity remained around 0.27 unit/mg (the initial specific activity was 0.5 unit/mg). The higher the dilution with β-cyclodextrin, the more dilute the final solution and the lower the activity. Results from these trials are shown in Fig. 2.

Effect of Dilution Ratios

Because the original dilution was seven parts foamate to three parts β-cyclodextrin, 700 µL of foamate solution from 200 mg/L cellulase and 100 mg/L CTAB (aerated at an air-flow rate of 12 mL/min) was diluted with three separate 100 µL β-cyclodextrin volumes. The results are displayed in Fig. 3.

Cellulase activity per unit volume of solution increased as soon as the first dilution was made. However, the activity dropped slightly after the second dilution with β-cyclodextrin. When the final dilution was made, the activity increased to 0.11, which was slightly above the initial activity. However, it was still lower than the activity resulting from a single large dilution (the 7:3 ratio in Fig. 2). This could be owing to enzyme degradation, because the solution was left on the lab counter during the 3 h of the experiment.

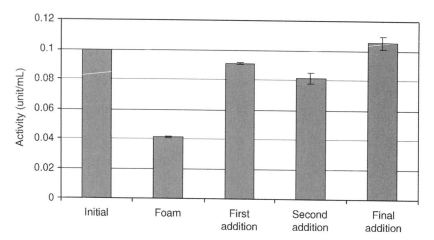

Fig. 3. Incremental addition to refold denatured cellulase, with activity measured before and after foaming and at each step of the dilution process. The foamate was collected at an air-flow rate of 12 mL/min from a mixture of 200 mg/L cellulase and 100 mg/L CTAB. β-cyclodextrin was added in three separate 100 µL volumes.

Effect of Standing Time on Refolding

The reference protocol (23) for investigation of refolding proteins over time suggested leaving the solution overnight to restore the protein activity. As (the physical) protein denaturation in foam fractionation is different than the chemically denatured protein studied in the reference, the time for this process to come to completion might be shorter or longer than specified. Foam fractionation was conducted at an air-flow rate of 8 mL/min for a 200 mg/L cellulase and 100 mg/L CTAB solution. Then, the foamate was diluted with 13 mM β-cyclodextrin (seven parts foamate to three parts β-cyclodextrin). The activity was measured as soon as the dilution was made, then after 4, 8, 12, and 24 h (the filter paper assay took 1 h to complete; therefore, there was effectively a 1-h delay in time measurement past the sampling time). These temperature variation experiments were also run at three different temperatures −4, 25, and 50°C. The results are shown in Fig. 4.

The recovered cellulase activity in the foamate was around 0.04 unit/mL before dilution with β-cyclodextrin, as seen in Fig. 3. After dilution with β-cyclodextrin, a dramatic increase in activity was observed. The recovered activity was highest at around (0.16) for a quadrupling of activity immediately following dilution. Because it takes about 1 h to complete the filter paper test for cellulase activity, refolding may have taken place before the filter paper test was completed (within 1 h) or even "instantaneously." The decrease in cellulase activity in time following the initial data point was probably owing to an unsuitable liquid environment which caused cellulase degradation. Degradation was fastest at 50°C. Keeping the solution in the refrigerator or at room temperature had a similar affect on the change in activity, so it

Effect of β-Cyclodextrin

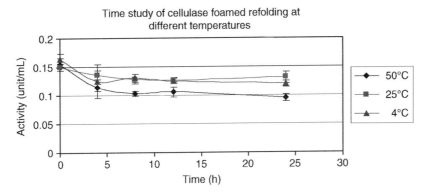

Fig. 4. Refolding of cellulase in the foamate as a function of standing time at three different temperatures. The foamate was collected from 12 mL of a 200 mg/L cellulase and 100 mg/L CTAB mixture aerated at 8 mL/min.

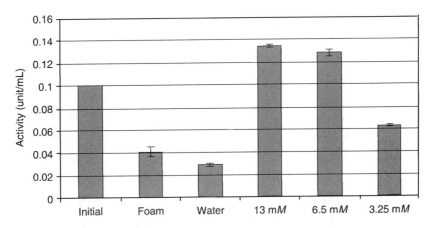

Fig. 5. Effect of β-cyclodextrin concentration on refolding of cellulase in the foamate. The foamate was collected from 12 mL of a 200 mg/L cellulase and 100 mg/L CTAB mixture aerated at 8 mL/min, 300 μL of β-cyclodextrin was added.

would be easier to run this process at room temperature on an industrial scale because it would save money on cooling cost.

Effect of β-Cyclodextrin Concentration

If the refolding β-cyclodextrin concentration used in the refolding process could be reduced, it would lower the cost of this process of concentrating cellulase. The β-cyclodextrin concentration normally used was 13 mM. The foamate solution from a 200 mg/L cellulase and 100 mg/L CTAB mixture that was aerated at 8 mL/min was individually diluted by 300 μL of β-cyclodextrin with different concentrations (13, 6.5, and 3.25 mM) or by deionized water. The dilution ratio was seven parts foamate and three parts β-cyclodextrin, with the results shown in Fig. 5.

Diluting the foamate with deionized water did not refold any of the cellulase, as expected in this control experiment. The activity decreased by the same ratio that the foamate was diluted with deionized water. As for β-cyclodextrin, a concentration of 13 mM worked best, but a 6.5 mM solution worked almost as well, 3.25 mM β-cyclodextrin renatured cellulase but not nearly as well as the other two concentrations. To save on chemical costs, it seems to be feasible to operate at 6.5 mM β-cyclodextrin instead of 13 mM. For industrial purposes, even corn dextrin might be used since it is inexpensive and as linear dextrin is known to successfully refold chemically denatured proteins *(24)*.

Conclusions

Three types of cyclodextrin were studied, and α-cyclodextrin was found to be as effective as β-cyclodextrin. The dilution ratio of cellulase-detergent to β-cyclodextrin was altered, but the specific activity of cellulase did not change for all three dilution ratios investigated. The concentration of β-cyclodextrin could be reduced by half without affecting the efficiency of renaturing the denatured cellulase. Pulse renaturation was applied in the dilution stage of cellulase foam fractionation. A single dilution volume was more effective than three separate dilution volumes (each 1/3 of the single one) for this process. Cellulase refolding was found to exhibit the highest activity immediately after β-cyclodextrin addition and only decreased with time. The refolding process was determined to occur optimally at room temperature.

Acknowledgment

The authors thank the USDA (grant no. 2001-52104-11476) for funding this research.

References

1. Suzuki, A., Yasuhara, K., Seki, H., and Maruyama, H. (2002), *J. Colloid Interface Sci.* **253(2)**:402–408.
2. London, M., Cohen, M., and Hudson, P. B. (1954), *Biochim. et Biophys. Acta* **13**:111–120.
3. Schnepf, R. W. and Gaden, E. L. (1959), *J. Biochem. Microbiol. Technol. Eng.* **1(1)**:1–8.
4. Lockwood, C. E., Bummer, P. M., and Jay, M. (1997), *Pharmaceutical Res.* **14(11)**: 1511–1515.
5. Brown, A. K., Kaul, A., and Varley, J. (1999), *Biotechnol. Bioeng.* **62(3)**:278–290.
6. Davis, D. A., Lynch, H. C., and Varley, J. (2001), *Enzyme Microbial. Technol.* **28(4–5)**: 346–354.
7. Noah, K. S., Fox, S. L., Bruhn, D. F., Thompson, D. N., and Bala, G. A. (2002), *Appl. Biochem. Biotechnol.* **98**:803–813.
8. Uraizee, F. and Narsimhan, G. (1990), *Enzyme Microbial. Technol.* **12(3)**:232–233.
9. Graham, D. E. and Phillips, M. C. (1979), *J. Colloid Interface Sci.* **70(3)**:403–414.
10. Phillips, M. C. (1981), *Food Technol.* **35(1)**:50.
11. Clarkson, J. R., Cui, Z. F., and Darton, R. C. (1999), *J. Colloid Interface Sci.* **215(2)**:323–332.
12. Phillips, L. G., Hawks, S. E., and German, J. B. (1995), *J. Agricul. Food Chem.* **43(3)**: 613–619.

13. Wilde, P. (2000), *Curr. Opin. Colloid Interface Sci.* **5(3-4)**:176–181.
14. Clarkson, J. R., Cui, Z. F., and Darton, R. C. (1999), *J. Colloid Interface Sci.* **215(2)**:333–338.
15. Clark, D. C., Smith, L. J., and Wilson, D. R. (1988), *J. Colloid Interface Sci.* **121(1)**:136–147.
16. Caessens, P., De Jongh, H. H. J., Norde, W., and Gruppen, H. (1999), *Biochim. et Biophys. Acta-Protein Struct. Mol. Enzymol.* **1430(1)**:73–83.
17. Clarkson, J. R., Cui, Z. F., and Darton, R. C. (2000), *Biochem. Eng. J.* **4(2)**:107–114.
18. Burapatana, V., Prokop, A., and Tanner, R. D. (2004), *Appl. Biochem.Biotechnol.* **113-16**: 619–625.
19. Smith, P. K., Krohn, R. I., Hermanson, G. T., et al. (1985), *Analytical Biochem.* **150(1)**: 76–85.
20. Griffin, H. L. (1973), *Analytical Biochem.* **56(2)**:621–625.
21. Rivers, D. B., Gracheck, S. J., Woodford, L. C., and Emert, G. H. (1984), *Biotechnol. Bioeng.* **26(7)**:800–802.
22. Kurganov, B. I. and Topchieva, I. N. (1998), *Biochem. Moscow* **63(4)**:413–419.
23. Rozema, D. and Gellman, S. H. J. (1995), *Am. Chem. Soc.* **117(8)**:2373–2374.
24. Sundari, C. S., Raman, B., and Balasubramanian, D. (1999), *FEBS Lett.* **443(2)**:215–219.

RSM Analysis of the Effects of the Oxygen Transfer Coefficient and Inoculum Size on the Xylitol Production by *Candida guilliermondii*

Mariana Peñuela Vásquez, Maurício Bezerra De Souza, Jr., and Nei Pereira, Jr.*

Departamento de Engenharia Bioquímica, Universidade Federal do Rio de Janeiro, Centro de Tecnologia, Bloco E, Rio de Janeiro, RJ, CEP 21.949-900, Brasil, E-mail: nei@eq.ufrj.br

Abstract

Biotechnology production of xylitol is an excellent alternative to the industrial chemical process for the production of this polyalcohol. In this work the behavior of *Candida guilliermondii* yeast was studied when crucial process variables were modified. The K_La (between 18 and 40/h) and the initial cell mass (between 4 and 10 g) were considered as control variables. A response surface methodology was applied to the experimental design to study the resulting effect when the control variables were modified. A regression model was developed and used to determine an optimal value that was further validated experimentally. The optimal values determined for K_La and X_0 were 32.85/h and 9.86 g, respectively, leading to maximum values for productivity (1.628 g/h) and xylitol yield (0.708 g/g).

Index Entries: Xylitol; response surface methodology analysis; xylose.

Introduction

Xylitol has been the subject of several studies owing to its important physicochemical properties, sweetening power, and anticariogenicity, among other advantages over other synthetic and natural sweeteners *(1)*.

The industrial production of xylitol basically follows a chemical route, which involves many different expensive steps, making unfavorable the competition of xylitol with other sweeteners as sorbitol or sucrose (the cost of production of xylitol is approx 10 times higher than the one for producing sucrose or sorbitol) *(1,2)*. For several reasons, the biotechnological process may be considered a viable alternative when compared with the industrial chemical route for the production of xylitol. It does not demand previous purification of xylose and specific microorganisms can

---

*Author to whom all correspondence and reprint requests should be addressed.

be used to act directly in the conversion of xylose to xylitol, requiring only simple purification steps and, consequently, a reduction of the production costs.

The conversion of xylose to xylitol by *Candida guilliermondii* occurs through a sequence of oxidation–reduction enzymatic reactions. First, xylose is reduced to xylitol by nicotinamide adenine dinucleotide phosphate (NADPH)-linked xylose reductase; then, xylitol can be excreted from the cell or oxidized to xylulose by NAD^+-linked xylitol dehydrogenase. The activity of these enzymes depends on their respective coenzyme concentration: NADPH is regenerated through the pentose-phospate route and NAD^+ is regenerated in the respiratory chain, being directly influenced by the availability of oxygen. Thus, a high concentration of oxygen leads to a higher reoxidation rate of NADH, favoring the actuation of the xylitol dehydrogenase and, consequently, cell growth. However, a condition of unavailability of oxygen would halt the metabolism of the yeast, causing its death. On the other hand, a low concentration of oxygen unbalances the redox reaction and generates the accumulation of xylitol (2–4).

The productivity of the bioconversion of xylitol from xylose using *C. guilliermondii* may be considered low, owing to the long times involved. Hence, studies aiming at the increase of productivity are necessary for this bioprocess to become an industrial reality. Such studies can be developed using experimental design statistical techniques for model analysis and determination of optimal conditions.

The optimization of the xylitol biological production is very important to the process scale-up. The classical method of optimization involves varying the level of one parameter at a time over a certain range, whereas holding the other test variables constant. This strategy is generally time consuming and requires a large number of experiments to be carried out (5). Experimental designs have been employed for the optimization of the cultivation process as they offer the possibility of studying several variables with a reduced number of experiments (6–8).

This work aimed the study of the kinetics of the production of xylitol by *C. guilliermondii* yeast. Experiments were planned and carried out in bioreactor, varying the aeration rates and the initial cell mass. The experimental conditions that lead to maximum productivity and yield were also determined.

Methods

Microorganism

The yeast *C. guilliermondii* IM/UFRJ 50088 was obtained from The Institute of Microbiology of The Federal University of Rio de Janeiro, Brazil. The strain was maintained at 4°C on xylose agar in a medium with the same composition, as follows.

Media Composition

Growth medium: 20 g/L D-xylose, 1.25 g/L urea, 1.1 g/L KH_2PO_4, 1.5 g/L yeast extract. Cultivation medium: 50 g/L D-xylose, 1.25 g/L urea, 1.1 g/L KH_2PO_4, 0.5 g/L yeast extract. Salt and citric acid solution (40 mL/L) were added to both growth and cultivation media (9). The initial pH was adjusted to 6 with HCl or NaOH. D-xylose was sterilized separately from other components to prevent damage to the nutritional qualities of the medium. The sterilization condition, in both cases, was 0.5 atm/15 min.

Inoculum Conditions

Cells were previously activated in a 500-mL conical flask containing 200 mL of working volume of growth medium, which was incubated at 30°C in a rotary shaker at 200 rpm (throw = 5 cm) for 24 h. The inoculum was prepared in 500-mL conical flask with 200 mL of working volume of growth medium, inoculated with 20 mL of activation culture and incubated in similar conditions for 32 h. After cell quantification, the volume required to achieve the inoculum concentration was centrifuged at 8000g for 20 min.

Cultivation Process

Experiments were carried out in a batch bioreactor (1.5 L; BIOFLO III, New Brunswick, USA) with 1.2 L working volume, at 30°C and controlled pH of 6.0. For each initial cell mass value (X_0) different oxygen transfer coefficient (K_La) values were evaluated. The pH and temperature were maintained constant throughout the cultivation. The cultivation was brought to an end when xylose in the medium was exhausted.

Analytical Methods

The samples were withdrawn, each at 3 h. The cells were measured for absorbance at 570 nm and a calibration curve, obtained by the dry weight method, was used.

D-xylose and xylitol were measured by high-performance liquid chromatography–Waters using a Shodex SC1011 ion-exchange column for sugars (300 × 8 mm^2) at 75°C and degasified Milli-Q water as the mobile phase at a flow rate of 0.8 mL/min (7,13).

The oxygen transfer coefficient (K_La) was determined through a correlation (Eq. 1) between K_La (/h), agitation (rpm), and air flow (L/min) developed for the system employed (10), the gassing-out method was used, as suggested by Moser (11).

$$K_La = [0.014 \times Log(Q) + 0.094] \times N \qquad (1)$$

Experimental Design

The K_La and initial cell mass (X_0) were chosen as the most important variables to improve the bioconversion performance. Based on previous studies for the process the others variables were kept constant (3,12,13).

Table 1
Central Composite Design and Experimental Results

Order of experiments	Original variables		Codified variables		$Y_{P/S}$ (g/g)	Q_P (g/h)
	Initial cell mass (g)	K_La (/h)	X_1	X_2		
F1	4	18	−1	−1	0.614	0.523
F2C	7	29	0	0	0.756	1.167
F3C	7	29	0	0	0.761	1.184
F4	4	40	−1	1	0.335	0.731
F5	10	40	1	1	0.514	1.572
F6	10	18	1	−1	0.632	0.442
F7	7	13.4	0	−1.414	0.518	0.391
F8	2.8	29	−1.414	0	0.879	1.428
F9	7	44.5	0	1.414	0.476	1.291
F10	11.2	29	1.414	0	0.639	1.213

C, central point; K_La, volumetric oxygen transfer coefficient; X_1, normalized initial cell mass; X_2, normalized volumetric oxygen transfer coefficient.

The yield of xylitol on xylose consumed ($Y_{P/S}$) and mass productivity (Q_P) are important factors to evaluate the viability to any process. For that reason, these factors were selected to analyze the process and to determine K_La and X_0 conditions that would permit to maximize the bioconversion process within the operational interval experimentally evaluated.

A central composite design was developed. High, intermediate, and low levels of the factors (K_La and X_0) were considered. The experimental design matrix is showed in Table 1 (14). The statistical analysis was performed using the program STATISTICA 6 (15).

Results and Discussion

Table 1 presents the experimental results obtained for each response variable (Q_P and $Y_{P/S}$). The yield of xylitol on xylose consumed ($Y_{P/S}$) is defined as the fraction of xylose that was converted to xylitol (g/g), and mass productivity (Q_P) is defined as xylitol amount produced in time unit (g/h). These values were calculated using the values obtained for xylitol and xylose at the end of each bioconversion.

Modifications in the studied variables (K_La and initial cell mass) induce important changes in the kinetic behavior of the yeast, as pointed out previously. The yeast performance is altered when oxygen supply is high. In this condition it uses substrate only for the cell growth, i.e., the regeneration of NAD$^+$ is increased in the respiratory chain in which oxygen is used as final electron receptor. In this circumstance, the xylitol dehydrogenase activity is stimulated allowing the enhancement in the oxidation rate of xylitol (2,3,12).

When oxygen supply is reduced, the cell growth is limited; xylose is consumed essentially for energy maintenance and xylitol production. The

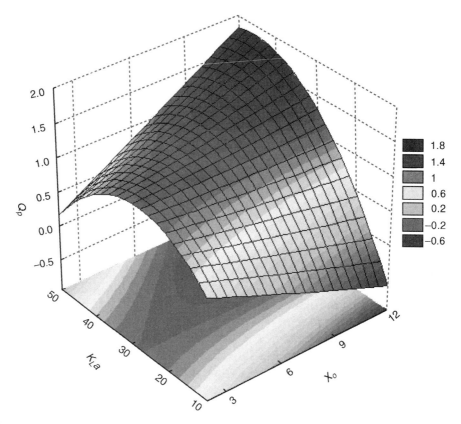

Fig. 1. Response surface for xylitol productivity (Q_P) using $K_L a$ and X_0 as controlled variables.

low-cell-growth rate is owing to a decrease of the regeneration rate of NAD^+ in the respiratory chain, allowing the accumulation of xylitol (2). On the other hand, bioconversion times are longer, lowering the productivity values (F1, F6, and F7).

In the bioconversion coded by F2C, F3C, F5, F8, F9, and F10, an equilibrium between cell mass and available oxygen was observed, allowing a low reoxidation of NADH and consequently, an accumulation of xylitol.

For the best productivity value (1.572 g/h) obtained in F5, the product yield was among the lowest one (0.514 g/g). It is possible to conclude that the amount of the dissolved oxygen for cell (in this cultivation) is still high, making the xylitol oxidation fast and facilitating the cell growth. This cannot be considered an appropriate condition to the synthesis of xylitol because the product yield is still low.

Table 1 shows that, for bioconversions with the same $K_L a$ and different inoculum levels, the yield coefficients do not present substantial changes. For instance, for a $K_L a$ of 29/h and initial cell mass variations between 2.76 and 11.24 g, the productivity and product yield values were approx 1 g/h and 0.7 g/g, respectively. It is possible to conclude that the behavior of the yeast was controlled basically by $K_L a$.

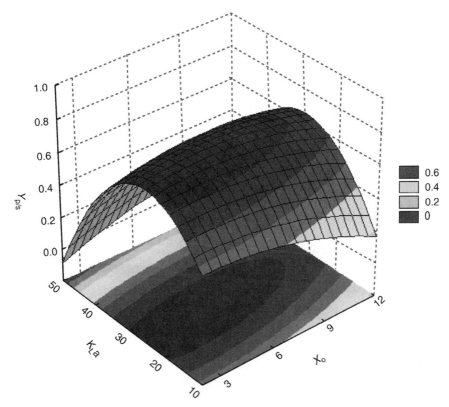

Fig. 2. Response surface for xylitol yield on xylose consumed ($Y_{P/S}$) using $K_L a$ and X_0 as controlled variables.

The following models were obtained for the relation between productivity (Eq. 2) and product yield (Eq. 3) with the controlled values:

$$Q_P = 1.183 + 0.057X_2 + 0.326X_1 - 0.236X_1^2 + 0.230X_1X_2 \qquad (2)$$

$$Y_{P/S} = 0.759 - 0.018X_2 - 0.026X_2^2 - 0.057X_1 - 0.157X_1^2 + 0.040X_2X_1 \qquad (3)$$

where all the coefficients in the equations above present statistical significance (p-level < 0.05). The pure errors were 1.4×10^{-4} and 1.1×10^{-5} for Q_P and $Y_{P/S}$, respectively *(10)*.

When the productivity was the sole response variable considered within the evaluated interval, an ascending profile was found and the maximum value was at the upper extreme point of the interval (Fig. 1).

On the other hand, when the product yield was the only variable considered, an optimal point was determined (0.771 g/g) for $K_L a$ of 26.25/h and an initial cell mass of 5.41 g (Fig. 2). However, it was important to consider the optimization of both response variables together through the *desirability* function *(14,15)*. This function allowed the superposition of product yield and productivity response surfaces ($Y_{P/S}$ and Q_P) and permitted the determination of the optimal process point.

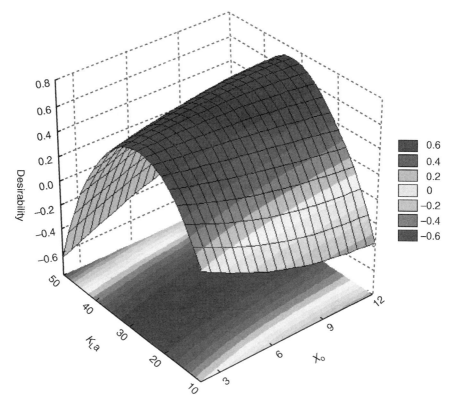

Fig. 3. Response surface for desirability function using $K_L a$ and X_0 as controlled variables.

Statistical Optimization

The multiobjective optimization resources of STATISTICA 6 were used to conjugate the response surface of each response variable and the determination of values which render the optimal point for both productivity and product yield. Figure 3 shows the superposition of product yield and productivity response surface. It displays in the dark region the maximum values of desirability function *(14,15)* (function that represents the superposition of two response variables will be maximized) and $Y_{P/S}$ and Q_P condition which lead to this optimal point.

The optimal point (desirability function = 0.748) corresponds to cell mass value of 9.86 g and $K_L a$ value of 32.85/h. In these conditions the model predicts productivity of 1.40 ± 0.09 g/h and a product yield of 0.70 ± 0.02 g/g, with a confidence level of 95%.

A new experiment was performed assuming the conditions established by the desirability function ($K_L a = 32.85$/h and $X_0 = 9.86$ g) with the aim to validate them. The results obtained are showed in the Fig. 4. The values of productivity (1.628 g/h) and product yield (0.708 g/g) achieved in this cultivation were closed to the ones predicted by desirability function.

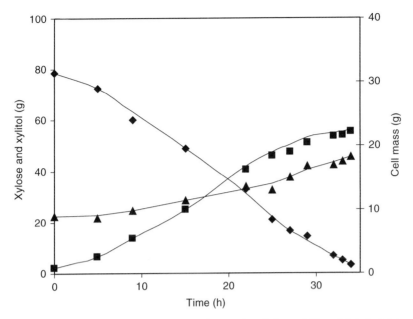

Fig. 4. Kinetics at the optimal condition ($K_L a = 32.85/h$ and $X_0 = 9.86$ g); (-♦-) xylose, (-■-) xylitol, and (-▲-) cell mass.

Conclusion

Experimental verification of optimal point leads to the conclusion that the statistical model can predict efficiently the bioconversion performance by *C. guilliermondii* in the bioreactor at the study conditions. The insertion of statistical tools into the realm of the experimental investigation of the bioconversion of xylose by *C. guilliermondii* allowed a systematic approach, which made possible the optimization of the process.

Acknowledgment

This work was supported by the Brazilian Council for Research and Graduate Studies (CNPq).

References

1. Mussatto, I. S. and e Roberto, I. C. (2002), *Biotecnl. Ciência e Desenvolvimento* **28**, 24–39.
2. Parajó, J. C., Dominguez, H., and Dominguez, J. M. (1998), *Bioresour. Technol.* **65**, 191–201.
3. Gimenes, M. A. P., Carlos, L. C. S., Faria, L. F. F., and Pereira, Jr., N. (2002), *Appl. Biochem. Biotechnol.* **98–100**, 1049–1059.
4. Nolleau, V., Preziosi-Belloy, L., and Navarro J. M. (1995), *Biotechnol. Lett.* **4**, 417–422.
5. Sen, R. and Swaminathan, T. (1997), *Appl. Microbiol. Biotechnol.* **47**, 358–363.
6. Sampaio, F. C., De Faveri, D., Mantovani, H. C., Passos, F. L., Perego, P., and Converti, A. (2005), *J. Food Eng.*, in press.
7. Rodrigues, D. C. G. A., Silva S. S., and Felipe, M. G. A. (1998), *J. Biotechnol.* **62**, 73–77.
8. Silva, C. J. S. M. and Roberto, I. C. (2001), *Proc. Biochem.* **36**, 1119–1124.

9. Du Preez, J. C. and Van Der Walt, J. P. (1983), *Biotechnol. Lett.* **6,** 395–400.
10. Vásquez, M. P. (2004), *M.Sc. Dissertation*. School of Chemistry of the, Federal University of Rio de Janeiro, Rio de Janeiro, Brazil.
11. Moser, A. (1988), *Bioprocess Technol—Kinetics and Reactors*, Springer-Verlag, NY.
12. Aguiar, Jr., W. B., Faria, L. F. F., Couto, M. A. P. G., Araujo, O. Q. F., and Pereira, Jr., N. (2002), *Biochem. Eng. J.* **12,** 49–59.
13. Faria, L. F. F., Gimenes, M. A. P. G., Nobrega, R., and Pereira, Jr., N. (2002), *Appl. Biochem. Biotechnol.* **98–100,** 449–158.
14. Montgomery, D. C. (2001), *Desing Analysis Experiments*. 5th Edition, Wiley, New York.
15. StatSoft, Inc. (2001). STATISTICA (Data analysis software system), version 6, www.statsoft.com.

Enzymatic Synthesis of Sorbitan Methacrylate According to Acyl Donors

GWI-TAEK JEONG,[1] HYE-JIN LEE,[1] HAE-SUNG KIM,[2] AND DON-HEE PARK*,[1,3-6]

[1]*School of Biological Sciences and Technology, Chonnam National University, Gwangju 500-757, Korea, E-mail: dhpark@chonnam.ac.kr;* [2]*Department of Chemical Engineering, Myongji University, Yongin 449-728, Korea;* [3]*Faculty of Applied Chemical Engineering, Chonnam National University, Gwangju 500-757, Korea;* [4]*Research Institute for Catalysis, Chonnam National University, Gwangju 500-757, Korea;* [5]*Biotechnology Research Institute, Chonnam National University, Gwangju 500-757, Korea;* [6]*Institute of Bioindustrial Technology, Chonnam National University, Gwangju 500-757, Korea*

Abstract

Recently, sugar polymers have been considered for use as biomaterials in medical applications. These biomaterials are already used extensively in burn dressings, artificial membranes, and contact lenses. In this study, we investigated the optimum conditions under which the enzymatic synthesis of sorbitan methacrylate can be affected using Novozym 435 in *t*-butanol from sorbitan and several acyl donors (ethyl methacrylate, methyl methacrylate, and vinyl methacrylate). The enzymatic synthesis of sorbitan methacrylate, catalyzed by Novozym 435 in *t*-butanol, reached an approx 68% conversion yield at 50 g/L of 1,4-sorbitan, 5% (w/v) of enzyme content, and a 1:5 molar ratio of sorbitan to ethyl methacrylate, with a reaction time of 36 h. Using methyl methacrylate as the acyl donor, we achieved a conversion yield of approx 78% at 50 g/L of 1,4-sorbitan, 7% (w/v) of enzyme content, at a 1:5 molar ratio, with a reaction time of 36 h. Sorbitan methacrylate synthesis using vinyl methacrylate as the acyl donor was expected to result in a superior conversion yield at 3% (w/v) of enzyme content, and at a molar ratio greater than 1:2.5. Higher molar ratios of acyl donor resulted in more rapid conversion rates. Vinyl methacrylate can be applied to obtain higher yields than are realized when using ethyl methacrylate or methyl methacrylate as acyl donors in esterification reactions catalyzed by Novozym 435 in organic solvents. Enzyme recycling resulted in a drastic reduction in conversion yields.

Index Entries: Immobilized enzyme; bioconversion; optimization; biocatalysis; sorbitan; esterification.

---

*Author to whom all correspondence and reprint requests should be addressed.

Introduction

"Biomaterials" is a generic term, basically referring to materials that are used for a medical purpose, which maintain direct contact with living tissues. Thus, any biomaterial must be carefully and microscopically fabricated, in order to best adjust to a living organism, regarding both functionality and structure. Biomaterials are also used as essential materials in the manufacture of catheters, contact lenses, and artificial human hearts (1,2). Recently, many researchers have noted the myriad of possible applications for sugar-containing polymeric materials, which can be synthesized from sugar esters (3–6). Sugars constitute an attractive group of multifunctional compounds. They are biologically relevant, and they harbor multiple hydroxyl groups. Sugar esters or esters that contain sugar molecules, have been receiving increasing interest, and are already being utilized in a variety of application fields. They have proven their advantage in a host of industries, and have been employed in flavorings, emulsifiers, lubricants, detergents, and cosmetic additives. Sugar esters are biodegradable, biocompatible, and nontoxic (3,7).

Mild reaction conditions and excellent selectivity associated with lipase-catalyzed reactions permit the generation of pure materials by more efficient and environmentally friendly processes than conventional chemical methods (5). Esterification is the principal process by which sugar esters are synthesized. This process has been studied extensively by both chemical and enzymatic processes. The chemical process is associated with low regioselectivity, resulting in poor selectivity, undesirable side reactions, and low yields. However, the enzymatic process can be applied to the regioselective transformations of mono- and disaccharides, and does not result in any undue complications (8). In this study, we opted to use alcoholysis as the applied esterification process (Fig. 1). Alcoholysis is the esterification of an ester (e.g., vinyl methacrylate [VMA]) with the hydroxyl group in the acyl acceptor (e.g., 1,4-sorbitan), and generating another ester and an alcohol as byproducts.

In the enzymatic process used to synthesize sorbitan methacrylate from 1,4-sorbitan, several factors can affect both the conversion yield and the rate of glycosylation. These factors include the reaction solvent, reaction temperature, the type and concentration of the acyl donor, enzyme content, and initial substrate concentration (5,9). The difficulty inherent to the dissolution of both hydrophobic and hydrophilic substrates in a common reaction solvent of low toxicity has constituted the primary limitation of biological synthesis (10). Reaction solvents used for the synthesis of glycosyl methacrylate must be selected for maximal rates and yields of glycosylation, enzyme activity, and the separation/purification of the product. Thus, t-butanol was used as the reaction solvent in all of our experiments, owing primarily to its high degree of substrate solubility, remarkable affinity for glycoside (acyl acceptor), and the ease inherent to the separation/purification of the products, which

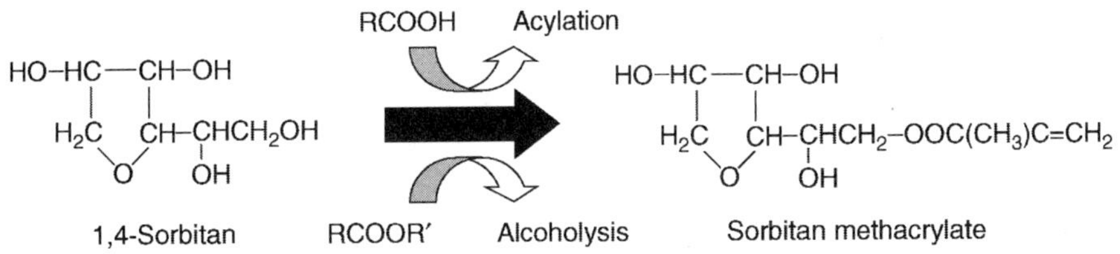

Fig. 1. Enzymatic synthesis of sorbitan methacrylate by acylation and alcoholysis.

is attributable to its low boiling point. In addition, *t*-butanol exhibits a regioselective effect with glycoside (acyl acceptor) *(5,9)*. In our article, *t*-butanol was selected for use in all of our experiments, owing to its ability to achieve a high degree of conversion, as well as the stability of the enzyme as described in previous reports *(5,6)*. Reaction temperature also influences the esterification rate. High reaction temperatures tend to induce enzyme inactivation. The optimum active temperature for Novozym 435 is between 40°C and 60°C. In the sorbitan acrylate *(5)* and β-methlyglucoside acrylate/methacrylate *(6)* esterification experiments, the optimum temperature was determined to be 50°C. In our study, the reaction temperature was set to 50°C in all experiments. We conducted the enzymatic synthesis of sorbitan methacrylate using Novozym 435 (derived from *Candida antarctica*), which is a well-known nonspecific lipase. Novozym 435 facilitates reactions between a wide range of alcohols and vinyl esters, and is a remarkably heat-tolerant enzyme *(11)*.

The objective of this study was to investigate the processes involved with the chemical and enzymatic synthesis of sorbitan methacrylate, which is the basic material used in biocompatible hydrogels.

Materials and Methods

Chemicals

Novozym 435 (lipase B from *C. antarctica*, EC 3.1.1.3), a nonspecific lipase immobilized on a macroporous acrylic resin, with 1–2% water content, and 10,000 propyl laurate units/g, was purchased from the Sigma-Aldrich Chemical Co. (St. Louis, MO). D-Sorbitol and VMA were obtained from the Sigma-Aldrich Chemical Co., USA. Methyl methacrylate (MMA) and ethyl methacrylate (EMA) were commercially obtained from Junsei Chemicals (Kyoto, Japan) and Acros Organics (NJ), respectively. Acetonitrile and methanol were supplied by Fisher Scientific Korea Ltd. (Korea). All other chemicals were of analytical grade, and the solvent used was dried with molecular sieves for 1 d before use.

1,4-Sorbitan Preparation

All dehydration reactions (sorbitol cyclization) for the synthesis of 1,4-sorbitan using *p*-toluenesulfonic acid (*p*-TSA) in a solvent-free process

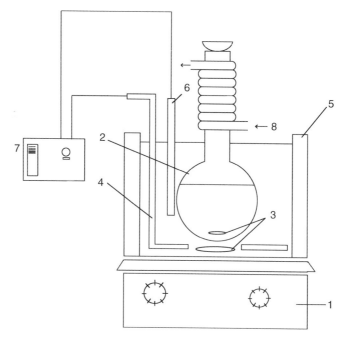

Fig. 2. Schematic diagram of experimental apparatus for the synthesis of 1,4-sorbitan esters. Magnetic stirrer, reactor, magnetic stirrer bar, heating coil, water bath, thermocouple, PID temperature controller, and condenser.

were conducted as previously reported (5). The dehydration reactions were carried out for 2 h at 130 ± 1°C, under 200 mmHg reduced pressure. The reactor volume was 50 mL. The reaction temperature was controlled with an oil bath equipped with a PID (proportional-integral-derivative) temperature controller. Agitation was conducted with a magnetic bar, spinning at approx 200 rpm.

Enzymatic Esterification

In this study, we applied esterification via alcoholysis. Esterification for the synthesis of 1,4-sorbitan esters with immobilized lipase (Novozym 435) was conducted using the apparatus displayed in Fig. 2. Reaction temperature was controlled with a water bath which had been equipped with a PID temperature controller. Mixing was conducted using a magnetic stirrer, spinning at approx 200 rpm. The condenser prevented reactant (t-butanol) evaporation. Results are expressed as the mean value of at least two independent measurements.

In order to determine optimal condition of enzymatic esterification, three parameters—molar ratio, enzyme amount, and initial sorbitan concentration—were investigated as follows: Prepared 1,4-sorbitan was added to the bottle, and either EMA, MMA, or VMA was subsequently added, in order to achieve the desired molar ratios (1,4-sorbitan : acyl donor) and enzyme amounts at 50°C, as shown in Table 1. t-Butanol was also added to the mixtures, in order to ensure a total volume of 20 mL.

Table 1
Experimental Condition of Enzymatic Esterification of Sorbitan Methacrylate

Reaction parameter	Kind of acyl donor	Reaction temperature (°C)	Molar ratio (acyl acceptor/acyl donor)	Enzyme amount (% [w/v])	Initial acyl acceptor concentration (g/L)	Reaction time (h)
Molar ratio (acyl acceptor/acyl donor)	EMA, MMA, VMA	50	1:0.5–1:5	3	50	48
Enzyme amount (% [w/v])	EMA, MMA, VMA	50	1:5	1–7	50	48
Initial acyl acceptor concentration (g/L)	EMA, MMA, VMA	50	1:5	5	30–100	48

During the reactions, 0.2 mL of the samples were withdrawn at set intervals, and monitored via high-performance liquid chromatography (HPLC).

In order to determine the effects of repeated enzyme usage on conversion yield, we added 50 g/L of prepared 1,4-sorbitan to the bottle, and VMA was subsequently added up to a 1:5 molar ratio. Reactions were then initiated by the addition of 5% (w/v) Novozym 435 at 50°C for 48 h. Each experiment for repeated enzyme usage was performed with the same enzyme for 48 h. During this experiment, 0.2 mL of samples were withdrawn at set intervals, and monitored via HPLC.

At least a trace of water is essential for bioconversion via enzyme action. Enzymes exhibit different reactions and selectivity in organic solvents than in water. In our experiments, there was no set initial water amount, with the exception of the water content of the enzymes.

Quantitative Analysis

Enzymatic reactions were monitored via analysis of the conversion yield of 1,4-sorbitan with the selected acyl donors (MMA, EMA, and VMA). Acyl donors were measured by HPLC with an ODS2 column (octadecyl [C_{18}], 5 μm, 120 Å, 250 × 4.6 mm, waters), which was constantly maintained at 35°C. A mixture of acetonitrile, methanol, and water (55:40:5 [v/v/v]) was used as a mobile phase, at a flow rate of 1 mL/min, to measure the acyl donors; 0.2 mL of samples were extracted from the reaction mixture at set intervals during the reaction. Enzymes were removed by sample filtering and appropriate dilution. Then we injected 20 μL of the prepared samples. Detection was conducted at 220 nm with MMA, EMA, and VMA as calibration standards.

Results and Discussion

In comparison with conventional chemical processes, the enzymatic synthesis process is accomplished under milder conditions. Therefore, it carries the advantages of high material stability, low energy cost, high selectivity, and low purification cost. The enzymatic process appears to constitute a favorable alternative synthesis process (4,5). In the enzymatic synthesis of sorbitan methacrylate from 1,4-sorbitan, several factors, including the reaction solvent, reaction temperature, the type and concentration of acyl donor, enzyme content, and initial substrate concentration, can impact both the conversion yield and the glycosylation rate (5).

The chemical synthesis of sorbitan, which was used to affect the enzymatic synthesis of sorbitan methacrylate, was accomplished via sorbitol dehydration. This resulted in a fine product, with minimized byproduct formation at 130°C and 200 mmHg reduced pressure, with 1% (w/w) p-TSA used as a catalyst, and 120 min of reaction time.

Enzymatic Synthesis of Sorbitan Methacrylate

In this study, we performed alcoholysis for the esterification of several acyl donors (EMA, MMA, or VMA) with the hydroxyl group in 1,4-sorbitan (acyl acceptor). The process was designed to generate sorbitan methacrylate and alcohol as byproducts. Esterifications for the enzymatic synthesis of 1,4-sorbitan ester using immobilized lipase (Novozym 435) were performed using MMA, EMA, and VMA as acyl donors, all in *t*-butanol as the organic solvent.

Effect of Molar Ratio

In the glycosylation of sorbitan, which harbors one primary hydroxyl group, only monoacrylate was synthesized, owing to the regioselective effects of the solvent (*t*-butanol). The theoretical molar ratio of acyl donor to sorbitan is 1:1, but in order to maintain a high reaction velocity and in consideration of the reversibility of glycosylation, we believed that the theoretical molar ratio would not prove to be sufficiently low *(5,9)*.

As shown in Fig. 3A, the enzymatic synthesis of sorbitan methacrylate catalyzed by Novozym 435 in *t*-butanol reached a conversion yield of approx 49% at 50 g/L of initial 1,4-sorbitan concentration, 3% (w/v) of enzyme content, a 1:5 sorbitan to EMA molar ratio, a reaction temperature of 50°C, and duration of 36 h, using EMA as the acyl donor. Using MMA as acyl donor (Fig. 3B) sorbitan methacrylate was synthesized at a conversion rate of approx 55%, at 50 g/L of initial 1,4-sorbitan concentration, 3% (w/v) of enzyme content, a 1:5 sorbitan to MMA molar ratio, a reaction temperature of 50°C, and a duration of 36 h. The degree to which esterification occurred was observed to have increased directly with the molar ratio in both cases.

Effect of Enzyme Amount

For an enzymatic process to be economically comparable with the conventional chemical synthesis process the amount of the added enzyme must be minimized. Higher enzyme loadings result in shorter reaction periods but increase operation costs. The process must be optimized concerning both productivity and enzyme loading. Above optimum enzyme concentrations, the equilibrated conversion yields tend to be reduced as the enzyme begins to form complex compounds not only with the acyl donor, which causes glycosylation, but also with sorbitan methacrylate resulting in the hydrolysis of the desired product *(5,9)*.

Enzyme loading were determined on the conversion of 1,4-sorbitan with EMA (or MMA) to sorbitan methacrylate using Novozym 435. As shown in Fig. 4A, the conversion of 1,4-sorbitan to sorbitan methacrylate was 68% at 36 h, and a 5% (w/v) enzyme amount using EMA as the acyl donor. Enzyme amounts were found to slightly influence conversion yields.

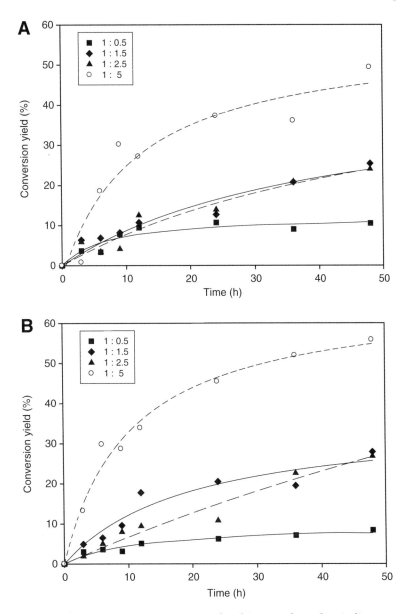

Fig. 3. Effect of molar ratio on conversion of sorbitan methacrylate in lipase-catalyzed glycosylation using **(A)** ethyl methacrylate and **(B)** methyl methacrylate as acyl donor.

However, high enzyme amounts can also result in economic problems. Higher enzyme contents tend to reduce reaction time but the final conversion appears to be independent of the amount of enzyme added. As shown in Fig. 4B, when MMA is used as the acyl donor, sorbitan methacrylate is synthesized at a conversion rate of approx 78%, at 50 g/L of initial 1,4-sorbitan concentration, 7% (w/v) of enzyme content, a sorbitan to MMA molar ratio of 1:5, and a reaction temperature of 50°C, with a reaction duration of 36 h.

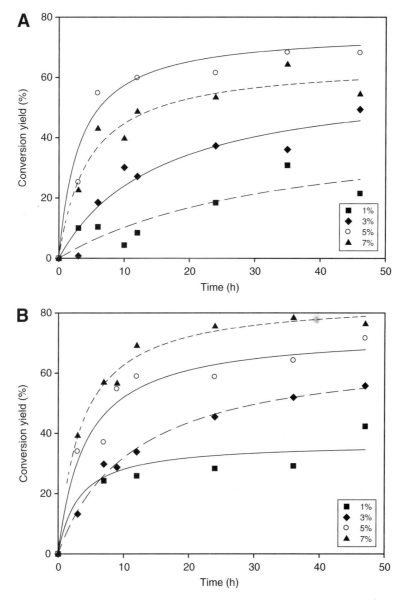

Fig. 4. Effect of enzyme content on conversion of sorbitan methacrylate in lipase-catalyzed glycosylation using **(A)** ethyl methacrylate and **(B)** methyl methacrylate as acyl donor.

Effect of Initial Sorbitan Concentration

As a result of the fact that enzymes exhibit lower activity than do chemical catalysts, reactions such as those conducted in this study often have longer reaction times, and often result in a relatively low conversion yield. This phenomenon can be circumvented by using high initial glycoside concentrations, but may result in an increase in the viscosity. An increase in the viscosity of the reactant at high concentrations of the initial substrate

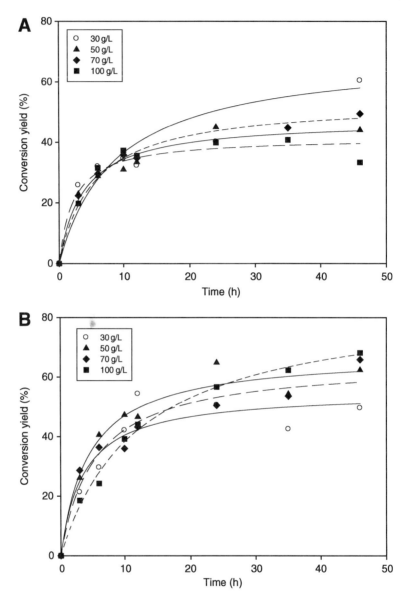

Fig. 5. Effect of initial 1,4-sorbitan concentration on conversion of sorbitan methacrylate in lipase-catalyzed glycosylation using **(A)** ethyl methacrylate and **(B)** methyl methacrylate.

results in lower conversion yields, and increases in the cost of both the raw and purified materials. Therefore, it was necessary to optimize the initial substrate concentration. In a previous experiment involving 1,4-sorbitan acrylate synthesis (5), the initial 1,4-sorbitan concentration was optimized at 50 g/L.

The effects of initial sorbitan concentration on the conversion of 1,4-sorbitan with EMA (or MMA) to sorbitan methacrylate were investigated using Novozym 435. As shown in Fig. 5A, during the reaction periods, the

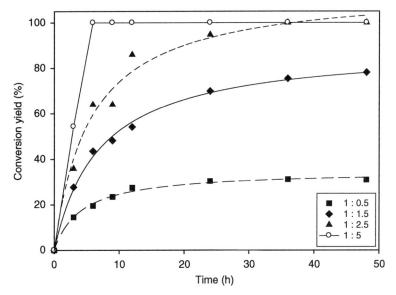

Fig. 6. Effect of molar ratio on conversion of sorbitan methacrylate in lipase-catalyzed glycosylation using vinyl methacrylate.

conversion of 1,4-sorbitan to sorbitan methacrylate was approx 60% at 46 h and a sorbitan concentration of 30 g/L. High initial sorbitan concentrations resulted in low conversion yields. As shown in Fig. 5B, when using MMA as the acyl donor, sorbitan methacrylate was synthesized at a conversion rate of approx 68% at the following conditions: 100 g/L of initial 1,4-sorbitan concentration, 5% (w/v) of enzyme content, 1:5 of molar ratio of sorbitan to MMA, a reaction temperature of 50°C, and a reaction time of 46 h. In contrast to the equilibrated final conversion yield, the initial conversion velocity was found to be higher at 50 g/L sorbitan within the first 12 h than with 100 g/L sorbitan. These results were different from the results of the MMA experiment and the previous sorbitan acrylate synthesis (1).

Enzymatic Esterification With VMA as an Acyl Donor

The synthesis of sorbitan methacrylate using VMA as an acyl donor was expected to result in a superior conversion yield under the following conditions: 3% (w/v) of enzyme content, 1:0.5–1:5 molar ratio, and 50 g/L of initial 1,4-sorbitan concentration. As shown in Fig. 6, during the reaction periods, the conversion of 1,4-sorbitan to sorbitan methacrylate was completed at 46 h and at a molar ratio above 1:2.5. Higher molar ratios of acyl donor resulted in more rapid conversion. The application of VMA to the reaction resulted in a higher conversion yield than is observed with EMA and MMA being used as acyl donors in enzymatic esterification catalyzed by Novozym 435 in organic solvents. In cases in which VMA is employed as an acyl donor, during the glycosylation periods, vinyl alcohol is produced. Simultaneously, tautomeric isomerization to acetaldehyde occurs.

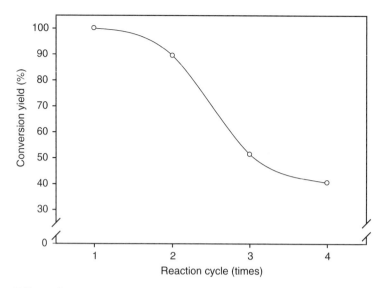

Fig. 7. Effect of enzyme recycling time on conversion of sorbitan methacrylate in lipase-catalyzed glycosylation using vinyl methacrylate.

Therefore, we can conclude that this process facilitates the irreversibility of glycosylation, resulting in higher conversion yields. In experiments involving β-methylglucoside acrylate/methacrylate esterification, higher conversion yields were obtained with the usage of vinyl acrylate and VMA as the acyl donors than when any other acyl donors, including acrylic acid and methacrylic acid, were used *(6)*.

Repeated Usage of Enzyme

In order for the enzymatic process to prove economically feasible as compared with the chemical synthesis process, the amount of added enzyme must be minimized, and must also be recycled in the reaction, as enzymes tend to be fairly expensive. As shown in Fig. 7, sorbitan methacrylate synthesis using VMA as acyl donor was expected to result in a superior conversion yield under the following conditions: 5% (w/v) of enzyme content, a molar ratio of 1:5, and 50 g/L of initial 1,4-sorbitan concentration. During the repeated reaction periods, we observed a reduction in the conversion yield of 1,4-sorbitan to sorbitan methacrylate. On the first reaction, the conversion was completed within 6 h, but during the repeated reactions, the conversion yield decreased dramatically, to approx 40%.

Conclusions

In this study, we attempted to determine the optimum conditions for the enzymatic synthesis of sorbitan methacrylate, using Novozym 435 in *t*-butanol from sorbitan and several acyl donors (EMA, MMA, and VMA). As we achieved an approximate conversion yield of sorbitan methacrylate catalyzed by Novozym 435 in *t*-butanol of 68% under the following conditions: 50 g/L initial 1,4-sorbitan concentration, 5% (w/v) enzyme content, a molar ratio of sorbitan to EMA of 1:5, and reaction time of 36 h. When MMA was utilized as an acyl donor, sorbitan methacrylate was synthesized at a conversion yield of approx 78% under the following conditions: 50 g/L initial 1,4-sorbitan concentration, 7% (w/v) enzyme content, a molar ratio of 1:5, and a reaction time of 36 h. Sorbitan methacrylate synthesis using VMA as the acyl donor was expected to result in a superior conversion yield at 3% (w/v) of enzyme content, and at a molar ratio of more than 1:2.5. Higher molar ratios of acyl donor resulted in faster conversion rates. VMA can be applied in order to obtain yields higher than those associated with EMA or MMA as acyl donors in our esterification reaction. Enzyme recycling resulted in low conversion yields, depending on recycling time.

Acknowledgment

This work was supported by the Korea Research Foundation Grant funded by the Korean Government (D00250) (R05-2004-000-11185-0).

References

1. Griffith, L. G. (2000), *Acta Mater.* **48,** 263–277.
2. Ladmiral, V., Melia, E., and Haddleton, D. M. (2004), *Eur. Polym. J.* **40,** 431–449.
3. Park, S. C., Chang, W. J., Lee, S. M., Kim, Y. J., and Koo, Y. M. (2005), *Biotechnol. Bioprocess Eng.* **10(1),** 99–102.
4. Park, O. J., Kim, D. Y., and Dordick, J. S. (2000), *Biotechnol. Bioeng.* **70,** 208–216.
5. Park, D. H., Lim, G. G., Jeong, G. T., et al. (2003), *Korean J. Biotechnol. Bioeng.* **18,** 222–228.
6. Park, D. W., Haam, S., Ahn, I. S., Lee, T. G., Kim, H. S., and Kim, W. S. (2004), *J. Biotechnol.* **107,** 151–160.
7. Torres, C. and Otero, C. (2001), *Enzyme Microb. Technol.* **29,** 3–12.
8. Park, H. G., Do, J. H., and Chang, H. N. (2003), *Biotechnol. Bioprocess Eng.* **8,** 1–8.
9. Park, D. H. and Kim, H. S. (2001), *Korean J. Biotechnol. Bioeng.* **16(1),** 82–86.
10. Castillo, E., Pezzotti, F., Navarro, A., and López-Munguía, A. (2003), *J. Biotechnol.* **102,** 251–259.
11. Vitro, C. and Adlercreutz, P. (2000), *Enzyme Microb. Technol.* **26,** 630–635.

Effect of Inhibitors Released During Steam-Explosion Pretreatment of Barley Straw on Enzymatic Hydrolysis

Mª Prado García-Aparicio, Ignacio Ballesteros, Alberto González, José Miguel Oliva, Mercedes Ballesteros, and Mª José Negro*

Renewable Energies Department—CIEMAT, Avenida Complutense, 22 28040-Madrid, Spain, E-mail: mariajose.negro@ciemat.es

Abstract

The influence of the liquid fraction (prehydrolysate) generated during steam-explosion pretreatment (210°C, 15 min) of barley straw on the enzymatic hydrolysis was determined. Prehydrolysate was analyzed for degradation compounds and sugars' content and used as a medium for enzymatic hydrolysis tests after pH adjusting to 4.8. Our results show that the presence of the compounds contained in the prehydrolysate strongly affects the hydrolysis step (a 25% decrease in cellulose conversion compared with control). Sugars are shown to be more potent inhibitors of enzymatic hydrolysis than degradation products.

Index Entries: Enzymatic hydrolysis; barley straw; inhibition; steam explosion.

Introduction

Current research and development on bioethanol production are being directed toward substituting higher-cost sugars and starch feedstocks with lower-cost lignocellulosic biomass as a way of reducing the cost of ethanol. Barley straw, an important residue from grain industry in Spain, is a promising substrate for microbial ethanol production.

Among biomass-to-ethanol processes, those based on enzymatic hydrolysis make out to be promising. However, there are physicochemical structural, and compositional factors that hinder the enzymatic digestibility of cellulose present in lignocellulose biomass. Unlike starch, which contains homogeneous and easily hydrolysed polymers, lignocellulose plant matter contains cellulose, hemicellulose, polyphenolic lignin, and other extractable components. These complex polymers must be broken down into low-molecular-weight sugars before microorganisms can complete the conversion. The native cellulose fraction of lignocellulosic biomass is recalcitrant

*Author to whom all correspondence and reprint requests should be addressed.

to enzymatic hydrolysis; so a pretreatment step is required to obtain high cellulose-to-glucose bioconversion yields.

Steam explosion is one of the most attractive pretreatment processes owing to its low use of chemicals and energy consumption (1). Steam explosion disrupts the lignin barrier and make cellulose more available to enzymatic attack by removing hemicellulose in order to increase the accessible surface area (2). In spite of these large advantages, there are also some limitations: steam explosion pretreatment, at least partially, degrades hemicellulose-derived sugars and solubilizes and transforms the lignin compounds to chemicals that can inhibit downstream process (3). It is also probable that the solubilization of extractives during the pretreatment step produce potent inhibitors in low concentrations.

Several authors have investigated the nature of the inhibitors present in diluted acid hydrolysates and steam explosion pretreated biomass (4–8). The nature and concentration of the final inhibitory compounds vary greatly with the pretreatment conditions (severity factor that consider both temperature and residence time), the raw material used (hardwood, softwood, or herbaceous plants) and the presence of acid catalyst. These inhibitors can be classified according to their chemical structure. They include weak acids (mainly acetic acid), furans (degradation product of hemicellulose sugars such as furfural, dehydration product of pentoses; and 5-hydroxymethylfurfural [HMF], a dehydration product of hexoses), and phenolic compounds from lignin (aromatic acids, alcohols such as catechol, and aldehydes such as 4-hydroxybenzaldehyde and vanillin). Some of the compounds formed during the degradation of hemicellulose and lignin, contaminate the water-phase product of the steam-explosion process, whereas others become embedded in the biomass and are released during successive bioconversion (9). Such inhibitors can affect enzymes in the hydrolysis step, reduce glucose conversion during fermentation, and depress the rate of ethanol formation at the end of the biomass-to-ethanol process (10). Consequently, the pretreated material should be filtered and washed to remove them. In fact, in most investigations, the slurry obtained after pretreatment is separated in a solid fraction (cellulose and lignin) and a liquid fraction or prehydrolysate (hemicellulose-derived sugars, sugar and lignin degradation products, acetic acid, and other compounds), and washed before enzymatic hydrolysis (10). However, from an economical and environmental point of view, it is preferable to include the prehydrolysate in the enzymatic hydrolysis step because it increases the concentration of fermentable sugars, and potentially provide a higher ethanol concentration in the fermentation step. The handling of pretreated material is facilitated and the capital cost of filtration and washing steps can be excluded. The high concentration of inhibitors in fermentation with high dry matter content might be overcome by applying a fed-batch technique in simultaneous saccharification and fermentation process (11).

Although a considerable amount of research has been carried out to study the effect of toxic components produced during pretreatment in both enzymatic and fermentation steps of hardwood and softwood (9–13), scarce references have been found in relation to agricultural residues such as barley straw.

The aim of this study is to investigate the influence of inhibitory compounds present in the liquid fraction (prehydrolysate) obtained after steam-explosion pretreatment of barley straw on cellulose conversion in the enzymatic hydrolysis step. It could reduce water consumption and residual water generated in an industrial scale process, as well as to enhance the overall sugar concentration in the hydrolysate before conversion to ethanol.

The pretreatment conditions, selected in a previous work, were 210°C and 5 min. After pretreatment the slurry was fractionated in a solid fraction and prehydrolysate by filtration. This prehydrolysate was analyzed for degradation compounds and sugars' content, and was used as a medium for enzymatic hydrolysis tests of the solid fraction. The effect of the two main fractions of the prehydrolysate, divided into hemicellulosic-derived sugars and degradation products, on cellulose conversion was also measured.

Materials and Methods

Biomass Pretreatment

Ground barley straw biomass (5% moisture) was supplied by Biocarburantes de Castilla y León (Salamanca, Spain). The composition of barley straw was 33% glucan, 20% xylan, 3.8% arabinan, 1% galactan, 16.1% lignin, 7.6% ash, and 13.8% extractives. The barley straw was pretreated in a small batch plant described in a previous work (14). After pretreatment, the material was recovered in a cyclone, and the slurry (solid/liquid fraction about 1/10 w/v) was cooled to about 40°C and then filtered for solid and liquid fraction (prehydrolysate) recovery. The solids fraction was thoroughly washed with water and dried at 45°C. Pretreatment conditions (210°C, 5 min) had been previously selected as the most adequate in terms of hemicellulose-derived sugar recovery in the liquid fraction, cellulose recovery in the solid fraction, and enzymatic hydrolysis yield.

The composition of native and steam-explosion pretreated biomass was determined by using standard methods developed at the National Renewable Energy Laboratory (NREL) (15). The prehydrolysate was analyzed regarding solubilized sugars and potentially inhibiting compounds (degradation products).

Enzymatic Hydrolysis Experiments

Enzymatic hydrolysis (EH) experiments were performed in 250-mL Erlenmeyer flasks containing 100 mL of 0.05 M citrate buffer (pH 4.8) at

5% w/v substrate (steam-exploded barley straw) loading, 50°C, 150 rpm, and 168 h. Periodically, 2.5-mL samples of the hydrolysis media were withdrawn and centrifuged at 9300g for 10 min. Sugar content was analyzed by high-performance liquid chromatography (HPLC) as decribed later.

The enzyme mixture of Celluclast 1.5L FG at 15 FPU (filter paper unit)/g cellulose and Novozym 188 at 15 IU β-glucosidase/g cellulose was employed. Enzymes were supplied by Novozymes A/S (Bagsvaerd, Denmark). Celluclast 1.5 L FG is provided as a liquid with a density of 1.2 g/mL; measured enzyme activities were 80 FPU/mL and 1.4 UI (β-glucosidase/mL. Novozyme 188 (β-glucosidase) has a density of 1.18 g/mL and β-glucosidase activity of 700 UI/mL. Measurement of the enzyme activities was performed as recommended by the International Union of Pure Applied Chemistry (16).

To test the effect of compounds produced during steam explosion pretreatment on enzymatic hydrolysis, experiments using the prehydrolysate (previously adjusted at pH 4.8) as enzymatic hydrolysis broth were carried out. Original prehydrolysate, twofold concentrate prehydrolysate, and 1 : 1 diluted prehydrolysate (DP) were used.

To study the effect of sugars and degradation compounds produced in steam explosion pretreatment on enzymatic hydrolysis, enzymatic hydrolysis experiments were performed in synthetic solutions (in buffer citrate 0.05 M) containing sugars or degradation products: (i) in concentration as in original prehydrolysate and (ii) in twofold concentration of original prehydrolysate.

Determination of the Inhibition Degree Caused by Sugar Monomers on Enzymatic Hydrolysis

Determination of inhibition was based on the ratio of the hydrolysis rates with and without the presence of supplemented sugars (glucose, xylose, arabinose, galactose, or mannose). Inhibition degree was expressed as V_I/V, in which V_I is the amount of glucose (g) produced from the substrate in the presence of supplemented sugars per 30 min, and V is the amount of glucose (g) produced from the substrate without sugar supplementation per 30 min. The hydrolysis rates were determined after 30 min of hydrolysis. The short hydrolysis period was selected to minimize the inhibitory effects of the released sugars (17).

Analytical Procedures

Analysis of HMF, furfural, vanillin, syringaldehyde, 4-hydroxybenzaldehyde, catechol, guaiacol, 4-hydroxybenzoic acid, syringic acid, vanillic acid, ferulic acid, coumaric acid, acetic acid, levulinic acid, and formic acid analysis were performed on a HPLC system as described previously (5).

The carbohydrate content of the liquid fraction after pretreatment was determined by performing a mild acid hydrolysis (3% [v/v] H_2SO_4 for

Table 1
Composition of Liquid Fraction Obtained After Steam
Explosion Pretreatment of Barley Straw

Compound	Concentration (g/L)	g/100 g Raw material	Compound	Concentration (mg/L)	mg/100 g Raw material
Xylose	17.4	7.1	4-Hydroxybenzaldehyde	7	2.8
Glucose	4.6	1.9	4-Hydroxyybenzoic acid	4	1.6
Arabinose	1.9	0.8	Catechol	42	17
Galactose	1.3	0.5	Syringaldehyde	31	13
Acetic acid	2.1	0.9	Syringic acid	6	2.4
Formic acid	0.8	0.33	Vanillin	63	25
Furfural	0.7	0.28	Vanillic acid	11	4.4
HMF	0.2	0.08	Ferulic acid	25	10
			Coumaric acid	44	18

120°C and 30 min) and measuring glucose, xylose, arabinose, galactose, and mannose concentration by Waters HPLC in a refractive index detector. Sugars released during enzymatic hydrolysis were also measured by HPLC as above (18).

Chemicals

All chemicals were of analytical grade and obtained from Sigma (St. Louis, MO).

Results and Discussion

Characterization of the Exploded Barley Straw and Composition of Hydrolysate

The composition of solid fraction obtained after steam explosion of barley straw was 60% glucan, 4.7% xylan, 1.4% arabinan, and 30% lignin. After the steam explosion (SE), the biomass composition changes because of the thermal degradation, mainly of the hemicellulose components. The chemical composition confirms that the matter loss primarily occurs at the expense of the hemicellulose, being the component more thermally degradable. Barley straw pretreatment resulted in a solid fraction enriched in cellulose (60%) and lignin (30%).

The sugar composition (expressed in g/L) as well as the degradation compounds from the liquid fraction are shown in Table 1. Results are also expressed as g/100 raw material. Using saturated water steam at high temperature, SE causes autohydrolysis reactions in which part of hemicellulose and lignin are converted into soluble compounds. The prehydrolysate of steam exploded barley straw consisted of a mixture of hydrolysable sugars

(25.2 g/L) and degradation products, for example, carboxylic acids (2.95 g/L), phenols (0.23 g/L), and furans (0.89 g/L). Regarding carbohydrates the major sugar released was the xylose being in a concentration of 17 g/L.

All degradation products that were found in prehydrolysate obtained from steam explosion pretreatment of barley straw biomass have been previously identified in other herbaceous biomass *(19)*. Acetic acid (2.14 g/L), formic acid (0.81 g/L), and furfural (0.69 g/L), from pentose degradation, were the main degradation products present in the prehydrolysate. Acetic acid from hydrolysis of hemicellulose and furfural from degradation of xylose were obtained as a consequence of the high xylan content in herbaceous biomass. The quantification of furfural can hardly explain hemicellulose losses during pretreatment. It is likely that hemicellulose were lost through volatilization of furfural. Formic acid is a product from sugar degradation *(13)*.

The presence of cinnamic acids reported to be present of herbaceous angiosperms is remarkable. The *p*-coumaric and the ferulic acids are major noncore lignin monomers that link hemicelluloses and core lignin *(20)*. It is worth to notice that vanillin and vanillic acid concentration, both formed by degradation of guaiacyl propane (G) units of lignin, are significantly higher than syringaldehyde and syringic acid, both produced by degradation of syringylpropane (S) units of lignin. This fact is consistent with the G/S ratio in herbaceous biomass *(13)*.

Effect of Prehydrolysate on the Enzymatic Hydrolysis

An efficient utilization of the water-soluble hemicellulose components is required to make the biorefinery approach feasible. Previous research has indicated that the biomass-to-ethanol process could be more economical by incorporating hemicellulose rich water-soluble fraction to the enzymatic hydrolysis of the solid fraction *(10)*. The influence of the prehydrolysate on the enzymatic hydrolysis of steam exploded barley straw was investigated. In order to test the effect of the prehydrolysate at different concentrations, three cases have been considered: (i) using the original prehydrolysate (P) obtained in the CIEMAT pilot plant (whose composition is shown in Table 1), (ii) using a 1 : 1 (v/v) diluted prehydrolysate (DP), and (iii) using twofold concentrated prehydrolysate (CP) considering that in a commercial plant the slurry produced would have increased solid content with consequent high loading of inhibitors. The time-course of sugar production was monitored and cellulose conversion determined. Cellulose conversion was calculated based on the amount of cellulose supplied to enzymatic hydrolysis step, which is converted into glucose. The effect of prehydrolysate on the cellulose conversion in the enzymatic hydrolysis step is shown in Fig. 1. The highest cellulose conversion (88% at 168 h) was obtained when the enzymatic hydrolysis of the steam-exploded biomass was assayed on citrate buffer (C). A decrease in cellulose conversion was observed in experiments using prehydrolysate instead of buffer as EH medium. When

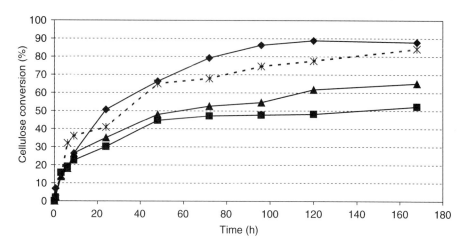

Fig. 1. Effect of the prehydrolysate on cellulose conversion (%) of solid fraction of steam-exploded barley straw at 5% loading (w/v). C (◆), OP (▲), CP (■), and DP (✶).

twofold CP was used, the lowest conversion was obtained (52% at 168 h), corresponding to 59% of the cellulose conversion obtained respect to the control (C). When the original hydrolysate (P) obtained after pretreatment was used as enzymatic hydrolysis medium, a reduction of 25% in the cellulose conversion occurred. The diluted hydrolysate produced 17% and 10% decrease with respect to the control at 48 h and at the end of the hydrolysis step, respectively. Similar findings were obtained by other authors *(10)* who reported cellulose conversion reduction up to 36% when a prehydrolysate of spruce impregnated with SO_2 and steam pretreated at 215°C for 3 min was used in EH.

As previously stated, the purpose of including the prehydrolysate in the enzymatic hydrolysis was also to enhance the overall sugar concentration before conversion to ethanol. When comparing with the control, the total sugar concentrations measured (glucose, cellobiose, and xylose) present in supplemented hydrolysates were higher. As expected, the higher total sugar concentrations in the enzymatic hydrolysis media was obtained by supplementing with the twofold concentrated hydrolysate. At these conditions, a concentration of 50 g/L after 96 h of hydrolysis was obtained. Results from the original liquid fraction and the diluted liquid fraction were quite similar (37 g/L), whereas when enzymatic hydrolysis was carried out in buffer (control) the concentration of sugars obtained was 30g/L. The proportion of different sugars present in the hydrolysate obtained by enzymatic hydrolysis was different. When the original filtrate was used as enzymatic hydrolysis medium the glucose/xylose ratio was 1.6, whereas with diluted prehydrolysate the ratio was 2.6 at 96 h.

Celluclast 1.5 L contains cellulase as the main activity, but also gives high xylose yields. Xylose was released from both prehydrolysate and steam exploded solids, which is consistent with the fact that Celluclast

Effect of Inhibitors

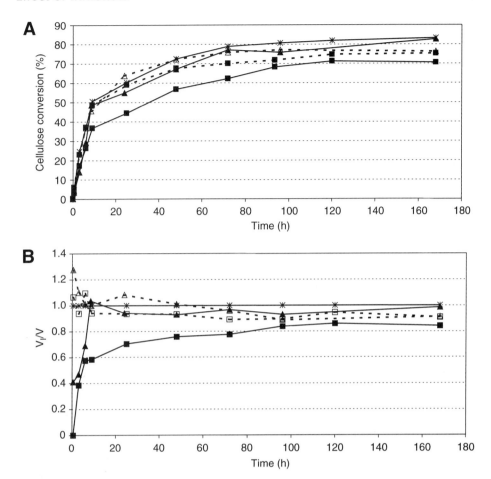

Fig. 2. Effect of supplementation of sugars and degradation product on **(A)** cellulose conversion percentage and **(B)** degree of inhibition (V_I/V) on hydrolysis of 5% steam-exploded barley straw: control (—✻—), twofold (—■—), and original (—▲—) sugars' concentration solution, and twofold (- -□- -) and original (- -△- -) degradation products.

1.5 L has also β-xylosidase activities, capable of catalyzing the hydrolysis of xylobiose and xylotriose to xylose *(21)*. This is an advantageous feature from the point of view of utilizing all sugars present in biomass. Recently, several strains of yeasts have been genetically engineered to effectively coferment glucose and xylose in hydrolysates from different cellulosic biomass to ethanol *(22)*.

Effect of Sugars and Degradation Compounds on the Enzymatic Hydrolysis

To distinguish the effect of sugars present in the prehydrolysate from the effect of other substances (degradation product), four experiments were performed adding to a buffer solution, separately, sugars and degradation products (both at the same concentration as found in original liquid and at twofold concentration). Figure 2A shows the results of the influence of the two fractions (sugars and degradation compounds) in the enzymatic hydrolysis. The sugar fraction was shown to have a greater inhibitory effect on enzymatic hydrolysis than did the degradation compounds. Cellulose conversion from the sugar-supplemented hydrolysis was lower than the control (without any sugar addition) over 120-h incubation time. The presence of hemicelluloses derived sugars at the twofold concentration of the original hydrolysate, decreased the cellulose conversion by 15% at the end of hydrolysis step (Fig. 2A). The degradation products were responsible for a minor part of the inhibition of enzymatic hydrolysis. However, the decrease observed in the original prehydrolysate was higher than the sum of the effect from the supplemented fraction (sugars and degradation product), which could be owing to another component not identified or the synergistic inhibition by the inhibitors.

The degree of inhibition (calculated as the ratio of the hydrolysis rates with and without the presence of supplemented sugars or degradation product), measured over the hydrolysis period, is shown in Fig. 2B. When media were supplemented with degradation products, a slight inhibitory effect was observed. The presence of hemicellulosic-derived sugars at the same concentration as the original prehydrolysate decreased the hydrolysis rate (in the first 3 h of hydrolysis) by 53% and 60% at twofold concentration. As the hydrolysis proceeded, digestion curves approached the control levels. This is probably because the inhibitory effects of the glucose during enzymatic hydrolysis surpassed the inhibitory effect of the supplemented sugars *(17)*.

A higher accumulation of cellobiose during the first 9 h of hydrolysis was observed when the sugar fraction was added to the run hydrolysis (Fig. 3). These results indicate that sugars may play an important role in inhibiting β-glucosidase in the early phase of hydrolysis step.

Degree of Inhibition on Enzymatic Hydrolysis by Monosaccharides

From the previous results, it can be deduced that the sugar fraction has a greater influence in the diminution of the cellulose conversion. Inhibitory effects of glucose, xylose, galactose, and arabinose as the major monosaccharides formed in the liquid fraction of steam-exploded herbaceous biomass were determined. Mannose has also been included although it was not found in barley straw composition. The degree of inhibition was studied supplementing glucose and xylose at 0–100 g/L and galactose, arabinose, and mannose at 0–30 g/L to hydrolysis broth in experiments with 10% (w/v) barley steam exploded as substrate.

As expected, the addition of glucose resulted in on the inhibition of the hydrolysis rate. Hydrolysis rates decreased by 80% after supplementation with glucose at 15 g/L. This effect is well documented in the literature. Xiao et al. *(17)*, performing studies of degree of sugar inhibition using Avicel as substrate and supplementing with 100 g/L of glucose, found an 80% of reduction of cellulase activity. Hemicellulose-derived sugars have

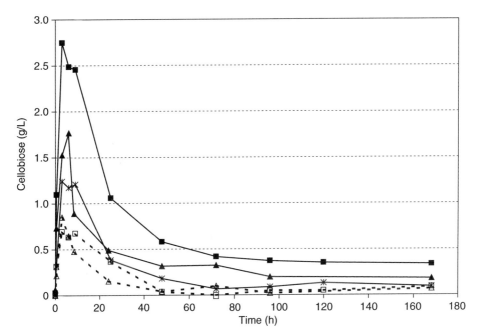

Fig. 3. Effect of supplementation of sugars and degradation product on cellobiose release in the enzymatic hydrolysis of 5% steam-exploded barley straw: (—✳—) control, (—■—) twofold and (—▲—) original sugars' concentration solution, (- -□- -) twofold and (- -△- -) original degradation products.

also been shown to have a direct inhibitory effect on the cellulase enzymes although less significant than glucose. Hydrolysis rates decrease by 35%, 13%, 11.5%, and 5% after supplementation with 20 g/L of xylose, arabinose, galactose, and mannose, respectively.

Conclusions

The presence of the compounds contained in the prehydrolysate from steam explosion of barley straw strongly affects the enzymatic hydrolysis of washed solid fraction from pretreatment. In enzymatic hydrolysis experiments performed in media supplemented with prehydrolysate at different concentrations a decrease in the cellulose conversion of 25% and 40% was obtained with original and twofold CP, respectively, compared with control tests.

Enzymatic hydrolysis conducted in media supplemented with the two major components contained in prehydrolysate (hemicellulose derived sugars and degradation compounds) showed that sugars were more potent inhibitors of enzymatic hydrolysis than degradations products. The presence of hemicellulose-derived sugars, at the same concentration than the prehydrolysate, decreases the hydrolysis rate by 53% in the first 3 h in comparison to control, whereas degradation products' components were responsible for a minor part of the inhibition of enzymatic hydrolysis. However, the inhibitory effect produced by prehydrolysate itself was higher than the sum of effects originated by the supplemented fraction (sugars and degradation product). This could be owing to other components not identified or/and the synergistic inhibition effect among the studied compounds.

The level of cellulase activity inhibition caused by individual sugars showed that glucose exerts strong inhibitory effect on hydrolysis rate (80% decrease after supplementation with glucose at 15 g/L). Xylose, the major hemicellulosic sugar, was also shown to produce a significant inhibitory effect.

References

1. Pereira Ramos, L. (2003), *Quim. Nova* **26**, 863–871.
2. Nathan, M., Wyman, C., Dale, B., et al. (2005), *Bioresour. Technol.* **96**, 673–686.
3. Palmqvist, E. and Hahn-Hägerdal, B. (2000), *Bioresour. Technol.* **74**, 25–33.
4. Ando, S., Arai, I., Kiyoto, K., and Hanai, S. (1986), *J. Ferment. Technol.* **64**, 567–570.
5. Oliva, J. M., Sáez, F., Ballesteros, I., et al. (2003), *Appl. Biochem. Biotechnol.* **105–108**, 141–153.
6. Jönsson, L., Palmqvist, E., Nivelbrant, N. O., and Hahn-Hägerdal, B. (1998), *Appl. Microbiol. Biotechnol.* **49**, 691–697.
7. Luo, C., Brink, D. L., and Blanch, H. W. (2002), *Biomass Bioenergy* **22**, 125–138.
8. Martínez, A., Rodríguez, M. E., Wells, M. L., York, S. W., Preston, J. F., and Ingram, L. O. (2001), *Biotechnol. Progr.* **17**, 287–293.
9. Cantarella, M., Cantarella, L., Gallifuoco, A., Spera, A., and Alfani, F. (2004), *Biotechnol. Progr.* **20**, 200–206.
10. Tengborg, C., Galbe, M., and Zacchi, G. (2001), *Enzyme Microb. Technol.* **28**, 835–844.
11. Rudolf, A., Alkasrawi, M., Zacchi, G., and Lidén, G. (2005), *Enzyme Microb. Technol.* **37**, 195–204.
12. Palmqvist, E., Hahn-Hägerdal, B., Galbe, M., and Zacchi, G. (1996), *Enzyme Microb. Technol.* **19**, 470–476.
13. Klinke, H. B., Thomsem, A. B., and Ahring, B. K. (2004), *Appl. Microbiol. Biotechnol.* **66**, 10–26.
14. Carrasco, J. E., Martínez, J. M., Negro, M. J., et al. (1989), In: *5th EC Conference on Biomass for Energy and Industry*, Vol. 2, Grassi, G., Gosse, G., and Dos Santos, G. (eds.), Elsevier, Essex, England, pp. 38–44.
15. National Renewable Energy Laboratory (NREL). Chemical analysis and testing laboratory analytical procedures: LAP-001 (1996), LAP-002 (1995), LAP-003 (1996), LAP-005 (1994), LAP-010 (1994), and LAP 017 (1998). NREL, Golden, CO, USA. http//www.eere.energy.gov/biomass/analytical_procedures.html.
16. Ghose, T. K. (1987), *Pure Appl. Chem.* **59**, 257–268.
17. Xiao, Z., Zhang, X., Greff, D. J., and Saddler, J. N. (2004), *Appl. Biochem. Biotechnol.* **113–116**, 1115–1126.
18. Negro, M. J., Manzanares, P., Ballesteros, I., Oliva, J. M., Cabañas, A., and Ballesteros, M. (2003), *Appl. Biochem. Biotechnol.* **105–108**, 87–100.
19. Martín, C., Galbe, M., Nilvebrant, N. O., and Jonson, L. J. (2002), *Appl. Microbiol. Biotechnol.* **98–100**, 699–716.
20. Jung, D. P. (1989), *Agron. J.* **81**, 33–38.
21. Sorensen, H. R., Pedersen, S., Vikso-Nielsen, A., and Meyer, A. S. (2005), *Enzyme Microb. Technol.* **36**, 773–784.
22. Sedlak, M. and Ho, N. W. (2004), *Appl. Biochem. Biotechnol.* **113–116**, 403–416.

Purification and Characterization of Two Xylanases From Alkalophilic and Thermophilic *Bacillus licheniformis* 77-2

VALQUIRIA B. DAMIANO,[1,2] RICHARD WARD,[3] ELENI GOMES,[1] HELOIZA FERREIRA ALVES-PRADO,[1,2] AND ROBERTO DA SILVA*,[1]

[1]*UNESP - State University of São Paulo, Biochemistry and Applied Microbiology Laboratory, Rua Cristóvão Colombo n 2265, 15054-000, São José do Rio Preto, SP, Brazil, E-mail: dasilva@ibilce.unesp.br;*
[2]*UNESP - State University of São Paulo, Biology Institute, Rio Claro, SP;*
and [3]*USP - University of São Paulo, Chemistry Institute, Ribeirão Preto, SP*

Abstract

The alkalophilic bacteria *Bacillus licheniformis* 77-2 produces significant quantities of thermostable cellulase-free xylanases. The crude xylanase was purified to apparent homogeneity by gel filtration (G-75) and ionic exchange chromatography (carboxymethyl sephadex, Q sepharose, and Mono Q), resulting in the isolation of two xylanases. The molecular masses of the enzymes were estimated to be 17 kDa (X-I) and 40 kDa (X-II), as determined by SDS-PAGE. The K_m and V_{max} values were 1.8 mg/mL and 7.05 U/mg protein (X-I), and 1.05 mg/mL and 9.1 U/mg protein (X-II). The xylanases demonstrated optimum activity at pH 7.0 and 8.0–10.0 for xylanase X-I and X-II, respectively, and, retained more than 75% of hydrolytic activity up to pH 11.0. The purified enzymes were most active at 70 and 75°C for X-I and X-II, respectively, and, retained more than 90% of hydrolytic activity after 1 h of heating at 50°C and 60°C for X-I and X-II, respectively. The predominant products of xylan hydrolysates indicated that these enzymes were endoxylanases.

Index Entries: Xylanase; *Bacillus licheniformis*; xylanase purification; alkalophilic bacteria; xylanase characterization.

Introduction

There is a strong interest in xylanolytic enzymes of microbial origin, owing to their numerous possibilities for industrial applications. Xylanases are typically produced as a mixture of different isoenzymes that act on xylan, degrading its backbone into small oligomers. The systems of cellulase-free xylanases have been receiving more attention, especially in systems in which the degradation of xylan without breaking cellulose fiber is required, like in the paper and pulp industries *(1)*. A new and promising

---

*Author to whom all correspondence and reprint requests should be addressed.

field for xylanase application is the biorefinery. A biorefinery is a facility that integrates biomass conversion processes and equipment to produce fuels, power, and chemicals from biomass *(2,3)*. Hemicelluloses and cellulose represent more than 50% of the dry weight of agricultural residues; they can be converted into soluble sugars either by acid or enzymatic hydrolysis, so they can be used as a plentiful and cheap source of renewable energy in the world. Enzymatic hydrolysis has been preferred as it is environmentally friendly. Xylan is the main component of the hemicellulose fraction, which is degraded by xylanases, thus if complete degradation of biomass feedstocks is required, xylanase must be present.

In many microorganisms, xylanase production is accompanied by the cellulase enzymes, which disables the application of the crude enzyme in the pulp and paper industry. The most economical form of obtaining cellulase-free xylanase is the isolation of microorganisms which only produce xylanase. Genetic manipulation can also be used to obtain cellulase-free mutants through the removal of the cellulase gene from a potent xylanase-producing microorganism. Another way to obtain cellulase-free xylanase is the development of a purification process to recover the pure xylanase from the fermentation broth *(4)*.

Many xylanolytic microorganisms produce more than one xylanase, which usually differ in their physicochemical and biochemical properties, allowing them to be separated, based on their molecular mass, electrical charge, and isoelectric point *(5)*. Enzymatic purification is an important stage for acquiring insight into the xylanolytic system of the microorganism and to understand its mechanism of action on the substrate, with the aim of optimizing its industrial application. Working directly with the crude enzyme makes it impossible to know whether the hydrolyzation of the xylan molecule was owing to the action of a single enzyme or to an enzymatic system acting synergistically.

In a previous work, *Bacillus licheniformis* 77-2 was isolated because it is a good producer of extracellular cellulase-free xylanase. The crude enzyme presented great activity at pH 6.0 and 70°C and pH stability in the range of 5.0–10.0, with 100% stability at 40°C for 1 h *(6)*. The crude xylanase was used in the biobleaching of eucalyptus Kraft pulp, allowing for a reduction of 33% in the ClO_2 load used for pulp delignification (κ number) or a 44% reduction to achieve the same level of brightness, in comparison with pulp that did not undergo the enzymatic treatment *(1)*. To the best of our knowledge, only one report has been published describing the purification of the predominant form of xylanase from this microorganism and its successful application in the delignification process *(7)*. In view of the scarcity of reports on xylanases from *B. licheniformis*, we believe that it was in the interest of certain industries to purify and characterize its isoenzymes in order to gain insight into its xylanolytic system with the aim of understanding its action during the biobleaching process.

Material and Methods

Microorganism and Growth Conditions

B. licheniformis 77-2 was isolated from decaying wood (6). The bacterium was grown in alkaline medium, pH 9.0 containing 1% beef extract powder, 1% peptone, 1% NaCl, 0.1% KH_2PO_4, 0.5% Na_2CO_3 (sterilized separately), and 2% Birchwood xylan. A 20 mL of the medium, in 125-mL Erlenmeyer flasks, were inoculated with 1 mL (10^7 cells) of a 16-h seed culture and incubated at 50°C with shaking at 150 rev/min. After 48 h, the bacteria were harvested by centrifugation at 10,000g at 4°C for 20 min. The cell-free supernatant was used as crude enzyme solution.

Enzyme and Protein Assays

Xylanase was assayed by incubating 0.1 mL of appropriately diluted enzyme, to ensure its initial rate of activity, with 0.9 mL of a suspension containing 1% of Birchwood xylan (Sigma, St. Louis, MO) in 0.1 M acetate buffer at pH 6.0. After incubation at 60°C for 10 min, the reducing substances released were assayed by 3,5-dinitrosalicylic acid (8). Controls were prepared with enzyme added after 10 min of boiling. One U of activity was defined as 1 µmol of xylose equivalent released/min under the above assay conditions, using a xylose standard curve. Protein content was determined according to Lowry using bovine serum albumin as the standard.

Xylanase Purification

Many steps were tried before those described here. The chromatographic steps were performed with AKTA purifier equipment (Amersham Pharmacia, Uppsala, Sweden). All steps were carried out at room temperature, except for the ethanol precipitation, which was performed at 18°C. Xylanase activity was measured after every purification step.

Ethanol Precipitation

The supernatant was cooled in an ice-bath slowly adding ethanol at –18°C (70% saturation, V/V). The mixture was incubated at 4°C overnight and the precipitate collected by centrifugation (10,000g, 10 min, at 4°C).

Gel Filtration Chromatography

The precipitate was suspended in Tris-HCl, 0.05 mol/L, pH 7.5, dialyzed and loaded on a Sephadex G-75 column (20 × 1000 mm^2) (Pharmacia) previously equilibrated with the same buffer. The flow rate was 12 mL/h. The fractions containing xylanase activity were pooled, two peaks corresponding to X-I and X-II enzymes.

Ion Exchange Chromatography

The pool containing xylanase X-I, was loaded on CM Sephadex C-50 (Pharmacia) previously equilibrated with 0.02 mol/L sodium acetate buffer (pH 5.0). The column was washed exhaustively with equilibrating buffer to wash off unbound protein and the bound proteins were eluted with linear saline gradient (0.01–1 mol/L NaCl in elution buffer) with a flow rate of 0.06 mL/min. The levels of protein were monitored at 280 nm.

Anion Exchange Chromatography—Q-Sepharose

After the elution from G-75, the pool containing X-II xylanase was concentrated using Centriprep-10 (Millipore), dialyzed overnight against 0.05 mol/L sodium acetate buffer, pH 5.0, and loaded again on previously equilibrated Q-Sepharose, exactly as performed on Sephadex C-50. The proteins were eluted with a linear gradient of 0.1–1 mol/L NaCl in elution buffer.

Anion Exchange Chromatography—Mono Q HR

After the elution from Q-Sepharose the pool containing X-II xylanase was concentrated again using Centriprep-10 (Millipore, Beldford, MA), dialyzed overnight against Tris-HCl, 0.02 mol/L, pH 7.2, and loaded again on previously equilibrated Mono Q HR 5/5 with the same buffer, exactly as performed on Q-Sepharose. The proteins were eluted with a linear gradient of 0.1–1 mol/L NaCl in elution buffer. The active fractions were pooled, dialyzed to remove salt, concentrated by 10 kDa membrane (centricon, Millipore) and sterilized by 0.22-μm membrane filtration (Millipore).

Electrophoresis and Silver Staining

Electrophoresis was carried out on denaturing sodium dodecyl sulfate-polyacrylamide gel electrophoresis (SDS-PAGE) using 10% gel with 1.5 M Tris-HCl buffer (pH 8.8) as described by Laemmli (9). The gels were silver stained according to Meril, 1970 (10). The molecular weight markers used were albumin bovine (66 kDa), carbonic anhydrase (29 kDa), and trypsin (20 kDa).

pH and Temperature Effects

The optimal pH for pure xylanase activity was measured in a pH range 3.0–11.0 at 60°C for 10 min. The following 100 mM buffer systems were used: Sodium acetate buffer pH 3.0–5.6, Citrate-phosphate pH 5.5–7.0; Tris-HCl pH 6.0–8.5, and Gly-NaOH pH 8.5–11.0. Optimal pH was the pH in which the enzyme displayed its maximal activity, which was considered 100% activity. For optimum temperature activity the pure xylanase was assayed between 40°C and 90°C for 10 min, although the pH was maintained at 7.0 for X-I and 9.0 for X-II. Optimal temperature was the temperature in which the enzyme displayed its maximal activity, which was considered 100% activity. The pH stability was determined by incubating the enzyme preparation without substrate in the same buffer systems used above, for 24 h at room temperature. The remaining activity was assayed under standard conditions at 60°C. The temperature stability was checked by subjecting the enzyme to pH 7.0 for X-I and 9.0 for X-II, without substrate at various temperatures (40–90°C) for 1 h, and then cooling in ice before measuring the residual activity under standard conditions at 60°C.

Determination of the Kinetic Parameters

The enzymatic reaction for the kinetic study of the xylanases of *B. licheniformis* 77-2 was carried out at pH 6.0 and 60°C for 10 min, using the specific activity of 18 and 18.3 U/mg protein of XI and XII, respectively, with different Birchwood xylan concentrations (0.2–20 mg/mL). The kinetic constants (K_m and V_{max}) were determined from the Lineweaver–Burk plot.

Analysis of Hydrolytic Products

A 5 U of the purified enzyme was incubated at 60°C for 2 h with 0.9 mL of 2% Birchwood xylan (Sigma) in 50 mM, pH 6.0 Tris-HCl buffer, with a final volume of 1 mL. After the incubation period, the enzyme was inactivated by 10 min of boiling. The identification of the hydrolysis products was carried out by the method of descending paper chromatography (Watman N. 1), using as run solvent ethyl acetate, isopropanol and water, in the proportion 6:3:1, respectively. The revelation solvent was acetone and saturated silver nitrate, washed with alcoholic silver hydroxide to visualize the stains. The standards used were xylotriose, xylobiose, and xylose.

Results and Discussion

Summary of Purification

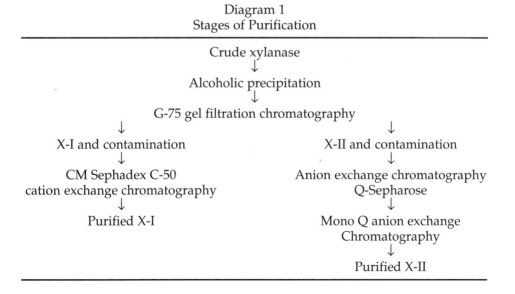

Diagram 1
Stages of Purification

Table 1
Summary of the Purification of the Xylanases Produced by B. licheniformis 77-2

Step		Volume (mL)	Total protein (mg)	Total activity (U)	Specific activity (U/mg)	Yield (%)	Fold purification
Supernatant		480	931	8640	9.3	100	1
Precipitation and dialysis		20	370	3700	10	42.8	1.1
Gel filtration (G-75)	X-I	118	12	1286	107	14.9	11.5
	X-II	136	22	2285	104	26.5	11.2
CM sephadex	X-I	78	0.78	562	720	6.5	77.4
Q-Sepharose	X-II	54	5.4	1350	250	15.6	27
Concentration centriprep	X-II	3.8	3.9	904	232	10.5	25
Mono Q HR5/5	X-II	4.5	0.27	99	367	1.2	40

Diagram 1 summarizes the purification stages realized. Starting from the crude supernatant of xylanase from *B. licheniformis* 77-2, two xylanases named X-I and X-II were separated.

The first purification stage was precipitation with alcohol. In this stage, the specific activity of the crude xylanase increased about 10 times and the total activity of xylanase was decreased from 8640 U to 3700 U, corresponding to a yield of 42%, as shown in Table 1. In the gel filtration process five protein peaks were observed and the xylanase activity appeared in two peaks that were designated X-I and II. X-I was separated from X-II with a yield of 15% and 26%, respectively. The purification of X-I by CM Sephadex C-50 raised the specific activity from 107 U/mg of protein to 720 U/mg of protein, resulting in an 80-fold purification factor and a yield of 6.5. The low yield of the X-I fraction may be explained on the basis of its low stability at pH 5.0 (Fig. 1A) used in the elution buffer and the fact that negative resins may have formed a chelate and eliminated certain metal ions important for enzyme activity *(11)*.

X-II did not bind to CM Sephadex C-50, thus it was applied on Q-Sepharose resin. The purification stage of xylanase II on the Q-Sepharose resin increased the purification factor from 11 to 28. Therefore, this stage was important in the purification of the enzyme, but it did not present sufficient resolution to separate it from all the contaminants.

With the application of X-II on the Mono Q HR 5/5 resin, the xylanase was completely separated. X-II was eluted with about 0.6 M of NaCl. In this stage, a low yield was obtained, but it was important for complete purification, as viewed in the electrophoresis stage (Fig. 2). The low yield of the X-II fraction may be also explained on the basis of its low stability in

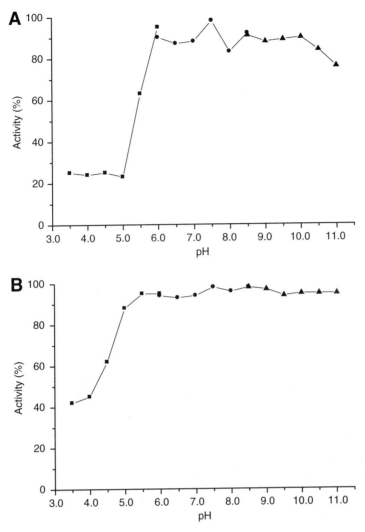

Fig. 1. Effects of pH on the stability of Xylanases from *B. licheniformis* **(A)** X-I and **(B)** X-II. The enzymes were incubated for 24 h at room temperature in the following 0.1 M buffer system: sodium acetate (pH 3.0–5.5); citrate-sodium phosphate (pH 5.5–7.0); Tris-HCl (pH 7.0–8.0) and Gly-NaOH (pH 8.5–10.0). The remaining activity was assayed at 60°C for 10 min at pH 6.0. A 100% activity was 2.1 U/mL for X-I and 6.6 U/mL for X-II.

the pH 5.0 buffer used for enzyme elution (Fig. 1B) and owing to the number steps used for complete purification of the enzyme. Generally, the recovery of proteins is inversely proportional to the number steps used to accomplish the purification procedure. Another reason for the low yields could be owing to fact that X-II showed evidence of glycosylation in results obtained on zymogram stained with Schiff reagent and in another separate experiment in which the xylanase bound to concanavalin A resin

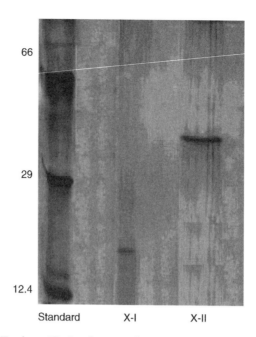

Fig. 2. SDS-PAGE of purified xylanases from *B. licheniformis*: The standards were β-amylase bovine serum albumin (66 kDa), carbonic anhydrase (29 kDa) and cytochrome C (12.4 kDa).

(data not shown). Glycoproteins display interactions with many chromatography gels *(12)*, which may result in lower recuperation. In this work, many modifications of the mobile phase were attempted, such as the use of high pH, decreased polarity and denatured condition, and so on, to reduce these interactions. In both xylanase purification procedures, no problems like protein precipitation were observed.

Determination of the Molecular Mass

The determination of the molecular mass in gel filtration is not an efficient methodology for many xylanases of the *Bacillus* spp., owing to their interaction with the resin. In such cases, SDS-PAGE is more suitable *(12)*. X-I and X-II fractions and standard proteins (bovine serum albumin, carbonic anhydrase, and cytochrome C) were applied to the 10% SDS-PAGE and proteins were stained with silver. Then, gel was dried between two sheets of cellophane. The molecular masses of the xylanases were determined (not digitized) by plotting the relative mobility (R_f value) of each protein against the log of its molecular weight using the standard protein (Fig. 2). X-I presented a molecular mass of 17 kDa and X-II presented a molecular mass of 40 kDa.

The xylanase from *B. amyloliquefaciens* displayed a molecular mass of 18 kDa *(12)*, which is very close to the value of X-I, and *Bacillus* TAR-1 showed a 40 kDa xylanase *(13)*, which was the same value presented by X-II. The molecular mass of xylanase from *B. licheniformis* A99 was 45 kDa.

Purified xylanases have been reported in the range of 5.5–85 kDa (7). Xylanases with low molecular mass are appropriate for the pulp and paper industry, because the size of xylanases is directly related to their accessibility to the fiber pore and the consequent removal of lignin (14).

Determination of the Kinetic Parameters

The kinetic study was carried out at 60°C for 10 min with different xylan concentrations (Birchwood, Sigma, St. Louis, MO). The results were analyzed according to the graphical procedure proposed by Lineweaver and Burk (data not shown). V_{max} was 7.05 and 9.1 U/mg of protein and K_m was 1.8 and 1.05 mg/mL, for the enzymes X-I and X-II, respectively. These values of K_m are in agreement with the reported range of K_m for xylanases (15), such as that of *Bacillus stearothermophilus* T-6 which produced a xylanase with a K_m of 1.63 mg/mL (16). These values are smaller than the K_m of 4.5 and 3.3 mg/mL of the xylanases from *B. amyloliquefaciens* (15) and *B. licheniformis* (7), respectively, and this difference is even greater when in comparison with the K_m of 10.14 mg/mL of the xylanase from the mushroom *Trichoderma longibrachiatum* (17).

Effect of pH on Activity and Stability of Xylanases

X-I presented maximum activity at pH 7.0, retaining about 80% of its activity at pH 8.0 and about 50% of its activity at pH 9.0 (Fig. 3A). Nath and Rao (18), and Archana and Satyanarayana (7) describe the enzymatic activity of Xyl II from *Bacillus* spp. (NCIM 59) and xylanase from *B. licheniformis* A99, respectively, with maximum activity in the pH range of 6.0–7.0. The high pH optimum of xylanase is a good property, as this is required for industrial pulping processes. The enzyme X-II presented an unusual pH-dependence behavior in the determination of its optimum pH (Fig. 3B). The bell-shaped curve was not observed, though a large range of optimum pH was observed, from pH above 7.0 until pH 11.0; the upper limit used in this paper. Both xylanases were very stable for 24 h over a large range of pH. X-I presented an average of about 90% stability from pH 6.0 to 10.0, and at pH 11.0 it still retained about 75% of its enzymatic activity. The enzyme X-II was similarly stable in a range from pH 5.5 to 11.0 (Fig. 1A,B). These results are also interesting for the application of these enzymes in the paper industry.

Effect of Temperature on Activity and Stability of Xylanases

The two enzymes presented optimal activities at 70–75°C. X-I presented 50% activity at 90°C (Fig. 4A), and X-II presented 80% activity at the same temperature (Fig. 4B). The xylanases of *Bacillus* spp. and *B. licheniformis* A99 showed a maximum activity at 55°C and 60°C respectively. *B. amiloliquefaciens* produced a xylanase with an optimum temperature of 80°C and retained 90% of its activity at 90°C (11). High temperatures of action for xylanases are also quite exciting from the point of view of their industrial application.

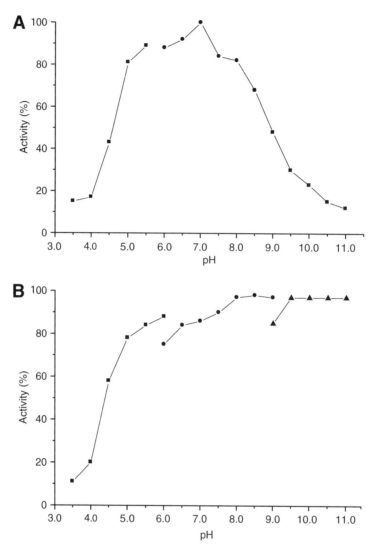

Fig. 3. Effects of pH on xylanases from *B. licheniformis* **(A)** X-I and **(B)** X-II. The effect of pH on xylanases was measured at 60°C for 10 min using Birchwood xylan as substrate in the following 0.1 M buffer system: sodium acetate (pH 3.0–5.5); citrate-sodium phosphate (pH 5.5–7.0); Tris-HCl (pH 7.0–8.5) and Gly-NaOH (pH 8.5–10.0). The 100% activity was 7 U/mL for X-I and 3.6 U/mL for X-II.

Both xylanases were 100% stable after one hour at 40°C. At 55°C X-I retained about 40% of its activity after 1 h of treatment, and that percentile remained up to 90°C (Fig. 5A). The enzyme X-II retained about 85% of its activity for temperatures up to 65°C. Approx 30% of the original activity was retained between 75 and 90°C (Fig. 5B).

Identification of the Hydrolysis Products

The hydrolysis products released from Birchwood (Sigma) by *B. licheniformis* X-I and X-II were analyzed by paper chromatography (Fig. 6).

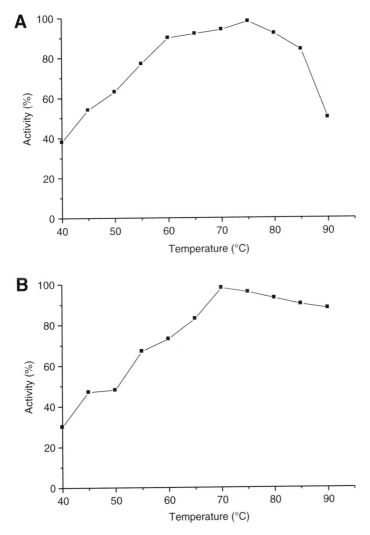

Fig. 4. Effects of temperature on xylanases from B. licheniformis (A) X-I and (B) X-II. Activity was measured at various temperatures for 10 min using Birchwood xylan as substrate in 0.1 M sodium acetate, pH 5.0.

X-I released mainly tetraose and larger xylooligosaccharides. In contrast, X-II liberated mainly xylotetraose, xylotriose, and xylobiose, whereas xylose was observed in smaller amounts. According to Biely et al. (3), enzymes with larger substrate binding sites hydrolyze a small extension of the xylan generating larger products. The results obtained with X-I xylanase are in accordance with this description (19).

Summary of the Characterization

Table 2 displays the characterization of the xylanases produced by B. licheniformis 77-2. X-I presented a molecular mass and hydrolysis products quite different from X-II, besides presenting smaller range of temperature

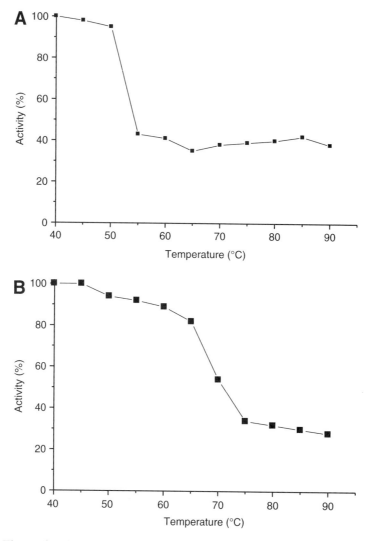

Fig. 5. Thermal stability of xylanases from *B. licheniformis*. **(A)** X-I and **(B)** X-II. The enzyme was maintained for 1 h at temperatures from 20°C to 90°C. The remaining activity was measured under standard assay conditions.

stability. *B. licheniformis* 77-2 studied here produces two different xylanase isoforms. The occurrence of more than one xylanase in *Bacillus* spp is quite common (7). These enzymes can be the products of distinct genes (20) or they can be secreted by the same gene, but undergo different posttranslational modifications, such as limited proteolysis and/or glycosylation (7,21).

X-I presented a low molecular mass in comparison with X-II, which can be an indication that the latter might have suffered differentiated glycosylation. The glycosylation of the isoenzyme X-II is reinforced by its greater stability at higher temperatures, when in comparison with X-I. Glycosylation has been reported as having an important role in stabilizing protein structure (22–29). The behavior of the xylanases concerning pH

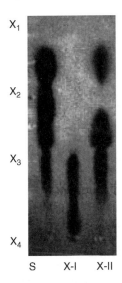

Fig. 6. Hydrolysis products of Birchwood by xylanases from *B. licheniformis* Standard (s): X_1 = xylose; X_2 = xylobiose; X_3 = xylotriose; X_4 = xylotetraose.

Table 2
Summary of the Characterization of Xylanases produced by *B. licheniformis* 77-2

	X-I	X-II
Molecular weight (kDa)	17	40
K_m (mg/mL)	1.8	1.05
V_{max} (U/mg)	7.05	9.1
Optimum pH	7.0	5.5–11.0
pH of stability	6.0–11.0	5.5–11.0
Optimum temperature (°C)	70–75	70–75
Temperature of stability (°C)	50	65
Products of hydrolyzation	$>X_4$ e X_4	X_4, X_3, X_2 e X_1

was also quite different, presenting a typical pH curve (bell-shaped), with optimum activity at pH 7.0. The atypical results found when analyzing the optimum pH of X-II may also be explained by glycosylation, because in the same way that groups of sugars linked to the protein contribute to the increased thermostability of the enzyme, they can also contribute to stabilizing the enzyme regarding changes in pH, which would explain its activity in the large range of alkaline pH. This is in agreement with Hamada et al. (27), who reported the stabilizing effect of glycosylation on the pH stability of cellulase (Ex-2). Furthermore, Wang et al. (25) demonstrated that, in addition to the apparent stabilizing effect, glycosylation may protect the enzyme against aggregation during heat treatment under acidic condition. Both xylanases acted as endoxylanases, but the hydrolysis products of xylan for X-I were quite different from that produced by X-II, indicating that these enzymes act differently on the xylan molecule.

Acknowledgment

The authors are grateful to the FAPESP and CNPq for their financial support.

References

1. Damiano, V. B., Bocchini, D. A., Gomes, E., and Da Silva, R. (2003), *World J. Microbiol. Biotechnol.* **19,** 139–144.
2. Ohara, H. (2003), *Appl. Microbiol. Biotechnol.* **62,** 474–477.
3. Mosier, N., Wyman, C., Dale, B., et al. (2005). *Bioresour. Technol.* **96,** 673–686
4. Alam, M., Gomes, I., Mohiuddin, G., and Hoq, M. M. (1994), *Enzyme Microb. Technol.* **16,** 298–302.
5. Biely, P., Vrsanslá, M., and Kucar, S. (1992), In: *Xylans and Xylanases*, Visser, J., ed., Elsevier, Wageningen, The Netherlands, pp. 80–95.
6. Tavares, V. B., Gomes, E., and Da Silva, R. (1997), *J. Brazilian Soc. Microbiol.* **28,** 179–182.
7. Archana, A. and Satyanarayana, T. (2003), *World J. Microbiol. Biotechnol.* **19,** 53–57.
8. Miller, G. L. (1959), *Anal. Chem.* **31,** 426–428.
9. Laemmli, U. K. (1970), *Nature* **227,** 680–685.
10. Meril, C. R. (1990), In: *Methods in Enzymology*, Deutscher, M. P. ed., Academic Press, New York, pp. 477–480.
11. McCroskery, P. A., Richards, J. F., and Harris, E. D. Jr. (1975), *Biochem. J.* **152,** 131–142.
12. Breccia, J. D., Siñeriz, F., Baigorí, M. D., Castro, G. R., and Haiti-Kaul, R. (1998), *Enzyme Microb. Technol.* **22,** 42–49.
13. Nakamura, S., Ishiguro, Y., Nakai, R., Wakabayashi, K., Aono, R., and Horikoshi, K. (1995), *J. Mol. Cat. B. Enz.* **1,** 7–15.
14. Maiti, B. and Whitmire, D. (1997), *Chem. Eng. Commun.* **162,** 169–175.
15. Beg, Q. K., Kapoor, M., Mahajan, L., and Hoondal, G. S. (2001), *Appl. Microbiol. Biotechnol.* **56,** 326–338.
16. Khasin, A., Alchanati, I., and Shoham, Y. (1993), *Appl. Environ. Microbiol.* **59,** 1725–1730.
17. Chen, C., Chen, J. L., and Lin, T. Y. (1997), *Enzyme Microb. Technol.* **21,** 91–96.
18. Nath, D. and Rao, M. (2001), *Enz. Microb. Technol.* **28,** 397–403.
19. Marques, S., Alves, L., Ribeiro, S., Girio, F. M., and Amaral Collaco, M. T. (1998), *Appl. Biochem. Biotechnol.* **73,** 59–72.
20. Honda, H., Kudo, T., Ikura, Y., and Horikoshi, K. (1985), *Canadian J. Microbiol.* **31,** 538–542.
21. Bastawde, K. B. (1992), *World J. Microbiol. Biotechnol.* **8,** 353–368.
22. Devine, P. L., Warren, J. A., and Layton, G. T. (1990), *Biotechniques* **8,** 354–356.
23. Dixon, B. (1991), *Biotechnology* **9,** 418–418.
24. Christov, L. P. and Prior, B. A. (1993), *Enzyme Microbiol. Technol.* **15,** 460–475.
25. Rudd, P. M., Joao. H. C., Coghill, E., et al. (1994), *Biochemistry* **33,** 17–22.
26. Wang, C., Eufemi, M., Turano, C., and Giartosio, A. (1996), *Biochemistry* **35,** 7299–7307.
27. Hamada, N., Ishikawa, K., and Fuse, N. J. (1999), *Biosc. Bioeng.* **87,** 442–451.
28. Li, P., Gao, X. G., and Arellano, R. O. (2001), *Protein Exp. Purification* **22,** 369–380.
29. Davis, B. G. (2002), *Chem. Rev.* **102,** 579–601.

Oxidation Capacity of Laccases and Peroxidases as Reflected in Experiments With Methoxy-Substituted Benzyl Alcohols

FENG HONG,[†,1] LEIF J. JÖNSSON,*,[1] KNUT LUNDQUIST,[2] AND YIJUN WEI[2]

[1]Biochemistry, Division for Chemistry, Karlstad University, SE-651 88, Karlstad, Sweden, E-mail: Leif.Jonsson@kau.se; and [2]Department of Organic Chemistry, Chalmers University of Technology, SE-412 96, Göteborg, Sweden

Abstract

A set of methoxy-substituted benzyl alcohol (MBA) congeners were examined regarding susceptibility to oxidation by *Trametes versicolor* laccase, *T. versicolor* lignin peroxidase and horseradish peroxidase: 2,4,5-trimethoxybenzyl alcohol (TMBA), 3,4,5-TMBA, 2,3,4-TMBA, 2,5-dimethoxybenzyl alcohol (DMBA), 3,4-DMBA, and 2,3-DMBA. The corresponding methoxy-substituted benzaldehydes were strongly predominant as products on enzymic oxidation. This together with different reaction rates and redox potentials makes the MBAs suitable as substrates in the characterization of ligninolytic enzymes. For fungal laccase, the reaction rate order was: 2,4,5-TMBA >> 2,5-DMBA > 3,4-DMBA > 3,4,5-TMBA ~ 2,3,4-TMBA ~ 2,3-DMBA. Horseradish peroxidase displayed a similar reactivity order. Oxidation of some of the MBAs with laccase and horseradish peroxidase was only observed when the reactions were carried out at low pH and with relatively high-substrate concentration. 3,4-DMBA (veratryl alcohol) was the best substrate for lignin peroxidase and the reaction rate order was: 3,4-DMBA > 2,4,5-TMBA ~ 3,4,5-TMBA > 2,5-DMBA > 2,3,4-TMBA ~ 2,3-DMBA. The oxidation experiments with different MBAs elucidate the potential of the enzymes as oxidants in various applications.

Index Entries: Lignin peroxidase; horseradish peroxidase; laccase; methoxy-substituted benzyl alcohols.

Introduction

Laccases (benzenediol: oxygen oxidoreductase, EC 1.10.3.2) are glycosylated blue copper-containing phenol oxidases that are widely distributed in higher plants and fungi (*1,2*). Laccase catalyzes the one-electron

---

*Author to whom all correspondence and reprint requests should be addressed.
†Present address: Institute of Biological Sciences and Biotechnology, Donghua University, 200051, Shanghai, P. R. China

oxidation of a variety of phenolic compounds, as well as diamines and hexacyanoferrate, concomitantly with the four-electron reduction of molecular oxygen to water. The catalytic ability of laccases has been related to the difference in redox potential between the substrate and the enzyme *(3,4)*. The action of laccase is usually limited to the types of substrates mentioned earlier and the enzyme is often described as not being active with nonphenolic lignin model compounds. However, in the presence of low-molecular mass mediators, laccase can be employed for the oxidation of a variety of nonphenolic aromatic compounds *(5)*. Thus, laccases have very broad substrate specificity regarding the reducing substrate and have an interesting potential as industrial enzymes. They are attracting considerable interest for a variety of biotechnological applications, such as organic synthesis/transformation *(6)*, textile processing, delignification of pulp, manufacture of new materials, construction of biosensors and biofuel cells, and detoxification *(7–9)*.

Laccases produced by the white-rot fungus *Trametes* (*Coriolus, Polyporus*) *versicolor* have a relatively high-redox potential (0.7–0.8 V) *(1)*. The catalytic site of the enzyme comprises four copper ions, one type 1 (T1), one type 2 (T2), and two type 3 (T3) copper ions. The T1 copper oxidizes compounds via one-electron abstraction and transfers electrons to the T2 and T3 copper ions. The T2 and T3 copper ions form a trinuclear cluster responsible for oxygen binding and reduction *(1,10)*. The catalytic mechanism of laccase is still not fully clear. Lignin peroxidase (LP; EC 1.11.1.14) and horseradish peroxidase (HRP; EC 1.11.1.7) are heme-containing enzymes displaying a similar catalytic cycle. A single two-electron oxidation of the peroxidase by hydrogen peroxide is followed by two single-electron oxidations of the reducing substrate by the peroxidase *(11)*. A comparison based on the oxidation of methoxy-substituted benzenes suggested that LP is a stronger oxidant than HRP, which in turn is a stronger oxidant than *T. versicolor* laccase *(3)*. Laccase oxidized only 1,2,4,5-tetramethoxybenzene, which has a relatively low half-wave potential *(12)*.

Methoxyl-substituted aromatic substrates are of interest considering that lignin is formed mainly from two methoxylated precursors, coniferyl alcohol and sinapyl alcohol *(13)*. In addition, 3,4-dimethoxybenzyl alcohol (DMBA; veratryl alcohol) is produced as a secondary metabolite by lignin-degrading fungi *(14)*. The investigation of the oxidation of methoxy-substituted benzyl alcohols (MBAs) is therefore relevant for the elucidation of lignin biodegradation in nature. The oxidation of veratryl alcohol has been examined using laccase-mediator systems *(5,15,16)*, HRP *(17,18)*, and LP *(19–21)*. Valli et al. *(22)* and Koduri and Tien *(23)* have studied the oxidation of 4-methoxybenzyl alcohol and 3,4,5-trimethoxybenzyl alcohol (TMBA) by LP. The oxidation of 3,4,5-TMBA with syringaldehyde as mediator in reactions catalyzed by laccase *(16)* and the oxidation of 2,4-DMBA with 2,2'-azino-*bis*(3-ethylbenzothiazoline-6-sulfonic acid) (ABTS) and 1-hydroxybenzotriazole (HBT) as mediators *(24)* have also been reported.

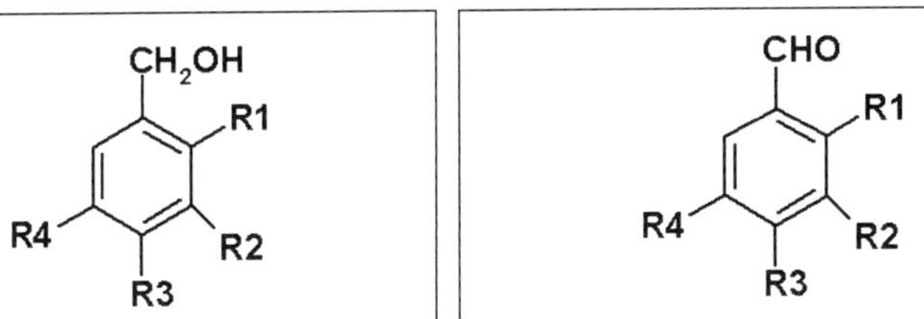

Oxidation Capacity of Laccases

Fig. 1. Methoxysubstituted benzyl alcohols used as substrates (MBAs) and their predominating conversion products on enzymic oxidation (MBDs).

	R1	R2	R3	R4
2,3-DMBA	OCH$_3$	OCH$_3$	H	H
2,5-DMBA	OCH$_3$	H	H	OCH$_3$
3,4-DMBA	H	OCH$_3$	OCH$_3$	H
2,3,4-TMBA	OCH$_3$	OCH$_3$	OCH$_3$	H
2,4,5-TMBA	OCH$_3$	H	OCH$_3$	OCH$_3$
3,4,5-TMBA	H	OCH$_3$	OCH$_3$	OCH$_3$

	R1	R2	R3	R4
2,3-DMBD	OCH$_3$	OCH$_3$	H	H
2,5-DMBD	OCH$_3$	H	H	OCH$_3$
3,4-DMBD	H	OCH$_3$	OCH$_3$	H
2,3,4-TMBD	OCH$_3$	OCH$_3$	OCH$_3$	H
2,4,5-TMBD	OCH$_3$	H	OCH$_3$	OCH$_3$
3,4,5-TMBD	H	OCH$_3$	OCH$_3$	OCH$_3$

We have investigated a set of MBAs (Fig. 1) as substrates with respect to oxidation by LP, HRP, and laccase. In contrast to substrates such as 1,2,4,5-tetramethoxybenzene *(3)*, the MBAs are advantageous in the respect that the methoxy-substituted benzaldehydes (MBDs) are produced as predominant products. This makes the MBAs suitable for kinetic studies. As expected based on theoretical considerations, it was found that one of the MBAs studied, 2,4,5-TMBA, was particularly well suited as substrate. The results from the oxidation experiments with MBAs provide a basis for judgment of the capacity of the enzymes as oxidants in various applications.

Methods

Chemicals

MBAs (Fig. 1), MBDs (Fig. 1), catechol (1,2-dihydroxybenzene), hydroquinone (1,4-dihydroxybenzene), ABTS, and HBT were obtained from Sigma-Aldrich (Steinheim, Germany). Procedures for the preparation of 1-(4-hydroxy-3,5-dimethoxyphenyl)-1-propanol *(25)* (mp 94.9–96.4°C; mp 96–97°C is reported in Pew and Connors *[25]*) and 1-(4-hydroxy-3,5-dimethoxyphenyl)-1-propanone *(26)* (mp 109.8–110.3°C; mp 109–110°C is reported in Pew and Connors *[26]*) have been described in the literature.

The preparation of bis(2,4,5-trimethoxyphenyl)methane has been described by von Usslar and Preuss *(27)* and Birch et al. *(28)*. We obtained *bis*(2,4,5-trimethoxyphenyl)methane by treatment of 2,4,5-TMBA with 0.06 M hydrochloric acid at room temperature. Precipitated *bis*(2,4,5-trimethoxyphenyl)methane was filtered off. Mp 99.5–100.5°C (mp 102–102.5°C is reported in *[27]*). ^{1}H NMR (400 MHz, CDCl$_3$, TMS, 20°C): δ 3.75 (6H, s, OCH$_3$), 3.81 (6H, s, OCH$_3$), 3.84 (2H, s, CH$_2$), 3.88 (6H, s, OCH$_3$), 6.58 (2H, s, H-Ar), and 6.66 (2H, s, H-Ar).

Table 1
TLC and UV-Visible Data for the MBAs and Their Oxidation Products

Compound	R_f	Color[a]	λ_{max} (nm)	ε
2,3-DMBA	0.40	Purple	274	1600
2,5-DMBA	0.44	Green	289	3000
3,4-DMBA	0.22	Pink	276	2800
2,3,4-TMBA	0.30	Red	270	900
2,4,5-TMBA	0.20	Blue	289	4400
3,4,5-TMBA	0.20	Brown	270	810
2,3-DMBD	0.77	Orange	260	9400
			322	2400
2,5-DMBD	0.78	Orange	258	8500
			355	4300
3,4-DMBD	0.58	Yellow	277	11000
			308	9200
2,3,4-TMBD	0.66	Yellow	285	14000
2,4,5-TMBD	0.42	Yellow	277	11000
			344	8500
3,4,5-TMBD	0.61	Yellow	286	11000

[a]Observed after spraying with formalin/H_2SO_4 solution.

Thin-Layer Chromatography Analyses

Thin layer chromatography (TLC) was performed using silica gel plates (Kieselgel 60F_{254}, Merck, Darmstadt, Germany). The eluent was toluene-ethyl acetate (2:1) (R_f values: MBAs and MBDs, see Table 1; 1-(4-hydroxy-3,5-dimethoxyphenyl)-1-propanol, 0.19; 1-(4-hydroxy-3,5- dimethoxyphenyl)-1-propanone, 0.42; bis(2,4,5-trimethoxyphenyl)-methane, 0.47). Spots were observed using ultraviolet (UV) light and by spraying with formalin/H_2SO_4 (1:9) and subsequent heating.

UV-Visible Spectroscopy

UV-visible spectra were recorded with a UV-1601PC (Shimadzu, Tokyo, Japan), unless otherwise stated. The compounds were first dissolved in 95% ethanol (to 60 mM) and then diluted to a suitable concentration with water. The λ_{max} and ε_{max} values are: MBAs and MBDs, see Table 1; catechol, λ_{max} 275 nm, ε_{max} 2400; hydroquinone, λ_{max} 289 nm, ε_{max} 2600. UV-visible spectra for 1-(4-hydroxy-3,5-dimethoxyphenyl)-1-propanol (λ_{max} 272 nm, ε_{max} 1800) and 1-(4-hydroxy-3,5-dimethoxyphenyl)-1-propanone (λ_{max} 299 nm, ε_{max} 10,000) were recorded using a Cary 4 UV-visible spectrophotometer (Varian, Palo Alto, CA).

Enzyme Preparations

T. versicolor laccase was obtained from Jülich Fine Chemicals GmbH (Jülich, Germany). The fungal laccase was purified to homogeneity using anion-exchange chromatography and 0.1 M phosphate buffer (pH 6.0), as reported previously (29), except that DEAE Sepharose Fast Flow resin (Amersham Biosciences, Uppsala, Sweden) was used instead of DEAE A50 Sephadex. The purity of the preparation was judged from sodium dodecyl sulfate-polyacrylamide gel electrophoresis and subsequent staining with Coomassie brilliant blue. The laccase activity was determined by using ABTS and the reactions were monitored at 414 nm ($\varepsilon = 3.6 \times 10^4$) (30). The reaction medium was 50 mM acetate buffer, pH 5.2, and the reaction temperature was 23°C. One unit equals the formation of 1 µmol ABTS radical cation per min. The protein concentration was determined with the dye assay (Bio-Rad, Hercules, CA) and bovine-serum albumin was used as the standard.

Lacquer tree (*Rhus vernicifera*) laccase and HRP VI (Reinheitzahl [RZ, A_{403}/A_{275}] 3) were obtained from Sigma Chemical Co. (St. Louis, MO). The HRP activity was determined in the same way as the laccase activity except that H_2O_2 was added to a concentration of 0.4 mM.

LP was produced by *T. versicolor* PRL572 and purified to homogeneity as previously described (31). The activity was determined using a 2 mM veratryl alcohol solution and the veratraldehyde (3,4-dimethoxybenzaldehyde [DMBD]) formation was monitored based on the absorbance at 310 nm as described previously (32). One unit equals the formation of 1 µmol veratraldehyde/min.

Screening for Acid Instability of MBAs

Solutions (10 mM) of the MBAs (Fig. 1) in 0.2 M HCl were kept at 23°C overnight. TLC analysis did not show any product formation, except for 2,4,5-TMBA. Treatment of 2,4,5-TMBA with 0.2 M HCl resulted in the formation of *bis*(2,4,5-trimethoxyphenyl)methane (*cf.* reports by von Usslar and Preuss [27] and Birch et al. [28]). Therefore, the effect of acidic conditions on 2,4,5-TMBA was investigated further. 2,4,5-TMBA (40 mM solution) was treated with citrate-phosphate buffers at pH 2.2, 3.0, 4.0, and 5.0. Samples for TLC analysis were taken after 15 min, 30 min, 1 h, 3 h, and 24 h. Treatment at pH 2.2 resulted in dimer formation after 15 min, whereas dimer formation at pH 3.0 was detected only after 1 h or longer. At pH 4.0 and 5.0, dimer formation was detected in samples taken after 3 and 24 h, respectively. Thus, dimer formation occurs even under weakly acidic conditions. However, both TLC and spectrophotometric analyses indicated that the rate of dimer formation at pH 3.0–5.0 was so slow that the results of the enzymic oxidations were not affected.

Examination of Oxidation Reactions of MBAs Using UV-Visible Spectroscopy and TLC

The enzymic reactions were monitored at 23°C by UV-visible spectrophotometry using wavelength scans (230–450 nm). The reactions were performed with a substrate concentration of 0.5 mM at pH 3.0, 5.2, and 6.0.

The concentrations of LP and HRP were 4 and 120 µg/mL, respectively, and the reactions were initiated by adding enzyme to the reaction mixture, which was 0.4 mM with respect to H_2O_2. For fungal laccase, the reaction medium was air-saturated and the enzyme concentration was 330 µg/mL. Control samples (no enzyme added) were examined with and without H_2O_2. The cuvets were sealed to avoid evaporation of water. Spectra were recorded every 2nd h.

In a second series of experiments, TLC was employed to verify the results obtained by UV-visible spectrophotometry. The enzymic reactions were carried out with 20 mM substrate concentration at pH 4.0 (50 mM acetate buffer) and 23°C overnight. For LP and HRP, enzyme and H_2O_2 were added in portions to decrease the effect of enzyme inactivation and to obtain enough product for TLC analysis. For fungal laccase, 3,4-DMBD was detected by TLC as a product from 3,4-DMBA in an experiment with a 3,4-DMBA concentration of 40 mM and an enzyme concentration of 1 mg/mL at pH 3.0 and at 23°C for 4 d. After the reaction, the mixture was extracted with 0.5 volume of ethyl acetate and a sample of the extract was applied to a TLC plate. Controls that did not contain any enzyme were examined as well.

The oxidation of 2,4,5-TMBA with tree laccase (*R. vernicifera*) was also examined. The reaction medium was 50 mM acetate buffer (pH 4.0). The substrate concentration was 2 mM and laccase concentrations up to 0.025 units/mL (as determined with ABTS at pH 5.2) were tested. The samples were monitored for 48 h, but no reaction was observed (UV spectroscopy).

Comparison of the Oxidation Rates of MBAs

The oxidation rates of the different MBAs were monitored based on the formation of MBDs using UV-visible spectrometry. The increase in absorbance was recorded at 322 nm for 2,3-DMBA ($\varepsilon = 2400$), 355 nm for 2,5-DMBA ($\varepsilon = 4300$), and 344 nm for 2,4,5-TMBA ($\varepsilon = 8500$). 3,4-DMBA ($\varepsilon = 9200$), 2,3,4-TMBA ($\varepsilon = 5500$), and 3,4,5-TMBA ($\varepsilon = 6300$) were monitored at 310 nm. The initial absorbance change was used to calculate the rate of product formation (µmol/min). Oxidation by LP was studied using reaction mixtures containing 1 mM substrate, 50 mM citrate-phosphate buffer (pH 3.0), 5 µg/mL LP, and 0.4 mM H_2O_2. HRP was investigated in the same manner as LP, except that the concentration of substrate was 10 mM and that the enzyme concentration (HRP) was 160 µg/mL. Laccase reaction mixtures contained 28 mM substrate, 50 mM citrate-phosphate buffer (pH 3.0) and 1 mg/mL fungal laccase. The highest reaction rate observed for each enzyme was arbitrarily set to 100.

Determination of Kinetic Constants With 2,4,5-TMBA and 3,4-DMBA as Substrates

The oxidation rates of 2,4,5-TMBA and 3,4-DMBA catalyzed by LP, HRP, and fungal laccase were determined at 23°C by UV-visible spectrophotometry using 50 mM acetate buffer (pH 4.0) as the reaction medium. The H_2O_2 concentration in the peroxidase-catalyzed reactions was 0.4 mM. The initial apparent reaction rates (V_0, μM/min) were calculated based on the initial increase of the absorbance at 344 nm (ε = 8500) for 2,4,5-TMBA and at 308 nm (ε = 9200) for 3,4-DMBA. The curves were of Michaelis-Menten type and the steady-state apparent Michaelis constant, K_m, was determined by fitting V_0 and the substrate concentration to the Michaelis-Menten equation with the Prizm program of GraphPad Software (San Diego, CA). All the experiments were repeated in duplicate and the standard deviations of the determination of the apparent k_{cat} and K_m values are given.

With LP, the 2,4,5-TMBA concentration was 0.01–1.2 mM and the enzyme concentration was 1.6 μg/mL. With HRP, the 2,4,5-TMBA concentration was 1–26 mM and the enzyme concentration was 0.1 μg/mL. For both peroxidases, the reactions were monitored for 5 min. With fungal laccase, the 2,4,5-TMBA concentration was 1–40 mM and the enzyme concentration 0.1 μg/mL. All the experiments with laccase were carried out in air-saturated solutions and the reactions were monitored for 60 min.

With LP, the 3,4-DMBA concentration was 0.01–1.4 mM and the enzyme concentration was 1.6 μg/mL. The reactions were monitored for 2 min. With HRP, the 3,4-DMBA concentration was 1-32 mM and the enzyme concentration was 100 μg/mL. The reactions were monitored for 15 min. With fungal laccase, the 3,4-DMBA concentration was 4–60 mM, the enzyme concentration was 100 μg/mL, and the reactions were monitored for 60 min.

pH Profiles

The initial reaction rates at 23°C were monitored using UV-visible spectrophotometry. The reaction media were 50 mM citrate-phosphate buffers and the pH range was 2.2–7.0. For LP and HRP, the concentration of H_2O_2 was 0.4 mM. The pH profiles with catechol (based on the decrease in absorbance at 275 nm), hydroquinone (based on the decrease in absorbance at 289 nm), 1-(4-hydroxy-3,5-dimethoxyphenyl)-1-propanol (based on the increase in absorbance at 295 nm [formation of 1-{4-hydroxy-3,5-dimethoxyphenyl}-1-propanone]), and ABTS (based on the increase in absorbance at 414 nm [formation of ABTS radical cation]) as substrates were determined for LP, HRP, and fungal laccase. The pH profile with ABTS as the substrate was also determined for tree laccase. In all cases mentioned earlier, the substrate concentration was 0.5 mM. The pH profiles with 2,4,5-TMBA (based on the increase in absorbance at 344 nm [formation of 2,4,5-trimethoxybenzaldehyde {TMBD}]) as the substrate were determined for LP, HRP, and fungal laccase. The pH profile with 3,4,5-TMBA (based on the increase in absorbance at 300 nm [formation of 3,4,5-TMBD]) as substrate was determined for LP. In reactions with 2,4,5-TMBA and 3,4,5-TMBA, the substrate concentration was 1 mM. The pH profiles for 2,5-DMBA (based on the increase in absorbance at 355 nm [formation

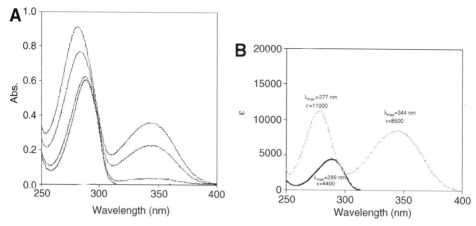

Fig. 2. (A) UV-visible spectral changes for a reaction with 2,4,5-trimethoxybenzyl alcohol (TMBA) catalyzed by *T. versicolor* laccase. After initiation of the reaction by addition of laccase (the initial spectrum [λ_{max} 289 nm] is shown as a black line [bottom]), the absorbance gradually increased (the gray lines represent spectra recorded after 1.5 h [second from bottom], 15 h [second from top] and 22 h [top]). The reaction was performed in 50 mM acetate buffer, pH 5.2, with 0.14 mM 2,4,5-TMBA and 56 µg/mL laccase in the reaction mixture. The yield of 2,4,5-TMBD was 3% after 1.5 h, 19% after 15 h, and 30% after 22 h. **(B)** UV-visible spectra for 2,4,5-TMBA (black line) and 2,4,5-TMBD (gray line).

of 2,5-DMBD]) and 3,4-DMBA (based on the increase in absorbance at 310 nm [formation of 3,4-DMBD]) as substrates (in a concentration of 10 mM) were determined for HRP. The concentration of LP was 0.4–2 µg/mL. For HRP, the enzyme concentration was 0.02–2 µg/mL, except when 2,5-DMBA and 3,4-DMBA were substrates. In the latter two cases, the enzyme concentration was 200 µg/mL. The laccase concentration was 0.07–50 µg/mL.

Results

To investigate if the MBA congeners (Fig. 1) served as substrates for LP, HRP, and laccase, the UV-visible spectral changes of the reaction mixtures obtained in experiments with the different MBAs were examined. UV-visible spectroscopy provides conclusive results as reactants and products differ distinctly from each other as concerns UV-visible properties (data are given in Table 1 and an example is provided in Fig. 2). Product identification was in general complemented by TLC examinations (data regarding R_f values and spot colors are given in Table 1). No formation of oxidation products was observed unless enzymes were added to the reaction mixture. Reactions were found to occur for all combinations of the six MBAs and LP, HRP and fungal laccase, but the oxidation of some of the MBAs by HRP and laccase was extremely slow. A low pH and a relatively high concentration of the MBAs had in general a beneficial impact on the reactions. In reaction mixtures containing HRP and 2,3-DMBA, 2,3,4-TMBA

Table 2
Relative Rate of Oxidation of Methoxy-Substituted Benzyl Alcohols by LP, HRP, and *T. versicolor* Laccase Determined by UV Spectroscopic Measurements[a]

Substrate	Relative rate of oxidation		
	LP/H_2O_2	HRP/H_2O_2	Laccase/O_2
2,3-DMBA	13[b]	≤0.02	≤0.01
2,5-DMBA	24[b]	0.05[b]	0.05[b]
3,4-DMBA	100[b]	0.03[b]	0.02[b]
2,3,4-TMBA	13[b]	≤0.02	≤0.01
2,4,5-TMBA	40[b]	100[b]	100[b]
3,4,5-TMBA	36[b]	≤0.02	≤0.01

[a]The rate of product formation with the best substrate is arbitrarily set to 100 for each enzyme. For reaction conditions, see Methods.
[b]The identity of the product was verified by TLC analysis.

or 3,4,5-TMBA, spectral changes were not observed within 24 h at pH 5.2 or 6.0, but reactions (very slow) were observed at pH 3.0. Similarly, in reaction mixtures containing laccase and 2,3-DMBA, 3,4-DMBA, 2,3,4-TMBA, or 3,4,5-TMBA, spectral changes were not observed within 24 h at pH 5.2 or 6.0, but reactions (very slow) were observed at pH 3.0. The spectral changes observed with fungal laccase and the best substrate, 2,4,5-TMBA, are illustrated in Fig. 2. Oxidation caused by LP and HRP led to similar spectral changes. Figure 2 shows that the enzymic oxidation caused a shift of λ_{max} from 289 to 277 nm and a new peak appeared at 344 nm. UV-visible and TLC examinations showed that the sole predominating oxidation product was 2,4,5-TMBD. With respect to product formation, the other MBA congeners exhibited analogous results with LP, HRP, and *T. versicolor* laccase; the products generated were the corresponding MBDs (UV-visible and TLC examinations).

When the laccase-catalyzed oxidation of 2,4,5-TMBA was allowed to proceed more extensively (the reaction mixture contained 1 m*M* substrate, 220 µg/mL fungal laccase and 50 m*M* acetate buffer, pH 4.0, and was incubated at 23°C for 24 h), the yield of 2,4,5-TMBD was about 80% (UV spectrometry). TLC analysis still showed only one product: 2,4,5-TMBD. When the reaction was allowed to continue for further 24 h, there was only a slight increase in the yield of 2,4,5-TMBD.

LP could oxidize all of the MBAs at a substantial rate (Table 2). With LP, the reaction rate order was: 3,4-DMBA > 2,4,5-TMBA ~ 3,4,5-TMBA > 2,5-DMBA > 2,3,4-TMBA ~ 2,3-DMBA. The reaction rates at pH 3 were 23 nmol/min with 2,5-DMBA and 94 nmol/min with 3,4-DMBA, when the substrate concentration was 1 m*M* and the concentration of LP was 5 µg/mL.

With HRP, the reaction rate order was 2,4,5-TMBA >> 2,5-DMBA > 3,4-DMBA > 2,3-DMBA ~ 3,4,5-TMBA ~ 2,3,4-TMBA (Table 2). The rates of

Table 3
Comparison of Apparent Kinetic Parameters for LP, HRP, and *T. versicolor*
Laccase With 2,4,5-TMBA and 3,4-DMBA as Substrates at pH 4.0

Enzyme	2,4,5-TMBA			3,4-DMBA		
	k_{cat} (/min)	K_m (mM)	k_{cat}/K_m (/mM/min)	k_{cat} (/min)	K_m (mM)	k_{cat}/K_m (/mM/min)
LP	135 ± 2	0.047 ± 0.003	2900	260 ± 3	0.115 ± 0.005	2300
HRP	2700 ± 200	8 ± 1	340	5 ± 0.1	26 ± 1	0.2
laccase	600 ± 200	73 ± 9	8	<1	>100	–

HRP-catalyzed oxidations at pH 3.0 were 7 nmol/min with 2,5-DMBA and 4 nmol/min with 3,4-DMBA, when the substrate concentration was 10 mM and the enzyme concentration was 160 μg/mL. An oxidized intermediate of HRP interfered with the determination of 2,3-DMBD. This could possibly lead to an overestimate of the oxidation rate of 2,3-DMBA by HRP.

With fungal laccase, the reaction rate order was: 2,4,5-TMBA >> 2,5-DMBA > 3,4-DMBA > 3,4,5-TMBA ~ 2,3,4-TMBA ~ 2,3-DMBA (Table 2). The reaction rates for fungal laccase at pH 3.0 were 0.7 nmol/min with 2,5-DMBA and 0.2 nmol/min with 3,4-DMBA, when the substrate concentration was 28 mM and the enzyme concentration was 1 mg/mL. The laccase-catalyzed oxidation of 2,4,5-TMBA proceeded at a 15-fold higher rate when the substrate concentration was raised from 1 to 16 mM, indicating a nearly linear relationship with the concentration change. With 3,4-DMBA as substrate for fungal laccase, the oxidation rate at pH 3.0 was 54 times higher when the substrate concentration was raised from 0.4 to 28 mM. Spectral changes in experiments with 2,4,5-TMBA as substrate were not observed with *R. vernicifera* laccase.

ABTS and HBT were studied as mediators for *T. versicolor* laccase in the oxidation of MBAs. The concentrations of ABTS and HBT were 1 and 0.8 mM, respectively. The pH range studied was 4.0–5.2. All the six MBAs (Fig. 1) were oxidized by the laccase-ABTS system and TLC analysis showed that the products were the corresponding MBDs. The effect of HBT on the oxidation of 2,4,5-TMBA to 2,4,5-TMBD was also studied. The oxidation rate was 87-fold faster when HBT was present in a reaction mixture in which the concentration of 2,4,5-TMBA was 1 mM.

Based on the oxidation experiments with the MBAs (Table 2), 2,4,5-TMBA and 3,4-DMBA were chosen as substrates in further kinetic studies and the results from the determination of apparent kinetic constants (k_{cat}, K_m and k_{cat}/K_m) are collected in Table 3. It is notable that for both 2,4,5-TMBA and 3,4-DMBA, LP had a comparatively low K_m. With all the three enzymes, the K_m value for 2,4,5-TMBA was lower than for 3,4-DMBA.

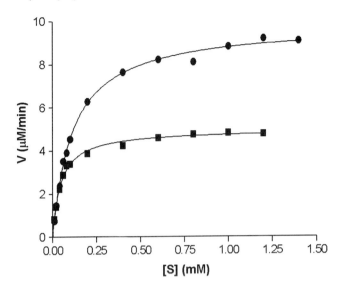

Fig. 3. LP-catalyzed oxidation of 2,4,5-TMBA (■) and 3,4-DMBA (●) at pH 4.

HRP and laccase had several 100-folds higher k_{cat} for 2,4,5-TMBA than for 3,4-DMBA. In contrast, LP had higher k_{cat} for 3,4-DMBA than for 2,4,5-TMBA. The clear difference in this regard is shown by the Michaelis-Menten plots for LP with 2,4,5-TMBA and 3,4-DMBA as substrates (Fig. 3).

The effect of pH on the initial oxidation rates of nonphenolic (Fig. 4A–C) and phenolic substrates (Fig. 4D–F) catalyzed by LP, HRP, and laccase was studied. With LP, an optimum at pH 3.0 was always observed for the initial reaction rates, regardless whether the substrate was nonphenolic (Fig. 4A) or phenolic (Fig. 4D).

With HRP, 3,4-DMBA, 2,5-DMBA, and 2,4,5-TMBA showed monotonic pH profiles; the rates decreased with increasing pH (Fig. 4B) (data for 2,5-DMBA are not shown as they are practically identical to those obtained with 3,4-DMBA). ABTS exhibited a bell-shaped pH profile in the pH range 2.2–7.0 and displayed an optimum at pH 4.0 (Fig. 4B). All phenolic substrates (Fig. 4E) showed pH profiles with optima in the range 6.0–7.0.

With *T. versicolor* laccase, the two nonphenolic substrates 2,4,5-TMBA and ABTS showed monotonic pH profiles; the rates decreased as the pH increased (Fig. 4C). With *R. vernicifera* laccase, ABTS showed an optimum at pH 3.0 (Fig. 4C). The three phenols exhibited bell-shaped pH profiles with an optimum at pH 4.0 (Fig. 4F).

Discussion

Enzymic oxidation of MBAs invariably resulted in formation of MBDs as the prevalent products. Even though LP, HRP, and *T. versicolor* laccase could catalyze the oxidation of all the different MBA congeners

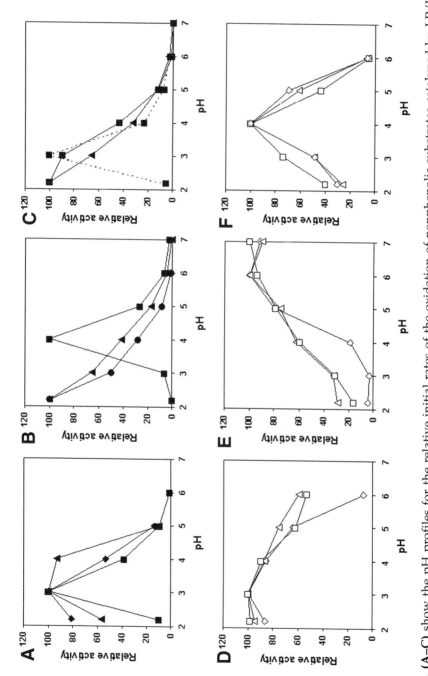

Fig. 4. (A–C) show the pH profiles for the relative initial rates of the oxidation of nonphenolic substrates catalyzed by LP/H$_2$O$_2$ (A), HRP/H$_2$O$_2$ (B), T. versicolor laccase/O$_2$ (C, solid line), and R. vernicifera laccase (C, dashed line). The substrates are 2,4,5-TMBA (▲), ABTS (■), 3,4,5-TMBA (♦), and 3,4-DMBA (●) (D–F) show the pH profiles for the relative initial rates of oxidation of phenolic substrates catalyzed by LP/H$_2$O$_2$ (D), HRP/H$_2$O$_2$ (E), and T. versicolor laccase/O$_2$ (F). The substrates are catechol (△), hydroquinone (□), and 1-(4-hydroxy-3,5-dimethoxyphenyl)-1-propanol (◇).

studied (Fig. 1), the rate of oxidation was very different (Table 2). For HRP and laccase, fastidious reaction conditions, such as low pH (3.0) and a relatively high concentration of substrate (10–28 mM), were required to obtain even very slow reactions with some of the MBAs.

The finding that a high-substrate concentration was required to obtain a detectable reaction rate for laccase and HRP may be explained in terms of high K_m values for these two enzymes compared with LP (Table 3). It may be possible to modify HRP and laccase by means of site-directed mutagenesis to obtain enzymes with lower K_m. Significant changes of the K_m and k_{cat} of laccases modified by site-directed mutations have been reported (33).

HRP concurred with *T. versicolor* laccase in the respect that 2,4,5-TMBA was by far the best substrate, whereas LP catalyzed the oxidation of 3,4-DMBA with higher rate. The catalytic efficiency of LP, discussed in terms of k_{cat}/K_m values (Table 3), was of the same order of magnitude with both 2,4,5-TMBA and 3,4-DMBA as substrates. In contrast, the k_{cat} and the catalytic efficiency for HRP and fungal laccase were several order of magnitudes higher with 2,4,5-TMBA than with 3,4-DMBA as substrate. Thus, there seems to be a fundamental difference regarding oxidation of the MBAs between HRP and fungal laccase on one hand, and LP on the other hand.

In contrast to laccase from *T. versicolor*, equivalent or even twofold higher doses of *R. vernicifera* laccase did not catalyze product formation from 2,4,5-TMBA. This is in all probability related to the difference in redox potential between *R. vernicifera* (around 0.4 V) and *T. versicolor* laccase (around 0.8 V) (1). Because MBAs are neither phenolic compounds nor diamines, they are not conventionally regarded as substrates for laccase. Laccases are normally used in combination with mediators in oxidization experiments with nonphenolic compounds (5).

Kersten et al. (3) observed enzymic oxidation of several compounds in a set of methoxy-substituted benzene congeners with *Phanerochaete chrysosporium* LP. HRP and *T. versicolor* laccase were found to catalyze the oxidation of few (HRP) or only one (laccase) of the methoxy-substituted benzenes studied. The enzymic oxidation of the methoxy-substituted benzenes was performed at low pH (3.0) and with relatively low substrate concentration (0.2 mM). With 1,2,4,5-tetramethoxybenzene, which was found to serve as substrate for all three enzymes, the K_m for LP was found to be considerably lower than for HRP and laccase (3), which is analogous to our results with 2,4,5-TMBA and 3,4-DMBA. That 1,2,4,5-tetramethoxybenzene served as substrate for all three enzymes was attributed to its comparatively low half-wave potential (0.81 V [12]) compared with, for example, methoxybenzene (1.76 V [12]), which was resistant to oxidation by all three enzymes (3). The catalytic action of a set of fungal laccases (4) has also been discussed in terms of the redox potential displayed by the enzymes in relation to the redox potential determined for the reducing substrate.

Benzyl alcohol has been reported to possess a redox potential of 2.4 V, whereas the redox potential of MBAs was found to be lower, for example, 1.7 V for 4-methoxybenzyl alcohol (34). The redox potential of three of the DMBAs used in this study has been determined (35): 2,5-DMBA, 1.33 V (vs normal hydrogen electrode); 3,4-DMBA, 1.36 V; 2,3-DMBA, 1.39 V. The present study showed that the oxidation rate order of these DMBAs with *T. versicolor* laccase and HRP is the following one: 2,5-DMBA > 3,4-DMBA > 2,3-DMBA. However, for LP, 3,4-DMBA was the best substrate (Table 2). Increasing rates of the oxidation of DMBAs by HRP and laccase follow decreasing redox potential, whereas the oxidation experiments indicate a special relationship between LP and 3,4-DMBA (veratryl alcohol). This compound is a secondary metabolite of *P. chrysosporium* (14). There are different hypotheses regarding the role of veratryl alcohol in lignin biodegradation (36). Veratryl alcohol may protect LP from inactivation by hydrogen peroxide by prevention of the formation of the inactive form Compound III. Veratryl alcohol may serve as the reducing substrate for Compound II to complete the catalytic cycle back to the native form of the enzyme. Furthermore, veratryl alcohol may act as a one-electron redox mediator in lignin biodegradation. Interestingly, site-directed mutagenesis of LP has suggested the existence of an electron-transfer pathway between Trp171 and the heme group, and that Trp171 is important for the oxidation of veratryl alcohol but not all reducing substrates (37). Site-directed mutagenesis of manganese peroxidase supported the role of Trp171 in the oxidation of veratryl alcohol by LP (38). Multiple alignments of peroxidase sequences early on indicated that LPs specifically contained this conserved Trp, located near the proximal His in the F helix (39). Thus, LP is known to possess a conserved amino-acid residue that may be responsible for the interaction with veratryl alcohol.

The oxidation of the MBAs was favored by low pH (2.0–3.0) for all the three enzymes. The oxidation of veratryl alcohol by LP has previously been shown to be favored by low pH (21,37,40). However, LP will eventually become instable if the pH is too low. Regarding laccase, Bourbonnais and Paice (5) reported that veratryl alcohol was not oxidized at pH 5.0. In the present study, a pH of 3.0 rather than 5.2 or 6.0 was found to be required in order to observe any laccase-catalyzed oxidation of veratryl alcohol (3,4-DMBA). HRP showed an optimum at higher pH with ABTS than with 2,4,5-TMBA, 3,4-DMBA, or 2,5-DMBA as substrate (Fig. 4B), which suggests that the optimum for ABTS oxidation observed at pH 4.0 is not a consequence of enzyme instability at low pH (2.0–3.0).

The pH optima for LP-catalyzed oxidation of phenolic substrates did not differ from those of nonphenolic substrates, whereas the optima for HRP-catalyzed and *T. versicolor* laccase-catalyzed oxidation of phenolic substrates were observed at higher pH than those for nonphenolic substrates (Fig. 4). The pH profiles obtained with fungal laccase and phenolic and nonphenolic substrates were studied by Xu (41), who proposed that the pH activity profiles of fungal laccase was the result of two opposing effects, namely the redox potential difference among the reducing substrate and the T1 copper of laccase, which would be favored by higher pH for a phenolic substrate, and the binding of a hydroxide ion to the T2/T3 copper center of laccase, which would result in inhibition and decreased activity at higher pH. The changes in E^0 with pH of fungal laccases were found to be rather modest (41). The contribution of hydroxide inhibition would consequently result in a monotonic pH activity profile for a nonphenolic substrate, as observed for *T. versicolor* laccase-catalyzed oxidation of ABTS and 2,4,5-TMBA (Fig. 4C).

The fact that the enzymes studied display different pH profiles for the same substrate, as observed for the oxidation of ABTS or phenolic substrates (Fig. 4), suggests that not only substrate-specific but also enzyme-specific factors need to be taken into account to explain the pH profiles. Indeed, site-specific mutagenesis of LP resulted in a novel form that displayed a shift in the pH/activity profile, at least regarding one substrate (37). More research is required to elucidate the detailed mechanisms behind the pH profiles.

1-(4-Hydroxy-3,5-dimethoxyphenyl)-1-propanol was found to serve as a substrate for LP, HRP, and laccase (Fig. 4). In a series of experiments with HRP at different pH (4.5, 5.0, and 7.0), 1-(4-hydroxy-3,5-dimethoxyphenyl)-1-propanone was obtained in about 80% yield. HRP-catalyzed dehydrogenation of 1-(4-hydroxy-3,5-dimethoxyphenyl)-1-propanol has previously been reported to produce 1-(4-hydroxy-3,5-dimethoxyphenyl)-1-propanone in high yield (26). Dehydrogenation of the structurally closely related compound 1-(4-hydroxy-3-methoxyphenyl)-1-propanol resulted in the formation of a dioxepin as a major product (25).

It appears from the examples discussed earlier that the MBA congeners are suitable in studies of oxidation with different enzymes or combinations of enzymes and mediators. One reason for this is that the oxidation of the MBAs produces the corresponding MBDs as the sole predominating products. Second, the MBAs exhibit characteristic UV properties that make it possible to follow the progress of the oxidation reactions by UV measurements. Third, the set of MBAs selected represents different requirements as concerns oxidation potential and this makes it possible to do comparative studies of the capability of the different enzymes as oxidation catalysts. For example, 2,4,5-TMBA could be used to discriminate among laccases from different sources, such as *T. versicolor* and *R. vernicifera*. We think that the MBAs are useful as substrates not only in studies of enzyme properties but also in connection with the design of processes based on enzymic catalysis aiming at detoxification of waste effluents (7), pulp bleaching (42), and production of chemicals (43). The comparisons of the different enzymes show that LP/H_2O_2 is an efficient catalyst as judged from the oxidation experiments with MBAs. However, in certain applications selectivity is of importance and in such cases HRP/H_2O_2 and laccase/O_2 may serve as suitable oxidants.

Acknowledgment

This work was supported by grants from the Swedish Research Council.

References

1. Reinhammar, B. (1984), In: *Copper Proteins and Copper Enzymes.* Lontie, R. (ed.), CRC Press, Boca Raton, FL, vol. 3, pp. 1–35.
2. Reinhammar, B. (1997), In: *Multi-Copper Oxidases.* Messerschmidt, A. (ed.), World Scientific, Singapore, pp. 167–200.
3. Kersten, P. J., Kalyanaraman, B., Hammel, K. E., Reinhammar, B., and Kirk, T. K. (1990), *Biochem. J.* **268**, 475–480.
4. Xu, F., Shin, W., Brown, S. H., Wahleithner, J. A., Sundaram, U. M., and Solomon, E. I. (1996), *Biochim. Biophys. Acta* **1292**, 303–311.
5. Bourbonnais, R. and Paice, M. G. (1990), *FEBS Lett.* **267**, 99–102.
6. Potthast, A., Rosenau, T., Chen, C. L., and Gratzl, J. S. (1996), *J. Mol. Catal. A: Chemical* **108**, 5–9.
7. Duran, N. and Esposito, E. (2000), *Appl. Catal. B: Environ.* **28**, 83–99.
8. Mayer, A. M. and Staples, R. C. (2002), *Phytochemistry* **60**, 551–565.
9. Smith, M., Thurston, C. F., and Wood, D. A. (1997), In: *Multi-Copper Oxidases.* Messerschmidt, A. (ed.), World Scientific, Singapore, pp. 201–224.
10. Farver, O. and Pecht, I. (1997), In: *Multi-Copper Oxidases.* Messerschmidt, A. (ed.), World Scientific, Singapore, pp. 355–389.
11. Dunford, H. B. (1991), In: *Peroxidases in Chemistry and Biology.* Everse, J., Everse, K. E., and Grisham, M. B. (eds.), CRC Press, Boca Raton, FL, vol. II, pp. 1–24.
12. Zweig, A., Hodgson, W. G., and Jura, W. H. (1964), *J. Am. Chem. Soc.* **86**, 4124–4129.
13. Brunow, G., Lundquist, K., and Gellerstedt, G. (1999), In: *Analytical Methods in Wood Chemistry, Pulping, and Papermaking.* Sjöström, E. and Alén, R. (eds.) Springer, Berlin, pp. 77–124.
14. Lundquist, K. and Kirk, T. K. (1978), *Phytochemistry* **17**, 1676.
15. Fabbrini, M., Galli, C., Gentili, P., and Macchitella, D. (2001), *Tetrahedron Lett.* **42**, 7551–7553.
16. Kawai, S., Umezawa, T., and Higuchi, T. (1989), *Wood Res.* **76**, 10–16.
17. Duran, N., Bromberg, N., and Kunz, A. (2001), *J. Inorg. Biochem.* **84**, 279–286.
18. McEldoon, J. P., Pokora, A. R., and Dordick, J. S. (1995), *Enzyme Microb. Technol.* **17**, 359–365.
19. Harvey, P. J., Schoemaker, H. E., and Palmer, J. M. (1986), *FEBS Lett.* **195**, 242–246.
20. Khindaria, A., Yamazaki, I., and Aust, S. D. (1995), *Biochemistry* **34**, 16,860–16,869.
21. Tien, M., Kirk, T. K., Bull, C., and Fee, J. A. (1986), *J. Biol. Chem.* **261**, 1687–1693.
22. Valli, K., Wariishi, H., and Gold, M. H. (1990), *Biochemistry* **29**, 8535–8539.
23. Koduri, R. S. and Tien, M. (1994), *Biochemistry* **33**, 4225–4230.
24. Potthast, A., Rosenau, T., Kosma, P., and Fischer, K. (1999), In: *Proceedings of 10th ISWPC.* Yokohama, Japan, vol. I, pp. 596–601.
25. Pew, J. C. and Connors, W. J. (1969), *J. Org. Chem.* **34**, 580–584.
26. Pew, J. C. and Connors, W. J. (1969), *J. Org. Chem.* **34**, 585–589.
27. von Usslar, U. and Preuss, F. R. (1970), *Archiv der Pharmazie* **303**, 842–850.
28. Birch, A. J., Jackson, A. H., Shannon, P. V. R., and Stewart, G. W. (1975), *J. Chem. Soc. Perkin I* 2492–2501.
29. Fåhraeus, G. and Reinhammar, B. (1967), *Acta Chem. Scand.* **21**, 2367–2378.
30. Childs, R. E. and Bardsley, W. G. (1975), *Biochem. J.* **145**, 93–103.
31. Jönsson, L., Johansson, T., Sjöström, K., and Nyman, P. O. (1987), *Acta Chem. Scand.* **B41**, 766–769.
32. Tien, M. and Kirk, T. K. (1988), *Methods Enzymol.* **161B**, 238–249.
33. Xu, F., Berka, R. M., Wahleithner, B. A., et al. (1998), *Biochem. J.* **334**, 63–70.

34. Fabbrini, M., Galli, C., and Gentili, P. (2002), *J. Mol. Catal. B: Enzymatic* **16,** 231–240.
35. Baciocchi, E., Gerini, M. F., Lanzalunga, O., and Mancinelli, S. (2002), *Tetrahedron* **58,** 8087–8093.
36. Zapanta, L. S. and Tien, M. (1997), *J. Biotechnol.* **53,** 93–102.
37. Doyle, W. A., Blodig, W., Veitch, N. C., Piontek, K., and Smith, A. T. (1998), *Biochemistry* **37,** 15,097–15,105.
38. Timofeevski, S. L., Nie, G., Reading, N. S., and Aust, S. D. (1999), *Biochem. Biophys. Res. Commun.* **256,** 500–504.
39. Jönsson, L. (1994), Ph.D. Thesis, Lund University, Lund, Sweden.
40. Marquez, L., Wariishi, H., Dunford, H. B., and Gold, M. H. (1988), *J. Biol. Chem.* **263,** 10,549–10,552.
41. Xu, F. (1997), *J. Biol. Chem.* **272,** 924–928.
42. Bajpai, P. (2004), *Crit. Rev. Biotechnol.* **24,** 1–58.
43. Arends, I. W. C. E. and Sheldon, R. A. (2004), In: *Modern Oxidation Methods.* Bäckvall, J. E. (ed.), Wiley-VCH, Weinheim, pp. 83–118.

Obtainment of Chelating Agents Through the Enzymatic Oxidation of Lignins by Phenol Oxidase

Gabriela M. M. Calabria and Adilson R. Gonçalves*

Biotechnology Department DEBIQ–FAENQUIL, CP 116, CEP 12600-970, Lorena—SP, Brazil; E-mail: adilson@debiq.faenquil.br

Abstract

Oxidation of lignin obtained from acetosolv and ethanol/water pulping of sugarcane bagasse was performed by phenol oxidases: tyrosinase (TYR) and laccase (LAC), to increase the number of carbonyl and hydroxyl groups in lignin, and to improve its chelating capacity. The chelating properties of the original and oxidized lignins were compared by monitoring the amount of Cu^{2+} bound to lignin by gel permeation chromatography. The Acetosolv lignin oxidized with TYR was 16.8% and with LAC 21% higher than that of the original lignin. For ethanol/water lignin oxidized with TYR was 17.2% and with LAC 18% higher than that of the original lignin.

Index Entries: Lignin; enzymatic oxidation; polyphenoloxidase; chelating agents.

Introduction

In the past years there was an increased interest in the use of agricultural residues, like as the sugarcane bagasse, seeking less harmful techniques to the environment as well as the technological development, to obtain products of higher aggregated value (1).

The lignin has many aromatic rings with high electronic density. Having the capacity to complex with metallic ions from aqueous solution this property can be used in the decontamination of industrial effluents.

The chemical or enzymatic oxidation of the lignin can promote the introduction of carbonyl groups and other functional group that have high electronic density, increasing the chelating power of this macromolecule (2).

Enzymatic oxidation is more selective and in this work phenol oxidases (PO) were used. PO are oxidoreductase enzymes which catalyze reactions involving direct activation of oxygen; those enzymes do not need any other cellular component act, being easier to be used. The two major groups of phenol oxidases are the tyrosinases and the laccases. Tyrosinases are produced by several types of organisms, such as mushrooms, bacteria

---

*Author to whom all correspondence and reprint requests should be addressed.

and superior plants being linked to the external membrane of the cells or soluble inside of the cellular medium *(3)*. This enzyme catalyzes two reactions: The insert of one oxygen in the orto position to the phenolic hydroxyl, and the oxidation of o-dihydroxyphenols to o-quinones *(2)*.

The laccase enzyme is produced by mushrooms and it can catalyze the oxidation of phenolic compounds to the correspondent form of free radical that is highly reactive, making possible the entry of oxygen molecules *(3)*.

In this work, lignins were oxidized by tyrosinase extracted from potatoes and by commercial laccase.

Materials and Methods

The enzyme tyrosinase (TYR) was extracted from potatoes and the laccase (LAC), was obtained from extracts of mushrooms (commercial enzyme NOVOZYM 51003, used as received). The potato extract was obtained by the grinding of 65 g of potatoes with 100 mL of 0.05 mol/L phosphate buffer pH 7.6. The homogeneous solution was filtered and centrifuged and the solution was immersed in an ice-bath and agitated, adding slowly 5 g of ammonium sulfate. The precipitate obtained was filtered and dried under a stream of nitrogen by 12 h and finally dissolved in 10 mL of phosphate buffer.

Lignin isolated from Acetosolv and ethanol/water pulping of sugarcane bagasse was oxidized by TYR and LAC using O_2 and glycerol in the homogenous phase (3:1 [v/v] 0.05 mol/L phosphate buffer: dioxane solution at pH 7.6, measured after mixing) for 5 h. Samples were taken every 30 min and analyzed by UV in a Cintra 200 spectrometer. Oxydized lignin was recovered after precipitation with HCl.

Chelated complexes were quantified using gel permeation chromatography with a Sephadex G-10 and the mobile phase was composed of 0.58 g NaCl, 1.21 g hydroxymethylaminomethane (Tris-buffer) and 0.128 g of $CuCl_2 \cdot 2H_2O$ in 800 mL and the pH was adjusted to 8.0. Samples were prepared with 17 mg of original or oxydized lignins

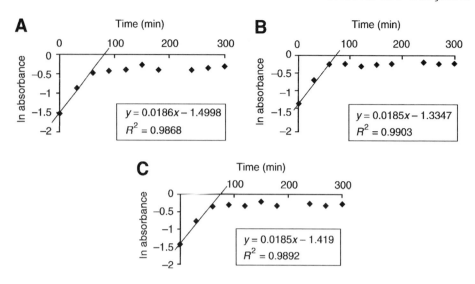

Fig. 1. ln (absorbance) in function of the time at 265 (**A**), 280 (**B**) and 320 nm (**C**) for the oxidation of lignin with TYR.

in 10 mL of a 0.04 mol/L NaOH. Samples were applied to the top of the 35-cm column (1 cm internal diameter), using 0.5 mL of the solution. Fractions were collected during the elution of about 1 mL every 5 min. A volume of 0.5 mL of each fraction was added to 5 mL of sodium diethyldithiocarbamate (0.1019 g in 100 mL) and the mixture was completed to 25 mL with distilled water. The absorbance of the complex obtained was measured at 450 nm in a Cintra 200 spectrometer. A blank experiment was performed using 0.5 mL of 0.04 mol/L NaOH and the negative area obtained was considered in the presented calculations. For the calibration curve 5 standard Cu^{2+} solutions were used. Calibration method was validated with 95% confidence level and the data presented are significative also in 95% confidence level.

Results and Discussion

Variation of absorbance values was adjusted to a pseudo-first order kinetics for the oxidation and graphics of ln (absorbance) in function of the time were obtained. Figure 1 shows graphics for the enzymatic oxidation of the lignin using TYR enzyme and Fig. 2 graphics for LAC enzyme. It was verified that the coefficients found for the reaction catalyzed by the TYR enzyme were four times higher and also more uniform in relation to that found for the reaction catalyzed with LAC enzyme. The mean for TYR was 0.0185/min and for LAC the means was 0.0044/min.

For the chromatographic determination of copper ions retained for the lignin chromatogram as showed in Fig. 3 was obtained. The lignin is a larger molecule, and therefore its speed is faster than the mobile phase. The lignin is complexed with copper ions and at the end of the column,

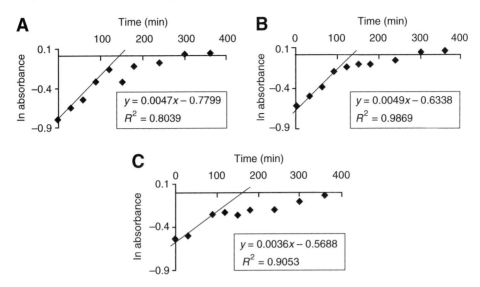

Fig. 2. ln (absorbance) in function of the time at 265 (**A**), 280 (**B**), and 320 nm (**C**) for the oxidation of lignin with LAC.

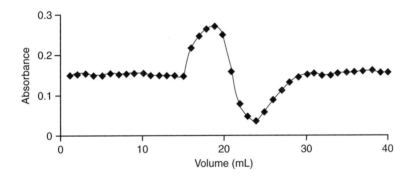

Fig. 3. Chromatogram for the elution of lignin with cupper ions.

a larger concentration of this ion will exist (positive area), and consequently a smaller concentration after the total passage of the lignin (negative area).

Theoretically the areas (positive and negative) should have the same value, and arithmetic mean the two areas was used, discounting the negative area due to the blank with NaOH.

Concentration at Cu^{2+} complexed by lignins Acetosolv original (LANO), oxidized by LAC (LAON) and by the TYR enzyme (LAOB) are displayed in Table 1. Through the Table 1 it was possible to observe that LAON presented an increase of 16.8% in the chelating power in all of the replicates in relation to LANO. The larger chelating power of 21% higher was obtained by the LAOB. In spite of both enzymes treatment presented an increase of chelating power, it was observed that the values obtained by

Table 1
Cu^{2+} Complexed by Acetosolv Lignin

Lignin		Area (absorbance units times volume [mL])	Corresponding Cu^{2+} mass (g)	mg Cu^{2+}/mg lignin Replicates	Mean
LANO	1	0.5121	0.1692	9.95	
	2	0.5986	0.2203	12.96	11.9 ± 1.7
	3	0.5973	0.2195	12.91	
LAON	1	0.6101	0.2271	13.36	
	2	0.6516	0.2516	14.8	13.9 ± 0.8
	3	0.6183	0.232	13.65	
LAOB	1	0.5798	0.2092	12.31	
	2	0.6499	0.2506	14.74	14.4 ± 1.9
	3	0.6873	0.2727	16.04	

Table 2
Cu^{2+} Complexed by Ethanol/Water Lignin

Lignin		Area (absorbance units times volume [mL])	Corresponding Cu^{2+} mass (g)	mg Cu^{2+}/mg lignin Replicates	Medium
LENO	1	0.6646	0.2409	14.17	
	2	0.5149	0.1738	10.22	12.2 ± 2.0
	3	0.5782	0.2082	12.25	
LEON	1	0.6531	0.2525	14.85	
	2	0.6206	0.2333	13.72	14.3 ± 0.6
	3	0.6387	0.2440	14.35	
LEOB	1	0.6863	0.2721	16.01	
	2	0.8801	0.3867	not used	14.4 ± 2.3
	3	0.5949	0.2181	12.83	

LAON were more uniform, as it can be observed in the value of the standard of mean values. Concentration at Cu^{2+} complexed by lignins ethanol/water original (LENO), oxidized by LAC (LEON) and by the TYR enzyme (LEOB) are displayed in Table 2. Through the Table 2 it was possible to observe again that LEON presented an increase of 17.2% in the chelating power in all of the replicates in relation to LENO. A similar chelating power (18%) was also obtained by the LEOB. The data presented confirm previous results with acetosolv lignins oxidized by TYR *(1,2)*.

Conclusions

The lignins were oxidized with success by TYR and LAC considering that the measure of absorbance was, theoretically, the increase of carbonyls and hydroxyls in the lignin. In both cases, the oxidize lignins were tested as for the chelating power, which increased due to the oxidation, and the best results of retention of copper ions were obtained by the lignin oxidized by the extracted enzyme of potatoes (tyrosinase). However, the results of the lignin oxidized by the commercial enzyme NOVOZYM 51003 (laccase) were more uniform. Comparing Acetosolv and ethanol/water lignins, the second one had the higher chelating power.

Acknowledgments

The authors wish to thank the Brazilian agencies Fundação de Amparo à Pesquisa do Estado de São Paulo (FAPESP) and Conselho Nacional do Desenvolvimento Científico e Tecnológico (CNPq).

References

1. Gonçalves, A. R., Benar, P., Costa, S. M., et al. (2005), *Appl. Biochem. Biotechnol.* **121–124,** 821–826.
2. Gonçalves, A. R. and Soto-Oviedo, M. A. (2002), *Appl. Biochem. Biotechnol.* **98–100,** 365–371.
3. Burton, S. (2003), G Laccases and Phenol Oxidases in organic synthesis: a review. *Curr. Organic Chem.* **7,** 1317–1331.

Reuse of the Xylanase Enzyme in the Biobleaching Process of the Sugarcane Bagasse Acetosolv Pulp

Luís R. M. Oliveira, Regina Y. Moriya, and Adilson R. Gonçalves*

Biotechnology Department (DEBIQ)—FAENQUIL, CP 116, CEP 12600-970, Lorena—SP, Brazil; E-mail: adilson@debiq.faenquil.br

Abstract

In this work, pretreatment-enzymatic series of the bagasse-sugarcane pulp and alkaline extraction of enzyme treated pulp were carried out. In the pretreatment an enzyme dose was utilized and acetosolv pulp suspension of 3% (w/v) with different solvents (distilled water, 0.05 mol/L acetate buffer pH 5.5 and 0.05 mol/L phosphate buffer pH 7.25) stirred at 85 rpm for 2 or 4 h. The enzymes used were pulpzyme and cartazyme, both commercial. The accompaniment of the enzymatic activity was carried out through measurement in initial and finish of each enzymatic pretreatment. The xylanase-treated pulps and xylanase-alkaline-extracted pulps were analyzed regarding kappa number and viscosity. Pulpzyme recovery was better in phosphate buffered medium (84, 46, and 23% for first, second, and third enzymatic treatment, respectively) although in aqueous medium reached only 2% for every treatments. However, the improvement of pulp properties was evidenced only in aqueous medium for pulpzyme. Cartazyme recovery was similar for both solvents (water and acetate buffer), reaching values around 19% for first enzymatic treatment and 9% for second one. Nevertheless, the pulp properties increased only in acetate buffered medium.

Index Entries: Enzymatic pretreatment; xylanase; acetosolv pulping; sugarcane bagasse.

Introduction

Brazil is the largest worldwide producer of sugarcane. At least 10% of the agricultural area, 5.5×10^6 ha of cane, are planted corresponding to an annual production of 357.5×10^6 t of cane *(1)*. The sugarcane bagasse is the principal byproduct of the alcohol industry, used as fuel in mills and its production surplus constitutes an environmentally problem and at the same time would be a renewable font of chemical resources.

---

*Author to whom all correspondence and reprint requests should be addressed.

Nowadays the studies of process that utilize agricultural residues for obtaining chemical products are growing. Pulping using organic solvents in aqueous solution known as organosolv is one of these processes. Chlorine utilized in the traditional bleaching sequences reacts with lignin and forms soluble chlorlignins in the basic medium. The low-molecular-weight-compounds (AOX) formed in high amounts own toxic mutagenic and/or carcinogen character (2,3). Environmental laws have put restrictions in using of the chlorine in the bleaching process in paper industry. Then modifications in pulping process and developing alternative ways in bleaching without use of the elementary chlorine were improved, like the bleaching-totally-chlorine-free methods. Hydrogen peroxide is one of the more widely used oxygen-based bleaching chemicals in the pulp and paper industry (4,5).

Enzymes have shown to be a biotechnological alternative in bleaching of pulps, used together with conventional-bleaching-chemical sequences (6–7). The main motive in the use of enzyme is the decrease of the chorine consumed in bleaching, reaching 34% (2–8), decreasing consequently the organochlorine-compound charge produced (8).

The mechanisms of the enzyme action in the pulp are the removal of the redeposited xylan on fibers and the rupture of the lignin–carbohydrate complex. The redeposited xylan on pulp-surface covers the residual lignin becoming it inaccessible to the bleaching reagents. The xylanase hydrolyzes part of this xylan permitting better access of the bleaching reagents to the residual lignin, becoming the removal of this lignin is easier (7,9,10). The lignin–carbohydrate complex theory assumes that there is a union between lignin and polyoses in pulp that restricts the removal of residual lignin. The xylan bond cleavage by xylanase separates the lignin-carbohydrate linkages improving the access of the bleaching reagents and facilitating the lignin removal in the subsequent bleaching chemical sequences (7–10).

The cost of enzyme production has been admitted as the greatest difficulty, impeding economical feasibility of the majority of biomass conversion processes (11). Several strategies have been used to increase the efficiency and reduce the cost of the process. A strategy of much success has been to enhance the enzyme generation productivity for mutation fungus and the culture conduction optimization. The increase of specific activity of the enzymes and enzyme reuse in the process are also strategies to be considered (12). In ethanol production process from lignocellulosic residue, the enzyme recycle lead to reduce the ethanol production cost. Ramos and Saddler (13) showed that recovery and recycle of the enzyme β-Glucosidase can reach approx 70% of the original protein added in the first hydrolysis reaction after seven hydrolysis cycles. Greeg et al. (14) showed that enzyme recycling using hydrolysis reactors for two cycles decreased the ethanol production cost in 12%. Despite several studies in this area, the reuse of the enzyme xylanase in pulp biobleaching process with intention of decreasing the cost this enzyme in the paper and cellulose industry was not yet investigated.

Materials and Methods

Acetosolv Pulping

The acetosolv sugarcane bagasse pulping was carried out with acetic acid 93% (v/v), HCl as catalyst, according to Benar (15). The bagasse/solvent ratio was 1 : 14 (w/v). The temperature of pulping was 120°C (temperature of solvent mixture) for 2 or 4 h. The pulp was washed with acetic acid 93% (v/v) and thoroughly washed with water until the neutral pH. The pulp utilized for cartazyme enzyme was the pulp obtained after 2 h, although for pulpzyme enzyme, the pulp was obtained after 4 h because bagasse did not turn in pulp with 2 h pulping.

Enzyme Reuse in the Pulp Pretreatment With Xylanase

Samples of acetosolv pulp with 3% (w/v) consistency were incubated in Erlenmeyer flasks in a shaker at 50°C with 85 rpm stirring for 10 min. Thereon a enzyme quantity, according to Ruzene and Gonçalves (16), 36 IU/g dry pulp for both enzymes, cartazyme and pulpzyme, was incubated in each Erlenmeyer and left to react for 2–4 h. After this time, the pulp was filtered on a Büchner funnel and washed thoroughly with distilled water.

The pulp retained in filter was transferred to a flask and 10 mL of solvent was added. Thereon the pulp was shaked for releasing of adsorbed enzymes. This mix was filtered and the obtained filtrate was blended with the filtrate of the enzymatic pretreatment and utilized in other pulp sample, one other pretreatment was carried out with the same reaction conditions described in the previous paragraph, except the addition of the enzyme. An aliquot of the filtrate was reserved for determination of the enzymatic activity. The procedure of enzyme reuse was repeated once for cartazyme with treatment duration of 4 h and twice for pulpzyme with treatment time of 2 h, decreasing the treatment time for this enzyme. Also the solvent was altered: distilled water and sodium acetate buffer with pH 5.5 and 0.05 mol/L for cartazyme and distilled water and sodium phosphate butter with pH 7.25 and 0.05 mol/L for pulpzyme. A control treatment was carried out with the same reaction conditions described above, except the enzyme addition.

Alkaline Extraction of the Pulp

Samples of acetosolv pulp with 3% (w/v) consistency were treated with sodium hydroxide solution 5% (w/w) for 1 h at 65°C. After, the alkaine-extracted pulp was filtered and thoroughly washed with water until the pH was neutral.

Determination of the Xylanase Enzymatic Activity

The two commercial xylanases, pulpzyme (Novozyme, Bagsvaerd, Denmark) and cartazyme HS (Sandoz, Leed, UK), were used. The cartazyme has pH and temperature optima of 3.0–5.0 and 35–55°C, respectively, with a molecular mass of 21 kDa. The pulpzyme was kindly furnished by Novozyme with batch number CKN00044 and has pH and temperature optima of 6.5–8.0 and 45–60°C, respectively.

The enzymatic activity was determined by measuring the reducing sugar quantity, released hereby Birchwood xylan breakage, according to Bailey et al. *(17)*. A diluted-xylanase solution with buffer was incubated in a substrate solution of Birchwood xylan for 5 min at 50°C. Reducing sugars were dosed by 3,5-dinitrosalicylic acid (DNS) *(18)*. The buffers used in the enzyme dilution were: sodium acetate buffer with pH 5.5 and concentration 0.05 mol/L for cartazyme and sodium phosphate butter with pH 7.25 and concentration 0.05 mol/L for pulpzyme.

Estimation of Kappa Number

Samples of pulp were submitted to the action of 0.1 eq/L $KMnO_4$ at 25°C for 10 min. Adding KI solution in excess stopped the reaction and the $KMnO_4$ consumed was determined from the results of back titrating of the liberated iodine with standard 0.1 eq/L sodium thiosulphate solution. Kappa number was calculated using the volume in mL of 0.1 N $KMnO_4$ consumed per gram of pulp *(19)*. Kappa number was determined for xylanase treated pulp and xylanase-alkaline-extracted pulp.

Determination of Viscosity

Viscosity was determined, through an Ostwald viscometer, by steeping of pulp in 12.5 mL of distilled water for 15 min and by dissolving pulp in 12.5 mL of cupriethylenediamine (Cu^{+2} 1 mol/L) for 30 min *(20)*. Viscosity was determined for xylanase treated pulp and xylanase-alkaline-extracted pulp.

Results and Discussion

Acetosolv Pulp Treated With Cartazyme Enzyme

In both solvents, with analysis of viscosity and kappa number ratio of xylanase treatment (X), only in first treatment a improvement in pulp properties was noted, whereas in the other one there was a slight decrease. Therefore with enzymatic treatment followed by alkaline extraction (XE) an improvement was noticed only in the treatment carried out in the buffered medium, because in aqueous medium the values decreased (Fig. 1, Table 1). The cartazyme effect was just enhanced in pulp viscosity analysis, whereas X stage as much as in XE stage, the viscosity suffered a rising. Because the kappa number analysis for XE stage presented similar values,

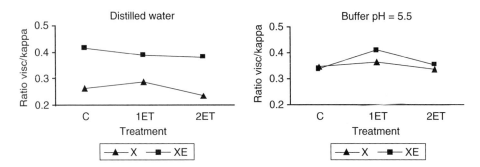

Fig. 1. Graphics of the ratio between viscosity and kappa number of cartazyme treated pulps and cartazyme-alkaline-extracted pulps. ET, enzymatic treatment; C, control treatment (without enzyme); X, xylanase; XE, xylanase and alkaline extraction.

Table 1
Values of Kappa Number and Viscosity of Cartazyme Treated Pulps and Cartazyme-Alkaline-Extracted Pulps

Medium	Sample	Kappa number X	Kappa number XE	Viscosity X (cP)	Viscosity XE (cP)	R X	R XE
Water	Control	28	18	7.3	7.3	0.261	0.415
	1st ET	34 ± 2	20 ± 1	9.8 ± 0.2	7.6 ± 0.6	0.286	0.387
	2nd ET	33 ± 1	18.4 ± 0.5	7.6 ± 0.8	7.0 ± 0.1	0.233	0.381
Buffer pH 5.5	Control	29	16	10.2	5.53	0.347	0.340
	1st ET	31 ± 1	18 ± 1	11.1 ± 2	7.4 ± 0.2	0.363	0.411
	2nd ET	33 ± 5	19.0 ± 0.2	11.2 ± 3	6.7 ± 0.2	0.335	0.352

ET, enzymatic treatment; X, xylanase; XE, xylanase plus alkaline extraction; R, ratio between viscosity and kappa.

that could be explained as bleaching agent was used in excess (NaOH solution 5% w/w), reaching bleaching saturation for this reagent. In such case the enzyme effect could not be checked by this analysis. Besides, in both solvents the enzymatic activity fall were similar, reveling that the use of buffer did not show any advantage regarding distilled water (Fig. 2). The cartazyme presented an enzyme recovery, in both solvents, around 19% and 9% for first and second enzymatic treatment, respectively. This recovery, equal for both solvents, is owing to acidity pulp, turning the aqueous medium slightly acid (similar enzymatic treatment condition in pH 5.5 buffer medium).

Acetosolv Pulp Treated With Pulpzyme Enzyme

In the Pulpzyme enzymatic treatment a significant improving in pulp properties was noted for treatment in aqueous medium, whereas in buffered medium the pulp properties were kept for X stage and reduced for XE stage (Table 2, Fig. 3). Also an increase in pulp viscosity was evidenced in

Reuse of the Xylanase Enzyme

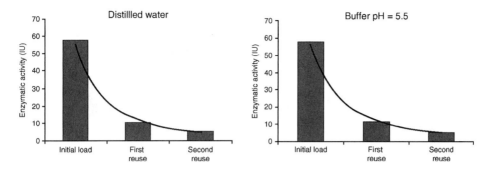

Fig. 2. Graphics of the enzymatic activity fall for the treatments enzymatic of cartazyme enzyme. The first column indicates initial enzymatic load and the others show the enzyme load of first and second reuses.

Table 2
Values of Kappa Number and Viscosity of Pulpzyme Treated Pulps and Pulpzyme-Alkaline-Extracted Pulps

Medium	Sample	Kappa number X	Kappa number XE	Viscosity X (cP)	Viscosity XE (cP)	R X	R XE
Water	Control	25	9.5	13.6	8.7	0.544	0.918
	1st ET	11 ± 2	9 ± 2	13 ± 2	9.4 ± 0.6	1.175	1.106
	2nd ET	22.2 ± 0.5	9.1 ± 0.2	12 ± 2	8.7 ± 0.4	0.559	0.922
	3rd ET	20.9 ± 0.5	9.1 ± 0.6	14 ± 1	9.1 ± 0.6	0.687	1.001
Buffer pH 7.25	Control	37	11	11.2	16.2	0.302	1.496
	1st ET	36.7 ± 0.6	10.7 ± 0.4	10 ± 1	13 ± 1	0.278	1.217
	2nd ET	35.7 ± 0.3	10.6 ± 0.1	12 ± 2	17 ± 3	0.336	1.555
	3rd ET	34.5 ± 0.5	10.2 ± 0.3	12 ± 2	16 ± 1	0.360	1.549

ET, enzymatic treatment; X, xylanase; XE, xylanse plus alkaline extraction; R, ratio between viscosity and kappa.

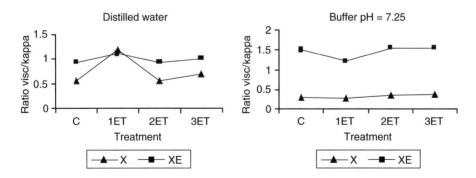

Fig. 3. Graphics of the ratio between viscosity and kappa number of pulpzyme treated pulps and pulpzyme-alkaline-extracted pulps. ET, enzymatic treatment; C, control treatment (without enzyme); X, xylanase; XE, xylanase and alkaline extraction.

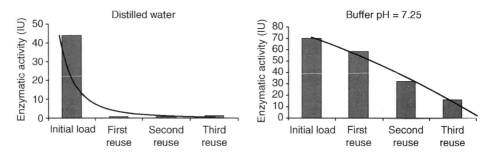

Fig. 4. Graphics of the enzymatic activity fall for the treatments enzymatic of pulpzyme enzyme. The first column indicates initial enzymatic load and the others show the enzyme load of first, second, and third reuses.

both stages and a bleaching saturation (kappa numbers similar) in XE stage. However, the use of buffer showed advantages regarding distilled water in the enzyme recovery. The enzymatic activity fall was very different (Fig. 4), in aqueous medium the enzyme recovery was low (approx 2% for every treatments) whereas in buffered medium the fall was more bland (84%, 46%, and 23% for first, second and third enzymatic treatment respectively).

Conclusions

With these results, it was concluded that pulpzyme enzyme would be more appropriate for use only in one batch in aqueous medium, whereas this enzyme achieves to increase the pulp properties (reduced enough the kappa number and increased the viscosity) and owing to this fact, it has small enzyme recovery in aqueous medium. The pulpzyme enzyme in buffered medium reached values of enzyme recovery higher than in aqueous medium, however the use of this enzyme, in buffered medium, did not revel improving in pulp properties in the studied time (2 h), so suggesting a larger time to evidence an effect of this enzyme. Regarding cartazyme, in buffered medium, this enzyme works better, whereas the increase of viscosity and kappa number ratios were higher than that in aqueous medium regarding the control. The enzyme recovery in both solvents was similar.

Acknowledgment

The authors wish to thank Fundação de Apoio à Pesquisa do Estado de São Paulo (FAPESP) and Conselho Nacional do Desenvolvimento Científico e Tecnológico (CNPq) (Brazilian agencies).

References

1. Unica. Available in http://www.unica.com.br, accessed in 19/04/2005.
2. Viikari, L. (1991), *Paperi ja Puu and Timber* **73**, 384–389.
3. Wang, S. H., Ferguson, J. F., and McCarthy, J. L. (1992), *Holzforschung* **46**, 219–223.
4. Shatalov, A. A. and Pereira, H. (2005), *Biores. Technol.* **96**, 865–872.

5. Chauvet, J. M., Comtat, J., and Noe, P. (1987) In: *International Symposium on Wood and Pulping Chemistry Notes*, ATIP, Paris, pp. 325.
6. Minor, J. L. (1996) In: *Pulp Bleaching: Principles and Practice*, Dence, C. W. and Reeve, D. W., eds., Tappi Press, Atlanta, pp. 363–377.
7. Wong, K. K. Y., Jong, E. D., Saddler, J. N., and Allison, R. W. (1997), *Appita J.* **50**, 415–422.
8. Paige, M. G., Bernier, R. J., and Jurasek, L. (1998), *Biotech. Bioeng.* **32**, 235–239.
9. Farrel, R. L., Viikari, L., and Senior, D. (1996), In: *Pulp Bleaching: Principles and Practice*, Dence, C. W. and Reeve, D. W., eds., TAPPI Press, Atlanta, pp. 363–377.
10. Young, R. A. and Akthar, M. (1998), In: *Environmentally Friendly Technologies for the Pulp and Paper Industry*, Young, R. A. and Akthar, M. eds., John Wiley & Sons, New York, pp. 5–69.
11. Nguyen, Q. A. and Saddler, J. N. (1991), *Biores. Technol.* **35**, 275–282.
12. Greeg, D. J. and Saddler, J. N. (1996), *Biotechnol. Bioeng.* **51**, 375–383.
13. Ramos, L. P. and Saddler, J. N. (1994), *Appl. Biochem. Biotechnol.* **45, 46,** 193–207.
14. Greeg, D. J., Boussaid, A., and Saddler, J. N. (1998), *Biores. Technol.* **63,** 7–12.
15. Benar, P. (1992), UNICAMP/Instituto de Química, Master Dissertation, Campinas.
16. Ruzene, D. S. and Gonçalves, A. R. (2003), *Appl. Biochem. Biotechnol.* **105–108,** 769–774.
17. Bailey, M. J., Biley, P. T., and Bountaen, K. J. (1992), Biotechnology **28,** 257–270.
18. Miller, G. L. (1956), *Anal. Chem.* **31,** 426–428.
19. Tappi standard methods. (1982), T 236.
20. Tappi standard methods. (1982), T 230.

Detection of Nisin Expression by *Lactococcus lactis* Using Two Susceptible Bacteria to Associate the Effects of Nisin With EDTA

THEREZA CHRISTINA VESSONI PENNA,*,[1] ANGELA FAUSTINO JOZALA,[1] THOMAS RODOLFO GENTILLE,[1] ADALBERTO PESSOA, JR.,[1] AND OLIVIA CHOLEWA[2]

[1]*Department of Pharmaceutical Technology, School of Pharmaceutical Science, University of São Paulo, SP, Brazil, E-mail: tcvpenna@usp.br; and* [2]*Molecular Probes, Inc. Eugene, OR, USA 97402, E-mail: olivia.cholewa@probes.com*

Abstract

Nisin, a bacteriocin produced during the exponential growth phase of *Lactococcus lactis* ATCC 11454, inhibits the growth of a broad range of Gram-positive bacteria. Gram-negative bacteria can also be inhibited by nisin with EDTA. In this study, nisin production was assayed by the agar diffusion method using *Lactobacillus sake* ATCC 15521 and a recombinant *Escherichia coli* DH5-α expressing the recombinant green fluorescent protein as the nisin-susceptible test organisms. The titers of nisin expressed and released in culture media were quantified and expressed in arbitrary units (AU/mL of medium) and converted to standard nisin concentration (Nisaplin®, 25 mg of pure nisin with an activity of 1×10^6 AU/mL). The expression and release of nisin by *L. lactis* in skimmed milk (9.09% total solids) with Man Rugosa Shepeer-Bacto Lactobacilli broth (1:1) was monitored in a 5 L New Brunswick fermentor. Combining EDTA with nisin increased the bactericidal effect of nisin on the bacteria examined. The presence of EDTA was necessary to inhibit *E. coli* growth with nisin. *L. sake* was shown to be a good indicator for the evaluation of nisin release in the culture media, including with the addition of EDTA.

Index Entries: Nisin; *Lactococcus lactis*; *Lactobacillus sake*; EDTA; *Escherichia coli*; recombinant green fluorescent protein.

Introduction

Nisin, a naturally occurring antimicrobial polypeptide discovered in 1928 *(1,2)*, is a monomeric pentacyclic subtype A lantibiotic peptide (3,353 Da) synthesized by *Lactococcus lactis* ssp. *lactis (3,4)* during exponential

---

*Author to whom all correspondence and reprint requests should be addressed.

growth (5,6). Nisin and other subtype A lantibiotics are characterized by a strong bactericidal activity by inhibiting spore germination (7) and the growth of many strains of Gram-positive bacteria (8). Nisin is used as a natural preservative in food and dairy industries and is approved by the US Food and Drug Administration and has GRAS (generally regarded as safe) status (9). This bacteriocin has been also applied in dental care products (10), and in pharmaceutical products for the potential use as a contraceptive and for the treatment of stomach ulcers and colon infections (11–13). Nisin solubility and stability increase substantially with increasing acidity. Nisin is stable at pH 2.0 and can be autoclaved at 121°C (14). The liberation of nisin from cells into the propagating medium is dependent on the pH of the medium. At pH < 6.0, at least 80% of expressed nisin is released into the growth medium. On the other hand, when the bacterium *L. lactis* is cultivated in pH > 6.0, the majority of the nisin is retained within the cell membrane or intracellularly (15). Nisin is not generally active against Gram-negative bacteria, yeasts and fungi. The outer membrane of Gram-negative bacteria prevents nisin from reaching the site of action. Outer membrane permeability can be altered by treatment with chelators, such as EDTA (disodium ethylenediamine tetraacetate), or high hydrostatic pressure, resulting in increased sensitivity toward nisin. (16–22). The recombinant green fluorescent protein (GFPuv), expressed by *Escherichia coli* DH5-α, is widely utilized as a genetic protein marker. GFPuv in the cells can be easily visualized by a UV hand lamp making this a versatile tool for a variety of biotechnological applications and as a potential biological indicator in the preservation of manufactured and processed products (23). In a system combining different antimicrobials, treatment with nisin/EDTA or nisin/potassium sorbate at 10°C exhibited significant population growth inhibition of *E. coli* O157:H7 compared with samples treated with nisin, EDTA or potassium sorbate alone (19,24). Vessoni Penna and Moraes (2002) (6) studied the production of nisin by *L. lactis* ATCC 11454 in Man Rugosa Shepeer-Bacto Lactobacilli (MRS) broth. Varying concentrations of sucrose (5–12.5 g/L), asparagine (7.5–75 g/L), potassium phosphate (6–18 g/L), and Tween-80 (1–6.6 g/L) were added to MRS broth to determine which additive was most influential on growth rates, nisin expression and release into the media.

The authors observed a positive correlation between nisin production and biomass of *L. lactis* dependent on the balance of sucrose and asparagine, but independent of potassium phosphate. In a previous study (25), culture media with milk was observed to provide better conditions for *L. lactis* growth and its concomitant expression of nisin. Milk alone (9.09% total solids) favored nisin expression and release into the media for all five transfers, from 0.1 to 0.4 g/L, similar to that attained at the first transfer for both 25% MRS plus 25% milk and 25% M17 plus 25% milk, with the latter media providing the highest nisin concentration, 3.6 g/L, after the fifth transfer. The effects of a milk-based medium on nisin expression by *L. lactis* were investigated further to analyze growth conditions in a fermentor, compared with batch culture conditions.

The inhibitory activity of nisin on Gram-negative organisms can be improved by combining nisin with EDTA in culture media *(26)*. The detection of expressed nisin was evaluated by the diffusion method in agar using two susceptible bacteria, *Lactobacillus sake* (Gram-positive), and *E. coli* (Gram-negative). To evaluate the bactericidal activity of nisin on Gram-negative bacteria, EDTA was added to the samples or standard nisin solution.

Material and Methods

The nisin-producing strain of *L. lactis* ATCC 11454, the nisin-susceptible indicator strains of *L. sake* ATCC 15521 and a recombinant *E. coli* ATCC DH5-α were used in this study. The cultures of *L. lactis* and *L. sake* were maintained at –80°C in MRS broth (Difco®, Detroit, MI) with 50% (v/v) of glycerol. *E. coli* was grown (24 h/100 rpm/37°C) in Luria Bertani broth (LB) (Difco) supplemented with 100 μg/mL ampicillin (amp) to maintain a plasmid, with cultures stored at –80°C with 50% (v/v) of glycerol, following previously published procedures *(23,25)*.

Microorganism and Inoculum Preparation

Before inoculating experimental media, 100 μL of the stock culture of *L. lactis* were added to 50 mL of MRS broth (Difco) in 250-mL flasks (preinoculum) and incubated for 36 h/100 rpm/30°C. From the preinoculum, 5 mL aliquots of the cell suspension were transferred into 50 mL of experimental medium (MRS broth plus skimmed milk, 1:1) in 250-mL flasks (first transfer) and incubated for another 36 h (100 rpm/30°C). Cultures were transferred five times (100 rpm/30°C/36 h) using 5 mL aliquots of broth culture for each new volume of the experimental medium in accordance with previous studies *(25)*; only the preinoculum and the first and second transfers into MRS plus skimmed milk (1:1) were subsequently used for the fermentor cultures in this work.

Single Batch Fermentation

A 15 mL aliquot of the first transfer culture was mixed into 150 mL of culture medium (75 mL MRS + 75 mL skimmed milk) and incubated at 36 h/100 rpm/30°C (second transfer culture). The entire 150 mL of this second transfer culture was poured into 1.5 L of the same medium (MRS plus skimmed milk, 1:1, pH = 6.25–6.31) in a 5 L bench-scale fermentor (NBS-MF 105, New Brunswick Scientific, New Brunswick, NJ). The initial cell concentration in the fermentor was 0.58 ± 0.10 g/L. The total incubation time was 16 h at 30°C to observe variations of nisin expression associated with growth conditions. Foaming was controlled as needed by adding 0.5 mL of dimethylpolysiloxane (Sigma-Aldrich, St. Louis, MO). Agitation and aeration were 200 rpm and 0.5 vvm, respectively. The airflow was measured by an on-line rotameter and set using a needle valve. The pH of the medium during cultivation was measured by an electrode (Ingold, Woburn, MA). Before the addition of inoculum to the fermentor, the propeller speed, aeration rate, and the temperature (30°C) were adjusted.

Analytical Procedures

After every 36 h incubation period, 10 mL of cell suspensions were aseptically withdrawn from the flasks and tested for pH, cellular density, colony number, and nisin concentrations. For this study, each fermentor culture was performed in triplicate. The pH measurements were performed in 10 mL of culture suspension with a Accumet AR20 pH/mV/conductivity meter (Fisher Scientific, Hampton, NH) calibrated with standard buffer solutions (Merck; Whitehouse Station, NJ) pH = 4.0, 7.0, 10.0, at 25°C. The cellular biomass concentration, expressed in mg of dried cellular weight per liter of broth (mg DCW/L), was determined from the optical density at 660 nm (OD_{660}) in a 1 cm path length quartz cuvet in a spectrophotometer (Beckman DU-600, CA, USA). The OD_{660} readings were calibrated against a standard dried cellular concentration curve of *L. lactis*, which was obtained by the gravimetric method of the biomass (mg/L) held on the surface of a 0.22 µm membrane (Millipore®, SP, Br). The equation for the calibration curve ($R^2 = 0.998$) was given by:

$$OD_{660} = 0.0145 + 0.0022 \times DCW \text{ or}$$
$$DCW (mg/L) = [(OD_{660}) - 0.0145]/(0.0022)$$

Culture populations were assayed by the plate count method expressed in colony forming units (CFU) per mL of broth (CFU/mL) in plate count agar (Difco) at 30°C for 24 h. Cell numbers were related to the reference curve associating OD_{660} to dry cell weight (mg DCW/L) of the same suspension, where $OD_{660} = 0.01$ was equivalent to 10^4 CFU/mL.

Nisin Activity Detection

As done in previous work *(25)*, the pH of the cell suspensions of *L. lactis* was adjusted from pH = 6.0 to pH = 4.0 and from pH = 4.0 to pH = 2.5 with 0.2 N HCl. The pH adjustment, before centrifugation, was done to coagulate the milk, releasing nisin into a clear supernatant (after centrifugation) to facilitate nisin detection.

For nisin activity detection, following methods of previous works *(6,24,25)*, the titers of nisin expressed and released in culture media were quantified and expressed in arbitrary units (AU/mL of medium) by the agar diffusion assay utilizing *L. sake* ATCC 15521 (Gram-positive) and a recombinant *E. coli* ATC DH5-α (Gram-negative) as nisin-susceptible indicator strains.

L. sake was grown in 50 mL of MRS broth in 250-mL flasks and incubated (24 h/100 rpm/30°C). A 1.5 mL aliquot of the suspension ($OD_{660} = 0.4$) was transferred and mixed with 250 mL of soft agar (MRS broth with 0.8%

of bacteriologic grade agar) in 500-mL flasks. Each 20 mL of inoculated medium was transferred to Petri plates (100 mm dia).

E. coli was grown in 50 mL of LB in 250-mL flasks and incubated (24 h/100 rpm/37°C) supplemented with 100 µg/mL amp. An aliquot of the suspension was transferred onto the surface of LB-amp agar-soft with isopropyl-β-D-thiogalactopyranoside (IPTG) added to a final concentration of 0.5 mM (w/v) and incubated at 37°C. Another aliquot of the suspension was serially diluted in physiologic buffer into 1:1000 dilution corresponding to an OD_{660} = 0.005. A 0.6 mL aliquot of this diluted suspension (OD_{660} = 0.005) was transferred and mixed with 100 mL of soft agar (LB broth with 100 µg/mL amp, 375 µL of IPTG and 0.8% of bacteriologic grade agar) in 250-mL flasks. Each 20 mL of inoculated medium was transferred to Petri plates. The recombinant *E. coli* was grown with amp and IPTG added to the media to maintain the plasmid and express GFP, conditions that will be used in future studies utilizing fluorescence detection.

From every fermentor culture, 50 µL of culture supernatant from centrifuged *L. lactis* suspension, with or without 7.45 mg/mL EDTA, was transferred into each of four wells on the surface of the *L. sake* inoculated agar. With culture supernatant mixed previously with 7.45 mg/mL EDTA or without, 50 µL of this mixture was transferred into each of four wells on the surface of the *E. coli* inoculated agar.

The relation between (AU/mL) and international units (IU/mL) was determined by using Nisaplin (a commercial purified nisin preparation containing 25 mg of nisin per gram of Nisaplin, corresponding to 10^6 IU/g Nisaplin; Aplin & Barret, Sigma Chemical); and a standard solution of nisin (Nisaplin), containing 250 µg of nisin per mL corresponded to 10^4 AU/mL. IU conversions are included in this work to correlate this work with past references *(24,25)*.

With *E. coli* as the susceptible bacteria, standard nisin was diluted in either deionized water or in 0.02N HCl, with a final concentration of 0.75% (w/v) NaCl, and 7.45 mg/mL added EDTA. For comparison with the sensitivity of *L. sake* to nisin mixed with EDTA, another standard curve was generated.

The activity of nisin expressed in AU/mL was converted for nisin in milligrams per milliliters (mg/mL), through the relation:

$$\text{Nisin (mg/mL)} = \frac{(z \times 0.025)}{1000}, \text{ where } z = \text{AU/mL}$$

Results and Discussion

Susceptible Bacteria and the Association of Nisin With EDTA

Using *E. coli* to evaluate the activity of nisin in solutions with 7.45 mg/mL EDTA (Table 1), the zones of inhibition ranged from 10 mm to 14.75 mm, which correlated to 10 to 10^5 AU/mL of nisin in acidified solution

Table 1
Standard Curves and Conversion Equations Associating the Inhibition Halos (H, mm, Zones Lacking *E. coli* or *L. sake* Growth) With the Standard Nisin Arbitrary Units ($10–10^5$ AU/mL)

Strains	Solutions	pH	Standard nisin activity (AU/mL) Diameters of inhibition halos (SD = ±0.5 mm)					Conversion equation		R^2
			10^{-1}	10^{-2}	10^{-3}	10^{-4}	10^{-5}	AU/mL		
E. coli	Nisin plus EDTA in 0.02N HCl	2.5	10	12.5	13.25	14.5	14.75	$10^{(0.783 \times H - 7.1011)}$		0.94
	Nisin plus EDTA in H$_2$O	4.0	13.3	14.8	15.7	16.25	17	$10^{(1.0693 \times H - 13.456)}$		0.96
L. sake	Nisin plus EDTA in 0.02N HCl	2.5	8.5	11.5	14.75	16.75	17.5	$10^{(0.4102 \times H - 2.6612)}$		0.95
	Nisin plus EDTA in H$_2$O	4.0	13	14	16.25	17.25	18.25	$10^{(0.7097 \times H - 8.1774)}$		0.97
	Nisin in skimmed milk	4.0	10	15.25	19.25	24.5	26.75	$10^{(0.2307 \times H - 1.4174)}$		0.98
	Nisin in skimmed milk plus EDTA	4.0	5	17	19.8	22.31	23.56	$10^{(0.2639 \times H - 0.3641)}$		0.98
	Nisin in skimmed milk	6.0	10.75	14	18.5	21	22.25	$10^{(0.3211 \times H - 2.5553)}$		0.96
	Nisin in skimmed milk plus EDTA	6.0	8.75	14.5	16.75	20.5	20.75	$10^{(0.3065 \times H - 1.9808)}$		0.92

Observation: the calibration curves between AU/mL and IU/mL, 1.09 ± 0.17 AU corresponded to 1.0 IU (40 IU = 1 µg of pure nisin A).

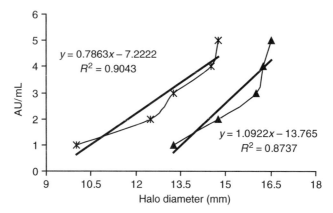

Fig. 1. Standard curve relating the action of nisin plus EDTA in [-✱-] acidified solution with HCl (pH 2.5) and [-▲-] solutions (pH 4.0) utilizing *E. coli* cells.

(pH 2.5). Nisin plus EDTA in aqueous solution (pH 4.0) gave zones of 13.3–17 mm corresponding the nisin activity of 10 to 10^5 AU/mL equivalent a concentration of 0.025 µg/mL to 2500 µg/mL. The relation between both slopes (1.0693 and 0.783) of the standard curves provided 37% higher arbitrary units on the nisin activity when nisin was diluted and mixed with EDTA in aqueous solution (pH 4.0) compared with the solution (pH 2.5) adjusted with 0.02N HCl. The acidified solution reduced the nisin activity on the susceptibility of the *E. coli* growth in agar (Fig. 1)

Using cells of *L. sake* to evaluate the activity of standard nisin in solutions of 7.45 mg/mL EDTA, (i) the zones of inhibition ranged from 8.5 mm to 17.5 mm correlated the nisin activity of 10 to 10^5 AU/mL, equivalent a concentration of 0.025 µg/mL to 2500 µg/mL from a conversion equation: Log AU/mL = 0.4102 × H – 2.6612, relating the action of nisin and EDTA in acidified solution with HCl (pH 2.5); (ii) the zones of inhibition ranged from 13 mm to 18.25 mm correlated the nisin activity of 10 to 10^5 AU/mL, equivalent a concentration of 0.025 µg/mL to 2500 µg/mL from a conversion equation: Log AU/mL = 0.7097 × H – 8.1774, relating the action of nisin plus EDTA in water (pH 4.0). The relation between both slope (0.4102 and 0.7097) of the standard curves provided 73% higher arbitrary units for nisin when diluted and mixed with EDTA in water (pH 4.0) compared with the acidified solution to pH 2.5 with 0.02N HCl.

The low pH in acidified solution (pH 2.5) lowered nisin susceptibility of *E. coli* and *L. sake*, and was verified previously by the authors (25), but growth was not inhibited by a nisin-free HCl solution (pH 2.5) added to the wells for *L. sake*, *E. coli*, and *L. sake* susceptibility in nisin solutions with EDTA adjusted to pH 2.5 exhibited higher variability, around 20% among halos, for the same AU/mL than with nisin with EDTA at pH 4.0 with variability around 4% (Table 2). Examining Tables 1 and 2, it can be noted that the association of EDTA and nisin has shown to increase the bactericidal effect on *E. coli*, the Gram-negative nisin-susceptible bacteria. *L. sake* has been

Table 2
Inhibitions Halos (mm), Zones Lacking *L. sake* or *E. coli* Growth, Converted to Nisin Arbitrary Units Per Milliliter (AU/mL) From the Placement of 50 µL of Either Supernatant (Centrifuged Samples From the Fermentor) or Standard Nisin Solutions, Mixed With 7.45 mg/mL EDTA Diluted in Water or in 0.02N HCl, Plus 0.75% NaCl Into the Wells on the Surface of Inoculated Agar (Diffusion Assay)

Supernatant from fermentor	pH	*L. sake* Halos Mm	Sensitive strain for nisin activity Conversion equation AU/mL	AU/mL	$\log_{10}$
With EDTA in water (1:1)	4.0	16	$10^{(0.7097 \times H - 8.1774)}$	1505	3.18
With EDTA in 0.02N HCl (1:1)	2.0	11.5	$10^{(0.4102 \times H - 2.6612)}$	114	2.06
With EDTA	4.8	13.75	$10^{(0.7097 \times H - 8.1774)}$	38	1.58
Untreated (without EDTA)	4.8	12.25	$10^{(0.33211 \times H - 2.5553)}$	33	1.51

Supernatant from fermentor	pH	*E. coli* Halos Mm	Sensitive strain for nisin activity Conversion equation AU/mL	AU/mL	$\log_{10}$
With EDTA in water (1:1)	4.0	16	$10^{(1.0693 \times H - 13.456)}$	4495	3.65
With EDTA in HCl (1:1)	2.0	0	$10^{(0.783 \times H - 7.1011)}$	nd	
With EDTA	4.8	13	$10^{(0.783 \times H - 7.1011)}$	11	1.05
Untreated (without EDTA)	4.8	0	$10^{(0.783 \times H - 7.1011)}$	nd	

Standard nisin solution	pH	*E. coli* Halos mm	Sensitive strain for nisin activity Conversion equation AU/mL	AU/mL	Log_{10}
50 µg/mL nisin (2 × 10³ AU/mL) In water	4.0	15.75	$10^{(1.0693 \times H - 13.456)}$	2429	3.39
50 µg/mL nisin (2 × 10³ AU/mL) in HCl	2.0	13	$10^{(0.783 \times H - 7.1011)}$	1197	3.08
50 µg/mL nisin (2 × 10³ AU/mL) In HCl : water (1:1)	2.0	13	$10^{(0.783 \times H - 7.1011)}$	1197	3.08
250 µg/mL nisin (1 × 10⁴ AU/mL) In water	4.0	17	$10^{(1.0693 \times H - 13.456)}$	52,735	4.72
250 µg/mL nisin (1 × 10⁴ AU/mL) In HCl	2.0	13.5	$10^{(0.783 \times H - 7.1011)}$	2947	3.47
250 µg/mL nisin (1 × 10⁴ AU/mL) In HCl : water (1:1)	2.0	14.8	$10^{(0.783 \times H - 7.1011)}$	30,711	4.49

nd, none detected.

confirmed to be a sensitive and reliable Gram-positive biological indicator for the detection of nisin, even without the addition of EDTA.

It was also observed that the formation of a greater inhibition zone of L. sake growth, around 45%, occurred with nisin plus EDTA in water rather than in acidified solution (pH 2.5), with a nisin activity to 10 AU/mL equivalent a concentration of 0.025 µg/mL (Table 1). On the other hand, for E. coli with nisin plus EDTA, the inhibition zone was 30% greater in water (nisin activity to 10 AU/mL equivalent a concentration of 0.025 µg/mL) than in acidified solution. The pH between 4.0 and 6.0 of skimmed milk showed a variation in the susceptibility of L. sake to nisin of around 20% (Table 1). Treatment with nisin solution without EDTA had no effect on the growth of E. coli. Only the supernatant, from centrifuged culture samples, treated with 7.45 mg/mL EDTA and diluted (1:1) in water were significantly different ($p < 0.05$) when compared with the untreated samples (Table 2). On the other hand, EDTA showed no influence on L. sake growth inhibition. For both strains E. coli and L. sake, mixing EDTA in water and then adding to the supernatant (1:1) seemed to enhance cell sensitivity and result in larger inhibition zones compared with wells applied with EDTA directly mixed into the supernatant. The mixing of EDTA in water and then adding to the supernatant (1:1) showed equal sized halos for L. sake and E. coli on the inoculated agar, corresponding to nisin activity of 1.51×10^3 AU/mL and 4.5×10^3 AU/mL, equivalent to a nisin concentration of 38 µg/mL and 113 µg/mL, respectively. For EDTA mixed directly into the supernatant, the inhibition halos of E. coli and L. sake growth were also similar, 13.75 mm and 13 mm, respectively, corresponding to 11 AU/mL and 38 AU/mL, with similar activity to 33 AU/mL nisin in the untreated supernatant.

The influence of EDTA on E. coli susceptibility was confirmed and diluting EDTA in water before mixing to the supernatant was significant. This susceptibility performance has been also confirmed by E. coli cells to evaluate the standard nisin activity at concentrations of 50 and 250 µg/mL (2×10^3 AU/mL and 10^4 AU/mL) in solutions with added EDTA. The zones of inhibition were greater for solutions mixed with EDTA diluted prior to mixing than directly added to the standard nisin solution and ranged from 15.75 mm to 17 mm, corresponding in nisin activity to 2.4×10^3 AU/mL and 5.27×10^4 AU/mL, equivalent to a nisin concentration of 60 µg/mL and 1318 µg/mL and dropped to 13 mm and 13.5 mm, respectively corresponding in nisin activity to 1.2×10^3 AU/mL and 2.9×10^3 AU/mL equivalent to a nisin concentration of 20 µg/mL and 73 µg/mL, 10 times lower than the former values obtained with EDTA diluted in water before adding to the nisin solution. Further work needs to be done to determine the basis of this discrepancy caused by mixing or direct addition of EDTA; this may be related to differences in the solubilization of EDTA under these conditions.

It has been claimed that the bactericidal effect of nisin activity on Gram-negative organisms can be improved when used in combination with other antimicrobial agents such as chelators (19,25). Cutter and

Siragusa (1995) *(25)* reported that nisin together with chelators exhibit a better antimicrobial activity on Gram-negative pathogens in culture media. However, this combination was found to be less effective when applied to real food systems, compared with culture media. Fang and Tsai (2003) *(19)* studied the inhibitory effect of antimicrobials and the combination of antimicrobials with EDTA over *E. coli* O157:H7 in ground beef at 10°C. They observed that the association of nisin (10^3 IU/mL) and EDTA (7.45 mg/mL) mixed to the ground beef samples (added with or without 1.5% $CaCl_2$) showed significant growth inhibition of *E.coli* compared with the samples treated with nisin or EDTA alone. The effect of the association of EDTA-nisin was similar in our experiments and confirmed our results.

We verified that the presence of EDTA was essential to improve nisin activity on *E. coli* growth inhibition. However, *L. sake* was more sensitive for the evaluation of nisin release in the culture media, even with nisin and EDTA in the supernatant. The mechanism of growth inhibition by EDTA is not fully understood, but generally attributed to its chelating activity. EDTA binds primarily divalent cations *(27)* that are present in the supernatant obtained from the growth media. Therefore, we can imply that washing the cells should be enough to extract the majority of salts from the culture media, in order not to limit the action of EDTA to destabilize the membrane of some Gram-negative strains by chelating calcium and magnesium, which are necessary for LPS to bind to the cell wall *(27–29)*. It is possible that the addition of EDTA can cause calcium and magnesium deprivation of the cells or EDTA may be toxic to the cells by some other mechanism.

It was observed that the formation of a larger inhibitory halos occurred when EDTA and nisin were diluted in water before mixing, at pH = 4.3 ± 0.5. In all the samples, for both sensitive bacteria, the dilution in aqueous solution with HCl (typically used to dissolve nisin) to pH = 2.5 provided smaller halo diameters when compared with the halos formed using water plus EDTA. This may be related to the pKa of EDTA and its effectiveness as a chelator.

Branen and Davidson (2004) *(30)* confirmed that EDTA enhanced the synergetic activity of nisin, monolaurim, and lysozyme in tryptic soy broth against two enterohemorrhagic *E. coli* strains, but not against *Salmonella enteritidis* or *P. fluorescens*. However, the authors observed that none of these antimicrobial combinations with EDTA was shown particularly effective in ultra high temperature milk held at 25°C against the same strains of *E. coli*, demonstrating that the activity of antimicrobials in a food system may be affected by a number of factors, as well as fat, protein, salt concentration, and storage temperature of the samples. Gill and Holley (2003) *(23)* found a probability of 89%, 73%, and 43% of two-agent inhibitory (EDTA plus chrisin) interactions against, respectively, *S. typhimurium*, *E. coli, and Serratia grimessi* on growing cells in nutrient broth with NaCl and nitrite added. Those findings reinforce the association of EDTA-nisin in water exhibiting significant synergetic activity on *E. coli* cells than in the

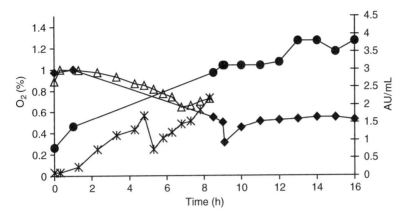

Fig. 2. Oxygen consumption and nisin activity of the first and second fermentations (●1° inoculum AU/mL; ♦1° inoculum %O_2; ✱ 2° inoculum AU/mL; Δ 2° inoculum %O_2). For media with 50% skimmed milk plus 50% MRS, fermentor cultures was started from the first transfer cultures, the relation between AU/mL and inhibition halo (H, mm) was calculated through the equation: $AU/mL = 10^{(0.2307 \times H - 1.4174)}$; and for the second transfer culture, the relation between AU/mL and inhibition halo (H, mm) was through the equation: $AU/mL = 10^{(0.3211 \times H - 2.5553)}$.

MRS plus skimmed milk (1:1), suggesting no interference of excess salts on the EDTA-nisin interaction.

Single Batch Fermentation

For 50% skimmed milk mixed with 50% MRS, extending the incubation to 16 h of fermentation growth for *L. lactis* resulted in higher expression of nisin, with titers of 5934 AU/mL, four times greater than the nisin titer of 1377 AU/mL obtained with 8 h fermentation, with a constant 50–55% O_2 demand and pH between 4.5 ± 0.5 (Fig. 2). Up to now, experiments were carried out with inoculum from the first and second transfer cultures for 8 h and 16 h fermentations, respectively. The data obtained from fermentation conditions were 22 times lower than the titer values obtained from *L. lactis* nisin expression grown in 25% milk and 25% MRS (with the fourth transfer culture as inoculum) in rotary shaker (100 rpm) for 36 h (at 30°C), observed from a previous study (24). These results (24) indicate that the fermentation growth conditions could be improved with culture medium (25% milk and 25% MRS) inoculated with the third, fourth, and fifth transfer cultures of *L. lactis*. Although the expression of nisin was observed throughout fermentation, the constant increase of nisin expression observed with 8 h fermentation was coincident to the stabilization of O_2 consumed, in which the O_2 concentrations were adjusted to cell growth.

Acknowledgment

The authors thank the Brazilian Committees for the Scientific Technology Research (CNPq, FAPESP, and CAPES) for financial support and scholarship; and biologist Irene A. Machoshvili for technical support.

References

1. Hurst, M. (1981), *Appl. Microbiol.* 27, 85–123.
2. Cleveland, J., Montville, T. J., Nes, I. F., and Chikindas, M. L. (2001), *Int. J. Food Microbiol.* **71,** 1–20.
3. Jung, G. (1991), *Angew. Chem. Int. Ed. Engl.* **30,** 1051–1192.
4. de Vuyst, L. and Vandamme, E. J. (1992), *J. Gen. Microbiol.* **138,** 571–578.
5. Buchman, G. W., Banerjee, S., and Hansen, J. N. (1988), *J. Biol. Chem.* **263,** 16,260–16,266.
6. Vessoni Penna, T. C. and Moraes, D. A. (2002), *Appl. Biochem. Biotech.* **98–100,** 775–789.
7. Vessoni Penna, T. C, Moraes, D. A., and Fajardo, D. N. (2001), *J. Food. Protect.* **65,** 419–422.
8. De Vuyst, L. and Vandamme, E. J. (1994), In: *Bacteriocins of Lactic Acid Bacteria.* De Vuyst, L and Vandamme, E. J. (eds.) Chapman & Hall, Glasgow, pp. 1 12.
9. Hansen, J. N. (1994), *Crit Rev Food Sci. Nutr.* **34,** 69–93.
10. Turner, S. R., Love, R. M., and Lyons, K. M. (2004), *Int. Endodontic J.* **37,** 664–671.
11. Aranha, C., Gupta, S., and Reddy, K. V. R. (2004), *Contraception* **69,** 333–338.
12. Dubois, A. (1995), *EID Digestive Diseases Division* **1(3),** 79–88.
13. Sakamoto I., Igarashi, M., and Kimura, K. (2001), *J. Antimicob. Chemother.* **47,** 709–710.
14. Biswas, S. R., Ray, P., Johnson, M. C., and Ray, B. (1991), *Appl. Environ. Microbiol.* **57,** 1265–1267.
15. Hurst, A. and Kruse, H. (1972), *Antimicrob. Agents Chemother.* **1,** 277–279.
16. Stevens, K. A., Sheldon, B. W., Klapes, N. A., and Klaenhammer T. R. (1991), *J. Food Protection* **55,** 763–766.
17. Ganzle, M. G., Hertel, C., and Hammes, W. P. (1999), *J. Food Microbiol.* **48,** 37–50.
18. Thomas, L. V., Clarkson, M., and Delves-Broughton, J. (2000), In: *Natural Food antimicrobial systems.* Naidu, A. S. (ed.), CRC Press, Washington DC, pp. 463–524.
19. Fang, T. J. and Hung-Chi, Tsai (2004), *Food Microbiol.* **20,** 243–253.
20. Ukuku, D. O. and Fett, W. (2004), *J. Food Protection.* **67(10),** 2143–2150.
21. Vaara, M. (1992), *Microbiol. Rev.* **56,** 395–411.
22. Hauben, K. J. A., Wuytack, E. Y., Soontjens, C. C. F., Michiels, C. W. (1996), *J. Food Prot.* **59,** 350–355.
23. Vessoni Penna, T. C., Ishii, M., Pessoa, Jr., A., Nascimento, L. O., A., Souza, L. C., and Cholewa, O. (2003), *Appl. Biochem. Biotechnol.* **113–116,** 453–468.
24. Gill, A. O. and Holley, R. A. (2003), *Int. J. Food Microbiol.* **80,** 251–259.
25. Vessoni Penna, T. C., Jozala, A. F., Novaes, L. C. L, Pessoa-Jr., A., and Cholewa, O. (2005), *Appl. Biochem. Biotech.* **121–124,** 1–20.
26. Cutter, C. N. and Siragusa, G. R. (1996), *Food Microbiol.* **13(1),** 23–33.
27. Shelef, L. A. and Seiter, J. (1993), In: *Antimicrobial in Foods.* Davidson, P. M., Branen, A. L. (ed.), Marcel Dekker, New York, pp. 539–569.
28. Gray, G. W. and Wilkson, S. G. (1965), *J. Appl. Microbiol.* **28,** 153.
29. Leive, L. (1965), *Biochem, biophys. Res. Commun.* **21,** 290–296.
30. Branen, J. K. and Davidson, P. M. (2004), *Int. J. Food Microbiol.* **90,** 63–74.

SESSION 2
Today's Biorefineries

Introduction to Session 2

PARIS TSOBANAKIS

Cargill, Inc., P.O. Box 5699, Minneapolis, MN 55440-5699

Session 2 focused on the transition from the biorefineries of today to the ones of the future. The various presenters provided an international perspective, offering several visions about how future biorefineries may look, what technologies they will require, and what products they will produce. One recurring high-level theme was the need to logically capitalize on the learning being accrued in our existing grain, sugar, pulp, and other plant biomass processing mills, which represent today's biorefineries. Such facilities are now being used to produce a large number of products, including starches, sugars, oils, fuels, organic acids, polymers, and power. Ethanol and biodiesel production, in particular, are experiencing tremendous growth and are benefiting from increased "critical mass" and significant economies of scale. Biorefineries based on ethanol and biodiesel production provide routes for dramatically increasing the use of renewable resources to supply fuels and chemicals for the global economy.

The first two articles in this session of these proceedings make the case that biorefineries of the future are likely to combine the use of both biochemically and thermochemically based conversion technologies, as well as apply various process intensification strategies, to develop integrated systems that produce fuels and a variety of bio-based chemical and energy products with very high overall efficiency. Kochergin and Kearney discuss the value of incorporating engineered fractal distributors into applicable biorefining separations to dramatically increase their efficiency and throughput. Eggeman and Verser highlight the important role that utility systems play in existing and prospective biorefinery operations, especially in determining facilities' fossil fuel requirements and efficiencies. The three subsequent articles focus on specific technologies that are likely to be implemented in future biorefineries. Ekenseair and coworkers address the opportunities to extract high value products such as flavonoids from biomass before it is processed into fuels and chemicals to help to improve overall process profitability. Etoc and colleagues report on the development of simplified procedures for screening antifoams for their efficacy in controlling foam during gas sparged fermentations (e.g., aerobic submerged cultivations). Finally, de Lima da Silva and coworkers report on optimizing production of biodiesel from castor oil.

Existing Biorefinery Operations That Benefit From Fractal-Based Process Intensification

VADIM KOCHERGIN* AND MIKE KEARNEY

*Amalgamated Research Inc., PO Box 228, Twin Falls, ID 83301,
E-mail: vkochergin@arifractal.com*

Abstract

Ion exchange, adsorption, and chromatography are examples of separation processes frequently used in today's biorefineries. The particular tasks for which these technologies can be successfully applied are highly influenced by capital cost and efficiency. There exists a potential for significantly increasing the efficiency of these processes whereas simultaneously decreasing their size and capital cost. This potential for process intensification can be realized with the use of engineered fractal equipment. The cost savings potential and the possibilities for broadening the use of fractal-based separation technologies in future biorefinery concepts is illustrated by examples of full-scale implementation in the sugar and sweetener industries.

Index Entries: Biorefinery; fractal; liquid separations; chromatography; ion exchange.

Introduction

In the past, agriculture was viewed as an industry providing single commodities such as grain, sugar, seed, and so on. For the past several decades agriculture has been slowly transforming into an industry processing agricultural feedstock into a wide range of value-added fuels and chemicals. An analogy can be made with petrochemical refining, in which oil is converted into various chemical products. Conventional corn processing can be used as an example of a biorefinery producing ethanol as a fuel, dextrose as base chemical, fructose syrup as a food sweetener, oils, vitamins, and complex animal feed. Global market conditions and competitive pressure are stimulating diversification in other agriculture-based industries, such as sugar, forestry, pulp, and paper. Additionally, a new industry is evolving, in which agricultural waste products, such as straw or other sources of inexpensive biomass are being used as raw materials for obtaining fuels and chemicals. The multicomponent nature of agricultural feedstock and the variability of its composition related to harvest and storage conditions make the task of creating a universal processing technology quite challenging. However, many lessons can be learned from

---

*Author to whom all correspondence and reprint requests should be addressed.

operation of existing biorefineries. Because certain similarities exist between plant-derived feedstocks, analysis of unit operations involved in current processing technologies provides a good start for evaluating process and equipment design for the emerging bioindustry.

Innovative concepts, such as process intensification, will favorably impact the economics of future biorefining. The term "process intensification" was first introduced by ICI (UK) in the late 1970s. Creation of new types of equipment with much higher efficiencies and smaller footprint compared with conventional designs is the main focus of process intensification efforts. Resulting reduction of capital and operating cost allows reconsidering use of technologies currently deemed marginally feasible for certain applications. Henderson (1) indicates, however, that the advantages extend beyond capital cost issues and additionally provide improved process performance, lower safety risks, and improved manufacturing capabilities.

Considering the complex nature of plant-derived materials, it is likely that affinity-based separation technologies capable of isolating and purifying components with similar physico-chemical characteristics will be used extensively in the future biorefineries. With the emphasis on the fact that separation processes account for 40–70% of both capital and operating costs in the existing industry Humphrey (2) introduced a term "maturity" of a separation technology to measure the relative levels of fundamental knowledge and practical applications. The maturity values for affinity-based technologies, such as ion exchange, adsorption, and industrial chromatography range were estimated at 20–30%, whereas for more conventional crystallization and distillation processes the values exceed 65%. Therefore, there is a definite need for technology development efforts on all levels starting with a better understanding of fundamentals and optimization of materials, equipment, and modes of operation. The purpose of this article is to illustrate how the concept of engineered fractals as used in today's biorefineries can benefit the emerging renewable feedstock processing industry.

The Importance of Liquid Separation Technologies

A generalized sequence of unit operations common to existing biorefineries is shown in Fig. 1. Although the flow diagram does not include multiple recycle streams and handling of waste streams, it clearly demonstrates the importance of separation technologies (especially liquid separations). The purpose of the following discussion is to illustrate the similarities between two existing biorefinery operations—corn wet milling and sugar processing.

Technologies recovering valuable components from an agricultural feedstock always include a step for cleaning and preparation, in which a large uniform surface area is created to facilitate subsequent chemical/biochemical reactions or mass transfer. Grinding corn or sugar cane stalks and sugar beet slicing are examples of such preparation steps. The prepared biomass is then subjected to initial fractionation or extraction of major

Existing Biorefinery Operations

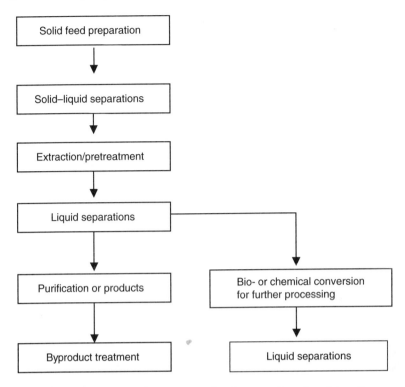

Fig. 1. A generalized sequence of unit operations in a biorefinery.

components, such as starch/gluten separation in the corn industry or a diffusion step in the cane or beet industry. Biochemical conversion into a new product in the corn wet milling process or chemical purification of sugars in beet processing are typically followed by ion exchange processes such as demineralization or softening, respectively. Examples of liquid separation technologies (adsorption, large-scale chromatography, and ion exchange) for downstream isolation and purification of products include enrichment of fructose syrup and polishing/decolorization of high-fructose corn syrup (HFCS) in the corn wet milling process and recovery of sugar from molasses and decolorization in cane refineries. These examples demonstrate the dominating role of liquid separation technologies in existing biorefineries. Out of 15 steps of HFCS production in a wet milling process five steps utilize ion exchange resin technologies for adsorption, fractionation, and purification (3). This is not surprising, considering that the described methods are very efficient for separating sugars from organic acids, salts, and other impurities. Even separation of various sugars with similar properties is relatively easy to accomplish. Separation of glucose and fructose using simulated bed chromatography is a good example of a commercially accepted process that achieves high purities and recoveries of both products (4).

Future biorefinery applications will take advantage of the technical solutions already developed in these example industries. It is anticipated,

Fig. 2. An example of a conventional distributor.

however, that additional developmental efforts will be needed, including in the area of process intensification, to improve feasibility. An example of such process intensification in the area of fluid distribution in the industrial equipment is considered later.

Distributors Based on Fractal Geometry

Most common distributors used in chromatographic, softener, decolorization applications are quite simple in design. The major drawback of common types of distributors roots in the lack of symmetry, which makes a scale-up task very complicated. An example of a conventional lateral pipe distributor is shown in Fig. 2. Typically the centrally located inlet is connected to a series of interconnected pipes with outlet holes. The residence time from the inlet to each hole varies depending on the distance from the inlet. For the same reason the pressure drop is different for each path from an inlet pipe to an outlet hole. Because the concentration of dissolved components often changes continuously, different residence time leads to spreading or "smearing" of the concentration front and therefore to loss of separation efficiency.

Conventional practice of distributor design based on high-pressure drop or variable outlet hole size leads to another set of problems. Among them is dependence of distribution quality on the feed rate. This problem is sometimes addressed as a turndown ratio. The ability of a distributor to convert fluid from a feed pipe into a uniform two-dimensional surface inside the column is extremely important for many industrial applications. For example in simulated moving bed (SMB) chromatography recirculation flow varies significantly during an operation cycle. Obviously, variable distribution quality in each step results in deviation from plug flow conditions and, hence, reduction of overall process efficiency.

A new generation of fluid distributors based on fractal geometry has been described extensively in the recent literature (5,6). Fractals can be defined in a number of ways. The following simple qualitative definition can be used to easily recognize such structures—"Fractals are self-similar objects whose pieces are smaller duplications of the whole object." Fractals can be "geometric" such that each iteration results in smaller and exact reproduction of the geometry (e.g., shapes in Fig. 3). In a separate category, "statistical" fractals are only self-similar in a statistical sense. An example is a tree in Fig. 3 (branches similar to the trunk, twigs similar to branches, and so on). The idea of application of engineered fractal for process applications has resulted in a number of patents on fluid distributors and

Fig. 3. Examples of geometrical (shapes) and statistical (tree) fractals.

engineered fractal cascades *(7,8)*. Fractal distributors originally installed in large-scale chromatographic columns (up to 7 m in diameter), have been modified for ion-exchange and decolorization applications, air distribution in conditioning silos, gas–liquid systems, and so on. A sample illustration of a fluid distributor for industrial chromatographic column up to 7 m in diameter is shown in Fig. 4. Fluid enters the predistributor designed to provide hydraulically equivalent flow to the fractal plates that are located on the top or bottom of the vessel. The final outlets provide *I* nearly ideal distribution quality. Each final outlet point is geometrically and hydraulically equivalent to the others, ensuring very uniform fluid distribution across column cross-section. Seemingly complicated, the distributors are rather easy to manufacture and install. The original distributors have been in industrial operation for more than 9 yr. Their cost compares favorably with conventional designs.

Process Intensification Using Fractal Technology

Related to fluid scaling and distribution methods evolved in nature (such as the human circulation system), engineered fractals exhibit unique features beneficial to the efficient operation of process equipment. They are especially efficient in the technologies that are sensitive to maintaining the narrow concentration front such as adsorption, ion exchange, distillation, sparging, mixing, and so on *(9)*. Engineered fractals were first applied in the industrial SMB chromatographic columns in 1992. Most important features of fractal devices are discussed later.

1. Fractals provide uniform fluid distribution to a surface or within a volume. This improves process efficiency in reaction vessels or mass transfer devices. As one example, fractals provide plug flow in

Fig. 4. Fractal distributor layout.

large-scale ion exchange or chromatographic columns. This results in significant savings in both capital and operating costs.

2. Fractals can be reliably scaled from small pilot equipment to large industrial equipment. The key to reliable scale-up is providing the same density and size of fluid inlets/outlets in both the pilot and subsequent industrial equipment. Thus, the hydrodynamic conditions for the fractals can be exactly reproduced. Construction of larger equipment is accomplished by adding larger fractal channeling to the initial pilot scale layout. Such scaling procedures have been validated for commercialization of large-scale chromatographic processes with columns up to 7 m in diameter. In these cases, the original pilot testing was performed using 3-in. diameter columns.

3. Fractals exhibit a very low sensitivity to changes in feed flow rate (high turndown). As an example, the quality of fluid distribution from a fractal distributor designed for an absorption/distillation process did not

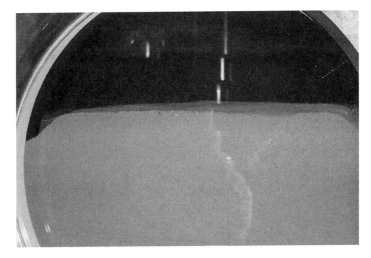

Fig. 5. Interface between water and 15% DS sugar solution above the resin bed.

change within a turndown ratio of 10 *(10)*. The primary reason is that the flow uniformity of fractals is based on symmetries and hydraulic equivalence of flow pathways rather than on the pressure drop criteria used for design of most conventional distributors. The typical very low-pressure drop exhibited by fractals provides additional opportunities for energy savings and novel equipment design.

New Concepts—From Retrofitting to Fractal Design

Selection of inlet/outlet point density and channel size within fractal distributors provides flexibility and control over fluid flow characteristics. Increasing the number of points allows narrowing of the distribution of certain parameters of fluid, such as momentum, concentration, particle size, velocity, and so on. A useful application of this design feature for ion exchange is discussed later.

Introduction of liquid through a fractal distributor with high exit density (several hundred outlets per square foot) leads to a reduction of the fluid momentum exiting each outlet. This inturn reduces the scale of turbulence, minimizing axial mixing. Figure 5 is a snapshot of a sight glass on a commercial ion exchanger during the rinse phase. Water moving downwards displaces colored sugar solution from the resin bed at a very high flow rate—500 bed volumes/h. Although the density difference between the two fully miscible water phases is less than 0.05 g/mL, the presence of a uniform fluid interface demonstrates the absence of turbulence in the headspace of the vessel. In addition to uniform plug flow, this feature of fractal distributors is important because turbulent eddies do not disturb the resin bed. In a conventional distributor design a short bed depth would be difficult to use owing to a low density of distribution points and, hence, high-liquid velocity and momentum.

Conventional ion exchangers utilize resin beds that are relatively deep regardless of process kinetics. In many instances increased bed depth is used to compensate for imperfections of the initial fluid distribution. Our experiments show that contrary to expectations, resin does not distribute liquid radially, and poor initial distribution usually leads to an overall reduction in efficiency. Additionally, increasing bed depth results in higher-pressure drop, which in turn limits equipment throughput. Use of fractal distributors allows constructing equipment with very shallow beds (just a few in. deep). In this case special resins with smaller beads or compressible resins can be used. Jensen et al. (11) have indicated that the use of smaller resin beads will increase adsorption kinetics significantly. Smaller diameter beads reduce the diffusion path in and out of the bead resulting in faster kinetics. However, engineering challenges related to pressure drop and fluid distribution were anticipated. As a second case, higher diffusion rates are expected in resins having lower crosslinking (hence, higher water content). In both cases pressure drop across the resin bed will limit equipment throughput. High quality liquid distribution is crucial for improving resin utilization by reducing the chance of premature breakthrough. Fractal equipment provides new opportunities for using these difficult media and for reducing resin inventories.

Applications of Fractal-Based Technologies

Several examples of fractal-based technologies that have found commercial application along with a few technologies under development are discussed later. The results illustrate that dramatic improvements can be made when equipment is designed specifically to take advantage of fractal characteristics.

Juice Softening in the Sugar Beet Industry

Industrial implementation of a new generation of fractal ion exchangers with shallow resin beds has opened new opportunities for the softening process in beet sugar factories (Fig. 6). Near ideal fluid distribution allows utilization of very small resin beds resulting in low capital and operating expenses (9,12). The fractal softener is quite compact (dimensions are $1.5 \times 1.5 \times 0.3$ m^3); however, it operates at the very high flow rate of 500 bed volumes per hour (BV/h).

(Table 1) illustrates the advantages of fractal weak acid softening systems compared with conventional weak acid softening systems. Regardless of high flow rates the pressure drop through the entire system (including two distributors and the bed of resin) does not exceed 0.25 bars, which allows construction of ion exchange cells without the requirement of pressure-rated vessels. Overhead tanks for gravity feed are sufficient for providing feed flow to the cells.

Existing Biorefinery Operations

Fig. 6. Fractal softener (an ion exchange vessel with fractal distributors installed on the top and bottom).

Table 1
Comparison of Conventional Weak Cation Juice Softening

Parameter	Conventional (lateral orifice distributor)	Fractal flat bed
Resin bed depth (m)	1.0	0.15
Exhaustion flow rate (BV/h)	50	500
Bed pressure drop (bar)	3.5–5.6	0.1
Regeneration flow rate (BV/h)	30	150

As a result of the introduction of this particular fractal implementation, overall capital cost of the weak cation juice softening process has been reduced to 35–40% of the conventional technology. Energy requirements for pumping have been reduced significantly. Peripheral tanks are smaller in size, which has had a positive influence on installation cost. A similar approach is being evaluated for adsorptive decolorization of sugar solutions. In the initial phase of an ongoing project a threefold reduction of resin bed depth has been demonstrated.

Molasses Desugarization

Molasses desugarization is a process, which has been used commercially in the beet sugar industry for the last 20 yr. The process is based on SMB chromatography and is principally similar to the SMB process used in corn wet milling for fructose purification. Being a final product of sugar processing, molasses still contains about 50% sucrose (based on dry substance), significant amounts of inorganic salts, amino acids, other sugars, and so on. Because of the large variation in molecular weight, size, and affinity of molasses components, separation of sucrose in a relatively pure form presents a challenge. Recovery of 92% pure sucrose with 90% yield has been long considered an industry standard. In a conventional system, an increase in molasses loading per unit of resin typically causes a significant reduction in performance. Introduction of fractal distributors coupled with the use of smaller resin has allowed a reduction of the inventory of ion exchange resin used as separation media by a factor of two without any loss of performance.

Arkenol Process

The potential of fractal technology can be further illustrated with reference to an application in the Arkenol process (13). In this process concentrated sulfuric acid is used to pretreat biomass. The task of acid recovery can be solved using SMB chromatography. A feasibility study performed by Arkenol indicated that acid recovery by chromatography should exceed 98%, and the recovery of sugars in the product fraction should be higher than 98%. Additionally, eluent use should be minimized to reduce dilution (and subsequent energy use for acid reconcentration). A project, in part addressing this problem, was financed by DOE grant no. DE-FC36-01ID-14016. A key accomplishment of the project was the reduction of SMB size by 75% with a simultaneous decrease of product dilution while maintaining required purities and recoveries (14). A short bed design allowed the use of small bead compressible separation media and thus, an improvement in process kinetics. The results would have been difficult (even if possible) with the use of conventional equipment. Japan Gas Corporation (Japan) is presently testing Arkenol technology using fractal SMB systems supplied by Amalgamated Research Inc. (Twin Falls, ID).

Fractal SMB Chromatography on a Model Biomass Conversion Stream

In the initial phase of the above DOE—funded research raw beet juice was considered to be a good model stream for biomass hydrolysate in regards to separation of sugars from impurities. Beet juice is derived from plant material and has many components; some of them poorly identified. The separation task was mainly isolating the groups of components from each other. Despite a high solids loading, it was possible to reduce the SMB system size by a factor of two compared to conventional equipment. The finding has had practical value for a new SMB separation process developed for the sugar industry.

The results of process intensification using fractal technology are summarized in Table 2. The size reduction factor is calculated by dividing the resin volume used in a conventional system by the resin volume in the

Table 2
Process Intensification With Fractals

Technology	Process	System size reduction factor
Juice softening (fast kinetics)	Ion exchange	10
Adsorptive decolorization (slow kinetics)	Ion exchange	3
Molasses desugarization	SMB chromatography	2
Acid recycle from biomass hydrolysate (Arkenol)	SMB chromatography	3
Model biomass stream (high solids loading)	SMB chromatography	2

fractal-based system. The equipment size reductions are quite impressive and may be improved further with more efficient separation media. Another way to present the system size reduction factors is to use them as multipliers for calculating the reduction in energy use and increase of specific loading per unit of separation media.

Existing biorefinery applications are not listed in Table 2 (e.g., HFCS production) may also benefit from fractal technology either via retrofitting or with new equipment.

Conclusions

It has been demonstrated that fractal-based technologies are used effectively in existing biorefineries. The equipment size and energy requirements can be significantly reduced owing to improvement of process efficiencies. Risk factors owing to implementation of fractal technology on a large scale are reduced dramatically because of the inherent scalability of fractal-based equipment. A high level of collaboration is needed to validate fractal concepts applied to process streams expected from future biorefineries. Manufacturers of resins and other separation media should consider the benefits of providing materials that take advantage of the possible synergistic relationship between media and fractal-based equipment.

References

1. Henderson, I. (2003), Topical Conference on Process Intensification, AIChE Spring National Meeting, March 30–April 3, 2003 New Orleans, LA, pp.110–114.
2. Humphrey, J. and Keller, G., II. (1997), Separation Process Technologies, McGraw-Hill.
3. Corn Sweetener refining and Ion exchange resins. (1990), Purolite Corporation technical publication, Bala Cynwyd, PA.

4. Rearick, D. E., Kearney, M., and Costesso, D. (1997), ChemTech, September 1997, pp. 36–40.
5. Kearney, M. (1999), *Chem. Eng. Comm.* **173**, 43–52.
6. Kearney, M. (1999), Engineered fractal cascades for fluid control applications, *Fractals in Engineering*, Institut National de Recherche en Informatique et en Automatique, Arcachon, France.
7. Kearney, M. (1994), US Patent 5,354,460, Fluid Transfer System With Uniform Fluid Distribution.
8. Kearney, M. (1999), US Patent 5,938,333, Fractal Cascade as an Alternative to Inter-Fluid Turbulence.
9. Kearney, M. (2003), Topical Conference on Process Intensification, AIChE Spring National Meeting, March 30–April 3, 2003 New Orleans, LA, pp.134–139.
10. Kochergin, V., Kearney, M., Kroon, M., Olujic, Z. (1997), Proc. AIChE Annual Meeting, Los Angeles, CA, pp. 69–74.
11. Jensen, C., Pillay, M., Rossiter, G., and Monnonen, H. (2004), Proceedings of IEX 2004, July 4–7, 2004, Cambridge University, UK, pp. 235.
12. Kearney, M., Velasquez, L., Petersen, K., Mumm, M., and Jacob, W. (2001), The Fractal Softener - Proc. of the 31st General Meeting of the American Soceity Sugar Beet Technologists, Orlando, FL, pp. 110.
13. Farone, W. A. and Cuzens, J. (1998), US Patent No. 5,820,687.
14. Industrial Membrane Filtration and Short-Bed Fractal Separation Systems for Separating Monomers from Heterogeneous Plant Material. Final technical report (2005), DOE Award No. DE-FC36-01ID-14016. http://www.arifractal.com/ARi%202004%20DOE%20biomass%20processing%20report.pdf (last accessed 12/29/05).

Copyright © 2006 by Humana Press Inc.
All rights of any nature whatsoever reserved.
0273-2289/06/129–132/361–381/$30.00

The Importance of Utility Systems in Today's Biorefineries and a Vision for Tomorrow

TIM EGGEMAN* AND DAN VERSER

*ZeaChem Inc., 2319 S. Ellis Ct., Lakewood, CO 80228;
E-mail: time@zeachem.com*

Abstract

Heat and power systems commonly found in today's corn processing facilities, sugar mills, and pulp and paper mills will be reviewed. We will also examine concepts for biorefineries of the future. We will show that energy ratio, defined as the ratio of renewable energy produced divided by the fossil energy input, can vary widely from near unity to values greater than 12. Renewable-based utility systems combined with low-fossil input agricultural systems lead to high-energy ratios.

Index Entries: Ethanol; indirect ethanol process; green power.

Introduction

Fossil resources are the major source of energy and chemicals in our society today. Among all renewable energy options, biomass is unique in its ability to potentially supply liquid transportation fuels and feed stocks for chemicals production. We believe this sector of our economy will eventually be transformed from today's fossil resource dominated system into a more sustainable biomass-based system.

The economics of production, energy security, sustainability, and rural economic development are four important policy issues for evaluating the performance of biorefineries. In this article, we discuss how utility systems affect the policy level performance for today's and tomorrow's biorefineries. Energy ratio, defined as the ratio of the renewable energy produced divided by the amount of primary fossil fuel consumed, is a useful proxy for comparing performance with respect to energy security and sustainability. We show that the energy ratio is strongly influenced by the utility system configuration and the fossil input required for growth and transport of biorefinery feed stocks. We also briefly discuss today's electrical power industry in the United States to provide background on green power markets. Our findings are used to guide performance requirements for the biorefineries of tomorrow.

An example showing the influence of utility system design on the performance of an advanced biorefinery concept for the production of

*Author to whom all correspondence and reprint requests should be addressed.

ethanol via an indirect fermentation route is discussed. This advanced biorefinery concept has the potential to give very high-energy ratios when combined with fossil fuel efficient agricultural systems. A further advantage is that implementation is not limited to tropical or subtropical climates; temperate climates such as those found in much of the United States could support facilities based on this advanced biorefinery concept.

Today's Biorefineries

Corn processing, sugar mills, and pulp and paper mills are examples of well-established biorefineries. The United States is a dominant player in both the corn processing and the pulp and paper industries; Brazil is a dominant player in sugar cane processing. Each of these industries will be discussed using supporting data from either the United States or Brazilian industry.

Corn Processing

Corn is processed through three types of mills. The two most common are dry mills designed to produce fuel ethanol and distiller's dried grains and solubles (DDGS), and wet mill complexes designed to produce a slate of starch derivatives and animal feed products. A third type of mill, also known as a dry mill, produces corn grits, corn meals, and corn flours. The volume of corn processed through this third type of mill is small in comparison to the other two types of mills, so our discussion will focus on the first two types.

Total capacity for the United States dry mill industry today is 2590 MMgal/yr (1). Individual plant capacities range from 30 to 100 MMgal (denatured)/yr of ethanol production. Capital cost for a modern grassroots facility is $1–1.50/annual gallon (denatured) of capacity (2). The steam and power systems are usually very simple and are responsible for only a small portion of the capital investment.

Figure 1 is a simplified block flow diagram of a typical corn dry mill. Corn is received, cleaned, and hammer milled. The milled corn is mixed with water and enzymes then cooked. Cooking has the dual function of pasteurizing the feed and liquefying/saccharifying the starch. The whole grain mash is fed to fermentation where the dextrose liberated from the starch during liquefaction/saccharification is converted into ethanol via yeast-based fermentation. The ethanol/water mixture is distilled to a near azeotropic mixture and molecular sieve technology is used to further dehydrate the ethanol to anhydrous ethanol specification. The whole stillage from distillation is centrifuged to separate the solids (i.e., distillers wet grains) from the liquid (i.e., thin stillage). The thin stillage is evaporated, the residual solubles mixed with the wet grains, and the mixture is dried to produce DDGS—an animal feed coproduct.

The power requirement for a conventional dry mill is about 1.09 kWh/gal (3) all of which is usually purchased from the grid. A typical 45

The Importance of Utility Systems

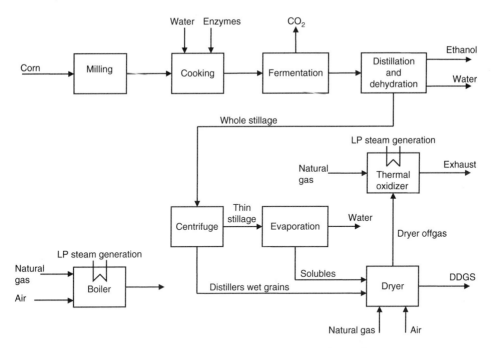

Fig. 1. Block flow diagram for a corn dry mill.

MMgal/yr ethanol plant will require 5–6 MW of base load electrical power; electrical power purchases will be roughly $1.9–2.4 MM/yr and contribute $0.04–0.05/gallon to the plant operating costs.

Thermal requirements for a dry mill are about 36,000 Btu/gal *(3)*. Roughly two-thirds is needed as steam for cooking, distillation, and thin stillage evaporation; the other third is typically supplied as direct fired heat for DDGS drying *(2)*. Historically, low pressure natural gas fired boilers were used to raise the necessary process steam. More recently, steam generation has been tied into the thermal oxidizer of the DDGS dryer emission control system. Natural gas is combusted to provide heat for DDGS drying, the dryer offgas is sent to a thermal oxidizer fired with supplemental natural gas as needed, and the hot gases from the thermal oxidizer are used to raise process steam. With natural gas valued at $6/MMBtu, a typical 45 MMgal/yr ethanol plant will purchase about $11 MM/yr of natural gas, contributing about $0.25/gallon to the plant operating cost. Natural gas purchases are the second largest operating cost in a typical dry mill, second only to corn. Financial instruments are often used to manage operating risk with respect to both corn and natural gas prices.

Figure 2 is a simplified block flow diagram for a corn wet mill facility. Wet mill facilities are technically more complex than corn dry mills. The front-end wet mill physically fractionates corn into its constitutive parts (i.e., starch, fiber, protein, and germ/oil). The fiber and protein fractions are sold as animal feeds; oil is extracted from the germ and is refined for human consumption. The starch slurry produced in the mill house is

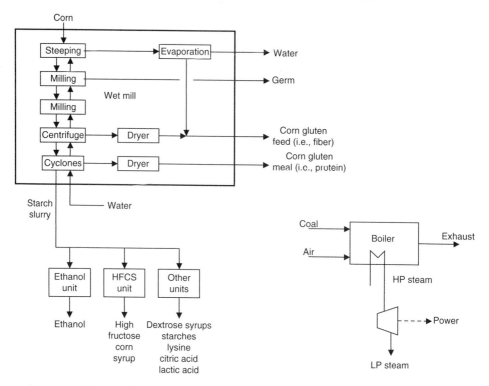

Fig. 2. Block flow diagram for a corn wet milling complex.

further processed by downstream units into a myriad of products including ethanol and high-fructose corn syrup. Current total wet mill capacity for ethanol production in the United States is 1109 MMgal/yr *(1)*, or slightly more than one-quarter of the industry's total grind capacity *(4)*.

Wet mill facilities are typically larger than dry mills. Grind capacities for an individual mill range from 50,000 to 500,000 bu/d of corn, with a typical mid-sized wet mill grinding about 200,000 bu/d or roughly 4300 mt (dry)/d. As a comparison, a 45 MM gal/yr (denatured) dry mill grinds about 47,000 bu/d or roughly 1000 mt (dry)/d. Capital costs for a wet mill are difficult to estimate because of all the possible configurations, but "rule-of-thumb" values are around $2.50–3/gall of capacity if all the starch were converted to ethanol.

Steam and power systems are also typically more complex. Large wet-mills often use coal fired boilers to produce high-pressure steam which is then let down across a steam turbine to produce power and low-pressure steam for process heating. Natural gas fired combined heat and power systems are found more frequently in small to mid-sized facilities. Electrical power and thermal requirements for a wet mill complex producing ethanol are approximately the same as for a dry mill when compared on a per bushel grind basis. Thus, the power to heat ratio for both types of facilities is roughly 0.10–0.15 depending on whether superheated steam or fired heat is used for the drying operations. This relatively low ratio suggests that most cogeneration systems were designed primarily for steam production. Galitsky et al. (5) estimates that in 1998 the United States wet mill industry, as a whole, produced only 21% of its electrical needs internally and purchased the balance of power from the grid. As in a dry mill, fuel costs are typically the second largest operating cost in a wet mill, second only to corn.

There has been much discussion in the literature over how to evaluate the energy producing capabilities of renewable energy systems. A life cycle assessment is a comprehensive accounting of all the inputs and outputs required to produce a good or service, and should conform to the standards set forth by the International Standards Organization (6). Although life cycle assessments are a rigorous means of evaluating renewable energy systems, conducting such an analysis is time consuming. In this article we use the short cut analysis method known as "net energy value." A net energy value analysis estimates the amount of fossil fuel required to produce a biobased fuel such as ethanol. The energy ratio is defined as the ratio of the renewable energy produced divided by the amount of fossil fuel used in production. Values greater than one imply success at harvesting sunlight and converting it into fuel. A net energy value analysis is less comprehensive than a life cycle assessment. For example, a net energy value analysis ignores the impact of ethanol on the performance of light duty vehicles whereas the boundaries for a life cycle assessment would likely be drawn to include this in a "farm-to-wheels" assessment. Although less comprehensive than a true life cycle assessment, the net energy value shortcut method has the benefit of being easy to calculate and the results can be used to qualitatively compare the energy producing capabilities of various ethanol production systems.

Workers at the United States Department of Agriculture (USDA) have authored a series of net energy value analyses for fuel ethanol production using domestic corn as the primary feedstock (3,7–8). Table 1 reproduces the USDA net energy analysis for the data reported in ref. 3. They give an industry composite energy ratio of 1.34, meaning that a gallon of ethanol produced by the industry only contains 34% more energy than the fossil fuels used in its production. Although the energy ratio is greater than one, it is does not exceed one by a large amount. As most of the fossil fuels are consumed during conversion of grain to ethanol, today's ethanol industry can be viewed as a means to convert coal and natural gas (and to a lesser extent petroleum) into a liquid transportation fuel.

Historically, ethanol has sold in the United States at a premium to unleaded gasoline even after adjusting for the effect of the excise tax exemption. The ethanol industry's capacity was relatively small and ethanol tended to compete in the fuel pool as either an oxygenate or octane blending component. The market situation has changed significantly over the past few years. Crude oil prices have risen from $20–25/barrel to over $50/barrel, but ethanol prices have remained relatively stable. Today

Table 1
Energy Ratios for Ethanol Plants

Values are in Btu/gal EtOH (neat)[a]

	Standard dry mill	Standard wet mill	Dry mill w/landfill gas	Dry mill w/gasifier	Cane mill + distillery, average	Cane mill + distillery, best practices	Transitional indirect w/ combined cycle	Advanced indirect
Feed stocks								
Process								
Utilities	Corn kernels Natural gas and coal	Corn kernels Natural gas and coal	Corn kernels Landfill gas and coal	Corn kernels Wood wastes	Cane juice Bagasse	Cane Juice Bagasse	Corn kernels Stover, natural gas and coal[d]	Lignocellulose Lignocellulose
Fossil energy use								
Crop production	21,803	21,430	21,803	21,803	6627	6076	15,566	2339
Crop transport	2284	2246	2284	2284	1792	1427	1631	2886
Other feed stocks	0	0	1369	2085	0	0	16,573	0
Ethanol conversion	48,772	54,239	11,482	0	2061	1556	20,833	0
Ethanol distribution	1588	1588	1588	1588	0	0	1588	1588
Subtotal	74,447	79,503	38,526	27,760	10,480	9059	56,191	6813
Coproduct credits								
Mill coproducts	13,115	14,804	13,115	13,115	0	0	10,670	0
Fuel or power export	0	0	0	0	7038	12,368	57,425	0
Subtotal	13,115	14,804	13,115	13,115	7038	12,368	68,095	0
Net energy value[b]	22,629	19,262	58,550	69,316	76,734	83,486	95,865	77,148
Energy ratio[c]	1.37	1.30	3.30	5.73	8.32	10.22	3.11	12.32

[a]Multiply by 0.2787163 to obtain MJ/m³. The cane mill entries are on a lower heating value basis and the calculations use Macedo's value for the lower heating value of ethanol (80,176 Btu/gal). The other entries are on a higher heating value basis and use the USDA's value for the higher heating value of ethanol (83, 961 Btu/gal).
[b]Net energy value heating = value of ethanol – (fossil energy use – coproduct credits).
[c]Energy ratio = (Heating value of ethanol + fuel or power export)/(fossil energy – mill coproducts).
[d]This case considers adding an indirect ethanol process to an exisitng wet mill. Corn stover is used as the main fuel for the indirect unit's utilities; natural gas and coal are the main fuels for the wet mill's utilities.

ethanol sells at roughly two-thirds the price of unleaded gasoline after adjusting for the effects of the $0.51/gallon excise tax exemption. This suggests that ethanol's role in the fuel pool has changed from an oxygenate or octane blending component to an energy source *(9)*.

US ethanol production is still relatively small when compared with demand for liquid transportation fuels. In 2003, the United States transportation sector consumed 26.8 Quads in the form of gasoline, diesel, and other fuels *(10)*. In that same year, ethanol supplied 0.239 Quads, or about 1% of our country's consumption of transportation fuels. Thus, despite recent growth of the industry, ethanol does not currently make significant contributions to our country's energy security.

US natural gas prices tend to track crude oil prices. The recent run-up in crude oil prices have been accompanied by a rise in natural gas prices from historic levels of $2–3/MMBtu to today's levels of $6–7/MMBtu. The high price of natural gas makes it difficult for corn processors to justify addition of natural gas fired cogeneration facilities. Some recently proposed dry mills have included coal fired boilers *(11)*. Although coal is a much less expensive energy source, coal fired utilities increase dry mill capital costs by 40–60% and have poor environmental performance when compared with natural gas fired utilities. There are two other niche modifications to utility systems being pursued today:

1. Some processors are substituting landfill gas for a portion of their natural gas needs. Successful projects require that landfill gas is locally available and can be purchased at a discount to natural gas on an energy content basis. Landfill gas substitution leads to improvements in the net energy value performance for the facility. Table 1 presents a net energy value analysis for a conventional dry mill that purchases electrical power from the grid and uses landfill gas to replace 100% of its natural gas requirement. Landfill gas is considered a renewable energy source, so the energy content of the gas itself is not included in the analysis; the other feed stocks category is an estimate of the fossil energy required for compression and processing at the landfill site. The energy ratio jumps to 3.3 when the entire natural gas requirement is replaced by landfill gas, a considerable improvement over a conventional dry mill.
2. Central Minnesota Ethanol Cooperative is pursuing a waste wood gasification project *(12)*. The syngas will be used to fire the thermal oxidizer/boiler at their existing corn dry mill. The heat released will be used to raise high-pressure steam, which in turn will be used to produce power and low-pressure steam for process needs. Capital cost is projected to be $15 MM; the existing plant capacity is about 20 MM gal/yr, so when implemented this will be a significant capital project. The benefit will be a complete elimination of natural gas and electricity purchases. Table 1 presents an order-of-magnitude net energy value analysis under the assumption that the fossil energy

consumed to produce and transport wood waste is 2085 Btu/gal. The energy ratio increases dramatically to 5.73.

Both examples show that changes to the utility system can improve plant economics while improving sustainability and maintaining rural economies. It is important to note that "green-ness," as indicated by an energy ratio much larger than one, does not necessarily imply a lignocellulosic-based fermentation. Starch fermentations can provide reasonably high-energy ratios when the utilities for the plant are provided by renewable resources. The long-term driver for lignocellulosic fermentation R&D is that lignocellulosics, unlike corn starch, are projected to be capable of providing enough feed stock to make a significant contribution to our country's energy supply (13).

Sugar Mills

Sugar is produced from either sugar beets or sugar cane. Sugar beets are grown in temperate climates of North America and Europe. Tropical and subtropical climates support sugar cane production. The beet sugar industry supplies roughly 30% of global sugar demand; the cane sugar industry provides the balance of demand (14).

The slicing season for sugar beets is about 100 d in Europe and can extend to as long as 250 d in North America. Beet processors operate integrated milling and refining operations. Thick juice storage/processing and molasses processing strategies are sometimes used to extend the operating season. Exhausted beet pulp is usually processed and sold into animal feed markets, thus it is not available as fuel for the factory. The utilities for beet factories are usually derived from fossil resources. Combined heat and power systems using a boiler/steam turbine cycle are common. Power and steam requirements for an energy efficient mill are 3.1 kWh/100 kg beet and 17–22 kg/100 kg beet (14). This gives a power to steam ratio of 0.23–0.30, at the high end of what can be provided by a simple boiler/turbine cycle.

The crushing season for sugar cane spans 5–8 mo. Unlike the beet sugar industry, most cane mills are not integrated with a refinery. Sugar refineries operate year-round to produce white sugars from the raw sugar produced by the mills. Colocating a mill with a refinery has the advantage of improving capital utilization as utilities and other infrastructure can be shared.

Brazil is by far the largest producer of raw sugar in the world. Its industry is unique in that significant capacity has been installed for both raw sugar and ethanol production, allowing integrated factories to swing production between the two products based on market conditions. In 1997, the total amount of gasoline, ethanol and diesel consumed in Brazil for transportation was 81.2×10^9 L, 16% of which was derived from domestically produced ethanol. Reference (15) gives further details of the policy level performance of the Brazilian industry.

Figure 3 is a simplified block flow diagram for a typical Brazilian cane mill. The harvested cane is received, cleaned, chopped if harvested as whole

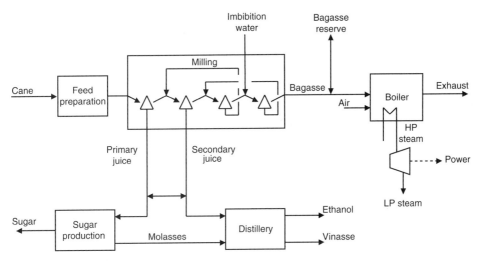

Fig. 3. Block flow diagram for a Brazilian cane mill.

cane, shredded then passed through the first of four to six milling tandems. The juice from the first mill, called the primary juice, is used mostly for sugar production. The bagasse (i.e., the fibrous remains of the cane plant after extraction of the sugar juices) from the first mill is further extracted in a series of three to five mills to produce a secondary juice and a final mill bagasse that is nearly exhausted of soluble sugars. Imbibition water is added at the back end of the milling train and the expressed juice from each mill is fed to the previous mill to provide countercurrent extraction. The secondary juice has lower concentration of sucrose and higher levels of nonsugar impurities when compared to the primary juice. The secondary juice is split between sugar production and ethanol production based on management's targets for sugar and ethanol. The distillery includes the fermentation, distillation, and dehydration steps used to produce commercial fuel alcohol. The carbohydrate feed to the distillery is a mixture of secondary juice and molasses from sugar production (i.e., the residual solubles after crystalline sugar has been recovered). In addition to ethanol, the distillery also produces vinasse, which contains the solubles not converted to ethanol during fermentation. Vinasse is recycled back to the cane fields as a fertilizer.

The bagasse from the last milling tandem is combusted in the plant boilers and the heat released is used to produce process steam, mechanical power to drive the milling equipment, and electricity to drive the rest of the rotating equipment in the factory. The three main factors affecting the mill energy balance are:

1. the fiber content of the incoming cane,
2. the energy efficiency of the juice evaporation and syrup crystallization steps, and
3. the energy efficiencies of the bagasse boiler and associated steam and power generation systems.

Most mills manage their operations so that bagasse provides all of the fuel to the mill's boilers during the crushing season, and the mill is usually operated so there is no net import or export of steam and power. Usually a slight excess of bagasse is held in reserve to cover fuel needs during start-up or shut down of the factory. Any bagasse produced in excess of the reserve needs is sold as fuel to other industrial facilities (e.g., orange juice processors), sold as a coproduct (e.g., source of pulp for papermaking, and various agricultural applications) or otherwise disposed. Coal, fuel oil, or another fossil fuel source is used to fuel the boilers only when the bagasse supply has been eliminated. Combined heat and power systems based on boilers/steam turbines are also common throughout the industry.

A current trend in Brazilian mill modernization is to improve electricity production. Historically, electricity prices in Brazil were very low because there was an oversupply of generating capacity. Hydroelectric plants supply 82–87% of the grid electricity (16). With poor market conditions for electricity, it was difficult for mills to justify investments to increase electricity production. In fact, up until 1985 most Brazilian mills were not self-sufficient with respect to electricity. Low thermal efficiencies in both the boiler and process operations were acceptable, perhaps even desired, as this was a way to dispose of bagasse. The oversupply in hydroelectric generating capacity has slowly been worked off over the past two decades. Today, market conditions are favorable for export electricity sales. Cane sugar mills currently provide about 6.1% of the grid power consumed in the service area for Paulista Company of Light and Power, a power utility in São Paulo state (16).

The experience of the Brazilian industry with respect to export electricity has been repeated throughout the cane sugar industry. Mills have installed energy efficiency projects that reduce their steam usage, resulting in extended bagasse supplies and/or higher power export revenues. Boiler house improvements focus on increasing the pressure of steam generation and improved steam turbine efficiencies. High-efficiency swirl-burner combustion systems have been installed in Australia and Cuba, leading to boiler combustion efficiencies of 81–89.6% vs 58.8–67.1% for older designs (17). The literature also contains much discussion about whether cane trash collection from the fields (i.e., leaves and tops) and gasification projects are worthwhile prospects for increasing export electricity production.

Macedo has published a series of net energy value analyses for the Brazilian ethanol industry (18–20). Table 1 presents the industry average case from Macedo's analysis of the 2001/2002 season (18). His reference plant is a seasonally operated mill + distillery that only produces ethanol (i.e., Macedo has simplified the analysis by excluding sugar and export electricity production). Power consumption is reported at 28.9 kWh/t cane (electrical: 12.9 kWh/t, mechanical: 16 kWh/t) and steam consumption ranges from 220 to 330 kWh/t cane, giving a process power:steam ratio requirement of 0.09–0.13. However, these utilities need not be counted in the net energy analysis as they are supplied by combustion of bagasse—a renewable resource. In fact, Macedo takes a credit for the bagasse in excess of that needed to operate the facility. He reports an energy ratio of 8.32 for the industry average and 10.22 for the best practices case.

Quantitative comparison with the USDA corn ethanol analysis is perilous because the underlying assumptions differ among the two analyses. For example, there are differences as whether higher or lower heating values are used for the basis, differences on whether to include energy embodied in buildings and equipment, and differences on whether to include energy for ethanol distribution. However, after adjustments to reconcile the analyses, the Brazilian ethanol industry appears to enjoy a significantly higher energy ratio than the United States industry. The use of green utilities in the ethanol conversion step is the chief reason for the difference. Reduced fossil fuel input to the agricultural system (e.g., lower fertilizer application rates) and bagasse fuel/green power export are also contributing factors.

Pulp and Paper Mill

A modern pulp and paper mill is a highly evolved and integrated system. The industry is by far the largest practitioner of lignocellulosic pretreatment processes to extract valuable components from biomass. We restrict our coverage to Kraft pulp mills because this is the predominant pulping technology used in North America.

A typical Kraft mill supplies all of its steam requirements and a substantial portion of its electrical needs internally. As shown in Fig. 4, a common configuration uses a combined heat and power system with two boilers raising high-pressure steam and a noncondensing steam turbine to produce power and low-pressure steam for process needs. A majority of the high-pressure steam is generated by the Tomlinson black liquor recovery boiler. This boiler has the dual function of regenerating process chemicals and producing high-pressure steam. It is fueled by the soluble lignin and hemicellulose components of black liquor, originally extracted from the wood chips during digestion. The hog fuel boiler provides the balance of steam and is fueled by bark and other waste wood. Roughly 65% of the mill's electricity requirement is produced internally; the balance is purchased from the grid. A mill also uses some fuel oil or natural gas to operate the lime kiln, part of the chemical cycle in the Kraft process.

We will not present a net energy analysis for the pulp and paper industry because few facilities currently produce ethanol. However, one would expect fairly high energy ratios would result as renewable fuels would be used to generate most of the steam and power utilities. Furthermore, the fossil inputs required for forestry operations are a fraction of those required for corn production (21).

Although Kraft pulping technology is very mature, some new technology is being considered to enhance the energy integration, yield

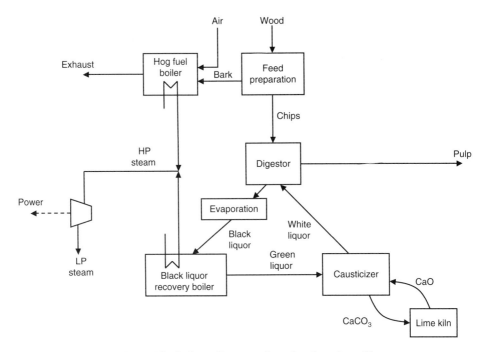

Fig. 4. Block flow diagram for a kraft pulp mill.

and flexibility of the process. Black liquor gasification is currently being tested at several demonstration scale industrial facilities. When implemented at full scale, black liquor gasification will replaced the functionality currently provided by the Tomlinson black liquor recovery boiler. Implementation of this technology will also enable polysulfide pulping, leading to 4–6% higher pulp yield for the mill.

The syngas produced by gasification can be used to replace the fossil fuel currently used to fire the lime kilns and also as a fuel for cogeneration of electricity and steam. The reference plant in (22), representative of a modern integrated Kraft pulp and paper mill located in the Southeastern United States, shows the current Tomlinson-type configuration requires a process power to steam ratio of 0.47. This high ratio cannot be met by a simple boiler plus noncondensing steam turbine cycle. Black liquor gasification produces a gaseous fuel that can be used to fuel a gas turbine-based combined cycle unit. Combined cycle cogeneration units are capable of meeting high ratios of power to steam. Energy balance calculations show that a full-scale replacement of the Tomlinson boiler with black liquor gasification plus combined cycle cogeneration will augment power generation to the point that the mill will become a net exporter of power while still maintaining its steam balance (22). In fact, the economic analysis in (22) suggests that mills may wish to purchase and gasify additional wood residuals to further boost export electricity production beyond that attainable with normal processing of pulp logs.

Today's US Electrical Power Industry

The US electric power industry was highly regulated for many decades. Industry goals were very simple: reliable low-cost supply of power. The result was the construction and operation of massive coal fired and nuclear power plants. These plants had high capital costs but very low operating costs. The high capital costs were not viewed as risky as the industry was regulated and capital charges could be amortized into the rate base at a low discount rate over a very long time period.

Ever increasing environmental standards and industry deregulation brought many changes to the US power industry in the 1990s. Lower capital cost projects, based both on combined heat and power installations at existing industrial facilities or as stand-alone gas fired combined cycle plants, dramatically improved the efficiency of power generation and brought the promise of low cost generation. The new players in the industry were much more responsive to market forces. Along with deregulation came new market forces such as green power programs. Green power, originally introduced as a competitive marketing tool in deregulated environments, was viewed as one way that new producers could compete against older utilities.

Today, green power programs are available at over 350 utility companies operating in 33 states (23). Approximately 1% of US electricity sales are through green power programs. Typical green power premiums are approx $0.025–0.03/kWh. Green generation capacity is mostly from wind, landfill gas, solar, or small scale hydroelectric power. Co-firing of biomass in existing coal plants and stand-alone biomass fired is also practiced (24).

The details of green power program implementation vary depending on each state's regulatory environment. Certain states have passed legislation requiring utilities to produce a certain percentage of their power from renewable resources. In fully deregulated markets, the customer may be required to switch from their default power provider to a provider that offers a green power package. In regulated markets, the utility offers green pricing programs to offset the higher production costs for green electricity. A third alternative, usually pursued by industrial rather than residential consumers, is to participate in green power trading programs through the purchase of renewable energy certificates (REC). REC also known as green tags, are also used by the utilities to support their green power programs.

An important benefit of green power generation is that it reduces the amount of NO_X and SO_X emissions produced per megawatt of generation in a utility's portfolio. NO_X and SO_X credits are separable from the green tags; they can be sold into the existing NO_X and SO_X trading systems. Old line utilities are becoming more interested in green power generation because these trading systems can be used to delay the large capital expenditures needed to modernize existing coal fired power plants.

Recent statistics on US power generation by fuel source are: coal–51%; nuclear–20%; natural gas–16%; hydroelectric–7%; others–6% *(10)*. Most coal and nuclear generating capacity is for base load power; today's high price of natural gas relegates its use to intermediate and peaking plants. It is very difficult to compete head-to-head with base load power plants without some type incentive such as green power credits. As was found in Brazil, we believe that in the long-term, biomass-based US biorefineries will eventually compete in electricity markets. Companies will decide that it makes more sense to invest in biorefinery cogeneration projects rather than massive new base load facilities to satisfy the incremental base load power demand caused by natural expansion of the economy.

In the short-term though, it is uncertain whether biorefineries will compete in electricity markets. Although petroleum and natural gas may become scarce, US coal reserves are quite large, so base load electricity prices should stay low in the foreseeable future. Green power programs may not be available at every biorefinery site. Production of export power drives-up capital projections for future biorefineries, further exacerbating risk for first-of-a-kind installations. This suggests that the technology developed for future biorefineries should provide flexibility with respect to business decisions on whether electrical power is to be exported from the facility.

Tomorrow's Biorefineries

Most experts agree that lignocellulosic feed stocks will eventually be required if renewable fuels are to make a significant contribution to our country's energy security. For example, the US DOE/USDA recently issued a joint report evaluating the land resources required to displace about 30% of our petroleum consumption. Their projections are that starch-based grains will contribute only 6% of the required feed stock; lignocellulosics will provide the balance *(13)*. Modern breeding tools can be used to create lignocellulosic crops with desirable agronomic and bioprocessing properties. It is not difficult to imagine a future in which dedicated farms grow energy crops designed specifically for biorefinery operations.

The technology used in biorefining operations will also likely evolve. One view of the future for ethanol production suggests that current direct fermentation technology will be merely augmented with front-end processes for lignocellulosic pretreatment and hydrolysis. The resulting hydrolyzate will then be converted into ethanol using direct fermentation technology, using a route similar to todays' with certain adaptations to account for differences between biomass and corn starch hydrolyzates. Although technically viable, this viewpoint fails to appreciate the poor match between the capabilities of direct fermentation technology and the distribution of energy among the fractions of the lignocellulosic feed stock.

Consider processing a typical lignocellulosic biomass into ethanol via direct fermentation of the biomass hydrolyzate. Roughly one-third of the energy content of the feed is present in the form of cellulose. Cellulose can be converted into dextrose with appropriate pretreatment and hydrolysis of the feedstock and then fermented with traditional direct fermentation yeasts or similar micro-organisms. Lignin and other nonfermentable materials account for about 40% of the energy content of the feed. The balance of the feed energy is in the form of hemicellulose, which produces a mixture of five and six carbon sugars upon pretreatment and hydrolysis of the biomass feedstock. Genetically engineered micro-organisms that produce ethanol from both five and six carbon sugars must be utilized to obtain high yield from the hemicellulose derived sugars as no wild-type micro-organisms exist that are capable of converting mixed sugars. Thus 40–60% of the chemical energy of the starting material (i.e., all of the lignin and nonfermentables fraction plus any unfermented materials from the cellulose and hemicellulose fractions of the biomass) is not available for ethanol production via direct fermentation.

So what can be done with the nonfermentable matter? One option could be to produce lignin-derived chemicals, such as adhesives, but these markets will become quickly oversaturated. Another option is to burn the nonfermentables, but very low process efficiencies result unless the heat released is used to generate steam and/or electricity. The amount of heat available is beyond the needs of the process, so the facility is forced to either compete for sales in electricity markets or accept low overall process efficiency.

Another view suggests radically different technologies will be used in the biorefineries of tomorrow. Concepts for advanced biorefineries often include biomass gasification/syngas production, bioprocessing, or combinations of the two. These technologies provide a better match between the energy distribution of the feed and the process technology capabilities. A common feature of syngas fermentation (25), catalytic production of ethanol or mixed alcohols from syngas (25), and the indirect fermentation process (26,27) is that gasification technology is used to access the chemical energy stored in lignin and other nonfermentable fractions of lignocellulosic feed stocks. Some benefits of these advanced concepts include: high overall process efficiency, good economics, and a good alignment with the other policy objectives mentioned in the Introduction. A downside of these advanced concepts is their high complexity, which increases risk for first-of-a-kind developments.

ZeaChem Inc. (Lakewood, CO) is engaged in the development of the indirect fermentation route for ethanol production. Our core chemistry can be broken down into three steps, as illustrated in Fig. 5 for the case of dextrose as the fermentable carbohydrate. In the first step, a homoacetogenic fermentation is used to produce acetic acid from carbohydrates at near 100% carbon yield. The acetic acid is then esterified with an alcohol to produce an ester. The ester undergoes hydrogenolysis to produce the desired ethanol

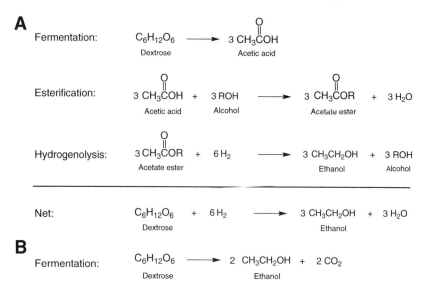

Fig. 5. Comparison of process chemistries. **(A)** Indirect ethanol route and **(B)** direct fermentation.

product and the recycle alcohol for the esterification step. The net result of the indirect route for ethanol production is a 50% improvement in molar yield compared to direct fermentation technology (i.e., three moles of ethanol per mole of six carbon sugar vs two moles ethanol per mole of six carbon sugar).

The energy for the third mole of ethanol is supplied by hydrogen. Biomass gasification is a particularly attractive means of hydrogen production as it converts the chemical energy stored in lignin and other non-fermentables into hydrogen, which in turn can be converted into the chemical energy stored in the ethanol product. This, combined with the fact that many homoacetogens metabolize both five and six carbon sugars, means that the chemical energy of all three major biomass fractions are converted into ethanol at high overall energy efficiency.

Figure 6 is a block flow diagram of one way to implement the indirect route during the current transitional period between starch and lignocellulosic feeds. Fermentable carbohydrate is supplied by a corn wet mill. Corn stover is gasified and further processed to recover hydrogen for the hydrogenolysis step. A fraction of the syngas is diverted for steam and power production. We recently completed a techno-economic study of the transitional implementation shown in Fig. 6, in which the indirect plant was assumed to be a grassroots facility located "across-the-fence" from an existing corn wet mill (26). The corn wet mill provided starch hydrolyzate to the ethanol facility; the ethanol facility produced its own utilities.

Rigorous material and energy balances were calculated using HYSYS 3.2, a commercial process simulator. A factored capital cost estimated was assembled, operating costs and revenues were estimated using the material balance as the basis, and discounted cash flow calculations were done to

The Importance of Utility Systems

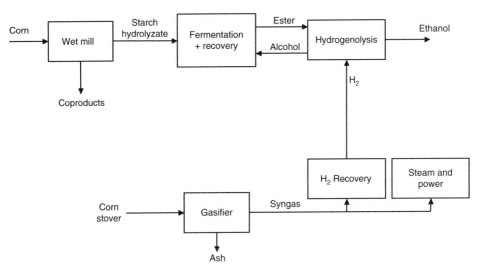

Fig. 6. Block flow diagram for a transitional indirect ethanol facility.

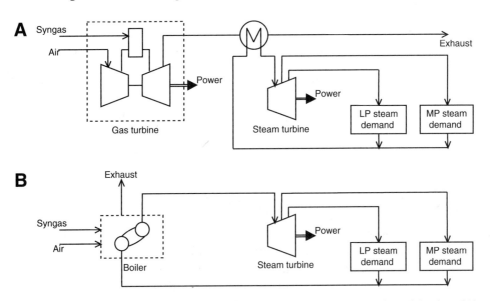

Fig. 7. Utility system configuration for the transitional indirect ethanol facility. (**A**) Combined cycle and (**B**) boiler/steam cycle.

combine the capital, operating and revenue estimates into standard performance measures. Both single parameter and Monte Carlo simulation methods were used to understand the sensitivities to various model assumptions.

One of the sensitivity studies was a comparison between two different implementations of the steam and power block. One implementation, shown in Fig. 7A, assumes a gas turbine-based combined cycle system with a noncondensing steam turbine. The combined cycle system provides the process steam requirements, all of the process electrical requirements,

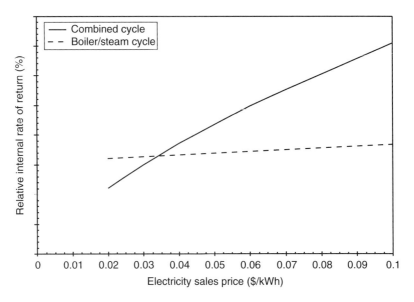

Fig. 8. Effect of utility system configuration on projected internal rates of return.

and a significant amount of export electricity for sales on the grid. The other implementation, shown in Fig. 7B, assumes the syngas is combusted in a boiler and the heat released used to produce high pressure steam, which in turn is expanded through a noncondensing steam turbine to produce electricity and process steam requirements. In this second case, the ethanol plant is nearly in power balance. The process steam requirement between the two cases is essentially identical. Because the combined cycle case produces a significant amount of export electricity, the combined cycle case has a higher stover feed rate and a larger gasification facility when compared with the simple boiler/steam cycle case.

Figure 8 shows the sensitivity of the internal rate of return for the two cases with respect to variations in the assumed selling price for export electricity. The two curves intersect at an electricity selling price of about $0.034/kWh. If the local value of base load green power is less than $0.034/kWh, the analysis suggests the boiler/steam cycle should be installed. If the local value of base load green power is greater than $0.034/kWh, the combined cycle system should be installed. A further risk analysis would probably lead to the conclusion that slightly higher electricity prices are needed to fully justify investment for the combined cycle option because it is more capital intensive. Additional creative business strategies, such as separating ownership of the utility system from ownership of the process plant, may also be needed to mitigate risk.

Table 1 presents an energy ratio analysis for the combined cycle case. The other feed stocks category reflects the fossil inputs required for stover production, harvesting and transport under the assumption that 1.69 MJ of nonrenewable energy is required per kg of harvested corn stover (28). An

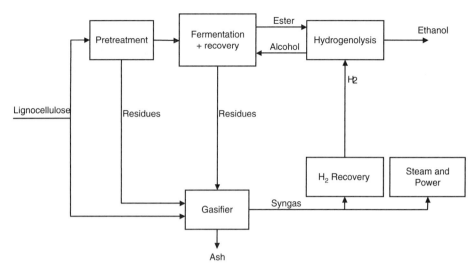

Fig. 9. Block flow diagram for an advanced indirect ethanol facility.

energy ratio of 3.11 is projected. Although the ethanol plant itself produces all of its steam and power from renewable resources, this modest value for the energy ratio is a result of the fact that fossil fuel consumption for the ethanol conversion step is still significant since the wet mill utilities are assumed to be supplied from fossil resources. The agricultural system used to produce corn kernels and stover is also a significant consumer of fossil fuels.

Many other implementations of the indirect process could be envisioned. For example, an all lignocellulosic case could be assembled by replacing the wet mill with biomass pretreatment and hydrolysis steps (see Fig. 9). Lignin rich fermentation residues would replace most of the stover used as feed for the gasifier. Significant improvements in the energy ratio would result if this advanced biorefinery concept were combined with improved agricultural systems for production and delivery of lignocellulosic feed stocks. The Advanced Indirect entry of Table 1 was derived from the values in (21) for fossil inputs required for lignocellulosic crops and an assumed chemical energy efficiency in ethanol conversion of 65% with the balance of feedstock energy used for utility production with no export/import of steam or power. An energy ratio of 12.32 is projected. Further improvements in energy ratio would result if the assumptions were changed to allow export of excess electrical power. Biorefineries with such high values for the energy ratio are not limited to tropical or subtropical climates; temperate climates throughout much of the US could support such a system.

Discussion

We have assumed the four main policy drivers affecting future biorefineries are economics, energy security, sustainability, and rural economic development. We showed that biorefinery utility systems are an important factor in policy level performance. Petroleum and natural gas fired systems look like nonstarters on all four policy level measures; coal fired systems only satisfy the first two policy objectives; biomass fired utility systems have potential to satisfy all four policy objectives.

We have outlined several desirable features for the process technology used in the biorefineries of tomorrow. Ethanol production technology should have flexibility with respect to feed stock to aid the transition from starch to lignocellulosic-based feed stocks. Ethanol production technology must be able to efficiently handle lignin and other non-fermentables; gasification based technologies appear to have a significant role in future biorefineries. Lastly, ethanol production technology needs to provide flexibility with respect to business decisions on power export.

Improved agricultural systems are also a likely feature of the biorefineries of tomorrow. Modern breeding tools can be used to create lignocellulosic crops with desirable agronomic and bioprocessing properties. Reduction in fossil fuel consumption for growing and harvesting plus ease of processing in biorefining operations will likely become important traits. Year round supply, or ability to store seasonally harvested crops, is another desirable feature. Seasonal operation of sugar mills results in low capital utilization, which makes it hard to justify capital investments to modernize existing facilities.

Finally, we suggest that the biorefineries of tomorrow will likely be capital intensive. The low capital intensity of a corn dry mill is a consequence of an easy to process feed stock and the low complexity of the supporting utility system. However, we have shown that today's conventional corn dry mill has a relatively low energy ratio and thus has poor policy level performance. We believe that future biorefineries will process lignocellulosic feed stocks and will derive their utilities from renewable resources. Both of these factors lead to higher capital intensity. The risk for an nth plant will be easy to quantify; new facilities will be justified provided the financial returns are adequate. However, high capital intensity will slow adoption of advanced biorefinery concepts.

Acknowledgments

This work was partially funded by the US Department of Energy—Inventions and Innovation Program (Grant Number DE-FG36-03GO13010). Support by US DOE does not constitute an endorsement by US DOE of the views expressed in this article.

References

1. Renewable Fuels Association (2005), http://www.ethanolrfa.org. Date accessed: May 2005.
2. BBI International (2003), *Ethanol Plant Development Handbook*, 4th edition, BBI International, Cotopaxi, CO.

3. Shapouri, H., Duffield, J. A., and Wang, M. (2002), U.S. Department of Agriculture, Agriculture Economic Report No. 814. http://www.usda.gov/oce/oepnu/. Date accessed: May 2005.
4. National Corn Growers Association (2005), *World of Corn*, Chesterfield, MO. http://www.ncga.com. Date accessed: May 2005.
5. Galitsky, C., Worrell, E., and Ruth, M. (2003), Lawrence Berkeley National Laboratory, LBNL-52307, Berkeley, CA.
6. International Organization for Standardization (1997), ISO 14040 Environmental Management–Life Cycle Assessment–Principles and Framework, International Organization for Standardization, Geneva. Also see related standards: ISO 14001, 14004, 14041, 14043, and 14049.
7. Shapouri, H., Duffield, J., McAloon, A., and Wang, M. (2004), Corn Utilization and Technology Conference, Indianapolis, IN. http://www.usda.gov/oce/oepnu/. Date accessed: May 2005.
8. Shapouri, H., Duffield, J. A., and Graboski, M. S. (1995), U.S. Department of Agriculture, Agriculture Economic Report No. 721. http://www.usda.gov/oce/oepnu/. Date accessed: May 2005.
9. Doering, O. and Seetin, M. (2005), *White Paper on Implications of Ethanol Pricing*, in press.
10. Energy Information Administration (2004), Annual Energy Review 2003, US Department of Energy, Washington, DC, DOE/EIA-0384(2003). http://www.eia.doe.gov/aer/. Date accessed: May 2005.
11. Nilles, D. (2004), *Ethanol Producer Mag.* **10(6)**, 28–52.
12. Nilles, D. (2004), *Ethanol Producer Mag.* **10(7)**, 32–34.
13. Perlack, R. D., Wright, L. L., Turhollow, A., et al. (2005), *Oak Ridge National Laboratory*, ORNL/TM-2005/66, Oak Ridge, TN.
14. Schröder, D. (1998), In: *Sugar Technology: Beet and Cane Sugar Manufacture*, van der Poel, P. W., Schiweck, H., Schwartz, T. (ed.), Bartens, Berlin, pp. 1067–1083.
15. Zanin, G. M., Santana, C. C., Bon, E., et al. (2000), Appl. Biochem. Biotechnol. **84–86**, 1147–1161.
16. Lamonica, H. M., Fioraneli, A., Linero, F. A. B., Leal, M. R. L. V. (2005), *Evolution of Surplus Power Generation in Brazilian Sugar/Ethanol Mills*, ISSCT 25th Congress, Guatemala, Paper CO2. See: http://issct.intnet.mu. Date accessed: May 2005.
17. Delavier, H. J. (1998), In: *Sugar Technology: Beet and Cane Sugar Manufacture*, van der Poel, P. W., Schiweck, H., Schwartz, T. (ed.), Bartens, Berlin, pp. 451–478.
18. Macedo, I. C., Leal, M., Ramos da Silva, J. (2004), Government of the State of São Paulo, Brazil. http://www.unica.com.br/i_pages/files/pdf_ingles.pdf. Date accessed: May 2005.
19. Macedo, I. C. (1998), *Biomass Bioenergy* **14(1)**, 77–81.
20. Macedo, I. C. (1992), *Biomass Bioenergy* **3(2)**, 77–80.
21. Lynd, L. R. and Wang, M. Q. (2004), *J. Ind. Ecol.* **7(3–4)**, 17–32.
22. Larson, E. D., Consonni, S., and Katofsky, R. E. (2003), Princeton University. http://www.princeton.edu/~energy/publications/texts.html. Date accessed: May 2005.
23. Bird, L. and Swezey, B. (2003), *National Renewable Energy Laboratory*, NREL/TP-620-35119, Golden, CO.
24. Wiltsee, G. (2000), *National Renewable Energy Laboratory*, NREL/SR-570-26949, Golden, CO.
25. Spath, P. and Dayton, D. (2003), *National Renewable Energy Laboratory*, NREL/TP-510-34929, Golden, CO.
26. Eggeman, T., Verser, D., and Weber, E. (2005), *An Indirect Route for Ethanol Production*, US Department of Energy, DE-FG36-03GO13010.
26. Verser, D. and Eggeman, T. (2003), US Patent 6 509 180.
27. Sheehan, J., Aden, A., Riley, C., et al. (2002), *Is Ethanol From Corn Stover Sustainable?* National Renewable Energy Laboratory, in press.

Extraction of Hyperoside and Quercitrin From Mimosa (*Albizia julibrissin*) Foliage

ADAM K. EKENSEAIR,[1] LIJAN DUAN,[1] DANIELLE JULIE CARRIER,[2] DAVID I. BRANSBY,[3] AND EDGAR C. CLAUSEN*,[1]

[1]*Department of Chemical Engineering, University of Arkansas, 3202 Bell Engineering Center, Fayetteville, AR 72701 E-mail: eclause@engr.uark.edu;* [2]*Department of Biological and Agricultural Engineering, University of Arkansas, 203 Engineering Hall, Fayetteville, AR 72701; and* [3]*Department of Agronomy and Soils, Auburn University, 202 Funchess Hall, Auburn, AL 36849*

Abstract

Mimosa, an excellent energy crop candidate because of its high growth yield, also contains, on a dry basis, 0.83% hyperoside and 0.90% quercitrin. Hyperoside has been documented as having anti-inflammatory and diuretic properties, whereas quercitrin may play a role in intestinal repair following chronic mucosal injury. Thus, mimosa might first be extracted for important antioxidant compounds and then used as a feedstock for energy production. This article presents results from studies aimed at determining the effect of three extraction parameters (temperature, solvent composition, and time) on the yield of these important quercetin compounds. Conditions are sought which maximize yield and concentration, whereas complementing subsequent biomass pretreatment, hydrolysis and fermentation.

Index Entries: Biomass pretreatment; energy crops; hyperoside; mimosa; quercitrin; value-added compounds.

Introduction

Although the US produced a record 3.4 billion gal (1.3×10^{10} L) of ethanol from grain in 2004 *(1)*, it is estimated that 150 billion gal (5.7×10^{11} L) of liquid fuel are required annually to make this nation energy independent. Lignocellulosic feedstocks, including crop and forestry wastes, municipal solid waste, and energy crops, are the logical feedstock alternative which could help to fill the gap. Depending on the conversion technology, $10–25 \times 10^6$ t of dry biomass are required to produce one billion gal of ethanol (2.4–6.0 kg/L). When this large feedstock requirement is coupled with the regional aspects and low bulk density of biomass, which makes shipping difficult and expensive, it seems prudent to utilize a host of biomass feedstock alternatives. In the southeastern United States, energy crops are a

*Author to whom all correspondence and reprint requests should be addressed.

viable source of lignocellulosic material because they can often be cultivated on marginal land, and thus serve as a source of additional revenue to farmers. Switchgrass is one such energy crop, with biomass yields of 6–9 dry t/acre/yr (1.3–2.0 kg/m^2/yr) *(2)*. A number of other energy crops have been proposed, including hybrid poplar, sericea lespedeza, arundo, and mimosa.

There are inherent difficulties in developing energy crops as biomass feedstocks for ethanol production. In addition to transportation costs, the farmer must receive compensation for producing and harvesting the energy crop, a fact, which increases the cost of the raw material for conversion. Lynd et al. *(3)* note that feedstock costs (at $40/t) account for 52% of the cost of producing sugar from hybrid poplar and up to 75% of the cost of ethanol. One way that feedstock costs can be offset is to select energy crops, which also contain compounds, which may be used in pharmaceutical and health care products. One such crop is mimosa, which showed forage yields of 6–7.5 dry t/acre/yr (1.3–1.7 kg/m^2/yr) in Alabama over a 5-yr test. Recent studies performed at the University of Arkansas showed that two documented antioxidant flavonoids, hyperoside (quercetin-3-O-galactoside) and quercitrin (quercetin-3-O-rhamnoside) constitute 2.3% of dry mimosa foliage *(4)*. Hyperoside has been documented as having anti-inflammatory and diurectic properties, whereas quercitrin may play a role in intestinal repair following chronic mucosal injury. Worldwide, about 50% of all drugs in clinical use are derived from natural products, and at least 25% of all prescription drugs contain ingredients extracted from higher plants *(5)*. These products can be very profitable. Each new drug is estimated to be worth an average of $94 million to a private drug company and $449 million to society as a whole *(6)*. Furthermore, pharmaceutical chemicals including conventional drugs and medicinals, vitamins, minerals, and herbal extracts constitute an $11.4 billion market in the United States and a $9.3 billion market in Europe *(7)*.

One key to effectively and economically extracting high value compounds from energy crops is to maximize the selectivity of the target compound in order to simplify downstream purification requirements. Another key is the ability to couple extraction with biomass conversion to ethanol, the major focus of biomass conversion. Although a number of technologies exist for the conversion of biomass to sugars (hydrolysis with dilute acid or concentrated acid, enzymatic hydrolysis), biomass pretreatment with either dilute acid or steam explosion followed by enzymatic hydrolysis is receiving the most attention because of its ability to develop a highly fermentable sugar stream. If high value products could be effectively extracted from energy crops prior to hydrolysis, a simple unit operation could be added to the existing biomass conversion technology to generate revenue from compounds that could be used in human and animal health care.

The purpose of this article is to investigate the feasibility of coupling antioxidant extraction and biomass conversion to ethanol by employing dilute acid, dilute base, or water as nonspecific extraction solvents. Simple batch extraction experiments were performed to measure hyperoside and quercitrin yields as a function of solvent composition and temperature. Most often flavonoids are extracted with acidic ethanol or methanol solutions *(8)*, and indeed, methanol is a very effective solvent for the extraction of hyperoside and quercitrin from mimosa *(9)*. However, an aqueous based extraction process would be more desirable, since the use of organic solvents for flavonoid extraction would require solvent recycle and residues containing organic solvent would not interface well with existing biomass conversion technologies.

Materials and Methods

Plant Material

Samples of fresh mimosa (*A. julibrissin*) foliage, ranging from 1 to 90 d in age, were harvested by hand in August, 2004, at Auburn, Alabama. The foliage included the petiole, branches and leaflets of the entire compound leaf. Foliage samples were placed in a forced air oven at 65°C within 1 h of harvesting, dried to constant weight, ground to a 0.32 mm particle size (as measured by ANSI/ASAE S319.3) and stored in sealed plastic bags at room temperature. A voucher specimen has been deposited at the Department of Chemical Engineering, University of Arkansas (Fayetteville, AR).

Chemicals

Hyperoside, quercitrin, and rutin (quercetin-3-*O*-rutinoside) were purchased from Indofine Chemical Company, Inc (Somerville, NJ). HPLC grade acetonitrile and methanol were obtained from EMD Chemicals Inc. (Gibbstown, NJ), formic acid was purchased from EM Science (Gibbstown, NJ), and H_2SO_4 and NaOH were purchased from VWR International (West Chester, PA).

Extraction of Mimosa Foliage

In extracting mimosa foliage, 2 g of dried biomass were extracted with 60 mL of solvent (either 100% methanol, 60% aqueous methanol, water, 0.1% NaOH, or 0.1% H_2SO_4) at 50, 70, and 85°C by blending the mixture in an insulated household blender for 10 min or by shaking the mixture in 280 mL amber bottles in a Precision shaking water bath (Winchester, VA) at 150 rpm. All experiments were performed at least in triplicate. Samples of biomass/solvent were removed with time in the shaking bath experiments. To separate the supernatant from the solids, the resulting mixture of solvent and solids was filtered through cheesecloth and then centrifuged at 12,000*g* for 30 min in an induction drive centrifuge (Beckman Coulter, Fullerton, CA). The crude extracts were collected and stored at 4°C for subsequent analysis. Aliquots (1 mL) of the extractions were dried under vacuum using a SpeedVac Plus (Savant Instruments, Holbrook, NY) without heat. After drying, the samples were redissolved in 1 mL of methanol for subsequent flavonoids analysis. The samples were then filtered through a 0.45 μm syringe filter (VWR International, West Chester, PA).

Quantitative Flavonoids Analysis by HPLC

The high-performance liquid chromatography (HPLC) analysis of flavonoids was conducted on a Waters Instrument, equipped with a 2996 photodiode array detector and a 2695 separations module controlled with Empower software. A 10 μL sample was injected into the Symmetry® C_{18} (150 mm × 4.6 mm) column (Waters, Milford, MA). The mobile phases used for the gradient consisted of solvent A (0.1% formic acid in water) and solvent B (0.1% formic acid in acetonitrile). The gradient program was initiated with 85:15 solvent A: solvent B, and maintained for 5 min. The gradient was linearly increased to 80:20 solvent A: solvent B over 20 min, and then increased to 20:80 solvent A: solvent B over 1 min. The gradient was then increased to 10:90 solvent A: solvent B over 10 min, and finally decreased to 85:15 solvent A: solvent B over 1 min, and then maintained for 3 min. This process was followed by a re-equilibration of the column at the final operating condition for 10 min. The flow rate was 0.75 mL/min, and the column temperature was 30°C. Each of the compounds was detected by the photodiode array detector at 360 nm. The authenticity of the peaks was verified as described by Lau et al. (4).

A typical HPLC chromatogram of a mimosa extract is shown in Fig. 1. Rutin was added as an internal standard (two parts sample to 1 part rutin, at a concentration of 1.0 mg/mL in methanol) and elutes at 13.1 min. Quercitrin elutes at 14.3 min and hyperoside elutes at 21.6 min. Quercetin-rhamnosylgalactoside, a third flavonoid found in mimosa extracts that has antioxidant activity (4), elutes at 9.9 min. Other compounds found in the extract have not been identified at this time.

Results and Discussion

Blender Experiments

An initial set of 10 min experiments was carried out with five solvents at three temperatures in the household blender to quickly evaluate the suitability of the solvents in extracting hyperoside, quercitrin, and quercetin-rhamnosylgalactoside from mimosa foliage. The household blender has the advantage of very good contact between solid and liquid in a short period of time, but suffers from the problem of maintaining constant temperature, despite the addition of insulation around the blender and restricting the experiment duration to 10 min. Results from these initial experiments are shown in Table 1, whereas the yields of flavonoids in ppm (mg flavonoid per kg of dried mimosa foliage) are shown for the five extraction solvents at three extraction temperatures. No standards are available for quercetin-rhamnosylgalactoside, so the yields of this quercetin

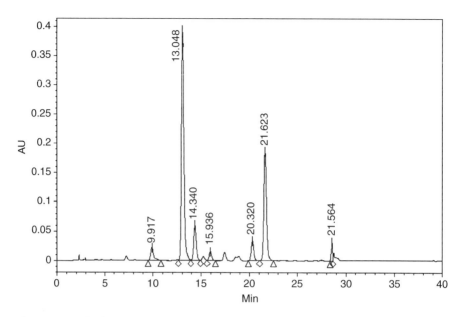

Fig. 1. HPLC chromatogram of mimosa foliage extracted with 50°C water for 10 min in a household blender. The retention times of quercetin-rhamnosylgalactoside, rutin (internal standard), quercitrin and hyperoside are 9.9, 13.1, 14.3, and 21.6 min, respectively.

Table 1
Extraction of Hyperoside and Quercitrin From Mimosa Foliage in a Household Blender Using Various Solvents

Temperature (°C)	Solvent				
	100% Methanol	60% Methanol	Water	0.1% NaOH	0.1% H_2SO_4
			ppm Hyperoside		
50	1896 ± 268	2551 ± 95	1701 ± 160	615 ± 109	2036 ± 36
70	N.A.	1903 ± 35	1708 ± 212	943 ± 288	1258 ± 308
85	N.A.	2542 ± 44	1596 ± 153	N.A.	1683 ± 38
			ppm, Quercitrin		
50	5431 ± 744	7520 ± 256	4954 ± 310	2853 ± 543	5589 ± 158
70	N.A.	5916 ± 95	5480 ± 608	4213 ± 1142	3709 ± 881
85	N.A.	7753 ± 123	4227 ± 1443	N.A.	4882 ± 11
			Estimated ppm Quercetin-rhamnosylgalactoside[a]		
50	552 ± 90	824 ± 15	645 ± 59	180 ± 37	742 ± 15
70	N.A.	655 ± 15	711 ± 58	367 ± 175	475 ± 128
85	N.A.	856 ± 19	624 ± 94	N.A.	638 ± 4

Not applicable, data were not taken at this condition.
[a]Estimated from the quercetrin calibration curve, because standards are no available.

compound were *estimated* using the calibration curve for quercitrin, a compound having a similar structure.

In analyzing the data, it is seen that 60% methanol is the best solvent for extracting mimosa flavonoids, followed by 0.1% H_2SO_4, 100% methanol and water. 0.1% NaOH was ineffective in flavonol extraction. In fact, 0.1% H_2SO_4 appears to be a very effective alternative to 60% methanol. The use of 60% methanol was far superior to 100% methanol, such that experiments at 70 and 85°C were not run for 100% methanol, since neither 100% methanol or 60% methanol would be a good candidate for linking extraction to biomass conversion to energy. Yields were generally expected to increase or remain constant with increasing temperature (little compound degradation was expected at these low extraction temperatures), but temperature variation in the blender during extraction may have affected the yields. Although the yields in water were lower than those obtained from 0.1% H_2SO_4 and 60% methanol, water could still be an interesting and effective solvent system when it is realized that the yield in extracting a high-value, relatively low demand compound from a high volume energy crop is relatively unimportant.

NaOH was not a good extraction solvent. The yields of hyperoside and quercitrin with 0.1% NaOH were one third to one half the yields from the other solvents. Furthermore, whereas the ratio of extracted hyperoside to extracted quercitrin was 0.31–0.38 for all of the other solvents, the hyperoside to quercitrin ratio was only 0.22 for NaOH. These negative extraction results with NaOH were somewhat expected, at least in a qualitative sense, since acidified solvents have been shown to be more effective in extracting flavonoids *(8)*.

Water Bath Experiments

Because of the inherent problems of maintaining a constant temperature in the blender experiments, time-course extraction experiments were carried out in a shaking water bath with 0.1% H_2SO_4 and water as the extraction solvents at 50 and 85°C. Thus, in these experiments, extraction efficiency was sacrificed for the ability to maintain a constant extraction temperature. Once again, at least three replicate experiments were performed at each extraction condition. Extraction times were expected to be longer in the water bath experiments with agitation at 150 rpm, so the first experiment in each set was carried out for 2–3 hr, whereas, after viewing the initial extraction results, subsequent experiments had a maximum extraction time of 60 min.

Water Extraction

Figure 2 shows the extraction yields (mg flavonoid per kg of dried mimosa foliage) of hyperoside, quercitrin, and quercetin-rhamnosylgalactoside from mimosa foliage in the water bath with water as the extraction solvent at 50 and 85°C. Once again, since standards are not available for quercetin-rhamnosylgalactoside, the yields of this quercetin compound

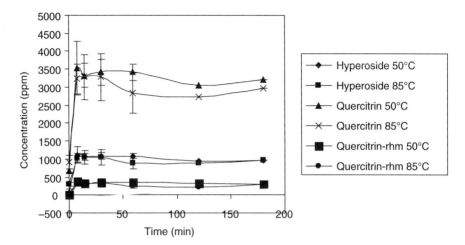

Fig. 2. Extraction of mimosa foliage with water at 50 and 85°C in a shaker bath to yield hyperoside, quercitrin, and quercetin-rhamnosylgalactoside.

were estimated using the calibration curve for quercitrin. As is noted in the figure, the maximum yield of each of the flavonoid compounds was attained in about 10 min, with perhaps a small but detectable decrease in yield observed after 30–60 min, possibly owing to compound degradation. Temperature had a minimal effect on extraction, with slightly higher yields at 50°C. The maximum yields of hyperoside, quercitrin and quercetin-rhamnosylgalactoside at 50°C were about 1100, 3500, and 400 mg/kg of dried mimosa foliage, respectively. Thus, the ratio of hyperoside to quercitrin after 10 min of extraction was 0.31 and the ratio of hyperoside to quercetin-rhamnosylgalactoside was 2.75, both quite similar to the values obtained in the blender experiments. Finally, the yields in the water bath experiments were only 60–80% of the yields in the blender (even at long extraction times), indicating that the flavonoids may be bound inside the mimosa cell matrix and are thus not accessible with water as the solvent without further breaking of the finely ground mimosa.

Extraction With 0.1% H_2SO_4

The extraction of mimosa foliage with 0.1% H_2SO_4 gave similar results, as presented in Fig. 3. Once again, the maximum yield of each of the flavonoid compounds was attained in about 10 min, and then leveled off. In contrast to the experiments with water, the maximum yields of the primary flavonoids with 0.1% H_2SO_4 (1500 mg hyperoside and 4500 mg quercitrin per kg of dried mimosa foliage) were obtained at 85°C. The ratio of hyperoside to quercitrin after 10 min of extraction at 85°C was 0.33 and the ratio of hyperoside to quercetin-rhamnosylgalactoside was 3.7, once again quite similar to the values obtained in the blender experiments. Finally, the yields in the water bath at 85°C were essentially the same as obtained in the blender experiments.

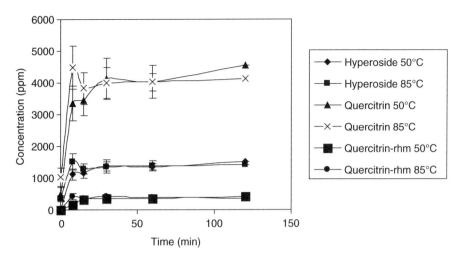

Fig. 3. Extraction of mimosa foliage with 0.1% H_2SO_4 at 50 and 85°C in a shaker bath to yield hyperoside, quercitrin and quercetin-rhamnosylgalactoside.

Conceptual Design

Figure 4 shows a conceptual design for coupling flavonoid extraction with biomass conversion to ethanol using enzymatic hydrolysis and fermentation. Milled biomass (mimosa or similar energy crop) is sent to extraction in which dilute acid (likely 0.1% H_2SO_4 or less) or water is used to extract flavonoids at a temperature of less than 100°C. Although experimental data have not yet been obtained for flavonoid extraction from mimosa foliage at temperatures higher than 85°C, there is likely an upper temperature limit in which compound degradation will become a problem. As an example of the potential for degradation at increased temperature, Duan et al. *(10)* saw significant silymarin degradation in extracting milk thistle with subcritical water at temperatures at or above 100°C. The dilute acid or water will be recycled in a continuous extraction process to maximize the flavonoid concentrations in the extract. It should be remembered that the large quantity of biomass required for ethanol production will generate an abundance of flavonoids so that only a portion of the biomass stream may be used in extraction. After extraction, the concentrated flavonoid stream is sent to flavonoid recovery to produce either an extract (solid mixture of mixed flavonoids) or the individual flavonoid compounds, which may be used as health beneficial compounds for humans or other animals.

Dilute acid is used to pretreat the biomass at 160–200°C before enzymatic hydrolysis and fermentation. Schell et al. *(11)*, for example, used 0.5–1.41% H_2SO_4 at 165–183°C for the pretreatment of corn stover in NREL pilot studies, and others have effectively used even lower acid concentrations. Fresh acid may be used for pretreatment, although an opportunity exists to use dilute acid from flavonoid extraction. Following pretreatment, the pretreated solids are available for subsequent hydrolysis and fermentation.

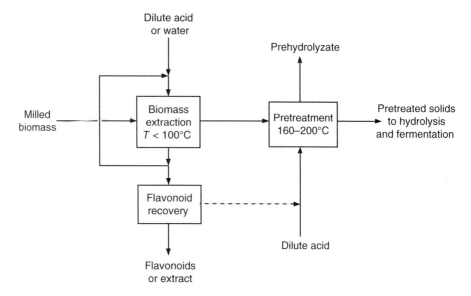

Fig. 4. Conceptual design for coupling flavonoid extraction with biomass pretreatment, hydrolysis and fermentation.

Conclusions

Water and 0.1% H_2SO_4 are very good solvents for the extraction of the flavonoids hyperoside, quercitrin and quercetin-rhamnosylgalactoside from mimosa foliage. By contrast, 0.1% NaOH was not effective as a solvent. When extracting mimosa foliage with 0.1% H_2SO_4 at 50°C in a household blender, flavonoid yields were 70–80% of the yields obtained when extracting with 60% methanol, the preferred solvent for flavonoid extraction. In using water as the extraction solvent, the yields fell to 60–70%, but water is still judged to be an effective solvent in extracting a high-value, relatively low demand compound from a high volume energy crop. Time-course experiments performed in a shaking water bath showed that the flavonoids could be extracted in about 10 min, and were not particularly sensitive to temperature between 50 and 85°C. A conceptual design shows that flavonoid extraction could be easily coupled with biomass pretreatment prior to enzymatic hydrolysis and fermentation.

References

1. Renewable Fuels Association (2005), Homegrown for the homeland, ethanol industry outlook 2005. www.ethanolrfn.org/outlook2005.html.
2. Bransby, D. (2001), Development of optimal establishment and cultural practices for switchgrass as an energy crop. Annual Report for US Dept. of Energy contract No. 19X-SY164C, Oak Ridge National Laboratory, Oak Ridge, TN.
3. Lynd, L., Wyman, C., and Gerngross, T. (1999), *Biotechnol. Prog.* **15(5)**, 777–793.
4. Lau, C., Carrier, D. J., Howard, L., Lay, J., Liyanage, R., and Clausen, E. C. (2006), *Bioresour. Technol.*, in press.

5. Carr, T., Pedersen, H., and Ramaswamy, S. (1993), *Environment* **35,** 14–38.
6. Mendelsohn, R. and Balick, M. J. (1995), *Econ. Bot.* **49,** 223–228.
7. Freedonia Group (1998), Cleveland, OH, http://freedoniagroup.imrmall.com.
8. Giusti, M. and Wrolstad, R. (2001), Characterization and measurement of anthocyanins by UV-visible spectroscopy. In *Current Protocols in Food Analytical Chemistry*, Wrolstda R (ed), John Wiley & Sons: New York.
9. Lau, C. S., Carrier, D. J., Howard, L. R., Lay, J. O. Jr., Archambault, J. A., and Clausen, E. C. (2004), *Appl. Biochem. Biotechnol.* **114,** 569–583.
10. Duan, L., Lovelady, J. K., Carrier, D. J., and Clausen, E. C. (2005), *J. Agri. Food Chem.*, in press.
11. Schell, D. J., Farmer, J., Newman, M., and McMillan, J. D. (2003), *Appl. Biochem. Biotechnol.* **105–108,** 69–85.

Foam Control in Fermentation Bioprocess

From Simple Aeration Tests to Bioreactor

A. Etoc,[1] F. Delvigne,[2] J. P. Lecomte,*,[1] and P. Thonart[2]

[1]Dow Corning s.a., Rue Jules Bordet, Parc Industriel - Zone C, 7180 Seneffe, Belgium, E-mail: j.lecomte@dowcorning.com; and [2]Centre Wallon de Biologie Industrielle, Unité de Bio-industries, Faculté des Sciences Agronomiques de Gembloux, 2 Passage des Déportés, 5030 Gembloux, Belgium

Abstract

In this article, we describe the development of a simple laboratory test for the effective screening of foam control agents on a selected fermentation system, the mass production of *Yarrowia lipolytica*. Aeration testing is based on sparging air in the foaming medium allowing partial reproduction of the gas–liquid hydrodynamic encountered in bioreactors. "Dynamic sparge test," for which measurements are made during foam formation, was used to compare the capacity of three antifoams, based on different technologies, to control the foam produced in the fermentation broth. The selected foam control agents were: (1) an organic antifoam (TEGO AFKS911), (2) a silicone-based emulsion containing in situ treated silica (DC-1520) and (3) a silicone/organic blend silica-free formulation. The testing results demonstrated dramatic differences among them and showed that the capacity of TEGO AFKS911 and DC-1520 to control the foam generated in the fermentation broth decreases as a function of fermentation time. This occurred to a much lesser extent for the silicone/organic blend formulation. These results were correlated with the change of the foam nature and the increase of foam stability of the fermentation broth with culture time. The increase in protein content as a function of growth time was correlated with an increase in foam stability and antifoam consumption. A "synthetic fermentation broth" was also developed, by adding both proteins and microorganism to the culture medium. This allowed us to mimic the fermentation broth, shown by the similar antifoams behaviour, and is therefore a simple methodology useful for the selection of appropriate antifoams.

Index Entries: Antifoam; aeration; bioreactor; silicone.

Introduction

Fermentation is often accompanied by foam formation because of the high foaming tendency of solutions containing biomaterials such as

*Author to whom all correspondence and reprint requests should be addressed.

proteins *(1)*. These proteins can act as surfactants owing to their amphiphilic structure. They can adsorb at the interface, partially unfold and form strong intermolecular interactions. This produces a visco-elastic, irreversibly adsorbed layer at the air/liquid surface, which stabilises the foam *(2)*. These kinds of films are not easily breakable. Restraining the formation of foam in a bioreactor is a crucial point for two reasons: on one hand it allows the control of the fermentation itself and on the other hand the fermentation equipments can be optimised and it therefore minimises the production costs. Screening of a range of antifoams during complete fermentation batches is one way to select the most appropriate product to control the foam produced during a specific process. This methodology presents major disadvantages such as the time it takes, and then the number of materials that can thus be tested is very limited. The selection of the proper antifoam will then often be limited to the availability of the material and on previous experiences.

The aim of this study was to develop a laboratory test for the effective screening of foam control agents. Previous work with the same aim used either a synthetic system made of bovine serum albumin *(3)* or a dynamic surface tension measurement *(4)*. However, at the beginning of the fermentation, the foam stabilisation will be caused by the proteins already present in the growing medium, whereas at the end of the fermentation, by the proteins produced by the microorganisms *(5)*. It was important to assess whether the overall protein type and content, as well as the microorganism content could modify the relative antifoam efficiency obtained in this laboratory test. Biomass production of *Yarrowia lipolytica* was selected as a fermentation system, as we experienced it to be especially foamy *(6)*. This specific system was used to assess the effect of the yeast mass production on the absolute and relative antifoam efficacy.

In this article, we describe the development of a laboratory test, which enabled us to compare the efficiency of three antifoams based on different technologies. The mechanism of action of antifoam was the subject of recent studies, which often concentrated on synthetic surfactant solutions *(7)*. Antifoams are often divided into two categories, depending whether they are based on soluble or insoluble oils *(8)*. The "soluble" antifoams are often based on surfactant containing polyethyleneoxyde or polypropylene oxide moieties. Having a lower surface tension than the solution, the antifoam molecules adsorb at the interface, forming a low viscosity film, which is weaker than the proteins visco-elastic film because of lower intermolecular interactions. The second type of antifoams is based on insoluble oils such as polydimethyl siloxane or mineral oil. They are most usually formulated with hydrophobic particles, which help the small antifoam droplet to enter the solution surface *(9,10)*. It has been already described that silicone-based antifoams based on simple polydimethyl siloxane formulated with silica have reduced efficacy when very viscoelastic protein films are the cause of the foam formation *(11)*. In this work, we compared the relative efficacy of:

1. an organic antifoam: a soluble antifoam based on polyglycol moieties (TEGO AFKS911 from Goldschmidt). This foam control agent is currently used to control the foam produced during *Y. lipolytica* mass production in our laboratory;
2. a silicone-based foam control emulsion: an emulsion of polydimethyl siloxane-based antifoam formulated with *in situ* treated silica (a solid hydrophobic particle), DC-1520 from Dow Corning. In the formulation, the active is 20% meaning that for the same quantity of material, this product contains five times less active that the two other formulations used in this work. We have however selected the approach to work at equivalent antifoam formulation volumes instead of working at constant antifoam active content, as it is more likely to be the practice used during antifoam selection;
3. a silicone/organic blend: a custom blend of polyglycol enriched with silicone materials having surface-active properties. The formulation contains 25% of Pluronic PE 6100 (BASF), 25% polypropylene glycol (Mw2000, as P2000 from BASF), 25% of a low Mw hydroxy-terminated PDMS (PA Fluid from Dow Corning) and 25% of a silicone polyether (FS 1270 from Dow Corning). This material was found especially efficient for the control of the foam during the fermentation of sugar cane molasses for the production of fuel ethanol *(12)*. This formulation, although based on silicone, does not contain silica and was especially interesting to add to this study, to cover a wider range of antifoam materials.

Materials and Methods

Strain

Yarrowia lipolytica strain 6481 was supplied by CWBI *(6)* and was stored at –80°C before utilization.

Culture Medium

The culture medium used for studying biomass production contained glucose (2%), peptone (1%), yeast extract (1%), and chloramphenicol (0.05%).

Evaluation of the Microbial Population

Microbial populations were evaluated by counting the yeast colonies on a jellified medium. After appropriate dilution (in sterile tubes containing a solution of 0.5% NaCl, 0.1% of peptone, and 0.2% of Tween-80), the cells suspension was spreaded onto Petri dishes containing a jellified growing medium made of glucose (2%), peptone (1%), yeast extract (1%), and Agar (1.4%). Petri dishes were incubated at 30°C for 48 h.

Bioreactor

A 20-L Biolafitte bioreactor fitted with two RDT4 (rushton disk turbine with four blades) impellers was employed for studying biomass production. The bioreactor was filled with 12 L of culture medium and then sterilized at 121°C for 20 min. A regulation system was used to control temperature at 30°C throughout the experiment after sterilization. A pH of 7.0 was maintained constant during the process using acidic (H3PO4 85% diluted twice) and basic (KOH 50%) regulation. The airflow rate was set to 1 vvm and the stirring speed was fixed at 200 rpm. A mechanical foam breaker was used when no antifoam addition was required for certain experiments. The bioreactor was inoculated with a 8% vol of inoculum under aseptic conditions. Precultures were made in 1-L Erlenmeyer flask containing glucose (2%), yeast extract (1%), and peptone (1%) and were inoculated 24 h before the beginning of bioreactor tests. During this time, preculture flasks were shook on an orbital agitator and maintained at 30°C.

Protein Assay

Protein assay was determined using the method developed by Lowry et al. *(13)* using the Folin-Ciocalteus reagent. The assay was performed on the precipitate formed by the addition of trichloroacetic acid to the solution to be analysed. The calibration line straight was completed with BSA (Albumin Fraction V from Merck) in concentrations varying from 0 to 200 µg/mL.

Sparge Test

Sparge tests were performed in a glass column (diameter, 9.2 cm; height, 30 cm). Airflow was supplied at the base of the column at a rate of 3 L/min. Fine bubble dispersion was obtained by forcing the air through a porous diffuser. The volume of foaming medium was 400 g. All experiments were performed at room temperature on "fresh" foaming medium (maximum 24 h storage at 4°C) or synthetic foaming medium.

Dynamic Sparge Test

In each experiment, testing was performed using the same amount of antifoam (for DC-1520, the amount of active antifoam is five times less than in TEGO AFKS911 and in the silicone/organic blend) added at the first stage of the test and during further aliquot addition. Aliquot volumes were however adapted from experiment to experiment, but they are mentioned in the legends accompanying the figures. An initial quantity of antifoam was deposited on the inner side of the graduated sparge column with a micropipette (Nichirya, Model 800) before the foaming medium was poured into the column. The mixture was stirred for 30 s before the gasflow was flowed in the liquid and the rise of foam height monitored. Under these conditions, very fast generation of stable foam was obtained. When the foam reaches the top of the column (arbitrarily set at 30 cm), an additional antifoam drop was added on top of the column. The collapse time of the foam was then recorded. The test ran for 10 min and the overall quantity of antifoam required to control the foam, during the experiment enabled comparison of antifoam efficiency.

Static Sparge Test

The foaming medium (400 g) was poured into the column before the airflow was started. Once the foam reaches the top of the column, the airflow was stopped. The foam level was recorded as a function of time.

Results and Discussion

Test Protocols

In laboratory studies, foam can be generated in different ways (2,14). Several air incorporation systems can be tested: sparging, whipping, shaking, or pouring. In the frame of this study, we selected the sparging of air, a simple aeration test, which allows partial reproduction of the gas–liquid hydrodynamic encountered in bioreactors. This method, commonly used as a research tool, involves forcing gas through the liquid via a sparger to create bubbles and hence foam, once surface-active materials are present in the liquid. It enables the delivery of controlled volumes and flow rates of a chosen gas to form a foam of a specific volume or for a specific time.

Foaming methods can be divided between the static and the dynamic methods. In static methods measurements are made after the foam has been formed whereas in dynamic methods measurements are taken during foam formation. Foam volume is often measured in static methods whilst time is often taken as the variable in dynamic measurements. Both types of methods were used in this work. Foam stability can be estimated by measuring the foam evolution as a function of time, after a given volume of foam is produced. The gasflow is stopped when the foam reaches the top of the column (arbitrarily set at 30 cm in our study) and the foam height progression is recorded during a definite time. This is referred as the "static sparge test" in this document. On the opposite, in the dynamic method, bubbling is maintained during the time of the experiment. Foam level is recorded as a function of sparge time, and the effect of different variables (for instance, level and type of antifoam, type of foaming medium) can be assessed by comparing what we will call the "foam profile." In a typical experiment, before the beginning of the measure, an initial quantity of antifoam was deposited inside the column. The liquid was stirred for half a minute to ensure proper dispersion of the antifoam. Once the flow of air started, foam was generated. Owing to the addition of antifoam, the foam level raised much more slowly than in its absence. Once the antifoam became progressively

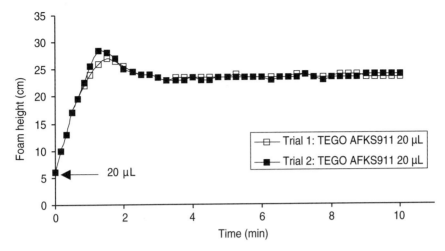

Fig. 1. TEGO AFKS911 ability to control the foam as a function of aparge time: reproducibility test—foam height as a function of sparge time (1 drop–arrow = 20 µL).

deactivated (15), the foam reached the top of the column, arbitrarily set at 30 cm. An additional quantity of antifoam was then added in top of the foam column, which enabled the assessment of the so-called "knock down" capacity of the antifoam. The test was stopped after 10 min and the total amount of antifoam needed to control the foam during this interval allowed comparison of the different antifoams efficiency. This will be referred as the "dynamic sparge test" in this document. The test was found to be reproducible, as presented in Fig. 1, when the same, fresh, foaming medium was used.

Foaming Tendency of Fermentation Broth

Before making any experiment in the presence of antifoam, it was important to assess the foaming behavior of the selected fermentation system in order to characterize the pattern of foaming during fermentation (5). Owing to the foaming tendency of the fermentation broth, only a static test could be selected. Mass production of *Y. lipolytica* was realized in flask culture and growth was quenched by cooling after different time periods. The fermentation broth samples were then tested by the "static sparge test" to assess the foaming tendency and foam stability after different stages of mass production.

The foam levels are plotted as a function of time in Fig. 2, which allows the determination of the influence of growth time on foam stability. The early parts of the curves (when air is still flowing) are all similar, showing that increasing culture time (i.e., yeast population) had no effect on medium foamability, foam was produced rapidly within a few seconds, independently of the fermentation medium composition. Foam capacity (16,17), which is related to the volume of foam normalized by the volume of air injected at the same time, was already high at the early stage of the culture and was not modified by the culture time. The foam stability was not

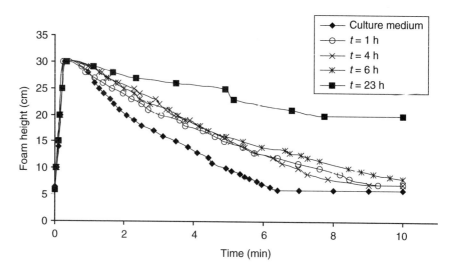

Fig. 2. Evolution of foam stability at different stages of *Y. lipolytica* mass production (culture medium, t = 1, 4, 6, and 23 h)—foam height as a function of time.

identical between the samples although the foam capacity was very similar, which is not uncommon for protein-stabilized foam *(18)*. Foam stability was assessed when air was stopped and by looking at the decay of foam as a function of time. Figure 2 clearly shows that the foam became more stable when culture time increased. During the lag phase in which no increase of the yeast population was recorded, the foam stability was constant but started to increase after 6 h of culture. These results highlight the fact that the nature of the foam changed with culture time and that yeast mass production significantly increased the foam stability of the fermentation broth. The foam is usually stabilised by proteins or other surface-active molecules present in the culture medium. The foaming capacity of the culture medium clearly showed this (*see* Fig. 3 as well). Proteins are furthermore produced during the process, *(19, 20)*, leading to more stable foam. The foaming pattern that this specific mass culture showed was then relatively simple to explain, as the foam stability had a tendency to increase as a function of fermentation time.

Efficiency of Antifoams in Fermentation Broth

In order to compare the efficiency of the three antifoams in the fermentation broth, the "dynamic sparge test" was used. Yeast mass production was run in a 20-L bioreactor using a mechanical foam breaker in order to enable the production of "antifoam-free" samples of fermentation broth after different times of culture. Samples of the culture broth were taken directly from the tank in a sterile manner at regular intervals. "Dynamic sparge tests" were performed on the different samples, directly after collection from the tank.

Figure 3 shows the foam profile obtained with the three antifoams, when tested in the culture medium (with no microorganisms). The foam

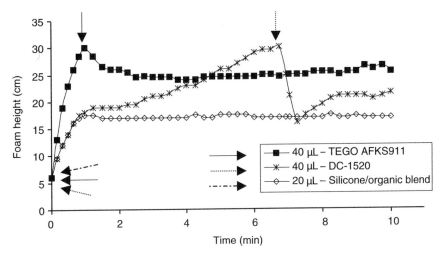

Fig. 3. Antifoam efficiency of selected materials on culture medium—foam height as a function of sparge time (1 drop–arrow = 20 µL).

profile enabled the differentiation of the antifoam efficiency of the different materials, and showed how a specific antifoam performed in the system. The silicone/organic blend showed excellent foam control. The foam remained low and controlled for the duration of the experiment. DC-1520 showed a gradual loss of efficiency, which lead to the need of an additional aliquot addition once the foam reached the top of the column. The consumption of antifoam required to control the foam during 10 min was then higher, reflecting poorer antifoam persistency. TEGO AFKS911 showed some difficulty to control the foam at the early beginning of the experiment, leading to the addition of an aliquot very early. Foam was then very well controlled during the rest of the time. This illustrated more a slower dispersion of the antifoam at the beginning of the experiment.

Figure 4 shows the foam profiles obtained with TEGO AFKS911 in fermentation broth of different culture times. Yeast density started at 2×10^6 cells/mL and reached 1.2×10^8 cells/mL after 24 h of culture. We can see that only one addition of TEGO AFKS911 (20 µL of antifoam) was required to control the foam of the fermentation broth at the early stage of the culture (0 and 6 h) for 10 min. The maximum foam height reached a limit of 25 cm. It is interesting to note that the addition of the micro-organism and the autoclave step seemingly improved the dispersion of the TEGO antifoam, leading to a more progressive foam profile (compared with the foam profile of TEGO in the culture medium, in which an early second addition was required). As the microorganism population started to rise, the amount of antifoam used to control the foam during the experimental time increased from 20 µL ($t = 6$ h) to 40 µL ($t = 9$ h) and reached 160 µL at the end of the culture ($t = 24$ h). The same trend was observed with the silicone emulsion DC-1520 (i.e., the quantity of antifoam required to control

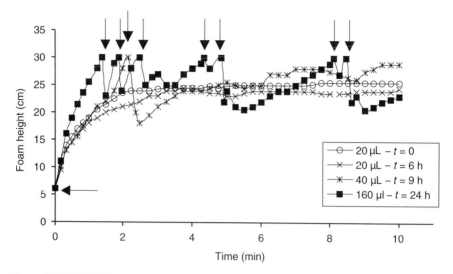

Fig. 4. TEGO AFKS911 ability to control the foam as a function of culture time ($t = 0$, 6, 9, and 23 h)—foam height as a function of sparge time (1 drop–arrow = 20 μL).

the foam increased as a function of growth time) but for the silicone/organic blend the quantity of antifoam required was constant and was not as strongly dependent on the yeast growth (results not shown). These results show that the yeast mass production decreased the capacity of the antifoam material (TEGO AFKS911 and DC-1520) to control the foam. Screening a wide range of antifoam materials should ideally be made with fermentation broth collected after 24 h of growth, as this corresponds to the more stringent testing conditions.

This kind of screening was exemplified with the three selected materials. Their ability to control the foam produced in the *Y. lipolytica* fermentation broth after 24 h of culture is illustrated in Fig. 5. The fermentation broth was produced in a 20 L bioreactor using a mechanical foam breaker to control the foam during the fermentation (this allows to obtain "antifoam-free" fermentation broth). After 24 h of culture, cell density reached 2.9×10^8 cells/mL.

Figure 5 demonstrates the significant difference between antifoam efficiencies. It should be noted that the antifoam aliquot used here was 100 μL compared with the 20 μL used for establishing antifoam testing protocol. A different volume of the aliquot was selected, which was used to discriminate efficiency of different materials, as foam profiles have to be obtained in appropriate conditions. If too small or too large aliquots were used, meaningless data could have been obtained (such as flat foam profiles or little to no foam control with each of the materials). With the commercial silicone-based emulsion DC-1520, an overflow was observed. Under these experimental conditions this material was not able to control the foam during the time of the experiment. The silicone/organic blend formulation was obviously more efficient than the TEGO AFKS911 material: 100 μL required

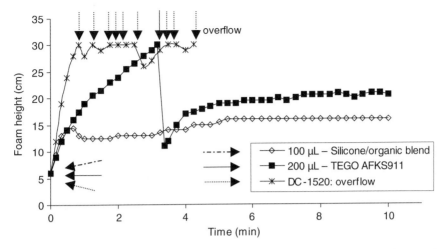

Fig. 5. Antifoam efficiency of selected materials on culture broth after 24 h of culture—foam height as a function of sparge time (1 drop–arrow = 100 µL).

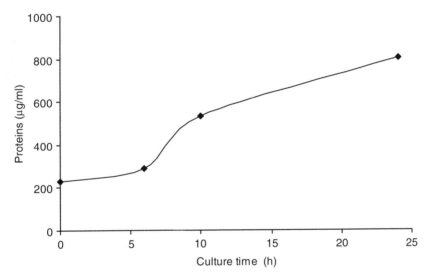

Fig. 6. Evolution of the protein concentration in the culture medium as a function of culture time.

to control the foam for 10 min for the former compared with 200 µL for the latter antifoam. Moreover, the foam profile obtained with the silicone/organic blend was flatter, suggesting longer lasting antifoam efficiency. The following antifoam efficiency ranking can be established: DC-1520 << TEGO AFKS911 << silicone/organic blend.

Foamability and Protein Content

Concentrations of protein in the fermentation broth were assessed by the method of Folin (13) and protein concentrations are plotted as a function of culture time in Fig. 6. Protein content increased as a function of

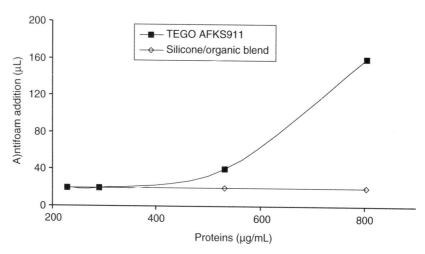

Fig. 7. Quantity of TEGO AFKS911 and silicone/organic blend needed to control the foam as a function of the protein concentration during *Y. lipolytica* mass production.

time, which correlated with an increase in foam stability and a decrease in efficiency of antifoam. Figure 7 shows that the increase of protein concentration in the broth is correlated with a lowering of the foam control efficiency (which was translated by an increased quantity of antifoam required to control the foam for 10 min) and an increase in antifoam consumption. The relation is obviously not linear. At some point, further increase of the extracellular concentration lead to a more significant decrease of antifoam efficiency and to a much higher antifoam consumption. Interestingly, the shape of the curve obtained with the TEGO AFKS911 and the silicone/organic blend was very different, showing that the silicone/organic blend formulation was much less sensitive to the protein content in the foaming medium.

Sparge Test on the Fermentation Broth vs Simple Synthetic Medium

We have shown that a simple dynamic sparge test can help to discriminate between antifoams of various efficiencies. This was exemplified with the *Y. lipolytica* mass production. This methodology is useful for comparative testing of antifoams but it was questioned if a simpler test could be designed. Other systems, which would mimic fermentation broth, were investigated to elaborate a simpler methodology useful for the selection of appropriate antifoams. Figure 3 shows the foam profiles obtained with the three antifoams for the culture medium free of microorganisms. The three antifoams were found to efficiently control the foam of the culture medium. The foam produced in the culture medium was less difficult to control than when produced in the fermentation broth after 24 h and the ranking of the antifoam efficiency was also different. The ranking in the culture medium is TEGO AFKS911 ± DC-1520 < silicone/organic blend, which is different from the one in the 24-h fermentation broth: DC-1520 << TEGO

Foam Control in Fermentation Bioprocess

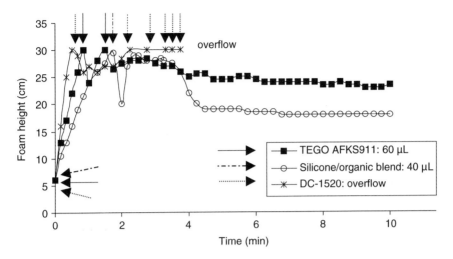

Fig. 8. Anitfoam efficiency of selected materials on culture medium containing additives (0.5% albumin and 13% baker yeast)—foam height as a function of sparge time (1 drop–arrow = 20 µL).

AFKS911 << silicone/organic blend. Simple culture medium was thus not representative of the fermentation broth after 24 h.

Enrichment of the starting culture medium was carried out by protein addition (albumin) and microorganism addition (baker's yeast). The ability of the three antifoams to control the foam of the culture medium was decreased by the addition of both proteins and yeast.

Adjustment of the protein and yeast content lead to a "synthetic fermentation broth" containing 0.5 % albumin and 13% baker yeast. The foam profiles obtained with the three antifoams in this "synthetic fermentation broth" are illustrated in Fig. 8. It is interesting to see that the ranking of antifoam performance is then similar to the one obtained with the 24 h-old fermentation broth: DC-1520 << TEGO AFKS911 << silicone/organic blend [20].

Conclusions

The relative antifoam efficiency of various materials can be assessed by using a simple sparge test, reproducing the aeration conditions of bioreactors. The capacity of antifoams to control the foam generated in culture medium or fermentation broth, decreases as a function of fermentation time. A fermentation broth obtained after 24 h of mass production is then the best foaming system to use to perform an antifoam screening exercise. The volume of antifoam required to control the foam during the test duration is then used as a measure of the antifoam efficiency. Efficiency of these antifoam was found to be very different and the silicone/organic blend formulation shows a longer lasting performance, i.e., a lower dependence of the antifoam efficiency on the proteins content.

A "synthetic fermentation broth" was developed, by adding both protein and baker's yeast to the culture medium. The behavior of the antifoams in this "synthetic fermentation broth" was approaching that of the fermentation broth after 24 h of culture.

Acknowledgment

The authors thank the Région Wallonne for financial support, in the frame of a fund for the technological and scientific research.

References

1. Wilde, P. and Clark, D. (1996), In: *Methods of Testing Protein Functionality*, Hall, G., ed., Blackie Academic & Professional, London, pp. 110–152.
2. Wilde, P. (2000), *Curr. Opinion Colloid Interval Sci.* **5**, 176–181.
3. Sie, T. and Schügerl, K. (1983), *Eur. J. Appl. Microbiol. Biotech.* **17**, 221–226.
4. Abribat, B., Molitor, J. -P., De Haut, C., Merlet, S., and Claessens, P. Patent WO 03/040699A1, Cognis, France S. A.
5. Hall, M., Dickinson, S., Pritchard, B., and Evans, J. (1973), *Prog. Industr. Microbiol.* **12**, 169–234.
6. Destain, J., Roblain, D., and Thonart, P. (1997), *Biotech. Let.* **19**, 105–107.
7. Garrett, P. (1993), In: *Defoaming: Theory and Industrial Applications*, Garrett, P., ed., Marcel Dekker, New York.
8. Lee, J. and Tynan, K. (1988), *2nd International Conference on Bioreactor Fluid Dynamics*, 353–377.
9. Wasan, D. and Christiano, S. (1997), In: *Handbook of Surface and Colloid Chemistry*, Birdi, K. S., ed., CRC Press, New York.
10. Denkov, N. (2004), *Langmuir* **20**, 9463–9505.
11. Christiano, S. and Fey, K. (2003), *J. Ind. Microbiol. Biotech.* **30**, 13–21.
12. Lecomte, J. -P., Cazaroto, M., Rivs, M., and Mivrat, T. (2004), Patent PCT/2004/035370 filed on October 21, 2004.
13. Lowry, O., Rosebrough, N., Farr, A., and Randall, R. (1951), *J. Biol. Chem.* **193**, 265–275.
14. Domingo, X., Fiquet, L., and Meijer, H. (1992), *Tenside Surf. Det.* **29**, 16–22.
15. Denkov, N., Marinova, K., Hristova, H., Hadjiiski, A., and Cooper, P. (2000), *Langmuir* **16**, 2515–2528.
16. Baniel, A., Fains, A., and Popineau, Y. (1997), *J. Food Sci.* **62**, 1–5.
17. Razafindralambo, H., Paquot, M., Baniel, A., et al. (1997), *Food Hydrocolloids* **11**, 59–62.
18. Popineau, Y., Loisel, W., Bertrand, D., and Gueguen, J. (1993), *Traitement Industriel des Fluides Alimentaires Non-Newtoniens* **27–29**, 72–81.
19. Ghildya, N., Lonsane, B., and Karanth, N. (1988), *Adv. Appl. Microbiol.* **33**, 173–222.
20. Schügerl, B. (1985), *Proc. Biochem.* **20**, 122, 123.

Optimization of Biodiesel Production From Castor Oil

NIVEA DE LIMA DA SILVA,* MARIA REGINA WOLF MACIEL, CÉSAR BENEDITO BATISTELLA, AND RUBENS MACIEL FILHO

Laboratory of Separation Process Development (LDPS), School of Chemical Engineering, State University of Campinas, São Paulo, Brazil, CP 6066, CEP 13081-970; E-mail: niveals@feq.unicamp.br

Abstract

The transesterification of castor oil with ethanol in the presence of sodium ethoxide as catalyst is an exceptional option for the Brazilian biodiesel production, because the castor nut is quite available in the country. Chemically, its oil contains about 90% of ricinoleic acid that gives to the oil some beneficial characteristics such as its alcohol solubility at 30°C. The transesterification variables studied in this work were reaction temperature, catalyst concentration and alcohol oil molar ratio. Through a star configuration experimental design with central points, this study shows that it is possible to achieve the same conversion of esters carrying out the transesterification reaction with a smaller alcohol quantity, and a new methodology was developed to obtain high purity biodiesel.

Index Entries: Biodiesel; castor oil; ethanolysis; transesterification; alkaline catalyst.

Introduction

Biodiesel is an alternative biodegradable and nontoxic fuel, which is essentially free of sulfur and aromatics. It is usually produced by a transesterification reaction of vegetable or waste oil with a low-molecular weight alcohol, such as ethanol or methanol. Industrially, the most common method for biodiesel production is a basic homogeneous reaction.

The demand for alternative energy sources is frequent, because there is a progressive decrease of the world's petroleum. Vegetable oil fuel or biodiesel is a potential substitute for diesel fuel because it is made from renewable resources. The American Society for Testing and Materials defines biodiesel fuel as monoalkyl esters of long chain fatty acids derived from a renewable lipid feedstock, such as vegetable oil or animal fat. Among the biodiesel advantages, it can be cited: biodegradability, no toxicity, renewable, reduction in greenhouse gas emission in line with the Kyoto

---

*Author to whom all correspondence and reprint requests should be addressed.

Protocol agreement, high viscosity and flash point, comparing with conventional diesel.

Biodiesel is made by transesterification reaction of vegetable oils and animal fats preferentially with alcohol of low-molecular weight. Ethanol is an alternative to methanol, because it allows production of entirely renewable fuel. This reaction can be carried out in the presence of alkaline, acid, and enzyme catalysts or using supercritical alcohol, pyrolysis, and microemulsions.

This work presents the transesterification of castor oil (CO) using sodium ethoxide. This vegetable oil has peculiar characteristics, such as its fatty acid composition (90% of ricinoleic acid). This acid has 18 carbon atoms with one hydroxyl in carbon 12, therefore it contains more oxygen than other oils and, for that, it is more soluble in ethanol (1). Its viscosity, is more than 100 times higher than the no. 2 diesel fuel. Other vegetable oils possess viscosities ranging from 10 to 20 times higher (2).

High-performance size-exclusion chromatography (HPSEC), also called gel-permeation chromatography, was used to evaluate the influence of different variables affecting the transesterification. HPSEC is based on the selective retention of the molecules according to their size, when they enter into the pores of the column polymer matrix. This chromatographic technique permits the separation of glyceride groups present, according to the molecule size (3). The sample for injection does not need preparation other than dilution in tetrahydrofuran. This method supplies the exact lipid composition of the reaction medium at any time and performs accurate kinetic measurements (4).

The transesterification reaction occurs in three consecutive and balanced stages, totaling six different rate constants. Triglycerides react with ethanol to produce diglycerides, and then diglycerides react to produce monoglycerides (5). Finally, monoglycerides react with alcohol to give glycerol as a byproduct, as following:

Overall reaction:

$$TG\,(triglycerides) + 3ROH \xleftrightarrow{catalyst} 3RCO_2R\,(ester) + GL\,(glycerol)$$

Stepwise Reactions:
1. $TG + ROH \leftrightarrow DG\,(diglycerides) + RCO_2R$
2. $DG + ROH \leftrightarrow MG\,(monoglycerides) + RCO_2R$
3. $MG + ROH \leftrightarrow RCO_2R + GL$

Different variables affect the castor oil transesterification such as reaction temperature (T), ethanol:castor oil molar ratio (A:O), catalyst concentration (C), level of agitation and reaction time. In this study, the level of agitation and the reaction time were kept constant, at 600 rpm and 30 min, respectively. Response-surface methodology (RSM) or response surface analysis was used, because it allows the simultaneous consideration of many variables at different levels and the interactions between those variables, using a smaller number of observations than conventional procedures. Furthermore, statistical-interference technique can be used to assess the importance of individual factors, the appropriateness of their functional form and the sensitivity of the response of each factor (6,7). The simplified RSM equations take into account only the significant coefficients. The coefficient b_o is the outcome (response) at the central point and the other coefficients measure the main effects and the interactions of the coded variable X_i on the response Y (8).

Experimental Procedures

Materials and Apparatus

Castor oil was donated by Aboissa (Brazilian Oil Company). Anhydrous ethanol and sodium ethoxide 95% pure were used.

The experiments were carried out in a 250 mL flask connected with a condenser. The reaction mixture was agitated by magnetic stirrer. After reaction, a Rota-Evaporator was used in order to recover the ethanol excess and the mixture is placed in a separation funnel. The water content was determined by Karl Fisher.

Reaction

Initially, the oil was added and preheated to the desired temperature. After that, the catalyst was prepared by dissolving the sodium ethoxide in the desired amount of ethanol. This ethanolic solution was added to the castor oil and the reaction initiated at this time. The system was kept at room pressure and the experiments carried out at constant temperature. The agitation was kept constant at 600 rpm to maintain uniform the mass transfer in the system. The reaction time was 30 min, and, during this time, samples were collected, diluted in tetrahydrofuran,cooled instantaneously and analyzed in the HPSEC, evaluating the triglycerides, diglycerides, monoglycerides, ethyl esters, and glycerol contents, according to the methodology used by Filléres et al. (4). The conversion was obtained analyzing the corresponding sample at 30 min of reaction in the HPSEC. The conversion (Y) was calculated by dividing the peak area of the ester by the sum of the peak area of all components.

$$\left(Y = \frac{A_{EE}}{A_{TG} + A_{DG} + A_{MG} + A_{EE}} 100 \right)$$

A typical chromatogram is presented in Fig. 1. After the reaction, the mixture is evaporated in the Rota-Evaporator under vacuum and then poured into a separatory funnel. After that, two layers are formed: the upper one made of ester and the lower one made of glycerol mixed with catalyst and some impurities. Later, the ester phase was separated (9).

Experimental Design

The RSM was chosen to study the optimization of three selected factors, temperature (T), ethanol:castor oil molar ratio (A:O) and catalyst

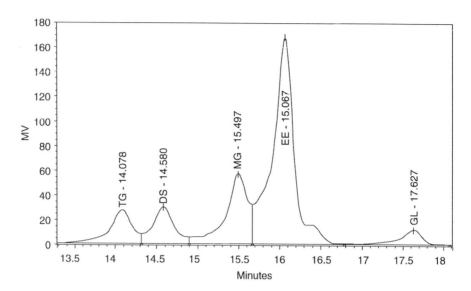

Fig. 1. Typical chromatogram of ethanolysis reaction using the HPSEC method: TG, triglycerides; DG, diglicerides; MG, monoglicerydes; EE, ethyl ester; GL, glycerol.

Table 1
Decoding Values of Independent Variables Used in the Experimental Design

Level	Temperature (°C)	Catalyst (wt%)	Ratio: ethanol/CO
$-\alpha$	30	0.5	12:1
-1	40	0.7	13.6:1
0	55	1	16:1
1	70	1.3	18.88:1
$+\alpha$	80	1.5	20:1

concentration (C). The experiments were carried out according to a 2^3 complete factorial design plus three central points and four axial points, called star points. The distance of the star points from the center point is given by $\alpha = (2^n)^{1/4}$, where n is the number of independent variables, for three factors $\alpha = 1.68$ *(10)*. Table 1 shows the decoding values for T, C, and $A:O$ and a total of 17 experiments were obtained as shown in Table 2. The repeatability of the experiments is available with the center points (runs 15, 16, and 17; Table 3). In this table, it is possible to observe that the repeatability is suitable. Typically, the central point is replicated several times to provide an independent estimate of experimental error *(11)*.

The factors (designated as X_i) were set independently of each other within the following limits: temperature (X_1): $30°C \leq T \leq 80°C$, catalyst concentration (X_2): $0.5\% \leq C \leq 1.5\%$, ethanol:castor oil molar ratio (X_3): $12:1 \leq A:O \leq 20:1$. Factorial level was chosen by considering the properties

Table 2
The Design of Castor Oil Transesterification Using RSM

	Experiment (run)	Coded variables			Conversion (%)
		X_1	X_2	X_3	
Factorial design	1	−1	−1	−1	52.34
	2	1	−1	−1	58.6
	3	−1	1	−1	90.21
	4	1	1	−1	93.69
	5	−1	−1	1	84.83
	6	1	−1	1	74.83
	7	−1	1	1	91.7
	8	1	1	1	90.11
Other points $\alpha = \pm 1.68$	9	−1.68	0	0	93.07
	10	1.68	0	0	93.97
	11	0	−1.68	0	46.88
	12	0	1.68	0	90.82
	13	0	0	−1.68	78.42
	14	0	0	1.68	90.14
Central point	15	0	0	0	90.96
	16	0	0	0	92.04
	17	0	0	0	78.01

Table 3
Quadratic Regression Coefficients

Term	Coefficient	p
Mean	87.16*	0
X1 (l)	−0.02	0.98
X1 (q)	1.78	0.28
X2 (l)	12.38*	0
X2 (q)	−6.96*	0
X3 (l)	4.86*	0.01
X3 (q)	−1.50	0.35
X1 × X2	0.70	0.70
X1 × X3	−2.67	0.18
X2 × X3	−6.35*	0.01

l, linear; q, quadratic
*Significant at the 95% confidence interval.

of the reactants (12). The upper temperature level, 80°C, was determined by the boiling point of ethanol; the lower level, 30°C, was the room temperature. Catalyst concentration level was varied from 0.5% to 1.5%, according to literature data (13) for other case studies.

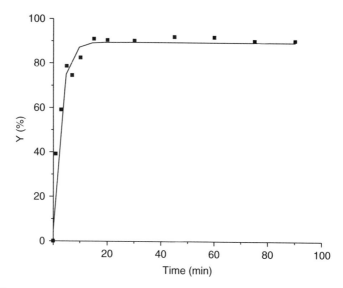

Fig. 2. Conversion to ethyl ester vs time. Catalyst: NaOET; T, 30°C; C:1 wt%; Ethanol:oil molar ratio – 19:1.

The complete RSM equation describes the contributions of the various factors on the outcome (response) of the ester conversion. The coefficient b_o is the outcome at central point and the other coefficients measure the mean effects and the interactions of the coded variables X_i on the response Y:

$$Y = b_0 + \sum b_i X_i + \sum b_{ii} X_i^2 + \sum b_{ij} X_i X_j \quad (i,,j) \tag{1}$$

As shown in Fig. 2, a plot of conversion to ethyl ester vs time, the reaction is very fast, conversions larger than 90% are reached within 15 min.

Results and Discussion

Table 3 shows the quadratic regression coefficients. As can be seen by the coefficients of the respective terms, only five terms are significant based on p-values. Then, the simplified RSM equation takes into account only these significant coefficients. Only the coefficients b_o (ester conversion at the central point), b_2 and b_{22} (catalyst content), b_{33} (ethanol:castor oil molar ratio), and b_{13} (interaction between temperature and ethanol:castor oil molar ratio) have statistically significant values. Table 4 shows the significant coefficients that were represented in a quadratic equation. Using this reduced model (Eq. 2), the response surfaces and the contour curves (Figs. 3–5) were constructed.

$$Y = 87.44 + 12.38 \times X_2 - 7.03 \times X_2^2 + 4.86 \times X_3 - 6.35 \times X_2 \times X_3 \tag{2}$$

Table 4
Significant Regression Coefficients

Term	Coefficient	p
Mean	87.44	0.00
X2 (l)	12.38	0.00
X2 (q)	−7.03	0.00
X3 (l)	4.86	0.00
X2 × X3	−6.35	0.00

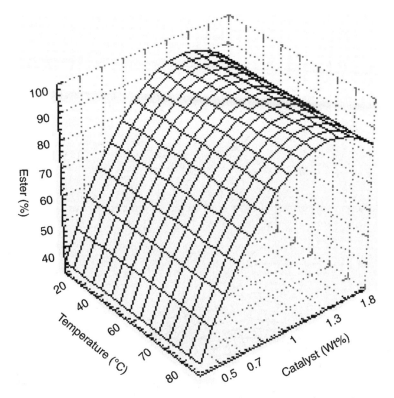

Fig. 3. Response surface of ester conversion vs catalyst concentration and temperature.

Table 5 shows the ANOVA for the full quadratic model. F_{calc} can be calculated $\left(F_{calc} = \frac{mean\ square\ regression}{mean\ square\ residual}\right)$, and was compared with the listed F. The reduced model may predict the castor oil transesterification, because the F_{calc} should be nine times higher than F_{listed}, at 95% of confidence.

Figure 3 shows how the catalyst content and temperature affect the esters conversion. The response surface indicates that, independently of the reaction temperature, the conversion of ethyl esters increases increasing catalyst concentration. Maximum ester conversions, more than 89.25%

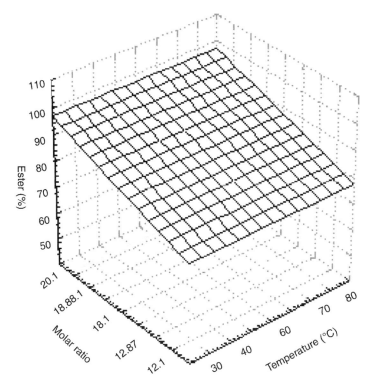

Fig. 4. Response surface of ester conversion vs temperature and molar ratio.

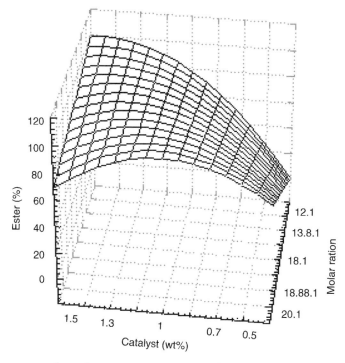

Fig. 5. Response surface of ester conversion vs catalyst concentration and molar ratio.

Table 5
ANOVA for the Full Quadratic Model Plus Three Central Points

Source of variation	Sum quadratic	Degrees of freedom	Mean quadratic (MQ)	F_{calc}	F_{test} F_{listed}
Regression	3377.35	4	844.34	31.11	3.26(F0.95,4,12)
Residual	325.71	12	27.14		
Lack of fit	203.80	10	20.38		
Pure error	121.90	2	60.95		
Total	3703.05	16			

F_{listed} values are significant at the 95% confidence level. Regression coefficient $R = 0.96$

were obtained for catalyst concentration from 1 to 1.5 wt%. Catalyst concentration is an important factor during transesterification, and its effect is positive.

In Figure 4, it can be seen how the temperature and the ethanol:castor oil molar ratio affect ester conversions. The response surface indicates that, independently of the reaction temperature, the conversion of ethyl esters increases increasing ethanol:castor oil molar ratio. The castor oil needs a large ethanol excess for the alkaline ethanolysis in order to achieve good conversion. Maximum ester conversions more than 95.95% were obtained for ethanol:castor oil molar ratio upper to 20:1. This is a very promising result, because Parente (14) has affirmed that an even larger excess of ethanol is necessary (650% or A:O = 39:1).

Figure 5 shows how the catalyst concentration and ethanol : castor oil molar ratio affect ester conversions. The maximum ester conversions more than 93.78% were obtained for catalyst concentrations from 0.8 to 1.2 wt% and for ethanol:castor oil molar ratio higher than 19:1. It is also observed that these conversions (more than 93.78%) happen for catalyst concentration up to 1.3 wt% at lower ethanol:castor oil molar ratios. The ester conversion has a moderate increase with ethanol:castor oil molar ratio, because this effect is positive and smaller than that catalyst effect.

Analysing these results, higher ethanol:castor oil ratio or higher catalyst concentration, it can be more interesting in economical terms to use the first condition. Furthermore, in this case, it is still possible to reuse the excess ethanol.

Conclusions

The methodologies of factorial design and response-surface analysis were useful for understanding the behavior of castor oil transesterification reaction. Among the three parameters studied, the temperature (X_1) does not influence the reaction, because the castor oil is soluble in ethanol at room temperature. In Table 2, small differences in esters conversions in runs 9 and 10, in the temperature level 30°C or 80°C can be seen. Catalyst concentration

(X_2) and ethanol:castor oil molar ratio (X_3) increase the ester conversion, because they have positive influence on the response. Catalyst concentration is the most important transesterification factor, because it has the highest effect. Higher ester conversion is obtained at 30°C, with large catalyst content, up to 1.3% wt and lower ethanol: castor oil molar ratio or with catalyst content from 0.8% wt to 1.2% wt, with large ethanol:castor oil molar ratio, up to 19:1. The statistical model generated can predict the castor oil ethanolysis, because it describes the experimental range adequately.

Acknowledgment

The authors are grateful to CAPES, CNPQ and FAPESP for the financial support.

References

1. Beltrão, N. E. M. www.tierramerica.net/2003/0526/ianalisis.shtml acess in 29/09/05.
2. Dermibas, A. (2003), *Energy Sources* **25**, 721–728.
3. Schoenfelder, W. (2003), *Eur. J. Lipid Sci. Technol.* **105**, 45–48.
4. Fillères, R., Benjelloum-Mlayah, B., and Delmas, M. (1995), *J. Am. Oil Chem. Soc.* **72**, 427–432.
5. Freedman, B., Butterfield, R. O., and Pryde, E. H. (1986), *J. Am. Oil Chem. Soc.* **63**, 1375–1380.
6. Mason, R. L., Gunst, R. F., and Hess, J. L.(1989), In: *Statistic Design and Analysis of Experiments: with Applications to Enginieesing and Science*, 2nd ed., John Wiley & Sons, New York.
7. Box, G. E. P., Hunter, W. G., and Hunter, J. S.(1978), In: *Statistics for Experimenters. An Introduction to Design, Data Analysis and Model Building*, 2nd ed., John Wiley & Sons, New York.
8. Hénon, G., Vigneron, P. Y., Stoclin, B., and Calgnez, J. (2001), *Eur. J. Lipid Sci. Tecnol.* **103**, 467–477.
9. Lima Silva, N. (2006), MSc Thesis, in preparation, School of Chemical Engineering, State University of Campinas, São Paulo, Brazil.
10. Neto, B. B., Scarminio, I. S., and Bruns, R. E. (2003), In: *Como fazer Experimentos*, 2nd ed., Unicamp, Brazil.
11. Kamimura, E. S., Medieta, O., Rodrigues, M. I., and Maugeri, F. (2001), *Biotech. Appl. Biochem.* **33**, 153–159.
12. Vicente, G., Coteron, A., Martinez, M., and Aracil, J. (1997), *Industrial Crops and Products* **8**, 29–35.
13. Freedman, B., Pryde, E.H., and Mounts, T.L. (1984), *J. Am. Oil Chem. Soc.* **61**, 1638–1643.
14. Parente, E. J. S. (2003), *Biodiesel: Uma Aventura Tecnológica num País Engraçado*, http://www.tecbio.com.br/Downloads/Livro%20Biodiesel.pdf. Acessed in 29/09/05.

SESSION 3A
Plant Biotechnology and Feedstock Genomics

Manipulating the Phenolic Acid Content and Digestibility of Italian Ryegrass (*Lolium multiflorum*) by Vacuolar-Targeted Expression of a Fungal Ferulic Acid Esterase

MARCIA M. DE O. BUANAFINA,* TIM LANGDON, BARBARA HAUCK, SUE J. DALTON, AND PHIL MORRIS

Plant Animal and Microbial Science, Department, Institute of Grassland and Environmental Research, Plas Gogerddan, Aberystwyth, SY23 3EB, Wales, UK

Abstract

In grass cell walls, ferulic acid esters linked to arabinosyl residues in arabinoxylans play a key role in crosslinking hemicellulose. Although such crosslinks have a number of important roles in the cell wall, they also hinder the rate and extent of cell wall degradation by ruminant microbes and by fungal glycohydrolyase enzymes. Ferulic acid esterase (FAE) can release both monomeric and dimeric ferulic acids from arabinoxylans making the cell wall more susceptible to further enzymatic attack. Transgenic plants of *Lolium multiflorum* expressing a ferulic acid esterase gene from *Aspergillus niger*, targeted to the vacuole under a constitutive rice actin promoter, have been produced following microprojectile bombardment of embryogenic cell cultures. The level of FAE activity was found to vary with leaf age and was highest in young leaves. FAE expression resulted in the release of monomeric and dimeric ferulic acids from cell walls on cell death and this was enhanced severalfold by the addition of exogenous β-1,4-endoxylanase. We also show that a number of plants expressing FAE had reduced levels of cell wall esterified monomeric and dimeric ferulates and increased in vitro dry-matter digestibility compared with nontransformed plants.

Index Entries: Constitutive vacuolar-targeted expression; digestibility; ferulic acid esterase; *L. multiflorum*; transgenic grasses.

Introduction

Cellulose and hemicellulose make up the bulk of plant material providing a large and renewable source of biomass with a high potential for degradation for generating fermentable sugars as feedstock for industry or for the production of biofuels. They also provide the major feed for the world's livestock, especially for ruminants, and could in fact meet their

---

*Author to whom all correspondence and reprint requests should be addressed.

energy requirements. However, in grasses this potential is not usually achieved because of cell wall lignification and the crosslinking of cell wall polysaccharides with phenolic residues *(1)* resulting in a low rate of cell wall digestion. For industrial uses, the degradation of the cell wall components require a combination of degrading enzymes and the cost of enzymatic hydrolysis of biomass is one of the major factors limiting the economic feasibility of the process.

Ferulic acid (4-hydroxy-3-methoxy-cinnamic acid) is the major hydroxycinnamic acid (HCA) identified in grass cell walls and is attached to the cell wall polymers via an ester linkage to the arabinose side chain of arabinoxylan *(2)* and can also be ether-linked to lignin monomers *(3)*. These HCAs are important components of the cell because they can be oxidatively coupled to form a variety of dehydrodiferulate dimers, crosslinking hemicellulose polysaccharides chains *(4)*; however, they also limit plant cell wall hydrolysis, hindering the rate and extent of cell wall degradation by ruminant microbes *(5)* or by fungal glycohydrolyases *(6)*.

We are interested in producing transgenic plants that can efficiently synthesize hydrolytic cell wall degrading enzymes so the plant cell wall can be broken down during biomass conversion or during processing of plants for foodstuffs and so will be more efficiently used not only for ruminant nutrition but also as a biomass resource.

In the present study, we test the targeted expression of a ferulic acid esterase in order to control the level of phenolic acids in *Lolium multiflorum* cell walls available for crosslink cell wall polysaccharides. A ferulic acid esterase (FAE) gene has been cloned from *Aspergillus niger (7)* and it has been shown that the recombinant enzyme releases ferulic acid and diferulate dimers from grass cell walls. Here, we test the vacuole targeting of FAE expressed using a constitutive actin promoter and delivered by microprojectile bombardment. We expect reduced levels of phenolic acids in pre-existing cell walls by release of FAE on cell death, which will increase the rates of cell wall digestion and carbohydrate availability. The efficiency of the gene targeting strategies, the levels of enzyme activity and the effects on cell wall composition and the release of esterified ferulates on cell death following release of vacuolar-targeted FAE of transgenic *L. multiflorum* is reported together with effects on end point digestibility.

Material and Methods

Vector Construction and Targeting

The pUA1K3 vector (Fig. 1) was constructed in a pUC plasmid with FAE from *A. niger (7)* provided as a genomic clone by Dr. Ben Bower (Genencor Inc., Palo Alto, CA). Modifications to the core open reading frame (ORF) of FAE were made to enhance expression in plants with a codon choice based on published barley preferences. The N-terminus of the protein contained an intact barley aleurain vacuolar targeting sequence *(8)* and was

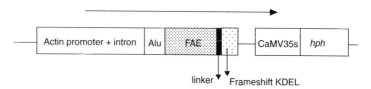

Fig. 1. Vector plasmid (pUA1-K3) used for plant transformation to target FAE to the vacuole under control of a rice actin promoter and intron.

driven by a rice actin promoter and first intron derived from pCOR105 *(9)*. Transient expression of FAE activity following microprojectile bombardment of grass cell cultures was found to be lower with the native *Aspergillus* C-terminal end (CTW) than when a linker sequence was added (CTW-PVAAA), suggesting that the native *Aspergillus* C-terminal end probably has deleterious effects in grasses. This linker and a mutated ER KDEL retention signal (ETTEG) formed the final C-terminal sequence. In order to verify that the targeting sequences were effective in delivering FAE to the vacuole, parallel vectors containing a monocot optimized green fluorescent protein (sgfp) *(10)* were constructed and the cellular distribution of gfp fluorescence was determined by confocal microscopy of single cells in transient expression studies of embryogenic cell cultures. The cointegrative vector pUA1K3 also contained a HindIII casset containing the hygromycin resistance gene (*hph*) under the control of a CaMV 35S promoter *(11)*. Further details of vector construction can be found in Dunn-Coleman et al. *(12)*.

Tissue Culture and Plant Transformation

Eight- to ten-week-old embryogenic suspension cultures of a responsive *L. multiflorum* genotype were bombarded with pUA1K3 using a Particle Inflow Gun as described by Dalton et al. *(13)*. Transformants were selected with hygromycin (25–50 mg/L) over a 10- to 12-wk period at 25°C under continuous white fluoresvcent light (60 µmol/m²/s) and plants were regenerated via somatic embryogenesis. Regenerated plants were screened for FAE activity on transfer to soil and expressing plants were grown to maturity in a containment growth room at 18°C under 16 h fluorescent lights (350 µmol/m²/s). Mature plants (6- to 8-wk old) were reassayed for FAE activity and fresh tissue harvested for self-digestion analysis. The remaining tissue was freeze-dried and powdered for cell wall phenolic analysis, and for in vitro dry-matter digestibility (IVDMD) determinations.

Determination of FAE Activity

The FAE activity was determined in soluble extracts of fresh (or frozen at –70°C) leaves (0.5 g fresh wt extracted with 0.1 M sodium acetated, pH 5.0 buffer). Extracts were incubated with 24 mM ethyl ferulate (EF ethyl-4-hydroxy-3-methoxy cinnamate) as substrate, at 28°C for 24 h and FAE activity calculated as the amount of ferulic acid released in which 1 unit FAE activity equals 1 µg ferulic acid released from EF in 24 h

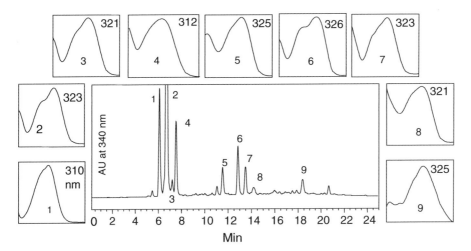

Fig. 2. HPLC and UV/visible spectra of esterified monomeric and dimeric ferulates of cell walls of mature *Lolium* leaves. (1) Trans *p*-coumaric acid. (2) Trans-ferulic acid, (3) 8-5′ diferulic acid, (4) *cis*-ferulic acid, (5) 5-5′ diferulic acid, (6) 8-O-4′ diferulic acid, (7) 8-5′ diferulic acid benzofuran form, (8) N, (9) K-unknown ferulate dimers.

at 28°C. FAE activity was also determined by measuring the release of monomeric and dimeric ferulates from "self-digested" leaf samples. Leaves (0.5 g fresh wt) were ground in 0.1 M sodium acetate extraction buffer pH 5.0, in the presence and absence of GC140 β-1,4-endoxylanase (1000 U/sample Genencor Inc.), without additional substrate, and incubated at 28°C for 72 h. Following centrifugation, soluble extracts were loaded onto a reverse phase C_{18} µNova Sep-Pak column (Waters Inc.), eluted with 100% MeOH and analyzed by high-performance liquid chromatography (HPLC).

Cell Wall Phenolic Analysis

Following exhaustive extraction of soluble phenolics and chlorophyll pigments with aqueous methanol, ester-bound compounds were extracted from the cell walls of freeze-dried powdered leaf material (50 mg), with 1 M NaOH (5 mL) followed by incubation at 25°C for 23 h, in the dark under N_2. After centrifugation, acidification and precipitation of solubilized carbohydrates with MeOH at 4°C, the extracted phenolics in the aqueous phase were loaded onto an activated reverse phase C_{18} µNova Sep-Pak column and eluted with 100% MeOH, and analyzed by HPLC.

HPLC analysis was carried out on a µNova Pak C_{18} 8 × 10 RCM (Waters Inc.) in 100% methanol: 5% acetic acid, either with a linear 35–65% MeOH gradient in 15 min (FAE assay) or with a 20–70% MeOH gradient in 25 min (for monomer and dimer cell wall components) (Fig. 2), at a flow rate of 2 mL/min. Hydroxycinnamic acids were monitored and quantified at 340 nm with a Waters 996 photo-diode array detector with UV/visible spectra collected at 240–400 nm, and analyzed with Millennium software (Waters Inc.) against authentic monomer standards,

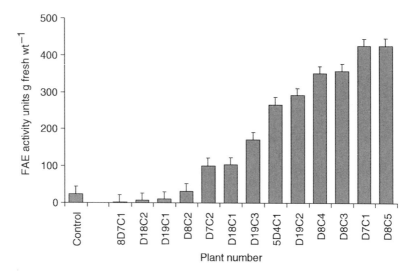

Fig. 3. Levels of FAE activity of mature leaves of 13 *L. multiflorum* plants transformed with pUA1-K3. One unit FAE activity equals 1 µg ferulic acid released from ethyl ferulate in 24 h at 28°C.

or using response factors for the various dehydrodiferulate dimers reported by Waldron et al. *(14)*.

End Point Digestion

IVDMD was estimated on 1 g dry weight of powdered leaf tissue of FAE expressing and control plants, using the pepsin/cellulase method of Jones and Hayward *(15)*.

Results

FAE Expression in Transgenic Plants

Mean FAE activity in the leaves of 13 transgenic plants was found to vary nearly 10-fold, between independent transgenic events (e.g., plants D8C2 and D7C1, Fig. 3). However, FAE activity was also found to vary with leaf age, with the highest activities in young leaves (Fig. 4) and FAE activity was also higher in young plants when directly removed from tissue culture and fell to a lower level as the plants matured (data not shown). This may partially explain some, but not all, of the observed variation in FAE activity shown in Fig. 3.

Cell Wall Composition of Transgenic Plants

The levels of esterified cell wall monomeric and dimeric hydroxycinnamic acid constituents of leaves of transformed plants expressing vacuolar-targeted FAE were also compared with tissue-cultured, nontransformed control plants. In some plants (e.g., D19C2, D8C4, D8C3, D7C1, and D8C5),

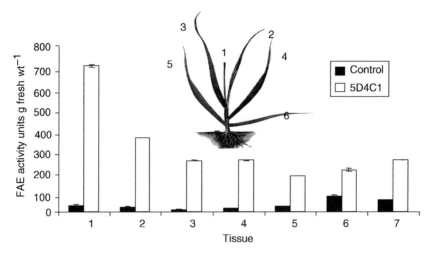

Fig. 4. FAE activities in leaves of different ages and leaf bases of plant 5D4C1 transformed with pUA1-K3, compared with a nontransformed control plant. One unit FAE activity equals 1 µg ferulic acid released from ethyl ferulate in 24 h at 28°C. Mean ± sem ($n = 3$).

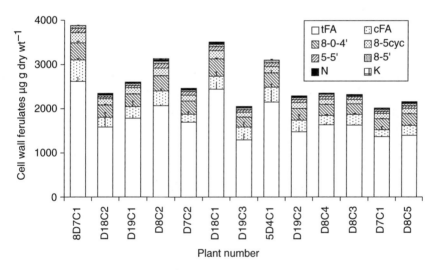

Fig. 5. Levels of ester-bound monomeric and dimeric ferulates in leaves of mature L. multiflorum plants transformed with pUA1-K3. tFA = trans-ferulic acid; cFA = cis-ferulic acid; 8-5cyc = diferulic acid benzofuran form, 8-0-4' diferulic acid, 5-5' diferulic acid, 8-5 diferulic acid, N and K = unknown ferulate dimers quantified as for ferulic acid. Mean ± sem ($n = 3$).

levels of both monomeric and dimeric ferulates were significantly lower than in control plants (Fig. 5) and this correlated with the highest levels of FAE expression (Fig. 3). However, other plants (e.g., 5D4C1) showed similar levels of FAE activity but with no significant changes in wall phenolics, whereas some plants with low measurable FAE activity (e.g., D18C2, and D19C1) showed significantly reduced levels of wall phenolics. Such

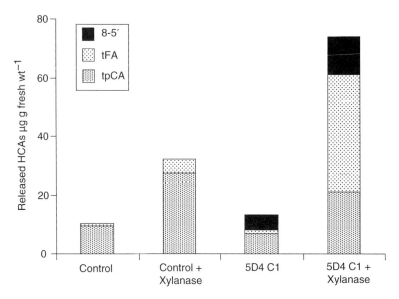

Fig. 6. Release of monomeric and dimeric hydroxycinnamic acids (HCAs) and effect of exogenous β-1,4-endo-xylanase following homogenization of leaves of plant 5D4C1 transformed with pUA1-K3 expressing FAE. tpCA = trans *p*-coumaric acid; tFA = trans ferulic acid; 8-5′ = diferulic acid.

anomalies may have arisen as a result of multiple gene insertions (evident in some plants following Southern blot analysis, data not shown) resulting in disrupted vacuolar targeting in which low levels of FAE resulted in disproportionate effects on cell wall feruoylation.

Effect of FAE Expression on Cell Death

We have demonstrated the potential of vacuole targeted FAE to aid "self-digestion" of the hemicellulose component of the cell wall after death in FAE-expressing plants. This "self-digestion" resulted in the release of small but significant quantities of esterified monomeric and dimeric ferulic acids as well as substantial amounts of *p*-coumaric acid (Fig. 6). In all cases this release was greatly stimulated by the exogenous addition of recombinant β-1,4-endo-xylanase (Fig. 6), which is known to act synergically with FAE in cell wall digestion by rumen microorganisms. This xylanase-enhanced release of HCAs was specific for FAE-expressing plants and was significant even in plants with very low levels of FAE activity. The kinetics of FAE-induced, xylanase-enhanced release of HCAs from constitutively expressed vacuolar-targeted FAE showed maximal release after 24 h incubation at 28°C, and after 8 h incubation at rumen temperatures of 40°C, amounting to some 70–80% release of the cell wall monomers and dimers (data not shown). It would appear therefore that, as with exogenously supplied FAE, xylanase activity is required in order to release arabinoxylans and to allow substrate access for FAE.

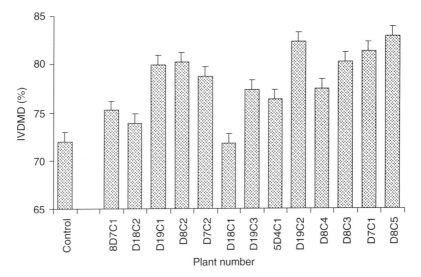

Fig. 7. In vitro dry-matter digestibility (IVDMD) of mature leaves of 13 *L. multiflorum* plants transformed with pUA1-K3.

Effects of FAE Expression on Cell Wall Digestion

End point digestibility, as determined by IVDMD using the pepsin/cellulase method of Jones and Hayward (15), was in some cases found to be higher in leaf tissues of some transformed plants expressing FAE compared to untransformed control plant (Fig. 7). However, this was not consistent, and control levels in other experiments were more variable and were much higher than in clonal plants of the parental genotype, suggesting some effects of in vitro-induced somaclonal variation. Lack of effects on end point digestion, but significant increases in the initial rates of fermentation and reductions in the time to reach maximal rates of fermentation under rumen-like conditions were found with *Festuca arundinacea* plants expressing FAE using in vitro gas production measurements with data fitted to the model of France et al. (16).

Discussion

Recent advances in plant biotechnology have given plant breeders new opportunities, not available using conventional plant breeding techniques, to transfer foreign genes from different genera, and kingdoms and to produce plants expressing these novel genes. Increasing the digestibility index of grasses has been a plant-breeding objective for many years, but owing to difficulties in fixing natural variation in the synthetic varieties derived from these outbreeding species, progress has been slow.

Grass cell walls, in particular, are characterized by the presence of a large amount of esterified ferulates. Removing labile phenolics by chemical treatment with alkali is known to increase biodegradability and the nutritional value of low-quality feed. Reducing the level of crosslinking of cell wall carbohydrate is therefore a predictable way of improving the rate of digestion of grass.

A ferulic acid esterase gene from *A. niger* has been cloned and sequenced *(7)* and the recombinant enzyme shown to release monomeric and dimeric ferulates from plant cell walls *(17)*. In the present investigation this gene has been modified to make it more compatible with monocot codon usage, and we have produced *L. multiflorum* plants expressing this gene driven by a constitutive promoter with the enzyme targeted to the vacuole. The level of FAE enzyme activity was found to vary between independent transformants. This variability may simply reflect the differing proportions of young and old leaves in different plants growing at different rates (as FAE activity was found to decline on leaf maturation) or may be because of typical differences in transgene expression found in different transformation events resulting from different copy numbers *(18)* and/or integration sites *(19)*.

It was expected that there would be no significant changes in wall phenolics with FAE targeted to the vacuole until cell death, yet some plants with the highest level of FAE expression, showed reduced levels of cell wall monomeric and dimeric ferulates. This may be due either to disrupted vacuole targeting in some lines or alternatively to reduced levels of feruoylation of the arabinoxylan in the ER/Golgi intramembrane system, prior to incorporation into the cell wall, during transit, and processing of the enzyme in the secretory pathway to the vacuole. Multiple gene insertions with sequence deletions and rearrangements of some copies are well known to occur following transformation by microprojectile bombardment, whereas disrupted feruoylation was only considered likely with FAE specifically targeted to the ER or Golgi.

We have also demonstrated that expression of FAE in *Lolium* leads to an increase in in vitro dry-matter digestibility. Whether this is because of the reduced levels of esterified cell wall phenolics or as a result of the release of vacuolar localized FAE on cell death, has yet to be determined. However, increased IVDMD could lead to significant financial benefits to the agricultural community, as Casler and Vogel *(20)* have estimated that for beef cattle, 1% unit increase in dry-matter digestibility can lead to a 3.2% increase in average daily live weight gain.

Although there are a number of reports of the expression of cell wall degrading enzymes in dicot plants, mainly cellulases and endo-xylanases, aimed at the possibility of using plant systems for the large-scale, cost-efficient production of heterologous enzymes, little attention has been given to alterations in cell wall chemistry of monocots aimed at increasing their digestibility. For example Ziegelhoffer et al. *(21)* reported the expression of thermostable cellulases from *Thermomonospora fusca* in Medicago, potato, and tobacco at levels as high as 0.1% soluble protein with no effects on plant phenotypes. However, Armstrong et al. *(22)* looked at the expression of a *Fibrobacter succinogenes* 1,3-1,4-β-glucanase in potato with the glucanase mature peptide-coding region, inserted in the sense orientation relative to a constitutive promoter (CaMV 35S). Analysis of the transformed plants revealed low levels of expression and defects in cell wall morphology within the transgenic potato lines, with the greatest effects detected in those plants exhibiting the highest levels of expression. Herbers et al. *(23)* also reported that expression of a thermostable xylanase from *Clostridium thermocellum* expressed at high levels in the apoplast of transgenic tobacco had no detrimental effects on plant development. Recently Kimura et al. *(24)* reported the constitutive expression of a xylanase domain from *C. thermocellum* in rice, and no phenotypic effects were noted.

In conclusion, we have been able to demonstrate for the first time that it is possible to genetically modify the phenolic composition and digestibility of monocot cell walls by vacuolar targeting of a fungal ferulic acid esterase in plants of *L. multiflorum*. To our knowledge this is the first report of genetically engineering plants to express a gene encoding a fungal esterase, and indicates the potential of using the vacuole as a reservoir of esterase, which can be released on cell death to aid cell wall degradation. This strategy could prove to be useful in other species in which FAE activity is desired to increase the rate of breakdown of plant cell walls more effectively and inexpensively than exogenous application of these enzymes, such as during biomass conversion to fermentable sugars, or during processing of plants for foodstuffs. In this respect we have also targeted FAE to the apoplast, ER, Golgi, and vacuole under both constitutive and inducible promoters in *Festuca arundiancea* which is less digestible, but more well adapted than *Lolium*. The results of this work will be reported elsewhere.

Acknowledgments

IGER is grant aided by the BBSRC and this work was supported by a BBSRC research grant and by Genencor Inc.

References

1. Hatfield, R. D., Ralph, J., and Grabber, J. H. (1999), *J. Sci. Food Agric.* **79**, 403–407.
2. Hartley, R. D. and Ford, C. W. (1989), *Am. Chem. Soc.* **9**, 137–145.
3. Scalbert, A., Monties, B., Lallemand, J. Y., Guittet, E., and Rolando, C. (1985), *Phytochemistry* **24**, 1359–1362.
4. Ralph, J., Quideau, S., Grabber, J. H., and Hatfield, R. D. (1994), *J. Chem. Soc.* **1**, 3485–3498.
5. Eraso, F. and Hartley, R. D. (1990), *J. Sci. Food Agric.* **51**, 163–170.
6. Grabber, J. H., Hatfield, J. R., and Ralph, J. (1998), *J. Sci. Food Agric.* **77**, 193–200.
7. de Vries, R. P., Michelsen, B., Poulsen, C. H., et al. (1997), *Appl. Environ. Microbiol.* **63**, 4638–4644.
8. Rogers, J. C., Dean, D., and Heck, G. R. (1985), *Proc. Natl. Acad. Sci. USA* **82**, 6512–6516.
9. McElroy, D., Zhang, W., Cao, J., and Wu, R. (1990), *Plant Cell* **2**, 163–171.
10. Heim, R., Prasher, D. C., and Tsien, R. Y. (1994), *Proc. Natl. Acad. Sci. USA* **91**, 12,501–12,504.
11. Bilang, R., Iida, S., Peterhans, A., Potrykus, I., and Paskowski, J. (1991), *Gene* **100**, 247–250.
12. Dunn-Coleman, N., Langdon, T., and Morris, P. (2001), *USA Patent Application No.* 20030024009.

13. Dalton, S. J., Bettany, A. J. E., Timms, E., and Morris, P. (1999), *Plant Cell Rep.* **18,** 721–726.
14. Waldron, K. W., Parr, A. J., Ng, A., and Ralph, J. (1996), *Phytochem. Anal.* **7,** 305–312.
15. Jones, D. I. H. and Hayward, M. V. (1975), *J. S. Food Agric.* **26,** 711–718.
16. France, J., Dhanoa, M. S., Theodorou, M. K., Lister, S. J., Davies, D. R., and Isac, D. (1993), *J. Theor. Biol.* **163,** 99, 100.
17. Bartolome, B., Faulds, C. B., Kroon, P. A., et al. (1997), *Appl. Environ. Microbiol.* **63,** 208–212.
18. Hobbs, S. L. A., Kpodar, P., and Delong, C. M. O. (1990), *Plant Mol. Biol.* **15,** 851–864.
19. Pröls, F. and Meyer, P. (1992), *Plant J.* **2,** 465–475.
20. Casler, M. D. and Vogel, K. P. (1999), *Crop Sci.* **39,** 12–20.
21. Ziegelhoffer, J., Will, J., and Austin-Phillips, S. (1999), *Mol. Breed.* **5,** 309–318.
22. Armstrong, J., Inglis, G., Kawchuk, L., et al. (2002), *Am. J. Potato Res.* **79,** 39–48.
23. Herbers, K., Wilke, I., and Sonnewald, U. (1995), *Biotechnology* **13,** 63–66.
24. Kimura, T., Mizutani, T., Tanaka, T., Koyama, T., Sakka, K., and Ohmiya, K. (2003), *Appl. Microbiol. Biotechnol.* **62,** 374–379.

Variation of S/G Ratio and Lignin Content in a *Populus* Family Influences the Release of Xylose by Dilute Acid Hydrolysis

BRIAN H. DAVISON,*,[1] SADIE R. DRESCHER,[1] GERALD A. TUSKAN,[1] MARK F. DAVIS,[2] AND NHUAN P. NGHIEM[3]

[1]Oak Ridge National Laboratory, Oak Ridge, TN 37831-6124, E-mail: davisonbh@ornl.gov; [2]National Renewable Energy Laboratory, Golden, CO 80401; and [3]Martek Biosciences Corporation, Winchester, KY 40391

Abstract

Wood samples from a second generation *Populus* cross were shown to have different lignin contents and S/G ratios (S: syringyl-like lignin structures; G: guaiacyl-like lignin structures). The lignin contents varied from 22.7% to 25.8% and the S/G ratio from 1.8 to 2.3. Selected samples spanning these ranges were hydrolyzed with dilute (1%) sulfuric acid to release fermentable sugars. The conditions were chosen for partial hydrolysis of the hemicellulosic fraction to maximize the expression of variation among samples. The results indicated that both lignin contents and S/G ratio significantly affected the yield of xylose. For example, the xylose yield of the 25.8% lignin and 2.3 S/G (high lignin, high S/G) sample produced 30% of the theoretical yield, whereas the xylose yield of the 22.7% lignin and 1.8 S/G (low lignin, low S/G) was 55% of the theoretical value. These results indicate that lignin content and composition among genetic variants within a single species can influence the hydrolyzability of the biomass.

Index Entries: Cell wall chemistry; genetic variation; hybrid poplar; hydrolysis; lignin.

Introduction

Production of fuels and chemicals from biomass crops is limited by the recalcitrance of lignocellulose to hydrolysis into its component pentose and hexose sugars. Two consecutive steps are generally involved in several proposed hydrolysis procedures *(1)*. In the first step, dilute acid hydrolysis is used. In the second, the nondissolved solids then are subjected to an enzymatic hydrolysis. Dilute acid hydrolysis serves two purposes. The first one is to generate five-carbon sugars from the hemicellulose fraction; the second, to

---

*Author to whom all correspondence and reprint requests should be addressed.

open the structure of the residual cellulose fraction to enhance enzymatic hydrolysis, which generates six-carbon sugars. The extent of dilute acid hydrolysis has strong influence on the efficiency of the overall process.

Plants have varying cell-wall compositions *(2)*. This variation between species (i.e., hardwoods or softwoods) has an influence on their conversion into other feed streams such as paper pulp or sugars. Total lignin content is of interest, as well as the composition of the lignin. One measure of lignin composition is the guaiacyl (G) and syringyl (S) contents. These aromatic subunits determine the type and number of crosslinks. Guaiacyl units can covalently crosslink with up to three other units whereas syringyl units may link to only two *(3)*. Vinzant et al. *(4)* showed that increased lignin content across 15 tree species decreased the total ethanol fermentation production via dilute acid hydrolysis followed by simultaneous saccharification by enzymes and fermentation (SSF). Chiang and Funaoka *(5)* showed that the S/G content between native hardwoods and softwoods correlated to their ease during Kraft delignification, with the S units facilitating cleavage of crosslinked bonds. The broad hypothesis is that variations in biomass composition can be controlled through genetic manipulation or breeding, and that these compositional changes will improve conversion of biomass. Conversion improvements can be from direct mass balance considerations (lower lignin may result in higher cellulose and leads to more fermentable hydrolysis sugars), as well as from more subtle composition influences, such as a shift in the S/G ratio. It may also allow milder pretreatment with production of a lower amount of inhibitory byproducts.

Transgenic and recombinant DNA methods have been used to modify lignin content and composition *(6)*. This is achieved by up or down regulating the pathways that lead to the formation of S, G, and other subunits. Suppression of a key lignin pathway in a recombinant *Populus* clone resulted in 45% less lignin *(7)*, faster growth *(6)*, and improved delignification during Kraft pulping *(8)*. Genetic engineering is also beginning to be considered as a means of changing the lignin composition of biomass feedstocks *(2)*. Much of the current work focuses on pulping or hydrolysis differences among different species. Several recombinant plants have shown changes in lignin content and composition including *Populus (2,9)*. Pilate et al. *(10)* indicate that genetically altered composition may not impact tree growth or fitness.

The experiments reported here were part of a larger project in which destructively sampled, clonally replicated field trials provided evidence that carbon allocation is genetically controlled *(11)*. Map-based identification of separate stem and root cell wall chemistry quantitative trait loci supported the hypothesis that carbon partitioning is also genetically controlled. Here, we report on a hydrolysis study of selected members of the above-cited *Populus* family to confirm the hypothesis that lignin composition and lignin content influence hydrolysis sugars production.

Materials and Methods

Biomass Samples

Approximately 300 progeny from a single F_2 segregating hybrid poplar family, Family 331, were established in three clonal replicates at Wallula, WA, and allowed to grow over the course of one growing season in a common garden. Each clonal replicate was completely removed from the site and sampled for component biomass allocation. Subsamples from several stems were preserved and used for the hydrolysis tests. After bark removal, the remaining subsample mass averaged approx 3 g DW. Samples were finely ground in a Thomas–Wiley mill (Thomas Scientific, Swedesboro, NJ) and screened through a 20-mesh screen. Several additional subsamples were obtained from the parental clones (F_1) and were used for establishing and testing the experimental conditions.

Wood Chemistry Analysis

Previously, increment cores were removed from each genotype from pairwise stem and root tissues in an 8-yr-old clonal replicate of Family 331 grown in Clatskanie, OR. These wood samples were subjected to pyrolysis molecular beam mass spectroscopy (pyMBMS) for determination of cell wall lignin content and S/G ratio by the National Renewable Energy Laboratory (12). Lignin content in the stem varied from 20.3 to 27.1% (by weight), with a mean of 24.9%. Root lignin content varied from 14.4 to 25.1%. The simple phenotypic correlation across all genotypes between stem lignin content and root lignin content was nonsignificant ($r = 0.24$). Similar results were obtained for S/G ratio. Stem S/G ratio varied from 1.6 to 2.5; root S/G ratio was lower and varied from 1.1 to 2. Again, the phenotypic correlation between stem and root S/G ratio was nonsignificant ($r = 0.21$). The theoretical xylose yield was calculated based on measured xylan contents from sample composition determined by pyMBMS. Xylan content varied between 15.5 and 18.8% in stems.

Dilute Acid Hydrolysis

Dilute acid hydrolysis was studied in stainless steel reactors. Five identical stainless steel reactors were constructed (316 stainless steel [SS], 1.8 cm ID, 1.9 mm OD, 12 cm length, 30 mL volume with Swagelock fittings). A thermocouple was inserted through one end of the reactor to monitor the temperature inside the biomass–acid mixture. A hot oil bath at 175°C was used to provide heating.

The tested biomass was ground to a uniform size (20-mesh sieve) and loaded into the reactor. A 1% (w/w) solution of sulfuric acid then was added to give a solid content of 1% (w/w). Sample size was 0.75 g.

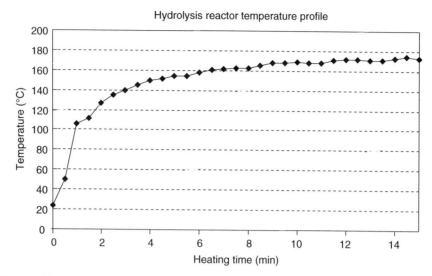

Fig. 1. Typical temperature profile for hydrolysis reactor used in a dilute acid hydrolysis experiment to test xylose yield from hybrid poplar stem samples with varying lignin content and S/G ratios.

The biomass was allowed to soak in this solution for 20 min. The reactors were placed in the preheated oil bath (175°C) for an empirically determined period (12 min). At the end of this period the reactors were quickly removed and placed in ice to quench the reaction. Figure 1 shows a typical temperature profile inside the reactors. The biomass and the liquid were removed from the reactors for analysis. The unreacted biomass solids were separated from the free liquid (larger samples were centrifuged; 0.75 g samples were squeezed dry). The solids were then dried at 90°C to determine the final residual mass and close the mass balance. The liquid samples were filtered and analyzed for xylose and other sugars by high-performance liquid chromatography (HPLC) (RMH Monosaccharide column (Phenomenex), 5 mM H_2SO_4 eluant; Waters 2410 RI detector, 10 µL sample). The total mass loss and the quantities of free sugars produced determined the extent of hydrolysis. Each experiment contained a single replicate, as it was not possible to subsample during the dilute acid treatment. Experiments were performed in triplicate.

Statistical Analysis

The raw data was analyzed using ANOVA and JMP statistical packages. To test the statistical significance of the data a simple linear model was used. The model took the form

$$X = a + b(L\%) + c(S/G) + d(L\%)(S/G) \tag{1}$$

where L% is the percent lignin in stem wood, and S/G is the syringyl to guaiacyl lignin ratio.

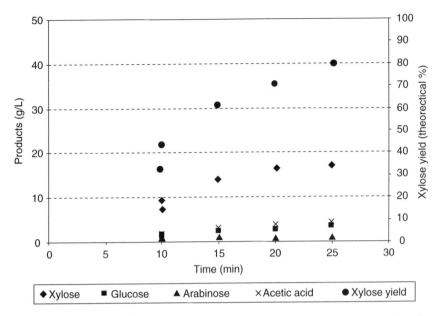

Fig. 2. Hydrolysis sugar production and xylose yield (as % of theoretical) with time for the test parental F_1 clone, 242.

Results and Discussion

Preliminary Experiments

The preliminary experiments to establish experimental conditions were performed on one of the F_1 parental clones (no. 242) in which there was sufficient biomass to use 1.5 g per test. Figure 2 shows these tests with the resulting sugar production and the calculated xylose yield at different hydrolysis times. The xylose yield was calculated using only the xylose produced and based on the total xylan content of the original specific sample. Because the objective was to study the effect of lignin composition on the efficiency of dilute acid hydrolysis, the experimental conditions were chosen for incomplete hydrolysis of the hemicellulosic fraction in order to maximize the expression of variation among samples. From this data we chose a hydrolysis time of 12 min for the comparative experiments.

Comparative Experiments

Dilute acid hydrolysis was performed on a subset of the F_2 *Populus* clones to determine the amount of fermentable sugars that are available from feedstocks of differing cell-wall composition. The selected samples spanning these ranges were hydrolyzed with dilute sulfuric acid (1% by weight) to release fermentable sugars (1% by-weight solids, 175°C maximum, 12-min hydrolysis time). The clones had been analyzed by pyMBMS for lignin content and for S/G ratio. A small subset (five) of the available clones was chosen across the distribution of these two values. The available samples were

Table 1
Variation in Cell-Wall Composition Among the Untreated Samples Used
in the Dilute Acid Hydrolysis Experiments and the Resulting Xylose Yield
(% theoretical) After Partial Hydrolysis

Sample No.	Factorial design	Lignin (%)	S/G ratio	Xylose yield (%)	Standard deviation
242	Center	24.6	1.9	44.5	0.057
1093	High–high	25.8	2.3	30.1	0.036
1640	High–low	24.8	1.8	39.5	0.056
1910	Low–high	22.7	2.1	28	0.043
1642	Low–low	22.7	1.8	54.9	0.01

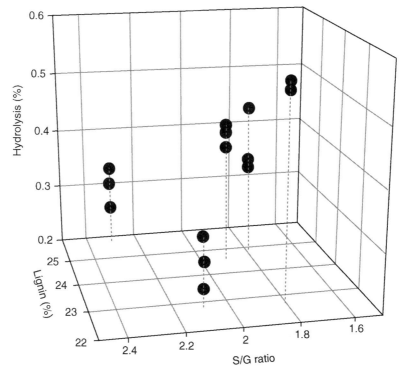

Fig. 3. Effect of lignin content and S/G ratio on xylose yields using a dilute acid hydrolysis. All data points are shown from triplicate runs.

small (most were <3 g) so the comparative hydrolysis experiments were downscaled to consume only 0.75 g of the milled wood chip per experiment, and were performed in triplicate on all five selected clones. Table 1 shows the compositions of the selected clones and the hydrolysis results. The change in hydrolysis results was greater than the replicate error of 5–10%. This was consistent with estimated measurement error of 9% at this small scale. The mass balance closure was greater than 90%. This data is also represented in Fig. 3. The data shows an effect of composition greater than the experimental

Table 2
ANOVA Analysis of Hydrolysis using Lignin% by S/G Ratio Factorial Experiment

	Variance in factorial experiment F ratio = 19.87, $p > F$ is 0.0002		
Source	DF	Sum of squares	Mean square
Model	3	0.1065	0.03551
Error	10	0.01788	0.00179
C. Total	13	0.1244	
	Lack of fit test F ratio = 3.098, $p > F$ is 0.112, max R^2 = 0.893		
Lack of fit	1	0.004578	0.00458
Pure error	9	0.0133	0.00148
Total error	10	0.01788	

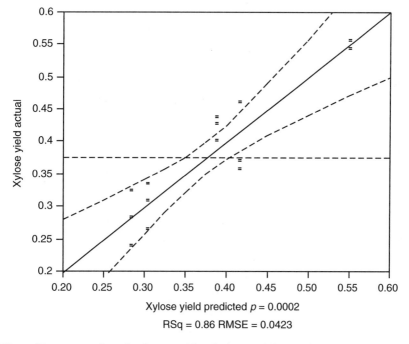

Fig. 4. Response plot of xylose yield–whole model actual vs predicted plot.

variability. The yield was highest at low–low combination of % lignin-S/G ratio and lowest at low–high combination of % lignin-S/G ratio.

Statistical Analysis

The ANOVA analysis presented in Table 2 indicates that the model is significant and that there is no "Lack of Fit" in the model. Figure 4 plots the actual vs predicted xylose yields with the 95% confidence limits. The increment of S/G ratio had a statistically significant negative impact on xylose yield. The effect of lignin percentage alone is not significant, but the interaction effect of lignin percentage and S/G ratio is significant.

Despite the lack of statistical significance, the magnitude and direction of the lignin percentage effect (decreased xylose yield with increased lignin) is consistent for the mass balance effect. Here the small increase in lignin decreases the potential xylan and the potential xylose yield by a few percent. Vinzant et al. *(4)* showed a significant effect of lignin content on ethanol yield across a variety of hardwood species. This effect was much greater than could be accounted for by the decrease in theoretically available hydrolysis sugars because of increased lignin. This supports the concept that lignin content alone is not the potentially most important factor. The S/G effect and the interaction effect are more interesting and indicate the importance of crosslinking on the ease of degradation.

Conclusions

A small decrease in S/G ratio resulted in statistically significant improvement in the rate of dilute acid hydrolysis. The significance of lignin content alone was not as clear. More data points with higher precision will be needed for a definite conclusion. The combined effect of lignin contents and S/G ratio on the rate of dilute acid hydrolysis; however was significant. These results were obtained despite imperfect distribution of independent variables and significant measurement error because of small sample size.

Interestingly, these results were opposite from our initial intuition. A lower S/G ratio implies more potential for covalent crosslinking and thus would be expected to result in less hydrolysis, not more hydrolysis as seen here. In paper pulping studies, alkali methods for digestion of wood with higher S content gave higher pulp yields. In these poplar samples, acid hydrolysis may break the G bonds more readily. However, a similar effect has been observed in animal digestibility studies *(13)*. Here various forage maize were tested with different compositions and lower S content correlated with more milk and meat production. They speculated that less S residues gave a more crosslinked but thinner cell wall that was easier to degrade. Following these studies, Fontaine et al. *(14)* again showed lower S content in maize associated with improved degradability, and hypothesized that lower S content is typical of a less mature, less lignified cell wall that is more accessible to chemical or enzymatic penetration. However, Reddy et al. *(15)* showed a increased lignin decreased alfalfa digestibility in transgenic lines but that lignin composition did not have an effect.

These results show that natural variation in poplar, whereas relatively narrow, can have a measurable effect on dilute acid hydrolysis. To the best of our knowledge, these results have not been observed for clones within a single species. Taken together, these results imply that genetic engineering or traditional breeding to enhance hydrolysis for release of fermentable sugars would be successful in modifying the composition of biomass crops.

Carbon allocation and carbon partitioning among and within plant tissues ultimately determine the growth and utility of plant materials for various applications including biochemical conversion to fuels and bio-based products as well as options for carbon sequestration. We proposed that cell wall chemistry variations within a single family within a single species could be manipulated to impact potential biomass use. Further work is needed to test if these hydrolysis increases also occur in a subsequent cellulose hydrolysis step. The proof-of-principle hydrolysis experiment reported here demonstrated that changes in cell wall chemistry could significantly enhance the conversion efficiency of the derived biomass feedstocks.

Acknowledgments

Lee Gunter, of Oak Ridge National Laboratory (ORNL) assisted in the sample selection and milling. Catherine Cheng, (now at Eli Lilly) performed the statistical analysis of the data set. We are grateful to Wilfred Vermerris, of Purdue for additional references on animal digestibility. This work was supported by the ORNL's Laboratory Director's Research and Development Fund. ORNL is operated under contract for the US Department of Energy under contract no. DE-AC05-00OR22725.

References

1. Aden, A., Ruth, M., Ibsen, K., et al. (2002), NREL/TP-510-32438, National Renewable Energy Laboratory (NREL), Golden, C. O.
2. Dinus, R. J., Payne, P., Sewell, M. M., Chiang, V. L., and Tuskan, G. A. (2000), *Crit. Rev. Plant Sci.* **20**, 51–69.
3. Boudet, A. M., Goffner, D., Marque, C., Teulieres, C., and Grima-Pettenati, J. (1998), *Ag. Biotech. News Inf.* **10**, 295–304.
4. Vinzant, T. B., Ehrman, C. I., Adney, W. S., Thomas, S. R., and Himmel, M. E. (1997), *Appl. Biochem. Biotechnol.* **62**, 99–104.
5. Chiang, V. L. and Funaoka, M. (1990), *Holzforschung* **44**, 309–313.
6. Hu, W. -J., Harding, S. A., Lung, J., et al. (1999), *Nat. Biotechnol.* **17**, 808–812.
7. Jouanin, L., Goujon, T., de Nadai, V., et al. (2000), *Plant Physiol.* **123**, 1363–1374.
8. LaPierre, C., Pollet, B., Petit-Conil, M., et al. (1999), *Plant Physiol.* **119**, 153–164.
9. Pena, L. and Seguin, A. (2001), *Trends Biotechnol.* **19**, 500–506.
10. Pilate, G., Guiney, E., Holt, K., et al. (2002), *Nat. Biotechnol.* **20**, 607–612.
11. Wullschleger, S., Yin, T. M., DiFazio, et al. (2005), *Can J. Forest Res.* **35**, 1779–1789.
12. Tuskan, G. A., West, D., Bradshaw, H. D., et al. (1999), *Appl. Biochem. Biotech.* **77–79**, 1–11.
13. Jung, H. G. and Deetz, D. A. (1993), In: *Cell Wall Lignification and Degradability*, Jung, H. G. Buxton, D. R,. Hatfield, R. D., and Ralph, J., eds., Madison, W. I., pp. 315–346.
14. Fontaine, A. S., Bout, S., Barriere, Y., and Vermerris, W. (2003), *J. Agric. Food Chem.* **51**, 8080–8087.
15. Reddy, M. S. S., Chen, F., Shadle, G., Jackson, L., Aljoe, H., and Dixon, R. A., (2005), *PNAS*, (in print).

Copyright © 2006 by Humana Press Inc.
All rights of any nature whatsoever reserved.
0273-2289/06/129–132/436–446/$30.00

Enhanced Secondary Metabolite Biosynthesis by Elicitation in Transformed Plant Root System

Effect of Abiotic Elicitors

GWI-TAEK JEONG[1] AND DON-HEE PARK*,[1–5]

[1]*School of Biological Sciences and Technology,* [2]*Faculty of Applied Chemical Engineering,* [3]*Research Institute for Catalysis,* [4]*Biotechnology Research Institute, and* [5]*Institute of BioIndustrial Technology, Chonnam National University, Gwangju 500-757, Korea, E-mail: dhpark@chonnam.ac.kr*

Abstract

Plants generally produce secondary metabolites in nature as a defense mechanism against pathogenic and insect attack. In this study, we applied several abiotic elicitors in order to enhance growth and ginseng saponin biosynthesis in the hairy roots of *Panax ginseng*. Generally, elicitor treatments were found to inhibit the growth of the hairy roots, although simultaneously enhancing ginseng saponin biosynthesis. Tannic acid profoundly inhibited the hairy root growth during growth period. Also, ginseng saponin content was not significantly different from that of the control. The addition of selenium at inoculum time did not significantly affect ginseng saponin biosynthesis. However, when 0.5 mM selenium was added as an elicitor after 21 d of culture, ginseng saponin content and productivity increased to about 1.31 and 1.33 times control levels, respectively. Also, the addition of 20 µM NiSO$_4$ resulted in an increase in ginseng saponin content and productivity, to about 1.20 and 1.23 times control levels, respectively, and also did not inhibit the growth of the roots. Sodium chloride treatment inhibited hairy root growth, except at a concentration of 0.3% (w/v). Increases in the amounts of synthesized ginseng saponin were observed at all concentrations of added sodium chloride. At 0.1% (w/v) sodium chloride, ginseng saponin content and productivity were increased to approx 1.15 and 1.13 times control values, respectively. These results suggest that processing time for the generation of ginseng saponin in a hairy root culture can be reduced via the application of an elicitor.

Index Entries: Elicitor; selenium; nickel; ginseng saponin; hairy roots.

Introduction

Plants constitute a large source of valuable compounds. About 100,000 compounds have been isolated from plant sources, with about 4000 new

*Author to whom all correspondence and reprint requests should be addressed.

compounds being discovered every year. These compounds not only are connected with important traits of the plant itself, including color, fragrance, taste, and resistance to pests and diseases, but also may prove useful in the manufacture of a host of useful products, including drugs, antioxidants, flavorings, fragrances, pigments, insecticides, and many other important industrial and medicinal raw materials *(1,2)*. Plant cell and tissue cultures are an attractive alternative to the cultivation of whole plants for production of valuable secondary metabolites *(3)*.

The hairy roots induced by *Rhizobium rhizogenes* constitute a valuable and promising source of root-derived phytochemicals. These transformed hairy roots can synthesize fairly robust amounts of several metabolites, at levels similar to, or even greater than, the amounts that would be generated by the original plants *(4)*. Hairy roots are characterized by high growth rates, high metabolite productivity, and inherent genetic stability *(5,6)*. In the near future, the plant-derived transgenic hairy root-based production approach may become a widely used method, allowing for the commercial production of a myriad of useful compounds *(7)*.

In general, plants produce secondary metabolites in nature as a defense mechanism against pathogenic and insect attack. In recent research into in vitro culture systems, a wide variety of elicitors have been employed in order to modify cell metabolism. These modifications are designed to enhance the productivity of useful metabolites in the cultures of the plant cells/tissues. The cultivation period, in particular, can be reduced by the application of elicitors, although maintaining high concentrations of product *(8,9)*. Elicitation strategies are compounds or treatments which cause plants to synthesize elevated levels of phytoalexins *(8)*. The active mechanisms employed by elicitors are complex and distinctive. As little is known regarding the biosynthetic pathways of most secondary plant metabolites, the effects of elicitation on a plant cell/tissue culture are difficult to predict. Therefore, elicitation approaches tend to be empirical steps. The effects of elicitors rely on a host of factors, including the concentration of the elicitor, the growth stage of the culture at the time of elicitation, and the contact duration of elicitation *(10)*. Both biotic and abiotic elicitors can be used to stimulate secondary metabolite biosynthesis in plant cell/tissue cultures, thereby reducing the processing time necessary for high product yields. Elicitors of nonbiological origin (abiotic elicitors), such as heavy metals and ultraviolet light, which induce phytoalexin synthesis, are generally designated as abiotic stresses *(8,10)*.

In previous researches, tannic acid, selenium, and $NiSO_4$ despite its functionality were not introduced to elicitation process for enhancing of secondary metabolites in plant cell/tissue cultures. Tannic acid, a commercial form of tannin, is not actually a true acid, but rather an acid-like substance, called a polyphenol. Tannic acid occurs naturally in tea, coffee, oak, and sumac bark. Tannin molecules crosslink proteins, rendering the tissues more resistant to both bacterial and fungal attacks. Many tannins are known to be cytotoxic to cell cultures. Tannins may also exhibit antibiotic activity that appears to be resulting from the precipitation of pathogen-produced extracellular enzymes, or interference with the metabolism of the pathogen *(11)*. For many years, selenium (Se) has been recognized as an essential trace element for both humans and livestock. Selenium has also been recognized to be an important cofactor in several selenoproteins. In order to increase selenium intake using natural sources, selenium content must be increased in cultivated crops. Supplementation is normally accomplished by the addition of either selenite or selenate to fertilizers, the spraying of crops with selenium salts, or treatment of the seeds with aqueous selenium. However, selenium is not an essential nutrient for plants. Also, the effects of selenium on plant growth and metabolite biosynthesis have yet to be thoroughly investigated *(12)*. Nickel (Ni) has been identified as an essential element for plant growth. Ni is required for the activation of functional urease, because it is a cofactor. To the best of our knowledge, the activation of urease is the sole proven function of Ni in higher plants. There have also, however, been some reports regarding the stimulatory effects of Ni on plant growth *(13,14)*. Salinity may affect a variety of metabolic processes, including photosynthesis, protein synthesis, respiration, nitrogen assimilation, and the turnover of plant growth regulators. Cytokinin synthesis in the roots and cytokinin transport to the shoots is also influenced by salinity, but abscisic acid synthesis is promoted *(15)*.

Panax ginseng C. A. Meyer, a member of the Araliaceae family, is one of the better known oriental medicinal plants and is widely distributed throughout the Korean peninsula and China. Ginseng plants have been reported to exert a variety of beneficial bioactive effects on human health including hemostatic qualities. Ginseng plants also appear to exhibit properties that promote blood circulation, relieve pain, stanch bleeding wounds and alleviate trauma, relieve stress, and improve immune function. A great deal of chemical, biochemical, and pharmacological research into the properties of ginseng plants has been recently carried out *(16,17)*. The principal compounds involved in the pharmaceutical effects of ginseng have been isolated and identified in a host of previous studies. These compounds include ginseng saponin (ginsenosides), polysaccharides, antioxidants, peptides, fatty acids, alcohols, vitamins, and phenolic compounds. The pharmacological characteristics and effects of the ginsenosides Rb_1 and Rg_1 are distinct, and are sometimes antagonistic. Ginsenoside Rb_1 manifests sedative, anticonvulsive, analgesic, antipyretic, anti-inflammatory, and antipsychotic properties, and also has been shown to improve gastrointestinal motility. Ginsenoside Rg_1 exerts stimulant and antifatigue effects, and also appears to enhance motor ability. In recent years, ginseng polysaccharides have been identified as potentially useful compounds that appear to exert important pharmacological effects. Ginseng polysaccharides also apparently possess immune stimulation, antitumor, antihepatitis, and mitogenic and hypoglycemic properties *(5,16,17)*.

It is known to that the ginseng saponin of *P. ginseng* was generated at the latter culture period. This biosynthesis pattern, nongrowth associated process, is well known regarding the formation of secondary metabolites in plant cell/tissue cultures. In this study, *P. ginseng* C.A. Meyer hairy root cultures, generated by infection with *R. rhizogenes* KCTC 2744, were employed in order to enhance the biosynthesis of a secondary metabolite, via the application of abiotic elicitors.

Materials and Methods

Hairy Roots and Culture

The hairy roots of *P. ginseng* C. A. Meyer were induced and maintained, as previously described *(18)*. In all of the experiments in this study, the hairy roots were cultivated in liquid hormone-free 1/2 MS medium that contained 30 g/L sucrose. The pH of the medium was adjusted to 5.8 with 2 N NaOH, and the medium was autoclaved at 121°C for 15 min before use. The cultures were then incubated at 23 ± 1°C in darkness in a 250 mL Erlenmeyer flask, on a rotary shaking incubator at 80 rpm.

Experimental Procedures

In order to determine the effects of several elicitors on hairy root growth and ginseng saponin biosynthesis in *P. ginseng* hairy roots in 250 mL flask cultures, tannic acid (0–10 mM), selenium (as selenite; 0–10 mg/L), nickel (as $NiSO_4$; 0–50 μM), NaCl (0–0.3% [w/v]), and some other abiotic elicitors (ascorbic acid 100 mg/L; salicylic acid [SA] 400 μM; H_2O_2 0.012% [v/v]; $CuSO_4$ 2 mM; $CaCl_2$ 20 mM; SA 400 μM + $CaCl_2$ 20 mM; SA 400 μM + H_2O_2 0.012% [v/v]) were applied. 0.2 mL of elicitors was added to each sample flasks at inoculum time. The pH of the prepared elicitor solutions was adjusted to 5.8, using either 1 N NaOH or 1 N HCl. At each experiment, the control was incubated to same culture condition for same period.

For selenium elicitation experiments, after 21 d of culture which did not added any elicitor (first culture), different concentrations of the prepared selenium elicitors (as selenite; 0–0.5 mM) were added to culture media. On day 3 of elicitation (second culture), the hairy roots were harvested, and the biomass and ginseng saponin content were measured. To compare with the elicitation effect of elicitor, the control was incubated to same culture condition for 24 d. All experiments are performed at least two independent experiments.

Analytical Methods

In order to determine the weight of the biomass, the hairy roots were harvested and rinsed with distilled water, and blotted to remove extra water and weighed. Dry weight was measured gravimetrically after drying the roots for 24 h at 60°C. Results are expressed as the mean value of at least two independent measurements.

Extraction and Analysis of Ginseng Saponin

In order to quantify the total amount of biosynthesized ginseng saponin, 100 mg of powdered dry hairy roots were soaked in 5 mL of distilled water-saturated n-butanol stored at 4°C for 24 h, sonicated in an ultrasonic cleaning bath for 1 h, and centrifuged twice at 5030g for 10 min. The collected supernatants were then used to determine the total amount of ginseng saponin that had been synthesized. Ginseng saponin content was measured via Vanillin-H_2SO_4 colorimetry (19). A calibration curve was established using a ginsenoside Re standard. Authentic ginsenoside Re was purchased from Sigma-Aldrich Co., Ltd (St. Louis, MO).

Calculations of Ginseng Saponin Content and Productivity

The ginseng saponin content of the *P. ginseng* hairy roots was calculated as according to the following formula:

Ginseng saponin content (mg/g) = ginseng saponin concentration of supernatant sample (mg/L) × total supernatant sample volume (L)/ dry weight (g) of sample hairy root

The ginseng saponin productivity of the *P. ginseng* hairy roots was calculated according to the following formula:

Ginseng saponin productivity (mg/L) = ginseng saponin content (mg/g) × harvested hairy root dry weight (g)/volume of culture medium (L)

Results and Discussion

Effect of Tannic Acid, Selenium, $NiSO_4$, and NaCl on Growth and Secondary Metabolite Production

Growth and ginseng saponin formation of *P. ginseng* hairy roots was monitored after tannic acid addition to cultures at inoculum time. Tannic acid was found to exert a potent inhibitory effect on hairy root growth (Fig. 1). When tannic acid was added in excess of 0.1 mM, serious inhibition of the growth of the hairy roots resulted. The addition of 0.01 mM tannic acid resulted in a ginseng saponin content (60.2 ± 4.3 mg/g) not significantly different from that of the control (60.3 ± 1.7 mg/g).

In order to investigate the effect of selenium on *P. ginseng* biomass and ginseng saponin biosynthesis, prepared selenium was added to medium at inoculum time. Selenium was found to strongly inhibit hairy root growth, but caused a slight increase in ginseng saponin content, as is shown in Fig. 2. The addition of 0.1 mM selenium resulted in a ginseng saponin content of 80.5 ± 4.4 mg/g (1.07 times control level), and a productivity of 402.3 mg/L (1.04 times control level). These results indicate that the addition of selenium at inoculum time do not greatly improve the productivity of ginseng saponin in *P. ginseng* hairy root cultures. Carvalho et al. (12) reported similar results that the effects of selenium supplementation

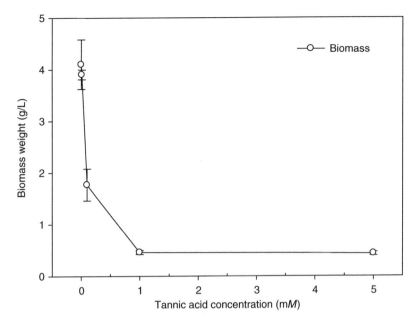

Fig. 1. Effect of tannic acid on growth of hairy roots.

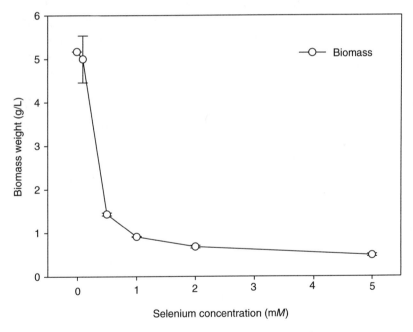

Fig. 2. Effect of selenium on growth of hairy roots.

on germination and plant growth and at sufficiently high concentrations, it can inhibit both the growth and germination of seeds.

As shown in Fig. 3, elicitation via the addition of 0.1 mM Se resulted in inhibited biomass growth. However, when 0.5 mM Se was added, biomass

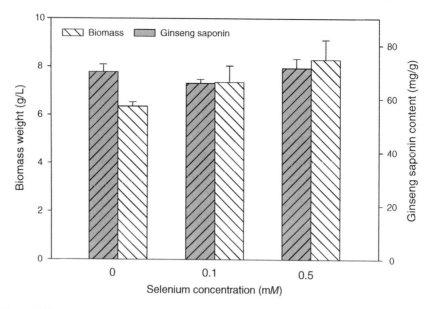

Fig. 3. Effect of selenium elicitor on growth and secondary metabolite accumulation of hairy roots.

was slightly higher than that observed in the control sample. The ginseng saponin content of hairy roots increased directly with increases in the added amount. As a result of the addition of 0.5 mM Se, ginseng saponin content was 74.7 ± 7.3 mg/g (1.31 times control value), and the productivity was 594 mg/L (1.33 times control value). Treatment with selenium as an elicitor is associated with several advantages in terms of both economy and operation compared with other treatments such as yeast elicitor treatments. Chief among these advantages are the low preparation cost and the ease of the preparation process. Preparation of a selenium elicitor is fairly simple, because the elicitor is applied as a solution, whereas the preparation of a yeast elicitor requires several processing steps, including repeated ethanol precipitation and purification steps. Also, the time required for preparation of a selenium elicitor is much less than that required for the preparation of yeast elicitor; the preparation of a yeast elicitor takes approx 1 wk. Therefore, application of selenium as an elicitor may dramatically reduce the costs associated with the large-scale production of ginseng saponin (ginsenosides).

In order to determine the influence of $NiSO_4$ on *P. ginseng* hairy root growth and ginseng saponin biosynthesis, different concentrations of $NiSO_4$ were introduced to the medium at inoculum time. Biomass growth, based on the dry weight of the hairy roots, was found to occur properly with the addition of 10 μM $NiSO_4$ as is shown in Fig. 4. Across the entire range of concentration, we observed no serious inhibition of hairy root growth. At less than 10 mM, the ginseng saponin content fluctuated. The addition of 20 μM $NiSO_4$ resulted in a ginseng saponin content of 91.6 ± 9.8 mg/g (1.20 times control value) and a productivity of 597 mg/L (1.23 times control value), with no inhibitory effects on hairy root growth.

Enhanced Secondary Metabolite Biosynthesis

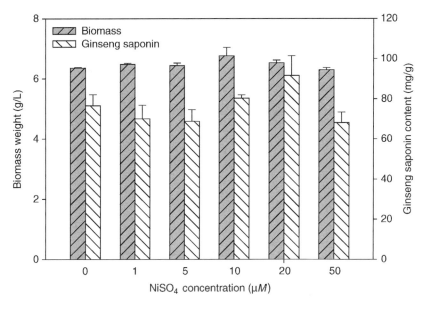

Fig. 4. Effect of NiSO$_4$ on growth and secondary metabolite production of hairy roots.

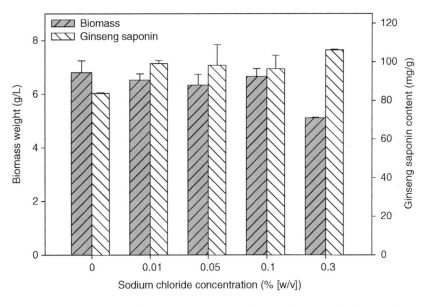

Fig. 5. Effect of sodium chloride on growth and secondary metabolite production of hairy roots.

In order to characterize the effects of sodium chloride on growth and ginseng saponin production in the *P. ginseng* hairy roots, 0–0.3% (w/v) sodium chloride was added to the medium at inoculum time. As shown in Fig. 5, sodium chloride was found to inhibit the hairy root growth, at a concentration of 0.3% (w/v). The addition of sodium chloride less than

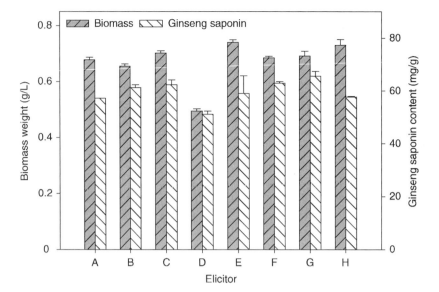

A: Ascorbic acid 100 mg/L; B: Salicylic acid (SA) 400 µM; C: H_2O_2 0.012% (v/v);
D: $CuSO_4$ 2 mM; E: $CaCl_2$ 20 mM; F: SA 400 µM+$CaCl_2$ 20 mM;
G: SA 400 µM + H_2O_2 0.012% (v/v); H: Control (no addition)

Fig. 6. Effect of several elicitors on growth and secondary metabolite accumulation of hairy roots.

0.1% (w/v) enhanced ginseng saponin content. The addition of 0.1% (w/v) sodium chloride resulted in a ginseng saponin content of 96.4 ± 6.9 mg/g (1.15 times control value) and a productivity of 642 mg/L (1.13 times control value), exerting no observable inhibitory effects on the growth of the hairy roots.

Effects of Some Other Abiotic Elicitors on Growth and Secondary Metabolite Production

As shown in Fig. 6, elicitation was performed with several abiotic compounds added to medium at inoculum time in flask cultures. Ascorbic acid had no enhancing effect on biomass accumulation, but did not inhibit accumulation of ginseng saponin. Other treatments, including 400 µM SA, 0.012% (v/v) H_2O_2, 400 µM SA + 20 mM $CaCl_2$, and 400 µM SA + 0.012% (v/v) H_2O_2 treatments resulted in slight increases in ginseng saponin accumulation, as compared with control values. Elicitation with 20 mM $CaCl_2$ resulted in slight enhancing in biomass growth and ginseng saponin formation. As a result, salicylic acid, hydrogen peroxide, and calcium chloride can apply to enhancement of secondary metabolite formation in plant hairy root cultures. In the case of 2 mM $CuSO_4$ addition, biomass and ginseng saponin accumulation was declined.

Conclusions

Plant hairy root culture appears to constitute a feasible technique for the in vitro production of valuable secondary metabolites. In this article, we attempted to determine the effects of several elicitors on growth and ginseng saponin biosynthesis of *P. ginseng* hairy root cultures. Some of the attempted elicitor treatments inhibited the growth of hairy roots, but increased ginseng saponin levels. The application of tannic acid profoundly inhibited the hairy root growth during growth period. The addition of selenium during the first culture period did not significantly improve ginseng saponin productivitys. However, at second culture period, on the addition of 0.5 mM Se as an elicitor, ginseng saponin content and productivity were increased to 1.31 and 1.33 times the control values, respectively. Throughout the entire range of added $NiSO_4$ concentrations, no profound inhibitory effects on hairy root growth resulted. The addition of 20 µM $NiSO_4$ resulted in a maximum ginseng saponin content and productivity of 1.20 and 1.23 times control values, respectively, but did not inhibit the growth of the cultures. Sodium chloride did not inhibit hairy root growth, except at a concentration of 0.3% (w/v). Ginseng saponin content increased as the result of the addition of sodium chloride, and these increases were observed at all concentrations. The addition of 0.1% (w/v) sodium chloride resulted in a ginseng saponin content and productivity of 1.15 and 1.13 times control values, respectively, but did not inhibit the growth of the hairy roots. The results of our study indicate that cultivation period can be reduced by the use of an elicitor and hairy root culture, making possible the large-scale production of useful metabolites in a much shorter time, and at a lower cost, than is currently possible with cultivation-based methods. However, it is suggested that subsequent studies be conducted to elucidate the enhancement of secondary metabolite productivity.

References

1. Verpoorte, R., van der Heijden, R., ten Hoopen, H. J. G., and Memelink, J. (1999), *Biotechnol. Lett.* **21**, 467–479.
2. Lee, J. H., Loc, N. H., Kwon, T. H., and Yang, M. S. (2004), *Biotech. Bioprocess Eng.* **9(1)**, 12–16.
3. Banthorpe, D.V. (1994), *Nat. Prod. Rep.* **11(3)**, 303–328.
4. Canto-Canche and Loyola-Vargas, V. M. (1999), In: *Chemicals Via Higher Plant Bioengineering (Advances in Experimental Medicine and Biology, 464)*, Kluwer, pp. 235–275.
5. Jeong, G. T., Park, D. H., Ryu, H. W., Hwang, B., and Woo, J. C. (2004), *Appl. Biochem. Biotechnol.* **116**, 1193–1203.
6. Furuya, T., Kojima, H., Syono, K., and Nishio, M. (1973), *Planta Med.* **47**, 183–187.
7. Giria, A. and Narasu, M. L. (2000), *Biotechnol. Adv.* **18**, 1–22.
8. Ramachandra, R. S. and Ravishankar, G. A. (2002), *Biotechnol. Adv.* **20(2)**, 101–153.
9. Akimoto, C., Aoyagi, H., and Tanaka, H. (1999), *Appl. Microbiol. Biotechnol.* **52**, 429–436.
10. Lu, M. B., Wong, H. L., and Teng, W. L. (2001), *Plant Cell Rep.* **20**, 674–677.
11. Seigler, D. S. (1995), In: *Plant Secondary Metabolism*, Kluwer, London.
12. Lopez-Bucio, J., Cruz-Ramirez, A., and Herrera-Estrella, L. (2003), *Curr. Opin. Plant Biol.* **6**, 280–287.

13. Witte, C. -P., Tiller, A. A., Taylor, M. A., and Davies, H. V. (2002), *Plant Cell Tiss. Org.* **68,** 103–104.
14. Gerendas, J., Polacco, J. C., Freyermuth, S. K., and Sattelmacher, B. (1999), *J. Plant Nutr. Soil Sci.* **162,** 241–256.
15. Arshi, A., Abdin, M. Z., and Iqbal, M. (2002), *Biol. Plantarum* **45(2),** 295–298.
16. Wu, J. and Zhong, J. J. (1999), *J. Biotechnol.* **68,** 89–99.
17. Jung, N. P. and Jin, S. H. (1996), *Korean J. Ginseng Sci.* **20(4),** 431–471.
18. Jeong, G. T. and Park, D. H. (2005), *Biotech. Bioprocess Eng.* **10,** 73–77.
19. Jeong, G. T. (2004), PhD Dissertation, Chonnam National University, Gwangju, Korea.

SESSION 3B
Biomass Pretreatment and Hydrolysis

Preliminary Results on Optimization of Pilot Scale Pretreatment of Wheat Straw Used in Coproduction of Bioethanol and Electricity

METTE HEDEGAARD THOMSEN,*,[1] ANDERS THYGESEN,[1]
HENNING JØRGENSEN,[2] JAN LARSEN,[3]
BØRGE HOLM CHRISTENSEN,[4] AND ANNE BELINDA THOMSEN[1]

[1]Biosystems Department, Risø National Laboratory, POB 49, DK-4000 Roskilde, Denmark, E-mail: mette.hedegaard.thomsen@risoe.dk; [2]Danish Center for Forest, Landscape and Planning, KVL, The Royal Veterinary and Agricultural, University, Højbakkegård Allé 1, DK-2630 Taastrup, Denmark; [3]Elsam A/S, Overgade 45, DK-7000 Fredericia, Denmark; and [4]Sicco K/S, Odinshøjvej 116, DK-3140 Aalsgaarde, Denmark

Abstract

The overall objective in this European Union-project is to develop cost and energy effective production systems for coproduction of bioethanol and electricity based on integrated biomass utilization. A pilot plan reactor for hydrothermal pretreatment (including weak acid hydrolysis, wet oxidation, and steam pretreatment) with a capacity of 100 kg/h was constructed and tested for pretreatment of wheat straw for ethanol production. Highest hemicellulose (C5 sugar) recovery and extraction of hemicellulose sugars was obtained at 190°C whereas highest C6 sugar yield was obtained at 200°C. Lowest toxicity of hydrolysates was observed at 190°C; however, addition of H_2O_2 improved the fermentability and sugar recoveries at the higher temperatures. The estimated total ethanol production was 223 kg/t straw assuming utilisation of both C6 and C5 during fermentation, and 0.5 g ethanol/g sugar.

Index Entries: Lignocellulose; hydrothermal; pretreatment; pilot plant; SSF; bioethanol.

Introduction

The recent years extensive research in bioethanol processes from lignocellulosic materials has brought the process closer to commercialization. A number of pretreatment methods are yet available—dilute acid pretreatment *(1)*, H_2SO_4 or SO_2-catalyzed steam explosion *(2,3)*,

*Author to whom all correspondence and reprint requests should be addressed.

and wet oxidation (4). Furthermore, the cost of employing enzymatic hydrolysis has during the past 3 yr been significantly reduced owing to the research grants from the US Department of Energy to Novozymes Biotech. and Genencore International. However, much of the research is still performed at laboratory scale and many of the promising results needs to be validated in pilot and commercial scale. Larger scale equipment and equipment for running in continuous mode also need to be tested.

Since the year 2000, it has been mandatory for Danish power plants to utilize a total of 1.2×10^6 t of biomass yearly, mainly from straw, for heat and power production. However, owing to the high content of especially potassium in straw, this type of biomass is not well suited for combustion. In a joint EU-project "coproduction biofuels," the largest Danish power producer Elsam A/S, The energy research centre—Risø National Laboratory, The Royal Veterinary and Agricultural University, and two small enter-prices—Sicco K/S and TMO Biotec (Ltd.), are exploring the possibility from combinations of lignocellulosic and starch/sugar feedstocks of merging the production of ethanol with the production of heat and power at combined heat and power plants. The objective is to develop a system that removes the potassium salts and a substantial part of the carbohydrates (mainly hemicellulose) from the straw, to be used as feedstock for ethanol fermentation with TMO's C5 and C6 fermenting thermophiles under development. The solid fraction is essentially free of potassium and therefore well suited for combustion. This cellulose-rich fraction containing the lignin can be burned for electricity or used for ethanol production. Byproducts from fermentation processes will be concentrated and used for animal feed or fertilizer (Fig. 1).

The bioethanol process will be placed at the power plants, giving advantages such as easy access to cheap supply of electricity and steam from the power plants, and to efficient electricity production from residual cellulose and lignin.

The main body of the project is the construction and testing of a pilot scale two-step pretreatment reactor system with a planned capacity of 1000 kg of biomass/h (Fig. 2). The reactor system is designed to be flexible, facilitating the handling of high concentrations of dry matter (exceeding 50%), large particles (e.g., straw more than 5 cm in length), temperatures up to 230°C and different pretreatment methods, e.g., counter-current extraction, diluted acid pretreatment, steam pretreatment, and wet oxidation. It will thus be possible to verify the many results reported from laboratory scale experiments in pilot scale.

This article describes the concept of the pilot plant and some results from the optimisation trials obtained during 2004 and 2005 with a one-step pretreatment reactor (Fig. 3) that will be integrated into the final two-step pretreatment system. The reactor has been tested with a

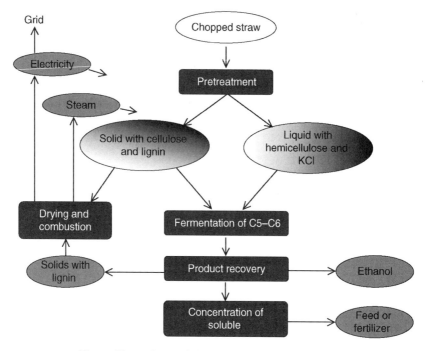

Fig. 1. Flow sheet of the straw-to-ethanol process.

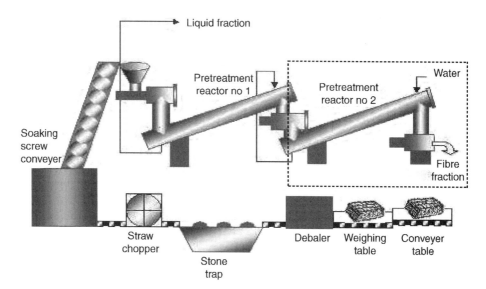

Fig. 2. Flow sheet of the process, as it will appear when the second pretreatment step has been added. The existing plant is a one step pretreatment plant. The liquid fraction from the reactors might either be mixed with the fiber fraction for enzymatic hydrolysis (SHF) or go directly to the fermentation step (SSF).

continuous flow of straw of 50 kg/h (6-min residence time). Several trials were made with varying parameters of water level, chemical addition and flow in the reactor.

Coproduction of Bioethanol and Electricity

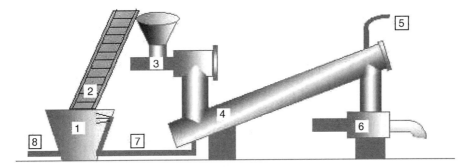

1. Intake of biomass and soaking tank with hot water recycled from reactor
2. Dosing belt conveyor with dewatering screen bottom
3. Particle pumb no.1 for introduction of biomass into the reactor
4. Reactor zone with double screw conveyor system
5. Water inlet into reactor
6. Particle pumb no. 2 for the extraction of biomass out of the reactor
7. Recycling system for process water
8. Drain

Fig. 3. Flow sheet of one-step test reactor.

Description of Pilot Plant

The final pilot plant is based on Sicco's proprietary design and will consist of a complete system for pretreating up to 1000 kg/h of straw and testing of this plant started September 2005 (Fig. 2).

A total of 3–4 big bales (approx 500 kg each) can be placed on the up front conveyer table thereby enabling automatic feeding of straw for up to 2 h. The flow of straw to the plant is controlled by the speed of the dosing belt into the de-baler unit. After the de-baler unit the straw goes through a stone trap to remove sand, stones, and metal parts before the chopping into approx 5 cm pieces in a commercial straw chopper. The cut straw goes through a presoaking step in a 10 m long screw conveyer. The temperature in this step can be up to 100°C and residence time can be up to 30 min. The straw is fed into the first reactor by means of a particle pump designed especially for introduction of biomass into the reactor against high pressure and temperature (5). The first reactor will be a screw conveyor system operated under pressure and temperatures up to 200°C. The conveyor moves the straw up through the reactor and water from the reactor 2 is passed counter currently to obtain leaching of the straw. From the first reactor the straw is let into reactor 2 also by the use of a particle pump. Reactor 2 is similar to reactor 1 but is able to operate at higher pressure and temperatures (up to 32 bar and 230°C). Biomass is removed from reactor 2 by another particle pump and can be used for production of boiler fuel or ethanol. The liquid extracted from reactor 1, which contains compounds such as monomeric sugars, salts, hemicellulose, lignin, and degradation products, can be used for bio-ethanol production by enzymatic hydrolysis and subsequent fermentation.

Table 1
Chemical Composition of Untreated Wheat Straw

	Composition straw [g/100 g DM]	
	Straw batch 1	Straw batch 2
Cellulose	30.4	33.9
Xylan	18.4	20.3
Arabinan	2.9	2.7
Total hemicellulose	21.3	23
Klason lignin	19.4	19.1
Ash	5.9	5.3
Residual	23	18.7

Two different batches of straw was used in the experiments.

Materials and Methods

Raw Materials

Wheat straw (*Triticum aestivum* L.) was grown and harvested after a drying period in Denmark during summer 2003. The straw was cut into 1–6 cm pieces on the field by a New Holland FX375 forage harvester and stored in containers at ambient temperature. The dry matter content was 90–92%. Table 1 shows the chemical composition of the untreated wheat straw.

Pretreatment of Wheat Straw

For these preliminary experiments a pretreatment pilot plant reactor with a capacity of 100 kg/h was used (Fig. 3). The straw was fed to a presoaking vessel continuously at a rate of 50 kg straw/h. The temperature in the presoaking vessel was 80–90°C, and the residence time was around 6 min. After presoaking the biomass was transported to the reactor inlet by a dosing belt conveyor. This conveyor would drain much of the free water from the biomass thereby securing a counter current flow of water against the biomass inside the reactor. Water was introduced at the top of the reactor giving a counter current flow. The pretreated biomass was transported out of the reactor by particle pump no. 2 following the same principle as particle pump no. 1. The pretreated biomass was weighed and collected in containers. Samples were collected and kept at –5°C until analysis. Processed water was recycled in the presoaking vessel and let to a drain. Samples of the liquid fraction were collected from the drain in the bottom of the reactor. Liquid samples were kept at –5°C until analysis. In this series of optimization trials on the pilot plant several experiments were run with varying parameters of water flow (0 L/h (steam), 100 L/h, 250 L/h, and 500 L/h), chemical addition (H_2SO_4, Na_2CO_3, NH_3, and H_2O_2 (wet oxidation), and temperature in the reactor (190°C, 195°C, and 200°C).

Analysis Methods

Analysis of Ash and Dry Matter Content

The ash content in the solid fraction was determined by incineration of around 0.5 g of dried sample at 550°C for 3 h. The dry matter content in the liquid extract was determined by drying 5 mL of sample overnight at 105°C. Dry matter content of the raw material and solid fractions was determined by drying and weighing in a mettler Toledo moisture analyzer HR83, Halogen.

Analysis of Carbohydrates in Solid Fractions

The composition of the raw and pretreated straw fibers was measured by strong acid hydrolysis of the carbohydrates. Dried and milled samples (160 mg) were treated with 72% (w/w) H_2SO_4 (1.5 mL) at 30°C for 1 h. The solutions were diluted with 42 mL of water and autoclaved at 121°C for 1 h. The hydrolysates were filtered, and the Klason lignin content was determined as the weight of the filter cake subtracted the ash content. The filtrates (5 mL) were mixed with 0.5 g $Ba(OH)_2 \cdot 8H_2O$ and after 5 min, the samples were centrifuged at approx $3000g$ for 5 min. The supernatant was analysed for sugars on high-performance liquid chromatography (HPLC). The recovery of D-glucose, D-xylose, and L-arabinose was determined by standard addition of sugars to samples before autoclavation. The sugars were determined after separation on an HPLC-system (Shimadzu) with a Rezex ROA column (Phenomenex) at 63°C using 4 mM H_2SO_4 as eluent and a flow rate of 0.6 mL/min. Detection was done by a refractive index detector (Shimadzu Corp., Kyoto, Japan). Conversion factors for dehydration on polymerization was 162/180 for glucose and was 132/150 for xylose and arabinose.

Analysis of Carbohydrates in Liquid Fractions

Carbohydrates in the liquid (filtrate) after pretreatment were both polymers and monomers, thus the samples were hydrolyzed using 4% (w/w) H_2SO_4 at 121°C for 10 min to determine the total glucose, xylose, and arabinose concentration. The sulphate anions in 10 mL acidic filtrate were precipitated by 0.5 g $Ba(OH)_2 \cdot 8H_2O$ and the supernatant was diluted 1 : 1 with 4 mM H_2SO_4. Glucose, xylose, arabinose, acetic acid, and ethanol were quantified by HPLC as described above.

Enzymatic Hydrolysis of Solid Fraction

The enzymatic conversion of the solid fraction was evaluated at 2% DM in 50 mM sodium acetate pH 4.8 using Cellubrix L added to obtain 30 FPU/g DM. The solid fibers were enzymatically hydrolyzed as milled samples. After 24 h at 50°C, the reaction was stopped by centrifugation at approx $3000g$ for 10 min. The concentration of glucose, xylose, and arabinose was measured by HPLC as described earlier.

Simultaneous Saccharification and Fermentation

Prehydrolysis (liquefaction) and, simultaneous saccharification and fermentation (SSF) was performed in 200-mL fermentation flasks. Eight grams of the dried solid fiber fraction were mixed with 60 mL of a 0.2 M acetate buffer (pH 4.8) or 60 mL of the pH adjusted (pH 4.8) filtrate originated from the same pretreatment (13% DM). Prehydrolysis of the WO solids was performed at 50°C for 24 h at an enzyme loading of 10 FPU/g DM filter cake, using Cellubrix L. After liquefaction, the fermentation flasks were supplemented with a second batch of enzymes (Cellubrix L) at an enzyme loading of 10 FPU/g DM, added 0.2 mL of a sterile filtered urea (24%), and inoculated with 0.2 g yeast after having cooled down to room temperature. The flasks were sealed with a yeast lock filled with glycerol and incubated at 32°C for 6–8 d. The cellulose to ethanol conversion was monitored by CO_2 loss, determined by weighing of the flasks at regular intervals. The final ethanol concentration was also determined by HPLC analysis as described above.

Calculations

Recoveries were calculated according to Eq. 1. Yields were calculated as percent of theoretical (in g/g original cellulose or hemicellulose in raw material) (Eqs. 2 and 3). Total yields were calculated as the total yield of hemicellulose/glucose in the liquid fraction and after enzymatic hydrolysis of the solid fraction (Eq. 4). The theoretical ethanol production based on the pretreatment and hydrolysis yields was calculated according to Eq. 5. The ethanol yield in SSF experiments was calculated as percentage of theoretical based on cellulose content of the fiber fraction and glucose in the filtrates (Eq. 6).

$$\text{Recovery} = \frac{(\text{sugar in filtrate (g / 100 g)} + \text{sugar in solid (g / 100 g)})}{(\text{sugar in raw material (g / 100 g)})} 100\% \quad (1)$$

$$\text{Yield}_{\text{hemicellulose}} = \frac{(\text{mass}_{\text{hemicellulose}} \text{ in filtrate})}{(\text{mass}_{\text{hemicellulose}} \text{ in raw material})} 100\% \quad (2)$$

$$\text{Hydrolysis yield}_{\text{glucose}} = \frac{(\text{mass}_{\text{glucose}} \text{ after enzymatic hydrolysis} \times 0{,}9)}{(\text{mass}_{\text{cellulose}} \text{ in raw material})} 100\% \quad (3)$$

Total yield$_{\text{sugar}}$

$$= \frac{(\text{mass}_{\text{sugar}} \text{ in filtrate}) + (\text{mass}_{\text{sugar}} \text{ after enzymatic hydrolysis})}{(\text{mass}_{\text{sugar}} \text{ in raw material})} 100\% \quad (4)$$

$$\text{Theoretical ethanol production} = \text{TSC*} \times 0.51 + \text{TSH*} \times 0.5 \quad (5)$$

*TSC = Total sugar from cellulose (after pretreatment and enzymatic hydrolysis) (g/100 g raw material)
*TSH = Total sugar from hemicellulose (after pretreatment and enzymatic hydrolysis) (g/100 g raw material)

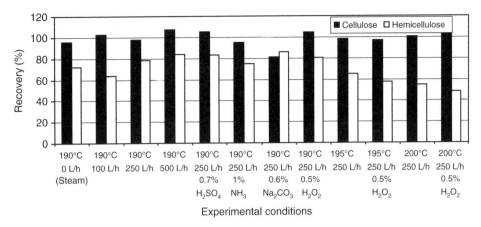

Fig. 4. Recovery of cellulose and hemicellulose in optimization experiments.

$$\text{EtOH yield} = \frac{\text{EtOH}_{\text{Gravimetric/HPLC}}}{\text{glucose in solid} \times 0.51 + \text{glucose in filtrate} \times 0.51} \times 100 \qquad (6)$$

Results and Discussion

The different pretreatments were evaluated with respect to recovery of sugars, depolymerization and extraction of hemicellulose (C5) sugars and cellulose convertibility (C6 sugars). The fermentability was evaluated in SSF fermentations with Baker's yeast. SSF of the cellulose-fraction was performed both in a buffer medium and in the hydrolysate to examine the inhibitory effect of the hydrolysates.

Recovery of Sugars

Very good cellulose recoveries (80–100%) were found in all trials (Fig. 4). The highest hemicellulose recovery (86%) was found in the experiment with Na_2CO_3 addition, but also the experiments with H_2SO_4, high flow (500 L/h), and H_2O_2 (wet oxidation) showed hemicellulose recoveries above 80%. In the experiments with low flow a substantial part of the hemicellulose sugars were lost (36%), and increasing the temperature to 195°C and 200°C also caused significant degradation of the hemicellulose sugars (40–50%) (Fig. 4).

Depolymerisation of Hemicellulose and Extraction of Sugars Into the Liquid

Figure 5 shows the yield of hemicellulose and glucose (in the liquid) after the pretreatment process. Especially, the experiment with H_2SO_4-addition stands out regarding extraction of hemicellulose sugars. In this experiment approx 50% of the hemicellulose sugars were extracted into the liquid. In the next best trials with high flow (500 L/h) and H_2O_2 addition around 30% of the hemicellulose sugars were extracted. The lowest

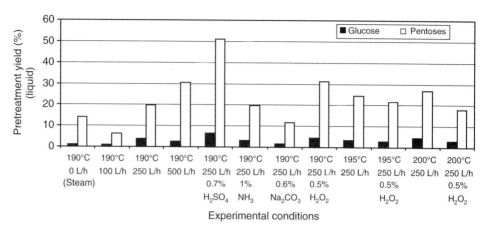

Fig. 5. Yield of glucose and pentoses after pretreatment.

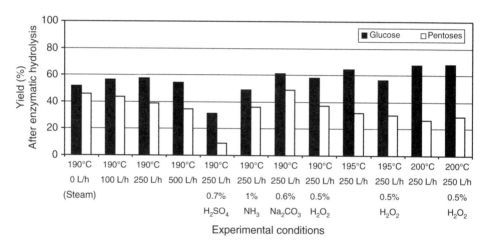

Fig. 6. Yield of glucose and pentoses after enzymatic hydrolysis of the solid fraction.

extraction of hemicellulose sugars was found in the experiment with low flow (100 L/h). By washing the solid (straw) fraction from the experiment with low flow in warm water (80°C), the hemicellulose yield was increased fourfold (data not shown). This showed that washing of the straw in the reactor was insufficient at low water flow and can be improved at higher flow.

Enzymatic Hydrolysis of the Solid (Straw) Fraction

One objective in this process is to use part of/or the entire solid fraction for bioethanol production, so the accessibility of the solid fraction for enzymatic hydrolysis is an important parameter. Figure 6 shows the yields of glucose and hemicellulose sugars after enzymatic hydrolysis of the solid fraction. In the majority of the optimization experiments performed at 190°C the yield of glucose is 50–60%. In the experiment with addition of H_2SO_4 the glucose yield is significantly lower (32%) than in the other experiments, despite the high extraction of hemicellulose sugars found

Table 2
Total Yield of Glucose and Pentoses After Pretreatment and Enzymatic Hydrolysis

Conditions			Yields	
Temperature (°C)	Flow (L/h)	Chemicals (%)	Glucose yield (%)	Pentoses yield (%)
190	0 (steam)	–	52.9	60.5
190	100	–	57.3	49.3
190	250	–	61.4	57.9
190	500	–	56	64.6
190	250	0.7% H_2SO_4	37.8	60
190	250	1% NH_3	52.1	55.5
190	250	0.6% Na_2CO_3	61.8	60.8
190	250	0.5% H_2O_2	62.0	68.3
195	250	–	67.8	56.2
195	250	0.5% H_2O_2	59.2	51.7
200	250	–	71.4	53.9
200	250	0.5% H_2O_2	70.5	47.2

in this experiment. Increasing the reaction temperature improves the convertibility of the cellulose, and the best hydrolysis of the cellulose fraction is found in the experiments performed at 200°C, in which the glucose yield is 67–68% (after 24 h of hydrolysis at 50°C). Also a part of the hemicellulose (left in the straw) was hydrolyzed by the enzymes. This conversion was most significant in the experiment with Na_2CO_3-addition, in which 49% of the hemicellulose was hydrolyzed in the enzymatic process.

Total Sugar Yields

The total yield of glucose was 50–60% in most of the experiments performed at 190°C (Table 2), but was significantly lower in the experiment with H_2SO_4-addition (38%) owing to the poor cellulose convertibility in this experiment. The best glucose yield was obtained in the experiments performed at 200°C. In these experiments the cellulose yield was improved by 10% in comparison with the best result obtained at 190°C.

In the best experiments a total hemicellulose yield of 60–68% was obtained, these are the experiments with H_2O_2, Na_2CO_3, high water flow (500 L/h), and H_2SO_4. It is also in these experiments the best hemicellulose recovery was found (Fig. 4). In the experiment with Na_2CO_3-addition most C5 sugars was obtained after enzymatic hydrolysis of the solid, whereas in the experiment with high water flow almost equal amounts of C5 sugars were obtained by pretreatment (in the liquid) and enzymatic hydrolysis (of the solid). The highest total yield of C5 sugars (68%) was found in the experiment performed at 190°C with H_2O_2 addition (wet oxidation).

Theoretical Ethanol Production

The theoretical ethanol production was calculated based on the total yields of C6 and C5 sugars (Eq. 5, Fig. 7). Because two different batches

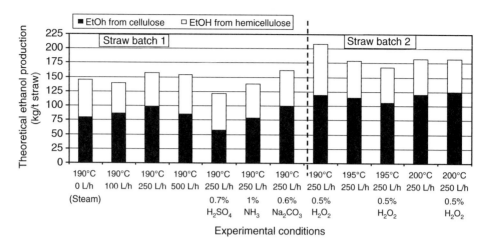

Fig. 7. Theoretical ethanol production based on the yields of C6 and C5 at different pretreatment conditions (Table 2). Two different batches of straw (with different sugar-content) were used in these trials.

of straw (with different sugar contents) were used, the results were not directly comparable. In the best experiment using straw no. 1 (pretreatment at 190°C with Na_2CO_3-addition) the theoretical ethanol production was 162 kg/t straw. Higher theoretical ethanol production was found in the experiments using straw no. 2. owing to the higher sugar content of this batch of straw. The highest theoretical ethanol production of 208 kg/t straw was obtained at 190°C with addition of H_2O_2 (wet-oxidation). For an optimal two-step process as described earlier, using the maximum obtained sugar yields from these trials (68% for hemicellulose and 71% for glucose, Table 2) and straw no. 2, the total theoretical ethanol production has been calculated to 223 kg/t straw.

SSF-Experiments

SSF of the cellulose with Baker's yeast has been performed on material from selected trials to examine the fermentability of the pretreated material (Fig. 8). The SSF were performed both in a buffer medium and in the liquid from the pretreatments (hydrolysates) to test the liquid for possible inhibitors.

In experiments with no addition of H_2O_2 (Fig. 8A), fermentation in the buffer medium takes place after a small lag phase, and very high yields are obtained (close to 100%), showing that under these conditions all of the cellulose can be converted to glucose and fermented to ethanol. Initially, some kind of lag phase is seen, however, 70% the ethanol produced was obtained within the first 50 h, after which the productivity decreased significantly (Fig. 8A). SSF in hydrolysate (instead of buffer medium) showed almost no ethanol production, indicating that the hydrolysates from these experiments contained too many inhibitors for fermentation to take place. A most reasonable explanation is degradation

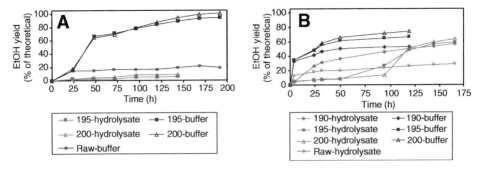

Fig. 8. SSF of solid fraction from pretreatments performed at temperatures ranging from 190–195°C in buffer and hydrolysate media. **(A)** is experiments performed with-out addition of H_2O_2 and **(B)** is experiments performed with addition of H_2O_2 (wet-oxidation).

of C5 sugars to carboxylic acids and furfurals, that are known inhibitors for most microorganisms, but also lignin degradation product have severe inhibition effects *(6)*. Analysis of specific inhibitors was not made in this study.

Using material from experiments with H_2O_2 addition the lag phase in buffer medium was avoided, and the initial productivity of the yeast was faster. Growth starts immediately in hydrolysate treated at 190°C, after approx 50 h in hydrolysate treated at 195°C, and after 94 h in hydrolysate treated at 200°C. The yeast probably uses these long lag phases to detoxify the medium *(6)*, and the more toxic the medium the longer the lag phase. The end yield is highest in fermentation of material pretreated at 200°C, which is owing to the better cellulose convertibility of these materials.

Conclusion and Future Experiments

Optimal depolymerisation and extraction of hemicellulose sugars was obtained at 190°C but improved cellulose conversion of the solid fraction was achieved at 200°C. A water flow of minimum 250 L/h through the reactor was of great importance to obtain optimal extraction of hemicellulose sugars. Addition of Na_2CO_3 and H_2O_2 both improved recoveries of sugars, giving a theoretical ethanol production of 162 kg/t straw with Na_2CO_3 and 208 kg/t straw with H_2O_2, when performed at 190°C. It was demonstrated that H_2O_2 improved the fermentability of the pretreated material and for further experiments a combination of Na_2CO_3 and H_2O_2 should be tested. For an optimal two-step process as suggested earlier, based on the maximum obtained sugar yields from these trials (68% for hemicellulose and 71% for glucose), the expected total ethanol production (theoretical) is 223 kg/t straw.

Acknowledgments

This work is financially supported by the EU contract ENK6-CT-2002-00650. We thank for the technical assistance from Tomas Fernqvist and

Ingelis Larsen from Risø National Laboratory and Joanna Møller Nielsen From the Royal Veterinary and Agricultural University. A special thanks to Erik Hedahl Frank from Elsam for daily operation and maintenance of the pilot plant.

References

1. Schell, D. J., Farmer, J., Newman, M., and McMillan, J. D. (2003), *Appl. Biochem. Biotechnol.* **105–108,** 69–85.
2. Söderström, J., Pilcher, L., Galbe, M., and Zacchi, G. (2002), *Appl. Biochem. Biotechnol.* **98–100,** 5–21.
3. Tengborg, C., Stenberg, K., Galbe, M., et al. (1998), *Appl. Biochem. Biotechnol.* **70–72,** 3–15.
4. Palonen, H., Thomsen, A. B., Tenkanen, M., Schmidt, A. S., and Viikari, U. (2004), *Appl. Biochem. Biotechnol.* **117,** 1–17.
5. Patent WO 03/013714 A1.
6. Klinke, H. B. (2001), PhD. thesis, Technical University of Denmark, Kgs. Lyngby, Denmark.

The Combined Effects of Acetic Acid, Formic Acid, and Hydroquinone on *Debaryomyces hansenii* Physiology

Luís C. Duarte, Florbela Carvalheiro, Joana Tadeu, and Francisco M. Gírio*

INETI, Departamento de Biotecnologia, Estrada do Paço do Lumiar, 22, 1649-038 Lisboa, Portugal; E-mail: francisco.girio@ineti.pt

Abstract

The combined effects of inhibitors present in lignocellulosic hydrolysates was studied using a multivariate statistical approach. Acetic acid (0–6 g/L), formic acid (0–4.6 g/L), and hydroquinone (0–3 g/L) were tested as model inhibitors in synthetic media containing a mixture of glucose, xylose, and arabinose simulating concentrated hemicellulosic hydrolysates. Inhibitors were consumed sequentially (acetic acid, formic acid, and hydroquinone), alongside to the monosaccharides (glucose, xylose, and arabinose). Xylitol was always the main metabolic product. Additionally, glycerol, ethanol, and arabitol were also obtained.

The inhibitory action of acetic acid on growth, on glucose consumption and on all product formation rates was found to be significant ($p \leq 0.05$), as well as formic acid inhibition on xylose consumption and biomass production. Hydroquinone negatively affected biomass productivity and yield, but it significantly increased xylose consumption and xylitol productivity. Hydroquinone interactions, either with acetic or formic acid or with both, are also statistically significant. Hydroquinone seems to partially lessen the acetic acid and amplify formic acid effects. The results clearly indicate that the interaction effects play an important role on the xylitol bioprocess.

Index Entries: Lignocellulosic byproducts; xylitol; interaction effects; *Debaryomyces hansenii*; inhibition.

Introduction

The xylitol bioproduction is a possible valuable alternative for upgrading the pentose-rich hemicellulosic hydrolysate stream arising from lignocellulosic materials pretreatments. However, the xylitol yield and productivity by yeasts (most noteworthy, some *Candida* spp. [1–4] and *Debaryomyces hansenii* [5–7]) are affected by several factors, for example, oxygen availability, and the initial concentrations of biomass, hexoses, pentoses, and inhibitors in the hydrolysates (8). Among those factors,

---

*Author to whom all correspondence and reprint requests should be addressed.

the inhibitors influence is still the less understood, conversely to what happens for the bioethanol process, for which the inhibitor effects are currently under extensive study both in bacteria, for example, *Escherichia coli (9–11)*, *Klebsiella oxytoca (11)*, or *Zymomonas mobilis (12)* and yeasts, for example, *Saccharomyces cerevisiae (13–15)*, *Pichia stipitis (16)*, or *Kluyveromyces marxianus (17,18)*. Unfortunately, much of the data concerning the effect of hydrolysis byproducts on bioethanol production do not fully apply to xylitol bioproduction, not only because different microorganisms are involved and tolerance is strain specific, but also owing to both the hydrolysates and operational conditions are relatively different between the two processes.

The potential inhibitors present in hemicellulosic hydrolysates can be divided into three major categories, aliphatic acids (e.g., acetic and formic acid), phenolic compounds (mostly with low molecular weight) and furan derivatives (furfural and HMF). The xylitol bioprocess requires concentrated hydrolysates because xylitol production is favored at high xylose concentrations. Therefore, a concentration (e.g., evaporation) step before the fermentation is usually performed *(8)*. This step may also act as a detoxification process and it is particularly effective in the reduction of furan derivatives levels *(19,20)*. Although phenolic compounds and, at some extent, aliphatic acids may also be partially evaporated, they usually have their levels increased thus becoming, in quantitative terms, the most important inhibitors present in concentrated hydrolysates.

The study of the inhibitors impact in the xylitol production is usually achieved using two approaches, experiments carried out in hydrolysate media, comparing between hydrolysates with different degrees/processes of detoxification *(19,21–23)* and experiments carried out in synthetic media using single inhibitors selected from the mentioned categories *(4,24–26)*. The first approach is by far the most used and gives relevant information relating the technological impact of the inhibitors. The second approach is more useful to quantify and understand the specific impact of the inhibitors on the microbial physiology. However, it is generally accepted the assumption that the toxicity also depends on the interaction of the inhibitors *(8,20)*, as it has been established for ethanol production *(10,15,17,18)*. Nevertheless, this has not yet been clearly shown for the xylitol bioproduction.

To identify the possible combination effects, a multivariate statistical approach is the best strategy because it enables to easily estimate both the main and the interaction effects *(27)*. Among the many possibilities, we applied a modified central composite design, that, in spite of some drawbacks (e.g., it is a nonorthogonal and nonrotatable design), has already been successfully applied to study the inhibitor interaction problem for the bioethanol process.

In this work, we studied the interaction effects induced by selected hydrolysis byproducts in *D. hansenii* semi-aerobic growth and metabolism. The compounds were chosen to be representative of the potential microbial inhibitors usually present in concentrated hydrolysates, namely, from the aliphatic acids, acetic, and formic acid, and from the phenolic compounds, hydroquinone. Their maximal concentrations were chosen based on the range of concentrations previously found in hemicellulosic hydrolysates.

Methods

Microorganism and Growth Conditions

Microorganism

D. hansenii CCMI 941 *(28)* obtained from Colecção de Culturas de Microrganismos Industriais (CCMI, Portugal), was used in all experiments. To increase inocula reproducibility a yeast stock culture was prepared as follows: a 24 h grown slant was inoculated in a synthetic medium *(26)* containing 10 g/L of xylose as sole carbon and energy source and incubated in the same conditions as described as follows. At late exponential growth phase sterile glycerol was added to a final concentration of 15% (vol/vol). 5 mL aliquots were then preserved in sterile vials at –70°C.

Media

Complete chemically defined media were used in all experiments. Inoculum medium contained 5 g/L D-glucose, 15 g/L D-xylose, and 5 g/L L-arabinose, as carbon sources and no inhibitors, as described elsewhere *(26)*.

In order to determine the main and interaction effects induced by acetic acid, hydroquinone, and formic acid, a central composite design *(15,18)* was used. The codified values for the inhibitor concentrations used are shown in Table 1. The experimental domain range between 0 and 6 g/L, 3 g/L and 4.6 g/L for acetic acid (Merck, Darmstadt, Germany), hydroquinone (Sigma Chemical Co., St. Louis, MO), and formic acid (Riedel-de Haën, Seelze, Germany), respectively.

This design was extended to incorporate replicates for the reference fermentations (fermentations without inhibitors) in each block, and also replicates of the central condition.

The growth media sugar composition aims to simulate the composition of concentrated pentoses-rich hemicellulosic acid hydrolysates. The reference medium contained per liter: 20 g D-glucose, 60 g D-xylose, 20 g L-arabinose, 24 g $(NH_4)H_2PO_4$, 8 g $(NH_4)_2HPO_4$, 4 g KH_2PO_4, and trace elements and vitamins in a double concentration of that described previously *(26)*. The phosphate sources were slightly modified *(26)* in order to increase the medium buffering capacity. Media pH was set to 5.5 before filter sterilization with 0.22 μm Gelman membrane filters (Ann Arbor, MI). To obtain the initial concentrations of inhibitors established by the experimental design, three similar media, supplemented either with 15 g/L acetic acid, 7.5 g/L hydroquinone, or 17 g/L sodium formate (11.5 g/L formic acid), freshly

Table 1
Codified Levels for the Modified Central Composite Design for the Factors
Acetic Acid (HAc), Hydroquinone (HQ), and Formic Acid (HF)

Block	Experiment	Factors		
		HAc	HQ	HF
I	1	−1	−1	−1
	2	−1	−1	−1
	3	−0.6	−0.6	−0.6
	4	0.6	−0.6	−0.6
	5	−0.6	−0.6	0.6
	6	0.6	−0.6	0.6
	7	−0.6	0.6	−0.6
	8	0.6	0.6	−0.6
	9	−0.6	0.6	0.6
	10	0.6	0.6	0.6
	11	0	0	0
II	1	−1	−1	−1
	2	−1	−1	−1
	3	−1	0	0
	4	1	0	0
	5	0	0	−1
	6	0	0	1
	7	0	−1	0
	8	0	1	0
	9	0	0	0
	10	0	0	0
	11	0	0	0

prepared, were mixed with the reference medium in appropriated volumes under sterile conditions.

Growth Conditions

The inoculum was prepared in 1000-mL baffled Erlenmeyer flasks capped with cotton wool stoppers containing 100 mL of inoculum medium and incubated in an Infors® Unitron (Bottmingen, Switzerland) orbital incubator set at 30°C and 150 rpm. Each flask was seeded with 0.5 mL of yeast stock suspension. After 17 h the culture was centrifuged at 8600g under sterile conditions in a Sartorius Sigma 2-16K centrifuge (Götingen, Germany). The cell mass was used to inoculate 55 mL of the different media to an initial cell dry weight of about 2.5 g/L. The experiments were carried out aerobically in 500-mL Erlenmeyer flasks also capped with cotton wool stoppers and incubated in the same conditions as the inoculum. The yeast growth was followed for 168 h for each condition established by the modified central composite design. Samples were taken at regular intervals. Experiments within the blocks were carried out simultaneously.

Analytical Methods

D-xylose, D-glucose, L-arabinose, formic and acetic acids, ethanol, and hydroquinone were analyzed by high-performance liquid chromatography (HPLC) using an Aminex HPX-87H column from Bio-Rad (Hercules, CA) in the same conditions and equipment described previously (26). Owing to the partial overlap of arabinose, xylitol, and arabitol, samples were also analyzed by HPLC using a Waters Sugar Pak 1 column (Millfort, MA) as described earlier (26). All samples were filtered by 0.45 µm Gelman membrane filters before analysis.

Biomass dry weight was quantified gravimetrically by centrifugation of 2 mL of culture broth at 20,600g in a Sartorius Sigma 2-16K centrifuge (Göttingen, Germany), using predried Eppendorf tubes. The resulting cell pellet was washed with an equal amount of filtered deionized water and dried overnight at 100°C to constant weight.

Calculations and Statistical Analysis

The volumetric substrate consumption rates (g/[Lh]), were based on grams of substrate consumed per liter of culture medium/h. The biomass and product productivities (g/[Lh]), were based on grams of biomass or product produced/liter of culture medium/h. The biomass and product yield (g/g), were calculated as the ratio between the productivity and the volumetric consumption rate for the relevant substrates. The values were calculated for the duration of maximum relevant monosaccharide consumption rate, i.e., parameters related to glycerol, ethanol, and biomass were calculated for the sample with higher glucose consumption rate. For xylitol and arabitol parameters the calculations were performed for the sample with higher xylose and arabitol consumption rate, respectively. The specific growth rate (μ, h^{-1}) was calculated by linear regression of the $\ln(OD/OD_i)$ vs time for the exponential growth phase.

In order to describe the influence of the factors on the dependent variables (the response variables), a second order polynomial model was used.

$$Y = \beta_0 + \beta_1 x_1 + \beta_2 x_2 + \beta_3 x_3 + \beta_{12} x_1 x_2 + \beta_{13} x_1 x_3 + \beta_{23} x_2 x_3 \\ + \beta_{11} x_1^2 + \beta_{22} x_2^2 + \beta_{33} x_3^2 + \beta_{123} x_1 x_2 x_3 + \varepsilon \qquad (1)$$

where Y is the response, the subscripts 1, 2, and 3 are referred to acetic acid, hydroquinone, and formic acid, respectively), β_0 is the specific intercept; x_1, x_2, and x_3 are the concentrations (g/L) of the factors 1, 2, and 3, respectively; β_1, β_2, and β_3 are the coefficients of the factors 1, 2, and 3, respectively; β_{12}, β_{13}, and β_{23} are the coefficients for the interactions between the factors 1 and 2; 1 and 3; and 2 and 3, respectively; β_{11}, β_{22}, and β_{33} are the coefficients for the quadratic effects of the factors 1, 2, and 3, respectively; and β_{123} is the coefficient for the interaction between the factors 1, 2, and 3; ε are independent random errors, assumed to be normally and independently distributed with expectation 0 and variance σ^2. The model was built on the assumption that there is no interaction between the regressor variables and the block effects. This means that all parameters are common to both blocks. Because the design used is not orthogonal, the regression model was fitted by the forward stepwise procedure implemented in STATISTICA (data analysis software system, version 6) from StatSoft, Inc. (2002).

Results and Discussion

Fermentation Kinetics of D. hansenii

Hemicellulosic hydrolysates used for xylitol production contain a mixture of hexoses and pentoses, which induce quite different microbial performances as in comparison with single substrate cultures *(29–32)*. Therefore, a synthetic culture medium that simulates the hydrolysate sugars composition must be used to investigate the effect of the inhibitor compounds. Otherwise, the true inhibitors impact on the monosaccharides metabolism might be unnoticed or misapprehended.

As examples of the fermentation kinetics obtained, the biomass, pH, monosaccharides, and metabolic product profiles during the fermentation time course for the reference medium and the medium containing all inhibitors in their central concentrations are presented in Figs. 1 and 2, respectively. Because no significant differences between the replicates were found, only the results for one of the assays is presented.

In the reference condition (Fig. 1), glucose was the preferred monosaccharide, being depleted from the culture medium in less than 24 h. Biomass productivity is high during glucose metabolism and after its depletion significantly decreases. As a consequence of glucose metabolism, ethanol, and some glycerol were produced, together with residual amounts (up to 0.4 g/L) of acetic acid. These products are typical for sub-optimum glucose metabolism and after glucose depletion, they were consumed by the yeast together with xylose *(33)*. Xylose utilization began after a short lag phase and it is consumed simultaneously with glucose, albeit at a slower rate. Xylose consumption induced the production of xylitol, arabitol, and residual amounts of acetate. Xylitol was the main product, presenting an higher production rate, and its accumulation occurs up to xylose depletion. Arabinose consumption slowly starts after xylose concentration decreases to ca. 10 g/L but increased substantially after xylose depletion. During late arabinose consumption, xylitol was also consumed. During this period, arabitol production rate increased being accumulated up to arabinose depletion, after which it is also metabolized.

In the medium containing inhibitors (Fig. 2), the overall monosaccharides consumption and products formation profiles were similar to the reference fermentations. Nevertheless, some specific changes occurred. Glucose and xylose consumption rates decreased, but arabinose consumption rate increased. The biomass and the broth pH profiles were considerably different to the reference ones. Growth rate was slower, although growth

Inhibitors Interaction on D. hansenii Physiology

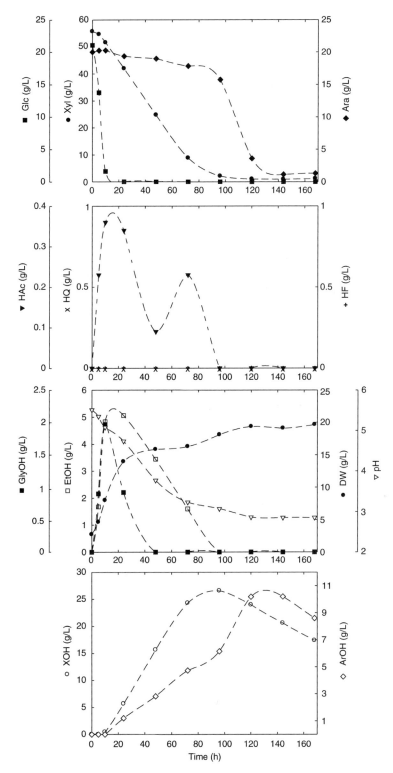

Fig. 1. Time-course of pH, biomass, monosaccharides, and metabolic products during the growth of *D. hansenii* CCMI 941 in the reference medium (Exp. I.1).

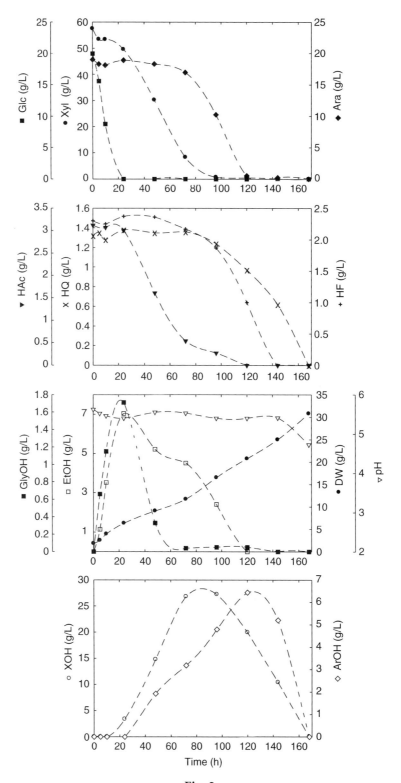

Fig. 2.

was sustained for a longer period, so that final biomass concentrations were higher than in the reference fermentation. This may be a consequence of the broth pH, that does not decrease so sharply, being maintained in values greater than 4.5 for the entire experiment.

After glucose depletion, acetic acid was consumed simultaneously with xylose. Formic acid consumption began by the end of acetic acid consumption, and its consumption rate increased after acetic acid depletion. Hydroquinone was metabolized simultaneously with formic acid, but at a lower rate, being the last inhibitor to be depleted.

These kinetic profiles show similar trends as described for this yeast when grown in brewery's spent grains hemicellulosic hydrolysates *(19,33)*.

The overall results for the modified central composite design as a function of the real inhibitor concentrations used in each assay are summarized in Table 2, in which the calculated monosaccharide and inhibitor consumption rates, product productivities and yields are presented, together with the specific growth rate for exponential phase and the broth pH at the maximum xylitol concentration.

Regardless the concentration, the inhibitors always induced a marked effect on the initial fermentation period, reducing the glucose metabolism and the rates of glucose consumption (Q_{glc}), product formation (Q_{glyOH}, Q_{etOH}), and the carbon assimilation into biomass (μ, Q_x, Y_x), clearly indicating a change in the carbon flux to maintenance processes. Regarding pentose metabolism, the inhibitor effects on xylose consumption are more complex. Although some mixtures did not seem to affect, (some even favor), xylose consumption, the majority tends to reduce it. Both biomass formation and yield from xylose were also decreased. Conversely, a different trend was observed for xylitol production and yield, that are stimulated in many of the inhibitor mixtures assayed, clearly indicating that a carbon flux shift from biomass to xylitol production occured. Similar effects, although not so marked, can also be found for arabinose and arabitol metabolism. Because arabitol is also produced during xylose metabolism, the reported values for arabitol yield from arabinose may have been slightly overestimated. The inhibitor consumption rates were much lower than monosaccharide consumption rates. Amongst the inhibitors, acetic acid always displayed the highest consumption rate and hydroquinone the lowest.

The Influence of Acetic Acid, Hydroquinone, and Formic Acid on D. hansenii Physiology

As stated earlier no significant differences were found among the replicates, both in the reference fermentations and in the central point.

Fig. 2. *(Opposite page)* Time-course of pH, biomass, monosaccharides, and metabolic products during the growth of *D. hansenii* CCMI 941 in a medium containing the central concentrations of all inhibitors (Exp. II.10).

Table 2
Experimental Matrix and Kinetic and Stoichiometric Parameters for *D. hansenii* Growth in Synthetic Media According to the Modified Central Composite Design[a]

Block	Exp.	Factors HAc (g/L)	HQ (g/L)	HF (g/L)	μ (h⁻¹)	pH	Q_{glc}	Q_{xyl}	Q_{ara}	Q_x	Q_{glyOH}	Q_{etOH} (g/[Lh])	Q_{xOH}	Q_{x_xOH}	Q_{arOH}	Q_{HAc}	Q_{HQ}	Q_{HF}	Y_x	Y_{glyOH}	Y_{etOH} (g/g)	Y_{xOH}	Y_{x_xOH}	Y_{arOH}
I	1	0.00	0.00	0.00	0.14	3.23	1.94	0.65	0.13	0.53	0.20	0.48	0.34	0.19	0.07[b]	0.00	0.00	0.00	0.27	0.10	0.25	0.52	0.19	0.54
	2	0.00	0.00	0.00	0.14	3.16	2.00	0.66	0.03[b]	0.64	0.21	0.51	0.34	0.20	0.05	0.00	0.00	0.00	0.32	0.11	0.26	0.53	0.21	1.79[b]
	3	1.11	0.64	0.97	0.12	4.71	1.72	0.68	0.16	0.38	0.16	0.45	0.41	0.19	0.10	0.01	0.00	0.01	0.22	0.09	0.26	0.60	0.20	0.62
	4	5.02	0.64	0.96	0.07	5.47	1.53	0.67	0.16	0.17	0.07	0.30	0.32	0.13	0.10	0.05	0.00	0.01	0.11	0.05	0.19	0.48	0.13	0.63
	5	1.29	0.64	3.50	0.11	5.33	1.72	0.59	0.16	0.28	0.12	0.41	0.35	0.15	0.10	0.01	0.00	0.02	0.16	0.07	0.24	0.60	0.17	0.61
	6	5.10	0.65	3.55	0.06	5.67	0.92	0.56	0.15	0.11	0.09	0.25	0.30	0.13	0.08	0.05	0.00	0.02	0.12	0.09	0.28	0.54	0.14	0.54
	7	1.26	2.40	0.97	0.09	4.99	1.63	0.72	0.15	0.20	0.14	0.51	0.45	0.17	0.03	0.02	0.01	0.01	0.12	0.09	0.31	0.62	0.14	0.21
	8	5.67	1.65	1.09	0.08	5.58	0.91	0.73	0.19	0.14	0.05	0.28	0.38	0.16	0.09	0.07	0.01	0.01	0.15	0.05	0.31	0.52	0.15	0.47
	9	1.31	2.37	3.40	0.08	5.19	1.16	0.62	0.13	0.14	0.10	0.34	0.34	0.11	0.03	0.01	0.00	0.02	0.12	0.09	0.29	0.54	0.12	0.22
	10	5.08	2.45	3.53	0.06	5.60	1.35[b]	0.55	0.13	0.13	0.10	0.40	0.25	0.09	0.04	0.04	0.01	0.02	0.10	0.08	0.30	0.49	0.10	0.29
	11	3.20	1.50	2.25	0.09	5.34	1.21	0.70	0.16	0.13[b]	0.09	0.32	0.40	0.14	0.07	0.04	0.01	0.02	0.11[b]	0.08	0.26	0.57	0.13	0.42
II	1	0.00	0.00	0.00	0.15	3.15	1.81	0.71	0.13	0.51	0.24	0.57	0.31	0.22	0.05	0.00	0.00	0.00	0.28	0.13	0.32	0.43	0.22	0.41
	2	0.00	0.00	0.00	0.15	3.03	1.79	0.73	0.13	0.56	0.23	0.54	0.31	0.23	0.04	0.00	0.00	0.00	0.31	0.13	0.30	0.43	0.22	0.32
	3	0.00	1.69	2.85	0.10	5.07	1.53	0.79	0.15	0.34	0.18	0.45	0.39	0.16	0.05	0.00	0.01	0.02	0.22	0.11	0.29	0.49	0.12	0.30
	4	6.50	1.43	2.38	0.06	5.79	0.85	0.63	0.13	0.18	0.06	0.32	0.33	0.12	0.08	0.06	0.01	0.01	0.18	0.08	0.38	0.53	0.13	0.52
	5	3.76	1.68	0.00	0.07	5.56	1.46	0.91	0.27	0.24	0.16	0.38	0.46	0.19	0.07	0.06	0.01	0.00	0.16	0.11	0.26	0.50	0.13	0.32
	6	3.14	1.38	4.57	0.08	5.71	0.89	0.59	0.13	0.17	0.10	0.29	0.34	0.14	0.06	0.03	0.01	0.03	0.17	0.11	0.33	0.57	0.16	0.35
	7	3.79	0.00	2.81	0.09	5.79	1.35	0.60	0.17	0.29	0.13	0.37	0.22	0.20	0.08	0.04	0.00	0.02	0.21	0.10	0.27	0.37	0.22	0.52
	8	3.03	2.74	2.21	0.05[b]	5.51	0.82	0.62	0.13	0.15	0.07	0.30	0.27	0.11	0.03	0.03	0.01	0.01	0.15	0.08	0.37	0.44	0.12	0.30
	9	3.17	1.44	2.31	0.09	5.49	1.17	0.71	0.15	0.20	0.13	0.37	0.38	0.15	0.05	0.04	0.01	0.02	0.17	0.11	0.32	0.54	0.15	0.33
	10	3.12	1.35	2.30	0.08	5.53	1.12	0.68	0.15	0.22	0.11	0.35	0.37	0.14	0.05	0.04	0.01	0.02	0.19	0.10	0.31	0.55	0.14	0.35
	11	3.17	1.42	2.35	0.08	5.51	1.12	0.68	0.15	0.21	0.12	0.36	0.35	0.15	0.07	0.04	0.01	0.02	0.19	0.11	0.32	0.51	0.16	0.47

[a]Values were calculated for the duration of maximum relevant monosaccharide consumption rate.
[b]Value was omitted from analysis.

μ, specific growth rate; pH, broth pH at the maximum xylitol concentration; Q_{glc}, volumetric glucose consumption rate; Q_{xyl}, volumetric xylose consumption rate; Q_{ara}, volumetric arabinose consumption rate; Q_x, volumetric biomass production rate; Q_{glyOH}, volumetric glycerol production rate; Q_{etOH}, volumetric ethanol production rate; Q_{xOH}, volumetric xylitol production rate; Q_{x_xOH}, volumetric biomass production rate at maximum xylitol productivity; Q_{arOH}, volumetric arabitol production rate; Q_{HAc}, volumetric acetic acid consumption rate; Q_{HQ}, volumetric hydroquinone consumption rate; Q_{HF}, volumetric formic acid consumption rate; Y_x, biomass yield on consumed monosaccharides at maximum xylitol productivity; Y_{arOH}, arabitol yield on arabinose.

Parameter values differ typically less than 10% independently of the block. Therefore, the statistical treatment was simplified by assuming that the intercept (β_0) is equal in all the blocks. There were minor differences between the values established by the design and the real inhibitor concentrations, but when using the correct natural variables *(15,18)* instead of the codified ones, the possible impacts that this may impose on the regression is negligible.

Statistical validation of the polynomial equations was made by analysis of variance. Model analysis by the coefficient of multiple determination (R^2) was also performed. Table 3 shows the regression results for the physiological parameters for which a significant regression could be found. ($p \leq 0.05$ and $R^2 > 0.8$). For simplicity, only statistically significant regression coefficients estimates ($p \leq 0.05$) are presented. The pure quadratic terms (β_{HAcHAc}, β_{HQHQ}, and β_{HFHF}), even if statistically significant, were omitted both from the table and the discussion. It is important to state that the forward stepwise procedure used for the regression is an iterative algorithm, that does not give unique definite solutions for the estimates. Thus, it is preferable to interpret the factors inclusion/exclusion from the model and their signal, but not their magnitude. For the 20 kinetic and stoichiometric parameters studied, only for arabitol productivity and product yields, it was not possible to identify significant effects of the assayed compounds, because it was not possible to find a significant regression model for these parameters. For those parameters for which it was possible to establish a model, a good agreement between experimental data for the reference fermentations (*see* Table 2) and the β_0 estimates was found, together with high correlation coefficients.

The inhibitory effects observed on the initial fermentation period can be partially explained by the presence of acetic acid. Essentially, this aliphatic acid negatively affected glucose consumption rate, specific growth rate, and product formation, both for products deriving from glucose metabolism and xylitol. Arabinose metabolism was not significantly affected, which may be explained by the fact that acetic acid is usually depleted when arabinose metabolism sets on. The negative impact of acetic acid on xylitol bioproduction in semisynthetic medium had already been described, although acetic acid could also improve both the xylitol yield and productivity for concentrations up to 1 g/L *(24)*, a value below the lower concentration assayed in this work. Recently, the acetic acid direct impact on intracellular xylose metabolism was evaluated *(25)* and it was found that it does not have a direct influence on the xylose metabolism specific enzymes. These, together with the results presented here, that suggest an unspecific cellular target for acetic acid, support the assumption that acetic acid inhibition occurs by means of a combined effect of a intracellular pH decrease and anions accumulation arising by the undissociated acid form diffusion across cell membrane *(34)*.

Table 3
Statistically Significant Regression Coefficients Estimates (± Standard Error) for the Polynomial Model for the Specific Growth Rate, Broth pH at Maximum Xylitol Productivity and Kinetic Parameters for *D. hansenii* Growth in Synthetic Media Containing Inhibitors According to the Modified Central Composite Design ($p \leq 0.05$)[a]

	β_0	β_{HAc}	β_{HQ}	β_{HF}	$\beta_{HAc\,HQ}$	$\beta_{HAc\,HF}$	$\beta_{HQ\,HF}$	$\beta_{HAc\,HQ\,HF}$	R^2
μ	0.149 ± 0.003	−0.018 ± 0.002	−0.042 ± 0.008		0.006 ± 0.001				0.98
pH	3.150 ± 0.118	0.454 ± 0.090	0.960 ± 0.236	0.511 ± 0.120					0.98
Q_{glc}	1.824 ± 0.054	−0.092 ± 0.024					−0.050 ± 0.022		0.89
Q_{xyl}	0.702 ± 0.026		0.197 ± 0.052	−0.078 ± 0.026				−0.006 ± 0.002	0.84
Q_{ara}	0.131 ± 0.008			−0.018 ± 0.007	0.015 ± 0.004			−0.007 ± 0.002	0.89
Q_x	0.538 ± 0.022	−0.082 ± 0.015	−0.109 ± 0.019						0.95
Q_{glyOH}	0.221 ± 0.013	−0.027 ± 0.004					−0.013 ± 0.004		0.86
Q_{etOH}	0.550 ± 0.015	−0.075 ± 0.011					−0.037 ± 0.009	0.007 ± 0.002	0.96
Q_{xOH}	0.312 ± 0.010	−0.020 ± 0.003	0.204 ± 0.020				−0.024 ± 0.004		0.95
Q_{x_xOH}	0.218 ± 0.010	−0.012 ± 0.003					−0.013 ± 0.003		0.81
Y_{x_xOH}	0.226 ± 0.009		−0.092 ± 0.019		0.009 ± 0.003				0.86
Q_{HAc}		0.013 ± 0.001			0.003 ± 0.001			−0.001 ± 0.000	0.99
Q_{HQ}			0.011 ± 0.002						0.87
Q_{HF}				0.009 ± 0.001					0.99

[a]β_{HAcHAc}, β_{HQHQ} and β_{HFHF}, even if statistically significant, were omitted both from the table and the discussion. Missing values correspond to statistically nonsignificant coefficients.

Formic acid had a quite diverse inhibitory pattern, negatively affecting xylose consumption and the initial biomass production. Conversely to previous reports (26,35), in which formic acid was described to enhance xylitol production in chemically defined media and at similar initial pH, in this work, formic acid per se had no statistically significant effect on xylitol productivity. This may be owing to the observed dissociation between xylose and formic acid consumption, suggesting that formic acid effect on xylose consumption may be owing to an indirect effect of acid inhibition on biomass production. Furthermore, an unspecific cell uptake and its related effects in a similar way as described for acetic acid, does not seem to occur, because formic acid concentrations remained almost unchanged for long periods and only decreased after acetic acid depletion. This is probably owing to the lower concentrations of the undissociated formic acid form at the fermentation pH.

Both aliphatic acids had a positive impact on broth pH, which is probably a direct consequence of buffering capacities, as consequence of their weak acid nature. However, the coefficient for the acids interaction was not statistically significant to explain any response, given that this factor was never incorporated in the explanatory model for any parameter (Table 3). Hydroquinone inhibited both the specific growth rate and the initial biomass productivity and statistically it did not significantly influence glucose or arabinose metabolism. Conversely, for values around 1.5 g/L hydroquinone specifically increased both xylose consumption, and xylitol productivity. Hydroquinone and acetic acid interaction had a significant positive impact on growth, arabinose consumption, and biomass yield. The interactions between hydroquinone and formic acid had a negative impact on glucose metabolism and all productivities. Hydroquinone seems to partially lessen the acetic acid and amplify formic acid effects. The effects of phenolic compounds on xylitol production is still poorly known, being reported that a phenolic content reduction in hydrolysates could increase both xylitol yield and productivity (20). On the contrary, their effect on yeast ethanol production from xylose has been studied and these compounds have been described to inhibit ethanol production, mainly owing to growth inhibition (10). Nevertheless, some evidences have been reported that lower xylitol productivities and/or yields can be found for some detoxification treatments that exhibited a higher removal of such compounds (19,36). The inhibitory mechanism of the phenolic compounds is considered to be related to a partition and loss of integrity of biological membranes, thereby affecting their ability to serve as selective barriers and enzyme matrices. However, these effects can be reverted if an energy source is available (37). This may explain why xylose consumption increased, but not its assimilation (hydroquinone negatively affects biomass yield). The observed increase in xylitol production would then be a consequence of overflow metabolism occurring at limited oxygen conditions. Furthermore, this membrane dysfunction might be helpful to diminish the stress induced by aliphatic acids. Formic acid and hydroquinone consumption are mainly dependent on their concentration. Conversely, acetic acid consumption rate is increased by hydroquinone, highlighting the possible mechanism involved in the positive interaction that hydroquinone exerts on lessen acetic acid inhibition.

Conclusions

Although the main effects induced by aliphatic acids and hydroquinone are important, the results obtained in this study highlight that the interactions among these compounds are essential to explain the overall performance, specifically the interactions involving hydroquinone. Furthermore, it was observed that the impacts induced by these compounds are not only in the induction of inhibitory effects. Remarkably, hydroquinone per se induced some positive effects, namely, on xylose consumption and xylitol production. Furthermore, hydroquinone and acetic acid combinations seem to limit the inhibitory impact of acetic acid. Finally, this work confirmed the existence of interaction effects between phenolic compounds and aliphatic acids, which overlay the need of further studies on the nature of the inhibitors interaction and on the microbial response mechanisms involved.

Acknowledgments

The authors wish to thank Amélia Marques, Carlos Barata, and Céu Penedo for their technical support, and Francisco Saias and J. C. Roseiro for their helpful discussions.

References

1. Roberto, I. C., Felipe, M. G. A., Lacis, L. S., Silva, S. S., and Mancilha, I. M. (1991), *Bioresour. Technol.* **36**, 271–275.
2. Vandeska, E., Amartey, S., Kuzmanova, S., and Jeffries, T. W. (1995), *World J. Microbiol. Biotechnol.* **11**, 213–218.
3. Oh, D. K., Kim, S. Y., and Kim, J. H. (1998), *Biotechnol. Bioeng.* **58**, 440–444.
4. Saha, B. C. and Bothast, R. J. (1999), *J. Ind. Microbiol. Biotechnol.* **22**, 633–636.
5. Roseiro, J. C., Peito, M. A., Gírio, F. M., and Amaral-Collaço, M. T. (1991), *Arch. Microbiol.* **156**, 484–490.
6. Domínguez, J. M. (1998), *Biotechnol. Lett.* **20**, 53–56.
7. Converti, A. and Domínguez, J. M. (2001), *Biotechnol. Bioeng.* **75**, 39–45.
8. Parajó, J. C., Domínguez, H., and Domínguez, J. M. (1998), *Bioresour. Technol.* **66**, 25–40.
9. Zaldivar, J. and Ingram, L. O. (1999), *Biotechnol. Bioeng.* **66**, 203–210.
10. Zaldivar, J., Martínez, A., and Ingram, L. O. (2000), *Biotechnol. Bioeng.* **68**, 524–530.
11. Gutierrez, T., Buszko, M. L., Ingram, L. O., and Preston, J. F. (2002), *Appl. Biochem. Biotechnol.* **98–100**, 327–340.
12. Ranatunga, T. D., Jervis, J., Helm, R. F., McMillan, J. D., and Hatzis, C. (1997), *Appl. Biochem. Biotechnol.* **67**, 185–198.
13. Larsson, S., Palmqvist, E., Hahn-Hägerdal, B., et al. (1999), *Enzyme Microbiol. Technol.* **24**, 151–159.

14. Larsson, S., Quintana-Sáinz, A., Reimann, A., Nilvebrant, N. O., and Jönsson, L. J. (2000), *Appl. Biochem. Biotechnol.* **84–6**, 617–632.
15. Palmqvist, E., Grage, H., Meinander, N. Q., and Hahn-Hägerdal, B. (1999), *Biotechnol. Bioeng.* **63**, 46–55.
16. Fenske, J. J., Griffin, D. A., and Penner, M. H. (1998), *J. Ind. Microbiol. Biotechnol.* **20**, 364–368.
17. Oliva, J. M., Ballesteros, I., Negro, M. J., Manzanares, P., Cabanas, A., and Ballesteros, M. (2004), *Biotechnol. Prog.* **20**, 715–720.
18. Oliva, J. M., Ballesteros, I., Negro, M. J., Manzanares, P., and Ballesteros, M. (2004), In: *26th Symposium on Biotechnology for Fuels and Chemicals*, Finkelstein, M. and Davison, B., eds., pp. 163.
19. Carvalheiro, F., Duarte, L. C., Lopes, S., Parajó, J. C., Pereira, H., and Gírio, F. M. (2005), *Process Biochem.* **40**, 1215–1223.
20. Mussatto, S. I. and Roberto, I. C. (2004), *Bioresour. Technol.* **93**, 1–10.
21. Mussatto, S. I., Santos, J. C., and Roberto, I. C. (2004), *J. Chem. Technol. Biotechnol.* **79**, 590–596.
22. Rodrigues, R. C. L. B., Felipe, M. G. A., Silva, J. B. A. E., and Vitolo, M. (2003), *Process Biochem.* **38**, 1231–1237.
23. Mancilha, I. M. and Karim, M. N. (2003), *Biotechnol. Prog.* **19**, 1837–1841.
24. Felipe, M. G. A., Vieira, D. C., Vitolo, M., Silva, S. S., Roberto, I. C., and Mancilha, I. M. (1995), *J. Basic Microbiol.* **35**, 171–177.
25. Lima, L. H. A., Felipe, M. G. A., Vitolo, M., and Torres, F. A. G. (2004), *Appl. Microbiol. Biotechnol.* **65**, 734–738.
26. Duarte, L. C., Carvalheiro, F., Neves, I., and Gírio, F. M. (2005), *Appl. Biochem Biotechnol.* **121**, 413–425.
27. Montgomery, D. C. (1997), Design and analysis of experiments, John Wiley & Sons, Inc., New York.
28. Amaral-Collaço, M. T., Gírio, F. M., and Peito, M. A. (1989), In: *Enzyme Systems for Lignocellulosic Degradation*, Coughlan, M. P., ed., Elsevier Applied Science, London, pp. 221–230.
29. Oh, D. K. and Kim, S. Y. (1998), *Appl. Microbiol. Biotechnol.* **50**, 419–425.
30. Tavares, J. M., Duarte, L. C., Amaral-Collaço, M. T., and Gírio, F. M. (2000), *Enzyme Microbiol. Technol.* **26**, 743–747.
31. Gírio, F. M., Amaro, C., Azinheira, H., Pelica, F., and Amaral-Collaço, M. T. (2000), *Bioresour. Technol.* **71**, 245–251.
32. Sánchez, S., Bravo, V., Castro, E., Moya, A. J., and Camacho, F. (2002), *J. Chem. Technol. Biotechnol.* **77**, 641–648.
33. Duarte, L. C., Carvalheiro, F., Lopes, S., Marques, S., Parajó, J. C., and Gírio, F. M. (2004), *Appl. Biochem. Biotechnol.* **113–116**, 1041–1058.
34. Pampulha, M. E. and Loureiro-Dias, M. C. (1989), *Appl. Microbiol. Biotechnol.* **31**, 547–550.
35. Granström, T. and Leisola, M. (2002), *Appl. Microbiol. Biotechnol.* **58**, 511–516.
36. Mussatto, S. I. and Roberto, I. C. (2004), *Biotechnol. Prog.* **20**, 134–139.
37. Heipieper, H. J., Keweloh, H., and Rehm, H. J. (1991), *Appl. Environ. Microbiol.* **57**, 1213–1217.

Bioethanol From Cellulose With Supercritical Water Treatment Followed by Enzymatic Hydrolysis

Toshiki Nakata, Hisashi Miyafuji, and Shiro Saka*

*Graduate School of Energy Science, Kyoto University, Japan,
E-mail: saka@energy.kyoto-u.ac.jp*

Abstract

The water-soluble portion and precipitates obtained by supercritical (SC) water treatment of microcrystalline cellulose (Avicel) were enzymatically hydrolyzed. Glucose could be produced easily from both substrates, compared with the Avicel. Therefore, SC water treatment was found to be effective for enhancing the productivity of glucose from cellulose by the enzymatic hydrolysis. It is also found that alkaline treatment or wood charcoal treatment reduced inhibitory effects by various decomposed compounds of cellulose on the enzymatic hydrolysis to achieve higher glucose yields. Furthermore, glucose obtained by SC water treatment followed by the enzymatic hydrolysis of cellulose could be converted to ethanol by fermentation without any inhibition.

Index Entries: Lignocellulosics; supercritical water; enzymatic hydrolysis; inhibitor; ethanol.

Introduction

Global warming owing to the use of the fossil resources is becoming a progressively more serious issue. Much attention has been focused on ethanol from lignocellulosic biomass as an alternative to fossil fuels because of its low green house gas emission and overall environmental friendliness. To obtain sugars from lignocellulosics for ethanol production, numerous studies have been performed on the hydrolysis of lignocellulosics by acid catalysis *(1,2)*, steam explosion *(3)*, liquid hot water *(4)*, and enzymatic saccharification *(5)*.

Recently, on the other hand, supercritical (SC) fluid technology has been applied for the conversion of lignocellulosics to fuels and chemicals *(6–16)*. Ehara and Saka *(13)* reported that cellulose can be converted to sugars such as polysaccharides, oligosaccharides, and monosaccharides by SC water treatment of cellulose for 0.5 s in a flow-type system. They also reported that some other decomposition compounds of cellulose were obtained simultaneously such as levoglucosan, 5-hydroxymethyl furfural,

---

*Author to whom all correspondence and reprint requests should be addressed.

erythrose, methylglyoxal, glycolaldehyde, and dihydroxyacetone. Sasaki et al. *(17)* reported that polysaccharides and oligosaccharides obtained by SC water treatment of cellulose can be effectively converted to cellobiose and glucose with cellulase. However, the ethanol fermentability of the obtained hydrolysate has not been studied. Furthermore, an influence of various decomposition compounds of cellulose obtained by SC water treatment on enzymatic hydrolysis and fermentation has not been studied.

The purpose of this study is, therefore, to establish the whole process for ethanol production from cellulose with SC water treatment followed by enzymatic hydrolysis. Thus, in this study, cellulase and β-glucosidase were applied after its SC water treatment to obtain glucose from cellulose for ethanol production. The effects of the various decomposition compounds of cellulose on enzyme activity and fermentation were also studied. In the previous studies, it is reported that alkaline treatment with wood ash *(18)* or wood charcoal treatment *(19)* is effective for removing inhibitors for ethanol fermentation in the wood hydrolysate obtained by dilute sulfuric acid hydrolysis of wood and improving its ethanol fermentability. Therefore, in this article, the treatment with alkali or wood charcoal was applied to SC water-treated decomposition compounds for enhancing enzyme activity as well as ethanol fermentability to attain high efficiency of ethanol production.

Materials and Methods

SC Water Treatment

Microcrystalline cellulose (Avicel PH-101) was treated by water at SC condition using the flow-type system described elsewhere *(13)*. A volume of the slurried Avicel (2 wt%) at room temperature was mixed with 10-fold volume of the SC water at 380°C and 40 MPa pressure. The reaction time was set at 0.12 s. Furthermore, the reactant was immediately cooled down to 150°C by introducing a 2.5-hold volume of cold water at ambient temperature into the reaction mixture directly and further cooled below 50°C by an external cooler.

The treated samples were filtered into a SC water-soluble (WS) portion and a water-insoluble portion. After settling for 12 h, the former was separated into a WS portion and a water-insoluble portion (precipitates), which are precipitated resulting from the change of dielectric constant of water from the SC state to ambient conditions. The precipitates were separated from the WS portion by filtration. The WS portion obtained was condensed 20 times *in vacuo*. (This condensed WS portion is described as "WS portion" later on in this article.) The separated precipitates were added to the same amount of the ultra pure water (Nacalai Tesque, Inc., Kyoto, Japan) as the WS portion. The obtained WS portion and the medium contained with precipitates were used for the enzymatic hydrolysis.

Preparation of the Wood Charcoal

Western red cedar (*Thuja plicata* D. Don) flours were treated under nitrogen at a heating rate of 4°C/min and maintained at 700°C for 1 h in a rotary furnace to prepare the wood charcoal.

Alkaline Treatment or Wood Charcoal Treatment on the WS Portion

For alkaline treatment, the WS portion was adjusted at pH 12.0 with solid $Ca(OH)_2$, and then stirred for 30 min. For wood charcoal treatment, the wood charcoal was added to the WS portion at a loading of 1 wt% the WS portion. The WS portion with the wood charcoal was then stirred for 10 min. Subsequently, the wood charcoal was separated from the WS portion by filtration.

Enzymatic Hydrolysis for the WS Portion and Precipitates

Before enzymatic hydrolysis, the media used in this study was adjusted at pH 5.0 with HCl aq. or solid $Ca(OH)_2$. Subsequently, cellulase (Wako Pure Chemical Industries, Ltd.), β-glucosidase (Oriental Yeast Co., Ltd.), and $(NH_4)_2HPO_4$ were mixed at concentrations of 1, 0.5, and 0.5 g/L, respectively. Enzymatic hydrolysis was carried out in a 50-mL glass vial containing 4 mL of the media for 24 h at 37°C with gently magnetic stirring.

Enzymatic Hydrolysis Followed by Ethanol Fermentation

The enzymatic hydrolysis was also performed on the mixture of the WS portion and precipitates. After pH adjustment at 5.0, the mixtures untreated or treated with alkali were hydrolyzed enzymatically by the same procedure as described earlier.

For evaluating the fermentability of the media after enzymatic hydrolysis, the fermentation was carried out after enzymatic hydrolysis for 24 h. To prepare the inoculum for ethanol fermentation, a yeast, *Saccharomyces cerevisiae* (IFO 233), was grown in a 300-mL Erlenmeyer-flask capped with a silicone plug, which contained 50 mL of sterile medium. The medium composition was as follows; 3 g/L yeast extract, 3 g/L malt extract, 5 g/L peptone, and 10 g/L glucose. The flask was set in a shaking water bath for 24 h at 28°C. Just after enzymatic hydrolysis of the mixture of the WS portion and precipitates, the inoculum, peptone, yeast extract, and $MgSO_4·7H_2O$ were added at concentrations of 10, 10, 10, and 0.5 g/L, respectively. The fermentation medium was then incubated at 37°C for 24 h with gently magnetic stirring.

Enzymatic Hydrolysis of Cellohexaose With Decomposed Compounds of Cellulose

Medium was prepared with 1.0 g/L cellulase, 0.5 g/L β-glucosidase, and 0.5 g/L $(NH_4)_2HPO_4$. To this mixture, various compounds listed in Table 1 were added at a concentration of 0.01 g/L after pH adjustment

Table 1
Concentrations of Various Compounds Obtained in the WS Portion
by Supercritical Water Treatment of Cellulose

	Compounds	Concentration (mg/L)		
		Untreated	Alkaline treatment	Wood charcoal treatment
Precipitates		3020	–	–
WS portion	Oligosaccharides	2247	2298.7	2188.2
	Glucose	157.5	156.4	140.4
	Fructose	30.6	29.3	26
	Methylglyoxal	9.4	13.3	7.9
	Glycolaldehyde	70.1	67.2	60.4
	Dihydroxyacetone	7.2	1.9	6.9
	Erythrose	4.7	1.9	3.1
	Levoglucosan	34.1	26.5	23.3
	5-Hydroxymethylfurfural	21.2	14.8	0
	Formic acid	18.5	0	17.3
	Glycolic acid	2.7	2.7	1.7
	Acetic acid	5.5	1.6	1.7
	Lactic acid	3.5	0	2.5

at 5.0 with HCl aq. Cellohexaose (Yaizu Suisankagaku Industry Co., Ltd., Shizuoka, Japan) was then mixed at a concentration of 1 g/L. Enzymatic hydrolysis on cellohexaose was carried out in a 50-mL glass vial containing 4 mL of the media for 24 h at 37°C with gently magnetic stirring.

Analytical Methods

Analysis of the WS portion was done by high-performance liquid chromatography (HPLC) (Shimadzu, Kyoto, Japan, LC-10A) equipped with an Ultron PS-80P column (Shinwa Chem. Ind. Co., Kyoto, Japan) and a refractive index detector (Shimadzu, RID-10A) or an ultraviolet detector (Shimadzu, SPD-10A) set at 280 nm. Ultra pure water (Nacalai Tesque, Inc., Kyoto, Japan) was used as mobile phase at a flow rate of 1.0 mL/min. The column oven temperature was set at 80°C.

For determining organic acids in the WS portion, capillary electrophoresis (CE) analysis was conducted with HP3D CE systems (Agilent Technologies, Palo Alto, CA). A fused-silica capillary (75 µm in inner diameter, 104 cm in total length, 95.5 cm in effective length) from Agilent Technologies was used. Indirect UV detection at 270 nm (reference wavelength at 350 nm) was performed using organic acid buffer (Agilent Technologies). Injection was carried out by pressure at the cathodic end at 50 mbar for 4 s. The applied voltage was set at –30 kV, with a capillary temperature at 15°C.

For determining glucose produced during enzymatic hydrolysis and ethanol produced during fermentation, the hydrolysis or fermentation medium were filtered through a 0.45-µm filter to separate the enzyme or yeast. The obtained filtrate was then analyzed by HPLC with the same conditions as in the analysis method for WS portion described earlier.

Results and Discussion

Chemical Composition of the Treated Cellulose With SC Water

After fractionation of treated cellulose samples with SC water, only the SC WS portion could be obtained without the SC water-insoluble residue. WS portion and precipitates were obtained from the SC water soluble portion. Oligosaccharides, glucose, fructose, methylglyoxal, glycolaldehyde, dihydroxyacetone, erythrose, levoglucosan, furfural, 5-hydroxymethylfurfural, acetic acid, formic acid, lactic acid, and glycolic acid could be identified by HPLC and CE analyses on the WS portion. The concentrations of these compounds in the WS portion and precipitates are quantified as shown in Table 1 (described as "Untreated"). In our previous study on the SC water treatment of cellulose, it was found that the oligosaccharides in the WS portion were made up of cello-oligosaccharides up to cello-dodecaose (13). It was also reported that the precipitates had noncrystalline structure and contained polysaccharides whose degree of polymerization is in a range between 13 and 100. Therefore, these oligosaccharides in the WS portion and polysaccharides can be the substrates for enzymatic hydrolysis to obtain glucose.

Enzymatic Hydrolysis of the WS Portion and Precipitates

To elucidate the enzyme activity on the WS portion and precipitates, enzymatic hydrolysis for these substrates were carried out separately, and the results are shown in Fig. 1. For comparison, results for untreated Avicel are also shown. Both substrates were found to be hydrolyzed to glucose more easily than Avicel, consistent with previous findings (17). Because of the lower degree of polymerization and noncrystalline structure of the oligosaccharide and precipitates, higher accessibility of enzymes must be obtained. A closer inspection indicated that glucose is produced from WS portion before the precipitates. The dissolved oligosaccharides in the WS portion are thought to be hydrolyzed easier, compared with the precipitates as the polysaccharides with the aggregated structure. These results clearly show that SC water treatment of cellulose can enhance enzyme activity for depolymerization and decrystallization of cellulose. However, the glucose concentration obtained after enzymatic hydrolysis was found to be lower than theoretical values of 2.2 g/L in the WS portion or 3 g/L in the precipitates approximately. Thus, it is assumed that various decomposed compounds of cellulose obtained by SC water treatment as shown in Table 1 affect on the enzymatic hydrolysis.

Effects of Decomposed Compounds of Cellulose on Enzymatic Hydrolysis

To evaluate the effects of various decomposed compounds of cellulose listed in Table 1 on the enzyme activity, a model media containing

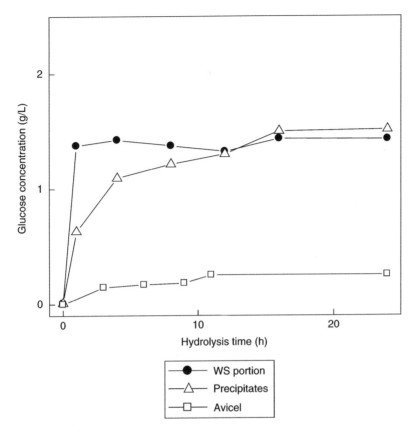

Fig. 1. Changes in glucose concentration during enzymatic hydrolysis of the WS portion and precipitates.

cellohexaose at 1 g/L with those compounds at 0.01 g/L was hydrolyzed enzymatically. Figure 2 shows the glucose concentrations obtained in the enzymatic hydrolysis of cellohexaose with various compounds after 9 h. Glucose concentrations were found to be lower in all media with the additives, compared with the medium with no additives. These results indicate that various compounds obtained by SC water treatment of cellulose have some inhibitory effects on enzymatic hydrolysis and these could account for the lower glucose yield after the enzymatic hydrolysis of the WS portion and precipitates in Fig. 1.

Alkaline or Wood Charcoal Treatments for Detoxification

To increase glucose yields, the enzymatic hydrolysis inhibitors in Fig. 2 should be detoxified or removed. In the previous studies, alkaline treatment with wood ash *(18)* or wood charcoal treatment *(19)* was found to be effective on improving ethanol fermentability of wood hydrolysate obtained by acid hydrolysis. In this study, therefore, we applied alkaline treatment or wood charcoal treatment to the enzymatic hydrolysis of the WS portion. Figure 3 shows the changes of the glucose concentrations for the WS portion

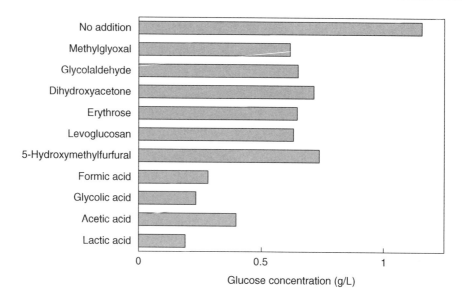

Fig. 2. The effect of various compounds on glucose release by enzymatic hydrolysis of cellohexaose for 9 h.

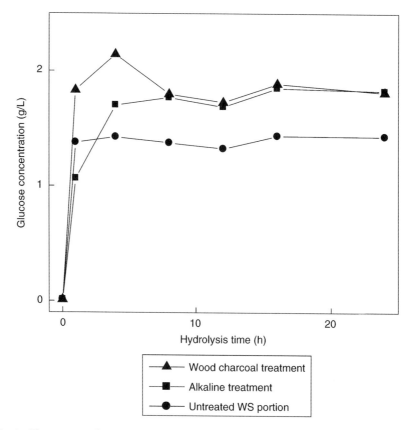

Fig. 3. Changes in glucose concentration for enzymatic hydrolysis of the WS portion without treatment and after treated with alkali or wood charcoal.

after the alkaline treatment or wood charcoal treatment in the enzymatic hydrolyses. Compared with the untreated WS portion, the glucose yields increased in the WS portion after the alkaline treatment or wood charcoal treatment. The concentrations of various compounds in the WS portion after alkaline or wood charcoal treatment are also shown in Table 1. For wood charcoal treatment, the concentrations of many compounds decreased probably owing to the adsorption by the wood charcoal, higher glucose yield were achieved as observed in Fig. 3. For alkaline treatment, on the other hand, the concentrations of many compounds listed in Table 1 also decreased and might be converted to other compounds which have less inhibitory effects. Although the details of the mechanism for reduction of inhibitors by alkaline treatment are unknown, these results showed that alkaline and wood charcoal treatment are effective in detoxifying the WS portion for the enzymatic hydrolysis.

Enzymatic Hydrolysis Followed by Ethanol Fermentation

For efficient glucose production from the WS portion and precipitates, it is preferred to hydrolyze both substrates together without separation. Furthermore, the medium after enzymatic hydrolysis which contains various decomposed compounds of cellulose should be evaluated on its fermentability to establish the process of ethanol production from cellulose with SC water treatment followed by enzymatic hydrolysis. Therefore, enzymatic hydrolysis of the mixture of the WS portion and precipitates was followed by ethanol fermentation as shown in Fig. 4. Alkaline treatment was applied to detoxify the mixture of the WS portion and precipitates. Unfortunately, it was difficult to apply wood charcoal treatment because the wood charcoal could not be separated selectively from the mixture after treatment. In the untreated medium some glucose was produced and fermented to ethanol. Considering that the maximum ethanol concentration was around half the consumed sugar concentration, all the produced glucose during enzymatic hydrolysis was thought to be converted to ethanol. Although the various inhibitory compounds in Fig. 2 were contained in the WS portion, inhibition of ethanol fermentation by those compounds was not observed. In the alkaline treatment, on the other hand, higher glucose yields could be attained in enzymatic hydrolysis, and the glucose obtained was fermented to ethanol without inhibition. These results reveal that SC water treatment of cellulose followed by alkaline treatment could achieve high ethanol yields from cellulose.

Conclusion

The oligosaccharides in the WS portion and polysaccharides as precipitates obtained by the SC water treatment of cellulose can be easily

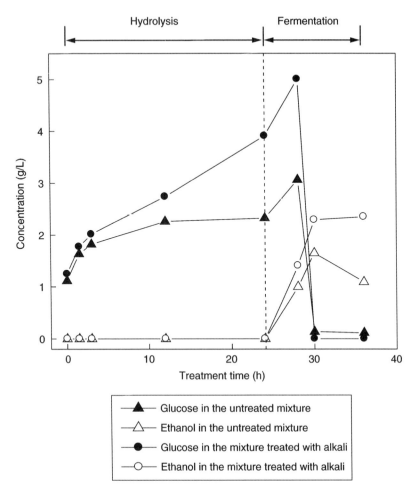

Fig. 4. Changes in glucose and ethanol concentration for enzymatic hydrolysis followed by fermentation of a mixture of WS portion and precipitates without treatment and after treated with alkali.

hydrolyzed to glucose by cellulose enzyme. Furthermore, the glucose obtained was fermented to ethanol without inhibition. Therefore, SC water treatment is a promising method for ethanol production from cellulose. Although various cellulose decomposition compounds from SC water treatment have some inhibitory effects on enzymatic hydrolysis, inhibition can be reduced by alkaline or wood charcoal treatment.

Acknowledgment

This work was carried out at the Kyoto University 21COE program of "Establishment of COE on Sustainable-Energy System" supported by the Ministry of Education, Science, Sports and Culture, Japan.

References

1. Yu, Z. and Zhang, H. (2004), *Biores. Technol.* **93,** 199–204.
2. Söderström, M., Pilcher, L., Galbe, M., and Zacchi, G. (2003), *Biomass. Bioenergy* **24,** 475–486.
3. Cantarella, M., Cantarella, L., Gallifuoco, A., Spera, A., and Alfani, F. (2003), *Biotechnol. Prog.* **20,** 200–206.
4. Laser, M., Schulman, D., Allen, S. G., Lichwa, J., Antal Jr., M. J., and Lynd, L. R. (2002), *Biores. Technol.* **81,** 33–44.
5. Ortega, N., Busto, M. D., and Perez-Maeteos, M. (2002), *Biomass Bioenergy* **47,** 7–14.
6. Ueno, T. and Saka, S. (1999), *Cellulose* **6,** 177–191.
7. Saka, S. and Konishi, R. (2001), In: *Progress in Thermochemical Biomass Conversion*, Bridgwater, A.V., ed., Blackwell Science, Oxford, pp. 1338–1348.
8. Iahikawa, Y. and Saka, S. (2001), *Cellulose* **8,** 189–195.
9. Ehara, K., Saka, S., and Kawamoto, H. (2002), *J. Wood Sci.* **48,** 320–325.
10. Tsujino, J., Kawamoto, H., and Saka, S. (2003), *Wood Sci. Technol.* **37,** 299–307.
11. Minami, E. and Saka, S. (2003), *J. Wood Sci.* **49,** 73–78.
12. Minami, E., Kawamoto, H., and Saka, S. (2003), *J. Wood Sci.* **49,** 158–165.
13. Ehara, K. and Saka, S. (2002), *Cellulose* **9,** 301–311.
14. Takada, D., Ehara, K., and Saka, S. (2004), *J. Wood Sci.* **50,** 253–259.
15. Yoshida, K., Ehara, K., and Saka, S. (2005), *Appl. Biochem. Biotechnol.* **121–124,** 795–806.
16. Miyafuji, H., Nakata, T., Ehara, K., and Saka, S. (2005), *Appl. Biochem. Biotechnol.* **121–124,** 963–971.
17. Sasaki, M., Iwasaki, K., Hamaya, T., Adschiri, T., and Arai, K. (2001), *Koubunshi Ronbunshu* **58,** 527–532.
18. Miyafuji, H., Danner, H., Neureiter, M., Thomasser, C., and Braun, R. (2003), *Biotechnol. Bioeng.* **84,** 390–393.
19. Miyafuji, H., Danner, H., Neureiter, M., et al. (2003), *Enzyme Microb. Technol.* **32,** 396–400.

Enhancement of the Enzymatic Digestibility of Waste Newspaper Using Tween

Sung Bae Kim,* Hyun Joo Kim, and Chang Joon Kim

Department of Chemical and Biological Engineering and ERI, Gyeongsang National University, Jinju 660-701 Korea; E-mail: sb_kim@gsnu.ac.kr

Abstract

Methods of increasing the enzymatic digestibility of waste newspaper by adding Tween (TW)-20 and 80 surfactants were investigated. Tween-series surfactants were selected because these surfactants increase cellulase activity during enzymatic hydrolysis and do not inhibit cell growth in downstream fermentation processes. When surfactant was used in a pretreatment, a benefic effect was expected in the enzymatic hydrolysis stage owing to surfactant carry-over from the pretreatment stage immediately upstream of the hydrolysis. However, because it was necessary to wash the pretreated substrate with water to remove inhibitors produced during pretreatment, no added benefit was obtained. When surfactant was used in the pretreatment only, it was found that it had a marked effect on digestibility and that this effect was higher at lower enzyme loadings. Also, TW-80 was found to be more effective than TW-20, and the addition of enzyme and TW-80 to substrate at the beginning of enzyme reaction was found to most effectively increase digestibility. When TW-80 was added into either the pretreatment stage or the hydrolysis stage the digestibilities of untreated sample increased by approx 40%, whereas an increase of only 45% was observed when TW-80 was added to both stages. These results show that the addition of surfactant to either the pretreatment or the enzymatic hydrolysis stage is sufficient to increase digestibility.

Index Entries: Pretreatment; newspaper; surfactant; hydrolysis; enzymatic digestibility.

Introduction

Energy supply problems owing to abrupt increases in petroleum prices have accelerated developments on alternative energy sources. One of the promising energy sources is bioethanol, which can be produced from renewable lignocellulosic materials such as forest and agricultural residues, and municipal solid wastes. Municipal solid wastes, such as waste paper, can provide cheap feedstocks for ethanol and meet targets of resource reutilization and environmental compatibility (1). Korea produces annually about 1.3 and 3.4×10^6 t of paper sludge and unrecycled waste

---

*Author to whom all correspondence and reprint requests should be addressed.

paper, respectively (2). In the case of organic sludge, landfilling was prohibited by law in 2003. Incineration involves high dewatering costs owing to the high moisture contents of sludge, and also produces secondary air pollution. Wastepaper constitutes half of municipal solid waste, and newspaper, which represents 14% of the waste, is recovered relatively easily (3). As in comparison with wood, waste cellulosic materials have very low associated costs and some of them contain high levels of cellulose, which can provide good feedstocks for ethanol production.

Newspaper is mostly derived from softwoods and exhibits low enzymatic digestibility because of its high lignin content and dense structure (4). However, because newspaper has already received considerable chemical and/or physical treatment, it does not require the extensive pretreatment required for woody and herbaceous materials. But ink and some additives added during the paper-making process can interfere with the enzymatic hydrolysis of wastepaper. The pretreatment methods developed for wastepaper were similar to those developed for woody and herbaceous materials. Of numerous studies on pretreatment, there have been limited studies on wastepaper pretreatment using electron beam irradiation (5), carbon dioxide explosion (6), and ammonia–hydrogen peroxide (7–9). These methods appear to be uneconomical because of high energy and/or chemical cost. To reduce energy consumption, we previously pretreated wastepaper with surfactant at temperature less than 50°C (10,11). Surfactant added to a pretreatment stage can remove many of the components that hinder enzyme access to substrate, and increase digestibility (8–11). Also, surfactant added during enzymatic hydrolysis can increase enzyme stability or positively affect enzyme–substrate interactions, and again have a positive effect on digestibility (12–15). Therefore, two types of experiments were carried out in this study. First, raw newspaper was pretreated using a surfactant to determine pretreatment performance, and the digestibility of this pretreated newspaper was measured without the addition of surfactant in hydrolysis stage. Second, pretreated newspaper was hydrolyzed in the presence of surfactant and again digestibility was measured to determine the effect of surfactant on enzymatic hydrolysis.

Materials and Methods

Materials

A mixture of Korean newspapers was used as substrate. Newspaper was cut into approx 0.5 × 0.5 cm pieces. Its moisture content was 7.2 wt% and it had the following composition on a dry substrate basis: 59 wt% glucan, 16.2 wt% xylan + mannan + galactan, 12.4 wt% klason lignin, and 6 wt% ash. Tween-series surfactants (Sigma Chemical Co., St. Louis, MO) were used in this study as listed in Table 1. Commercial cellulase and β-glucosidase (Novo Nordisk, Bagvard, Denmark) were supplied from Novozymes Korea Ltd., (Seoul, Korea), and a mixture of Celluclast (80 filter

Table 1
Nonionic Surfactants

	Chemical name	HLB[a]
TW-20	Polyoxyethylene sorbitan monolaurate	16.7
TW-80	Polyoxyethylene sorbitan monooleate	15

[a]HLB, hydrophile-lypophile balance; TW, Tween.

paper units [FPU]/ml) and Novozym 188 (792 cellobiase units [CBU]/mL) was used in the ratio of 4 FPU of Celluclast/CBU Novozym to alleviate end-product inhibition by cellobiose.

Pretreatment

Ten grams of substrate was added to a 500 mL round flask containing 200 g of deionized water. Then, 0–2 wt% of a surfactant was added to this solution and the contents were agitated at 400 rpm and 40°C for 1 h. The concentration of the surfactant was calculated as wt% based on the 10 g of dry substrate. After pretreatment, the wet solid material was washed with 1 L of deionized water, filtered to a moisture content of 70–80%, and then separated into two portions. One was oven dried at 105°C overnight to determine moisture content and weight loss during pretreatment, and analyzed for composition. The other was stored in a refrigerator until required for enzymatic digestibility testing.

Enzymatic Digestibility Test

Enzymatic digestibility of pretreated substrate was performed in duplicate according to National Renewable Energy Laboratory (NREL) standard procedure no. 009. The NREL standard procedures can be obtained from the following website: www.ott.doe.gov/biofuels/analytical_methods.html. An amount of solid required to produce 0.5 g glucan in 50 mL was added to a 250-mL flask. The buffer solution was 0.05 M citrate, pH 4.8. Cellulase was then loaded at 30 FPU/g glucan, and 0.5% of the dry substrate weight of surfactant, when needed, was added. The contents of the flask were preheated to 50°C before the enzyme was added. The flask was then placed on a shaking bath at 50°C and 90 strokes/min. Samples were taken periodically and analyzed for glucose using high-performance liquid chromatography (HPLC). The glucose contents after 24, 48, and 72 h of hydrolysis were used to calculate the enzymatic digestibility.

Analytical Methods

Solid biomass sample was analyzed for sugars, klason lignin, and ash according to NREL standard procedures nos. 002, 003, and 005. Sugars concentrations were measured by HPLC (Thermo Separation Products)

Enhancement of Enzymatic Digestibility

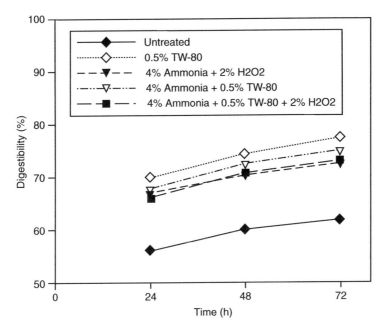

Fig. 1. Effect of composition of pretreatment solution on the enzymatic digestibility of newspaper.

using a Bio-Rad HPX-87H column (condition; 0.6 mL/min, 65°C, 0.005 M H_2SO_4). Because this column did not resolve xylose, mannose, and galactose, their combined concentrations were used.

Results and Discussion

Effect of Surfactant on Pretreatment

A mixture of ammonia and hydrogen peroxide was previously found to be highly effective at pretreating lignocellulosic biomasses, such as, corn cobs/stover mixture *(16)*. And the combined use of ammonia, hydrogen peroxide, and surfactant was found to be effective at pretreating wastepaper before hydrolysis *(8,9)*. These studies revealed that newspaper should not be pretreated using the methods developed for wood and herbaceous biomass, because paper is produced from wood using many processing stages, and a multitude of additives and ink are added during the paper producing and printing processes. Thus, wastepaper was pretreated at near room temperatures, not at the high temperatures and pressures usually required to treat other lignocellulosic biomasses. Although ammonia and hydrogen peroxide help to swell cellulosic fibers and remove ink, we investigated methods that require minimal amount of chemicals in the present study, because of the environmental aspects and because these chemicals represent a major fraction of the total pretreatment cost.

As shown in Fig. 1, in this study we examined combinations of ammonia, hydrogen peroxide, and/or surfactant to pretreat newspaper.

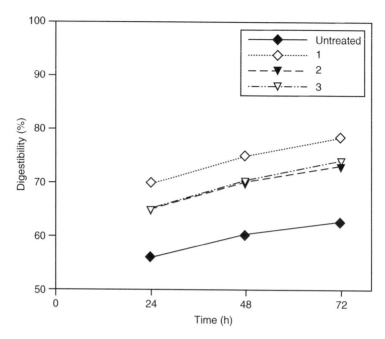

Fig. 2. Effect of treatment method after pretreatment on the enzymatic digestibility of newspaper.

Method	Pretreatment	→	Hydrolysis
1	0.5% TW-80	Washing	No surfactant
2	"	Filtering	No surfactant
3	"	Filtering	Compensate TW-80 to make 0.5% in hydrolysis solution

The digestibility of materials obtained using any combination of these three chemicals was much higher than that of untreated sample. Here, untreated sample is defined as a substrate without any form of treatment, because the digestibility of untreated feedstock at 72 h is only 2% lower than that of feedstock pretreated by soaking in water for an hour (data not shown). The results obtained showed that pretreatment with surfactant only (i.e., containing no ammonia and hydrogen peroxide) proved best. This finding was unexpected, but it does mean that pretreatment using surfactant offers processing economies and environmental benefits.

We selected Tween surfactants because these surfactants increase cellulase activity during enzymatic hydrolysis and do not inhibit cell growth during subsequent fermentation processes *(12,14,15)*. This suggests that Tween can be used in both the pretreatment and hydrolysis processes. When Tween (TW)-80 is used in the pretreatment, the pretreatment solution can be partially removed by filtration without water washing. Because the surfactant left in the pretreated substrate could have a positive effect on enzymatic digestibility, three schemes were devised as shown in Fig. 2. The results

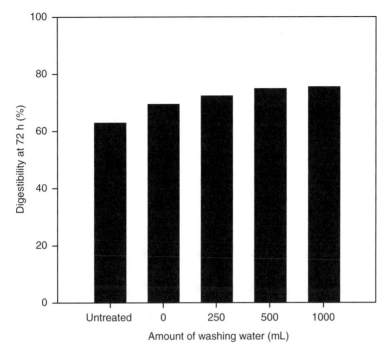

Fig. 3. Effect of amount of washing water used after pretreatment on enzymatic digestibility (pretreatment: 0.5% TW-80).

showed that method 1, in which the pretreated sample was washed with water, had 5% higher digestibility than method 2, in which the pretreatment solution was partly removed by filtration, and than method 3, in which the surfactant was added (to a concentration of 0.5% in the hydrolysis stage) to material obtained using method 2. These results showed that a washing step is required between the pretreatment and hydrolysis stages to obtain high digestibility. This is probably because ink components and other additives detached from paper fibers directly hinder the enzymatic hydrolysis or reattach to the fiber surface and reduce digestibility.

Figure 3 shows the effect of the amount of washing water on 72 h digestibility. The digestibility of the unwashed material was 6% higher than that of untreated sample and digestibility increased with the amount of used water, but this effect was negligible at more than 500 mL. This means that the components that hinder hydrolysis are effectively removed by washing pretreated substrate with 500 mL of water. Figure 4 shows the effect of TW-80 loading in the pretreatment stage on 72 h digestibility. Digestibility increased with surfactant loading, but this increase was negligible above 0.5%, thus 0.5% was used in further experiments.

Table 2 shows the effect of surfactant type and enzyme loading on 72 h digestibility after newspaper was pretreated with 0.5% surfactant. In comparison with digestibility at 30 FPU, digestibilities at 15 FPU, were 14.9%, 7.2%, and 6.4% lower for untreated sample, and TW-20 and TW-80-pretreated

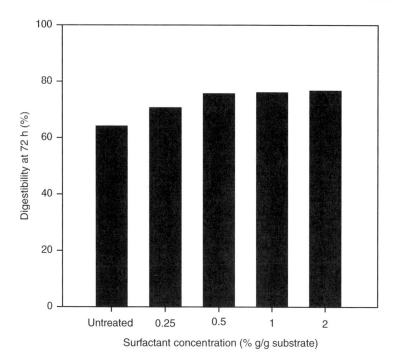

Fig. 4. Effect of TW-80 loading in the pretreatment stage on the enzymatic digestibility of newspaper.

Table 2
Effect of Surfactant Type in the Pretreatment Stage
and Enzyme Loading on 72 h Enzymatic Digestibility
of Newspaper (Surfactant Concentration = 0.5%)

Enzyme loading (FPU/g glucan)	Surfactant		
	Untreated	TW-20	TW-80
15	49.1	62.9	68.1
30	64	70.1	74.5

samples, respectively. When the enzyme loading was halved, the digestibility of untreated sample was maximally reduced, but surfactants significantly attenuated this reduction. This indicates that surfactants have a greater effect on digestibility when enzyme loading is lowered *(10,13)*. The digestibility of TW-80-pretreated sample at 15 FPU was about 4% higher than that of untreated sample at 30 FPU. Because TW-80 increased digestibility by 4% vs TW-20, we concluded that it is probably more suitable for the pretreatment of newspaper.

Effect of Surfactant on Enzymatic Hydrolysis

Enzyme activity reduces as the reaction proceeds, and this deactivation can be reduced by the addition of surfactants. The mechanism underlying

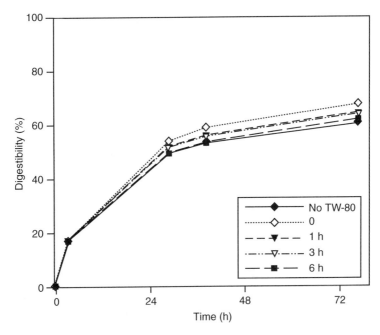

Fig. 5. Effect of time of surfactant addition on enzymatic digestibility (hydrolysis: 0.5% TW-80).

this phenomenon has not been elucidated, but one of the most promising explanations is that surfactant causes cellulase to desorb easily from cellulose surface after hydrolysis *(12–15)*. In order to convert solid cellulose to soluble sugars, enzymes must be adsorbed onto the cellulose surface. Because surfactant could prevent the adsorption of enzyme onto a cellulose surface, we considered that enzyme should be added first and that the surfactant be added later. Figure 5 shows the effect of time of surfactant (TW-80) addition on the digestibility of newspaper. Digestibility was maximized when surfactant was present from the beginning of the hydrolysis and was reduced if the surfactant addition was postponed. This implies that surfactant does not prevent binding between enzyme and substrate. Also this result supports the explanation that the surfactant increases the hydrophilicity of the substrate, and facilitates enzymes access *(13)*, or that it prevents unproductive enzyme adsorption to the lignin part of the substrate during the initial stage of the hydrolysis *(12)*.

The digestibility of newspaper has been reported to be dramatically increased when TW-80 or other surfactants are added to the hydrolysis reaction *(11–15)*. Figure 6 shows the effect of surfactant in the pretreatment and/or in the hydrolysis stage at an enzyme loading of 15 FPU/g glucan. Process 1, in which the surfactant was present in the pretreatment stage only, and process 2, in which surfactant was present in the hydrolysis stage only, showed 20% higher digestibility than the untreated sample. Thus, it can be concluded that the effect of surfactant on either process is similar and

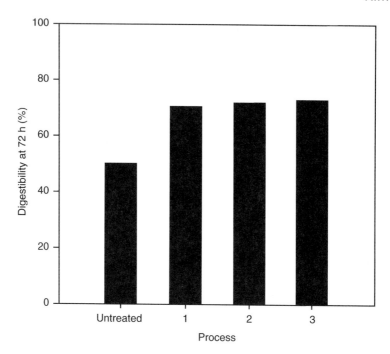

Fig. 6. Effect of process on the enzymatic digestibility of newspaper (enzyme loading = 15 FPU/g glucan).

Process no.	Pretreatment	→	Hydrolysis
1	0.5% TW-80		No surfactant
2	No surfactant	Washing	0.5% TW-80
3	0.5% TW-80		0.5% TW-80

significant. Process 3, in which surfactant was present in both stages, was found to have a 2–3% higher digestibility than processes 1 or 2. This implies that surfactant does not affect the hydrolysis if newspaper is pretreated with surfactant, and that surfactant is needed only once, in the either pretreatment or hydrolysis stage. Surfactant in the pretreatment stage removes components that retard enzymatic hydrolysis, and surfactant in the hydrolysis stage prevents enzyme deactivation and improves enzyme-substrate interactions, both of which would lead to enhanced digestibility. However, we cannot explain why the effect of surfactant on either pretreatment or hydrolysis is significant, whereas the effect of surfactant on the hydrolysis of surfactant-pretreated substrate is almost negligible.

Conclusion

We investigated the possibility of enhancing the enzymatic digestibility of waste newspaper using nonionic surfactants. In order to increase the digestibility of newspaper, pretreated substrate containing surfactant should be washed with water before hydrolysis. Therefore, a surfactant used in the pretreatment stage cannot affect the hydrolysis stage, and thus it is not necessary to use the same surfactant in both stages. Moreover, because surfactant in the hydrolysis stage may inhibit cell growth in a subsequent fermentation, surfactant selection is limited, whereas any surfactant can be used in the pretreatment stage, because of the intervening washing stage before the hydrolysis. Also surfactant and enzyme must be added from the beginning of the hydrolysis reaction to obtain high digestibility. One addition of TW-80 to either the pretreatment or hydrolysis stage was found to be sufficient, because the effect of surfactant addition to both stages found to be only marginally higher than its effect when added to either of the two stages.

References

1. Bergeron, P. W. and Riley, C. J. (1991), *Wastepaper as a Feedstock for Ethanol Production*, NREL/TP-232-4237, National Renewable Energy Laboratory, Golden, Colorado, USA.
2. Korea Paper Manufacture's Association (2003), *Report for Production and Treatment of Paper Mill Waste*, Seoul, Korea.
3. Scott, C. D., Davison, B. H., Scott, T. C., Woodward, J., Dees, C. and Rothrock, D. S. (1994), *Appl. Biochem. Biotechnol.* **45/46,** 641–653.
4. Holtzapple, M. T., Lundeen, J. E., Sturgis, R., Lewis, J. E., and Dale, B. E. (1992), *Appl. Biochem. Biotechnol.* **34/35,** 5–21.
5. Khan, A. W., Labrie, J., and McKeown, J. (1987), *Radiat. Phys. Chem.* **29,** 117–120.
6. Zheng, Y., Lin, H., and Tsao, G. T. (1998), *Biotechnol. Prog.* **14,** 890–896.
7. Kim, J. S., Lee, Y. Y., and Park, S. C. (2000), *Appl. Biochem. Biotechnol.* **84/86,** 129–139.
8. Moon, N. K. and Kim, S. B. (2001), *Korean J. Biotechnol. Bioeng.* **16(5),** 446–451.
9. Kim, S. B. and Moon, N. K. (2003), *Appl. Biochem. Biotechnol.* **105/108,** 365–373.
10. Kim, S. B. and Chun, J. W. (2004), *Appl. Biochem. Biotechnol.* **113/116,** 1023–1031.
11. Kim, S. B. and Chun, J. W. (2004), In: *Lignocellulosic Biodegradation,* Saha, B. C. and Hayashi, K., eds., ACS Books, Washington, USA, pp. 36–48.
12. Eriksson, T., Börjesson, J., and Tjerneld, F. (2002), *Enzyme Micro. Technol.* **6112,** 1–12.
13. Helle, S. S., Duff, S. J. B., and Cooper, D. G. (1993), *Biotechnol. Bioeng.* **42,** 611–617.
14. Karr, W. E. and Holtzapple, M. T. (1998), *Biotechnol. Bioeng.* **59,** 419–427.
15. Wu, J. and Ju, L. (1998), *Biotechnol. Prog.* **14,** 649–652.
16. Kim, S. B. and Lee, Y. Y. (1996), *Appl. Biochem. Biotechnol.* **57/58,** 146–156.

Ethanol Production From Steam-Explosion Pretreated Wheat Straw

Ignacio Ballesteros, Mª José Negro, José Miguel Oliva, Araceli Cabañas, Paloma Manzanares,* and Mercedes Ballesteros

CIEMAT-Renewable Energies Division, Av. Complutense, 22, 28040-Madrid-Spain; E-mail: p.manzanares@ciemat.es

Abstract

Bioconversion of cereal straw to bioethanol is becoming an attractive alternative to conventional fuel ethanol production from grains. In this work, the best operational conditions for steam-explosion pretreatment of wheat straw for ethanol production by a simultaneous saccharification and fermentation process were studied, using diluted acid [H_2SO_4 0.9 % (w/w)] and water as preimpregnation agents. Acid- or water-impregnated biomass was steam-exploded at different temperatures (160–200°C) and residence times (5, 10, and 20 min). Composition of solid and filtrate obtained after pretreatment, enzymatic digestibility and ethanol production of pretreated wheat straw at different experimental conditions was analyzed. The best pretreatment conditions to obtain high conversion yield to ethanol (approx 80% of theoretical) of cellulose-rich residue after steam-explosion were 190°C and 10 min or 200°C and 5 min, in acid-impregnated straw. However, 180°C for 10 min in acid-impregnated biomass provided the highest ethanol yield referred to raw material (140 L/t wheat straw), and sugars recovery yield in the filtrate (300 g/kg wheat straw).

Index Entries: Wheat straw; ethanol; diluted acid pretreatment; steam-explosion.

Introduction

During the past century, world energy consumption has mostly depended on the utilization of fossil fuels, which has led to harmful changes in our climate and increased the amount of greenhouse gases in the atmosphere. According to the World energy, technology and climate policy outlook report published in 2003 by the European Union *(1)*, given the continued dominance of fossil fuels, world CO_2 emissions are expected to increase more rapidly than the energy consumption (2.1%/yr on an average). In 2030, world CO_2 emissions are expected to be more than twice the level of 1990. This scenario is a clear encouragement for developing

---

*Author to whom all correspondence and reprint requests should be addressed.

alternative sources that mitigate the detrimental environmental effects of fossil fuels use.

With the search for alternative renewable energy sources, bioethanol is fast becoming a viable solution, as it is a nonfossil fuel from a renewable source that may result in a cost-efficient way to reduce greenhouse gases and gasoline use in transport provided that it is produced in an efficient conversion process. Although conventional fuel ethanol is derived from grains such as corn and wheat competing as a food source for humans, the cereal industry produces vast amounts of residue that has little use today. According to Kim and Dale (2), under the 60% ground cover practice, about 354 millions of tons of wheat straw could be available globally and could produce 104 GL of bioethanol. Europe production would account for about 38% of this world bioethanol capacity. In Spain grain industry generates important amounts of wheat straw, a part of which is used as bedding straw and the remainder is burned or left on the land to fertilize the soil. Bioconversion of this residue to fuel ethanol would provide an attractive possibility to boost the development of biofuels in our country in a sustainable way.

The lignocellulosic nature of wheat straw makes the pretreatment an essential step because the physical and chemical barriers caused by the close association of main components greatly limits the suceptibility to bioprocesses such as simultaneous saccharification and fermentation (SSF). Among processes developed to pretreat lignocellulosic biomass, steam-explosion (SE) has been extensively studied and claimed as one of the most successful techniques for fractionating biomass and enhancing the accessibility of cellulose to enzymes. SE has been proved to be effective in a great variety of lignocellulosic biomass, including hardwoods (3,4), softwoods (5,6), and herbaceous residues such as corn stover (7), sugarcane bagasse (8), and wheat straw (9,10). As a way to further improve the effectiveness of SE pretreatment, the addition of an impregnation agent before pretreatment has been shown to be an effective method to increase cellulose digestibility of pretreated substrates and solubilize a significant portion of the hemicellulosic component. Preimpregnation of biomass with acid catalyst such as dilute SO_2 and H_2SO_4 has been shown to decrease both temperature and time requirements whereas achieving optima fractionation, sugar recovery, and enzymatic hydrolysis (EH) of steam-pretreated samples (11). The efficiency of dilute H_2SO_4 pretreatment in some agricultural residues as corn stover has been already demonstrated (12,13). Neverthless, few references are found about the effect of acid addition in agricultural residues as wheat straw.

In this work, the best operational conditions for SE pretreatment of wheat straw for ethanol production by a SSF process were studied, using diluted acid (H_2SO_4 0.9 % [w/w]) or water as impregnation agents before pretreatment. Acid- or water-impregnated biomass was steam-exploded at different temperatures (160–200°C) and residence times (5, 10, and 20 min). The effectiveness of SE was evaluated in terms of cellulose recovery in the water-insoluble solids (WIS) fraction, hemicellulose-derived sugar

(HS) recovery in the filtrate, and EH yield of WIS fraction. Finally, pretreated wheat straw was tested in SSF process with the thermotolerant yeast strain *Kluyveromyces marxianus* CECT 10875.

Materials and Methods

Raw Material

Wheat straw (6% moisture content) was provided by Ecocarburantes de Castilla y León (Salamanca, Spain). Biomass was coarsely crushed using a laboratory hammer mill (Retsch GmbH & Co. KG, Germany), homogenised and stored until used.

The chemical composition of raw material and WIS fraction was determined using the standard laboratory analytical procedures for biomass analysis provided by the National Renewable Energy Laboratory (Colorado) *(14)*. The chemical analysis of raw material showed the following compostion (% dry weight): cellulose, 30.2; hemicellulose, 22.3 (xylan, 18.7; arabinan, 2.8; and galactan, 0.8); acid insoluble lignin, 15.3; acid soluble lignin, 1.7; acetyl groups, 2.6; ash, 4.7; and extractives, 14.7 (total, 91.5%).

SE Pretreatment

Pretreatment assays were performed by applying Masonite technology in a 2L-SE pilot unit as described in a previous work *(3)*. Before pretreatment, wheat straw was soaked for 18 h at 45°C in 0.9% w/w diluted sulphuric acid solution or water (solid–liquid ratio: 1/10). The soaked material was vacuum filtered to approx 20% solids content and then steam-exploded. Temperature pretreatment ranged from 160°C to 200°C and time from 5 to 20 min, depending on temperature. Pretreatment experiments on wheat straw preimpregnated with water were performed at all tempertaures and selected times, except for the lowest temperature of 160°C where only acid impregnated biomass was tested. A summary of the operation conditions assayed are shown in Table 1. After pretreatment, the material was recovered in a cyclone, cooled to approx 40°C and filtered to recover two fractions: (1) the WIS fraction and (2) the filtrate or prehydrolyzate. After separating the filtrate, WIS fraction was throughly washed with water, weighted and dried at 45°C for storage. Solid recovery yield was then calculated as dry weight of WIS remaining after pretreatment referred to 100 g of raw material.

WIS fraction was analyzed for carbohydrates and acid-insoluble lignin content, and used as substrate in EH and SSF tests. Sugars, furfural, and hydroxymethylfurfural (HMF) content of the filtrate were also analyzed.

Enzymatic Hydrolysis Tests

The washed WIS fraction after pretreatment was used as substrate for EH experiments. EH tests were performed in 100-mL Erlenmeyer flasks,

Table 1
Conditions for SE Pretreatment of Wheat Straw

Temperature (°C)	Time	Impregnation agent
160	20	Catalyst[a]
170	20	H_2O
	5	Catalyst
	10	Catalyst
180	10	H_2O
	5	Catalyst
	10	Catalyst
190	10	H_2O
	5	Catalyst
	10	Catalyst
200	10	H_2O
	5	Catalyst

[a]0.9% (w/w) H_2SO_4.

each containing 25 mL of 0.1 M sodium acetate buffer (pH 4.8), 10% (w/v) dry WIS loading, at 50°C for 72 h. Enzyme loading of 15 FPU/g dry WIS of Celluclast 1.5 L and 12.6 IU/g dry WIS of β-glucosidase Novozyme 188 was employed. Enzymes were a gift from Novozymes A/S (Bagsvaerd, Denmark). After EH assays completion, glucose was analyzed by HPLC as decribed earlier. Experiments were performed in duplicate. EH yield was calculated as the ratio of g glucose in the EH/100 g potential glucose in WIS. Glucose yield in the EH referred to untreated initial material was also calculated by taking into account the solid recovery yield attained in each experiment.

Microorganisms and Growth Conditions

K. marxianus CECT 10875, purchased in the Spanish collection of type cultures, was used in SSF experiments. Active cultures for inoculation were prepared by growing the organism on a rotary shaker at 150 rpm and 42°C for 16 h, in a growth medium containing (g/L): yeast extract, 5; NH_4Cl, 2; KH_2PO_4, 1; $MgSO_4·7H_2O$, 0.3; and glucose, 30. All chemicals were from Sigma (Sigma-Aldrich Inc., St. Louis, MO).

SSF Tests

SSF experiments were carried out under no sterile conditions in 100-mL Erlenmeyer flasks, each containing 50 mL of the fermentation medium (without glucose) as described earlier, and were incubated at 150 rpm and 42°C for 72 h. No contamination was detected at the end of the experiments. The WIS fraction obtained after pretreatment was used as substrate at 10% (w/v) concentration. Enzymatic complex and loading used in SSF experiments were the same as that for EH tests. Flasks were inoculated with 0.2 g/L yeast culture and periodically analyzed for ethanol and glucose. Experiments were performed in duplicate.

SSF results are reported in percentage of the theoretical yield. The theoretical SSF yield was calculated by assuming that all the potential glucose in the WIS fraction is available for fermentation, and a fermentation yield of 0.51 g ethanol/g glucose. Results are also reported as ethanol yield based on the initial untreated material, taking into account the solid recovery yield.

Analytical Methods

The carbohydrate content of the prehydrolyzate after pretreatment was measured by performing a mild acid hydrolysis (3% [v/v] H_2SO_4, 120°C and 30 min) and measuring glucose, xylose, arabinose, galactose, and mannose concentration by HPLC in a Waters 2695 liquid chromatograph with refractive index detector. An AMINEX HPX-87P carbohydrate analysis column (Bio-Rad, Hercules, CA) operating at 85°C with deionized water as mobile-phase (0.6 mL/min) was used. Likewise, glucose concentration after completion of EH tests was measured in EH media following this method. HPLC was also used to analyze the prehydrolyzate for furfural and HMF as described previously (3).

Ethanol was measured by gas chromatography, using a HP 5890 Series II apparatus equipped with an Agilent 6890 series injector, a flame ionization detector and a column of Carbowax 20 *M* at 85°C. The injector and detector temperature was maintained at 150°C.

Results and Discussion

SE Pretreatment

Results of solid recovery yield and composition of WIS fraction after SE at different process conditions are shown in Table 2. Recovery yields ranged from 42 to 60%, depending on pretreatment conditions. Higher solubilization was obtained at harsher conditions, showing great effect of residence time and acid addition on solids recovery. Regarding WIS composition, cellulose content increased in relation to untreated material (30.2%) in all conditions tested. Cellulose concentration ranged from 50% to approx 64%, depending on the pretreatment conditions. The maximum cellulose content (approx 64%) was obtained in acid-impregnated biomass at 180°C and 10 min. Higher temperatures produced somewhat lower cellulose content in pretreated materials, specially in acid-impregnated biomass. It indicates that SE at harsh conditions causes an initial breakdown of the cellulose fiber and a consequent loss of glucose in solid residue (10). Acid-insoluble lignin was

Table 2
Dry Matter Recovery and Composition of Pretreated Wheat Straw
(WIS Fraction) After SE at Different Process Conditions

Conditions			Solid recovery (%)	Component (% dry WIS)		
Temperature (°C)	Time (min)	Impregnation agent		Cellulose	Hemicellulose	Lignin
160	20	Acid[a]	60	50.3	15	25.4
170	10	Water	55.2	50	13.6	20.8
	5	Acid	51.8	54.1	8.6	25.5
	10	Acid	51.7	58.5	5.9	27
180	10	Water	48.6	60.2	7.5	25.9
	5	Acid	47.4	62.7	5.2	27.5
	10	Acid	46.8	63.5	1.3	32.6
190	10	Water	45.6	59.6	7.6	23.5
	5	Acid	43.4	54.8	1.8	28.3
	10	Acid	43.1	55.8	1	33.6
200	10	Water	45.1	61.9	4.5	27.9
	5	Acid	41.6	55.2	1.2	33.1
Untreated				30.2	22.3	15.3

[a] 0.9% (w/w) H_2SO_4.

considerably concentrated in comparison with untreated material (15.3%), reaching values up to 33% at the most severe conditions. Referred to raw material content, lignin losses accounted for 1–5% of initial lignin values, depending on the conditions.

As expected, results show a substantial removal of hemicelluloses during pretreatment. The most severe conditions of 180°C and 190°C for 10 min and 200°C for 5 min, in acid-impregnated biomass, led to almost complete dissolution of hemicellulose component, remaining only 1.5% in WIS fraction (<1% referred to raw material content). At lower temperature of 160°C, the hemicellulose content remaining in the solid accounted for up to 15%, corresponding to approx 50% of the content in raw material. Experiments performed in acid-impregnated biomass led to higher hemicelluloses removal in comparison with water-impregnated biomass in all temperatures tested.

Enhancing effect of acid-impregnation of biomass in solubilization of hemicelluloses during pretreatment has been previously reported in other residues as corn stover. Varga et al. (7) found that impregnation of corn stover with 0.5 and 2% (w/w) H_2SO_4 had greater effect in decreasing hemicellulose fraction in the solid residue than pretreatment temperature. SE pretreatment at 210°C, 5 min and 2% H_2SO_4 gave rise to almost complete hemicellulose removal from solid fraction. However, it is well-known that a high sugar solubilization from the solid fraction does not always correlate with good sugar recoveries in prehydrolyzate at harsh

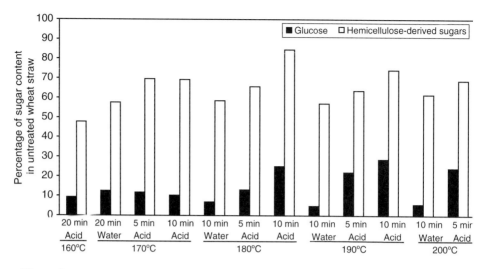

Fig. 1. Sugar recovery yield in prehydrolyzates from SE of wheat straw at different pretreatment conditions. Results reported as percentage of sugar content in untreated raw material.

pretreatment conditions, owing to sugar losses by degradation in acidic conditions *(15,16)*. So, the filtrate obtained after SE experiments must be analyzed for total sugar recovery assessment. Results of glucose and HS recovery yields in prehydrolyzate (expressed as percentage of glucose and HS content in untreated raw material) after pretreatments are shown in Fig. 1. Comparing HS recovery yields for water and acid impregnated biomass experiments, it can be observed that yields in water-impregnated biomass reached similar values (57–61%) in all temperatures tested and that the addition of acid resulted in increased values up to 65–85%, depending on temperature. A maximum HS recovery of 85% was achieved in experiments performed with acid-impregnated biomass at 180°C and 10 min. This value corresponds to 21 g/100 g raw material of which xylose is the major component (82%), followed by arabinose (13%), and low amounts of galactose (5%). Because at this condition HS content in WIS fraction only accounted for <1% of content in raw material, the remaining 15% HS must be attributed to sugar degradation. Higher temperatures resulted in decreased HS recovery yields in filtrate of approx 70%, showing increased sugar losses. However, high recovery values >90% have been obtained in other biomass residues such as corn stover at 190°C SE pretreatment of dilute-acid (1% H_2SO_4) impregnated material, but at shorter residence time of 90–110 sec *(12)*. In our study, the combination of high temperature of 190–200°C with longer times of 5–10 min in acid-impregnated biomass provided much too severe conditions for hemicellulose recovery.

Regarding glucose recovery in filtrate, acid-impregnation before pretreatment at the most severe conditions produced high glucose recovery

Table 3
Degradation Products Content and pH Values of Prehydrolyzate From SE of Wheat Straw in Different Process Conditions

| Conditions | | | | HMF | Furfural |
Temperature (°C)	Time (min)	Impregnation agent	pH	(g/100 g raw material)	(g/100 g raw material)
160	20	Acid[a]	1.8	0.03	0.03
170	10	Water	3.8	0.01	0.01
	5	Acid	1.8	0.04	0.02
	10	Acid	1.7	0.05	0.02
180	10	Water	3.8	0.03	0.01
	5	Acid	1.9	0.25	0.07
	10	Acid	2.0	0.32	0.07
190	10	Water	3.8	0.06	0.02
	5	Acid	1.9	0.51	0.08
	10	Acid	1.8	0.75	0.14
200	10	Water	3.5	0.04	0.16
	5	Acid	2.0	1.51	0.24

[a]0.9% (w/w) H_2SO_4.

values from 25 to 30%, corresponding to 8–10 g glucose/100 g raw material (Fig. 1). Considering the total yield of monomers in the filtrate after pretreatment (glucose and HS), a maximum total release of about 30 g sugars/100 g raw material was found in acid-impregnated biomass at 180°C and 10 min. This value is significantly higher than that reported by Palmarola-Adrados et al. (17) who obtained a maximum yield of 6.4 g/100 g raw material working with starch-free wheat fiber steam-pretreated at 200°C and 10 min, without acid addition. Our results are in accordance with the fact that the addition of acid before pretreatment allows for lower pretreatment temperatures whereas achieving good recovery yields.

Concerning degradation products and pH values in prehydrolyzate (Table 3), acid-impregnation of biomass in experiments at elevated residence time gave rise to lower pH and higher values of furans, mainly HMF, in all temperatures tested. Nevertheless, the quantification of furans can hardly explain hemicellulose losses in pretreatment, particularly at 190–200°C, where high xylose degradation was found to occur. It is possible that hemicelluloses were lost through high volatilization of furfural (very low amounts were detected in all experiments) and recondensation reactions.

Enzymatic Hydrolysis and SSF

To assess the effect of different pretreatment conditions tested on the digestibility of pretreated wheat straw (WIS fraction), EH, and SSF tests were performed. EH yield, expressed as percentage of glucose released

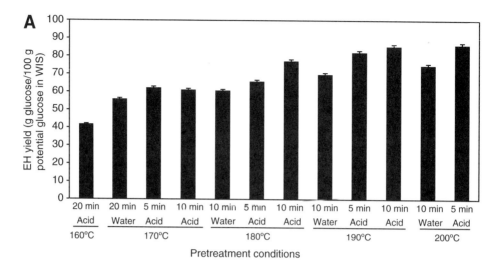

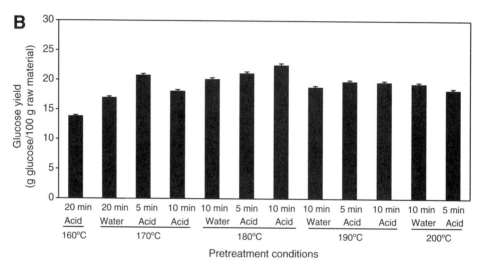

Fig. 2. Enzymatic hydrolysis yield of steam-exploded wheat straw at different pretreatment conditions, reported as g glucose obtained in EH/100 g potential glucose in WIS **(A)**, and glucose yield referred to untreated wheat straw (g/100 g raw material) **(B)**.

during hydrolysis at 72 h in relation to potential glucose in WIS fraction, together with glucose yield based on untreated raw material, are illustrated in Fig. 2. Results regarding the digestibility of pretreated straw (panel A) show that it is markedly dependent on temperature pretreatment, rising from 40% at 160°C to maximum yields of approx 85% at 190°C and 200°C in acid impregnated biomass at 10 and 5 min, respectively. It can be considered a fairly good result if we take into account that EH tests were performed at high substrate loading of 10% (w/v), which may involve end product inhibition and mixing problems. Palmarola et al. *(17)* reported increased saccharification yield of 92.4% in 190°C, 10 min pretreatment of wheat straw using similar cellulase loading of 15 FPU/g subtrate, but at 5% substrate loading. The difficulties of performing EH at high substrate loading have been also disccused by Alfani et al. *(10)* in bioconversion of steam-exploded wheat straw.

The increasing tendency of EH yield found in water-biomass samples might suggest that temperatures over 200°C could be tested to improve cellulose digestibility without acid addition. Nevertheless, experiments carried out by the authors in steam-pretreated wheat straw at 210–230°C and shorter time (data not shown) did not produce enhanced cellulose digestibilities. Beltrame et al. *(18)* working at high temperatures of 230°C in steam-exploded wheat straw did not achieve either saccharification yields exceeding 75%. Only when steam-exploded biomass was next extracted with dioxane/water was EH increased up to 93%.

In all temperatures tested, impregnation with acid resulted in higher degree of digestibility of WIS fraction compared to experiments in water-impregnated biomass. EH yields exceeded values found in water pretreated samples by 7% for experiments with acid at 170°C, and in 15–17% at higher temperatures of 180°C to 200°C in acid pretreated samples. Within acid-impregnated biomass experiments, higher residence times of 10 min also produced elevated digestibilities in comparison with 5 min experiments, except in 170°C tests, in which similar values were found among different times. The effect of harsh pretreatment conditions on enhancing the enzymatic digestibilty of pretreated substrates has been reported to be owing to a greater fraction of hemicellulose being dissolved, so allowing the enzyme to have greater access to cellulose. In experiments performed in this work, the lower HS removal from the 160°C to 180°C pretreated samples (Table 2) resulted in decreased EH yields of 40–65%, supporting the idea of incomplete hemicellulose hydrolysis lowering cellulose digestibilty. Um et al. *(13)* have described a linear relationship between cellulose digestibility and percentage of xylan removal from the solid in dilute-acid pretreated corn stover.

EH yield referred to potential glucose in WIS fraction is a unquestionable useful tool to valuate the effectiveness of pretreatment on saccharification performance. However, in establishing optimum conditions it is also valuable to calculate glucose produced by EH per gram of untreated raw material (Fig. 2B). This calculation takes into account solid recovery values in the pretreatment step and so, it provides a more representative data of overall process efficiency. In fact, in our experiments a maximum glucose yield on untreated material basis of about 23 g/100 g raw material was obtained at 180°C, 10 min in acid-impregnated biomass, a slightly higher value than that calculated for 190°C, 10 min (19 g/100 g raw material), which provided maximum EH yield reported on a WIS basis.

Figure 3 illustrates SSF yields in steam-exploded wheat straw as percentage of theoretical yield on WIS basis (panel A) and ethanol yields based on untreated raw material (panel B). Similarly to EH performance, yields reported on WIS basis increased as temperature rose, reaching maximum values of 80% of theoretical in acid-impregnated biomass at 190°C and 200°C

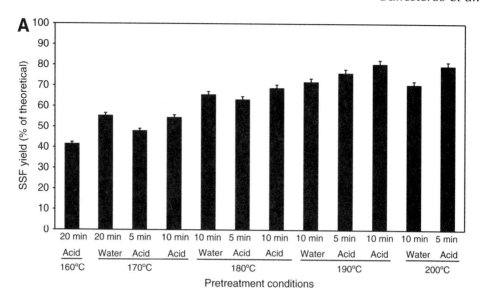

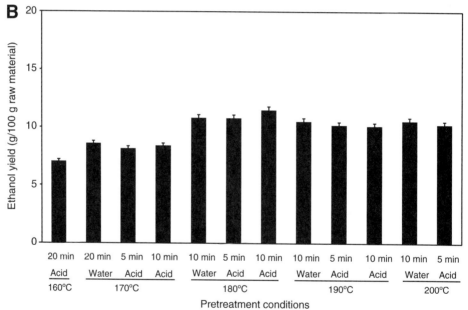

Fig. 3. SSF yield of steam-exploded wheat straw at different pretreatment conditions, calculated as the ethanol produced/potential glucose in WIS and reported as percent of theoretical **(A)**, and ethanol yield based on untreated wheat straw (g/100 g raw material) **(B)**.

during 10 and 5 min, respectively. At 180°C, values decreased up to 69% in acid-impregnated biomass pretreated for 10 min. Tucker et al. *(12)* described good SSF performance using similar cellulase and substrate loading than in our work, and *Saccharomyces cerevisiae* as fermenting yeast in WIS fraction from acid steam-pretreated corn stover at 190°C (90% SSF yield). Lowering the temperature to 180°C or 160°C resulted also in a drop in SSF of up to 65%.

However, if ethanol yields are based on untreated material, a highest value of 120 g ethanol/kg wheat straw is again found at 180°C, 10 min in acid-impregnated biomass. It is about 70% of the maximum theoretical amount of ethanol that could be attained taking into account the glucose content in raw material. At harsher conditions of 190°C and 200°C, some cellulose losses occur during pretreatment (Table 2), which result in slightly lower yields when referring to untreated material.

Considering our results, the evaluation of pretreatment effectiveness should consider not only the enhancement of cellulose hydrolysis but the overall process from cellulose to ethanol as well.

Conclusions

This study shows that dilute acid-impregnation of biomass is an efficient method to increase the EH and SSF of cellulose to ethanol of wheat straw, compared with water-impregnated biomass. Addition of acid enhances solubilization of hemicelluloses from raw material, although at high temperature and residence times tested, some release of glucose may also occur. The best pretreatment conditions to obtain high conversion yield to ethanol of cellulose-rich residue after SE were 190°C for 10 min, or 200°C during 5 min in straw impregnated with 0.9% (w/w) H_2SO_4 when reported in WIS basis. However, when calculating glucose and ethanol yields referred to initial raw material before pretreatment, 180°C for 10 min in acid-impregnated biomass gave better results than that obtained at 190°C. On the other hand, hemisugars recovery yield, considered also as a key parameter in the assessment of pretreatment efficiency was maximized in experiments performed with acid-impregnated biomass at 180°C and 10 min. So, considering that the overall utilization of the carbohydrate components in the feedstock is essential to decrease ethanol production costs, and that a lower temperature would be always advantageous since it results in saving energy cost, 180°C and 10 min would be the most adequate pretreatment conditions of acid-impregnated wheat straw for ethanol production, in the operational conditions used in this work.

Acknowledgment

This work has been supported by the European Community under the Energy, Environment and Sustainable Development Programme (contract no. NNE5/2001/685).

References

1. Directorate-General for Energy Research. European Commission. (2003), Publication EUR 20366: World energy, technology and climate policy outlook 2030 (WETO). Luxembourg, Office for Official Publications of the European Communities, 137 pp.
2. Kim, S. and Dale, B. (2004), *Biomass Bioener.* **26**, 361–375.

3. Negro, M. J., Manzanares, P., Ballesteros, I., Oliva, J. M., Cabañas, A., and Ballesteros, M. (2003), *Appl. Biochem. Biotechnol.* **105–108,** 87–100.
4. Emmel, A., Mathias, A., Wypych, F., and Ramos, L. P. (2003), *Bioresour. Technol.* **86,** 105–115.
5. Galbe, M. and Zacchi, G. (2002), *Appl. Microbiol. Biotechnol.* **59,** 618–628.
6. Saddler, J. N., Ramos, L. P., and Breuil, C. (1993), *Bioconversion of Forest and Agricultural Plant Wastes.* Saddler, J. N. (ed.), CAB International, London.
7. Varga, E., Réczey, K, and Zacchi, G. (2004), *Appl. Biochem. Biotechnol.* **113–116,** 509–523.
8. Glasser, W. G. and Wright, R. S. (1998), *Biomass Bioenergy* **14(3),** 219–235.
9. Montane, D., Farriol, X., Salvadó, J., Jollez, P., and Chornet, E. (1998), *J. Wood Chem. Technol.* **18,** 171–191.
10. Alfani, F., Gallifuoco, A., Saporosi, A., Spera, A., and Cantarella, M. (2000), *J. Ind. Microbiol. Biotechnol.* **25,** 184–192.
11. Ramos, L. P. (2003), *Quim. Nova,* **26,** 863–871.
12. Tucker, M. P., Kim, K. H., Newman, M. M., and Quang, A. N. (2003), *Appl. Biochem. Biotechnol.* **105–108,** 165–177.
13. Um, B. H., Karim, M. N., and Henk, L. L. (2003), *Appl. Biochem. Biotechnol.* **105–108,** 115–125.
14. National Renewable Energy Laboratory (NREL). Chemical Analysis and Testing Laboratory Analytical Procedures: LAP-001 to LAP-005, LAP-010 and LAP-017. NREL, Golden, CO. www. ott.doe.gov/biofuels/analytical_methods.html.
15. Ramos, L. P. and Saddler, J. N. (1994), *Enzymatic Conversion of Biomass for Fuels Production.* Himmel, M. E., Baker, J. D., Overend, R. P. (eds.), American Chemical Society Symposium Series 566, Washington.
16. Boussaid, A., Robinson, J., Cai, Y., Gregg, D. J., and Saddler, J. N. (1999), *Biotechnol. Bioeng.* **64(3),** 284–289.
17. Palmarola-Adrados, B., Glabe, M., and Zacchi, G. (2004), *Appl. Biochem. Biotechnol.* **113–116,** 989–1002.
18. Beltrame, F. and Marzetti, A. (1992), *Bioresour.Technol.* **39,** 165–171.

Catalyst Transport in Corn Stover Internodes

Elucidating Transport Mechanisms Using Direct Blue-I

Sridhar Viamajala,* Michael J. Selig, Todd B. Vinzant, Melvin P. Tucker, Michael E. Himmel, James D. McMillan, and Stephen R. Decker

National Bioenergy Center, National Renewable Energy Laboratory, 1617 Cole Boulevard, Golden, CO 80401, E-mail: Sridhar_viamajala@nrel.gov

Abstract

The transport of catalysts (chemicals and enzymes) within plant biomass is believed to be a major bottleneck during thermochemical pretreatment and enzymatic conversion of lignocellulose. Subjecting biomass to size reduction and mechanical homogenization can reduce catalyst transport limitations; however, such processing adds complexity and cost to the overall process. Using high-resolution light microscopy, we have monitored the transport of an aqueous solution of Direct Blue-I (DB-I) dye through intact corn internodes under a variety of impregnation conditions. DB-I is a hydrophilic anionic dye with affinity for cellulose. This model system has enabled us to visualize likely barriers and mechanisms of catalyst transport in corn stems. Microscopic images were compared with calculated degrees of saturation (i.e., volume fraction of internode void space occupied by dye solution) to correlate impregnation strategies with dye distribution and transport mechanisms. Results show the waxy rind exterior and air trapped within individual cells to be the major barriers to dye transport, whereas the vascular bundles, apoplastic continuum (i.e., the intercellular void space at cell junctions), and fissures formed during the drying process provided the most utilized pathways for transport. Although representing only 20–30% of the internode volume, complete saturation of the apoplast and vascular bundles by fluid allowed dye contact with a majority of the cells in the internode interior.

Index Entries: Biomass conversion; internode transport; dilute acid pretreatment; direct blue-I; biomass recalcitrance.

Introduction

Conversion of biomass to ethanol involves mechanical and catalytic steps to disrupt the complex lignocellulosic structure, exposing polysaccharides that can then be hydrolyzed to fermentable sugars. Catalysts, such as dilute sulfuric acid are used to condition or "pretreat" biomass

*Author to whom all correspondence and reprint requests should be addressed.

to increase its susceptibility to enzymatic hydrolysis *(1,2)*. Uneven distribution of catalyst within the biomass tissue results in heterogeneous pretreatment zones, with only a fraction of the biomass exposed to optimal conditions *(3)*. This situation can be especially problematic with dilute acid catalysis, as areas of excessive pretreatment lead to the formation of byproducts, such as furfural, which inhibit fermentation of downstream sugars and decrease conversion yields *(2)*, whereas incomplete pretreatment zones result in lowered enzymatic digestibility *(4)*.

Within the pulp and paper industries, efficient transport of catalyst into biomass has long been recognized as an important step in the pulping process. Unit operations to uniformly impregnate wood chips are often part of the process design and operation *(5)*. In the biomass-to-ethanol process literature, the effects of catalyst transport have generally been neglected, although this issue has been recognized as one of the factors that could contribute to a decrease in xylan hydrolysis rates during dilute acid pretreatments *(6)*. The lack of concern toward catalyst transport mechanisms results from several contributing factors, including the tendency to use milled biomass in batch laboratory experiments *(6–8)* and the use of compression screw feeders in larger scale studies, which exert high shear and cause biomass size reduction; examples are the Sunds reactor *(9)* or the StakeTech reactor *(4)*. However, size reduction, either as a stand-alone unit operation or in a compression screw feeder system, can lead to significant increases in equipment and operating costs *(2,10)*. In some cases, compression of the feed stock may in fact be detrimental to pretreatment, causing biomass pore structure collapse and leading to uneven heat and mass transfer during pretreatment *(11)*. In an effort to understand the effect of catalyst impregnation on biomass conversion in this study, several methods of impregnation were employed, including soaking at various standard temperatures and atmospheric pressure, soaking under vacuum, and preheating of biomass in air before submersion. It was conceived that this latter approach would mimic the steam preheating, common in the pulping industry, and allow catalyst penetration as the hot entrained air contracted after submersion.

Corn stover is available in significant quantities for conversion to ethanol and other bio-based products *(9)*. Corn stover included the leaves, sheaths, stalks, cobs, and husks, with stalks contributing to about 60% of the dry mass and at least 50% of the total stover carbohydrate *(1)*. As shown in Fig. 1, the stalks consist of internodes and nodes, with internodes containing two-thirds of the total stalk dry- and carbohydrate-mass *(1)*.

A typical internode cross-section depicts several morphologically and functionally different cell types (Fig. 1) with the dominant structures being pith, vascular bundles, and rind. The pith, characterized by relatively large, thin-walled parenchyma cells, comprises a volumetrically large portion of the internode. Parenchyma cells, with a thin primary cell wall and a *de minimus* secondary cell wall, function primarily as storage reservoirs in the

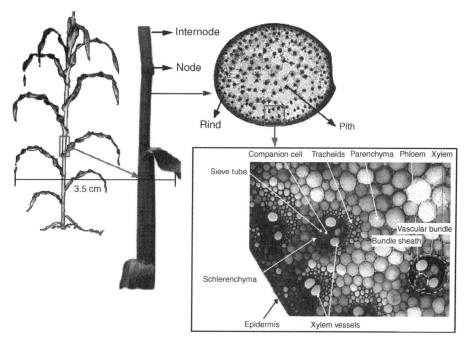

Fig. 1. Corn stover stem anatomy; adapted from Thomas et al. *(1)*, with permission.

internode. Although comprising a very large volumetric percentage of the plant, they included only about 11–12% of the mass in the internode *(12)*.

Distributed throughout the pith are vascular bundles, which run along the length of the corn stem. Mature vascular bundles are made up of an array of vascular tissues, primarily xylem, and phloem. The majority of the mass is sclerenchyma-derived xylem vessels and xylem tracheid fibers containing thick secondary cell walls. In the living plant these bundles serve as a means to transport water and nutrients along the length of the plant stem and leaves. Despite their relatively low contribution to the cross-sectional volume, the thick secondary cell walls of the vascular bundles increase their relative density, allowing the pith vascular bundles to make up approx 11–12% of the internode mass, about the same mass as the more volumetrically abundant parenchyma cells *(12)*.

The rind as a tissue type makes up about 75% of the mass of the internode and thus is a critical component for biomass conversion *(12)*. The rind layer has dense vascular bundles as well as collenchyma, parenchyma, and epidermal cells and is coated by an exterior waxy cuticle. The collenchyma cells are distinct from parenchyma in having thickened primary cell walls. The dense and water resistant nature of this protective layer can pose a significant barrier to catalyst transport.

Transport among nonvascular cells in a living plant is carried out through pits, areas of thin primary cell wall devoid of secondary cell wall among adjacent cells. Pits contain plasmodesmata, which are direct cytoplasmic connections among living cells. Pits also facilitate movement of intracellular components between vascular cells and surrounding tissues. This plant-wide contiguous cytoplasm is referred to as the symplast. In contrast to the intracellular symplast, the apoplast is the intercellular space exterior to the cell membranes, providing a route for transport throughout the plant outside of the symplast *(13)*. In dry senesced plants, the vascular bundles, pits, and apoplast may serve as conduits for catalyst transport.

Understanding how plant structure affects catalyst transport is a key to overcoming a number of pretreatment performance barriers traditionally addressed through costly process modifications. In this study, we have attempted to characterize the bulk transport characteristics of catalyst through dry intact corn stover using solutions of direct blue-I (DB-I) dye to simulate generic catalyst behavior. The affinity of DB-I for cellulose may mimic the binding of catalytic enzymes to cellulose; however, we are making no attempt to qualify or quantify this phenomenon. It should also be noted that the molecular weight of DB-I is about 10 times that of H_2SO_4, placing limitations on the routes of penetration used by DB-I in comparison with H_2SO_4. These differences, however, are likely to be undetectable as both limits are far below the resolving capacity of light microscopy. Using corn stem internodes to provide a simplified and reproducible model system, we have observed dye transport over a range of impregnation conditions. Through light microscopy and subsequent image analysis, we have observed in the internode matrix, a number of possible catalyst transport barriers and pathways.

Methods

Corn Stover

Whole corn stover, hand-cut from a single field at the Gustafson Farm in Weld County, CO, was derived from Round-Up Ready Pioneer corn hybrid, 36N18. The plants, cut in mid-January of 2005, had completely senesced in the field. At cutting, the average moisture content of the corn stalks was roughly 20%, which was reduced by air-drying to about 10% moisture content before any further preparations. For this study, the third, fourth, and fifth internodes from the base of the corn stalk were cut from the dried stalks using a size 3/0 jeweler's saw blade. All nodes were removed to provide a relatively uniform structure from which smaller segments could be cut.

Measurement of Internode Densities and Void Fraction

The bulk density (ρ_{bulk}), tissue density (ρ_{abs}), and void fraction (ε) were calculated for the internodes used in this study. This information was used to estimate the degree of impregnation (% saturation of the internode void space) that occurred during each impregnation scenario. Analysis of the saturation data was conducted in conjunction with microscopic inspection of the dye internodes in order to understand how the distribution of catalyst (dye) throughout the internode correlated with the degree of impregnation under different conditions. For this study, bulk density is defined as the ratio of the internode mass to the volume of the intact structure. Saturation is defined as the potential amount of dye that can fill all spaces in the biomass, both empty air spaces inside cells and vessels, as well as any gaps between cell wall components. It is estimated from the experimentally determined difference in bulk densities of intact and finely milled internodes. The volume, in this case, is the sum of the volume of the plant tissue and the air trapped within the internode. The absolute density is defined as the average density of the solid tissue alone.

To measure average bulk density, approx 60 internode segments, each 3–5 cm in length, were cut from larger internode sections. The air-dried mass of each segment was recorded to the nearest 0.1 mg. Segment volume was measured to the nearest 0.1 mL by volumetric displacement and then corrected for any water absorption that occurred during the measurements by differential mass measurements. Segments were then oven dried at 105°C overnight in order to obtain a dry mass value. The average internode bulk density was then calculated (m/V) on the basis of both the air-dried and oven-dried masses.

Absolute density was measured by milling a number of internodes to pass a 20-mesh screen. A known amount of water was added to a measured mass of milled stover, vortexed briefly, and centrifuged for 10 min at 1500g. After centrifugation, the difference between the total volume and the liquid volume was taken as the solids volume and used to calculate ρ_{abs}. Using both the absolute and bulk densities, the average internode void fraction (Eq. 1) was calculated.

$$\text{Void Fraction} = \varepsilon = 1 - (\rho_{bulk}/\rho_{abs}) \qquad (1)$$

Transport Studies With DB-I

Powdered DB-I was obtained from Pylam Products Company Inc., Tempe, Arizona. In general, direct dyes are anionic and have a planar aromatic structure and a strong affinity for cellulose. DB-I, often referred to as pontamine sky blue, is a tetrasulfonated dye with a molecular weight of 993 and a molecular diameter of about 1 nm. Solutions of 1% and 0.5% (w/v) were prepared in water from the powdered dye for use throughout this study [14].

Preparation of Internodes

Segments measuring 3.5 cm ± 0.1 cm in length, were cut from the larger internode sections and the ends were briefly exposed to moderate vacuum (approx 10 inch Hg) in order to remove debris from the cutting process that may have been lodged in any open-ended structures, such as the vascular bundles. The air-dried mass (m_{dry}) of each internode segment was recorded to the nearest 0.1 mg. The internode volume (V_{tot}) and void volume (V_{void}) were approximated for each segment using the following equations:

$$V_{tot} = m_{dry}/\rho_{bulk} \qquad (2)$$

$$V_{void} = V_{tot}\varepsilon \qquad (3)$$

Impregnation of Internode Segments With DB-I

Dye transport through internode sections was studied under three impregnation conditions:

1. submersion in dye solution at atmospheric pressure and varying temperatures,
2. submersion under varying vacuum pressure, and
3. preheating segments in air before submersion into room temperature dye solution.

At atmospheric pressure, segments were impregnated at room temperature (approx 22°C) for a range of exposure times between 10 s and 24 h. Segments were also impregnated at atmospheric pressure with dye solutions heated to 50, 60, 70, 80, and 90°C over a range of exposure times between 30 s and 2 h. To explore the effect of vacuum pressure, segments were impregnated under a 22.5 inch Hg vacuum for exposure times between 30 s and 24 h. In addition, the effect of lower vacuum pressures (5, 10, 15, and 20 inch Hg) was also tested for shorter exposure times (1–30 min). The effect of preheating on impregnation was studied by heating segments in an oven that was set to the desired temperature. After fixed exposure times, the hot segments were rapidly transferred into room temperature dye solution and submerged for 5 min. In this study, segments were oven heated for 5, 10, and 20 min at the following temperatures: 50, 60, 85, 105, and 120°C.

For all impregnations, internode segments were completely submerged in DB-I solution and either held stationary with metal tongs for short duration exposures or weighed down with stainless steel hose clamps for long-term exposures. Triplicate segments were dyed for all time points under all conditions. On removal from the dye solution, any excess dye was wiped from the exterior of each segment and the mass was recorded to the nearest 0.1 mg. This "wet" mass was then used to calculate an approximate value for the percent saturation (Eq. 4) of the segments void space.

$$\% \text{ Saturation} = 100*((m_{wet} - m_{dry})/\rho_{dye})/(V_{void}) \qquad (4)$$

Microscopy: Sectioning and Image Analysis

After impregnation, internode segments were air-dried in a laminar flow hood before any manipulation. Dry segments were laterally bisected by hand using Performa Super double-edged razor blades from the

Table 1
Corn Stover Density Measurements From Literature

Authors	Bulk density (g/cm^3)	Tissue density (g/cm^3)
This study	0.13	1.13
Mani et al. (15) Set 1	0.132	1.14
Mani et al. (15) Set 2	0.156	1.171
Mani et al. (15) Set 3	0.158	1.179
Thomas et al. (1) Set 1	0.11	NA
Thomas et al. (1) Set 2	0.15	NA

American Safety Razor Company in Verona, VA. In order to evaluate dye movement along the length of the internode, images of the internal face of the bisected segments were captured using a Zeiss Stemi 2000-C stereomicroscope. The microscope was attached with a Sony DSP 3 CCD color video camera. A few wet segments were laterally bisected to check for dye movement that may have occurred during drying. After stereomicroscope imaging, lateral and transverse thin sections (approx 0.5 mm) were cut from the bisected halves. Before making transverse sections, the rind was removed to facilitate easy and uniform thickness cuts. Thin sections were imaged using an Olympus DP70 Microscope Digital Camera (maximum resolution of approx 12.5 million pixels) attached to an Olympus IX71 Inverted Microscope.

Results

Internode Bulk Properties

The average bulk density was measured on the basis of both air-dried and oven-dried internodes. The average moisture content of the air-dried internodes was measured to be 9.3 ± 0.7% ($N = 56$). The average bulk density of air-dried internodes was measured to be 0.130 ± 0.007 g/cm^3 ($N = 56$) and the average bulk density of the internodes on an oven-dried basis (0% moisture) was 0.142 ± 0.009 g/cm^3 ($N = 56$). The air-dried bulk density was used for the remainder of the calculations in this study. The average absolute internode tissue density (air-dried basis) was determined to be 1.13 ± 0.05 g/cm^3 ($N = 8$). These measured corn stover densities were in agreement with previous measurements reported by Mani et al (15) and Thomas et al. (1) (Table 1).

Using the values for bulk density and absolute density, the void fraction of the air-dried internodes was estimated to be 0.89 ± 0.01. This void fraction was used to estimate the void volume of an internode segment (Eq. 3), which represents the volume of an air-dried internode that can be occupied by a catalyst solution. The void volume was further used to calculate the degree of saturation (Eq. 4) after impregnation, to provide an estimate of the volume occupied by dye solution as a fraction of the total

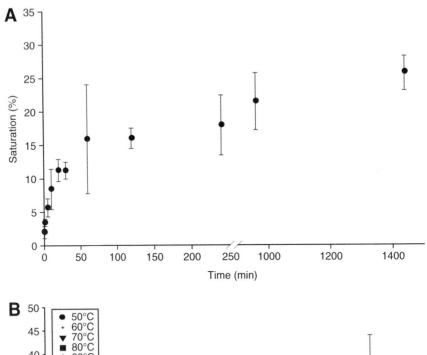

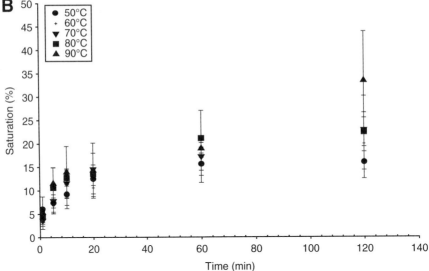

Fig. 2. Saturation of internode void space vs time after submerging in DB-I solution under atmospheric pressure at **(A)** room temperature and **(B)** varying dye temperatures.

available volume. This calculated degree of saturation was used to compare uptake of dye solution under the tested conditions.

Effect of Temperature and Pressure on Dye Uptake and Distribution

Impregnation at Atmospheric Pressure

Dye impregnation was very slow at room temperature and atmospheric pressure. Even after 24 h submerged in dye solution, a saturation of only 20–25% was achieved (Fig. 2A). When the segments were laterally

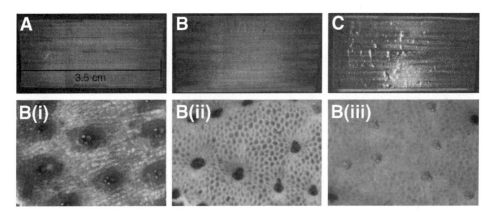

Fig. 3. Internode segments impregnated with DB-I at room temperature and atmospheric pressure. **(A)** lateral section after 5 min, **(B)** lateral section after 2 h; (i) 1 mm, (ii) 10 mm, and (iii) approx 175 mm from end; **(C)** lateral section after 24 h.

Fig. 4. Lateral sections of internodes impregnated with DB-I at atmospheric pressure and elevated temperatures: **(A)** 50°C for 2 h, **(B)** 70°C for 2 h, and **(C)** 90°C for 2 h.

bisected, dye movement was observed only along the length (Fig. 3) indicating that dye entered laterally through segment ends and not through the rind. After 24 h, although dye penetrated throughout the length of the piece, there were zones along the rind in which dye did not appear to reach. Increasing the temperature of the dye solution did not significantly improve dye uptake into the internodes. After 2 h of exposure to dye solution at higher temperatures, the saturation was only approx 25% and dye movement was restricted to internode ends. (Figs. 2B and 4). As can be seen in Fig. 4, the lateral dye movement profiles of internodes that were impregnated for 2 h were very similar throughout the entire temperature range. After 2 h of impregnation, dye was present both in the vascular bundles and the apoplast in sections 2 mm from the end. However, in sections that were 1 cm away from the end, dye was present only in the vascular bundles, whereas cross-sections cut through the middle of the internodes did not show any dye.

Vacuum Impregnation

Saturation of 25–40% of the void space was achieved immediately (<30 s) when segments were impregnated under 22.5 inch Hg vacuum (Fig. 5A). Under this condition, dye moved throughout the entire internode

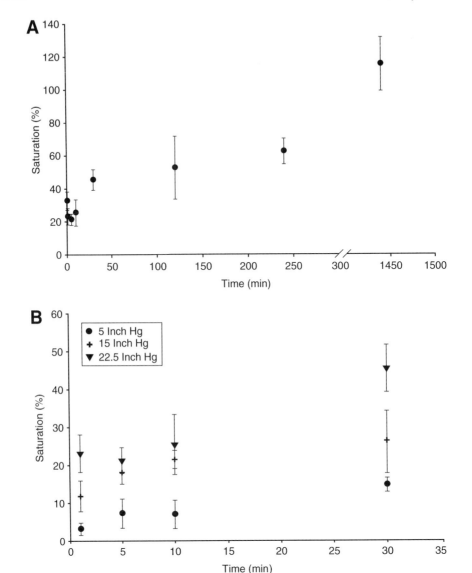

Fig. 5. Vacuum saturation of the internode void space after impregnation by submerging in DB-I under **(A)** 22.5 inch Hg and **(B)** varying vacuum pressures.

length and microscopic observation of cross-sections showed the presence of dye within and along the vascular bundles, as well as throughout the entire apoplastic continuum (Fig. 6). After the initial rapid saturation, further increase in moisture content in the internodes occurred at a much slower rate, and a long period under vacuum was required (>12 h) to achieve 100% saturation. At lower vacuum strengths, more time was required to achieve greater than 25% saturation (Fig. 5B). In cases in which internodes were less than 25% saturated, uniform dye distribution was not achieved (data not shown).

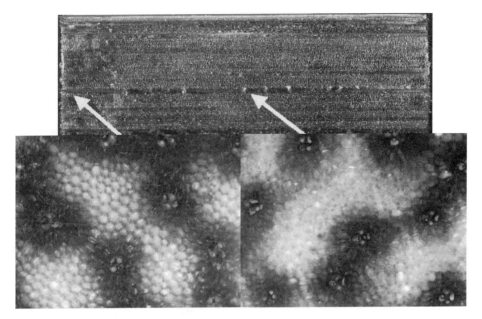

Fig. 6. Impregnating for 30 s under 22.5 inch Hg vacuum allowed for 30% saturation of the available void space. The lateral section shows uniform distribution of dye over the length of the internode whereas transverse sections at the ends (left) and the interior (right) show dye movement into the vascular bundle–parenchyma cell interface area.

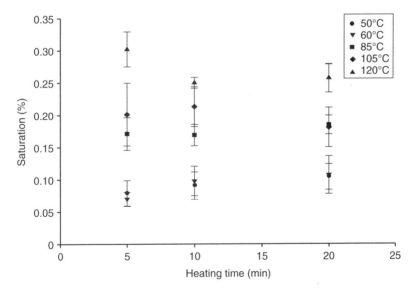

Fig. 7. Saturation of internode void space after preheating internodes segments and submerging in room temperature DB-I solution.

Dry Heat Impregnation of Internodes

Preheating dry internodes and then rapidly submerging them in dye solution was studied as another method of dye transport into internodes. As can be seen from Fig. 7, saturations levels of 20–30% could be achieved

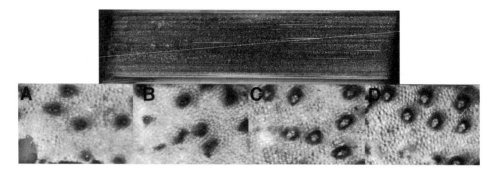

Fig. 8. Lateral and transverse sections of an internode preheated at 120°C for 15 min before submersion in DB-I solution. Cross-sections were taken from **(A)** 2 mm from left end, **(B)** 10 mm from left end, **(C)** at the internode midpoint, and **(D)** 5 mm from the right end.

by preheating pieces above 100°C for only 5 min and longer heating times did not appear to significantly enhance dye uptake (Fig. 7). Examination of lateral sections showed dye movement throughout the internode length at these saturation levels (Fig. 8). Closer examination of cross-sections showed that dye was present in most of the apoplastic space. Although dye was present both within and around the vascular bundles near the ends, it was present only in the apoplast around the vascular bundles in the center of the pieces, approx 175 mm from the cut end.

Discussion

Our experiments with DB-I revealed a number of pathways and barriers to the transport of dye into the model corn internode system. In general, the primary entry points were at the internode ends. Dye movement from this point occurred via four main pathways:

1. through the intercellular space surrounding parenchyma cells (the apoplastic continuum),
2. through vascular bundles,
3. along the parenchyma-vascular bundle interface, and
4. along the inner rind–pith interface.

Radial movement of dye into the internode through the rind was not observed, suggesting that the waxy outer surface of the rind acts as a barrier to penetration (just as nature intended). The mechanisms driving fluid into these pathways in dry corn stems are most likely to be capillary action, varying pressure differentials or a combination of both. The primary barriers to dye movement were the waxy outer surface of the rind and the air trapped within individual cells.

Pathways of Catalyst Distribution

The apoplastic continuum and vascular bundles are the major routes for dye transport into the internode. Whereas movement through the vascular bundles is not surprising, because they form hollow channels in senesced plants, the extent and rapidity of liquid transport through the parenchymal apoplastic space was not anticipated. Although transport of dye through vascular bundles is obvious, it appears to be limited in extent compared to transport through the apoplast. Dye penetration through the vascular bundles under atmospheric conditions does not completely penetrate the entire bundle, presumably owing to the inability to displace the entrained air and through restriction by cellular junctions, perforation plates, and sieve plates.

Apoplast

Virtually all dye transport through the volumetric space occupied by thin-walled parenchyma tissue occurred via the intercellular space among cells, referred to here as the apoplastic continuum. The bulk of movement through this space presumably occurs at the junctures in which three or more cells come together and small air channels form during senescence. Although small, there appear to be several key factors that enhance the apoplast's role in catalyst distribution. First, the pathways with the apoplastic space are continuous and unimpeded. Unlike the vascular bundle components, in which transport is impeded by cell junctions, sieve plates, and perforation plates, the apoplastic continuum's main limitation is its narrow size. Second, the air passages within the apoplast, allow direct access of the catalyst to cell wall material, without the need to fill up the entire cell volume or to penetrate any residual membrane material. Examination of high magnification (×200) images (Fig. 9A,B) reveals clear ovular shaped disks formed by the abutment of cell walls from two adjacent parenchyma cells. From these images it was unclear whether or not dye penetrates into the middle lamella. How rapidly and extensively catalysts, specifically the hydronium ion and enzymes, can penetrate radially from the apoplastic continuum through the middle lamella and into the innermost layer of cell wall is still unanswered.

Because enzymes have specific affinities for biomass, much like the affinity of DB-I for cellulose, the potential exists for a titration effect of catalyst transported by the apoplast. We see indirect evidence of this in the fading of dye levels in parenchyma cells as you move radially out from the vascular bundles, but we do not have any direct evidence to support this idea. Images in Fig. 10 display distinct layers of cells surrounding the vascular bundles in which the cells walls appear to be dyed in decreasing hues of blue as they move outward. In these cases only a distinct layer of cells have their walls dyed blue. It is possible that high pressure differentials could drive the dye solution from the interior of vascular cells through pits and plasmodesmata and into adjacent parenchyma cells. The pressure differential would decrease as fluid moved into each subsequent layer of cells until it was not strong enough to push dye through into the next layer, and movement would essentially

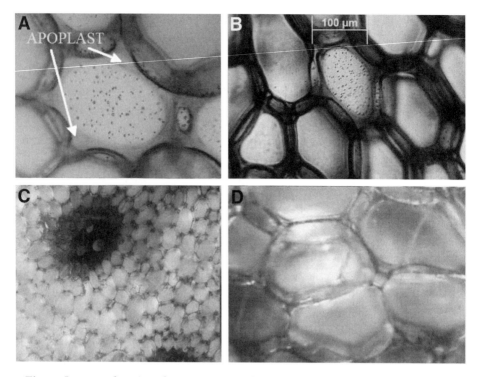

Fig. 9. Images showing dye movement through the intracellular spaces between parenchyma cells. Pits, cell wall junctions, and the apoplastic continuum framework are clearly visible in images taken at ×200 (**A,B**) with peripheral light source and at ×40 (**C**) and ×40 plus digital zooming (**D**) without peripheral light illumination.

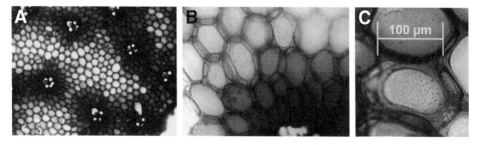

Fig. 10. Images of distinct layers of dyed cells surrounding the vascular bundles showing fading parenchymal dye levels distal to the vascular bundles. A, B, and C show increasing detail with higher magnification.

stop. More experimentation will be necessary to test this hypothesis through higher resolution, microscopy techniques.

Vascular Bundles

Vascular bundles provide the primary bulk fluid transport in living plants. These vessels provide convenient channels in dried biomass, but are limited in their ability to distribute catalyst radially into the parenchyma cells under any condition (see Fig. 10) and even laterally under

Fig. 11. Lateral cross-section of a dyed internode depicting dye movement from the cut ends of the internode into the interior and no movement through the exterior.

atmospheric pressure impregnation conditions (Fig. 3Biii). When impregnated under vacuum, the extent of transport in the vascular bundles is much greater, suggesting that the requirement for air displacement is a limitation (Fig. 6). When heating and submersion was used to impregnate the biomass, the vascular bundles did not show the same extent of interior transport as when vacuum was applied, however, the area immediately adjacent to the bundles was extensively saturated with dye (Fig. 8).

Vascular Bundle–Parenchyma Cell and Rind–Pith Interfaces

The vascular bundle–parenchyma cell interface, the result of differential shrinkage of the two tissue types during drying, appears to rapidly and extensively move dye the length of the internode. This transport route has the additional benefit of allowing direct access of catalyst to the apoplastic continuum along the length of the internode, without the need to penetrate the cell walls of vascular bundle cells. Presumably, a similar transport mechanism exists along the rind–pith interface. Dye movement in the vicinity of the rind was readily observed along the inner surface of the rind, through what we hypothesized to be cracks and fissures at the rind–parechyma interface. In some cases, this movement allowed for dye contact with the entire inner surface of the rind. Dye movement was also observed through the rind vascular bundles in a number of cases; however, the exterior of the rind completely blocked any movement of dye through the rind to the interior internode (Fig. 11). In all cases, dye transport occurred through the open ends of the internode. Because of this fact, it is probable that impregnation would be difficult with the scenario of a whole internode-node system in which the internode ends are not exposed. This observation indicates that at least a small degree of mechanical crushing or size reduction of corn stems is necessary to allow access of catalyst to the parenchyma though the cracks in the rind, as well as by creating cracks in the parenchyma. However, excessive mechanical shear and crushing could lead to collapse of the pore structure and result in poor catalyst penetration *(11)*.

Impregnation Barriers

Entrained Air

Although apoplastic movement of minerals is common in plants tissues, the air-trapped within parenchyma cells appears to be difficult to displace. The slow and incomplete movement of dye into the parenchyma cells under atmospheric pressure is likely owing to slow removal of the air trapped within cells. Under vacuum, this entrained air is removable, allowing for displacement by liquid. When the air is blocked from exiting however, such as by submersion in liquid, movement of liquid into the parenchyma intracellular space is prevented and movement in the apoplastic continuum is limited. The primary escape route for the entrapped air is likely through the plasmodesmata in pit fields that connect parenchyma cells. However, these pits are very small (approx 20 nm) and likely blocked owing to cell wall drying (data not shown). Note that the surface tension of water would prevent movement through these narrow pores into a dry space, especially if air displacement were limited. From these images it is unclear whether any dye actually penetrates into the cell wall beyond the apoplastic air gap. The air trapped within individual cells appears to be a primary barrier to dye movement through the internode.

Although Kim and Lee *(16)* have previously proposed diffusion into lignocellulosic biomass as the mass transfer mechanism, their experiments were performed with milled corn stover with a maximum particle size of 14 mesh. It is likely that at such small particle sizes, the cell structure is destroyed or disrupted and air contained within cells provides little if any barrier to being displaced by liquid. Whereas particle size reduction could result in rapid acid movement and contact with biomass *(16)*, milling biomass to a fine size is impractical on a large scale owing to the high-energy input required *(2)*. We hypothesized that heating intact internode sections would cause the entrained air to expand and on cooling by submersion in dye, the contracting air would be displaced by dye. This appeared to be the case, as heated samples rapidly took up dye on submersion and the dye distribution appeared to be rapid and extensive throughout the apoplast. Air removal techniques, such as vacuum or heat displacement of intracellular air, resulted in significantly improved dye penetration compared to diffusion enhancing mechanisms, such as submersion in liquids at high temperature.

The primary mechanism of liquid movement into and through intact corn internodes appears to be by bulk transport of liquid, which occurs as the air entrained within cells gets displaced by liquid. Similar fluid penetration mechanisms have previously been hypothesized for penetration of chemicals into wood during wood pulping operations *(5)* and removal of air has been accomplished by "penetration aid" techniques, such as presteaming, purging, and evacuation. Our method of preheating the segments before impregnation by immersion is similar to the common practice of presteaming of woodchips before liquor impregnation during the paper pulping process. Elevated temperature causes entrained air to expand, driving the air out and reducing the pressure within the void space. When heated segments are submerged in dye solution, the pressure differential generated by the contracting air acts to quickly draw liquid into the void space (5). Under these conditions, cross-sectional segments distal to the internode ends showed the presence of dye in the apoplast surrounding the vascular bundles, but the lack of dye inside the vascular tissue suggests that dye movement occurs directly through the apoplast, specifically along the exterior of the vascular bundles. A proposed mechanism for this result lies in the differential contraction of vascular tissue and parenchyma cells during senescence and drying. As the parenchyma cells contract to a higher extent than vascular tissue (presumably owing to the much higher structural strength of vascular tissue), fissures and cracks are generated along the parenchyma-vascular bundle interface. This structural gap could allow much more rapid and extensive transport of liquids by capillary action, as the gap lacks the physical barriers present in vascular tissue, such as perforation and sieve plates. In addition, because this gap occurs at the parenchyma-vascular bundle interface, the liquid has direct access to the apoplastic continuum, without needing to rely on dye transport out of the vascular bundles and into the apoplast.

Dye transport was slow and uniform distribution was difficult to achieve when impregnation was conducted at atmospheric pressure. Quick and uniform distribution of dye was easily achieved when internodes were preheated before impregnation or impregnated under high vacuum. In both cases, uniform distribution throughout most of the observed pathways mentioned above occurred when the internode void space was 25–35% saturated. Presumably, the remainder of unsaturated volume was the interior void space of the bulk of the parenchyma cells. We speculate that dye may move to the interior of cells immediately surrounding the vascular bundles, but have questioned whether transport occurs into the middle lamella and cell wall layers when the apoplast is fully saturated. If catalyst from the apoplast can be absorbed into the cell walls, then catalyst could be in contact with each hydrolyzable site without the need to saturate the entire parenchyma intracellular space. Complete impregnation of cell walls with catalyst could be achieved with a moisture content of only approx 30%. If our speculations are correct, then one could theoretically achieve catalyst contact with 100% of the cell wall surfaces by saturating only 30% of the internode void volume. This could, in turn, increase the solids content in pretreatment reactors up to about 70%, significantly higher than current solid loadings of about 30–50% (9,17). Higher solids content during pretreatment would result in lower use of process water, reduction in volume of catalyst, and higher sugar concentrations; all of which would reduce downstream processing costs.

Conclusions

The primary observed pathways for dye transport through our corn internode models were:

1. through the vascular bundles,
2. along the vascular bundle–parenchyma cell interface,
3. throughout the apoplastic continuum, and
4. along the rind–parenchyma interface.

Transport via all routes was initiated at the open ends of the internodes, whereas virtually no transport occurred through the impermeable rind exterior. Air trapped inside cells, particularly the parenchyma, proved to be a difficult barrier, blocking access to the interior cell wall surfaces. Dye transport was slow and a uniform distribution was difficult to achieve when impregnation was conducted at atmospheric pressure. Quick and uniform distribution of dye was achieved when internodes were preheated before impregnation or impregnated under high vacuum. In both cases uniform distribution throughout most of the observed impregnation pathways mentioned above occurred when the internode void space was 25–35% saturated. We speculated that dye may move to the interior of cells surrounding the vascular bundles through pit membranes and question whether transport occurs into the middle lamella when the apoplast is fully saturated. Currently, more experimental work is necessary to bring light to these issues, but if our speculations are correct then one could theoretically achieve catalyst contact with 100% of the cell wall biomass by saturating only 30% of the internode void volume.

In continuation of this work, we will extend our simple internode model studies to include more complex anatomical structures, such as the nodes, and employ higher magnification microscopy tools, such as SEM, TEM, and AFM. We have also initiated experiments aimed at determining how effectively whole internode segments can be pretreated with varying levels of substrate impregnation. Besides providing us with information about catalyst transport in bulk corn stover, this work will also enhance experimental effort to understand interactions occurring in biomass at the ultrastructural and molecular level during both pretreatment and enzymatic hydrolysis.

Acknowledgments

This work was supported by the US Department of Energy (DOE), Office of the Biomass Program. The authors would like to thank Donald Gustafson for supplying the stover used in this study, and Stephanie

Porter, Steven Thomas and Rick Elander for their advice and assistance during preparation of this article.

References

1. Thomas, S., Porter, S., Jurich, J., et al. (2004), *Technical Report FY04-561*, National Renewable Energy Laboratory, Golden, CO.
2. McMillan, J. D. (1994), *ACS Symposium Series* **566**, 292–324.
3. Kazi, K. M. F., Jollez, P., and Chornet, E. (1998), *Biomass Bioenergy* **15**, 125–141.
4. Ramos, L. P. (2003), *Quimica Nova* **26**, 863–871.
5. Malkov, S., Tikka, P., and Gullichsen, J. (2001), *Paper Timber* **83**, 468–473.
6. Esteghlalian, A., Hashimoto, A. G., Fenske, J. J., and Penner, M. H. (1997), *Biores. Technol.* **59**, 129–136.
7. Soderstrom, J., Pilcher, L., Galbe, M., and Zacchi, G. (2003), *Biomass and Bioenergy* **24**, 475–486.
8. Zhu, Y., Lee, Y. Y., and Elander, R. T. (2004), *Appl. Biochem. Biotechnol.* **117**, 103–114
9. Schell, D. J., Farmer, J., Newman, M., and McMillan, J. D. (2003), *Appl. Biochem Biotechnol.* **105/108**, 69–85.
10. Aden, A., Ruth, M., Ibsen, K., et al. (2002), *Technical Report NREL/TP-510-32438*, National Renewable Energy Laboratory, Golden, CO.
11. Kim, K. H., Tucker, M. P., and Nguyen, Q. A. (2002), *Biotechnol. Prog.* **18**, 489–494.
12. Decker, S. and Vinzant, T. (2004), *Office of Biomass Program E Milestone*, ID No. FY04-602.
13. Raven, P. H., Evert, R. F., and Eichhorn, S. E. (1992), *Biology of Plants, 5th edition*, Worth Publishers, Inc., New York.
14. Horobin, R. W. and Kiernan, J. A. (2002), *Conn's Biological Stains*, 10th edition, Biological Stain Commission, Oxford.
15. Mani, S., Tabil, L. G., and Sokhansanj, S. (2004), *Canadian Biosys. Eng.* **46**, 55–61.
16. Kim, S. B. and Lee, Y. Y. (2002), *Biores. Technol.* **83**, 165–171.
17. Tucker, M. P., Kim, K. H., Newman, M. M., and Nguyen, Q. A. (2003), *Appl. Biochem. Biotechnol.* **105**, 165–177.

Evaluation of Cellulase Preparations for Hydrolysis of Hardwood Substrates

Alex Berlin,*,[1] Neil Gilkes,[1] Douglas Kilburn,[1] Vera Maximenko,[1] Renata Bura,[1] Alexander Markov,[2] Anton Skomarovsky,[2] Alexander Gusakov,[2] Arkady Sinitsyn,[2] Oleg Okunev,[3] Irina Solovieva,[3] and John N. Saddler[1]

[1]*Forest Products Biotechnology, Faculty of Forestry, The University of British Columbia, Vancouver, BC V6T 1Z4, Canada, E-mail: alex.berlin@ubc.ca;*
[2]*Department of Chemical Enzymology, Faculty of Chemistry, M. V. Lomonosov Moscow State University, Vorobyevy Gory, Moscow 119899, Russian Federation; and* [3]*Institute of Biochemistry and Physiology of Microorganisms, Russian Academy of Sciences, Pushchino, Moscow Region 142292, Russian Federation*

Abstract

Seven cellulase preparations from *Penicillium* and *Trichoderma* spp. were evaluated for their ability to hydrolyze the cellulose fraction of hardwoods (yellow poplar and red maple) pretreated by organosolv extraction, as well as model cellulosic substrates such as filter paper. There was no significant correlation among hydrolytic performance on pretreated hardwood, based on glucose release, and filter paper activity. However, performance on pretreated hardwood showed significant correlations to the levels of endogenous β-glucosidase and xylanase activities in the cellulase preparation. Accordingly, differences in performance were reduced or eliminated following supplementation with a crude β-glucosidase preparation containing both activities. These results complement a previous investigation using softwoods pretreated by either organosolv extraction or steam explosion. Cellulase preparations that performed best on hardwood also showed superior performance on the softwood substrates.

Index Entries: Cellulase; xylanase; hemicellulose; lignocellulose; bioconversion.

Introduction

Concerns about diminishing resources, national energy security, and the excessive production of greenhouse gases continue to motivate the search for alternatives to petroleum. Lignocellulosic biomass contains large

---

*Author to whom all correspondence and reprint requests should be addressed.

amounts of polymeric carbohydrates that represent an attractive source of sugars to produce alternative fuels and other chemical commodities. Potential feedstocks include agricultural residues such as corn stover, "purpose-grown" energy crops such as hybrid poplar, and hard- or softwood wastes from the forest industry.

One of the bioconversion schemes currently under active investigation involves enzymatic hydrolysis of the carbohydrate fraction, included largely of cellulose and hemicellulose, to produce glucose and other simple sugars for fermentation to fuel-grade ethanol. However, this approach is problematic because cellulose is inherently resistant to enzyme attack and because both cellulose and hemicellulose are protected by the surrounding matrix of lignin. Consequently, lignocellulosic biomass requires pretreatment to disrupt cellulose and lignin in order to improve enzyme accessibility.

Typically, pretreatment produces an enriched cellulose fraction containing residual hemicellulose and lignin, although the composition of pretreated material varies considerably, according to the type of feedstock, the pretreatment technology employed, and the process parameters that affect pretreatment severity. Various pretreatment technologies are now being optimized in attempts to produce appropriate substrates for hydrolysis at realistic cost (1). Concurrently, enzyme manufacturers are investigating ways to reduce production costs and improve the specific activities of the enzyme complexes required for the hydrolysis of pretreated feedstocks (2).

Efficient cellulose hydrolysis requires the concerted action of several endo- and exoglucanases (3). These cellulases are prime targets in attempts to improve enzyme activity. A further strategy involves optimization of so-called "accessory enzymes" that hydrolyze the complex array of glycosidic bonds in hemicellulose. Efficient hemicellulose hydrolysis is important, not only for recovery of sugars from residual hemicellulose, but also because hemicellulose appears to hinder the access of cellulases to cellulose fibers. In some feedstocks, similar considerations may apply to residual pectin.

In previous research, we examined the hydrolysis of several softwood substrates by a panel of seven cellulase preparations in order to evaluate their hydrolytic performance (4). Substrates were pretreated by SO_2-catalyzed steam explosion or ethanol organosolv extraction. We demonstrated that evaluation of enzyme performance using the target substrate is essential because the ability to hydrolyze model cellulosic substrates, such as filter paper, provides a poor estimate of activity on pretreated softwood. We also presented indirect evidence that the activity of a cellulase preparation is related to the endogenous levels of two activities: β-glucosidase (cellobiase) and xylanase. In this article, we present complementary data for two hardwood substrates prepared by organosolv pretreatment. This is relevant because hardwoods and softwoods contain different types of hemicellulose and lignin (5,6), factors that may influence enzyme performance.

The flexibility to process a broad range of lignocellulosics would benefit commercial bioconversion process economics, so versatile cellulase preparations offer significant advantages.

Materials and Methods

Substrate and Pretreatment

Representative samples of yellow poplar (*Liriodendron tulipifera*), and red maple (*Acer rubra*), were collected in Eastern Canada. Samples were chipped to approx 2 × 2 × 0.5 cm after removal of bark, screened for uniformity, and equilibrated at 4°C in sealed plastic bags to approx 9% (w/w) moisture content before pretreatment.

Organosolv pretreatment of poplar and maple was carried out in a 1 L-stainless steel pressure reactor (Parr Instrument Co., Moline, IL) using 50% (w/w) ethanol, adjusted to pH 2.4 with 10% (v/v) sulfuric acid, at 195°C and approx 3.2 MPa (460 psi). The solvent:wood ratio was 7:1 (w:w). The pretreatment time was 40 min for both substrates. The time required to reach the target cooking temperature was approx 53 min, in all cases. After cooking, the reactor was quenched in ice until the inside temperature was ≤55°C and the spent liquor removed by decantation. The solids were homogenized for 5 min in 70% (v/v) ethanol at 70°C (solids: ethanol approx 9:1) in a British disintegrator (TMI, Montreal, Canada), then washed three times with 1 L of warm 70% ethanol and rinsed extensively with water. Pretreated solids were then separated by filtration and stored in sealed plastic bags at 4°C.

Chemical Analysis of Untreated and Pretreated Hardwoods

The carbohydrate composition and lignin content of untreated and pretreated hardwood samples was determined using a modified Klason lignin method derived from the TAPPI standard method T222 om-88, as previously described (7). Monosaccharides were analyzed by HPLC with fucose as internal standard according to the procedure described elsewhere (7).

Cellulase Preparations

Three commercial *Trichoderma reesei* cellulase preparations and four laboratory preparations produced by mutant strains of *Penicillium* spp. and *Trichoderma* spp. (see Table 2) were evaluated. In some assays, cellulase preparations were supplemented with Novozym 188 (Novozymes), a commercial β-glucosidase preparation from *Aspergillus niger* containing 340 cellobiose units (CBU)/mL, as described later.

Batch Hydrolysis of Pretreated Hardwood and Data Analysis

Hydrolysis experiments were performed in triplicate in 100-mL flasks, at 50°C and shaken at 250 rpm. The reaction mixture contained

0.1M acetate buffer, pH 5.0, 5% (w/v) substrate and 10 filter paper units (FPU) of cellulase activity per gram dry substrate, in a total volume of 10 mL. In experiments involving supplementation with Novozym 188, the FPU:CBU ratio was 1 : 2. Samples were taken at 1, 3, 6, and 12 h. Glucose concentrations were determined using the glucose oxidase-peroxidase method *(8)*.

Two indices, specific conversion (SC) and mean specific rate (MSR) were calculated to compare the various hydrolysis progress curves obtained for pretreated hardwoods, as described in the Results and Discussion.

Statistical Analysis

Statistical analyses were performed using Origin 6.0 software (Microcal Software, Inc.). Analyses of variance were performed using Origin's one-way ANOVA test.

Enzyme Assays

Enzymes activities on model substrates were performed as previously described *(4)*, according to the recommendations of the International Union of Pure and Applied Chemistry *(8,9)*. Xylanase activity was determined by monitoring the release of reducing sugars from birchwood xylan (Sigma) by the Somogyi-Nelson method *(10)*, as previously described *(4)*. β-glucanase activity was measured using the xylanase assay procedure, with barley β-glucan (Sigma) replacing xylan. Pectinase and mannanase activities were measured using polygalacturonic acid and galactomannan, respectively, as previously described *(11,12)*. The protein concentration in enzyme preparations was determined by the Lowry method *(13)*, following precipitation with trichloroacetic acid.

Results and Discussion

Carbohydrate and Lignin Composition of Pretreated Hardwoods

The compositions of untreated and pretreated hardwood samples (% dry weight) are shown in Table 1. Untreated poplar and maple contained approx 45% and 42% cellulose (assuming all glucose represents cellulose) respectively. The percentage of cellulose content increased to approx 80% and 72%, respectively, as a result of pretreatment, reflecting partial extraction of lignin and hemicellulose during organosolv extraction. The xylan content of both hardwood samples decreased following organosolv extraction whereas the mannan content increased, suggesting preferential extraction of xylan. Typical hardwood hemicelluloses have a high-xylan content and low-mannan content, relative to softwoods. However, the xylan and mannan contents of organosolv pretreated hardwoods (Table 1) and softwoods *(4)* show no particular trend, demonstrating that the hemicellulosic sugar content of pretreated lignocellulosic substrates does not necessarily reflect that of the untreated feedstock.

Table 1
Carbohydrate and Lignin Composition of Untreated and Pretreated Hardwoods (% From Dry Weight)

	Arabinan	Galactan	Glucan	Xylan	Mannan	Klason lignin	Acid-soluble lignin
Untreated yellow poplar	0.34 ± 0.06	0.40 ± 0.03	44.65 ± 0.99	17.06 ± 0.32	1.82 ± 0.07	21.14 ± 0.19	2.74 ± 0.06
Organosolv-pretreated yellow poplar	0	0.02 ± 0.00	79.57 ± 2.04	7.05 ± 0.17	2.52 ± 0.08	5.35 ± 0.02	1.32 ± 0.32
Untreated red maple	0.29 ± 0.02	0.60 ± 0.03	41.86 ± 0.49	6.22 ± 0.11	1.76 ± 0.10	28.13 ± 0.19	2.06 ± 0.09
Organosolv-pretreated red maple	0	0	72.15 ± 0.28	2.83 ± 0.03	2.01 ± 0.03	17.50 ± 0.32	0.95 ± 0.02

Activities of Cellulase Preparations on Model Carbohydrate Substrates

The protein content of the seven cellulase preparations, and their specific hydrolytic activities against a panel of model cellulosic and hemicellulosic substrates and related glycans, are shown in Table 2. The preparations showed similar specific filter paper activity (0.7–1.0 FPU/mg), CMCase activity (14.1–24.4 U/mg), and Avicelase activity (1.6–2.5 U/mg). However, the preparations demonstrated significant differences in their levels of specific β-glucosidase activity (0.15–1.16 U/mg) and in their levels of xylanase, mannanase, and pectinase activities. All preparations contained similar levels of β-glucanase activity.

Analysis of Hydrolysis of Pretreated Hardwood Substrates

Two indices, MSR and SC, were used to evaluate the hydrolysis of cellulose by the various preparations for the two pretreated hardwood substrates, as previously described for softwood (4). The MSR index (g glucose/L/h/mg) estimates the average rate of cellulose hydrolysis during the first 12 h of hydrolysis, normalized for total protein. The method of calculation, involving curve triangulation is illustrated in Fig. 1A. To validate this method, all hydrolytic progress curves were fitted to an arbitrary hyperbolic function ($G = [k_1 t]/[k_2 + t]$; where G = glucose concentration (g/L), t = time (h), and k_1 and k_2 are constants). Regression analysis showed that this function produced a good fit to all hardwood hydrolysis data described below ($\chi^2 \leq 0.015$; $r^2 = 0.99$). Analysis of variance (Fig. 1B) was then used to demonstrate that MSR values calculated by curve triangulation did not differ significantly from values calculated using the first derivative of the fitted curves ($F = 0.02–0.16$; $p = 0.67–0.95$). The SC index (%/mg) describes the % of total cellulose in the sample hydrolyzed to glucose in the 12 h incubation period, normalized for total protein.

Activities of Cellulase Preparations on Pretreated Hardwood Substrates

Data for the hydrolysis of pretreated poplar and maple by the seven different cellulase preparations are shown in Figs. 2A and 3A, respectively. Analysis of these data (Table 3) shows that the rate (MSR) and extent (SC) of cellulose hydrolysis by MSUBC1 (*Penicillium* sp. cellulase preparation, *see* Table 2) were significantly greater than seen for the other preparations tested, as previously reported for a panel of pretreated softwood substrates (4). Also, as previously reported for softwood substrates, the MSR and SC indices for all the cellulase preparations show a poor correlation with activities determined using filter paper ($r_{MSR} = 0.428$, $p = 0.127$; $r_{SC} = 0.508$, $p = 0.06$), CMC ($r_{MSR} = -0.203$, $p = 0.486$; $r_{SC} = -0.124$, $p = 0.674$) or Avicel ($r_{MSR} = -0.462$, $p = 0.097$; $r_{SC} = -0.517$, $p = 0.059$). This result emphasizes the previous conclusion (4) that filter paper activity does not provide a reliable prediction

Table 2
Enzyme Activities in Cellulase Preparations

Cellulase preparation and source	Protein concentration[a]	Cellulase activities (U/mg protein)				Other enzyme activities (U/mg protein)			
		FPA	CMC	Avicelase	β-Glucosidase	β-Glucanase	Xylanase	Pectinase	Mannanase
Laboratory preparations									
MSUBC1 (*Penicillium* spp.)	838	0.9	16.9	2	1.16	17	39.1	0.64	0.3
MSUBC2 (*Trichoderma* spp.)	556	0.7	19.2	2.4	0.16	14.6	9.3	0.25	0.26
MSUBC3 (*Trichoderma* spp.)	771	0.8	17.7	1.9	0.15	15.6	3.8	1.56	0.18
MSUBC4 (*Trichoderma* spp.)	460	0.8	24.4	2.5	0.2	12.2	6.1	1.07	0.43
Commercial preparations									
TR1 (*Trichoderma* spp.)	129	0.9	14.1	2.2	0.19	16.8	3.5	0.07	0.11
TR2 (*Trichoderma* spp.)	130	1	20.7	1.6	0.66	19.2	11.8	0.05	0.06
TR3 (*Trichoderma* spp.)	149	0.9	21.6	2.5	0.28	14.2	13.1	0.03	0.03

[a]Protein concentration, — mg/g, except TR1 and TR2 (mg/mL).

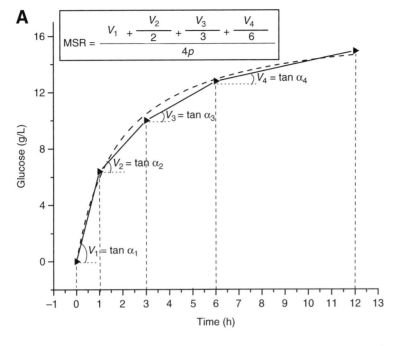
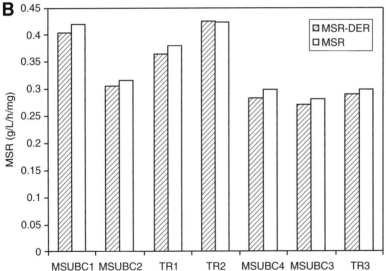

Fig. 1. Calculation of the MSR of hydrolysis by curve triangulation. **(A)** Hydrolysis of organosolv-pretreated poplar by the MSUBC1 cellulase with exogenous β-glucosidase supplementation is illustrated as an example. p—total protein loaded (mg) and **(B)** one-way ANOVA. MSR-DER calculated from the first derivative of the fitting function; MSR-calculated by curve triangulation.

of the ability of a cellulase preparation to hydrolyze cellulose into glucose in complex lignocellulosic substrates, despite suggestions to the contrary (14).

However, linear regression analysis revealed a significant correlation ($r \geq 0.80$, $p < 0.0001$) between the efficiency of cellulose hydrolysis (MSR

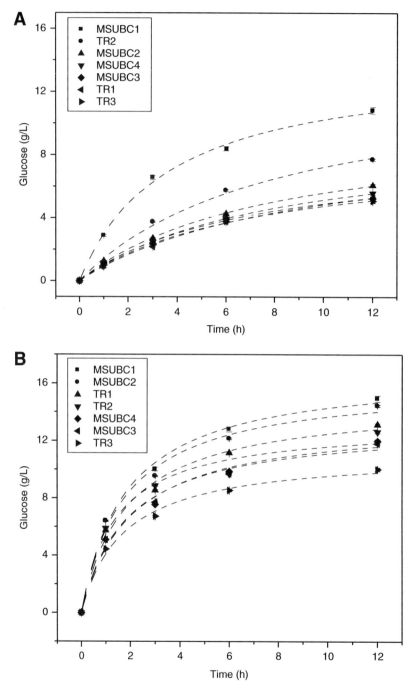

Fig. 2. Hydrolysis of organosolv-pretreated poplar by cellulase preparations. **(A)** Without β-glucosidase supplementation and **(B)** with β-glucosidase supplementation (FPU:CBU 1:2). Dashed lines are fitted curves.

or SC) in pretreated hardwood and the levels of endogenous β-glucosidase and xylanase activities (Figs. 4A, 5A, 6A, and 7A). This result provides indirect evidence for the hypothesis that differences in the endogenous levels of

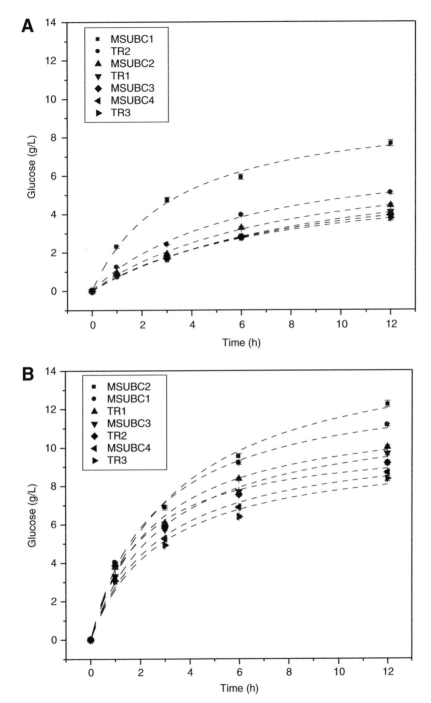

Fig. 3. Hydrolysis of organosolv-pretreated maple by cellulase preparations. **(A)** Without β-glucosidase supplementation and **(B)** with β-glucosidase supplementation (FPU:CBU—1:2). Dashed lines are fitted curves.

these two activities are at least partially responsible for the differences in cellulase performance seen on hardwood substrates, as also reported for softwoods *(4)*. This evidence is supported by the demonstration (Figs. 2B and 3B)

Table 3
Mean Specific Rates (MSR) and SCs for Hydrolysis of Hardwood Samples by Cellulase Preparations With and Without Exogenous β-Glucosidase (β-G) Supplementation

Enzyme	Protein load per assay (mg)	Organosolv-pretreated yellow poplar		Organosolv-pretreated yellow poplar with exogenous β-G		Organosolv-pretreated red maple		Organosolv-pretreated red maple with exogenous β-G	
		MSR[a]	SC[b]	MSR[a]	SC[b]	MSR[a]	SC[b]	MSR[a]	SC[b]
MSUBC1	5.66	0.25	4.32 (24.48)	0.42	5.96 (33.75)	0.19	3.4 (19.26)	0.29	4.22 (30.74)
MSUBC2	7.28	0.1	1.89 (13.72)	0.32	4.48 (32.63)	0.07	1.52 (11.07)	0.23	4.56 (25)
MSUBC3	6.57	0.09	1.81 (11.87)	0.28	4.06 (26.7)	0.07	1.5 (9.84)	0.21	4.76 (22.95)
MSUBC4	6.22	0.1	2.03 (12.61)	0.3	4.35 (27.07)	0.07	1.79 (9.84)	0.2	3.49 (21.72)
TR1	5.48	0.11	2.17 (11.87)	0.38	5.41 (29.67)	0.08	1.65 (10.25)	0.27	3.68 (24.18)
TR2	4.82	0.17	3.62 (17.43)	0.42	5.92 (28.56)	0.13	2.64 (12.71)	0.3	4.92 (27.87)
TR3	5.39	0.11	2.13 (11.5)	0.3	4.20 (22.62)	0.08	1.75 (9.43)	0.22	3.88 (20.9)

[a] Average rate of cellulose hydrolysis from 0 to 12 h/mg protein (g glucose/L/h/mg).
[b] Percentage of total cellulose converted to glucose in 12 h/mg protein (%/mg); in parentheses: cellulose conversion (%) from 0 to 12 h.
[c] β-G (Novozym 188—a β-glucosidase preparation from *Aspergillus niger*).

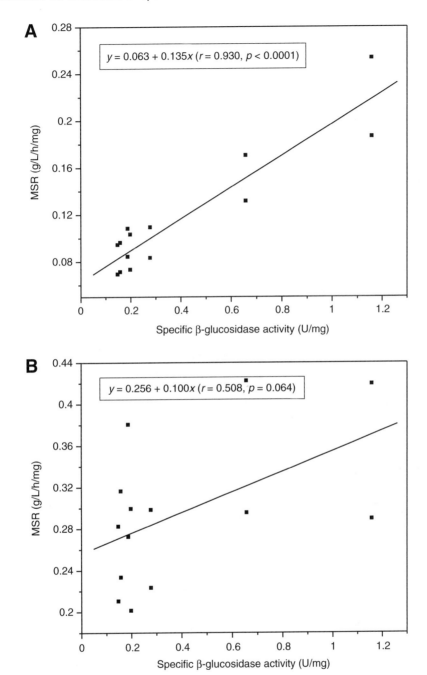

Fig. 4. Regression analyses of MSR for hydrolysis of hardwood samples vs specific β-glucosidase activity without **(A)** and with **(B)** β-glucosidase supplementation (FPU:CBU—1:2). MSR—mean specific rate.

that differences in the efficiencies of cellulose hydrolysis by the various cellulase preparations on hardwood substrates were reduced or eliminated after supplementation with Novozym 188, a commercial β-glucosidase

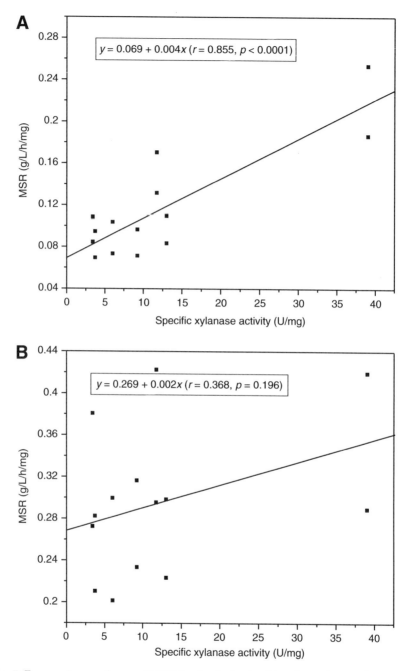

Fig. 5. Regression analyses of MSR for hydrolysis of hardwood samples vs specific xylanase activity without **(A)** and with **(B)** β-glucosidase supplementation (FPU: CBU—1:2). MSR—mean specific rate.

preparation commonly used to improve cellulase performance. Consequently, the correlation between cellulose hydrolysis in pretreated hardwood and level of endogenous β-glucosidase or xylanase activity was markedly reduced following supplementation (Figs. 4B and 6B). It should

Evaluation of Cellulase Preparations

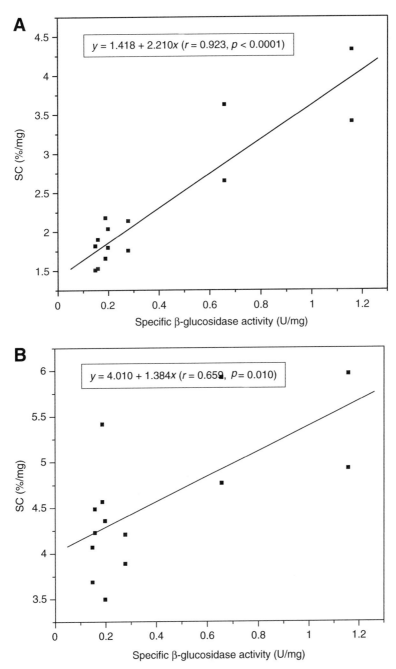

Fig. 6. Regression analyses of SC for hydrolysis of hardwood samples vs specific β-glucosidase activity without **(A)** and with **(B)** β-glucosidase supplementation (FPU: CBU—1:2). SC—specific conversion.

be noted that Novozym 188 contains significant xylanase activity (0.58 U/mg protein, based on hydrolysis of birchwood xylan) (4). Therefore, it appears that deficiencies in the levels of both activities are compensated

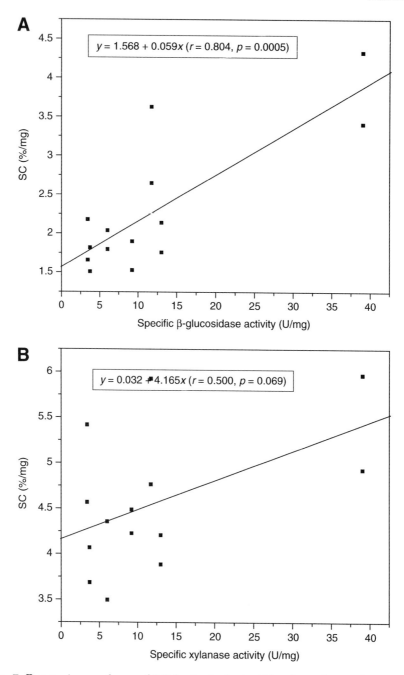

Fig. 7. Regression analyses of SC for hydrolysis of hardwood samples vs specific xylanase activity without (A) and with (B) β-glucosidase supplementation (FPU: CBU—1:2). SC—specific conversion.

by supplementation with corresponding activities present in the β-glucosidase preparation during the hydrolysis of both softwood and hardwood substrates. Further experiments using defined enzymes are required to determine the relative importance of these two activities.

The role of β-glucosidase in relieving end product inhibition caused by accumulation of cellobiose is well documented (15). Presumably, xylanases improve cellulose hydrolysis by removing hemicellulose on fiber surfaces, thereby increasing the accessibility of cellulose to cellulases. Although the xylan content of pretreated hardwood samples is low (≤ 7.1%; Table 1), it is probable that a fraction of hemicellulose is solubilized during pretreatment and redeposited on fiber surfaces during the late stages of pretreatment, as in kraft pulping (16); consequently, steric hindrance owing to hemicellulose may be significant. Xylanases may also increase cellulose accessibility indirectly by facilitating lignin removal (16,17). In contrast to xylanase, no significant correlation was seen between cellulose hydrolysis in pretreated hardwoods and levels of endogenous mannanase (r_{MSR} = 0.064, p = 0.827; r_{SC} = 0.030, p = 0.918), pectinase (r_{MSR} = –0.227, p = 0.435; r_{SC} = –0.280, p = 0.333) or β-glucanase activity (r_{MSR} = 0.536, p = 0.046; r_{SC} = 0.577, p = 0.031). Mannans are not major components of hardwood hemicelluloses, in contrast to softwoods (6); however, a similar lack of correlation between endogenous mannanase activity and cellulose hydrolysis was also reported for softwood substrates (4). These results suggest that the residual mannan does not significantly restrict access to cellulose in pretreated woody substrates, or that mannan hydrolysis is limited by other factors. It is noted that significant differences in the rate and extent of cellulose hydrolysis by the various cellulase preparations remain after supplementation, and that supplementation of cellulase preparations with Novozym 188 did not improve the correlation between cellulase hydrolysis in hardwood substrates and hydrolysis of filter paper (r_{MSR} = 0.465, p = 0.094; r_{SC} = 0.373, p = 0.189), CMCase (r_{MSR} = –0.290, p = 0.315; r_{SC} = –0.260, p = 0.370) or Avicelase (r_{MSR} = –0.420, p = 0.135; r_{SC} = –0.571, p = 0.033) indicating that the various preparations are distinguished by additional differences in enzyme properties.

These results are relevant to current attempts to reduce the cost of cellulase preparations for bioconversion of lignocellulosic substrates. First, they demonstrate that discovery of novel enzyme complexes, coupled with mutagenesis, is a viable empirical method to significantly improve cellulase activity on lignocellulosic substrates. The MSUBC strains used to provide cellulase preparations in this study were derived by reiterative strain selection and random mutagenesis (18). Screening for improved enzyme complexes should involve the target substrate, because activity on filter paper, or other model cellulosic substrates, provides a poor indication of the ability to hydrolyze cellulose in lignocellulose. Secondly, they support the concept that further improvements in performance can be achieved by supplementation of cellulase preparations with accessory enzymes, such as xylanases, that facilitate the removal of noncellulosic components.

The results presented here for hardwoods, and previously for softwoods (4), provide no indication that the improvements in cellulose hydrolysis produced by both these approaches are restricted to particular

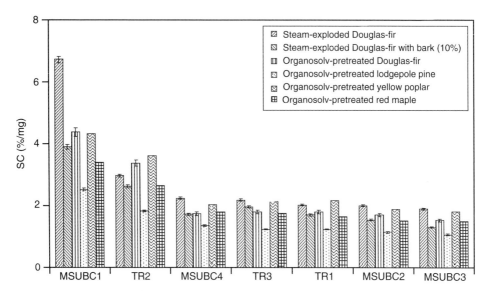

Fig. 8. Comparison of enzyme performance (SC) on a range of substrates produced by SO_2-catalyzed steam explosion or organosolv pretreatment of hardwoods and softwoods.

classes of substrates or pretreatment methodologies: MSUBC1 and TR2 show superior performance on all substrates examined so far (Fig. 8), although further experiments using a broader range of substrates are required to substantiate any conclusion. Robust enzyme preparations (i.e., those that perform well on a broad range of substrates) should simplify enzyme production and biomass conversion processes and reduce costs, and the flexibility to process a range of feedstocks would mitigate potential problems with feedstock supply that may arise from reliance on one feedstock. Nevertheless, it is reasonable to expect that incremental increases in hydrolytic performance could be achieved by systematically fine tuning the composition of enzyme complexes for particular substrates; for example, by addition of further enzymes to existing cellulase preparations to produce specific "cocktails." If all relevant enzymes are assumed to have approximately equal production costs (and stabilities), a simplistic model suggests that this strategy is cost effective only when the improvement, per unit weight of protein added, exceeds that achieved by simply increasing the loading of unsupplemented cellulase preparation. Similarly, it is still unclear whether use of enzyme supplementation to improve the performance of cellulase preparations such as TR2 offers any economic advantage over use of unsupplemented preparations like MSUBC1 (Table 3).

Acknowledgments

This research was supported by the Natural Science and Engineering Research Council of Canada and Natural Resources Canada. We thank Novozymes for providing samples of Novozym 188.

References

1. Wyman, C. E., Dale, B. E., Elander, R. T., Holtzapple, M., Ladisch, M. R., and Lee, Y. Y. (2005), *Bioresour. Technol.* **96,** 1959–1966.
2. US Department of Energy (ed) (2004), *Biomass Program—Cellulase Enzyme Research.* http://www.eere.energy.gov/biomass/cellulase_enzyme.html.
3. Rabinovich, M. L., Melnik, M. S., and Bolobova, A. V. (2002), *Appl. Biochem. Microbiol.* **38,** 355–373.
4. Berlin, A., Gilkes, N., Kilburn, D., et al. (2005), *Enzyme Microb. Technol.* **37,** 175–184.
5. McMillan, J. D. (1993), In: *Enzymatic conversion of biomass for fuel production* Himmel, M. F., Baker, J. O., and Overend, R. P. (eds.), American Chemical Society, Washington, D.C., pp. 292–232.
6. Sjöström, E. (1993), *Wood Chemistry. Fundamentals and Applications*, Academic Press, New York.
7. Bura, R., Mansfield, S. D., Saddler, J. N., and Bothast, R. J. (2002), *Appl. Biochem. Biotechnol.* **98–100,** 59–72.
8. Wood, T. M. and Bhat, K. M. (1988), In: *Methods in Enzymology*, Wood, T. M. and Kellogg, S. T. (eds.), vol. 160, Academic Press Inc., London.
9. Ghose, T. K. (1987), *Pure Appl. Chem.* **59,** 257–268.
10. Somogyi, M. (1952), *J. Biol. Chem.* **195,** 19–23.
11. Semenova, M. V., Grishutin, S. G., Gusakov, A. V., Okunev, O. N., and Sinitsyn, A. P. (2003), *Biochemistry (Moscow)* **68,** 559–569.
12. Baraznenok, V. A., Becker, E. G., Ankudimova, N. V., and Okunev, O. N. (1999), *Enzyme Microb. Technol.* **25,** 651–659.
13. Lowry, O. H., Rosebrough, N. J., Farr, A. L., and Randall, R. J. (1951), *J. Biol. Chem.* **193,** 265–275.
14. Xiao, Z., Storms, R., and Tsang, A. (2004), *Biotechnol. Bioeng.* **88,** 832–837.
15. Coughlan, M. P. (1985), *Biotechnol. Genet. Eng. Rev.* **2,** 39–109.
16. Buchert, J., Carlsson, G., Viikari, L., and Ström, G. (1996), *Holzforschung* **50,** 69–74.
17. Mansfield, S. D. and Esteghlaglian, A. R. (2003), In: *Applications of enzymes to lignocellulosics*, Mansfield, S. D. and Saddler, J. N. (eds.), vol. ACS Symposium Series 855, American Chemical Society, Washington, D.C., pp. 2–29.
18. Solovieva, I.V., Okunev, D. N., Velkov, V. V., et al. (2005), *Mikrobiologiia* **74(2),** 172–178. Russian.

Steam Pretreatment of Acid-Sprayed and Acid-Soaked Barley Straw for Production of Ethanol

Marie Linde, Mats Galbe, and Guido Zacchi*

Department of Chemical Engineering, Lund University, P. O. Box 124, S-221 00 Lund, Sweden; E-mail: Guido.Zacchi@chemeng.lth.se

Abstract

Barley is an abundant crop in Europe, which makes its straw residues an interesting cellulose source for ethanol production. Steam pretreatment of the straw followed by enzymatic hydrolysis converts the cellulose to fermentable sugars. Prior to pretreatment the material is impregnated with a catalyst, for example, H_2SO_4, to enhance enzymatic digestibility of the pretreated straw. Different impregnation techniques can be applied. In this study, soaking and spraying were investigated and compared at the same pretreatment condition in terms of overall yield of glucose and xylose. The overall yield includes the soluble sugars in the liquid from pretreatment, including soluble oligomers, and monomer sugars obtained in the enzymatic hydrolysis. The yields obtained differed for the impregnation techniques. Acid-soaked barley straw gave the highest overall yield of glucose, regardless of impregnation time (10 or 30 min) or acid concentration (0.2 or 1.0 wt%). For xylose, soaking gave the highest overall yield at 0.2 wt% H_2SO_4. An increase in acid concentration resulted in a decrease in xylose yield for both acid-soaked and acid-sprayed barley straw. Optimization of the pretreatment conditions for acid-sprayed barley straw was performed to obtain yields using spraying that were as high as those with soaking. For acid-sprayed barley straw the optimum pretreatment condition for glucose, 1.0 wt% H_2SO_4 and 220°C for 5 min, gave an overall glucose yield of 92% of theoretical based on the composition of the raw material. Pretreatment with 0.2 wt% H_2SO_4 at 190°C for 5 min resulted in the highest overall xylose yield, 67% of theoretical based on the composition of the raw material.

Index Entries: Barley straw; pretreatment; enzymatic hydrolysis; H_2SO_4; ethanol.

Introduction

Interest in alternative, renewable fuels is increasing world-wide owing to alarming reports on environmental issues as well as the limited oil resources. Within the EU, the goal is to replace 5.75%, calculated on the basis of energy content, of all petrol and diesel for transport purposes with

---

*Author to whom all correspondence and reprint requests should be addressed.

biofuels by the year 2010 (1). Transport fuels alone account for 32% of total EU energy consumption (2). Replacing fossil-based transport fuels with biofuels is therefore an essential step in decreasing fossil energy consumption and the emission of greenhouse gases. One alternative fuel that has been found to be well suited to this purpose is ethanol produced from biomass. It can be produced from various lignocellulosic materials, is easy to introduce in the present infrastructure, and results in low net contribution of carbon dioxide to the atmosphere (3). Several varieties of biomass, such as agricultural residues and wood residues (3–5) have been evaluated for ethanol production. The choice of biomass in a specific region depends on its availability. In Europe, barley is an abundant crop, which makes barley straw a potential lignocellulosic source for ethanol production (6,7).

The main constituents of straw are cellulose, hemicellulose, and lignin. Cellulose is a linear, polymeric chain of glucose units (8,9). Hemicellulose consists mainly of a polymeric chain of xylose that acts as a backbone with glucuronic acid, arabinose, and acetyl side groups (10). By pretreating the straw at high temperature and pressure and then subjecting it to enzymatic hydrolysis, the polymeric chains are hydrolyzed into monomeric sugar units (3,11).

Addition of sulphuric acid prior to the pretreatment step has been shown to increase the enzymatic digestibility of biomass (12,13). However, the technique of adding the acid to the biomass can vary. A technique common at the laboratory scale is to immerse the material in a large volume of dilute acid and press the wet straw to a desired dry-matter (DM) content. This technique is referred to as soaking. A previous study on pretreatment of acid-soaked barley straw showed that the highest overall yield of glucose, 38.3 g/100 g raw material, was obtained under the condition of 170°C with 1 wt% H_2SO_4 for 5 min. However, use of soaking in an industrial plant is not realistic as it requires large volumes of impregnation liquid and consumes more chemicals. To avoid this, spraying the acid onto the raw material is considered a more feasible alternative. In this study the impregnation techniques of soaking and spraying have been evaluated comparatively to see if the overall yield of fermentable sugars differs.

For ethanol to become more attractive as an alternative fuel, its production cost must be competitive with that of gasoline. The most important factor for the economic outcome of the bio-ethanol process is the overall ethanol yield (14,15). As a consequence, it is important to maximize the overall sugar yield in the process, that is, obtain high yields of both glucose and hemicellulosic sugars. In this study pretreatment of acid-sprayed barley straw was optimized in terms of overall yields of glucose and xylose in order to attain the same as for pretreatment of acid-soaked barley straw.

Another factor affecting the production cost is the substrate loading in enzymatic hydrolysis (11,15,16). Increased substrate loading results in decreased flow rate of streams for downstream processing, and thus,

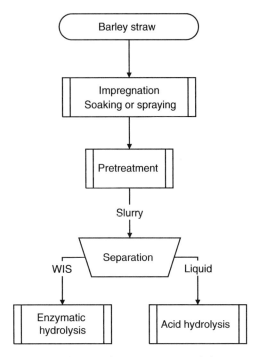

Fig. 1. Experimental setup for assessment of the pretreatment.

reduced energy demand and capital cost in the distillation and the evaporation steps. To achieve this, the amount of DM in the pretreatment has to be increased. The effect of increased substrate loading in the pretreatment step was therefore also investigated.

Methods

The experimental procedure is outlined in Fig. 1. The barley straw was impregnated, either by soaking or spraying, before steam pretreatment. Each impregnation was performed in duplicate. The water-insoluble solids (WIS) in the slurry from pretreatment were separated from the liquid by filtration. The liquid was analyzed for its content of monomeric sugars and acid hydrolysis was performed to determine the content of soluble oligomeric sugars. The filter cake was washed with deionized water to remove all water-soluble solids (WS), from the WIS and then filtrated. An enzymatic hydrolysis was performed on the WIS to assay the pretreatment.

Raw Material

Three batches of barley straw were kindly donated by Abengoa Bioenergy, Spain, in 2004. They were delivered as pieces of approx 50–100 mm and had a DM content of 92%. The straw was cut into smaller sections in a hammer mill, sieved to pieces of 10–30 mm, and stored at room temperature.

Impregnation

Prior to pretreatment the raw material was impregnated with H_2SO_4 solution. Two impregnation techniques were used: either soaking or spraying. Each impregnation was performed in duplicate at room temperature. When evaluating the two different impregnation techniques, the impregnation times were either 10 or 30 min and the acid concentrations in the liquid after impregnation were 0.2 or 1.0 wt%. After comparing the impregnation techniques the pretreatment of acid-sprayed barley straw was optimized and the impregnation time was then 60 min and the acid concentrations in the liquid were 0.2, 0.5, 1.0, or 2.0 wt%.

Soaking

The barley straw was immersed in a liquid containing H_2SO_4 (20 g liquid/g dry straw) and stored in a sealed bucket. After the impregnation time was reached, the wet straw was pressed in a 3-L cylindrical container with a sieve plate in the bottom, which let the liquid pass through but retained the straw. The pressing cylinder was attached to a long screw stem, and was screwed down with manual power onto the wet straw. The straw was pressed to a DM content of 40 wt% and then directly steam pretreated.

Spraying

Sulphuric acid was sprayed over the barley straw through a nozzle creating a mist. The straw was agitated in a cement mixer lined with stainless steel whereas being sprayed to ensure an even layer of acid. When the acid addition techniques were compared, the DM content of the straw after impregnation was 40 wt%. In the optimization of the pretreatment step of the acid-sprayed barley straw the DM content of the straw after impregnation was 60 wt%. Spraying took approx 5 min to complete. The acid-sprayed straw was stored in a sealed bucket to prevent evaporation until reaching the desired impregnation time and was then directly steam pretreated.

Pretreatment

Barley straw was pretreated in a steam pretreatment unit, comprising a 10-L reactor, which has been described elsewhere (17). The temperature was maintained using saturated steam. After the desired pretreatment time of 5 min was reached, the pressure was released and the material collected in a flash tank. The WIS in the slurry from the pretreatment were separated from the liquid by filtration.

The substrate loading into the pretreatment unit was 400 g of DM when the impregnation techniques were compared. The pretreatment temperature was set to 200°C as this temperature gave a high yield of glucose when soaking was optimized in a previous study. For the optimization of the pretreatment of acid-sprayed barley straw the substrate loading was 500 g of DM. The pretreatment conditions are presented in Table 1.

Table 1
Conditions in the Pretreatment Step for the Optimization of Pretreatment
of Acid-Sprayed Barley Straw

Time (min)	5												
H$_2$SO$_4$ (wt%)	0.2				0.5				1.0				2.0
Temperature (°C)	190	200	210	220	190	200	210	220	190	200	210	220	190 200

Acid Hydrolysis for Oligosaccharide Determination

The liquid part of the slurry of pretreated material was analyzed for its content of oligosaccharides using the National Renewable Energy Laboratory (NREL) dilute acid hydrolysis procedure for determination of total sugars in the liquid fraction of process samples LAP-014 *(18)*. The oligosaccharide concentration was determined as the difference in monomer sugar concentration before and after acid hydrolysis of the oligosaccharides to monomeric sugars.

Enzymatic Hydrolysis

The effects of the impregnation and the pretreatment conditions on the digestibility were assessed by enzymatic hydrolysis of washed WIS from the pretreatment. The WIS were washed with deionized water to remove all WS from the WIS and then filtrated. All filtrations were performed through a Munktell filter paper, grade 5. Enzymatic hydrolysis was performed in 500-g batches with 2 wt% WIS in 0.1 M sodium acetate buffer at a temperature of 40°C for 96 h. The following enzymes were added: 2.52-g *Celluclast 1.5L* (65 FPU/g and 17 β-glucosidase IU/g) and 0.52-g *Novozyme 188* (376 β-glucosidase IU/g), both kindly donated by Novozymes A/S (Bagsværd, Denmark). Samples were withdrawn after 0, 2, 4, 6, 8, 24, 48, 72, and 96 h to monitor hydrolysis. Duplicates were run for each sample.

Titration

The buffering capacity of barley straw was investigated by titration; 50 g of unwashed barley straw in pieces of 10–30 mm was immersed in 1000-g deionized water at a temperature of 80°C for 30 min and then filtered to remove the straw. The resulting barley water was titrated with 0.01 M H$_2$SO$_4$ (0.1 wt% H$_2$SO$_4$) with a GP Titrino 736 from Metrohm Ion analysis (Switzerland). Deionized water was used as reference.

Increasing Substrate Loadings

The effect of increased substrate loading on the concentration of WIS in the slurry from the pretreatment step was investigated. In the study on soaking vs spraying, 1000 g of impregnated straw with a DM content of 40 wt%

was pretreated in a 10-L reactor at 200°C for 5 min with 0.2 wt% H_2SO_4. The acid concentration corresponds to 0.30-g H_2SO_4/100-g dry straw. This pretreatment condition was compared with one performed in the optimization of the pretreatment step, which resulted in approximately the same acid : dry straw ratio but with a different substrate loading. The loading of acid-sprayed barley straw was then 833 g of impregnated straw with a DM content of 60 wt% and pretreated in the 10-L reactor at 200°C for 5 min with 0.5 wt% H_2SO_4. The pretreatment condition corresponded to an acid : dry straw ratio of 0.33-g H_2SO_4/100-g dry straw.

Analysis

DM contents were determined by drying samples in an oven at 105°C until constant weight was obtained. The composition of barley straw in batches 1 and 2, as well as the WIS after pretreatment, was determined according to LAP-002, LAP-003, LAP-004, and LAP-005 from NREL *(19–22)*. The composition of barley straw in batch 3 was analyzed after removing all starch present by hydrolyzing to monomeric sugar using *Termamyl 120L* and *AMG 300L*, kindly donated by Novozymes A/S (Bagsværd, Denmark). The straw to water ratio was 1:10, and the starch-removal procedure was divided into two steps. In the first, the starch was liquefied by thermostable α-amylases (24-mL termamyl 120 L/kg dry straw) for 4 h at 80°C and pH 6.0 to catalyze the hydrolysis of α-1,4 linkages. In the second step, saccharification of the liquefied starch was performed with amyloglucosidase (72 mL AMG 300 L/kg straw) at 55°C and pH 5.0 for 48 h *(23)*. The composition of batch 3 and the starch-free straw, as well as the WIS after pretreatment of batch 3, was determined according to the NREL procedure for determination of structural carbohydrates and lignin in biomass *(24)*.

The liquids from the determination of carbohydrates in biomass, the filtrate from the pretreated material, and the liquids after acid hydrolysis for oligosaccharide determination were analyzed for their content of monomeric sugars using HPLC (Shimadzu, Kyoto, Japan) equipped with a refractive index detector (Shimadzu). The column used was an Aminex HPX-87P (Bio-Rad, Hercules, CA) at 85°C with an eluent flow rate of 0.5 mL/min for separation of glucose, xylose, galactose, arabinose, and mannose. The resulting liquids from the filtration of the pretreated materials were also analyzed for their content of byproducts, HMF, and furfural, using an Aminex HPX-87H column (Bio-Rad, Hercules, CA) at 65°C, with 5 mM H_2SO_4 as eluent, at a flow rate of 0.5 mL/min. All samples were filtered through a 0.2-µm filter before analysis to remove particles.

Results and Discussion

All yields are expressed as g/100 g raw material unless otherwise stated. The overall yield includes the soluble sugars in the liquid from

Table 2
Composition of Barley Straw

Percentage	Batch 1	Batch 2	Batch 3	
	B	B	B	A
Glucan	37.1	38.2	39.6	40.5
Starch	n.d.[a]	n.d.[a]	n.d.[a]	2.4
Xylan	21.4	21.7	16.4	24.2
Galactan	BDL[b]	BDL[b]	1.1	0.4
Arabinan	3.1	2.6	3.9	3.0
Lignin	19.5	19.6	23.9	19.8
Ash	2.7	2.6	7.2	3.8
Others	16.2	15.3	7.9	5.9

[a]Not determined.
[b]Below detectable level.
"A" and "B" are determined with and without starch removal, respectively.

pretreatment, including soluble oligomers, and monomer sugars obtained in the enzymatic hydrolysis. The recovery after pretreatment is defined as the sum of monomeric, oligomeric, and polymeric glucose or xylose in the liquid and WIS as the percentage of the theoretical value for either sugar, based on the raw material.

Soaking Vs Spraying

The comparison of the impregnation techniques was performed for two different concentrations of sulphuric acid: 0.2 and 1 wt%. Raw materials were from batches 1 and 2, respectively, see Table 2. The starch content was not measured or removed before the analysis of these batches. In batch 3 the composition was determined with and without starch removal, and the result differed between the analyses, see Table 2. The starch did not only have an effect on the glucose content but also on measurement of xylose, probably owing to interference from starch in the raw material analysis. Batches 1 and 2 resulted in overall sugar yields above what is theoretically possible. This is most probably owing to the presence of starch in these batches as well but the starch content was not analyzed in these raw materials as all material was consumed before the problem was realized. However, it is still possible to compare the impregnation techniques when the yield is expressed in g/100 g raw material, as this is not affected by the determination of the composition of the raw material.

The slurry after pretreatment contains mainly WIS and oligomeric and monomeric sugars from hydrolyzed hemicellulose and cellulose. Figure 2 shows the yield, including oligomers, of glucose and xylose in the liquid from the slurry of the pretreated material after pretreatment. There is no noticeable difference in glucose yield between the impregnation techniques at an acid concentration 0.2 wt%. At the higher acid concentration,

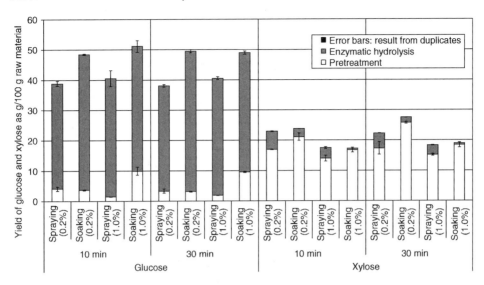

Fig. 2. Yield of glucose and xylose in pretreatment, enzymatic hydrolysis, and overall for various impregnation and pretreatment conditions. The maximum yield of glucose is 40.8 and 42 g/100 g straw for 0.2% and 1.0% H_2SO_4, respectively. The maximum yield of xylose is 37.1 and 38.2 g/100 g straw for 0.2% and 1.0% H_2SO_4, respectively.

1.0 wt%, there was a considerable difference in glucose yield between the impregnation techniques in which soaking resulted in the highest yield, 9.6 and 10.0 g/100 g raw material at 10 and 30 min impregnation time, respectively. Spraying resulted in glucose yields below 1.4 and 1.9 g/100 g raw material, respectively. The xylose yield after pretreatment was higher for soaking at both acid concentrations. Soaking using 0.2 wt% H_2SO_4 for 30 min resulted in the highest xylose yield after pretreatment, 25.8 g/100 g raw material. The highest xylose yield after the pretreatment for acid-sprayed barley straw, 17.3 g/100 g raw material, was also obtained using 0.2 wt% H_2SO_4.

There was a decrease in glucose yield over the pretreatment step for acid-sprayed barley straw when the acid concentration was increased from 0.2 to 1.0 wt%. If the glucose in the liquid were solely a product of hydrolyzed cellulose, one would expect an increase in glucose yield when the severity was increased. However, the drop in glucose amount supports the assumption that the material contained starch. The decrease of glucose can then be explained if starch was released at 0.2 wt% H_2SO_4 and further degraded to HMF at 1.0 wt% H_2SO_4, consistent with the observed increase in HMF production shown in Fig. 3. The production of degradation products, HMF and furfural, increased when the acid concentration increased owing to higher severity in the pretreatment step *(17,25–27)*. The production of degradation products was also higher for the acid-soaked material than the acid-sprayed material. This shows that spraying resulted in less severe conditions in pretreatment.

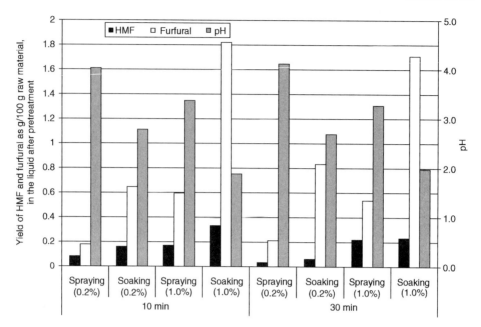

Fig. 3. Yield of HMF and furfural as g/100 g raw material, and pH, in the liquid after pretreatment.

Even though soaking and spraying were performed at the same acid concentrations, soaking produced a liquid after pretreatment with a lower pH than spraying, *see* Fig. 3. At an acid concentration of 0.2 wt%, soaking gave a pH of approx 2.7, whereas spraying gave a pH of 4.0. At an acid concentration of 1.0 wt%, the pH was 1.9 and 3.3 for soaking and spraying, respectively, i.e., 1.0 wt% H_2SO_4-sprayed barley straw had a higher pH than 0.2 wt% H_2SO_4-soaked barley straw. This too shows that soaking resulted in more severe conditions for barley straw than spraying.

The difference in pH is probably owing to the buffering capacity of barley straw, which was overcome by the large volume of impregnation liquid when soaking was applied, thus increasing the severity of the pretreatment. Figure 4 shows the titration curve of the barley water and of deionized water. A decrease in pH from 6.0 to 3.0 in pure water resulted in consumption of approx 3 mL, 0.01 M H_2SO_4, whereas the barley water consumed approx 35 mL, 0.01 M H_2SO_4. This shows that soaking of barley straw in water extracts compounds that buffer, for example, ash components (28). As impregnation with soaking is performed with an excess of dilute acid, the buffering is overcome, whereas this is not the case for spraying.

Enzymatic hydrolysis was performed to assess the effect of the pretreatment on digestibility, with Fig. 2 showing the yields of glucose and xylose for enzymatic hydrolysis. Soaking gave a higher yield of glucose than spraying for both 0.2 and 1.0 wt% H_2SO_4 in the enzymatic hydrolysis, but the difference in yield between the impregnation techniques decreased at 1.0 wt%. This was owing not only to an increase in yield for the sprayed

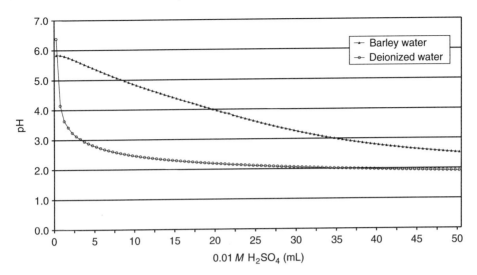

Fig. 4. Titration curves with 0.01 M H_2SO_4 (0.1 wt%) for water from a barley-straw and water slurry and deionized water.

material but also to a decrease in yield for the soaked material caused by an increase in released glucose in the pretreatment at 1.0 wt% H_2SO_4.

The xylose yield for enzymatic hydrolysis was lower for the soaked material, which is owing to the fact that almost all the xylan was already hydrolyzed in the pretreatment step. The impregnation time had a negligible effect on both the glucose and xylose yield in the enzymatic hydrolysis.

Soaking resulted in a higher overall sugar yield than spraying when compared at the same acid concentration and pretreatment condition, see Fig. 2. Soaking in 1.0 wt% sulphuric acid for 10 min gave the highest overall yield of glucose, 51.1 g/100 g raw material (122% of theoretical). Increased impregnation time did not have any effect on the overall yield, i.e., it was not possible to make up for the lower glucose yield obtained with spraying by increasing the impregnation time from 10 to 30 min. The increase in acid concentration from 0.2 to 1.0 wt% H_2SO_4 using spraying resulted in a lower glucose yield than with soaking at 0.2 wt%. The overall glucose yield when soaking at 0.2 wt% was 48.9 g/100 g raw material, whereas spraying at 1.0 wt% only reached an overall glucose yield of 40.5 g/100 g raw material in spite of the fact that the acid concentration was five times higher.

The highest overall yield of xylose, 27.6 g/100 g raw material (115% of theoretical), was obtained for soaking in 0.2 wt% H_2SO_4 for 30 min. The higher acid concentration, 1.0 wt%, had a negative effect on the overall yield of xylose for both impregnation techniques. For acid-soaked barley straw, the xylose yield in pretreatment was higher than for the acid-sprayed barley straw. In the enzymatic hydrolysis step it was the opposite; i.e., acid-sprayed barley straw had a higher xylose yield in the enzymatic hydrolysis step than acid-soaked barley straw. The overall xylose yield therefore did not markedly differ between the impregnation techniques.

Table 3
Recovery of Glucose and Xylose in the WIS and in the Liquid
After the Pretreatment Step as % of Theoretical Value Based on the Glucan
and Xylan Content in the Raw Material

H_2SO_4 (wt%)	°C	WIS (%)[a]	Glucose (%)			Xylose (%)		
			Total	WIS	Liquid	Total	WIS	Liquid
0.2	190	80	102	97	5	86	50	36
	200	64	96	92	4	63	26	38
	210	71	86	84	2	81	39	42
	220	72	92	89	3	57	34	22
0.5	190	70	91	88	3	82	45	37
	200	62	94	90	4	72	24	48
	210	64	101	97	4	48	11	37
	220	58	91	87	4	22	5	17
1	190	68	98	95	3	76	40	36
	200	61	99	94	5	71	23	48
	210	58	95	91	4	41	9	32
	220	62	97	93	4	20	5	15
2	190	63	99	94	5	71	20	51
	200	58	96	90	6	57	9	48

[a]Recovery of WIS after pretreatment.

Optimization of Steam Pretreatment of Acid-Sprayed Barley Straw

In this study with unwashed barley straw, the yield as well as the response to increased acid concentration differed between the impregnation techniques. As spraying is a more feasible impregnation alternative than soaking, in terms of use in an industrial plant, it was of interest to study the possibility of attaining as high a yield with spraying as with soaking. Optimization was therefore performed on the pretreatment step using acid-sprayed barley straw to find the maximum yield of glucose and xylose for acid-sprayed barley straw.

Batch 3, see Table 2, was used as raw material when optimizing the pretreatment step for acid-soaked barley straw. The total glucose content was 47.1 g/100 g raw material, of which 2.6 g/100 g raw material was from starch. The xylose content was 27.1 g/100 g raw material.

Table 1 shows the pretreatment conditions, whereas Table 3 shows the recovery of glucose and xylose after pretreatment. When the temperature was increased from 190°C to 220°C, xylose recovery decreased from 86% to 57% for 0.2 wt% H_2SO_4 and from 76% to 20% for 1.0 wt% H_2SO_4. This is owing to degradation of the xylose during the pretreatment, which can be seen as an increase of furfural in the liquid from the pretreatment, see Fig. 5. The production of degradation compounds increased with increased temperature and acid concentration. Figures 6 and 7 show the yields including oligomers, of glucose and xylose in the liquid from the slurry of the pretreated

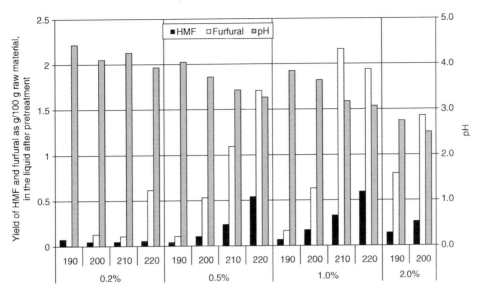

Fig. 5. Yield of HMF and furfural as g/100 g raw material, and pH, in the liquid after pretreatment.

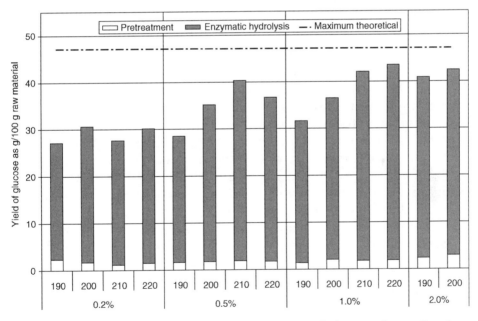

Fig. 6. Yield of glucose in pretreatment, enzymatic hydrolysis, and overall as function of the pretreatment conditions.

material and in the liquid after enzymatic hydrolysis. The maximum yield of xylose in the liquid from the pretreatment step was 14 g/100 g raw material (52% of theoretical) at 190°C with 2.0 wt% H_2SO_4.

Enzymatic hydrolysis was performed on washed WIS after pretreatment to assess the effect of the pretreatment step on the digestibility. Yields

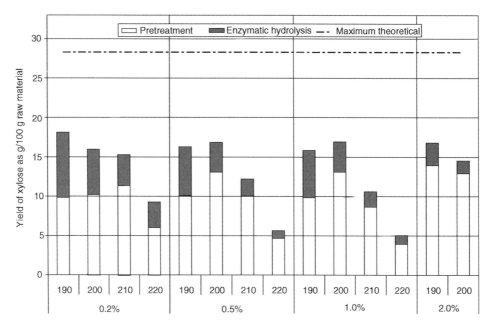

Fig. 7. Yield of xylose in pretreatment, enzymatic hydrolysis, and overall as function of pretreatment conditions.

were calculated from the 96-h sample taken during enzymatic hydrolysis. Figures 6 and 7 show the yield of glucose and xylose, respectively, for enzymatic hydrolysis. The glucose yield in the pretreatment step was low and approximately the same for all pretreatment conditions. Thus, the pretreatment conditions that gave the highest glucose yield in the enzymatic hydrolysis also resulted in the highest overall glucose yield. The highest overall glucose yield was 43.5 g/100 g raw material (which corresponds to a glucose yield of 92% based on the glucose content in the raw material including starch), obtained at the pretreatment condition of 1.0 wt% H_2SO_4 and 220°C for 5 min, see Fig. 6.

Severe pretreatment conditions hydrolyze more of the hemicellulose in the straw to monomeric sugars, and the xylose yield in the enzymatic hydrolysis will be low owing to less xylan in the material. However, if the pretreatment conditions are too severe degradation of xylose to furfural will occur in the pretreatment step and result in a low overall xylose yield. An example is the pretreatment conditions, 190°C and 200°C at 0.5 wt% H_2SO_4 (Fig. 7), in which the xylose yield in the pretreatment step was higher at 200°C. However, the xylose yield in the enzymatic hydrolysis was higher for 190°C, resulting in approximately the same overall xylose yield at 190°C as at 200°C, owing to less degradation of xylose to furfural in the pretreatment step (Fig. 5). The maximum overall yield of xylose, 18.1 g/100 g raw material (which corresponds to a xylose yield of 67% based on the xylose content in the raw material), was obtained at 0.2 wt% H_2SO_4 and 190°C for 5 min, see Fig. 7.

Table 4
Overall Yield of Glucose and Xylose at Pretreatment Temperature of 200°C for Different Substrate Loadings as g/100 g Raw Material

	1000 g wet straw	833 g wet straw	
	40 wt% DM	60 wt% DM	
	0.2 wt% H_2SO_4	0.2 wt% H_2SO_4	0.5 wt% H_2SO_4
Acid : dry-straw ratio	0.30	0.13	0.33
Glucose (g/100 g)	38.5 (40.8)[a]	30.6 (47.1)	35.2 (47.1)
Xylose (g/100 g)	22.6 (24.0)[a]	15.9 (27.1)	16.8 (27.1)

[a]Starch content not analyzed and removed before raw material analysis.
The values in bracket are the maximum (theoretical) values based on the composition of the raw material.

The two different optima for glucose and xylose are near the two extremes in pretreatment conditions evaluated with acid-sprayed barley straw. The overall yield of xylose is more affected by an increase in temperature from 190°C to 220°C than by an increase in acid concentration from 0.2 to 2.0 wt%. The best total overall yield of glucose and xylose, 57.6 g/100 g raw material (which corresponds to 78% based on the glucose and xylose content in the raw material), was obtained at 2.0 wt% H_2SO_4 and 190°C for 5 min.

One objective of the study was to increase the concentration of WIS in the slurry from the pretreatment step in order to decrease the flow rate of the stream going to the downstream processing and increase the concentration of ethanol *(15)*. The concentration of WIS and total DM (including WS) obtained in the slurry after the pretreatment of spray-impregnated barley straw at 200°C and acid:dry-straw ratio 0.3 was measured for the substrate loadings of 1000 g with 40 wt% DM and 833 g with 60 wt% DM. Concentrations of WIS obtained in the pretreated material were 6.9–8.1 and 13.9 g/100 g slurry for 40 and 60 wt% DM, respectively. DM concentrations obtained in the pretreated material were 10.4–12.5 and 20.4 g/100 g slurry for 40 and 60 wt% DM, respectively. The increase in amount of dry-straw and DM concentration in the pretreatment step increased the WIS and DM content in the pretreated material by approx 100%. This shows that the substrate loading have an evident effect on the WIS and DM content after pretreatment and thus the ethanol concentration.

The overall yield of glucose for spray-impregnated barley straw at pretreatment condition 200°C with 0.2 wt% H_2SO_4 and a substrate loading of 1000 g straw with 40 wt% DM was 38.5 g/100 g raw material, *see* Table 4. For the same conditions but with 833-g straw and 60 wt% DM the overall glucose yield was lower, 30.6 g/100 g raw material. The overall xylose yields with 0.2 wt% H_2SO_4 were 22.6 and 15.9 g/100 g raw material at 40 and 60 wt% DM, respectively. However, the overall yield of glucose at 60 wt% DM and the same acid:dry-straw ratio as with 40 wt% DM, i.e., 0.33, gave an overall glucose yield of 35.2 g/100 g raw material. Thus, an increase in dry-straw loading and DM concentration in the pretreatment at the same acid : dry-straw ratio still decreased the yield of glucose, although the decrease was not as marked. The overall xylose yield at acid: dry-straw ratio of 0.33 and 60 wt% DM was 16.8 g/100 g. The low overall yield of xylose at an increased dry-straw loading and DM concentration was owing to an increase in xylose concentration in the pretreatment liquid, which probably increased the rate of degradation *(29)*.

Conclusions

Pretreatment of acid-soaked and acid-sprayed barley straw under the same conditions followed by enzymatic hydrolysis resulted in different overall yields of glucose and xylose. Acid-soaked barley straw gave the highest overall yield of glucose, regardless of impregnation time and acid concentration. An increased acid concentration from 0.2 to 1.0 wt% did not increase the yield from the acid-sprayed barley straw to the level of 0.2 wt% acid-soaked barley straw. For xylose, an increase in acid concentration resulted in a decrease in yield for both acid-soaked and acid-sprayed barley straw. In this study, the raw material was unwashed barley straw. A similar study with washed barley straw should be performed. As a result of washing the barley straw, its buffering capacity would probably decrease and the difference in severity in the pretreatment of acid-sprayed and acid-soaked barley straw would diminish thus yielding the same amount of fermentable sugars for the same acid concentration.

For acid-sprayed barley straw the pretreatment optimum for glucose and xylose are close to the two extremes of pretreatment conditions evaluated in this study. Thus, to obtain a high yield of glucose, a severe condition in the pretreatment is preferable when using spraying as an impregnation technique. However, a high yield of xylose is an important aspect for a future process when a pentose-fermenting microorganism is available. This suggests that a two-step steam pretreatment *(30)* where the xylose is released in the first step at a low severity would be a better option. The WIS is then pretreated again under a more severe condition, improving the enzymatic hydrolysis and thus achieving high overall yields of both glucose and xylose.

The concentration of DM and WIS in the pretreated material increased when the dry-straw loading and DM concentration were increased. However, the overall yield of both glucose and xylose decreased and partially counteracted the positive effect of an increase in DM concentration, i.e., an increase in ethanol concentration. The decrease in overall glucose yield, on the other hand, was significantly reduced when the acid:dry-straw ratio was kept the same for the different substrate loadings, whereas the decrease in overall xylose yield remained unchanged.

The pretreatment optimum for acid-soaked barley straw was not the same as the optimum for acid-sprayed barley straw. Studies on acid-soaked barley straw should be used with caution for design of processes utilizing acid-sprayed barley straw, and the buffering capacity of barley straw should not be overlooked as it has a significant effect on the severity in pretreatment.

Acknowledgments

The authors are grateful to Carl-Gustav Mårtensson and Gustav Sassner for their assistance in the experimental work. This study was financed by the European Commission Framework V, Contract no. NNE5-2001-00685.

References

1. Directive 2003/30/EC of the European parliament and of the council of 8 May 2003 on the promotion of the use of biofuels or other renewable fuels for transport, Official Journal of the European Union.
2. European Communities, 2004, Promoting biofuels in Europe, European Commission, Directorate-General for Energy and Transport, B-1049 Brussels.
3. Galbe, M. and Zacchi, G. (2002), *Appl. Microbiol. Biotechnol.* **59**, 618–628.
4. Palmarola-Adrados, B., Galbe, M., and Zacchi, G. (2005), *J. Chem. Technol. Biotechnol.* **80(1)**, 85–91.
5. Söderström, J., Pilcher, L., Galbe, M., and Zacchi, G. (2003), *Appl. Biochem. Biotechnol.* **105–108**, 127–140.
6. Bettio, M., Bos, S., Bruyas, P., Cross, D., Weiler, F., and Zampogna, F. (2003), Eurostat yearbook 2003: the statistical guide to Europe: data 1991–2001, 8th ed., Office for official publications of the European communities, Luxembourg.
7. Kim, S. and Dale, B. E. (2004), *Biomed. Bioener.* **26**, 361–375.
8. Kadam, K. L. (1996), *Handbook on Bioethanol: Production and Utilization*, Taylor & Francis, Washington DC, pp. 213–252.
9. Sun, R. C., Fang, J. M., Rowlands, P., and Bolton, J. (1998), *J. Agric. Food Chem.* **46**, 2804–2809.
10. Glasser, W., Kaar, W. E., Jain, R. K., and Sealey, J. S. (2000), *Cellulose* **7**, 299–317.
11. Sun, Y. and Cheng, J. (2002), *Biores. Technol.* **83**, 1–11.
12. Tengborg, C., Stenberg, K., Galbe, M., et al. (1998), *Appl. Biochem. Biotechnol.* **70–72**, 3–15.
13. Nguyen, Q. A., Tucker, M. P., Boynton, B. L., Keller, F. A., and Schell, D. J. (1998), *Appl. Biochem. Biotechnol.* **70–72**, 77–87.
14. von Sivers, M. and Zacchi, G. (1996), *Biores. Technol.* **56**, 131–140.
15. Wingren, A., Galbe, M., and Zacchi, G. (2003), *Biotechnol. Progr.* **19(4)**, 1109–1117.
16. Nguyen, Q. A. and Saddler, J. N. (1991), *Biores. Technol.* **35**, 275–282.
17. Palmqvist, E., Hahn-Hägerdal, B., Galbe, M., et al. (1996), *Biores. Technol.* **58(2)**, 171–179.
18. Ruiz, R. and Ehrman, T. (1996), Dilute Acid Hydrolysis Procedure for Determination of Total Sugars in the Liquid Fraction of Process Samples; Laboratory Analytical Procedure-014, National Renewable Energy Laboratory, Golden, CO.
19. Ehrman, T. (1994), Standard Method for Ash in Biomass, Laboratory Analytical Procedure-005, National Renewable Energy Laboratory, Golden, CO.
20. Ehrman, T. (1996), Determination of Acid-Soluble Lignin in Biomass, Laboratory Analytical Procedure-004, National Renewable Energy Laboratory, Golden, CO.
21. Ruiz, R. and Ehrman, T. (1996), Determination of Carbohydrates in Biomass by High Performance Liquid Chromatography, Laboratory Analytical Procedure-002, National Renewable Energy Laboratory, Golden, CO.

22. Templeton, D. and Ehrman, T. (1995), Determination of Acid-Insoluble Lignin in Biomass, Laboratory Analytical Procedure-003, National Renewable Energy Laboratory, Golden, CO.
23. Palmarola-Adrados, B., Chotĭborská, P., Galbe, M., and Zacchi, G. (2005), *Biores. Technol.* **96(7)**, 843–850.
24. Sluiter, A., Hames, B., Ruiz, R., Scarlata, C., Sluiter, J., and Templeton, D. (2004), Determination of Structural Carbohydrates and Lignin in Biomass, NREL, Golden, CO.
25. Baugh, K. D. and McCarty, P. L. (1988), *Biotechnol. Bioeng.* **31(1)**, 50–61.
26. Clark, T. A. and Mackie, K. L. (1984), *J. Chem. Technol. Biotechnol.* **34B**, 101–110.
27. Palmqvist, E. and Hahn-Hägerdal, B. (2000), *Biores. Technol.* **74**, 25–33.
28. Grohmann, K., Torget, R., and Himmel, M. (1986), *Biotech. Bioeng. Symp.* **17**, 135–151.
29. Sassner, P., Galbe, M., and Zacchi, G. (2005), *Appl. Biochem. Biotechnol.* **121–124**, 1101–1117.
30. Söderström, J., Pilcher, L., Galbe, M., and Zacchi, G. (2003), *Biom. Bioen.* **24**, 475–486.

Reaction Kinetics of Stover Liquefaction in Recycled Stover Polyol

FEI YU,[1] ROGER RUAN,*,[1,2] XIANGYANG LIN,[2] YUHUAN LIU,[2] RONG FU,[1] YUHONG LI,[1] PAUL CHEN,[1] AND YINYU GAO[2]

[1]Center for Biorefining and Biosystems and Agricultural Engineering, University of Minnesota, 1390 Eckles Avenue, St Paul, MN 55108; and [2]MOE Key Laboratory of Food Science, Nanchang University, Jiangxi 330047, China, E-mail: ruanx001@umn.edu

Abstract

The purpose of this research was to study the kinetics of liquefaction of crop residues. The liquefaction of corn stover in the presence of ethylene glycol and ethylene carbonate using sulfuric acid as a catalyst was studied. It was found that the liquefaction yield was a function of ratio of solvent to corn stover, temperature, residence time, and amount of catalyst. Liquefaction of corn stover was conducted over a range of conditions encompassing residence times of 0–2.5 h, temperatures of 150–170°C, sulfuric acid concentrations of 2–4% (w/w), and liquefaction reagent/corn stover ratio of 1–3. The liquefaction rate constants for individual sets of conditions were examined using a first-order reaction model. Rate constant increased with the increasing of liquefaction temperature, catalyst content, and liquefaction reagent/corn stover ratio. Reuse of liquefied biomass as liquefying agent was also evaluated. When using recycled liquefied biomass instead of fresh liquefaction reagent, the conversion is reduced. It appeared that 82% of liquefaction yield was achieved after two times of reuse.

Index Entries: Corn stover; ethylene carbonate; ethylene glycol; liquefaction; polyol.

Introduction

About 80 million acres of corn are planted each year in the United States resulting in an estimated 120 million dry tons of corn stover. Additional agricultural byproducts/waste products available across the United States each year include an estimated 95×10^6 t of agricultural waste and $100–280 \times 10^6$ t of forest waste. Corn stover consists of the stalks, leaves, and cobs after the corn kernels are harvested. More than 90% of the stover is left in the fields. Less than 1% of corn stover is collected for industrial processing. About 5% is baled for animal feed and bedding. Much of the remaining 90+% must be plowed. The plowing operation can cause organic carbon and nitrogen

---

*Author to whom all correspondence and reprint requests should be addressed.

losses because of oxidation, and increase the amount of fertilizer chemicals that need to be applied. The decay of the stover increases the release of CO_2, a greenhouse gas. Although some residue is required to protect the soil from erosion, most residues can be safely taken out of the fields for valuable utilization, which has the potential to be a win-win situation for the producer, processor, and the environment.

A number of studies have been performed on the effective utilization of crop residues and byproducts (1–4). Liquefaction of biomass, also called solvolytic reaction, is one of the effective ways to produce valuable fuels and polymer materials. In the liquefaction process, crop residues are liquefied in the acidic conditions with liquefying reagents, such as ethylene glycol and ethylene carbonate. As a result, hydroxyl groups are introduced into the product. Some polyester, polyurethane or fuels have been prepared from the liquefied polyol product (5,6). Yamada (7) reported that liquefaction of lignocellulosic biomass can be done in the presence of ethylene carbonate using acid catalysts at temperatures of 120–180°C. Heitz (8) have investigated the solvolytic power of some organic solvents such as aliphatic alcohol, polyols, phenol, lactic acid, and ethanolamine within the liquefaction process. Yao (9) focused his work on hydrolytic reactions of depolymerization in the cellulosic chain in phenol and phenol derivatives. In the present article, the influence of various reaction conditions on the liquefaction of corn stover such as liquefaction time, liquefaction temperature, sulfuric acid concentration, and the corn stover to liquefying reagent ratio was investigated. The solubility limit was also evaluated in order to understand the role of temperature and time in complete solvolysis of corn stovers. A first-order reaction model was then used to fit the kinetic of corn stover liquefaction.

Material and Methods

Materials

Air dried corn stover provided by Agricultural Utilization Research Institute was milled and then screened, and only the fraction less than 1 mm sieve was used in this research. Ethylene carbonate and ethylene glycol (Sigma, Minneapolis, MN) were used as liquefying reagents. Sulfuric acid (Sigma, Minneapolis, MN) was used as the catalyst.

Liquefaction Procedures

The liquefaction reaction was carried out in a 500-mL separable three-branch flask equipped with a stirring system and a reflux condenser in a mantle heater. After the desired reaction time, the heater was turned off, and the stirrer kept running until the mixture cooled down. At the end of liquefaction, the reactant was diluted by 200-mL dioxane-water solution (4/1, v/v) and then filtrated through filter paper under vacuum. The residue was dried to a constant weight at 105°C. The liquefaction yield was calculated by the following equation:

$$\text{Liquefaction yield} = \left(1 - \frac{\text{weight of dried residue}}{\text{weight of starting biomass}}\right) \times 100$$

The effects of following variables on liquefaction were investigated: (1) type of solvent, (2) temperature, (3) time, (4) amount of catalyst, (5) ratio of biomass solid to solvent, and (6) number of recycling liquefied biomass. All experiments and analysis were performed in triplicate.

Results and Discussion

Effect of Organic Solvent

Liquefaction of corn stover varied substantially with the two different solvents used in this study (Fig. 1). Liquefaction yield for ethylene carbonate approached maximum level in 0.5 h whereas liquefaction using ethylene glycol was very slow and 15% the corn stover still remained unliquefied after 2.5 h. Many bubbles were observed during the liquefaction with ethylene carbonate, which can be attributed to carbon dioxide gas emitted as a result of reactions between ethylene carbonate and biomass. This would be undesirable if it gets out of control in large-scale production. It also decreases the liquefaction yield because of the carbon loss through carbon dioxide emission. In the subsequent experiments, mixed solvents containing 90% of ethylene glycol and 10% of ethylene carbonate was used to achieve a reasonable liquefaction yield in an acceptable time frame.

The difference in liquefaction efficiency between ethylene carbonate and ethylene glycol may be related to their dielectric value (7). In our liquefaction process, the catalyst (sulfuric acid) was dissolved in organic solvents. The acid potential thus depends on the solvent's dielectric value. Because ethylene carbonate has a higher dielectric value than ethylene glycol, it is reasonable that liquefaction of corn stover was faster with ethylene carbonate than with ethylene glycol.

Effect of Liquefaction Time and Temperature

Figure 2 shows the effect of liquefaction time and temperature on liquefaction yield. It demonstrates that both time and temperature have a great influence on the liquefaction yield. Liquefaction of corn stover took place rapidly in the initial period and gradually leveled off. The liquefaction yield increased with increasing temperature and time. When the temperature is 160°C, the liquefaction yield became relatively high (91%) at 2 h, and a further increase in temperature and time resulted in a depressed increase in liquefaction yield. Additionally, the liquefaction curves exhibited an exponential trend. This reaction could therefore be regarded as first-order reaction. Furthermore, this figure shows that a satisfied liquefaction result can be obtained at 2 h and 160°C.

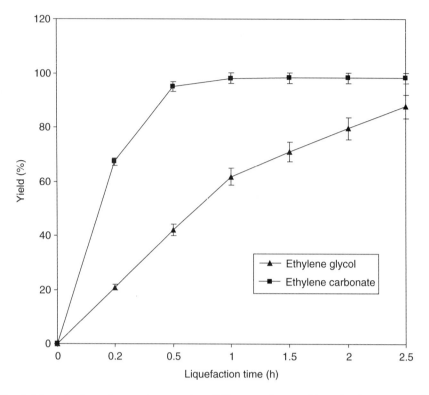

Fig. 1. Liquefaction of corn stover using different solvents. Liquefaction reagent/corn stover ratio, 3; catalyst content, 3% (w/w, on liquefaction reagent); temperature, 150°C; ▲, ethylene glycol; ■, ethylene carbonate. Error bars represent standard deviations calculated from the data obtained from three duplicated experiments.

Effect of Catalyst Content

The effect of the catalyst content on liquefaction is shown in Fig. 3. It is observed that the liquefaction yield increased with the increasing of sulfuric acid content. 92% liquefaction yield was obtained at 3% sulfuric acid in 2 h. We also attempted to substitute hydrochloric acid for sulfuric acid for economic considerations. After 2 h reaction at 6% hydrochloric acid, the liquefaction yield was only 54%, which is lower than the value obtained by the sulfuric acid catalyzed method. It indicated that sulfuric acid was better than hydrochloric acid, which may be because sulfuric acid is higher in acidity than hydrochloric acid. However, the use of sulfuric acid will cause condensation of degraded residues and increase the viscosity of the liquefied materials.

Effect of Liquefaction Reagent/Corn Stover Ratio

Figure 4 shows the effect of the liquefaction reagent/corn stover ratio on the liquefaction yield. The slope of the curve for ratio 3 is greater than it for ratios 1 and 2, indicating that the liquefaction reagent/corn stover ratio has a significant influence on the liquefaction yield. When the liquefaction

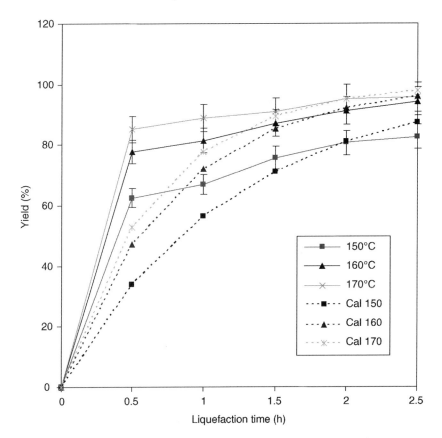

Fig. 2. Effect of time and temperature on the liquefaction of corn stover. Ethylene glycol 90% and 10% of ethylene carbonate acts as liquefaction reagent; liquefaction reagent/corn stover ratio, 3; catalyst content, 3%; liquefaction temperature: ■, 150°C; ▲, 160°C; ×170°C. Solid and dot curve represent the experimental and predicted curve respectively. Error bars represent standard deviations calculated from the data obtained from three duplicated experiments.

reagent/corn stover ratio reaches 3, the liquefaction yield is 91% after 2 h. A higher yield could be expected at even lower biomass loading but at the cost of a lower productivity. A ratio around 3 appears to be practical in our laboratory setting. A lower ratio may be possible and more economically viable in industrial scales.

Using Liquefied Biomass as Solvent

It would be economically important to recycle organic solvents in biomass liquefaction (10,11). However, it is expensive to separate the solvents from the liquefied biomass. In this study, we tested the feasibility of using liquefied biomass as solvent, this based on the fact that the liquefied biomass is rich in hydroxyl groups, which is a characteristic of organic solvents for biomass liquefaction. Figure 5 shows the liquefaction yields as a function of number of times the liquefaction solvent was reused. Before

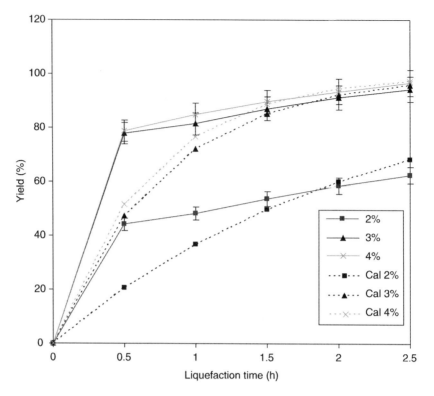

Fig. 3. Effect of the catalyst content on the liquefaction of corn stover. Ninety percentage of ethylene glycol and 10% of ethylene carbonate as liquefaction reagent; liquefaction reagent/corn stover ratio, 3; liquefaction temperature, 160°C; liquefaction time, 2.5 h; catalyst content: ■, 2%; ▲, 3%; x, 4%. Solid and dot curve represent the experimental and predicted curve respectively. Error bars represent standard deviations calculated from the data obtained from three duplicated experiments.

the reuse, the acidity of the liquefied biomass was readjusted to a sulfuric acid concentration of 3%. The result indicated that the liquefaction yield decreased with increasing number of reuse. The decrease in the yield may be attributed to: (1) decreased solvolytic capability of liquefied biomass, and (2) dramatically increased viscosity, which would limit reaction rates and present a challenge to mixing. We determined that 82% of liquefaction yield was achieved after two times of reuse.

Solubility Limit

Solubility limit (S_L) was used to describe the maximum concentration of corn stover dissolvable in liquefaction reagent. It had shown that S_L was 5% at 80°C and 12% at 100°C after 2 h liquefaction at 3% sulfuric acid (Fig. 6). The solubility limit is a construct that is valid only for steady-state conditions. It is likely to be a strong function of the ability of the cellulose to uncoil from its tight rigid rod configuration, which would explain the behavior near the glass transition temperature of approx 30°C. This uncoiling takes time, and as such the stovers require between 5 and 7 min becoming

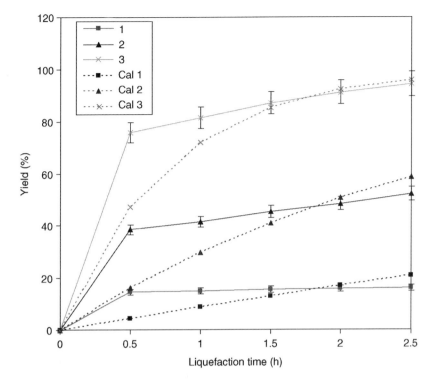

Fig. 4. Effect of the liquefaction reagent/corn stover ratio on the liquefaction of corn stover. Ninety percent of ethylene glycol and 10% of ethylene carbonate as liquefaction reagent; liquefaction temperature: 160°C; liquefaction time: 2.5 h; catalyst content, 3%; liquefaction reagent/corn stover ratio: ■, 1; ▲, 2; x, 3. Solid and dot curve represent the experimental and predicted curve respectively. Error bars represent standard deviations calculated from the data obtained from three duplicated experiments.

fully soluble in the polyol. At temperatures above 80°C, the S_L appears to increase almost linearly through the range studied.

The results presented in Fig. 2 suggest that 160°C was a desirable operational temperature, based on the high liquefaction yield and low unliquefied residue. For this, a starting point of 30% stovers in solution was tried. This is obviously far above what can be expected to react, as it quickly formed a solid mass. At 13 min, the system became more liquid-like, and at around 17 min was a stirred liquid system. This was likely the solubility limit, as shown in Fig. 7. A solubility limit S_L of 20% should be taken as maximum that can be expected to be in solution at any given time, and therefore the maximum that can react. For a fed system, in which the concentration is held constant, the feed rate would be set correspondingly. It was observed that slightly more stover can be present in the system, but at a risk of overloading, which could lead to interruption.

Solubility itself presents unique challenges. Solubility of the stovers in the polyol does not begin until approximately more than 80°C, in which a saturated solution forms a gel-like liquid–solid equilibrium. Cooling this quickly to less than approx 40°C causes the system to solidify, and slow

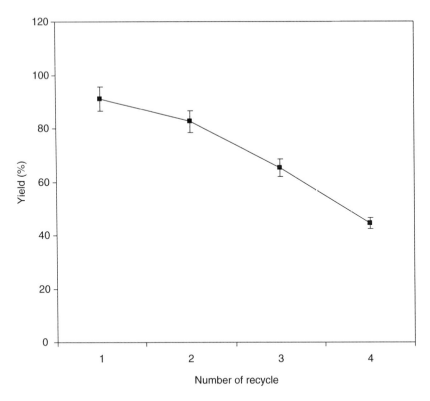

Fig. 5. Liquefaction rate of corn stover as a function of the recycle number. Liquefaction temperature, 160°C; liquefaction time, 2.5 h; catalyst content, 3%; recycled stover polyol/corn stover ratio, 3; Error bars represent standard deviations calculated from the data obtained from three duplicated experiments.

cooling allows the cellulose fraction to recrystalize out of solution. This may be a technique for purifying the cellulose/hemicellulose fraction.

First-Order Reaction Kinetics of Corn Stover Liquefaction

The liquefaction curves appeared to follow the first-order reaction. In a first-order reaction, the rate is proportional to the concentration of reactant of interest, which can be written as *(12)*

$$-\frac{d[S]}{[S]} = k dt \qquad (1)$$

where S is the mass of corn stover, t the reaction time, and k a rate constant. Integrating Eq. 1 yields

$$\ln[S] = -kt + C \qquad (2)$$

The constant of integration C can be evaluated by using boundary conditions.

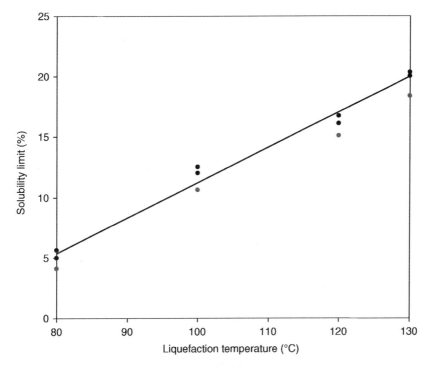

Fig. 6. Solubility limit of corn stover as a function of temperature. Ninety percent of ethylene glycol and 10% of ethylene carbonate as liquefaction reagent; liquefaction time, 2 h; catalyst content, 3%; corn stover content, 30%. Dots and solid line represent the experimental data and the predicted line, respectively.

When $t = 0$, $[S] = [S_0]$ where $[S_0]$ is the initial mass of reactant. Substituting into Eq. 2 gives the value of the constant of integration.

Finally an equation for the first-order reaction can be written as

$$\ln \frac{[S]}{[S_0]} = -kt \quad \text{or} \quad [S] = [S_0]e^{-kt} \qquad (3)$$

The value of $[S]$ at different time t could be obtained from the experimental curves (Figs. 2–4). The rate constant k values for various parameters of the liquefaction processes were then calculated using Eq. 3 and shown in Table 1. The predicted curves for different liquefaction temperature, catalyst content and liquefaction reagent/corn stover ratio are presented in Figs. 2–4. Rate constant k increased with the increasing of liquefaction temperature, catalyst content and liquefaction reagent/corn stover ratio.

The average deviation is defined as

$$\text{deviation } (\%) = 100 \frac{\sqrt{\sum ([S]_i^{\text{exp}} - [S]_i^{\text{cal}})^2 / N}}{[S]_{\text{max}}}$$

where N is the number of data points and $[S]_{\text{max}}$ is the maximum value of experiment data, subscript i refers to the data points used, $[S]^{\text{exp}}$ represents

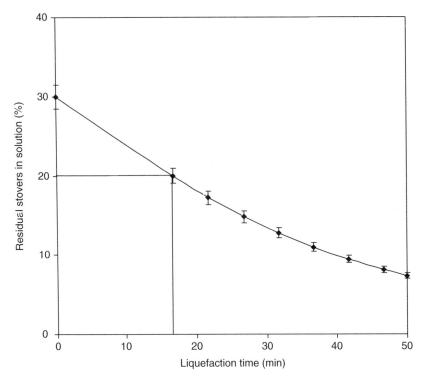

Fig. 7. Solubility limit of corn stover as a function of time. Ninety percent of ethylene glycol and 10% of ethylene carbonate as liquefaction reagent; liquefaction temperature: 160°C; catalyst content, 3%; corn stover content, 30%. Error bars represent standard deviations calculated from the data obtained from three duplicated experiments.

Table 1
k-Value for Liquefaction Process

		k (/h)	Deviation (%)
Temperature (catalyst = 3% sulfuric acid and ratio of solvent to biomass = 3)	150°C	0.83	15.4
	160°C	1.27	13.9
	170°C	1.50	14.6
Catalyst (temperature = 160°C, ratio of solvent to biomass = 3)	2%	0.46	17.8
	3%	1.27	13.8
	4%	1.45	12.0
Liquefaction reagent/corn stover ratio (temperature = 160°C, catalyst = 3% sulfuric acid)	1	0.09	32.7
	2	0.35	20.8
	3	1.27	13.1

the observed values and $[S]^{cal}$ represents those calculated by Eq. 3 for a given set of parameters. The experimental results were then described by the first-order reaction with an average deviation about 12–33% calculated in Table 1.

Conclusions

Liquefaction of corn stover was investigated in this study. The liquefaction yields varied with type of solvent, ratio of biomass solid to solvent, temperature, and type and amount of catalyst. Our research indicated that ethylene carbonate was better than ethylene glycol. Liquefaction yield increased with increasing ratio of solvent to biomass solid. A ratio around 3 was preferred for productivity and economical reasons. Liquefaction yield also increased with increasing temperature in the range from 150°C to 170°C. It was found that sulfuric acid was a better catalyst than hydrochloric acid for the liquefaction. Almost 3% of sulfuric acid was considered a practical catalyst concentration for our systems. The use of liquefied biomass as solvent was studied. It was found the liquefied biomass was not a good solvent; however 82% of liquefaction yield was achieved after two times of reuse. A first-order reaction model was used to explain the kinetic of corn stover's liquefaction with 12–33% deviation of prediction.

References

1. Rustamov, V. R., Abdullayev, K. M., and Samedov, E. A. (1998), *Energy Conversion Manag.* **39**, 869–875.
2. Flachowsky, G., Ochrimenko, W. I., Schneider, M., and Richter, G. H. (1996), *Animal Feed Sci. Technol.* **60**, 117–130.
3. Gorhmann, K., Torget, R., and Himmel, M. (1985), *Biotechnol. Bioeng. Symp.* **15**, 59–80.
4. Jackson, M. G. (1977), *Animal Feed Sci. Technol.* **2**, 105–130.
5. Karunanandaa, K. and Varga, G. A. (1996), *Animal Feed Sci. Technol.* **61**, 1–16.
6. Montane, D., Farriol, X., Salvado, J., Jollez, P., and Chornet, E. (1998), *J. Wood Chem. Technol.* **18**, 171–191.
7. Yamada, T. and Ono, H. (1999), *Biores. Technol.* **70**, 61–67.
8. Heitz, M., Brown, A., and Chornet, E. (1994), *Can. J. Chem. Eng.* **72**, 1021–1027.
9. Yao, Y., Yoshioka, M., and Shiraishi, N. (1994), *J. Appl. Polym. Science.* **52**, 1629–1636.
10. Rezzoug, S. A. and Capart, R. (2002), *Appl. Energy* **72**, 631–644.
11. Rezzoug, S. A. and Capart, R. (2003), *Energy Conversion Manag.* **44**, 783–794.
12. Sharma, R. K., Yang, J., Zondlo, J. W., and Dadyburjor, D. B. (1998), *Catalysis Today* **40**, 307–320.

Liquefaction of Corn Stover and Preparation of Polyester From the Liquefied Polyol

Fei Yu,[1] Yuhuan Liu,[2] Xuejun Pan,[3] Xiangyang Lin,[2] Chengmei Liu,[2] Paul Chen,[1] and Roger Ruan*,[1,2]

[1]Center for Biorefining and Biosystems and Agricultural Engineering, University of Minnesota, 1390 Eckles Avenue, St. Paul, MN 55108; [2]MOE Key Laboratory of Food Science, Nanchang University, Jiangxi 330047, China, E-mail: ruanx001@umn.edu; and [3]Forest Products Biotechnology, University of British Columbia, 2424 Main Mall, Vancouver, BC, V6T 1Z4, Canada

Abstract

This research investigated a novel process to prepare polyester from corn stover through liquefaction and crosslinking processes. First, corn stover was liquefied in organic solvents (90 wt% ethylene glycol and 10 wt% ethylene carbonate) with catalysts at moderate temperature under atmospheric pressure. The effect of liquefaction temperature, biomass content, and type of catalyst, such as H_2SO_4, HCl, H_3PO_4, and $ZnCl_2$, was evaluated. Higher liquefaction yield was achieved in 2 wt% sulfuric acid, 1/4 (w/w) stover to liquefying reagent ratio; 160°C temperature, in 2 h. The liquefied corn stover was rich in polyols, which can be directly used as feedstock for making polymers without further separation or purification. Second, polyester was made from the liquefied corn stover by crosslinking with multifunctional carboxylic acids and/or cyclic acid anhydrides. The tensile strength of polyester is about 5 MPa and the elongation is around 35%. The polyester is stable in cold water and organic solvents and readily biodegradable as indicated by 82% weight loss when buried in damp soil for 10 mo. The results indicate that this novel polyester could be used for the biodegradable garden mulch film production.

Index Entries: Biodegradability; liquefaction; polyester; strength; solubility.

Introduction

Most synthetic polymers used in daily life and industries are derived from petroleum. The main shortcomings are (1) dependence on the petroleum resource that is not renewable and (2) poor biodegradability and consequent negative impact on the environment. Numerous attempts have been made to develop biodegradable materials from renewable resource to replace petrochemical-based ones.

*Author to whom all correspondence and reprint requests should be addressed.

Crop residues are naturally occurring polymers. For example, lignocellulosic materials are made up of mainly of cellulose, hemicellulose, and lignin, which are rich hydroxyl groups. It is the hydroxyl group that makes it possible to convert crop residues into biopolymers. Liquefaction is an effective way to convert lignocellulosic materials into intermediates rich in hydroxyl groups. Japanese researchers investigated liquefaction of wood with polyethylene glycol and phenols, and prepared polyurethane foams from liquefied materials [1,2]. Yamada and Ono [3] reported the production of phenol–formaldehyde resin from liquefied wood. Kurimoto and Doi synthesized polyurethane-like foams obtained by reacting liquefied wood with polyisocyanate [4]. New resin systems have also been illustrated by reacting with multifunctional epoxy compounds by Kobayashi and his colleagues [5]. However, no study of polyester synthesis from liquefied biomass has been reported.

This research was focused on the development of a novel process to convert crop residues such as corn stovers into bio-polyester. The process involves two steps: (1) converting corn stover to bio-polyols by means of chemical liquefaction, in which, the corn stover experienced partial degradation and chemical reactions, and becomes a homogeneous liquid. (2) Preparing biodegradable polyesters from the bio-polyols through crosslinking with carboxyl acids and acid anhydrides.

Material and Methods

Raw Materials and Procedures of Liquefaction

The biomass samples used was corn stover taken from Agricultural Utilization Research Institute in Minnesota. The dried samples were ground in a Wiley mill to pass a screen of 1 mm aperture. The ground sample and liquefaction reagent (90 wt% ethylene glycol and 10 wt% ethylene carbonate) containing the appropriate amount of catalyst were put into a round bottom distilling flask with three cylindrical standard ground joint necks. To one neck of the flask was attached a condenser. The temperature of liquefaction was measured with a thermometer (upper limit of 300°C), which was attached to one neck of the distilling flask. The required reaction time for the liquefaction, as determined, was about 2.5 h. A stirrer was attached to the top neck of the distilling flask. Liquefaction was completed with continuous stirring under atmospheric pressure. After reaction, the stirrer kept running until the mixture cooled down to room temperature. Collect the liquefied mixture for later use. All experiments and analysis were performed in five duplicates.

Cyclic carbonates, such as ethylene carbonate and propylene carbonate, have been found to be novel reagents that can liquefy lignocellulosic materials rapidly at medium temperature under atmospheric pressure, although reaction mechanisms involved have not been well established [3].

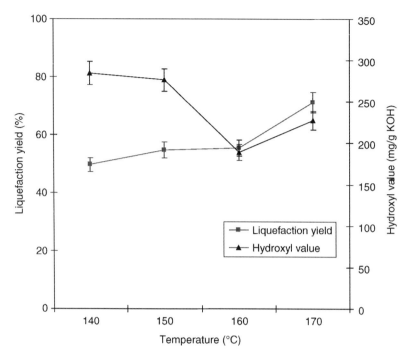

Fig. 1. Effect of temperature on liquefaction of corn stover. Liquefaction reagent/ corn stover ratio: 4; catalyst content: 2% (w/w, on liquefaction reagent); liquefaction time 2 h. Error bars represent standard deviations calculated from the data obtained from five duplicated experiments.

In this article, mixed solvents containing 90% of ethylene glycol and 10% of ethylene carbonate was used as liquefaction reagent. Figure 1 shows the effect of temperature on liquefaction yield and hydroxyl value. Liquefaction yield increased rapidly with increasing temperature. Temperatures higher than 160°C are required to achieve a high liquefaction yield. On the other hand, hydroxyl value decreased with increasing temperature. This could be considered that hydroxyl group-rich carbohydrate were decomposed and converted to low-molecular weight products, such as organic acids, at high temperature.

Corn stover meal is a porous and low-density material. Its liquid absorption capability is very high. To get a satisfactory liquefaction, it is very important to mix uniformly the stover meal with liquefying reagent. An effective stirring is definitely necessary. Otherwise, carbonization may occur because of incomplete heating. Table 1 shows the liquefaction results at various loading level of corn stover. When the ratio of stover to liquefaction reagent was below 30/100, liquefaction was smooth, and a uniform liquid was obtained. A mild stirring was enough. With increasing ratio, violent stirring became very important. The viscosity of the liquefied product increased significantly. For example, corn stover could be completely liquefied at a ratio of 40/100, but the viscosity was higher. When the ratio was over

Table 1
Possibility of Liquefaction at High Corn Stover Load

Stover/liquefying reagent	Stirring method	Result
30/100	Magnetic stirrer	Uniform liquefaction, thin liquid
40/100	Mechanical stirrer	Uniform liquefaction, thicker liquid
50/100	Mechanical stirrer	Partial carbonization

Conditions were liquefaction temperature 170°C, liquefaction time 4 h, catalyst content 5% (w/w, on liquefaction reagent).

50/100, the liquefaction was unable to complete because the liquefaction reagent was too little to maintain a uniform reaction. Carbonization occurred at this ratio.

In the presence of catalyst sulfuric acid, corn stover was subjected to a partial chemical degradation as well as reacted with ethylene carbonate and ethylene glycol to form a variety of glucosides. Bubbles were observed during the liquefaction, which could be attributed to the carbon dioxide produced during the reaction.

Preparation of Polyester From the Stover Polyols

The stover polyols used in this study for polyester were prepared under this condition: sulfuric acid (96%) at 2% on liquefying reagents; ratio of stover to liquefying reagent at 1/4 (w/w); temperature at 160°C; time in 2 h. The polyols had a hydroxyl value of 200 mg KOH/g and a viscosity of 2×10^5 MPa·s.

Polyester sheets were prepared by crosslinking the liquefied stover with carboxylic acids or anhydrides to form a network of polyester. First, the weighed liquefied stover, crosslinking chemicals, and other additives were mixed and heated together with stirring, and then the obtained homogeneous mixture was molded into a container or cast on a polished plate to form a uniform layer of the mixture (0.15–0.75 mm). The polyester sheet was obtained after curing by heating the mixture in an oven. Curing temperature and time varied (1–5 h, 140–180°C) depending on the formulation and thickness of the sheets.

Determination of Crosslinking Extent of Polyester

Because completely crosslinked network of polyester is almost insoluble in mixture of dioxane and water (4/1, v/v) (note: uncrosslinked polyol is soluble in the solvent), we used the percentage of insoluble residue in dioxane/water to evaluate, indirectly, the crosslinking extent of polyester. Weigh sample of crosslinked polyester, put it into a flask containing dioxane/water mixture (4/1, v/v) for 24 h at room temperature with stirring. After that, solvent was filtered out. The sample was washed with the solvent until no visible black color, and then the insoluble residue was dried overnight in an oven at 105°C to determine its weight. The crosslinking extent of polyester was calculated using the following equation:

$$\text{Crosslinking extent (\%)} = (W_r / W_s) \times 100$$

where W_s is the weight of initial sample and W_r is the weight of insoluble residue.

Solubility of Polyester

Polyester sheets were cut into small pieces of about 3 × 3 mm². Weigh the polyester pieces, put them into flasks containing water or selected solvents, and keep for 8 d at room temperature. After that, water or solvent was filtrated out. The samples were washed with corresponding water or solvent, and then dried overnight in oven at 105°C to determine weight loss. For the solubility of polyester sheets in hot water, alkaline and acid solution, the samples were refluxed for 6 h in water, 2 M NaOH or 2 M H_2SO_4, and then washed with water, dried to determine weight loss. The solubility of polyester sheets was calculated using the following formula.

$$\text{Solubility (\%)} = [(W_i - W_f) / W_i] \times 100$$

where W_i is the initial weight of sample and W_f is the final weight of sample after treatment with water, solvents, alkali, and acid solution.

Tensile Strength of Polyester Sheets

Tensile strength of the polyester sheets was evaluated with a material testing system (Model APEX-T1000, Satec Systems, Inc., Grove City, Pennsylvania). Width of testing sample was 25 mm. Loading speed was 5 mm/min. Before testing, the thickness and width of the sample and initial length of the sample between grippers were entered into the control system. After testing, the tensile strength, elongation and strain–stress curve of the sample were produced automatically by the computer.

Biodegradability of Polyester

To evaluate the biodegradability of polyester, natural degradation of polymers in soil was simulated in our laboratory. The soil used in this research was a merchandised potting soil. To guarantee the existence of desired microorganisms, about twofold garden soil collected from yard was mixed with the potting soil. The test samples were buried in the potting soil in flowerpots. The flowerpots were kept at 25°C in a cultivating room. The samples were watered twice a week. Every month, three samples were taken out, washed with water, dried in oven at 105°C to determine weight loss of the samples. The rate of biodegradation was indicated by the weight loss.

Liquefaction of Corn Stover

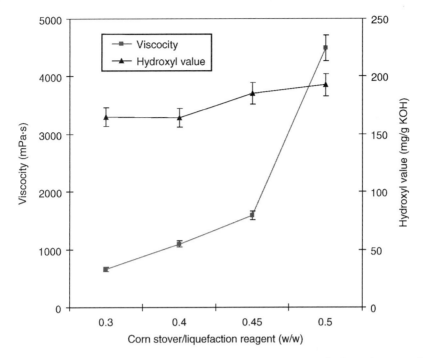

Fig. 2. Effect of the corn stover/liquefaction reagent ratio on the liquefaction of corn stover. Liquefaction temperature: 160°C; liquefaction time: 2 h; catalyst content: 2% (w/w, on liquefaction reagent). Error bars represent standard deviations calculated from the data obtained from five duplicated experiments.

Results and Discussion

Preparation of Bio-Polyols From Crop Residues

Figure 2 shows the liquefaction result of corn stover under various ratio of corn stover to liquefying reagent. With increasing solid content, the viscosity of liquefaction product increased sharply. The ratio of 45/100 was a critical point. A ratio higher than 45/100 will result in a very high viscosity, even a failure of liquefaction. The slight increase in hydroxyl value was because of high corn stover content.

Sulfuric acid is an ideal catalyst for the liquefaction. Usually, 1–3% of sulfuric acid on corn stover can make a good liquefaction. We also investigated other catalyst, such as hydrochloric acid, phosphoric acid, and zinc chloride. As shown in Table 2, these catalysts did not work for the liquefaction of corn stover. After 4-h reaction at 170°C with these catalysts, corn stover still maintained their original state (particles).

Preparation of Polyester From Polyol

The liquefied corn stover consisted of degraded stover fragments (oligosaccharides), glucosides, and residual and decomposed liquefying reagents, all of which contained two or more hydroxyl groups. Therefore,

Table 2
Liquefaction of Corn Stover With Different Catalysts

Catalyst	Dosage of catalyst % on straw	Straw/liquefaction reagent	Result
H_2SO_4	1.5	30/100	Completely liquefied
HCl	2.0	20/100	Unable to liquefy
H_3PO_4	6.0	20/100	Unable to liquefy
$ZnCl_2$	3.0	30/100	Unable to liquefy

Conditions were liquefaction temperature 170°C, liquefaction time 4 h.

Table 3
Crosslinking Chemicals for Stover-Based Polyols

Crosslinking chemicals	Formula	Result and observation
Adipic acid	$HOOC(CH_2)_4COOH$	Fair sheet, bubbles
Sebacid acid	$HOOC(CH_2)_8COOH$	Not well crosslinked, sticky, bubbles
Terephthalic acid	$HOOC(C_6H_4)COOH$	Weak sheet, brittle, bubbles
Terephthaloyl chloride	$ClOC(C_6H_4)COCl$	Many bubbles, like foam; brittle
Citric acid	$HOOCCH_2(HOOC)C(OH)CH_2COOH$	Brittle, bubbles
Maleic anhydride	$C_4H_2O_3$	Fair sheet, stiff
Succinic anhydride	$C_4H_4O_3$	Good sheet, flexible

Condition was OH/COOH = 1/1, sheets were cured at 140°C for 4 h.

it is feasible to crosslink the polyols into a network of polyester through esterification reaction between the hydroxyl groups of the polyols and carboxyl groups in crosslinking chemicals having two or more functional groups, such as dicarboxylic acids or cyclic acid anhydrides. Meanwhile, residual ethylene carbonate underwent ester interchange reaction with diacids (or anhydrides) to become polymer.

To crosslink the polyols, a variety of dicarboxylic acid, diacid chloride, and cyclic anhydride as crosslinking reagents were tested in this study, as shown in Table 3. The results show that although they all were able to react with the polyols to form polyester network, they varied in affecting the properties of formed polyester. A common problem of dicarboxylic acids is the bubbles caused by the water vapor, a byproduct of esterification. The chain length of diacid has an effect on crosslinking. It seems that a longer chain might result in a poorer crosslinking. Benzene ring makes terephthalic acid molecules stiff and bulky, which retarded the reaction between hydroxyl and carboxyl groups. Therefore, the formed polyesters were weak and brittle. Although diacid chloride is much more reactive than diacid, it did not give a good result. Because the

Table 4
Typical Formula of Polyester From BioPolyols

Carboxyl group/hydroxyl group	1.1–1.2/1
Carboxyl contributor (wt% in total carboxyl)	Succinic anhydride, 95% Citric acid, 5%
Hydroxyl contributor (wt% in total hydroxyl)	Polyol, 90% Polyethylene glycol 400, 5% Hexanediol, 2% Glycerol, 3%

reaction between diacid chloride and the polyols were so fast that many bubbles were formed in a short time, resulting from HCl gas produced as a byproduct during the esterification reaction. The results show that cyclic anhydrides, such as succinic and maleic anhydride, are ideal crosslinking reagent for the biopolyols. Because the amount of water produced by the anhydrides was only half of that by diacids, the polyester crosslinked by anhydrides had less and smaller bubbles. Compared with succinic anhydride, maleic anhydride produced stiffer and less flexible polyester, because the double bond in maleric anhydride makes its molecular not as flexible as succinic anhydride. Although citric acid has three carboxyl groups and one hydroxyl group, it is not a good crosslinker as expected. The polyester crosslinked by citric acid appears brittle and foamy *(6,7)*.

Effect of Additives

We found that a small amount of polyhydric alcohols, such as polyethylene glycol, hexanediol, and glycerol, can greatly improve the strength of the polyester. Succinic anhydride alone can react with the biopolyols into polyester, but it was weak and less flexible. The long-chain ones (polyethylene glycol and hexanediol) improve the flexibility of polyester, although glycerol (with more hydroxyl groups) is helpful to increase the crosslinking density. Glycerol also functions as a plasticizer to improve the flexibility of polyester. A small amount of citric acid is positive to the strength of the polyester owing to a higher crosslinking density.

Table 4 shows the formula tested in this study. The polyester samples used in next section to evaluate the properties of polyester sheet were prepared using this formula.

In addition to crosslinking reagents, the curing reaction of the polyol is dependent greatly on curing temperature and time. Figure 3 shows crosslinking results of biopolyols as a function of curing temperatures and time. It was noted that at the same temperature, crosslinking extend increased with increasing reaction time and high temperature enhanced the curing reaction. For example, when the sheet was cured at 140°C for

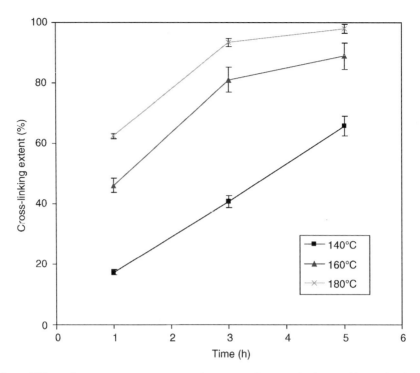

Fig. 3. Effect of curing temperature and time on the crosslinking of biopolyols. Error bars represent standard deviations calculated from the data obtained from five duplicated experiments.

5 h, the crosslinking extend was less than 65%, although if the sheet was cured at 180°C for 5 h, the curing extent was near 100%.

Properties of the Polyester Sheet

Solubility

Figure 4 shows that polyester was quite stable in cold organic solvents and cold water. Only about 8–13% was dissolved after 8 d soaking, depending on the solvents. The dissolved contents should represent those of unreacted crosslinking chemicals, residual liquefying reagents, and small pieces not completely crosslinked in the polyester. The solubility of polyester in hot water was 3–4 times higher than that in cold water. Because of the trace of sulfuric acid in the polyester, when it was refluxed in boiling water, the hydrolysis of the polyester was catalyzed by the sulfuric acid, which could be the main reason for the higher solubility of polyester in hot water than in cold water.

It is well known that ester can be catalytically hydrolyzed under either alkaline or acidic condition. Therefore, when the polyester was refluxed in 2 M NaOH and 2 M H_2SO_4 for 6 h, they were significantly soluble. The solubility was 66% in 2 M H_2SO_4 and 86% in 2 M NaOH, respectively. The polyester was more soluble in alkaline solution than in acidic one.

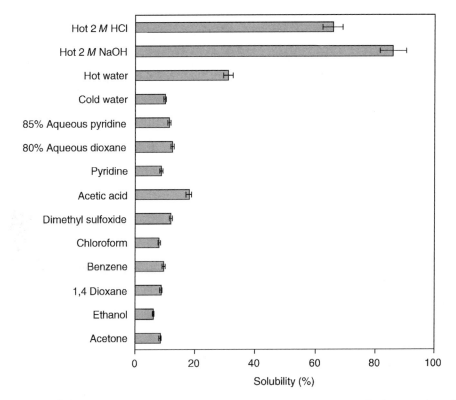

Fig. 4. Solubility of polyester sheet in organic solvents, water, alkaline, and acidic solutions. Sheets were cured at 140°C for 4 h. Error bars represent standard deviations calculated from the data obtained from five duplicated experiments.

Tensile Strength and Elongation

The tensile strength and elongation of polyester was shown in Fig. 5. Two samples were evaluated. One was prepared from the biopolyols produced with ethylene carbonate/ethylene glycol (9/1) (sample 1), and to compare, another was made from the biopolyols produced with 100% ethylene glycol (sample 2). Two samples had similar strength and elongation. Although the sheets were not very strong (about 5 MPa), the strength is acceptable in some cases, such as garden mulch film. We are making more effort to improve the strength of the biopolyester *(8)*.

Biodegradability

The biodegradability of polyester films from corn stover biopolyols was shown in Fig. 6. It was found that the polyester films lost 82% of their initial weight during 10 mo. There is a certain amount of chemicals uncrosslinked or partially crosslinked remaining in the polyester films, which might be less resistant to microbial attacks than fully crosslinked materials, which may contribute to the great weight loss during the 10 mo. Microorganism spots or strains were observable on the surface of films

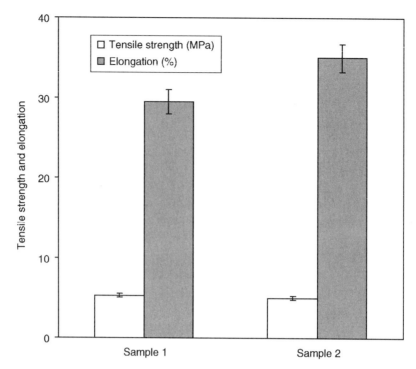

Fig. 5. Tensile strength and elongation of polyester sheets. Sheets were cured at 140°C for 4 h. Error bars represent standard deviations calculated from the data obtained from five duplicated experiments.

and in the surrounding soil, suggesting that there were nutrients for the microorganisms around the films, which were degraded chemicals from the bio-polyester films *(9,10)*.

Conclusions

This study demonstrates that biopolyols produced from corn stover through an organ-solvent liquefaction procedure at moderate temperatures and atmosphere pressure can be directly used as chemical building blocks for making biodegradable polymers without separation and purification. The biopolyols are rich in hydroxyl groups that can react and form crosslink with carboxyl groups of some suitable chemicals to produce biopolyester. A variety of dicarboxylic acids, diacid chloride, and cyclic anhydrides as crosslinking reagents were tested in this study. Although they all were able to react with the stover biopolyols to form polyester, they varied in affecting the properties of the formed polyester. It was found that succinic anhydride combined with a small amount of citric acid, glycol, hexanediol, and polyethylene glycol is a good crosslinking reagent for making biopolyester from stover biopolyols. In general, the formed polyester were stable in cold water and solvents, had acceptable mechanical strength, and readily biodegradable. Biopolyesters made from

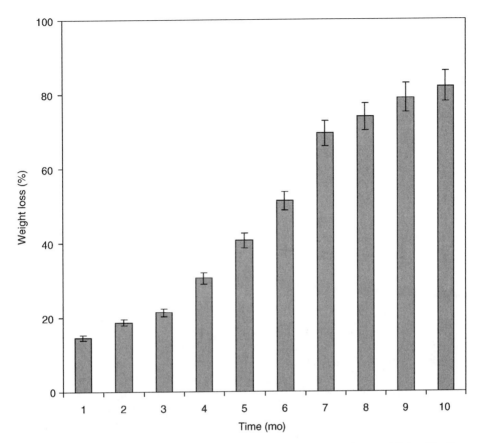

Fig. 6. Biodegradability of polyester films. Sheets were cured at 140°C for 4 h. Error bars represent standard deviations calculated from the data obtained from five duplicated experiments.

corn stover may be processed into sheets, films (e.g., mulching film), and fibers, which may be used in agriculture, gardens, and packaging and textile industries.

References

1. Yao, Y., Yoshioka, M., and Shiraishi, N. (1995), *Mokuzai Gakkaishi* **41,** 659–668.
2. Yao, Y., Yoshioka, M., and Shiraishi, N. (1996), *J. Appl. Polym. Sci.* **60,** 1939–1949.
3. Yamada, T. and Ono, H. (1999), *Bioresour. Technol.* **70,** 61–67.
4. Kurimoto, Y. and Doi, S. (2001), *Biomass. Bioenergy* **21,** 381–390.
5. Kobayashi, M., Tukamoto, K., and Tomita, B. (1999), *Holzforschung* **53,** 617–622.
6. Fang, Q. and Hanna, M. A. (2001), *Bioresour. Technol.* **78,** 115–122.
7. Gomes, M. E., Ribeiro, A. S., Malafaya, P. B., Reis, R. L., and Cunha, A. M. (2001), *Biomaterials* **22,** 883–889.
8. Wang, W., Flores, R. A., and Huang, C. T. (1995), *Cereal Chem.* **72,** 38–41.
9. Marques, A. P., Reis, R. L., and Hunt, J. A. (2002), *Biomaterials* **23,** 1471–1478.
10. Tudorachi, N., Cascaval, C. N., Rusu, M., and Pruteanu, M. (2000), *Polym. Test.* **19,** 785–799.

Enzymatic Production of Xylooligosaccharides From Corn Stover and Corn Cobs Treated With Aqueous Ammonia

Yongming Zhu,[1] Tae Hyun Kim,[1] Y. Y. Lee,*,[1] Rongfu Chen,[2] and Richard T. Elander[3]

[1]Department of Chemical Engineering, Auburn University, AL 36849, E-mail: yylee@eng.auburn.edu; [2]SNC Lavalin GDS, Inc. Houston, TX 77096; and [3]National Bioenergy Center, National Renewable Energy Laboratory, Golden, CO 80401

Abstract

A novel method of producing food-grade xylooligosaccharides from corn stover and corn cobs was investigated. The process starts with pretreatment of feedstock in aqueous ammonia, which results delignified and xylan-rich substrate. The pretreated substrates are subjected to enzymatic hydrolysis of xylan using endoxylanase for production of xylooligosaccharides. The conventional enzyme-based method involves extraction of xylan with a strong alkaline solution to form a liquid intermediate containing soluble xylan. This intermediate is heavily contaminated with various extraneous components. A costly purification step is therefore required before enzymatic hydrolysis. In the present method, xylan is obtained in solid form after pretreatment. Water-washing is all that is required for enzymatic hydrolysis of this material. The complex step of purifying soluble xylan from contaminant is essentially eliminated.

Refining of xylooligosaccharides to food-grade is accomplished by charcoal adsorption followed by ethanol elution. Xylanlytic hydrolysis of the pretreated corn stover yielded glucan-rich residue that is easily digestible by cellulase enzyme. The digestibility of the residue reached 86% with enzyme loading of 10 filter paper units/g-glucan. As a feedstock for xylooligosaccharides production, corn cobs are superior to corn stover because of high xylan content and high packing density. The high packing density of corn cobs reduces water input and eventually raises the product concentration.

Index Entries: Corn stover; corn cobs; xylooligosaccharides; xylan; aqueous ammonia; pretreatment; endoxylanase.

Introduction

Xylooligosaccharides (XOS) with a low degree of polymerization (DP) have been proven to promote proliferation of bifidobacteria, beneficial microorganisms in human intestine *(1–3)*. The demand of XOS as a food additive has shown rapid growth over the last two decades *(4,5)*. XOS are

*Author to whom all correspondence and reprint requests should be addressed.

not foreign for human consumption as they exist in natural plants and food substances including bamboo shoots, fruits, vegetables, milk, and honey. XOS can be produced by autohydrolysis of hemicellulose in lignocellulosic biomass *(5–7)*. In this case, relatively mild conditions are applied because XOS can easily be converted to monomer (xylose). The hydrolyzates from those processes contain a variety of undesirable components, such as soluble lignin, lignin- and sugar-degradation products, organic acids, and ash. Extensive downstream purification is, therefore, required *(8)*.

Alternatively, XOS can be produced by enzymatic hydrolysis. Of the three major types of xylanases (endoxylanase, exoxylanase, and xylosidase), endoxylanase and exoxylanase are the ones responsible for production of XOS. Certain natural plant materials can be digested directly by endoxylanase *(9)*, but such feedstock is scarce, not available in large enough quantities for commercial production. In most natural lignocellulosic biomass, xylan exists mainly as xylan–lignin complex *(10–12)*, and becomes resistant to enzyme attack. For this reason, current commercial processes are carried out in two stages: alkaline extraction of xylan from lignocellulosic biomass followed by enzymatic hydrolysis of the dissolved xylan *(2,3)*. Because the alkaline extract is heavily contaminated, a complex purification process must be applied to the crude xylan before the enzymatic conversion. The situation is about the same as for the xylan produced by autohydrolysis. Recently, wood pulp has been investigated as feedstock for enzymatic production of XOS *(13)*. In this process, the washed wood pulp was hydrolyzed with hemicellulase or xylanase to give a XOS-lignin complex, which is further treated with acid or heat to generate XOS.

A pretreatment method based on soaking in aqueous ammonia (SAA) under moderate severity reaction condition for extended period (e.g., 10–15% NH_3, 65–90°C, 12–24 h) has been investigated in our laboratory. This method was found to be very efficient as a pretreatment for corn stover. The unique features of the SAA are that most of the lignin is removed and all of the cellulose and most of the xylan is retained in solid after treatment. A considerable portion of the ash is also dissolved in this treatment. The SAA-treated corn stover is therefore clean, carbohydrate-rich, and amenable for enzymatic digestion. It is well suitable for enzymatic production of XOS. The primary goal of this research is to assess the feasibility of producing low-DP XOS by enzymatic conversion of the SAA-treated corn stover and corn cobs. It was also of our interest to evaluate the reacted glucan-rich residue, a byproduct, as a feedstock for enzymatic saccharification.

Experimental Methods

Materials

Corn stover supplied by NREL was stored at 5°C. The moisture content was 9–14%. Corn cob was a kind gift of Andersons, Inc, Maumee, Ohio. It has a supplier's coding of 2040 WC, which is "milled woody portion of corn cobs." It has a moisture content of 8–10 %. The composition of 2040 WC as determined by the supplier is: 47.1% cellulose, 37.3% hemicellulose, 6.8% lignin, and 1.2% ash. Endoxylanase, extracted from *Thermomyces lanuginosus*, was purchased from Sigma-Aldrich (cat. no. X2753, lot no. 100K1359). It has manufacturer's nominal activity of approx 2500 xylanase units/g. The cellulase Spezyme CP (Genencor) was supplied by NREL. The cellulolytic activity as determined by NREL was 30 filter paper units/mL. Activated carbon powder was purchased from Sigma-Aldrich (cat. no. C7606, batch no. 073K0037).

Aqueous Ammonia Treatment

The treatment was conducted in a 250-mL autoclave. Heating and temperature controls were done in a GC oven (Varian Model 3700). Twenty grams of dry corn stover and 200 g of 15% ammonia solution (or in the case of corn cobs, 28 g dry corn cob particles and 75 g of 15% ammonia solution) were placed into autoclave for each treatment. The treated materials were washed with deionized water until the pH became neutral. The washed corn stover and corn cobs were then subjected to composition analysis and used as the substrate for XOS production.

Enzymatic Hydrolysis

Enzymatic hydrolysis was conducted in 250-mL Erlenmeyer flasks, which were placed in a laboratory shaking incubator (150 rpm) with temperature control. Solid substrates (treated or untreated corn stover/corn cobs and α-cellulose) were mixed with citrate buffer to reach a total volume of 100 mL. The substrates and buffer were sterilized at 121°C for 30 min before being placed in the incubator. Enzyme was added after the flask content reached the desired temperature. Addition of cellulase enzyme (Spezyme CP) was done on the basis of the glucan content of the substrate, whereas the addition of endoxylanase enzyme was done on the basis of dry weight of substrate.

Purification of XOS

The hydrolyzate from the xylanolytic hydrolysis was centrifuged at $3800g$ for 10 min. Activated carbon powder was added into supernatant liquid with loadings varying in the range of 1–10% of the liquid weight. The flask was placed in a room-temperature incubator shaker at 200 rpm for 30 min to stabilize the carbohydrate-carbon adsorption. The mixture was suction filtered with a 50-mL Pyrex crucible filter and washed with 4×50 mL distilled and deionized water. The XOS-enriched carbon cake thus obtained was eluted by ethanol twice in succession: 2×50 mL 15% ethanol and 2×50 mL 30% ethanol. Where necessary, the third elution was applied with 50% ethanol. The eluates were concentrated by vacuum rotoevaporation and freeze-dried to recover the product in solid form.

Table 1
Composition of Untreated and SAA-Treated Corn Stover[a] (% [w/w], Dry Basis of Samples) and Recovery of Sugars and Solids After Washing

	Untreated corn stover	Ammonia-treated corn stover	Recovery (%)
Glucan	36.8	54.4	96.1
Xylan	22	24.9	73.6
Galactan	0.68	1	95.6
Arabinan	3.5	3.1	57.6
Mannan	0.7	0.6	55.7
Insoluble lignin	17.2	6.1	23
Acid-soluble lignin	3.2	1.6	32.5
Acetyl	3.2	0	0
Ash	7.5	7.9	68.5
Total	94.8	99.6	65[b]

[a]SAA-treatment conditions: 15% ammonia, L/S = 10, 90°C, 24 h.
[b]Solid remaining.

Analysis

The liquid samples were analyzed for sugars by HPLC operated with Bio-Rad Aminex HPX-87P column and a refractive index detector. The HPLC was operated at 85°C with a flow rate of 0.55 mL/min. Sugar oligomers in the hydrolyzate were determined by the increase of sugar monomers after secondary hydrolysis (4% H_2SO_4, 121°C, 1 h) following the NREL Standard Analytical Procedure No. 014 *(14)*. The sugar, acetyl, and lignin contents in the solids were measured by NREL Standard Analytical Procedure No. 002 *(14)*. The cellulose digestibility of the treated biomass was determined by NREL Standard Analytical Procedure No. 009 *(14)*.

Results and Discussion

Aqueous Ammonia Treatment

The details of the SAA-treatment procedure and the results on corn stover were reported by Kim and Lee *(15)*. Table 1 shows the effect of SAA treatment on the composition of corn stover. It is clearly seen that the lignin content (both acid-soluble and -insoluble lignin) decreased significantly after treatment, whereas the sugars were largely retained (96.1% of glucan and 73.6% of xylan being recovered in treated and washed solids). These data support the idea of using the SAA-treated corn stover as a substrate for enzymatic production of XOS.

Optimal Conditions for Enzyme Activity and Hydrolysis

A series of experiments were carried out to determine the optimal pH and temperature for the enzymatic activity. The SAA-treated corn stover

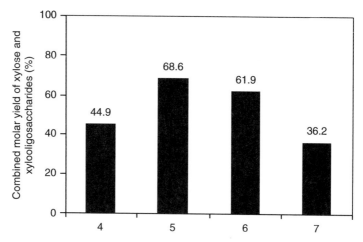

Fig. 1. Effect of pH on the combined molar yield of xylose and XOS from hydrolysis of SAA-treated corn stover at 50°C after 96 h. Enzyme complex: endoxylanase X2753; enzyme loading: 0.04 g/g solids; SAA-treatment conditions: 15% ammonia, 90°C, L/S = 10, 24 h.

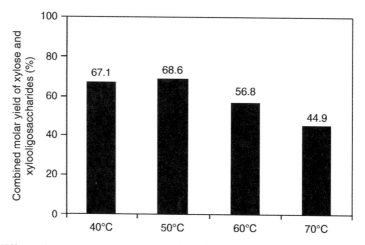

Fig. 2. Effect of temperature on the combined yield of xylose and XOS from hydrolysis of SAA-treated corn stover at pH = 5.0 after 96 h. Enzyme complex: endoxylanase X2753; enzyme loading: 0.04 g/g solids; SAA-treatment conditions: 15% ammonia, 90°C, L/S = 10, 24 h.

used as the substrate. The enzyme activity was expressed as the combined molar yields of xylose and XOS. From the results shown in Figs. 1 and 2, pH 5.0 and 40–50°C were the optimal range applicable for the endoxylanase enzyme (X2753). These findings are in agreement with the optimal ranges reported for endo- and exo-xylanase enzyme complexes *(6,16–18)*.

In order to evaluate the efficiency of the enzyme complex for the digestion of xylan, three enzyme loadings (w/w) were applied at pH 5.0 and 50°C (Fig. 3). The combined molar yield of xylose and XOS increased with the increase of enzyme loading within the range of 0.01–0.16. However, the

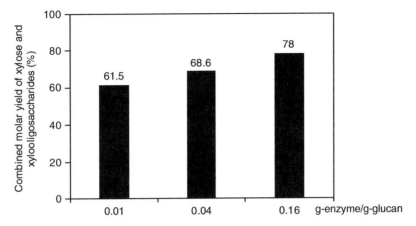

Fig. 3. Effect of enzyme loading on the combine molar yield of xylose and XOS from hydrolysis of SAA-treated corn stover at pH = 5.0 and 50°C after 96 h. Enzyme complex: endoxylanase X2753; SAA-treatment conditions: 15% ammonia, 90°C, L/S = 10, 24 h.

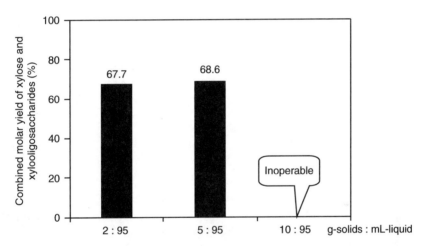

Fig. 4. Effect of substrate loading on the combined molar yield of xylose and XOS from hydrolysis of SAA-treated corn stover at pH = 5.0 and 50°C after 96 h. Enzyme complex: endoxylanase X2753; enzyme loading: 0.04g enzyme/g solids; SAA-treatment conditions: 15% ammonia, 90°C, L/S = 10, 24 h.

efficiency of the enzyme (sugar yield/enzyme loading) decreased with the increase of enzyme loading. Considering the cost of enzyme, we have selected the enzyme/solids ratio of 0.04 (w/w) in subsequent experiments.

Figure 4 presents the effect of substrate concentration on the total molar yield of xylose and XOS. We found no significant change of yield within the solids concentration of 2–5% (w/w). Attempts to further increase the solid concentration to 10% (w/w) failed because of the difficulty of agitating the medium. Obviously the solids concentration appropriate for digestion of SAA-treated corn stover with xylanase is less than 10% (w/w). Usually a fed-batch operation is a solution for this type of

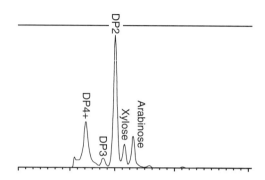

Fig. 5. XOS from enzymatic hydrolysis of SAA-treated corn stover. Hydrolysi conditions: 5% (w/w) solids, pH = 5.0, 50°C, 150 rpm, 72 h. Enzyme complex: Endoxylanase X2753; Enzyme loading: 0.04 g/g-solids; SAA-treatment conditions: 15% ammonia, 90°C, L/S = 10, 24 h.

Table 2
Composition of Xylanlytic Hydrolyzate From SAA-Treated Corn Stover (g/L)

	Monosaccharide	Oligosaccharides	Total
Glucose	–	–	1.488
Xylose	1.147	8.754	9.901
Arabinose	0.078	1.072	1.15

Hydrolysis conditions: 5% (w/w) solids, pH = 5.0, 50°C, and 72 h. Enzyme complex: Endoxylanase X2753; Enzyme loading: 0.04 g/g-solids; SAA-treatment conditions: 15% ammonia, 90°C, L/S = 10, 24 h.

problem. However, it is questionable whether it will work in this case because hydrolysis of xylan does not alter the structure of cellulose fibers, which constitute the framework of the lignocellulosic biomass. As a proof, we noticed that the shape and rigidity of the SAA-treated corn stover solids were not changed significantly after 96 h of digestion. In subsequent experiments, the enzymatic hydrolysis of SAA-treated corn stover for XOS production was conducted under pH = 5.0, 50°C, 0.04 g X2753 endoxylanase/g-solids, solids loading 5% (w/w).

Characterization of XOS

Hydrolysis of xylan in SAA-treated corn stover with endoxylanase generated a liquid product containing XOS with a wide distribution of DP. As shown in the chromatogram of Fig. 5, the hydrolyzate is made mainly of XOS with DP2 and DP4+. Small amount of xylose and arabinose was also detected. It appears that portion of the XOS of DP4+ exists in the form of heteropolysaccharides (i.e., arabinoxylan), as indicated by appearance of sugars other than xylose in the secondary hydrolyzates (Table 2). Nonetheless, no negative effects of these heteropolysaccharides were reported pertaining to their use as food additives.

Table 3
Xylan Digestibility of Untreated and SAA-Treated Corn Stover

	Untreated corn stover	SAA-treated corn stover
Xylan digestibility (%)	14	66.1

Note: Xylan digestibility is expressed as the formation of xylose and XOS. Hydrolysis conditions: 5% (w/w) solids, pH = 5.0, 50°C, and 72 h. Enzyme complex: Endoxylanase X2753; Enzyme loading: 0.04 g/g-solids; SAA-treatment conditions: 15% ammonia, 90°C, L/S = 10, 24 h.

Improvement of Xylan Digestibility by SAA Treatment

The positive effect of delignification on the cellulose digestibility is well known. However, very little study, if any, was focused on the influence of delignification on xylan digestibility. This is primarily because few of the previous pretreatment methods lead to the concurrence of extensive lignin removal and large xylan retention. In other words, the selectivity toward delignification was usually very low in most pretreatment methods. However, the SAA pretreatment provided a unique substrate for this study. As shown in Table 1, when the corn stover was treated with 15% ammonia at 90°C for 24 h, the Klason lignin was reduced from 17.2% to 8.2%, whereas the percentage of xylan increased slightly from 22% to 24.96%. The increase in percent xylan is owing to the weight decrease after pretreatment.

Table 3 compares the digestibilities of corn stover xylan with and without delignification. The SAA treatment increased the xylan digestibility from 14% to 66.1%. This improvement is in line with the increase of cellulose digestibility on delignification. However, the fates of cellulose and xylan in the SAA treatment are different. In lignocellulosic substrates, cellulose exists as homopolymers, which are surrounded by a complex of lignin and hemicellulose. The removal of lignin facilitates the transport of cellulase enzyme to the cellulose surface, but leaves the cellulose chains intact. However, in the ammonia treatment, some of the hemicellulose-lignin linkages are disrupted, and some of the side chains attached to the xylose backbone are removed. Ammonia treatment brings about significant changes to the hemicellulose structure. Despite these differences, the increased substrate accessibility by enzymes remains as the common factor responsible for increased digestibility in both cases.

Refining and Fractionation of XOS

Hydrolysis of SAA-pretreated corn stover with endoxylanase produces hydrolyzates made mainly of XOS (Table 2). It is necessary to remove the impurities such as soluble lignin and ash in the hydrolyzate to bring it to food-grade. For this purpose, the XOS was treated with charcoal (carbon) adsorption and ethanol elution. The carbon adsorption takes advantage of the different interactions between the XOS and carbon (19). The higher the

Table 4
Fates of XOS in Carbon-Ethanol Refining[a]

Carbon : Hydrolyzate (w/w)	0.01	0.05	0.1
XOS loss by water washing	Significant	Moderate	Small
XOS yield from 15% ethanol elution	5.9%	21.3%	34.5%
XOS yield from 30% ethanol elution	2.5%	15.6%	15.9%
XOS yield from 50% ethanol elution	Not done	Not done	4.4%
Total XOS yield from ethanol elution	8.4%	36.9%	54.8%

[a]Yields are molar yields on the basis of xylan in SAA-treated corn stover.

DP the stronger the adsorption to charcoal. For example, xylose (DP1) has little interaction with carbon showing no adsorption on it. It is therefore removed from the XOS-carbon complex by water washing. Small amount of XOS are also lost in water-washing stage before the ethanol elution, the actual amount depending on DP and carbon loading. As shown in Table 4, significant loss of XOS occurs with carbon-to-hydrolyzate ratio of 0.01 (w/w). Increasing the carbon to hydrolyzate ratio to 0.05 (w/w) markedly reduced the XOS loss thus increasing the yield of it. Further increase of the carbon-to-hydrolyzate ratio to 0.1 (w/w) essentially eliminated XOS loss as evidenced by negligible amount of XOS being detected in the water eluate. Most of the XOS adsorbed on charcoal were recovered with 15% ethanol elution. The remaining XOS were recovered by 30% ethanol elution. Although very low in quantity, the products recovered from the 50% ethanol elution included only high-DP XOS. In all practical sense, 30% ethanol elution is sufficient for recovery of the XOS of interest. Knowing that the xylan digestibility is 66.1% (Table 3), the recovery of XOS from the solubilized xylan is estimated to be more than 70% when 30% ethanol elution is applied with carbon-to-hydrolyzate ratio 0.1 (w/w). These data also indicate that successive elution with different ethanol levels may be used as a tool, if needed, to further fractionate the XOS into products of different DP. The final XOS retained pure-white color after freeze-drying.

Glucan Digestibility After Xylanase Treatment

Corn stover, after the SAA treatment, turned into a carbohydrate-rich product. Subsequent hydrolysis of the SAA-treated corn stover by xylanase generated a glucan-rich substrate. As shown in Table 5, the glucan content in the corn stover rose to 69.6% after hydrolysis by endoxylanase, which is almost twice that of untreated corn stover. The residue remaining after XOS production is a byproduct suitable for further bioconversion. We have measured the enzymatic digestibility of this substrate (Fig. 6). For the digestibility test, the moisture of the treated solids was reduced (to 79.2%) by squeezing. The solids, however, were not sterilized so that endoxylanase enzyme remains active along with the cellulase enzyme. It is conceivable that the combined action of cellulase and xylanase may not deteriorate the

Table 5
Composition of SAA-Treated Corn Stover After Xylanolytic
Hydrolysis (% [w/w], Dry Basis of Sample)

	SAA-treated corn cobs
Glucan	69.6
Xylan	11.9
Arabinan	1.4
Mannan	0.7
Acid-insoluble lignin	6.4
Acid-soluble lignin	1.3
Acetyl	0

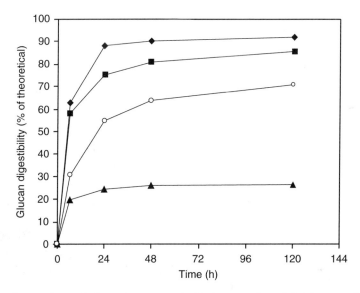

Fig. 6. Glucan digestibility of untreated, SAA-treated and SAA-treated xylanase-digested corn stover and α-cellulose. All the substrates except the xylanase-digested SAA-treated corn stover were sterilized at 121°C for 30 min before enzyme addition. Digestibility test conditions: 1 wt% glucan loading, 10 FPU/g-glucan, pH = 4.8, 50°C, 150 rpm. ▲, untreated corn stover; ♦, SAA-treated corn stover; ■, SAA-treated xylanase-digested corn stover; ○, α-cellulose.

digestibility of glucan. As shown by the curves in Fig. 6, the glucan digestibility of the SAA-treated corn stover maintained at a considerably high level (86% after 120 h) after xylan removal by endoxylanase. This means that the xylanase-treated corn stover can continue to serve as an excellent substrate for further saccharification.

XOS Production From Corn Cobs

Corn cobs were also evaluated as a substrate for XOS production. The procedures were almost identical to those of corn stover; treated with aqueous ammonia and hydrolyzed by endoxylanase. The SAA-treatment conditions

Table 6
Composition of SAA-Treated Corn Cobs[a]

	Dry basis (%)
Glucan	50.4
Xylan	37
Galactan	1.2
Arabinan	3.9
Total sugars	92.5
Acetyl	0
Acid-insoluble lignin	3.6
Acid-soluble lignin	2
Ash	1
Total components	99.1

[a]SAA-treatment conditions: 15% aqueous ammonia, L/S = 2.8, 60°C, 48 h.

Table 7
XOS Production Using Untreated and SAA-Treated Corn Cobs

	Untreated	Treated
Concentration in hydrolyzate (g/L)	4.11	15.74
Molar yield (% of xylan)	21.1	80.5

Note: Xylanlytic hydrolysis conditions: 5% solids loading, 0.04 g-xylanase/g-solids, pH = 5.0, 50°C, 150 rpm, 72 h. SAA-treatment conditions: 15% aqueous ammonia, L/S = 2.8, 60°C, 48 h.

were 15% ammonia, 60°C, L/S 2.8, and 48 h. One noticeable difference was that a significantly lower liquid-to-solid ratio was applied for the corn cobs (only 2.8) because it has a much higher bulk density. High solid loading is beneficial in that it lowers energy input as well as wastewater.

SAA treatment of corn cobs yielded a superior substrate for saccharification. As shown in Table 6, the total sugar content reached 92.5%, the total lignin level went down to 5.6%, and the ash content was reduced to 1%. The following hydrolysis of the treated and washed corn cobs using endoxylanase gave a fairly clear hydrolyzate. Table 7 shows the concentrations and yields of XOS in the hydrolyzate for both untreated and SAA-treated corn cobs. The yield increased by a factor of four after SAA treatment and the XOS concentration reached 1.57%. The SAA-treated corn cobs can be loaded more densely into the hydrolysis reactor than corn stover (Table 8). The XOS concentration rose to 4.7% with 15% solids loading. The product inhibition by XOS to the xylanase appears to be insignificant as the XOS yield decreased slightly with increase of solid loading from 5% to 15%. The HPLC chromatogram of XOS from the hydrolysis of SAA-treated corn cobs is very similar to that from the SAA-treated corn stover (data not shown). Xylose accounted for approx 10% (w/w) of the hydrolyzed xylan, the rest being XOS.

Table 8
Enzymatic Production of XOS From SAA-Treated Corn Cobs
Using Different Solids Loadings

Solids loading	5%	10%	15%
Concentration in liquor (g/L)	15.74	30.02	47.18
Molar yield (%)	80.5	72.7	71.8

Note: Enzymatic digestion conditions: 0.04 g-xylanase/g-solids, pH = 5.0, 50°C, 150 rpm, 72 h. SAA-treatment conditions: 15% aqueous ammonia, L/S = 2.8, 60°C, 48 h.

Conclusions

SAA treatment of corn stover and corn cobs resulted in delignified and xylan-enriched substrates. They are highly susceptible to enzymatic hydrolysis by endoxylanase. The hydrolyzates contained predominantly XOS with a small amount of xylose and other components. Refining of the XOS can be accomplished by carbon adsorption of oligomers followed by ethanol elution. Under optimal-treatment conditions, xylose was totally removed and the XOS were obtained in good yield. Removal of xylan from the SAA-treated corn stover caused a slight decrease in the digestibility of the remaining glucan, but still retained a level above 85% with 10 FPU/g-glucan. For XOS production, corn cobs are a more suitable feedstock than corn stover because corn cobs possess higher xylan content and greater bulk density.

Acknowledgments

The financial support for this research was provided by US Environmental Protection Agency through a grant (Technology for a Sustaining Environment-Grant No. EPA-RD-83164501). A cost-share was provided to this project by Auburn University.

References

1. Modler, H. W. (1994), *Int. Dairy J.* **4**, 383–407.
2. Suwa, Y., Koga, K., Fujikawa, S., Okazaki, M., Irie, T., and Nakada, T. (1999), US 5939309.
3. Yu, S., Yong, Q., and Xu, Y. (2002), *Faming Zhuanli Shengqing Gongkai Shuomingshu*, CN 1364911 A, 7.
4. Crittenden, R. G. and Playne, M. J. (1996), *Trends Food Sci. Technol.* **7**, 353–361.
5. Vazquez, M. J., Alonso, J. L., Dongminguez, H., and Parajo, J. C. (2000), *Trends Food Sci.Technol.* **11**, 387–393.
6. Vazquez, M. J., Alonso, J. L., Dongminguez, H., and Parajo, J. C. (2001), *World J. Microbiol. Biotechnol.* **17**, 817–822.
7. Kabel, M. A., Carvalheiro, F., Garrote, G., et al. (2002), *Carbohydrate Polymers* **50**, 47–56.
8. Carvalheiro, F., Esteves, M. P., Parajo, J. C., Pereira, H., and Girio, F. M. (2003), *Bioresource Technol.* **91**, 93–100.
9. Takao, Y. and Yoshio, I. (1996), *Japanese Patent JP* 5194241.
10. Watanabe, T., Karina, M., Sudiyani, Y., Koshijima, T., and Kuwahara, M., (1993), *Wood Res.* **79**, 13–22.

11. Gubitz, G. M., Csom, G., Johansson, C. I., and Saddler, J. N. (1998), *International Conference on Biotechnology in the Pulp and Paper Industry, 7th, Vancouver, B. C., June 16–19, CC167–C170*. Canadian Pulp and Paper Association, Technical Section, Montreal, Que.
12. Yang, R., Xu, S., and Wang, Z. (2001), *Zhongguo Liangyou Xuebao* **16(6)**, 43–46.
13. Izumi, Y., Sugiura, J., Kagawa, H., and Azumi, N. (2002), *US patent application* 2002195213 A1, 16 pp.
14. Laboratory Analytical Procedures (2004), NREL, Golden, CO.
15. Kim, T. H. and Lee, Y. Y. (2004), *App. Biochem. Biotechnol.* **121–124,** 1119–1131.
16. Frederick, M. M., Frederick, J. R., Fratzke, A. R., and Reilly, P. J. (1981), *Carbohydrate Res.* **97(1),** 87–103.
17. Cai, J. M., Wu, K., Pan, R. R., et al. (2000), *Chinese patent* CN 1266633.
18. Mao, L. -S., Xu, Y., Song, X. -Y., Yong, Q., Yu, and S. -Y. (2001), *Linchan Huaxue Yu Gongye* **21(4),** 33–38.
19. Pellerin, P., Gosselin, M., Lepoutre, J. -P., Samain, E., and Debeire, P. (1991), *Enzyme Microb. Technol.* **13,** 617–621.

Optimal Conditions for Alkaline Detoxification of Dilute-Acid Lignocellulose Hydrolysates

Björn Alriksson,*,[1] Anders Sjöde,[1] Nils-Olof Nilvebrant,[2] and Leif J. Jönsson[1]

[1]Biochemistry, Division for Chemistry, Karlstad University, SE-651 88 Karlstad, Sweden, E-mail: Bjorn.Alriksson@kau.se; and [2]STFI-Packforsk, PO Box 5604, SE-114 86 Stockholm, Sweden

Abstract

Alkaline detoxification strongly improves the fermentability of dilute-acid hydrolysates in the production of bioethanol from lignocellulose with *Saccharomyces cerevisiae*. New experiments were performed with NH_4OH and NaOH to define optimal conditions for detoxification and make a comparison with $Ca(OH)_2$ treatment feasible. As too harsh conditions lead to sugar degradation, the detoxification treatments were evaluated through the balanced ethanol yield, which takes both the ethanol production and the loss of fermentable sugars into account. The optimization treatments were performed as factorial experiments with 3-h duration and varying pH and temperature. Optimal conditions were found roughly in an area around pH 9.0/60°C for NH_4OH treatment and in a narrow area stretching from pH 9.0/80°C to pH 12.0/30°C for NaOH treatment. By optimizing treatment with NH_4OH, NaOH, and $Ca(OH)_2$, it was possible to find conditions that resulted in a fermentability that was equal or better than that of a reference fermentation of a synthetic sugar solution without inhibitors, regardless of the type of alkali used. The considerable difference in the amount of precipitate generated after treatment with different types of alkali appears critical for industrial implementation.

Index Entries: Ethanol; lignocellulose; detoxification; alkali; inhibitor.

Introduction

In the production of fuel ethanol by fermentation of dilute-acid hydrolysates, fermentation inhibitors such as furan aldehydes, phenols, and aliphatic acids can cause a considerable decrease in fermentability. Treatment of the hydrolysate with $Ca(OH)_2$ prior to the fermentation, overliming, is a well established method to improve the fermentability. It is one of the most efficient detoxification methods known *(1)* and it has been in the focus of attention of several studies *(2–6)*. Optimization of the

*Author to whom all correspondence and reprint requests should be addressed.

conditions for overliming of a dilute-acid hydrolysate of spruce has shown that an ethanol productivity and an ethanol yield equal to or even superior to that of a reference fermentation of a synthetic sugar solution can be achieved *(6)*.

One drawback with overliming is the formation of a calcium sulfate precipitate (gypsum). Another drawback is that if the treatment is done under too harsh conditions (high pH and temperature), a considerable degradation of fermentable sugars occurs *(3,5,7)*. Chemical analysis of overliming combined with fermentation experiments suggest that it is difficult to find conditions that separate degradation of fermentable sugars from degradation of inhibiting furan aldehydes *(6)*. Even though very harsh conditions lead to quantitative degradation of some inhibitors, such as furan aldehydes, the fermentability does not increase further, which can be attributed to formation of other inhibitors, such as phenols and formic acid *(6)*. Thus, the treatment has to be optimized to meet the objectives of combining the highest improvement in fermentability with the lowest sugar degradation. A tool for optimizing detoxification procedures is the balanced ethanol yield (ψ_{EtOH}), which takes both the ethanol yield in relation to a reference fermentation of a synthetic sugar solution and the sugar degradation into account *(6)*. Although the problem with sugar degradation can be minimized using this approach, the problem with gypsum formation persists. Other attempts to optimize overliming include the use of titration with NaOH to predict optimal lime addition for detoxification of sugarcane bagasse hydrolysates *(3)*.

A potential approach to overcome the problems associated with overliming is to use another form of alkali. Ammonium does not form poorly soluble salts and previous results suggest that treatment with ammonium hydroxide compares favorable with overliming *(4,8)*. Sodium hydroxide would be another option, but in comparisons performed under similar conditions sodium hydroxide treatment has so far been less efficient than overliming (*see* e.g., refs. *1* and *8*). Novel studies of overliming *(6)* suggest that the conditions for the treatments would need to be carefully optimized to make comparisons been different forms of alkali truly meaningful. To address this and to find an alternative to overliming, factorial designed experiments for treatment of a dilute-acid hydrolysate with ammonium hydroxide and sodium hydroxide were performed to find the optimal treatment conditions. In order to further clarify the role of ammonia in alkaline treatments and to investigate the importance of pH adjustment during detoxification, an experimental series was carried out in which a combination of ammonium hydroxide and sodium hydroxide was used. In addition, a simultaneous comparison among treatments with calcium hydroxide, ammonium hydroxide and sodium hydroxide was carried out on basis of the results of the optimizations. The benefits and drawbacks of each of the methods are discussed.

Methods

Dilute-Acid Hydrolysate From Spruce

The dilute-acid hydrolysate used was prepared for the Swedish Ethanol Research Program and supplied by Dr. R. Eklund (Midsweden University, Örnsköldsvik, Sweden). The hydrolysate was prepared by a two-step hydrolysis in a 250-L batch reactor. Chipped Norway spruce (*Picea abies*) was impregnated with H_2SO_4 (0.5% w/v). Steam was loaded to 12×10^5 Pa (190°C) and the pressure was kept for 10 min. The liquid and solid fractions were separated by filtration. The solid fraction was washed with water and reimpregnated with H_2SO_4 and loaded into the reactor again. Steam was added to 21×10^5 Pa (215°C) and the pressure was kept for 10 min. The liquid fraction was recovered by filtration and pooled with the liquid fraction from the first step. This solution, referred to as the hydrolysate, had a pH of 1.9. The hydrolysate contained 17.3 g/L glucose, 13.3 g/L mannose, 3.1 g/L 5-hydroxymethylfurfural (HMF), 0.8 g/L furfural, 1.7 g/L acetic acid and 1 g/L formic acid. The total concentration of phenols was 3 g/L (calculated as vanillin equivalents).

NH_4OH Detoxification

A multilevel factorial experiment was designed for treatment of the spruce hydrolysate with NH_4OH at different pH values and temperatures using MODDE 7.0 software (Umetrics AB, Umeå, Sweden). Based on previous experience with NH_4OH (8), $Ca(OH)_2$ (6) and NaOH treatments (9,10), different pH-conditions (pH 8.0, 9.0, and 10.0) and different temperatures (5, 30, 42.5, 55, and 80°C) were investigated in different combinations during 3 h. Treatments were not performed over pH 10.0 owing to the pK_a of ammonium (9.25 at 25°C). The 13 different conditions are shown in Table 1. The treatments were performed using a TIM 900 Titration Manager (Radiometer Analytical, Copenhagen, Denmark) in pH-static mode. The pH-meter was corrected against the temperature before use, as the pH is influenced by the temperature and incorrect pH values otherwise will result at temperatures higher than 25°C. After the different treatments, the pH was adjusted to 5.5 by addition of H_2SO_4. Because all the samples had been treated in different ways, water was added to give a final dilution of 5% for all hydrolysate samples.

NaOH Detoxification

A reduced multilevel factorial experiment was designed for treatment of the spruce hydrolysate at different combinations of pH and temperature using the MODDE 7.0 software (Umetrics). The different combinations of pH (9.0, 10.0, 11.0, and 12.0) and temperature (30, 55, and 80°C) used for treatment during 3 h are shown in Table 2. The treatments were performed in pH-static mode as described above. After the different treatments, the

Table 1
Concentrations of Inhibitors and Sugars After NH_4OH Detoxification
and Assessment of Fermentability of the Treated Hydrolysates

Treatment conditions Temperature (°C)	pH	Inhibitors and sugars (% of conc. in untreated sample)					Assessment of fermentability ψ_{EtOH} (% of reference fermentation)
		Furan aldehydes		Phenols	Sugars		
		HMF	Furfural		Glucose	Mannose	
5	8	95	92	91	93	94	33
5	8	97	96	94	97	99	35
30	8	89	91	83	96	98	54
55	8	83	73	93	93	94	109
80	8	69	49	128	95	92	97
5	9	75	81	104	94	94	60
30	9	77	66	90	99	102	77
42.5	9	67	68	100	93	95	103
42.5	9	63	65	102	90	91	109
55	9	68	66	87	93	94	120
80	9	57	44	109	88	90	111
5	10	78	80	82	95	96	42
30	10	68	69	95	94	96	44
55	10	50	43	88	92	93	102
80	10	11	7	126	75	84	99
80	10	4	2	112	73	86	102
Untreated		100	100	100	100	100	7

pH was adjusted to 5.5 by addition of HCl and the volume was adjusted with water so that the total dilution was always 5%.

Combined $NaOH/NH_4OH$ Detoxification

To investigate the role of ammonium and the importance of pH adjustment during detoxification, a combined $NaOH/NH_4OH$ treatment was performed. Different doses of NH_4OH (2.5, 15, 30, 45, and 55 mmol) were added to 114 mL of the hydrolysate. Then, NaOH was added until the pH reached 10.15. The sample to which 55 mmol of NH_4OH was added reached pH 10.15 directly and in that case no addition of NaOH was made. In addition, one hydrolysate sample was treated with only NaOH at 10.15. All samples were treated at 55°C for 3 h. After the treatment, the pH was adjusted to 5.5 by addition of HCl. The volume of all samples was adjusted with water so that the final dilution was 5%.

Simultaneous Comparison of Detoxification With Different Types of Alkali

As the optimization experiments with NH_4OH and NaOH (this study) and $Ca(OH)_2$ (6) were conducted at separate occasions and using separate

Table 2
Concentrations of Inhibitors and Sugars After NaOH Detoxification and Assessment of Fermentability of the Treated Hydrolysates

Treatment conditions		Inhibitors and sugars (% of concentration in untreated sample)							Assessment of fermentability
		Furan aldehydes		Phenols	Aliphatic acids		Sugars		Ψ_{EtOH} (% of reference fermentation)
Temp.(°C) / pH		HMF	Furfural		Acetic	Formic	Glucose	Mannose	
55	9	78	99	96	90	82	94	93	19
55	9	79	102	96	96	85	96	93	27
80	9	71	63	88	98	93	94	89	111
30	10	63	93	97	97	93	95	92	18
55	10	52	49	105	105	104	103	97	77
80	10	65	36	107	107	117	72	72	74
30	11	30	32	94	94	89	92	88	80
55	11	<1	<1	127	102	108	75	88	82
80	11	2	<1	156	115	130	51	61	74
30	12	5	5	114	101	103	84	95	65
30	12	5	6	111	99	101	89	100	80
55	12	0	0	168	133	166	35	34	63
Untreated		100	100	100	100	100	100	100	6

inocula, an experimental series was performed in which selected conditions for treatment with $Ca(OH)_2$, NH_4OH, and NaOH were applied and the detoxified samples were fermented in parallel. The samples were treated for 3 h using the pH-stat as previously described. The conditions were: NaOH, 80°C and pH 9.0; NH_4OH, 55°C and pH 9.0; $Ca(OH)_2$, 30°C and pH 11.0. The total amounts of added alkali to achieve and maintain the conditions described were approx 20 mmol of NaOH, 50 mmol of NH_4OH, and 15 mmol of $Ca(OH)_2$. The subsequent adjustment of the pH to 5.5 required an addition of HCl of 1.5 mmol for the NaOH-treated sample, 20 mmol for the NH_4OH-treated sample, and 10 mmol for the $Ca(OH)_2$-treated sample.

Yeast Strains and Growth Conditions

The fermentations were carried out using *Saccharomyces cerevisiae* (Jästbolaget AB, Rotebro, Sweden). Agar plates with yeast extract peptone dextrose (YEPD) medium (2% yeast extract, 1% peptone, 2% D-glucose, and 2% agar) were used to maintain the strain. Cultures for preparing inocula were grown in 2000 mL cotton-plugged Erlenmeyer flasks containing 1200 mL YEPD medium. The flasks were incubated with agitation at 30°C for approx 12 h. Cells were harvested in the exponential phase by centrifugation (Sorvall RC26 Plus, Dupont) at $1500g$ and 4°C for 5 min. Thereafter, the cells were washed with a sodium chloride solution (9 g/L) and centrifuged as before. To determine the dry weight of the inoculum, a membrane filter (0.45 μm HA filter, Millipore, Milford, MA) was dried in a microwave oven (Husqvarna Micronett, Sweden) set at a power scale of 3 for 15 min and thereafter placed in a desiccator. After 2 h, the filter was taken from the desiccator and weighed on an analytical scale. One and a half milliliters of the yeast suspension were then filtered through the dried filter under the influence of suction. The filter was washed with 5 mL water, dried as previously described, and weighed.

The hydrolysate samples were filtered before fermentation to remove precipitate and the pH was 5.5. The hydrolysate sample (47.5 mL) (or, alternatively, 47.5 mL of a synthetic sugar solution in water for reference fermentations) was mixed with 1 mL of a nutrient solution (consisting of 50 g/L yeast extract, 25 g/L $(NH_4)_2HPO_4$, 1.25 g/L $MgSO_4$ 7 H_2O, 79.4 g/L NaH_2PO_4 H_2O) and 1.5 mL of the inoculum (adjusted to give an initial biomass concentration of 2 g/L dry weight [DW]) in each fermentation vessel. The fermentation vessels were equipped with magnets for stirring and sealed with rubber stoppers with cannulas for outlet of CO_2. The vessels were then placed in an incubator at 30°C with magnetic stirring. The glucose levels during the fermentation were monitored by using a glucometer (Glucometer Elite XL, Bayer AG, Leverkusen, Germany). Samples (0.2 mL) taken from the vessels were diluted with water (1.8 mL) and filtered through an HPLC filter (0.45 μm GHP Acrodisc 13 mm syringe filter, Pall, Ann Arbor, MI) and stored at −20°C until analyzed.

Analysis of Hydrolysates

The concentrations of glucose and mannose were determined as previously described (11) using high-performance anion-exchange chromatography (HPAEC) with a DX 500 chromatography system (Dionex, Sunnyvale, CA) equipped with a CarboPac PA-1 column (Dionex).

HMF and furfural were analyzed using HPLC with a 2690 separation module from Waters (Milford, MA) equipped with a binary pump, an auto injector, and a photo diode array detector (PDAD) set at 282 nm. The separation was performed on an ODS-AL column (50 × 3 mm, 120 Å, and 5-μm particles [Waters]) using conditions previously described (11).

Quantification of acetic acid and formic acid was done using ion chromatography (IC). Before analysis, all samples were filtered through a 0.45 μm Acrodisc syringe filter (Pall Gelman Laboratory, Ann Arbor, MI) and diluted with Milli-Q water. The analysis was done by HPAEC with a Dionex ICS-2000 chromatography system with a conductivity detector, using an IonPac AS 15 (4 × 250 mm) analytical column and an IonPac AG15 (4 × 50) guard column (all from Dionex). An isocratic eluent concentration of 35 mM sodium hydroxide and a flow rate of 1.2 mL/min was used for separation of the anions. External calibration curves were used for the quantification. The total run time of the analysis was 16 min.

The total concentration of phenols was determined using a spectrophotometric method (12), which is based on Folin and Ciocalteu's reagent (Sigma, Steinheim, Germany). Vanillin was used as the standard.

Analysis of Fermentation Samples

The concentrations of glucose and mannose were determined by HPAEC as previously described. The concentration of ethanol was measured using an HP 5890 gas chromatograph (Hewlett-Packard, Palo Alto, CA) equipped with a BP-20 column (film thickness of 1 μm) (SGE, Austin, TX) and a flame ionization detector (11).

Results

The conditions and results of the NH_4OH optimization experiments are shown in Table 1. The software MODDE 7.0 was used for the evaluation of the effect on sugars, inhibitors and precipitate formation. The degradation of fermentable sugars was generally moderate, as shown in the graph for mannose in Fig. 1A. In contrast, the concentrations of HMF and furfural decreased rapidly under conditions with high pH and temperature, as indicated in Fig. 1C,D. The total concentration of phenols decreased slightly except for the treatments performed at 80°C where it increased. The concentrations of acetic and formic acid were only measured for the samples 5°C/pH 8.0 and 80°C/pH 10.0. No significant difference between the two samples was detected. The balanced ethanol yield is graphically

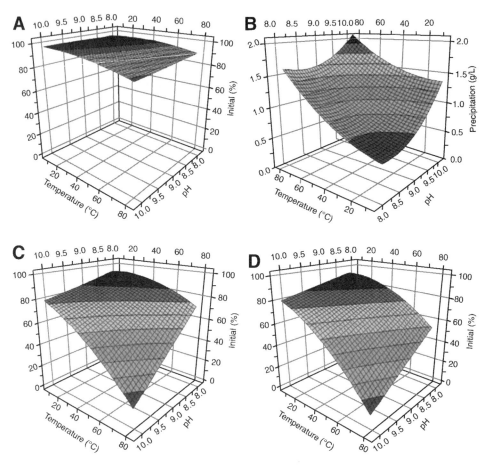

Fig. 1. Response surfaces of NH$_4$OH detoxification (generated by MODDE 7.0) for **(A)** mannose, **(B)** formation of precipitate, **(C)** HMF, and **(D)** furfural.

illustrated, using the software MatLab 7.0.4, in Fig. 2A. All treated samples displayed higher balanced ethanol yield than the untreated sample. The highest balanced ethanol yield was obtained after treatment at intermediate pH (approx 9.0) and intermediate-high temperature (approx 55°C) (Fig. 2A) and reached 120% of the value of the reference fermentation. During the treatments a brownish precipitate formed. The amount of precipitate increased with increasing pH and temperature (Fig. 1B).

The conditions and data used for finding optimal conditions for NaOH treatment are shown in Table 2. As expected *(10)*, the degradation of glucose and mannose was substantial under harsh conditions. Increase of temperature and pH within the intervals studied affected the degradation of the sugars similarly. The concentrations of HMF and furfural decreased under harsher conditions in a similar way. In the sample treated at 55°C/pH 12.0, no detectable levels of HMF or furfural were found. The total concentration of phenols decreased slightly under milder conditions but increased under harsh conditions. The concentrations of acetic and formic acid increased

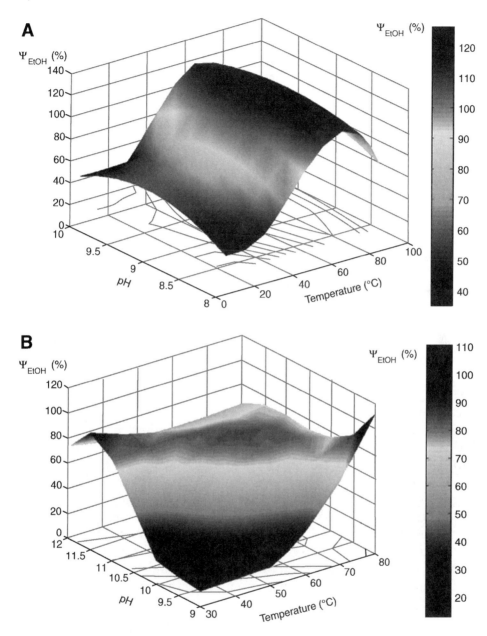

Fig. 2. The balanced ethanol yield (ΨEtOH) for **(A)** NH_4OH and **(B)** NaOH treatments, given in percent of the reference fermentation and plotted using MatLab 7.0.4.

under harsh conditions. The balanced ethanol yield is shown in Fig. 2B. All treatment conditions studied resulted in a higher balanced ethanol yield than for the untreated sample. The best conditions were high temperature in combination with moderate pH or moderate temperature in combination with high pH resulting in a ridge between these conditions, as displayed in Fig. 2B. The highest balanced ethanol yield corresponded to approx 110%

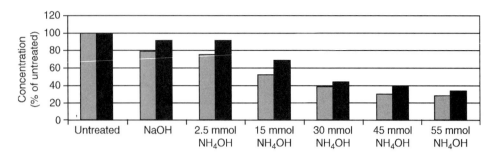

Fig. 3. The degradation of HMF (gray bars) and furfural (black bars) after treatment with NH_4OH, NaOH, or combinations thereof.

of the value of the reference fermentation. No significant amount of precipitate was formed during the NaOH treatments.

The combined experiment was carried out without the use of a pH-stat. This resulted in a decrease in pH for all samples during the 3 h treatment time. The largest drop in pH, from 10.15 to 7.8, was monitored in the sample that was treated only with NaOH. The smallest drop in pH, from 10.15 to 10.1, was monitored in the sample that was treated with 55 mmol of NH_4OH. The drop in pH for the rest of the samples increased with the decreasing amount of NH_4OH added. None of the samples to which NH_4OH was added dropped below pH 9.5. The concentrations of HMF and furfural decreased with the increasing amount of added NH_4OH (Fig. 3). The degradation ranged from 20% of the HMF and 8% of the furfural in the sample treated with only NaOH up to 70% of the HMF and 64% of the furfural in the sample treated with 55 mmol of NH_4OH. The degradation of phenols followed a similar pattern as that observed for furan aldehyde degradation, although it was less extensive, reaching a level of 12% in the sample treated with 55 mmol of NH_4OH. The concentrations of aliphatic acids were not measured. The fermentability was evaluated by comparing the glucose consumption for each sample during fermentation. For the samples with an NH_4OH addition of 15, 30, 45, and 55 mmol, all glucose was consumed within 4 h. For the sample with an NH_4OH addition of 2.5 mmol, all glucose was consumed within 6 h. In contrast, it took 28 h for all glucose to be consumed in the NaOH-treated sample.

Detoxification with $Ca(OH)_2$, NH_4OH, and NaOH was compared in a parallel fermentation experiment. The balanced ethanol yield for all detoxifications was approximately equal to or even better than that of the reference fermentation (Fig. 4). In contrast, the balanced ethanol yield of the untreated sample reached only 14% of that of the reference fermentation.

Discussion

Although overliming is an efficient detoxification method, there are several types of alkali that tentatively could be used as alternatives, especially

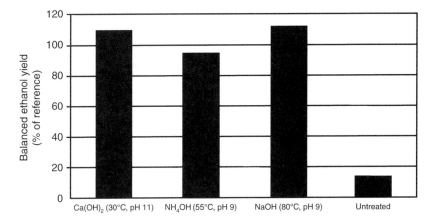

Fig. 4. Balanced ethanol yield in the experimental series in which hydrolysate samples detoxified with Ca(OH)$_2$, NH$_4$OH, and NaOH were fermented in parallel.

as overliming may be associated with problems that are not acceptable for industrial implementation, such as simultaneous degradation of fermentable sugars, resulting in poor ethanol yield, and formation of gypsum. As expected from previous results with NH$_4$OH (4,8), it was possible to achieve excellent fermentability combined with low sugar degradation (Fig. 2A). The ethanol productivity and ethanol yield of some of the NH$_4$OH-treated samples even exceeded the values obtained for the reference fermentation, which has previously also been reported after optimization of overliming (6). This can be explained by the presence of aliphatic and aromatic acids that in proper concentrations stimulate ethanol production at the expense of biomass formation (13–16). Under the relatively mild conditions that were used for detoxification with NH$_4$OH, only small differences in the concentrations of aliphatic acids were observed, which can be attributed to the importance of sugars as precursors to aliphatic acids in the alkaline sugar degradation (17). In these experiments only minor sugar degradation occurred (Fig. 1A). If the sugar degradation and the levels of aliphatic acid are compared at the same temperature and pH (this work and [6]), there are no large differences between NH$_4$OH, NaOH, and Ca(OH)$_2$ treatment. The chemical composition of the precipitate formed in the experiments with NH$_4$OH was not established. The amount of furan aldehydes decreased whereas the amount of precipitate increased, which suggests that a reaction between furan aldehydes and ammonia may have occurred. It has previously been suggested that the furan aldehydes may be involved in alkali-assisted aldol condensation reactions (6). Advantages with NH$_4$OH include that the amount of precipitate formed using NH$_4$OH is lower than the amount of gypsum formed using Ca(OH)$_2$, relatively mild conditions can be used with NH$_4$OH, and the removal of furan aldehydes and phenols is relatively extensive.

Detoxification with NaOH did not give rise to any precipitate as in detoxification with Ca(OH)$_2$ and NH$_4$OH. No significant amount of precipitate was observed regardless whether H_2SO_4 or HCl was used for the pH adjustments. The absence of precipitation might be owing to several factors, one being the low degree of removal of furan aldehydes. Precipitation of toxic factors has been suggested to be an explanation for good detoxification results achieved with calcium salts (18). Other studies suggest that chemical conversions rather than coprecipitation are important for removal of inhibitors (4). Our study shows that precipitation is not prerequisite for efficient detoxification.

The experimental series in which NH_4OH and NaOH was combined shows that one of the benefits of NH_4OH is the buffering effect which keeps the pH on a high and stable level during the detoxification. Even the addition of an amount of NH_4OH corresponding to only 21 mM (the 2.5 mmol experiment) resulted in a much more stable pH. This result can explain that NH_4OH generally performs well if comparisons between different forms of alkali are performed without using pH-stat. The experiment also illustrates the importance of pH adjustment during detoxification with NaOH. Despite the differences in HMF and furfural levels observed in the different samples (Fig. 3), the fermentabilty of the four samples with a NH_4OH addition between 15 and 55 mmol was equally good. This indicates that neither of the furan aldehydes are important as inhibitors in the particular hydrolysate studied, which is also evident from the modest initial concentrations. Spruce hydrolysates may contain considerably higher concentrations of furan aldehydes (1,19). Although the total levels of phenols did not change very much, there might be separate phenols in the hydrolysate that are more toxic to the yeast than others and who are affected differently by different kinds of alkali. Further studies are needed to answer this hypothesis.

The simultaneous comparison of samples detoxified using $Ca(OH)_2$, NH_4OH, or NaOH (Fig. 4) verified the results from the three separate optimization experiments (NH_4OH and NaOH, this work; $Ca(OH)_2$, [6]) in the sense that either type of alkali can be used to achieve high balanced ethanol yield, as long as optimal or near-optimal conditions are applied.

In conclusion, the results show that excellent fermentability combined with low sugar degradation can be achieved regardless of whether $Ca(OH)_2$, NH_4OH, or NaOH is used for alkali detoxification. NH_4OH treatment can be performed under mild conditions, gives a good buffering effect, low sugar degradation and extensive removal of inhibitors, particularly of furan aldehydes. NH_4OH gives rise to a precipitate, which was not observed for samples treated with NaOH. The choice of detoxification method is also dependent of other factors, such as the cost of different forms of alkali and possibly also costs for heating to treatment temperature. To avoid costs for heating, it should be attractive to perform alkaline treatment directly after dilute-acid hydrolysis, before cooling. The optimization experiment shows that treatment at low temperature is a feasible option. Further studies are needed to elucidate the composition of the precipitate generated by NH_4OH treatment, the role of separate phenolic compounds as fermentation inhibitors, and the interconversion between different phenolic compounds that takes place during alkali treatment. The results obtained are an important step toward the development of an efficient detoxification procedure that is suitable for industrial implementation.

Acknowledgments

This work was supported by grants from the Swedish National Energy Administration. We thank Dr. Robert Eklund for providing the hydrolysate.

References

1. Larsson, S., Reimann, A., Nilvebrant, N. -O., and Jönsson, L. J. (1999), *Appl. Biochem. Biotechnol.* **77–79**, 91–103.
2. Martinez, A., Rodriguez, M. E., York, S. W., Preston, J. F., and Ingram, L. -O. (2000), *Biotechnol. Bioeng.* **69**, 526–536.
3. Martinez, A., Rodriguez, M. E., Wells, M. L., York, S. W., Preston, J. F., and Ingram, L. O. (2001), *Biotechnol. Prog.* **17**, 287–93.
4. Persson, P., Andersson, J., Gorton, L., Larsson, S., Nilvebrant, N. -O., and Jönsson, L. J. (2002), *J. Agric. Food Chem.* **50**, 5318–5325.
5. Millati, R., Niklasson, C., and Taherzadeh, M. J. (2002), *Process Biochem.* **38**, 515–522.
6. Sárvári Horváth, I., Sjöde, A., Alriksson, B., Jönsson, L. J., and Nilvebrant, N. -O. (2005), *Appl. Biochem. Biotechnol.* **121–124**, 1031–1044.
7. Nigam, J. N. (2001), *J. Biotechnol.* **87**, 17–27.
8. Alriksson, B., Sárvári Horváth, I., Sjöde, A., Nilvebrant, N. -O., and Jönsson, L. J. (2005), *Appl. Biochem. Biotechnol.* **121–124**, 911–922.
9. Nilvebrant, N. -O., Reimann, A., Larsson, S., and Jönsson, L. J. (2001), *Appl. Biochem. Biotechnol.* **91–93**, 35–49.
10. Nilvebrant, N. -O., Persson, P., Reimann, A., de Sousa, F., Gorton, L., and Jönsson, L. J. (2003), *Appl. Biochem. Biotechnol.* **105–108**, 615–628.
11. Sárvári Horváth, I., Sjöde, A., Nilvebrant, N. -O., Zagorodni, A., and Jönsson, L. J. (2004), *Appl. Biochem. Biotechnol.* **114**, 525–538.
12. Singleton, V. L., Orhofer, R., and Lamuela-Raventos, R. M. (1999), *Methods Enzymol.* **299**, 152–178.
13. Larsson, S., Palmqvist, E., Hahn-Hägerdal, B., et al. (1999), *Enzyme Microb. Technol.* **24**, 151–159.
14. Verduyn, C., Postma, E., Scheffer, W. A., and van Dijken, J. P. (1992), *Yeast* **8**, 501–517.
15. Taherzadeh, M. J., Niklasson, C., and Lidén, G. (1997), *Chem. Eng. Sci.* **52**, 2653–2659.
16. Lawford, H. G. and Rousseau J. D. (2003), *Appl. Biochem. Biotechnol.* **106**, 457–470.
17. de Bruijn, J. M., Kieboom, A. P. G., van Bekkum, H., and van der Poel, P. W. (1986), *Sugar Technol. Rev.* **13**, 21–52.
18. Van Zyl, C., Prior, B. A., and Du Preez, J. C. (1988), *Appl. Biochem. Biotechnol.* **17**, 357–369.
19. Taherzadeh, M. J., Eklund, R., Gustafsson, L. C., and Lidén, G. (1997) *Ind. Eng. Chem. Res.* **36**, 4659–4665.

Reintroduced Solids Increase Inhibitor Levels in a Pretreated Corn Stover Hydrolysate

R. Eric Berson,*,[1] John S. Young,[1] and Thomas R. Hanley[2]

[1]Department of Chemical Engineering, University of Louisville, Louisville, Kentucky 40292, E-mail: eric.berson@louisville.edu; and [2]Auburn University, Auburn, AL 36849

Abstract

Following detoxification of the liquid hydrolysate produced in a corn stover pretreatment process, inhibitor levels are seen to increase with the re-addition of solids for the ensuing hydrolysis and fermentation processes. The solids that were separated from the slurry before detoxification of the liquor contain approx 60% (w/w) moisture, and contamination occurs owing to the diffusion of inhibitors from the moisture entrained in the porous structure of the corn stover solids into the bulk fluid. This evidence suggests the need for additional separation and detoxification steps to purge residual inhibitors entrained in the moisture in the solids. An overliming process to remove furans from the hydrolysate failed to reduce total organic acids concentration, so acids were removed by treatment with an activated carbon powder. Smaller carbon doses proved more efficient in removing organic acids in terms of grams of acid removed per gram of carbon powder. Sugar adsorption by the activated carbon powder was minimal.

Index Entries: Activated carbon; detoxification; organic acids; overliming; pretreated corn stover hydrolyzate.

Introduction

Cellulosic crops and other sources of biomass harvested for the conversion to fuels and chemicals do not contain significant quantities of readily fermentable simple sugars. The development of economically viable processes for the extraction of simple sugars from biomass is ongoing and there is room for improvement in all phases of the conversion process, especially the hydrolysis of hemicellulose and cellulose and the subsequent removal of undesirable byproducts. Acid treatments that hydrolyze hemicellulose and prepare the cellulose for subsequent enzymatic hydrolysis have been studied for some time. Dilute acid hydrolysis is

*Author to whom all correspondence and reprint requests should be addressed.

currently the most common form of biomass pretreatment. Often in a dilute acid hydrolysis process, the biomass is first ground into small pellets and then soaked in a dilute acid, usually sulfuric acid *(1)*. The pretreatment of most biomass systems causes degradation of sugars and lignin resulting in the formation of byproducts such as furfural, hydroxy methyl furfural, phenolic compounds, acetic acid, and other organic acids. Although these chemicals are not always produced in great quantities, they can have a toxic effect on the fermentative ability of ethanologenic organisms, especially bacteria *(2)*. Concentrations of acetic acid found in pretreated softwoods may reach 10 g/L, completely inhibiting the fermentative ability of Zymomonas mobilis *(3)*.

Several detoxification methods have been investigated to overcome the toxic effects of these byproducts *(4–9)*. Overliming is a common and effective process in which the pH of hot hydrolysate is adjusted with lime resulting in the formation of insoluble salts *(4–6)*. Although common and effective, overliming does not affect acetic acid levels *(5)*, which are known to inhibit ethanol production above 2 g/L *(10)*. Detoxification by activated carbon adsorption has proven to effectively remove acetic acid from a synthetic hydrolysate solution formulated to mimic the composition of an actual hydrolysate *(11)* and an actual corn stover hydrolysate *(12)*. Carbon has been used as an adsorbent for hundreds of years because it was discovered that it could be used to purify drinking water and remove color impurities and is still commonly used today in wastewater, drinking water, refinery waste, and chemical clarification applications *(13)*. Activated carbons are regenerated easily with steam, and the stripped components may be recovered and marketed. A number of studies have qualitatively reported using activated carbons on acid-hydrolyzed wood and sugar cane substrates before fermentation with varying degrees of success *(14–17)*.

Overliming and carbon adsorption procedures are performed in liquid hydrolysates from which biomass solids are necessarily first separated to prevent interference during the detoxification procedure. The cellulose containing solids are added back to the conditioned liquor for the enzymatic hydrolysis of cellulose and fermentation. In this study, the effect of reintroducing corn stover solids on inhibitor levels in conditioned hydrolysates is investigated. Conditioning is performed by overliming and carbon adsorption methods. Furans and organic acids are measured before detoxification, following detoxification, and following the reintroduction of solids. The effect of carbon loading on adsorption efficiency is also examined.

Materials and Methods

A corn stover slurry pretreated with dilute sulfuric acid (190°C, 1.6% acid, 30% solids) was provided by the National Renewable Energy Laboratory. The slurry viscosity is on the order of 25,000 cP or higher at room temperature and a shear rate of 1/s, and because of the impracticality of dispersing carbon or $Ca(OH)_2$ solids in this viscous material, liquid hydrolysate is necessarily first separated from the slurry. The solids are later returned to the detoxified liquid. Vacuum filtration is employed for the separation process. After maximum liquid separation from the vacuum filtration, the solids were found to contain 60% moisture. Overliming was performed by raising the liquor temperature to 50°C, manually adding $Ca(OH)_2$ until the pH reached a value of 10.0, and maintaining these conditions for 30 min. Total furans, before and after overliming, were measured using the UV spectra method described by Martinez et al. *(18)* in a Genesys 20 spectrophotometer from ThermoSpectronic (Waltham, MA). The UV spectra method provides a reasonable estimation of relative concentrations based on the difference in absorbance readings between 284 and 320 nm, so instead of quantifying absolute furans concentrations, results are reported in terms of ΔA ($A_{284}-A_{320}$).

Calgon BL activated carbon powder provided by Calgon Carbon Corporation (Pittsburgh, PA) was used for organic acid adsorption testing. Tests with carbon concentrations equal or less than 80 g/L were run in an Innova Model 4230 benchtop refrigerated incubator/shaker from New Brunswick Scientific (Edison, NJ). The incubator contains an Erlenmeyer flask platform capable of holding 25 × 250 mL flasks. Flasks were loaded with 100 mL of liquid and carbon powder concentrations of 20, 40, or 80 g/L. The shaker was operated at 250 rpm and 30°C. Flasks were covered with parafilm to prevent evaporation. Tests at higher carbon concentrations were mixed with a Lightnin Mixer fitted with a R100 impeller to prevent carbon aggregation. After 30 min of carbon treatment in the hydrolysate, carbon was separated from the liquid by centrifugation followed by passing the remaining liquid through a 0.1 μm filter (diameter of carbon particles is 150 μm). Initial and equilibrium total organic acids concentrations were measured by titration with NaOH. To determine the extent of sugar adsorption by the carbon, glucose, and xylose compositions were measured using an Alltech HPLC equipped with a Biorad Aminex HPX-87H column and ELSD detector.

Results and Discussion

Previously, attempts at fermenting a broth containing detoxified hydrolysate with cellulosic rich corn stover solids returned to the liquor resulted in lower ethanol production than expected *(19)* based on fermentations in synthetic liquid broths containing varying levels of inhibitors *(10)*. The moisture (60%) in the returning solids should contain inhibitors similar to the preconditioned bulk fluid, so it was postulated that diffusion from the moisture in the pores of the solid back into the conditioned bulk fluid would cause a rise in bulk fluid inhibitor levels.

Total furans were measured before an overliming process, following an overliming process, and following the addition of varying levels of corn

Table 1
Effect of Reintroduced Solids on Total Furans
Concentration After Overliming

Stage	Total furans in hydrolysate ($A_{284}-A_{320}$)
Initial	0.393
Overlimed	0.001
10% Solids added	0.139
15% Solids added	0.164
17% Solids added	0.175
20% Solids added	0.205

stover solids after overliming appear in Table 1. The solids added back to the hydrolsate were allowed to equilibrate for 2 h. Furans (after dilute acid pretreatment) were reduced from an initial ΔA of 0.393 to 0.001 following the overliming process. After 100 g/L corn stover solids were reintroduced into the hydrolysate, ΔA increased to 0.139. After reintroduction of 150 g/L, 170 g/L, and 200 g/L solids, ΔA measured 0.164, 0.175, and 0.205, respectively. Although the actual furans concentrations are not quantified, the increase owing to the reintroduction of solids is clearly illustrated, a 17.6% increase in reintroduction of solids (170 to 200 g/L) results in a 17.1% increase (0.175 to 0.205) in relative furans concentration.

Total organic acid levels increased as well when filtered corn stover solids were returned to the conditioned hydrolysate. Results presented in Fig. 1 show the effect of reintroduced solids concentration on the new equilibrium concentration of total organic acids in the bulk liquid. Initial acid concentration (approx 18 g/L) was reduced to 6 g/L through a series of activated carbon treatments using 80 g/L carbon per step. After the carbon was separated from the hydrolysate, corn stover solids were reintroduced and allowed to equilibrate for 2 h. Tests were performed using 100, 150, and 200 g/L corn stover solids. The solids were filtered from the hydrolysate to allow titration for organic acids concentration.

The vertical lines after the fourth treatment step represent the increase in organic acids following the reintroduction of solids. The reintroduction of 100 g/L solids increased the organic acids concentration from 6 to 8.3 g/L. The reintroduction of 150 g/L solids increased the organic acids concentration from 6 to 10.3 g/L. The reintroduction of 200 g/L solids increased the organic acids concentration from 6 to 11.9 g/L. The correlation between organic acids increase and mass of solids added is approximately linear. A 50% increase from 10 to 15% solids resulted in a 53% increase in organic acids (4.3 g/L vs 2.3 g/L), and a 33% increase from 15 to 20% solids resulted in a 37% increase in organic acids (5.9 g/L vs 4.3 g/L). The concentration gradient between the liquid in the pores of the reintroduced solids and the bulk liquid will determine the new equilibrium concentration of a component in the bulk liquid, with material diffusing from

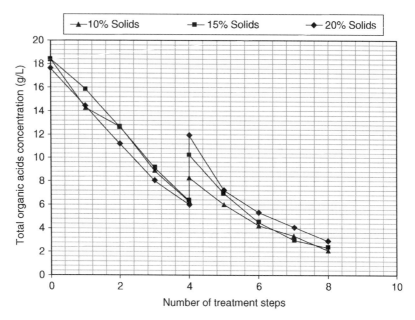

Fig. 1. Effect of reintroduced solids on bulk fluid total organic acids concentration.

Table 2
Effect of Initial Organic Acids Concentration on Final Organic Acids Concentration After the Reintroduction of 20% Solids

Initial total organic acids (g/L)	Final total organic acids (g/L)	Change (g/L)
11.6	13.1	1.5
10.9	13.2	2.3
6	11.9	5.9

the pores of the solids in which untreated liquid resides into the bulk liquid in which the inhibitor concentrations have been reduced. Therefore, it would be expected that inhibitor levels will increase more when solids are added to hydrolysate with lower initial inhibitor levels. In this context, "initial" refers to the time immediately before the reintroduction of solids.

Table 2 shows the change in organic acids levels after the reintroduction of 200 g/L solids into hydrolysate with varying initial organic acids concentrations. Initial organic acids concentrations were established by altering the number of carbon treatment repetitions; fewer repetitions removed less acid resulting in higher initial acid concentrations. Data in the table shows the expected trend: for initial organic acids concentrations of 11.6, 10.9, and 6.0 g/L, final equilibrium concentrations increased by 1.5, 2.3, and 5.9 g/L. Inhibitors are removed from hydrolysate for the benefit of the downstream fermentation and the increase in inhibitors following the reintroduction of solids indicates the need for additional detoxification steps, or a means of

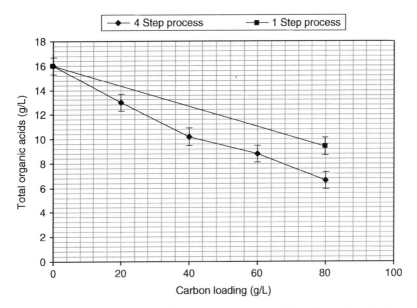

Fig. 2. Carbon treatment efficiency comparison (80 g/L total carbon).

purging inhibitors directly from the solids through a washing process. If additional detoxification steps are favored, the process of detoxifying and reintroducing solids may need to be repeated until inhibitor levels stabilize.

Previously reported adsorption isotherm data (12) suggests that lower activated carbon concentration may be more efficient at removing organic acids in terms of acid removed per gram of carbon. This was verified by performing a single step detoxification with a given carbon loading and comparing the final organic acids concentration to the final concentration in a test in which an equivalent amount of total carbon was divided into four batches and four consecutive detoxifications were performed. The hydrolysate used in these two tests originated from the same batch of pretreated corn stover so the initial compositions were equivalent.

Results in Fig. 2 indicate organic acids concentration was lowered from 16 to 9.4 g/L using 80 g/L carbon in a single step and from 16 to 6.6 g/L using 20 g/L carbon in four consecutive steps, an improvement of 42% adsorption. The improvement was even greater in tests with lower total carbon loading. When 10 g/L carbon in four consecutive steps is in comparison with a single 40 g/L step, organic acids removal improved by 55% (7 g/L removed vs 4.5 g/L), and when 5 g/L carbon in four consecutive steps is in comparison with a single 20 g/L step, organic acids removal improved by 117% (3.5 g/L removed vs 7.6 g/L). The final organic acids levels are above known critical limits for *Z. mobilis* growth, but the data is intended to illustrate the differences in efficiency for varying carbon concentration levels.

With fewer carbon particles present in the liquid at lower carbon loadings, a higher concentration gradient exists between organic acids in

Table 3
Carbon Treatment Efficiency Comparison—High Shear Environment

100 g/L carbon/stage × 2 stages		50 g/L carbon/stage × 4 stages		25 g/L carbon/stage × 8 stages	
Stage	Acid (g/L)	Stage	Acid (g/L)	Stage	Acid (g/L)
Initial	16.2	Initial	16.2	Initial	16.2
1	9.8	1	13.3	1	14.7
2	4.6	2	9.7	2	12.4
		3	6.1	3	11.1
		4	3.9	4	9.8
		5	7.9		
		6	6.4		
		7	4.8		
		8	3.6		

the bulk fluid and each carbon particle, driving more acid into each particle. This may be evidence that the rate limiting step of adsorption of organic acids by the carbon particles is diffusion across the bulk liquid-particle surface interface rather than diffusion within the particle pores.

If carbon particles aggregated while mixing in the shaker flasks during the adsorption treatment process, which would be more likely at higher carbon loadings, the total surface area of carbon will be reduced, which would reduce the efficiency at higher carbon loadings. To address this, a test was performed in which carbon was mixed in the hydrolysate with a high-shear R100 impeller at 250 rpm instead of mixed in shaker flasks to prevent aggregation. Results in Table 3 for this high-shear system are consistent with shaker flask tests; organic acids adsorption occurs with higher efficiency when equivalent total carbon loading is distributed among more treatment steps, ruling out aggregation of carbon particles as a cause for reduced efficiency with higher carbon concentration.

Because the removal of organic acids by the carbon treatment process will not be considered useful if significant amounts of sugars are removed, xylose and glucose concentrations were measured before and after treatment to determine the extent of their adsorption. Xylose is present in the acid-treated hydrolysate at 30.5 g/L and glucose is present at 3.7 g/L. Following hydrolysate conditioning by carbon treatment, xylose, and glucose concentrations were remeasured (Table 4). Depending on carbon loading, xylose concentration was reduced by 1.3 to 3.9 g/L and glucose concentration was reduced by 0.3 to 0.5 g/L. The loss of xylose and glucose from hydrolysate is comparable to losses reported owing to overliming (6). The carbon's affinity for organic acids and lack of affinity for xylose and glucose may be attributed to the size of the molecules. Although the size of the pore opening is unknown, the molecular weights of xylose, 150 g/mol, and glucose, 180 g/mol, are

Table 4
Adsorption of Sugars During Carbon Treatment

Carbon load (g/L)	Xylose (g/L)	Glucose (g/L)
Initial	29.7	4.1
80	25.8	3.6
40	28.4	3.8
20	26.9	3.7

two and a half and three times the size of the molecular weight of acetic acid, 60 g/mol.

Conclusions

Following the reintroduction of corn stover solids into detoxified hydrolysate, total furans and organic acids levels increased significantly enough to require additional treatment before fermentation. The amount of increase depended on the amount of solids returned and the existing concentration of the inhibitor. Carbon efficiency for removing organic acids was higher for lower carbon loadings. Adsorption of both xylose and glucose by the carbon was minimal.

Acknowledgment

This work was funded by the US Department of Energy's National Renewable Energy Laboratory, subcontract: XCO-1-31016-01.

References

1. Nguyen, Q. A., Tucker, M. P., Keller, F. A., Beaty, D. A., Connors, K. M., and Eddy, F. P. (1999), *Appl. Biochem. Biotechnol.* **77–79,** 133–142.
2. Larsson, S., Palmqvist, E., Hahn-Hagerdal, B., et al. (1999), *Enzyme Microbiol. Technol.* **24,** 151–159.
3. Ranatunga, T. D., Jervis, J., Helm, R. F., McMillan, J. D., and Hatzis, C. (1997), *Appl. Biochem. Biotechnol.* **67,** 185–198.
4. Larsson, S., Reimann, A., Nilvebrant, N. O., and Jonsson, L. J. (1999), *Appl. Biochem. Biotechnol.* **77–79,** 91–103.
5. Martinez, A., Rodrigues, M. E., Wells, M. L., York, S. W., Preston, J. F., and Ingram, L. O. (2001), *Biotechnol. Prog.* **17,** 287–293.
6. Mohagheghi, A., Ruth, M., and Schell, D. (2004), Tracking the fate of calcium and sulfur through the overliming process used to condition hydrolysates produced by dilute sulfuric-acid pretreatment of lignocellulosic biomass, presented at the 26th Symposium on Biotechnology for Fuels and Chemicals, Chattanooga, TN.
7. Lee, W. G., Lee, J. S., Shin, C. S., Park, S. C., Chang, H. N., and Chaik, Y. K. (1999), *Appl. Biochem. Biotechnol.* **77–79,** 547–559.
8. Rivard, C. J., Engel, R. E., Hayard, T. K., Nagle, N. J., Hatzis, C., and Philippidis, G. P. (1996), *Appl. Biochem. Biotechnol.* **57–58,** 183–191.
9. Jonsson, L. J., Palmqvist, E., Nilvebrant, N. O., and Hahn-Hagerdal, B. (1998), *Appl. Microbiol. Biotechnol.* **49,** 691–697.
10. Priddy, S. A. (2002), PhD Dissertation, University of Louisville, Louisville, Kentucky.

11. Priddy S. A. and Hanley, T. R. (2003), *Appl. Biochem. Biotechnol.* **105–108,** 353–364.
12. Berson, R. E., Young, J. S., Kamer, S. N., and Hanley, T. R. (2005), *Appl. Biochem. Biotechnol.* in press.
13. Morresi, A. C. and Cheremisinoff, P. N. (1978), In: *Carbon Adsorption Handbook,* Cheremisinoff, P. N and Ellerbusch, F., eds., Ann Arbor Science Publishers, Ann Arbor, MI, pp. 1–54.
14. Fein, E. F., Tallim, S. R., and Lawford, G. R. (1984), *Can. J. Microb.* **30,** 682–690.
15. Frazer, F. R. and McCaskey, T. A. (1989), *Biomass.* **18,** 31–42.
16. Roberto, I. C., Lacis, L. S., Barbosa, M. F. S., and de Mancilha, I. M. (1991), *Process Biochem.* **26,** 15–21.
17. Parajo, J. C., Dominguez, H., and Dominguez, J. M. (1997), *Enzyme Microb. Technol.* **21,** 18–24.
18. Martinez, A., Rodrigues, M. E., York, S. W., Preston, J. F., and Ingram, L. O. (2000), *Biotechnol. Prog.* **16,** 637–641.
19. Kamer, S. (2004), Masters thesis, University of Louisville, Louisville, Kentucky.

Modeling of a Continuous Pretreatment Reactor Using Computational Fluid Dynamics

R. Eric Berson,*,[1] Rajesh K. Dasari,[1] and Thomas R. Hanley[2]

[1]*Department of Chemical Engineering, University of Louisville, Louisville, KY 40292, E-mail: eric.berson@ louisville.edu; and* [2]*Auburn University, Auburn, AL 36849*

Abstract

Computational fluid dynamic simulations are employed to predict flow characteristics in a continuous auger driven reactor designed for the dilute acid pretreatment of biomass. Slurry containing a high concentration of biomass solids exhibits a high viscosity, which poses unique mixing issues within the reactor. The viscosity increases significantly with a small increase in solids concentration and also varies with temperature. A well-mixed slurry is desirable to evenly distribute acid on biomass, prevent buildup on the walls of the reactor, and provides an uniform final product. Simulations provide flow patterns obtained over a wide range of viscosities and pressure distributions, which may affect reaction rates. Results provide a tool for analyzing sources of inconsistencies in product quality and insight into future design and operating parameters.

Index Entries: Auger; biomass; CFD simulations; pretreatment reactor; screw conveyor.

Introduction

Methods such as acid pretreatments and enzymatic hydrolysis for converting biomass into high yields of pentose and hexose sugars to be used in fermentations are the subject of continuing investigations. Dilute acid hydrolysis, the most commonly used form of pretreatment, is a thermo-chemical process in which the biomass is in contact with a dilute acid, typically 2% or less sulfuric acid, at temperatures greater than 140°C for short residence times, usually on the order of 10 min or less (1–5). The process produces hydrolysate liquor rich in pentose sugars and opens the ligno-cellulose pore structure to increase the susceptibility of cellulose to enzymatic attack (6). Less than optimal processing conditions result in low sugar concentration and sugar degradation into substances toxic to a fermenting organism.

---

*Author to whom all correspondence and reprint requests should be addressed.

Other acid hydrolysis strategies are also presented in the literature: Nguyen et al. *(7)* recommended soaking in a 0.4–0.7 % sulfuric acid solution at 60°C for 4 h. Teixeira et al. *(8)* used a 60% solution of peracetic acid for 7 d. Choi and Matthews used a 2% solution of sulfuric acid at 132°C for 40 min followed by a 15% sulfuric acid solution at 132°C for 70 min *(9)*. Taherzadeh et al. *(10)* soaked different wood species in 5 g/L of sulfuric acid for 7 min between 188°C and 234°C *(10)*. Pretreatment at pilot or production scale must be performed as a continuous process to be economical. A number of successful continuous dilute acid processes have been reported *(7,11–16)*, but interest in processing at high solids concentrations has created unique processing challenges owing to high slurry viscosities. Screw augers are typically used for conveying viscous fluids but may not provide the optimal incorporation of acid and steam to produce a consistent pretreated biomass product. Over or under exposure of biomass to the acid, buildup on reactor walls, and poor heat transfer result from poor mixing and will lead to inconsistencies in the final product. Higher solids loading will provide more sugar yield if processed efficiently, but the increased viscosity may present a trade-off in terms of yield vs efficiency.

Berson and Hanley *(17)* presented flow patterns in screw auger pretreatment reactors created using computational fluid dynamics (CFD). Results showed distinct flow differences because of the auger design for a given viscosity. At solids loading in the 15–25% range, increasing the concentration of solids by 1% can increase the viscosity on the order of thousands of centipoise, significantly affecting flow conditions in the reactor. The objective of this study is to compare conditions in a screw auger pretreatment reactor for varying viscosities associated with varying solids loading and temperature using FLUENT 6.2, a commercial CFD flow simulation package. The reactor consists of a tubular shell around a screw auger used to convey the biomass slurry. Flow patterns and pressure distributions are revealed from simulations performed for slurries with solids concentrations ranging from 10% to 25%.

Materials and Methods

Biomass Slurry Preparation

Corn stover slurry pretreated with dilute sulfuric acid (190°C, 1.6% acid, 30% solids, 5 min residence time) was provided by the National Renewable Energy Laboratory. To prepare slurry for viscosity measurements, liquid hydrolysate and solids were separated by vacuum filtration. An appropriate amount of separated solids, containing 60% moisture after the filtration, were returned to the hydrolysate to obtain the desired solids concentration.

Viscosity Measurements

Biomass slurries containing suspended solids require a viscosity determination technique that maintains the solids in a uniform suspension to prevent phase separation during measurement. A helical ribbon impeller fitted to Brookfield RVDV III (7187 dyne cm maximum torque) and HBDV III (57496 dyne cm maximum torque) rheometers is used to maintain particles in suspension for homogeneous viscosity determinations as suggested by Allen and Robinson (18). This method assumes an average shear rate for the complex flow field created by the helical impeller that is proportional to the impeller speed:

$$\gamma_{avg} = kN \tag{1}$$

and assumes the Reynolds number is inversely proportional to the dimensionless power number for Newtonian fluids in laminar flow:

$$c/\mathrm{Re} = \frac{2\pi M}{\rho N^2 D^5} \tag{2}$$

with:

$$\mathrm{Re} = \frac{\rho N D^2}{\eta}$$

where k and c are empirically determined constants. Solving for viscosity yields:

$$\eta = \frac{2\pi M}{cND^3} \tag{3}$$

or:

$$\eta = \frac{2\pi k M}{\gamma_{avg} c D^3} \tag{4}$$

Newtonian calibration fluids are used to calculate c in Eq. 3 using torque and rotational speed measurements from the viscometer. With a known value for c, k for the helical impeller is determined from Eq. 4. With known parameters, c and k, torque and rotational rate data from the helical fitted viscometer can be used to determine biomass slurry viscosities.

CFD Modeling

The Gambit 1.6 preprocessor is used for creating a geometric rendering representing the flow domain in the reactor with the domain divided into 261,773 tetrahedral shaped discrete control volumes that comprise the computational grid. Solutions are obtained with the FLUENT 6.2 solver, a commercial CFD software package that models fluid flow in any geometry. The solver employs a finite-volume discretization process to numerically solve the governing equations for conservation of mass and momentum. Algebraic equations are constructed from the governing differential equations and integrated into each control volume. The equations are linearized and

Table 1
Viscosity as a Function of Solids Concentration
($25°C$, $\gamma = 1/s$)

Solids (%)	Viscosity (cP)
10	680
15	16,000
17	28,600
25	41,700

a solution of the equation system gives updated values for the dependent variables, such as velocities, pressures, and so on. The general form of the mass conservation equation is written as:

$$\frac{\partial \rho}{\partial t} + \nabla \cdot (\rho V) = S_m \qquad (5)$$

and the momentum conservation equation is written as:

$$\frac{\partial}{\partial t}(\rho V) + \nabla \cdot (\rho V V) = -\nabla P + \rho g + \nabla(\eta \nabla V) \qquad (6)$$

that includes terms to account for pressure, gravitational, and viscous forces.

The reactor dimensions are 1.52 m in length and 0.15 m in diameter. A screw shaft along the length of the reactor rotates at 5 rpm. Material enters at one of the horizontal reactor and for a boundary condition, the pressure at the exit is assumed to be atmospheric. Gravitational forces are neglected. All simulations are performed using the laminar flow model: Reynolds numbers ranged between 5.6 at 1000 cP and 112.5 at 20,000 cP. Motion of the shaft is generated with the rotating reference frame model that provides a steady state solution and is commonly used to model flow in rotating equipment (19).

Results and Discussion

Table 1 shows a significant increase in viscosity for a given temperature, in this case 25°C, as the concentration of biomass solids increases from 10% to 25%. The increase is particularly significant in the 10% through 17% range in which an increase of 1% solids results in an average increase of 4000 cP. Above 17%, the consistency becomes closer to a wet solid rather than a suspension of particles. Viscosity still increases considerably with a small increase in solids between 17% and 25%, but the average increase drops to 1600 cP per 1% increase in solids.

If the material is heated to the reactor's typical operating temperature, 190°C, evaporation of the liquid will result in inaccurate viscosity measurements. Instead, viscosities were measured at temperatures from

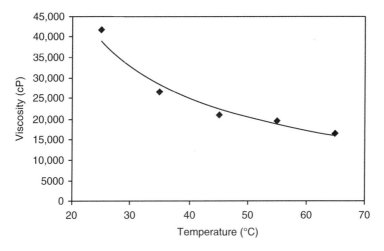

Fig. 1. Viscosity as a function of temperature (25% solids, $\gamma = 1/s$).

Fig. 2. Reactor shaft with location of cross-sectional plane.

25°C to 65°C, the maximum temperature before liquid losses begin to affect torque measurements by the viscometer. It is inadvisable to extrapolate viscosity to 190°C and, fortunately, not necessary since the decrease in viscosity begins to level off by 65°C. Figure 1 shows the viscosity–temperature relationship for 25% solids concentration between 25°C and 65°C. The viscosity decreases from 41,700 cP at 25°C to less than 20,000 cP at 65°C. The consistency of the lowest solids concentration considered here, 10%, is much thinner, and temperature has less effect on viscosity. Torque at 25°C, corresponding to a viscosity of 680 cP, is less than 3% of the viscometer's maximum. At this low range, the viscometer did not register differences in torque measurements for different temperatures.

Based on viscosity–temperature data for 10% and 25% solids, viscosities = 1000, 5000, 10,000, and 20,000 cP were chosen for simulations to compare flow characteristics over this range of concentrations. Because the actual viscosity is not known at 190°C, the four viscosities do not correlate directly to specific solids concentrations and temperatures, but are chosen

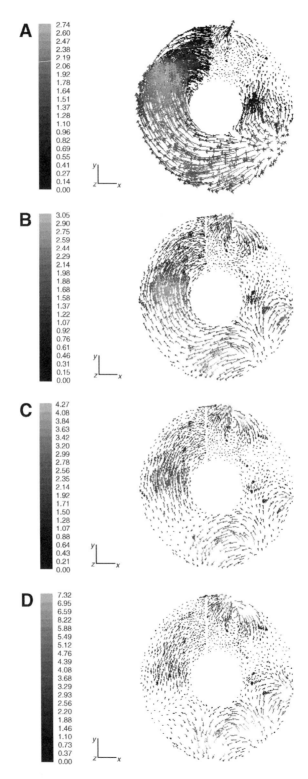

Fig. 3. Velocity vectors in a cross-sectional mixing plane (scale in m/s). **(A)** 1000 cP b, **(B)** 5000 cP, **(C)** 10,000 cP, **(D)** 20,000 cP.

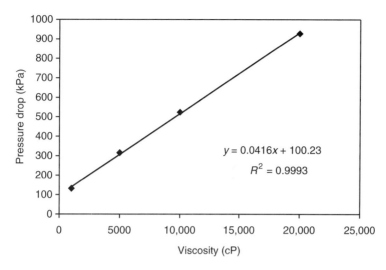

Fig. 4. Pressure drop across reactor as a function of viscosity.

to provide a qualitative comparison covering the approximate range of viscosities that likely exists within the reactor.

Velocity vectors provide a qualitative means of comparing flow characteristics between the four cases. Figure 2 shows the location of a cross-sectional plane 1 m from in which the material enters the reactor. In Fig. 3A through 3D, a perpendicular view shows vectors in this plane for each of the four cases. At the lowest viscosity, a more organized circular pattern is observed (Fig. 3A). As the viscosity increases, the velocity vectors become less pronounced in the tangential direction as the rotation of the shaft has less of a rotational impact on fluid away from the surface of each screw flight. The greater flow resistance exhibited at higher viscosities results in less tangential and more axial flow between screw flights.

The spiraling nature of the fluid movement, as it circulates while proceeding axially, is evident at any viscosity as some areas within a plane contain longer tangential vectors, and some areas contain smaller vectors that appear as dots, indicating motion perpendicular to the plane. The flow patterns indicate better mixing is likely occurring at lower viscosities, which is expected, because there is more tangential motion in addition to the axial motion. However, the higher viscosity fluids do exhibit some localized regions of high intensity circulation that provides some mixing benefits.

Higher solids loading is desirable to help reduce the size of production scale reactors, but these results indicate a trade-off may be necessary if poor mixing at higher viscosities results in inconsistencies in the final product stream. Even if the solids are soaked in the liquid phase before entering the reactor, good mixing is still needed for a uniform temperature distribution. The inclusion of vertical baffles located between screw flights should significantly enhance mixing.

Pressure drop appears to vary linearly with viscosity (Fig. 4). Poiseuille's equation,

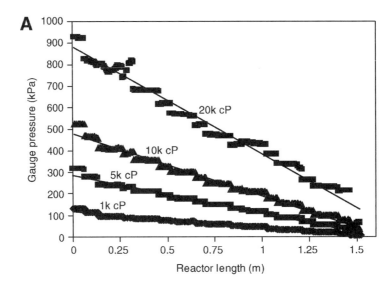

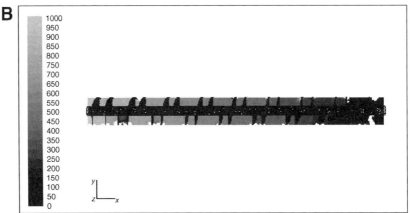

Fig. 5. Pressure distribution along length of reactor. **(A)** Pressure distribution along length of reactor (linear regressions, 1k cP = 0.971, 5k cP = 0.980, 10k cP = 0.982, 20k cP = 0.975). **(B)** Contours of pressure distribution along length of reactor (20k cP).

$$\Delta P = \frac{F 8 \eta L}{\pi r^4} \tag{7}$$

predicts this for laminar flow in a straight pipe, but interestingly, this relationship is valid even with the flow altered owing to the rotating shaft. Although the relationship applies at this slow rotation rate, 5 rpm, it is uncertain if it will apply at higher rotation rates having different flow characteristics.

The pressure drop across the length of the reactor increases about sevenfold as the viscosity increases from 1000 cP to 20,000 cP. Figure 5A shows total pressure distribution along the length of the reactor for each of the four cases. Higher pressure may result in higher hydrolysis reaction rates that can have adverse implications regarding sugar degradation into toxic byproducts. The 132 kPa pressure drop for a 1000 cP fluid provides a more uniform pressure distribution throughout the length of the reactor, and likely more uniform local reaction rates, compared with the higher viscosity fluids were the pressure drop is as high as 927 kPa for a 20,000 cP fluid.

Pressure appears to vary linearly with reactor length, which is also predicted by Poiseuille's equation. The data segments that appear to "stair-step" in Fig. 5A coincide with the shaded pressure contours in between consecutive screw flights in Fig. 5B (only 20k cP case is shown), with each segment beginning and ending at a screw flight.

Conclusions

Viscosity increases by approx 4000 cP for each 1% increase in solids concentration between 10% and 17%, and by approx 1600 cP for each 1% increase between 17% and 25%. The tangential component of flow is more pronounced at lower viscosities, indicating the likelihood of increased contact between the dilute acid liquid and biomass solids. Pressure drop increases linearly with viscosity, and is seven times that for 20,000 cP than for 1000 cP. Results provide a tool for analyzing sources of inconsistencies in product quality and insight into future design and operating parameters.

Acknowledgment

This work was funded by the US Department of Energy's National Renewable Energy Laboratory, subcontract: XCO-1-31016-01.

Nomenclature

- c   Newtonian proportionality constant (dimensionless)
- D   Diameter (m)
- F   Volumetric flow rate (m^3/s)
- k   Shear rate constant (dimensionless)
- L   Length (m)
- M   Torque (Nm)
- N   Rotation rate (rps)
- P   Pressure (Pa)
- r   Radius (m)
- Re Reynolds number (dimensionless)
- V   Velocity (m/s)
- γ Shear rate (/s)
- η Viscosity (kg/m·s)
- ρ Density (kg/m^3)

References

1. Schell, D. J., Walter, P. J., and Johnson, D. K. (1992), *Appl. Biochem. Biotechnol.* **34–35,** 659–665.
2. Schell, D. J., Farmer, J., Newman, M., and McMillan, J. D. (2003), *Appl. Biochem. Biotechnol.* **105–108,** 69–85.
3. Zhu, Y., Lee, Y. Y., and Elander, R. T. (2005), *Appl. Biochem. Biotechnol.* **121–124,** 1045–1054.
4. Torget, R. W., Walter, P. J., Himmel, M., and Grohmann, K. (1991), *Appl. Biochem. Biotechnol.* **28–29,** 75–86.
5. Wang, S. S. and Converse, A. O. (1992), *Appl. Biochem. Biotechnol.* **34–35,** 61–75.
6. McMillan, J. D. (1997), *Renewable Energy* **10(2/3),** 295–302.
7. Nguyen, Q. A., Keller, F. A., Tucker, M. P., et al. (1999), *Appl. Biochem. Biotechnol.* **77–79,** 455–472.
8. Teixeira, L. C., Linden, J. C., and Schroeder, H. A. (1999), *Appl. Biochem. Biotechnol.* **77–79,** 19–34.
9. Choi, C. H. and Matthews, A. P. (1996), *Bioresour. Technol.* **58(2),** 101–106.
10. Taherzadeh, M. J., Elkund, R., Gustafsson, L., Nicklasson, C., and Liden, G. (1997), *Ind. Eng. Chem. Res.* **36(11),** 4659–4665.
11. Tucker, M. P., Farmer, J. D., Keller, F. A., Schell, D. J., and Nguyen, Q. A. (1998), *Appl. Biochem. Biotechnol.* **70–72,** 25–35.
12. Nguyen, Q. A., Tucker, M. P., Keller, F. A., and Eddy, F. P. (2000), *Appl. Biochem. Biotechnol.* **84–86,** 561–576.
13. Torget, R. W., Hayward, T. K., and Elander, R. (1997), presented at the *19th Symposium on Biotechnology for Fuels and Chemicals*, Colorado Springs, CO.
14. Chen, R., Wu, Z., and Lee, Y. Y. (1998), *Appl. Biochem. Biotechnol.* **70–72,** 37–49.
15. Lee, Y. Y., Wu, Z., and Torget, R. W. (2000), *Bioresour. Technol.* **71,** 29–39.
16. Converse, A. O. (2002), *Bioresour. Technol.* **81,** 109–116.
17. Berson, R. E. and Hanley, T. R. (2005), *Appl. Biochem. Biotechnol.* **121–124,** 935–945.
18. Allen, D. G. and Robinson, C. W. (1990), *Chem. Eng. Sci.* **45(1),** 37–48.
19. Fluent, Inc. (2003) FLUENT 6.1 User's Guide, Fluent, Inc., Lebanon, NH.

Ethanol Production From Pretreated Olive Tree Wood and Sunflower Stalks by an SSF Process

ENCARNACIÓN RUIZ,[1] CRISTÓBAL CARA,[1]
MERCEDES BALLESTEROS,[2] PALOMA MANZANARES,[2]
IGNACIO BALLESTEROS,[2] AND EULOGIO CASTRO*,[1]

[1]Department of Chemical, Environmental and Materials Engineering,
University of Jaén, Campus Las Lagunillas, 23071 Jaén, Spain,
E-mail: ecastro@ujaen.es; and [2]DER-CIEMAT,
Avenida Complutense 22, 28040 Madrid, Spain

Abstract

Olive tree wood and sunflower stalks are agricultural residues largely available at low cost in Mediterranean countries. As renewable lignocellulosic materials, their bioconversion may allow both obtaining a value-added product, for fuel ethanol, and facilitating their elimination. In this work, the ethanol production from olive tree wood and sunflower stalks by a simultaneous saccharification and fermentation (SSF) process is studied. As a pretreatment, steam explosion at different temperatures was applied. The water insoluble fractions of steam-pretreated sunflower stalks and steamed, delignified olive tree wood were used as substrates at 10% w/v concentration for an SSF process by a cellulolytic commercial complex and *Saccharomyces cerevisiae*. After 72-h fermentation, ethanol concentrations up to 30 g/L were obtained in delignified steam-pretreated olive tree wood at 230°C and 5 min. Sunflower stalks pretretated at 220°C and 5 min gave maximum ethanol concentrations of 21 g/L in SSF experiments.

Index Entries: Ethanol; olive tree wood; sunflower stalks; SSF; pretreatment.

Introduction

Olive tree wood and sunflower stalks are among the main agricultural residues found in Mediterranean countries whose conversion into fuel ethanol has considerable advantages. On the one side, these materials must be eliminated, thus preventing propagation of vegetal diseases and keeping fields clear; till date, disposal methods include burning or grinding and scattering, with associated costs and no economical alternatives. On the other side, the huge amount annually generated and their lignocellulosic

*Author to whom all correspondence and reprint requests should be addressed.

nature make both residues a major low-cost renewable source of sugars that may be converted into valuable products, for example, fuel ethanol.

More than 8×10^6 ha of olive trees are cultivated in the world, most of them in Mediterranean countries (1). Pruning is an essential operation in olive tree cultivation to eliminate less productive branches and prepare trees for the next crop. This action generates an annual volume of lignocellulosic residues estimated at 3000 kg/ha (2). Olive tree pruning is composed of leaves, thin branches, and wood in different proportions, depending on culture conditions, production, and local uses. A typical olive tree pruning lot includes 30% of wood that is separated and put to domestic use as firewood. Sunflower is the fourth largest source of oil-seeds worldwide, representing around 23×10^6 ha of cultivated land (1). Stalks, heads, and leaves account for 3–7 t of dry matter/ha (3) and are left in the fields after harvesting for elimination.

The utilization of lignocellulosic residues of these types in a bioconversion process requires pretreatment of the raw material to improve the release of sugars from both the hemicellulose and the cellulose fractions. In steam-explosion pretreatment, biomass is exposed to pressurized steam followed by rapid reduction in pressure. The treatment results in substantial breakdown of the lignocellulosic structure, hydrolysis of the hemicellulosic fraction, depolymerization of the lignin components, and defibration. Therefore, the accessibility of the cellulose components to degradation by enzymes is greatly increased (4).

The production of ethanol from pretreated material may be accomplished either by sequential hydrolysis and fermentation or by a simultaneous saccharification and fermentation (SSF) process. In the SSF process, end product inhibition can be overcome as glucose is fermented as it is formed; another advantage is that a single reactor is used for both saccharification and fermentation. On the contrary, the enzymatic reaction in SSF process is operated at a temperature lower than its optimum level owing to the mismatch in optimum temperatures for hydrolysis (approx 50°C) and fermentation (approx 30°C) (5).

In previous works (6,7), we analyzed the influence of steam-explosion pretreatment on sugar recoveries in both solids and liquids and the enzymatic hydrolysis of solids from sunflower stalks and olive tree wood, as a first step in the residue-to-ethanol bioconversion process. The objective of this work is to determine the feasibility of the production of ethanol by the SSF of steam-exploded sunflower stalks and olive tree wood.

Methods

Raw Materials

Olive tree wood and sunflower stalks were collected locally after crop harvesting, air-dried at room temperature to equilibrium moisture content (approx 10% for olive tree wood and approx 8.2% for sunflower stalks), milled using a laboratory hammer mill (Retsch, Germany) to a particle size smaller than 10 mm, homogenized in a single lot, and stored at room temperature until used.

Steam Explosion Pretreatment

Steam explosion of raw material was carried out in a batch pilot unit based on Masonite technology using a 2-L reaction vessel designed to reach a maximum operating pressure of 4.12 MPa *(8)*. The reactor was charged with 200 (olive tree wood chips) or 150 (sunflower stalks) g of feedstock (dry matter) per batch and heated to the desired temperature with saturated steam for 5 min. Assayed steam temperatures were 190, 210, 230, and 240°C for olive tree wood and 180, 190, 200, 210, 220, and 230°C in the case of sunflower stalks. The exploded material was recovered in a cyclone and after cooling to about 40°C, filtered for liquid and solid recovery. Dried solids were weighed, analyzed for sugars and lignin composition, and used in enzymatic hydrolysis and SSF tests.

Alkaline-Peroxide Delignification

The water-insoluble fiber from steam-explosion pretreatment of olive tree wood was delignified using a hot alkaline peroxide treatment protocol adapted from Yang et al. *(9)* in 1% (w/v) H_2O_2 solution in 4% (w/v) solids concentration. The pH was adjusted to 11.5 using 4 M NaOH. The treatment lasted for 45 min at 80°C, and then the suspension was filtered and water-washed until neutral pH. Delignified, dried solid was weighed and analyzed for carbohydrates and lignin composition.

Enzymatic Hydrolysis

The washed water-insoluble residues from steam-pretreated sunflower stalks and those from steam-pretreated, delignified olive tree wood were enzymatically hydrolyzed by a cellulolytic complex (Celluclast 1.5 L) kindly provided Novozymes A/S (Bagsvaerd, Denmark). Cellulase enzyme loading was 15 FPU/g substrate. Fungal β-glucosidase (Novozyme 188, Novozymes A/S) was used to supplement the β-glucosidase activity with an enzyme loading of 12.6 international unit (IU)/g substrate. Enzymatic hydrolysis was performed in 0.05M sodium citrate buffer (pH 4.8) at 50°C on a rotary shaker (Certomat-R, B-Braun, Germany) at 150 rpm for 72 h and at 10% (w/v) pretreated material concentration. Samples were taken every 24 h for glucose concentration determination. All enzymatic hydrolysis experiments were performed in duplicate and average results are given.

Simultaneous Saccharification and Fermentation

The washed water-insoluble residues from steam-pretreated sunflower stalks and that from steam-pretreated, delignified olive tree wood were submitted to an SSF process. The same cellulolytic complex (15 FPU/g substrate enzyme loading) supplemented by β-glucosidase (12.6 IU/g substrate activity) as in the enzymatic hydrolysis was used for saccharification. Fermentation was performed by *Saccharomyces cerevisiae* (DER-CIEMAT culture collection no. 1701).

SSF experiments were carried out in 100-mL Erlenmeyer flasks, each containing 25 mL of fermentation medium in 0.05M sodium citrate buffer (pH 4.8) at 35°C on a rotary shaker (Certomat-R, B-Braun, Germany) at 150 rpm for 72 h and at 10% (w/v) pretreated material concentration. The fermentation medium contained 5 g/L yeast extract, 2 g/L NH_4Cl, 1 g/L KH_2PO_4, and 0.3 g/L $MgSO_4·7H_2O$. Flasks were inoculated with 4% (v/v) *S. cerevisiae* culture obtained by growing the organism on a rotary shaker at 150 rpm for 20 h at 35°C in the same growth medium with glucose (30 g/L). Samples were taken every 24 h and analyzed for glucose and ethanol production. Average results of duplicate experiments are shown.

Analytical Methods

The composition of solid materials was determined according to the National Renewable Energy Laboratory (NREL, Golden, CO) analytical methods for biomass *(10)*. Cellulose and hemicellulose content were determined by HPLC with a Waters 2695 liquid chromatograph with refractive index detector. An AMINEX HPX-87P carbohydrate analysis column (Bio-Rad, Hercules, CA) operating at 85°C with deionized water as the mobile-phase (0.6 mL/min) was used. Glucose concentration from enzymatic hydrolysis samples was measured by an enzymatic determination glucose assay kit (Sigma GAHK-20).

Ethanol was measured by gas chromatography, using a HP 5890 Series II apparatus equipped with an Agilent 6890 series injector, a flame ionization detector, and a column of Carbowax 20 M at 85°C. The injector and detector temperature was maintained at 150°C. All analytical determinations were performed in duplicate and average results are shown. Relative standard deviations were in all cases below 5%.

Results and Discussion

Raw Material Composition

Table 1 summarizes the composition of raw materials. Although olive tree wood and sunflower stalks are quite different, the cellulose and hemicellulose contents are similar. Xylose is the main sugar of the hemicellulosic fraction in both materials (approx 80%). Lignin content is smaller for sunflower stalks. The main differences are extractives and ash contents.

Pretreatments and Enzymatic Hydrolysis Assays

Both raw materials were subjected to steam-explosion pretreatment for 5 min at different temperatures. In addition, the water-insoluble fiber obtained from steam explosion of olive tree wood was further submitted to

Table 1
Raw Material Composition (% Dry Matter)

Composition	Olive tree wood	Sunflower stalks
Cellulose as glucose	34.4	33.8
Hemicellulosic sugars	20.3	20.2
Xylose	16	16.1
Mannose	1.4	1.7
Galactose	1	1.4
Arabinose	1.9	1
AIL	18	14.6
Acid-soluble lignin (ASL)	2.4	2.7
Acetyl groups	1.8	2.5
Extractives	15.4	6.9
Ash	1.7	9.6

Table 2
Composition of Water-Insoluble Fiber (%) Resulting From Steam Explosion of Sunflower Stalks and Olive Tree Wood at Different Pretreatment Temperatures and Composition of Exploded, Delignified Olive Tree Wood (%)

Steam-explosion temperature (°C)	Olive tree wood				Sunflower stalks					
	190	210	230	240	180	190	200	210	220	230
Steamed water-insoluble fiber										
Total gravimetric recovery	54.2	44.1	44.3	40.4	65.1	55.3	45.8	43.8	38.3	35.5
Glucose	50.5	55.1	64.4	56.2	45.5	52	56.8	58.1	60.6	60.1
Hemicellulosic sugars	13.8	6.8	2.9	2.6	21.5	18.1	11.6	6.4	5.9	2.7
Acid-insoluble lignin	31.4	34.4	35.5	38.2	22.1	24.7	27.6	30.1	34	35.4
Steamed, delignified solid										
Delignified solid recovery	69.1	66.9	57.8	58.9						
Glucose	64	77.6	90.3	85.8						
Hemicellulosic sugars	11.4	6.9	2.8	3.6						
Acid-insoluble lignin	20.7	16.7	11.1	12.5						

an alkaline peroxide delignification step. The complete studies of the influence of pretreatments on sugar recoveries in both solids and liquids and the enzymatic hydrolysis of solids are reported elsewhere (6,7). As a starting point for the research on ethanol production from both residues, the main results of enzymatic hydrolysis performance are briefly presented here.

Table 2 shows, as a function of pretreatment temperature, the composition of water-insoluble fibers obtained after steam explosion of sunflower stalks and olive tree wood. For both lignocellulosic residues, a decrease

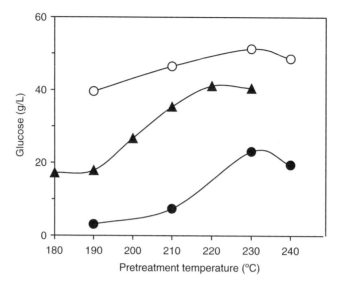

Fig. 1. Glucose concentration (g/L) obtained by enzymatic hydrolysis (72 h, 10% substrate loading) of steam-exploded sunflower stalks (▲), steam-exploded olive tree wood (●), and steam-exploded and delignified olive tree wood (○) at different temperature conditions for steam explosion.

of total gravimetric recovery (solids remaining after pretreatment divided by original oven-dried weight) was detected as the pretreatment temperature increased. This is mainly attributed to the solubilization of hemicellulosic fraction; the hemicellulosic-derived sugars recovery in the water-insoluble fiber decreased as long as the steaming temperature increased. The contents in acid-insoluble lignin (AIL) of the pretreated residue showed slight solubilization increasing, in general, with pretreatment temperature. The cellulose content in the solid increased with pretreatment temperature except for the highest one. The maximum cellulose content for olive tree wood (64.4%) was attained at 230°C and for sunflower stalks (60.6%) at 220°C. Partial solubilization of cellulose was detected for both materials, and hence the cellulose recovery diminished in general as pretreatment temperature was increased.

Enzymatic hydrolysis assays were performed on steam-pretreated residues with a cellulase complex (Celluclast 1.5L) supplemented with β-glucosidase (Novozyme 188). Figure 1 shows glucose concentration obtained after a 72-h period of enzymatic action. Enzymatic hydrolysis yields (determined from glucose obtained in the enzymatic hydrolysis divided by the potential glucose in the pretreated material) depended on the temperature of pretreatment. For sunflower stalks, yields increased as the temperature was raised up to 220°C (67.8%) with a light decrease at 230°C. Regarding steam-pretreated olive tree wood, the enzymatic hydrolysis results were quite poor. The maximum enzymatic yield was just 35.9% for a steam-pretreatment temperature of 230°C. In order to improve the enzymatic hydrolysis of olive tree wood, a delignification step on the steamed solid residue was done. The composition of the insoluble fiber resulting after alkaline-peroxide delignification is shown in Table 2.

As can be observed in Fig. 1, results corresponding to enzymatic hydrolysis on exploded and delignified olive tree wood were much better than those obtained with just steam pretreatment. For all steaming temperatures, considerable improvements in the susceptibility to enzymatic hydrolysis were evidenced. Nevertheless, glucose inhibition was detected in solids pretreated at the highest temperatures in which cellulose content was elevated.

Simultaneous Saccharification and Fermentation

Solid residues from steam-exploded sunflower stalks and steam-exploded, delignified olive tree wood were further submitted to an SSF process by *S. cerevisiae* at a temperature of 35°C and 10% substrate concentration. In the SSF process, the glucose released by the enzymatic attack is simultaneously converted into ethanol by the yeast, thus reducing enzyme inhibition from glucose. Figure 2 shows ethanol concentrations determined every 24 h in a 72-h SSF period for both lignocellulose materials. Culture samples were also analyzed for glucose content (data not shown); in all cases very low glucose concentrations were obtained, showing a good yeast fermentation performance.

In both residues, the concentration of ethanol increased with time, regardless the pretreatment temperature. At a given time, the concentration of ethanol generally increased with pretreatment temperature until the maximum is reached. Maximum ethanol concentrations of 21 and 29.4 g/L were obtained for sunflower stalks pretreated at 220°C and olive tree wood pretreated at 230°C and delignified, respectively. The higher ethanol concentrations obtained in olive tree wood agreed with the higher glucose content of this solid remaining after the delignification step. This fact can also be confirmed in Table 3, in which ethanol productivity values determined at each sample time are shown. A maximum ethanol productivity at 72 h of 0.291 g/L·h was calculated for SSF-sunflower stalks experiment from 220°C pretreatment temperature. In the case of olive tree wood, ethanol productivities were higher, the maximum value (0.409 g/L·h) being found for the SSF experiment performed with the solid pretreated at 230°C. Comparison of the ethanol productivities calculated after 72-h fermentation time shows a decrease to the half of those obtained after 24 h. Kádár et al. *(11)* found ethanol productivities after 72 h of 0.197 and 0.125 g/L·h when applying SSF process by *S. cerevisiae* to paper-industry wastes. Our results compare with the work of Alfani et al. *(12)*, who reported 0.837 g ethanol/L·h after 30 h SSF time using steam-exploded wheat straw as substrate; these authors remark that SSF ethanol productivities are much higher than ethanol productivities obtained by separate hydrolysis and fermentation, making SSF process more favorable from an industrial point of view.

Although the highest ethanol productivities are obtained after 24 h, high ethanol yields are also essential in order to get a more efficient use

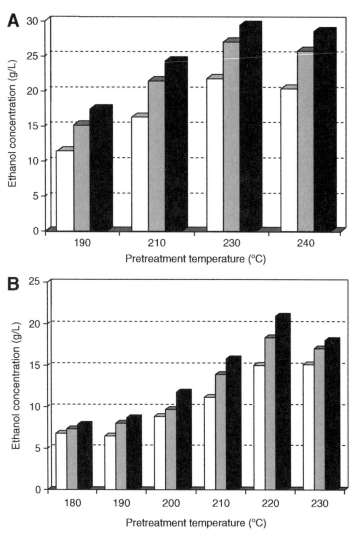

Fig. 2. Concentration of ethanol solutions (g/L) obtained in the SSF process after 24 (white bar), 48 (gray bar), and 72 h (black bar). **(A)** Steam-exploded, delignified olive tree wood. **(B)** Steam-exploded sunflower stalks.

of raw material. Ethanol yields obtained in the SSF process, calculated from ethanol concentration data, are shown in Tables 4 and 5. $Y_{p/s}$ stands for mass of ethanol obtained at different times divided by 100 g of cellulose in pretreated raw material. SSF-ethanol yields were calculated as a percentage of the maximum theoretical ethanol yield by assuming that all the potential glucose in the starting material is available for fermentation. Maximum values of 65.2% and 67.7% were reached for olive tree wood and sunflower stalks, respectively. These results are in the order of those reported by Alfani et al. *(12)*; SSF–ethanol yields close to 68% of theoretical were achieved with steam-exploded (217°C, 3 min) wheat straw washed with NaOH solutions. Stenberg et al. *(13)* in a study on the SSF

Table 3
Ethanol Productivities (g/L·h) at Different Times for SSF Experiments
With Steam-Pretreated Sunflower Stalks and Steam-Pretreated,
Delignified Olive Tree Wood

Pretreatment	Sunflower stalks		
Temperature (°C)	24 h	48 h	72 h
180	0.283	0.153	0.109
190	0.270	0.167	0.121
200	0.368	0.201	0.164
210	0.465	0.290	0.220
220	0.626	0.382	0.291
230	0.630	0.355	0.250
	Olive tree wood		
	24 h	48 h	72 h
190	0.478	0.315	0.242
210	0.679	0.447	0.338
230	0.908	0.563	0.409
240	0.848	0.536	0.397

Table 4
SSF-Ethanol and Enzymatic Hydrolysis Yields After 24, 48, and 72 h
for Steam-Pretreated, Delignified Olive Tree Wood at Different
Pretreatment Temperatures

Pretreatment temperature (°C)	Y p/s[a]			SSF-ethanol yield[b]			Enzymatic hydrolysis yield[c]		
	24 h	48 h	72 h	24 h	48 h	72 h	24 h	48 h	72 h
190	17.9	23.7	27.3	35	46.4	53.4	41.2	52	61.8
210	21	27.7	31.3	41.1	54.2	61.2	42.9	52.8	60
230	24.1	29.9	32.6	47.2	58.5	63.8	36.1	50.5	56.8
240	23.7	30	33.3	46.4	58.7	65.2	38.5	51.3	56.6

[a]Yield product/substrate, g ethanol/100 g cellulose (expressed as potential glucose) in pretreated raw material.
[b]As percentage of theoretical ethanol yield (0.511 g ethanol/g glucose).
[c]As percentage of glucose obtained by enzymatic hydrolysis at 50°C from total potential glucose in the pretreated material.

of steam-pretreated softwood, obtained, as a maximum ethanol yield in the SSF step by *S. cerevisiae*, 82% of theoretical using an enzyme loading of 32 FPU/g substrate and 5% (w/v) substrate concentration, instead of 15 FPU/g and 10% (w/v) substrate concentration used in our work.

The feasibility of using 10% (w/v) substrate concentration in SSF is considered to be relevant, since earlier studies on this process have reported the limiting effect of elevated substrate concentrations owing to

Table 5
SSF–Ethanol and Enzymatic Hydrolysis Yields After 24, 48, and 72 h
for Steam-Pretreated Sunflower Stalks at Different Pretreatment Temperatures

Pretreatment temperature (°C)	Y p/s[a]			SSF-ethanol yield[b]			Enzymatic hydrolysis yield[c]		
	24 h	48 h	72 h	24 h	48 h	72 h	24 h	48 h	72 h
180	14.9	16.2	17.3	29.2	31.7	33.8	33.3	35.6	37.9
190	12.5	15.4	16.7	24.5	30.1	32.7	31.4	33	34.5
200	15.5	17	20.7	30.3	33.3	40.5	38.9	43.8	46.9
210	19.2	24	27.2	37.6	47	53.2	49.7	57.6	60.9
220	24.8	30.2	34.6	48.5	59.1	67.7	55	60.4	67.8
230	25.1	28.4	30	49.1	55.6	58.7	57.1	64.7	67.3

[a]Yield product/substrate, g ethanol/g cellulose (expressed as potential glucose) in pretreated raw material.
[b]As percentage of theoretical ethanol yield (0.511 g ethanol/g glucose).
[c]As percentage of glucose obtained by enzymatic hydrolysis at 50°C from total potential glucose in the pretreated material.

difficulties in stirring the material, mass transfer problems, or high ethanol inhibiting concentration *(11,14)*. In fact, in most SSF experiments reported in the literature, substrate concentrations lower than 10% (w/v) are employed. On the other hand, low concentrations of substrate would increase the capital cost of equipment and would yield low concentrations of ethanol for distillation *(15)*. As far as the enzyme loading is concerned, high enzyme concentrations can increase conversion yields *(13)*; increasing cellulase loading from 15 to 45 FPU/g substrate in SSF process for ethanol production from paper material resulted in improved ethanol yields coming up from 56.4% to 70.4% *(15)*. Nevertheless, the economic implications of higher cellulase concentrations must be carefully considered.

In an attempt to overcome the problem of the difference in optimum temperature of the cellulases and the fermenting microorganisms, thermotolerant yeast strains, such as *Kluyveromyces marxianus*, have been assayed for ethanol production by SSF process. Nevertheless, Kádár et al. *(11)* found that there was no significant difference between *S. cerevisiae* and *K. marxianus* when comparing ethanol yields in SSF conversion of industrial paper wastes. The SSF–ethanol yields were in the range of 0.31–0.34 g/g for both strains used. Ballesteros et al. *(8)* studied the steam explosion of several herbaceous residues and woods and the subsequent SSF-ethanol production by *K. marxianus* reporting maximum ethanol yields in the range 0.31–0.36 g/g (60.9–71.2% of the theoretical) with maximum ethanol contents from 16 to 19 g/L in fermentation media. Our maximum ethanol yields reported here, obtained at the same conditions of enzyme loading and substrate concentration, are in the same range, but we attained higher ethanol concentrations (21 and 29.4 g/L for sunflower stalks and olive tree wood, respectively) because of a higher cellulose content in pretreated materials.

For comparison purposes, Tables 4 and 5 show enzymatic hydrolysis yields, determined from glucose concentration values (Fig. 1) and the potential glucose concentration in the pretreated material (Table 2) for olive tree wood and sunflower stalks, respectively. Results for olive tree wood indicate that cellulose conversion yields in SSF are higher than enzymatic hydrolysis yields for all pretreatment temperatures except the lowest one. The improvement in SSF-cellulose conversion is greater for 230 and 240°C pretreatment temperatures at which glucose inhibition in the enzymatic hydrolysis for a 10% substrate concentration had been detected. Enzymatic hydrolysis yields at 2% substrate loading (data not shown) resulted in similar values as those in SSF, confirming that the consumption of glucose by the yeast as it is formed alleviates the inhibiting effect of cellulase activity *(16)*. In comparison to enzymatic hydrolysis, pretreated sunflower stalks submitted to SSF led to lower cellulose conversion except at the pretreatment temperature of 220°C, in which cellulose conversion was similar by both procedures. It must be pointed out that, even in the case of equivalent yields, the SSF process is preferable to the separate hydrolysis and fermentation process (SHF) because using a single bioreactor reduces investment and operation costs *(12)*. Another advantage is that the SSF process leads to a significantly increased ethanol productivity *(18)*. Low sugar is also available for invasing organisms.

Overall Process Considerations

The main objective of the present study was to evaluate ethanol production from pretreated sunflower stalks and olive tree wood by an SSF process. According to the compositions shown in Table 1, the maximum ethanol yield attainable if all sugars present in the raw material were converted into ethanol is 17.6 g/100 g olive tree wood and 17.3 g/100 g sunflower stalks, equivalent to 223.1 L/t and 219.3 L/t, respectively. In order to determine the amount of ethanol actually obtainable under the assayed operational conditions, first, the glucose recovery must be taken into account in the steam-explosion-pretreated solid; next, and only in the case of olive tree wood, the glucose material recovery after delignification; and, finally, the ethanol yield attained in the SSF process. From data in Table 2, the values of glucose content in the pretreated material are multiplied by the total gravimetric recovery of pretreatment step (and by the delignified solid recovery in the case of olive tree wood) in order to refer results to initial raw material; finally, the obtained values are multiplied by SSF yields ($Y_{p/s}$ in Tables 4 and 5), leading to the overall ethanol yield that may be obtained from 100 g of each raw material. The results are shown in Fig. 3. From 100 g of sunflower stalks, submitted to steam-explosion pretreatment at 220°C and further SSF by *S. cerevisiae* on the solid-pretreated residue, a maximum of 8 g of ethanol is obtained, representing 46% of the theoretical referred to raw material. Increasing the temperature of steam pretreatment

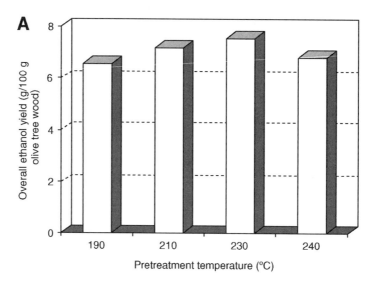

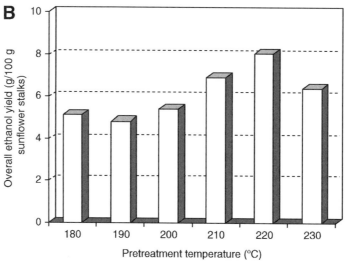

Fig. 3. Overall ethanol yield referred to raw material at different steam-explosion temperatures. **(A)** Olive tree wood (maximum yield attainable: 17.6 g ethanol/100 g raw material). **(B)** Sunflower stalks (maximum yield attainable: 17.3 g ethanol/100 g raw material).

from 180°C to 220°C resulted in an increase in overall ethanol yield because of improvement in the susceptibilty of pretreated material to enzymatic action. Beyond 220°C, the decline in overall ethanol yield is attributed to lower recovery of cellulose.

From 100 g of olive tree wood, steam-explosion pretreatment at 230°C, further alkaline-peroxide delignification on the solid-pretreated residue, and submission to SSF process, 7.5 g of ethanol is obtained, which corresponds to 43% of theoretical if all glucose in the raw material was available for fermentation. Slight differences in the overall ethanol yield are found for other assayed steam temperatures.

Comparing the global process for both residues, process yields are a little better for sunflower stalks and the global scheme implies only two steps (steam explosion and SSF) instead of three for olive tree wood. On the contrary, more concentrated ethanol solutions are obtained for the latter as a consequence of a higher cellulose content in the pretreated solid after delignification. Thus, 1 L of ethanol may be obtained from 2.7 kg of steamed, delignified olive tree residue, comparing to 3.8 kg of steam-exploded sunflower stalks. This must be taken into account for an eventual scale-up of the process. Referred to a raw material basis, 95.1 L ethanol/t olive tree wood and 101.4 L ethanol/t sunflower stalks may be obtained.

As a conclusion, SSF process is an interesting option for producing ethanol from both agricultural residues. Nevertheless, it is assumed that yields obtained are all relatively low for industrial ethanol production processes and that further improvements in terms of increased ethanol yields are necessary to achieve an economical process. Research on advanced reactor configuration or on the utilization of the whole slurry generated in the pretreatment step appear to be a promising means to increase final ethanol yieds in SSF process.

References

1. FAOSTAT (2004), http://www.fao.org/es/ess.
2. Sánchez, S., Moya, A. J., Moya, M., et al. (2002), *Ing. Quím.* **34**, 194–202.
3. Marechal, V. and Rigal, L. (1999), *Ind. Crops Prod.* **10**, 185–200.
4. Moniruzzaman, M. (1996), *Appl. Biochem. Biotechnol.* **59**, 283–297.
5. Wu, Z. and Lee, Y. Y. (1998), *Appl. Biochem. Biotechnol.* **70–72**, 479–492.
6. Ruiz, E., Cara, C., Manzanares, P., Ballesteros, M., and Castro, E. (2005), *Enzyme Microb. Technol.*, submitted.
7. Cara, C., Ruiz, E., Ballesteros, I., Negro, M. J., and Castro, E. (2005), *Process Biochem.*, in press.
8. Ballesteros, M., Oliva, J. M., Negro, M. J., Manzanares, P., and Ballesteros, I. (2004), *Process Biochem.* **39**, 1843–1848.
9. Yang, B., Boussaid, A., Mansfield, S. D., Gregg, D. J., and Saddler, J. N. (2002), *Biotechnol. Bioeng.* **77**, 678–684.
10. National Renewable Energy Laboratory (NREL). Chemical analysis and testing laboratory analytical procedures: LAP-002 (1996), LAP-003 (1995), LAP-004 (1996), LAP-005 (1994), LAP-010 (1994), and LAP 017 (1998). NREL, Golden, CO, USA. http://www.eere.energy.gov/biomass/analytical_procedures.html.
11. Kádár, Zs., Szengyel, Zs., and Réczey, K. (2004), *Ind. Crops Prod.* **20**, 103–110.
12. Alfani, F., Gallifuoco, A., Saporosi, A., Spera, A., and Cantarella, M. (2000), *J. Ind. Microb. Biotechnol.* **25**, 184–192.
13. Stenberg, K., Bollók, M., Réczey, K., Galbe, M., and Zacchi, G. (2000), *Biotechnol. Bioeng.* **68**, 204–210.
14. Mohagheghi, A., Tucker, M., Grohman, K., and Wyman, C. (1992), *Appl. Biochem. Biotechnol.* **33**, 67–81.
15. Ballesteros, M., Oliva, J. M., Manzanares, P., Negro, M. J., and Ballesteros, I. (2002), *World J. Microbiol. Biotechnol.* **18**, 559–561.
16. Cantarella, M., Cantarella, L., Gallifuoco, A., Spera, A., and Alfani, F. (2004), *Biotechnol. Prog.* **20**, 200–206.
17. Cantarella, M., Cantarella, L., Gallifuoco, A., Spera, A., and Alfani, F. (2004), *Process Biochem.* **39**, 1533–1542.

SESSION 4
Industrial Biobased Products

The Development of Cement and Concrete Additive

Based on Xylonic Acid Derived Via Bioconversion of Xylose[‡]

BYONG-WA CHUN,[*,1] **BENITA DAIR,**[†,1]
PATRICK J. MACUCH,[2] **DEBBIE WIEBE,**[2]
CHARLOTTE PORTENEUVE,[1] **AND ARA JEKNAVORIAN**[1]

[1]Grace Performance Chemicals, W.R. Grace & Co., 62 Whittemore Avenue, Cambridge, MA 02140, E-mail: byong-wa.chun@grace.com; and
[2]Altran Corporation, Boston, MA 02210

Abstract

The present work attempted to utilize xylose by converting it to an aldonic acid. In the present study, xylose was converted to xylonic acid by using commercial glucose oxidase enzyme, palladium catalysis, and microbial bioconversion. The enzyme conversion was successfully done using a commercial glucose oxidase. The microbial conversion with *Gluconobactor oxydans* proceeded even with the presence of a large amount of lignosulfonate. Thus obtained xylonic acid products were evaluated as a cement dispersing agent in cement and concrete tests. It was found that xylonic acid is approximately twice as effective as lignosulfonate. Xylonic acid can be effectively utilized in concrete water reducer application.

Index Entries: Xylose utilization; biochemical oxidation; xylonic acid; cement; concrete.

Introduction

Concrete based on Portland cement is probably one of the most important construction and building materials to support human activity. Important urban infrastructures, such as bridges, highways, and buildings cannot be built without concrete. Various admixture chemicals are used in concrete today to improve its physical properties. One of the most important classes of concrete admixture is called water reducer. The water reducer is fundamentally a cement-dispersing chemical that deflocculates cement grains so that the cement paste becomes more fluid. The fluidity translates into the improved placement ability of the concrete. Alternatively, the action

---

*Author to whom all correspondence and reprint requests should be addressed.
†Present address: FDA, Rockville, MD 20852.
‡A part of the work was presented at American Chemical Society Annual Meeting at Anaheim in 2004.

of water reducer allows less amounts of water in concrete without sacrificing workability or placement ability, which results in the greater strength and better durability of the concrete. The most widely used water-reducing chemical is lignosulfonate contained in spent sulfite liquor (SSL) from a sulfite pulp mill. Meanwhile, aldonic acid, particularly gluconic acid, is known to be a superior inorganic-dispersing chemical. Sodium gluconate is used in cement and concrete industries as a cement-dispersing and a retarding agent today, but its use as a general water-reducing chemical is somewhat limited as a result of its cement set retarding tendency.

In the present study, the authors attempted to convert xylose to xylonic acid and evaluate its potential as a water-reducing chemical in concrete. Xylose was selected as a starting aldose because, although it is abundant in various agricultural waste streams, the pentose is significantly underutilized because of its poor fermentatability. Xylonic acid is expected to be a cement dispersant because of its similarity to gluconic acid. However, little or no research has been done on evaluating xylonic acid as a dispersant.

Materials and Methods

Bioconversion of Xylose With Commercial Glucose Oxidase

Glucose oxidase enzymes used are OxyGo® 1500 (Genencor International, Rochester, NY) and Novozym 771® (Novozymes North America, Franklinton, NC) and catalase used was Fermacolase® 1000 (Genencor International). The enzymes were kindly supplied by the manufacturers. The enzyme conversion experiments were done according to Lantero et al. (1) except that 1 atm back-pressure was not applied. A preliminary experiment showed that the OxyGo sustained its activity longer than Novozym, especially at the elevated temperature; therefore, most of the experiments were done with OxyGo. Typical reaction conditions are as follows: OxyGo (2.25 mL) and Fermcolase (1.55 mL) were added to 125 g of a sugar (10–30 wt% concentration). Air was bubbled at 3 SCFH rate. The pH controller was set at 5.0, and the temperature was kept at either 40 or 60°C. A 0.1 N NaOH solution was used for pH adjustment. The conversion rate was measured primarily by the alkaline consumption. The aldonic acid formation was confirmed by high-performance liquid chromatography (HPLC) method described later.

Catalytic Oxidation of Xylose

The catalytic oxidation was performed using 5% palladium on activated carbon (Aldrich, Milwaukee, WI). Typical conditions of aldose conversion with palladium catalyst is described elsewhere (2). Twenty grams of sugar were dissolved in 170 mL of water, and 4 g of wet 5% palladium on carbon catalyst (2 g dry powder) were added. Air was bubbled through sintered glass at 3 SCFH. Temperature was held at 35 or 50°C. The pH was adjusted at 10.0 with using 25 wt% NaOH solution. The consumption of the sodium hydroxide solution was monitored to measure the conversion rate.

Microbial Oxidation of Xylose in SSL

Microbial oxidation of xylose in SSL was carried out using *Gluconobactor oxydans suboxydans* (ATCC 621). Substrate used was concentrated hardwood (birch) SSL from Fraser Paper (Park Falls, WI), which contains approx 30 wt% of xylose on dry solid basis.

Fermentation experiments were done based on the studies by Buchert et al. *(3,4)*. Inoculums was prepared by a shake flask culture of *G. oxydans* using pure xylose basal media according to Buchert et al. *(3,4)*; the basal media was inoculated (10^5 cells/mL) and incubated (50 mL in 250 mL) at 25°C. The pH, OD (550 nm), and viable counts were checked for the cell growth. Typically after 5 d (over 10^7 cells/mL) of cultivation, although *G. oxydans* is under exponential growth phase, the cell was spun down and used for the bioconversion test. The bioconversion of xylose in SSL was performed with sterilized 500 mL fermentation setup. Four hundred milliliters of SSL substrate solution at 15 wt% was inoculated at the cell density of either 1.02 or 4.59 (g dc/L), which corresponded to the viable count of 2×10^7 and 6×10^7 (CFU). Filtered air was supplied to the reactor and pH was controlled constant at 5.5. Quantification of the xylonic acid as well as the xylose/xylonic acid ratio were measured by HPLC using an organic acid column (Aminex HPX-87H, Bio-Rad, Hercules, CA) in combination with both ultraviolet (UV) detector and refractive index (RI) detector. By using both UV and RI responses of pure xylose and xylonic acid standards, the calibration factors matrix could be established so that each concentration of xylose and xylonic acid in the mixture could be accurately determined even though the two peaks were overlapped.

Cement Paste Test

A cement paste test was conducted according to the procedure described elsewhere *(5,6)*. Xylonic acid was also tested in a cement heat calorimeter, 3114/3236 TAM Air Isothermal Calorimeter (Thermometric AB, Sweden), to study its setting time behavior. The initial set-time was determined by the onset of the main heat peak. Two commercial Type I/II ordinary Portland cements were used for this calorimetric testing, referred to as Cement B and C. Cement B had a lower soluble alkali content than Cement C.

Mortar Flow Test

Mortar flow test was performed based on JIS A 5201 and A 1173. Mortar was mixed using Hobart mixer. Admixture chemical was added 1 min after mortar was mixed with a mixing water ("delayed" addition). The delayed addition is known to represent actual field concrete performance better than adding a chemical to the mixing water before the mortar preparation. Mortar mix proportions were either cement/sand/water ratio of 460/1350/235 g (w/c = 0.511) or 384/1350/230 g (w/c = 0.599). Cement used was Type I/II ordinary Portland cement. Sand was ISO standard EN-sand. Both mortar slump and flow were measured, and workability = (slump + flow) − 100, was calculated. No tests were made on hardened mortar specimens.

Concrete Test

Concrete testing was performed according to ASTM C192 (specimen preparation), C143 (slump test) and C39 (compressive strength) with the xylonic acid prepared above and compared with performance of the desugared calcium lignosulfonate (Fraser Paper, WI) and sodium gluconate. The starting material, D-xylose, was also tested for slump for a comparison. The concentrated SSL before desugaring process was also obtained from Fraser Paper. The SSL samples before and after the microbial conversion were tested in concrete along with the desugared SSL. The cement factors applied were either 564 lb/yd^3 concrete (334.6 kg/m^3) or 517 lb/yd^3 concrete (306.7 kg/m^3). Coarse aggregate amount was 1800 lb/yd^3 concrete (1068 kg/m^3). The water-to-cement ratios were between 0.567 and 0.617. Varied dosages of the samples were added to concrete and 9-min slump, air content, initial set-time, and 3, 7, and 28-d compressive strength were measured according to ASTM.

Results and Discussion

Bioconversion of Xylose With Commercial Glucose Oxidase

Figure 1 shows the initial enzymatic conversion rate of glucose and xylose at 40°C. As seen in the figure, xylose was successfully converted to xylonic acid although the conversion rate was approx 1/5 of glucose. Interestingly, when the temperature raised to 60°C with all the other variables constant, the conversion rate of xylose increased dramatically and became comparable to that of glucose at 40°C (Fig. 2). The results imply that the oxygen supply might become the rate-determining step for glucose conversion, although the other step was rate determining in the case of xylose conversion (7).

Table 1 summarizes the initial conversion rate of xylose compared to glucose. Under the present conditions, the impact on the xylose conversion rate with commercial GOx appears the following order: temperature > enzyme dosage = substrate concentration > catalase dosage. It should be noted that OxyGo already contains some catalase. It seems that the enzymatic oxidation of xylose may be practical if the temperature tolerance of the GOx enzyme is further improved.

Development of Cement and Concrete Additive

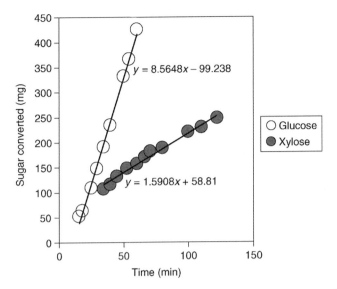

Fig. 1. Initial conversion rate of glucose and xylose using glucose oxidase (OxyGo 1500 at 2.25 mL/125 g substrate) and catalase (fermacolase at 1.55 mL/125 g substrate) at 40°C. Substrate concentration is 25%.

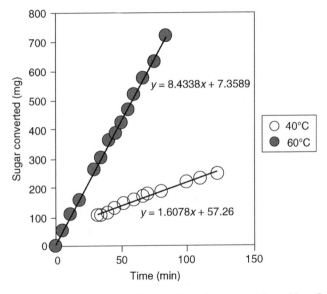

Fig. 2. Initial conversion rate of xylose using glucose oxidase (OxyGo 1500 at 2.25 mL/125 g substrate) and catalase (Fermacolase at 1.55 mL/125 g substrate) at 40°C and 60°C. Substrate concentration is 25%.

Catalytic Oxidation of Xylose

Palladium catalyst conversions was done in a similar way to the enzymatic conversion except that the pH was adjusted to 10.0. Figure 3 shows the catalytic conversion results of glucose and xylose at 35°C with

Table 1
Initial Conversion Rate of Xylose and Glucose Using Glucose Oxidase
(Oxygo 1500) and Catalase (Fermacolase)

Substrate (125 g)	Substrate concentration (%)	Temperature (°C)	OxyGo (mL)	Fermcolase (mL)	Initial rate (mg/min)
Glucose	25	40	2.25	1.55	9.1
Glucose	10	60	2.25	1.55	11.8
Xylose	25	40	2.25	1.55	1.6
Xylose	25	60	2.25	1.55	8.4
Xylose	10	60	2.25	1.55	5.1
Xylose	10	60	2.25	0.8	3.7
Xylose	10	60	4.5	1.55	8.7
Xylose	10	60	4.5	0.8	9.5
Xylose	10	60	4.5	3.2	9.6
Xylose	10	63.5	2.25	0	3.4
Xylose	10	63.5	2.25	0.4	3.1
Xylose	10	63.5	1.15	0.4	1.2
Xylose	10	63.5	1.15	0.2	1.7

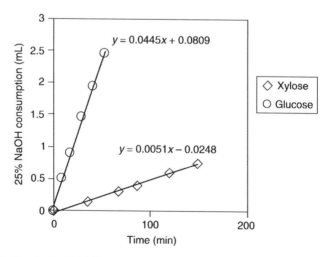

Fig. 3. Catalytic (Pd/C) oxidation rates of xylose and glucose at 35°C.

25 wt% substrate concentration. The conversion rate of xylose turns out to be approx 1/8 of glucose. When the temperature raised to 50°C, the conversion rate of xylose became comparable with glucose at 35°C (Fig. 4). The observed temperature effect is similar to the enzymatic conversion case. Mannose and galactose were even slower than xylose (Fig. 4).

Microbial Oxidation of Xylose in SSL

During the bioconversion, viable count remained relatively constant. The microbe was not growing but held its viability and enzyme activity.

Development of Cement and Concrete Additive

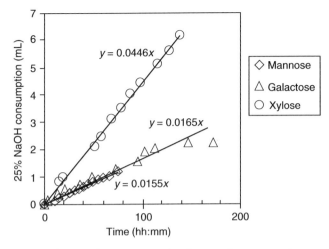

Fig. 4. Catalytic (Pd/C) oxidation rates of xylose, galactose, and mannose at 50°C.

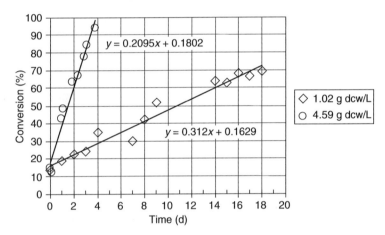

Fig. 5. Microbial bioconversion of xylose in hardwood (birch) spent sulfite liquor to xylonic acid using *Gluconobactor oxydans suboxydans* (ATCC 621).

Figure 5 shows the two bioconversion tests results with different amount of cell density. The conversion rate appears to be linear to the cell density in this range. With the higher cell density, the conversion was completed within 4 d. The high-cell density result, 4 d turn around, implies that the conversion could potentially be done without sterilized condition. Thus obtained product was tested in cement paste, mortar, and concrete to estimate the performance improvement by the bioconversion.

Cement Paste Calorimetric Test

Figure 6A and B shows the cement paste calorimety results of xylonic acid and sodium gluconate in two different cements. Both cements showed considerable set-time retardation with gluconate, although xylonic acid increases the set-time in a more linear manner. The tendency is generally

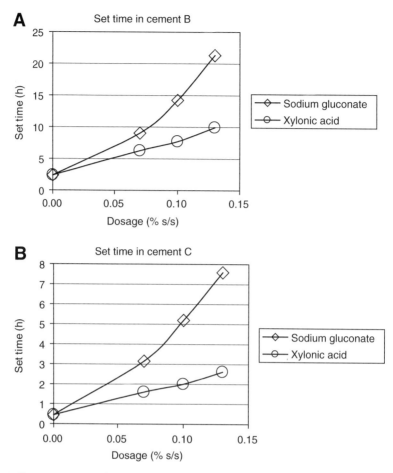

Fig. 6. Cement paste calorimetry results of xylonic acid and sodium gluconate with **(A)** regular soluble alkali cement, **(B)** higher soluble alkali cement.

favorable for water-reducing chemical application because the set-time is more predictable. The cement retardation mechanism of gluconic acid is still unclear, but, in general, it is believed that the precipitation of gluconate on the hydrating cement surface prohibits further hydration of cement. This tendency can be seen in many compounds having α-hydroxy carbonyl functionality *(8)* although there are several exceptions, such as lactic acid *(9)*.

Mortar and Concrete Test Results of Xylonic Acid

Table 2 shows the mortar test results on xylonic acid. The workability result confirmed that xylose becomes an active cement dispersant by converting it to the corresponding aldonic acid. The dispersion effect was comparable to that of gluconate while the retardation was not as large as gluconate.

Figures 7 and 8 show the concrete slump and set-time test results of xylonic acid as a function of applied dosage (wt% on cement weight), respectively. The concrete test results also demonstrate that the xylonic

Development of Cement and Concrete Additive

Table 2
Mortar Test Results of Xylonic Acid

	Workability index (mm)	Setting time (hh:mm)
Blank	83	4:21
Sodium gluconate	182	13:53
Xylose	55	7:04
Xylonic acid	174	8:52

The cement/sand/water ratio was 460/1350/235 (w/c = 0.51). Chemicals' dosage fixed at 0.1 wt% on cement.

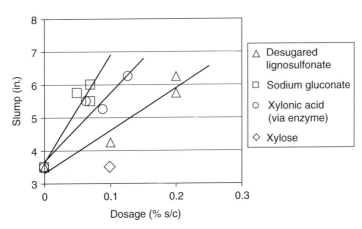

Fig. 7. Concrete slump test results of xylonic acid, sodium gluconate, and desugared lignosulfonate, cement factor = 564 lb/yd^3 (334.6 kg/m^3), coarse aggregate = 1800 lb/yd^3 (1068 kg/m^3), and w/c = 0.567.

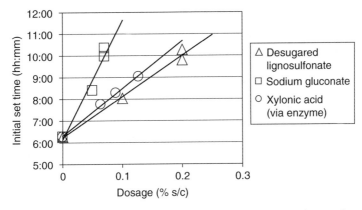

Fig. 8. Concrete initial set-time results of xylonic acid, sodium gluconate, and desugared lignosulfonate, cement factor = 564 lb/yd^3 (334.6 kg/m^3), coarse aggregate = 1800 lb/yd^3 (1068 kg/m^3), and w/c = 0.567.

acid effectiveness as a cement dispersant. The results indicates that the xylonic acid is twice as effective as lignosulfonate and slightly inferior to gluconate. The concrete set-time result also in agreement with the paste

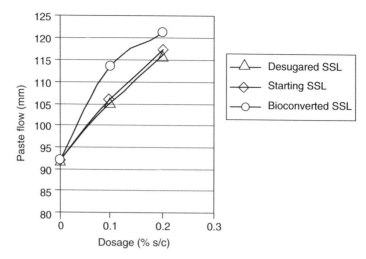

Fig. 9. Cement paste flow results of the bioconverted hardwood (birch) spent sulfite liquor (SSL) along with the starting SSL and the corresponding desugared SSL. The water-to-cement ratio is 0.5. The data points at 0.1% dosage was average of duplicate tests.

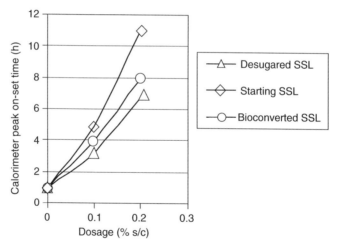

Fig. 10. Cement paste set-time results of the bioconverted hardwood (birch) spent sulfite liquor (SSL) along with the starting SSL and the corresponding desugared SSL. The water-to-cement ratio is 0.5. The data points at 0.1% dosage was average of duplicate tests. Set-time was determined by calorimetry.

and mortar results. It should be noted that the significantly shorter set-time of xylonic acid than gluconate may be somewhat exaggerated by residual chloride salt originated from the enzyme solution because chloride is a known set acceleration chemical.

Cement Paste, Mortar, and Concrete Test Results of Bioconverted SSL

Figures 9 and 10 show the cement paste tests results of bioconverted SSL along with the starting SSL containing about 30 wt% of xylose and the

Development of Cement and Concrete Additive

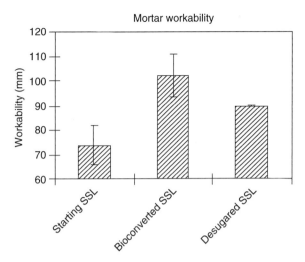

Fig. 11. Mortar workability test results of the bioconverted hardwood (birch) spent sulfite liquor (SSL) along with the starting SSL and the corresponding desugared SSL. The cement/sand/water ratio was 384/1350/230 (w/c = 0.598). Dosage was fixed at 0.1% on cement. Average of duplicate tests and the bars in the chart show the range of the data.

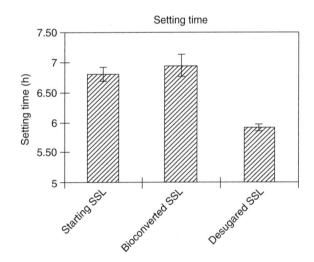

Fig. 12. Mortar set-time test results of the bioconverted hardwood (birch) spent sulfite liquor (SSL) along with the starting SSL and the corresponding desugared SSL. The cement/sand/water ratio was 384/1350/230 (w/c = 0.598). Dosage was fixed at 0.1% on cement. The set-time was determined by penetrometer method. Average of duplicate tests and the bars in the chart show the range of the data.

desugared lignosulfonate from the same source. It can be seen in Fig. 9 that the paste flow performance is improved by the bioconversion. Considering the xylose content in the SSL is about 30%, the improvement by the bioconverted seen in the result is in good agreement with the above straight xylonic acid results. Meanwhile the retardation of SSL was moderately increased by the bioconversion. Figures 11 and 12 shows the mortar results.

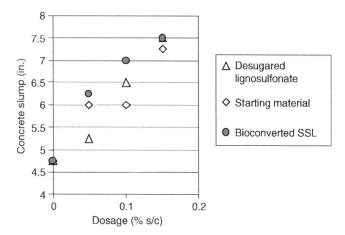

Fig. 13. Concrete slump test results of the bioconverted hardwood (birch) spent sulfite liquor (SSL) along with the starting SSL and the corresponding desugared SSL. Cement factor was 517 lb/yd³ (306.7 kg/m³). Coarse aggregate amount was 1800 lb/yd³ (1068 kg/m³). The water-to-cement ratios was 0.617.

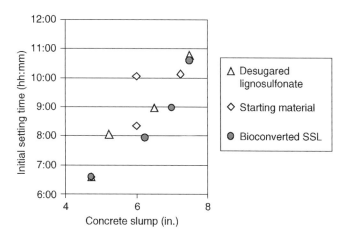

Fig. 14. Concrete initial set-time results of the bioconverted hardwood (birch) spent sulfite liquor (SSL) along with the starting SSL and the corresponding desugared SSL, plotted against respective slump value. Cement factor was 517 lb/yd³ (306.7 kg/m³). Coarse aggregate amount was 1800 lb/yd³ (1068 kg/m³). The water-to-cement ratios was 0.617.

The dosage is fixed at 0.1% solid on cement weight. The data are the average of two results. The mortar results are essentially the same as the above paste results. Figure 13 shows the concrete slump result of the sample. Although the data is rather scattered, the bioconverted SSL demonstrated superior slump performance against the benchmark desugared SSL. In Fig. 14, the setting times of the samples are plotted against the slump data so that set-time can be compared at the equivalent slump point. The plot shows that the bioconverted SSL has a shorter setting time at the same slump value, which is desirable for the water-reducing chemical application.

Table 3
Concrete Test Results (Two Sets) of Bioconverted SSL

	Dosage[a] (wt%)	Slump (cm)	Plastic air (%)	Set-time (hh:mm) Initial	Set-time (hh:mm) Final	Strength (MPa) 3 d	Strength (MPa) 7 d	Strength (MPa) 28 d
Blank	–	n/a[b]	2	n/a[b]	n/a[b]	21.7	25.6	31.6
Desugared lignosulfonate	0.15	17.1	2	6:21	7:58	21.9	28.6	33
Starting material	0.15	16.5	2.2	7:01	9:19	23.4	28.8	34.7
Bioconverted SSL	0.15	18.4	2.2	6:29	8:18	23.4	28.2	33.1
Blank	–	14	2.2	4:55	7:09	17	24.1	32.9
Desugared lignosulfonate	0.1	17.8	1.8	5:55	8:02	18.2	24.3	34.7
Starting material	0.1	18.4	1.7	6:30	8:25	18.5	25.1	36.9
Bioconverted SSL	0.1	19.1	1.7	6:25	8:34	17.4	23.8	34.5

[a]wt% of solid chemicals on cement weight.
[b]not available.
Cement factor was 517 lb/yd^3 (306.7 kg/m^3). Coarse aggregate amount was 1800 lb/yd^3 (1068 kg/m^3). The water-to-cement ratios were 0.603 and 0.600. Blank concrete has the same w/c ratio as concretes with the admixtures.

Table 3 shows the strength results of the bioconverted SSL. All the strength results of the admixture containing concrete are comparable and better than the blank concretes that have the same w/c ratio as the admixture concretes. Considering the better strength in addition to the clearly superior slump improvement of the bioconverted SSL, there is no doubt that the bioconverted SSL will be qualified as ASTM Type-A water reducer even without further formulation which is always done in a commercial water reducer products. Application and ASTM Type-F and/or Type-D may require some formulation for controlling setting time as well as potential air entrainment at higher dosage level.

All the performance test results of the bioconverted SSL indicate that the dosage performance of starting SSL was improved by approx 30%, which is in good agreement with the straight xylonic acid results; 30 wt% of nonactive xylose in the SSL was converted to the active xylonic acid that is twice as effective as lignosulfonate.

Conclusion

It was demonstrated that the xylonic acid can be derived either by enzymatic, microbial, or catalytic conversion. Enzymatic xylose conversion using a commercial glucose oxidase proceeded at the reasonable rate when higher temperature is applied. Microbial conversion seems to have an advantage if a large amount of impurities exist. Xylonic acid demonstrated that it can be effective cement dispersant having superior performance over lignosulfonate or sodium gluconate. In summary, xylose can be effectively utilized by converting to its corresponding aldonic acid. As shown here, it cannot only replace gluconate in the existing applications, but can potentially replace other materials, such as lignosulfonate. It is expected that further investigation may provide more unique applications of xylonic acid.

Acknowledgments

The authors are indebted to Dr. Marc Mittelman and Dr. Roy Pang for their advice and guidance on microbiological tests and fermentation. The authors also thank Mr. Gregory S. Freeman and Dr. David F. Myers for their continuous support and guidance during the project.

References

1. Lantero, O. J. and Shetty, J. K. (1999), US Patent 5,998,179.
2. Hattori, K., Miya, B., Matsuda, M., Ishii, M., Watanabe, H., and Takizawa, H. (1978), US Patent 4,108,891.
3. Buchert, J., Puls, J., and Poutanen, K. (1988), *Appl. Microbiol. Biotechnol.* **28(4/5),** 367–372.
4. Buchert, J. and Viikari L. (1988), *Appl. Microbiol. Biotechnol.* **29(4),** 375–379.
5. Chun, B. (2001), *Cem. Concrete Res.* **31(6),** 959–963.
6. Kantro, D.L. (1980), *Cem. Conc. Aggr.* **2,** 95–102.
7. Keilin, D. and Hartree, E. F. (1952), *Biochem. J.* **50,** 331–341.
8. Ithoh, M. and Takeuchi, T. (1989), Chi-en oyobi Chou-Chi-en no Mekanizumu, Semento/Konkuriito Kagaku to Sono Ouyou, ISBN 4-88175-020-8 C3043, Semento Kyoukai (chapter 14).
9. Singh, N. B., Singh, S. P., and Singh, A. K. (1986), *Cem. Concrete Res.* **16,** 545–553.

Production of *Bacillus sphaericus* Entomopathogenic Biomass Using Brewery Residues

Cristiane Darco Cruz Martins, Paula Fernandes De Aguiar, and Eliana Flavia Camporese Sérvulo*

Universidade Federal do Rio de Janeiro, Escola de Química, Centro de Tecnologia, Bloco E, Ilha do Fundão, 21949-900 Rio de Janeiro, RJ, Brazil; E-mail: servulo@eq.ufrj.br

Abstract

The use of brewery residues—yeast and trub—has been evaluated aiming to minimize the costs of the industrial production of *Bacillus sphaericus*-based bioinsecticide. Both brewery residues promoted growth and sporulation of the three *B. sphaericus* strains that were isolated from Brazilian soils (S1, S2, and S20). However, distinct growth and sporulation behaviors were observed in relation to the different nutritional conditions and strain used. The maximum sporulation percentage was obtained through the cultivation of S20 strain in brewery residual yeast. In general, the entomopathogenic biomasses produced showed good results for toxicity to *Culex* larvae. The minimum values of larvae population (LC50) were observed for the S20 strain grown on yeast brewery residue-containing media. After fermentation, a considerable decrease in the organic material of alternative media was verified, although the residual values were still higher than that considered appropriate for effluent discharge.

Index Entries: *Bacillus sphaericus*; bioinsecticide production; industrial residues; entomopathogenic biomass; spore/crystal toxins.

Introduction

The continuous and indiscriminating use of chemical pesticides to control insects and other vectors during several decades, in many cases, has resulted in environmental imbalance, particularly owing to the lack of specificity of the chemicals used *(1,2)*. Among other deleterious effects, the application of pesticides, with long toxic residual action, may cause the development of resistant insect populations *(2)*. For these reasons, the use of biological products is an attractive alternative, owing to their low toxicity and insect specificity. Furthermore, the use of biological insecticides should grow in the future as environmental protection agencies become

---

*Author to whom all correspondence and reprint requests should be addressed.

more rigorous, in particular, those regulating developing countries, and also by launching of new, more efficient, and less expensive formulation products.

In the past few years, an increasing interest has been focused toward biological products with the ability to control tropical diseases. In Brazil, the goal is to eliminate specifically *Culex* (filariases vector), *Anopheles* (malaria vector), *Aedes* (dengue vector), and *Simulium* (oncocercosis vector) *(3)*. Recently in Brazil, commercial microbial agents, such as *Bacillus thuringiensis* and *B. sphaericus*, are being applied to kill *Aedes/Simulium* and *Aedes* mosquitoes, respectively. Among those two bacilli, *B. sphaericus* toxins present several advantages such as high and long larvicidal activity and spore persistence, even in polluted environments. However, *B. sphaericus*, distinctly from the other species, is not able to utilize carbohydrates as carbon sources.

Considerable progress has been achieved in microbial insecticide production, either concerning the isolation of strains with high larvicidal activity against target organisms or through the development of new formulations, more stable, with longer environmental persistence *(2)*. However, there is a need to reduce microbial insecticides' production costs, which are higher than that of synthetic insecticides. Hopefully, use of locally available and low-cost raw materials could make a bioinsecticide large-scale production more economically feasible *(4,5)*. In Brazil, an industrial scale brewery normally produces 2.8 L residual yeast/hL beer with a standard production of 700,000 hL/mo, so the destination of such a waste is a problem of great magnitude to the industry. In addition, for each 1500 L of beer produced, approx 30–40 L of trub are accumulated, generating another problem for the company. Technically, trub is defined as the sludge consisting of proteins and hops that precipitate out of wort during the boiling and chilling processes of brewing production. This work aims to evaluate the substitution of the medium normally used for *B. sphaericus* entomopathogenic biomass production by brewery residues, attempting to make the biopesticide competitive with chemical pesticides production.

Methods

Microorganism

B. sphaericus S1 was isolated from a river sample from Vitoria (Espirito Santo, Brazil), *B. sphaericus* S2 from a soil sample from Brasilia (Distrito Federal, Brazil), and *B. sphaericus* S20 from a soil sample from Pantanal (Mato Grosso do Sul, Brazil).

Media

Culture growth has been performed on standard medium (medium S) containing: 1 g/L yeast extract, 3 g/L meat extract, 5 g/L peptone, 1 g/L KH_2PO_4, 0.15 g/L $CaCl_2$, 0.10 g/L $MgSO_4 \cdot 7H_2O$, 0.01 g/L $FeSO_4 \cdot 7H_2O$, 0.01 g/L $MnSO_4 \cdot H_2O$, and 0.01 g/L $ZnSO_4 \cdot 7H_2O$.

Table 1
Alternative Media Composition

Production medium	Residues (% [v/v])	
	Residual yeast	Trub
T	–	38.5
YT	46	19
Y	92	–

Table 2
Chemical Composition of Both Brewery Residues Used

Residue	Protein (g/L)	Fat (g/L)	Total carbon (g/L)
Residual yeast	7.6	0.7	36.9
Trub	18.2	2.3	32.2

Three other formulations consisting of the brewery residues (yeast and trub), isolated or in mixture, were tested for entomopathogenic biomass production (Table 1). The chemical compositions of both brewery residues are presented in Table 2. The protein content of the alternative media was established to assure an initial protein concentration of 7 g/L, according to previous studies (6). Before use, the brewer's yeast residue has been heated to 60°C for 12 h in order to promote cell autolysis, filtered in qualitative paper no. 4 (Whatman plc, UK) for solid removal, and then stored at –20°C. The trub residue was filtered and stored using the same procedure.

Inoculum

A loopful of bacterial growth from nutrient agar (Merck 1.05450.0500) was used to inoculate a 500-mL Erlenmeyer flask containing 100 mL of medium S. After incubation of the flask on a rotatory shaker (New Brunswick, US) at 250 rpm and 30°C for 24 h, volumes corresponding to approx 0.15 g/L of exponential-phase cells were used as inoculum for production media. The cell concentration was determined by dry weight (7).

Bioinsectide Production

Batch cultures were performed in triplicate on 500-mL Erlenmeyer flasks containing 100 mL of production medium. After 48 h of incubation, at 29 ± 1°C and 250 rpm, samples of fermented media were withdrawn aseptically for microbiological analysis. Other samples were centrifuged for cell removal, lyophilized, and the toxicity of spore/crystals produced was determined by bioassays. Supernatants were submitted to chemical analysis.

Analytical

Total viable cell was quantified by serial plating technique as described by Moraes and Alves *(8)*. For spore count, samples were heated at 80°C for 12 min before plating. Total organic carbon quantity was measured in media, before inoculation, and after cultivation process, using a Shimatzu TOC 5000 analyzer. Total protein and lipid content was measured in media before inoculation and was determined according to the micro-Kjeldahl *(9)* and the Bligh and Dyer methods *(10)*, respectively. The pH values were determined using a potenciometer (DMPH-1 DA Digimed). In order to determine the entomotoxic activity of *B. sphaericus* strains, the biomasses produced were assayed against third-instar larvae of *Culex quinquefasciatus* according to the protocol of Monnerat et al. *(11)*. The lethal concentration necessary to reduce 50% of larvae population (LC50) was calculated by Probit analysis *(12)*. All samples were taken in duplicate and the counts were averaged.

Results and Discussion

The tested brewery residues—yeast and trub—were both able to enhance the growth and sporogenesis of the three selected strains (Fig. 1). However, distinct growth and sporulation behaviors were observed related both to the strain and to the composition media used. In most cases, these results were superior to those found for the production of *B. sphaericus* bioinsecticide using the medium S. In general, the cultivation of the S1 strain produced the largest growth in the tested alternative media. However, regarding the sporulation percentage, the S20 strain proved to be, overall, the most efficient of the studied strains. It must be emphasized that all presented data are linked to a fermentation time of 48 h, as longer periods did not lead to best results (data not shown).

Among all media, that which constituted exclusively of trub (medium T) was the most appropriate for *B. sphaericus* strain growth and S1 in particular. On the other hand, in the case of S1 strain in medium T, the sporulation percentages were not so satisfactory in comparison with the cultures grown in the other alternative media. The medium consisting of both brewery residues (medium YT) promoted a different growth profile for each of the studied strains. In particular, S2 was the only one to present maximum sporulation in this medium, which was even greater than that observed in medium S. Mainly, in the media containing residual brewery yeast, media Y and YT, S20 achieved the maximum sporulation frequencies of 92% and 66%, respectively. Those results were even higher than those evidenced for the other strains. According to Pelczar et al. *(13)*, autolyzed yeast extract solution is a recommendable nitrogen source. This product is added to culture media to stimulate microbial growth, as it contains vitamins, in particular B-complex vitamins, as well as amino

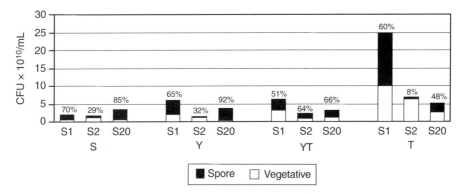

Fig. 1. Vegetative and spore counts (CFU) and sporulation percentage of *Bacillus sphaericus* strains (S1, S2, and S20) on different media (S, standard medium; Y, 100% residual yeast; YT, 50% residual yeast plus 50% trub; and T, 100% trub) after 48-h incubation.

acids and other stimulating substances that are important to the metabolic activity of these microorganisms. Trub residue also has high protein content, but its polyphenol content (11–26%) and lack of growth factors can interfere with bacteria metabolism *(14)*.

The results obtained are in agreement with those reported by other authors. According to Dias *(15)* the growth and sporulation of *B. sphaericus* 2362 were different as a function of the composition medium. The author obtained maximum growth and sporulation of 9.1×10^{10} cell/mL and 84.7%, respectively, through the use of brain heart infusion (BHI) medium *(16)*. A similar sporulation frequency was obtained in NYSM (a peptone-based broth) medium *(17)* although, under this condition, a considerably lower cell concentration (1.4×10^8 cell/mL) was determined *(18)*.

The spores/crystals of *B. sphaericus* strains produced from tested media and bioassayed against third-instar larvae of *C. quinquefasciatus* are given in Fig. 2. Our results show LC50 values for alternative media (Y, T, and YT), varying from 8 to 7815 µg/L. It should be highlighted that the smaller the LC50 value, the greater the lethal power of the toxin produced by the microorganism. Again, the dependence of larvicidal activity on the nutritional condition was confirmed: *B. sphaericus* S2 has shown greater entomotoxic activity, of 88 and 87 µg/L, when cultivated on media T and YT, respectively. On the other hand, the S1 strain has shown the poorest lethal larvae activity when cultivated on media YT. The toxicity of *B. sphaericus* S20 proved to be, overall, the most efficient of the studied strains. This strain showed good larvicidal activity in all alternative production media in comparison with medium S cultives. The best LC50 values of 8 and 10 were obtained for S20 on media Y and YT, respectively.

The fermentation of *B. sphaericus* 2362 on protein-based media supplemented with glycerol, a byproduct of yeast metabolism, led to an almost

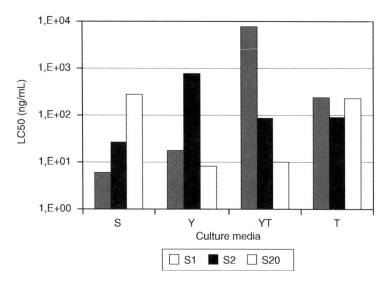

Fig. 2. Effect of growth of different strains (S1, S2, and S20) on different media (S, standard medium; Y, 100% residual yeast; YT, 50% residual yeast plus 50% trub; and T, 100% trub) after 48-h incubation in entomotoxic analysis (LC50, lethal concentration necessary to kill 50% of larvae).

complete sporulation and a considerably improved larvicide production (19). Also, Nigerian agricultural products provided good sporulation (53–80%) of *B. sphaericus* 1593, whose powders were effective against *C. quinquefasciatus* larvae with LC50 of 193 µg/L (20). The culture of this same strain on defatted groundnut cake and defatted milk powder led to 85% sporulation and a LC50 value of 52.4 µg/L (21).

Distinct behaviors concerning the pH were observed depending on the strain and the alternative media used for bioinsecticide production (Fig. 3). In some conditions, a slight increase of the pH was observed. All fermented media by the *B. sphaericus* S20 strain presented this profile. This is probably a reflection of the protein consumption, mainly as a carbon source. According to Arcas (22), the pH gradually increases during the fermentation process, reaching values varying from 8.0 to 9.0, and it is possibly owing to the consumption of amino acids as the carbon source.

According to Yousten (23), the need to control the pH during the fermentation process seems to be related to the *B. sphaericus* strain used. On the other hand, Yousten and Wallis (17), in later studies, did not observe any significant difference for spore count and insecticidal activity for *B. sphaericus* 2362 when the pH was controlled between 7.2 and 7.3.

The trub-containing media (T and YT) showed a decrease in the pH values after fermentation by S1 and S2 strains, suggesting the formation of metabolic acids. The oxidative metabolism of the bacilli occurs through the Embden-Meyerhof-Parnas and/or Entner-Doudoroff pathway, yielding pyruvate, which is subsequently metabolized by the tricarboxylic acid

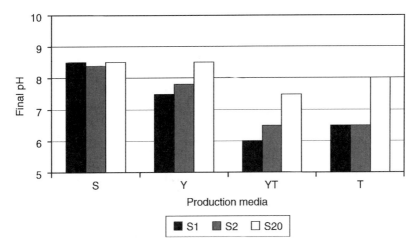

Fig. 3. Effect of growth of different strains (S1, S2, and S20) on different media (S, standard medium; Y, 100% residual yeast; YT, 50% residual yeast plus 50% trub; and T, 100% trub) after 48-h incubation in pH values.

(TCA) cycle. According to Russel et al. *(24)*, the pH decreases during vegetative growth, as the glucose consumption leads to the accumulation of acetate, until a depression of the TCA enzymes occurs; then, the acetate can be metabolized by this pathway, as observed for *B. cereus*. At the end of the exponential phase, which coincides with the transition of the cell from the vegetative to the sporulated form, an increase in the pH is observed as a consequence of the oxidation of organic acids, which are intermediate compounds of this metabolic pathway. However, owing to the incapacity of *B. sphaericus* to assimilate carbohydrates, it requires alternative carbon sources for its metabolism, such as protein material *(24,25)*. The presence of protein materials in the culture medium as carbon and nitrogen sources results in a greater assimilation of carbon in comparison with nitrogen, causing an accumulation of ammonia ions and a subsequent increase of the pH.

Concerning the organic material present in the alternative media, it can be observed that they contain a much higher quantity of organic carbon than the medium S (Fig. 4). After fermentation, a reduction in the organic material of both residues was verified, especially for the S20 culture in medium Y. The results obtained indicate the use of brewery residues for bioinsecticide production as another possible strategy in pollution treatment.

The protein uptake in the different media by the three *B. sphaericus* strains is shown in Fig. 5. The major consumptions are observed when strains S1 and S2 were cultivated in the medium T. The protein uptake by the S20 strain showed similar behaviors under all culture conditions tested.

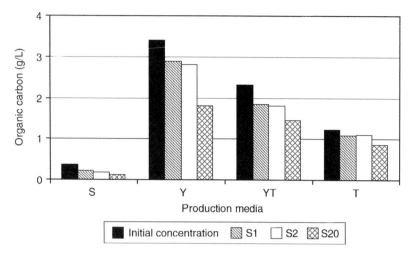

Fig. 4. Total organic carbon uptake by *Bacillus sphaericus* strains after 48-h incubation in different media (S, standard medium; Y, 100% residual yeast; YT, 50% residual yeast plus 50% trub; and T, 100% trub).

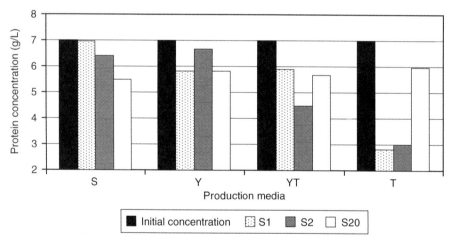

Fig. 5. Protein uptake by *Bacillus sphaericus* strains after 48-h incubation in different media (S, standard medium; Y, 100% residual yeast; YT, 50% residual yeast plus 50% trub; and T, 100% trub).

Conclusions

The Brazilian isolated strains of *B. sphaericus* S1, S2, and S20 were able to grow and sporulate when cultivated in brewery residues. Trub promoted the growth of the three strains, especially of S1. Maximum sporulation frequencies of most bacilli were achieved by cultivation in the brewery yeast residue. The spore forming of the bacilli in the alternative media was superior to that obtained in the medium S. However, different growth and sporulation patterns have been evidenced regarding the constituents of the industrial residues. The spore/crystals produced by *B. sphaericus* strains on alternative media showed good entomotoxic activity against *C. quinquefasciatus*, especially S20. The results obtained recommend the use of brewery residues for *B. sphaericus* insecticidal toxin production because of their wide availability and low cost in Brazil and other countries. Considering the study, we can estimate a cost of 1 USD to prepare 100 L of medium according to data provided by the brewing industry. The organic matter was reduced after the fermentative process, mainly by the S20 strain cultivated in yeast brewery residue, thus reducing total carbon in the effluent stream and the cost owing to its treatment discharge.

References

1. Federici, B. A. (1995), *J. Am. Mosquito Control Assoc.* **11(2)**, 260–268.
2. Capalbo, D. M. F. (1995), *Mem. Inst. Oswaldo Cruz* **90(1)**, 249–253.
3. Fridlender, B., Keren-Zur, M., Hofstein, R., et al. (1998), *Mem. Inst. Oswaldo Cruz* **84**, 123–127.
4. Obeta, J. A. N. and Okafor, N. (1983), *Can. J. Microbiol.* **29**, 704–709.
5. Hoti, S. L. and Balaraman, K. (1986), *Ind. J. Med. Res.* **83**, 166–170.
6. Roberg, R. A. P. (2000), PhD Thesis, Departmento de Engenharia Bioquímica, Escola de Química, Universidade Federal do Rio de Janeiro, Rio de Janeiro, Brasil.
7. Reis, F. A. S. L., Servulo, E. F. C., and e de França, F. P. (2004), *Appl. Biochem. Biotechnol.* **113–116**, 899–912.
8. Moraes, S. A. and Alves, S. B. (1986), In: *Controle Microbiano de Insetos, Quantificação de Inóculo de Patógenos de Insetos*. Alves, S. B. (ed.), Manole Ltda., Brasil, pp. 278–288.
9. Association of Official Anaytical Chemists (AOAC) (1984), *Official Methods of Analysis*, 14th edition, Virginia, p. 1141.
10. Bligh, E. G. and Dyer, W. J. (1959), *Can. J. Biochem. Phisiol.* **37**, 911–917.
11. Monnerat, R. G., Dias, D. G. S., Silva, S. F., et al. (2005), *Pesq. Agropec. Bras.* **40(2)**, 103–106.
12. Finney, D. J. (1971), In: *Probit Analysis*. Cambridge University Press, Cambridge, England.
13. Pelczar, D. J., Cham, E. C. S., and Krieng, N. R. (1993), In: *Microbiology. Concepts and Applications*, 5th edition, McGraw-Hill, New York.
14. Barchet, R. (1994), *Brewing Techniq.* **2(2)**, 1–10.
15. Dias, J. M. C. S. (1992), *Pesq. Agropec. Bras.* **27**, 59–76.
16. Rowan, N. J., Deans, K., Anderson, J. G., Gemmell, C. G., Hunter, I. S., and Chaithong, T. (2001), *Appl. Environ. Microbiol.* **67(9)**, 3873–3881.
17. Yousten, A. A. and Wallis, D. A. (1987), *J. Ind. Microbiol.* **2**, 277–283.
18. Andreev, J., Dibrov, P. A., Braun, D., and Klein, S. B. (1994), *FEBS Lett.* **349**, 416–419.
19. Klein, D., Yanai, P., Hofstein, R., Fridlender, B., and Braun, S. (1989), *Appl. Microbiol. Biotechnol.* **30**, 580–584.
20. Obeta, J. A. N. and Okafor, N. (1983), *Can. J. Microbiol.* **29**, 704–709.
21. Desai, S. Y. and Shethna, Y. I. (1991), *Ind. J. Med. Res.* **93**, 318–323.
22. Arcas, J. A. (1996), In: *Producción de Bacterias Entomopatogenicas*. Leucona, R. (ed.), Miroorganismos Patogenos Empleados en el Control Microbiano de Insectos Plagas, Buenos Aires, vol. 1, pp. 208–222.
23. Yousten, A. A. (1984), *Adv. Biotech. Proc.* **3**, 315–343.
24. Russel, B. L., Jelley, S. A., and Yousten, A. A. (1989), *Appl. Environ. Microbiol.* **55(2)**, 294–297.
25. Yousten, A. A., Wallis, D. A., and Singer, S. (1984), *Curr. Microbiol.* **11**, 175–178.

Batch (One- and Two-Stage) Production of Biodiesel Fuel From Rapeseed Oil

GWI-TAEK JEONG[1] AND DON-HEE PARK*,[1–5]

[1]School of Biological Sciences and Technology, Chonnam National University, Gwangju 500-757, Korea, E-mail: dhpark@chonnam.ac.kr; [2]Faculty of Applied Chemical Engineering, Chonnam National University, Gwangju 500-757, Korea; [3]Research Institute for Catalysis, Chonnam National University, Gwangju 500-757, Korea; [4]Biotechnology Research Institute, Chonnam National University, Gwangju 500-757, Korea; and [5]Institute of Bioindustrial Technology, Chonnam National University, Gwangju 500-757, Korea

Abstract

Biodiesel fuel is an alternative and renewable energy source, which may help to reduce air pollution, as well as our dependence on petroleum for energy. Several processes have already been developed for the production of biodiesel. Alkali-catalyzed transesterification with short-chain alcohols, for example, generates high yields of methyl esters in short reaction times. In this study, we have evaluated the efficacy of batch (one- and two-stage) transesterification of rapeseed oil in the production of rapeseed methyl ester. The conversion of rapeseed oil exhibited similar reaction patterns and yields in 30- and 1-L reaction systems. Approximately 98% of the rapeseed oil was converted at 400 rpm within 20 min, under the following conditions: 1% (w/w) KOH, 1:10 methanol molar ratio, and at 60°C. In the 30-L, two-stage transesterification process, approx 98.5% of the rapeseed oil was converted at a 1:4.5 molar ratio and 1% (w/w) KOH at 60°C for 30 min (first reaction condition), and at a 1:1 molar ratio and 0.2% (w/w) KOH at 60°C for 30 min (second reaction condition).

Index Entries: Biodiesel fuel; transesterification; rapeseed oil; two stage.

Introduction

The methyl esters of the higher fatty acids C_{14}–C_{22}, which can be derived from vegetable oils and animal fats via transesterification with alcohol, may be usable as an alternative fuel for diesel engines. This compound is also referred to as a biodiesel fuel. However, it also holds tremendous promise in a wide range of industrial purpose, either in direct form (e.g., paint-stripper, graffiti remover, cleaning solvent, emulsifier, and absorbent of volatile organic pollutants) or as starting materials for the

---

*Author to whom all correspondence and reprint requests should be addressed.

production of other compounds (e.g., alkanolamides, isopropyl esters, fatty alcohols, fatty acid α-sulpho-esters, synthetic esters, and sugar polymers) *(1–3)*. Biodiesel fuel is a renewable alternative energy source, which may help to reduce air pollution, as well as our dependence on petroleum for energy. The ratio of fatty acid methyl ester (FAME) in mixed biodiesel fuel ranges between 5 and 30 wt%, but is usually between 20 and 30 wt%. Mixed fuels are characterized by good emissions profiles, and also tend to minimize the problems generally associated with biodiesel fuels *(3)*. Biodiesel is compatible with petrodiesel in compression–ignition engines, and mixtures of biodiesel and petrodiesel have been used without any need for engine modification *(4)*. These fuels also exhibit low viscosities, similar to those of petrodiesel. Also, many of the characteristics of biodiesel fuels, including volumetric heating value, cetane number, and flash point, are fairly similar to those of petrodiesel *(5)*. The use of biodiesel fuel is also associated with several unique advantages. It may be used as an alternative renewable energy source, or a biodegradable nontoxic fuel. Biodiesel fuels are generally associated with the generation of a low degree of air pollution from particulates, carbon monoxide, SO_x emissions, and CO_2 recycling over short periods *(6)*. Also, biodiesel exhaust gases contain a relatively low amount of polycyclic aromatic hydrocarbons, which exert proven carcinogenic and mutagenic effects *(3)*.

Despite the fact that biodiesel is being increasingly eyed as an attractive alternative fuel, problems remain regarding the costs of its production. Two factors are extremely relevant to the cost of biodiesel production: raw materials' costs (oil and alcohol), and process operation costs. The cost of raw material accounts for approx 60–75% of the total biodiesel production cost *(7)*. Whereas in the United States, soybean oil has been primarily used in the development of biodiesel, rapeseed oil has become the dominant feedstock for biodiesel production in Europe *(8)*. Several processes have been developed for the production of biodiesel fuel via acid-, alkali-, and enzyme-catalyzed transesterification reactions *(9–11)*. Alkali-catalyzed transesterification proceeds approx 4000 times faster in the presence of an alkaline catalyst than when catalyzed by an equal amount of an acid catalyst *(12)*. Transesterification, also commonly referred to as alcoholysis, is the displacement of alcohol from an ester by another alcohol, in a process which is fairly reminiscent of hydrolysis. Transesterification involves a number of consecutive and reversible reactions. During the reaction step, triglycerides are converted to diglycerides. The reaction step is followed by the conversion of diglycerides to monoglycerides, and then of monoglycerides to glyceride *(13,14)*. The main factors which affect transesterification include the molar ratio of vegetable oil to alcohol, the type and amount of catalyst, the reaction temperature and reaction time, and the contents of free fatty acid and water in the oil sources used *(10)*. In alkali-catalyzed transesterification, the constitutent oils and alcohol must be largely anhydrous, as water can cause saponification when mixed with oil. Another significant factor which affects conversion yield is the molar ratio of alcohol to vegetable oil. The stoichiometry of the transesterification reaction requires 3 mol alcohol per mole of triglyceride, in order to yield 3 mol fatty esters and 1 mol of glycerol. Higher oil to alcohol molar ratios tend to result in greater conversion yields in a shorter reaction time. The recommended amount of alkali catalyst for effective transesterification is somewhere between 0.1% and 1% (w/w) of oils and fats. In general, the reaction temperature is set at near the boiling point of alcohol. Higher reaction temperatures tend to facilitate the reaction, and shorten reaction time *(2,10)*. The other important factor in this process is agitation speed, which plays a vital role in the transesterification process. The degree of homogeneity of alcohol with the triglyceride is critical to the success of the transesterification process *(12)*.

In this article, we have attempted to determine the optimal transesterification conditions inherent to the production of biodiesel via batch (one- and two-stage) transesterification, using rapeseed oil with methanol and potassium hydroxide.

Materials and Methods

Materials

Refined and bleached rapeseed oil was obtained from the Onbio Co., Ltd. (Pucheon-Si, Korea). Table 1 illustrates the fatty acid composition and characteristics of the rapeseed oil. Reference standards of FAMEs, including palmitic, stearic, linolenic, linoleic, and oleic methyl ester, all of >99% purity, were purchased from the Sigma Chemical Co. (St. Louis, MO). Methanol and potassium hydroxide were analytical grade.

Apparatus

1-, 5-, and 30-L reaction systems were applied to the transesterification process. The 1-L reactor system was composed of a 1-L four-necked reactor, which was equipped with a reflux condenser, a thermometer, and a sampling port. The reactor was immersed in a constant-temperature water bath which was controlled by a PID temperature controller, capable of controlling the temperature to within ±0.2°C of the set point. Mixing was done by an electrical motor equipped with a propeller-type impeller *(2)*. The 5-L reaction system consisted of a 5-L reactor, and an impeller with six flat blades. The 30-L reaction system (KobioTec, Korea) consisted of a 30-L reactor and three impellers with six flat blades. During the experiments, samples were withdrawn at preset time intervals with a 1-mL glass pipet through a sampling port in the reactor.

Reaction Procedures and Conditions

For the batch transesterification of the rapeseed oil, the reactor was initially charged with a given amount (400 g in the 1-L reaction system, 3.5 L

Table 1
Fatty Acid Composition and Characteristics
of Rapeseed Oil

Characteristics	Content (%)
Specific gravity	0.917
Moisture content	0.01%
Free fatty acid	0.018%
Unsaponifiable matter	0.39%
Fatty acid % (w/w)	
Palmitic acid ($C_{16:0}$)	5.7%
Stearic acid ($C_{18:0}$)	2.2%
Oleic acid ($C_{18:1}$)	58.5%
Linoleic acid ($C_{18:2}$)	24.5%
Linolenic acid ($C_{18:3}$)	9.1%

in the 5-L system, and 18 L in the 30-L system) of rapeseed oil, and then heated to the set temperature along with agitation. The catalyst was prepared by dissolution in the required amount of methanol. After the set temperature of oil and methanol was achieved, a methanolic catalyst was added to the base of the reactor, in order to prevent the methanol from evaporating. The reaction was timed, beginning immediately after the addition of the methanol and the catalyst.

For the two-stage transesterification of the rapeseed oil, the first reaction was performed under the following conditions: 1:4.5 molar ratio, 1% (w/w) potassium hydroxide, and at 60°C. After the first reaction ceased, the supernatant was rereacted by adding potassium hydroxide and methanol.

Sample Preparation and Analysis

Samples were drawn at preset time intervals. Approximately 1 mL of the sample mixtures were collected in 10-mL test tubes, to which 1 N hydrochloric acid was immediately added and vortexed, in order to neutralize the catalyst and halt the reaction. The pretreated samples were evaporated in order to remove any nonreacted methanol, and then centrifuged at low temperature to remove the glycerin. The supernatant was then evaporated under a vacuum, and diluted with methanol for high-performance liquid chromatography (HPLC) analysis.

Prepared samples were analyzed for FAMEs, using HPLC. An HPLC (Waters Korea Ltd., Korea) was equipped with a model Waters 1525 binary HPLC pump and a ultraviolet detector (Waters 2487 dual λ absorbance detector, 205 nm). A Waters Spherisorb ODS2 column (4.6 × 250 mm² with 5 µm particle size) was used for the separation. The mobile phase consisted of a 48:48:4 volumetric mixture of acetonitrile, acetone, and water. The mobile phase was degassed by sonication for 1 h. The pump was operated at a 1 mL/min constant flow rate, and column temperature was maintained in the column chamber at 35°C. The sampling injection volume used was 20 µL, and peak identification was conducted by comparing retention times between the sample and the standard materials. The conversion yields were finally calculated using calibration curves for FAMEs.

Results and Discussion

In order to perform the alkali-catalyzed transesterification process with the rapeseed oil, we applied several reaction systems. In the alkali-catalyzed transesterification, the amount of free fatty acid was proposed to be less than 0.5% on the basis of oil weight, in order to ensure a high conversion yield (15). As it exhibits a high acid value, the activity of the catalyst was diminished during the transesterification reaction. As reported in Table 1, the fatty acid content of the rapeseed oil used in this experiment was 0.018%, which was lower than the proposed value (below 0.5%). In our previous report (2), we optimized rapeseed FAME (biodiesel) production by the alkali-catalyzed transesterification reaction, using anhydrous methanol and potassium hydroxide. The optimized conditions for alkali-catalyzed transesterification using KOH were determined to be the following: 1:8 to 1:10 molar proportion of oil to methanol, 1% (w/w) by oil weight of KOH catalyst; 60°C reaction temperature, and 30 min reaction time. Under the given conditions, the conversion yield was approx 98%. From the refined product (rapeseed FAME, biodiesel), the purity of the product was found to be more than 99% through posttreatment, which included washing and centrifugation.

Batch Production System of Rapeseed Methyl Ester

Agitation speed appears to play a vital role in the transesterification process. The degree to which the alcohol is homogeneous with the triglycerides is also relevant to the transesterification process (12). We assessed the effects of agitation speed on the biodiesel conversion yield in the 1-L reaction system. In this reaction system, the agitation was provided by an electrical motor, which was equipped with a propeller-type impeller with a diameter of 50 mm, and three blades. As shown in Fig. 1, we compared conversion yields at different agitation speeds, from 400 to 1800 rpm, under the following conditions: 1% (w/w) potassium hydroxide, a methanol molar ratio of 1:6, and 60°C after a reaction time of 20 min. Below 600 rpm, the rapeseed oil and methanol were visually confirmed to mix rather poorly. The conversion yields occurring at 600 and 1800 rpm were approx 75% and 90%, respectively. Above 1400 rpm, production yield was unaffected by agitation speed.

We also attempted to determine the effects of agitation speed on the production of FAME in the 5-L reaction system. We organized the 5-L reaction system, which consisted of an impeller with six flat blades. The conversion

Batch Production of Biodiesel Fuel

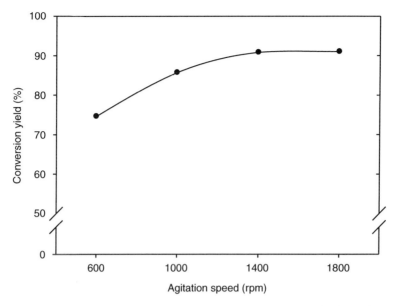

Fig. 1. Effect of agitation speed on conversion yield at 1% (w/w) of potassium hydroxide (1:6 methanol molar ratio and 60°C) after 20 min in 1-L reaction system.

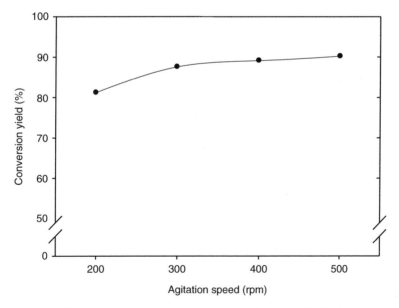

Fig. 2. Effect of agitation speed on conversion yield at 1% (w/w) of potassium hydroxide (1:6 methanol molar ratio and 60°C) after 30 min in 5-L reaction system.

yields were compared at different agitation speeds from 200 to 500 rpm, under the following conditions: 1% (w/w) KOH, a methanol molar ratio of 1:6, and a temperature of 60°C after 30-min reaction. Figure 2 shows the effects of agitation speed on the conversion of rapeseed oil to biodiesel.

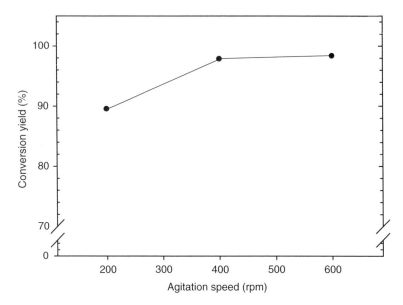

Fig. 3. Effect of agitation speed on conversion yield at 1% (w/w) of potassium hydroxide (1:10 methanol molar ratio and 60°C) after 30 min in 30-L reaction system.

The rapeseed oil and methanol were observed to mix poorly at less than 200 rpm. The conversion yields at 200 and 600 rpm were approx 80% and 90%, respectively. Above 400 rpm, the production yields of rapeseed FAME were unaffected by agitation speed.

We also organized a 30-L reaction system, which consisted of a 30-L reactor and three impellers with six flat blades, which were used to agitate the reactants. Figure 3 shows the effects of agitation speed on the conversion of rapeseed oil to biodiesel at different agitation speeds, ranging from 200 to 600 rpm, in 1% (w/w) KOH, at a methanol molar ratio of 1:10, and at 60°C, after reacting for 30 min. At less than 200 rpm, the rapeseed oil and methanol did not mix well. Approximately 89% of the rapeseed oil was converted at 200 rpm within 20 min, and approx 98% was converted at 400 rpm under the same conditions. Above 400 rpm, the production yield of the rapeseed FAME was unaffected by agitation speed.

Regarding the operating parameters for this process, reaction time constitutes one of the most important factors in the determination of production cost. Shorter reaction times with stable conversion are associated with less costly operation processes. Figure 4 shows the time-course of rapeseed oil transesterification in the 30-L reaction system, using different methanol-to-oil molar ratios. The conversion yield of the rapeseed oil in the 30-L reaction system provided a similar reaction pattern and yield, as was obtained in the 1-L reaction system (2). Within 30 min, approx 91.2% of the rapeseed oil was converted, at a methanol molar ratio of 1:6. Approximately 98.5% of the rapeseed oil was converted in 30 min at a molar ratio of 1:10. However, both conversions achieved equilibrium after 5 min.

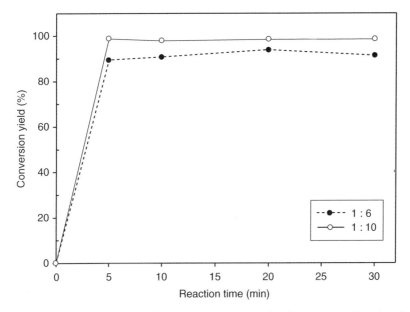

Fig. 4. Effect of methanol molar ratio on rapeseed oil transesterification in 30-L reaction system.

Several researchers have reported that the conversion of vegetable oils to FAME was above 80% within 5 min when a sufficient molar ratio was used *(2,16,17)*. Kim and Kang *(16)* reported that the optimum molar ratio was 1:6 at a reaction temperature of more than 60°C, in an experiment involving soybean oil with high unsaturated fatty acid content. However, rapeseed oil with high oleic acid content was found to have an optimal molar proportion of 1:6 to 1:8, and a reaction temperature below 60°C. Zmudzinska-Zurek and Buzdygan *(17)* reported that the complete factorial design of the KOH-catalyzed transesterification of refined rapeseed oil could be optimized at 50°C, 0.9% KOH, and 6:1 MeOH to oil ratio; the conversion yield was 80.3%.

Two-Stage Transesterification of Rapeseed Oil

Several processes have been developed for the production of biodiesel fuel via two-step transesterification *(9,18–20)*. Tanaka et al. *(18)* performed two-step transesterification using 6:1 to 30:1 molar ratios with alkali catalysis, in order to achieve a 99.5% conversion yield in his two-step reaction of oils and fats, including tallow, coconut oil, and palm oil. The first reaction was conducted at or near the boiling temperature of the lower alcohol, for a period of 0.5–2 h. The crude ester layer was then employed in a second reaction, which used 8–20% alcohol and 0.2–0.5% alkali catalyst, and was conducted for 5–60 min. Zhong *(19)* conducted a conversion with edible beef tallow containing 0.27% free fatty acids, using a 6:1 molar ratio of methanol to tallow, 1% NaOH, and a temperature of 60°C, for approx 30 min. After the separation of glycerol, the ester layer

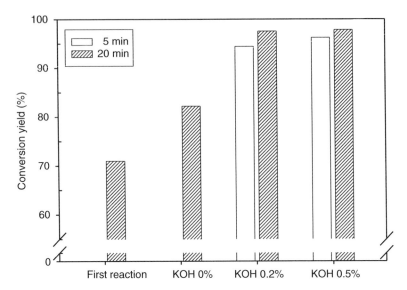

Fig. 5. Effect of reaction time and KOH concentration on second transesterification with addition of 1:1 molar ratio in 1-L reaction system.

was rereacted using 0.2% NaOH and 20% methanol at 60°C, for approx 1 h. Cvengroš and Cvengrošová (20) conducted a two-step reaction in one reactor, in which the oil to MeOH molar ratio was 1:3.5 for step I, and 1:0.95 for step II, with each step lasting 1.5–2 h. Also, Ahn et al. (9) carried out the transesterification of two steps in the VN process for the production of biodiesel, resulting in a reported 99% conversion rate.

In our previously mentioned trials, successful yields for the batch transesterification process of rapeseed oil were achieved under the following conditions: methanol molar ratio of 1:10, potassium hydroxide concentration of 1% (w/w), reaction temperature of 60°C, and a reaction time of within 30 min. This process consumed a large amount of methanol, which increased the manufacturing cost of the biodiesel, and also required a great deal of energy to recover the nonreacted methanol. To save the required amount of methanol and obtain a high conversion yield, we conducted two-stage transesterification of rapeseed oil with different methanol molar ratios and KOH concentrations. Figure 5 shows the effects of reaction time in the second reaction, using a molar ratio of 1:1 and different KOH concentrations (0–0.5% [w/w]). The first reaction was performed for 20 min with a 1:4.5 molar ratio and 1% (w/w) KOH, and resulted in a conversion yield of approx 71%. The second reaction was conducted with a 1:1 molar ratio and different KOH concentrations (0–0.5% [w/w]). The conversion yields were 82%, 97%, and 98% in 0%, 0.2%, and 0.5% (w/w) KOH after a reaction time of 20 min, respectively. Within 5–20 min, rapeseed FAME was produced at a concentration of 94.4–97.5%.

Figure 6 shows the effects of added KOH and methanol amounts on the conversion yields of rapeseed oil under the second transesterification

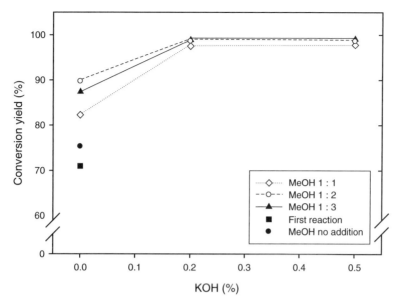

Fig. 6. Effect of added KOH on added MeOH molar ratio in second transesterification in 1-L reactor system.

conditions. The first reaction was performed with a 1:4.5 molar ratio and 1% (w/w) KOH, and achieved a conversion yield of approx 71%. In the second reaction without added KOH, the enhanced conversion of the rapeseed oil was observed to added amount of 1:2 molar ratio. Conversion yields were not significantly affected by amounts of added KOH of more than 0.2% (w/w) with different methanol molar ratios. Above 0.2% (w/w) addition of KOH, more than 98% of the rapeseed oil was converted, at methanol molar proportion of 1:1 to 1:3. According to our results, the second reaction scale-up condition in the two-stage transesterification of rapeseed oil was set to a methanol molar ratio of 1:1, and a 0.2% (w/w) amount of added KOH.

Figure 7 shows the conversion yields associated with the two-stage transesterification of rapeseed oil in the 30-L reaction system. In this system, the first reaction was characterized by a 1:4.5 molar ratio and 1% (w/w) of added KOH, and the second reaction was characterized by a 1:1 molar ratio and 0.2% (w/w) of added KOH. In the first reaction, approx 72% of the rapeseed oil was converted within 5 min, and was equilibrated to approx 73% for 30 min. After 30 min, during the first reaction, the reaction was discontinued and the upper product (mixture of FAME, nonreacted rapeseed oil, and methanol) was separated. The nonfinished product was then rereacted with methanol at a 1:1 molar ratio with 0.2% (w/w) added KOH at 60°C for 30 min. The conversion of rapeseed oil ultimately generated a yield of approx 98.5% under the second reaction conditions.

From our analysis of the refined products (rapeseed FAME; biodiesel) of the batch (one- and two-stage) transesterifications, we found that the

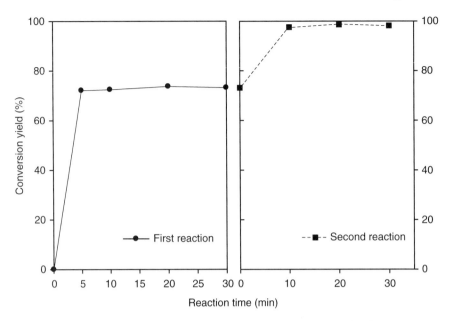

Fig. 7. Time-course of two-stage transesterification of rapeseed oil in 30-L reaction system. First reaction: 1:4.5 molar ratio and 1% (w/w) KOH; second reaction: 1:1 molar ratio and 0.2% (w/w) KOH.

purity of the products obtained was more than 99% after posttreatment, which included washing and centrifugation.

Conclusions

In this article, we have attempted to evaluate the efficacy of a method for the production of methyl ester of rapeseed oil via alkali-catalyzed batch (one- and two-stage) transesterifications. Approximately 89% and 98% of the rapeseed oil was converted at 200 and 400 rpm within 20 min, respectively, under the following conditions: 1% (w/w) potassium hydroxide, a methanol molar ratio of 1:10, and a reaction temperature of 60°C after 30 min of reaction in a 30-L reaction system. The rapeseed oil conversion yield showed a similar reaction pattern and yield in the 30-L reaction system, as was seen in the 1-L reaction system. For the two-stage transesterification of rapeseed oil in the 1-L reaction system, the first reaction was performed using a 1:4.5 molar ratio and 1% (w/w) of added KOH, and a conversion yield of 71% was achieved. In the second reaction, the conversion yield was not significantly affected by the amounts of added KOH of more than 0.2% (w/w) with different methanol molar ratios. With the addition of 0.2% (w/w) KOH, over 98% of the rapeseed oil was converted using methanol molar proportion of 1:1 to 1:3. In the 30-L, two-stage transesterification reaction system, 72% of the rapeseed oil was converted within 5 min, and was equilibrated to approx 73% for 30 min in the first reaction. The conversion of rapeseed oil was discontinued at a yield of above

98.5% in the second reaction condition, involving a 1:1 methanol molar ratio and 0.2% (w/w) of added KOH at 60°C for 30 min. According to our analysis of the final refined FAME product (biodiesel), the purity obtained was found to be more than 99% after posttreatment, which involved washing, centrifugation, and drying.

Acknowledgment

The authors would like to acknowledge the Korea Energy Management Corporation for their financial support of this study.

References

1. Kusdiana, D. and Saka, S. (2004), *Appl. Biochem. Biotechnol.* **113–116,** 781–791.
2. Jeong, G. T., Park, D. H., Kang, C. H., et al. (2004), *Appl. Biochem. Biotechnol.* **113–116,** 747–758.
3. Cvengroš, J. and Cvengrošová, Z. (2004), *Biomass Bioener.* **27,** 173–181.
4. Cardone, M., Mazzoncini, M., Menini, S., et al. (2003), *Biomass Bioener.* **25,** 623–636.
5. Lang, X., Dalai, A. K., Bakhshi, N. N., Reaney, M. J., and Hertz, P. B. (2001), *Biores. Technol.* **80,** 53–62.
6. Graboski, M. S. and McCormick, R. L. (1998), *Prog. Ener. Combust. Sci.* **24,** 125–164.
7. Krawczyk, T. (1996), *INFORM* **7,** 801–829.
8. Raheman, H. and Phadatare, A. G. (2004), *Biomass Bioener.* **27,** 393–397.
9. Ahn, E., Mittelbach, M., and Marr, R. (1995), *Sep. Sci. Technol.* **30(7–9),** 2021–2033.
10. Freedman, B., Pryde, E. H., and Mounts, T. L. (1984), *JAOCS* **61(10),** 1638–1643.
11. Nelson, L. A., Foglia, T. A., and Marmer, W. N. (1996), *JAOCS* **73(8),** 1191–1195.
12. De Oliveira, D., Di Luccio, M., Faccio, C., et al. (2004), *Appl. Biochem. Biotechnol.* **113–116,** 771–780.
13. Darnoko, D. and Cheryan, M. (2000), *JAOCS* **77(12),** 1263–1267.
14. Noureddini, H. and Zhu, D. (1997), *JAOCS* **74(11),** 1457–1463.
15. Ma, F., Clements, L. D., and Hanna, M. A. (1998), *Ind. Eng. Chem. Res.* **37,** 3768–3771.
16. Kim, H. S. and Kang, Y. M. (2001), *J. Korean Oil Chem. Soc.* **18(4),** 298–305.
17. Zmudzinska-Zurek, B. and Buzdygan, S. (2002), *Przemysl Chemiczny* **81(10),** 656–658.
18. Tanaka, Y., Okabe, A., and Ando, S. (1981), Method for the preparation of a lower alkyl ester of fatty acids, US Patent 4, 303–590.
19. Zhong, D. (1994), Master thesis, Food Science and Technology, University of Nebraska-Lincoln, USA.
20. Cvengroš, J. and Cvengrošová, Z. (1994), *J. Am. Oil Chem. Soc.* **71,** 1349–1352.

Optimization of Distilled Monoglycerides Production

LEONARDO VASCONCELOS FREGOLENTE,* CÉSAR BENEDITO BATISTELLA, RUBENS MACIEL FILHO, AND MARIA REGINA WOLF MACIEL

Laboratory of Separation Process Development (LDPS) School of Chemical Engineering, UNICAMP CP 6066-CEP 13081-970-Campinas, SP, Brazil; E-mail: leonardo@feq.unicamp.br

Abstract

Monoglycerides (MG) are emulsifiers widely used in food and pharmaceutical industries. Current industrial processes for MG production consist of the interesterification of triglycerides with glycerol (GL), in the presence of inorganic catalysts at high temperatures (>200°C). This reaction is known as glycerolysis and produces a mixture of approx 50% of MG. This level of concentration is suitable for many applications, although, for some specific uses like margarine, shortening, icing, and cream filling, require distilled MGs, which are purified MG (min. 90%) obtained by the molecular distillation process. Therefore, in this work, a 2^3 factorial design was employed to evaluate the effects of reaction parameters in the MG content after the interesterification reaction of refined soybean oil with GL in the presence of sodium hydroxide as catalyst. After that, the MG content in the reaction product was enhanced through the molecular distillation process in order to obtain distilled MG.

Index Entries: Glycerolysis; soybean oil; distilled monoglycerides; molecular distillation; short path distillation; factorial design.

Introduction

Many different types of lipid-based emulsifier can be applied to the food, cosmetic, and pharmaceutical industries. The manufacturer must select the one that is most suitable for each particular product, considering the physicochemical properties of the final product, cost, and availability of the emulsifier and its compatibility with other ingredients (1).

Monoglycerides (MG) are the predominant type of emulsifier, representing about 70% of the synthetic emulsifiers produced (2). Some researchers have developed three lipase-catalyzed routes to MG: (1) hydrolysis or alcoholysis of triglyceride (TG), (2) glycerolysis of TG, and (3) esterification or transesterification of glycerol (GL), considering the mild condition

---

*Author to whom all correspondence and reprint requests should be addressed.

requirements of the lipases (low temperatures and near neutral pH). Furthermore, they can explore the lipase fatty acid selectivity and regioselectivity for the primary vs secondary positions in the GL *(3)*.

However, industrially, MG are manufactured by batch or continuous reactions at temperatures greater than 200°C, using inorganic catalysts. Products with around 50% of MG content are achieved by direct esterification reaction, when the starting material is fatty acid or, by interesterification when the starting material is TG. As the dominant part of the manufacturing cost is the price of the feed material, interesterification is preferred, because acids are more expensive than TG. Direct esterification is frequently used when MG with a specified acid distribution is required.

The MG content obtained in these chemical reactions is suitable for many applications, although for some specific uses such as cake or icing, the mouth-melt of the product is critical. The use of a commercially prepared MG emulsifier could impact texture or mouthfeel of the product *(4)*, so that distilled MG are required, which are purified MG (min. 90%), normally obtained through the molecular distillation process.

Also known as short-path distillation, molecular distillation is characterized by a short exposure of the distilled liquid to the operating temperature and high vacuum *(5,6)*. Because of these features, besides the concentration of MG, this process has been widely applied in lipid areas. Some of these applications include the recovering of carotenoids from palm oil *(7)*, recovering of tocopherol from crude deodorizer distillate of soya oil *(8)*, purification and deodorization of structured lipids *(9)*, and the preparation of purified concentrates of polyunsaturated fatty acid *(10)*.

The aim of this study is to find the best conditions for the glycerolysis of refined soybean oil in a batch reactor, carried out at relatively low temperatures (190–210°C), and also to obtain distilled MG from the reaction products using molecular distillation process.

Methods

Determination of TG, Diglycerides, MG, Free Fatty Acids, and GL

For the determination of TG, diglycerides (DG), MG, free fatty acids, and GL high-performance size exclusion chromatography (HPSEC) was used according to Schoenfelder (2003) *(11)*. The chromatographic system consists of an isocratic pump, model 515 high-performance liquid chromatography (HPLC) Pump (Waters Inc., Milford, MA), a differential refractometer detector model 2410 (Waters), and an oven for columns thermostatted at 40°C by a temperature control module (Waters). The samples were injected using a manual injector, model rheodyne 7725i with a 20-µL sample loop. Two HPSEC columns Styragel HR 1 and HR 2 (Waters), with dimensions of 7.8 × 300 mm and particle size of 5 µm packed with styrene-divinylbenzene copolymer were connected in series. The mobile phase used was HPLC-grade tetrahydrofuran from Tedia Inc. (Fairfield, OH) and the flow rate was 1 mL/min. The typical pressure at this flow rate was 470 psi. All the standards were obtained from Supelco (Supelco, Bellefonte, PA). The data processing was done by the Millenium software 2010 Chromatography Manager Software from Waters Inc.

Analysis of Fatty Acid Composition

The fatty acid composition of the reaction products was determined by gas–liquid chromatography (GLC). Acylglycerols were converted into fatty acid methyl esters (FAME) according to Hartman (1973) *(12)*. The FAME mixture was analyzed by a Varian gas chromatograph model STAR 3600CX (Lexington, MA) equipped with a flame ionization detector and with a DB 23 column (30 m × 0.53 mm, J&W Scientific, Folsom, CA). Injector and detector temperatures were set at 250°C and 300°C, respectively. The carrier gas used was helium at 46 mL/min. Air and hydrogen flow rates were 334 and 34 mL/min, respectively. The program of the oven temperature was as follows: starting at 50°C for 2 min; from 50°C to 180°C at 10°C/min; 180°C was held for 5 min; from 180°C to 240°C at 5°C/min. Identification of different FAME was based on a reference standard mixture F.A.M.E. Mix C4-C24 (Supelco, PA).

Glycerolysis Reaction

In this work, to evaluate the effect of the mass ratio of GL to refined oil (GL/TG), and also the effects of the reaction temperature (T) and of the amount of catalyst (NaOH), a 2^3 factorial design with three central points was planned, as shown in Table 1. Experiments were carried out in a 250-mL glass-stoppered volumetric flask of 250 mL.

Initially, 50 g of reactants consisting of commercial refined soybean oil and GL (Labsynth, SP, Brazil) were fed in the batch reactor. Then, the mixture of reactants was heated by a stirring hot plate (Fisatom, SP, Brazil) to a defined temperature, in nitrogen atmosphere, under agitation. When the defined temperature was reached, a known amount of catalyst (NaOH, LabSynth, Brazil) was added to the reactor (reaction time = 0). In order to monitor the conversion of TG into MG and also to verify whether the reaction reached the equilibrium condition, samples were withdrawn from the reactor at different times. The mass ratio (GL/TG) of each run, as well as the amount of catalyst (amount of NaOH) and the reaction temperatures (T) studied are shown in Table 1.

Manufacturers avoid oils rich in unsaturated fatty acids, such as soybean oil, because at temperatures >200°C, they burn or polymerize causing a dark color, off-odor and burnt taste; so, usually, they use partially or fully hydrogenated fat. For these reasons, the experimental range of the reaction temperature was studied at relatively low levels, between 190°C and 210°C, and the fatty acid composition of the reaction products was accompanied. In case of significant degradation, the fatty acids composition of the reaction products may vary during the reaction time, what can

Table 1
Coded Levels and Real Levels (In Parentheses) of the Variables Studied and the Results Obtained in the Glycerolysis Reaction

Runs	NaOH (g)	T (°C)	GL/TG	% of MG[a] (t = 90 min)
1	−1 (0.07)	−1 (190)	−1 (0.18)	41.0
2	1 (0.13)	−1 (190)	−1 (0.18)	41.8
3	−1 (0.07)	1 (210)	−1 (0.18)	41.8
4	1 (0.13)	1 (210)	−1 (0.18)	40.7
5	−1 (0.07)	−1 (190)	1 (0.30)	48.3
6	1 (0.13)	−1 (190)	1 (0.30)	51.8
7	−1 (0.07)	1 (210)	1 (0.30)	53.8
8	1 (0.13)	1 (210)	1 (0.30)	54.5
9	0 (0.10)	0 (200)	0 (0.24)	48.6
10	0 (0.10)	0 (200)	0 (0.24)	48.6
11	0 (0.10)	0 (200)	0 (0.24)	49.6

[a]Normalized peak area.

be detected by the GLC analyses. Furthermore, if significant polymerization occurs, the molecules of high-molecular weight formed in the reaction may be detected by the HPSEC analysis, as this method of analysis identifies components by their sizes.

Usually, in the factorial design, the variables levels are denoted by coded numbers. The superior levels are represented by +1 and the inferior levels by −1. The central points, which are used for the error estimation, are denoted by 0. They are replicates in the center of the experimental range that permit the evaluation of the repeatability of the experiments. The variation between them reflects the variability of whole design (13). The response analyzed was the MG concentration at the equilibrium condition.

As GL is not totally soluble in the mixture, even in a system with good agitation, the percentage of GL varies considerably in different points of the reactor, causing sampling problem. To avoid this problem, the percentage of MG (in the equilibrium condition) reported in Table 1 was obtained by normalizing the peak areas of TG, DG, and MG. Furthermore, this procedure permits the comparison among runs, with different initial GL concentration.

Molecular Distillation

Molecular distillation, also known as short path distillation, is a separation process characterized by a short exposure of its distilled liquid to the operating temperature, high vacuum in the distillation space (distance between the evaporator and condenser), and a small distance between these elements. The separation principle is the vacuum (enabling the molecules to evaporate from the evaporator to the condenser) and, in case of a centrifugal distillator, the centrifuge force (promoting a thin film on the evaporator

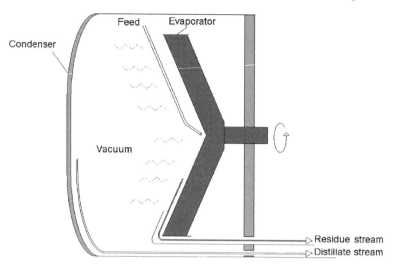

Fig. 1. Simplified scheme of a centrifugal molecular distillator.

surface). Two product streams are generated: Distillate (rich in the molecules that escape from the evaporator and reach the condenser) and Residue (rich in the heavier molecules that remain in the evaporator).

In this work, the equipment used was a centrifugal distillator from Myers Vacuum Inc. (Kittanning, PA), with an evaporator area of 0.0046 m². The system was operated at a pressure of 1.6×10^{-4} bar. The starting material was fed at 60°C in the center of the evaporator (as indicated in Fig. 1), with a defined flow rate (Q) and evaporator temperatures (TEV). The condenser temperature is fixed at 55°C.

Results and Discussion

Glycerolysis Optimization

As can be seen in Fig. 2, after 1 h, the system reached the equilibrium condition, for all the runs. It can be noted that the higher reaction yields were obtained in the runs with GL/TG = 0.30 (runs 5–8). Analyzing the three central points (runs 9–11), the good repeatability of the experiments can be seen, as their results are very close.

Figure 3A shows the time-course of the glycerolysis for run 2, representing the runs carried out with GL/TG = 0.18 and Fig. 3B shows the time-course of the glycerolysis for run 7, representing the runs carried out with GL/TG = 0.30.

It can be seen that, in the experiments carried out with the mass ratio GL/TG = 0.30, the difference between the MG and DG concentrations at the equilibrium condition is approx 20%, whereas in the experiments carried out with GL/TG = 0.18, these concentration values are very close. Furthermore, it can be noted that the TG equilibrium concentration in the

Optimization of Distilled MG Production

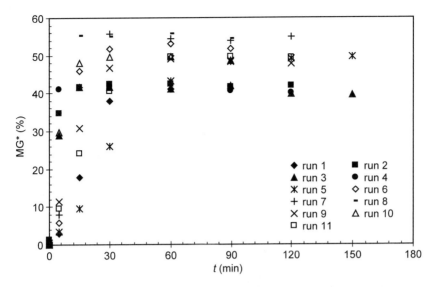

Fig. 2. MG concentration during the glycerolysis reaction. *Normalized peak areas of TG, DG, and MG.

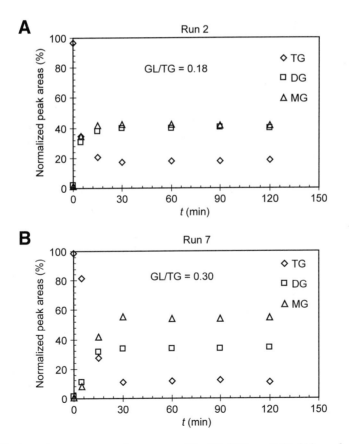

Fig. 3. Comparison among the runs with GL/TG = 0.18 (**A**) and runs with GL/TG = 0.30 (**B**).

Table 2
Main Effects of GL/TG, T, Amount of NaOH, and the Binary Interaction Effect Involving These Factors

	Effect (%)	Standard error	t (2)	p-value
(1)NaOH	0.96	0.204	2.388	0.1395
(2)T^a	1.97	0.204	4.838	0.0401
(3)GL/TG^a	10.77	0.204	26.393	0.0014
1 by 2	−1.20	0.204	−2.878	0.1025
1 by 3	1.17	0.204	2.756	0.1103
2 by 3^a	2.13	0.204	5.205	0.0350

^{a}Significant effects at 95% of confidence.

runs with GL/TG = 0.18 is approximately twice the TG equilibrium concentration when GL/TG = 0.30. This confirms that, at the experimental conditions studied, the higher yields were obtained with the GL/TG = 0.30.

The main effects of the three variables studied and the binary interaction effect involving these factors are shown in Table 2. It can be seen that, at 95% of confidence, GL/TG, T, and the interaction between T and GL/TG presented significant effects (p-value < 0.05). As shown before, the effect of the GL/TG is very relevant in the experimental condition studied, presenting an effect of 10.77%, what means that, on an average, the percentage of MG increases 10.77% when the level of the GL/TG is increased from −1 to +1.

The effect of T is positive and also significant, but much lower than the GL/TG effect. This may be owing to the narrow T experimental range studied. The interaction effect between T and GL/TG is significant at 95% of confidence, but also lower than the GL/TG effect. As expected, the main effect of the amount of catalyst (NaOH) is not significant at this level of confidence, as the catalyst concentration does not displace the equilibrium condition toward any side (product or reagent sides).

Because all the variables studied showed positive main effects, new experiments were carried out in order to explore a new experimental range, confirming the results obtained in the experimental design. Thus, higher levels of the mass ratio GL/TG was studied (Fig. 4) as its effect was significant at 95% of confidence level (the highest effect among the studied variables). The temperature was kept at 210°C (the highest studied level in the experimental design). Although its effect is positive and significant at 95% of confidence level, in the experimental range studied, temperatures >210°C were not studied because it may cause considerable polymerization reactions. The amount of catalyst (NaOH) added in the system did not affect significantly the equilibrium condition, therefore it was fixed in 0.13 g for 50 g of reactants.

Figure 4 shows the MG concentration (normalized peak areas of TG, DG, and MG) in the equilibrium condition as a function of the GL/TG ratio. It can be noted that, in this new experimental range explored

Optimization of Distilled MG Production

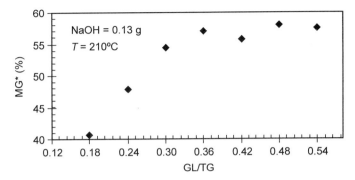

Fig. 4. MG concentration in the equilibrium condition as a function of the GL/TG ratio. *Normalized peak areas of TG, DG, and MG.

Table 3
Fatty Acids Composition of Acylglycerols

t (min)	Fatty acid composition (%)[a]								
	16:0	18:0	18:1	18:2	18:3	20:0	20:1	22:0	24:0
0	12.1	3.3	21.8	54.2	5.4	0.4	0.2	0.5	0.2
30	12.4	3.8	22.3	53.2	5.2	0.4	0.2	0.5	0.2
60	12.6	3.4	22.5	53.2	5.1	0.4	0.2	0.5	0.2
90	12.8	3.7	22.2	52.4	5.1	0.4	0.2	0.5	0.2

[a]16:0, palmitic acid; 18:0, stearic acid; 18:1, oleic acid; 18:2, linoleic acid; 18:3, α-linolenic acid; 20:0, arachidic acid; 20:1, eicosenoic acid; 22:0, behenic acid; 24:0, lignoceric acid. GL/TG = 0.48, T = 210°C, NaOH = 0.13 g, mass of reagents = 50 g.

(GL/TG values >0.30), there is a small increase in the equilibrium MG concentration. However, it seems that for GL/TG values greater than 0.36, the equilibrium MG concentration remains the same. The maximum MG concentration obtained (as normalized peak area) in the studied experimental range is around 57%.

Table 3 contains experimental data on fatty acid composition of acylglycerols at different times of reaction. One can see that, at the experimental conditions studied, the fatty acid composition changes lightly with time, indicating just a moderate occurrence of polymerization. The concentrations of linoleic acid and α-linolenic acid, both polyunsaturated, decrease because they are more susceptible to degradation. This partial degradation causes a change in the color of the oil, which becomes darker as well as a change in the odor. Despite the moderate degradation of the polyunsaturated, HPSEC analysis did not detect polymers formed.

Molecular Distillation Optimization

Fregolente et al. (2005) *(14)* have shown a study to find optimized operating conditions of the molecular distillation in order to enrich the distillate

Table 4
Properties of the Commercial MG Used as Starting Material

Product name	Source	Acid value (mg KOH/g)	Iodine value (cg I 2/g)	MG content (%)	Physical form
BRAWAX MGS C	Vegetable	1.3	58.3	43.6	Waxy solid

Table 5
Central Composite Design Carried Out to Optimize the Molecular Distillation Process

TEV (°C)	Q (mL/min)	MGD (%)	MGR (%)
−1(215)	−1(6.5)	78.1	37.5
+1(285)	−1(6.5)	70.9	25.7
−1(215)	+1(13.5)	56.3	41.8
+1(285)	+1(13.5)	77.7	36.1
−1.41(200)	0(10)	56.9	41.5
1.41(300)	0(10)	65.2	29.3
0(250)	−1.41(5)	74.8	26.8
0(250)	1.41(15)	68.8	41.0
0(250)	0(10)	75.2	37.4
0(250)	0(10)	77.5	36.6

stream in MG, starting from a commercial MG. Its properties are shown in Table 4. The composition of this commercial MG is similar to the composition of the products obtained in the glycerolysis products (40–50%).

The variables studied were the evaporator temperature (TEV) and the feed flow rate (Q), because they are very important process variables in the molecular distillation process. The experimental range of these variables was chosen according to previous experience. Values of feed flow rate <4 mL/min may not be high enough to form a uniform thin film on the evaporator surface. This uniform thin film promotes efficient mass and energy transfers (15). For feed flow rate values greater than 15 mL/min, it was noted that the system operated with low effectiveness, as the residence time of the molecules on the evaporator is too low. Therefore, the feed flow rate varied from 5 to 15 mL/min. The variation range of TEV was from 200 to 300°C (Table 5).

To find the optimized operating conditions, response surface methodology (RSM) was applied, that is a set of mathematical and statistical methods developed for modeling phenomena and finding combinations of a number of experimental factors that will lead to optimum responses (16). Usually, in the RSM, simple coded models such as linear and quadratic are fitted. In this work, independent and dependent variables were fitted to a second-degree polynomial equation (Eq. 1), where y is the estimated response (MG concentration in the distillate stream [MGD], or MG concentration in the residue stream [MGR]), b_0 is a constant, b_{ij} are the coefficients for each term and x_i are the independent factors in coded values (x_1 corresponds to the coded value for TEV and x_2 corresponds to the coded value for Q).

$$y = b_0 + b_1 x_1 + b_2 x_2 + b_{11} x_1^2 + b_{22} x_2^2 + b_{12} x_1 x_2 \qquad (1)$$

Analyzing Eq. 1, which is a quadratic model with two variables, it can be seen that it contains six parameters, so that the number of combinations of the independent variable levels must be >6, as it is not possible to predict values when the number of equation parameters is higher than the number of independent variable levels. Thus, a factorial design consisting of 2^2 trials plus a star configuration (four axial points) with three central points was carried out. The distance of the axial points from the central point is calculated from the equation, $\alpha = (2^n)^{1/4}$, where α is the distance of the axial points and n is the number of independent variables (17). This kind of factorial design, also known as central composite design, is suitable for the fit of Eq. 1, because for two independent variables, it contains nine different combinations of the independent variable levels.

The quality of the fitted models was evaluated by the analysis of variance (ANOVA), based on F-test (18) and on the percentage of explained variance, which provides a measurement of how much of the variability in the observed response values could be explained by the experimental factors and their interactions (19).

The fitted coded models for the MGD and MGR are shown in Eqs. 2 and 3, respectively, in which, all the coefficients of Eq. 1 were considered.

$$\text{MGD} = 75.27 + 3.25 \times \text{TEV} - 6.04 \times \text{TEV}^2 - 2.94 \times Q \\ - 0.66 \times Q^2 + 7.13 \times \text{TEV} \times Q \qquad (2)$$

$$\text{MGR} = 36.88 - 4.37 \times \text{TEV} - 0.59 \times \text{TEV}^2 + 4.35 \times Q \\ - 1.32 \times Q^2 + 1.51 \times \text{TEV} \times Q \qquad (3)$$

Through the ANOVA, shown in Table 6, it can be concluded that there is no evidence of lack of fit for the fitted models, since the calculated F-value (lack of fit/pure error) are lower than the critical F-value ($F_{0.95,3,2} = 19.16$) at 95% confidence, for both models. Furthermore, the results show that the model for the MG concentration in the residue (MGR), Eq. 3, is predictive in the experimental conditions studied, because the percent of explained variance is high (98.53%) and the calculated F-value (regression/residual) is more than 13 times higher than the critical F-value at 95% of confidence ($F_{0.95,5,5} = 5.05$). As a practical rule, the regression can be considered useful to predict values when the calculated F-value (regression/residual) is more than 10 times higher than the critical F-value (20).

As can be seen in Fig. 5, obtained from Eq. 3, at low feed flow rate, the MGD increases up to a maximum of 80% approx, and starts decreasing at TEV higher that 250°C owing to the significant amount of DG that leave the distillator in the distillate stream at these conditions. The feed flow rate (Q) is important, because at high Q, the residence time of the molecules on

Table 6
Analysis of Variance for the Fitted Models[a]

Source of variation	Model	Sum of square	Degrees of freedom	Mean square	F-ratio
Regression	Eq. 2 (MGD)	570.002	5	114.000	10.57[b]
	Eq. 3 (MGR)	323.592	5	64.718	66.93[b]
Residual	Eq. 2 (MGD)	53.919	5	10.783	–
	Eq. 3 (MGR)	4.833	5	0.967	–
Lack of fit	Eq. 2 (MGD)	44.262	3	14.754	3.06[c]
	Eq. 3 (MGR)	4.373	3	1.458	6.34[c]
Pure error	Eq. 2 (MGD)	9.657	2	4.829	–
	Eq. 3 (MGR)	0.460	2	0.230	–
Total	Eq. 2 (MGD)	623.921	10		
	Eq. 3 (MGR)	328.425	10		

[a]MGD, percent of explained variance, 91.36; percent of explicable variance, 98.45; MGR, percent of explained variance, 98.53; percent of explicable variance, 99.86
[b]F-ratio (regression/residual)
[c]F-ratio (lack of fit/pure error).

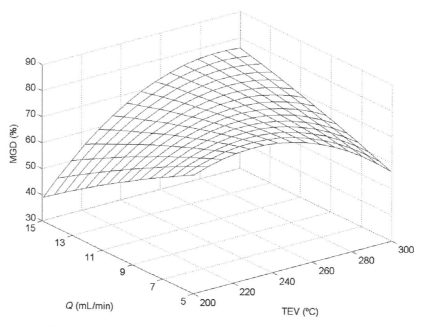

Fig. 5. Response surface for the MG concentration in the distillate stream.

the evaporator surface is low, so that the DG evaporation may not be high enough to dilute the MGD.

Distillation of the Glycerolysis Reaction Product

To obtain enough material for the distillation of the glycerolysis product in a optimized operating condition, the 50 g experiments was scaled up to 500 g with the following conditions: GL/TG = 0.48, T = 210°C and

Table 7
Composition of the Two Phases Separated, After the Glycerolysis Reaction in the 500 g System and the Compositions of the Streams Obtained in the Molecular Distillations

Material analyzed	TG (%)	DG (%)	MG (%)	FFA (%)	GL (%)
Superior phase (acylglycerols)	8.22	30.99	50.13	0.85	9.81
Inferior phase (GL)		0.34	1.84	0.9	96.91
Distillate stream of the first distillation		3.11	80.81	0.91	15.17
Residue stream of the second distillation		3.78	90.88	0.81	4.53

NaOH = 1.3 g. When the system reached the equilibrium condition, the reaction mixture was transferred to a separatory funnel in order to separate the insoluble GL. The composition of the two phases formed is shown in Table 7.

It is clear that the MG losses were small, as the MG concentration in the inferior phase is low (1.84%). Furthermore, it can be seen that the MG concentration level reached is 50.13, a good value when it is compared with reactions carried out at higher temperatures.

The superior phase was submitted to a first distillation at TEV = 250°C and Q = 5 mL/min, because according to Fig. 5, this is a suitable condition for the concentration of MG in the distillate stream. The MG concentration obtained in the distillate stream was 80.81 as shown in Table 7, with a MG recovery of 56%. It can be noted that the GL concentration in the distillate is high (15.17%). Therefore, a second distillation was carried out at TEV = 170°C and Q = 5 mL/min in order to remove the GL from the distillate stream. The MG concentration obtained in the residue stream of the second distillation is 90.88%, which can already be considered a commercial distilled monoglyceride (min. 90%).

Besides the MG content, the color of the product, as well as odor and taste, are essential properties to be considered. Therefore, it can be seen that the change in the color of the reaction products during the glycerolysis does not compromise the quality of the distilled MG obtained, as it is practically colorless. Furthermore, one can note that the odor was effectively removed by the molecular distillation process, as verified by Kuhrt et al. (1950) *(21)*.

Concluding Remarks

This work represents an important contribution for the manufacturers of MG, as it shows an optimization study of the MG production from soybean oil and also a study of the usual process for MG concentration.

After 60 min, the glycerolysis of the soybean oil reached the equilibrium condition for all the runs. Among the three variables studied (T, GL/TG, and NaOH), the GL/TG is the most relevant, in the experimental range studied. Its main effect is 10.77%. The maximum of MG concentration obtained in the glycerolysis reaction was around 50%, when the independent variables T and NaOH were fixed in the their superior levels studied and the GL/TG ratio was greater than 0.36.

For effective separation of the reaction products, careful attention must be given to feed flow rates (Q) and evaporator temperature (TEV). According to the results obtained in this work, with only one distillation step, the max. MGD was 80.81%, at evaporator temperature = 250°C and feed flow rate = 5 mL/min. From this stream, a second distillation step was carried out in order to remove GL of the distillate stream (evaporator temperature = 170°C and feed flow rate = 5 mL/min), obtaining a colorless product with 90.88% of MG, which has the desired characteristics of Distilled MG.

Acknowledgment

The authors are grateful to the financial support of CAPES, CNPq, and FAPESP.

References

1. Akoh, C. C. and Min, D. B. (1998), *Food Lipids: Chemistry, Nutrition, and Biotechnology*, Marcel Dekker, New York.
2. Ferreira-Dias, S., Correia, A. C., Baptista, F. O., and Fonseca, M. M. R. (2001), *J. Mol. Catal B: Enzymatic.* **11**, 669.
3. Bornscheuer, U. T. and Kazlauskas, R. J. (1999), Hydrolases in Organic Synthesis: Regio- and Stereoselective Biotransformations, Wiley-VCH, Weinheim.
4. Bailey, A. E. (1996), *Bailey's Industrial Oil & Fat Products*, 5th ed., Edited by Y. H. Hui, John Wiley, New York, pp. 569–601.
5. Mikov, M., Lutisan, J., and Cvengros, J. (1997), *Sep. Sci. Technol.* **32**, 3051–3066.
6. Batistella, C. B., Maciel, M. R. W., and Maciel Filho, R. (2000), *Comput. Chem. Eng.* **24**, 1309–1315.
7. Batistella, C. B., Moraes, E. B., Maciel Filho, R., and Maciel, M. R. W. (2002), *Appl. Biochem. Biotechnol.* **98**, 1149–1159.
8. Moraes, E. B., Batistella, C. B., Alvarez, M. E. T., Maciel Filho, R., and Maciel, M. R. W. (2004), *Appl. Biochem. Biotechnol.* **113**, 689–711.
9. Xu, X., Jacobsen, C., Nielsen, N. S., Heinrich, M. T., and Zhou, D. (2002), *Eur. J. Lip. Sci. Technol.* **104**, 745–755.
10. Breivik, H., Haraldsson, G. G., and Kristinsson, B. (1997), *J. Am. Oil Chem. Soc.* **74**, 1425–1429.
11. Schoenfelder, W. (2003), *Eur. J. Lip. Sci. Technol.* **105**, 45–48.
12. Hartman, L. and Lago, R. C. A. (1973), *Lab. Practice* **22**, 475–476.
13. Carvalho, C. M. L., Serralheiro, M. L. M., Cabral, J. M. S., and Aires-Barros, M. R. (1997), *Enzyme Microb. Technol.* **21**, 117–123.
14. Fregolente, L. V., Batistella, C. B., Maciel Filho, R., and Maciel, M. R. W. (2005), *J. Am. Oil Chem. Soc.* **82**, 673–678.
15. Batistella, C. B. and Maciel, M. R. W. (1998), *Comput. Chem. Eng.* **22**, S53–S60.
16. Ferreira-Dias, S., Correia, A. C., Baptista, F. O., and Fonseca, M. M. R. (2001), *J. Mol. Catal B: Enzymatic.* **11**, 669–711.
17. Khuri, A. I. and Cornell, J. A. (1987), *Response Surface Design and Analyses*, Marcel Dekker, New York.

18. Box, G. E. P., Hunter, W. G., and Hunter, J. S. (1978), *Statistics for Experimenters*, Wiley, New York, pp. 510–539.
19. Burkert, J. F. M., Maugeri, F., and Rodrigues, M. I. (2004), *Bioresour. Technol.* **91,** 77–84.
20. Barros Neto, B., Scarminio, I. S., and Bruns, R. E. (2003), *Como Fazer Experimentos*, 2nd ed., Editora da UNICAMP, São Paulo, Brazil, pp. 251–266 (in Portuguese).
21. Kuhrt, N. H., Welch, E. A., and Kovarik, F. J. (1950), *J. Am. Oil Chem. Soc.* **27,** 310–313.

Production of Lactic Acid From Cheese Whey by Batch and Repeated Batch Cultures of *Lactobacillus* sp. RKY2

HYANG-OK KIM,[1] YOUNG-JUNG WEE,[2] JIN-NAM KIM,[1] JONG-SUN YUN,[3] AND HWA-WON RYU*,[2]

[1]*Department of Material Chemical and Biochemical Engineering, Chonnam National University, Gwangju 500-757, Korea;* [2]*School of Biological Sciences and Technology, Chonnam National University, Gwangju 500-757, Korea, E-mail: hwryu@chonnam.ac.kr; and* [3]*Biohelix, Naju, Jeonnam 502-811, Korea*

Abstract

The fermentative production of lactic acid from cheese whey and corn steep liquor (CSL) as cheap raw materials was investigated by using *Lactobacillus* sp. RKY2 in order to develop a cost-effective fermentation medium. Lactic acid yields based on consumed lactose were obtained at more than 0.98 g/g from the medium containing whey lactose. Lactic acid productivities and yields obtained from whey lactose medium were slightly higher than those obtained from pure lactose medium. The lactic acid productivity gradually decreased with increase in substrate concentration owing to substrate and product inhibitions. The fermentation efficiencies were improved by the addition of more CSL to the medium. Moreover, through the cell-recycle repeated batch fermentation, lactic acid productivity was maximized to 6.34 g/L/h, which was 6.2 times higher than that of the batch fermentation.

Index Entries: Corn steep liquor; lactic acid; *Lactobacillus*; lactose; whey.

Introduction

Lactic acid has numerous applications in food, chemical, textile, pharmaceutical, and other industries *(1)*. Recently, there has been a great demand for lactic acid, because it can be used as a monomer for the production of biodegradable polymer, polylactic acid (PLA), which can be alternative to synthetic polymers derived from petroleum resources *(2)*. In 1999, the annual world production of lactic acid was estimated to be approx 80,000 t produced both by chemical synthesis and by biological fermentation processes *(1)*. However, a number of applications of lactic acid currently resulted in a significant increase in its demand. While the

*Author to whom all correspondence and reprint requests should be addressed.

only racemic DL-lactic acid is produced through a chemical synthesis, a desired stereoisomer, i.e., an optically pure L(+)- or D(-)-lactic acid, could be produced through a fermentative production from renewable resources, if the proper microorganisms would be chosen for lactic acid fermentation *(3)*. The optical purity of lactic acid is important to the physical properties of PLA. Especially, L(+)-lactic acid of high purity is polymerized to a high crystal polymer, that is suitable for fiber and oriented film. Also as an optically active material, L(+)-lactic acid is expected to be useful for the production of liquid crystals *(4)*. However, lactic acid fermentation process generally requires a complex basal medium, which may result in an increased production cost. Nevertheless, many studies on the production of lactic acid by the lactic acid bacteria were mainly focused on producing from the dextrose-based media and/or expensive nutrients such as yeast extract *(3,5)*. Therefore, the studies on alternative and low-cost media for lactic acid fermentation will be needed owing to its industrial feasibility and an economic consideration. The economics of lactic acid fermentation would be typically improved by using cheap raw materials.

Whey is a major byproduct of dairy industry, and it contains approx 60–65% (w/v) of lactose and some moieties of protein, fat, and mineral salts. The worldwide production of whey is approx 120×10^6 t/yr and its greater portion remains unutilized, which causes an environmental pollution as a result of high biochemical oxygen demand (BOD, 40,000–60,000 ppm) and chemical oxygen demand (COD, 50,000–80,000 ppm). More than 90% of whey BOD is caused by lactose moiety in the whey. In order to reduce the BOD level and to acquire some useful compounds, this nutrient-rich whey can be utilized for the production of lactic acid by bacteria as a cheap carbohydrate source *(6–12)*. However, lactic acid bacteria have a complex nutrient requirement because they have a limited capacity to synthesize B-vitamin and amino acids *(1)*. Therefore, for complete conversion of lactose to lactic acid, the supplementation of nitrogen sources such as yeast extract or corn steep liquor (CSL) is needed *(13)*.

CSL is a byproduct from corn milling industry, which has been used as an inexpensive nutrient source for several fermentations. In the production of lactic acid, though CSL seems to negatively affect the separation and purification of the produced lactic acid and to reduce the productivity, it should be an attractive source for the economical production of lactic acid as mentioned before *(14,15)*. This study was mainly focused on the utilization of whey lactose as a substrate for the production of lactic acid by batch culture of *Lactobacillus* sp. RKY2. In addition, the effects of CSL on lactic acid fermentation using cheese whey were also investigated. The cell-recycle repeated batch production of lactic acid using cheese whey and CSL as raw materials was tried in order to further enhance the productivity of lactic acid.

Materials and Methods

Microorganism

Lactobacillus sp. RKY2, which is stocked in the Korea Collection for Type Cultures (KCTC) with the accession number KCTC 10353BP *(16–18)*, was used throughout this study. The stock cultures were maintained at –20°C in 5-mL vial containing 50% (v/v) glycerol and culture medium composed of 30 g glucose, 10 g yeast extract, 2 g $(NH_4)_2HPO_4$, and 0.1 g $MnSO_4$ per 1 L of deionized water. After three consecutive transfers to 20- mL vial containing the fresh medium, the final culture was transferred to 50-mL vial with 40 mL of the fresh medium, which was then used for the inoculation after incubation for 12 h.

Preparation of Medium

Whey powder containing 60–65% (w/v) lactose was obtained from Samick Dairy Industry (Gimje, Korea). It was dissolved in deionized water to attain the desired lactose concentration, and the pH was adjusted to 4.0 by adding 10 M HCl, which was then heated to 100°C for 10 min and cooled to the room temperature *(19)*. The resulting whey solution was centrifuged at 15,540g to remove a solid, and the supernatant was used for the fermentation. Unless otherwise indicated, whey solution was supplemented with 30 g/L CSL, 1 g/L yeast extract, 2 g/L $(NH_4)_2HPO_4$, and 0.1 g/L $MnSO_4$. Whey solution and other components were autoclaved separately at 121°C for 15 min.

Batch Fermentation

The batch fermentations were carried out in a 2.5-L jar-fermenter (KF-2.5 L; Kobiotech, Incheon, Korea) with 1-L working volume. The culture temperature was controlled at 36°C and the agitation speed was adjusted to 300 rpm. The culture pH was maintained at 6.0 by automatically adding 10 M NaOH.

Cell-Recycle Repeated Batch Fermentation

The repeated batch fermentation with cell-recycle was performed in the same fermenter stated above in order to improve the lactic acid productivity and nutrient economics. A hollow fiber unit (HUF 1010-BPN30, Chemicore Inc., Daejeon, Korea) was used for cell recycling, and it contained 100 polysulfone hollow fiber membranes. The internal diameter and length of the module were 32 and 300 mm, respectively. A nominal molecular weight cutoff of the membranes was 300 kDa and the total filtration area was 0.07 m^2. A peristaltic pump was used for recycling the culture broth through the hollow fiber filtration unit with a flow rate of approx 100 mL/min. The hollow fiber filtration unit was sterilized with 200 ppm sodium hypochlorite for 12 h, and then washed with sterile water

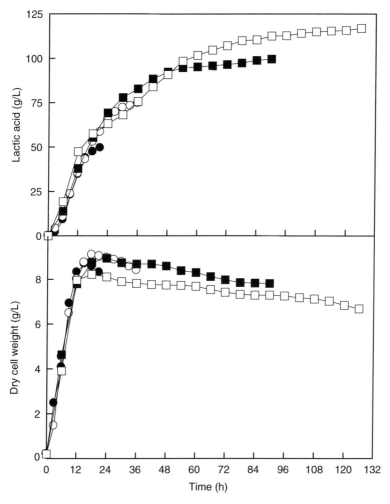

Fig. 1. Lactic acid production and cell growth from various concentrations of whey lactose by batch culture of *Lactobacillus* sp. RKY2. Symbols: -●-, whey lactose 50 g/L; -○-, whey lactose 75 g/L; -■-, whey lactose 100 g/L; -□-, whey lactose 125 g/L.

to remove the remainders of sodium hypochlorite before each fermentation experiment. In cell-recycle repeated batch fermentations, once the sugar was completely consumed at the former batch run, 90% (v/v) of the culture broth was taken out of the fermenter through the hollow fiber filtration unit. Then, the same volume of fresh medium was fed into the fermenter. The subsequent batch fermentations were performed by the same manner as described earlier.

Analyses

Cell concentration was determined turbidimetrically by measuring the optical density at 660 nm using a spectrophotometer (UV-Vis 1700, Shimadzu, Kyoto, Japan). The optical density was converted to dry cell weight through an appropriate calibration curve. The samples obtained at different time intervals were centrifuged at 17,620g. The resulting supernatants were used

Table 1
Effect of Lactose Concentrations on Lactic Acid Produced, Residual Lactose,
Maximal Dry Cell Weight, Yield, and Productivity in Batch Culture
of Lactobacillus sp. RKY2

Substrate	Initial lactose (g/L)	Residual lactose (g/L)	Lactic acid (g/L)	Max. dry cell weight (g/L)	Yield (g/g)[a]	Productivity (g/L/h)
Whey lactose	50	0	49.45	8.69	0.989	2.355
	75	0	74.51	9.09	0.988	2.07
	100	0	99.35	8.94	0.987	1.106
	125	6.3	116.92	8.22	0.985	0.928
Pure lactose	50	0	49.4	7.71	0.987	2.352
	75	0	74.32	8.18	0.985	2.064
	100	3.2	95.5	8.25	0.978	1.061
	125	9.9	109.96	7.76	0.955	0.873

[a]The yield of lactic acid was calculated on the basis of the amount of consumed lactose in whey.

for the analysis of lactic acid and lactose. Lactose concentration was measured by DNSA methods using lactose as a standard (20). Lactic acid was analyzed by a high-performance liquid chromatography equipped with an Aminex HPX-87H ion-exclusion column (300 × 7.8 mm; Bio-Rad, CA) under the following conditions: column temperature, 35°C; mobile phase, 5 mM H_2SO_4; flow rate, 0.6 mL/min; detection, UV 210 nm.

Results and Discussion

Influence of Initial Whey Lactose Concentration

In an attempt to evaluate the influence of whey lactose concentrations on lactic acid fermentation by batch culture of Lactobacillus sp. RKY2, the medium containing 50, 75, 100, and 125 g/L of whey lactose was tested. The fermentations were conducted in a 2.5-L jar-fermenter with 1-L working volume at 36°C and 200 rpm. As shown in Fig. 1, lactic acid production curves followed the similar pattern and the final lactic acid produced increased with increasing the initial whey lactose concentration. The maximal lactic acid (116.92 g/L) was obtained after 126 h of fermentation at an initial whey lactose concentration of 125 g/L. The lactic acid yields based on consumed lactose in cheese whey were 0.989, 0.988, 0.987, and 0.985 g/g at 50, 75, 100, and 125 g/L of whey lactose, respectively (Table 1). The lactic acid productivities usually decreased with increasing the initial whey lactose concentration. Table 1 shows also the fermentation parameters, such as final lactic acid produced, lactic acid yields, and volumetric productivity, at various concentrations of whey lactose. However, when the medium was supplemented with 125 g/L of whey lactose, more than 10 g/L of whey lactose remained even after 126 h of fermentation without being used, which is probably owing to substrate and product inhibitions. In addition, when the medium was supplemented with 100 g/L of whey lactose, the fermentation time was severely prolonged. This was expected because there was also a rapid decrease in dry cell weight. Dry cell weight usually increased with increasing the initial whey lactose concentration up to 75 g/L, but then decreased beyond this value. Although Büyükkileci and Harsa (6) reported that there was no substrate inhibition in lactic acid production from whey by batch culture of *Lactobacillus casei*, *Lactobacillus* sp. RKY2 used in this study was slightly inhibited by whey lactose above 100 g/L.

Influence of Initial Pure Lactose Concentration

To investigate the effect of initial pure lactose concentration on lactic acid fermentation by batch culture of *Lactobacillus* sp. RKY2, the medium containing 50, 75, 100, and 125 g/L of pure lactose was tested. The fermentation of pure lactose showed similar profiles of lactic acid production and cell growth as compared with the fermentation of whey lactose (Fig. 2). In cases of lactic acid yields and productivities, these values were almost similar when 50–75 g/L of whey lactose and pure lactose were used, but lactic acid yields and productivities are slightly higher when more than 100 g/L of whey lactose was used as a substrate (Table 1). As shown in Fig. 2, the final lactic acid produced increased with increasing the initial pure lactose concentration. The maximal lactic acid (109.96 g/L) was obtained after 126 h of fermentation at an initial pure lactose concentration of 125 g/L. The lactic acid yields based on consumed lactose were obtained to 0.987, 0.985, 0.987, and 0.955 g/g at 50, 75, 100, and 125 g/L of pure lactose, respectively (Table 1). In all experiments, however, a nutrient supplement is needed for complete conversion of lactose to lactic acid. Otherwise, it may result in incomplete utilization of lactose and prolonged fermentation.

Influence of CSL Concentration

Lactic acid bacteria are generally fastidious organisms, which require complex nutrients such as amino acids and vitamins for cell growth. To investigate the effect of CSL as a cheaper nitrogen source on lactic acid fermentation, 15–60 g/L of CSL was supplemented to whey lactose medium. Figure 3 shows the time-course of lactic acid production and cell growth during the fermentation at various concentrations of CSL. As can be seen in Fig. 3, both the lactic acid production rate and cell growth increased with increasing the CSL concentrations. The maximal cell growth (10.36 g/L of dry cell weight) was obtained with the medium supplemented with 60 g/L of CSL. Although a little lactose remained unconverted after the end of fermentation, most of the lactose was nearly converted to lactic acid (Table 2).

Table 2 summarizes the fermentation parameters of the earlier experiments at various concentrations of CSL, such as lactic acid produced, lactic acid yields, residual lactose, and productivities. From 30 to 60 g/L

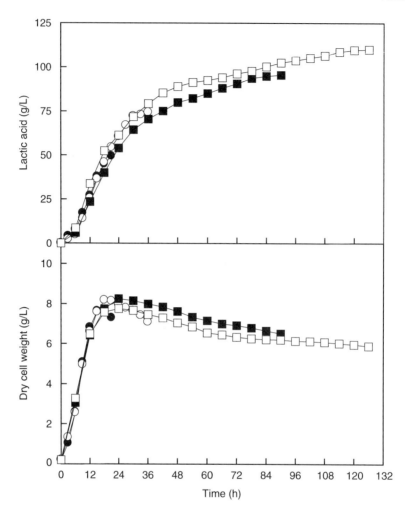

Fig. 2. Lactic acid production and cell growth from various concentrations of pure lactose by batch culture of *Lactobacillus* sp. RKY2. Symbols: -●-, lactose 50 g/L; -○-, lactose 75 g/L; -■-, lactose 100 g/L; -□-, lactose 125 g/L.

of CSL concentrations, lactic acid yields based on total lactose consumed was maintained above 0.98 g/g. In general, lactic acid fermentation using *Lactobacillus* sp. RKY2 should be severely affected by CSL concentration as a nitrogen source added to the medium, which is well agreed with the previous reports by Hujanen and Linko (21) who investigated the effect of nitrogen sources on lactic acid production by *Lactobacillus casei*.

Lactic Acid Production Through Cell-Recycle Repeated Batch Fermentation

Because lactic acid is a high-volume and low-price chemical, it is necessary to reduce the manufacturing cost. Numerous studies on reducing the manufacturing cost of lactic acid were carried out by several groups. According to Leudeking and Piret (22), lactic acid production rates could

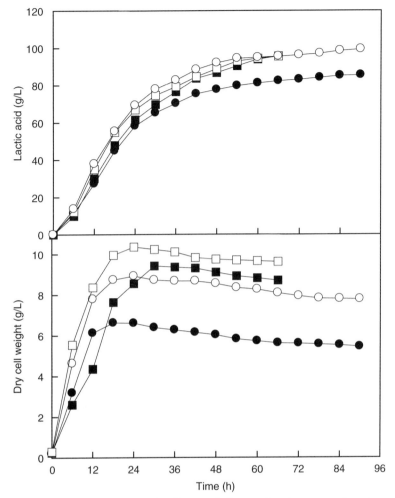

Fig. 3. Lactic acid production and cell growth from whey lactose and various concentrations of CSL by batch culture of *Lactobacillus* sp. RKY2. Symbols: -●-, CSL 15 g/L; -○-, CSL 30 g/L; -■-, CSL 45 g/L; -□-, CSL 60 g/L.

be increased by increase of cell densities in the bioreactor. In order to achieve this, the cell-recycle repeated batch fermentation was conducted by using a hollow fiber membrane. For economical fermentation of lactic acid, cost of fermentation medium as well as purification cost should be reduced. In addition, fermentation method such as batch, repeated batch, and continuous fermentations should be considered to improve productivity of final product. From this point of view, the cell-recycle repeated batch system might be a potential process that can maintain a reasonable productivity and reuse of the former culture as a seed.

The cell-recycle repeated batch fermentation by *Lactobacillus* sp. RKY2 was conducted to improve volumetric productivity. Figure 4 shows the results of the cell-recycle repeated batch fermentation. The medium for the first batch run contained 100 g/L of whey lactose, 30 g/L of CSL, 1 g/L of

Table 2
Effect of CSL Concentrations on Lactic Acid Produced, Residual Lactose, Maximal Dry Cell Weight, Yield, and Productivity in Batch Culture of *Lactobacillus* sp. RKY2

CSL concentration (g/L)	Residual lactose (g/L)	Lactic acid (g/L)	Max. dry cell weight (g/L)	Yield (g/g)[a]	Productivity (g/L/h)
15	3.4	85.32	6.63	0.88	0.95
30	0	99.15	8.94	0.983	1.1
45	3.9	95.5	9.43	0.988	1.44
60	3.8	95.38	10.36	0.989	1.45

[a]The yield of lactic acid was calculated on the basis of the amount of consumed lactose in whey.

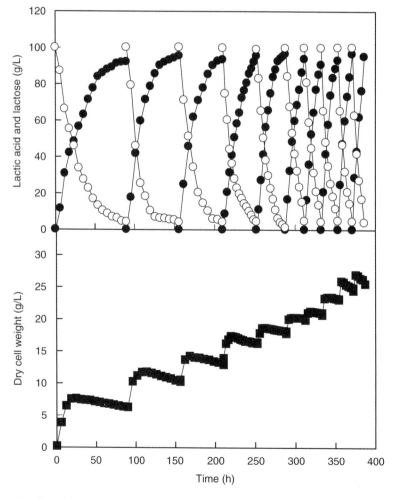

Fig. 4. Lactic acid production and cell growth from whey lactose and CSL by cell-recycle repeated batch culture of *Lactobacillus* sp. RKY2. Symbols: -●-, lactic acid; -○-, lactose; -■-, dry cell weight.

Table 3
Fermentation Results of the Cell-Recycle Repeated Batch Culture
of *Lactobacillus* sp. RKY2

Batch number	Lactic acid (g/L)	Max. dry cell weight (g/L)	Yield (g/g)[a]	Productivity (g/L/h)
1	94.06	7.64	0.98	1.06
2	95.47	11.82	0.99	1.45
3	93.38	14.28	0.97	1.72
4	94.61	17.4	0.99	2.25
5	96.53	18.7	0.98	2.68
6	93.79	20.29	0.98	3.91
7	90.61	21.26	0.95	4.31
8	93.03	23.46	0.98	4.42
9	94.84	25.97	0.99	5.27
10	95.11	26.98	0.99	6.34

[a]The yield of lactic acid was calculated on the basis of the amount of consumed lactose in whey.

yeast extract, 2 g/L of $(NH_4)_2HPO_4$, and 0.1 g/L of $MnSO_4$, and the medium for subsequent batches contained 100 g/L of whey lactose, 30 g/L of CSL, 1 g/L of yeast extract, and 2 g/L of $(NH_4)_2HPO_4$. The second batch run was completed 36 h faster than the first batch, which resulted in the improvement of lactic acid productivity from 1.02 g/L/h to 1.45 g/L/h. During 10 repeated batch runs, the volumetric productivities ranged from 1.02 to 6.34 g/L/h (Table 3). In other words, lactic acid productivity of the 10th batch run was improved by 6.2-fold higher value compared with that of the first batch run. The cell concentration maximized to 26.98 g/L at the 10th batch run, which corresponded to 3.53 times higher value than that of the first batch run. During the cell-recycle repeated batch operations, lactic acid productivity increased proportionally to the increase in cell growth. Therefore, the cell-recycle system seems to be a potential tool for the production of lactic acid from whey and CSL as cheap raw materials.

Conclusions

The biological production of lactic acid from cheese whey and CSL by *Lactobacillus* sp. RKY2 was investigated. In batch lactic acid fermentation, lactic acid yields based on consumed lactose were obtained at more than 0.98 g/g from the medium containing whey lactose or pure lactose. However, lactic acid productivities and yields obtained from whey lactose were slightly higher than those obtained from the pure lactose. The final concentration of lactic acid increased with increasing whey lactose concentration. Lactic acid productivity and cell growth in cheese whey and CSL were improved by the addition of more CSL to the medium. When the cell-recycle repeated batch fermentation was performed with 100 g/L of lactose in cheese whey, 30 g/L of CSL, and 1 g/L of yeast extract, lactic acid productivity was maximized to 6.34 g/L/h, which was 6.2 times higher than that of the batch fermentation.

References

1. Hofvendahl, K. and Hahn-Hägerdal, B. (2002), *Enzyme Microb. Technol.* **26,** 87–107.
2. Datta, R., Tsai, S. P., Bonsignore, P., Moon, S. H., and Frank, J. R. (1995), *FEMS Microbiol. Rev.* **16,** 221–231.
3. Yun, J. S., Wee, Y. J., and Ryu, H. W. (2003), *Enzyme Microb. Technol.* **33,** 416–423.
4. Ohara, H. and Yahata, M. (1996), *J. Ferment. Bioeng.* **81,** 272–274.
5. Fitzpatrick, J. J., Murphy, C., Mota, F. M., and Pauli, T. (2003), *Int. Dairy J.* **13,** 575–580.
6. Büyükkileci, A. O. and Harsa, S. (2004), *J. Chem. Technol. Biotechnol.* **79,** 1036–1040.
7. Irvine, D. M. and Hill, A. R. (1985), In: *Comprehensive Biotechnology*, Vol. 3, Moo-Young, M., (Ed.), Pergamon Press, New York, pp. 523–566.
8. Amrane, A. and Prigent, Y. (1994), *Appl. Microbiol. Biotechnol.* **40,** 644–649.
9. Eliseo, C. U., Alma Rosa, N. M., Francisco, J. M. R., Cleotilde, J. R., Nora, R. O., and Juvencio, G. M. (2000), *Process Biochem.* **35,** 649–657.
10. Ben-Hassan, R. M. and Ghaly, A. E. (1994), *Appl. Biochem. Biotechnol.* **47,** 89–105.
11. Fournier, D., Schwitzguebel, J. P., and Peringer, P. (1993), *Biotechnol. Lett.* **15,** 627–632.
12. Kisaalita, W. S., Lo, K. V., and Pinder, K. L. (1990), *Biotechnol. Bioeng.* **36,** 642–646.
13. Stanier, R. Y., Ingraham, J. L., Wheelis, J. L., and Painter, P. R. (1987), *General Microbiology*, 5th edition, Prentice-Hall, NJ.
14. Amrane, A. and Prigent, Y. (1993), *Biotechnol. Lett.* **15(3),** 239–244.
15. Parekh, M., Formanek, J., and Blascheck, H. P. (1999), *Appl. Microbiol. Biotechnol.* **51,** 152–157.
16. Wee, Y. J., Kim, J. N., Yun, J. S., and Ryu, H. W. (2005), *Biotechnol. Bioprocess Eng.* **10,** 23–28.
17. Wee, Y. J., Yun, J. S., Park, D. H., and Ryu, H. W. (2004), *Biotechnol. Bioprocess Eng.* **9,** 303–308.
18. Lee, J. H., Choi, M. H., Park, J. Y., et al. (2004), *Biotechnol. Bioprocess Eng.* **9,** 318–322.
19. Dlamini, A. M. and Peiris, P. S. (1997), *Biotechnol. Lett.* **19(2),** 127–130.
20. Miller, G. L. (1959), *Anal. Chem.* **31,** 426–428.
21. Hujanen, M. and Linko, Y. Y. (1996), *Appl. Microbiol. Biotechnol.* **45,** 307–313.
22. Luedeking, R. and Piret, E. L. (1959), *J. Biochem. Microbiol. Technol.* **1,** 393–412.

Production of Bacterial Cellulose by *Gluconacetobacter* sp. RKY5 Isolated From Persimmon Vinegar

SOO-YEON KIM,[1] JIN-NAM KIM,[1] YOUNG-JUNG WEE,[2] DON-HEE PARK,[2] AND HWA-WON RYU*,[2]

[1]*Department of Material Chemical and Biochemical Engineering, Chonnam National University, Gwangju 500-757, Korea; and* [2]*School of Biological Sciences and Technology, Chonnam National University, Gwangju 500-757, Korea, E-mail: hwryu@chonnam.ac.kr*

Abstract

The optimum fermentation medium for the production of bacterial cellulose (BC) by a newly isolated *Gluconacetobacter* sp. RKY5 was investigated. The optimized medium composition for cellulose production was determined to be 15 g/L glycerol, 8 g/L yeast extract, 3 g/L K_2HPO_4, and 3 g/L acetic acid. Under these optimized culture medium, *Gluconacetobacter* sp. RKY5 produced 5.63 g/L of BC after 144 h of shaken culture, although 4.59 g/L of BC was produced after 144 h of static culture. The amount of BC produced by *Gluconacetobacter* sp. RKY5 was more than 2 times in the optimized medium found in this study than in a standard Hestrin and Shramm medium, which was generally used for the cultivation of BC-producing organisms.

Index Entries: Bacterial cellulose; fermentation; *Gluconacetobacter*; optimization; persimmon vinegar.

Introduction

Cellulose (poly-β-1,4-glucose) is the most abundant biological macromolecules in the world, with an estimated production of 10^{11} t/yr. It forms a structural matrix of cell walls of several fungi, algae, and nearly all plants, in which the cellulose forms semi-crystalline microfibrils of several nm in diameter *(1,2)*. In general, most of the plants generate thinner microfibrils with lower cellulose crystallinity, and the microfibrils are often intimately associated with the encrusted materials including hemicellulose *(3)*. The bacterium, *Acetobacter xylinum*, is known to be able to produce an extracellular cellulose, which is called bacterial cellulose (BC) that was first described by Brown in 1886 *(4,5)*. The cellulose synthesized by *A. xylinum* is structurally identical to that made by plants or algae. The microbial

*Author to whom all correspondence and reprint requests should be addressed.

cellulose synthesis may offer a more cost-effective means of supply, because the microorganism secretes cellulose which is pure and free of lignin, pectin, hemicellulose, and phytate found in plant cellulose as a crystalline pellicle, making its recovery simple and relatively inexpensive (6). Because of high-tensile strength and water-holding capacity of BC, it has been used as a raw material for manufacturing of high-fidelity acoustic speaker, high-quality paper, and dietary foods.

It is well known that the cellulose-producing bacteria are subject to spontaneous mutation yielding cellulose-negative mutant, Cel^-, when exposed to strong shear stress fields or serial cultivation. Many strategies for the production strains capable of producing cellulose with a high yield have been attempted including chemical mutation, gene modification, development of improved fermentation technique, and improved components in culture broth. However, the productivity of BC is still quite low, which makes the production costs extremely high. Therefore, it is necessary to isolate a novel microorganism for stable production of BC and to establish a production medium for mass production of BC.

In this study, we presented an optimized medium composition for the efficient production of BC by batch fermentation of *Gluconacetobacter* sp. RKY5. Furthermore, under the optimized culture medium, we compare the fermentation efficiencies between static and agitated cultures of *Gluconacetobacter* sp. RKY5.

Materials and Methods

Microorganism

The microorganism used in this study was *Gluconacetobacter* sp. RKY5 KCTC 10683BP, which was previously isolated from persimmon vinegar in Korea. Based on the Bergey's Manual of Determinative Bacteriology (7), several biochemical tests were carried out such as gram staining, ability to oxidize ethanol, acetate and lactate, and ketogenic activity for glycerol. In addition, morphological test was carried out using scanning electron microscope and the produced BC was identified by X-ray diffractometer (4,8–10). According to the results from the above tests, this isolate was suggested to be *Acetobacter* sp. and eventually identified as *Gluconacetobacter* sp. through 16s rDNA sequence analysis (11). It was maintained by monthly transfer on Hestrin and Shramm medium (HS medium) and kept at –20°C with 50% (v/v) glycerol.

Cultivation Conditions

The standard medium used in this study was HS medium, consisting of 20 g/L glucose, 5 g/L yeast extract, 5 g/L peptone, 2.7 g/L Na_2HPO_4, and 1.15 g/L citric acid monohydrate (12). For seed culture, 1% (v/v) of the stock culture was inoculated into 50 mL of HS medium in a 250-mL Erlenmeyer flask, which was then cultivated at 30°C and 150 rpm for 48 h in a shaking incubator (KMC-8480SF, Vision Scientific, Daejeon, Korea). After that, the culture broth was vigorously shaken in order to release the cells from the BC pellicle. The resulting cell suspension was then filtered through 12 layers of sterilized gauze, and 2% (v/v) of the filtrate was inoculated into 50 mL of fermentation medium in a 250-mL Erlenmeyer flask for static and agitated culture.

Optimization of Culture Medium

Based on HS medium as a standard medium, several carbon, nitrogen, phosphate sources, and secondary substrates were tested in sequence to determine the optimal composition of culture medium for the production of BC. After one component was chosen, the test for its optimum concentration followed. After all components were selected and its concentrations were determined, these were used for further experiment. The amounts of BC produced were measured in static and agitated culture conditions under the optimized medium.

Analytical Methods

Cell growth was evaluated by measuring the optical density at 660 nm using a UV-1700 spectrophotometer (Shimadzu, Kyoto, Japan), after the culture broth was treated with 0.1% (v/v) cellulase (Celluclast 1.5 L, Novozymes A/S, Bagsvaerd, Denmark) at 50°C with shaking at 150 rpm. The dry cell weight (g/L) was then calculated by using a predetermined calibration curve.

In order to measure the amount of BC produced, the gelatinous membrane of BC that was formed on the surface of the culture was picked up with tweezers, and the pellets of the agitated culture broth were centrifuged at 9940g for 15 min (8). After the separation of BC from the culture broth, the pellicles or pellets were washed with distilled water to eliminate the medium components, which was treated with 0.1 N NaOH at 80°C for 20 min in order to dissolve the cells (13). After the neutralization with 0.1 M acetate buffer, the BC was rinsed again with distilled water, and then filtered. The purified BC was dried at 80°C under a vacuum until a constant weight was obtained. For each investigation, two independent experiments were carried out and each sample was analyzed in three replicates, from which we obtained mean values.

Results and Discussion

Effect of Carbon Sources on BC Production

To investigate the effect of carbon sources on the production of BC, 2% (w/v) of several carbon sources were added to the glucose-free standard medium. As shown in Fig. 1A, a high level of BC production was observed when glycerol, fructose, or sucrose was used as a carbon source.

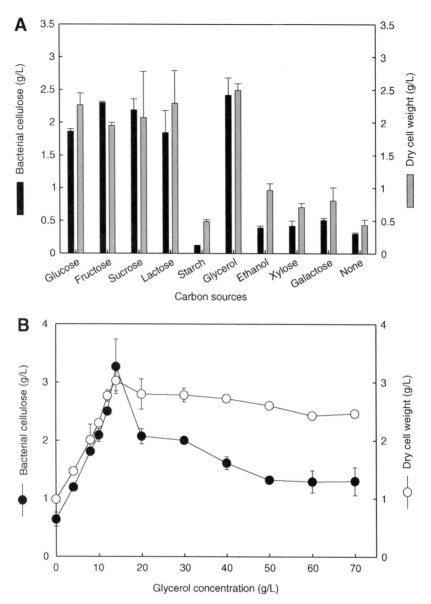

Fig. 1. Effect of various carbon sources (A) and glycerol concentrations (B) on the production of bacterial cellulose. The cultivation medium was composed of 20 g/L carbon source, 5 g/L yeast extract, 5 g/L peptone, 2.7 g/L Na$_2$HPO$_4$, and 1.15 g/L citric acid monohydrate. Data are presented as the means of two replicates and error bars indicate the standard deviation.

However, when starch, ethanol, xylose, or galactose was used as a carbon source, poor cellulose production was observed. It might be expected that, when the glycerol was used as a carbon source, cell growth was much higher than that of other carbon sources because the culture pH during the fermentation using glycerol as a carbon source did not significantly drop. Therefore, the amount of BC produced from glycerol as a carbon source was highest. Although BC-producing bacteria generally metabolize glucose or fructose for the production of BC, glycerol was found to be the best carbon source for cultivation of *Gluconacetobacter* sp. RKY5. Therefore, it is supposed that *Gluconacetobacter* sp. RKY5 has another metabolic pathway to synthesize cellulose from C_3 component, whereas *Acetobacter* species synthesize cellulose via such pathway as conversion of glucose to uridine diphosphate glucose (UDP-glucose) *(14)*. Therefore, the glycerol was selected as an optimum carbon source for BC production, and the effects of glycerol concentration on the production of BC was investigated. As shown in Fig. 1B, the amount of BC produced increased with increases in glycerol concentrations up to 14 g/L, and decreased beyond this value. The maximum BC production (3.27 g/L) was observed at 14 g/L of glycerol. Yields might be improved in batch fermentation if the cultures are initiated with a low glycerol concentration, because high initial glycerol concentrations resulted in low amount of BC production.

Effect of Nitrogen Sources on BC Production

Several nitrogen sources were added to the standard medium at a level of 1% (w/v) to investigate the effect of nitrogen sources on the production of BC. As shown in Fig. 2A, when yeast extract was added to the medium, the highest amount of BC (4.49 g/L) was obtained. Tryptone and corn steep liquor (CSL) were also tested for BC production and yielded 3 g/L and 2.16 g/L of BC, respectively. The other nitrogen sources resulted in poor BC production. The addition of yeast extract considerably stimulated BC production, probably owing to its abundant nutrients as well as several growth factors. To find the optimum concentration of yeast extract, various concentrations ranging from 0 to 20 g/L were tested. As shown in Fig. 2B, the BC production increased gradually with increases in yeast extract concentrations up to 8 g/L, and remained constant beyond this value. Therefore, 8 g/L of yeast extract was selected as an optimum nitrogen source concentration for the production of BC by batch fermentation of *Gluconacetobacter* sp. RKY5.

Effect of Phosphate Sources and Secondary Substrates on BC Production

To determine the effect of phosphate sources on the production of BC, various phosphate sources were tested by addition to the standard medium in the same range of concentration. All phosphate sources except $(NH_4)_2HPO_4$ exhibited similar BC production and cell growth, but K_2HPO_4 was found to be the best (Fig. 3A). To assess the effect of K_2HPO_4 concentrations on BC production, various concentrations of K_2HPO_4

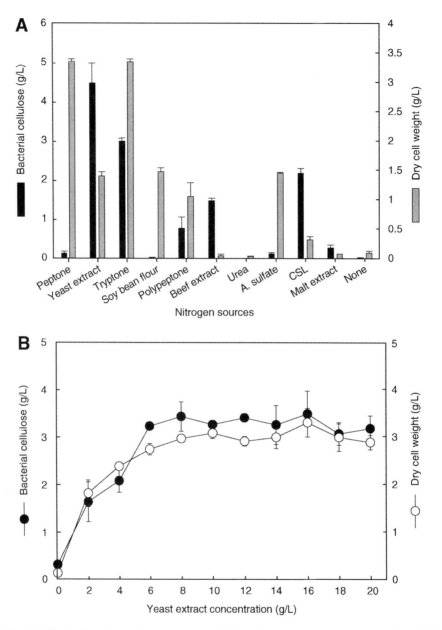

Fig. 2. Effect of various nitrogen sources **(A)** and yeast extract concentrations **(B)** on the production of bacterial cellulose. The cultivation medium was composed of 15 g/L glycerol, 10 g/L nitrogen source, 2.7 g/L Na_2HPO_4, and 1.15 g/L citric acid monohydrate. Data are presented as the means of two replicates and error bars indicate the standard deviation.

ranging from 0 to 5 g/L were added to the medium. As shown in Fig. 3B, the maximum yield of BC (3.21 g/L) was observed in the medium containing 3 g/L of K_2HPO_4, but the production of BC decreased slightly beyond this value.

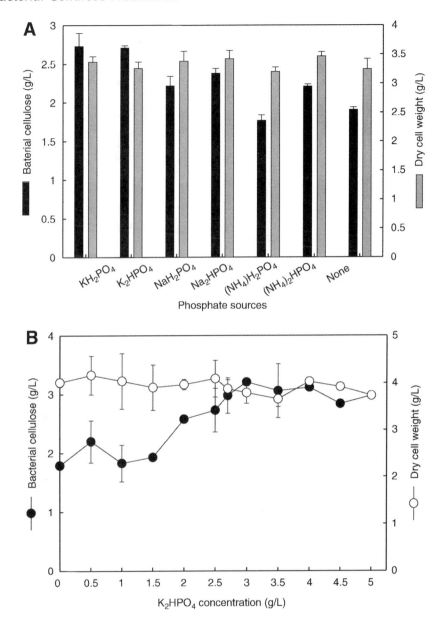

Fig. 3. Effect of various phosphate sources **(A)** and K_2HPO_4 concentrations **(B)** on the production of bacterial cellulose. The cultivation medium was composed of 15 g/L glycerol, 8 g/L yeast extract, 2.7 g/L phosphate source, and 1.15 g/L citric acid monohydrate. Data are presented as the means of two replicates and error bars indicate the standard deviation.

It is well known that the addition of secondary substrate, including some organic acids or ethanol, to the fermentation medium stimulates the production of BC. Therefore, several organic acids and ethanol were tested for the investigation of their effects on the production of BC by

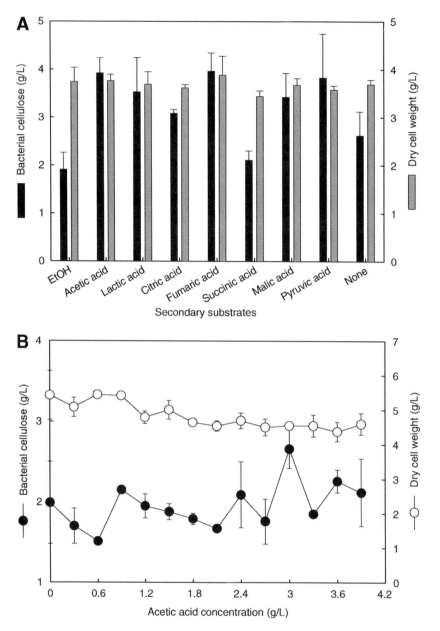

Fig. 4. Effect of various secondary substrates **(A)** and acetic acid concentrations **(B)** on the production of bacterial cellulose. The cultivation medium was composed of 15 g/L glycerol, 8 g/L yeast extract, 3 g/L K_2HPO_4, and 1.15 g/L secondary substrate. Data are presented as the means of two replicates and error bars indicate the standard deviation.

Gluconacetobacter sp. RKY5. As can be seen in Fig. 4A, all the compounds added except ethanol and succinic acid enhanced the production of BC. Among the compounds tested, acetic and fumaric acids were the best for BC production. According to the previous reports, the addition of lactic acid to the medium stimulated the production of BC by *A. xylinum* subsp. *sucrofermentans* BPR 2001 *(5)*, and Son et al. *(15)* improved the production of BC by adding the ethanol to the medium in batch culture of *Acetobacter* sp. A9. However, for the production of BC, *Gluconacetobacter* sp. RKY5 used in this study preferred acetic acid and fumaric acid to other compounds tested, which is a distinguishing characteristic of *Gluconacetobacter* sp. RKY5 from the other strains. To investigate the effect of acetic acid concentrations on BC production, various concentrations of acetic acid ranging from 0 to 3.9 g/L were added to the medium. As shown in Fig. 4B, the maximum yield of BC (3.92 g/L) was observed in the medium containing 3 g/L of acetic acid, but acetic acid concentrations below or above this value resulted in low amount of BC. However, the yields of cellulose were quite variable.

BC Production in Optimal Medium

As investigated earlier, the optimized culture medium for BC production by *Gluconacetobacter* sp. RKY5 was found to be 15 g/L glycerol, 8 g/L yeast extract, 3 g/L K_2HPO_4, and 3 g/L acetic acid. This media composition was designated as modified HS (MHS) medium. Figure 5 shows the time-course of BC production and cell growth during static (Fig. 5A) and agitated (150 rpm, Fig. 5B) cultivations of *Gluconacetobacter* sp. RKY5 in the MHS medium. In a static culture condition, the amount of BC produced gradually increased, and the maximum amount of BC (4.59 g/L) was obtained after 144 h of cultivation. In case of an agitated culture condition, the BC produced was poor before 48 h of cultivation, but it sharply increased after 48 h of cultivation. When mean values were compared, the maximum amount of BC produced (5.63 g/L) was observed at 144 h of cultivation. Because the production of BC started during exponential cell growth, BC production by *Gluconacetobacter* sp. RKY5 might be initially associated with cell growth. However, maximum production of BC occurred in stationary growth phase. Because the agitated culture condition has more shear stress than the static culture condition, this might cause the delay in production. On the other hand, in the static culture condition, the mobility of the microorganism decreases compared to those in the agitated culture condition. Therefore, the uptake of carbon source may be dependent on diffusion permitting more glycerol to go to cellulose-synthesizing enzyme. However, for the production of BC, the agitated culture condition was better than the static culture condition in this study. It seemed that the cellulose-synthesizing mechanism of *Gluconacetobacter* sp. RKY5 yielded more BC in MHS medium for agitated culture, probably as a result of better aeration. In conclusion, the amount of BC produced was approximately more than 2 times in the optimized medium established in this study than that in a standard HS medium.

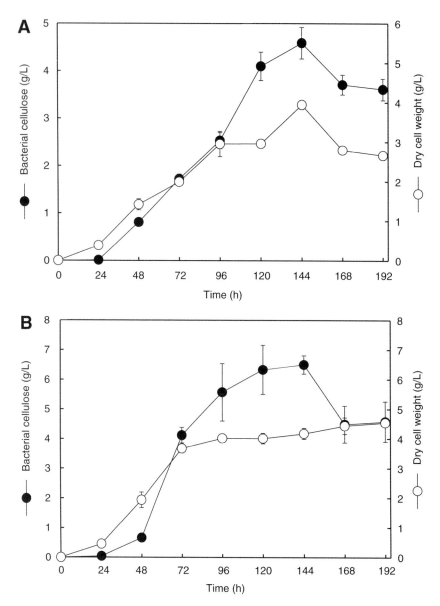

Fig. 5. Time-course of bacterial cellulose production and cell growth using MHS medium. The cultivation medium was composed of 15 g/L glycerol, 8 g/L yeast extract, 3 g/L K_2HPO_4, and 3 g/L acetic acid. **(A)** static culture condition; **(B)** agitated culture condition. Data are presented as the means of two replicates and error bars indicate the standard deviation.

References

1. Astley, O. M., Chanliaud, E., Danald, A. M., and Gidley, M. J. (2001), *Int. J. Biol. Macromol.* **29,** 193–202.
2. Jonas, R. and Farah, L. F. (1998), *Polym. Degrad. Stabil.* **59,** 101–106.
3. Hult, E. L., Yamanaka, S., Ishihara, M., and Sugiyama, J. (2003), *Carbohydr. Polym.* **53,** 9–14.

4. Brown, A. J. (1886), *J. Chem. Soc.* **49,** 432–439.
5. Matsuoka, M., Tsuchida, T., Matsushita, K., Adachi, O. A., and Yoshinaga, F. (1996), *Biosci. Biotechnol. Biochem.* **60(4),** 575–579.
6. Wulf, P. D., Joris, K., and Vandamme, E. J. (1996), *J. Chem. Techchol. Biotechnol.* **67,** 376–380.
7. Buchanan, R. E. and Gibbons, N. E. (1974), *Bergey's Manual of Determinative Bacteriology*, 8th edition, Williams and Wilkins Co., Baltimore.
8. Park, J. K., Park, Y. H., and Jung, J. Y. (2003), *Biotechnol. Bioprocess Eng.* **8,** 83–88.
9. Toyosaki, H., Naritomi, T., Seto, A., Matsuoka, M., Tsuchida, T., and Yoshinaga, F. (1995), *Biosci. Biotech. Biochem.* **59,** 1498–1502.
10. Yamada, Y., Hoshino, K., and Ishikawa, T. (1997), *Biosci. Biotech. Biochem.* **61,** 1244–1251.
11. Ryu, H. W., Wee, Y. J., Kim, S. Y., Kim, J. N., and Yun, J. S. (2004), *Korea Patent Application No.* 10-2004-0104790.
12. Schramm, M. and Hestrin, S. (1954), *Biochem. J.* **56,** 163–166.
13. Son, C., Chung, S., Lee, J., and Kim, S. (2002), *J. Microbiol. Biotechnol.* **12(5),** 722–728.
14. Tonouchi, N., Tsuchida, T., Yoshinaga, F., Beppu, T., and Horinouchi, S. (1996), *Biosci. Biotechnol. Biochem.* **60(8),** 1377–1379.
15. Son, H. J., Heo, M. S., Kim, Y. G., and Lee, S. J. (2001), *Biotechnol. Appl. Biochem.* **33,** 1–5.

Natural Compounds Obtained Through Centrifugal Molecular Distillation

Vanessa Mayumi Ito, Patricia Fazzio Martins, César Benedito Batistella, Rubens Maciel Filho, and Maria Regina Wolf Maciel*

Separation Process Development Laboratory, School of Chemical Engineering, State University of Campinas, PO Box 6066, 13081-970, Campinas, Brazil, E-mail: wolf@feq.unicamp.br and vanessa@lopca.feq.unicamp.br

Abstract

Soybean oil deodorized distillate (SODD) is a byproduct from refining edible soybean oil; however, the deodorization process removes unsaponifiable materials, such as sterols and tocopherols. Tocopherols are highly added value materials. Molecular distillation has large potential to be used in order to concentrate tocopherols, because it uses very low levels of temperatures because of the high vacuum and short operating time for separation and, also, it does not use solvents. However, nowadays, the conventional way to recover tocopherols is carrying out chemical reactions prior to molecular distillation, making the process not so suitable to deal with natural products. The purpose of this work is to use only molecular distillation in order to recover tocopherols from SODD. Experiments were performed in the range of 140–220°C. The feed flow rate varied from 5 to 15 g/min. The objective of this study was to remove the maximum amount of free fatty acids (FFA) and, so, to increase the tocopherol concentration without add any extra component to the system. The percentage of FFA in the distillate stream of the molecular still is larger at low feed flow rates and low evaporator temperatures, avoiding thermal decomposition effects.

Index Entries: Centrifugal distillation; molecular distillation; natural products; tocopherol.

Introduction

Soybean oil is the most consumed vegetable oil in the world, representing 54% of the total world production (1). Brazil is the second largest producer of soybeans. The refining process has several steps: degumming, refining, bleaching, hydrogenation, and deodorization. Deodorization removes compounds that give odor and flavor to the oil. Soybean oil deodorized distillate (SODD) is produced in the deodorization step of the

---

*Author to whom all correspondence and reprint requests should be addressed.

oil refining *(2)*, corresponding from 0.1 to 0.4 wt% of the crude oil weight, and contains between 0.8 and 10 wt% of tocopherols *(3)*. SODD is a complex mixture made up of free fatty acids (FFA), sterols, tocopherols, sterol esters, hydrocarbons, breakdown products of fatty acids, aldehydes, ketones, and acylglycerol species *(4)*, and, therefore, it can be a good raw material for producing vitamin E and sterols *(5)*. Tocopherols are natural antioxidants and have vitaminic activity. Sterols are important in the production of hormones and in the artificial production of other vitamins *(6)*, for instance, vitamin D.

FFA are different according to the number of carbons and saturations. SODD contains high levels of FFA (25–75%) and acylglycerols (3–56%), *(7)*, depending on the raw material and on the type (physical or chemical way) and conditions of the refining process *(4)*. FFA have molecular weights in the range 180–300 g/gmol, and higher vapor pressure than tocopherols.

Tocopherols, which are substances physiologically active as vitamin E, are important natural antioxidants and find extensive applications in food, cosmetic, and pharmaceutical industries *(7)*. Preparing high-purity concentrates of tocopherols normally involves a series of physical and chemical treatment steps in conventional processes *(8)*. Because deodorized distillate is a complex mixture and the properties of its components are very similar, it is difficult to recover high-quality concentrates of tocopherols with good yield *(9)*.

Several processes can be used in order to separate FFA from tocopherols and phytosterols. Some of them use molecular distillation together chemical reactions, what mischaracterizes them as natural processes.

Ramamurthi et al. *(10)* studied the lipase-catalyzed esterification of FFA. This process used methanol to esterify the FFA from canola oil deodorized distillate.

Chang et al. *(11)* recovered tocopherols and FFA, investigating a supercritical fluid CO_2 extraction process of SODD. Good recovery is achievable, but it is necessary to employ high pressure. Lee et al. *(12)* studied the lipase-catalyzed esterification, transesterification to form methyl esters, followed by supercritical fluid CO_2.

Buczenko et al. *(13)* performed the separation of tocopherols from sterols using a liquefied petroleum gas extraction. First, saponification step is used to separate fatty acids from the raw material. Then, the separation was carried out in the liquefied petroleum gas extractor.

Chu et al. *(14,15)* recovered tocopherols from palm distillate by batch adsorption using an orbital shaker. The raw material was preconcentrated through neutralization and so, extracted with hexane, centrifuged, and dried. Then, the preconcentrate mixture is carried out at batch adsorption equipment.

Chu et al. *(16)* studied the enzymatic hydrolysis using a commercial immobilized *Candida antarctica* lipase, followed by hydrolysis and washing to remove FFA salts. Hirota et al. *(2)* distillated the SODD by molecular

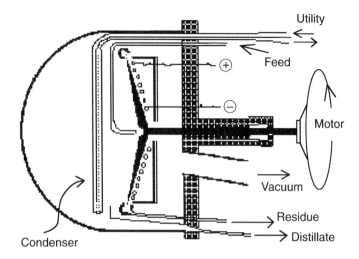

Fig. 1. Centrifugal molecular still schematic.

distillation, followed by a lipase-catalyzed hydrolysis and another molecular distillation step to separate FFA from steryl esters.

Ghosh and Bhattacharyya (17) recovered tocopherol by lipase-catalyzed hydrolysis and esterification reactions, followed by fractional distillation of the derived ester product. Owing to the similar volatility of sterols, tocopherols, and fatty acids, it is quite difficult to separate tocopherols and sterols from FFA using fractional distillation. Exposure to high temperatures may degrade tocopherols.

The processes above have various operating steps and are laborious. They need attention to prevent decomposition of tocopherols because of the exposure to molecular oxygen, to light, and to high temperatures.

Molecular or short-path distillation is, generally, accepted as the most suitable distillation method to separate and to purify both thermosensitive and high-molecular-weight compounds (18). Two main types of molecular stills are available: falling film and centrifugal (19). The centrifugal molecular still scheme, used in this work, is shown in Fig. 1. Figure 2 shows the material stream throughout the molecular distillator (20). The feed stream is introduced in the center of the equipment. The liquid flows, by centrifugal force, uniformly around the evaporator until the border of the rotor in a thin film (21). The light compounds are volatilized and condensed (distillate stream) and the heavy compounds are collected as residue stream. The process is characterized by high vacuum in the distillation space, and a small distance between the evaporator and the condenser, resulting in a short exposure of the distilled liquid to the operating temperature, on the order of a few seconds, which is guaranteed by distributing the liquid in the form of a thin film (22). Under these conditions, i.e., short residence time and low temperature, distillation of heat-sensitive materials is accompanied by only negligible thermal decomposition (21).

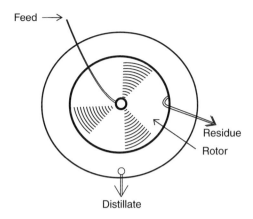

Fig. 2. Centrifugal molecular still material streams.

Molecular distillation is applied at several areas, such as monoglycerides concentration (23), carotenoids recovery from palm oil (24), heavy petroleum characterization (25), and herbicides. The present study aims the recovery of tocopherols and FFA from SODD just using molecular distillation.

Materials and Methods

Materials

SODD was provided by Bunge Ltda., São Paulo, Brazil. All samples were stored in the refrigerator at 4°C until analysis. All solvents and reactants for the analyses were of analytical grade. A tocopherol kit consisting of α-, β-, γ-, and δ-tocopherol (purity ≥95%) was purchased from Calbiochem (San Diego, CA) and used as reference standards for tocopherols analysis. The reactants used for the free fatty acid analysis were ethyl alcohol, phenolphetalein and sodium hydroxide. The solvents, hexane, and isopropanol for tocopherol analysis were of high-performance liquid chromatography (HPLC) grade from Tedia Company, Inc. (Fairfield, OH).

Methods

Centrifugal Molecular Distillation

The FFA distillation was performed using a centrifugal molecular still. The process conditions were maintained at 13.3 Pa, the feed temperature at 50°C, and the condenser temperature at 50°C. The molecular distillation experiments were conducted according to the following procedure: a sample of SODD was homogenized before feeding the equipment. The distillation was performed at 140–220°C. The feed flow rate varied in the range 5–15 g/min. For each molecular distillation run, samples of both streams (distillate and residue) were collected and submitted to FFA and tocopherols analyses.

Table 1
Raw Material Characteristics

Analysis	SODD
FFA (wt% as oleic acid)	53.04 ± 1.1
α-tocopherol	1.41 ± 0.25
β-tocopherol	0.12 ± 0.05
γ-tocopherol	1.95 ± 0.15
δ-tocopherol	0.58 ± 0.09
Tocopherol total	4.06 ± 0.54

Analytical Methods

The FFA analysis: FFA content were determined according to Method AOCS Ca 5 a-40 (26). This method used titration with a standard alkali, NaOH. The FFA concentration is expressed as percentage of oleic acid ($C_{18:1}$). The expression is:

$$\%FFA \text{ as oleic acid} = \frac{\text{alkali volume (mL)} \cdot \text{alkali normality} \cdot 28.2}{\text{sample weight (g)}}$$

Tocopherols analysis: the method AOCS Ce 8-89 (26) was used to determine the α-, β-, γ-, and δ-tocopherol contents. A known amount of the sample was dissolved in hexane (approx 1 mg/mL) and 20 µL of the solution was injected into a HPLC modular equipment composed by Waters 515 HPLC pump (Mildford, MA), equipped with a fluorescence detector (Waters model 2475 multifluorescence, Mildford, MA). The flow rate of the mobile phase (hexane:isopropanol, 99:1 v/v) was set at 1 mL/min. The separation was conducted in a µporasil column 125 Å, with particle size of 10 µm and 3.9 × 300 mm of dimension (Waters, Ireland). The tocopherols detected in the chromatograms were identified comparing the retention time of the compounds with the retention time of standard solutions. Quantification of each type of tocopherols was done using calibration curves. The data processing was carried out through the Millennium software (Waters, Mildford, MA).

Results and Discussion

The SODD was analyzed in relation to the FFA and tocopherols contents. Table 1 shows the characteristics of the soybean oil deodorizer distillate. The SODD used is brownish and semisolid at room temperature.

The feed flow rate and the evaporator temperature are the main variables in the molecular distillation process (23,24), so the experiments were performed from 140 to 220°C and the feed flow rate varied from 5 to 15 g/min. The procedure consists of keeping the evaporator temperature fixed and the feed flow rate is varied. For all distillation runs, the condenser temperature was maintained at 50°C (to avoid solidification of distillate stream on the condenser) and the feed temperate at 50°C.

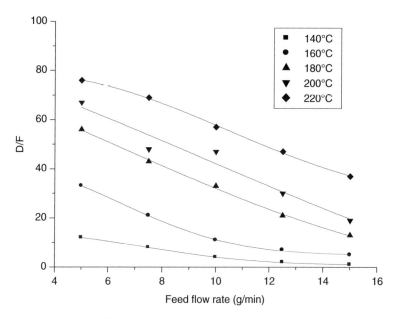

Fig. 3. Distillation profile for the ratio D/F in function of feed flow rate at different values of evaporator temperature.

The evaporation ratio D/F (mass of distillate/mass of feed) is a useful parameter to evaluate the molecular distillation process. Figure 3 shows the profile of evaporation ratio D/F as a function of feed flow rate (F) at different evaporator temperatures. The results show that the feed flow rate affects significantly the evaporation ratio D/F. Increasing the feed flow rate, the material, that it will be distillate, increases. The evaporator efficiency is lower because of the poor contact between the material and the heated surface, so the ratio D/F declines. At constant temperature, the ratio D/F decreases increasing the feed flow rate. On the other hand, increasing the evaporator temperature and maintaining the same feed flow rate, the evaporation ratio D/F increases. At feed flow rate of 5 g/min and 220°C is obtained the maximum distillate amount. Figure 4 shows the profile of residue flow rate of the centrifugal molecular still. The residue flow rate is directly proportional to the feed flow rate and inversely to the evaporator temperature. As the vaporization rate was constant at one temperature; the residue percentage rose with increased feed flow rate.

The percentage of FFA of SODD (before the centrifugal molecular distillation) was 53.04%. FFA tends to concentrate in the distillate stream because they are lighter molecules than tocopherols. As it is known, the split ratio D/R (mass of distillate/mass of residue) is another important parameter to evaluate the molecular distillation process (27). Figure 5 shows the free fatty acid recovery in the distillate stream in function of the split ratio D/R. It can be obtained a recovery higher than 90% using a split ratio higher than 2.

Figure 6 shows the FFA content in the residue streams. The loss of FFA occurs in the residue stream of the centrifugal molecular still with the

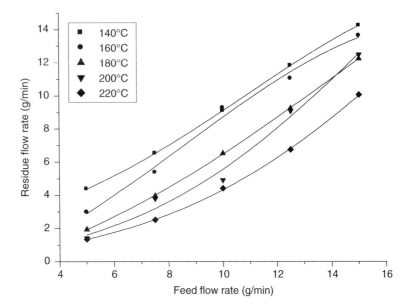

Fig. 4. Residue stream from centrifugal molecular still.

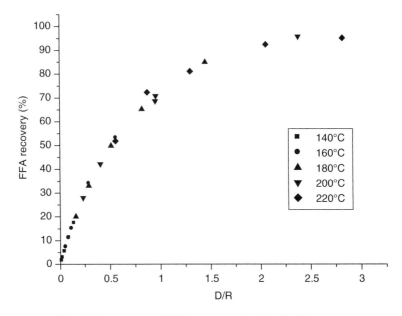

Fig. 5. Percentage of FFA recovery in the distillate stream.

increase of feed flow rate and the decrease of the evaporator temperature. It is important to note that the percentage of FFA in the residue stream depends on the evaporator temperature and feed flow rate. At temperatures of 140°C and 160°C, there is a great loss of FFA in the residue stream. Then, it is necessary to operate the molecular distillation conditions at 5 g/min and evaporator temperature greater than 180°C.

Centrifugal Molecular Distillation

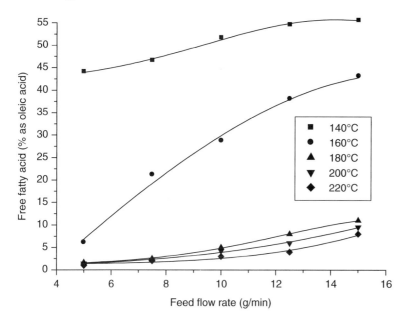

Fig. 6. Free Fatty Acid content in the residue stream.

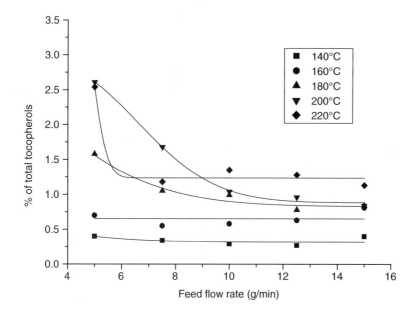

Fig. 7. Total tocopherols content vs feed flow rate in the distillate stream.

Figure 7 shows the total tocopherols content in the distillate stream (loss of tocopherols, they are to be recovered in the residue stream). At 140°C, the percentage of total tocopherols was lower than 0.5% for all the feed flow rates. At higher temperatures, the loss of tocopherols increases.

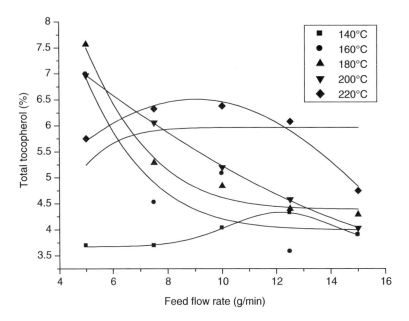

Fig. 8. Total tocopherols content vs feed flow rate in the residue stream.

The increase of feed flow rate decreases the loss of tocopherols. The optimum operating conditions are low evaporator temperatures (140°C) and high feed flow rate (12.5 g/min).

The tocopherols content in the residue stream is affected by the evaporator temperature and feed flow rate. It was observed in Fig. 8 that at 180°C and 5 g/min the product reached 7.7% of total tocopherols. In this case, the tocopherol enrichment was 1.9 for just one molecular distillation step.

The profile of tocopherol recovery evaluating the split ratio was observed in Fig. 9. It is possible to obtain a recovery higher than 90% using a split ratio lower than 0.25. The increase of tocopherols recovery implies a loss of FFA in the residue stream. The FFA loss decreases with increasing the split ratio D/R.

It can be seen in Fig. 10 that the intersection of the tocopherol and FFA recovery curves is the optimum ratio D/R. The optimum ratio D/R value is 0.96. In this operating condition, 73% of total tocopherol and FFA are recovered in its respective streams.

Concluding Remarks

In this work, the elimination of FFA and concentration of tocopherols present in SODD through the centrifugal molecular distillation was studied. The initial composition of FFA in SODD was around 53%. The percentages of distillate and residue compositions depend on operating variables (evaporator temperature and feed flow rate). High concentration of FFA is obtained using a split ratio higher than 2. It can be

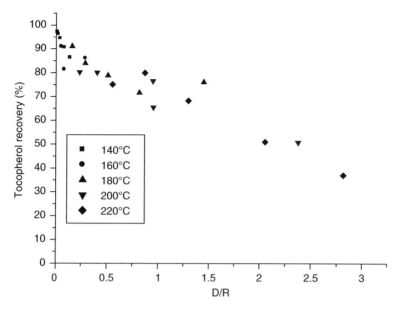

Fig. 9. Tocopherol recovery in the residue stream.

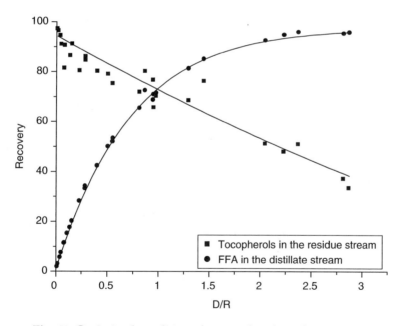

Fig. 10. Optimized conditions for tocopherols and FFA recovery.

used different evaporator temperature and feed flow rate to reach the same split ratio. To concentrate tocopherols, it is necessary to work at a low evaporator temperature and feed flow rate. Tocopherols were concentrated twice in relation to the raw material. Then, the split ratio used to reach high recovery of tocopherols is low. At D/R split ratio of 0.96, it is obtained a recovery of 73% of tocopherols and FFA.

Acknowledgments

The authors are grateful to Fapesp (Fundação de Amparo à Pesquisa do Estado de São Paulo), CAPES (Coordenação de Aperfeiçoamento de Pessoal de Nível Superior) and CNPq (Conselho Nacional de Desenvolvimento Científico e Tecnológico) for the financial support.

References

1. Mendes, M. F., Pessoa, F. L. P., and Uller, A. M. C. (2002), *J. Supercritical Fluids* **23**, 257–262.
2. Hirota, Y., Nagao, T., Watanabe, Y., et al. (2003), *J. Am. Oil Chem. Soc.* **80**, 341–346.
3. Augusto, M. M. M. (1988), M. Sc. Thesis, State University of Campinas, Brazil.
4. Ramamurthi, S., and McCurdy, A. R. (1993), *J. Am. Oil Chem. Soc.* **70(3)**, 287–295.
5. Moraes, E. B., Batistella, C. B., Torres Alvarez, M. E., Maciel Filho, R., and Wolf Maciel, M. R. (2004), *Appl. Biochem. Biotechnol.* **113**, 689–711.
6. Bondioli, P., Mariani, C., Lanzani, A., Fedeli, E., and Muller, A. (1993), *J. Am. Oil Chem. Soc.* **70(8)**, 763.
7. Chu, B. S., Baharin, B. S., and Quek, S. Y. (2002), *Food Chem.* **79(1)**, 55–59.
8. Nagesha, G. K., Subramanian, R., and Sankar, K. U. (2003), *J. Am. Oil Chem. Soc.* **80**, 397.
9. Lin, K. M. (2002), National Chung Hsing University, M.S., Texas A & M University.
10. Ramamurthi, S., Bhirud, P. R., and McCurdy, A. R. (1991), *J. Am. Oil Chem. Soc.* **68(12)**, 970–975.
11. Chang, C. J., Chang, Y., Lee, H., Lin, J., and Yang, P. (2000), *Ind. Eng. Chem. Res.* **39**, 4521–4525.
12. Lee, H., Chung, B. H., and Park, Y. H. (1991), Concentration of tocopherols from soybean sludge by supercritical carbon-dioxide, *J. Am. Oil Chem. Soc.* **68(8)**, 571–573.
13. Buczenko, G. M., Oliveira, J. S., and von Meien, O. F. (2003), *Eur. J. Lipid Sci. Technol.* **105**, 668–671.
14. Chu, B. S., Baharin, B. S., Che Man Y. B., and Quek, S. Y. (2004), *J. Food Eng.* **62(1)**, 97.
15. Chu, B. S., Baharin, B. S., Che Man Y. B., and Quek, S. Y. (2004), *J. Food Eng.* **62(1)**, 105.
16. Chu, B. S., Quek, S. Y., and Baharin, B. S. (2003), *Food Chem.* **80(3)**, 295–302.
17. Ghosh, S., and Bhattacharyya, D. K. (1996), *J. Am. Oil Chem. Soc.* **73(10)**, 1271–1274.
18. Cvengros, J., Mrazik, M., and Kmetty, G. (1999), *Chemical papers-chemicke Zvesti*, **53(2)**, 102–106.
19. Batistella, C. B. and Maciel, M. R. W. (1996), *Comp. Chem. Eng.* **20**, S-19–S-24.
20. Batistella, C. B., Moraes, E. B., Maciel Filho, R., and Wolf Maciel, M. R. (2002), *Appl. Biochem. Biotechnol.* **98–100**, 1187–1206.
21. Batistella, C. B. and Wolf Maciel, M. R. (1998), *Com. Chem. Eng.* **22**, S53–S60.
22. Batistella, C. B., Maciel, M. R. W., and Maciel Filho, R. (2000), *Comput. Chem. Eng.* **24**, 1309–1315.
23. Fregolente, L. V., Batistella, C. B., Maciel Filho, R., and Wolf Maciel, M. R. (2005), *J. Am. Oil Chem. Soc.* **82(9)**, 673–678.
24. Batistella, C. B., Moraes, E. B., Maciel Filho, R., and Wolf Maciel, M. R. (2002), *Appl. Biochem. Biotechnol.* **98–100**, 1149–1159.
25. Sbaite, P., Vasconcelos, C. J. G., and Martins, P. F., et al. (2003), 4th European Congress in Chemical Engineering (ECCE4), Granada, Spain.
26. AOCS (1997), *Official Methods and Recommended Practices of the American Oil Chemists Society*. 5th ed., Champaign, IL.
27. Martins, P. F., Ito, V. M., Batistella, C. B., and Maciel, M. R. W. (2006), *Sep. Purification Technol.* **48**, 78–84.
28. Cvengros, J., Lutisan, J., and Micov, M. (2000), *Chem. Eng. J.* **78**, 61.

Biosurfactants Production by *Pseudomonas aeruginosa* FR Using Palm Oil

Fernando J. S. Oliveira, Leonardo Vazquez, Norberto P. de Campos, and Francisca P. de França*

Departamento de Engenharia Bioquímica, Escola de Química, Centro de Tecnologia, Universidade Federal do Rio de Janeiro, Ilha do Fundão, Rio de Janeiro, RJ, CEP 21949-900, Brazil; E-mail: fpfranca@eq.ufrj.br

Abstract

Biosurfactants production by a strain of *Pseudomonas aeruginosa* using palm oil as a sole carbon source was investigated. The experiments were carried out in 500-mL conical flasks containing 100 mL of mineral media supplemented with palm oil as the sole carbon source. The *P. aeruginosa* FR strain was able to reduce surface tension of three tested inorganic media. Rotation velocities from 100 to 150 rpm provided free-cell fermented media with the lowest surface tension of approx 33 mN/m. Emulsification index results of even 100% were achieved when diesel was used as oil phase. Eight surface-active compounds produced by the bacterium were identified by mass spectrometry.

Index Entries: Biosurfactants production; rhamnolipids; *Pseudomonas aeruginosa*; palm oil; *Ellaus guineensis* fruit; mass spectra.

Introduction

Surfactants are a wide class of amphiphilic molecules capable of reducing surface and/or interfacial tension between gasses, liquids, and solids. Therefore, these substances have been used in a great variety of products and processes in various chemical industries *(1,2)*.

The high toxicity, low biodegradability, and effectiveness in a narrow pH and temperature range of synthetic surface-active compounds raised the interest in surfactants produced by microorganisms, also called biosurfactants. These substances present ecological acceptance and effectiveness in a wide range of pH and temperature values *(2,3)*. Beside the surface and interfacial activities, some of the biosurfactants presented antifungal, antiviral, and soil clean-up processes, for example, metal sorption and enhancement of oil biodegradation *(3–5)*. Most biosurfactants are a complex mixture of molecules, comprised of various chemical structures, such as fatty acids, glycolipids, peptides, polysaccharides, and proteins *(1,6)*, that is, a function of the organism, raw matter, and process condition,

---

*Author to whom all correspondence and reprints request should be addressed.

increasing the interest in its composition, characterization, and production optimization (7,8).

Aiming at the final biosurfactant cost reduction, the development of economical alternatives for its production has been investigated. Thus, the use of low-cost raw matter appears as a natural choice to generate an overall economy. In this perspective, some substrates have been used for microbial surfactants production, such as agro-industrial wastes, hydrocarbon mixtures, and vegetable oil. Brazil is one of the world's largest producers of vegetable oil, for example, soybean oil, corn oil, babacu oil, and palm oil. Palm oil is extracted from *Ellaus guineensis* fruit, easily available in our country (Brazil). This oil is widely used in Brazilian foods because of its particular taste, odor, and thermal stability. Recently, various studies have been published on biosurfactant production by various bacterial genus. However, to our knowledge no reports have been published on the biosurfactants production from palm oil by *Pseudomonas aeruginosa* bacteria. The scope of this investigation was the production and characterization of biosurfactants synthesized by a *P. aeruginosa* FR strain growing on mineral media using palm oil as the sole carbon source.

Material and Methods

Microorganism

The bacterium strain employed in this study was isolated from soil contaminated with crude oil (Galeão Beach, Rio de Janeiro, Brazil). The microorganism was identified according to Palleronni (9). The strain was maintained on nutrient agar slant medium (Difco laboratories, 0001, Detroit, MI) at 4°C. The inocula were prepared using bacterial cells transferred from the storage culture to a test tube containing 10 mL of nutrient broth. After incubation at 29 ± 1°C for 24 h, the inoculum was propagated to a 500 mL flask containing 100 mL of the mineral medium with the following composition in (g/L): KH_2PO_4, 1.4; K_2HPO_4, 0.5; $CaCl_2 \cdot 2H_2O$, 0.02; $MnSO_4 \cdot 1H_2O$, 0.03; NH_4NO_3, 2.0. The pH of the medium was adjusted to 7.0 ± 0.2 using 1 N NaOH, supplemented with 0.5% (v/v) of palm oil and also incubated while agitating (150 rpm) for 24 h at 29 ± 1°C.

Growth on Mineral Media

Three inorganic media were used in this study on biosurfactants production containing distilled water, in g/L: (a) MM1: 1.0, $(NH_4)_3PO_4$; (b) MM2: 3.0, inorganic commercial fertilizer N:P:K (10:10:10), Ouro Verde Company (São Paulo, Brazil); and (c) MM3: 0.5, Na_2HPO_4; 4.5, KH_2PO_4, 2.0, NH_4Cl, 0.1, $MgSO_4 \cdot 7H_2O$. pH values of the media were adjusted to 7.0 ± 0.2 using 1 N NaOH, these media were sterilized autoclaving at 121°C for 15 min. Palm oil was sterilized separately at 111°C for 20 min and added aseptically (0.5% v/v) to the flasks containing the inorganic media, after cooling.

The bacterial concentration was monitored from 0 to 72 h of process. At programmed time intervals, colony forming unity (CFU) determinations were executed using the *pour plate* method, employing a nutrient agar medium. Petri dishes were incubated at 29 ± 1°C for 48 h, and the CFU enumeration was presented as a mean of three independent experiments.

Aliquots of 50 mL of fermented media were transferred to a conical separation flask for the removal of the palm oil. Aqueous phase was then centrifuged at 4000g for 15 min for cell removal. Supernatants were submitted to surface tension and the critical micelle dilution (CMD^{-1} and CMD^{-2}) was measured in a Sigma Tensiometer model 70 (Helsinki, Finland). The CMD^{-1} and CMD^{-2} were estimated by respective surface tension measurement of 10^{-1} and 10^{-2} diluted free-cell fermented medium in distilled water. All results were presented as a mean of three independent experiments.

Rotation Velocity Effect

The effect of rotation velocity on biosurfactant production was investigated using 500-mL conical flasks containing 100 mL of the MM3. The flasks were incubated using different rotary velocities: 50, 100, 150, and 200 rpm for 72 h at 29 ± 1°C. At programmed time intervals, aliquots of 80 mL of fermented media were transferred to conical separation flask for the removal of palm oil. Aqueous phase was then centrifuged at 4000g for 15 min for cell removal. Supernatants were submitted to surface tension, CMD^{-1}, CMD^{-2}, and emulsification index determinations. Emulsification capacities were determined by the addition, into crap tubes, of equal volumes of free-cell fermented media and selected hydrocarbon (hexane, jet fuel, and diesel). The tubes were mixed using a vortex for 2 min, and left to rest for 24, 48, and 72 h at 29 ± 1°C. The emulsification index was determined as the percentage of the height of the emulsion column.

Extraction and Characterization of the Biosurfactants

Rhamnolipids species were extracted from fermentation media using chloroform. The extract was dried under nitrogen atmosphere at 40°C and resuspended in methanol. This solution was submitted to high-performance liquid chromatography (HPLC) coupled to a quadrupole mass spectrometer in negative mode (Apployd 200 Varian San Francisco, CA). The chromatographic system was equipped with an automatic injector and a 250 × 4.6 mm² monocron C8 reverse phase column. Aliquots of 10 µL was injected into HPLC, and an acetronitrile–water gradient was used as mobile phase and the elution was started with 70% acetonitrile for 4 min, and subsequent acetonitrile concentration was raised to 70–100% for 40 min at 1.0 mL/min.

In a HPLC-electrospray mass spectrometer, postcolunm addition of acetone at 250 µL/min was executed using a syringe pump (Agilent 1100 thermo autosample, Paulo Alto, CA). Mobile phase and acetone were mixed in the system, and a split ratio of 1/10 was used to introduce the effluent into electrospray. The injection was performed in full scan mode with mass range from 280 to 750 mass spectrometer in centroid mode using a frequency of 1/s and a interscan time of 0.1 s. Mass spectrometry was carried out using multiple reaction monitoring, dry nitrogen as carrier gas at 80°C and 400 L/h. The capillary was held at 3.5 kV potential and the extraction voltage was –34 V.

Results and Discussion

Mineral Media Screening

In Fig. 1A–C, the results of the cell count and the surface tension are presented as a function of the fermentation time when applied to the media MM1, MM2, and MM3, supplemented with palm oil. These test aim to investigate the ability of the *P. aeruginosa* strain to grow in media containing different kinds of nitrogen sources and the possibility to reduce costs by utilizing low-cost nutrient sources. For all the tested media, cell growth can be verified, which indicates that the *P. aeruginosa* strain was able to utilize nitrogen from various sources for cell multiplication. These results corroborate the ones presented by Bonilla (8) and Abalos et al. (10) who reported a growth of the *P. aeruginosa* strains in cultures containing different nitrogen sources. As presented in Fig. 1A–C, it can be observed that the microbial strain presented the lowest cell multiplication rate when applied to the MM1 medium. In this case, the highest cell concentration was 1.2×10^7 CFU/mL. When the MM2 and MM3 media were applied, similar maximum cell concentrations were obtained of approx 4.2×10^7 and 4.7×10^7 CFU/mL, respectively. These results indicate that the use of the medium containing commercial N:P:K fertilizer (MM2) or inorganic MM3 medium could promoted similar the bacterium growth. In studies related to the bioremediation of soil contaminated with crude-oil applying a mixed culture containing *P. aeruginosa*, Oliveira and de Franca (11,12) verified similar behavior, that is, the microbial growth occurred irrespective of the use of a complex mineral medium or commercial fertilizer.

In Fig. 1A–C it can also be verified that the biosynthesis of the emulsifiers was initiated in the exponential phase of the growth of the microorganism, continuing in the stationary phase, irrespective of the inorganic medium that was utilized. In the 24-h period of the process, a marked reduction of the surface tension of the medium occurred. It is key to emphasize that the lowest values for surface tension were achieved applying the MM2 and MM3 media, presenting values of approx 35 mN/m.

CMD is an indirect means of measuring the surfactant production related to the range of the critical micelle concentration (13,14). The data presented in Fig. 1B and C demonstrate that the CMD^{-1} revealed a slight increase of the surface tension of the fermented medium, whereas the CMD^{-2} caused a remarkable increased of the surface tension, which is an indicator of the elevated quantity of surfactants in the respective media.

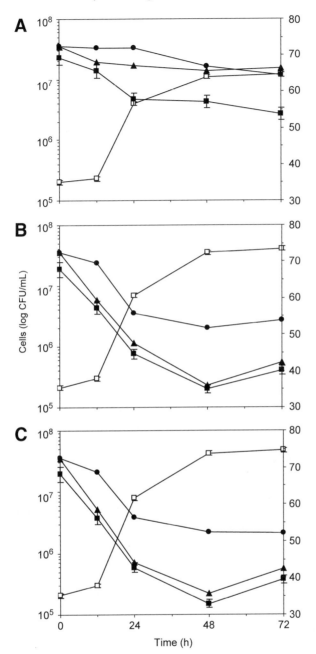

Fig. 1. Data of monitored parameters using three inorganic tested media: **(A)** MM1; **(B)** MM2; **(C)** MM3. (□), cells counting; (■), surface tension; (▲), CMD^{-1}; (●), CMD^{-2}.

When the MM1 medium was applied (*see* Fig. 1A), it was verified that both the CMD^{-1} and the CMD^{-2} presented high surface tension values, which was indicative of the low quantity of biosurfactants in the medium.

The selection of the mineral medium for the production of biosurfactants was based on the results of the surface tension of each of the three

Table 1
Surface Tension and CMD^{-1} Data Collected at 48 h of Bath Fermentations Using
P. aeruginosa FR Growing on Inorganic Media MM2 Enriched With Palm Oil

Rotary velocity (rpm)	Surface tension (mN/m)	CMD^{-1} (mN/m)
50	55 ± 1.0	65 ± 0.7
100	39 ± 1.2	44 ± 0.9
150	35 ± 1.2	40 ± 1.0
200	51 ± 1.1	63 ± 0.8

studied media and of its CMD^{-1} and CMD^{-2}. The results indicate that the use of the media MM2 or MM3 promoted a higher biosurfactant production when compared with the mineral medium MM1. Furthermore, from the presented results it is verifiable that palm oil can be applied as a low-cost substratum, renewable and adequate for the surfactant production by *P. aeruginosa*. Because the commercial fertilizer is cheaper than the MM3 medium components, the use of the MM2 medium was favored in the subsequent tests for the bioemulsifier production by the strain that was studied.

Effect of Rotation Velocity

The variation in the agitation speed of the cultures of 50–200 rpm influenced the reduction of the surface tension of the medium. As can be observed in Table 1, the values for surface tension of the cell-free medium were higher for tests carried out employing 50 and 200 rpm. During the 48 h of the bioprocess, it was verified that the surface tension of the cultures agitated at 50 and 200 rpm was superior to 50 mN/m and presented a pronounced increase of the values for surface tension when submitted to the CMD^{-1} measuring. These results indicate a lower production of surface-active compounds under these process conditions. When the medium was agitated at 100 or 150 rpm, the lowest values of surface tension, CMD^{-1} and CMD^{-2} were achieved, pointing to greater surface-active compounds biosynthesis. The results now presented corroborate those reported by Cunha et al. *(15)* who found greater biosurfactants production by *Pseudomonas* spp. using ethanol-blended gasoline as the sole carbon source at 100 rpm. *P. aeruginosa* is a facultative microorganism that may present growth in environments with low oxygen concentrations *(9)*; the production of surface-active compounds, however, involves stages of oxidation of the substrate as described by Maier and Sobéron-Chávez *(16)*. The agitation speed of the medium is a determining factor in the mixture of the aqueous and oily phase as well as in the oxygen mass transfer into the cultures using agitated flasks *(17)*. Thus, the agitation speed may have influenced the surfactant production by the applied strain, promoting, in some cases, a phase mixture and/or an adequate oxygen transfer rate for bioemulsifier effective production.

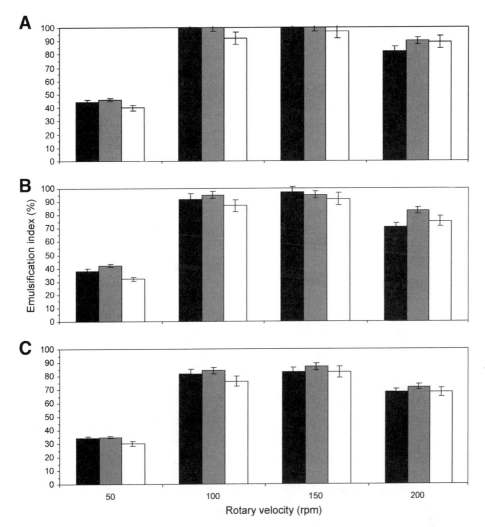

Fig. 2. Effect of rotary velocity on biosurfactant production at (■), 24 h; (■), 48 h; and (□), 72 h of the bioprocess, using different oil phases: **(A)** diesel; **(B)** jet fuel; **(C)** hexane.

Figure 2A–C represents the effect of the agitation on the emulsification index of the fermented cell-free MM2 medium obtained at 24, 48, and 72 h of process. It can be verified that rest time does not influenced emulsification index results, irrespective of agitation speeds of the cultures. It can be also verified that all the tested oily phases were emulsified. The use of 50 rpm furnished fermented media with lowest emulsification index, approx 30–45%. The cultures agitated at 200 rpm yielded free-cell fermented media with emulsification index from 55–80%. However, it is important to note that total emulsification was obtained with the fermented media from the cultures obtained at an agitation from 100 to 150 rpm. These results indicate that a higher concentration of surface-active compounds is obtained at rotation velocities from 100 to 150 rpm. Rahman et al. *(18)*

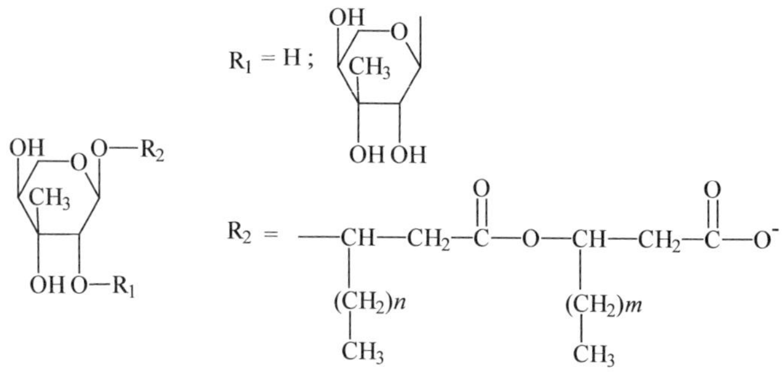

Fig. 3. Scheme of rhamnolipid general structure. $n = 4, 6, 8$ and $m = 4, 6, 8$ for length chain C_8, C_{10}, or C_{12} carbons atoms.

report on the surfactant production for two *P. aeruginosa* strains grown in a medium containing various carbon sources; also, they report an emulsification index from 25% to 90% for some oily phases such as hexane, diesel, and kerosene. In the present study, emulsification index of even 100% of these hydrocarbons was obtained with a stable time of up to 72 h, accounting for the good quality of the mixture of the surface-active compounds produced by the *P. aeruginosa* FR strain using palm oil as the sole carbon source.

Biosurfactant Characterization Using Mass Spectrometry

Some authors have been reporting on the characterization of surfactants produced by *P. aeruginosa* strains employing liquid chromatography coupled with mass spectrometry (7,18–20). Muligan (5) and Santos et al. (6) report that *Pseudomonas* strains produce rhamnolipids from carbon sources that are either soluble or nonsoluble in water. The applied extraction method was based on the solubility of the lipids in chloroform and solvent strip under nitrogen stream at 40°C was performed aiming to avoid molecules oxidation. The extract was applied into the column of HPLC equipment and the elution with acetonitrile/water yielded the separation of eight fractions that were submitted to electrospray ionization. Analysis of mass spectra data verified a typical structure between the compounds, with the presence of rhamnose and lipidic group. Figure 3 presents the standard molecular structure of the rhamnolipids (7), whereas Fig. 4 presents mass spectra of rhamnolipids mixture extracted from *P. aeruginosa* cultures on inorganic medium MM2 supplemented with palm oil as sole carbon source. Table 2 represents a synthesis of the data of the mass spectrometry of the each component of the rhamnolipids mixture.

The study of the mass spectrum of the rhamnolipids was initiated with the verification of the pseudomolecular ions and its fragments. The pseudomolecular ions with m/z 503 and 649 were most abundant in the samples, indicating a higher degree of biosynthesis, in agreement with the results reported by Déziel et al. (20). The greater production of Rha-$C_{10}C_{10}$

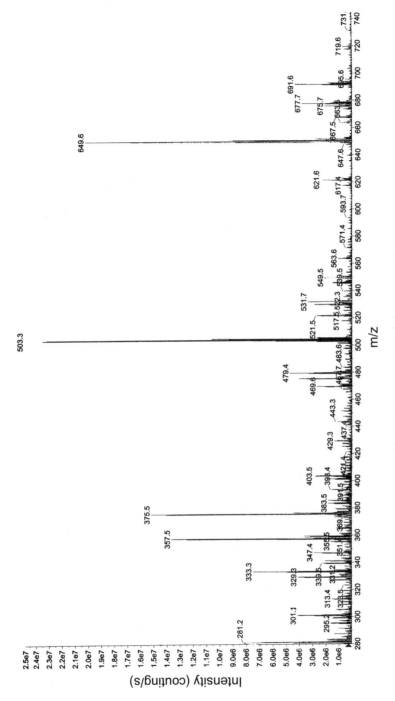

Fig. 4. Mass spectra of rhamnolipids.

Table 2
Mass Spectra Data of Isolated Rhamnolipds Produced by *P. aeruginosa* FR
Growing on Inorganic Media MM2 Supplemented With Palm Oil

Rhamnolipids structure	Pseudomolecular ion (m/z)	Ion fragments (m/z)
R_1 = Rha; R_2 = $C_{10}C_{12}$	677	163, 169, 187, 357, 503, 530, 650
R_1 = Rha; R_2 = $C_{10}C_{10}$	649	163, 169, 197, 311, 339, 479
R_1 = Rha; R_2 = C_8C_{10}	621	141, 163, 169, 311, 452
R_1 = H; R_2 = $C_{12}C_{10}$	531	161, 163, 169, 187, 333
R_1 = H; R_2 = $C_{10}C_{12:1}$	529	151, 163, 169, 187, 333
R_1 = H; R_2 = $C_{10}C_{10}$	503	103, 119, 163, 169, 333, 339
R_1 = H; R_2 = $C_{12:2}$	357	187, 333
R_1 = H; R_2 = $C_{8:2}$	301	131, 169, 187

and Rha-Rha-$C_{10}C_{10}$, m/z 503, 649, respectively, as compared with the other rhamnolipids species could be attributed to the culture conditions, carbon and nitrogen sources as well their ratio, as pointed by Santos et al. (6) and by Déziel (20).

The pseudomolecular ions with 677, 621, and 649 m/z presented rhamnose group as R_1, and the others, pseudomolecular ion with m/z of 531, 529, 503, 357, and 301, has hydrogen as R_1. Some rhamnolipids, having unsaturated fatty acid group at side chain as also found, corresponding to the ions m/z 301, 357, 529 and corroborates with the rhamnolipid characterization results described in the literature (19,20). The ion fragment at m/z 163 is related to the cleavage of the rhamnose group, and the fragments with the 119 and 103 m/z corresponding to its fragmentation (18–20).

These data clearly show the synthesis of the rhamnolipids by *P. aeruginosa* FR growing in low-cost mineral media and carbon sources.

Conclusion

The synthesis of surfactants by *P. aeruginosa* FR strain was verified. The bacterium could grow in three inorganic media using of palm oil as sole carbon source, producing surface-active compounds. It was also verified that nitrogen source in the mineral medium and rotation velocity influenced biosurfactants production. The biosurfactants produced by bacterium growing in palm oil presents emulsification index of even 100%, when diesel was used as oily phase and temporal (72 h) stable emulsions were obtained. HPLC-mass spectrometry data clearly shows the presence of eight rhamnolipids species in the cultures. The results permitted us to conclude that rhamnolipids can be synthesized by *P. aeruginosa* FR using a low cost and renewable carbon source, the palm oil.

Acknowledgments

We would like to thank Coordenação de Apoio a Pessoal de Nível Superior (CAPES), Conselho Nacional de Desenvolvimento Científico e Tecnológico (CNPq), and Financiadora de Projetos (FINEP) for financial support.

References

1. Puntus, I. F., Sakharovsky, V. G., Filinov, A. E., and Boronin, A. M. (2005), *Process Biochem.* **40,** 2643–2648.
2. Banat, I. M. (1995), *Bioresource Tech.* **51,** 1–12.
3. Bognolo, G. (1999), *Coll. Surf.—Part A* **152,** 41–52.
4. Abalos, A., Maximo, F., Manresa, M. A., and Baptista, J. (2002), *J. Chem. Tech. Biotech.* **77,** 777–784.
5. Mulligan, C. N. (2005), *Environ. Poll.* **133,** 183–198.
6. Santos, A. S., Sampaio A. P. W., Vasquez, G. S., Sant'Anna, L. M., Pereira, N., and Freire, D. M. G. (2002), *Appl. Biochem. Biotech.* **98,** 1025–1035.
7. Abalos, A., Maximo, F., Manresa, M. A., and Baptista, J. (2001), *Langmuir* **17,** 1367–1371.
8. Bonilla, M., Olivaro, C., Corona, M., and Soubes, M. (2005), *J. Appl. Microbiol.* **98,** 456–463.
9. Palleroni, J. (1984), In: *Bergey's Manual of Systematic Bacteriology*, Vol. 1, Krieg, N. R. ed., Williams & Wilkins, MD, USA, pp. 141–198.
10. Abalos, A., Deroncelé, V., Espuny, J., Bermúdez, R., and Manresa, A. (2000), *Rev. Cub. Quím.* **XII,** 24–29.
11. Oliveira, F. J. S. and de França, F. P. (2004), *Soil Rocks Lat. Am. Geotech. J.* **27,** 287–292.
12. Oliveira, F. J. S. and de França, F. P. (2005), *Appl. Biochem. Biotech.* **122,** 583–604.
13. Jain, D. K., Collins-Thompson, D. L., Lee, H., and Trevors, J. T. (1991), *J. Microbiol. Methods* **13,** 271–279.
14. Cassidy, D., Efendiev, S., and White, D. M. (2000), *Water Res.* **34,** 4333–4342.
15. Cunha, C. D., do Rosário, M., Rosado, A. S., and Leite, S. G. F. (2004), *Process Biochem.* **39,** 2277–2282.
16. Maier, R. M. and Soberó-Chávez, G. (2000), *Appl. Microbiol. Biotechnol.* **54,** 625–633.
17. Bailey, J. E. and Ollis, D. F. (1986), *Biochemical Engineering Fundamentals*, 2nd ed., McGraw-Hill Book Company, New York.
18. Rahman, K. S. M., Vasudevan, N., and Lakshmanaperumalsamy, P. (1999), *J. Environ. Poll.* **6,** 85–93.
19. Haba E., Pinazo A., Jauregui O., Espuny M. J., Infante M. R., and Manresa A. (2003), *Biotechnol. Bioeng.* **81,** 316–322.
20. Déziel, E., Lépine, F., Milot, S., and Villemur, R. (2000), *Biochem. Biophys. Acta.* 1485, 145–152.

Novel Approach of Corn Fiber Utilization

G. KÁLMÁN,*,1 K. RECSEG,2 M. GÁSPÁR,1 AND K. RÉCZEY1

[1]Budapest University of Technology and Economics,
Department of Agricultural Chemical Technology, H-1111 Budapest,
Szt. Gellért tér 4, E-mail: gergely_kalman@mkt.bme.hu;
and [2]Bunge Rt. H-1139 Budapest, Petneházy u. 2-4

Abstract

The corn wet milling process produces a 10% (w/w of the processed corn) byproduct called corn fiber, which is utilized worldwide as a low-value feedstock for cattle. The aim of this study was to find a higher value use of corn fiber. The main fractions of corn fiber are: 20% starch, 40% hemicellulose, 14% cellulose, and 14% protein. Extraction of the highly valuable, cholesterol-lowering corn fiber oil is not feasible owing to its low (2% w/w) concentration in the fiber. The developed technology is based on simple and inexpensive procedures, like washing with hot water, dilute acid hydrolysis at 120°C, enzymatic hydrolysis of cellulose, screening, drying, and extraction. The main fractions are sharply separated in the order of starch, hemicellulose, cellulose, lipoprotein, and lignin). The lipoprotein fraction adds up to 10% of the original dry corn fiber, and contains 45% corn fiber oil, thus yielding more oil than direct extraction of the fiber. It is concluded that the defined method makes the extraction of the corn fiber oil economically feasible. The fractionation process also significantly increases the yield of cholesterol-lowering substances (sterols and sterol-esters). At the same time clear and utilizable fractions of monosaccharides, protein, and lignin are produced.

Index Entries: Bioethanol; corn fiber hydrolysis; corn fiber oil; phytosterol.

Introduction

Corn fiber is an abundant and inexpensive byproduct of the corn wet milling process, comprising about 10% of the processed dry corn. Corn fiber is a main ingredient of corn gluten feed (CGF), which is a low-value feed for cattle. Some countries, like the US export the corn fiber overseas, as the demand of the CGF is lower than the supply. The worldwide increasing demand for bioethanol results in more corn processing, but at the same time the demand for animal feed—depending on the country—has remained stagnant or is decreasing. The value of the fiber in the EU-25 fluctuates around 0.1 $/kg. The aim of this study was to find a more economic way of utilizing corn fiber.

*Author to whom all correspondence and reprint requests should be addressed.

Opponents of bioethanol argue that its production is not economical. In fact, it is difficult to define the price of bioethanol because it significantly depends on the substrate from which it is made. It is necessary to define the raw material and the technology of production when economic feasibility of bioethanol is surveyed. Abundant and cheap substrates for bioethanol fermentation are still in demand. Such substrates could be agro-industrial byproducts like corn fiber or wheat bran. The American example shows that the significant increase in production of ethanol from corn leads to a surplus of corn fiber, so it is important to search for alternative options for utilizing the excess corn fiber.

Not only is the bioethanol production potential of the corn fiber valuable and noteworthy, but also the extractable lipid content of the fiber (called corn fiber oil) contains valuable cholesterol-lowering substances like phytosterols and phytosterol-esters (like ferulate-phytosterol-esters) *(1–15)*. According to our analysis, corn fiber oil contains phytosterols in the highest concentration (6–10%) ever reported in edible plant oils. Phytosterols are effective reducers of LDL-cholesterol levels in humans. A high LDL-cholesterol level is the main reason for heart and circulatory diseases. The oil also contains some tocopherols and tocotrienols, which are commonly known as vitamin-E. Ferulic acid, tocopherols, and tocotrienols are effective antioxidants. However, extraction of the highly valuable corn fiber oil is not economically feasible owing to its low (2–2.5% w/w) concentration in the fiber.

Dilute acid hydrolysis of lignocelluloses and corn residues has been widely discussed in the literature *(16–27)*. Dilute (0.5–5% w/w) acid pretreatment at the temperature of 120–180°C can effectively remove the hemicelluloses from the lignocellulose matrix. It has been shown that there is a direct relationship between dilute acid-mediated xylan removal and concomitant increased enzymatic breakdown of cellulose *(16)*. It seems clear that the efficiency of dilute acid hydrolysis—in terms of promoting enzymatic digestibility—depends on the hemicelluloses/cellulose ratio. Torget et al. *(16)* reported 90–100% digestibility of corn cob cellulose after dilute acid pretreatment, whereas corn stover, and wood samples did not reach this value. The hemicellulose/cellulose ratio of corn cob is 0.89, whereas at the other samples it varies between 0.43 and 0.64. Enzymatic hydrolysis of cellulose has been studied for a long time *(16,17)*. Pretreated celluloses usually need about 40 g FPU/g cellulose enzyme loading, plus 5–15 g IU/g cellulose β-glucosidase activity to achieve satisfactory glucose yield *(16)*.

Singh et al. studied the removal of nonlipid components of corn fiber, and thereby the increase of corn fiber oil and phytosterol concentration in the remaining fiber *(13–15)*. They examined the effect of dilute sulfuric acid and enzymatic pretreatments *(13)*. Depending on the treatment, the oil concentration of the fiber increased from 1.4% to 12.2% and the total phytosterol level increased from 2 to 14 mg/g fiber. This result was obtained by dilute sulfuric acid pretreatment at the temperature of 121°C. In this case the weight loss was 76.1%. The authors also determined that the typical composition of the native corn fiber is: approx 40% hemicellulose, 12% cellulose, 25% starch, 10% protein, 10% lignin and ash, and approx 3% oil. With this background information a unique fractionation process was developed for corn fiber.

Materials and Methods

Analysis of the Raw Material

The raw material was kindly donated by Hungrana Rt. (Szabadegyháza, Hungary). The dry material content of the native fiber was approx 40%, so the fiber was immediately dried at room temperature to avoid biological decay. All analytical and technological processes were performed using the dry fiber, except when the catalytic effect of the sulfurous acid content of the native fiber was tested.

The polysaccharide content was measured as follows. The fiber was milled and dried at room temperature (less then 40°C). Dry material content of the milled fiber was then measured by drying at 105°C until weight stability. Five mililiter of sulfuric acid (72% w/w) was added to 1 g milled fiber. The suspension was incubated at room temperature for 1.5 h. Then it was diluted with 145 mL distilled water, and the suspension was stirred and incubated at 120°C for 1.5 h. The suspension was then chilled, and the supernatant was sampled. The sample was measured for monosaccharides by high-performance liquid chromatography (HPLC). The glucan, xylan, and arabinan content of the raw material were calculated according to Eq. 1:

$$C = \frac{c \times V}{f \times m} \tag{1}$$

where:
C is the polysaccharide content (mass fraction);
c the monosaccharide concentration measured by HPLC (g/L);
V the volume of the suspension (L);
F the hydration factor (glucan: 180/162; xylan: 150/132; arabinan: 150/132); and
m the dry weight of the analyzed raw material (g).

The starch content was measured as follows. Twenty milliliters of dilute hydrochloric acid (ρ = 1.125 g/mL) and 250 mL distilled water (DW) was added to 3 g milled fiber, then the suspension was incubated at 120°C for 1.5 h. Then the suspension was chilled, and sample was taken from the supernatant for HPLC analysis. Starch content of the raw material was calculated according to Eq. 1. Cellulose content was calculated as the difference of the glucan and the starch content. The protein content was measured according to Dumas' method (28).

The oil content was measured as follows: 5–15 g milled fiber was placed into a paper vial, and extracted with hexane in a soxhlet-extractor until weight stability. Oil content was calculated as the yielded oil divided by the dry weight of the sample.

Ash content was measured by burning of 1–5 g dry fiber at 600°C for 6 h. Ash content was calculated as the yielded ash divided by the dry weight of the sample.

The HPLC analysis was performed using an Aminex HPX-87H column at 65°C. The eluent was 5 mM H_2SO_4 at a flow rate of 0.5 mL/min. Glucose, xylose, arabinose, cellobiose, acetic acid, and ethanol were detected and quantified by refractive index.

Fractionation

Step 1 (Destarching, Hydrolysis, and Fermentation of Supernatant)

Air-dried raw material was used except when the catalytic effect of the sulfurous acid content of the native fiber was tested. The dried fiber was diluted with DW so that the dry material concentration was 8% w/w. The suspension was incubated at 120°C for 2 h. At elevated temperatures the sulfurous acid catalyses the hydrolysis of starch by lowering the pH of the suspension. After cooling down the suspension was filtered through a 150 µm mesh nylon filter, thus fiber and supernatant were separated. The fiber was washed with 80°C DW three times to completely remove adsorbed substances. The fiber was dried at room temperature (less then 40°C). The supernatants were combined. The combined supernatant was diluted 1:1 with 8% (w/w) dilute sulfuric acid and then incubated at 120°C for 10 min to hydrolyze poly- and oligosaccharides. Then the solution was chilled and sample was taken for HPLC analysis.

Fermentability of starch containing supernatant was tested. The combined supernatant was not suitable for ethanol fermentation because of its low starch concentration, so the supernatant of the first separation was tested for fermentation. Before fermentation the starch was hydrolyzed with 0.2 g α-amylase/g starch (Termamyl Supra, Novozymes, Denmark) and 0.1 g glucoamylase/g starch (AMG 300L, Novozymes). The first supernatant was incubated at 120°C for 10 min at the beginning of the hydrolysis (0.1 g α-amylase addition), later temperature was reduced to 90°C (+0.1 g α-amylase addition), and then to 60°C, when the glucoamylase was added. The pH of the hydrolysis was 4.6. Fermentations were conducted using commercial *Saccharomyces cerevisiae* at the temperature of 35°C, pH was adjusted to 4.8 using acetate buffer.

For inoculum preparation fresh commercial compressed baker's yeast (*S. cerevisiae*, Budafok Yeast and Spirit Factory Ltd., Budapest, Hungary) was used. The inoculum was prepared in a 750 mL-Erlenmeyer-flask containing 150 ml sterile solution in which the concentration of nutrients in g/L were 50 glucose, 2.5 yeast extract, 5 peptone, 1 KH_2PO_4, 0.3 $MgSO_4$, and 2 NH_4Cl. The Erlenmeyer flask was incubated in a rotary shaker at 30°C and 300 rpm for 24 h.

To 90 mL of the first supernatant 5 mL of yeast inoculum and 5 mL acetate buffer (1M, pH = 5.5) were added and the flask was shaken at 300 rpm. Ethanol yield and glucose conversion were calculated in two different ways. Three flasks were sampled for HPLC analysis after 0, 3, 6, 24, and 48 h. Ethanol yield and glucose conversion were calculated using the concentration values provided by the HPLC before and after fermentation as:

$$Y = \frac{Y_{EtOH}}{Y_{EtOHmax}} \times 100\,(\%),$$

where:
Y is the ethanol yield;
Y_{EtOH} the yielded ethanol (g); and
$Y_{EtOH\ max}$ the maximal (theoretical) ethanol yield (grams), equals 0.51 times the consumed glucose.

$$C = \left(1 - \frac{C_{glucose}}{C_{glucose,\ start}}\right) \times 100\,(\%)$$

where:
C is the conversion
$C_{glucose}$ the concentration of glucose (g/L); and
$C_{glucose,\ start}$ the concentration of glucose (g/L) at the beginning of the fermentation.

Three other flasks were equipped with oneway valve plug. The weights of these flasks were measured before and after fermentation. Ethanol yield can be calculated from the weight difference, which is caused by carbon dioxide loss. The yielded ethanol is calculated as 92/88 times the weight of the carbon dioxide (weight difference).

Step 2 (Dilute Acid Hydrolysis)

After destarching, the fiber was suspended in 1% dilute sulfuric acid so that the dry material concentration of the suspension was set to 8% w/w. The suspension was incubated at 120°C for 2 h. Then the suspension was cooled down and filtered through a 150-μm mesh nylon filter, thus fiber and supernatant were separated. The fiber was washed with 80°C DW three times to completely remove adsorbed substances. The fiber was dried at room temperature (less then 40°C). The supernatants were combined. The combined supernatant was diluted 1:1 with 8% w/w dilute sulfuric acid and then incubated at 120°C for 10 min to hydrolyze poly- and oligosaccharides. After cooling down 3 mL of sample was taken for HPLC analysis.

Step 3 (Enzymatic Hydrolysis of Cellulose)

The dried fiber was suspended in $0.1M$ acetic acid/sodium-acetate buffer (pH = 4.8) so that the dry material concentration of the suspension was set to 3% w/w. The suspension was incubated at 50°C for 48 h and flasks were shaken at 200 rpm. Commercial enzymes, like Celluclast L 1.5 (Novozymes) and Novozyme 188 (Novozymes) were loaded. The total loaded enzyme was 5 FPU/g dry substrate plus 5 IU/g dry substrate β-glucosidase activity, added in four equal doses. These correspond to 10.2 IU/g cellulose enzymatic activity, respectively. After 48 h the fibrous material formed two sharply separating fractions: a white fine fiber with low-settling velocity and a brown coarse fiber with high-settling velocity. Thus the suspension was filtered in two steps. In the first step, the slowly settling white (fine) fiber fraction was decanted and filtered on a G3 glass funnel. In the second step, the quickly settling brown (coarse) fiber fraction was filtered through a 150-μm mesh nylon filter. Thus fine, coarse fibers and supernatant were separated. The fibers were dried at room temperature (less then 40°C). The supernatant was analyzed for carbohydrates by HPLC. Each fiber fractions of the above subscribed fractionation method were analyzed for dry material content, starch, cellulose, xylan, arabinan, acetate, protein, lipids, and ash using the defined methods.

Results and Discussion

Raw Material Analysis

Corn fibers of three seasons (2002–2004) and seven individual charges were analyzed. The starch content of the raw material was the most variable among the components. It varied from 14% to 26.3% w/w. This variability is caused by the quality of the raw corn, according to the processor (Hungrana Rt., Hungary). Average composition of the corn fiber can be seen in Table 1. It is noteworthy, that the 5% total lipid content is not directly extractable. Sixty percent of the lipids fraction is chemically bound as lipoproteins and lipopolisaccharides. These bonds must be cut in order to make total lipid content extractable. The high (37.1/15 = 2.5) hemicelluloses/cellulose ratio significantly differs from the value of most lignocelluloses (approx 25/50 = 0.5).

Composition of the dried fiber after destarching can be seen in Table 2. Components are indicated as g/100 g native (original) dry corn fiber. Table 2 shows that 100% of the starch could be extracted. It is noteworthy that no chemical reagents and no stirring was applied during extraction, thus this step might be an inexpensive, convenient process even at industrial scale. A significant part (23.6%) of the proteins was also dissolved, whereas only a slight (6.2%) loss of the hemicelluloses was detected. Cellulose and lipids were untouched during the first step of the fractionation. Therefore it can be concluded that this step can sharply separate the starch from other components.

Table 1
Composition of Corn Fiber

Component	%
Starch	21.3
Cellulose	15
Hemicellulose	37.1
Xylan	22
Arabinan	10.9
Glucan	2.2
Acetate	2
Protein	14
Total lipids	5
Extractable oil	2
Bound lipids	3
Ash	1
Sum	93.2

Table 2
Composition of the Fiber After Destarching (Step 1)

Component	g/100 g Native dry fiber	Change (%)
Dry weight	72.5	−27.5
Starch	n.d.[a]	−100
Cellulose	15	0
Hemicellulose	34.8	−6.2
Xylan	20.5	−6.8
Arabinan	10.1	−7.3
Glucan	2.2	0
Acetate	2	0
Protein	10.7	−23.6
Total lipids	5	0
Extractable oil	2	0
Bound lipids	3	0
Ash	1	0
Sum	66.3	−28.3

[a]n.d., not detectable.

Table 3 shows the composition of the liquid phase of the starch extraction. It can be seen that the starch was completely found in the liquid phase, whereas there is a slight difference between the extracted and the measured xylan and arabinan. This difference may be owing to the fact that the starch is extracted in the form of polysaccharides, whereas the extracted xylan and arabinan is probably in the form of mono- and oligosaccharides, which are subject to thermal destruction.

A slight change was made in the destarching process when the catalytic effect of the sulfuric dioxide absorbed in the native wet corn fiber

Table 3
Composition of the Liquid Phase After Destarching (Step 1)

Component	g/100 g Native dry fiber
Starch	21.3
Hemicellulose	1.9
Xylan	1.3
Arabinan	0.6
Glucan	n.d.
Acetate	n.d.
Sum	23.2

n.d., not detectable.

was tested. Fiber was not dried before starch extraction, and temperature was reduced from 120°C to 100°C, other parameters remained unchanged. pH of the suspension was 3.6 throughout the process. Results showed that only 55% of the starch was extracted, thus the reduction of the incubation temperature was not compensated with the catalytic effect of the sulfuric dioxide.

The supernatant was hydrolyzed with commercial amylases and then tested for ethanol fermentation. After 24 h of fermentation the glucose conversion was 100%, whereas the ethanol yield was 96.4–98.3%. Gravimetric method of tracing ethanol fermentation was also tested. However, gravimetric results were significantly different from the results given by HPLC. Ethanol yields measured by the gravimetric method were 6–16% lower, presumably because of the absorption of carbon dioxide in the fermentation liquor.

After destarching, the resulting fiber was dried and suspended in dilute sulfuric acid to extract hemicelluloses. According to Table 4 hydrolysis with dilute acid impacted nearly every components: 64% of the dry weight of the fiber gained after destarching was extracted in step 2. Ninety seven percent of the hemicelluloses were extracted, and 56% of the proteins were also dissolved. Arabinan and acetate components of the hemicellulose seem to be easier to hydrolyze, presumably because of that they form side chains or substitutes, which are easier to hydrolyze than the xylan backbone. A significant part (15.3%) of the cellulose was also hydrolyzed, which was not the goal of this step. Yield of corn fiber oil extraction increased significantly (75%) owing to the tearing of lipopolysaccharide and lipoprotein bonds. Ash content of the corn fiber is 100% soluble in dilute sulfuric acid. Table 5 shows the composition of the liquid phase of the dilute acid hydrolysis.

Several dilute acid hydrolyses were performed to optimize the concentration of the sulfuric acid. It is noteworthy that only 0.5% dilute sulfuric acid at the relatively low temperature of 120°C can hydrolyze 97% of the hemicelluloses. We have previously reported (17) that 0.5% dilute sulfuric

Table 4
Composition of the Fiber Fraction After Dilute Acid Hydrolysis (Step 2)

Component	g/100 g Native dry fiber	Change (%)
Dry weight	26	−64.1
Starch	n.d.	0
Cellulose	12.7	−15.3
Hemicellulose	0.9	−97.4
Xylan	0.7	−96.6
Arabinan	0.2	−98
Glucan	n.d.	−100
Acetate	n.d.	−100
Protein	4.7	−56.1
Total lipids	5	0
Extractable oil	3.5	75
Bound lipids	1.5	−50
Ash	n.d.	−100
Sum	23.3	−64.9

n.d., not detectable.

Table 5
Composition of the Liquid Phase After Dilute Acid Hydrolysis (Step 2)

Component	g/100 g Native dry fiber
Hemicellulose	32.1
Xylan	19.2
Arabinan	9.9
Glucan	1
Acetate	2
Sum	32.1

acid at the same temperature (120°C) and residence time (2 h) can extract only 27.4% of corn-stalk hemicelluloses. Only 2% dilute sulfuric acid can extract 92–94% of corn-stalk hemicelluloses, which shows that corn fiber is a relatively easy-to-hydrolyze substrate compared to other lignocellulosic byproducts. Dilute acid hydrolyses were run to set the optimal concentration of the sulfuric acid. Acid concentrations, pH of the supernatant after hydrolysis, dry weight and hemicellulose content of the fiber after hydrolysis are shown in Table 6. The results show that it is not rational to decrease the concentration of sulfuric acid below 0.5% w/w, as the mass of the residual fiber and the amount of hemicellulose in the fiber increases, thus the sharpness of separation decreases.

After dilute acid treatment, the fiber was air-dried and subjected to enzymatic hydrolysis. After enzymatic hydrolysis, two solid fractions were obtained. There was a white fine, and a brown coarse fiber. The brown fiber

Table 6
Results of Different Dilute Acid Hydrolyses

Acid (%)	pH	Dry weight of the fiber	Hemicellulose in the fiber
0.5	1.02	26	0.9
0.7	0.85	27.1	0.7
0.9	0.72	24.6	0.3
1.1	0.67	23.8	0.4
1.3	0.56	23.4	0.4
1.5	0.50	25	0.3

Table 7
Composition of the Fine Fiber After Enzymatic Hydrolysis (Step 3)

Component	Fine fiber g/100 g native dry fiber	Coarse fiber g/100 g native dry fiber	Total change (%)
Dry weight	10.1	1.7	−54.6
Cellulose	0.4	0.7	−91.3
Hemicellulose	n.d.	n.d.	−100
Protein	4.7	n.d.	0
Total lipids	4.6	0.4	0
Extractable oil	4.6	0.4	42.9
Bound lipids	0	0	−100
Sum	9.7	1.4	−52.4

n.d., not detectable.

settled faster then the white fiber, thus they could be separated via decantation. Results are seen in Table 7. It can be seen that even 4 IU/g DW enzymatic activity could hydrolyze more than 91% of the cellulose present in the fiber after dilute acid hydrolysis. This proves the superior digestibility of the corn fiber cellulose after pretreatment. As previously reported (17), dilute (2%) acid hydrolysis at the temperature of 120°C resulted in only 56% cellulose digestibility in corn stalk. We draw attention to the fact that we produced a solid fraction of corn fiber, which contains 46% corn fiber oil. This is significantly more than that reached by Singh et al. (13) (12.2%) through dilute sulfuric acid pretreatment. This novel technology yielded 5 g corn fiber oil/100 g dry corn fiber. In fact, this is the highest yield of corn fiber oil ever reported. During the sequential hydrolyzes lipopolysaccharides and lipoproteins degraded, thus bound lipids became extractable. Concentration of phytosterols and phytosterol-esters in the oil contents of each fiber fractions was also determined (Table 8). The results show that the yield of phytosterol-esters slightly decreased—presumably because of acid catalyzed hydrolysis—but the yield of phytosterols significantly increased. The total phytosterol yield increased from 1.45 to 1.84 g/kg CF, which corresponds to 27% increase.

Table 8
Yields of Phytosterols and Phytosterol-Esters

	Direct extraction	Fractionation
Free phytosterol yield (g/kg CF)	0.46	1.24
Phytosterol-ester yield (g/kg CF)	0.99	0.60
Sum	1.45	1.84

Table 9
Composition of the Liquid Phase After Enzymatic Hydrolysis (Step 3)

Component	g/100 g native dry fiber
Cellulose	11.6
Hemicellulose	0.7
Xylan	0.7
Arabinan	n.d.
Sum	12.3

n.d., not detectable.

Through the fractionation process it is proven that corn fiber contains no significant amounts of lignin. The fine and coarse fibers contain only 0.7 g undetermined ("others") fraction/100 g processed native fiber altogether, which means that this is the maximum (theoretical) amount of acid insoluble lignin present in the native fiber. Those results from Singh et al. (13–15) and other previous authors are inconsistent with these results. The applied Hägglund's method (29) does not seem to be suitable when the raw material contains significant amounts of unhydrolizable protein and lipids, which is the case at corn fiber. Those high values of lignin (7–12%) given by previous authors include protein and lipid fractions (and probably nucleic acids, as well). The real amount of acid insoluble lignin present in native corn fiber is less than 0.7 g/g fiber, but the exact determination of this value was not the goal of this study.

Table 9 shows the composition of the liquid phase after enzymatic hydrolysis. The cellulose fraction can be found in the liquid phase in the form of glucose monosaccharides. These conditions of enzymatic hydrolysis yielded no byproducts, so the glucose content of this solution is fermentable to ethanol at theoretical yields.

Conclusions

The novel technology described in this article implements the simultaneous, optimized production of bioethanol and corn fiber oil, which we consider the two most valuable products that can be made from corn fiber. This process is able to cleanly separate the fractions of native corn fiber. The first step removed all the starch content of the native corn fiber. The liquid phase of the first step—containing the extracted starch—was 96.4–98.3% fermentable to ethanol after enzymatic hydrolysis by commercial amylases. Only 0.5% w/w dilute sulfuric acid and the temperature of 120°C was needed to remove 97.4% of the hemicellulose and also to make remaining cellulose 91.3% enzymatically digestible.

The third step hydrolyzed the cellulose, which resulted in two solid fractions: a white, fine, and a brown, coarse fiber. These solid fractions could be separated via decantation or flotation. The fine fiber added up to 10% of the original CF, and contained 46% w/w corn fiber oil. This means that we produced 5 g corn fiber oil/100 g CF, which is the highest yield of CFO reported. It is noteworthy that only 40% of the lipid content of corn fiber is directly extractable, since 60% of the lipids are in the bound forms of lipoproteins and lipopolysaccharides. The original concentration of corn fiber oil in the fiber (2%) was 23 times increased in the residue. The process described herein liberated all of the lipids present in the corn fiber.

Another important observation of this study is that corn fiber contains no significant amount of lignin. It is likely that the values (7–12%) of lignin content given by previous papers may not be correct. The applied Hägglund's method (29) is not adequate for determining the lignin content of corn fiber owing to its significant fraction of unhydrolysable proteins and lipids. Our mass balances indicate that lignin can only be present in traces in the native corn fiber.

Acknowledgment

This work was supported by the Hungarian Ministry of Education (NKFP-OM 00231/2001) and the Hungarian Scientific Research Fund (OTKA-TS049849) project. The authors also want to thank Novozymes for supplying industrial enzymes and Hungrana Ltd for supplying corn fiber.

References

1. Hicks, K. B. and Moreau, R. A. (2001), *Food Technol.* **55**, 63–67.
2. Vanhanen, H. T., Kajander, J., Lehtovirta, H., and Miettinen, T. A. (1994), *Clin. Sci.* **87**, 61–67.
3. Singh, V., Moreau, R. A., and Cooke, P. H. (2001), *Cereal Chem.* **78**, 436–441.
4. Moreau, R. A., Hicks, K. B., and Powell, M. J. (1999), *J. Agric. Food Chem.* **47**, 2869–2871.
5. Moreau, R. A., Singh, V., Eckhoff, S. R., Powell, M. J., Hicks, K. B., and Norton, R. A. (1999), *Cereal Chem.* **76**, 449–451.
6. Singh, V., Moreau, R. A., Doner, L. W., Eckhoff, S. R., and Hicks, K. B. (1999), *Cereal Chem.* **76**, 868–872.
7. Ramjiganesh, T., Roy, S., Vega-Lopez, S., McIntyre, J., and Fernandez, M. L. (1999), *FASEB J.* **13**, 561.
8. Moreau, R. A., Powell, M. J., and Hicks, K. B. (1996), *J. Agric. Food Chem.* **44**, 2149–2154.
9. Singh, V., Moreau, R. A., Haken, A. E., Hicks, K. B., and Eckhoff, S. R. (2000), *Cereal Chem.* **77**, 665–668.
10. Taylor, S. L. and King, J. W. (2000), *J. Am. Oil Chem. Soc.* **77**, 687–688.
11. Wilson, T. A., DeSimone, A. P., Romano, C. A., and Nicolosi, R. J. (2000), *J. Nutr. Biochem.* **11**, 443–449.

12. Singh, V., Moreau, R. A., Haken, A. E., Eckhoff, S. R., and Hicks, K. B. (2000), *Cereal Chem.* **77,** 692–695.
13. Singh, V., Johnston, D. B., Moreau, R. A., Hicks, K. B., Dien, B. S., and Bothast, R. J. (2003), *Cereal Chem.* **80,** 118–122.
14. Singh, V. and Moreau, R. A. (2003), *Cereal Chem.* **80,** 123–125.
15. Singh, V., Moreau, R. A., and Hicks, K. B. (2003), *Cereal Chem.* **80,** 126–129.
16. Torget, R., Walter, P., and Himmel, M. (1991), *Appl. Biochem. Biotech.* **28–29,** 75–86.
17. Kálmán, G., Varga, E., and Réczey, K. (2001), *Chem. Biochem. Eng. Quart* **16,** 151–157.
18. Esteghlalian, A. and Hashimoto, A. G. (1997), *Bioresour. Technol.* **59,** 129–136.
19. Schell, D. J., Walter, P. J., and Johnson, D. K. (1992), *Appl. Biochem. Biotech.* **34–35,** 659–665.
20. Lee, Y. Y., Iyer, P., and Torget, R. W. (1999), *Adv. Biochem. Eng. Biotechnol.* **65,** 93–115.
21. Chen, R., Lee, Y. Y., and Torget, R. (1999), *Appl. Biochem. Biotech.* **57–58,** 133–146.
22. Torget, R. and Hsu, T. (1994), *Appl. Biochem. Biotech.* **45–46,** 5–21.
23. Torget, R., Himmel, M. E., and Grohmann, K. (1991), *Biores. Technol.* **35,** 239–246.
24. Chen, R., Wu, Z., and Lee, Y. Y. (1998), *Appl. Biochem. Biotech.* **70–72,** 37–49.
25. Hsu, T. A. and Himmel, M. (1996), *Appl. Biochem. Biotech.* **57–58,** 3–18.
26. Hsu, T. and Nguyen, Q. (1995), *Biotechnol. Tech.* **9,** 25–29.
27. Schell, D. J., Torget, R., Power, A., Walter, P. J., Grohmann, K., and Hinman, N. D. (1991), *Appl. Biochem. Biotech.* **28–29,** 87–97.
28. AOCS Official Method Ba 4e-93: Generic Combustion Method for Determination of Crude Protein.
29. Hägglund, E. (1951), *Chemistry of Wood*, Academic Press, New York.

Stimulation of Nisin Production From Whey by a Mixed Culture of *Lactococcus lactis* and *Saccharomyces cerevisiae*

CHUANBIN LIU, BO HU, YAN LIU, AND SHULIN CHEN*

Department of Biological Systems Engineering, Washington State University, Pullman, WA 99164-6120; E-mail: chens@wsu.edu

Abstract

The production of nisin, a natural food preservative, by *Lactococcus lactis* subsp. *lactis* (ATCC 11454) is associated with the simultaneous formation of lactic acid during fermentation in a whey-based medium. As a result of the low concentration and high separation cost of lactic acid, recovering lactic acid as a product may not be economical, but its removal from the fermentation broth is important because the accumulation of lactic acid inhibits nisin biosynthesis. In this study, lactic acid removal was accomplished by biological means. A mixed culture of *L. lactis* and *Saccharomyces cerevisiae* was established in order to stimulate the production of nisin via the *in situ* consumption of lactic acid by the yeast strain, which is capable of utilizing lactic acid as carbon source. The *S. cerevisiae* in the mixed culture did not compete with the nisin-producing bacteria because the yeast does not utilize lactose, the major carbohydrate in whey for bacterial growth and nisin production. The results showed that lactic acid produced by the bacteria was almost totally utilized by the yeast and the pH of the mixed culture could be maintained at around 6.0. Nisin production by the mixed culture system reached 150.3 mg/L, which was 0.85 times higher than that by a pure culture of *L. lactis*.

Index Entries: Nisin; whey; mixed culture; fermentation.

Introduction

Cheese whey is a byproduct of the dairy industry obtained by separating the coagulum from whole milk, cream, or skim milk and represents about 85–90% of the milk volume whereas retaining 55% of the milk nutrients [1]. About 30×10^6 t of liquid whey is produced annually in the United States alone with just a small portion being actively utilized to produce whey protein [2]. Whey permeate is the liquid that passes through the ultra-filtration membrane by which the whey protein is retained [1]. The majority of the nutrition in whey and whey permeate, which includes lactose, soluble proteins, lipids, and mineral salts, is not fully utilized because

---

*Author to whom all correspondence and reprint requests should be addressed.

of the low value and limited market of whey products. Because of its low concentration of milk constituents (e.g., lactose content is only 4.5–5%, [w/v]), whey and whey permeate are commonly considered wastes. Environmentally friendly and economic disposal of these byproducts is a great challenge to the dairy industry *(3)*. Biological processing of these cheese byproducts for production of value-added products is considered one of the most profitable utilization alternatives, with generation of high-value coproducts from this currently available waste stream being an example of "biorefinery" in action *(4)*.

Nisin is an antimicrobial peptide produced by certain *Lactococcus* bacteria *(5)*. The peptide has strong antimicrobial activity against almost all Gram-positive bacteria and their spores, especially several food-borne pathogens such as *Listeria monocytogenes*, *Staphylococcus aureus*, and psychrotrophic enterotoxigenic *Bacillus cereus (6–8)*. Therefore, nisin has been accepted as a safe and natural preservative in more than 50 countries and is widely used as an antimicrobial agent in the food industry *(9)*. The US Food and Drug Administration views nisin derived from *Lactococcus lactis* subsp. *lactis* to be a GRAS (generally recognized as safe) substance for use as an antimicrobial agent *(10)*. Direct addition of nisin to various types of foods, such as cheese, margarine, flavored milk, canned foods, and so on, is permitted *(11)*. In addition, nisin is also being considered for use in health and cosmetic products *(12)*.

Several studies *(13,14)* have indicated that cheese whey could be used as feedstock for the production of nisin, assuming supplementation of some essential nutrients within the fermentation process. Biosynthesis of nisin is coupled with the growth of lactic acid bacteria, and a significant amount of lactic acid is simultaneously formed alongside the nisin biosynthesis. Lactic acid is an important chemical for food processing. It can also be used as a raw material in the production of the biodegradable polymer poly(lactic) acid *(15)*. Presently, though, lactic acid is not recovered in the current industrial process for nisin production. However, previous work by the authors has revealed that it is feasible to produce nisin and lactic acid simultaneously by fermentation, because the optimal conditions for nisin biosynthesis and lactic acid formation by *L. lactis* using cheese byproducts as feedstock are almost the same *(16)*. Unfortunately, owing to the low concentration and high cost of separating and recovering lactic acid, the process may not be economical.

Considering that the accumulation of lactic acid in fermentation broth inhibits nisin biosynthesis *(17)*, this study is focused on stimulation of nisin production by the removal of lactic acid via use of a mixed culture system. A mixed culture of *L. lactis* and *Saccharomyces cerevisiae* (Fig. 1) was established in order to stimulate the production of nisin via the *in situ* consumption of lactic acid by the yeast strain, which is capable of utilizing lactic acid as its carbon source. The *S. cerevisiae* in the mixed culture does not compete with the nisin producing bacteria because the yeast does not

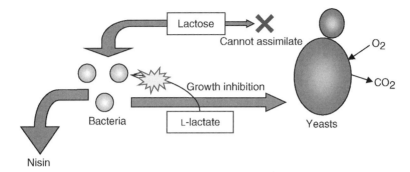

Fig. 1. The interaction between *S. cerevisiae* and *L. lactis* in a mixed culture system for nisin production.

utilize lactose, which is the major carbohydrate in whey for bacterial growth and nisin production. A pH control strategy without alkali addition was developed and nisin production in the mixed culture was compared with that in a pure culture of *L. lactis*.

Materials and Methods

Microorganisms and Media

L. lactis subsp. *lactis* (ATCC 11454) was the nisin-producing microorganism used in this work. *Micrococcus luteus* (ATCC 9341) was used as an indicating microorganism in the bioassay of nisin concentrations. The compositions of media used for the growth of these microorganisms are summarized as follows. Medium I, used for seed culture of *L. lactis* (pH 7.0), contained 5 g/L glucose, 5 g/L polypeptone, and 5 g/L yeast extract. Medium II, used for bioassay of nisin (pH 7.0), contained 10 g/L glucose, 5 g/L polypeptone, 5 g/L yeast extract, and 5 g/L NaCl. Medium III, used for the main fermentation, contained 20 g/L sweet whey powder (provided by WesternFarm Food Inc., Seattle, WA), 12 g/L yeast extract, 0.6 g/L KH_2PO_4, and 0.6 g/L $MgSO_4$.

Six *S. cerevisiae* strains, i.e., ATCC 4098, ATCC 4126, ATCC 8766, ATCC 9080, ATCC 9763, and ATCC 9841, were used in this work. The medium for yeast culture contained 1 g/L KH_2PO_4, 5 g/L $(NH_4)_2SO_4$, 1 g/L yeast extract, and 5 g/L carbon source. The pH was adjusted to 5.0 before sterilization at 121°C for 30 min. Lactose and lactic acid were applied as carbon sources, respectively, for the testing of the capabilities of the yeast strains growing on lactose and lactic acid. Glucose was used as carbon source when preparing yeast seed for further mixed culture study.

Mutagenic Treatment of *S. cerevisiae*

The suspension of the parental yeast strain was transferred to sterilized Petri plates. The Petri plates were then placed under a ultraviolet

(UV) lamp, (emitting the energy of 1.6×10 J/m/s) for 10 min. The plates were then placed in the incubator at 30°C for 3 d. The colonies showing a larger size on lactic acid containing agar as compared with the parental strain were picked up and subjected to natural selection. These mutant strains were further cultivated on lactic acid containing agar. The stable traits obtained were then compared for their capabilities for growing on lactose and lactic acid.

Analysis

Cell concentration of the pure cultures was measured as dry cell mass and optical density (OD). The viable cell concentrations of *L. lactis* and *S. cerevisiae* in mixed culture were determined as colony forming units (CFU) on selective media. Concentrations of L-lactic acid and acetic acid in the medium and fermentation broth were analyzed using a high-performance anion-exchange chromatography method *(18)*. Nisin concentration was measured by a bioassay method based on the method of Shimizu et al. *(17)*.

Cultivation Method

Before main cultivation was performed, culture size was scaled up by two steps in order to increase the amount of cells with high growth activity. Seed culture of *L. lactis* and *S. cerevisiae* was conducted in 125-mL Erlenmeyer flasks placed on an orbital shaker at 160 rpm and 30°C for 8 h. Main fermentations were performed in a 5 L Bioflo 110 fermentor (New Brunswick Scientific, Edison, NJ) equipped with temperature, pH, dissolved oxygen concentration and gas flow control systems. The working volume was 2 L. Air was supplied to the fermentor for aerobic cultivation conditions.

Results and Discussion

Nisin Production by the Pure Culture of L. lactis

The time-course of nisin production by the pure culture of *L. lactis* on a whey-based medium is shown in Fig. 2. The pH was controlled at 6.0 by the addition of NaOH. Biosynthesis of nisin increased rapidly in the first 8 h of fermentation, and reached a maximum concentration of 81.2 mg/L within the broth. Nisin concentration decreased slightly with further increase of time. Accompanied with nisin synthesis was the formation of a significant amount of lactic acid. The concentration of lactic acid increased rapidly to 10.1 g/L after 10-h fermentation, and reached 11.3 g/L in 24 h.

Lactic Acid Utilizing Yeast Strain Development

Six *S. cerevisiae* strains, ATCC 4098, ATCC 4126, ATCC 8766, ATCC 9080, ATCC 9763, and ATCC 9841, were compared for their capabilities of growth on lactic acid and lactose. The results of aerobic culture of these six yeast strains showed that none of them could use lactose as substrate. The

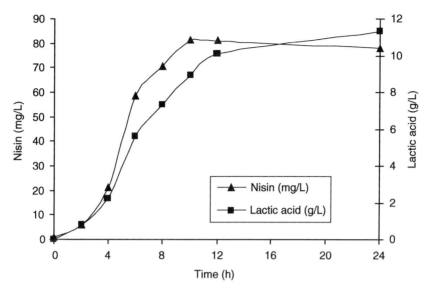

Fig. 2. Nisin production in a pure culture of *L. lactis* with pH control by alkali addition.

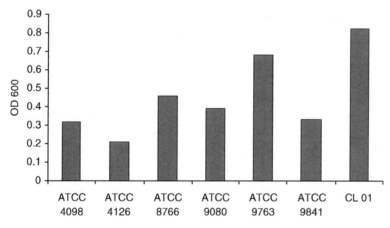

Fig. 3. The growth of different yeast strains on lactic acid-based media after 16 h aerobic culture.

growth of these six yeast strains on lactic acid was also compared. Each yeast strain was cultured on glucose-based media for 12 h, and then inoculated into lactic acid-based media. The cell growth was quantified by measuring OD_{600} after 16 h of aerobic culture. As shown in Fig. 3, the OD_{600} of ATCC 9763 was the highest among all of these six yeast strains. In order to obtain a good yeast strain for *in situ* removal of lactic acid in the mixed culture system, therefore, ATCC 9763 was mutated using UV treatment. After cultivating the mutants on lactic acid containing agar, five stable traits were obtained, and the mutant that showed the best performance for lactic acid assimilation was obtained and applied for the study of lactic acid removal in the mixed culture. The mutant was named *S. cerevisiae* CL01.

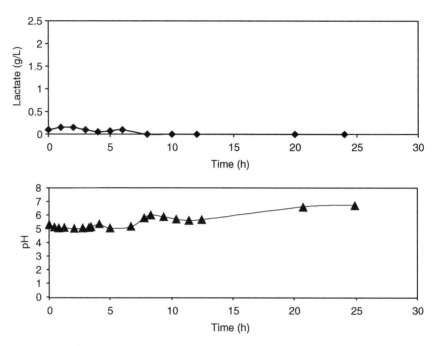

Fig. 4. Lactic acid accumulation and pH profile in a mixed culture with 3 h induction.

Effects of Induction Time on Lactic Acid Removal in the Mixed Culture

In order to overcome the inhibition of lactic acid to nisin biosynthesis, the lactic acid should be totally consumed by the yeast very quickly. In other words, there should be no accumulation of lactic acid in the fermentation broth. This objective can be achieved by activating the yeast in advance using lactic acid and providing sufficient yeast biomass in the mixed culture.

It is well known that the existence of glucose in medium will depress the capabilities of yeast to utilize other carbon source. As a result, an induction period is required to allow yeast cells to use up the glucose and shift their metabolic pathway from glycolysis to lactic acid assimilation. As shown in Fig. 4, lactic acid content in the mixed culture was well controlled at very low level (<0.2 g/L) after the yeast biomass were induced using 0.2 g/L of lactic acid for 3 h. In contrast, significant accumulation of lactic acid was observed (Fig. 5) without induction. Therefore, induction of yeast using lactic acid before mixed culture is essential for pH control. The appropriate induction time was 3 h.

Effects of Inoculum Ratio on Mixed Culture

Considering that the imbalance of the cell concentrations of the yeast and the bacteria in a mixed culture may cause the accumulation of lactic acid and thus inhibit the growth of *L. lactis* and the biosynthesis of nisin, the inoculum ratio of yeast biomass to bacteria population should be an

Stimulation of Nisin Production

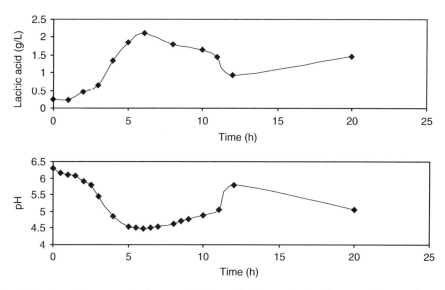

Fig. 5. Lactic acid accumulation and pH profile in a mixed culture without induction.

important parameter to achieve a stable pH control without alkali addition. Therefore, the effects of inoculum ratio on lactic acid removal were investigated. The time-course of three batches of mixed culture with different inoculum ratio of yeast biomass to bacteria population are shown in Figs. 6–8. The inoculum size of *L. lactis* in these three experiments was the same (2.4×10^6 CFU/mL), whereas the inoculum size of yeast in Figs. 6–8 was 0.8×10^6 CFU/mL, 2.2×10^6 CFU/mL, and 4.8×10^6 CFU/mL, respectively. The initial ratio of yeast biomass to bacteria population was proved to be very important for a stable pH of the mixed culture. The pH shown in Fig. 6 was not stable in the initial 8 h of fermentation, and thus caused the accumulation of lactic acid to the level of 0.1 g/L with nisin concentration reaching 135.2 mg/L. As shown in Figs. 7 and 8, with the increase of yeast inoculum, the pH of the mixed culture was well controlled around 6.0 in the first 12 h of fermentation, and nisin concentration was further increased to 150.3 mg/L. It was also noticed that the pH of the mixed culture increased gradually to 7.0. This fact could be owing to the fact that yeast assimilated not only lactic acid but also acetic acid. Analysis of the fermentation broth indicated that both lactic acid and acetic acid were removed.

The comparison of nisin production with different inoculum ratio (Figs. 6–8) indicates that sufficient inoculation of yeast is the key to a successful pH control in the mixed culture system. The inoculum ratio of the yeast to the bacteria should be 1:1. When inoculum size of *L. lactis* is greater than the expected value, or inoculum size of *S. cerevisiae* is less than the expected value, lactic acid will not be completely assimilated by *S. cerevisiae* and the pH will decrease. In such a case, growth of *L. lactis* will be inhibited. If *S. cerevisiae* grow and lactic acid concentration is decreased, both

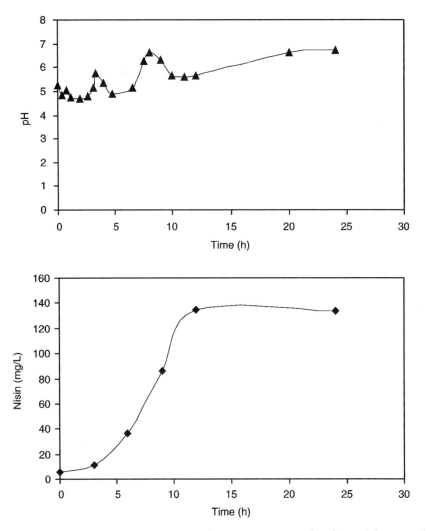

Fig. 6. The pH profile and nisin production in a mixed culture. The inoculum of yeast and bacteria was 0.8×10^6 and 2.4×10^6 CFU/mL respectively.

microorganisms are able to grow again. However, if the growth activity of *L. lactis* is lost under low pH for this period, growth of both microorganisms should be stopped *(19–22)*.

High Production of Nisin in the Mixed Culture System

By comparing the time-course of nisin production in the mixed culture shown in Figs. 7 and 8 to that in the pure culture shown in Fig. 2, nisin production was significantly stimulated in the mixed culture. Nisin concentration in the end reached 150.3 mg/L and it was 0.85 times greater than the nisin production shown in Fig. 2, which represents the production without mixed culture. Kim et al. *(23)* tried to increase nisin production

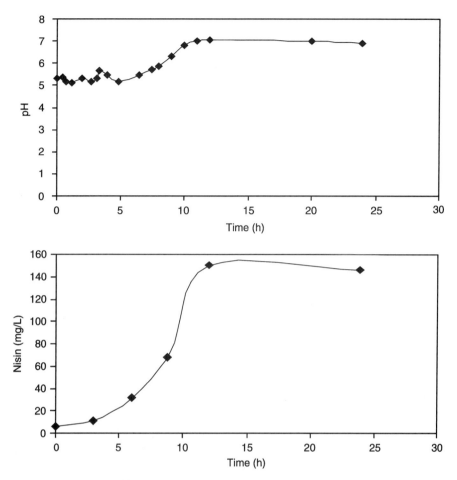

Fig. 7. The pH profile and nisin production in a mixed culture. The inoculum of yeast and bacteria was 2.2×10^6 and 2.4×10^6 CFU/mL respectively.

through genetic engineering, and they achieved 10% improvement by introducing foreign nisin immunity/resistance plasmid pND300 to *L. lactis*. The results of this study show that mixed culture improves nisin production when compared with the pure culture of *L. lactis* and is mainly limited by lactic acid inhibition rather than by the immunity/resistance of the bacteria to nisin.

Conclusions

A mixed culture of *L. lactis* and *S. cerevisiae* was established for the purpose of stimulating nisin production. The lactic acid produced by the bacteria was *in situ* utilized by the yeast and the pH of the mixed culture could be maintained at around 6.0 without alkali addition. Nisin concentration reached 150.3 mg/L and this value was 85% greater than that in a pure culture of the bacteria.

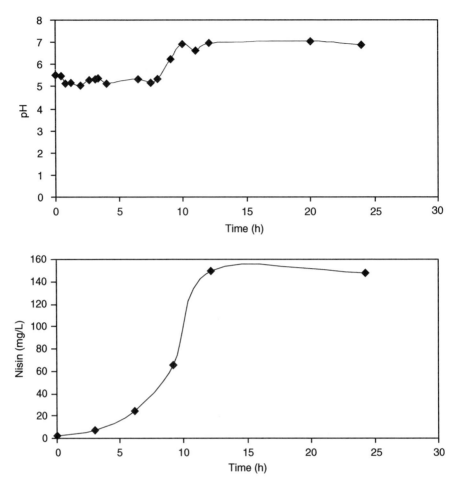

Fig. 8. The pH profile and nisin production in a mixed culture. The inoculum of yeast and bacteria was 4.8×10^6 and 2.4×10^6 CFU/mL, respectively.

Acknowledgments

This work was carried out under a grant from the Washington State Dairy Product Commission. Special thanks are given to WesternFarm Food Inc. (Seattle, WA) for providing cheese whey for the experiments.

References

1. Gonzalez, S. M. I. (1996), *Biores. Technol.* **57,** 1–11.
2. Yang, S. T. and Silva, E. M. (1995), *J. Dairy Sci.* **78,** 2541–2562.
3. Warwaha, S. S. and Kennedy, J. F. (1988), *Int. J. Food Sci. Technol.* **23,** 323–336.
4. Kamm, B. and Kamm, M. (2004), *Appl. Microbiol. Biotechnol.* **64,** 137–145.
5. Broughton, J. B. (1990), *Food Technol.* **44,** 100–117.
6. Montville, T. J. and Chen, Y. (1998), *Appl. Microbiol. Biotechnol.* **50,** 511–519.
7. Cleveland, J., Thomas, J., Montville, J. T., Nes, F. I., and Chikindas, L. M. (2001), *Int. J. Food Microbiol.* **71,** 1–20.
8. Parente, E. and Ricciardi, A. (1999), *Appl. Microbiol. Biotechnol.* **52,** 628–638.

9. Sablon, E., Contreras, B., and Vandamme, E. (2000), *Adv. Biochem. Eng. Biotechnol.* **68,** 21–59.
10. Food and Drug Administration (2001), GRAS Notice No. GRN 000065.
11. Sloan, A. E. (1998), *Food Technol.* **52,** 37–44.
12. Jack, R. W., Tagg, J. R., and Ray, B. (1995), *Microbiol. Rev.* **59,** 171–200.
13. Nelson, P. G. and Lorenzo, P. (2001), *Biotechnol. Lett.* **23,** 609–612.
14. Penna, T. C. V. and Moraes, D. A. (2002), *Appl. Biochem. Biotechnol.* **98–100,** 775–789.
15. Datta, R. and Tsai, S. P. (1997), In: *ACS Symposium Series 666*, pp. 224–236.
16. Liu, C., Liu, Y., Liao, W., Wen, Z., and Chen, S. (2004), *Appl. Biochem. Biotechnol.* **114,** 627–638.
17. Shimizu, H., Mizuguchi, T., Tanaka, E., and Shioya, S. (1999), *Appl. Environ. Microbiol.* **65,** 3134–3141.
18. Saccani, G., Gherardi, S., Trifirò, A., Soresi, B. C., Calza, M., and Freddi, C. (1995), *J. Chromatography A.* **706,** 395–403.
19. Kim, W. S., Hall, R. J., and Dunn, N. W. (1997), *Appl. Microbiol. Biotechnol.* **48,** 449–453.
20. De Vuyst, L. (1995), *J. Appl. Bacteriol.* **78,** 28–33.
21. De Vuyst, L. and Vandamme, E. J. (1993), *Appl. Microbiol. Biotechnol.* **40,** 17–22.
22. Meghrous, J., Huo, M., Quittelier, M., and Petitdemange, H. (1992), *Res. Microbiol.* **143,** 879–890.
23. Kim, W. S., Hall, R. J., and Dunn, N. W. (1998), *Appl. Microbiol. Biotechnol.* **50,** 429–433.

Biochar As a Precursor of Activated Carbon

R. Azargohar and A. K. Dalai*

*Catalysis and Chemical Reaction Engineering Laboratories,
Department of Chemical Engineering, University of Saskatchewan,
Saskatoon, SK, Canada S7N 5C5; E-mail: ajay.dalai@usask.ca*

Abstract

Biochar was evaluated as a precursor of activated carbon. This product was produced by chemical activation using potassium hydroxide. The effects of operating conditions of activation process, such as temperature, activating agent to biochar mass ratio, and nitrogen flow rate, on the textural and chemical properties of the product were investigated. Activated carbon produced by this method has internal surface area at least 50 times than that of the precursor and is highly microporous, which is also confirmed by scanning electron microscopy analysis. Fourier-transform infrared spectroscopy analysis showed development of aromatization in the structure of activated carbon. X-ray diffraction data indicated the formation of small, two-dimensional graphite-like structure at high temperatures. Thermogravimetric study showed that when potassium hydroxide to biochar mass ratio was more than one, the weight loss decreased.

Index Entries: Biochar; activated carbon; chemical activation; potassium hydroxide.

Introduction

Activated carbon is a highly porous material which has high surface area and exhibits good adsorptive capacities. It can be used as a catalyst *(1)* or as a catalyst support *(2,3)*. The high porosity and considerable catalytic and adsorptive properties are the results of activation process. There are two kinds of activation process, physical or chemical *(4)*. In chemical activation, carbon precursor reacts with a chemical reagent (dehydrating agent). After preparation of the mixture, it is heated, according to a specific heat treatment schedule, in an inert atmosphere and then the reaction product is washed to remove the activating agent. In a recovery unit, the chemical agent is recovered from solution. Carbon, after separation from the slurry, is dried and classified *(5)*. Commercial processes for making activated carbon use degraded and coalified plant matter (e.g., peat, lignite, and all types of coal) or of botanical origin (e.g., wood, coconut shells, and nut shells) as precursors *(6)*. Biochar is a product of fast pyrolysis process of biomass (including forest residues such as bark, sawdust, and shavings;

---

*Author to whom all correspondence and reprint requests should be addressed.

and agricultural wastes such as wheat straw and bagasse). The other products of this process are gases and a liquid mixture of organic compounds. The liquid phase can be converted to a high-value-added product, known as bio-oil. This oil can be used in diesel engines and furnaces. Biochar is a high-heating-value solid fuel that is commonly used in kilns and boilers. Owing to high demand for activated carbons, the pyrolytic solid product may be converted to activated carbon, a high-value end product. The intent of this work is to produce high-quality activated carbon from biochar and then to study the effects of chemical activation process by evaluating the internal surface area and surface chemistry of the product.

Experiment

Raw Material

Biochar, received from the Dynamotive Corporation (Vancouver, BC, Canada), was used as the starting material. The as-received char was sieved, and particles between –30 and +100 mesh (150–600 µm) were collected for activation.

Process

The schematic diagram of the experimental setup is shown in Fig. 1. A fixed bed, inconel tubular reactor (25.4-mm outer diameter, 22-mm internal diameter, and 870-mm length), was used to produce activated carbons from biochar. A furnace mounted vertically on a steel frame was installed to supply heat to the reactor. For chemical activation, 20 g of the char was placed on glass wool supported by a steel web welded to the wall of the reactor. Bed temperatures were recorded by placing a K-type thermocouple in the middle of the char bed. Nitrogen flow (as the carrier gas) into the reactor was controlled by a mass flow controller (Brooks Instrument, 5850S/B). Temperature of the furnace was controlled by a temperature controller (Eurotherm 2416). The system can be operated to a maximum temperature of 1200°C. An ice bath was used to condensate any trace of steam in outlet flow.

KOH, in the form of pellet, was mixed with the biochar at the desired ratios, and then 100 mL of water was added until all the activating agent was dissolved. This mixture was kept at room temperature for 2 h to ensure the access of KOH to the interior of the biochar and then the mixture was dried overnight at 120°C in an oven. The samples were placed in reactor and were heated from room temperature to 300°C at 3°C/min and then held at this temperature for 1 h. This step was introduced to prevent carbon loss through the direct attack of steam (7). The temperature was further increased to a final activation temperature at the same heating rate and held for 2 h before cooling down under nitrogen flow. Biochar was impregnated with 0.25–3 g KOH/g char and the heat treatment was performed to reach the final temperature in the range 550–800°C. After heat treatment, products were thoroughly washed with water, followed by

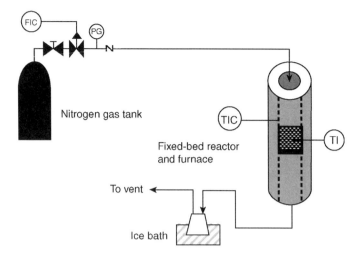

Fig. 1. Activation reactor setup.

washing with 0.1 M HCl, and finally with distilled water until the pH of the solution was between 6.0 and 7.0. Then the sample was dried at 110°C for 12 h and then characterized for its physical and chemical properties. Acid-washing step removes the soluble salts (5) and the potassium compounds, which results in the development of an accessible microporosity (8). Chlorine ions are also removed from the carbon by distilled water (9).

Design of Experiments

The activation process was studied with a standard response surface method (RSM) design called central composite design (CCD). It is suited for fitting a quadratic surface. This method helps to optimize the effective parameters with a minimum number of experiments, as well as to analyze the interaction effect between those parameters. Generally, the CCD consists of a 2^k factorial runs with $2k$ axial or star runs and n_c center runs (six replicates) (10). Activation temperatures (T), mass ratio of potassium hydroxide (KOH) to biochar (R), and flow rate of nitrogen (F) were three parameters investigated and their related ranges were 550–800°C, 0.25–3 g/g, and 80–250 cm^3/min, respectively. Six replications were performed at the center point in order to estimate the residual error. Therefore, the total number of experiments (N) required, for this investigation on activated carbon, is as follows:

$$N = 2^k + 2k + n_c = 2^3 + 2(3) + 6 = 20 \tag{1}$$

Characterization

Activated carbons were characterized by nitrogen gas adsorption, scanning electron microscopy (SEM), X-ray diffraction (XRD), Fourier-transform infrared spectrometry (FTIR), and thermogravimetry/differential thermal analysis (TG/DTA). To determine the surface area and pore structure of raw material and activated carbons, the nitrogen adsorption–desorption isotherms at 77 K were measured by an automated gas adsorption analyzer ASAP2000 (Micromeritics, Norcross, GA) with ±5% accuracy. SEM analysis of samples was performed by using a Phillips SEM-505 scanning electron microscope. The SEM instrument was operated at 300 kV/SE and 50°C inclination. Before analysis, all samples were gold-coated in a sputter-coating unit (Edwards Vacuum Components Ltd., Sussex, England) for electrical conduction. The micrographs were recorded using photographic techniques. XRD analysis was performed using Rigaku diffractometer (Rigaku, Tokyo, Japan) using Cu-K_α (λ = 1.5406 Å) radiation filtered by a graphic monochromator at a setting of 40 kV and 130 mA. The powdered catalyst samples were smeared on glass slide with methanol and dried at room temperature. The XRD analysis was carried out in the scanning angle (2θ) range of 3–85° at a scanning speed of 5°/min. FTIR spectra were obtained using a spectroscope (Spectrum GX; Perkin-Elmer, Norwalk, CT) at a resolution 4/cm. Undiluted activated carbons, in the powdered form, were scanned and recorded between 4000 and 450/cm. TG studies were carried out to study the effects of an activating agent on the product yield and thermal stability, by a pyres-diamond TG/DTA (Perkin-Elmer instruments) under the flow of argon gas. The pH of biochar and products were measured according to ASTM D 3838-80. Iodine number and ash content of activated carbons were measured according to ASTM D4607-94 and D2866-94, respectively.

Results and Discussion

Analysis of the Char

The biochar was analyzed for ultimate and bulk ash analysis by the Loring laboratory in Calgary (Canada). The ultimate analysis of biochar (moisture free) is 83.07 wt% carbon, 3.76 wt% hydrogen, 0.11 wt% nitrogen, 0.01 wt% sulphur, 9.6 wt% oxygen, and 3.44 wt% ash. The bulk ash analysis of biochar is 53.48 wt% SiO_2, 7.73 wt% Al_2O_3, 0.1 wt% TiO_2, 2.52 wt% Fe_2O_3, 17.98 wt% CaO, 4.2 wt% MgO, 2.07 wt% Na_2O, 6.93 wt% K_2O, 0.94 wt% P_2O_5, 1.06 wt% SO_3, and 2.99 wt% undetermined. The pH of biochar is 7.64 and its BET (Brunauer, Emmett, and Teller) surface area is 10 m^2/g.

Product Quality and Thermal Stability

Chemical activation by KOH is an effective method for developing highly microporous activated carbon (11). A typical isotherm of activated carbon prepared from chemical activation of biochar using KOH, at 537.5°C, KOH–biochar mass ratio of 1.63, and a nitrogen flow rate of 165 cm^3/min, is shown in Fig. 2, indicating the development of microporous structure in the product. This product has BET surface area of 585 m^2/g, micropore area of 470 m^2/g, total pore volume of 0.286 cm^3/g, micropore volume of 0.222 cm^3/g, and average pore diameter of 14.5°A. Therefore, micropores

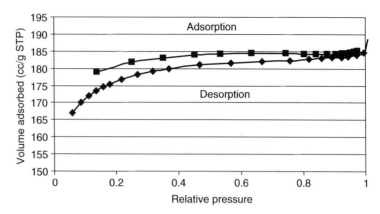

Fig. 2. Isotherm plot of activated carbon prepared at 537.5°C, KOH–biochar mass ratio of 1.63, and nitrogen flow rate of 165 cm^3/min.

Table 1
Effect of Temperature on the Porosity of Activated Carbon Prepared With KOH to Biochar Mass Ratio of 1.63 and Nitrogen Gas Flow Rate of 165 cm^3/min

Temperature (°C)	BET surface area (m^2/g)
537.5	585
675	927
812.5	1578

Table 2
Effect of Mass Ratio of KOH to Biochar on the Porosity of Activated Carbon Produced at 675°C and Nitrogen Flow Rate of 165 cm^3/min

KOH–biochar mass ratio	BET surface area (m^2/g)
0.11	248
1.63	927
3.14	659

account for 80.3% of the surface area and 77.6% of the pore volume of this activated carbon. Table 1 shows the effect of temperature on the porosity of activated carbon. It shows that by increasing the temperature from 537.5 to 812.5°C, BET surface area increased from 585 to 1578 m^2/g. Increasing the temperature increases the rate of reaction between char and KOH, thus creating new pores. Table 2 shows the effect of mass ratio of activating agent (KOH) to biochar on the porosity. It shows that the BET surface area increased from 248 to 927 m^2/g by increasing the ratio from 0.11 to 1.63 and then it decreased to 659 m^2/g. This trend has been observed for chemical activation of other precursors using KOH *(8,12)*. This is owing to the

Table 3
Effect of Nitrogen Flow Rate on the Porosity
of Activated Carbon Prepared at 675°C
and KOH–Biochar Mass Ratio of 1.63

Nitrogen flow rate (cm^3/min)	BET surface area (m^2/g)
71.5	582
165	927
258	1210

Table 4
Comparison of Properties of Chemically Activated Carbon Produced
From Biochar With Those of Three Commercial Activated Carbons

	Activated carbon from Envirotrol Inc., coconut shell EI-48S	Activated carbon from Fisher Scientific Co., coconut shell S-690-A	Norit FGD activated carbon	Chemically activated carbon from biochar
BET surface area (m^2/g)	960	640	556	1578
Micropore area (m^2/g)	830	450	246	1355
Total pore volume (cm^3/g)	0.46	0.31	0.54	0.75
Micropore volume (cm^3/g)	0.39	0.21	0.11	0.63
Average pore diameter (Å)	14	14	28	14
Iodine number (mg/g)	902	446	608	1576
Ash content (wt%)	0.6	27.70	26.67	0.5

widening of micropores in product by increasing the mass ratio. Table 3 shows the effect of nitrogen flow rate on the porosity of activated carbon. BET surface area increased from 582 to 1210 m^2/g by increasing the flow rate of gas from 71.5 to 258 cm^3/min. It can be related to the faster removal of gases evolved during the activation process at increased nitrogen flow and therefore shifting the equilibrium of the reaction to the further production of these gases and micropore materials (8). These results indicate that in order to develop a highly porous activated carbon with surface area more than 1000 m^2/g, temperature more than 675°C, KOH–biochar mass ratio equal to 1.63, and gas flow rate more than 165 cm^3/min are required. A sample prepared at 812.5°C, mass ratio of 1.63, and nitrogen flow rate of 165 cm^3/min has a BET surface area equal to 1578 m^2/g. In Table 4, the

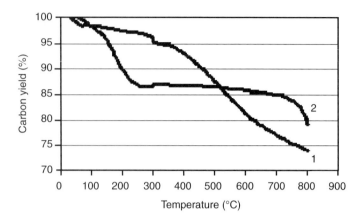

Fig. 3. TGA of biochar (1) and activated carbon (2) prepared at 675°C, KOH–biochar mass ratio of 1.63, and nitrogen flow rate of 165 cm³/min.

characteristics of this chemically activated carbon have been compared with three commercial activated carbons (activated carbon from Envirotrol Inc., Sewickley, PA [coconut shell EI-48S], activated carbon from Fisher scientific Co., Ottawa, ON, Canada [coconut shell S-690-A], and Norit FGD activated carbon, Marshall, TX). The average pore diameter of EI-48S and S-690-A are equal to that of chemically activated carbon. But the BET surface area, total pore volume, micropore surface area, and micropore volume of the product from biochar are more than those of these two commercial activated carbons. Norit FGD has a bigger average pore diameter and lower porous characteristics than chemically activated carbon. The iodine number, as a relative indicator of porosity in activated carbon, shows that the product from biochar has high porous structure. The ash content of this product, as a result of low ash content of its precursor, is obviously less than the ash contents of these commercial activated carbons. It shows that by using biomass as the main precursor, activated carbon can be produced with highly porous characteristics and low ash content. Therefore, biochar produced in fast pyrolysis of biomass, instead of using as solid fuel, can be converted to an attractive byproduct that can be used as a sorbent for liquids and gases or as catalyst or catalyst support.

Figure 3 shows the results in the TG analyses carried out examining the thermal stability of materials during activation process, under an inert (Ar) atmosphere for biochar and activated carbon, prepared at 675°C, KOH-biochar mass ratio of 1.63, and nitrogen flow rate of 165 cm³/min. It can be obviously seen that addition of an activating agent increases the carbon yield. Table 5 shows the amount of weight loss of sample for different mass ratios of KOH to biochar. According to this table, KOH acts as dehydrating agent, influences the pyrolytic decomposition, inhibits tar formation, and increases carbon yield (14). In order to minimize the weight loss of activated carbon, the ratio of the activating agent to biochar should be maintained more than one.

Table 5
Effect of Mass Ratio on the Weight Loss of Precursor During
Activation Process for Activated Carbon Prepared at 675°C
and Nitrogen Flow Rate of 165 cm³/min

KOH/biochar (g/g)	Total weight loss %
0	26.4
0.25	36.4
1	28.4
1.63	21.5
3	19.7

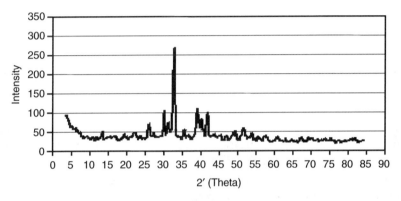

Fig. 4. X-ray diffraction (XRD) of activated carbon prepared at 675°C, KOH–biochar mass ratio of 1.63, and nitrogen flow rate of 165 cm³/min, before acid-washing.

Crystallinity and Surface Chemistry of Activated Carbon

Figure 4 shows the XRD plot of activated carbon prepared at 675°C, KOH–biochar mass ratio of 1.63, and nitrogen flow rate of 165 cm³/min, before acid-washing. According to the observed peaks, KO_3 and $K_2CO_3 \cdot (1.5)H_2O$ were identified in the product of chemical activation before acid-washing (peaks at $2\theta = 32.5°$, $39.8°$, and $41.9°$ for KO_3 and peaks at $2\theta = 12.8°$, $29.8°$, $32.2°$, $32.5°$, $32.7°$, and $41.5°$ for $K_2CO_3 \cdot [1.5H_2O]$). More study is required to specify the other possible potassium compounds in the product before acid-washing. The products were then acid-washed and subjected to XRD analysis.

Figure 5 shows the XRD spectra for biochar and two activated carbons prepared at the same mass ratio and nitrogen flow rate but at different temperatures. It shows that by increasing activation temperature from 537.5°C to 675°C a small graphite-like structure forms on activated carbon (peaks at $2\theta = 26.6°$ and $44.5°$). This structure, known as turbostratic (11), can be of two-dimensional order created by parallel orientation of carbon layer planes. This also indicates that potassium species are completely removed from the activated carbon.

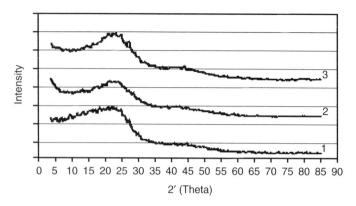

Fig. 5. X-ray diffraction (XRD) of biochar (1) and activated carbon prepared at 537.5°C (2) and 675°C (3) (at constant KOH–biochar mass ratio of 1.63 and nitrogen flow rate of 165 cm³/min for both activated carbons).

The scanning electron micrographs were taken for biochar, activated carbon prepared at 675°C, KOH–biochar mass ratio of 1.63, and nitrogen flow rate of 165 cm³/min before and after acid-washing to examine their microporsity (see Fig. 6A–C). According to these micrographs, biochar does not have a porous structure, as confirmed by very low BET surface area of it, and for activated carbons, before acid-washing pores are blocked by potassium and basic compounds, but washing with HCl/water removes them and exposes the porosity of the product.

Diffuse-reflectance FTIR is a powerful technique for IR analysis of fine particles and powders in the concentration range from undiluted to parts per thousand. In the IR technique, the interpretation of the spectra is complicated because each group indicates several bands at different wave numbers, therefore each band may include contributions from various groups (13). The IR bands which can be associated with aromatic hydrocarbons are as follows.

Peak between 900 and 675/cm (out-of-plane bending of the ring C–H bonds), peak between 1300 and 1000/cm (in-plane bending bands), peaks between 1600 and 1585/cm as well as 1500 and 1400/cm (skeletal vibrations involving carbon–carbon stretching within the ring), peak between 3100 and 3000/cm (aromatic C-H stretching bands), and peak between 2000 and 1650/cm (weak combination and overtone bands). Figure 7 shows the FTIR results for biochar, the product of heat treatment of biochar to the final temperature of 675°C without using an activating agent and activated carbon prepared at 675°C, KOH–biochar mass ratio of 1.63, and nitrogen flow rate of 165 cm³/min. The spectrum of activated carbon prepared by activating agent shows the effect of this agent in aromatization of the product.

Figure 7 shows that there are low levels of aromatic compounds in biochar, which have disappeared after heat treatment without any activating agent. However, use of activating agent with the same heat treatment

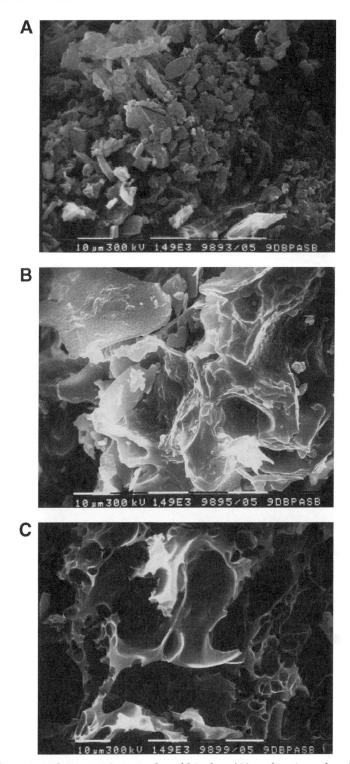

Fig. 6. Scanning electron micrographs of biochar **(A)** and activated carbon before acid-washing **(B)** and after acid-washing **(C)** (prepared at 675°C, KOH–biochar mass ratio of 1.63, and nitrogen flow rate of 165 cm³/min).

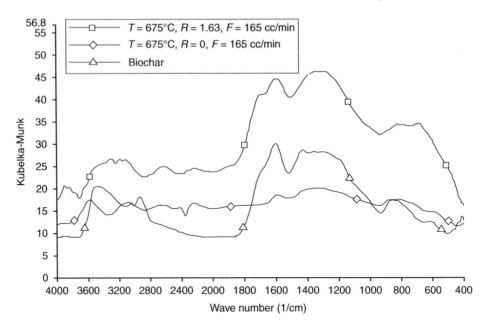

Fig. 7. Effect of KOH–biochar mass ratio on the surface chemistry.

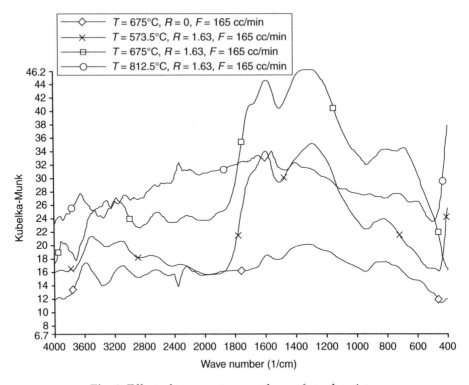

Fig. 8. Effect of temperature on the surface chemistry.

and gas flow rate has developed the aromatic structure. Figure 8 shows that the activation temperature up to 675°C increases apparent aromatization of structure of activated carbon, after which the aromaticity decreases.

Conclusions

Biochar, the solid product of fast pyrolysis of biomass, instead of using as fuel for energy generation, can be converted to activated carbon, which is a value-added product. Chemical activation of biochar, using KOH, produces activated carbon with high porous structure and low ash content, which is comparable with commercial, activated carbons. Some potassium compounds, such as KO_3 and $K_2CO_3 \cdot (1.5H_2O)$, were identified in the activated carbon before acid-washing, which were completely removed by acid-washing. BET surface area of activated carbon produced by this material can be as large as 1500 m^2/g. There was presence of small graphite-like structure in the activated carbon produced at higher temperatures. For KOH–biochar ratios of more than one, chemical activation increases the carbon yield. The activating agent increases apparent aromatization in the product.

References

1. Dalai, A. K., Majumdar, A., Chowdhury, A., and Tollefson, E. L. (1993), *Can. J. Chem. Eng.* **71,** 75.
2. Rodriguez-Reinoso, F. (1995), In: *Porosity in Carbons: Characterization and Applications,* Patrick, W., Jr., ed., Edward Arnold, London, Chapter 10.
3. Radovic, L. R. and Rodriguez-Reinoso, F. (1997), In: *Chemistry and Physics of Carbon,* Vol. 25, Thrower, P. A., ed., Marcel Dekker, New York, p. 243.
4. Kyotani, T. (2000), *Carbon* **38,** 269–286.
5. Rodriguez-Reinoso, F. In: *Handbook of Porous Solids,* Schuth, F., Sing, K. S. W., and Weitkamp, J., eds., Wiley-VCH Verlag Gmbh, Weinheim, Germany, 2002, p. 4.8.1.
6. El-Hendawy, A. A., Samra, S. E., and Girgis, B. S. (2000), *Colloids Surf. A: Physicochem. Eng. Aspects* **180,** 209–221.
7. Otowa, T., Nojima, Y., and Miyazaki, T. (1997), *Carbon* **35,** 1315.
8. Lozano-Castello, D., Lillo-Rodenas, M. A., Cazorla-Amoros, D., and Linares-Solano, A. (2001), *Carbon* **39,** 741–749.
9. Lillo-Rodenas, M. A., Lozano-Castello, D., Cazorla-Amoros, D., and Linares-Solano, A. (2001), *Carbon* **39,** 751–759.
10. Montgomery, D. C. (1997), *Design and Analysis of Experiments,* 4th ed., John Wiley & Sons, USA, Chapter 13.
11. McEnaney, B. (2002), In: *Handbook of Porous Solids,* Schuth, F., Sing, K. S. W., and Weitkamp, J., eds., Wiley-VCH Verlag Gmbh, Weinheim, Germany, p. 4.8.2.
12. Ghen, X. S. and McEnaney, B. (2001), In: Extended abstracts, 25th American carbon conference, Lexington, KY.
13. Byrne, J. F. and Marsh, H. (1995), In: *Porosity in Carbons: Characterization and Applications,* 1st ed., Patrick, W., Jr., ed., Halsted Press, London, Chapter 1.
14. Figuciredo, J. L., Pereira, M. F. R., Freitas, M. M. A., and Orfao, J. J. M. (1999), *Carbon* **37,** 1379–1389.

Moisture Sorption, Transport, and Hydrolytic Degradation in Polylactide

Richard A. Cairncross,*,[1] Jeffrey G. Becker,[2] Shri Ramaswamy,[2] and Ryan O'Connor[3]

[1]Chemical and Biological Engineering, Drexel University, Philadelphia, PA 19104, E-mail: cairncross@drexel.edu; [2]Bio-Based Products, University of Minnesota, St. Paul, MN 5510; and [3]NatureWorks, LLC, Minnetonka, MN 55345

Abstract

Management of moisture penetration and hydrolytic degradation of polylactide (PLA) is extremely important during the manufacturing, shipping, storage, and end-use of PLA products. Moisture transport, crystallization, and degradation in PLA have been measured through a variety of experimental techniques including size-exclusion chromatography, differential scanning calorimetry, and X-ray diffraction. Quartz crystal microbalance and dynamic vapor sorption experiments have also been used to measure moisture sorption isotherms in PLA films with varying crystallinity. A surprising result is that, within the accuracy of the experiments, crystalline and amorphous PLA films exhibit identical sorption isotherms.

Index Entries: PLA; biodegradable polymers; bio-based polymers; polylactic acid; diffusion; hydrolysis.

Introduction

Even though there is a significant growing interest in *bio-based polymers* derived from renewable resources, these polymers have had limited success competing against petroleum-based plastics in the commodity plastics markets owing to several reasons including economics, inferior properties, and environmental performance. One of the major technical challenges to widespread acceptance of bio-based polymers is difficulties achieving mechanical and barrier properties comparable with conventional synthetic polymers while maintaining biodegradability. The leading bio-based polymer is currently polylactide or polylactic acid (PLA), which is being produced commercially by NatureWorks (formerly Cargill Dow) in the United States and other companies worldwide (1,2). The main application areas for PLA are food packaging, bottles, films, and serviceware, with a number of PLA products now commercially available. Bio-based polymers

*Author to whom all correspondence and reprint requests should be addressed.

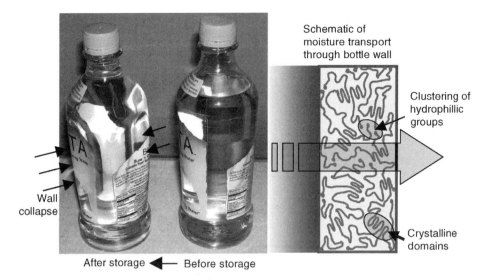

Fig. 1. Photo showing buckling of polylactide (PLA) water bottle owing to moisture loss through walls and schematic of moisture transport through a PLA film.

and biodegradable polymers are also popular for biomedical applications. Polylactides and copolymers of lactic acid with other monomers, such as glycolic acid, are leading candidates for degradable drug-delivery devices, tissue scaffolds, and other biomedical applications *(3)*.

A barrier to effective commercialization of bio-based polyesters is their inferior moisture barrier properties compared to those of synthetic polymers. Fundamental knowledge of moisture transport in bio-based polyesters will help the polymer's life cycle in the following aspects: drying of bio-based polymer pellets during production (melt-processed pellets are quenched in water), moisture regain (i.e., equilibrium moisture content) and degradation during processing, water losses through products (e.g., water bottles shown in Fig. 1), industrial composting and degradation kinetics, and life cycle assessment of bio-based polyester products.

Despite the importance of moisture penetration in evaluating the performance, degradation, and life cycle of polymer products, there have been few published papers on moisture transport in renewable polymers. In the published literature, there is some disagreement about the dominant mechanisms of moisture transport. Sharp et al. (2001) proposed that the equilibration time of glassy polymers such as PLA lead to non-Fickian diffusion effects, which is inferred by thickness-dependent diffusion coefficients *(4)*. Siparsky et al. (1997) suggested that for many samples (especially thick samples), it is not possible to separate the diffusion process from hydrolytic degradation, so accurate measurement of transport properties requires analyzing the simultaneous diffusion/reaction processes *(5)*. Auras et al. (2002) measured water vapor permeability in PLA at various temperatures and relative humidities; surprisingly, the permeability was independent of relative humidity and decreased with increasing temperature *(6,7)*. Shogren (1997) also reported that PLA permeability to water decreases with increasing temperature, although permeability to both CO_2 and O_2 increased with temperature *(8)*. Diffusion coefficients of water in PLA and other biopolymers were measured by gravimetric techniques and changes in solubility were correlated with hydrophobicity of biopolymers and related to formation of clusters of water molecules *(5,9)*.

The literature is also inconsistent about how crystallinity affects moisture sorption and moisture transport. One study suggests that the mass of water that absorbs into PLA at equilibrium is essentially independent of crystallinity *(5)*. Other studies show that the mass of water that absorbs into PLA at equilibrium scales the percent crystallinity *(8,9)*. In most studies on other semicrystalline polymers, crystallinity does affect water transport with lower solubility and lower diffusion coefficients at higher crystalline fraction *(10,11)*. Vert (1998) showed that changing the catalyst used for ring-opening polymerization of PLA can have a significant impact on moisture sorption, and this was attributed to transesterification reactions of impurities in the catalyst, leading to more hydrophobic end groups *(3)*. A systematic study of the influence of end groups (Fig. 2) on moisture sorption in bio-based polyesters is needed and will help to explain why the published data are inconsistent.

Degradation of bio-based polyesters is commonly understood to occur in two steps: (1) hydrolytic degradation of ester bonds into lower molecular weight polymers and oligomers followed by (2) biological degradation to carbon dioxide and water. Several researchers show that hydrolytic degradation must reduce molecular weight of PLA to 15,000–40,000 before biodegradation can take over *(2,12)*. Hydrolytic degradation of PLA can either occur randomly within the polymer or from the ends (unzipping); the ratio of these reactions is pH-dependent *(13–16)*. In semicrystalline PLA, degradation can lead to bimodal molecular weight distributions through hydrolytic cleavage of the polymer tie chains spanning among crystallites *(17)*. The kinetic studies on PLA have shown the reaction to be pseudo-first-order with an autocatalytic effect owing to production of acidic end groups by the hydrolytic reaction *(14,16)*. Autocatalysis also leads to faster degradation inside of PLA specimen immersed in water because soluble oligomers are trapped on the interior of the specimen and raise the interior pH relative to the surface *(3,16,18)*.

Experiment

Three different samples of polylactide were provided by NatureWorks LLC: 4032, 4060, and stereocomplex. The 4032 grade of PLA is synthesized from a high percentage of L-lactic acid and can be amorphous or crystalline, depending on heat treatment. The 4060 grade of PLA is synthesized from a mixture of lactic acid stereoisomers and does not crystallize.

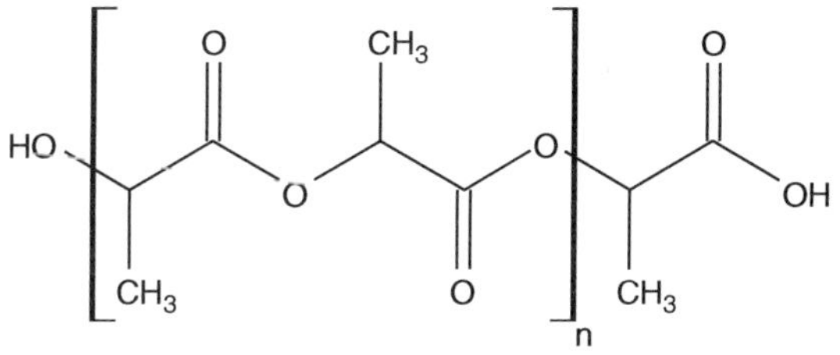

Polylactide

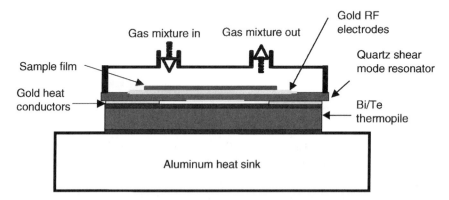

Fig. 2. Chemical structure of polylactide with hydrophilic acid and alcohol end groups.

Fig. 3. Schematic of quartz crystal microbalance heat conduction calorimeter (QCM/HCC) mass/heat flow sensor and gas sample chamber.

The stereocomplex samples are a 50/50 blend of PLLA and PDLA (homopolymers prepared from nearly pure L- or D-lactic acid); the stereocomplex sample crystallizes rapidly into a different, higher-melting-point crystalline structure than homopolymer PLA.

Quartz Crystal Microbalance Heat Conduction Calorimeter

The quartz crystal microbalance heat conduction calorimeter (QCM/HCC) combines a *quartz crystal microbalance* coated with a thin polymer film in intimate thermal contact with a *heat-flow sensor* mounted on a heat sink (Fig. 3). Using QCM/HCC, Smith and co-workers *(19–21)* have been able to measure simultaneously the change in mass per unit area (to ±2 ng/cm^2) and the resulting heat flows (to ±1 µW) when the polymer on the sample QCM surface takes up or releases solvent vapor. When the gas composition above a polymer film is changed, the resulting change in mass and the integrated thermal power signal gives the enthalpy of sorption, $\Delta_{sorp}H$, of the vapor in the polymer at a specific vapor activity. The combination of heat and mass measurements enables both $\Delta_{sorp}H$ and $\Delta_{sorp}G$ to be measured, so $\Delta_{sorp}S$ can also be determined. Thus, a single set of experiment at a constant temperature is sufficient to determine all three thermodynamic properties characterizing the polymer/solvent interaction. Recently, Masscal Co. (Chatham, MA) developed a commercial version of QCM/HCC, the Masscal G1™, which was used for the results in this article by special arrangement with Masscal. Humidity was controlled by combining a dry nitrogen gas stream and a water-saturated nitrogen gas stream at various ratios. The saturated stream was prepared by bubbling nitrogen through water, and the humidity produced by this system was verified using a Buck Research CR-4™ chilled mirror hygrometer.

PLA films were prepared for QCM by dissolving PLA in chloroform (stereocomplex sample was dissolved in hexafluoroisopropanol) and spin coating. Films were air-dried for 2 d at ambient conditions, and then dried at 60°C and 80°C for at least 1 d, each in an inert nitrogen atmosphere. Amorphous films were prepared by heating to above the melting temperature for 2 h under dry nitrogen and cooling rapidly to room temperature.

Dynamic Vapor Sorption

Dynamic vapor sorption (DVS) is a gravimetric sorption method that combines a highly sensitive (i.e., lower detection limit of 0.1 µg) electrobalance with a controlled-humidity gas handling system in controlled-temperature environment. DVS is an ideal system for measuring sorption isotherms (i.e., mass uptake vs relative humidity at constant temperature). For moderately hydrophobic bio-based aliphatic polyesters, sorption isotherms can be measured with sample sizes as small as 5 mg. The DVS measurements in this paper were measured using a surface-measurement-systems DVS instrument located in Gary Reineccius's laboratory in the Food Science and Nutrition Department at the University of Minnesota (St. Paul, MN).

PLA films were prepared for DVS by melt-pressing with a heated hydraulic press at 200°C (for 4032 and 4060) or 250°C (for stereocomplex) under dry nitrogen. PLA pellets were inserted between sheets of cleaned aluminum foil, connected to a supply of dry nitrogen, and inserted into the press. After pressing for approx 5 min at approx 100 lb of force, the sample was removed and cooled quickly by pouring water over the outside of the aluminum foil surrounding the sample.

Crystallinity Measurements

Crystallinity of PLA films was determined using differential scanning calorimetry (DSC) and X-ray diffraction (XRD). Wide-angle XRD was performed on the films prepared for QCM/HCC and DVS using a Bruker-AXS rapid XRD™ microdiffractometer in the Characterization Facility at the University of Minnesota. Sharp, crystalline, broad, and amorphous peaks were identified in the plots of intensity vs 2θ, and the areas under crystalline and amorphous peaks were calculated using the JADE™ software package to obtain a qualitative percent crystallinity.

DSC was performed on melt-pressed films using either a Perkin Elmer TGA 7™ in Ted Labuza's laboratory in the Food Science and Nutrition Department at the University of Minnesota (St. Paul, MN) or a TA Instrument

Q1000™ in the Polymer Characterization Lab in the Chemical Engineering and Materials Science Department at the University of Minnesota. Crystallinity was determined by integrating the areas under the crystallization and melting peaks, subtracting them, and dividing by the enthalpy of fusion for 100% crystalline PLA (taken to be 93 kJ/kg for 4032 PLA, and 135 kJ/kg for stereocomplex PLA).

Results

Quartz Crystal Microbalance Heat Conduction Calorimeter

Figures 4–6 display data measured by QCM/HCC for moisture sorption in PLA films. The raw QCM/HCC data in Fig. 4 show that sorption of water closely follows changes of water partial pressure (or relative humidity) within the sample chamber through two cycles of stepwise humidity changes from zero to approx 25% at 40°C. The frequency of the QCM decreases as water absorbs into the film, and the mass change is calculated by the Saurbrey equation. The thermal power (heat flow through HCC) exhibits spikes whenever the humidity changes and mass sorbs into or desorbs out of the film. The motional resistance is a measure of the damping properties of the film, and as water absorbs into PLA films, the motional resistance increases, which indicates that water acts as a plasticizer in PLA. The frequency and motional resistance are reproducible over the two cycles of humidity, indicating that degradation of PLA during the experiment is negligible.

Sorption isotherms in PLA films can be estimated from the mass absorption data from QCM (Fig. 5). Crystalline and amorphous films have been prepared and tested with little difference observed among their sorption properties. Figure 5 displays a comparison among moisture sorption into amorphous PLA films from 4032 PLA and stereocomplex PLA. The sorption isotherms are nearly linear and are fit well using the Flory-Huggins equation with a high interaction parameter. The stereocomplex PLA sample exhibited somewhat smaller moisture sorption, and even though both samples were amorphous; heat treatment did not affect the sorption isotherms.

Simultaneous measurement of mass changes and heat flow (thermal power) with QCM/HCC enables calculating a sorption enthalpy directly from the experimental data. Figure 6 shows measured values of sorption enthalpy vs moisture content for PLA films at 40°C. There is considerable scatter in the data because the low levels of sorption (less than 0.2%) lead to heat flows that are close to the limit of accuracy for the equipment. However, there is a general trend of decreasing sorption enthalpy with increasing moisture content, and the sorption enthalpy for 4032 PLA appears slightly lower than stereocomplexed PLA. All of the measured sorption enthalpies are higher than the heat of vaporization of pure water (approx 44 kJ/mol); such high sorption enthalpies have been related to formation of water clusters when water absorbs into a hydrophobic polymer (5,9).

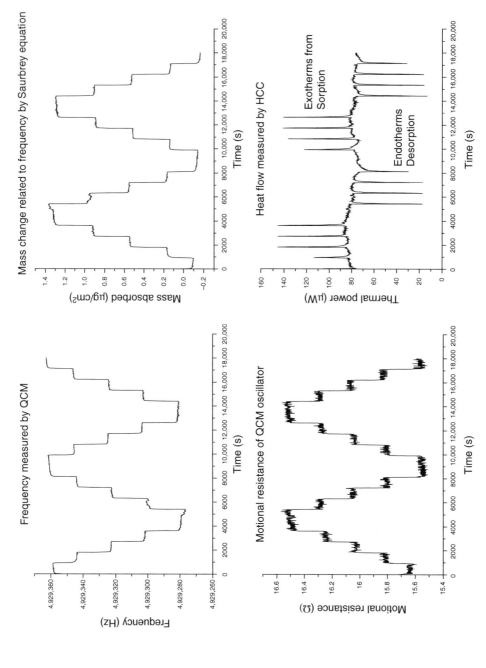

Fig. 4. Quartz crystal microbalance heat conduction calorimeter (QCM/HCC) measurements of frequency, mass of water absorbed, thermal power, and motional resistance for a PLA film subjected to a cycle of humidities at 40°C.

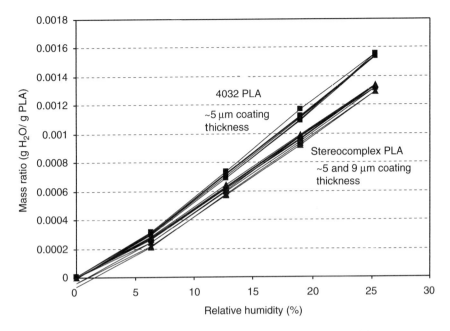

Fig. 5. Sorption isotherms for amorphous PLA films measured by quartz crystal microbalance heat conduction calorimeter (QCM/HCC) at 40°C.

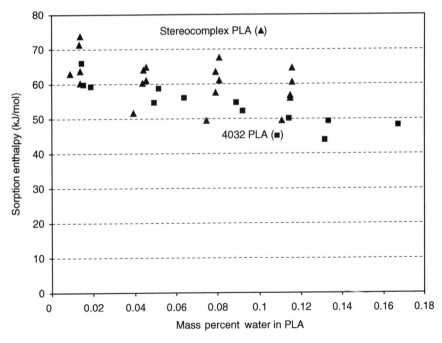

Fig. 6. Sorption enthalpies for amorphous polylactide (PLA) films measured by quartz crystal microbalance heat conduction calorimeter (QCM/HCC) at 40°C.

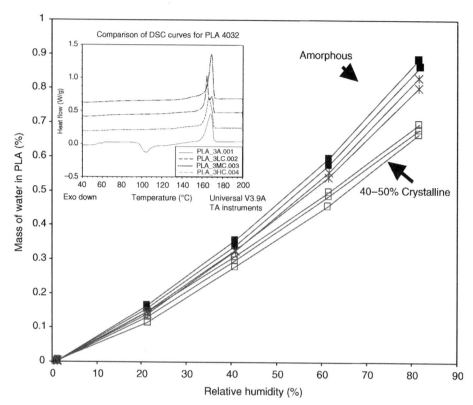

Fig. 7. Moisture sorption isotherm for polylactide (PLA) at 40°C at several different crystallinities. Inset shows differential scanning calorimetry (DSC) measurements used to determine crystallinity.

Dynamic Vapor Sorption

Sorption isotherms were also measured using DVS (Figs. 7–9). The sorption isotherms for amorphous and crystalline PLA plotted in Figs. 7 and 8 from DVS are comparable with those from QCM/HCC. The sorption isotherms for both 4032 PLA (Fig. 7) and stereocomplexed PLA (Fig. 8) show that crystallinity has a small influence on moisture sorption in PLA. For films that are 40–50% crystalline, the amount of water absorbed decreases by less than 20% compared with the amorphous films; this result is independent of sample thickness and the type of crystalline structure.

Degradation (and molecular weight) is known to have a significant effect on moisture sorption and transport in PLA. Measurements of sorption isotherms on partially degraded PLA films (of amorphous 4060 PLA) show a rapid increase in sorption with degradation. Films were exposed to high humidity (85%) and high temperature (80°C) before being tested with DVS. After 1 d degradation, the change in sorption is small; this may be owing to a time-lag for the sample chamber to rise to an elevated humidity or owing to the reported autocatalytic effect for PLA degradation. The

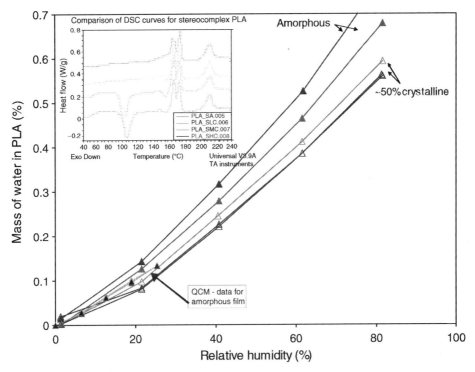

Fig. 8. Moisture sorption isotherms for a stereocomplex blend of polylactide (PLA) at 40°C measured by dynamic vapor sorption (DVS). Inset shows differential scanning calorimetry (DSC) measurements used to determine crystallinity.

amounts of moisture absorbed after 2 and 3 d of degradation are about twice and ten times as much, respectively. The increase in moisture sorption with degradation is consistent with the hypothesis of end groups controlling sorption because more end groups are present in a degraded sample.

Discussion

With the growing interest in bio-based and degradable polymers, there is a need for a more fundamental understanding of moisture transport within these polymers. A goal of the US Department of Energy and the US Department of Agriculture is to increase the fraction of chemicals and materials produced from biomass from the 2001 level of 5–12% in 2010 and 25% in 2020 (22). Achieving this goal requires the development of a number of enabling technologies for converting biomass into chemicals and materials and for processing materials into useful, competitive products.

The results presented in this article suggest that molecular structures of bio-based polymers affect moisture transport and barrier properties in a way not discussed previously. The insensitivity of sorption to crystallinity is unexpected and inconsistent with results for other semicrystalline polymers. For sorption to be independent of crystallinity in PLA, the sites for water sorption in PLA must be largely excluded from the crystalline

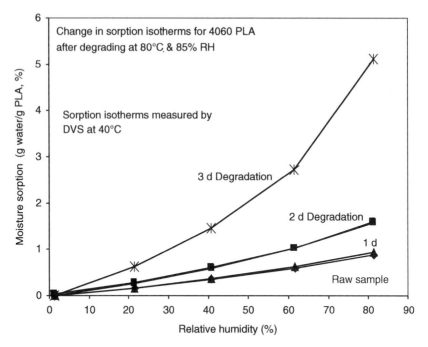

Fig. 9. Sorption isotherms for amorphous polylactide (PLA) films after being degraded at 80°C and 85% relative humidity. Isotherms measured by dynamic vapor sorption (DVS) at 40°C.

domains, because it is commonly thought that water cannot enter into the crystalline domains (and it has been established that crystallinity does not change during sorption/desorption). Based on these observations, we hypothesize that moisture sorption is controlled by hydrophilic end groups (and as indicated in Fig. 1) in PLA. Future experiments will investigate the effect of molecular weight and chemical modification of the end groups on sorption in PLA.

There are several open issues raised by research described in this article, such as why sorption and diffusion in hydrophobic polyesters are apparently insensitive to crystallinity, the role of the hydrophilic end groups on sorption, whether dynamic sorption follows Fickian kinetics, and whether water absorbs as clusters in bio-based polyesters. Also, the published literature on sorption and diffusion in PLA is inconsistent; so there is a need for systematic research focused on molecular and structural factors affecting sorption and diffusion.

Acknowledgments

The research conducted for this project was supported by the National Research Initiative of the USDA Cooperative State Research, Education and Extension Service, grant no. 2004-35504-14636, which funded the research of Cairncross during his sabbatical in the BioBased Products Department at the University of Minnesota. The research was also supported by seed grant funding from the University of Minnesota, Initiative for Renewable Energy and the Environment (IREE) and by funding and samples from NatureWorks, LLC. Access to experimental facilities in the laboratories of Ted Labuza and Gary Reineccius in the Food Science and Nutrition Department at the University of Minnesota is gratefully acknowledged. Allan Smith and Fletcher Smith aided in this research by allowing access to the Masscal G1™ and their expertise in its use. Discussions with Marc Hillmyer (University of Minnesota, Chemistry Department) and Roger Ruan (University of Minnesota, Biosystems and Agricultural Engineering Department) were also helpful in this research.

References

1. Crank, M., Patel, M., Marscheider-Weidmann, F., Schleich, J., Husing, B., and Angerer, G. (2004), Technical Report, European Commission's Institute for Prospective Studies, Seville, Spain.
2. Drumright, R. E., Gruber, P. R., and Henton, D. E. (2000), *Adv. Mater.* **12(23),** 1841.
3. Vert, M., Schwach, G., Engel, R., and Coudane, J. (1998), *J. Controlled Release* **53(1–3),** 85.
4. Sharp, J. S., Forrest, J. A., and Jones, R. A. L. (2001), *Macromolecules* **34(25),** 8752.
5. Siparsky, G. L., Voorhees, K. J., Dorgan, J. R., and Schilling, K. (1997), *J. Env. Poly. Deg.* **5(3),** 125.
6. Auras, R., Harte, B., and Selke, S. (2004), *J. Appl. Polym. Sci.* **92(3),** 1790.
7. Auras, R., Harte, B., and Selke, S. (2004), *Macromol. Biosci.* **4(9),** 835.
8. Shogren, R. (1997), *J. Environ. Polym. Degrad.* **5(2),** 91.
9. Yoon, J. S., Jung, H. W., Kim, M. N., and Park, E. S. (2000), *J. Appl. Polym. Sci.* **77(8),** 1716.
10. Iordanskii, A. L., Kamaev, P. P., Ol'khov, A. A., and Wasserman, A. M. (1999), *Desalination* **126(1–3),** 139.
11. Olkhov, A. A., Vlasov, S. V., Iordanskii, A. L., Zaikov, G. E., and Lobo, V. M. M. (2003), *J. Appl. Polym. Sci.* **90(6),** 1471.
12. Agarwal, M., Koelling, K. W., and Chalmers, J. J. (1998), *Biotechnol. Prog.* **14(3),** 517.
13. Shih, C. (1995), *Pharm. Res.* **12(12),** 2036.
14. Schliecker, G., Schmidt, C., Fuchs, S., Kissel, T. (2003), *Biomaterials* **24(21),** 3835.
15. de Jong, S. J., Arias, E. R., Rijkers, D. T. S., van Nostrum, C. F., Kettenes-van den Bosch, J. J., and Hennink, W. E. (2001), *Polymer* **42(7),** 2795.
16. Tsuji, H. and Ikada, Y. (2000), *Polym. Degrad. Stab.* **67(1),** 179.
17. Tsuji, H. and Miyauchi, S. (2001), *Polym. Degrad. Stab.* **71(3),** 415.
18. Grizzi, I., Garreau, H., Li, S., and Vert, M. (1995), *Biomaterials* **16(4),** 305.
19. Smith, A. L., Wadso, I., and Shirazi, H. (1998), *Abstr. Papers Am. Chem. Soc.* **216,** U714.
20. Smith, A. L., Shirazi, H. M., and Mulligan, S. R. (2002), *Biochim. Biophys. Acta Protein Struct. Mol. Enzymol.* **1594(1),** 150.
21. Smith, A. L., Mulligan, R. B., and Shirazi, H. M. (2004), *J. Polym. Sci. [Part B-Polym. Phys.]* **42(21),** 3893.
22. Perlack, R. D., Wright, L. L., Turhollow, A., Graham, R. L., Stokes, B., and Erbach, D. C. (2005), Oak Ridge National Laboratory Report ORNL/TM-2005/66 1.

SESSION 5
Microbial Catalysis and Metabolic Engineering

Zymomonas mobilis As Catalyst for the Biotechnological Production of Sorbitol and Gluconic Acid

GILMAR SIDNEY ERZINGER[1] AND MICHELE VITOLO*,[2]

[1]University of Joinville, Pharmacy School, Santa Catarina, SC, Brazil;
and [2]Biochemical and Pharmaceutical Technology Department, Faculdade
de Ciências Farmacêuticas, University of São Paulo,
Av. Prof. Lineu Prestes, 580, B.16. 05508-900, São Paulo, SP, Brazil,
E-mail: michenzi@usp.br

Abstract

The conversion of glucose and fructose into gluconic acid (GA) and sorbitol (SOR) was conducted in a batch reactor with free (CTAB-treated or not) or immobilized cells of *Zymomonas mobilis*. High yields (more than 90%) of gluconic acid and sorbitol were attained at initial substrate concentration of 600 g/L (glucose plus fructose at 1:1 ratio), using cells with glucose-fructose-oxidoreductase activity of 75 U/L. The concentration of the products varied hyperbolically with time according to the equations $(GA) = t\,(GA)_{max}/(W_{GA} + t)$, $(SOR) = t\,(SOR)_{max}/(W_{SOR} + t)$, $v_{GA} = [W_{GA}\,(GA)_{max}]/(W_{GA} + t)^2$ and $v_{SOR} = [W_{SOR}\,(SOR)_{max}]/(W_{SOR} + t)^2$. Taking the test carried out with free CTAB-treated cells as an example, the constant parameters were $(GA)_{max} = 541$ g/L, $(SOR)_{max} = 552$ g/L, $W_{GA} = 4.8$ h, $W_{SOR} = 4.9$ h, $v_{GA} = 112.7$ g/L·h and $v_{SOR} = 112.7$ g/L·h.

Index Entries: *Zymomonas mobilis*; sorbitol; gluconic acid.

Introduction

Sorbitol (SOR) and gluconic acid (GA) are compounds employed in food (SOR as humectant, softener, and texturizer, whereas GA as coagulant in tofu production and in the prevention of milkstone and beerstone in dairy and brewing products, respectively), pharmaceutical (SOR as flavor enhancer and a tablet excipient, whereas GA as a replenisher of magnesium, calcium, and ferrous ions) and chemical (SOR in the synthesis of sorbose, ascorbic acid, and propylene glycol, whereas GA in galvanoplasty and in the textile printing process) industry.

Zymomonas mobilis, an anaerobic, gram-negative, and nonpathogenic bacterium, has in the periplasmic space of the cell envelope the glucose-fructose-oxidoreductase (GFOR), which converts glucose (G) and fructose

*Author to whom all correspondence and reprint requests should be addressed.

(F) in GA and SOR, respectively *(1)*. The GFOR has tightly linked in its structure the NADP, which is reduced to $NADPH_2$ during the glucose oxidation to glucono-δ-lactone and oxidized again by the reduction of fructose to sorbitol *(1)*. The cyclic nature of GFOR catalysis is quite advantageous because the cofactor, generally an expensive reagent, is not consumed. Moreover, the enzymatic conversion is less polluting and health hazardous than the fructose/sorbitol nickel catalyzed hydrogenation- (carried out at 150°C and pressure of 40–50 atm) and more simple than the glucose/gluconic acid fermentative oxidation by fungi of the genus *Aspergillus*. Despite the technical advantages of the GFOR catalyzed bioconversion over the fermentation and chemical synthesis, the market prices for GA and SOR combined with the production overall costs have not been favorable to the biotechnological process worldwide *(1)*. However, according to Silveira and Jonas *(1)* the biotechnological production of sorbitol is economically possible in at least some countries. This aspect stimulates further studies about G-F/GA-SOR bioconversion *(2)*.

The conspicuous localization of GFOR in *Z. mobilis* cell, the establishment of GFOR catalytic mechanism and the large data available on the biochemistry of this bacterium has stimulated its use in the G-F/GA-SOR bioconversion carried out in reactors operated through batch, fed-batch, or continuous process. Yields greater than 90% were attained by using permeabilized or not, free or immobilized cells of *Z. mobilis* through continuous and discontinuous processes *(1)*.

From the data published in the literature it can be seen that the formation of GA and SOR along the reaction time follows a hyperbolic pattern when the bioconversion is carried out with an initial substrate concentration (glucose + fructose) higher than 300 g/L *(3–4)*. Based on this result a mathematical model might be developed aiming to quantify the G-F/GA-SOR bioconversion as far as nothing about this subject was found in the literature.

The present work aims to establish a quite simple mathematical model for estimating the maximum amount of GA and SOR that might be attained through the bioconversion catalyzed by free (permeabilized or not) and immobilized *Z. mobilis* cells.

Materials and Methods

Microorganism and Culture Conditions

Z. mobilis ATCC 29191 was maintained and cultivated anaerobically as described previously *(5)*. The standard medium (SM) contained (per L): 2 g $(NH_4)_2SO_4$, 1 g $MgSO_4 \cdot 7H_2O$, 0.01 g $FeSO_4$, 3.5 g KH_2PO_4, 5 g Bacto yeast extract, 0.2 g sodium citrate, and 150 g glucose. Batch fermentation was performed in 20-L fermentor (BIOSTAT ED, B. Braun Diessel Biotech, Melsungen, Germany) containing 14 L of sterilized SM and inoculated with 1 L of inoculum (4.5 g dry matter). The culture was carried out under anaerobic conditions at 30°C, impeller speed 400 rpm and pH 5.5 (adjusted with MES-buffer (2-[N-Morpholino]ethanesulfonic acid).

Cell Treatment With CTAB

The cell permeation using CTAB (cetyl trimethyl ammonium bromide) was accomplished as follows: 160 µL of CTAB solution (3 g/L) were added to a cell suspension of 9 g dry matter/L under agitation of 300 rpm at 4°C. After 10 min of stirring, the mixture was centrifuged (9000g; 10 min) and the supernatant discharged. The cake was suspended in 80 mL of distilled water and employed in the bioconversion test.

Cell Immobilization

The cell immobilization was carried out as described previously (6). Almost 4 g of sodium alginate (Satialgine S1100X, viscosity of 550 cp and mannuronic acid/guluronic acid ratio of 1:5, purchased from SKW-Biosystems, Trenton, NJ) were dissolved in 100 mL of distilled water, followed by the addition of 100 mL of cell suspension (30 g dry matter/L). The final suspension was left under agitation of 150 rpm for 2 h. Then, the cell suspension was totally dropped into 0.3 M $CaCl_2$ solution for obtaining calcium alginate beads (external diameter of 2 mm) with Z. mobilis cells entrapped. The cell-entrapped beads were maintained immersed in distilled water at 4°C till their use in the bioconversion test. Not less than 95% of the initial cell concentration was entrapped inside the calcium alginate beads, as determined by counting the free cells present in $CaCl_2$ solution through a Neubauer chamber (1/400 mm^2 × 0.1 mm).

Glucose-Fructose-Oxidoreductase Activity

The following assay conditions were defined: cell concentration, 30 g dry matter/L; substrate concentration, 180 g of glucose/L and 180 g of fructose/L; temperature, 39°C; pH, 6.4 (1.0 M phosphate/citrate buffer); reaction time, 20 min. A volume of 1.6 mL of the substrate solution (containing glucose and fructose) was mixed with 0.4 mL of the cell suspension. After 20 min the reaction was stopped by immersing the test tube in a boiling-water bath for 1 min, followed by centrifugation for 10 min at 8000g. The supernatant was collected and stored at 4°C for the determination of gluconic acid. The buffer solution without sugars was used as a blank. One GFOR unit (U) was defined as the amount of enzyme catalyzing the formation of 1 g gluconic acid/h at 39°C. Cells used in all tests had a GFOR activity around 75 U/L (2.5 U/g of dry matter) regarding gluconic acid.

Bioconversion Tests

In a 250-mL batch reactor were mixed the 600 g/L substrate solution (glucose and fructose, ratio 1:1, dissolved in 0.5 MES/MES.K buffer, pH 6.2) and the suspension containing free (CTAB-permeated or not cells) or Ca-alginate entrapped cells. The reaction was carried out at 39°C, 300 rpm and the pH maintained at 6.2 by the addition of 2 M NaOH. In all tests the suspension had a total cell concentration of 30 g dry matter/L. All tests were made in triplicate and the whole volume of samples taken for analytical purposes was lower than 5% of the reacting medium inside the reactor.

Assuming that one molecule of glucose reacts with one molecule of fructose giving one molecule of gluconic acid and one molecule of sorbitol, respectively, then the yield regarding gluconic acid (R_{GA}) and sorbitol (R_{SOR}) formed were calculated, respectively, through the following equations:

$$R_{GA} = (Y_{G/GA} / 1.09) \times 100 \quad (1)$$

$$R_{SOR} = (Y_{F/SOR} / 1.01) \times 100 \quad (2)$$

where $Y_{G/GA}$ (glucose/gluconic acid conversion factor) = (GA_{final} − $GA_{initial}$)/($G_{initial}$ − G_{final}); $Y_{F/SOR}$ (fructose/sorbitol conversion factor) = (SOR_{final} − $SOR_{initial}$)/($F_{initial}$ − F_{final}); (GA_{final}, G_{final}, SOR_{final} and F_{final}) and ($GA_{initial}$, $G_{initial}$, $SOR_{initial}$ and $F_{initial}$) are, respectively, the final and initial concentrations (g/L) of gluconic acid, glucose, sorbitol and fructose.

Analytical Techniques

The cell dry matter was determined according to Neves et al. *(7)*. Gluconic acid was assayed enzymatically using the test kit of Boehringer (Mannheim, Germany) *(8)*. Glucose, fructose, and sorbitol concentrations were analysed by using a Merck (Darmstadt, Germany) HPLC with a Eurokat-Pb (KANAUER, Munich, Germany) 300 × 8 column and a software (Merck D-6000) was used for data retrieval and analysis. Ethanol was determined by gas chromatography (Hewlett-Packard HP5890-II, Palo Alto, CA) with a HP-FFAP capillary column.

Results and Discussion

From Table 1, it can be seen that the Ca-alginate entrapped and free (CTAB-treated or not) cells converted glucose and fructose, respectively, in gluconic acid and sorbitol at a yield higher than 90% as well as no ethanol formation occurred. These results are in accordance with the data already published *(1)*. Figures 1–3 show that the curves related to the variation of SOR and GA concentrations along the reaction in all tests studied had a hyperbolic-like profile. Therefore, the following generic equations can be written:

$$(GA) = t\,(GA)_{max} / (W_{GA} + t) \quad (3)$$

$$(SOR) = t\,(SOR)_{max} / (W_{SOR} + t) \quad (4)$$

Table 1
Formation of Sorbitol (SOR), Ethanol (E), and Gluconic Acid (GA) in the Batch Bioconversion Catalyzed by Nonpermeabilized Free Cells (Test 01), CTAB-Permeabilized Free Cells (Test 02), and Ca-Alginate Immobilized Cells (Test 03)

Test parameters	01	02	03
S (g/L)[a]	600	600	600
(SOR) (g/L)	290	290	290
(GA) (g/L)	288	290	299
(E) (g/L)	0	0	0
R_{GA} (%)	95	92	90
R_{SOR} (%)	95.1	93	90

[a]Initial substrate concentration constituted by 50% (w/w) of glucose and fructose.

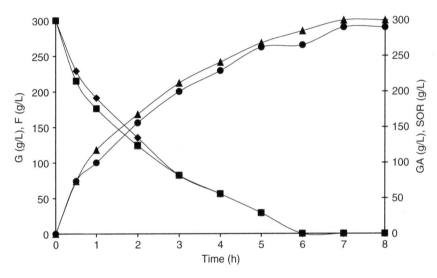

Fig. 1. Variation of glucose (G; ♦), fructose (F; ■), gluconic acid (GA; ▲) and sorbitol (SOR; •) concentrations in a biotransformation performed with whole cells of Z. mobilis at initial substrate concentration (glucose plus fructose at 1:1 ratio) equal to 600 g/L (Test 01). The linear equations related to (1/GA) × (1/t) and (1/SOR) × (1/t) are, respectively, 1/(GA) = 5.72 × 10^{-3}t^{-1} + 2.69 × 10^{-3} (r = 0.998) and 1/(SOR) = 5.65 × 10^{-3}t^{-1} + 3.24 × 10^{-3} (r = 0.99).

where (GA) = gluconic acid concentration (g/L) during the reaction time (h); (GA)$_{max}$ = maximum gluconic acid concentration (g/L); (SOR) = sorbitol concentration (g/L) during the reaction time (h); (SOR)$_{max}$ = maximum sorbitol concentration (g/L); W_{GA} and W_{SOR} are constants representing the time to which the gluconic acid and sorbitol concentration, respectively, correspond to the half of maximal concentration achieved for each of them. Moreover, an estimate of the GFOR activity regarding the production

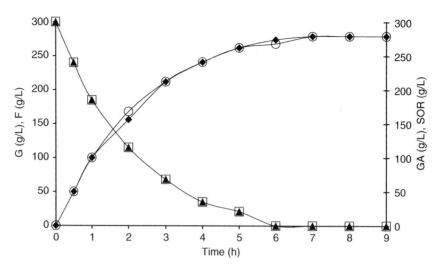

Fig. 2. Variation of glucose (G; ▲), fructose (F; O), gluconic acid (GA; ♦) and sorbitol (SOR; O) concentrations in a biotransformation performed with whole cells of *Z. mobilis* treated with CTAB. The initial substrate concentration (glucose plus fructose at 1:1 ratio)) was 600 g/L (Test 02). The linear equations related to $(1/GA) \times (1/t)$ and $(1/SOR) \times (1/t)$ are, respectively, $1/(GA) = 8.91 \times 10^{-3} t^{-1} + 1.85 \times 10^{-3}$ (r = 0.998) and $1/(SOR) = 8.91 \times 10^{-3} t^{-1} + 1.81 \times 10^{-3}$ (r = 0.997).

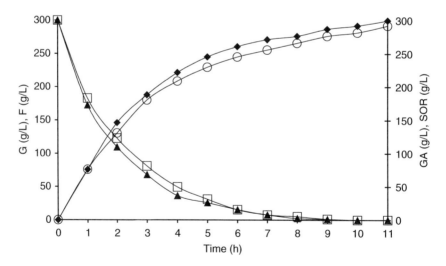

Fig. 3. Variation of glucose (G; ▲), fructose (F; O), gluconic acid (GA; ♦) and sorbitol (SOR; O) concentrations in a biotransformation performed with whole cells of *Z. mobilis* entrapped into calcium alginate beads. The initial substrate concentration (glucose plus fructose at 1:1 ratio)) was 600 g/L (Test 03). The linear equations related to $(1/GA) \times (1/t)$ and $(1/SOR) \times (1/t)$ are, respectively, $1/(GA) = 1.07 \times 10^{-2} t^{-1} + 2.09 \times 10^{-3}$ (r = 0.994) and $1/(SOR) = 1.07 \times 10^{-2} t^{-1} + 2.31 \times 10^{-3}$ (r = 0.998).

of gluconic acid and sorbitol along the reaction could be attained at each time through the derivatives of Eqs. 3 and 4, which are respectively

Table 2
Constant Parameters Related to the Formation of Gluconic Acid and Sorbitol From Glucose and Fructose, Respectively, in Bioconversions Catalyzed by Free and Immobilized Z. mobilis Cells

Test (no.)	$(GA)_{max}$ (g/L)	$(SOR)_{max}$ (g/L)	W_{GA} (h)	W_{SOR} (h)	$(GA)_{max}/(SOR)_{max}$	W_{GA}/W_{SOR}	v_{GA} (g/L·h)	v_{SOR} (g/L·h)
01	372	309	2.1	1.8	1.2	1.2	177.1	177.7
02[a]	541	552	4.8	4.9	0.98	0.98	112.7	112.7
03[b]	478	432	5.1	4.6	1.1	1.1	93.7	93.7

[a]The cells were treated with CTAB.
[b]The cells were entrapped into calcium alginate beads.

It were also presented the initial GFOR activity [calculated through Eqs. 5 and 6, which at t = 0 h become $v_{GA} = (GA)_{max}/W_{GA}$ and $v_{SOR} = (SOR)_{max}/W_{SOR}$, respectively] related to the production of gluconic acid (v_{GA}) and sorbitol (v_{SOR}).

$$v_{GA} = d(GA)/dt = [W_{GA} \cdot (GA)_{max}]/(W_{GA} + t)^2 \quad (5)$$

$$v_{SOR} = d(SOR)/dt = [W_{SOR} \cdot (SOR)_{max}]/(W_{SOR} + t)^2 \quad (6)$$

where v_{GA} and v_{SOR} are the GFOR activity related to gluconic acid and sorbitol, respectively.

Regarding Eqs. 5 and 6, two aspects might be borne out. First, the Eqs. 5 and 6 at t = 0 h become, respectively, $v_{GA} = (GA)_{max}/W_{GA}$ and $v_{SOR} = (SOR)_{max}/W_{SOR}$, which should represent the maximal initial GFOR activity presented by a particular biocatalyst (in the present work, CTAB nontreated, CTAB-treated and immobilized Z. mobilis cells) regarding the formation of both products (gluconic acid and sorbitol). So that, these ratios could allow attaining a first glance on the catalytic potential of the biocatalyst in bioconversions conducted under fixed initial substrate concentration (glucose plus fructose at 1:1 ratio), in the present case equal to 600 g/L. Accordingly, for tests 01, 02, and 03, v_{GA} is approx v_{SOR} and equal to 177.1, 112.7 and 93.7 g/L·h, respectively (Table 2). Second, coupling the v_{GA} and v_{SOR} values (calculated by Eqs. 5 and 6 and related to the GFOR activity needed) with the consumption of glucose and fructose at each time along the reaction (taken from Figs. 1–3, by subtracting G and F concentrations from 300 g/L at a desired time), it could be possible setting the suitable amount of substrate to be added into the bioreactor, when the bioconversion was planned to be conducted as a fed-batch process.

Through plots of $1/(GA) \times 1/t$ and $1/(SOR) \times 1/t$ the linear equations were established from which the parameters $(GA)_{max}$, W_{GA}, $(SOR)_{max}$ and W_{SOR} were calculated for all tests. The correspondent equations are presented in the captions of Figs. 1–3.

From Table 2 it can be seen that cells treated with CTAB (test 02) had $(GA)_{max}$, $(SOR)_{max}$, W_{GA} and W_{SOR}, as 31, 44, 56, and 64% higher than those of nontreated cells (test 01), respectively. Undoubtedly, the CTAB treatment increases the cell wall fluidity for substrates (glucose and fructose) and products (sorbitol and gluconic acid). It could also be considered that the CTAB treatment did not affect significantly the GFOR catalysis, as the ratios $(GA)_{max}/(SOR)_{max}$ and W_{GA}/W_{SOR} for test 02 were near one (Table 2). Moreover, for test 03 the ratios $(GA)_{max}/(SOR)_{max}$ and W_{GA}/W_{SOR} both equal to 1.1 were quite similar to those of tests 01 and 02 (Table 2), an indication, even so indirect, that the immobilization did not change significantly the GFOR catalytic pattern. Thereby, an enlargement of the feasibility of the G-F/GA-SOR bioconversion might be achieved by using immobilized cells because the reaction can be conducted through continuous processes using different kinds of reactors (continuous stirred tank reactor, membrane reactor, fluidized, or packed bed reactor) *(3,9)*. In addition, an immobilized system could facilitate a future scale-up of this bioconversion.

Finally, the constants W_{GA} and W_{SOR}, which have the dimension of time (h), correspond to the moment of the reaction in which $(GA) = (GA)_{max}/2$ and $(SOR) = (SOR)_{max}/2$, respectively. The times to which a particular GA and SOR concentrations are attained, could be considered as the starting points to set the residence time, when the bioconversion is planned to be realized through a continuous process. As two products are involved, so the average between W_{GA} and W_{SOR} would be a reasonable procedure to set the residence time. Accordingly, the average time for tests 01, 02, and 03 are 1.95, 4.85, and 4.85 h, respectively (Table 2).

Conclusions

Yields greater than 90% for the G-F/GA-SOR bioconversion were attained using either free (CTAB-treated or not) or Ca-alginate entrapped *Z. mobilis* cells. The profiles of the curves $(GA) \times t$ and $(SOR) \times t$ were described by hyperbolic equations, from which the maximum amount of GA and SOR formed might be calculated, provided that the initial substrate concentration (glucose plus fructose at 1:1 ratio) was not less than 600 g/L.

References

1. Silveira, M. M. and Jonas, R. (2002), *Appl. Microbiol. Biotechnol.* **59**, 400–408.
2. Jonas, R. and Silveira, M. M. (2004), *Appl. Biochem. Biotechnol.* **118**, 321–336.
3. Ferraz, H. C., Borges, P. C., and Alves, L. M. (2000), *Appl. Biochem. Biotechnol.* **89**, 43–53.
4. Shene, C. and Bravo, S. (2001), *Appl. Microbiol. Biotechnol.* **57**, 323–328.
5. Erzinger, G. S., Silveira, M. M., Vitolo, M., and Jonas, R. (1996), *World J. Microbiol. Biotechnol.* **12**, 22–24.
6. Carvalho, W., Silva, S. S., Converti, A., and Vitolo, M. (2002), *Biotechnol. Bioeng.* **79**, 1–10.
7. Das Neves, L. C. M., Pessoa A. Jr., and Vitolo, M. (2005), *Biotechnol. Prog.* **21**, 1136–1139.
8. Rehr, B., Wilhelm, C., and Sahm, H. (1991), *Appl. Microbiol. Biotechnol.* **35**, 144–148.
9. Tomotani, E. J., Das Neves, L. C. M., and Vitolo, M. (2005), *Appl. Biochem. Biotechnol.* **121**, 149–158.

Metabolic Engineering of *Saccharomyces cerevisiae* for Efficient Production of Pure L-(+)-Lactic Acid

Nobuhiro Ishida,*,[†,1] Satoshi Saitoh,[†,2] Toru Ohnishi,[2] Kenro Tokuhiro,[1] Eiji Nagamori,[1] Katsuhiko Kitamoto,[3] and Haruo Takahashi[1]

[1]*Biotechnology Laboratory, Toyota Central R&D Labs Inc., Aichi 480-1192, E-mail: e1168@mosk.tytlabs.co.jp;* [2]*Toyota Biotechnology & Afforestation Laboratory, Toyota Motor Co., Aichi 470-0201; and* [3]*Department of Biotechnology, The University of Tokyo, Bunkyo-ku, Tokyo 113-8657, Japan*

Abstract

We developed a metabolically engineered *Saccharomyces cerevisiae*, which produces optically pure L-lactic acid efficiently using cane juice-based medium. In this recombinant, the coding region of pyruvate decarboxylase *(PDC)1* was completely deleted, and six copies of the bovine L-lactate dehydrogenase *(L-LDH)* genes were introduced on the genome under the control of the *PDC1* promoter. To confirm optically pure lactate production in low-cost medium, cane juice-based medium was used in fermentation with neutralizing conditions. L-lactate production reached 122 g/L, with 61% of sugar being transformed into L-lactate finally. The optical purity of this L-lactate, that affects the physical characteristics of poly-L-lactic acid, was extremely high, 99.9% or over.

Index Entries: Cane juice-based medium; L-lactic acid production; optical purity; *Saccharomyces cerevisiae*.

Introduction

Poly lactic acid (PLA) is being developed as a renewable alternative for conventional petroleum-based plastics. The advancement of a sustainable society has created an urgent need for large-scale production of lactic acid, which is used as a monomer for polymerization into PLA. But it has been pointed out that this polymer was only thermostable up to approx 58°C *(1)*. The problem is that PLA is weak and therefore receives heat and has increased attention for the expanded use of this renewable plastic. Because it has been reported that optical purity affects the physical characteristics, such as crystallization, thermostability, the biodegradation rate,

---

*Author to whom all correspondence and reprint requests should be addressed.
[†]These two authors contributed equally to this work.

and performance *(2)*, it is important to establish a processing technology for L-lactic acid with high optical purity.

The goal of this study is to establish efficient production of optically pure L-lactic acid. L-Lactic acid is generally produced using lactic acid bacteria such as *Lactobacillus* spp. The optical purity of this monomer is approx 95% *(3)*, but this purity is not suitable for the high physical properties of PLA. To improve the purification of lactic acid, the separation of optical isomers through the crystallization has been reported *(4,5)*. But it is important to obtain L-lactic acid of extremely high optical purity during the fermentation stage. *Lactobacillus* spp. are not able to produce lactic acid of extremely high optical purity, because some of them have both L- and D-lactate dehydrogenase (L-, D-*LDH*) genes. Even *Lactobacillus sakei*, which has only an L-*LDH*, also produces D-lactic acid owing to its lactate racemase activity *(6)*. Recently, improvement of its optical purity with *Lactobacillus* spp. was reported *(7–9)*, but it was not enough for industrial use.

On the other hand, yeasts, including *Saccharomyces cerevisiae*, hardly produce either L- or D-lactic acid, because they do not have lactate racemase or L- and D-LDH. Therefore, new methods for producing lactic acid with genetically engineered yeast have been developed, and applied for large-scale production on a trial basis. Such transgenic yeasts were first reported by Dequin and Barre *(10)* and Porro et al. *(11)*, who showed that the recombinants yielded about 10–20 g of lactate/L. To increase the metabolic flow from pyruvic acid to lactic acid, a mutant strain, such as the *pdc1, pdc5 (12)*, or *adh1 (13)* mutant, was utilized as the genetic background for obtaining an L-*LDH* expressing yeast. However, a remarkable improvement in L-lactic acid production has not been observed, and the analysis of the optical purity was not mentioned in those reports.

In a previous study, we attained efficient production of L-lactic acid with a recombinant wine yeast *(14)*. To achieve mass production of L-lactic acid of extremely high optical purity by using *S. cerevisiae*, we have developed a more genetically modified yeast strain following the previous work. This new recombinant has six copies of the bovine L-*LDH* gene on the genome, and all the copies are expressed under the control of the *PDC1* promoter. In this study, we examined the optical purity of the lactic acid produced by this transgenic strain.

Additionally, fermentation analysis with an inexpensive medium, such as one including an unused resource, would also be significant for producing L-lactic acid of high purity on an industrial scale. Such examination with media containing cane juice *(15,16)*, corn steep liquor *(17)*, hydrolyzed sago starch *(18)*, and biological pretreatment corncob *(19)* has been reported. However, the reported production involving the use of a transgenic yeast involved YPD medium rather than an inexpensive medium. In this study, we also examined the lactic acid productivity under the cane juice-based medium. It is expected that the achievement of the

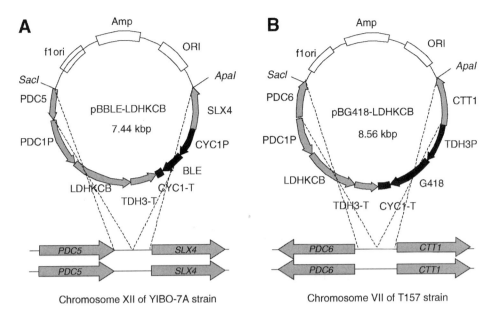

Fig. 1. Maps of the plasmid vectors and breeding of transgenic *S. cerevisiae*. **(A)** Construction of the T157 strain. The constructed DNA fragment, which was obtained by digesting the pBBLE-LDHKCB vector with *Sac*I and *Apa*I, was integrated into the *PDC5* downstream locus of chromosome XII in the YIBO-7A transgenic strain. **(B)** Construction of the T165 strain. The constructed DNA fragment, which was obtained by digesting the pBG418-LDHKCB vector with similar restriction enzymes, was integrated into the *PDC6* upstream locus of chromosome VII in the T157 transgenic strain.

efficient production of the lactic acid in an inexpensive medium contributes to low-cost production of PLA.

Materials and Methods

Strains and Media

The *Escherichia coli* strain used for molecular cloning was JM109 (Toyobo, Osaka, Japan). *E. coli* cultivation and the medium preparation were carried out by standard procedures *(20)*. The *S. cerevisiae* OC-2T strain (a/α *trp1/trp1*) was derived from the wine yeast IFO2260 strain *(21)*. The culture medium used for *S. cerevisiae* was YPD medium (1% bacto yeast extract, 2% bacto peptone, and 2% glucose, wt/vol).

Plasmid Construction

The plasmid vectors used in this study are shown in Fig. 1. The method used to modify the bovine L-*LDH* gene and the construction of the YIBO-7A recombinant strain were described in detail in the previous report *(14)*. Two genome integration vectors, pBBLE-LDHKCB (Fig. 1A) and pBG418-LDHKCB (Fig. 1B), were constructed using the pBluescript SKII⁺ vector (Stratagene, La Jolla, CA).

Table 1
Primers Used for Constructing pBBLE-PDC1P-LDHKCB
and pBG418G-PDC1P-LDHKCB

Primer	Sequence (5'–3')[a]	Restiction site
PDC5-U	ATATAT<u>GAGCTC</u>CATGATTAGATGGGGTTTGAAGCC	Sac I
PDC5-D	ATATAT<u>GCGGCCGC</u>CTGGAAGACAGGACAGAAAAGT	Not I
SLX4-U	ATATAT<u>GTCGAC</u>GGTTAAAGATTAGCTTCTAATA	Sal I
SLX4-D	ATATAT<u>GGGCCC</u>GGGCAACTGAACTACTGGTTATT	Apa I
PDC6-U	ATATAT<u>GAGCTC</u>GTTGGCAATATGTTTTTGC	Sac I
PDC6-D	ATATAT<u>GCGGCCGC</u>TTCCAAGCATCTTATAAACC	Not I
CTT1-U	ATATAT<u>GGGCCC</u>GATGTACGATCGCCTGCACTAT	Apa I
CTT1-D	ATATAT<u>GGTACC</u>GGGCAAGTAACGACAAGATTG	Kpn I

[a]Restrictions sites in the primer sequences are underlined.

In the pBBLE-LDHKCB vector (Fig. 1A), the phleomycin resistance gene casset was *Tn5 BLE* of bacterial transposon *Tn5 (22)*, which was fused downstream from the *S. cerevisiae* cytochrome-c (*CYC1*) promoter. The *PDC1* promoter, *PDC5* and *SLX4* gene fragments were isolated by PCR using the genomic DNA of the *S. cerevisiae* OC-2T strain as a template. Genomic DNA was prepared using a Fast DNA Kit (Q-Biogene, Carlsbad, CA), and the concentration was determined with an Ultro spec 300 spectral photometer (Pharmacia Biotech, Uppsala, Sweden). KOD DNA polymerase was used for PCR amplification, and the oligonucleotide sequences of the primers are shown in Table 1 (*see* the previous report for details of the primer sequence of the *PDC1* promoter fragment, *[14]*). The amplification fragments were treated with each restriction enzyme (Takara Bio, Otsu, Japan), and then ligated to a vector. The ligase reaction was performed with a Lig Fast Rapid DNA Ligation System (Promega, Madison, WI), and the competent cells used for transformation were of the *E. coli* JM109 strain (Toyobo). To confirm subcloning of the vector, the nucleotide sequence was determined with an ABI PRISM 310 Genetic Analyzer (Applied Biosystems, Foster City, CA).

In the pBG418-LDHKCB vector (Fig. 1B), the kanamycin (G418) resistance gene is the aminoglycoside phosphotransferase (*APT*) gene *(23)*, which confers geneticin resistance on yeasts, fused downstream from the *S. cerevisiae* glyceraldehyde-3-phosphate dehydrogenase 3 (*TDH3*) promoter. The *PDC6* and *CTT1* gene fragments were isolated by PCR using the genomic DNA of the *S. cerevisiae* OC-2T strain as a template, and the oligonucleotide sequences of the primers are shown in Table 1. The vector was constructed by a similar technique to that described earlier.

Breeding of Yeasts

S. cerevisiae transformation was performed by the lithium acetate procedure *(24)*, and each transformant was selected on YPD medium

Table 2
Transgenic Strains Used in this Study

Strain	Relevant genotype	Copy number of LDH gene	Reference
OC-2T (Host strain)	trp1/trp1	0 copies	21
YIBO-7A	pdc1/pdc1	2 copies	14
T157	pdc1/pdc1, phlemycin^r	4 copies	This study
T165	pdc1/pdc1, phlemycin^r, kanamycin^r	6 copies	This study
T165R	pdc1/pdc1, phlemycin^r, kanamycin^r, EMS mutagenized	6 copies	This study

containing 7.5 µg/mL phleomycin (Sigma, St. Louis, MO) or 150 µg/mL G418 (Calbiochem, San Diego, CA).

Table 2 shows the transgenic strains that were constructed in this study. First, the pBBLE-LDHKCB vector fragment, which had been digested with SacI and ApaI, was transformed into the YIBO-7A recombinant strain (Fig. 1A), which showed high lactic acid production in the previous study (12). Host strain OC-2T is a diploid and homotallic strain (21). After transformation, the L-LDH caset was usually located on one side of a pair of chromosomes. The heterologous gene on one side of a chromosome could be duplicated through spore formation. Spore formation was performed on sporulation plates (1% acetate, 0.1% D-glucose, 0.1% yeast extract, and 2% agar, wt/vol). Diploid formation was performed using the homothallic property, and tetrads were dissected under an optical microscope (Olympus, Tokyo, Japan) with a micro-manipulator (Narishige Science, Tokyo, Japan). After colonies had been isolated the target gene integration was confirmed by PCR. The primer sequences used were as follows; F2,5'-ACCAGCC-CATCTCAATCCATCT-3';R2,5'-ACACCCAATCTTTCACCCATCA-3'. The resulting recombinant yeast was named the T157 strain, which included four copies of the bovine L-LDH gene on the genome.

Second, the pBG418-LDHKCB vector fragment, which had been digested with similar restriction enzymes, was transformed into the T157 strain (Fig. 1B). The transgenic strain was constructed by a similar method to that described earlier. Target gene integration was confirmed by PCR. The primer sequences used were as follows; F3,5'-TCATTGGTGACG-GTTCTCTACA-3';R3,5'-CGATAGCAAGTAGATCAAGACA-3'. The resulting recombinant yeast was named the T165 strain, which included six copies of L-LDH gene on the genome.

Ethylmethane Sulfonate Mutagenesis

Stationary phase cells were harvested from independent cultures grown in YPD. The cells were resuspended in 200 mM sodium phosphate (pH 7.0) containing 50 µL ethylmethane sulfonate (EMS). The mixtures were vortexed, and then agitated gently on a rotary shaker for 1 h at room temperature. The cells were promptly washed in 10 mL sterile water. The final suspensions, which contained approx 10^6–10^7 viable cells/mL, were incubated at 30°C on YPD medium.

LDH Specific Activity

Cell extracts were prepared with a SONIFIER 250 (Branson, Danbury, CT) as described previously *(25)*. LDH-specific activity was determined in freshly prepared extracts as described by Minowa et al. *(26)*. Protein concentrations in cell extracts were determined with a DC protein assay kit (Bio-Rad, Richmond, CA), using bovine serum albumin (Sigma, St. Louis, MO) as a standard.

Fermentation

The fermentation experiment was performed at 30°C in a 100 mL flask with a working volume of 40 mL in YPD10 medium (1% bacto yeast extract, 2% bactopeptone, 10% D-glucose) containing 3% of sterilized calcium carbonate (wt/vol). The inoculum was prepared by transferring a strain from a stock culture to a flask containing 5 mL of YPD medium. The culture was performed for 24 h at 30°C on a shaker, followed by transfer to the fermentation medium at an inoculum size of 0.1% packed cell volume (PCV). Cells were inoculated into a 1-L jar-fermenter made by Biotto Corporation (Tokyo, Japan). The jar conditions were kept at 32°C and pH 5.2, with aeration at 0.15 L/min. Its agitation rate was controlled at 60 rpm, and NaOH was used for neutralization. The medium consisted of cane juice (sugar concentration, approx 20%) containing 0.3% yeast extract (wt/vol). Glucose, lactic acid, and ethanol concentrations were measured with a Biosensor BF-4 (Oji Keisoku Kiki, Amagasaki, Japan). The optical purity of L-lactic acid was calculated as follows. Each value of the following expression shows the quantity (%, wt/vol) of lactic acid.

$$\text{Optical purity of L-lactic acid (\%)} = \frac{\text{L-lactic acid} - \text{D-lactic acid}}{\text{Total lactic acid}} \times 100$$

Results

Construction of Strain Containing Multi-Copies of L-LDH Gene

A transgenic *S. cerevisiae* with an increased L-LDH gene copy number was constructed based on the YIBO-7A strain, which was constructed in previous study *(14)*. In the YIBO-7A strain, the coding region for *PDC1* on chromosome XII is substituted for that of the L-LDH through homologous recombination. The expression of mRNA for the genome-integrated L-LDH is regulated under the control of the native *PDC1* promoter, whereas *PDC1* is completely disrupted *(14)*. First, T157 strain, which has four copies of L-LDH gene, was constructed. The heterologous L-LDH was integrated

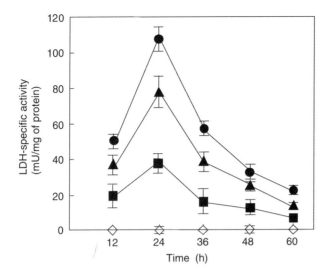

Fig. 2. Comparison of LDH specific activity with increasing gene copy number. The control strain OC-2T (□). YIBO-7A strain with two copies (■), T157 strain with four copies (▲), and T165 strain with six copies of the L-LDH gene (●). The average and deviation for three independent experiments are presented.

between PDC5 and SLX4 (Fig. 1A), but these two coding regions were not disrupted on homologous recombination, because it was reported that pdc1 and pdc5 double mutant strain shows suppression of growth rate (12). Second, the T165 strain, which has six copies of L-LDH, was constructed (Fig. 1B). The PDC6 and CTT1 genes were also not disrupted during the integration.

LDH-Specific Activity

LDH-specific activity of these three recombinant strains, YIBO-7A (L-LDH; two copies), T157 (four copies), and T165 (six copies), were measured. Host strain OC-2T (no L-LDH) was used as a control. As shown in Fig. 2, in every strain, the highest activity was observed at 24 h, and the activity decreased significantly after 36 h. The expression of PDC1 is strongly induced by glucose (27), and glucose responding elements in S. cerevisiae have already been reported (28). It was supposed that the time-course of the activity was correlated closely with the glucose concentration in the medium. Improvement of the LDH-specific activity was observed with an increasing L-LDH copy number on the genome. The T165 strain showed the highest LDH activity, 108.2 mU/mg of protein, at 24 h. This is an approx 2.8 times increase compared with for the YIBO-7A strain, which had two copies of the L-LDH.

EMS Mutagenesis of T165 Recombinant Strain

Despite high LDH activity, the T165 strain (L-LDH; six copies) exhibited remarkably suppressed of the growth rate (data not shown).

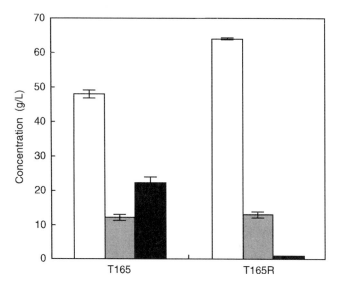

Fig. 3. Comparison of fermentation with the *S. cerevisiae* T165 strain and T165R strain (EMS mutagenesis) in YPD medium containing 100 g/L of glucose and 50 g/L of CaCO$_3$. L-Lactate (white bars), ethanol (gray bars), and glucose (black bars). Each strain was cultivated for 72 h under micro-anaerobic conditions at 30°C. The average and deviation for two independent experiments are presented.

Furthermore, this strain could not completely consume the glucose in fermentation, and lactate productivity decreased (Fig. 3). Following this, we selected the T165 recovered strain (T165R) by EMS mutagenesis. The growth of the T165R strain was recovered to the same level as for the T157 strain (four copies) and this strain completely consumed the glucose on fermentation (Fig. 3). In fermentation analysis, it was clarified that the micro-aerobic condition is optimum for the T165R strains (data not shown), whereas the efficient production of lactate was not confirmed in the anaerobic condition.

Fermentation of Recombinant Strain in YPD Medium

The lactate productivity of these recombinants was examined by cultivation in YPD medium containing 100 g/L of glucose. The T165R strain, with six copies of L-LDH gene being expressed from *PDC1* promoter, was observed to produce both L-lactate (68 g/L) and ethanol (9.8 g/L), with up to 68% of the glucose being transformed into L-lactic acid (Fig. 4). However, the T157 and T165R strains exhibited glucose consumption ability. The lactate productivity of T167R was improved by more than 1.28 times compared with that of the recombinant strain YIBO-7A in a previous study *(14)*. Improvement of the productivity was observed with increasing L-LDH copy number on the genome, as well as from the LDH-specific activity results (Fig. 2).

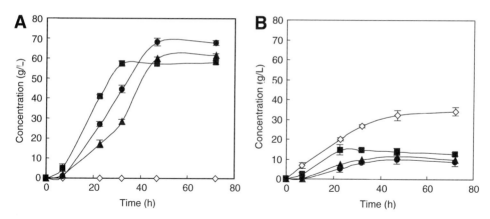

Fig. 4. Time courses of cultivation of four transgenic *S. cerivisiae* strains on a flask scale. **(A)** L-Lactate, **(B)** Ethanol. The control strain OC-2T (□). YIBO-7A strain with two copies (■), T157 strain with four copies (▲), and T165R strain with six copies of the *L-LDH* gene (●). The average and deviation for three independent experiments are presented.

Fermentation of T165R Strain in Cane Juice-Based Medium

To confirm lactate production in low-cost medium, we examined media other than YPD. Cane juice is obtained by squeezed sugar cane, and contains high concentration of glucose, sucrose, and many vitamins and minerals. To prepare cane-based medium, the cane juice was diluted into 20% of sugar concentration, and 0.3% yeast extract (wt/vol) was added. Using a 1-L jar-fermenter with pH control, L-lactate production of T165R strain reached 122 g/L, with up to 61% of the sugar being transformed into lactic acid (Fig. 5). The T165R strain showed production of high concentration of L-lactate, although yield on sugar was decreased compared with in the case of 10% YPD (68%, Fig. 4).

Optical Purity

The optical purity of L-lactic acid produced by the T165R strain was measured. The purity of L-lactate was at least 99.9% in both YPD and cane juice-based medium. This purity was obviously high compared with that of lactic acid produced by other lactic acid bacteria (Table 3).

Discussion

Our goals are to establish a means of mass production of l-lactic acid monomer, and to obtain L-lactic acid of high optical purity with transgenic wine yeast. In a previous study, we constructed recombinant yeast that expressed bovine *L-LDH* under the control of the *PDC1* promoter on the genome. Also, genome integration led to efficient lactic acid production compared with the YEp-multicopy method *(14)*. For mass production of lactic acid, we newly constructed recombinants with increased copy

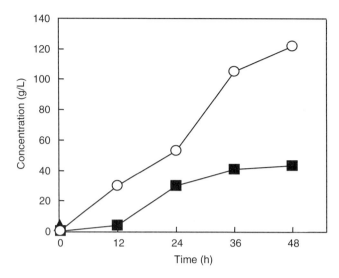

Fig. 5. Fermentation analysis with the *S. cerevisiae* T165R strain in the cane juice-based medium (sugar concentration, approx 20%). The pH was controlled at pH 5.2 with 1 *N* NaOH. o, L-lactate; ■, ethanol. Each strain was cultivated for 72 h under micro-anaerobic conditions at 30°C.

Table 3
Comparison of Optical Purity of L-Lactic Acid Between the Transgenic Yeast and Lactic Acid Bacteria

Strain	Optical purity %	Reference
Saccharomyces cericisiae OC2T T165R	>99.9 (D-lactic acid < 0.01)	This study
Lactobacillus amylophilus ATCC49845	93.0	35
Lactobacillus delbrueckii spp. *bulgaricus* ATCC11842	91.0–95.0	36
Lactobacillus delbrueckii spp. *delbrueckii* ATCC9649	94.0–95.0	36
Lactobacillus rhamnosus ATCC10863	95.0	37
Lactobacillus salivarius spp. *salivarius* ATCC11742	86.0–90.0	38
Streptococcus bovis 148	95.6	16

numbers of the L-LDH gene on the genome. However, a decrease in cell growth rate was observed with increasing copies of the L-LDH. In particular, transgenic strain T165 (L-LDH; six copies) exhibited remarkable suppression of the growth, reduced glucose consumption, and finally decreased lactate productivity, although the LDH-specific activity was increased. It was pointed out by Maris et al. *(29)* that intercellular ATP regeneration and the redox balance were important for increasing lactic acid productivity, and oxygen-limited chemostat cultures showed that lactic acid-producing *S. cerevisiae* strains require oxygen for the generation of ATP. To relieve intracellular stress, transgenic yeasts excrete intracellular lactic acid extracellularly, for which more energy is needed. The intracellular ATP decrease caused remarkable suppression of the growth. The T165R strain selected through EMS mutagenesis exhibited growth, and showed complete glucose consumption compared with the T165 unmutagenized strain. As shown in Fig. 4, the improved L-lactate productivity of the T165R strain was explained by the increased number of copies of the L-LDH, and 68.0 g/L of L-lactate was produced from 10% YPD medium in 72 h finally. In the case of the fermentation in a jar-fermenter, the microaerobic conditions (aeration at 0.15 L/min) led to higher proliferation than under anaerobic conditions. These results show that the improved L-lactic acid productivity of the T165R strain would be the cause of the recovery of ATP generation through the TCA cycle.

For industrial scale production of L-lactic acid, it is also important to use inexpensive media. Cane juice is one of the largest biomass sources available, more than 100×10^6 t of it being generated in Latin America and Tropical Asia (15). Cane juice, which contains more of mineral salts and sucrose, is a low-cost medium compared with YPD. As shown in Fig. 5, the T165R strain showed high lactic acid production even in cane juice-based medium (more than 120 g/L of lactate was produced from this medium in 48 h with a yield of 60% based on the initial sugar concentration). This yield was not greatly decreased compared with in the case of YPD medium. The host strain, OC-2T, was derived from the wine yeast IFO2260 strain, which has been used for producing wine on an industrial scale. Lactic acid is generally produced with lactic acid bacteria, such as *Lactobacillus* spp., which are hard to cultivate at high density and show high auxotrophy concerning growth (3). Considering fermentative production in inexpensive media, it is appropriate to use an industrial yeast, such as a wine yeast, for low cost production of PLA.

The optical purity of L-lactic acid affects the physical properties of poly L-lactic acid, such as the biodegradation rate, crystallization, and thermostability (2). The optical purity of L-lactic acid produced with lactic acid bacteria is generally about 95%, but this purity is insufficient for the high physical properties of PLA. To improve the optical purity, the separation of optical isomers through crystallization has been reported (5), however, this process increases the production cost. Lactic acid bacteria have D-LDH, and attempt to delete this gene have been reported (8). On the other hand, other hosts, i.e., fungi (19,30), genetically engineered *E. coli* (31–33), and genetically engineered yeasts (10–13), have been investigated regarding the production of L-lactic acid. Although several organisms producing lactic acid have been discussed about optical purity (33,34), genetically engineered yeasts have been not analyzed about it yet. In this study, we confirmed that a recombinant wine yeast was able to produce L-lactate of high optical purity (99.9%; Table 3). With recombinant strain T165R, the yield of lactic acid was low compared with those with lactic acid bacteria, because ethanol was still produced. However, it can be said that there are the following three advantages regarding the use of this recombinant yeast. First, L-lactic acid of high optical purity can be produced. Second, lactic acid can be produced even if one uses an inexpensive media, such as one based on cane juice. Last, because yeasts exhibit a low pH tolerance, free lactic acid production can be expected without neutralization (12). Regarding the mass production of L-lactic acid of high optical purity, these results indicate that the use of this transgenic wine yeast has several advantages. We expect this research will lead to further use of transgenic yeasts.

Acknowledgments

We wish to thank Osamu Saotome, Noriko Yasutani, Dr. Takashi Matsuyama, and Dr. Masana Hirai for the valuable discussions. We also thank Miyoko Imoto, Wakana Takase, and Keiko Uemura for the technical assistance.

References

1. Ozeki, E. (1996), *Shimadzu Rev.* **53**, 1–8.
2. Ohara, H. (1998), *Nippon Kagaku Kaishi* **6**, 323–331.
3. Hofvendahl, K. and Hahn-Hagerdal, B. (2000), *Enzyme Microb. Technol.* **26**, 87–107.
4. Benthin, S. (1995), *Appl. Microbiol. Biotechnol.* **42**, 826–829.
5. van Breugel, J., Van Krieken, J., Cerda Baro, A., Vidal Lancis, J. M., and Camprubi Vila, M. (2002), US Patent No. 6,630,603.
6. Hiyama, T., Fukui, S., and Kitahara, K. (1968), *J. Biochem.* **64**, 100–107.
7. Malleret, C., Lauret, R., Ehrlich, S. D., Morel-Deville, F., and Zagorec, M. (1998), *Microbiology* **144**, 3327–3333.
8. Lapierre, L., Germond, J. E., Ott, A., Delley, M., and Mollet, B. (1999), *Appl. Environ. Microbiol.* **65**, 4002–4007.
9. Kyla-Nikkila, K., Hujanen, M., Leisola, M., and Palva, A. (2000), *Appl. Environ. Microbiol.* **66**, 3835–3841.
10. Dequin, S. and Barre, P. (1994), *Bio/Technology* **12**, 173–177.
11. Porro, D., Barmbilla, L., Ranzi, B. M., Martegani, E., and Alberghina, L. (1995), *Biotechnol. Prog.* **11**, 294–298.
12. Adachi, E., Torigoe, M., Sugiyama, S., Nikawa, J., and Shimizu, K. (1998), *J. Ferment. Bioeng.* **86**, 284–289.
13. Skory, C. D. (2003), *J. Ind. Microbiol. Biotehnol.* **67**, 22–27.
14. Ishida, N., Saitoh, S., Tokuhiro, K., et al. (2005), *Appl. Environ. Microbiol.* **71**, 1964–1970.
15. Fontana, J. D., Guimaraes, M. F., Martins, N. T., Fontana, C. A., and Baron. M. (1996), *Appl. Biochem. Biotechnol.* **58**, 413–422.
16. Narita, J., Nakahara, S., Fukuda, H., and Kondo, A. (2004), *J. Biosci. Bioeng.* **97**, 423–425.
17. Ohara, H., Doi, U., Otsuka, H., Okuyama, H., and Okada, S. (2001), *Nippon Seibutukougaku Kaishi* **79**, 142–148.
18. Hipolto, C. N., Matsunaka, T., Kobayashi, G., Sonomoto, K., and Ishizaki, A. (2002), *J. Biosci. Bioeng.* **93**, 281–287.
19. Miura, S., Arimura, T., Itoda, N., et al. (2004), *J. Biosci. Bioeng.* **97**, 153–157.
20. Sambrook, J., Fritsh, E. F., and Maniatis, T. (1989), *Molecular Cloning: A Laboratory Manual*. 2nd ed., Cold Spring Harbor Laboratory Press, NY.
21. Saitoh, S., Mieno, Y., Nagashima, T., Kumagai, C., and Kitamoto, K. (1996), *J. Ferment. Bioeng.* **81**, 98–103.
22. Gatignol, A., Baron, M., and Tiraby, G. (1987), *Mol. Gen. Genet.* **207**, 342–348.

23. Hadfild, C., Jordan, B. E., Mount, R. C., Pretorius, G. H. J., and Burak, E. (1990), *Curr. Genet.* **18,** 303–313.
24. Ito, H., Fukuda, Y., Murata, K., and Kimura, H. (1983), *J. Bacteriol.* **153,** 163–168.
25. Pronk, T. J., De Steensma, H. Y., and van Dijiken, J. P. (1996), *Yeast* **12,** 1607–1633.
26. Minowa, T., Iwata, S., Sakai, H., and Ohata, T. (1989), *Gene* **85,** 161–168.
27. Kellermann, E. and Hollenberg, C. P. (1988), *Curr. Genet.* **14,** 337–344.
28. Butler, G. and Mc Connell, D. J. (1988), *Curr. Gent.* **14,** 405–412.
29. van Maris, A. J. A., Winkler, A. A., Porro, D., van Dijken, J. P., and Pronk, J. T. (2004), *Appl. Environ. Microbiol.* **70,** 2898–2905.
30. Skory, C. D. (2004), *Appl. Microbiol. Biotechnol.* **64,** 237–242.
31. Chang, D.E., Shin, S., Rhee, J., and Pan, J. (1999), *Appl. Environ. Microbiol.* **65,** 1384–1389.
32. Dien, B. S., Nichols, N. N., and Bothast, R. J. (2001), *J. Ind. Microbiol. Biotechnol.* **27,** 259–264.
33. Zhou, S., Causey, T. B., Hasona, A., Shanmugam, K. T., and Ingram, L. O. (2003), *Appl. Environ. Microbiol.* **69,** 399–407.
34. Zhou, S., Shanmugam, K. T., and Ingram, L. O. (2003), *Appl. Environ. Microbiol.* **69,** 2237–2244.
35. Yumoto, I. and Ikeda, K. (1995), *Biotechnol. Lett.* **17,** 543–546.
36. Hofvendahl, K. and Hahn-Hagerdal, B. (1997), *Enzyme Microb. Technol.* **20,** 301–307.
37. Olmos-Dichara, A., Ampe, F., Uribelarrea, J. L., Pareilleux, A., and Goma, G. (1997), *Biotechnol. Lett.* **19,** 709–714.
38. Siebold, M., von Frieling, P., Joppien, R., Rindfleisch, D., Schugerl, K., and Roper, H. (1995) *Process Biochem.* **30,** 81–95.

A Unique Feature of Hydrogen Recovery in Endogenous Starch-to-Alcohol Fermentation of the Marine Microalga, *Chlamydomonas perigranulata*

KOYU HON-NAMI

Engineering R&D Division, Tokyo Electric Power Co. Ltd., 1-1-3 Uchisaiwai-cho, Chiyoda-ku, Tokyo 100-8560, Japan, E-mail: koyu.honnami@tepco.co.jp

Abstract

A unicellular marine green alga, *Chlamydomonas perigranulata*, was demonstrated to synthesize starch through photosynthesis, store it in a cell, and ferment it under anaerobic conditions in the dark to produce ethanol, 2,3-butanediol (butanediol), acetic acid, and carbon dioxide (CO_2). Previous fermentation data of an algal biomass cultivated outdoors in a 50-L tubular photo-bioreactor showed good carbon (C) recovery in the fermentation balance, with a higher ratio to alcohols and, therefore, lower ratio to CO_2 in the C distribution of products than what would be expected from the Embden-Myerhof-Parnas pathway. These findings led to a proposed concept for a CO_2-ethanol conversion system (CDECS). The above data were evaluated in terms of hydrogen (H) recovery with the following results: C recovery at 105% was well balanced, although H recovery was as high as 139%, meaning an additional gain of H through fermentation. This finding was reproduced wholly in a set of experiments carried out in the same month of the following year, October, whereas another set of experiments was carried out in the following June provided ordinary fermentation results in terms of C and H recoveries with poor growth. Further analyses of these data revealed that butanediol is equal to ethanol as a product from a putative conversion system from CO_2 to the detected fermentation products, leading to the revision of the CDECS concept to a CO_2-alcohol conversion system (CDACS). The relevance of the CDACS will be discussed in relation to the cultivation conditions employed by chance.

Index Entries: Marine green microalga; *Chlamydomonas perigranulata*; endogenous starch fermentation; fuel alcohol; hydrogen recovery.

Introduction

Global warming and climate change are topics of world concern; one of their causes is recognized to be anthropogenic increases in the atmospheric levels of greenhouse gases, including carbon dioxide (CO_2) derived from fossil fuels through combustion. The reduction of such combustion and expansion of the CO_2 sink are essential principal measures to take in addition to the direct separation and sequestration of CO_2 at emission sites. An effective approach to mitigating climate change is the subject of debate *(1)*; biological measures, including the utilization microalgae, have been proposed as possible options *(2–4)*, and efforts to discover new organisms *(5)*, even genetic materials *(6)*, that might be of value for carbon sequestration and the synthesis of renewable fuels are continuing.

Among renewable fuels, bio-ethanol produced by fermenting microorganisms such as yeast, followed by distillation and dehydration if necessary, has gained wide use as an alternative motor fuel and has recently been the subject of increased attention, mainly because of unstable oil prices and the conviction to produce a major movable hydrogen source for fuel cells in the future *(7)*. In general, ethanol fermentation by yeast results in a carbon loss of one-third, as carbohydrate is converted to ethanol through enzymatic processes known as the Embden-Meyerhof-Parnas pathway. We showed previously that *Chlamydomonas perigranulata*, a unicellular marine green alga *(8)*, converts endogenous carbohydrate, intracellular starch stored through photosynthesis to valuable substances including fuels and chemicals such as ethanol and 2,3-butanediol (butanediol) *(9–11)*. In this fermentation, it was found that the whole metabolic reaction provides about 40% more alcohols and results in a higher recovery of carbon than expected from the consumed starch, a result that led us to propose the concept of a CO_2-ethanol conversion system (CDECS) *(11)*.

This article describes an evaluation of the previous results in terms of hydrogen (H) recovery in the fermentation balance, and presents a revised proposal for a CO_2-alcohol conversion system (CDACS). Cultivation conditions will be discussed in relation to the CDACS and the resultant efficient carbon (C) recovery in alcohol and an unexpected gain in hydrogen.

Materials and Methods

Microalga

The microalga used in this study was *C. perigranulata* *(8)*, which was isolated from a water sample obtained at Yambu, Saudi Arabia, and on the Red Sea.

Microalga Cultivation

The culture medium and growth conditions were described previously *(10)*. F/2 medium *(12,13)* was used with a slight modification: the concentrations of nitrate and phosphate were enriched four times. The F/2 medium contained (/L) 75 mg $NaNO_3$; 5 mg $NaH_2PO4·2H_2O$; 1 mL mineral solution; 1 mL vitamin solution; an artificial seawater agent, 16 g instant ocean; 22 g NaCl. The mineral solution concentrations (g/L) were as follows: 0.18 $MnCl_2 \cdot 4H_2O$; 0.022 $ZnSO_4 \cdot 5H_2O$; 0.01 $CuSO_4 \cdot 5H_2O$; 0.006 $Na_2MoO_4 \cdot 2H_2O$; 0.01 $CoCl_2 \cdot 6H_2O$; 3.15 $FeCl_3(6H_2O)$; 4.36 Na_2EDTA. The stock vitamin solution contained (mg/mL), 0.1 thiamine-HCl; 0.005 biotin; 0.0005 vitamin B12. The artificial seawater agent, Instant Ocean, was purchased from Aquarium Systems, Inc., Mentor, Ohio. Cells were pre-cultured in a 10-L carboy (working volume of 8 L) and supplied with 1% CO_2 in air at a rate of 300 mL/min. Illumination was provided by a fluorescence lamp, which surrounded the carboy and gave a light intensity of about 7 kLx at the outer surface of the vessel set in a room in which the temperature was controlled at 25°C. After about 7 d of culture, an algal concentration of 0.5 g/L was attained and used for inoculation.

Outdoor Cultivation in a 50-L Scale Tubular Bioreactor

Outdoor cultivations of the microalga were carried out between July and October, 1996 *(10)*, and between June and October, 1997, at the Hiroshima lab of Mitsubishi Heavy Industries, Ltd., in Hiroshima. Cultivation conditions were as described in previous papers *(10,11)*.

The F/2 medium was used with further modification for outdoor cultivation; the concentrations of nutrients, including nitrate, phosphate, metals, and vitamins, were enriched 8 times. A working volume of 50 L was used in a 53-L air-lifted tubular bioreactor. Tap water was used to prepare the medium after filtration through polypropylene-wound cartridge filters (TWC-1N-PPS, Advantec Tokyo Ltd., Tokyo). The reactor was sterilized with sodium hypochlorite solution (10 g/L) by overnight circulation before use *(14)*. Cultivation was started by the inoculation of 16 L of a preculture (about 30% of the working volume). Air containing 1.8% CO_2 was supplied at 1.7 L/min, which drove the medium circulation at a rate of 0.3 m/s. Illumination and temperature employed for cultivation were natural except for spraying water on the tubes if necessary to keep the temperature under 35°C. Illumination by sunlight was measured with a digital illuminometer (IM-3, Topkon Corporation, Tokyo). The reactor system consisted of a polyacryl tubular bioreactor, air compressor, flow meters for CO_2 and air, and separate temperature sensors for the culture medium and the atmosphere. The reactor specifications are as follows: two horizontal tubes (length of 2000 mm) and two vertical tubes (height of 3880 mm) with an inside diameter (ID) of 70 mm and an outside diameter of 76 mm, a tank (140 mm ID and 450 mm long), and a gas injection port. Cultures were harvested by centrifugation in a rotor (Qn rotor, Kokusan Co. Ltd., Tokyo) at 10 k rpm (9100*g*) and a flow rate of about 1 L/min. This system can yield about 100 g of cells in 50 L culture medium within 1 h. The packed cells were transferred immediately to a vessel as described later for fermentation.

Fermentation in the Dark

The fermentation procedures were described previously *(10,11)*. Cells were suspended in artificial sea water or an appropriate buffer to a suitable concentration (150–200 g of cells/L). The air was removed by flushing with N_2 (1 L/min) at the start of fermentation. The microalgal slurry was maintained automatically at 25°C in a water bath and at pH 7.0 by intermittent additions of 2M NaOH solution coupled with a pH sensor and a controller that moved a U-shaped vane at a constant 60 rpm. The vane's shaft was made of polytetrafluor-ethylene (70-mm long and 10-mm wide for the vertical part and 70-mm long and 15-mm wide for the horizontal part, 5 mm in thickness). The fermentation apparatus consisted of a fermenter, a cooler for volatile substances, and a gas collector (Tedlar Bag for 1 L, Iuchi Seieido Ltd., Osaka). The fermentation vessel (a cylindrical glass flask with a flat bottom, No 0582-500, Shibata Scientific Technology Ltd., Tokyo) had an inner diameter of 85 mm and a height of 115 mm with a removable cover with four necks (No 0580-4), a mechanical-rotation drive and automatic pH-controlling systems.

Analyses

The analytical procedures undertaken and their conditions were described previously *(10,11)*. Starch in cells was measured by coupled methods with perchloric acid treatment *(15)*: glucose produced by hydrolysis by perchloric acid was determined using a Glucose CII-test kit (Wako Pure Chemical Industries Ltd., Osaka). Ethanol, butanediol, and gas concentrations were analyzed by gas chromatography on a Shimadzu (Shimadzu Corporation, Kyoto, Japan) GC-8A instrument equipped with a data processor (Chromatopac C-R1A), a flame ionization detector (FID) and a thermal conductivity detector. For alcohol determination, a glass column (ID 3.2 × 2.1 m) of PEG-20M 20% Uniport HP 60/80 (GL Sciences Ltd., Tokyo) was used. The column temperature was maintained at 200°C, the carrier gas was N_2, and the flow rate was 35 mL/min. For gas measurements, a SUS column (ID 1.4 × 1.8 m) of WG-100 (GL Sciences Ltd., Tokyo) was used at a column temperature of 50°C, with N_2 as the carrier gas at a flow rate of 35 mL/min. The concentrations of organic acids, including acetate and lactate, in the fermented microalgal slurry were measured by gas chromatography on a Shimadzu GC-9A instrument equipped with a data processor (Chromatopac C-R3A), a column (ID 3.2 × 2.6 m) of Thermon 3000 (Shinwa Chemical Industries Ltd., Kyoto) and a FID. The column temperature was maintained at 160°C, the carrier gas was N_2, and the flow rate was 35 mL/min. Inorganic carbon was determined by a total organic carbon analyzer *(16)* using a Shimadzu TOC-5000 instrument. The sample was treated at 150°C and CO_2 produced was measured. Nitrate concentration in the culture medium was determined by the brucine absorption method *(17)*.

Calculations

The calculations employed in this study were based on the literature *(18)*: carbon recovery is defined as the ratio of mol of C in the fermentative end products to the mol of C in the starch metabolized; hydrogen recovery

Table 1
Summary of the Outdoor Cultivation of C. perigranulata
and the Subsequent Fermentation Volume

Cultivation series #	T33[a]	T35	T38
Cultivation date	Oct 18–30,1996	Jun 4–17,1997	Oct 7–20,1997
Concentration of biomass harvested (g/L)	1.53	0.55	1.55
Biomass harvested (g)	79	27	77
Starch content (%)	35	25	28
Fermentation volume (mL)	437	138	350

[a](See ref. 10)

is defined as the ratio of molecule of available H in the fermentative end products to the molecule of available H in the starch metabolized, in which the molecule of available H in a compound with a general formula of $C_aH_nO_z$ is calculated as $4a + n - 2z$; the O/R index is defined as the ratio of the weighed sum of the O/R values of the oxidized end products to the weighed sum of the O/R values of the reduced end products, in which the O/R value of a compound with general formula of $C_aH_nO_z$ is calculated as $z - n/2$.

Chemicals

All chemicals used in this study were reagent grade unless otherwise stated. Ethanol and 2,3-butanediol were purchased from Wako Pure Chemical Industries, Ltd., Osaka.

Results and Discussion

Re-Evaluation of Previous Fermentation Results

A summary of the outdoor cultivation of C. perigranulata reported previously (10) is listed in Table 1 with the results of similar cultivations carried out the following year. The fermentation results as described in a previous paper (10,11) are evaluated in terms of H recovery in the fermentation balance in addition to C recovery, as listed in Table 2. The data show that C recovery was almost balanced (105%), while H recovery was significantly higher (139%), and leaning toward the product side, with more H gained during fermentation, which is also shown by the O/R index being as low as 0.23. These findings suggest that some reducing power was added to a metabolite(s) to lead to the production of more alcohols, and that, on the whole, the fermentation products recovered were more reduced than the substrate in the case of this alga.

Revision of the CO_2–Ethanol Conversion System

In the CDECS mentioned previously, only CO_2 derived from the original butanediol fermentation pathway was counted tentatively as source

Table 2
Endogenous Starch Fermentation of *C. perigranulata* and Recovery of Carbon and Hydrogen

Product	T33[a] mmol	T33[a] mmol C	T33[a] mmol H	T35 mmol	T35 mmol C	T35 mmol H	T38 mmol	T38 mmol C	T38 mmol H
Starch stored	170	1022		45	268		131	784	
Starch consumed	−134	−807(100)	−3216(100)	−41	−246(100)	−984(100)	−102	−614(100)	−2448(100)
Ethanol	209	418(52)	2508	52.5	105(43)	624	187	375(61)	2244
Butanediol	88	352(44)	1936	19.5	78(32)	418	48	191(31)	1056
Acetate	2	4(0)	16	2	4(2)	16	16	32(5)	128
CO_2	43	77[b](10)	–	88	88(36)	–	44	44(7)	–
Inorganic C	33	–	–	0	–	–	0	–	–
Total C analyzed		851(105)			275(112)			642(105)	
Total H analyzed		4460(139)			1058(108)			3428(140)	
O/R index		0.23			1.09			0.17	

[a]A part of data was reported previously (refs. *10, 11*).
[b]Including inorganic C.

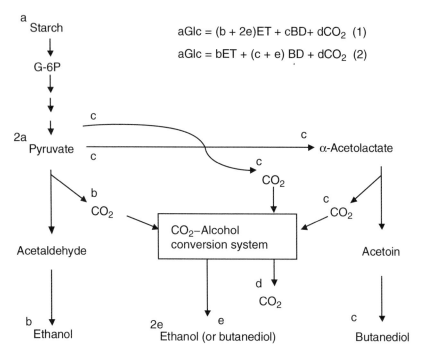

Fig. 1. A possible endogenous starch fermentation pathway in C. *perigranulata* and the CDACS. Molar amounts of substances in the pathway are presented as a–e for starch, ethanol, 2,3-butanediol, CO_2, and 2,3-butanediol produced through the CDACS, respectively. This figure indicates that alcohol obtainable from the CDACS is either ethanol or 2,3-butanediol. Equations 1 and 2 represent the stoichiometry for each case of ethanol and 2,3-butanediol as a product of the CDACS, respectively. Abbreviations: G-6P, glucose-6 phosphate; Glc, glucose; ET, ethanol; and BD, 2,3-butanediol. For other abbreviations *see* the text.

for the additional formation of ethanol through CDECS in order to explain the stoichiometry of the above novel fermentation, a higher ratio of alcohols to CO_2 than expected on the basis of the Embden-Meyerhof-Parnas pathway. In this article the CDECS is revised as follows along with its name. CO_2, which is formed at the decarboxylation step not only in the butanediol pathway, but also in the original ethanol pathway, is converted to ethanol and/or butanediol through a CDACS, as shown in Fig. 1. The revised system takes into account the fact that butanediol should not be excluded as an additional product produced through the CDACS, and that, therefore, this alcohol is equal to ethanol as a product through a putative CO_2 conversion system. CDECS is one type of such a putative CO_2 conversion system.

Considering a possible metabolic pathway from starch to alcohol via pyruvate and CDACS, acetic acid is assumed to be the same compound as ethanol, which was also a case with the previous CDECS. The basis of the above assumption is as follows: acetic acid (1) is a minor product in this fermentation; (2) is equivalent to ethanol in C number as well as in basic chemical structure, although different in redox level; and (3) has the same metabolic pathway until just before the final step in a series of reactions in yeast fermentation, which is probably also the case in this alga, because pyruvate decarboxylase activity has been demonstrated in vitro under anaerobic and dark conditions *(9)*, and so the final step from acetaldehyde to ethanol would be catalyzed by this enzyme as in yeast.

Stoichiometry in the Fermentation of the Microalgae

Some features of the fermentation results, including product pattern and the distribution of C and H among products, are listed in Table 2. The conversion pathway in the CDACS is explained in the following three cases, the formation of ethanol only, the formation butanediol only, and the simultaneous formation of both ethanol and butanediol.

Possible Ethanol Formation in the CDACS

On the basis of the determined amount of substrate and the products listed in Table 3 and shown in Fig. 2A, the validity of the possible pathway and CDACS is described as follows: assuming that all of the 88 mmol of butanediol detected was formed through the original fermentation pathway from pyruvate to butanediol via α-acetolactate, the original ethanol pathway will yield as much as 93 (= 269 – 176) mmol ethanol, an amount that is estimated by subtracting the 176 mmol pyruvate used for butanediol formation from the 269 mmol pyruvate metabolized by glycolysis derived through starch consumption. As 211 mmol ethanol was detected, the additional amount of ethanol required is 118 (= 211 – 93) mmol, which is obtained by subtracting the above mentioned 93 mmol ethanol from the 211 mmol ethanol that was determined. On the other hand, if the CDACS is active and converts CO_2 to ethanol, the expected amount of ethanol will be as much as 96 mmol, which corresponds to 192 (= 269 – 77) mmol as carbon; the latter is estimated by subtracting the detected 77 mmol CO_2 from the 269 mmol CO_2 that is expected to be formed through both the original ethanol and butanediol fermentation pathways.

The above amounts of 118 and 96 mmol ethanol were obtained by subtracting the expected amount of ethanol, which is obtained if no more butanediol is formed through the CDACS, from the amount of ethanol actually determined, and by subtracting the amount of CO_2 actually determined from the expected amount of CO_2 derived through each original pathway. It should be mentioned that the former amount was derived through product side estimation, and the latter amount through substrate side estimation. The difference, 22 mmol ethanol, is as much as 44 mmol in carbon equivalents, which corresponds to the difference in the moles carbon in the total fermentation products and the starch consumed as substrate (Table 2). The extra ethanol equivalent to 44 mmol C probably arose

Table 3
Endogenous Starch Fermentation and Its Products Via the CDACS

	Assumed product from CO_2			
	(Cultivation T33)		(Cultivation T38)	
Product (mmol)	Ethanol	Butanediol[a]	Ethanol	Butanediol[a]
Starch consumed	−134	−134	−102	−102
Ethanol[b]				
Observed	211	211	203	203
Produced from acetaldehyde(A)	93	211	109	203
Produced from CO_2 (B)	96	N/A	80.5	N/A
Calculated (A + B)	189	N/A	189.5	N/A
Calculated and observed difference	22	N/A	13.5	N/A
Butanediol[a]				
Observed	88	88	48	48
Produced from acetoin(C)	88	29	48	1
Produced from CO_2 (D)	N/A	48	N/A	40
Calculated(C + D)	N/A	77	N/A	41
Calculated and observed difference	N/A	11	N/A	7
CO_2	77[c]	77[c]	44	44

[a]2,3-butanediol.
[b]Including acetic acid.
[c]Including inorganic C.

from starch derivatives, such as intermediates that comprise the glycolysis and fermentation pathway from glucose to alcohol already present before the start of the fermentation experiment, including starch measurement. It can be concluded, therefore, that the CDACS is active.

Possible Butanediol Formation Via the CDACS

In the same manner, the validity of the above pathway is described for butanediol as a product formed via the CDACS, as shown in Table 3 and Fig. 2B. In this case the amount of butanediol required is more 59 (= 88 −29) mmol more than that obtained by subtracting 29 mmol butanediol from 88 mmol butanediol. The expected amount of butanediol from the CDACS, on the other hand, is as much as 48 mmol. The difference, 11 mmol butanediol, is again equivalent to as much as 44 mmol C, which indicates that butanediol production provides an alternative situation to the case of ethanol production as discussed in the previous section. Therefore, CDACS is also demonstrated to be active, and butanediol can also be expected as a product of the CDACS.

Effective Microalgal Alcohol Fermentation

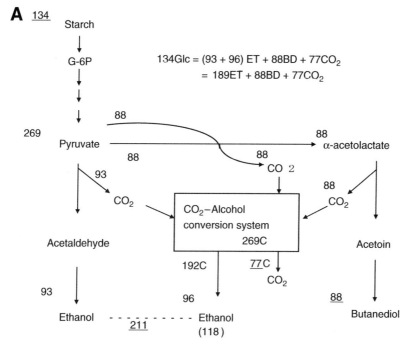

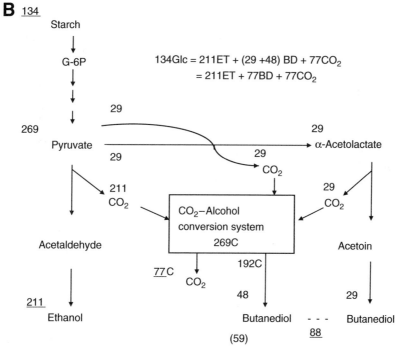

Fig. 2. Analysis of C flux in a possible endogenous starch fermentation pathway and CDACS in *C. perigranulata* cultivation T33. **(A)** All 2,3-butanediol analyzed is assumed to have been formed through the original 2,3-butanediol pathway shown in Fig. 1, and, therefore, extra CO_2 results in ethanol formation through CDACS. **(B)** All ethanol analyzed, in reverse, is also assumed to have been formed through the

Simultaneous Formation of Ethanol and 2,3-Butanediol via the CDACS

It should be mentioned that the simultaneous formation of ethanol and butanediol by the CDACS is highly probable. Ethanol formed via the CDACS is coupled with its formation in the original fermentation pathway as well as butanediol formation in both the original pathway and CDACS. Therefore, for example, a decrease in ethanol formed by the CDACS could be complemented by an increase in ethanol formed by the original pathway, which would lead to a decrease in the amount of butanediol formed by the original pathway and a compensating increase in the amount of butanediol formed in the CDACS. The above compensation would still range between 96 mmol and none for ethanol, and between none and 48 mmol for butanediol.

Reproduced Cultivation and Fermentation With Confirmation of the CDACS Concept

The following year, several sets of the experiments involving outdoor cultivation and subsequent fermentation were again carried out during the period from June to October, and their typical results are listed in Tables 1 and 2 and shown in Fig. 3. In the case of microalgal cells obtained in October (T38), the fermentation results reproduced almost completely the results obtained the previous year with a high C recovery and a significantly higher H recovery (140%), thus confirming the CDACS concept. The CDACS is assumed to account for 80.5 mmol ethanol or 40 mmol butanediol, a difference of as much as 13.5 mmol ethanol or 7 mmol butanediol from those analyzed on the product side (Table 3, Fig. 3).

Poor Cultivation and Resultant Inactive CDACS

Another set of experiments carried out in June, 1998 (T35), on the other hand, failed to reproduce the above results (Tables 1 and 2, and Fig. 4). In these experiments, the cultivation results were very poor: the biomass concentration at the stationary phase of growth was only about one-third that observed in the case of excellent growths, as in T33 and T38, and, therefore, the resultant biomass obtained was also only one-third of that obtained in T33 and T38 with a lower starch content (Table 1).

With respect to the fermentation results, as listed in Tables 2 and 4 and shown in Fig. 4, the C and H recoveries were 112% and 108%, respectively, values very close to each other. The O/R index was 1.09 (Table 2), and evolved CO_2 corresponded to one-third of the total mol C of the

---

Fig. 2. (*Continued*) original ethanol pathway shown in Fig. 1, and, therefore, extra CO_2 results in 2,3-butanediol formation through the CDACS. Numerical values represent the amount of each compound in mmol; values that are underlined and in parentheses represent analyzed values and the amounts required to conform with the analyzed values, respectively. For explanations, *see* Fig. 1.

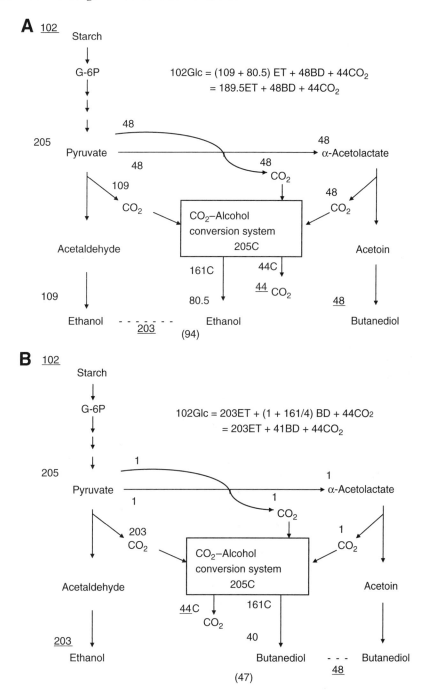

Fig. 3. Analysis of C flux in a possible endogenous starch fermentation pathway and CDACS in *C. perigranulata* cultivation T38. For explanations, *see* Figs. 1 and 2.

products. Taken together, although the C and H recoveries in this fermentation were increased as much as about 10% in stoichiometry, the fermentation itself in this case seems to have proceeded according to only the

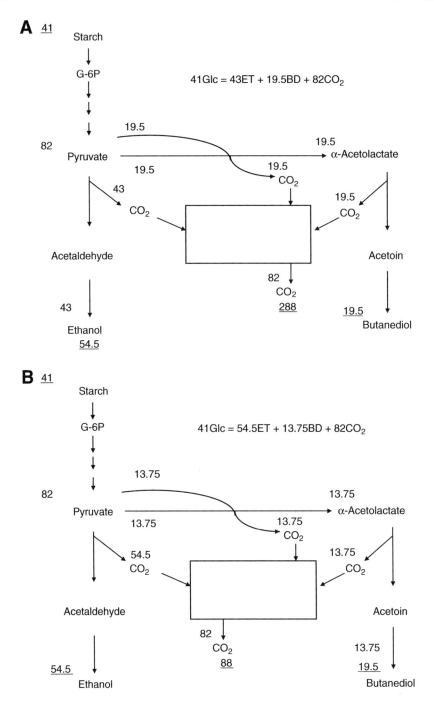

Fig. 4. Analysis of C flux in a possible endogenous starch fermentation pathway of *C. perigranulata* cultivation T35, a case where the CDACS was inactive and, therefore, no alcohol could come out through the CDACS. For explanations, *see* Figs. 1 and 2.

Embden-Meyerhof-Parnas pathway, which means the CDACS is not needed to explain the fermentation results.

Table 4
Endogenous Starch Fermentation and Its Products
When the CDACS Is Inactive (Cultivation T35)

Product[a] (mmol)	Butanediol[b]	Ethanol
Starch consumed	−41	−41
Ethanol		
Observed	54.5[c]	54.5[c]
Produced from acetaldehyde	43	54.5
Calculated and observed difference	11.5	0
Butanediol[b]		
Observed	19.5	19.5
Produced from acetoin	19.5	13.75
Calculated and observed difference	0	5.75
CO_2		
Produced from pyruvate to acetaldehyde and from pyruvate to acetoin	82	82
Calculated and observed difference	6	6

[a]Product: Observed amount of the product was assumed to be obtained through its original metabolic pathway.
[b]2,3-butanediol.
[c]Including acetic acid.

On the basis of the data in Fig. 4A, the validity of the CDACS being inactive is described as follows. In this system, according to Embden-Meyerhof-Parnas pathway, 41 mmol glucose-equivalents of starch were metabolized to 82 mmol pyruvate as an intermediate and converted finally to ethanol, butanediol, and CO_2. The resultant CO_2 formed is expected to equal the amount of that of pyruvate metabolized and to be as much as 82 mmol. The measured amount of CO_2, however, was 88 mmol, just slightly more than expected. Therefore, no additional CO_2 needed to enter the CDACS to be turned into alcohol. With respect to the amount of alcohol produced, assuming all of the 19.5 mmol butanediol determined was formed through the original pathway, the original ethanol pathway will produce 43 (= 82 (19.5 × 2) mmol ethanol, which is the amount obtained by subtracting 19.5 mmol pyruvate twice, the amount of butanediol determined from the 82 mmol pyruvate formed through glycolysis of the starch consumed. As 54.5 mmol ethanol was found, there was an ethanol shortage of as much as 11.5 (= 54.5 (43) mmol.

Taken together, because 41 mmol starch was estimated to be consumed as the starting substrate and 19.5 mmol butanediol was formed, 11.5 mmol ethanol, and 6 mmol CO_2 were missing, which corresponds to 29 mmol C, the amount of the difference between the substrate and total products (Table 4). Therefore, owing to the shortage of CO_2 in the fermentation system employing cultured T35, it is apparent that CO_2 did not enter the CDACS and be transformed into ethanol. The extra CO_2 and ethanol probably derived from metabolic intermediates that existed prior to the start of the fermentation experiments.

In a similar manner, the inactivity of the CDACS is confirmed by the other case in which the 54.5 mmol ethanol detected is assumed to have been formed through its original metabolic pathway (Fig. 4B). In this case, because 82 mmol pyruvate was considered to be metabolized and because 54.5 mmol ethanol was also considered to be formed, 5.75 mmol butanediol and 6 mmol CO_2, equivalent to 29 mmol C, were lacking as listed in Table 4. Because of a lack of CO_2 to enter the CDACS, the conversion of CO_2 to butanediol through the CDACS can be excluded. The extra CO_2 and butanediol probably derived from metabolic intermediates that existed prior to the start of the fermentation experiments. In conclusion, poor cultivation and subsequent fermentation resulted in the CDACS remaining inactive.

Conditions for CDACS to Be Active

In this section, conditions under which the CDACS becomes active are discussed in term of the cultivation and fermentation conditions. Typical profiles of the outdoor cultivation of *C. perigranulata* are shown in Fig. 5. Both the T33 and T38 cultivations were carried out in the same season in different years and resulted in excellent growth and fermentation. They showed the following features in common:

1. sufficient and stable illumination by sunlight;
2. biomass concentration reaching more than 1.5 g/L at the stationary phase of growth;
3. high-starch content near 30%;
4. relatively wide cultivation temperature from 30°C to 5°C; and
5. complete nitrate consumption and further cultivation for several days under nitrate starvation conditions.

The cultivation of T35, on the other hand, in which the CDACS was judged to be inactive, showed the following marked features:

1. biomass concentration of the medium at harvest as low as 0.5 g/L, about 1/3 of that observed for the excellent growth cases;
2. lower starch content of 25%; and
3. a narrow range of cultivation temperature relative to T33 and T38, which is owing to the temperature at night being higher at around 20°C.

The resultant failure of growth in the cultivation of T35 may be as a result of the strong illumination sunlight (over 200 Ly/d) during the initial phase of growth, although the illumination and cultivation temperature data for 3 d during the early stages of cultivation (5–7 June) could not be

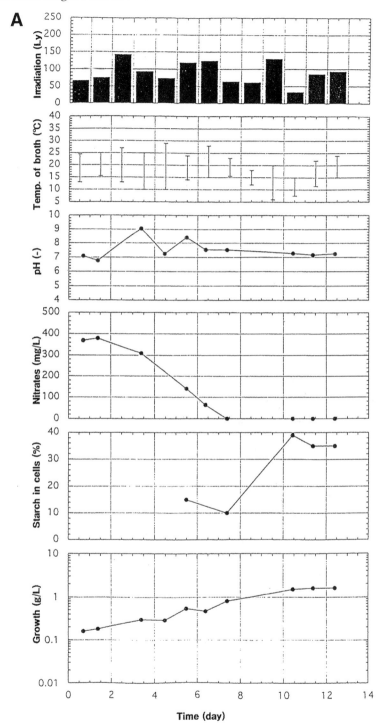

Fig. 5. Time courses of *C. perigranulata* growth and its parameters during outdoor cultivation. **(A)**, **(B)**, and **(C)** are cultivating series numbers T33, T35, and T38, respectively (*see* Table 1). Some results **(A)** have been reported previously *(10)*. In **(B)**, data for the illumination and cultivation temperature for 3 d in the early stages of cultivation (June 5–7) could not be obtained as a result of technical circumstances.

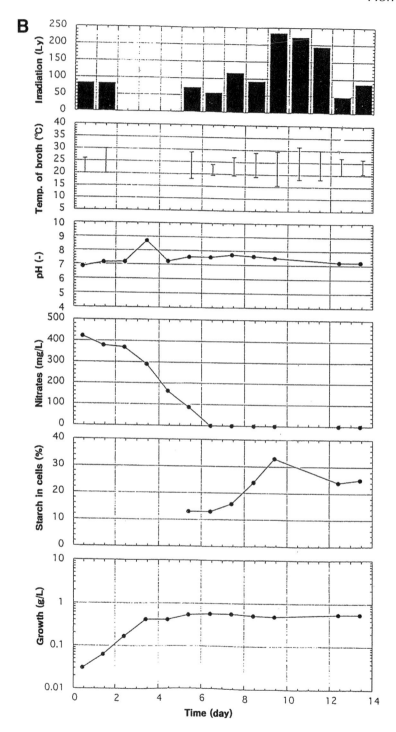

Fig. 5 *(Continued)*

obtained due to some technical difficulties. The former data were estimated to be 220, 120, and 250 Ly/d, respectively, on the basis of the values observed and reported at the Hiroshima Meteorological Observatory.

Effective Microalgal Alcohol Fermentation

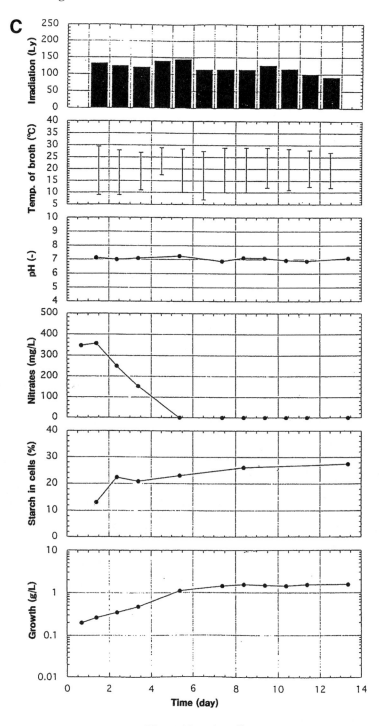

Fig. 5 (*Continued*)

With respect to the fermentation conditions, on the other hand, the algal slurry concentration employed for all three lots was adjusted to about 20%, and the volume of each fermentation mixture varied depending on the biomass obtained (Table 1) with no significant effects on the CDACS activity.

Taken together, some cultivation conditions are considered to be essential for the CDACS to be active. The remarkable differences between the excellent cases (T33 and T38) and the poor case (T35) offer suggestions as to favorable cultivation conditions:

1. stable sunlight illumination and excellent growth; and
2. a temperature range of the medium during cultivation varying from 30°C to 5°C.

Therefore, potential conditions under which the CDACS is active are considered to be as follows:

1. excellent growth under stable illumination; and
2. a culturing temperature providing suitable level of warmth during the daytime accompanied by cool at nights; the latter, especially, should be required during the stationary phase of growth under nitrate starvation.

It is well known that during the stationary phase of growth, continued cultivation and starvation of some nutrients leads to an increase in starch content *(19)*. If the night temperature is high enough to consume the accumulated starch, algal cells will begin to degrade it without any reducing power being accumulated. The CDACS is, therefore, considered to be no longer active. In conclusion, the accumulation of sufficient amounts of starch and reducing power are needed for the CDACS to be active. In order to maintain these accumulations in the culture medium during periods of low illumination, the temperatures should be maintained as low as possible.

Possible Metabolic Pathway of Alcohol Fermentation Via the CDACS

As described previously, the refixation of CO_2 formed through alcohol fermentation and its further conversion to alcohol are proposed in Fig. 1. In the previous proposal of the CDECS, a formic acid derivative was presented as a candidate for the initial product of CO_2 fixation *(11)*; here, another possibility using the existing photosynthetic system may be pointed out, a part of this photosynthetic system, especially the CO_2 fixation process including Rubisco, would be expected to work even under anaerobic and dark conditions. In addition, a possibility using the existing central carbon metabolic pathway and alcohol fermentation pathways should be also pointed out for the CDACS, after its fixation, CO_2 would be introduced into the central carbon metabolic pathway as an intermediate, and, finally, some of it would be converted to alcohol.

CO_2 fixation by photosynthesis stands always on a background of enriched energy by ATP storage and reducing power such as NADPH, reduced Fd and reduced a state of the Fd/thioredoxin system *(20,21)*. These substances or states in this microalga could also be produced and stored by photosynthesis during cultivation, especially during nitrate starvation where no proliferation can proceed *(22)*.

Recently, the refixation and utilization of formed CO_2 through lipid formation have been reported in green rape seeds under conditions of some illumination *(23)*. The CDACS that is active in this microalga is also a novel example of the re-fixation and utilization of formed CO_2, and it should be pointed out that this alga performs them under anaerobic and dark conditions. It is an important model system with which to elucidate the mechanisms of effective C and H utilization because it involves single cells and the reaction occurs after photosynthesis and can be separated from the light reaction of photosynthesis. Finally, it should be noted that this algal fermentation system was demonstrated to produce lactic acid as an additional product under some conditions of cultivation and fermentation *(24)*.

In conclusion, the potential for microalgae to contribute to CO2 mitigation might be higher than currently expected. Recently, a large report of a microalgae utilization R&D project undertaken over almost two decades has appeared *(25)*. In this report, it is noted that the use of some microalgal species results in the production of biodiesel, an alternative fossil fuel has the potential to become competitive depending on current petroleum costs. This present study is similar in that, although biodiesel was not produced, the bio-alcohol products are renewable. Many issues, including the physiology and applicability to mitigation remain to be elucidated *(3)*.

Acknowledgments

The author wishes to thank L. G. Ljungdahl for useful advice. He also wishes to thank A. Hirano, Y. Samejima, S. Kunito, S. Hirayama, R. Ueda, Y. Ogushi, M. Kaneko, T. Hamada, and H. Nakayama for useful discussion; S. Yoshihara and M. Hada for expert technical assistance. He also thanks M. M. Dooley for critical reading and editing of the manuscript and useful advice. He is grateful for the encouragement of M. Ishibashi, K. Goto, and P. D. Boyer in this study.

References

1. Claussen, E. (2004), *Science*, **306,** 816.
2. Benemann, J. R. (2001), In: *Photosynthetic Microorganisms in Environmental Biotechnoloy*, Kojima, H. and Lee, Y. K., eds., Springer-Verlag, Hong Kong, pp. 1–10.
3. Pedroni, P., Davison, J., Beckert, H., Bergman, P., and Benemann, J. (2001), *J. Energy Environ. Res.* **1,** 136–150.
4. Hon-Nami, K. (2001), In: *Photosynthetic Microorganisms in Environmental Biotechnoloy*. Kojima, H. and Lee, Y. K., eds., Springer-Verlag, Hong Kong, pp. 291–310.
5. Smith, H. O., Friedman, R., and Venter, J. C. (2003), *The BRIDGE*, Summer, pp. 36–40; available at http://www.princeton.edu/~seasplan/lifesciences/NAE%20Bridge.pdf. Accessed date: Feb. 23, 2004.
6. Venter, J. C., Remington, K., Heidelberg, J., (2004), *Science*, **304,** 66–77.
7. Deluga, G. A., Salge, J. R., Schmidt, L. D., and Verykios, X. E. (2004), *Science* **303,** 993–997.

8. Hirayama, S., Ueda, R Nakayama, T., and Inoulye, I. (2001), *Botanica Marina* **44,** 41–46.
9. Hirayama, S., Ueda, R., Ogushi, Y., Hirano, A., Hon-Nami, K., and Kunito, S. (1997), In: *Proceedings of the Annual Meeting of the Japanese Society for Marine Biotechnology*, Tokyo, June, pp. 78 (in Japanese).
10. Hirano, A., Samejima, Y., Hon-Nami, K., et al. (1997), In: *Making Business from Biomass in Energy, Environment, Chemicals, Fibers and Materials*, Overend, R. P. and Chornet E., eds., Pergamon, New York, pp. 1069–1076.
11. Hon-Nami, K., Hirano, A., Samejima, Y., et al. (1998), In: *Biomass for Energy and Industry*, Kopetz, H., Weber, T., Palz, W., Chartier, P., and Ferrero, G. L., eds., C.A.R.M.E.M., Rimpar, pp 602–605.
12. Ong, L. J., Glazer, A. N., and Waterbury, J. B. (1984), *Science* **224,** 80–83.
13. Castenholz, R. W. (1988), *Methods Enzymol.* **167,** 68–92.
14. Tatewaki, M. (1979), In: *Sourui-Kenkyu-Hou*, Nishizawa, K. and Chihara, M. eds., Kyoritsu Shuppan Ltd., Tokyo, pp. 69–87 (in Japanese).
15. Ohta, S., Miyamoto, K., and Miura, Y. (1987), *Plant Physiol.* **83,** 1022–1026.
16. JIS K0102 22 (1996), In: *JIS Handobukku Kankyousokutei*, Japanese Standard Association, Tokyo (in Japanese), pp. 1083–1085.
17. JIS K0102 43. 2. 4 (1996), In: *JIS Handobukku Kankyousokutei*, Japanese Standard Association, Tokyo (in Japanese), pp. 159.
18. Gfeller, R. P. and Gibbs, M. (1984), *Plant Physiol.* **75,** 212–218.
19. Martin, M. C. and Goodenough, U. W. (1975), *J. Cell Biol.* **67,** 587–605.
20. Buchanan, B. B. (1991), *Arch. Biochem. Biophys.* **288,** 1–9.
21. Huppe, H. C., Farr, T. J., and Turpin, D. H. (1994), *Plant Physiol.* **105,** 1043–1048.
22. Poolman, M. G., Fell, D. A., and Raines, C. A. (2003), *Eur. J. Biochem.* **270,** 430–439.
23. Schwender, J., Goffman, F., Ohlrogge, J. B., and Shachar-Hill, Y. (2004), *Nature* **432,** 779–782.
24. Hon-Nami, K. (2004), In: *Biotechnoloy of Lignocellulose Degradation and Biomass Utilization*, Ohmiya, K., Sakka, K., Karita, S., Kimura, T., Sakka, M., and Ohnishi, Y., eds., Uni Publishers Co., Ltd., Tokyo, pp. 746–754.
25. Sheehan, J., Dunahay, T., Benemann, J., and Roessler, P. (1998), *A look back at the US Department of Energy's aquatic species program-biodiesel from algae.* Prepared by the National Renewable Energy Laboratory, A national laboratory of the U.S. Department of Energy operated by Midwest Research Institute, Under Contract No. DE-AC36-83CH10093. http://205.168.79.26/docs/legosti/fy98/24190.pdf. Accessed date: Oct. 21, 2005.

Detailed Analysis of Modifications in Lignin After Treatment With Cultures Screened for Lignin Depolymerizing Agents

AARTI GIDH,*,[1] DINESH TALREJA,[2] TODD B. VINZANT,[3] TODD CLINT WILLIFORD,[1] AND ALFRED MIKELL[2]

[1]Departments of Chemical Engineering; [2]Biology,
University of Mississippi, University, MS 38677,
E-mail: agidh@olemiss.edu; and [3]National Bioenergy Center,
National Renewable Energy Laboratory, Golden, CO 80401

Abstract

Termites, beetles, and other arthropods can digest living and decaying wood plus other lignocellulosic plant litter. Microbial sources like other wood-eating insect guts and wastewater treatment sludge were screened for lignin depolymerization. Near infrared spectroscopy and atomic force microscopy (AFM) along with high-performance liquid chromatography (HPLC), were used to track changes in functional groups, size, shape, and molecular weight of lignin molecules during incubations. *Odontotaenius disjunctus* (Betsy beetle) guts dissected whole or separately as midgut, foregut, and hindgut, consumed corn stover but did not show lignin depolymerization. The sludge-treated lignin did show some reduction in molecular weight on the HPLC, particle size (350–650 nm initially to 135–220 nm by day 30) and particles per field on AFM. pH and the presence of nutrients had a substantial effect on the extent of depolymerization. Cultures in lignin and nutrients showed higher growth than cultures with lignin only. Colony characteristics within the beetle gut and the sludge were also evaluated.

Index Entries: Lignin; beetles; NIR; HPLC; AFM; depolymerization.

Introduction

Earth's most abundant biomass is produced by photosynthetic fixation of carbon dioxide in our biosphere, yielding approx 136×10^{15} g of dry plant material annually *(1)*. Lignin (20–30% dry weight) is conventionally defined as a complex hydrophobic network of phenyl propenoid units derived from the oxidative polymerization of one or more of three types of hydroxycinnamyl alcohol precursors—*p*-coumaryl, coniferyl, and sinapyl alcohols. Lignin can vary depending on the plant type or plant tissue, depending on the ratio of the monolignols and the degree of methoxylation *(2,3)*. Lignin is

---

*Author to whom all correspondence should be addressed.

particularly difficult to biodegrade, and it reduces the bioavailability of other cell-wall constituents including the predominant constituent, cellulose. Knowledge of these different constituents is essential for understanding the vastly different rates at which different plant materials decompose. Presently in bioethanol production, cellulose is separated from lignocellulose by dilute acidification, and then the cellulose is converted in two stages, enzymatic hydrolysis using cellulases and subsequent fermentation to ethanol. The acid hydrolysis and separation stages involve significant chemical use, generate waste streams, and require costly alloys, increasing capital costs.

With this method lignin cannot be isolated from lignocellulose without partial denaturation, because they are intimately linked with carbohydrate polymers and covalently linked to hemicelluloses (4). Lignin was discovered in 1839 by Payen as the fraction of cell walls insoluble in acids, and since then progress has been made in understanding the mechanisms of lignin biodegradation. Ligninase (lignin peroxidase) is a powerful oxidant, which initiates lignin degradation by one electron oxidation. Lignin is also degraded by peroxidase and laccases of white-rot fungi. Although degradation of lignin by fungi has been reported, most bacteria degrade only low-molecular weight components of lignin but not the higher ones (5). Most lignin loss reported from solid substrates is because of solubilization of lignin and not because of lignin degradation (1). Bacterial lignin degradation has, however, been reported to be more specific than with fungal systems, an advantage, potentially leading to many industrial applications like vanillin, adhesives, binder for laminated or composite wood products, and so on (6).

Few bacteria are able to degrade fully intact lignified wood cells on their own. Although, a large number of actinomycetes have been isolated from decayed wood, pure cultures cause only limited decay. Complete degradation of lignified cell walls probably requires a complex interaction and succession of a variety of bacteria and fungi. When successful, cell-wall erosion and delignification are some of the important processes that can be observed (7).

For these reasons, bioprospecting for effective and alternate sources of lignin modification is important. Little work has been done on prokaryotes, in particular those derived from herbivore guts. Lignin-degrading filamentous bacteria have been isolated from the guts of wood-eating termites (8). The occurrence of specific gut microbiota among the insects remains to be systematically studied, yet there is sufficient evidence of the presence of digestive symbionts for representative insect orders (1,8). The most prominent examples are among the Coleoptera and Diptera. Coleopterans have actively fermenting gut microbiota, including cellulolytic and hemicellulolytic bacteria (8). Wood infested by these beetles is usually well decomposed and falls apart readily (9).

Lignin conversion involves complex reactions involving bond scission and functional group alteration. It is also important to track and quantify these alterations in enzyme-mediated depolymerization studies. Because of complexity of the lignin structure, most work has involved the use of lignin model compounds. Lignin degradation by bacteria has been studied with various softwood, hardwood, or graminaceous lignin and lignocellulosic substrates *(6)*. The ^{14}C labeled lignins *(10,11)*, lignin model compounds *(12–14)*, and technical lignins like Kraft lignin *(15)* or lignosulfonates have also been studied.

Techniques employed to monitor lignin degradation have included ^{14}C-labeling *(16)*, dry weight analysis of solid substrates combined with the HPLC analysis of acid soublized substrate *(17)*, pyrolysis and Fourier transform infrared analysis *(18)*, microscopic examinations of substrates *(19)*, and protein and enzymatic activity assays *(20,21)*. Nuclear magnetic resonance (NMR) spectra *(22)* and gas chromatography-mass spectroscopy (GC-MS) *(23,24)* can be used for analysis of only lignin model compounds, their degradation products, and metabolites or to estimate bond linkages. The above listed techniques, however, do not give a comprehensive understanding of the changes to the parent lignin molecule.

Matrix-assisted Laser Desorption Ionization time-of-flight mass spectrometry has been used to detect lignin compounds up to 1.9 kDa with guaiacol as model substrate *(25)*. Although this is a useful method for compositional studies, this technique requires significant sample preparation and there are difficulties with ionization and desorption of higher molecular weight lignin molecules. Changes resulting from reactions during evaporation also cannot be ruled out *(26)*. In order to evaluate depolymerization of the parent lignins it is useful to monitor changes in molecular weight distributions. Molecular weight distributions for pure lignin substrates have been determined by means of various size exclusion columns using tetrahydrofuran or aqueous NaOH as eluents *(27)*.

This work had two objectives: investigating new bacteria for lignin depolymerization and new analytical protocols for measuring their effects. The source for new bacteria included consortia from wood-eating insect guts and some other sources for lignin depolymerization. The wood-eating insect in this initial investigation of choice was Betsy beetles, *Odontotaenius disjunctus*. (Order: Coleoptera). Dissection and serial plating techniques were optimized for beetles. We also explored the consortia within activated sludge. We have chosen to look at lignin from Kraft pulp black liquor, extracted from Jack Pine *(28)*. Lignin concentration in the liquor varies from 35.86% to 44.35% by weight. The high molecular weight, acid insoluble lignin that can be separated and is used in this work varies from 23.1% to 29.9% of the total liquor solids *(29)*. The smaller acid soluble fraction acts as a plasticizer for the larger lignin fraction. The kraft lignin purified is a very complex, highly chemically modified form of lignin which is very recalcitrant and thus more difficult to depolymerize then lignin purified by less stringent techniques. The screening incubations were analyzed only with HPLC for molecular weight shifts. The second objective involved development of protocols for atomic force microscopy (AFM) for monitoring molecular conformations of lignin during incubation, and near infrared spectroscopy (NIR) for functional group tracking along with high-performance liquid chromatography (HPLC). These techniques proved very effective in validating our depolymerization results, and give a better understanding of the changes to lignin.

Materials and Methods

Dissection Technique

Betsy beetles were treated with ethyl alcohol to surface sterilize them. Then they were treated with boiling KOH for 1 min to soften the cuticle. The specimen was pinned through the dorsal portion of the thorax, to a suitable dissection tray filled with beeswax under a dissecting microscope with magnification up to ×10. Invertebrate saline kept the tissues moist and maintained correct osmotic conditions. The elytra (wing covers) were first raised using both the fine point forceps and microdissection scissors, thereby exposing the membranous wings and the soft underlying abdominal cuticle. Membranous wings were then cut off at their base. The soft abdominal tissue was gently lifted and a small incision was made in the cuticle directly under the forceps to expose the gut. Then, a cut was made along the entire length of the abdomen in a posterior to anterior direction and the cuticle was gently pulled upward to locate the gut. The unwanted organs were cleared away and the gut was removed and immediately placed in sterile invertebrate saline. Using a sterile scalpel, the gut was divided into foregut, midgut, and hindgut separately and added to sterile 10-mL invertebrate saline for serial dilution and plating (Fig. 1).

Serial Dilution and Plating

The foregut, midgut, and hindguts were processed separately using serial dilution technique with invertebrate saline (0.425% NaCl) as indicated below and dilutions 10^{-3}–10^{-8} were plated on tryptate soy agar (TSA) agar media plates and incubated at 30°C for 36–48 h. Distinct colonies were isolated and refreshed on TSA agar media again and colony characteristics like morphology, shape, and color and Gram nature were accessed *(30)*.

Dilutions: gut + 10-mL invertebrate saline = 10^{-1}

1 mL of 10^{-1} + 9 mL invertebrate saline = 10^{-2} up to 10^{-8}.

Propagation of Wastewater Treatment Sludge

Cultures in the sludge obtained from a wastewater treatment facility were initially grown with corn stover (National Renewable Energy Laboratories) and soytone peptone as substrate in a 3-L fermenter. After a week, 1 g/L lignin was added daily for the second week to replenish the carbon

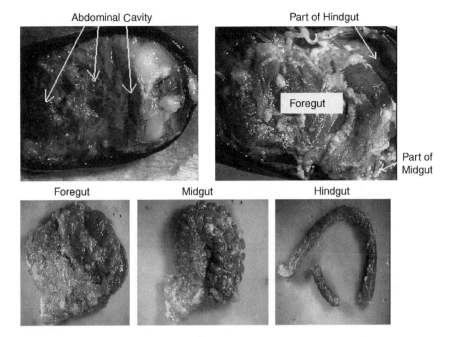

Fig. 1. Beetle dissection—abdominal cavity including the gut of Betsy beetle.

source and acclimatize the culture for a lignin substrate and the pH was maintained at 7.5 with HCl.

Screening Incubations

Batch screening incubation studies were conducted in duplicate. A 30-mL total reaction volume was used in 200-mL conical flasks covered with aluminum foil. Each liter of media consisted of 1 mL Cohen-Bazire's modified Hunters stock solution and 11 mL of 1:10 dilution Mineral base solution. Unlyophilized lignin (11.092 g/L) purified from Kraft black liquor was added to give a final concentration of 1 g/L. The media was prepared in bulk so that each flask would be the same containing 1 g/L lignin as the primary control (L). Lignin with just the nutrients and without the inoculum was used as the secondary control to ensure that the changes in the lignin profile were seen only because of the inoculum effects (M + L). Each flask, containing media, was then sterilized by autoclaving at 120°C for 45 min. The sterilized media was allowed to cool to ambient conditions for 1 d before inoculation. Propagated sludge and other inoculate sources were then added to make up 10% volume fraction (M + L + I). Beetles were inoculated in five forms as crushed whole, dissected with the whole gut, or separated as foregut, midgut, and hindgut. Inoculum was prepared in 30 mL of invertebrate saline. The other six sources investigated for lignin depolymerization were soil obtained from around rotting wood, worms, termites, deer dung, cow dung, and cow rumen. The solid inoculum was diluted with deionized water to slurry. The entire sets of experiments were performed in duplicates for data validation, with static aerobic incubation at 32°C for 30 d.

Secondary sets of experiments were repeated with beetle gut inoculation with two different oxygen environments. Aerobic conditions were maintained as explained earlier and a set of tubes were also incubated in a limited oxygen environment (candle jar). Centrifuge tubes (50 mL) were placed in a desiccator with a burning candle, to develop limited oxygen environment when the candle extinguished. Hydrogen peroxide (150 µL) was added to some flasks to test for improved lignin depolymerization. Inoculation was with dissected beetle gut—foregut, midgut, and hindgut.

Samples were withdrawn every 3–4 d for HPLC analysis and molecular weight distribution tracking. Sample analysis was conducted using HPLC. The size exclusion column used was Asahipak GS-320 (Shodex). A 100-µL sample loop was used. 0.1 M NaOH (pH 12.0) at a flow rate of 0.5 mL/min was the mobile phase used. The detector connected in series was Dionex AD20 Absorbance Detector to track changes in molecular weight.

Detailed Analysis

The screening results indicated that the wastewater sludge depolymerized lignin. A detailed study was then conducted with the sludge to get more comprehensive data. Batch studies (200 mL) were conducted in duplicate with this culture. The reaction mixture contained by volume 5% unlyophilized lignin (final concentration—0.55 g/L) and 85% Yeast Malt extract media. pH was adjusted to 8.5 with 0.1 N NaOH before autoclaving. Inoculation, incubation, and sample withdrawal was conducted as described earlier. Mixtures of 0.55 g/L lignin without nutrients (L) and lignin with nutrients (Ym + L), both without inoculum, were used as primary controls to ensure the change seen was because of inoculation. Lignin without nutrient (L + I) and nutrient broth without lignin (Ym + I) both inoculated, and just inoculation in deionized water (I) were used as secondary controls, to study the effect of the nutrient broth and lignin on microbial growth. The final set had all three components (Ym + L + I). All six sets were conducted in duplicate for data verification.

Analysis, along with HPLC for lignin tracking, involved Vecco Multi-Mode AFM and Bruker EQUINOX© 55/S (Bruker Optics, Model No. 502) FT-IR Spectrophotometer equipped with OPUS© 2.2 software (Opus Supplies Limited). The spectra were collected in the NIR mode using a tungsten lamp source and an air-cooled GE-Diode detector with a quartz beam splitter. Lignin did not adsorb very well on the mica surface tried initially for AFM studies. The final concentration selected for lignin was 0.01 g/L and the surface selected was graphite. Bacterial colonies were counted for the initial and final day samples with Axiovert 25 CFL Microscope.

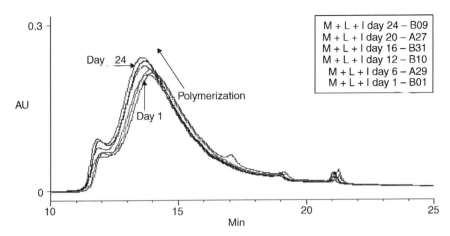

Fig. 2. Lignin polymerized with beetle gut inoculation. M + L + I indicates the flask contained nutrients, lignin, and inoculum.

Results and Discussion

Screening Incubations

Figure 1 shows the lignin profile overlays for flask inoculated with whole beetle gut. It is a representative diagram for the inoculum screening. Similar profiles were obtained for all different sources except the wastewater treatment sludge. Size exclusion columns elute the higher molecular fractions first followed by the lower molecular weight. Thus, a decrease in the earlier peaks with a consequent increase in the peaks eluting later, would indicate depolymerization. The six lignocellulose digester inoculum sources showed only polymerization with the tried fermentation conditions (Fig. 2). Lignin polymerized with beetle gut inoculation for both types of aerobic and limited oxygen environments. H_2O_2 seemed to bleach the samples from day 1 (Fig. 3). When beetle guts had earlier been applied to corn stover (10% dry solid content), a 7–8% total weight loss had indicated lignocellulose utilization and thus seemed a promising source for further investigation (data not shown). Morphology, shape, color, and Gram character of the bacterial, fungal, and yeast-like colonies isolated after beetle gut dissection are listed in Tables 1–3. Another batch was tried with the whole gut to see if the mutualistic behavior of the whole gut bacteria would improve depolymerization. The whole gut was able to achieve weight loss on corn stover incubations but when investigated with lignin showed only polymerization.

Figure 4 is a representative diagram for controls with only lignin during each set of experiments. Similar plots were obtained for lignin and nutrient (M + L) secondary controls. There was substantial overlap of the profiles over the entire incubation period without any change in the molecular weight distribution. This confirmed that any changes seen (polymerization or depolymerization) were because of the inoculum and

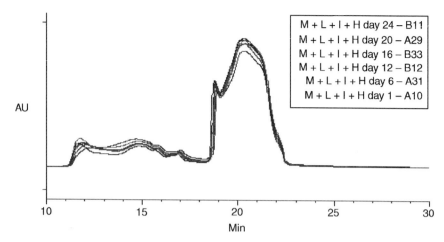

Fig. 3. Lignin bleached from day 1 in flasks supplemented with H_2O_2. M + L + I indicates the flask contained nutrients, lignin, and inoculum. H–H_2O_2.

Table 1
Colony Characteristics and Gram Character of Bacterial Colonies
From the Foregut of *Odontotaenius disjunctus* (Betsy beetle)

Legend	Gram	Morphology	Color/Shape
A	+	Rods	Cream/rhizoid, filamentous, flat
B	–	Cocci	Red/circular, entire, convex
C	Nd*	Nd*	Filamentous fungus
D	+	Rods	Big white/circular, entire, flat
E	+	Cocci	Yellow/circular, entire, raised
F	–	Cocci	Cream/circular, entire, raised
G	+	Cocci	Tiny white/circular, entire, convex

Shapes are in the order: whole colony forms, Margin edge forms, form of elevation.
*Nd, not determined.

Table 2
Colony Characteristics and Gram Character of Bacterial Colonies
From the Midgut of *Odontotaenius disjunctus* (Betsy beetle)

Legend	Gram	Morphology	Shape/Color
I	Nd*	Nd*	White/irregular, serrate, umbonate
II	–	Cocci	Red/circular, entire, convex
III	Nd*	Nd*	Filamentous fungus
IV	+	Rods	White/irregular, undulate, flat
V	+	Rods	Big white/circular, entire, flat

Shapes are in the order: whole colony forms, Margin edge forms, form of elevation.
*Nd, not determined.

that the lignin was stable. Lignin depolymerization was seen only with flasks inoculated with the wastewater sludge. Because of this a more comprehensive study was conducted on this inoculum source.

Table 3
Colony Characteristics and Gram Character of Bacterial Colonies From the Foregut of *Odontotaenius disjunctus* (Betsy beetle)

Legend	Gram	Morphology	Shape/Color
a	+	Rods	Translucent/circular, entire, flat
b	−	Rods	Yellow/irregular, undulate, raised
c	+	Cocci	Big off-white/circular, entire, convex
d	−	Cocci	Red/circular, entire, convex
e	+	Rods	Cream/rhizoid, filamentous, flat
f	−	Cocci	Tiny white/circular, entire, convex
g	Nd*	Nd*	Filamentous fungus

Shapes are in the order: whole colony forms, margin edge forms, and form of elevation.
*Nd, not determined.

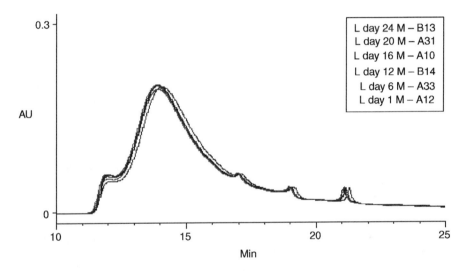

Fig. 4. Lignin primary control without any inoculum showed no molecular weight shifts. M indicates primary control from the midgut inoculum set.

Detailed Analysis for Sludge

Lignin depolymerization was conclusively seen only with the wastewater treatment sludge obtained from a brewery. Initially polymerization was observed as indicated by the increase in the higher molecular weight peaks eluting first (Fig. 4). But after 12 d depolymerization began. pH and the presence of yeast malt extract had a substantial effect on depolymerization. Lignin without yeast malt extract (L + I) showed only polymerization (Fig. 5), but lignin samples with yeast malt extract (Ym + L + I) showed depolymerization and degradation after initial polymerization (Fig. 4). The total area under the curve for the entire lignin profile also seemed to decrease with time indicating not only depolymerization but also complete decomposition of the lignin. Initial pH was adjusted to 8.5 with 0.1 N NaOH. pH declined

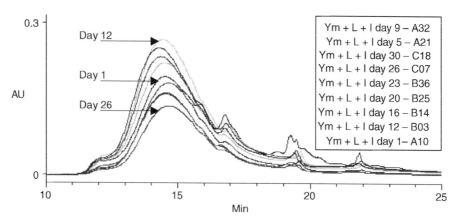

Fig. 5. Lignin polymerized and then depolymerized with wastewater treatment sludge.

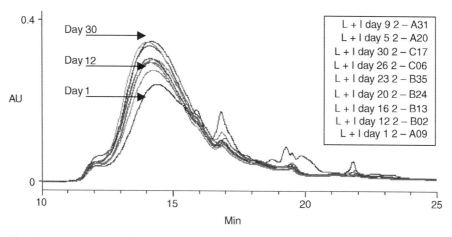

Fig. 6. Incubations without yeast malt extract did not show any depolymerization. Lignin only polymerized.

with incubation for flasks containing yeast malt extract (Ym). The final pH ranged from 4.0 to 4.5 in all the flasks containing Ym, whereas the pH in the other flasks maintained at 8–8.5. This might be another reason for no lignin depolymerization in flasks without Ym (Fig. 6).

An AFM protocol (never attempted before) was developed for lignin, as it was expected to give an idea of the molecular conformation of lignin. Mica was initially tried as a substrate, but a very weak adsorption was seen for lignin with the tip sweeping away the lignin. The substrate selected for AFM studies was graphite (hydrophobic). Graphite also gave a weak interaction with lignin in water or lignin with nutrients (M + L). But for a sample containing all three (lignin, nutrients, and inoculum), a very good, stable, and uniform picture was obtained which showed a monolayer of spherical lignin molecules at a concentration of 0.01 g/L (Fig. 7). The size range of the molecules (350–650 nm) seemed to match with the data (hydrodynamic and root mean square radius) that had been initially obtained for the same lignin with a light scattering detector characterization study (28).

Screening Cultures for Lignin Depolymerization

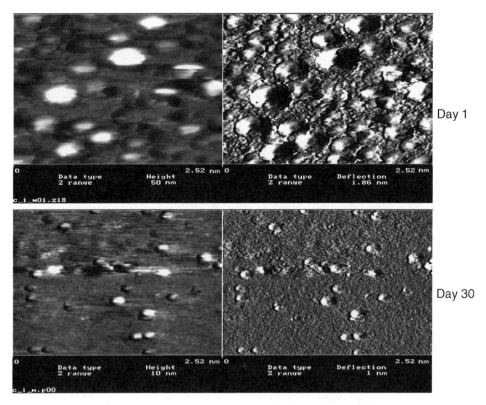

Fig. 7. AFM studies showed spherical lignin molecules.

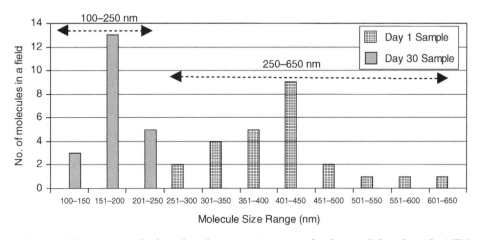

Fig. 8. Wastewater sludge depolymerization was further validated with AFM. Lignin depolymerized from a size ranging from 350 to 650 nm initially, to range from 135 to 220 nm by day 30.

Depolymerization of lignin for wastewater sludge seen on the HPLC was further confirmed with the AFM studies done on samples of days 1 and 30 (Figs. 7 and 8). The molecular size ranging initially from 250 to

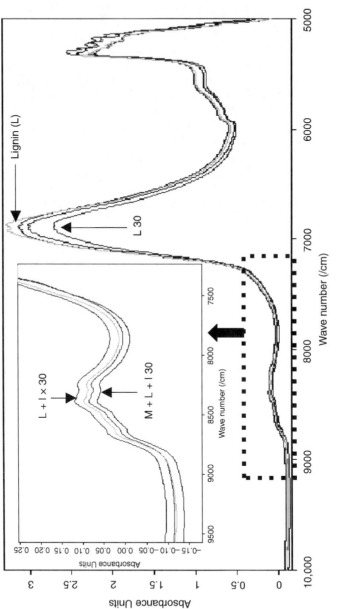

Fig. 9. NIR studies confirmed reduction in concentration of a range of functional groups. L, lignin; M, nutrient media; I, inoculum.

650 nm, reduced to range from 135 to 220 nm (Fig. 8). It can be seen that there is a reduction in not only the size but also the number of molecules, as was indicated with the HPLC results.

A NIR protocol was developed for lignin. A good R^2 of 0.99 was obtained for a range from Wave numbers 10,000.9–7497.8/cm. The functional groups that are covered in this wave number range are C—H, O—H, or N—H str. second overtone (CH, CH_2, CH_3, NH_2, HC=CH, OH, aromatic), 2X C—H str. + 2X C—C str. (Benzene, cyclopropane), 2X C—H str. + 2X C—H def. + $(CH_2)_n$—CH_2 (oil) and 2X N—H str. + 2X amide I (proteins). Figure 9 further validates the results seen on the HPLC and AFM that lignin was decomposed by the sludge. The calibrated zone (insert zoomed up) shows the curve for Ym + L + I is lower than the primary control L (day 1 and 30) indicating that there is an overall lowering in the functional groups.

The next step was to understand the characteristics of the colonies that depolymerized lignin. The colony count CFU for these flasks showed increased growth (20% higher) for inoculated flasks containing Ym + L + I, compared with flasks containing Ym + I or L + I only. There was no colony seen for dilutions of primary controls, indicating that there was no contamination in the flasks. Gram staining showed that there were only two bacterial species growing in all the flasks inoculated. They were white and off-white Gram positive rods. There were, however, slight modifications in both the species compared to the day 1 sample. The colonies from flasks containing only lignin (L + I) and both yeast malt extract and lignin (Ym + L + I) looked similar to those from day 1 inoculation (I). But the colonies from flasks containing only yeast malt extract (Ym + I) were slightly smaller compared to those from day 1 when seen thorough the microscope.

Conclusions

These studies conducted with lignin using analytical tools combining HPLC (UV detector), AFM, NIR, and Gram staining gave a more comprehensive idea regarding the changes that might be occurring to lignin during the incubations. Current studies with beetle guts and other lignin digester sources did not show depolymerization of lignin purified from kraft black liquor. One reason could be that this lignin has undergone many chemical modifications making it a high-molecular weight recalcitrant molecule.

Wastewater treatment sludge inoculum, however, did depolymerize and decompose even this form of lignin, reducing both the number and size of molecules as was validated on the HPLC, AFM, and NIR studies. AFM studies were very effective in studying molecular shape and also size reduction of lignin resulting from depolymerization. Lignin with inoculum gave a stable interaction and a very distinct picture of the lignin monolayer. Presence of nutrients or Ym was required for depolymerization. Lack of nutrients caused only polymerization. But this could also be because the initial pH was readjusted to range around 4.5 which might be more conducive for the growth of the two white and off-white (Gram positive rods) bacterial isolated.

Beetle gut is an ideal system for lignocellulose processing combining all aspects of size reduction, lignocellulose separation, hydrolysis, saccharification, and eventual fermentation of the cellulose through the series of guts. The lignin separated is in a low-molecular weight solubilized form, which could then have many applications (vanillin, adhesives, binder for laminated or composite wood products, automotive plastics, and so on) and thus a more value-added product than in the present process when it is just used as fuel in the burners. A replication of this ideal system would be an excellent process for lignocellulose conversion to bioethanol, as also seen through the weight loss studies conducted with consortia separated from beetle gut (publication in preparation).

References

1. Brezneck, J. A. and Brune, A. (1994), *Annu. Rev. Entomol.* **39,** 453–487.
2. Burlat, V., Ambert, K., and Ruel, K. (1997), *Plant Physiol. Biochem.* **35(8),** 645–654.
3. Lewis, N. G. and Paice, M. G. (1989), *Plant Cell Wall Polymers*, American Chemical Society, Washington, D. C.
4. McMullan, G. C., Meehan, C., Conneely, A., et al. (2001), *Appl. Microbiol. Biotechnol.* **56,** 81–87.
5. Kato, K., Kozaki, S., and Sakuranaga M. (1998), *Biotechnol. Lett.* **20(5),** 459–462.
6. Zimmermann, W. (1990), *J. Biotechnol.* **13,** 119–130.
7. Berrocal, M., Ball, A. D., Huerta, S., et al. (2000), *Appl. Microbiol. Biotechnol.* **54,** 764–771.
8. Brune, A. (2003), *Encyclopedia of Insects*, Academic Press, NY. pp. 1102–1107.
9. Drees, B. M. and Jackman, J. A. (1999), *A Field Guide to Common Texas Insects*, Gulf Publishing Company, Houston, TX.
10. Kaplan, D. L. and Hardenstein, R. (1987), *Soil Biol. Biochem.* **12,** 65–75.
11. Odier, E., Janin, G., and Monties, B. (1981), *Appl. Environ. Microbiol.* **41,** 337–341.
12. Gonzalez, K., Olave, I., Calderon, I., and Vicuna, R. (1988), *Arch. Microbiol.* **149,** 389–394.
13. Liaw, J. H. and Srinivasan, V. R. (1989), *Appl. Environ. Microbiol.* **55,** 2220–2225.
14. Kuhnigk, T. and Kong H. (1997), *J. Basic Microbiol.* **37,** 205–211.
15. Nutsubidize, N. N., Sarkanen, S., Elmer, L., and Shashikant, S. (1998), *Phytochemistry* **49(5),** 1203–1212.
16. Bernard-Vailhe, M. A., Besle, J. M., and Dore, J. (1995), *Appl. Environ. Micobiol.* **61(1),** 379–381.
17. Akin, D. E. and Benner, R. (1988), *Appl. Environ. Microbiol.* **54(5),** 1117–1125.
18. Bourdon, S., Laggoun-Defarge, F., Disnar, J. R., et al. (2000), *Org. Geochem.* **31,** 421–438.
19. Rhodes, T. L., Mikell, A. T., and Eley, M. H. (1995), *Can. J. Microbiol.* **41,** 592–600.
20. Moreira, M. T., Feijoo, G., Mester, T., Mayorga, P., Sierra-Alvarez, R., and Field, J. A. (1998), *Appl. Environ. Microbiol.* **64(7),** 2409–2417.
21. Rothschild, N., Levkowitz, A., Hadar, Y., and Dosoretz, C. G. (1999), *Appl. Environ. Microbiol.* **65(2),** 483–488.
22. Kajikawa, H., Kudo, H., Kondor, T., et al. (2000), *FEMS Microbiol. Lett.* **187,** 15–20.
23. Kato, K., Kozaki, S., and Sakuranaga, M. (1998), *Biotechnol. Lett.* **20(5),** 459–462.

24. Mester, T., Swarts, H. J., Romero i Sole, S., de Bont, J. A., and Field, J. A. (1997), *Appl. Environ. Microbiol.* **63(5),** 1987–1994.
25. Potthast, A., Rosenau, T., Koch, H., and Fischer, K. (1999), *Holzforschung* **53,** 175–180.
26. Rittstieg, K., Suurnakki, A., Suortti, T., Kruus, K., Guebitz, G., and Buchert, J. (2002), *Enzyme Microbiol. Technol.* **31,** 403–410.
27. Wong, K. K. Y. and Jong, E. (1996), *J. Chromatogr. B* **737,** 193–203.
28. Gidh A. V., Decker S., See C., Himmel M., and Williford C. (2006), *Anal. Chemica. Acta.* **555(2),** 250–258.
29. Dong, D. J. and Fricke A. L. (1995), *Polymer* **36,** 2075–2078.
30. Nester, A. E. W., Anderson, D., Roberts, C. E., Jr., Pearsall, N., and Nester, M. (1997), *Microbiology, A Human Perspective*, 4th ed., McGraw Hill, Boston.

Optimization of L-(+)-Lactic Acid Production Using Pelletized Filamentous *Rhizopus oryzae* NRRL 395

YAN LIU,* WEI LIAO, CHUANBIN LIU, AND SHULIN CHEN

Department of Biological Systems Engineering and Center for Multiphase Environmental Research, Washington State University, L. J. Smith 213, Pullman, WA 99164-6120, E-mail: yanliu@mail.wsu.edu

Abstract

Lactic acid is used as a food additive for flavor and preservation and a precursor in the development of poly-lactic acid, a product used to make biodegradable plastics and textiles. *Rhizopus oryzae* NRRL 395 is known to be a strain that produces optically pure L-(+)-lactic acid. The morphology of *Rhizopus* cultures is complex, forming filamentous, clumps, and pellet mycelia. Different morphology growth has significant effects on lactic acid production. In bioreactors, the filamentous or clump mycelia increase the viscosity of the medium, wrap around impellers, and block the nutrient transportation, leading to a decrease in production efficiency and bioreactor performance. Growing fungi in pellet form can significantly improve these problems. In this study, factors that affect lactic acid production in pelletized flask cultures using *R. oryzae* NRRL 395 were investigated in detail. Completely randomized designs were used to determine the influence of culture temperature, time, concentration of glucose, and inoculum size. Lactic acid fermentation using clump and pellet morphologies were performed in a 5 L fermentor at the optimal values obtained from flask culture. Finally, fed-batch culture was used to enhance the lactate concentration in broth. The final lactate concentration of fed-batch culture reached 92 g/L. The data presented in the article can provide useful information on optimizing lactic acid production using alternative source materials.

Index Entries: *Rhizopus oryzae*; lactic acid; pellet morphology.

Introduction

Lactic acid ($CH_3CHOHCOOH$) is a colorless compound that is important in both international and domestic markets, and in several industries. Lactic acid is currently widely used as an acidulant, flavoring, and preservative in food industries. There is also increased interest in its

---

*Author to whom all correspondence and reprint requests should be addressed.

application in the production of poly-lactic acid, which can be used as biodegradable plastic. Both the polymers and copolymers derived from lactic acid are especially attractive for biomedical application because of their biocompatibility, body absorption, and blood compatibility. The increased use of lactic acid in exciting applications and potential for use in biodegradable plastics has made producing lactic acid an attractive investment.

Rhizopus oryzae NRRL 395 is known to be a strain producing optically pure L-(+)-lactic acid *(1–5)*. However, there are some disadvantages in the fungal fermentative organic acids production. The fungi tend to form cotton-like mycelia in reactors, therefore the reactor is difficult to control under a homogeneous condition. Mass transfer and oxygen transfer through cotton-like mycelium are much more difficult than other forms of morphology. In addition, cotton-like morphology makes fungal biomass reuse impossible. All above factors ultimately lead to a low efficiency and yield of organic acid fermentation process *(6)*. Growing fungi in pellet form can significantly improve the mass transfer condition and benefits for lactic acid production.

In this article, the lactic acid production using palletized *R. oryzae* has been studied. The optimal conditions for the production of lactic acid with pellets were determined in flask culture. Enhanced production of lactic acid was conducted in a stirred fermentor using fed-batch culture.

Methods and Materials

Microorganism and Spore Culture Method

The fungus *R. oryzae* NRRL 395 (ATCC 9363) was obtained from the American type culture collection (Manassas, VA). The fungus was first grown on potato-dextrose agar (Difco, Sparks, MI) slants at 30°C for 7 d. For experimentation, the fungal spores in the slant were suspended in sterilized water maintained at 4°C. For storage, the spores were placed in 20% glycerol solution at –80°C.

Seed Culture

The composition of the seed medium was 24 g/L potato dextrose broth (PDB) (Difco, Sparks, MI) with 6 g/L $CaCO_3$. In terms of achieving pellet form, spore solution was inoculated into a 125-mL Erlenmeyer flask containing 50-mL seed medium with a spore concentration of 1×10^6 spore/mL and cultured at 27°C with 170 rpm shake speed for one day. The broth was used as the pellet seed for the following experiments. The diameter of seed pellet was 1.13 ± 0.22 mm (Fig. 1). As for the culture of clump seed used in the section of lactic acid production in stirred fermentor, culture conditions were same as the culture of pellet seed, except that the culture was in the incubator without shaking.

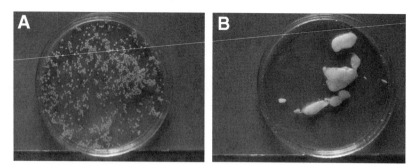

Fig. 1. The morphology difference of pellet and clump from seed culture. **(A)** is the seed pellets; **(B)** is the seed clumps.

Effects of Glucose Concentration, Reaction Time, and Temperature on Lactic Acid Production in Flask Culture

The experiment was carried out by a completely randomized design, with three replicates of 18 culture combinations. Three glucose concentrations (60, 100, and 120 g/L) and three culture durations (2, 3, and 4 d) were studied at two different temperatures (27°C and 30°C). The shake speed was 170 rpm. Calcium carbonate as the neutralizer was added into the flasks in order to maintain the pH value of approx 6.0 during the culture. The ratio of $CaCO_3$ to glucose was 1:2. The seed for all cultures was inoculated into the flasks at a fixed concentration of 0.45 g dry biomass/L. The cultures were performed in 250-mL flasks containing 100 mL of culture medium. The culture medium with varied glucose concentration (60, 100, and 120 g/L) was obtained by adding different amount of solid glucose into PDB medium (PDB contains 20 g/L glucose). All media were autoclaved at 121°C for 15 min before inoculation.

Effects of Inoculum Size on Lactic Acid Production in Flask Culture

Four seed concentrations (0.13, 0.23, 0.45, and 0.68 g dry biomass/L) were inoculated to the media which contained 24 g/L PDB, 100 g/L glucose and 60 g/L $CaCO_3$. The culture temperature was 27°C. And the culture duration was 4 d. Other culture conditions were same as described in previous section.

Lactic Acid Production in Stirred Fermentor

A 7-L (5.6-L effective volume) stirred tank fermentor (Bioflo 110 Modular Benchtop Fermentor, New Brunswick Scienific, NJ) equipped with a pH controller was used to carry out lactic acid fermentation. For batch culture, 2 L of deionized (DI) water, 48 g PDB and 200 g glucose were added into fermentor and autoclaved at 121°C for 15 min. The aeration rate and agitation speed were 1 vvm (volume/volume/minute) and 200 rpm, respectively. The pH was adjusted at 7.0 ± 0.1 using 20% calcium hydroxide. Two different seeds of clump and pellet were used to study the effects of morphology on lactic acid production. 0.23 g dry biomass/L was inoculated into fermentor and cultured at 27°C for 3 d. For fed-batch fermentation, culture conditions were the same as the batch fermentation except that the inoculum size was 0.45 g dry biomass/L, and extra 150 mL medium with 80 g glucose and 12 g PDB powder was fed into the fermentation on d 2.5 and 3.5.

Statistical Analysis

The effects of glucose concentration, reaction time, temperature, and inoculum size on lactic acid production in flask culture were analyzed by general lineal model using the statistical analysis system program 8 (SAS institute Inc. NC). Pair wise comparison and Tukey-Kramer multiple comparison were conducted to identify the difference of lactic acid production from different culture combinations.

Analytical Methods

A high-performance anion-exchange chromatography apparatus was used for the analyses. Because calcium carbonate was added to neutralize the pH, the lactate concentration in the broth was reported instead of the lactic acid concentration. The lactate in broth was analyzed using a Dionex DX-500 system (Sunnyvale, CA) including an AS11-HC (4 mm 10–32) column, a quaternary gradient pump (GP40), a CD20 conductivity detector, and an AS3500 auto-sampler (5). Glucose concentration was measured using the modified dinitrosalicylic acid method (7). Dry biomass was determined by washing the pellet mycelia with 6 N HCl and then washing to pH 6.0 with DI water. The washed biomass was dried at 100°C over night before weight analysis. The diameter of seed pellets was determined using an Olympus microphotograph (Tokyo, Japan).

Results and Discussion

Effects of Substrate Concentration, Temperature, and Fermentation Time for Lactic Acid Production by Pelletized R. oryzae NRRL 395 in Flask Culture

The production of lactic acid by pelletized R. *oryzae* with different medium glucose concentrations at 27°C and 30°C was shown in Figs. 2–4. At the culture duration of 2 d, lower glucose concentrations had higher lactate yields. 60 g/L glucose produced 33 g/L lactate (corresponding lactate yield of 55%) at both culture temperature of 27°C and 30°C; and 100 g/L and 120 g/L glucose at 30°C produced more lactate than that at 27°C (Figs. 2 and 3). Then, lactate yields from both 100 and 120 g/L glucose increased following the increase of culture duration, whereas the yields from 60 g/L glucose kept stable at approx 55% (Figs. 2 and 3). After 3 d of culture duration, the statistical analysis of pair wise comparison

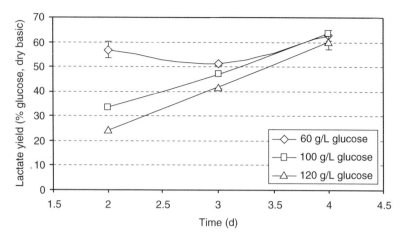

Fig. 2. Effects of glucose concentration on lactic acid yield. Fermentation was performed at 27°C with 0.45 g/L dry biomass inoculum size. Data was the average of triplicates with standard deviations ($n = 3$).

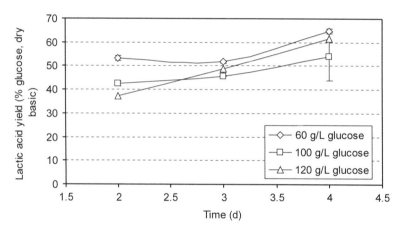

Fig. 3. Effects of glucose concentration on lactic acid yield. Fermentation was performed at 30°C with 0.45 g/L dry biomass inoculum size. Data was the average of triplicates with standard deviations ($n = 3$).

showed that there were no significant ($p > 0.05$) differences on lactate yields between each other of the three glucose concentrations, at each individual culture durations of 3 and 4 d (Figs. 2 and 3), in addition there were also no significant ($p > 0.05$) difference between culture temperatures of 27°C and 30°C at the culture duration of more than 3 d (Figs. 2 and 3). The highest lactate yield of 60% has been reached at 4 d of culture time, no matter what glucose concentration and culture temperature were. These results elucidated that the influence of temperature on lactate yield was diminished once the culture duration was increased, although it had significant ($p > 0.05$) influences at low glucose concentration in the short culture durations of 2 and 3 d. Glucose concentration had no significant

Optimization of L-(+)-Lactic Acid Production

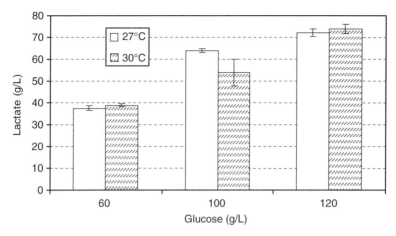

Fig. 4. Effect of temperature on lactate production. Fermentation was cultured for 4 d with 0.45 g/L dry biomass inoculum size. Data was the average of triplicates with standard deviations ($n = 3$).

influence on lactate yield at long culture durations, meanwhile, the higher glucose concentration produced more lactate at the same yield. Thus, the highest lactate concentration of 72 g/L was obtained at 4 d of culture duration with 120 g/L glucose concentration in the medium (Fig. 4). Therefore, 120 g/L glucose concentration and 4 d of fermentation at either temperature of 27°C and 30°C were chosen for the experiments in the following sections.

Effects of Inoculum Size on Lactic Acid Production in Flask Culture

Figure 5 and Table 1 showed that the lactate concentration varied with inoculum size. The inoculum size of 0.68 g/L dry biomass produced significantly less lactate than other inoculum sizes. The lowest lactate concentration of 37.3 g/L was obtained from the inoculum size of 0.68 g/L at 4 d of culture duration (Fig. 5). Meanwhile, the other three inoculum reached average lactate concentration of 78.5 g/L in the same culture duration. The Tukey-Kramer Multiple Comparison test confirmed that there was no significant ($p > 0.05$) difference on lactate concentration between each other of the inoculum sizes of 0.13, 0.23, and 0.45 g/L dry biomass (Table 1). This means that inoculum size between 0.13 and 0.45 g/L dry biomass had no significant influences on lactic acid production. Therefore, inoculum sizes in the range of 0.13–0.45 g/L dry biomass were the optimal ones for flask culture of lactic acid production.

Comparison of Lactic Acid Production Using Different Fungal Morphologies in a Stirred Tank Batch Culture

Lactic acid fermentation with clump and pellet morphologies was performed in a 5-L stirred fermentor (Fig. 6). The results demonstrated that

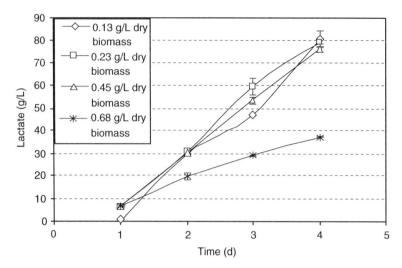

Fig. 5. Effects of inoculum size on lactic acid production. Fermentation was performed at 27°C with 120 g/L glucose in medium. Data was the average of triplicates with standard deviations ($n = 3$).

Table 1
Tukey-Kramer Multiple-Comparison Test of Lactate Concentration*

Group of inoculum size (g/L dry biomass)	Mean of lactate concentration (g/L)	Different from groups
0.68	37.3	0.45, 0.23, 0.13
0.45	76.3	0.68
0.23	79.5	0.68
0.13	80.6	0.68

*This report provides multiple comparison tests for all pair wise differences between the means at $\alpha = 0.05$. Fermentation was performed for 4 d at 27°C with 120 g/L glucose concentration.

the there were significant ($p < 0.05$) difference on lactic acid production between clump and pellet morphologies. The lactate concentration of clump fermentation reached to 33 g/L in 2.5 d of culture duration whereas the pellet fermentation produced 60 g/L. The data indicated that the lactic acid production was significant increased using pelletized fungal fermentation.

Enhancing Lactic Acid Production by Fed-Batch Culture in a Stirred Tank Fermentor

In order to improve lactic acid production, a fed-batch culture was performed under the optimal condition obtained from the previous section of the flask culture. Figure 7 showed that lactate concentration from the part of batch fermentation reached 50 g/L in 2 d, then two extra glucose solutions were fed into the fermentor at the culture durations of 2.5 and 3.5 d,

Optimization of L-(+)-Lactic Acid Production

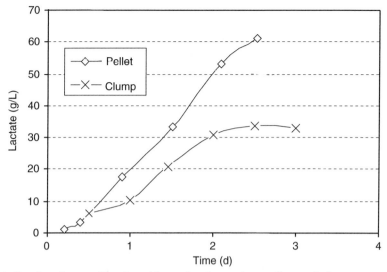

Fig. 6. Batch culture of lactic acid production using pellet and clump morphology in a 5-L stirred tank. The aeration rate and agitation speed were 1 vvm and 200 rpm, respectively. The pH was adjusted at 7.0 ± 0.1 with 20% $Ca(OH)_2$ and 0.23 g/L dry biomass seed was inoculated into fermentor and cultured at 27°C.

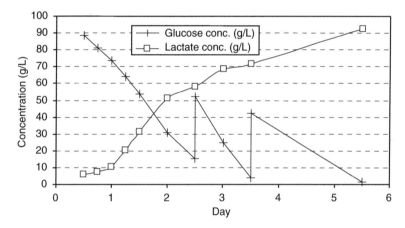

Fig. 7. Lactic acid production by fed-batch culture of *R. oryzae* NRRL 395 in a 5-L stirred tank fermentor with the aeration rate and agitation speed were 1 vvm and 200 rpm, respectively. The pH was adjusted at 7.0 ± 0.1 with 20% $Ca(OH)_2$ and 0.45 g/L dry biomass seed was inoculated into fermentor and cultured at 27°C for 5.5 d and 150-mL fed-medium with 80 g glucose and 12 g PDB powder was added at 2.5 and 3.5 d.

respectively. After feeding, the lactate concentration of fed-batch culture eventually reached 92 g/L in 5.5 d. The data also presented the difference of production rates on different phases of batch fermentation and fed-batch fermentation. The production rate after feeding was 0.53 g/L h, which was much slower than 1.02 g/L h of the rate before feeding (Fig. 7). The low production rate was mainly caused by the formation of calcium

Table 2
Comparison of Different Processes of Lactic Acid Production Using R. oryzae

Fermentation conditions*	Reactor	Lactate (g/L)	Yield (g/100g glucose)	Productivity (g/L h)	Reference
Cells immobilized on cotton cloth	Rotating bed	126	90	2.5	(8)
Pellets	Air-lift	86	–	1.07	(6)
Immobilized cells	Fluidized bed	73	65	1.6	(9)
Pellets	Flask	76–80	63–67	1.1	This study
Pellet (batch)	Stirred tank	60	66	1	This study
Pellets (Fed batch)	Stirred tank	92	60	0.7 (average)	This study

*Glucose was used as substrate for all fermentations.

lactate crystalline and substrate limitation. During fermentation process, the product of calcium lactate started crystallizing out once the lactate concentration reached 70 g/L. The calcium lactate crystals were attached on the surface of pellets and influenced the mass/oxygen transfer though pellets, and further reduced the production rate.

It is well known that production of fermentation processes are mainly controlled by both product inhibition and substrate limitation. In this particular case, it was apparent that both the substrate limitation and product inhibition played equal important roles on lactate production. Fed-batch fermentation can only minimize the substrate limitation, but it has nothing to do with production inhibition. In terms of further improving the fermentation performance, the product inhibition of calcium lactate crystalline has to be eliminated. Other alkalis such as sodium hydroxide and ammonia, producing soluble lactate during fermentation process, could be good alternatives of calcium hydroxide. However, high concentration of those cations could inhibit the lactic acid production. Thus, mixture of alkalis such as calcium hydroxide, sodium hydroxide, and ammonia could be used as neutralizers to control pH in fed-batch culture, which will result in low concentration of each cation and high concentration of soluble lactate.

Comparison of Different Processes of Lactic Acid Production Using R. oryzae

Several cell immobilization methods have been developed to control fungal morphology, to achieve higher content of fungal biomass and eliminate mass transfer limitations inside the fungal mycelia, to increase the total amount of final product and its productivity (Table 2). In these studies, fungal mycelia were either entrapped in a polymeric matrix or attached on a support surface. The performances were apparently improved. The best results obtained in these studies are: 126 g/L final lactate concentration in broth, 2.5 g/L h productivity, and 90% lactic acid yield from the immobilized fungal biomass on a cotton cloth (8). However, the systems with cell immobilization will add extra cost on lactic acid production. Thus, if the cell can directly form small pellets, the operation would be still highlys efficient and much more economical. Yin (6) formed small pellets in an air-lift fermentor to produce lactic acid. The lactic acid concentration from the air-lift fermentor was 86 g/L (Table 2). Compared with the pellet fermentation of air-lift system, our submerged pellet fermentation achieved a higher lactate concentration of 92 g/L. In addition, the average productivity and yield of the submerged pellet fermentation were 0.7 g/L h and 60%, respectively.

Conclusions

Pelletized morphology significantly increases lactic acid production. The optimal fermentation conditions of pelletized flask culture with *R. oryzae* NRRL 395 were 120 g/L of glucose concentration, 0.13–0.45 g/L dry biomass inoculum size, 27–30°C culture temperature and 3–4 d fermentation time. Under these conditions, the lactic acid productivity, yield, and lactate concentration reached 1.1 g/L h, 63%, and 76 g/L, respectively. Fed-batch culture can significantly increase lactate concentration in broth. The final lactate concentration reached 92 g/L after 5.5 d of fermentation. To enhance lactic acid concentration, productivity, and yield in broth, future work should focus on the fed-batch fermentation using mixed alkali to adjust pH.

Acknowledgments

This research was supported by the Washington State IMPACT Center (International Marketing Program for Agricultural Commodities and Trade) and the Agricultural Research Center of Washington State University. We are grateful to Craig Frear, Mike Richardson, and Dan Hardesty for editorial assistance.

References

1. Hang, Y. (1989), *Biotechnol. Lett.* **11,** 299–300.
2. Soccol, C. R., Martin, B., and Raimault, M. (1994), *J. Appl. Microbiol. Biotechnol.* **41,** 286–290.
3. Yin, P. M., Nishina, N., Kosakai, Y., Yahiro, K., Park, Y., and Okabe, M. (1997), *J. Ferment. Bioeng.* **84,** 249–253.
4. Oda, Y., Saito, K., Yamauchi, H., and Mori, M. (2002), *Curr. Microbiol.* **45,** 1–4.
5. Liu, Y., Wen, Z., Liao, W., Liu, C., and Chen, S. (2005), *Eng. Life Sci.* **5,** 343–349.
6. Yin, P. M., Yahiro, K., Ishigaki, T., Park, Y., and Okabe, M. (1998), *J. Ferment. Bioeng.* **85,** 96–100.
7. Miller, G. L. (1958), *Anal. Chem.* 426–428.
8. Tay, A. and Yang, S. (2002), *Biotechnol. Bioeng.* **80,** 1–12.
9. Hamamci, H. and Ryu, D. D. Y. (1994), *Appl. Biochem. Biotechnol.* **44,** 125–133.

Copyright © 2006 by Humana Press Inc.
All rights of any nature whatsoever reserved.
0273-2289/06/129–132/854–863/$30.00

A Simple Method to Generate Chromosomal Mutations in *Lactobacillus plantarum* Strain TF103 to Eliminate Undesired Fermentation Products

SIQING LIU

Bioproducts and Biocatalysis Research Unit, National Center for Agriculture Utilization Research, USDA,[†] ARS, 1815 N. University St., Peoria, IL 61604, E-mail: lius@ncaur.usda.gov

Abstract

Gram-positive bacteria have been explored to convert lignocellulosic biomass to biofuel and bioproducts. Our long-term goal is to create genetically engineered lactic acid bacteria (LAB) strains that convert agricultural biomass into ethanol and other value-added products. The immediate approaches toward this goal involve genetic manipulations by either introducing ethanol production pathway genes or inactivating pathways genes that lead to production of undesired byproducts. The widely studied species *Lactobacillus plantarum* is now considered a model for genetic manipulations of LAB. In this study, *L. plantarum* TF103 strain, in which two of the chromosomal L-*ldh* and D-*ldh* genes are inactivated, was used to introduce additional mutations on the chromosome to eliminate undesired fermentation products. We targeted the acetolactate synthase gene (*als*) that converts pyruvate to acetolactate, to eliminate the production of acetoin and 2,3-butanodial. A pBluescript derivative containing sections of the *als* coding region and an erythromycin resistance gene was directly introduced into *L. plantarum* TF103 cells to create mutations under selection pressure. The resulting erythromycin resistant (Em^r) TF103 strain appears to have chromosomal mutations of both the *als* and the adjacent *lysP* genes as revealed by polymerase chain reaction and Southern blot analyses. Mutations were thus generated via targeted homologous recombination using a Gram-negative cloning vector, eliminating the use of a shuttle vector. This method should facilitate research in targeted inactivation of other genes in LAB.

Index Entries: Lactic acid bacteria; *als* and *lysP* mutant; ethanol; acetoin; 2,3-butanodial; acetolactate synthase.

[†]Names are necessary to report factually on available data; however, the USDA neither guarantees nor warrants the standard of the product, and the use of the names by USDA implies no approval of the product to the exclusion of others that may also be suitable.

Introduction

Metabolic engineering often requires the modification of the bacterial genome by inactivation of undesired genes and products. The creation of a genetic knockout mutant usually involves the replacement of a chromosomal gene by a mutated version via homologous recombination. However, gene inactivation studies in Gram-positive bacteria have been limited as a result of the lack of mature recombinant DNA technology in these rather difficult hosts when compared with the well-developed systems in Gram-negative bacteria.

An example of a genetic knockout study in a Gram-positive Group B *Streptococcus* (GBS) involved a two-step procedure as described by Li *(1)*. The first step is a single crossover, resulting in recombinant Kanamycin and chloromphenical ($Km^r Cm^r$) cells after growth of transformed host cells at a nonpermissive temperature. The second step involved further growth under culture conditions without antibiotics, which allowed the second double crossover, resulting in $Km^r Cm^s$ colonies that carry a mutant copy of the targeted gene. This strategy requires multiple cloning and selection steps that lead to the double crossover event. First, the kanamycin gene was inserted into the target gene in a cloning vector. Then, the fusion DNA was subcloned into a specific shuttle vector containing the thermosensitive region (Ts) and chloromphenical resistance gene (Cm^r) and introduced into the GBS host cells. Under the nonpermissive growth temperature, the xeno-plasmid can be forcefully integrated into the host chromosome. The second recombination event, between the chromosomal target gene and its deleted copy, resulted in segregational loss of the excised vector, producing a chlorom phenical-sensitive phenotype. Therefore, the excision step can generate either the wild-type gene or a deletion of the target gene in the chromosome *(1)*. Targeted mutations of L-*ldh* and D-*ldh* genes were created in a similar fashion in the Gram-positive bacterium *Lactobacillus plantarum*, that resulted in Cm^r-Em^s TF103 strain *(2)*.

Another method is to introduce an internal fragment of the target gene into a suicide vector incapable of replicating in the organism of interest and then select for a single homologous recombination event. This results in the insertion of the entire plasmid into the target gene by Campbell-type integration, flanked by two truncated copies of the target gene. A single crossover that led to targeted gene inactivation has been reported in *Lactococcus lactis (3)*, *Lactobacillus helveticus (4)*, and *Lactobacillus pentosus (5)*.

The long-term goal of this project is to create genetically engineered LAB strains that convert agricultural biomass into ethanol and other value-added products *(6,7)*. *L. plantarum* is a lactic acid bacterium that ferments glucose to a variety of secondary end products including lactate, ethanol, acetoin, 2,3-butanediol, acetate, and mannitol *(8)*. Extensive genetic studies have been reported using *L. plantarum*, and therefore it is considered a

Table 1
Oligonucleotides Used in This Study

als2306	GA**CTCTAGA**TTCTGGTCAAGTTCAGCGTG
als3423	ACC**AAGCTT**GCCATCATTCCAGATCAAA
als3884	ACC**AAGCTT**GCAGCTAGAATAGTTGCGGG
als4922	CAA**CTCGAG**ACCCAGCCCGTTTAAAGACT
als4467	TAA**CTCGAG**AAAACGAGCGCAATCAGACT
als2669	ACG**AAGCTT**GCTTTCGTCACTTCTGGTG
als3140	ACG**AAGCTT**ACAACGGGTCAGTGATGACA

The bold nucleotides indicate corresponding restriction enzyme sites introduced in the primers.

model system for genetic engineering (9,10). The rationale for this project is based on the L-*ldh* and D-*ldh* double knock-out studies by Ferain (2,8); the mutant strain produced mainly acetoin, as well as other end products including ethanol, 2,3-butanediol, and mannitol (8). The inactivation of the *als* gene, which converts pyruvate to acetolactate, would block the down-stream production of acetoin and 2,3-butanediol (from acetolactate), and therefore might allow pyruvate to be rerouted toward ethanol and/or mannitol production when both L-*ldh* and D-*ldh* are inactivated (11). In this study, we report the generation of targeted mutations on the chromosome of *L. plantarum* TF103 via homologous recombination by using the nonreplicative, Gram-negative pBluescript cloning vector.

Materials and Methods

Bacterial Strains and Growth Conditions

Escherichia coli strain *Dh5α* cells were grown at 37°C in Luna Bertani (LB) medium supplemented with 100 µg/mL ampicillin and/or 35 µg/mL chloramphenicol when necessary. *L. plantarum* NCIMB8826 derivative strain TF103 was provided by Ferain (2). This strain, defective for both D- and L-lactate dehydrogenase activities, was grown in Deman Rogosa Sharpe (MRS) broth with chloramphenicol (10 µg/mL) at 110 rpm and 37°C.

Cloning of Deleted Copy of the als Gene Into pBluescript

The preparation of chromosomal DNA from *L. plantarum* strain TF103 was performed using a Bactozol Kit (Molecular Research Center, Inc., Cincinnati, OH) as described in the manufacture's protocol. Polymerase chain reaction (PCR) primers (Table 1) were designed according to the published *L. plantarum* sequences (10). Basic molecular biology techniques were performed as described (12). *L. plantarum* strain TF103 genomic DNA was used as template in following PCR amplifications. PCR primers als2306 and als3423 were used to amplify an 1117-bp fragment. Following digestion with XbaI/HindIII, the fragment was cloned into XbaI/HindIII sites of pBluescript to generate pBlue2306$_{3423}$. Similarly, the 1038 bp PCR products from primers als3884 and als4922 were cloned into the HindIII and XhoI sites of pBlue2306$_{3423}$ to generate pBlue2306$_{3423}$-3884$_{4922}$, with a 461 bp sequence deleted between als3424 and 3884. The pBlue2306$_{3423}$-3884$_{4922}$ was digested with NheI (cut at 2606) and HindIII. The resulting larger vector fragment was ligated with the 1 kb XbaI-HindIII erythromycin fragment from pEI2 (13) to generate pBlue2306$_{2606}$Em3884$_{4922}$, which has an erythromycin resistance gene inserted between als2306-2606 and als3884-4922 sequences. All the constructs were confirmed by sequencing using the ABI Prism 310 and the ABI Prism Dye Terminator Cycle Sequencing Ready Reaction Kit (Perkin-Elmer, Foster City, CA). Sequence analyses were performed with the SDSC biology workbench (http://www.sdsc.edu/Research/biology/) and NCBI (http://www.ncbi.nlm.nih.gov/).

Generation and Confirmation of the Em^r TF103 als Mutant

The genetic *als* deletional mutant was obtained by electroporating *L. plantarum* TF103 cells with the pBluescript derived plasmid pBlue2306$_{2606}$Em3884$_{4922}$ as described previously (14). The transformants were selected for erythromycin resistance on MRS plates supplemented with 5 µg/mL erythromycin. Five Em^r mutants were selected and genomic DNA was prepared as described (6). Genomic PCR of the mutant strains was performed using two primers (als2306 and als4667) located on each side of the deleted region of the *als* gene. The original pBlue2306$_{2606}$Em3884$_{4922}$ plasmid DNA and parent TF103 genomic DNA were used as templates in control reactions.

Southern Hybridization

Alkaline transfer of EcoRI digested genomic DNA on 0.8% agarose gels to nylon membranes was done as described in ref. (12). Probes were labeled by PCR using appropriate primers (see Table 1) and digoxigenin-dUTP of Roche Applied Science (Penzberg, Germany). As indicated in Fig. 1, probe 1 was labeled from the intentionally deleted 363 bp fragment amplified from TF103 genomic DNA by using primers als2306 and als2669, probe 2 was labeled from the 283 bp PCR fragment amplified using primers als3140, als3423; and probe3 was labeled from the 783 bp PCR product using primers als3884 and als4667. The DNA sequences of all three probes were confirmed by sequencing. Filters were hybridized and washed using the DIG Easy Hyb kit (Roche) as instructed by the manufacturer. The digoxigenin-labeled DNA molecular weight marker set III was purchased from Roche. Immunological detection of DIG-labeled nucleic acids was carried out using the DIG Luminescent Detection Kit (Roche Applied Science, Penzberg, Germany).

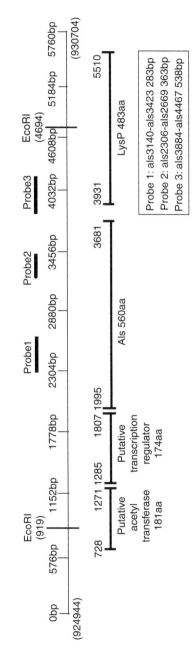

Fig. 1. Schematic diagram of the *L. plantarum* chromosomal regions around the *als* gene and its close vicinity including the *lysP* gene. The nucleotide positions are indicated and the relative positions and sizes of four ORFs including the Als and LysP proteins are drawn in the same scale. The positions of three probes used in Southern hybridization analyses are shown at the top. Probe sizes and PCR primers used to amplify these probes are indicated inside the box at the bottom of the diagram.

Results and Discussion

The chromosomal sequences of the L. plantarum als gene and vicinity were extracted from the published data of Kleerebezem (10). A total of 5760 bp corresponding to genomic positions 924,944–930,704 were analyzed in this study. As shown in Fig. 1, there are four open reading frames (ORFs) within the 5760 bp DNA fragment (http://www.ncbi.nlm.nih.gov/genomes/ framik.cgi? db=genome& gi=276). The relative positions of each ORF are indicated by nucleotide numbers relative to 0 (924,944). The als gene, with an ORF of 560 amino acids, is flanked on the left side by a putative transcription regulator (ORF of 174 amino acids) and on the right side by the lysP gene encoding a lysine transport protein consisting of 483 amino acids. There are two EcoRI sites at positions 919 and 4694. The positions of probes for Southern blotting and PCR primers are all marked by nucleotide numbers relative to the 0 point.

The construction of the pBluescript derivative pBlue2306_{2606}Em3884_{4922}, in which the ery gene was inserted into the als sequences, was done in E. coli Dh5a cells. The plasmid linear map is illustrated in Fig. 2 (the top bar). This plasmid was used to transform TF103 cells and the resulting Em^r colonies were further analyzed for chromosomal integrations.

Two specific primers (als2306 and als4667) located on each side of the deleted region would result in PCR amplification of a 1883 bp fragment from the control plasmid pBlue2306_{2606}Em3884_{4922} (the ery gene was inserted at the deleted region, Fig. 2), and of a 2161 bp fragment from the parent TF103 (from 2306 to 4667, Figs. 1 and 2). Total genomic DNA preps from Em^r candidates were subjected to PCR analyses to screen for clones bearing the deleted copy of the als gene.

As expected, the 1883 bp and 2161 bp fragments were detected when the control plasmid pBlue2306_{2606}Em3884_{4922} (Fig. 3, lane 1) and parent strain TF103 genomic DNA (Fig. 3, lane2) were used as templates, respectively. However, among five TF103 Em^r genomic DNA samples tested, all yielded a shorter 1683 bp PCR fragment (Fig. 3, lane 3, only one sample shown). The absence of the 2161 bp fragment in the mutant suggested that the mutant strain does not carry the original copy of the als gene. Rather, a deleted copy was substituted as reflected by a shorter 1683 bp PCR product. These data suggest that the mutant may result from DNA integration into the chromosome (Fig. 2).

Southern blotting was performed on total DNA from the parent TF103 and an als mutant, which was digested by EcoRI and hybridized with three different probes. The positions of each probe are indicated in Fig. 1. Probe 1 and probe 2 are located within the als coding region, whereas probe 3 is located at the end of als, extends to the beginning of lysP that encoding the N-terminal part of the LysP protein (Fig. 1).

Positive hybridization of about 3775 bp DNA fragment from EcoRI digested genomic DNA, which cuts at position 919 and 4694 (Fig. 1), was

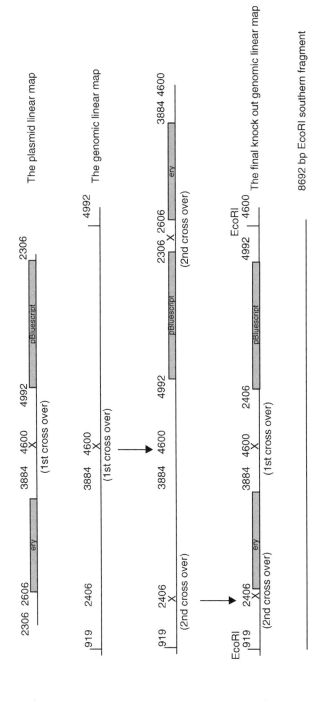

Fig. 2. Schematic representation of a system for generating targeted chromosomal replacement by a two-step integration procedure. Sequential homologous recombinations led to the construction of the chromosomal mutant TF103 Δ*als*Δ*lysP*. The approximate positions of two crossover and *Eco*RI digestion sites are shown. The approximate size of *Eco*RI fragment as detected by Southern and the expected PCR amplicon from the mutant are indicated.

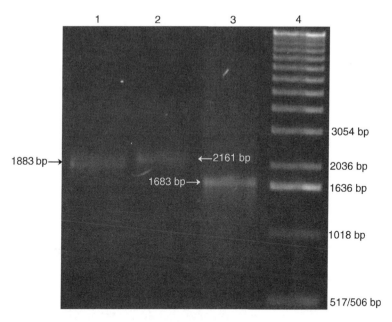

Fig. 3. Agarose (1%) gel electrophoresis of PCR products formed on chromosomal DNA of the new mutant strain TF103 Δ *als*Δ *lysP* using als2306 and als4467 primers. Lane 1, PCR product from control pBlue2306$_{2606}$Em3884$_{4922;}$ Lane 2, PCR product from control genomic DNA of TF103; Lane 3, PCR products from genomic DNA of the TF103 Δ *als*Δ *lysP* mutant; Lane 4, DNA molecular weight markers from invitrogen.

obtained with the parent strain TF103 regardless of the probes used (Fig. 4, the first lanes of panel A–C), and this 3775 bp fragment was not detected in the mutant with any of the three probes. These results confirmed that the mutant clone does not bear the original copies of wild type *als* and *lysP*. Instead, both genes were interrupted.

As illustrated in Fig. 1, probe 1 fragment covers the 2306–2669 region that is located in the deleted region of *als* (Fig. 2). Southern blotting results using this specific probe failed to detect any signals in the mutant genomic DNA, indicated that the intented deletion was successful (Fig. 4, panel B, the third lane). Meanwhile, a fragment of about 9 kb was detected with the 283 bp probe 2 specific for the als3140–3423 region and the 587 bp probe 3 specific for the als3884–4667 (Fig. 4, the third lanes of panels B and C).

The combined results of Southern blotting and PCR analyses suggest that the knockout is a result of two crossover events. A schematic diagram of how the mutant was generated is shown in Fig. 2. The first crossover occurred at approx 4600, and the second crossover occurred at approx 2400. After homologous recombination at two positions, both *als* and *lysP* were interrupted. The recombination events led to the hybridization to a 8692 bp fragment when using probes 1 and 3, larger than the 3775-bp wild-type gene fragment. As indicated in Fig. 2, the mutant includes the interruption of the *als* gene with the *ery* insertion in addition to an approx 200 bp

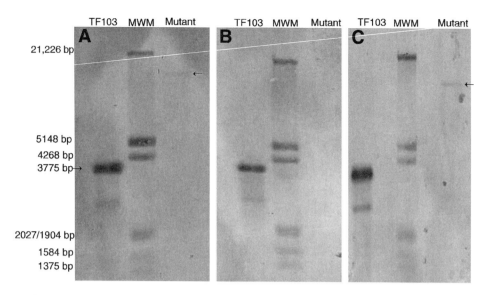

Fig. 4. Southern hybridization analysis of *Eco*RI digested chromosomal DNAs of parent TF103 and the TF103 Δ *als*Δ *lysP* mutant using three gene-specific probes. The relative positions of each probe used are indicated in Fig. 1. MWM: the Molecular Weight Marker III purchased from Roche, with the sizes of the digoxigenin-labeled DNA fragments are indicated on the left.

internal deletion from 2406–2606, and also the interruption of the *lysP* gene by the pBluescript vector insertion. This mutant is therefore designated as TF103 Δ *als*Δ *lysP*.

Acetolactate synthase (ALS: EC 4.1.3.18) is the first enzyme in the biosynthetic pathway of leucine, valine, and isoleucine. ALS catalyzes the formation of 2-acetolactate or 2-aceto-2-hydroxy-butanoate from pyruvate using thiamine pyrophosphate as a cofactor. Unlike *Lactococcus lactis* (15,16), which has three *als* genes, two acetolactate synthase large subunit (*ilvB*), and one acetolactate synthase small subunit (*ilvN*), *L. plantarium* WCFS1 genome has only one *als* gene (10), and therefore can not synthesize the branched-chain amino acids valine, leucine, and isoleucine. The *als* in *L. plantarum* is presumably used for the fermentative production of acetoin and 2, 3-butanodial from pyruvate. Detailed biochemical characterization and fermentation product analyses of the mutant are in progress.

The *lysP* gene encodes a membrane protein that is reported to be responsible for lysine transport in *E. coli* (17). The mutant will be made available for interested researchers in *lysP* related studies.

In summary, we have generated an insertional-deletion mutant TF103 Δ *als*Δ *lysP* with inactivated *als* and also its nearby *lysP* locus. The approach is simple, using only a Gram-negative cloning vector, and eliminating the tedious cloning procedures involved with many Gram-positive shuttle vectors. The results of this study suggest that the pBluescript vector can be used to introduce engineered DNA into the host chromosome. This opens potential applications of the pBluescript vector in engineering other Gram-positive hosts.

Acknowledgments

The author wish to thank Dr. Thierry Ferain (Laboratorire de Genetique Moleculaire, Universite Catholique de Louvain, Belgium) for providing *L. plantarium* TF103 strain and Theresa Holly for excellent technical assistance. The valuable review of this manuscript by Dr. Timothy D. Leathers is gratefully acknowledged.

References

1. Li, J., Kasper, D. L., Ausubel, F. M., Rosner, B., and Michel, J. L. (1997), *Proc. Natl. Acad. Sci. USA* **94**, 13,251–13,256.
2. Ferain, T., Hobbs, J. N., Jr., Richardson, J., et al. (1996), *J. Bacteriol.* **178**, 5431–5437.
3. Leenhouts, K. J., Kok, J., and Venema, G. (1989), *Appl. Environ. Microbiol.* **55**, 394–400.
4. Bhowmik, T. and Steele, J. L. (1993), *J. Gen. Microbiol.* **139**, 1433–1439.
5. Lokman, B. C., Heerikhuisen, M., Leer, R. J., et al. (1997), *J. Bacteriol.* **179**, 5391–5397.
6. Liu, S., Nichols, N. N., Dien, B. S., and Cotta, M. A. (2005), *J. Ind. Microbiol. Biotechnol.*, in press.
7. Bothast, R. J., Nichols, N. N., and Dien, B. S. (1999), *Biotechnol. Prog.* **15**, 867–875.
8. Ferain, T., Schanck, A. N., and Delcour, J. (1996), *J. Bacteriol.* **178**, 7311–7315.
9. Djordjevic, G. M. and Klaenhammer, T. R. (1996), *Plasmid* **35**, 37–45.
10. Kleerebezem, M., Boekhorst, J., van Kranenburg, R., et al. (2003), *Proc. Natl. Acad. Sci. USA* **100**, 1990–1995.
11. Neves, A. R., Ramos, A., Costa, H., et al. (2002), *Appl. Environ. Microbiol.* **68**, 6332–6342.
12. Sambrook, J., Fritsch, E.F., and Maniatis T. (1989), *Cold Spring Harbor Laboratory Press*, Cold Spring Marbor, NY.
13. Posno, M., Leer, R. J., van Luijk, N., et al. (1991), *Appl. Environ. Microbiol.* **57**, 1822–1828.
14. Liu, S., Dien, B. S., and Cotta, M. A. (2005), *Curr. Microbiol.* **50**, 1–6.
15. Godon, J. J., Delorme, C., Bardowski, J., Chopin, M. C., Ehrlich, S. D., and Renault, P. (1993), *J. Bacteriol.* **175**, 4383–4390.
16. Bolotin, A., Wincker, P., Mauger, S., et al. (2001), *Genome Res.* **11**, 731–753.
17. Neely, M. N., Dell, C. L., and Olson, E. R. (1994), *J. Bacteriol.* **176**, 3278–3285.

Production of Insoluble Exopolysaccharide of *Agrobacterium* sp. (ATCC 31749 and IFO 13140)

Márcia Portilho,*,[1] Graciette Matioli,[1] Gisella Maria Zanin,[2] Flávio Faria de Moraes,[2] and Adilma Regina Pippa Scamparini[3]

[1]*Pharmacy and Pharmacology Department, State University of Maringá, Avenida Colombo, 5790-BL P-02, 87020-900 Maringá, PR, Brazil, E-mail: mportilho@uem.br;* [2]*Chemical Engineering Department, State University of Maringá, Avenida Colombo, 5790-BL P-02, 87020-900 Maringá, PR, Brazil; and* [3]*Faculty of Food Engineering, Universidade Estadual de Campinas, Campinas, Brazil*

Abstract

Agrobacterium isolated from soil samples produced two extracellular polysaccharides: succinoglycan, an acidic soluble polymer, and curdlan gum, a neutral, insoluble polymer. Maize glucose, cassava glucose, and maize maltose were used in fermentation medium to produce insoluble polysaccharide. Two *Agrobacterium* sp. strains which were used (ATCC 31749 and IFO 13140) in the production of insoluble exopolysaccharide presented equal or superior yields compared to the literature. The strain ATCC 31749 yielded better production when using maize maltose, whose yield was 85%, whereas strain IFO 13140 produced more when fed maize glucose, producing a yield of 50% (on reducing sugars).

Index Entries: Microbial exopolysaccharides; *Agrobacterium* sp.; microbial gums; curdlan; fermentation.

Introduction

An insoluble and extracellular microbial polysaccharide composed completely of D-glucose residues linked by β-D-1,3, was isolated and identified through Gram-negative bacteria culture, *Alcaligenes faecalis, myxogenes* variety, mutant 10C3K, currently classified as ATCC of *Agrobacterium* sp. The exopolysaccharide (EPS) was tested and found to be soluble and form a gel in a warmed aqueous suspension and was named "curdlan" *(1–3)*. This gum was approved by the Food and Drug Administration (FDA) in 1996, being used in food industry owing to its ability of producing an excellent, hard, and resistant gel.

*Author to whom all correspondence and reprint requests should be addressed.

The production of curdlan can be done through continuous and batch systems, using a chemically defined medium, having a carbohydrate as the main source of carbon. The glucose is the most commonly used carbohydrate in different concentrations: varying from 4% to 8% (4–6), producing about 50% gum on a mass basis. Lee et al. (7) obtained excellent results using 10% of maltose (yield of 48%) and sucrose (yield of 47%) as carbon sources in the production medium. Using the same concentration of glucose in the evaluated conditions, those researchers reached a yield of 40%. They used fructose, galactose, lactose, and raffinose too, but with lower yield. According to Sutherland (8), the hydrolyzed starch and glucose are on a large scale accepted by industry for microbial transformation, although they are found in different levels of purity. They are available substrates worldwide and in general, have lower prices. This fact is valid for sugarcane molasses, which need a clarification process before use. So more, research to find a more economical carbon source is essential for industrial production of the polysaccharide.

The gum is easily separated from the fermentation medium owing to its solubility in the alkaline medium (9,10), eliminating the use of solvents very common in the separation of others polysaccharides. The carbon source used in fermentation is the most expensive part of the process.

Materials and Methods

Microorganism

It was acquired two lyophilized strains of bacterium *Agrobacterium* sp. from the American Type Culture Collection (ATCC 31749, United States) and from the Institute for Fermentation of Osaka (IFO 13140, Japan).

Carbon Sources

Cassava's glucose syrup: Commercial food product, furnished by Indemil Indústria e Comércio de Milho Ltda, Paranavaí, Pr, Brazil (commercial label: Manicandy® 4084). The composition declared by producer is presented in the Table 1. Maize's glucose syrup: commercial food product, furnished by Corn Products Brasil Ingredientes Industriais, Balsa Nova, Pr, Brazil (commercial label: Excell® 1040). The composition declared by producer is presented in the Table 1. Maize's high maltose syrup: commercial food product, furnished by Corn Products Brasil Ingredientes Industriais, Balsa Nova, Pr, Brazil (commercial label: Mor-sweet® 1557). The composition declared by producer is presented in the Table 1.

Production of EPS

The production of the insoluble EPS was carried out in flasks Schott. Cultivation at 30°C for 15 d at 120 rpm. A chemically defined medium was

Table 1
Characteristics of the Glucose and Maltose Syrups of Commercial Origin

Characteristic	Cassava's glucose syrup Manicandy®	Maize's glucose syrup Excell®	Maize's high-maltose syrup Mor-sweet®
Substance dries (%)	83–84.5	81–83	85
Equivalent dextrose (%)	37–41	38–40	n.c.
pH	4.5	4.5–5.5	4.5–5.5
SO_2 (ppm)	150	150	n.c.
Dextrose (%)	16.9	15	12
Maltose (%)	13.2	12	42
Other sugars (%)	69.9	73	46

n.c., not certain. The symbol ® stands for commercial products trademark.

used, described by Nakanishi et al. (4) (g/100 mL): Glucose, 4; $(NH_4)_2HPO_4$, 0.15; KH_2PO_4, 0.10; $MgSO_4$, 0.05; $Fe_2(SO4)_3 \cdot 7H_2O$, 0.005; $MnSO_4 \cdot nH_2O$, 0.002; $CoCl_2 \cdot 6H_2O$, 0.001; $ZnCl_2$, 0.001; $CaCO_3$, 0.30. The pH was adjusted to 7.0. Commercial products of this study substituted the glucose. The quantity syrups used was calculated based on total reducing sugar (TRS) concentration. The samples were collected at predetermined periods to determine TRSs.

Total Reducing Sugars

The technique described by Nelson (11) was used for TRS determination.

Polysaccharide Recovery

The production medium containing insoluble EPS was centrifuged and the sediment was solubilized in a solution of 1 N NaOH. After centrifugation the sediment (cells and insoluble solids) was discarded and the supernatant (solubilized EPS) was precipitated by neutralization, using a solution of 1 N HCl. The insoluble material was water washed and dried by lyophilization. The percent yield was calculated based on the EPS weight (g) obtained compared with the TRS weight (g) consumed in the process.

Results and Discussion

The two strains studied responded differently during production with commercial sources of carbon. Figures 1 and 2 show the TRS consumption of each commercial product and the laboratory glucose.

When using maize glucose ("Excell") as carbon source, the two strains showed a satisfactory production of EPS, considering yield of 55% for strain ATCC 31749 and 50% for strain IFO 13140 based on TRS. The strain IFO 13140 showed the best result for the three alternative sources researched.

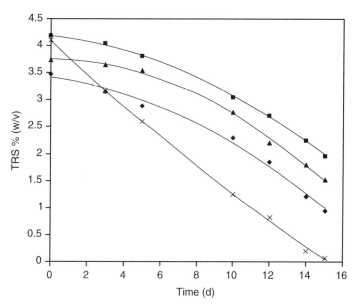

Fig. 1. Observation of the consumption of total reducing sugars (TRSs) during insoluble polysaccharide production by strain ATCC 31749 of *Agrobacterium* sp. in a medium made with different carbon sources. ◆, cassava glucose; ■, maize maltose; ▲, maize glucose; ×, laboratory glucose—control.

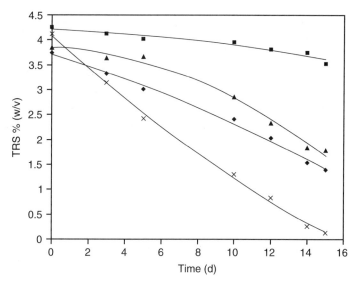

Fig. 2. Observation of the consumption of total reducing sugars (TRSs) during insoluble polysaccharide production by IFO 13140 de *Agrobacterium* sp. in a medium made with different carbon sources. ◆, cassava glucose; ■, maize maltose; ▲, maize glucose; ×, laboratory glucose—control.

The two strains were able to use the carbohydrates presented in the commercial product "Mor-sweet" (maize maltose). Strain ATCC 31749 was more efficient than the other ones, showing a yield of 85%, whereas the strain IFO presented a yield of 48% of insoluble EPS on the consumed TRS during fermentation.

The use of cassava glucose ("Manicandy") produced insoluble EPS with the two strains employed too, but it was indeed the commercial product which had lower yield results, as follows of 44% on the consumed TRS for strain ATCC 31749 and 45% for strain IFO 13140.

Conclusion

Although the commercial carbohydrates used did not have a highly defined chemical composition, they were useful in the production of insoluble EPS from *Agrobacterium* sp. However, the two strains of *Agrobacterium* sp. did not consume completely the present sugars in the commercial products. Lee et al. *(2)* cite monosaccharides (glucose, fructose, and galactose), disaccharides (maltose and sucrose), and a trisaccharide (raffinose) as usual carbon sources used by *Agrobacterium* sp. in the curdlan production. In the composition of the studied commercial products, some carbohydrates are not reported in literature as producers of the polysaccharide (*see* Table 1; "other sugars" are trisaccharide, oligosaccharides, and dextrins). These sugars were not metabolized in the medium and, therefore, they were not converted into polysaccharides. It is also possible that certain carbohydrates of the medium can control the curdlan biosynthesis (rarely explored in literature), which explains the low sugar consumption in the medium of production.

The commercial products used show relatively lower cost than the commercial glucose cost—50% Excell, 20% Mor-sweet, and 60% Manicandy—cheaper than the glucose. However, the amount used for producing culture medium resulted in a higher final cost of polysaccharide compared with the cost of glucose used as carbon source. The latter was completely metabolized by the strains used in this study and the commercial carbohydrates were not consumed completely by microorganism, which does not justify its use in the commercial processes.

Acknowledgments

The authors acknowledge the Fundação Coordenação de Aperfeiçoamento de Pessoal de nível Superior (CAPES), Universidade Estadual de Campinas, and Universidade Estadual de Maringá.

References

1. Harada, T., Misaki, A., and Saito, H. (1968), *Arch. Biochem. Biophys.* **124,** 192–298.
2. Lee, I. K., Seo, W. T., Kim, G. J. et al. (1997), *J. Industr. Microbiol. Biotechnol.* **18(40),** 255–259.
3. Lee, J. W., Yeomans, W. G., Allen, A. L., Gross, R. A., and Kaplan, D. L. (1997), *Biotechnol. Lett.* **19(12),** 1217–1221.
4. Nakanishi, N., Kimura, I., et al. (1976), *J. Gen. Appl. Microbiol.* **22(1),** 1–11.

5. Harada, T. In: *Studies on Bacterial Gel Forming β-1,3-Glucans (Curdlan-Type Polysaccharides) in Japan*. Terui and Gyozo (eds.). *Proc. IV IFS: Ferment. Technol. Today*. Society of Fermentation Technology, Osaka, pp. 603–607.
6. Harada, T. (1974), *Proc. Biochem.* **9(1),** 21–25.
7. Lee, J. W., Yeomans, A. L. A., Kaplan, D. L., Deng, F., and Gross, R. A. (1997), *Can. J. Microbiol.* **43,** 149–156.
8. Sutherland, I. W. (1996), *Int. Biodeter. Biodegrad.* **38(3,4),** 249–261.
9. Pace, W. and Righelato, C. (1980), *Adv. Biochem. Eng.* **15,** 41–70.
10. Phillips, K. and Lawford, H. G. (1983), *Prog. Ind. Microbiol.* **18,** 201–229.
11. Nelson, P. and Norton, L. B. (1944), *J. Biol. Chem.* **183,** 375–380.

Selective Utilization of Fructose to Glucose by *Candida magnoliae*, an Erythritol Producer

Ji-Hee Yu,[1] Dae-Hee Lee,[1] Yong-Joo Oh,[1] Ki-Cheol Han,[2] Yeon-Woo Ryu,[3] and Jin-Ho Seo*,[1]

[1]School of Agricultural Biotechnology and Center for Agricultural Biomaterials, Seoul National University, Seoul 151-742, Korea, E-mail: jhseo94@snu.ac.kr; [2]Korea Institute of Science and Technology, Seoul 130-650, Korea; and [3]Department of Molecular Science and Technology, Ajou University, Suwon 442-749, Korea

Abstract

Candida magnoliae isolated from honeycomb is an industrially important yeast with high erythritol-producing ability. Erythritol has been used as functional sugar substitute for various foods. In order to analyze the physiological properties of *C. magnoliae*, a study on sugar utilization pattern was carried out. The fermentation kinetics of glucose and fructose revealed that *C. magnoliae* has the discrepancy in glucose and fructose utilization when it produces erythritol. In contrast to most yeasts, *C. magnoliae* showed preference for fructose to glucose as a carbon source, deserving the designation of fructophilic yeast. Such a peculiar pattern of sugar utilization in *C. magnoliae* seems to be related to the evolutionary environment.

Index Entries: *Candida magnoliae*; fructophilic yeast; fructose utilization; erythritol.

Introduction

Erythritol is a four-carbon sugar alcohol used as a food ingredient with taste and mouthfeel-enhancing properties. It is a naturally occurring sweetener in various fruits and fermented foods including grape, wine, beer, and soy sauce *(1)*. Erythritol has 60–80% sweetness relative to sucrose and is a good low-calorie sweetener *(2)*. Interestingly, more than 90% of ingested erythritol is not metabolized by the human body and is excreted in the urine without changing blood glucose and insulin levels *(3)*. Therefore, the four-carbon sugar alcohol might be advantageously used as a functional sugar substitute in special foods for people with diabetes and obesity *(4)*. Erythritol has been approved in the United States and used as

*Author to whom all correspondence and reprint requests should be addressed.

a flavor enhancer, nutritive sweetener, and sugar substitute (5). In Japan, it has been used since 1990 as a sugar substitute for various foods such as candies, soft drinks, chewing gum, jams, and yogurt (6).

Candida magnoliae is a high erythritol-producing strain isolated from honeycomb (Deposition No.KFCC 11023). To improve the erythritol-producing ability, the parental wild strain was mutated by ultraviolet irradiation and nitrosoguanidine treatment to give a mutant strain (7,8). The optimized fed-batch fermentation using the mutant strain of C. magnoliae resulted in 200 g/L erythritol concentration, 1.2 g/L/h productivity, and 0.43 g/g yield (9). The genome of C. magnoliae was not sequenced yet and hence information was not sufficient to explore the erythritol production pathway at a gene level. In our previous reports, the expression levels of metabolic enzymes illustrated on two-dimensional electrophoresis (2-DE) served to infer the physiological properties of C. magnoliae at the protein level (10–12). C. magnoliae showed unique physiological characteristics compared with typical yeasts including Saccharomyces cerevisiae. It is able to grow in a wide range of pH values in the presence of high concentrations of sugars (8). Unlike S. cerevisiae, C. magnoliae does not produce ethanol, but produces erythritol and glycerol. The most interesting characteristic of C. magnoliae deserved by our attention is a peculiar preference of fructose to glucose as a carbon source. More investigations on the mode of sugar metabolism by C. magnoliae are desirable to use this strain for erythritol production in an industrial scale.

Not many studies on the preference for glucose or fructose in yeast strains have been performed. *Candida shellata* and *Zygosaccharomyces bailii* have a preference for fructose, whereas *S. cerevisiae* in general appears to be glucophilic (13,14). The aim of this work was to investigate in more detail the sugar metabolism of *C. magnoliae* by analyzing fermentation kinetics under various sugar-containing media. Specifically, growth rate, biomass yield, and production of metabolites such as erythritol and glycerol were measured and compared.

Materials and Methods

Yeast Strains and Culture Conditions

C. magnoliae KFCC 11023 wild type was used for this study. The strain was isolated from honeycomb and is osmotolerant to produce erythritol as a major product. *C. magnoliae* was propagated and cultured under laboratory conditions at 30°C in medium, which contained 300 g/L glucose, 10 g/L yeast extract, and 20 g/L bactopeptone. Solid medium was supplemented with 20 g/L agar.

Fermentation Media and Condition

Seed culture for fermentation was performed in a 500-mL baffled flask containing 50-mL growth medium at 30°C, 10*g*, and for 24 h. The growth medium was composed of 10 g/L yeast extract and 20 g/L bactopeptone supplemented with one of the various carbon sources (glucose, fructose, sucrose, and a mixture of glucose and fructose). Batch culture in a fermentor was performed with 2.5-L jar fermentors (KoBioTech, Incheon, Korea) containing 1-L fermentation medium at 30°C with agitation at $40g$ and aeration at 1 vvm. Concentrations of sugars used as carbon source were 300 g/L glucose, 300 g/L fructose, 300 g/L sucrose, and a mixture of 150 g/L glucose and 150 g/L fructose. All experiments were done in duplicate under the same conditions.

Analytical Methods

Dry cell weight was estimated using a calibration curve derived from the relationship between the absorbance at 600 nm and cell dry weight. The conversion factor of absorbance to dry cell weight was 0.26 for the *C. magnoliae* wild strain. Optical density was measured with a spectrophotometer (UltroSpec 2000, Pharmacia Biotech, NJ). Concentrations of glucose, fructose, sucrose, erythritol, and glycerol were determined by high-performance liquid chromatography (HPLC; Knauer, Berlin, Germany) equipped with the Aminex HPX-87H column (Bio-Rad, Richmond, CA) at 65°C. The flow rate of a mobile phase, $0.01N\ H_2SO_4$, was set at 0.6 mL/min. Detection was done using a differential reflective index detector (Knauer, Germany).

Results

Evaluation of Carbon Utilization Patterns

Experimental results for batch fermentations with initial glucose or fructose concentration of 300 g/L were illustrated in Figs. 1 and 2, respectively. Glucose concentration in the medium declined at a constant rate, 0.8 g/L/h, after the lag phase of cell growth. Cells grew exponentially until 45 h and then grew up to 42 g/L; 4 g/L erythritol was produced and glycerol was not formed when grown in the glucose medium. Cells in the fructose medium grew exponentially until fructose was consumed about 30% of the initial concentration and then the biomass slightly increased, actively producing erythritol and glycerol during the stationary phase (Fig. 2). Finally, the cells grew up to 54 g/L and erythritol was produced at a constant rate of 3.0 g/L/h from the early exponential phase of cell growth until 100% of fructose initially added was depleted. This value represents a 3.8-fold increase in fructose consumption rate compared with glucose consumption rate. The final concentration of erythritol in the 300 g/L fructose medium was 85 g/L, which is about 21 times higher than that produced with 300 g/L glucose. Glycerol was produced in proportion to the cell mass during the exponential phase of cell growth but was not consumed later as a carbons source. The final concentration of glycerol produced was 77 g/L, which is almost the same amount of erythritol. The residual concentrations of sugars after the end of the fermentation were

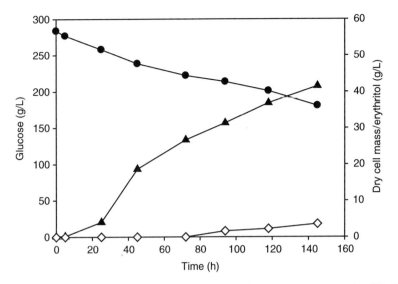

Fig. 1. Fermentation profiles of *C. magnoliae* wild type grown in 300 g/L glucose: dry cell mass (▲), glucose (●), and erythritol (◊).

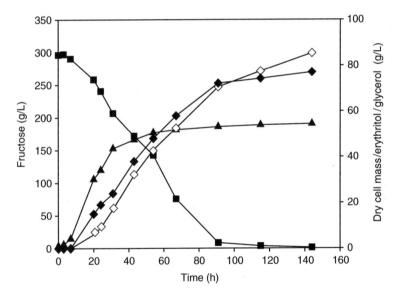

Fig. 2. Fermentation profiles of *C. magnoliae* wild type grown in 300 g/L fructose: dry cell mass (▲), fructose (■), erythritol (◊), and glycerol (♦).

181 g/L glucose for the glucose medium and 1 g/L fructose for the fructose medium in 144 h, indicating *C. magnoliae* was not able to utilize glucose completely, whereas it was able to use high concentration of fructose when it produced erythritol. Fermentation parameters obtained with 300 g/L glucose or fructose was summarized in Table 1. These results clearly showed that fructose favors erythritol production and cell growth in *C. magnoliae*.

Table 1
Summarized Results of Batch Fermentations by *C. magnoliae* Grown in Various Compositions of Sugars

Sugar	Concentration (g/L)	Maximum dry cell mass (g/L)	Maximum erythritol concentration (g/L)	Specific growth rate, μ (/h)	Erythritol productivity (g/L/h)	Erythritol yield (g/g sugar)	Sugar consumption rate (g/L/h)
Glucose	300	42	4	0.068	0.05	0.086	0.81
Fructose	300	54	85	0.097	1	0.29	3.24
Glucose + Fructose	150 + 150	50	50	0.19	1	0.29	2.96 (Fructose) 0.71 (Glucose)
Sucrose	300	63	65	0.10	1	0.21	3.21 (Fructose) 0.18 (Glucose)

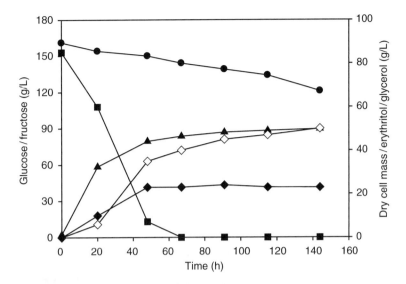

Fig. 3. Fermentation profiles of *C. magnoliae* wild type grown in a mixture of 150 g/L glucose and 150 g/L fructose: dry cell mass (▲), glucose (●), fructose (■), erythritol (◇), and glycerol (◆).

Selective Utilization of Fructose to Glucose

To confirm clearly the selective utilization of fructose by *C. magnoliae*, the batch fermentations were performed with 300 g/L sucrose or a mixture of 150 g/L glucose and 150 g/L fructose. A fermentation profile in the sugar mixture was shown in Fig. 3. Even though the same amount of glucose and fructose was added initially, the selective utilization of fructose compared with glucose led to a discrepancy between the amount of glucose and fructose consumed (GF discrepancy) during the linearly erythritol-producing period of the fermentation. Right after inoculation, a significant amount of fructose was taken up at a constant rate of 2.9 g/L/h. This initial rapid uptake was followed by a period of almost depletion of fructose. Toward the end of fermentation, when glucose became more limiting, the GF discrepancy decreased. When the fermentation finally ceased, glucose was found at significantly higher concentrations than fructose. The GF discrepancy increased almost linearly during the fermentation (Fig. 4). The GF discrepancy determined after depletion of fructose was disregarded and focused on the fermentation stages when fructose was linearly consumed. A fermentation pattern for 300 g/L sucrose was displayed in Fig. 5. The results showed the *C. magnoliae* had the ability to hydrolyze sucrose to glucose and fructose by its invertase. *C. magnoliae* produced mainly extracellular invertase like other sucrose-consuming yeasts during the fermentation (data not shown). After the lag phase of cell growth, fast hydrolysis of sucrose was observed, resulting in accumulation of glucose and fructose. In the beginning of the fermentation, fructose also started to accumulate because its production by

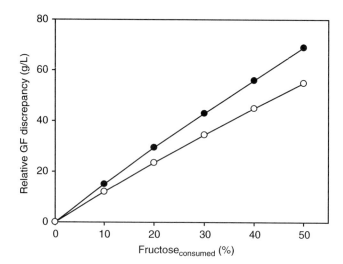

Fig. 4. Relative concentration of glucose to fructose consumed with fermentation time for the 300 g/L sucrose medium (○) and the mixture of 150 g/L glucose and 150 g/L fructose medium (●).

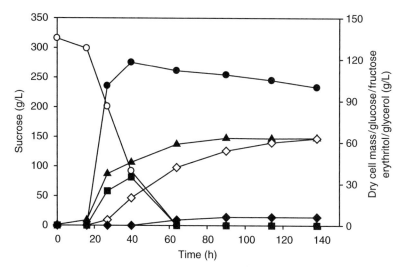

Fig. 5. Fermentation profiles of *C. magnoliae* wild type grown in 300 g/L sucrose: dry cell mass (▲), sucrose (○), glucose (●), fructose (■), erythritol (◊), and glycerol (◆).

hydrolysis of sucrose was faster than its consumption. However, glucose continuously accumulated up to approx 120 g/L until the residual fructose was almost consumed. After depletion of fructose, glucose was used and the fermentation profile is similar to that in the medium containing glucose alone. Erythritol increased to 65 g/L, with an erythritol yield of 0.21 g/g of fructose and glucose consumed. Glycerol was produced as well. Fermentation results of *C. magnoliae* in batch cultures were summarized in Table 1. The rate of fructose consumption was higher than that of glucose in both sucrose and a mixture of glucose and fructose. In both fermentations, fructose was first used up followed by consumption of glucose by C. magnoliae (Figs. 3 and 5). Interestingly, the diauxic growth was not observed in disagreement with the cases typically in which two carbon sources were present in the same growth medium. C. magnoliae showed preference to fructose when using a mixture of glucose, fructose, and sucrose as carbon source. The GF discrepancy in a mixture of sugars was also higher than that in the sucrose medium (Fig. 4). These results are consistent with fructophilic behavior described previously for other strains such as C. shellata and Z. baillii (13,14).

Discussion

The experimental results with C. magnoliae isolated from honeycomb have shown clearly the selective utilization of fructose relative to glucose, indicating C. magnoliae to be fructophilic. The exact reason for such an observation needs to be characterized. In S. cerevisiae, fructose is used concomitantly with glucose, the latter is used first, which gives rise to a discrepancy between glucose and fructose (15,16). The metabolic pathway of fructose utilization is very similar to the glucose assimilation pathway. The transporters are shared although their affinity for glucose is higher than for fructose, but V_{max} for the two sugars is similar. Hence, the transport step could be a reason for selective utilization of glucose in S. cerevisiae. For C. magnoliae, however, glucose was not consumed when fructose was present in the growth medium. Research efforts to characterize hexose transporters in yeasts usually deal with glucose as substrate, but most transporters characterized so far do not discriminate between glucose and fructose. Recently, the fructose-specific transporter of Z. bailii was cloned and characterized by functional complementation of S. cerevisiae incapable of growth on hexoses (17). Like C. magnoliae, Z. bailii consumed fructose faster than glucose (14). In Z. bailii, fructose promoted the inactivation of the glucose transporter, preventing the utilization of glucose when fructose was also available. Moreover, fructose crosses the plasma membrane through fructose-specific transporter at a higher rate than glucose and can be metabolized faster when fructose is present in high concentrations. After transport, glucose is phosphorylated by glucokinase and hexokinases 1 and 2, whereas fructose is only phosphorylated by the latter two enzymes in S. cerevisiae (18). The affinity of the hexokinases is higher for glucose. Hence, the phosphorylation step could be another reason for the cause of the discrepancy in glucose and fructose fermentation. Fructose is a ketose sugar, nearly 30% of which is present in the furanose form in solution (19), whereas glucose is an aldose, nearly 99.9% of which is present in the pyranose form (20). Because glucose and other sugars are transported in the pyranose form rather than in the furanose form, the actual transport-competent concentration of fructose is below its total concentration *(21)*. Differences in physicochemical properties like these may explain the lower affinity for fructose of the transport system *(22)* and the hexokinases *(23,24)*. After phosphorylation, fructose-6-phosphate readily enters glycolysis by conversion into fructose-1,6-bisphosphate, whereas glucose-6-phosphate still has to be converted first into fructose-6-phosphate by phosphogluco-isomerase. The cause of the GF discrepancy therefore appears to be located in the transport and/or phosphorylation steps of the fermentation pathway.

In conclusion, the selective utilization of fructose by *C. magnoliae* was observed especially when the yeast produced erythritol from sucrose or a mixture of glucose and fructose. More research has to be done to elucidate the fructophilic behavior of *C. magnoliae* at a gene or protein level.

Acknowledgments

This work was supported by the Ministry of Commerce, Industry, and Energy, and by the Ministry of Education and Human Resources Development through the BK21 program.

References

1. Goossen, J. and RÖper, H. (1994), *Confect. Prod.* **24**, 182–188.
2. De Cock, P. (1999), *World Rev. Nutr. Diab.* **85**, 110–116.
3. Noda, K. and Oku, T. (1992), *J. Nutr.* **122**, 1266–1272.
4. Hiele, M., Ghoos, Y., Rutgeerts, P., and Vantrappen, G. (1993), *Br. J. Nutr.* **69**, 169–176.
5. U.S. Food and Drug Administration, Center for Food Safety and Applied Nutrition, Office of Food Additives Safety: Agency Response Letter: GRAS Notice No. GRN 000076, September 2001.
6. Scientific Committee of European Commission (2003), Scientific opinion on erythritol.
7. Ryu, Y. W., Park, Y. C., Park, J. B., Kim, S. Y., and Seo, J. H. (2000), *J. Indust. Microbiol. Biotechnol.* **25**, 100–103.
8. Lee, K. H., Seo, J. H., and Ryu, Y. W. (2002), *Kor. J. Biotechnol. Bioeng.* **17**, 509–514.
9. Koh, E. S., Lee, T. H., Lee, D. Y., Kim, H. J., Ryu, Y. W., and Seo, J. H. (2003), *Biotechnol. Lett.* **25**, 2103–2105.
10. Park, Y. C., Lee, D. Y., Lee, D. H., Kim, H. J., Ryu, Y. W., and Seo, J. H. (2005), *J. Chromatogr. B*, **815**, 251–260.
11. Lee, D. Y., Park, Y. C., Kim, H. J., Ryu, Y. W., and Seo, J. H. (2003), *Proteomics* **3**, 2330–2338.
12. Kim, H. J., Lee, D. Y., Lee, D. H., et al. (2004), *Proteomics* **4**, 3588–3599.
13. Ciani, M., Ferraro, L., and Fatichenti, F. (2000), *Enzyme Microb. Technol.* **27**, 698–703.
14. Sousa-Dias, S., Goncalves, T., Leyva, J. S., Peinado, J. M., and Loureiro-Dias, M. C. (1996), *Microbiology* **142**, 1733–1738.
15. Trumbly, R. J. (1992), *Mol. Microbiol.* **6**, 5–21.
16. Ronne, H. (1995), *Trends Genet.* **11**, 12–17.
17. Pina, C., Goncalves, P., Prista, C., and Loureiro-Dias, M. C. (2004), *Microbiology* **150**, 2429–2433.
18. Barnett, J. A. (1997), Sugar Utilization by *Saccharomyces cerevisiae*. In: Zimmermann, F. K. and Entian, K. D. (eds.), *Yeast Sugar Metabolism*, Technomic Publishing, Switzerland, pp. 35–43.
19. Flood, A. E., Johns, M. R., and White, E. T. (1996), *Carb. Res.* **288**, 45–56.

20. Hyvöen, L., Varo, P., and Koivistoinen, P. (1977), *J. Food Sci.* **42,** 654–656.
21. Bisson, L. F. (1999), *Am. J. Enol. Vitic.* **50,** 107–119.
22. Reifenberger, E., Boles, E., and Ciriacy, M. (1997), *Eur. J. Biochem.* **245,** 324–333.
23. Sols, A., De La Fuente, G., Villar-Palasi, C., and Asensio, C. (1958), *Biochim. Biophys. Acta* **30,** 92–101.
24. Rose, M., Albig, W., and Entian, K. D. (1991), *Eur. J. Biochem.* **199,** 511–518.

Biosurfactant Production by *Rhodococcus erythropolis* Grown on Glycerol As Sole Carbon Source

Elisa M. P. Ciapina,[1] Walber C. Melo,[1] Lidia M. M. Santa Anna,[2] Alexandre S. Santos,[1] Denise M. G. Freire,[3] and Nei Pereira, Jr.*,[1]

[1]*Departamento de Bioquímica, Universidade Federal do Rio de Janeiro, Centro de Tecnologia, Bloco E, Rio de Janeiro, RJ, CEP 21949-900, Brasil, E-mail: nei@eq.ufrj.br;* [2]*Cenpes, Petrobrás, Gerência de Biotecnologia e Tratamentos Ambientais, Avenida Jequitibá 950, Ilha do Fundão, Rio de Janeiro, RJ, Brasil; and* [3]*Instituto de Química, Universidade Federal do Rio de Janeiro, Centro de Tecnologia, Bloco A, Rio de Janeiro, RJ, Brasil*

Abstract

The production of biosurfactant by *Rhodococcus erythropolis* during the growth on glycerol was investigated. The process was carried out at 28°C in a 1.5-L bioreactor using glycerol as carbon source. The bioprocess was monitored through measurements of biosurfactant concentration and glycerol consumption. After 51 h of cultivation, 1.7 g/L of biosurfactant, surface, and interfacial tensions values (with *n*-hexadecane) of 43 and 15 mN/m, respectively, 67% of Emulsifying Index (E_{24}), and 94% of oil removal were obtained. The use of glycerol rather than what happens with hydrophobic carbon source allowed the release of the biosurfactant, originally associated to the cell wall.

Index entries: Biosurfactant; *Rhodococcus*; oil removal.

Introduction

Biosurfactants are amphipathic molecules that can be divided in low-molecular-weight compounds such as glycolipids, phospholipids, and lipopeptides, which present lower interfacial tension and high-molecular-weight biosurfactants such as polysaccharides, proteins, lipoproteins, or complex of these biopolymers *(1)*. These high-molecular-weight compounds are associated with the production of stable emulsions, but the lowering of the surface tension or interfacial tension is not a usual trait of them and is called as bioemulsifiers *(2)*.

*Author to whom all correspondence and reprint requests should be addressed.

The increasing interest in the potential applications of microbial surface-active compounds is based on their broad range of functional properties that include wetting, foaming, emulsification, viscosity reduction, phase separation, and solubilization. There are many areas of industrial application in which chemical surfactants could be substituted by biosurfactants in several fields as industrial cleaning, agriculture, construction, food, paper and metal industries, textiles, cosmetics, pharmaceutical, and petroleum and petrochemical industries, including applications in environmental bioremediation. Biosurfactants have gained attention because of their biodegradability, low toxicity, ecological acceptability, and ability to be produced from renewable and cheaper substrates *(3)*.

Surfactants are produced by a variety of microbes (bacteria and fungi), secreted either extracellularly or attached to parts of cells and are often only produced when growing on *n*-alkanes or other water-immiscible substrates to facilitate adhesion to hydrophobic substrates. However, some microbial surfactants can be produced on water-soluble growth substrates *(4)*.

The genus *Rhodococcus* is a group of Gram-positive bacteria exhibiting a diverse range of metabolic activities. Some of them have the ability to degrade a variety of organic compounds including polychlorinated biphenyls and aliphatic and aromatic hydrocarbons. This property is often accompanied by the ability to produce biosurfactants *(5)*. These molecules are predominantly glycolipids but other types have also been reported such as polysaccharides associated to the cell wall *(6,7)*.

Many of the potential applications that have been considered for biosurfactants depend on whether these can be produced economically in commercial quantities. The parameters that affected the economics of biosurfactant manufacture include the choice of nutrients and strain *(4)*. The purpose of this study was to investigate the production of biosurfactant by *Rhodococcus erythropolis* on glycerol and its application in hydrocarbon removal from oily sludge.

Materials and Methods

Bacteria

R. erythropolis ATCC 4277 was obtained from bacteria collection of the Department of Microbiology, University of São Paulo, Brazil.

Media and Growth Conditions

Inoculum

The inoculum was grown in sterilized medium containing 0.3% yeast extract; 1.5% peptone; and 0.1% glycerol, at 28°C in a rotary shaker at 200 rpm (throw = 5 cm), during 24 h of cultivation. The cells were centrifuged (10,000g, 10 min) and used as inoculum for shake flask and bioreactor experiments.

Shake Flask Experiments

The basal medium used for all culture experiments contained distilled water in the following amounts: 2 g/L KH_2PO_4, 1 g/L KNO_3, 2 g/L K_2HPO_4, 2 g/L $(NH_4)_2SO_4$, 1 g/L NaCl, 0.2 g/L $MgSO_4$, 0.02 g/L $CaCl_2$ $2H_2O$, 0.01 g/L $FeCl_3$ $7H_2O$, and 0.1 g/L yeast extract. The pH was adjusted to 7.0. The medium was sterilized for 15 min at 121°C. Glycerol and hexadecane were investigated as carbon sources to a final concentration of 1.5% (w/v). The experiments were performed at 28°C, at 200 rpm and on an orbital shaker for 1 wk.

Batch Experiment in Bioreactor

The basal medium containing glycerol (1.5%) was used. The bioreactor was run at 28°C, with a constant dissolved oxygen level of 30% of saturation in the fermentation broth (maintained by automatic stirred control). The pH was maintained at 7.0 by automatic addition of acid and base solutions. The fermenter used was a Biostat® B 2.0 L model (B. Braun Biotech International, Germany).

Biomass Content

The biomass growth was monitored by measurements of the optical density at 600 nm of the cell suspension, followed by normalizing to the dry cell weight from a calibration plot.

Glycerol Concentration

The glycerol concentration was analyzed by enzymatic-colorimetric assay using a triglycerides kit (GPO/POD, CELM/Brazil).

Determination of Surfactant Production

The production of biosurfactant was estimated by the phenol-sulfuric method *(8)*. The choice of this method for surfactant quantitative determination was determined by the ability of *Rhodococcus* sp. to produce glycolipids and polysaccharides class surfactant (both with carbohydrates portion). The concentration of carbohydrate was reported as biosurfactant concentration.

Surface Characteristics

The whole broth of the bioreactor was sonicated (10 KHz, 10 min) before measurement for the release of the biosurfactant attached to the cell wall. The cells were removed by centrifugation at 10,000*g* for 15 min. The surface and interfacial tension of the cell-free spent medium was measured with a Du Nuoy ring tensiometer (Kruss Tensiometer, K9 model, Germany). Interfacial tension was measured with *n*-hexadecane as the oil phase. Emulsifying activity was estimated by the method of Cooper and

Table 1
Levels of Variables in Experimental Design to Hydrocarbons Removal Experiment

Variables	Low level (–)	Central point (0)	High level (+)
Oily sludge (g)	0.5	2	3.5
Agitation (rpm, throw = 5 cm)	100	200	300
Time (min)	30	75	120
Biosurfactant (g/L)	0.5	2.25	4

Goldenger (9). Two milliliters of a cell-free spent medium were added to 2 mL of n-hexadecane and vortex for 2 min and left to stand for 24 h. The Emulsification Index (E_{24}) was given as percentage of height of the emulsified layer (mm) divided by the total height of the liquid column (mm).

Evaluation of the Biosurfactant in Hydrocarbon Removal

The studies of submerged washing of oily sludge for hydrocarbon removal using the spent medium containing the biosurfactant were performed according to Urum (10). The experiments were carried out following the experimental design as shown in Table 1, in which the variables investigated were oily sludge amount, agitation speed, process time, and biosurfactant concentration. The software STATISTICA (version 5.0, Statsoft Inc., Tulsa, OK) was used to generate the experimental design and the results were analyzed utilizing as response variable the removal of total hydrocarbons.

Results and Discussion

Shake Flask Experiments

Several works have reported the production of biosurfactant by *Rhodococcus* sp. using n-alkanes as the carbon source (11–13), and observed the association of surfactants with the cell wall. According to Lang and Philp (7) only a minor portion of the produced surfactants is released.

In this study, a hydrophilic and hydrophobic carbon sources were evaluated in the production of biological active surface compounds. Biosurfactant concentrations of 0.4 and 0.1 g/L on glycerol and hexadecane, respectively, were obtained. In order to increase biosurfactant content, the effect of sonication was investigated. There was an increase of fourfold in surfactant release using sonication with the culture grown on hexadecane (0.4 g/L) and no significant difference (0.45 g/L) was observed with that culture grown on glycerol. Although production of surface-active lipids by *Rhodococcus* can be induced by the presence of n-akanes in culture medium (7), these results showed that glycerol was able to promote the production of biosurfactant and allowed its release.

Physiologically, the production of biosurfactant is associated with the assimilatory mechanism to hydrophobic substrates. This mechanism

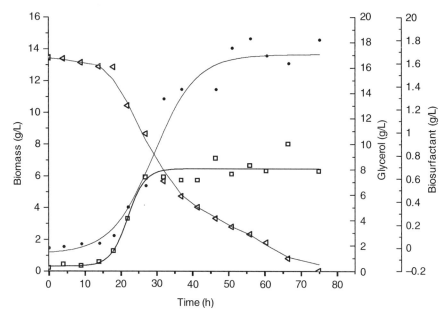

Fig. 1. Time-course of growth, biosurfactant production, and glycerol consumption during cultivation of *R. erythropolis* in bioreactor (□, biomass [g/L]; •, biosurfactant [g/L]; △, glycerol [g/L]).

would consist in direct contact of cells with large oil droplet with little or no emulsification, or the contact with fine oil droplet, culminating in emulsification. In the first, the biosurfactant is retained on the outer cell surface, facilitating the attachment and subsequent transport of hydrophobic compounds to the cell inside *(14)*. In the second case the free biosurfactant, released in the culture medium would form a hydrocarbon–surfactant complex that pseudosolubilize the substrates and hence increase availability to the cell *(15)*.

Batch Experiment in Bioreactor

Figure 1 shows the cell growth on glycerol and the synthesis of the biosurfactant. The exponential growth of *R. erythropolis* commenced at approx 14 h with a specific growth rate (μ_x) of 0.184/h, corresponding to a doubling time of 3.8 h. After 27 h, the culture achieved the stationary phase and the glycerol consumption rate was reduced from 0.64 to 0.15 g/L h. It is well known that the entrance in the stationary growth phase is associated with the depletion of any nutrient, even though carbon source is still available. The production of the biosurfactant started in early exponential growth phase and continued even when growth had ceased indicating a typical growth-semiassociated production. The biosurfactant concentration increased approximately threefold during stationary growth phase, accumulating 1.7 g/L after 51 h. In the end of the process, the volumetric productivity and the biosurfactant yield on cell growth were 0.033 g/L h

Table 2
Conditions and Results of the Experimental Design Method

Run	Oily sludge (g)	Agitation (rpm)	Time (min)	Biosurfactant (g/L)	Removal (%)
1	3.5	300	120	4	93.6 ± 3.1
2	3.5	300	120	0.5	57.7 ± 2.4
3	3.5	300	30	4	71.6 ± 2.2
4	3.5	300	30	0.5	22 ± 4.9
5	3.5	100	120	4	94.1 ± 2.9
6	3.5	100	120	0.5	24.9 ± 1.4
7	3.5	100	30	4	17.1 ± 1.3
8	3.5	100	30	0.5	88.9 ± 4.1
9	0.5	300	120	4	84 ± 3.2
10	0.5	300	120	0.5	87.8 ± 3.3
11	0.5	300	30	4	40 ± 2.4
12	0.5	300	30	0.5	65.1 ± 3.3
13	0.5	100	120	4	19.6 ± 0.9
14	0.5	100	120	0.5	32.6 ± 1.6
15	0.5	100	30	4	71.9 ± 1.9
16	0.5	100	30	0.5	63.8 ± 1.3
CP	2	200	75	2.25	64.3 ± 2.9

CP, central point.

and 0.28 g/g, respectively. The biosurfactant yield on glycerol consumed was 0.13 g/g.

The obtained production was equivalent to values reported for the production of biosurfactant by *Rhodococcus* sp. varying over a wide range (from 0.5 to 30 g/L) *(16)*. No excess of foam formation was observed. This is an advantage, when compared with other biosurfactant producing species, because the excessive foam formation in bioreactor is difficult to control *(17)*. In the end of the experiment, the minimum values of surface and interfacial tensions (with *n*-hexadecane) were 43 and 15 mN/m, respectively, comparable with values reported in the literature with glycerol as the carbon source *(18)*.

The E_{24} of 67% for a *n*-hexadecane-water binary system was obtained. It is a value comparable with results reported by other authors, working with *Rhodococcus* species *(12,16,19)*. This indicates that the biosurfactant produced in this study is a good emulsifier.

Evaluation of the Biosurfactant in Hydrocarbon Removal

In order to investigate the efficiency of the biosurfactant produced by this strain using glycerol as the carbon source, a preliminary experiment using the spent medium containing the surfactant was performed to verify the removal of hydrocarbons from oily sludge. The best condition pointed out by the experimental design is depicted in the experiment 5 (94.1%) (Table 2), confirming the good potential of this biosurfactant for oil recuperation and environmental remediation. After oily sludge washings, the oil phase was maintained dispersed in the water phase, characterizing the removal phenomenon associated to dispersion and dislocation processes. This effect suggests the successfully oil recuperation in conventional separation system oil/water to industrial reprocessing.

These studies demonstrated that *R. erythropolis* ATCC 4277 strain grown on glycerol produces mainly free biological surfactant with surface active properties and good ability for emulsification. Furthermore, the excellent result obtained in the application of the produced biosurfactant in the hydrocarbon removal from oily sludge lead to a subsequent research for its chemical characterization, production optimization, and applicability evaluation.

Acknowledgments

This work was financially supported by CAPES, CNPq, FAPERJ. We thank Prof. Vivian H. Pellizari, Environmental Microbial Laboratory, Biomedical Institute, University of São Paulo, USP, Brazil, for providing the strain of *R. erythropolis* ATCC 4277.

References

1. Rosenberg, E. and Ron, E. Z. (1990), *Appl. Microbiol. Biotechnol.* **52**, 154–162.
2. Bognolo, G. (1999), *Colloids Surf.* **152**, 41–52.
3. Banat, I. M., Makkar, R. S., and Cameotra, S. S. (2000), *Appl. Microbiol. Biotechnol.* **53**, 495–508.
4. Fiechter, A. (1992), *Trends Biotechnol.* **10**, 208–217.
5. Bell, K. S., Philp, J. C., Aw, K. J. W., and Christofi, N. (1998), *J. Appl. Microbiol.* **85**, 195–210.
6. Neu, T. R. and Poralla, K. (1990), *Appl. Microbiol. Biotechnol.* **32**, 521–525.
7. Lang, S. and Philp, J. C. (1998), *Antonie van leeuwenhoek* **74**, 59–70.
8. Dubois, M., Gilles, K. A., Hamilton, J. K., Rebers, P. A., and Smith, F. (1956), *Anal. Chem.* **28**, 350–356.
9. Cooper, D. G. and Goldenberg, B. G. (1987), *Appl. Environ. Microbiol.* **53**, 224–229.
10. Urum, K. and Pekdemir, T. (2004), Chemosphere **57**, 1139–1150.
11. Kim, J. S., Powalla, M., Lang, S., Wagner, F., Lünsdorf, H., and Wray, F. (1990), *J. Biotechnol.* **13**, 257–266.
12. Bicca, F. C., Fleck, L. C., and Ayub, M. Az. (1999), *Revista de Microbiol.* **30**, 231–236.
13. Philp, J. C., Kuyukina, M. S., Ivshina, I. B., et al. (2002), *Appl. Microbiol. Biotechnol.* **59**, 318–324.
14. Wagner, F. B., Behrendt, U., Bock, H., Kretschmer, A., Lang, S., and Syldatk, C. (1983), In: *Microbial Enhanced Oil Recorery*, Zajic, J. E., Cooper, D. G., Jack, T. R., and Kosaric, N. eds., Pennwell, Tulsa, Okla, pp. 55–60.
15. Beal, R. and Betts, W. B. (2000), *J. Appl. Microbiol.* **89**, 158–168.
16. Pirog, T. P., Shevchuk, T. A., Voloshina, I. N., and Karpenko, E. V. (2004), *Appl. Biochem. Microbiol.* **40**, 544–550.
17. Fiechter, A. (1992), *Pure Appl. Chem.* **64**, 1739–1743.
18. Espuny, M. J., Egido, S., Rodón, I., Manresa, A., and Mercadé, M. E. (1996), *Biotechnol. Lett.* **18**, 521–526.
19. Ivshina, I. B., Kuykina, M. S., Philp, J. C., and Christofi, N. (1998), *World J. Microbiol. Biotechnol.* **14**, 711–717.

Methane Production in a 100-L Upflow Bioreactor by Anaerobic Digestion of Farm Waste

ABHIJEET P. BOROLE,[1] K. THOMAS KLASSON,*,[2] WHITNEY RIDENOUR,[3] JUSTIN HOLLAND,[3] KHURSHEED KARIM,[4] AND MUTHANNA H. AL-DAHHAN[4]

[1]Oak Ridge National Laboratory, Oak Ridge, TN 37831-6226;
[2]Southern Regional Research Center, USDA-ARS, 1100 Robert E. Lee Blvd, New Orleans, LA 70124, E-mail: tklasson@srrc.ars.usda.gov; [3]Oak Ridge Institute for Science and Education, ORAU, Oak Ridge, TN 37830; and [4]Chemical Reaction Engineering Laboratory (CREL), Department of Chemical Engineering, Washington University, St. Louis, MO 63130

Abstract

Manure waste from dairy farms has been used for methane production for decades, however, problems such as digester failure are routine. The problem has been investigated in small scale (1–2 L) digesters in the laboratory; however, very little scale-up to intermediate scales are available. We report production of methane in a 100-L digester and the results of an investigation into the effect of partial mixing induced by gas upflow/recirculation in the digester. The digester was operated for a period of about 70 d (with 16-d hydraulic retention time) with and without the mixing induced by gas recirculation through an internal draft tube. The results show a clear effect of mixing on digester operation. Without any mixing, the digester performance deteriorated within 30–50 d, whereas with mixing continuous production of methane was observed. This study demonstrates the importance of mixing and its critical role in design of large scale anaerobic digesters.

Index Entries: Anaerobic digestion; animal manure; gas recirculation; mixing; biogas.

Introduction

Methane produced by animal wastes is a clean replacement for coal and other fossil fuels which negatively affect air quality. Animal wastes represent a large unused source of sustainable, affordable, and renewable energy. In the United States, at least one billion tons of animal wastes are generated annually which is equivalent to approx 100 Mt coal/yr (1).

---

*Author to whom all correspondence and reprint requests should be addressed.

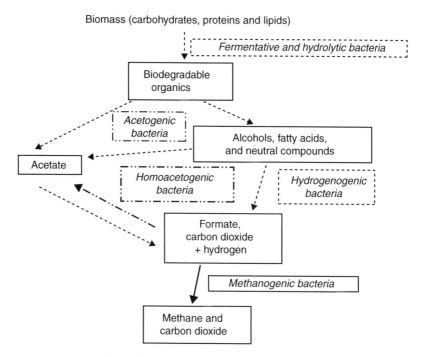

Fig. 1. Fermentation metabolic pathway *(3)*.

Methane is also a cleaner energy than traditional fossil fuels since it is compliant with policies of the Clean Air and Energy Policy Act representing a fuel source that can reduce SO_X emissions (biomass contains low amounts of sulfur), reduce NO_X emissions (biomass contains less nitrogen than coal), and reduce methane (formed in degradation of unused biomass) released into the atmosphere *(2)*. The complex organics in animal wastes produces methane through a process that includes four main microbial cultures which work together to break down the waste, producing fatty acids that are further broken down to produce methane. The fermentation process (Fig. 1) consists of two final steps in which 70% of the methane produced is metabolized from acetate and 30% from carbon dioxide reduction with hydrogen *(3)*.

Anaerobic digesters can be used to effectively process these wastes and collect the released methane; however, the effect of hydrodynamics and mixing in anaerobic digesters are not well studied. This could possibly contribute to the high-failure rate encountered in digester applications. The Department of Energy recently reviewed the history and performance of large anaerobic digesters implemented by the farming community for the treatment of animal wastes and the generation of methane for energy production *(4)*. A total of 94 digesters of various designs were investigated as part of the study. The designs were categorized as

1. Plug-flow digesters—These digesters are of simple design in form of a trough, and a slurry mixture is fed once a day to one end of the

digester. The dimensions are in the range of 1:5 (channel width to length), and the total size is determined by the size of the daily feed. An expandable cover is used to collect the biogas. The hydraulic retention time (HRT) is on the order of 20–30 d, and the solids concentration is 11–13%. The plug-flow digesters are sensitive to the amount of solids present in the feed, since the feeding of the solids to one end, provides the "pushing action" to drive the content towards the other end.

2. Complete-mix digesters—These digesters have internal mixing and are usually similar to a chemical reactor—tall, circular, heated, with good controls. They suffer from high capital and maintenance costs. Sizes range from 95,000–1,900,000 L. The concentration of the solids are 3–10% and the HRT is 10–20 d.

3. Slurry digesters—The slurry digester operates in the same solids regime as the plug flow and complete-mix digesters. They require no mechanical mixing and are often constructed in silo configurations where internal convection (from temperature gradients and gas evolution) provides mixing.

4. Covered lagoon digesters—The lagoon digesters are used to treat streams with low solids concentration (<3%). It is a popular method used for methane production, in which the manure cleaning is accomplished by a flushing mechanism, generating large volumes of low-solid waste. The HRT is on the order of 60 d, implying that the conversion rate is very slow. It can take a couple of years to reach steady-state conditions in the lagoon. A floating cover collects the methane. The cost of these digesters is low and they are not heated, making methane production very dependent on the weather conditions.

5. Miscellaneous—Other types of digesters include designs that are not yet commercialized for farm use, such as upflow sludge blanket reactors and sequencing batch reactors. These types have potential for faster processing rates, but rely on more complex designs and thus higher capital and operational costs.

Of the 94 digesters reviewed, only 74 had actually been constructed, whereas the others were still in the planning stages or were never built. Only 28 of the digesters were in operation, the others had been shut down, either as a result of operational difficulties of the digester or by termination of farm operations. Based on the data available, it was determined that the failure rate was 50% overall. In the case of plug-flow and complete-mix digesters, the failure rate was 63% and 70%, respectively. No failures were seen with the slurry digesters, but the report concluded that there were too few cases (seven in total), to get an accurate estimate of the failure rate. The majority of failures were attributed to poor designs and installation, improper equipment and incompatible choice of materials of construction, and incorrect operation and lack of maintenance. The conclusion

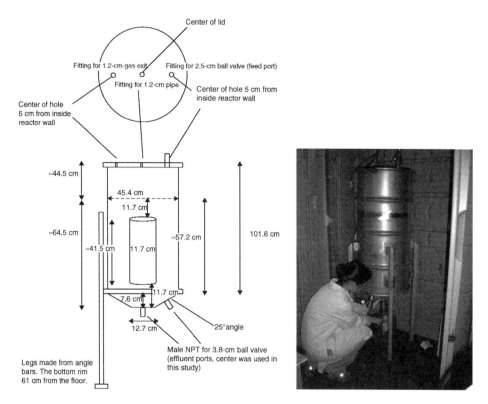

Fig. 2. Picture of the digester and collection of effluent.

of a poor design is usually indicative of inadequate mixing, resulting in plugging problems, problem obtaining desired pH balance, insufficient gas production, sand build-up, and so on. Improper choice of equipment and wrong choice of materials resulted in mechanical failures and in corrosion of materials from sulfur gases.

Thorough mixing of the substrate in the digester is regarded as essential in high-rate anaerobic digesters (5,6). The importance of mixing in achieving efficient substrate conversion has been reported by several other researchers (7–9), but the optimum mixing pattern is a subject of much debate. Some reports indicate that an intermediate degree of mixing appear to be optimal for substrate conversion (10).

In one of our previous studies (10), experiments with small-scale (4 L) digesters mixed via gas recirculation showed that the performance was independent of mixing rate and that unmixed digesters performed equally well. The purpose of the current investigation was to conduct larger-scale digester experiments and compare their performance to small-scale digesters.

Materials and Methods

The pilot-scale, stainless-steel digester held approx 96 L of bovine waste with 80 L of head space (Fig. 2). The digester was operated with a

16 d hydraulic retention time (HRT) and an average manure feed rate of 700 g total solids (TS) per day (containing approx 310 g total volatile solids [TVS]/d). These conditions were selected based on results from previous small-scale experiments conducted under different mixing conditions, amounts of TVS loaded, and hydraulic retention times (data not shown).

The digester was housed in a temperature-controlled (35°C) area (see Fig. 2). In order to provide mixing in the digester, gas was pumped from the top of the reactor by gas pumps and returned, from the top, to the digester at the lowest point of the draft tube insert. The cylindrical draft tube causes the gas to be directed through its interior which in turn causes the liquid to rise and mix. The gas recirculation rate through the digesters was approx 6 L/min. The digester was also operated under nonmixed conditions. In the mixed condition, the gas from the headspace of the digester was recirculated continuously through the digester for a period of 73 d. In the non-mixed condition, immediately following the previous phase, the digester was operated for a period of 66 d. On the 67th day, the mixing was reinitiated by starting the gas flow but other parameters remained unchanged. The digester was then operated in the mixed condition for an additional 16 d. The gas generated in the digesters was collected in Tedlar gas bags, and when these were full, the flow passed through an oil-filled wet gas meter (GSA/Precision Scientific, Chicago, Il) capable of determining cumulative gas volume produced. This allowed measurement of total gas generation between sampling events.

The digester was operated using bovine manure collected from a dairy farm in the Anderson County, TN area. The cow manure was obtained fresh (i.e., <7-d old) from grass-fed cows kept in a pasture under no antibiotic treatment (i.e., antibiotic treatment of cows limits the viability of methane generating microorganisms in the cow manure). The cow manure was then refrigerated at 4°C until use. Before feeding the digester, the manure was prepared by blending tap water and wet manure in a 1:3 ratio for 2 min with an impeller mixer and placed into a large bucket for the heavy solids (sand, and so on) to settle out. TVS were determined and the slurry was diluted as needed to an estimated 6.6% (w/v) and passed through a sieve with a 9.5-mm pore size.

Feeding events for the digester occurred every other day in which gas composition and cumulative gas production volume was determined. On the sampling day, 12 L of reactor content (effluent) was removed (see Fig. 2), and 10 L of feed and 2 L of tap water were then added to the top of the digester (see Fig. 3). Triplicate samples were collected from the feed and effluent for TS and TVS determinations. Samples were also collected for pH, total fatty acids, and alkalinity measurements after each feeding event. TS were determined by drying a known weight of slurry at 105°C overnight, and TVS by volatilization of a known weight of dried slurry at 540°C for a minimum of 45 min (11). Total alkalinity was assessed by titrating

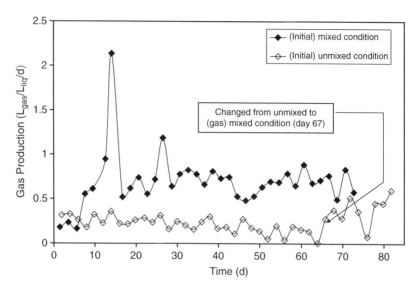

Fig. 3. Biogas productivity of the pilot-scale digester during different mixing conditions.

a known weight of manure slurry sample (10–20 g) with 0.1 M hydrochloric acid to a pH of 4.5 *(12)*.

Total fatty acids (TFA) analyses in the effluent and feed samples were performed by centrifugation and filtering samples through a 0.2-μm-pore-size filter, followed by injection of the filtrate into an HPLC. The mobile phase (filtered 5 mM H_2SO_4) of the HPLC was pumped at 0.6 mL/min through a 300 mm × 7.8 mm (8-μm-particle-size) RHM Monosaccharide column (Phenomenex, Torrance, CA) held at a temperature of 65°C to a refractive index detector (Model 2410, Waters Corporation, Miltford, MA) held at a temperature of 40°C. The sample injection volume was 10 μL and the resulting chromatograms were compared with injections of standards for acid concentration.

Gas samples (150 μL) were collected using a gas-tight syringe from a sampling port in the gas production line. They were injected in duplicate into a Hewlett Packard (Model 5890 Series II, Avondale, PA) gas chromatograph (GC) with a 30 m × 0.53 mm GS-Q phase capillary column (J&W Scientific, Folsom, CA). The injector, oven, and thermal conductivity detector (TCD) temperatures were 125, 50, and 150°C, respectively. The carrier gas (helium) flow rate through the column was 4 mL/min. The sample was injected in a split mode with approx 10% of the sample going through the column. The column make-up gas and the reference gas in the GC was helium. GC calibration of methane (CH_4), carbon dioxide (CO_2), and air gases was initially performed by injecting different volumes (50–200 μL) of pure gases. Later, periodic calibration was performed by injecting different amounts of air and using the relationship between TCD response factors *(13)*.

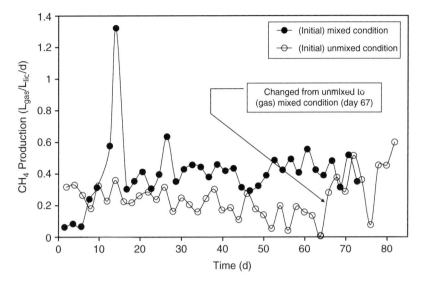

Fig. 4. Methane productivity of the pilot-scale digester during different mixing conditions.

Results and Discussion

The biogas (i.e., methane and carbon dioxide) and methane generation are shown in Figs. 3 and 4 for both the mixed (via recirculating gas) and unmixed operating conditions. It should be noted that the digester was operated continuously: 73 d mixed, 66 d unmixed, and 16 d mixed. Larger fluctuations were noted during the initial start-up phase, but the performance stabilized at approx 0.64 L biogas per L (or 0.4 L CH_4) digester slurry volume/d when the digester was mixed. After an initial phase of predominantly CO_2 production (probably owing to presence of residual oxygen), the CH_4 content of the biogas was approx 60% CH_4 and 40% CO_2, respectively (see Fig. 5).

In comparison with the mixed condition, the performance of the digester under the nonmixed condition was very different. The gas productivity of 0.64 $L_{gas}/L_{liq}/d$ observed under the mixed condition fell to more than half and, on certain days, there was negligible gas production under the nonmixed condition after 50 d of operation (Fig. 3). The methane production rate also decreased accordingly (Fig. 4) and the composition of the biogas became leaner (Fig. 5).

In order to confirm the effect of mixing, the gas recirculation was resumed on d 67. As observed from Figs. 3–5, the methane and biogas production picked up immediately upon starting the gas recirculation (indicated by an arrow).

The average amount of TS in the feed and effluent under the mixed condition at steady state were 115 and 83 g/L, respectively (see Table 1). At steady state the average amount of TVS in the feed and effluent were

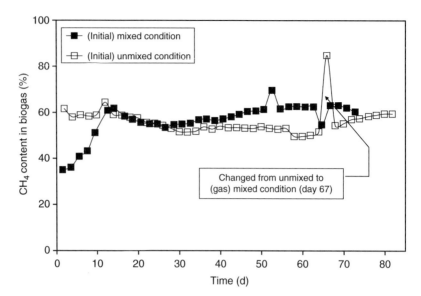

Fig. 5. Gas composition of biogas from the pilot-scale digester during different mixing conditions.

Table 1
Different Measured Parameters for Feed and Effluent at Steady State

	TS	TVS	TFA	CH_4 productivity ($L_{gas}/L_{liq}/d$)
Mixed condition				
Feed[a]	115	52	5.5	
				0.39
Effluent	83	44	0.1	
Unmixed condition				
Feed[a]	98	47	3.1	
				0.14
Effluent	113	53	5.5	

[a]The feed concentration has been corrected for dilution.

52 and 44 g/L (Table 1), while the TFA of the feed and effluent at steady state were 5.5 and 0.1 g/L (Table 1), respectively. During start up, there was an excess of propionic acid in the effluent (data not shown), but upon achieving steady state conditions there was little acid found in the effluent.

After the digester had been operated in an unmixed mode for 60–66 d, the TS in the feed and effluent were 98 and 113 g/L (see Table 1) and the average concentration of total volatile solids in the feed and effluent were 47 and 53 g/L, respectively (Table 1). The apparently higher concentration of TS and TVS in the effluent demonstrates the difficulty of obtaining accurate operating conditions for unmixed conditions. It is likely that significant settling occurs during unmixed operation. Consequently, effluent samples drawn from the bottom of the digester will not necessarily be indicative of the average composition of the digester. The incomplete conversion of TVS during unmixed operation corresponds to the poor methane productivity observed (Fig. 4). The TFA were higher in the effluent than in the feed (i.e., 5.5 vs 3.1 g/L). This suggests that the acetogenic bacteria are active as the fatty acids accumulated in (at least the bottom of) the digester, while the methanogenic bacteria were not. The alkalinity of the feed and effluent were essentially constant for both mixing conditions (data not shown).

The results obtained in this study contradicts the results obtained in our small-scale studies, where performance were unaffected by mixing conditions (10). In both studies, biogas recirculation were used as the mode of mixing, and it has been reported that biogas recirculation is the most efficient mode of agitation for anaerobic digesters (8,14,15). Although our current studies shows better performance under mixed conditions, higher methane production rates in unmixed digesters have been shown by Ghaly and Ben-Hassan (16); however, it has also been suggested that unmixed reactors perform worse, especially large reactors (17). It is possible that the size of digester is an important factor to consider, Ghaly and Ben-Hassan (16) found in their literature review that the digester units should have a diameter greater than 25 cm and a liquid depth of 20 cm, or greater, in order to provide reliable data that can be used for scale-up. The digester used in our pilot study fits this criterion, while the digesters used in our previous experiment did not.

Conclusions

Biogas production in a 100-L pilot-scale digester was evaluated and the effect of partial mixing induced by gas upflow/recirculation in the digester was studied. The digester was operated for a period of about 70 d (at a 16-d hydraulic retention time) with and without the mixing induced by gas upflow. A steady-state methane production of 0.4 $L_{gas}/L_{liq}/d$ was obtained. This result is slightly lower than that obtained in the small-scale study (10). However, in the pilot-scale study, the results show a dramatic effect of mixing on digester operation. Without any mixing, the digester performance deteriorates within 30–50 d, while with mixing, continuous and consistent production of methane is observed. This study demonstrates the importance of mixing and its critical role in design of large-scale anaerobic digesters. Future work on this project is targeted towards application to larger-scale designs as well as studies with small-scale digesters to determine the effect of scale-up. Further areas of interest will be in optimizing the mixing parameters and comparison with mechanical mixing.

Acknowledgments

The funding for this research was provided by US Department of Energy Office of Energy Efficiency and Renewable Energy. We would also like to thank Carl and Michelle Hoefer from the local dairy farm in Anderson County, TN, for their cooperation. Whitney Ridenour was supported through the Student Undergraduate Laboratory Internships program. The mention and use of firm names or trade products does not imply that they are endorsed or recommended by the US Department of Agriculture over other firms or similar products not mentioned.

References

1. Sheffield, J. (1999), *Summary Report of the Workshop on Opportunities to Improve and Benefit from the Management of Animal Waste*, JIEE Research Paper 99–01, Joint Institute for Energy and Environment, Knoxville, TN.
2. Robinson, A., Baxter, L., Junker, H., et al. (1988), *Fireside Issues Associated with Coal-Biomass Cofiring*, NREL/TP-570-25767, National Renewable Energy Laboratory, Golden, CO.
3. Hill, D. T. (1982), *Trans. ASAE* **25(5)**, 1374–1380.
4. Lusk, P. (1998), *Methane Recovery from Animal Manures: The Current Opportunities Casebook*, NREL/SR-580-2545, National Renewable Energy Laboratory, Golden, CO.
5. Sawyer, C. N. and Grumbling, A. M. (1960), *J. Sanitation Eng. Div. ASCE* **86**, 49–63.
6. Meynell, P. -J. (1976), *Methane: Planning a Digester*, Prism Press, London, pp. 55–57.
7. Casey, T. J. (1986), In: *Anaerobic Digestion of Sewage Sludge and Organic Agricultural Wastes*, Bruce, A. M., Kouzeli-Katsiri, A., and Newman, P. J., eds., Elsevier Applied Science Publisher, London, pp. 90–103.
8. Lee, S. R., Cho, N. K., and Maeng, W. J. (1995), *J. Ferment. Bioeng.* **80(4)**, 415–417.
9. Smith, L. C., Elliot, D. J., and James, A. (1996), *Water Res.* **30(12)**, 3061–3073.
10. Karim, K., Klasson, K. T., Hoffmann, R., Drescher, S. R., DePaoli, D. W., and Al-Dahhan, M. H. (2005), *Biores. Technol.* **96(16)**, 1771–1781.
11. Clesceri, L. S., Greenberg, A., and Trussell, R. (1989), *Standard Methods for the Examination of Water and Wastewater*, 17th edition, American Public Health Association, Washington DC, pp. 2-77.
12. Clesceri, L. S., Greenberg, A., and Trussell, R. (1989), *Standard Methods for the Examination of Water and Wastewater*, 17th edition, American Public Health Association, Washington DC, pp. 2-35–2-39.
13. Deitz, W. A. (1967), *J. Gas Chromatograph.* **5**, 68–71.
14. Morgan, P. F. and Neuspiel, P. J. (1958), In: *Biological Treatment of Sewage and Industrial Wastes*, Vol. 2., McCabe, J. and Eckenfelder, W. W., eds., Reinhold, New York, pp. 61–69.
15. Kontandt, H. G. and Roediger, A. G. (1977), In: *Microbial Energy Conversion*, Schlegel, H. G. and Barnea, J., eds., Pergamon Press, New York, pp. 379–392.
16. Ghaly, A. E. and Ben-Hassan, R. M. (1989), *Appl. Biochem. Biotechnol.* **20–21**, 541–559.
17. Bello-Mendoza, R. and Sharratt, P. N. (1998), *J. Chem. Technol. Biotechnol.* **71**, 121–130.

Copyright © 2006 by Humana Press Inc.
All rights of any nature whatsoever reserved.
0273-2289/06/129–132/897–908/$30.00

Biomodification of Coal to Remove Mercury*

K. Thomas Klasson,[†,1] Abhijeet P. Borole,[§,1] Catherine K. McKeown,[1] and Choo Y. Hamilton[2]

[1]*Oak Ridge National Laboratory, Oak Ridge, TN 37831-6226, E-mail: borolea@ornl.gov;* and [2]*The University of Tennessee, Knoxville, TN 37996*

Abstract

A biological process for removal of mercury from coal is under investigation. Iron and sulfur oxidizing bacteria have previously been used for desulfurization of coal and for mineral mining. We have shown that removal of mercury from coal is also possible via the same principles. Two pure cultures, *Leptospirillum ferrooxidans* and *Acidithiobacillus ferrooxidans* and four environmental consortium samples obtained from an acid mine drainage site were studied for mercury removal from coal. Four different coal samples were included in the study and the preliminary results have shown that up to 20% of the mercury can be removed in batch cultures compared to control. Additional parameters such as media composition and inoculum size were also studied. This is the first report demonstrating successful leaching of mercury from coal using biological treatment.

Index Entries: Bioleaching; mercury; coal; ferrooxidans.

Introduction

Emissions of mercury from coal-fired burners are in the range of 0.5 to 22 lb/10^{12} Btu and there is a plausible link between emissions and mercury bioaccumulation in the food chain *(1,2)*. Current operations of coal-fired power plants do not require dedicated mercury removal equipment and emissions control in the combustion of coal has traditionally been limited to the removal of mercury from off-gases. The power-generating industry emits about 50 t of mercury each year, about a third of the total manmade emissions *(3)*. In 2003, EPA suggested two approaches to reduce mercury emissions *(4)*. In the first approach, emissions would be reduced from 48 to 34 t/yr by 2007 using existing technology and a second approach to reduce emissions by 70% by 2018. Currently, there is a debate on the amount of mercury emission

---

*The submitted manuscript has been authored by a contractor of the US Government under contract No. DE-AC05-00OR22725. Accordingly, the US Government retains a nonexclusive, royalty-free license to publish or reproduce the published form of this contribution, or allow others to do so, for US Government purposes.
[†]Currently at USDA, ARS, Southern Regional Research Center, New Orleans, LA 70124
[§]Author to whom all correspondence and reprint requests should be addressed.

reduction, with stress by regulators to reduce emissions beyond what has been proposed by EPA. Needless to say, new technologies capable of reducing mercury emissions significantly will be needed in near future.

Mercury is naturally present in coal from different world sources and the concentration is typically in the range of 0.02 to 0.4 mg/kg (5). In the United States, coal from the Gulf Coast and Appalachian regions has the highest average concentration of mercury at 0.21–0.22 mg/kg (6). In a comprehensive review by Toole-O'Neil et al. (6), it was concluded that mercury in coal is most likely associated with the sulfur-containing iron compounds such as pyrite; however, a fraction of the mercury may be associated with the organic matter. It is expected that mercury and sulfur are closely associated in the coal as it is well known that mercury sulfide is a low-solubility inorganic salt (7).

Analysis of trace metals in coal is sometimes based on leaching the coal with a dilute nitric acid. In unpublished studies, as much as 75% of the mercury could be removed through nitric acid leaching (6) and published results summarizing data from commercial cleaning facilities suggest that 12–78% removal is possible when pyrite is removed from coal via froth floatation (8). Use of a two-step hydrochloric acid wash process has also been demonstrated to leach mercury from coal up to 77% (9).

It is well-known that pyrite in coal can be utilized by members of the bacteria *Acidithiobacillus* (formerly *Thiobacillus*) *ferrooxidans* (*A. ferrooxidans*) (10–12), and others which use both the reduced iron and sulfur in pyrite with the overall reaction

$$4\ FeS_2(s) + 15\ O_2(g) + 2\ H_2O(l) \Rightarrow 2\ Fe_2(SO_4)_3(aq) + 2\ H_2SO_4(aq).$$

This reaction is stepwise beginning with the interaction between the pyrite surface and soluble Fe(III) to liberate elemental sulfur, S(0), and Fe(II). Fe(II) and S(0) are oxidized by the bacteria, yielding the overall reaction above. The reactions are carried out by the bacterium *A. ferrooxidans* or by two bacteria (*A. ferrooxidans* and *A. thiooxidans*) working together. The generation of sulfuric acid in the process lowers the pH and helps with further dissolution of pyrite, and will aid in the dissolution of mercury-sulfur compounds. The optimal pH for iron removal from coal pyrite was determined by Torma and Olsen (13) to be pH 2.0 in experiments with *A. ferrooxidans*. This pH is naturally obtained through the release of sulfuric acid by the bacteria.

Another organism with good metal bioleaching capability is *Leptospirillium ferrooxidans* (*L. ferrooxidans*). This organism was found to comprise more than 50% population in microbial species habitating biotopes such as mines and surrounding dump sites (14) at temperatures above 20°C. Other reports also suggest the dominance of *Leptospirillum* genus in acid mine drainage environments (15). This is a strict chemolithoautotroph, metabolizing ferrous iron, and pyrite.

Microbial leaching for copper and uranium recovery has been used commercially for low-grade ore (16,17). Other metals including Ni, Cu, and

Table 1
ATCC 2039 Nutrient Medium and Fe(III) Medium for Experiments

Solution A	
$(NH_4)_2SO_4$	0.8 g
$MgSO_4 \cdot 7H_2O$	2 g
K_2HPO_4	0.4 g
Wolfe's mineral solution	5 mL
deionized water	795 mL
Adjust with 20% H_2SO_4 to pH 2.3.	
Solution B (prepared fresh)	
$FeSO_4 \cdot 7H_2O$	20 g
deionized water, pH 2.3 (acidified with H_2SO_4)	200 mL
Solution C (prepared fresh)	
$Fe_2(SO_4)_3 \cdot 1.5H_2O$	0.13 g
deionized water, pH 2.3 (acidified with H_2SO_4)	50 mL
ATCC 2039 [Fe (II) medium]	
Filter sterilize solutions A and B separately and combine 800 mL solution A with 200 mL Solution B.	
Fe (III) medium	
Combine 100 mL of filter-sterilized Solution A with 2 mL of filter sterilized Solution C.	

Pb have also been studied for bioleaching potential using the same organisms. The objective of this study was to investigate the feasibility of developing a biological coal modification technique based on bioleaching that will aid in the removal of mercury from coal before thermal processing.

Materials and Methods

The coals used in this study were obtained from Penn State Coal Bank (University Park, PA) and analyzed for mercury content. The coals obtained were PSOC-275 (Ohio No. 6A, Lower Freeport Seam) (0.31 mg/kg of Hg, slightly lower than published values [18,19]), PSOC-1286 (Ohio No. 5, Lower Kittanning Seam) (0.4 mg/kg of Hg), PSOC-1296 (Pennsylvania B, Lower Kittanning Seam) (0.26 mg/kg of Hg), PSOC-1368P (Weir-Pittsburg/Cherokee Seam, Missouri) (0.26 mg/kg of Hg), PSOC-1470 (Pratt Seam, Alabama) (0.4 mg/kg of Hg). These samples represented coals with a high-pyritic sulfur content and possibly mercury concentration based on coal seam data (6). The coal samples were obtained, packaged under an inert atmosphere and were maintained under those conditions until used.

A. ferrooxidans American type culture collection (ATCC) 19859 and *L. ferrooxidans* Markosian ATCC 53993 were obtained from ATCC and grown in 5 mL of ATCC medium 2039 (Table 1) in screw-cap test tubes placed in slanted configuration in a 30°C incubator. In addition to the pure cultures mentioned earlier, a consortia of bacteria was also used for mercury leaching. Acid mine drainage from an abandoned lead and zinc mine (the Valzinco mine in Spotsylvania county, Virginia) was sampled and transferred to the laboratory. The acid mine drainage from this abandoned mine coats the creek-banks orange (suggestive of iron-oxidizing bacteria) and downstream the acid mine drainage flows past an abandoned gold mine (the Mitchell Mine) in which mercury was used in the recovery of gold (20). Thus, there was a possibility of collecting microorganisms which have been exposed to (or not exposed to) mercury within one watershed. The microbial consortia were labeled VA No. 1, 2, 3, and 4. The VA No. 1 and VA No. 2 samples were collected from an area just downstream from the lead/zinc mine and VA No. 3 and VA No. 4 were collected downstream from the lead/zinc mine and the gold mine. The VA No. 1 and VA No. 2 were similar, except that they were collected from two different points in the same area. The same was the case with Samples VA No. 3 and VA No. 4. These natural environmental cultures were enriched by growing them in the laboratory using ATCC 2039 medium and then transferring them several times to obtain enrichment cultures. Coal biotransformation experiments were conducted using these four environmental enrichments as well.

The cultures were transferred to fresh medium every 4 wk with a 25% inoculum rate and, before an experiment, a larger (125 mL) screw-cap bottle with 20-mL ATCC medium 2039 was inoculated with 1 mL from a 4-wk-old culture.

Cell suspensions used for the studies were prepared by growing the culture in 1 L of ATCC medium 2039 in a stirred (150 rpm) batch tank reactor with pH and temperature (30°C) control. Aeration was provided via sparging with air containing 1–3% CO_2. Growth was usually noted after 24–48 h, as indicated by heavy iron(III) oxide/hydroxide precipitation. After 1 wk of incubation, the broth was centrifuged at 4000g for 30 min at 4°C and the supernatant removed via suction. The precipitate/cells were combined with approx 150 mL of sterile 4°C deionized water at pH 2.6 (acidified with 20% H_2SO_4) in a 250-mL media bottle. The contents were shaken vigorously for 2 min and stored at 4°C overnight. On the second day, the supernatant (containing cells) was transferred to a sterile 500-mL media bottle and store at 4°C. The precipitate (in the 250-mL bottle) was again shaken with approx 75 mL of sterile 4°C deionized water at pH 2.6 and stored at 4°C overnight. This precipitate washing protocol was repeated two more times. The supernatants collected were combined and centrifuged at 10,240g for 20 min at 4°C and the resulting cell pellet was resuspended with approx 15 mL of sterile 4°C deionized water at pH 3.5 (acidified with 20% H_2SO_4). The cell suspension was centrifuged at 12,180g for 20 min at 4°C and, finally, the cell pellet was resuspended with approx 15 mL of sterile 4°C deionized water at pH 3.5 (acidified with 20% H_2SO_4) and stored at 4°C until used. The OD_{600} of this culture suspension was approx 1. A similar procedure of cell harvesting has previously been reported (21). It should be noted that whereas most of the Fe(III) precipitate is removed as a result of the above procedure, it is impossible to remove all the iron particulates, as small iron precipitate particles are strongly attached to the bacteria.

Biotreatment experiments

The feasibility for biotreatment of coal was done in batch experiments by contacting 0.1 g of coal with 4.5 mL of (pH-adjusted) medium and 0.5 mL of concentrated cell culture in 15-mL (screw cap) test tubes. In controls, 0.5 mL of pH 3.5 sulfuric acid was added in place of the cultures. Two different media formulations were used, one containing Fe(II) and the other containing Fe(III) (*see* Table 1). The effect of mixing by intermittent shaking was also studied for certain experimental conditions. The tubes were shaken gently to mix the coal and the microbial culture and then incubated for 1–8 wk at 30°C.

Sampling and Analysis

The test tubes were sampled and analyzed for mercury by collecting 4 mL samples after the culture test tubes had been centrifuged. The samples were oxidized using bromine monochloride, followed by neutralization of excess oxidant with hydroxylamine hydrochloride and reduction of mercury using stannous chloride to produce elemental mercury, which was analyzed using single gold trap amalgamation and cold vapor atomic fluorescence spectroscopy (Hg analyzer, BrooksRand, Seattle, WA). The method closely followed US EPA Method 1631E and is a standard methodology which works for mercury at levels ranging from 10 ng/L to 2 µg/L. The instrument can measure sub- to low-picogram levels of mercury which allows very low levels of detection and analysis of small volumes of samples.

Results and Discussion

The goal of this work was to determine the feasibility for mercury removal under different conditions by different organisms. There was significant mercury removal in most of the experiments compared with control experiments. In the first experiment, two pure cultures *A. ferrooxidans* and *L. ferrooxidans* were compared when they were acting on the different coals. A typical result from one of those experiments are shown in Fig. 1. Duplicate analysis of a few samples was conducted and found to be within 10% of each other. As seen in the Fig. 1, *L. ferrooxidans* was found to remove more mercury from coal PSOC-1470 than *A. ferrooxidans* in a 4-wk period. Another experiment investigating an 8-wk-time period indicated that no additional mercury was removed under these conditions (data not shown). Tests were also performed with the other coal types using the two pure cultures and these tests indicated a similar or lower amount of mercury removal as compared with coal PSOC-1470. In most cases, *L. ferrooxidans* was found to remove more mercury from a particular coal than *A. ferrooxidans* after 4 wk. It is also important to note that the pH changes during the incubation in inoculated samples, dropping from an initial pH

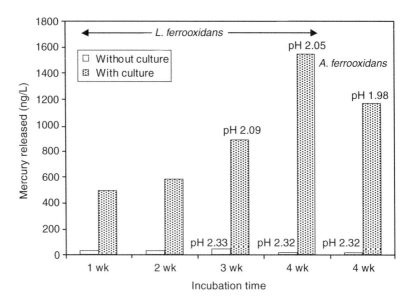

Fig. 1. Removal of mercury from coal PSOC-1470 using pure cultures at 30°C. The pH values given are at the end of the experiment. The starting pH was 2.3.

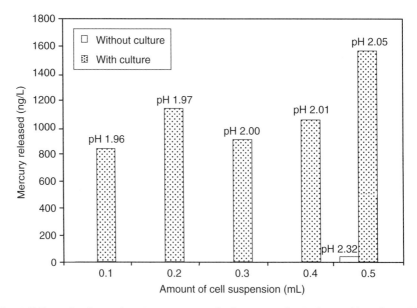

Fig. 2. Effect of culture density on removal of mercury by *L. ferrooxidans* from PSOC-1470 coal. The initial pH was 2.3.

of 2.3 to pH 2.0 or slightly below; the pH in control samples remained approximately constant during the incubation (Fig. 1).

The effect of culture density on mercury removal was studied by varying it between approx 2 and 10% by volume (0.1–0.5 mL/5 mL medium). The results are shown in Fig. 2—in general, there was little effect of culture density over the range studied. Thus, it was decided to use 0.5 mL of cell

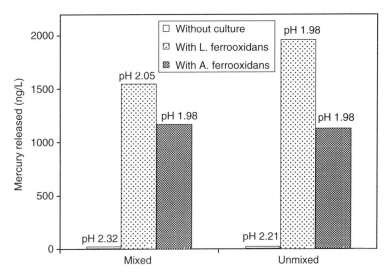

Fig. 3. Effect of mixing on the removal of mercury by *L. ferrooxidans* from PSOC-1470 coal after 4 wk at 30°C.

suspension per 5 mL coal sample slurry for all experiments. An experiment was conducted to study the effect of intermittent mixing (Fig. 3). The experiment was conducted over a 4-wk period and mixing was done on alternate days for a 5-min period via vortex mixing. The results show that there was less mercury removed under mixing conditions. The iron oxidizing bacteria are known to work by attaching themselves to ore surface and this may explain why static conditions may be better *(22)*. However, the experimental data set is insufficient to make a strong case for this observation. As the mixing appeared to have a negative effect on the release of mercury, further experiments were done without mixing.

Natural environmental cultures also proved to liberate more mercury from the coal compared with controls. As part of these studies, augmentation of different species of iron was studied. The significance of adding Fe(III) vs Fe(II) is as follows: the reduced form of iron (Fe[II]) is required for growth of microorganisms, whereas the oxidized form is the one which reacts with pyrite/FeS_2 in coal, thereby oxidizing the sulfide species known to bind mercury. This chemical reduction process also produces Fe(II) which is then available for organisms to grow on. It has been shown that Fe(III) can be helpful when it is present as the only iron species in the beginning of the chemical leaching process *(23,12)*. However, in a long-term process, it may not make a significant difference to add either.

Results from experiments with culture enrichments from the acid mine drainage sites are shown in Figs. 4–7. Removal of mercury was observed by all cultures/enrichments. In general, VA No. 1 and VA No. 2 performed better than the other two enrichment cultures. Up to 1300 ng/L of mercury was leached from the coal samples by the former enrichments. Among these two, VA No. 2 performed better under the Fe(III) media conditions.

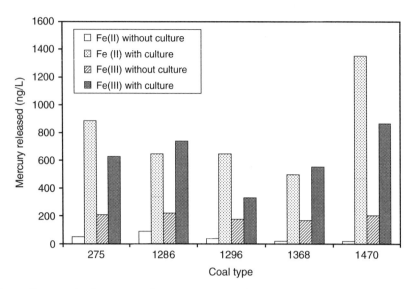

Fig. 4. Removal of mercury from various coals by environmental culture enrichment VA No. 1 with different species of iron augmentation.

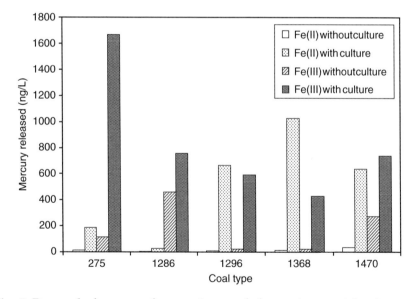

Fig. 5. Removal of mercury from various coals by environmental culture enrichment VA No. 2 with different species of iron augmentation.

A comparison of mercury removal by pure cultures vs environmental cultures can be done for coal PSOC-1470. For example, about 1548 ng/L Hg was released by *L. ferrooxidans* in 4-wk period (Fig. 1), whereas 1353, 637, 520, and 21 ng/L Hg was released by VA No. 1, 2, 3, and 4 cultures, respectively, under Fe(II) augmentation condition (Figs. 4–7). Although it may seem that the pure culture is able to remove higher level of mercury from coal PSOC-1470 than environmental cultures, this statement cannot

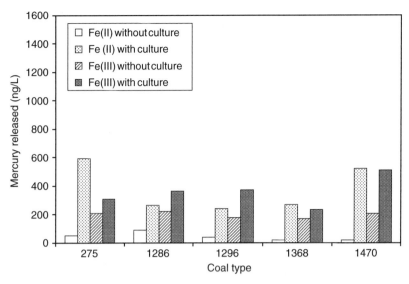

Fig. 6. Removal of mercury from various coals by environmental culture enrichment VA No. 3 with different species of iron augmentation.

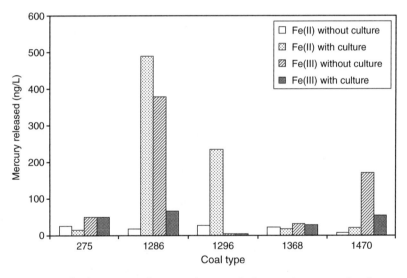

Fig. 7. Removal of mercury from various coals by environmental culture enrichment VA No. 4 with different species of iron augmentation.

be generalized. However, literature reports indicate that *L. ferrooxidans* is one of the most common organisms in acid mine *(24,25)* and metal bioleaching studies *(14)* and that it is a very effective organism for oxidation of metal sulfides such as pyrite.

Previous research has shown that mercury leaches out of coal under acidic conditions. Dronnen et al. *(9)* found that neither *A. ferrooxidans* nor *A. thiobacillus* improved the leachability of mercury from North Dakota lignite—the authors concluded that their growth medium with a pH of 2.0

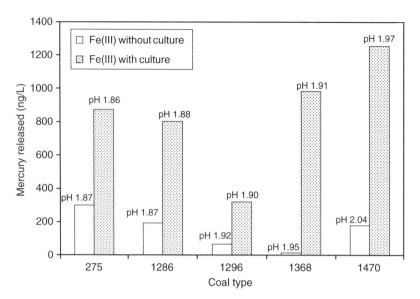

Fig. 8. Removal of mercury from various coal by environmental culture enrichment VA No. 2 with ferric iron, Fe(III), augmentation.

or 5.0 removed as much mercury as either of the organisms did when grown together with the lignite at pH 2.0 for 116 d. In our studies, the pH dropped during cultivation; consequently we performed additional studies in which the pH in the control experiments were initially reduced to a level we anticipated the inoculated experiments would reach after 4 wk of incubation. The result from such an experiment may be seen in Fig. 8. As is noted, the pH in some of the control test tube experiments ended up lower than in the inoculated test tubes and the opposite was found in other experiments. Regardless, mercury release from the coal was clearly better in the inoculated cases.

In our studies, the mercury release from the coal was detected by monitoring the concentration of mercury in the leaching liquid. In other studies, investigators have reported volatilization of mercury from soil slurries contaminated with mercury in the presence of a strain of *A. ferrooxidans* (26). In a limited number of experiments, vials with coal, medium, and pure strain organisms were incubated in a larger sealed glass vial with a Teflon-lined septum. After 4 wk of incubation, the gas phase was sampled for mercury but none detected. This suggests that the strains (both, *A. ferrooxidans* and *L. ferrooxidans*) used in these experiments do not volatilize mercury under the conditions studied, but solubilized the mercury and released it into the liquid medium.

The reproducibility of mercury removal by the bacterial strains is an important question. Although the biotreatment experiments were not run in duplicate, repeat experiments were conducted to assess mercury release from coal PSOC-1470 by *L. ferrooxidans* and have shown qualitatively similar results (data not shown). The differences observed in release of mercury may be owing to the differences in density of the cell culture used in the three experiments with *L. ferrooxidans* (Figs. 1–3). A quantitative analysis of the amount of mercury released (g inoculated cells) per day is being conducted and will be reported in a subsequent publication targeted at kinetics and mass transfer issues. Other issues such as effect of higher concentrations of mercury on bacteria, effect of iron species on growth, and microbial growth during leaching process are also under investigation. It was also difficult to assess the exact release of mercury in noninoculated controls. For example, mercury release in the control experiments for coal PSOC-1368P with Fe(III)-augmentation ranged between 21 and 170 ng/L (Figs. 4–7). It is not entirely clear what caused this variation. It should be noted that samples for analysis were taken after centrifugation but were unfiltered and any small quantity of carbon accidentally transferred from the culture test tube to the sample container would increase the apparent mercury concentration in the sample. Further, the control experiments were done under nonsterile conditions. Some bacteria are expected to be present associated with the coal. It is possible that these indigenous bacteria result in release of mercury in coal. Regardless of the cause, the variations of mercury release in control experiments are not significant enough to change our conclusions.

The amount of mercury released from the different coal types has been observed to vary significantly. The reason for this may be the difference in porosity, pyrite distribution, and sample heterogeneity of the coal samples. Overall, the greatest mercury removal was observed from coal PSOC-1470, which contained about 0.4 mg/kg of mercury. About 1600 ng/L of mercury was observed in the leachates from this coal, amounting to about 20% removal of mercury. A process capable of continuous mercury removal using mercury-specific sorbents can potentially improve the extent of mercury removed from coal.

Conclusions

Preliminary experiments have shown the capability of pure strains, as well as environmental enrichments, to assist in the leaching of mercury from coal. Mercury was released to a different extent from each coal type. Environmental enrichment samples did as well as pure cultures. The augmentation of ferric iron at low concentration help leaching under some conditions. Under the best conditions, 20% of the mercury initially present in the coal was leached into the liquid and it was mediated by the presence of bacteria.

References

1. Kotnik, J., Horvat, M., and Jereb, V. (2002), *Environ. Mod. Software* **17**, 593–611.
2. Meij, R., Vredenbregt, L. H. J., and Winkel, H. T. (2002), *J. Air Waste Mngt. Ass.* **52**, 912–917.
3. Brown, T. D., Smith, D. N., Hargis, R. A., and O'Dowd, W. J. (1999), *J. Air Waste Mngt. Ass.* **49**, 628–640.

4. Johnson, J. (2005), *Chem. Eng. News* **83**, 44, 45.
5. Richaud, R., Lachas, H., Collot, A. G., et al. (1998), *Fuel* **77**, 359–368.
6. Toole-O'Neil, B., Tewalt, S. J., Finkelman, R. B., and Akers, D. J. (1999), *Fuel* **78**, 47–54.
7. Dean, J. A. (1985), *Lange's Handbook of Chemistry*. McGraw-Hill, New York.
8. Akers, D. J. (1995), In: *Environmental Aspects of Trace Metals in Coal*, Swaine D. J. and Goodarzi F. (ed.). Kluwer Academic Publishers, Dordrecht, The Netherlands, p. 93–110.
9. Dronen, L. C., Moore, A. E., Kozliak, E. I., and Seames, W. S. (2004), *Fuel* **83**, 181–186.
10. Andrews, G. F. and Maczuga, J. (1982), *Biotech. Bioeng. Symp.* **12**, S337–S348.
11. Dugan, P. R. and Apel, W. A. (1978), In: *Metallurgical Applications of Bacterial Leaching and Related Microbial Phenomena*. Murr, L. E., Torma, A. E., and Brierly, J. A. (ed.), Academic Press, London, pp. 223–250.
12. Silverman, M. P., Rogoff, H., and Wender, I. (1963), *Fuel* **42**, 113–124.
13. Torma, A. E. and Olsen, T. M. (1988), *Appl. Biochem. Biotechnol.* **18**, 341–354.
14. Sand, W., Rohde, K., Sobotke, B., and Zenneck, C. (1992), *Appl. Environ. Microbiol.* **58**, 85–92.
15. Ram, R. N. Verberkmoes, M., Thelen, G, et al. (2005), *Science* **308**, 1915.
16. Berry, V. K., and Murry, L. E. (1978), In: *Metallurgical Applications of Bacterial Leaching and Related Microbial Phenomena*, Murr, L. E., Torma, A. E., and Brierly J. A. (ed.), Academic Press, London, pp. 103–136.
17. Bosecker, K., Neuschutz, D., and Scheffer, U. (1978), In: *Metallurgical Applications of Bacterial Leaching and Related Microbial Phenomena*, Murr, L. E., Torma, A. E., and Brierly J. A. (ed.), Academic Press, London, pp. 389–402.
18. Merdes, A. C., Keener, T. C., Kang, S.-J., and Jenkins, R. G. (1998), *Fuel* **77**, 1783–1792.
19. Wang, M., Keener, T. C., and Khang, S.-J. (2000), *Fuel Process Technol.* **67**, 147–161.
20. Seal II, R. R., Johnson, A. N., Hammarstrom, J. M., and Meier, A. L. (2002), *Geochemical Characterization of Drainage Prior to Reclamation at the Abandoned Valzinco Mine, Spotsylvania County, Virginia*, Open-File Report 02-360. US Geological Survey.
21. Silverman, M. P. and Lundgren D. G. (1959), *J. Bacteriol.* **78**, 326–331.
22. Rohwerder, T., Jozsa, P., Gehrke, T., and Sand, W. (2002), In: *Encyclopedia of Environmental Microbiology*, Bitton, G. (ed.), John Wiley & Sons, Inc., New York. Vol. 2, p. 632.
23. Sand, W., Gerke, T., Hallmann, R., and Schippers, A. (1995), *Appl. Microbiol. Biotechnol.* **43**, 961–966.
24. Gonzalez-Toril, E., Llobet-Brossa, E., Casamayor, E. O., Amann, R., and Amils, R. (2003), *Appl. Environ. Microbiol.* **69**, 4853–4865.
25. Hallberg, K. B. and Johnson, D. B. (2003), *Hydrometallurgy* **71**, 139–148.
26. Takeuchi, F., Iwahori, K., Kamimura, K., Negishi, A., Maeda, T., and Sugio, T. (2001), *Biosci. Biotechnol. Biochem.* **65**, 1981–1986.

SESSION 6
Bioprocess Research and Development

Introduction to Session 6

MICHAEL R. LADISCH

Laboratory of Renewable Resources Engineering, Potter Engineering Building, 500 Central Drive, Purdue University, West Lafayette, IN 47907-1295

Bioprocess research and development seeks to translate laboratory results into practical industrial processes. Fundamental issues that must be addressed include mass transfer, reactions at liquid gas interfaces, use of solid phase catalysts to carry out rapid conversion of oligosaccharides to fermentable sugars, and bioreactor design for the conversion of fermentable sugars to ethanol and other value-added products. This session addressed key developments in these areas, with presentations highlighting some of the critical challenges and exciting advances that are occurring to develop bioprocess-related approaches for transforming renewable resources into various fuels and chemicals products. Several new bioprocessing concepts were also proposed—such as enzymatic production of biodiesel—to better exploit the ever-increasing power of biocatalysis (both enzymatic and microbial).

This session's oral paper and poster presentations provided timely examples of how bioprocess engineering and bioprocess engineering principles are being applied across a wide range of applications. Although many of the papers and posters concerned processes to produce liquid transportation fuels products, most notably ethanol and biodiesel, a large number addressed issues related to producing or recovering a variety of non-fuel products, including: polyhydrolyalkonates (PHAs) and other biopolymers; hydrolytic enzymes; vitamins/vitamin precurcors; and organic acids, especially lactic acid. The preponderance of proceedings papers discussing non-fuel bioproducts reflects the importance of such products in today's marketplace, as well as the fact that such products will likely represent important value-added coproducts in future biorefineries in which fuels will be the primary product, at least on a volumetric basis. An emerging theme that came out of this session is that efficient separations technologies are required to be able to develop cost effective processes to produce (and recover) new bioproducts.

Enzymatic Conversion of Waste Cooking Oils Into Alternative Fuel—Biodiesel

GUANYI CHEN,* MING YING, AND WEIZHUN LI

Section of Bioenergy and Environment (Faculty of Environmental Science and Engineering)/State Key Lab of Internal Combustion Engine, Tianjin University, 300072, Tianjin, China; E-mail: chen@tju.edu.cn

Abstract

Production of biodiesel from pure oils through chemical conversion may not be applicable to waste oils/fats. Therefore, enzymatic conversion using immobilized lipase based on *Rhizopus orzyae* is considered in this article. This article studies this technological process, focusing on optimization of several process parameters, including the molar ratio of methanol to waste oils, biocatalyst load, and adding method, reaction temperature, and water content. The results indicate that methanol/oils ratio of 4, immobilized lipase/oils of 30 wt% and 40°C are suitable for waste oils under 1 atm. The irreversible inactivation of the lipase is presumed and a stepwise addition of methanol to reduce inactivation of immobilized lipases is proposed. Under the optimum conditions the yield of methyl esters is around 88–90%.

Index Entries: Biocatalyst; biodiesel; immobilized *R. oryzae* lipase; transesterification; waste oils.

Introduction

Methylesters (MEs) produced by primary or secondary alcoholysis of vegetable oils and animal fats, which are collectively named biodiesel, are viewed as a promising renewable and degradable energy source. Specifically, biodiesel is considered as an environment-friendly fuel as it leads to much lower levels of sulfur oxides, halogens, and soot in engine exhaust gas than fossil fuel, particularly carcinogenic polyaromatic hydrocarbons and nitric acid-containing compounds, which are usually reduced by 75 and 85% *(1)*. Biodiesel is therefore becoming one of most attractive clean alternative automobile fuels in the world, particularly in China, taking into account, on one hand, the significant imbalance in the amount of oil produced and consumed (in 2004, China imported approx 120×10^6 t of oils), and on the other hand heavy pollution related to transportation sector in most of the metropolises.

The commonly used method to produce biodiesel is transesterification of vegetable oils and animal fats by methanol with the assistance of catalyst.

*Author to whom all correspondence and reprint requests should be addressed.

The transesterification reaction can be carried out chemically or enzymatically. At present, biodiesel at demonstration scale or more is produced by chemical method, however, this has to be proceeded at a relatively high temperature and complex downstream operation and usually results in high yield of byproduct glycerol (2). The major disadvantage of both alkali and acid catalyzed process is that homogenous catalysts are removed together with the glycerol layer after the reaction and cannot be reused (3). In this case, the purification of glycerol as an additionally valuable product is getting more difficult as inorganic material carried out with glycerol has to be completely removed. The disadvantage caused by chemical catalysts can be simply prevented by the use of lipases which allows mild reaction conditions, easy recovery/reuse of glycerol without delicate purification, simple separation process of MEs from other compounds and almost no chemical waste liquid produced. So recently, enzymatic transesterification using lipase has become more attractive in the oleochemical industry.

Enzymatic transesterification of waste oils/fats to produce biodiesel refers to an enzyme-catalyzed reaction involving triacylglycerols and methanol to yield MEs and glycerol as shown in Fig. 1. Triacylglycerols, the main component of vegetable oil, consist of three long-chain fatty acids esterified to a glycerol backbone. When triacylglycerols react with the methanol, the three fatty acid chains are released from the glycerol skeleton and combine with the methanol to yield MEs. Glycerol is produced as a byproduct. In general, a large excess of methanol is used to shift the equilibrium far to the right in this reaction.

Recently, methanolysis of pure soybean oils or sunflower oil using lipases as catalyst has been reported (4,5). Shimada et al. (6) investigated methanolysis of pure vegetable oil by enzymatic alcoholysis reaction and reported that immobilized *Candida antarctica* lipase was most effective for the methanolysis of pure vegetable oil. Later they continued conducting the transesterification reaction by *Candida antarctica* lipases after pretreatment of pure plant oils (7).

The cost of biodiesel, however, is the main hurdle to its commercialization in the market. Pure vegetable oils as feedstock for biodiesel in China at this moment is not affordable, and therefore using zero or even negative cost of waste oils/fats as resources seems a good choice. It may not only significantly reduce the cost of biodiesel produced but also eliminate the environment and human health risk caused by large quantities of waste oils/fats directly released by restaurants, slaughterhouses, and oil-processing plants. Waste oils/fats amount to approx $5–6 \times 10^6$ t/yr in China (1,8,9). Currently, about 80% waste oils/fats in China are illegally recycled into market through simple processing thereby leading to health risks. However, only less than 1 wt% of waste oils/fats is being subjected to transesterification into biodiesel by alkali- or acid-catalyzed process.

The physical and chemical properties of waste oils/fats is quite complex, particularly in terms of content of free fatty acids, water, and impurities

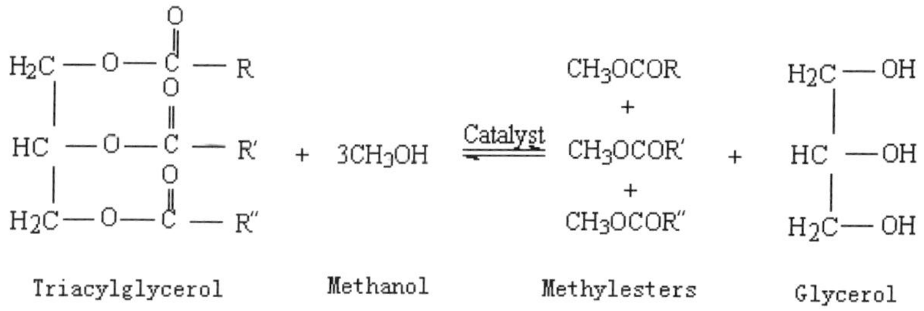

$$H_2C-O-\overset{O}{\underset{\|}{C}}-R$$
$$HC-O-\overset{O}{\underset{\|}{C}}-R' + 3CH_3OH \xrightleftharpoons{\text{Catalyst}} \begin{array}{c} CH_3OCOR \\ + \\ CH_3OCOR' \\ + \\ CH_3OCOR'' \end{array} + \begin{array}{c} H_2C-OH \\ | \\ HC-OH \\ | \\ H_2C-OH \end{array}$$
$$H_2C-O-\overset{O}{\underset{\|}{C}}-R''$$

Triacylglycerol   Methanol   Methylesters   Glycerol

Fig. 1. The mechanism of transesterification to produce MEs.

compared with pure vegetable oils. The technological process of converting them into biodiesel, is therefore different from that of pure vegetable oils. The aim of the present study is to investigate immobilized *Rhizopus orzyae*-based lipase-catalyzing methanolysis of waste cooking oils with a stepwise process, with reference to the impact of reaction parameters, such as temperature, substrate molar ratio, enzyme load, and others.

Experimental Methods

Chemical and Physical Properties of Waste Cooking Oils

The waste cooking oils were obtained from Yizhongyuan restaurant next to Tianjin University campus. Saponification value of the waste cooking oils was determined according to the method described by Yasuhiko (10). Viscosity of refined oils, waste cooking oils, and the corresponding products after methanolysis were measured at 20 and 40°C, respectively, with a viscometer (Model SYD-265D-I, Shanghai Changji Geological Apparatus Co. Ltd, China). Fatty acid composition was determined by gas chromatography. Identification of fatty acids contained in waste cooking oils was performed by comparison of retention time with fatty acids standard (Sigma Chemical Co., St. Louis, MO).

Methanolysis Reaction

For the methanolysis reactions conducted at stoichiometric molar ratio of waste cooking oil to methanol, 33.4 g oil and 3.613 g methanol were placed into the reaction flask, stirred by a magnetic stirrer at 220 rpm and heated to the reaction temperature. In subsequent experiments, in which the effect of molar ratio of oil/methanol was investigated, total weight of reacting compounds was always kept constant at approx 37 g. The appropriate amount of immobilized enzyme based on oil weight was added to the flask. After a certain period of time, the reaction was stopped and the enzyme was removed from the reaction mixture by filtration using 100 mL of hexane as an extracting agent. Methanol is separated from MEs and glycerol by delicate distillation. Quantitative analyses of MEs product are carried out by gas chromatography (GC) as described in later paragraphs. Transesterification reactions were carried out in duplicate. In order to determine the effect of lipase load, reactions were performed with stoichiometric oil/methanol molar ratio at 40°C with lipase load at 10, 15, 20, 25, 30, 35, 40, and 45% based on oil weight, respectively. The reaction time was kept constant at 7 h for all experiments, if not specifically indicated. Because the alcoholysis reaction is reversible, an increase in the amount of one of the reactants will result in higher ester yield and at least 3 M equivalents of methanol are required for the complete conversion of the oil to its corresponding MEs. The role of substrate molar ratio in methanolysis of waste cooking oils was also investigated at 40°C and 7 h, focusing on 1:1, 1:2, 1:3, 1:4, 1:5, and 1:6 oil/methanol molar ratios with the enzyme content of 30% based on oil weight. In order to investigate the effect of reaction temperature, the reactions were conducted at 30, 40, 50, 60, 70°C, respectively, under the following conditions: 30% of enzyme based on the oil weight, oil/methanol molar ratio 1:4, and reaction time 7 h. In all cases, samples (1 mL) were taken from the reaction mixture at specified times and centrifuged at 13,000 rpm for 1 min to separate lipase powder and obtain supernatant for GC analysis.

Lipase Activity

Immobilized lipase powder from *R. oryzae* was obtained from Amano Enzyme Inc. in Japan, which is 1,3-positional specificity lipase. Enzyme activity defined as the amount of enzyme that liberates 1 mmol of lauric acid/min was measured in our lab. The activity of the immobilized lipase was 150,000 u/g.

Analytical Methods of Methylester

The methylester content of the reaction mixture was quantified using an Agilent 6890N Series GC system (Agilent Technologied Corp.), and HP6890 Chemstation software was used for data analysis. The GC was equipped with a HP9091s-413 capillary column (300 μm × 30 m). For GC analysis, 500 μL of sample supernatant and 500 μL hexane as diluent were mixed in a 1.5-mL bottle. A 1-μL aliquot of the dilutes sample was injected into the gas chromatograph. A split injector was used with a split ratio of 20:1 and the temperatures of injector and detector were set at 300 and 260°C, respectively. The carrier gas was nitrogen with a flow rate of 20 mL/min. A flame ionization detector (FID) was used and the oven was initially held at 50°C, then elevated to 130°C at 20°C/min holding 5 min, and finally to 260°C at 2.5°C/min. The oven was held at this temperature for 10 min before returning to 50°C. Total run time for this method was about 70 min. Calibration of the GC method was carried out by analyzing standard solutions of methyl palmitate, linoleic acid methylester, methyl oleate, and stearic acid methyl ester. The standards were diluted in hexane like the reaction samples. ME yield was expressed as the percentage of MEs produced relative to the theoretical maximum based on the amount of original oils. In this article, ME yield is sometimes expressed as ME content or conversion.

Table 1
Comparison of Properties of Pure Vegetable Oils and Waste Cooking Oils

Properties	Pure vegetable oils			Waste cooking oils
	Soybean oil	Palm oil	Rapeseed oil	
Acid value mgKOH/g oil	0.24	0.28	0.26	5.96
Saponification value	193	180	181	188
Unsaponification (wt%)	1	1	1	4.5
Fatty acid (wt%)[a]	89.6	89.4	90.3	88.9
Viscosity (mm^2/s, 40°C)	4.7	4.3	4.1	156

[a]Fatty acid percentage was obtained after chemical hydrolysis of vegetable oils with 2.5% sulfuric acid-methanol solution.

Result and Discussion

Chemical Properties of Waste Cooking Oil

The waste cooking oils obtained from Yizhongyuan restaurant were used in our experiments to determine their physical and chemical properties, and pure vegetable oils were used for comparison. Some chemical properties of waste cooking oils are summarized in Table 1, and its compositions identified by GC and the corresponding fatty acid content (wt%) are showed in Table 2. It is observed from the table that unsaponification matter and acid value of waste cooking oils are significantly higher than that of pure vegetable oils and the saponification value of waste oils is slightly higher than that of pure oils. The higher content of unsaponification matter is mainly made up of pigments, residual gum, and oxidized materials (11). Waste cooking oils show higher viscosity than pure oils.

Lipase Amount

The impact of lipase amount on the methanolysis of waste cooking oils is presented in Fig. 2. From the figure, it can be seen that MEs yield continuously increases with the increase of lipase quantity up to 30 wt%. The highest ME formation (89 wt%) is observed for the case using 30% lipase based on oil weight. This result is similar to the data reported previously by Nelson et al. (12) who carried out the research using pure vegetable oils.

Effect of Substrate Molar Ratio

Results for the influence of substrate molar ratio are given in Table 3. The MEs yield at 1:1 waste cooking oils to methanol molar ratio is 30.15%, but increasing the molar equivalents of methanol up to four initially in the oil promotes the methanolysis. The highest ME yield (85.12%) is obtained at 1:4 oil/methanol molar ratio. The decrease in activity and MEs yield at high methanol concentrations (1:7 and 1:8) may reflect the ability of the methanol excess to distort the essential water layer that stabilizes the

Table 2
Composition of Fatty Acids in Waste Cooking Oils

Fatty acid compositions[a]	Content (wt%)	Fatty acid compositions	Content (wt%)
Oleic acid	34.28	Myristic acid	0.17
Stearic acid	5.21	Arachidic acid	0.35
Palitic acid	0.01	Tetracosanoic acid	0.35
Zoomaric acid	0.13	Pentadecanoic acid	0.06
Linoleic acid	40.69	Docosanoic acid	0.01
Daturic acid	0.09	Pentadecanoic acid	0.06

[a]The compositions of fatty acid in waste cooking oils were identified by gas chromatography.

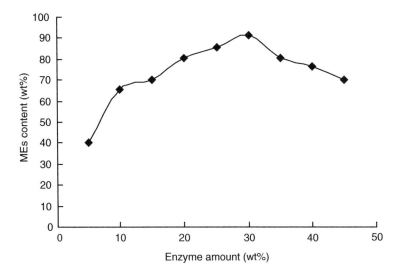

Fig. 2. Effect of enzyme load on methanolysis of waste cooking oils.

immobilized enzyme as reported by Soumanou and Bornscheuer (13). An interesting fact is found here that one-third oil/methanol of molar ratio at 9 h leads to much lower MEs content (75.36 wt%) than at 7 h (89 wt%). This could be attributed to the reversible reaction shifting to the left from right because methanol is used up completely during reaction time of 7 h. Further work is needed to support such a presumption.

Effect of Reaction Temperature

Experiments carried out at 40°C gives the highest MEs yield (87%) as shown in the Fig. 3. From the Fig. 3, it can be seen that at temperatures higher than 40°C, MEs yield decreases as a result of enzyme activity being negatively affected by high temperatures. This result is in agreement with the

Table 3
Effect of Oil/methanol Molar Ratio on Methanolysis of Waste Cooking Oils[a]

Oil/methanol (mol/mol)	MEs content[b] (wt%)	Oil/methanol (mol/mol)	MEs content[b] (wt%)
1/1	30.15	1/5	73.12
1/2	58.91	1/6	59.91
1/3	75.36	1/7	34.35
1/4	85.12	1/8	28.18

[a]Reaction were performed in duplicate at 40°C for 9 h in admixture of 37 g oil/methanol, 30% immobilized *R. orzyae* lipase based on oil weight.
[b]Methyl esters.

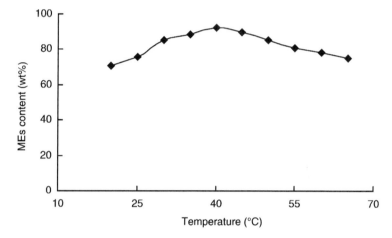

Fig. 3. Effect of temperature on methanolysis of waste cooking oils.

results obtained from methanolysis of vegetable oil using the same lipase (14). The Methyl ester yield at 8 h increases with increasing temperature until 40°C, and after that starts to decrease. The reason for an optimal reaction temperature of 40°C is still not clear and needs further investigation.

Effect of Water Content on Methanolysis of Waste Cooking Oils

The waste cooking oils is different significantly from pure vegetable oils in water content. The waste cooking oils used here contained 2008 ppm water. In order to study the effect of amount of water present in the reaction mixture on methanolysis of waste cooking oils, water content was varied from 15% to 100% by weight of substrate. The experimental results are illustrated in Fig. 4. It is clear that methyl ester production increases with increasing amount of water in the reaction mixture, however, an abrupt decrease is observed at water contents of 75% and 100%. ME yield reaches

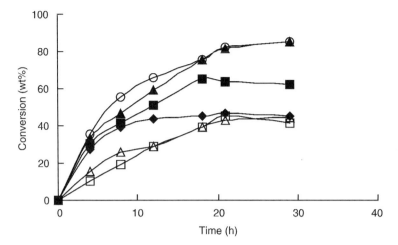

Fig. 4. Effect of water content on methanolysis of waste cooking oils. Oil to methanol molar ratio of 1:3 and 30% R. oryzae lipase used, water content (- □ -) 15%, (- ♦ -) 25%, (- ▲ -) 30%, (- ○ -) 50%, (- ■ -) 75%, and (- △ -) 100%.

the highest value of 60% (w/w) after 24 h of reaction with water content of 75%. It has been reported that significant water excess greatly reduces the amount of ester formed when vegetables oils are esterified with methanol (15,16). Water contents ranging between 4% and 30% are recommended for pure oils (17,18). As for waste cooking oils, owing to their greater viscosity, reaction carried out with 50% (w/w) water content was important to facilitate the good mixing of substrate and to guarantee a greater oil/water interface area at which R. oryzae lipase displays activity.

Three-Step Batch Methanolysis of Waste Cooking Oil

Because immobilized R. oryzae lipase is inactivated easily by higher methanol concentration (19), the lipase that had been used for five cycles was employed for three-step methanolysis of waste cooking oils. As shown in Fig. 5, the MEs yield reaches about 40% and 60% after the first- and second-step reactions, respectively. However, the conversion of waste oil reaches 92.5% after the three-step methanolysis. Such conversion is still less than that of pure vegetable oils. The difference may be attributed to the oxidized fatty acid compounds in waste oil. In general, when a vegetable oil is used for frying, some fatty acids are converted to epoxides, aldehydes, polymers, and so on, by oxidation or thermal polymerization (20). Because the lipase cannot recognize these oxidized compounds, the conversion of waste cooking oils decreases a little (21). To investigate the lipase stability, the three-step batchwise methanolysis was repeated by transferring the lipase to a fresh substrate: first-step, 10 h; second-step, 14 h; and third-step, more than 24 h. The results indicate that the conversion doesn't significantly decrease even after 40 cycles (80 d), showing that contaminants in waste oil do not affect the stability of the lipase.

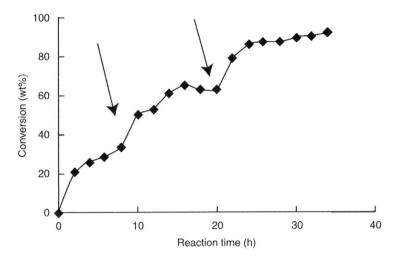

Fig. 5. Three-step batchwise methanolysis of waste cooking oils with R. oryzae lipase.

Physico-Chemical Properties of Biodiesel Under the Optimum Conditions

The sample obtained from waste cooking oils after immobilized R. orzyae lipase-catalyzed transesterification under the optimum processing conditions (i.e., methanol to oils molar ratio of 4:1, with immobilized lipase loaded at 30% based on the oil weight, at a temperature of reaction of 40°C, and using three stepwise addition of methanol) was analyzed both by our lab instrument and the Petroleum Products Analysis Institute of Tianjin. The results are shown in Table 4. It can be seen that the biodiesel sample obtained in the Lab is following within the European Biodiesel Standard DIN EN 14214 and the fuel properties of China diesel no. 0. There are slight differences in density and viscosity comparing with diesel no. 0 but the properties of the biodiesel are completely acceptable. The higher flashpoint of the biodiesel is beneficial from a safety aspect, and the low sulfur content is the reason for the extremely low SO_x emissions associated with its use as a fuel. The cetane number is higher than diesel no. 0 resulting in a smoother running of the engine with less noise. Biodiesel is an oxygenated fuel naturally with oxygen content about 10% which contributes to the favorable emission, but lead to a lower caloric value 32 MJ/kg compared with petro diesel. Cold Filter Pugging Point (CFPP) of the sample is very close to diesel no. 0 in summer and lower in winter thereby contributing to its winter operability.

Conclusions

Waste cooking oils were studied, concentrating on the effect of reaction parameters (the molar ratio of methanol to waste oils, biocatalyst

Table 4
Comparison of Physico-Chemical Properties
of Biodiesel Samples With Standard Data

Properties	European biodiesel standard DIN EN 14214	Biodiesel sample of this work	Chinese no. 0 diesel standard
CFPP[a]			
Winter °C	<0	−10	−5
Summer °C	<−10	−20	−20
Density at 15°C (g/mL)[b]	0.875 ~ 0.9	0.89	0.83
Viscosity at 40°C (mm^2/s)[b]	3.5 ~ 5	4.88	2–4
Flash point (°C)[b]	>120	171	60
Cetane[a]	>49	56.6	>49
Caloric value (MJ/kg)[a]	32.9	32	35
Carbon residue (%w/w)[b]	<0.05	0.2	0.03
Sulfur content (%w/w)[b]	<0.01	0.01	<0.2
Ash (%w/w)[b]	<0.03	0.01	n/a
Water content (ppm)[a]	<300	150	n/a
Oxygen content (%v/v)[a]	<10.9	10	0
Copper strip corrosion[b]	No. 1	No. 1	No. 1
Cloud point (°C)[a]	−4	0	−

[a]Analysis in Petroleum Products Analysis Institute of Tianjin.
[b]Analysis in our lab.

addition amount, reaction temperature and water content) on the yield of MEs. A three-step batch methanolysis of waste cooking oils with immobilized *R. oryzae* lipase was carried out. The results indicate that a methanol to oils molar ratio of 4, immobilized lipase to oils weight ratio of 30%, and temperature of 40°C are suitable for the production of biodiesel from waste cooking oils under 1 atm. The biodiesel produced is good in quality and its yield is satisfactory, up to about 92%. Further investigation is underway to lower the cost of production of biodiesel through developing the whole-cell catalyzed transesterification with waste cooking oils and designing continuous bioreaction process to stably produce biodiesel.

Acknowledgments

The financial support provided partly by Ministry of Science and Technology of China (contract no. 2005CCA06500), European Commission through Asia-Pro-Eco programme and partly by Tianjin Commission of Science and Technology is gratefully acknowledged. The authors also thank Dr. Jiuling Chen working in Faculty of Chemical Engineering and Technology, Tianjin University, for his assistance in analyzing properties of waste cooking oil.

References

1. Leung, D. Y. C. and Chen, G. (2000), In: *Report, Symposium on Energy Engineering*, Cheng P. (eds.), Hong Kong, China.
2. Zhang, Y., Dube, M. A., McLean, D. D., and Kates, M. (2003), *Biores. Technol.* **89,** 1–16.
3. Köse, Ö., Tüter, M., and Aksoy, A. H. (2002), *Biores. Technol.* **83,** 125–129.
4. Watanabe, Y., Shimada, Y., Sugihara, A., and Tominaga, Y. (2002), *J. Mol. Catalysis B: Enzymatic* **17,** 151–155.
5. Antolin, G., Tinaut, F. V., Briceno, Y., Castano, V., Perez, C., and Ramirez, A. I. (2002), *Biores. Technol.* **83,** 111–114.
6. Shimada, Y., Watanable Y., Samukawa T., et al. (1999), *J. Am. Oil Chem. Soc.* **76,** 789–793.
7. Samukawa T., Kaieda M., Matsumoto T., et al. (2000), *J. Biosci. Bioeng.* **90(2),** 180–183.
8. Zhang, Y., Dube, M. A., McLean, D. D., and Kates, M. (2003), *Biores. Technol.* **90,** 229–240.
9. Ji, X. and Hao, X. L. (2002), *Energy China* **5,** 16–18.
10. Yasuhiko, F. (1990), In: *Introduction to Lipid Assay*, 5th ed., Gakkai Publish Center, Japanese.
11. Shah, S., Sharma, S., and Gupta, M. N. (2004), *Energy Fuel* **18,** 154–159.
12. Nelson, L. A., Foglia, T. A., and Marner, W. N. (1996), *J. Am. Oil Chem. Soc.* **73,** 1191–1195.
13. Soumanou, M. M. and Bornscheue, U. T. (2003), *Enzyme Microbial Technol.* **33,** 97–103.
14. Kaieda, M., Samukawa, T., Matsumoto, T., and Fukuda, H. (1999), *Biosci. Bioeng.* **88(6),** 627–631.
15. Shieh, C. J., Liao, H. F., and Lee, C. C. (2003), *Biores. Technol.* **88,** 103–106.
16. Vicente, G., Coteron, A., Martinez, M., and Aracil (1998), *J. Indus. Crops Products* **8,** 29–35.
17. Ban, K., Kaieda, M., Matsumoto, T., Kondo, A., and Fukuda, H., (2001), *J. Biochem. Eng.* **8,** 39–43.
18. Park, S. B., Endo, Y., Maruyama, K., and Fujimoto, K. (2000), *Food Sci. Technol.* **6,** 192–196.
19. Watanabe, Y., Shimada, Y., Sugihara, A., Noda, H., Fukuda, H., and Tominaga, Y. J. (2000), *J. Am. Oil Chem. Soc.* **77,** 55–261.
20. Salah, A. B., Sayari, A., Verger, R., and Gargouri, Y. (2001), *Biochemie* **83,** 463–469.
21. Kikugawa, K. (2000), In: *The Handbook of Oil Chemistry-Lipids and Surfactants*, 4th ed., Tokyo Maruzen.

Inulin-Containing Biomass for Ethanol Production

Carbohydrate Extraction and Ethanol Fermentation

Mª José Negro, Ignacio Ballesteros, Paloma Manzanares, José Miguel Oliva, Felicia Sáez, and Mercedes Ballesteros*

Renewable Energies Division-CIEMAT, Avda. Complutense, 22 28040-MADRID, SPAIN; E-mail: m.ballesteros@ciemat.es

Abstract

The use of stalks instead of tubers as a source of carbohydrates for ethanol production has been investigated. The inulin present in the stalks of Jerusalem artichoke was extracted with water and the effect of solid–liquid ratio, temperature, and acid addition was studied and optimized in order to attain a high-fructose fermentable extract. The maximum extraction efficiency (corresponding to 35 g/L) of soluble sugars was obtained at 1/6 solid–liquid ratio.

Fermentations of hydrolyzed extracts by baker's yeast and direct fermentation by an inulinase activity yeast were also performed and the potential to use this feedstock for bioethanol production assessed. The results show that the carbohydrates derived from Jerusalem artichoke stalks can be converted efficiently to ethanol by acidic hydrolysis followed by fermentation with *Saccharomyces cerevisiae* or by direct fermentation of inulin using *Kluyveromyces marxianus* strains. In this last case about 30 h to complete fermentation was required in comparison with 8–9 h obtained in experiments with *S. cerevisiae* growth on acid extracted juices.

Index Entries: Ethanol; fermentation; Jerusalem artichoke; sugar extraction.

Introduction

Jerusalem artichoke (*Helianthus tuberosus* L.) has shown excellent potential as an alternative sugar crop (1). Jerusalem artichoke is a perennial herbaceous plant belonging to the sunflower family, which is well adapted to a wide variety of climates (2). It does not require soil fertility and develops underground stolons forming shaped tubers, which are similar to potatoes. Like sugar beet, Jerusalem artichoke produces sugars in the above ground and stores them in the roots and tubers. The tubers consist

---

*Author to whom all correspondence and reprint requests should be addressed.

of 75–79% water, 2–3% proteins, and 15–16% carbohydrates, of which the D-fructose polymer inulin can constitute 80% or more.

This D-fructose polymer, which is initially in high levels in the stems, is transferred to tubers by the end of the growing cycle and the majority of the sugar produced in the leaves does not enter the tuber until the plant has nearly reached the end of its productive life *(3,4)*. Traditionally, Jerusalem artichoke has been grown for the tubers utilization and it is harvested in late autumn when the migration of carbohydrates from the aerial part of the plant to underground tubers has been completed. According to its high solubility in hot water, techniques similar to those based on sugars diffusion for sugar beet have been used to extract the inulin.

Over the past decades, Jerusalem artichoke has received interest for the production of fructose syrups *(5,6)* and ethanol *(7,8)* from tubers but the high cost of the harvesting and the infesting nature of tubers left in the soil that has been a limiting factor to the expansion of the crop. However, an alternative to harvest and handle the tuber exists. The possibility to harvest the above-ground biomass before tubers development and use the inulin containing stalks as feedstock for ethanol production is now envisaged as an interesting option. Thus, it could be possible to harvest the Jerusalem artichoke crop when the sugar content in the stalk reaches a maximum, thereby avoiding the harvesting of the tubers. In this case, the harvesting equipment and procedures are essentially the same as for harvesting sweet sorghum or corn for ensilage, reducing operation costs.

Although various strains of yeasts (*Kluyveromyces*, *Candida*, and *Schizosaccharomyces*) have been efficiently used for the direct fermentation of extracts from Jerusalem artichoke tubers *(9–11)*, the process is slow, taking typically 30–40 h. As a consequence, for a commercial operation a hydrolysis step before fermentation with conventional yeast is preferable.

In this work the use of stalks instead of tubers as a source of carbohydrates for ethanol production has been investigated. The effect of solid–liquid ratio, temperature, and acid addition on the water extraction of inulin from fresh Jerusalem artichoke stalks has been studied and optimized in order to attain a high fructose fermentable extract. Furthermore, fermentation of hydrolyzed extracts by baker's yeast *Saccharomyces cerevisiae* and direct fermentation by yeast *Kluyveromyces marxianus* CECT 10875 having inulinase activity has been performed and the potential to use this feedstock for bioethanol production assessed.

Material and Methods

Material

Stalks from *Helianthus tuberosus* L. var. Violet de Rennes were kindly supplied by Agronomy School of Madrid. The harvest was performed at the end of September, before flowering. Fresh stalks, without leaves, were crushed with a Blixer 4 vv (Robot Cupe, UK) and frozen at –18°C for storage.

Carbohydrate Extraction Procedure

The aqueous extraction of free sugars, sucrose, and inulin from stalks was carried out at different temperatures and different solid–liquid ratios (1/6, 1/4, and 1/3 w/w).

In order to test the effect of water temperature on free sugars, sucrose, and inulin solubilization, 20 g of stalks, crushed into small pieces, were placed into 500-mL Erlenmeyer flasks containing 120 mL of deionized water at 60°C and 80°C and boiling water for 30 min. The extract then, was separated and the solids were resuspended with an equal amount of water for a second extraction. As very low sugar content (<2%) was achieved in the second extraction, just one-step extraction was used in further experiments.

After the extraction samples were taken, the liquid phase removed by filtration and an aliquot of this extract analyzed for glucose, fructose, and sucrose content. To determine inulin concentration, a part of the filtrate was treated with HCl 1 N at 60°C for 30 min for inulin hydrolysis and then analyzed for sugars content. The glucose and fructose concentrations released from inulin were calculated by the difference in both determinations. The inulin content, corrected for water loss during hydrolysis, was calculated as follows (12):

$$\text{Inulin} = k\,(G_i + F_i)$$

where: $k = [180 + 162\,(n-1)]/180n$; n is the average polymerization degree $n = [(F_i/G_i) + 1]$; G_i the glucose release from fructans; and F_i the fructose release from fructans.

$$\text{Considering that: } G_i = G_t - G_l \text{ and } F_i = F_t - F_l$$

where: G_t is the total glucose after hydrolysis; G_l the free glucose before hydrolysis and glucose from sucrose; F_t the total fructose after hydrolysis; and F_l the free fructose before hydrolysis and fructose from sucrose.

In order to test the effect of solid content on sugars solubilization, experiments at different solid–liquid ratios (1/6, 1/4, and 1/3) were carried out. So, 20, 30, and 40 g of stalks, crushed into small pieces, were placed into 500-mL Erlenmeyer flasks containing 120 mL of deionized water and boiled for 30 min. Samples were withdrawn at 10, 20, and 30 min and analyzed as described earlier. In order to test the direct extraction of monosaccharides (glucose and fructose) from inulin contained in Jerusalem artichoke stalks, clorhydric acid at concentrations of 0.05, 0.1, and 0.5 N instead of water were used. Extractions were performed at 1/6 solid–liquid ratio and boiled for 30 min. Samples were taken at 10, 20, and 30 min and analyzed.

Microorganisms and Growth Conditions

Saccharomyces cerevisiae baker's yeasts and *Kluyveromyces marxianus* CECT 10875 (a microorganism capable of fermenting both monomers and inulin) were used in fermentation experiments. Active cultures for inoculation were prepared by growing the organisms on a rotary shaker at 150 rpm for 16 h at 35°C in a growth medium containing 5 g/L yeast extract (Difco, East Molesley, UK), 5 g/L peptone (Oxoid, Hampshire, UK); 2 g/L NH_4Cl, 1 g/L KH_2PO_4, 0.3 g/L $MgSO_4·7H_2O$, and 30 g/L glucose or inulin.

Fermentation Assays

Fermentation experiments were carried out in 100-mL Erlenmeyer flasks, each containing 50 mL of the Jerusalem artichoke stalk extract, and were incubated at 35°C and 150 rpm. Flasks were inoculated with 4% (v/v) yeast cultures and periodically analyzed for ethanol and sugars. Fermentations by *S. cerevisiae* were performed on acid-extracted juice. *K. marxianus* was grown on extracted juice without any hydrolysis step. Fermentation results of ethanol production and sugars consumption were reported as percentage of the theoretical yield.

Analytical Procedures

Compositional analysis of the substrate was performed according to the Laboratory Analytical Procedures developed by the National Renewable Energy Laboratory (13).

The maximum amount of free sugars, sucrose, and inulin in the stalks were determined according to AOAC standard method (14) and used to calculate extraction efficiency.

Sugars were quantified by high-performance liquid chromatography (HPLC) in a 1081B Hewlett Packard (HP) apparatus with refractive index detector under the following conditions: column, Aminex HPX-87P (300 mm × 7.6 mm) (BioRad, Hercules, CA); temperature, 85°C; eluent, water at 0.6 mL/min.

Ethanol was measured by gas chromatography (GC), using a HP 5890 Series II apparatus (Palo Alto, CA), equipped with an Agilent 6890 automatic injector and a flame ionization detector (Palo Alto, CA). A column of Carbowax 20 *M* (2 m × 1/8 in.) using helium as carrier gas at 35 mL/min was used. The GC oven temperature was held at 85°C. The injector and detector temperature was maintained at 150°C.

Results and Discussion

Composition of Jerusalem Artichoke Stalks

Table 1 shows chemical composition of Jerusalem artichoke stalks. Stalks were made up of about 58.5% moisture, and the remaining 41.5% solids were made up of 58% organic and aqueous extractives, 17.2% of cellulose, 14.6% klason lignin, 6.5% hemicellulose (mainly xylans), and 3.7% total ash. The total water soluble sugars (both free glucose and fructose, sucrose, and inulin) content in Jerusalem artichoke stalks was 58.7% of dry

Table 1
Composition of Jerusalem Artichoke Stalks

Component	% Dry matter
Cellulose	17.2
Glucose	18.9 ± 0.16
Hemicellulose	6.5
Xylose	5.6 ± 0.55
Galactose	1.2 ± 0.01
Arabinose	0.6 ± 0.01
Klason lignin	14.6 ± 0.41
Extractives	58.0 ± 0.1
Total ash	3.7 ± 0.05
Total	100

Data are the mean value of three replications.

weight of which 13% were monosaccharides, 3% was sucrose and 84% was oligofructose consisted of polymers about 20 fructose units with one glucose unit at the end of the molecule.

It is important to note that the total potential sugar content (both structural and nonstructural carbohydrates) in stalks at this developmental stage of the plants is close to 80% of dry matter content.

Effect of Temperature, Solid–Liquid Ratio, and Acid on Water Extraction of Inulin

Different methods, such as grinding or cutting the tubers with sugar beet slicer, followed by water extraction (with both boiling or cold water) have been normally used for the extraction of inulin in Jerusalem artichoke (15,16). In this work, a similar method for sugar extraction was applied to Jerusalem artichoke stalks. The effect of different variables (water temperature, amount of water, and acid addition) on the extraction of inulin, sucrose, and simple sugars was studied.

To test the effect of water temperature on the extraction efficiency of carbohydrates (inulin, sucrose, and simple sugars) from Jerusalem artichoke stalks, water at 60°C and 80°C temperature and boiling water were employed (Fig. 1). Extraction efficiency was calculated as the ratio of the amount of sugars obtained to the theoretical maximum amount of sugars available (7.8, 1.4, and 49.5% dry-wt for free sugars, sucrose, and inulin, respectively). As it can be observed, water temperature had not effect on the extraction of free sugars and sucrose, obtaining 100% extraction efficiency for all tested temperatures. However, at 60 and 80°C water temperature only 80% of the potential inulin was recovered in the juice, whereas using boiling water inulin extraction of 100% of the theoretical was obtained. As a result, for the extraction of carbohydrates in further experiments boiling water were used.

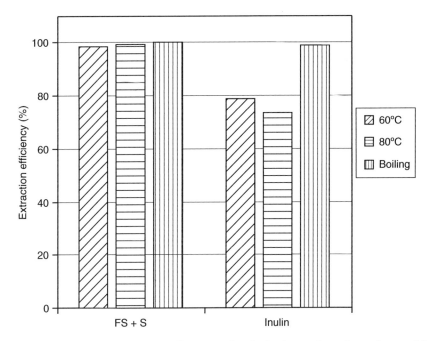

Fig. 1. Carbohydrate extraction efficiency of carbohydrates from Jerusalem artichoke stalks at different water temperatures. FS, free sugars; S, sucrose.

Figure 2A shows the sugar content in water extracts obtained from the extraction of Jerusalem artichoke stalks with boiling water at different solid–liquid ratio. Incubations longer than 10 min did not result in higher sugar concentration in the water extract. As expected, higher oligofructose concentration was obtained at higher substrate loading. Juices containing more than 56 g/L of total sugars were obtained at 1/3 solid–liquid ratio.

The effect of water amount on the extraction efficiency of soluble sugars was also analyzed (Fig. 2B). As can be observed, higher sugar extraction efficiencies were obtained at lower solid–liquid ratio. Free sugars and inulin efficiencies extraction about 90% were achieved at 1/3 and 1/4 solid – liquid ratio. The maximum extraction efficiency of 100% for free sugars and sucrose and 95% for inulin was attained at 1/6 solid–liquid ratio (corresponding of total carbohydrate concentration of 38.5 g/L); therefore these conditions were selected to further studies.

For ethanol production from carbohydrates (free sugars, sucrose, and inulin) contained in stalks of Jerusalem artichoke by *S. cerevisiae*, the inulin has to be first converted into simple sugars (glucose and fructose) by acidic or enzymatic hydrolysis. In order to examine the direct extraction of monosaccharides contained in stalks, boiling clorhydric acid at concentrations of 0.05, 0.1, and 0.5 N instead of water were used (Fig. 3). Carbohydrates were completely extracted and easily hydrolyzed in 0.05 N HCl within 10 min with no change thereafter. However, higher acid concentrations produced a negative effect on the sugar concentration owing to the

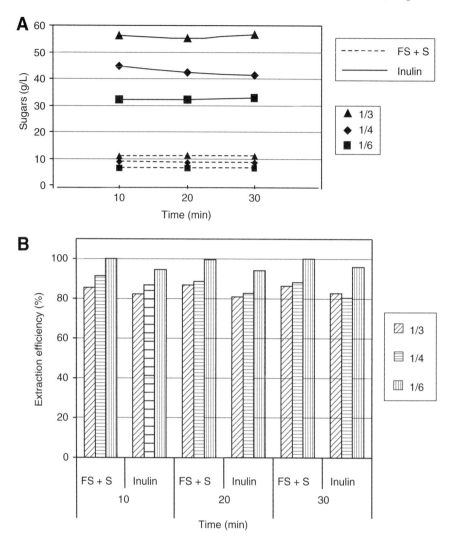

Fig. 2. Carbohydrate concentration (**A**) and extraction efficiency (**B**) in extracts obtained from the extraction of Jerusalem artichoke stalks using boiling water at different solid–liquid ratio (1/3, 1/4, and 1/6). FS, free sugars; S, sucrose.

degradation of fructose formed during depolymerization process. So, juices prepared by acid extraction using 0.05 N HCl were used for further fermentation experiments.

Ethanol Fermentation Assays

Sugars consumption and ethanol production results by *S. cerevisiae* of acid-extracted juices from Jerusalem artichoke stalks at different solid to liquid ratios are shown in Fig. 4A. Neither the rate of sugars utilization nor the final sugars concentration remaining in the media were affected by the initial sugar concentration in the range tested. Fermentation was completed

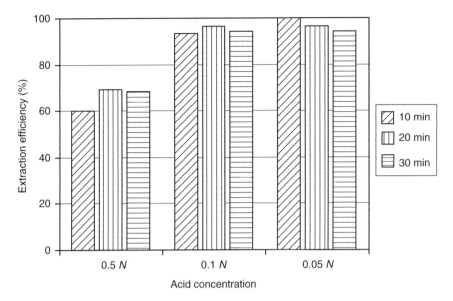

Fig. 3. Sugars extraction efficiency using boiling clorhydric acid at different concentrations.

in about 8–9 h from duration of inoculation. Using initial sugar concentrations of 38.3, 58, and 71.8 g/L (corresponding to 1/6, 1/4, and 1/3 solid–liquid ratio), final ethanol concentrations of 18, 27, and 32.5 g/L, respectively were obtained.

The direct conversion (without hydrolysis step) of inulin into ethanol using a selected *K. marxianus* strain growing on juices from Jerusalem artichoke stalks extracted at different solid–liquid ratio was also studied. (Fig. 4B). Direct conversion of inulin to ethanol by yeasts, which possesses the inulinase enzyme activity, was slower than fermentation of simple sugars by *S. cerevisiae*. Juices containing inulin could be completely fermented in 30 h regardless of initial carbohydrate concentration, although slight amounts of residual inulin were found at the end of fermentations at the highest initial sugar concentration.

As it can be seen, the rate of sugars utilization was not affected by initial inulin concentration. On the other hand, a delay in ethanol production was observed in comparison with inulin consumption. Thus, although inulin content in the media was negligible after 15 h from the onset of fermentation, the maximum ethanol concentration was found at 30 h. These results suggest that the microorganism has to hydrolyze the inulin first to monomeric sugars and subsequently convert them into ethanol. It should be noted that juices containing inulin used as fermentation broth, were prepared at a low inoculum loading and without any external assimilable nitrogenous sources. Long fermentation times required for the direct inulin conversion into ethanol could be reduced increasing cell population in the media. Thus, the use of higher inoculum size and

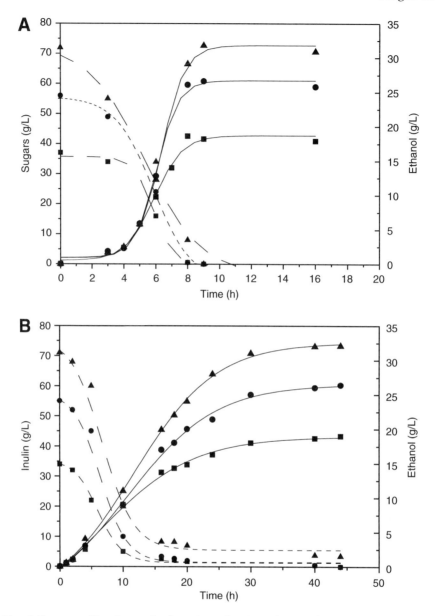

Fig. 4. Sugars utilization and ethanol production of juices from Jerusalem artichoke stalks extracted at different solid–liquid ratios (▲, 1/3; ●, 1/4; and ■, 1/6). **(A)** Free sugars fermentation by *S. cerevisiae* on acid-extracted juice; **(B)** direct inulin fermentation by *K. marxianus* on water-extracted juice.

nutrient supplementation of the media should be studied. Further investigation will be carried out in our laboratory.

Data in Fig. 4 were used to calculate ethanol yield and percentage of substrate utilization (Fig. 5). Percentages of substrate utilization in direct inulin fermentation were slightly affected by initial sugar concentration, ranging from 96% to 100%. Ethanol yield was found to decrease with

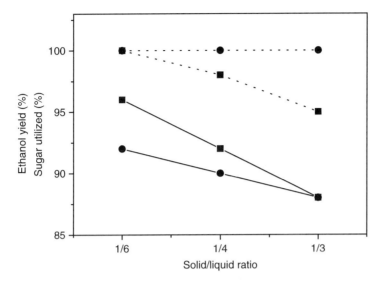

Fig. 5. Total sugar utilization and ethanol yield (expressed as percentage of the theoretical) of free sugars fermentation by *S. cerevisiae* (●) and direct inulin fermentation by *K. marxianus* (■) on juices obtained from Jerusalem artichoke stalks extracted at different solid–liquid ratios (1/3, 1/4, and 1/6).

increasing initial sugars content. Lower ethanol yields, between 88% and 92%, were obtained in fructose fermentations by *S. cerevisiae* in comparison to those obtained by *K. marxianus* growing on inulin. It has been stated that *S. cerevisiae* is glucophilic yeast, being less efficient in the utilization of fructose *(17,18)*. Ethanol yields obtained in this work were similar to those obtained by Bajpai and Margaritis *(19)* working with another strain of *K. marxianus* using inulin from an extract of Jerusalem artichoke tubers.

Conclusions

Results show that the carbohydrates derived from Jerusalem artichoke stalks can be converted efficiently to ethanol by acidic hydrolysis followed by fermentation with *S. cerevisiae* or by direct fermentation of inulin using *K. marxianus* strains, although in this last case about 30 h to complete fermentation were required in comparison with 8–9 h for experiments with *S. cerevisiae* growth on acid extracted juices. Studies to decrease fermentation time are being carried out.

On the basis of 4.5 t/ha produced annually from Jerusalem artichoke stalks reported in the literature *(20)* and the ethanol yield of 0.49 obtained in our study for *K. marxianus*, it is feasible to produce 2.2 t of ethanol/ha/yr. Moreover, the solid left after extraction of soluble sugars contains about 24% of structural carbohydrates (cellulose and hemicellulose) on dry-weight basis of stalks. They could also be converted into ethanol, which would increase significantly ethanol production from Jerusalem artichoke stalks. Work is underway in our laboratory for the enzymatic hydrolysis and fermentation of cellulose component of Jerusalem artichoke stalks.

Acknowledgment

This work has been supported by the Spanish Science and Education Ministry Contract No.

References

1. Gosse, G. (1988), In: *Topinambour (Jerusalem artichoke)*, EUR 11855, G. Grassi and G. Gosse, eds., Commission of the European Communities, pp. 3–14.
2. Kosarik, N., Consentino, G. P., and Wieczorek, A. (1984), *Biomass* **5**, 1–36.
3. Incoll, L. D. and Neales, T. F. (1970), *J. Exp. Bot.* **21(67)**, 469–476.
4. Ballesteros, M. (1989), PhD thesis, UAM University of Madrid, Spain.
5. Chubey, B. B. and Dorrel, D. G. (1974), *Can. Inst. Food Sci. Tech.* **7**, 98–100.
6. Kriestan, M. P. J. (1978), *Biotech. Bioeng.* **3**, 447–4503.
7. Margaritis, A., Bajpai, P., and Cannell, E. (1981), *Biotech. Lett.* **3**, 595–599.
8. Szambelan, K., Nowak, J., and Chrapkowska, K. J. (2004), *Acta Sci. Pol. Technol. Aliment.* **3(1)**, 45–53.
9. Guiraud, J. P., Daurelles, J., and Galzy, P. (1981), *Biotech. Bioeng.* **23**, 1401–1420.
10. Margaritis, A. and Bajpai, P. (1981), *Biotech. Lett.* **3**, 679–685.
11. Poncet, S., Jacob, F. H., Berton M. C., and Couble, A. (1985), *Ann. Inst. Pasteur. Microbiol.* **136B(1)**, 99–109.
12. Prosky, L. and Hoebregs, H. (1999), *J. Nutr.* **129**, 1418S–1423S.
13. National Renewable Energy Laboratory (NREL). Chemical Analysis and Testing Laboratory Analytical Procedures: LAP-001 to LAP-005, LAP-010 and LAP-017. NREL, Golden, CO. www.ott.doe.gov/biofuels/analytical_methods.html.
14. Hoebregs, H. (1997), *J. AOAC Int.* **80**, 1029–1037.
15. Byun, S. M. and Nahm, B. H. (1948), *J. Food Sci.* **43**, 1871–1879.
16. Margaritis, A., Bajpai, P., and Bajpai, P. K. (1983), *Develop. Ind. Microbiol.* **24**, 321–327.
17. Berthels, N. J., Cordero Otero, R. R., Bauer, F. F., Thevelein, J. M., and Pretorius, I. (2004), *FEMS Yeast Res.* **4**, 683–689.
18. Wang, D., Xu, Y., Hu, J., and Zhao, G. (2004), *J. Inst. Brew.* **110(4)**, 340–346.
19. Bajpai, P. and Margaritis, A. (1982), *App. Environ. Microbiol.* **44(6)**, 1325–1329.
20. Caserta, G. and Cervigni, T. (1991), *Biores. Technol.* **35**, 247–250.

Production of Medium-Chain-Length Polyhydroxyalkanoates by *Pseudomonas aeruginosa* With Fatty Acids and Alternative Carbon Sources

PUI-LING CHAN, VINCENT YU, LAM WAI, AND HOI-FU YU*

State Key Laboratory of Chinese Medicine and Molecular Pharmacology, Shenzhen, China, and Applied Biology and Chemical Technology, Hong Kong Polytechnic University, Hong Kong, China; E-mail: bcpyu@polyu.edu.hk

Abstract

In this study, medium-chain-length polyhydroxyalkanoates (mcl-PHAs) were produced by *Pseudomonas aeruginosa* using different carbon sources. Decanoic acid induced the highest (9.71% [±0.7]) mcl-PHAs accumulation in bacterial cells at 47 h. The cells preferred to accumulate and degrade the polyhydroxyoctanoate than polyhydroxydecanoate (PHD) during early stage and final stage of the growth, respectively. The production cost of mcl-PHAs can be reduced by using edible oils as the carbon source. The bacteria accumulated 6% (±0.7) of mcl-PHAs in the presence of olive oil. Besides, reused oil was another potential carbon source for the reduction of the production cost of mcl-PHAs. Overall, PHD was the major constituent in the accumulated mcl-PHAs.

Index Entries: Edible oils; fatty acids; medium-chain-length polyhydroxyalkanoates; *Pseudomonas aeruginosa*.

Introduction

Polyhydroxyalkanoates (PHAs) are polymers that are accumulated as energy storage materials in various microorganisms when cells have adequate supplies of carbon but are limited for another nutrient, such as nitrogen, phosphate, or oxygen *(1–3)*. In general, PHAs can be divided into two groups according to the number of carbon atoms in the monomer units. Short-chain-length PHAs (scl-PHAs) contain 3–5 carbon atoms and are commonly produced by *Ralstonia eutropha* and *Alcaligenes latus*. Medium-chain-length PHAs (mcl-PHAs) contain 6–14 carbon atoms and are typically produced by fluorescent pseudomonads *(4)*. The properties of PHAs follow in part from their bacterial origin. scl-PHAs are highly crystalline materials and are brittle in nature. Owing to their rigid property, they are not suitable

---

*Author to whom all correspondence and reprint requests should be addressed.

for making films for medical and special devices (5). Mcl-PHAs are elastomeric thermoplastics. They are rubbery and flexible materials with low crystallinity and can be used in a wide range of industrial and medical applications which cannot be fulfilled by scl-PHAs (6). The mode of biosynthesis of polyhydroxybutyrate (PHB) is well understood. However, the synthesis of mcl-PHAs has been studied in recent years (7). Like all PHAs, accumulation of mcl-PHAs is in the form of granules (8). It is believed that mcl-PHAs synthesis from fatty acids is via the fatty acid β-oxidation to yield acyl-CoAs intermediates such as enoyl-CoA, 3-Ketoacyl-CoA, and/or S-3-hydroxyalcyl-CoA, which are substrates for mcl-PHAs synthase when fatty acids are used as the sole carbon source (7). *Pseudomonas aeruginosa* is a Gram-negative, aerobic rod, belonging to the bacterial family *Pseudomonadaceae* and is reported to produce mcl-PHAs (9). Highly homologous genes have been isolated from *P. aeruginosa* in order to enhance the production of mcl-PHAs (10–12).

Owing to the promising future of mcl-PHAs in industrial and medical areas, the production of mcl-PHAs is highly desired. In this study, we studied mcl-PHAs synthesis by *P. aeruginosa* using different fatty acids. Moreover, edible oils from the local market were used as substitute carbon sources in order to reduce the production cost.

Materials and Methods

Bacterial Cultures and Growth Conditions

A strain of *P. aeruginosa* PAO1 was donated by Institute of Microbiology of Shandong University (11). The strain was stored at –20°C in the presence of 25% (v/v) glycerol and was maintained on a nutrient agar (Oxoid, Hampshire, England) slant at 4°C by monthly subculture. In the shake flask study, the strain was inoculated in 1-L conical flasks containing 200 mL of cultivation media. The cultivation medium is shown in Table 1. Glucose and fatty acids are obtained from International Laboratory USA, whereas edible oils are obtained from the local market. The pH was adjusted to 7.0, and the medium was autoclaved. The cells were cultivated in the above media for 48 h under aerobic conditions in a temperature-controlled shaker set at 250 rpm and at 30°C, harvested by centrifugation, washed and then lyophilized by freeze drier. Triplicates were done for each manipulation and the values (presented later) after the ± symbol in the parentheses indicated the standard deviation.

Extraction of Polymers

The extraction method (13) was modified. In brief, lyophilized cells were suspended in chloroform (30 mL/g) and extracted overnight at 60°C. The cell debris was then separated by filtration. The chloroform solution was concentrated by rotary evaporator under 60°C and the polymer was precipitated overnight on cold methanol (for better precipitation)

Table 1
The Cultivation Medium for the Production of mcl-PHAs

Component (/L)	Constituents	
E medium	3 g/L $(NH_4)_2HPO_4$	
	5.8 g/L K_2HPO_4	
	3.7 g/L KH_2PO_4	
100 mM $MgSO_4 \cdot 7H_2O$		
1 mL microelement		
solution (/L of 1 N HCl)	2.78 g/L $FeSO_4 \cdot 7H_2O$	
	1.98 g/L $MnCl \cdot_4 H_2O$	
	2.81 g/L $CoSO_4 \cdot 7H_2O$	
	1.67 g/L $CaCl_2 \cdot 2H_2O$	
	0.17 g/L $CuCl_2 \cdot 2H_2O$	
	0.29 g/L $ZnSO_4 \cdot 7H_2O$	
Carbon source	Glucose (20 g/L)	
	Fatty acids (8 g/L)	Citric acid
		Octanoic acid
		Nonanoic acid
		Decanoic acid
		Lauric acid
	Edible oils (8 g/L)	Corn oil
		Soybean oil
		Peanut oil
		Canola oil
		Olive oil
		Reused oil

(1:9 v/v chloroform: methanol). The precipitated polymer was collected by centrifugation (Beckman J2-21) at 9000g (Beckman, CA).

Preparation and Analysis of PHAs

Polyesters in dried cells (20 mg) or in purified forms (10 mg) were methyl-esterfied in a (1:1; v/v) mixture of chloroform and methanol-sulfuric acid (14). The lower chloroform solution was analyzed by gas chromatography (GC). A Hewlett Packard 5890 Series II Gas Chromatograph system (Hewlett Packard) was used, equipped with an AT-50 Alltech® capillary column (Alltech) and with a flame ionization detector. The operating program was set according to Ballistreri et al. (13). Cell concentration was defined as cell dry weight (g)/L of culture broth. In brief, the cell was centrifuged in order to remove the supernatant and oven dried. The cell weight was measured as the cell concentration. The PHA content (wt%) was defined as the percentage of the ratio of PHA concentration to cell concentration (i.e., g [%] mcl-PHAs/g cell mass).

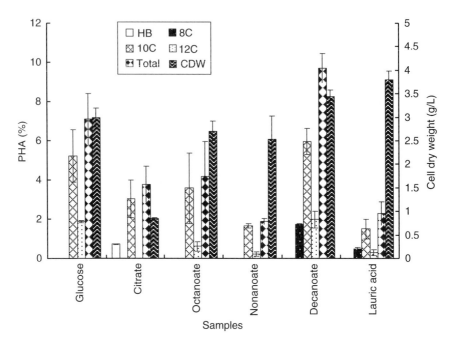

Fig. 1. The percentage of different mcl-PHAs accumulated by *P. aeruginosa* in the presence of glucose and other fatty acids and their relative cell dry weight. HB, hydroxybutyrate; 8C, hydroxyoctanoate (HO); 10C, hydroxydecanoate (HD); 12C, hydroxydodecanoate (HDD); total, total polyhydroxyalkanoates accumulated in the cell; CDW, cell dry weight.

Results

The results of the different mcl-PHAs accumulated by *P. aeruginosa* in the presence of glucose and other fatty acids and their relative cell dry weight are shown in Fig. 1 and Table 2. The results showed that the 3-hydroxydecanoate (HD) and 3-hydroxyoctanoate (HO) were the major types of mcl-PHAs accumulated in the bacterial cell. Polyhydroxydecanoate-co-Polyhydroxydodecanoate (PHD-co-PHDD) was accumulated in the cells in medium containing glucose, octanoic acid, and nonanoic acid. Polyhydroxybutyrate-co-Polyhydroxydecanoate (PHB-co-PHD) was accumulated in the cells cultured in citric acid medium. For the decanoic acid and lauric acid medium, the bacteria produced Polyhydroxyoctanoate-co-Polyhydroxydecanoate-co-Polyhydroxdodecanoate (PHO-co-PHD-co-PHDD), with Polyhydroxyoctanoate (PHO) and Polyhydroxydro-decanoate (PHD) being accumulated in greater proportion.

Figure 2 showed the growth curve of *P. aeruginosa* and time points for different mcl-PHAs accumulation using the decanoic acid as the carbon source. The results showed that *P. aeruginosa* in medium containing decaonic acid grew exponentially from 16–33 h. The cell produced more PHO than PHD between 22 and 60 h. However, the accumulation of PHD was higher after 60 h of fermentation. However, only a low percentage

Table 2
The PHA Accumulation in the Bacterial Cell and Their Cell Dry Weight
After 48 h Cultivation in Medium Containing Different Carbon Sources

Carbon sources	mcl-PHAs accumulated (%)	Cell dry weight (g/L)
Glucose	7.1 (±1.3)[a]	3.0 (±0.2)
Citric acid	3.78 (±0.9)	0.84 (±0.02)
Octanoic acid	4.19 (±1.8)	2.7 (±0.21)
Nonanoic acid	1.89 (±0.1)	2.5 (±0.5)
Decanoic acid	9.71 (±0.7)	3.4 (±0.14)
Lauric acid	2.28 (±0.6)	3.8 (±0.18)

[a]The values in the parentheses were standard deviations.

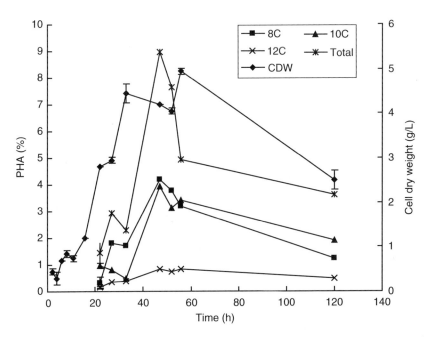

Fig. 2. The growth curve and the percentage of different mcl-PHAs accumulated by *P. aeruginosa* in 120 h shake flask fermentation by using decanoic acid as the sole carbon source. 8C, hydroxyoctanoate (HO); 10C, hydroxydecanoate (HD); 12C, hydroxydodecanoate (HDD); total, total polyhydroxyalkanoates accumulated in the cell; CDW, cell dry weight.

of Polyhydroxydodecanoate (PHDD) was accumulated in cell during the fermentation. The highest percentage of mcl-PHAs accumulation was up to 8.96 (±0.5)% at 47 h.

The effect of edible oils as carbon sources on the accumulation of mcl-PHAs by *P. aeruginosa* are shown in Table 3 and Fig. 3. The results showed that all edible oils used except corn oil could induce the PHO-co-PHD-co-PHDD accumulation in the bacterial cells. As in the fatty acid fermentation,

Table 3
The PHA Accumulation in the Bacterial Cell and Their Cell Dry Weight
Using Various Edible Oils as Carbon Source

Edible oil	Total PHA accumulated in cell (%)	Cell Dry Weight (g/L)
Corn oil	0.72 (±0.1)[a]	4.04 (±0.3)
Soybean oil	2.38 (±0.4)	2.55 (±0.1)
Peanut oil	3.84 (±0.5)	3.99 (±0.3)
Canola oil	3.31 (±0.5)	4.19 (±0.3)
Olive oil	6.0 (±0.7)	4.49 (±0.2)
Reused oil	5.72 (±1.0)	5.41 (±0.05)

[a]The values in the parentheses were standard deviations.

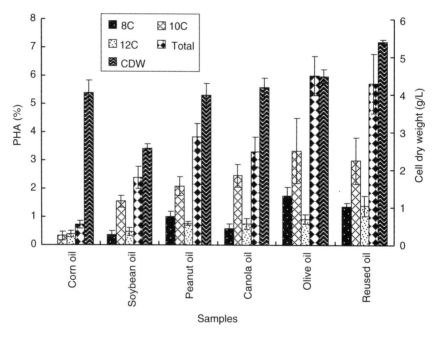

Fig. 3. The percentage of different mcl-PHAs accumulated by *P. aeruginosa* in the presence of different edible oils. HB, hydroxybutyrate; 8C, hydroxyoctanoate (HO); 10C, hydroxydecanoate (HD); 12C, hydroxydodecanoate (HDD); Total, total polyhydroxyalkanoates accumulated in the cell; CDW, cell dry weight.

the bacteria accumulated higher percentage of PHD in medium containing edible oils. The bacteria grew better in the medium containing reused oil. The highest amount of mcl-PHAs accumulated was achieved with the medium containing olive oil. The estimated costs for carbon source for producing mcl-PHAs are set out in Table 4. The costs are around $0.68, $1, $0.16, $0.18, and $0.35/g of mcl-PHA for corn oil, soybean oil, peanut oil, canola oil, and olive oil, respectively.

Table 4
The Estimated Carbon Source Costs
Used for Producing/L g of mcl-PHAs

Sample	Cost (USD)[a]
Decanoic acid	24.1
Corn oil	0.68
Soybean oil	1
Peanut oil	0.16
Canola oil	0.18
Olive oil	0.35

[a]The carbon source costs are calculated according to the prices of the carbon sources and their performance on the mcl-PHAs accumulation. Others costs (e.g., ingredients of medium, manpower, and so on) did not included in these costs.

Discussion

Mcl-PHAs produced by *P. aeruginosa* in medium containing different carbon sources were extracted and analyzed. The results in Fig. 1 and Table 2 demonstrated that besides related carbon sources (fatty acids), mcl-PHAs could also be produced from an unrelated carbon source, i.e., glucose. The results showed that the high cell mass will not contribute to the high PHA production. This is because the growth of the cell is not related to the production of PHAs. The growth of the cell only depends on the environmental conditions and richness of the nutrient presented in the medium. Once the environmental conditions are suitable and the nutrient is rich in the medium, the bacteria can grow well. However, the production of PHA is not related to the growth of the cell. The metabolic condition of the cell contributes to the accumulation of PHA as PHA is a secondary metabolite. The bacteria will accumulate the PHA as the carbon storage compound under limited nutrient, such as nitrogen, phosphate, or oxygen but excess carbon source. Also, when the carbon source is limited in the medium, the bacteria will breakdown the PHA as the carbon source. Therefore, the cell mass is not directly proportional to the PHA concentration. The results also demonstrated that HD was the major constituent in PHAs and HO being the second most common monomers in the polymer. The highest mcl-PHAs yield of 9.71 (±0.7)% was obtained with decanoic acid mineral medium.

Because the bacteria accumulated the highest percentage of mcl-PHAs in the presence of decanoic acid, therefore, decanoic acid was chosen for studying cell growth and the mcl-PHA accumulation between time intervals. The results demonstrated that the exponential phase of *P. aeruginosa* was between 16 h and 33 h. The results in Fig. 2 showed that the cells accumulated more 3-HO than 3-HD between 22 and 60 h and the reverse after

60 h. This demonstrated that the bacteria preferred to accumulate PHO than PHD when decanoic acid is still rich in the medium. After 60 h fermentation, the nutrient began to be depleted and the bacteria preferred to hydrolyze the PHO than PHD as the carbon source for growth. Therefore, the higher accumulation of PHD was found at the end of the fermentation. The bacteria accumulated up to 8.96 (±0.5)% mcl-PHAs at 47 h and started to decrease accumulation after 47 h. Therefore, the bacterial cells should be harvested at 47 h after fermentation in order to obtain the optimal percentage of mcl-PHAs.

Wide application of PHAs is mainly limited by the high cost, especially for the carbon sources, of producing PHA by bacterial fermentation. Therefore, synthesis of PHAs using cheaper carbon sources has been regarded as an attractive alternative for the production of PHA at low cost. Different edible oils purchased from local markets were used as alternative carbon sources for the production of mcl-PHAs. Edible oils are cheaper carbon sources compared with fatty acids. The estimated carbon source costs used for producing per gram of mcl-PHA are calculated in Table 4 based on the prices of different edible oils and their performance on inducing production of mcl-PHAs in this study. Compared with decanoic acid ($24.1/g), the estimated carbon source cost used for producing mcl-PHAs can be reduced by around 93.8–99%. Reused oils (after fried for several times) and maltwaste are even cheaper carbon sources that can further reduce the production cost. However, preliminary results (data not shown) demonstrated that maltwaste is not a suitable carbon source for the production of mcl-PHAs. The results in Fig. 3 showed that 3-HD content is still the major constituent in the polymer as the same in the fatty acid containing medium. Olive oil is the most promising edible oil for producing mcl-PHAs as it can induce the highest mcl-PHAs accumulation. Although bacteria only yielded 5.72 (±1)% of mcl-PHAs in reused oil medium compared with 6 (±0.7)% for olive oil medium, it is usually obtained free from restaurant. Therefore, it is also a good choice for reducing the production cost of mcl-PHAs.

Conclusions

The highest yield of up to 9.71 (±0.7)% mcl-PHAs by *P. aeruginosa* was obtained in decaonic acid mineral medium. The highest amount of mcl-PHAs was found to be accumulated at 47 h. The cells preferred to accumulate and degrade PHO at different time intervals during the cell growth. Among the edible oils, olive oil was found to be a good choice for inducing the production of mcl-PHAs in *P. aeruginosa*, with mcl-PHAs yields as high as 6 (±0.7)%. Reused oil from restaurant is another potential carbon sources that can further reduce the production cost of mcl-PHAs. Overall, PHD was the major constituent in the accumulated mcl-PHAs.

Acknowledgments

The authors would like to express their sincere gratitude to The Hong Kong Polytechnic University and the research Grant Council of the Hong Kong special administration region, CHINA (Polyu5272/01M, Polyu5257/02M and Polyu5403/03M).

References

1. Fidler, S. and Dennis, D. (1992), *FEMS Microbiol. Lett.* **103,** 231–235.
2. Gonzalez-Lopez, J., Pozo, C., Martinez-Toledo, M. V., Rodelas, B., and Salmeron, V. (1996), *Int. Biodeterior. Biodegrad.* **38,** 271–276.
3. Brandl, H., Gross, R. A., Lenz, R. W., and Fuller, R. C. (1990), *Adv. Biochem. Eng. Biotechnol.* **41,** 77–93.
4. Zhang, G., Wu, Q., Tain, W., et al. (1999), *Tsinghua Sci. Technol.* **4:3,** 1535–1538.
5. Sánchez, R. J., Schripsema, J., Silva, L. F. D., Taciro, M. K., Pradella, J. G. C., and Gomez, J. G. C. (2003), *Eur. Polymer J.* **39,** 1385–1394.
6. Quinteros, R., Goodwin, S., Lenz, R. W., and Park, W. H. (1999), *Int. J. Biol. Macromole.* **25,** 135–143.
7. Steinbuchel, A. and Hein, S. (2001), *Adv. Bioeng. Biotechnol.* **71,** 81–122.
8. Kroumova, A. B., Wagner, G. J., and Davies, H. M. (2002), *Arch. Biochem. Biophy.* **405,** 5–103.
9. Krieg, N. R., Holt, J. G., Murray, R. G. E., et al. (1984), *Bergey's Manual of Systematic Bacteriology* Volume 1. Williams & Wilkins, Baltimore. 964 p.
10. Hoffmann, N. and Rehm, B. H. A. (2004), *FEMS Microbiol. Lett.* **237,** 1–7.
11. Qi, Q., Rehm, B. H. A., and Steinbűchel, A. (1997), *FEMS Microbiol. Lett.* **157,** 155–162.
12. Campos-Garcia, J., Caro, A. D., Nájera, R., Miller-Maier, R. M., Al-Tahhan, R. A., and Soberón-Chávez, G. (1998), *J. Bacteriol.* **180,** 4442–4451.
13. Ballistreri, A., Giuffrida, M., Guglielmino, S. P. P., Carnazza, S., Ferreri, A., and Impallomeni, G. (2001), *Int. J. Biol. Macromole.* **29,** 107–114.
14. Braunegg, G., Sonnleitner, B., and Lafferty, R. M. (1978), *Eur. J. Appl. Microbiol. Biotechnol.* **6,** 29–37.

Production and Rheological Characterization of Biopolymer of *Sphingomonas capsulata* ATCC 14666 Using Conventional and Industrial Media

ANA LUIZA DA SILVA BERWANGER,[1] NATALIA MOLOSSI DOMINGUES,[1] LARISSA TONIAL VANZO,[1] MARCO DI LUCCIO,[1] HELEN TREICHEL,*,[1] FRANCINE FERREIRA PADILHA,[1] AND ADILMA REGINA PIPPA SCAMPARINI[2]

[1]*Department of Food Engineering, URI, Campus de Erechim, Av. Sete de Setembro, 1621-Erechim-RS, 99700-000, Brazil, E-mail: helen@uricer.edu.br; and* [2]*Department of Food Science, UNICAMP, CP6121 cep 13083-970, Campinas, SP, Brazil*

Abstract

This work was aimed at the production and rheological characterization of biopolymer by *Sphingomonas capsulata* ATCC 14666, using conventional and industrial media. The productivity reached the maximum of 0.038 g/L·h, at 208 rpm and 4% (w/v) of sucrose. For this condition, different concentrations of industrial medium were tested (2.66, 4, 6, and 8%). The best productivity was obtained using pretreated molasses 8% (w/v) (0.296 g/L·h), residue of textured soybean protein 6% (wt/v) (0.244 g/L·h) and crude molasses 8% (w/v) (0.192 g/L·h), respectively. Apparent viscosity presented similar results when compared with those in the literature for other biopolymers.

Index Entries: Agroindustry waste; biopolymers; experimental design; rheology; *Sphingomonas capsulata*.

Introduction

The production of microbial biopolymers, also known as extracellular polymeric substances, is an alternative to gums extracted from plants, because they present physico-chemical properties of high industrial interest, which are essential to define their applications *(1)*. In the past two decades significant progress was observed, concerning the identification, characterization, and utilization of microbial polysaccharides *(2)*. These

---

*Author to whom all correspondence and reprint request should be addressed.

compounds have the ability to form gels and viscous solutions in aqueous medium, even at low concentrations *(3)*.

Bacteria of the genus *Sphingomonas* produce biopolymers, like gelan, welan, ramsan, and diutan, which present gelation characteristics, high viscosity, and better thermal stability than other gums. Thus, such gums are interesting for the food, pharmaceutical, and petrochemical industries *(4–6)*.

Glucose and sucrose are preferentially used as carbon sources for biopolymer production. However, the use of low-cost substrates, such as agroindustry wastes and byproducts in fermentation processes, may favor the reduction of production costs and minimize environmental problems *(7)*. Some alternative sources have been suggested, such as sugarcane molasses, soybean industry wastes, cheese whey, and others *(8–11)*. Molasses is a byproduct of sugar production and a very economical carbon source in bioprocesses *(8,10)*. A byproduct of soybean processing is a protein rich source and is sold as a low-value residue. Depending on the process, it may contain a considerable amount of carbohydrates. Cheese whey has been studied as an alternative source of xanthan gum production *(12,13)*.

Industrial media are very complex, and some of their components may be responsible for inhibiting polysaccharide production or hindering its recovery and purification *(14)*. Contaminants, such as, heavy metals and specific inhibitors may be removed with special pretreatments *(10)*. These pretreatments may clarify the medium without affecting fermentation performance and then guarantee better product extraction and purification *(14)*.

The chemical structures of each biopolymer determine its rheological characteristics and thus, their applications *(1)*. Solutions of microbial polysaccharides present pseudoplastic behavior, i.e., the viscosity decreases with the increase in shear rate *(15)*. In this context, the aim of this work was to study the production and characterization of biopolymers produced by *Sphingomonas capsulata* ATCC 14666, using conventional and industrial media (raw and pretreated molasses and textured soybean protein waste).

Material and Methods

Microorganism

S. capsulata ATCC 14666 was maintained at 4°C, in Yeast Malt (YM) medium containing 3 g/L yeast extract, 3 g/L malt extract, 5 g/L peptone, 10 g/L glucose, and 20 g/L agar.

Fermentation

The production of cells was carried out in 50 mL liquid YM medium in 300-mL Erlenmeyer flask, in two steps. First a preinoculum was prepared, inoculating a loopful of stock culture in 50 mL of YM medium and incubating at 120 rpm, 28°C ± 2°C, for 17 to 19 h. The inoculum was prepared by the addition of 1 mL of preinoculum culture to 50 mL of YM

Table 1
Values of Coded Levels and Real Values Used in the Experimental
Design With Conventional Medium

Coded variable levels	−1.41	−1	0	+1	+1.41
Sucrose (%)	1.18	2	4	6	6.82
Stirring rate (rpm)	152	160	180	200	208

Table 2
Industrial Media

Industrial media	Concentrations
Molasses	4, 6, and 8%
Pretreated molasses	4, 6, and 8%
Aqueous extract of STP waste	2.66, 4, and 6%

medium incubated in an orbital shaker, at 120 rpm, 28°C ± 2°C, for 21 to 23h, when cell concentration reached 10^8 CFU·m/L.

The conventional medium contained sucrose in concentrations according to the experimental design, adding salts to the medium to reach the following concentrations, 1 g/L KNO_3, 0.2 g/L $MgSO_4·7H_2O$, 0.5 g/L K_2HPO_4, 0.1 g/L $CaSO_4$, 0.05 g/L $NaMoO_4$ (16). The production medium was added to the medium containing cells and incubated in an orbital shaker at 28°C ± 2°C, 72 h. The stirring rate was defined by the experimental design. After inoculation, work volume was 150 mL. Table 1 presents the factors and levels investigated in the complete experimental design 2^2, with two axial points and a central point. Productivity was the response variable and the results were statistically analyzed using Statistica 5.1® (StatSoft Inc).

Biopolymer production was carried out using raw and pretreated molasses and an aqueous extract of soybean textured protein waste (STP) (17) in different concentrations, according to Table 2. Production medium was added to the medium containing cells and incubated in an orbital shaker at 208 rpm, 28°C ± 2°C. Experimental runs were performed in triplicate, and productivity was used as the response. Results were statistically evaluated using Tukey's test.

Biopolymer Recovery

After fermentation, the broth was centrifuged at 4700g for 40 min at 4°C, for cell separation. The polysaccharide was precipitated from the supernatant with the addition of isopropyl alcohol (1:3), followed by refrigeration for 12 h. After, samples were centrifuged at 4700g for 40 min at 4°C. The precipitate was dried at constant weight at 50°C ± 5°C, dialyzed for 48 h against sterile Milli-Q® water, lyophilized, and stored in hermetical flasks until further analyses.

S. capsulata ATCC 14666

Table 3
Matrix of the Experimental Design With the Coded and Real Values
(in Parentheses) for the Response Productivity

Run	Sucrose (%)	Stirring rate (rpm)	Productivity (g/L·h)
1	−1 (2)	−1 (160)	0.003
2	+1 (6)	−1 (160)	0.004
3	−1 (2)	+1 (200)	0.004
4	+1 (6)	+1 (200)	0.015
5	−1.41 (1.18)	0 (180)	0.001
6	+1.41 (6.82)	0 (180)	0.007
7	0 (4)	−1.41 (152)	0.005
8	0 (4)	+1.41 (208)	0.038
9	0 (4)	0 (180)	0.019
10	0 (4)	0 (180)	0.016
11	0 (4)	0 (180)	0.015

Rheological Analysis

Aqueous solutions (3% w/v) of the gum were prepared for apparent viscosity analysis, at 25°C and 60°C. A digital rheometer (Brookfield, LVDV III+) coupled to a water bath (Brookfield, TC-502P) was used for viscosity determinations. Readings of viscosity and shear stress were taken at 10 s intervals, varying shear rate (0-264-0/s), based on each of the sample characteristics. Viscosity data was fitted for increasing and decreasing shear rate, using the Ostwald-de-Waele model ($\sigma = K\gamma^n$), where σ is shear stress, and γ is shear rate, K and n are the consistency index and power law exponent, respectively. For Newtonian flow n is equal to one and K equals the viscosity. For shear-thinning fluids n is less than one whereas it is bigger than one for shear-thickening fluids (8).

Results and Discussion

Experimental Design for Conventional Medium

Table 3 presents the matrix of the experimental design and the response in terms of productivity. The highest productivity (0.038 g/L·h) was obtained at 28°C ± 2°C, 208 rpm, in a medium containing 4% of sucrose at 72 h (run 8). Regression coefficients and standard deviations were calculated using the data of Table 3, and are presented in Table 4. Second order coefficient for sucrose concentration and the first order coefficient for stirring rate were statistically significant ($p < 0.05$). Nonsignificant parameters were added to lack of fit for analysis of variance (ANOVA) shown in Table 5.

The correlation (R) was 0.85 and F calculated was 2.43 times higher than F tabulated, what permits the formulation of a nonlinear coded second order model (Eq. 1). This model describes productivity as a function

Table 4
Regression Coefficients for Biopolymer Productivity

Parameter	Regression coefficients	Standard deviation
Mean	0.017[a]	0.001
Sucrose (Linear parameter)	0.005	0.001
Sucrose (Quadratic parameter)	−0.016[a]	0.002
Stirring rate (Linear parameter)	0.015[a]	0.001
Stirring rate (Quadratic parameter)	0.002	0.002
Interaction sucrose × Stirring rate	0.005	0.002

[a]Statistically significant for productivity at the 95% confidence level.

Table 5
Analysis of Variance to Biopolymer Productivity

Source of variation	Sum of squares	Degrees of freedom	Mean squares	F-test
Regression	0.00084	2	0.00042	10.85
Residual	0.00031	8	3.85×10^{-5}	
Lack of fit	0.0003	6		
Pure error	7.92×10^{-6}	2		
Total	0.00114	10		

Regression coefficient: R = 0.85.
$F_{0.95;2;8} = 4.46$.

of the independent variables (stirring rate and sucrose concentration), inside the investigated range.

$$P = 0.017 - 0.008 \cdot (C_s)^2 + 0.007 \cdot A \quad (1)$$

where P is the productivity; C_s the sucrose concentration; A the stirring rate.

The coded model was used to plot the surface presented in Fig. 1. The concentration of sucrose that resulted in higher productivity was 4%, which corresponds to the central point. Upper levels of stirring rate showed a tendency to increase gum yields.

To check if productivity would increase with stirring rate, as indicated by statistical data analysis, two experimental runs were carried out. The first run repeated the maximized experimental condition, which corresponds to run 8 of the experimental design. The second run was carried out in the same experimental conditions as before, except for the stirring rate, which was increased to 236 rpm in order to evaluate its influence on productivity.

Average productivity decreased from 0.041 g/L·h at 208 rpm to 0.009 g/L·h at 236 rpm. Tukey's test showed that these two results are statistically different ($p < 0.05$). Thus, the maximization of biopolymer production by *S. capsulata* ATCC 14666 was achieved, for conventional medium, inside

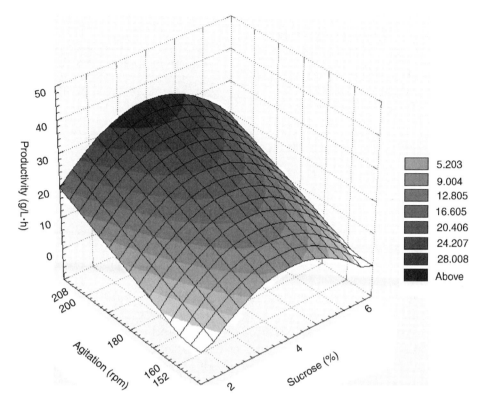

Fig. 1. Response surface for productivities of biopolymer.

the range investigated. It was also possible to note, through the repetition of run 8 of the first experimental design, that a good reproducibility is obtained.

Industrial Media

Tukey's test was carried out to compare the best results (that showed higher gum productivity) for each substrate, considering the different substrates (Table 6). The results for the three different concentrations of the STP waste extract are significantly ($p < 0.05$) different.

Gum productivity increases with the increase in the concentration of this substrate. The highest productivity was obtained with 6% of the STP extract. The productivities obtained with molasses (raw and pretreated) at 4% and 6% were not statistically different ($p < 0.05$), although the addition of 8% of this substrate showed a result significantly higher than the other two.

The statistical analysis of the highest productivity for each substrate showed that the STP waste extract at 6%, raw molasses at 8%, and pretreated molasses at 8%, yielded the same productivities ($p < 0.05$). However, pretreated molasses 8% showed higher productivity than raw molasses at 8% (Table 6).

Table 6
Productivities Obtained Using Industrial Media

Condition	Productivity and standard deviation
Aqueous extract of STP waste 6 %	$0.2437^a \pm 0.0158$
Molasses 8%	$0.1917^a \pm 0.048$
Pretreated molasses 8%	$0.2957^a \pm 0.0403$

^{a}Means with different letters in the same column differ significantly ($p \leq 0.05$) in Tukey's test.

The average biopolymer productivity obtained in this work was 0.24 g/L·h, using 6% of aqueous extract of STP waste, without any supplementation. In the production of gelan using soybean sauce residue Jin et al. *(9)* achieved 0.1 g/L·h, using a 7-L bioreactor and a medium containing 2% (w/v) glucose and 2% (w/v) soybean residue.

Using raw molasses the average productivity was 0.19 g/L·h, and when molasses was pretreated it increased to 0.29 g/L·h. In both cases, molasses was used as the sole source of carbon and nitrogen, without further supplementations. Kalogiannis et al. *(10)* obtained higher polymer productivity (2.21 g/L·h), but they used a concentration of molasses almost two times higher than in our work (175 g/L), and supplemented the medium with salts.

Biopolymer Rheology

Experimental results showed that aqueous solutions of the biopolymer produced in this work presented pseudoplastic behavior. Viscosity also decreases with a temperature increase. Table 7 presents the Ostwald-de-Waele model parameters (power law exponent and consistency index). All correlation coefficients were higher than 0.99, indicating a good fit of experimental data to the model. The power law exponents were less than one, indicating the pseudoplastic behavior. The slight tixotropic behavior was observed at lower shear rates at 60°C (data not shown).

Viscosity data obtained in our work are comparable to those reported in literature for other polymers. At 10/s we obtained 176 cP, for 3% solution of the biopolymer at 25°C. Navarrete and Shah *(6)* obtained apparent viscosity of 100 cP, for 1.4×10^{-4}% solutions of diutan, at 24°C. Ashtaputre and Shah *(16)* obtained 200 cP for 0.5% solution of biopolymer at 30°C.

The biopolymer produced by *S. capsulata* showed lower viscosity than those reported in literature, because we worked at higher concentrations. This biopolymer may find application when the product requires lower thickener ability. Interaction studies of this gum with other biopolymers should still be studied, aiming to induce synergetic increase in viscosity, as it occurs with xanthans and galactomannans.

Table 7
Ostwald-de-Waele Model Parameters (Behavior and Consistency Indexes)

Biopolymer obtained from	T (°C)	n	K	R^2
Sucrose 4%	25	0.894	2.2898	0.9995
	60	0.9023	0.7436	0.9999
Pretreated molasses 8%	25	0.8781	1.4973	0.9996
	60	0.8848	0.5299	0.9999
Aqueous extract of TSP waste 6%	25	0.8109	0.1955	0.9996
	60	0.8121	0.1148	0.9999

Conclusions

The best experimental condition for biopolymer production by *S. capsulata* ATCC 14666 was at 28°C ± 2°C, 208 rpm, in a medium containing 4% of sucrose for 72 h of fermentation, in which an average productivity of 0.038 g/L·h was obtained. It is possible to use industrial media like raw and pre-treated molasses and STP aqueous extracts to produce biopolymers using *S. capsulata*, which may reduce production costs and minimize environmental problems. The best productivities were achieved with pretreated molasses 8% (0.296 g/L·h), STP 6% (0.244 g/L·h), and raw molasses 8% (0.192 g/L·h). Aqueous solutions of the biopolymer presented pseudoplastic behavior, confirmed by the good fit of the experimental data to the Ostwald-de-Waele model.

Acknowledgment

The authors thank CAPES, FAPERGS, and FAPESP for the financial support of this work and scholarships.

References

1. Pace, G. W. (1991), Polímeros Microbianos, In: Bu'Lock, J. & Kristiansen, B. *Biotecnología Básica*. Editorial Acribia S. A., Espanha, pp. 449–462.
2. Padilha, F. F. (1997), MSc Thesis, Síntese e caracterização de biopolímeros por *Beijerinckia* sp 7070. Pelotas, Universidade Federal de Pelotas (UFPel).
3. Moreira, A. N., Del Pino, F. A. B., Vendruscolo, C. T. (2003), *Ciência e Tecnologia de Alimentos*. **23(2),** 300–305.
4. Maugeri, F. (2001), Produção de Polissacarídeos, In: Lima, U. A., Aquarone, E., Borzani, W., Schmidell, W. *Biotecnologia Industrial: Processos Fermentativos e Enzimáticos*. São Paulo: Editora Edgard Blucher Ltda., Brasil, v.3, pp.125–154.
5. Morris, E. R., Gothard, M. G. E., Hember, M. W. N., Manning, C. E., Robinson, G. (1996), *Carbohyd. Polym.* **30,** 165–175.
6. Navarrete, R. C. and Shah, S. N. (2001), *Soc. Pet. Eng.* 68487, Richardson, TX, USA.
7. Woiciechowski, A. L. (2001), PhD thesis, Desenvolvimento de bioprocesso para a produção de xantana a partir de resíduos agroindustriais de café e de mandioca. Universidade Federal do Paraná (UFPR) Curitiba.
8. Bae, S. and Shoda, M. (2004), *Biotech. Progr.* **20,** 1366–1371.
9. Jin, H., Lee, N. -K., Shin, M. -K., Kim, S. -K., Kaplan, D. L., and Lee, J. -W. (2003), *Biochem. Eng. J.* **16,** 357–360.

10. Kalogiannis, S., Iakovidou, G., Liakopoulou-Kyriakides, M., Kyriakidis, D. A., and Skaracis, G. N. (2003), *Process Biochem.* **39,** 249–256.
11. Nitschke, M., Rodrigues, V., and Schinatto, L. F. (2001), *Ciência e Tecnologia de alimentos.* **21(1),** 82–85.
12. Marsul Proteinas LTDA. (2005), Ficha Técnica de Proteína Texturizada de Soja.
13. Soy Protein Council. Soy protein products—characteristics, nutritional aspects and utilization, Technical Report, 1997.
14. Treichel, H. (2005), PhD Thesis, Estudo da otimização da produção de inulinase por *Kluyveromyces marxianus* NRRL Y-7571 em meios industriais pré-tratados, FEA/UNICAMP Campinas.
15. Padilha, F. F. (2003), PhD Thesis, Produção de biopolímeros sintetizados por microorganismos. FEA/UNICAMP, Campinas.
16. Ashtaputre, A. A. and Shah, A. K. (1995b), *Appl. Environ. Microb.* **61(3),** 1159–1162.
17. Berwanger, A. L. S. (2005), MSc Thesis, Produção e caracterização de biopolímero sintetizado por *Sphingomonas capsulata*. URI–Campus de Erechim, Erechim, 2005.

Inulinase Production by *Kluyveromyces marxianus* NRRL Y-7571 Using Solid State Fermentation

João Paulo Bender, Marcio Antônio Mazutti, Débora de Oliveira, Marco Di Luccio, and Helen Treichel*

Universidade Regional Integrada do Alto Uruguai e das Missões – URI Campus de Erechim – Departamento de Engenharia de Alimentos Av. Sete de Setembro 1621, 99700-000, Erechim – RS; E-mail: helen@uricer.edu.br

Abstract

Inulinase is an enzyme relevant to fructose production by enzymatic hydrolysis of inulin. This enzyme is also applied in the production of fructo-oligosaccharides that may be used as a new food functional ingredient. Commercial inulinase is currently obtained using inulin as substrate, which is a relatively expensive raw material. In Brazil, the production of this enzyme using residues of sugarcane and corn industry (sugarcane bagasse, molasses, and corn steep liquor) is economically attractive, owing to the high amount and low cost of such residues. In this context, the aim of this work was the assessment of inulinase production by solid state fermentation using by *Kluyveromyces marxianus* NRRL Y-7571. The solid medium consisted of sugar cane bagasse supplemented with molasses and corn steep liquor. The production of inulinase was carried out using experimental design technique. The effect of temperature, moisture, and supplements content were investigated. The enzymatic activity reached a maximum of 445 units of inulinase per gram of dry substrate.

Index Entries: Inulinase; solid state; *Kluyveromyces marxianus*.

Introduction

Inulinases (E.C. 3.2.1.7) are glycosidases that perform the endohydrolysis of 2,1-β-D-fructosidic linkages in inulin. These enzymes are potentially useful in the production of high fructose syrups, using inulin as substrate (1). The production of fructose from inulin is more advantageous than the conventional method that uses starch, given that the reaction using inulinase is simpler and yields products with fructose content higher than 95%. Inulinases have also been used for production of fructo-oligosaccharides (2). Researches involving fructo-oligosaccharides have been stimulated by

*Author to whom all correspondence and reprint requests should be addressed.

the increase in the demand for healthy foods *(3)*, because the ingestion of fructoligosaccharides increases the production of *Bifidobacterium*, which are the most important bacteria of intestinal microflora *(4)*.

Several microorganisms may be used to produce inulinase. The selection of such microorganisms depends on their physiological characteristics and the fulfillment of Generally Recognized as Safe (GRAS) and Food and Drug Administration (FDA) criteria for food products *(5)*. The gender *Kluyveromyces* satisfies these requirements, which makes this microorganism interesting for commercial applications. *Kluyveromyces marxianus* NRRL Y-7571 has shown excellent results in submerged medium in the experience we have gathered in collaboration with the Laboratory of Bioprocesses in UNICAMP (results not shown). Several studies about inulinase production can be found in literature. Most of them focus on submerged fermentation systems and other microorganisms *(6–10)*. However, few reports about inulinase production using solid state fermentation (SSF) can be found *(11)*. The use of low cost residues, higher productivities, low energy requirements, lower wastewater production, extended stability of products, and low production costs are some of the main advantages of SSF *(12,13)*. The use of SSF presents the possibility of using low cost substrate, like agroindustry wastes and bioproducts, then adding value to such materials *(14)*. This work examined the production of inulinase by *K. marxianus* by SSF, using sugar cane bagasse as substrate and sugar cane molasses (SCM) and corn steep liquor (CSL) as of carbon and nitrogen supplementary sources.

Material and Methods

Microorganism

K. marxianus NRRL Y-7571 obtained from NRRL (Northern Regional Research Laboratory, now the National Center For Agricultural Utilization Research, Peoria, IL was maintained in Agar slants at 5°C on YM Agar medium (Yeast Malt) containing (g/L) 3, 3, 5, 10, and 20 g/L of yeast extract, malt extract, peptone, glucose, and agar, respectively.

Preinoculum preparation was carried out inoculating 10 mL of liquid YM medium in a 50-mL tube with a loopful of stock culture, and incubating at 30°C, for 24 h. Inoculum medium contained: 20 g/L sucrose, 5 g/L yeast extract, 5 g/L K_2HPO_4, 1.5 g/L NH_4Cl, 1.15 g/L KCl, and 0.65 g/L $MgSO_4 \cdot 7H_2O$ at initial pH 6.8. All reagents and medium components were analytical grade and supplied by Vetec Ltd. (Rio de Janeiro, RJ, Brazil). Each tube with liquid YM medium was transferred to a 500 mL Erlenmeyer containing 100 mL of medium and then incubated at 30°C and 150 rpm for 24 h.

Solid State Fermentation

Sugar cane bagasse from a local industry was used as substrate for inulinase production. Fermentations were carried out in conical reactors using 5 g of dry bagasse, supplemented with CSL and SCM in the concentrations

Table 1
Coded Levels and Real Values Used in the First
Complete Factorial Design (2^4)

Coded variable levels	−1	0	+1
Temperature (°C)	30	35	40
Moisture (%)	50	65	80
Molasses (%)	0	5	10
Corn steep liquor (%)	0	5	10

Table 2
Coded Levels and Real Values Used in the Second Complete
Factorial Design (2^3)

Coded variable levels	−1.68	−1	0	+1	+1.68
Temperature (°C)	31.6	35	40	45	48.4
Moisture (%)	39.8	50	65	80	90.2
Corn steep liquor (%)	6.6	10	15	20	23.4

defined in the experimental design. All flasks were sterilized at 121°C for 20 min. Each reactor was inoculated with 3 mL of the inoculum cell suspension obtained previously and incubated for 72 h. Cultivation was performed in an incubator with humidified air injection. All experimental conditions were carried out in duplicate. Control reactors were also carried out to discount enzyme activity already present in the inoculum. In control runs the enzyme activity was determined right after inoculation. The effect of temperature of incubation, initial moisture of the substrate, and supplement concentrations on inulinase production was assessed by a complete 2^4 experimental design. Table 1 presents the range of the factors investigated. Details on experimental design technique are described by Rodrigues and Iemma (15). After analysis of the results of the first experimental design using a Experimental Design Analysis software tool (Statistica® 5.1 Statsoft Inc, Tulsa, OK), a second experiment was performed based on a central composite design with two factors and six axial points. The range of the factors that were investigated is presented in Table 2.

Enzyme Extraction

After incubation, enzyme was extracted adding 50 mL of sodium acetate buffer 0.1 M pH 4.5 to the fermented medium, following incubation at 30°C for 30 min at 150 rpm and filtration with Whatman qualitative paper, no. 3. Enzyme activity was assayed in the supernatant after filtration.

Inulinase Assay

Inulinase activity was assayed as follows: 1 mL enzyme solution, obtained after extraction, was mixed with 9 mL of 2% (w/v) sucrose on sodium acetate buffer 0.1 M and pH 4.5. The mixture was maintained at

Table 3
Matrix of the First Experimental Design With Coded Levels and Real Values
(in Parenthesis) for the Inulinase Activity

Run	Moisture (%)	Molasses (%)	CSL (%)	Temperature (°C)	Activity (U/g)
1	−1 (50)	−1 (0)	−1 (0)	−1 (30)	n.d.
2	+1 (80)	−1 (0)	−1 (0)	−1 (30)	45
3	−1 (50)	+1 (10)	−1 (0)	−1 (30)	9
4	+1 (80)	+1 (10)	−1 (0)	−1 (30)	n.d.
5	−1 (50)	−1 (0)	+1 (10)	−1 (30)	n.d.
6	+1 (80)	−1 (0)	+1 (10)	−1 (30)	29.6
7	−1 (50)	+1 (10)	+1 (10)	−1 (30)	n.d.
8	+1 (80)	+1 (10)	+1 (10)	−1 (30)	60.1
9	−1 (50)	−1 (0)	−1 (0)	+1 (40)	n.d.
10	+1 (80)	−1 (0)	−1 (0)	+1 (40)	n.d.
11	−1 (50)	+1 (10)	−1 (0)	+1 (40)	n.d.
12	+1 (80)	+1 (10)	−1 (0)	+1 (40)	n.d.
13	−1 (50)	−1 (0)	+1 (10)	+1 (40)	n.d.
14	+1 (80)	−1 (0)	+1 (10)	+1 (40)	113.8
15	−1 (50)	+1 (10)	+1 (10)	+1 (40)	11.1
16	+1 (80)	+1 (10)	+1 (10)	+1 (40)	78.6
17	0 (65)	0 (5)	0 (5)	0 (35)	n.d.
18	0 (65)	0 (5)	0 (5)	0 (35)	n.d.
19	0 (65)	0 (5)	0 (5)	0 (35)	0.5

n.d., not detected.

50°C and the rate of appearance of fructose was determined by the dinitrosalicylic acid (DNS) method *(16)*. One unit of inulinase activity is defined as the amount of enzyme that hydrolyses 1 µmol of sucrose per min, under the reaction conditions.

Results and Discussion

Table 3 presents the matrix of the first complete experimental design with central points and inulinase activity as response. The highest inulinase activity (114 U/g) was obtained using the substrate with 80% of moisture added of 0% of molasses, and 10% of CSL, and incubated at 40°C (run 14). We observed that the extracts present very good stability at room temperature (results not shown).

The data presented in Table 2 were statistically treated and the effects of the factors on inulinase activity are presented as the Pareto chart shown in Fig. 1. Moisture and CSL concentration were the factors that most influenced inulinase activity, with a positive significant effect ($p < 0.05$). The positive effect of moisture is probably related to higher water requirements of yeasts when compared with fungi, which are most commonly used in SSF. CSL is a complex supplement, which is a good source of nitrogen

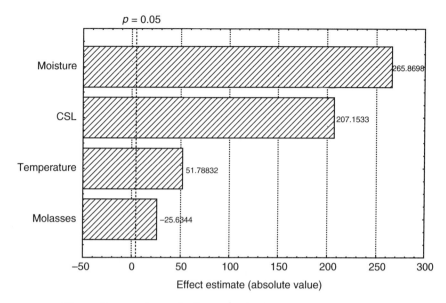

Fig. 1. Pareto chart of effects for the first experimental design.

and micronutrients. The positive effect of the concentration of this supplement on inulinase activity shows that the amount of CSL added in the substrate should be improved to satisfy the microorganism nutrient requirements.

The results obtained in the first experimental design show that the use of molasses as supplement is unnecessary, because this factor presented a negative significant effect. Other possibility is the presence of heavy metals in the molasses, which would inhibit enzyme production by the microorganism (6). Therefore, a second experimental design was carried out, aiming to optimize inulinase production. The ranges of the factors studied were adjusted based on the results of the first experimental design. Table 4 presents the results of the second complete experimental design. Maximum inulinase activity obtained in this work was 445 U/g, obtained at the axial point (40°C, 65% moisture, 23.4% CSL). An increase in inulinase production can be noticed when results of second experimental design are compared with the results of the first one. Coded models for inulinase activity as a function of the studied factors could not be validated for any of group of experimental data.

Statistical analysis of the second experimental design yielded the Pareto chart presented in Fig. 2. It is worth noting that CSL concentration still positively influences enzyme production, indicating that further increase in this factor may still increase inulinase yield.

The only study that investigated inulinase production by SSF is reported by Selvakumar et al. (11). They obtained about 108 U/g using *Staphylococus* spp. and 123 U/g with *K. marxianus* ATCC 52466 (11). These results are nearly fourfold lower when compared with inulinase activities

Table 4
Matrix of the Second Experimental Design With Coded Levels and Real Values (in Parenthesis) for the Inulinase Activity

Run	Moisture (%)	CSL (%)	Temperature (°C)	Activity (U/g)
1	−1 (50)	−1 (10)	−1 (35)	42.8
2	+1 (80)	−1 (10)	−1 (35)	105.6
3	−1 (50)	+1 (20)	−1 (35)	136.3
4	+1 (80)	+1 (20)	−1 (35)	217.9
5	−1 (50)	−1 (10)	+1 (45)	n.d.
6	+1 (80)	−1 (10)	+1 (45)	11.2
7	−1 (50)	+1 (20)	+1 (45)	11.9
8	+1 (80)	+1 (20)	+1 (45)	52.2
9	−1.68 (39.8)	0 (15)	0 (40)	58.6
10	+1.68 (90.2)	0 (15)	0 (40)	168.5
11	0 (65)	−1.68 (6.6)	0 (40)	90.2
12	0 (65)	+1.68 (23.4)	0 (40)	444.8
13	0 (65)	0 (15)	−1.68 (31.6)	343.4
14	0 (65)	0 (15)	+1.68 (48.4)	n.d.
15	0 (15)	0 (40)	0 (40)	223.3
16	0 (15)	0 (40)	0 (40)	196.1
17	0 (15)	0 (40)	0 (40)	265.7

n.d., not detected.

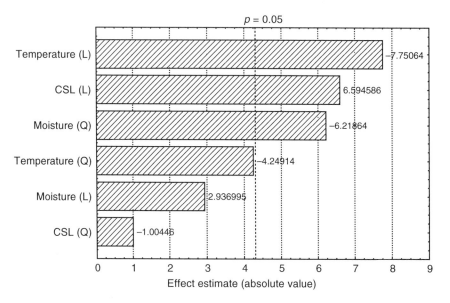

Fig. 2. Pareto chart of effects for the second experimental design.

obtained in our study. This suggests that sugar cane bagasse and CSL are good nutrient sources for inulinase production by *K. marxianus* NRRL Y-7571 using SSF. The moisture content and temperature of incubation that maximize inulinase production were 65% and 40°C, respectively, similar to those reported by Selvakumar et al. *(11)* where maximum inulinase production by K. *marxianus* was obtained at a moisture content of 65% and temperature of 37°C. A related study of product kinetics was carried out by our group and is described elsewhere *(17)*. Characterization of the inulinase produced by SSF in relation to specificity (endo or exoinulinase), optimum temperature, pH, and stability is currently under investigation.

Other studies concerning inulinase production by *Kluyveromyces* in submerged fermentations show that the amount of enzyme produced greatly vary with the strain and the kind of nutrient supplementation used. Gupta et al. *(18)* obtained maximum production of 7 U·m/L, using fructanes as carbon source for K. *fragilis*. Santos et al. *(10)* reported a production of 7.5 U/g using peptone, sucrose, and yeast extract as nutrient sources for K. *bulgarius*. Kalil et al. *(19)* used sucrose as carbon source and obtained 127 U·m/L of inulinase by K. *marxianus*. In a preliminary comparison, the results obtained in the present work show that production of inulinase by SSF may yield higher amounts of the enzyme. Economical feasibility of inulinase production by SSF is also evident, owing to the use of low cost substrates. An important factor that turns SSF practical for industrial purposes is the simple recovery and purification of the enzyme, because it is found concentrated in the extraction buffer *(12)*.

Conclusions

In this work the maximum inulinase production was 445 U·(g/dry substrate), which is almost fourfold higher than inulinase activities found in literature. The maximum activity was obtained by fermentation of sugar cane bagasse with 65% moisture at 40°C and supplemented with 23.4% of CSL. The results show the feasibility of using SSF for inulinase production by K. *marxianus* and suggest new experimental runs for optimization of enzyme production.

Acknowledgment

Authors are grateful to FAPERGS for financial support and to Laboratório de Engenharia de Bioprocessos do Departamento de Engenharia de Alimentos/FEA/UNICAMP for providing the microorganism used in this work.

References

1. Ettalibi, M. and Baratti, J. C. (1987), *Appl. Microbiol. Biotechnol.* **26**, 13–20.
2. Zhang, L., Zhao, C., Zhu, D., Otha, Y., and Wang, Y. (2004), *Protein Exp. Purif.* **35**, 272–275.
3. Chien, C., Lee, W., and Lin, T. (2001), *Enzyme Microbial. Technol.* **29**, 252–257.
4. Gibson, G. R. and Wang, X. (1994), *J. Appl. Bacteriol.* **77(4)**, 412–420.

5. Schneider, A. L. S. (1998), Estudo da Produção de Inulinase por *Kluyveromyces marxianus* ATCC 36907. Florianópolis, 102p. Dissertation (Mestrado em Engenharia Química) Departamento de Engenharia Química, Universidade Federal de Santa Catarina.
6. Treichel, H. (2004), Estudo da otimização da produção de inulinase por *Kluyveromyces marxianus* NRRL Y-7571 em meios industriais pré-tratados. Campinas, Thesis (Doutorado em Engenharia de Alimentos), FEA/UNICAMP.
7. Silva-Santisteban, B. O. Y. and Maugeri Filho, R. (2005), *Enzyme Microbial. Technol.* **36**, 717–724.
8. Cruz-Guerrero, A., Garcia-Peña, I., Barzana, E., Garcia-Garibay, M., and Gomez-Ruiz, L. (1995), *J. Fermentatio. Bioeng.* **80(2)**, 159–163.
9. Gill, P. K., Sharma, A. D., Harchand, R. K., and Singh, P. (2003), *Bioresour. Technol.* **87**, 359–362.
10. SANTOS, A. M. P. (1998), Produção de Oligossacarídeos por Inulinase de *Kluyveromyces bulgaricus*. Campinas. Dissertation (Mestrado em Engenharia de Alimentos), FEA/UNICAMP.
11. Selvakumar, P. and Pandey, A. (1999), *Process Biochem.* **34**, 851–855.
12. Pandey, A. (2003), *Biochem. Eng. J.* **13**, 81–84.
13. Holker, U. and Lenz, J. (2005), *Curr. Opin. Microbiol.* **8**, 301–306.
14. Pandey, A., Soccol, C. R., Nigam, P., and Soccol, V. T. (2000), *Biores. Technol.* **74**, 69–80.
15. Rodrigues, M. I. and Iemma, A. F. (2005), *Planejamento de Experimentos e Otimização de Processos*, Ed. Casa do Pão, Campinas – SP - Brasil.
16. Miller, G. L. (1959), *Anal. Chem.* **31**, 426–428.
17. Mazutti, M., Bender, J. P., Treichel, H., and Di Luccio, M. (2006), *Enzyme Microbial. Technol.* in press.
18. Gupta, A. N., Davinder, P. S., Kaur, N., and Singh, R. (1994), *J. Chem. Technol. Biotechnol.* **59**, 377–385.
19. Kalil, S. J., Suzan, R., Maugeri, F., and Rodrigues, M. I. (2001), *Appl. Biochem. Biotechnol.* **94**, 257–264.

Macroscopic Mass and Energy Balance of a Pilot Plant Anaerobic Bioreactor Operated Under Thermophilic Conditions

Teodoro Espinosa-Solares,*,[1,2] John Bombardiere,[2] Mark Chatfield,[2,3] Max Domaschko,[2] Michael Easter,[2] David A. Stafford,[2,3] Saul Castillo-Angeles,[1] and Nehemias Castellanos-Hernandez[1]

[1]*Agroindustrial Engineering Department, Autonomous University of Chapingo, Chapingo, 56230, Edo. de Mexico, MEXICO, E-mail: espinosa@correo.chapingo.mx; [2]West Virginia State University, Division of Agricultural, Consumer, Environmental, & Outreach Programs, Institute, WV 25112-1000; and [3]West Virginia State University, Department of Biology, Institute, WV 25112-1000*

Abstract

Intensive poultry production generates over 100,000 t of litter annually in West Virginia and 9×10^6 t nationwide. Current available technological alternatives based on thermophilic anaerobic digestion for residuals treatment are diverse. A modification of the typical continuous stirred tank reactor is a promising process being relatively stable and owing to its capability to manage considerable amounts of residuals at low operational cost. A 40-m³ pilot plant digester was used for performance evaluation considering energy input and methane production. Results suggest some changes to the pilot plant configuration are necessary to reduce power consumption although maximizing biodigester performance.

Index Entries: Biogas; livestock residual; methane; thermophilic anaerobic digestion.

Introduction

Livestock residuals have been becoming an environmental concern as the production of animals has increased throughout the world (1). In the United States alone, more than 350×10^6 t of livestock residuals must be disposed of annually (2). These residuals are often disposed of by application directly on land or are occasionally composted (3). In West Virginia, intensive poultry production generates more than 100,000 t of broiler litter annually.

Alternative processes for reducing the environmental impact are highly demanded. Since the 1960s and 1970s, several research groups have been

*Author to whom all correspondence and reprint requests should be addressed.

proposing anaerobic digestion as an alternative for pollution control (4). Currently the available technological alternatives based on this process are very diverse. A modification of the typical continuous stirred tank reactor is a promising process owing to its capability to manage considerable amounts of residuals at low operational cost and being a relatively stable process.

There are several factors that influence the anaerobic process. Some of the principal criteria that should be considered to design the processes are as follows: temperature, chemical and biochemical composition, hydraulic retention time (HRT), and mixing. As it has been explained in the literature, the thermophilic process has a better performance than mesophilic and psychrophilic processes (5). As an example, Kim et al. (6) reported a higher yield in the thermophilic process (716 mL/g_{VS}/d) compared with the mesophilic one (556 mL/g_{VS}/d). However, that study did not include an evaluation of the energetic cost of the yield increment. It has been reported that the type of manure determines biogas production. Güngör-Demirci and Demirer (7) evaluated anaerobic batch fermentation of manure. These authors, working with mixtures of broiler and cattle residuals, reported that the biogas yields decrease as the fraction of broiler manure increased. This behavior was attributed to ammonia inhibition. Regarding power consumption, de Pinho et al. (8) worked with anaerobic sequencing batch reactors containing granular or flocculent biomass for the treatment of piggery wastewater. They reported a volumetric power input (VPI) range of 0.18–0.31 kW/m^3 for a 4.5-L reactor operated at 30°C, the device was mechanically stirred by three typical Rushton turbines.

For full-scale treatment plants, these considerations take a key role in the economic feasibility of the technological alternatives. In that context and considering that energy evaluation of the anaerobic digestion process is limited in the published literature, the aim of this article was to evaluate the performance of an anaerobic process, using a mass and energy balance and chemical characterization as tools to analyze a pilot plant scale biodigester.

Materials and Methods

Experiments were carried out at the facilities of the Bioplex Project at West Virginia State University (WV). The anaerobic process is presented in Fig. 1. The dilution system included an axial mixer (M-101), a recirculation loop and settling tank, which was used for removing grit and wood. Diluted chicken litter slurry was fed automatically into a 40 m^3 tank. The biodigester was operated in a semicontinuous process, being fed with fresh slurry every 2 h using a pump (L-102). The feed slurry flow rate was defined to achieve a HRT of 10 d. During the experiments, the slurry volume inside the tank was kept at 27.43 ± 0.06 m^3. The fermentation media in the digester was bubbled. A gas blower (JB-207) extracted biogas from the top of the digester and recirculated the gas through a bubbling ring located at the bottom of the digester. The bubbling rate was fixed at 0.01 ± 0.0005 vvm (gas volume/liquid volume/min), operating for 5 min every

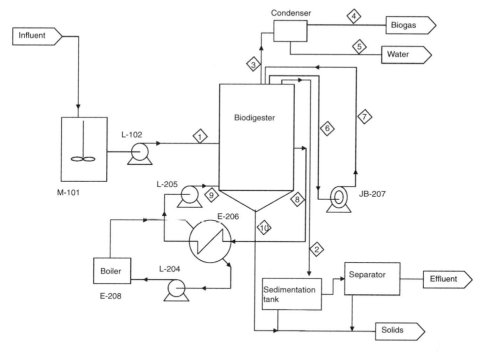

Fig. 1. Process diagram of the thermophilic anaerobic pilot plant.

30 min. The bioreactor operated under thermophilic conditions (56.6°C). The fermentation media was automatically heated when the temperature fell 0.1°C below target. A pump (L-204) recycled the work fluid (water and glycol mixture) for the heat exchanger (E-208). The digester liquid was pumped through an external heat exchanger and recirculated back into the digester using a pump (L-205). The digester effluent was discharged into a 5700-L sedimentation tank. The biogas flowed out by pressure differential through a flow meter and then to the flare for burning. Liquid overflowed onto a separator screen with 1000 mesh and then into a holding tank.

Mass and energy balance *(9)* was evaluated considering a steady state. For that purpose, 24 h of continuous operation was considered as an experimental unit, evaluated in three replicates. Motor power consumption of each device was evaluated at a sampling rate of 0.5 s, using Watt transducers (Omega, OM10 series, Stamford, CT). Signals were acquired and processed using a data acquisition system (National Instruments, NI-DAQ7, NI 4351, and NI 4350, Austin, TX) and a data logger (National Instruments, NI VI Logger, Austin, TX). Energy used by the devices was evaluated from a plot of the power as a function of time. The area under the curve, that corresponded to energy, was integrated using the SigmaPlot 2002 © software (Version 8.02a, Point Richmond, CA).

Liquid flow rate was evaluated using an ultrasonic device (Doppler, 301) although for biogas flow rate a coriolis meter (Emerson, no. CMF0-25M319NABAEZZZ, Boulder, CO) was used. Composition of the biogas was determined using a Drager Multiwarn II methane detector. Natural gas consumption by the hot water boiler used for the heat exchanger loop was recorded every 24-h period. The amount of energy used by this device was calculated based on the material used and the measured composition of the regional natural gas, i.e., 100% methane, and the corresponding heat of combustion (10).

Litter was obtained from a commercial poultry farm located in Moorefield, WV. The litter was taken from a broiler house using wood chips for bedding. Litter was reused for six flocks of grow out, which is approx 1 yr. Before digestion, litter was removed from the house and stored in a litter shed for 3 mo. Dry matter, hemicellulose, cellulose, lignin, crude fat, and crude protein were determined using the corresponding procedures of the Official Methods of Analysis (11). Starch was evaluated as described by Hall et al. (12). BOD (biochemical oxygen demand), ALK (alkalinity), TS (total solids), and VS (volatile solids) were determined using the Standard methods for the examination of water and wastewater (13). In the case of VA (volatile acids), COD (chemical oxygen demand), and AMM (ammonia), the methods 8196, 8000, and 10,031 reported in the Hach Water Analysis Handbook were used (14). Chemical composition of the feed and effluent were evaluated by three replicates for each experimental unit. Feed slurry samples were obtained from the feed inlet pipe during feeding events. Effluent samples were taken at the time of discharge from the digester tank, before sedimentation or separation. Dry litter samples (before dilution) were used for hemicellulose, cellulose, lignin, crude fat, and crude protein analysis of the feed. The digester biochemical parameters along with biogas and methane percent measurements were used to determine steady state. At similar hydraulic and carbon loading rates, when the biogas production fluctuated less than 10% and methane percent fluctuated less than 2% in a 24-h period, along with pH fluctuations within the digester of less than 0.1 and VA fluctuations inside the digester smaller than 10% over a 4-d period, the digester was judged to be at steady state.

The biodigester has been operating continuously for 4 yr, having activity reduction just during the winter months (mid-December until March 1). For the experiments presented here, the anaerobic reactor had an adaptation period of 120 d since last winter, which consisted of gradually increasing the feeding rate up to the normal operation conditions. The process was in a steady state for 4 d previous to data acquisition.

Results and Discussion

The poultry litter coming from the farm had a dry matter of 70.93%. The dry matter basis manure proximal analysis indicated the following composition: hemicellulose 12.96%, cellulose 23.75%, lignin 5.07%, starch 1.16%, crude fat 2.36%, and crude protein 21.06%. Chemical characteristics of the slurry used for feeding are presented in Table 1.

Table 1
Characteristics of (Diluted) Manure Used in Experiments

Parameter[a]	Minimum	Maximum	Mean[b] ± SD
TS (%)	5.01	5.73	5.46 ± 0.25
VS of TS (%)	73.3	76.6	74.7 ± 1.24
COD (mg/L)	44,366	64,433	49,626 ± 6054
BOD (mg/L)	8700	14,700	12,478 ± 2291
VA (mg/L)	2196	5546	3681 ± 1079
AMM (mg/L)	1223	1290	1237 ± 21.4
ALK (CaCO$_3$ mg/L)	4900	7490	6210 ± 719
pH	6.72	6.88	6.78 ± 0.06

[a]TS, total solids; VS, volatile solids; COD, carbon oxygen demand; BOD, biochemical oxygen demand; VA, volatile acids; AMM, ammonia; ALK, alkalinity; pH, hydrogen potential.
[b]Mean of three replicates.

The mass balance of the anaerobic process is reported in Table 2. Results indicated that the HRT achieved during the experiments was 9.91 ± 0.15 d. The steady state of the process was confirmed based on the low coefficient of variation registered for the feeding and effluent currents and also on the stabilization of chemical composition of the currents. It is important to note that the volume of digester slurry recycled by pumping was 5.95 ± 0.56 times larger than the slurry inside the tank; although the volume used for mixing by bubbling was 6.41 ± 0.26 times higher than the volume of the slurry inside the tank. The mixing strategy followed in this work indicated that the gas feeding was under the typical order of magnitude used for devices that only use gas to promote mixing. The 0.01 vvm used here is considerably lower than the 1.5 vvm used by bubble columns (15) and external-loop airlift reactors (16), and even lower than the 2.5 vvm reported for airlift reactors (17). Similar high gas feeding values (1 vvm) are reported for aerated stirred tanks (18). The low gas feeding value is more evident if it is considered that the bubbling is intermittent. Thus the process presented here suggests that pumping has a key role in mixing.

Table 3 presents the energy used by the process devices during a 24-h period of continuous operation. The main energy consumer is the heat exchanger (94% of the total). The VPI, considering heating, was 0.401 kW/m^3. This value is similar in order of magnitude to the one reported by de Pinho et al. (8). However, these authors only included the mechanical power input in the evaluation. For this particular case, a detailed analysis of the heating system should be considered in order to determine the alternatives to diminish the power consumption. On scale up, fermentative mass and heat transfer area of the vessel do not increase linearly. In fact, when the volume of the tank scales up, the area exposed to heat transfer, and thus the energy exchanged to ambient, increases in lower magnitude than

Table 2
Mass Balance of the Process During a 24-h Period

	Line									
	1	2	3	4	5	6	7	8	9	10
	Mass (kg)									
Trial 1	2891.7	2797.8	39.4	37	2.4	242.3	242.3	188,838.6	188,838.6	54.5
Trial 2	2779.7	2688	39.4	37	2.4	218.7	218.7	155,971	155,971	52.4
Trial 3	2799	2705.7	40.6	38.1	2.5	226	226	154,430.9	154,430.9	52.7
Average	2823.5	2730.5	39.8	37.4	2.4	229	229	166,413.5	166,413.5	53.2
Standard deviation	48.9	48.1	0.584	0.548	0.036	9.88	9.88	15,869	15,869	0.92
Coefficient of variation (%)	1.7	1.8	1.5	1.5	1.5	4.3	4.3	9.5	9.5	1.7

Line numbers correspond to the currents presented in Fig. 1.

Table 3
Energy Balance in a 24 h of Continuous Operation

Device or source	Minimum (kJ)	Maximum (kJ)	Mean[b] ± SD (kJ)
M-101	2780	3316	2993 ± 233
L-102	1196	1374	1316 ± 85
L-204	3214	3982	3492 ± 348
L-205	29,652	35,745	31,684 ± 2872
JB-207	16,344	17,868	16,950 ± 660
E-208[a]	866,010	942,143	894,560 ± 33,870
Biogas produced[a]	661,224	694,213	678,086 ± 13,478

[a]The energy was calculated based on a heat of combustion of 33 943 kJ/m^3.
[b]Mean of three replicates.

does volume. Therefore, in regional digester applications 100 to 1000 times larger than the pilot plant, the heat supply required diminishes considerably.

Exclusively evaluating the mechanical power input, the pump L-205 and the blower JB-207 consume, respectively, 56% and 30%. Both devices have the main purpose of promoting homogeneity inside the vessel in terms of temperature and composition. However, the higher power consumption of the pump L-205 over the blower JB-207 and the reduced gas feeding (0.01 vvm) used in the process confirm that the principal device used for mixing is the pump L-205.

Taking the mechanical energy used for the different devices reported in Table 3 and considering operation and nonoperation times, the total average power of the process was 0.653 ± 0.0015 kW. This value could be transformed taking into account the volume involved in the process, thus the VPI of the process was 0.024 ± 0.0015 kW/m^3, which is in the same order of magnitude of an anaerobic sewage sludge digestion process reported by Pierkiel and Lanting (19). These authors found a VPI ranging from 0.02 to 0.04 kW/m^3. It is important to stress that the VPI registered in the present work is similar to the 0.022–0.038 kW/m^3 range reported for an aerated stirred hybrid reactor (20). On the other hand, the mechanical energy used in the process reported here, represents between 7.7% and 13.2% of the energy used with an anaerobic batch biofilm reactor with mechanical stirring (8). In that report VPI ranged from 0.18 to 0.31 kW/m^3, depending on rotational speed. This showed that the claimed low power consumption using bubbling when is compared with stirred vessels, depends strongly in the kind of impeller. In this particular case, bubbling has similar power consumption like an hybrid geometry (20), but it requires a smaller power input than the Rushton turbines reported by de Pinho et al. (8). Consequently, new strategies for mixing should be explored in order to reduce power consumption.

When the operation time is evaluated, as presented in Table 4, it is easy to observe that in a 24-h period, both the L-204 and L-205 pumps worked for almost 11 h, followed by the blower JB-207 with around 3.5 h. Considering that, for the same period of time, the material inside the tank is recycled almost six times and the L-205 operated for 11 h, the average time that slurry was passing through the pump, and consequently exposed to mechanical stress, was close to 110 min. This long period of time could lead to a detriment in the performance of the process owing to the fact that consortia aggregates could be exposed to intermittent mechanical stress. Thus, the hydrodynamics imposed in the process could have an important influence on biodigester performance. As an example, Jin and Lant (21) working with bubble column, airlift, and aerated stirred reactors found that flow regime and hydrodynamics influence the floc size distribution. Further work is necessary in order to elucidate the effect of mechanical stress on microbial communities and thus optimize the biodigester performance.

Table 4
Operation Time in a 24 h Period

Device	Minimum (min)	Maximum (min)	Mean[a] ± SD (min)
M-101	70.4	75.4	73 ± 2.1
L-102	9.7	10.9	10.2 ± 0.5
L-204	606.6	741.8	653.7 ± 62.3
L-205	606.2	750.9	657.6 ± 66.1
JB-207	208.1	230.6	217.9 ± 9.4

[a]Mean of three replicates.

Table 5
Characteristics of the Effluent

Parameter[a]	Minimum	Maximum	Mean[a] ± SD
TS (%)	3.14	3.75	3.35 ± 0.2
VS of TS (%)	62.9	64.9	63.4 ± 0.7
COD (mg/L)	33,433	42,233	36,918 ± 2646
BOD (mg/L)	6200	9900	7322 ± 1212
VA (mg/L)	3546	6735	4875 ± 1142
AMM (mg/L)	1923	2027	2004 ± 32
ALK (as $CaCO_3$ mg/L)	9520	11,900	10,859 ± 719
pH	7.68	7.81	7.74 ± 0.05

[a]Mean of three replicates.

The effluent composition reported in Table 5 was obtained under steady state conditions. The pH in the effluent ranged from 7.68 to 7.81, although in feed ranged from 6.72 to 6.88. The increase in pH from the feed to the effluent is a result of metabolic reactions that occur with anaerobic digestion of poultry litter. Over the 4 yr, the digester has been fed solely poultry litter, pH in the digester has fluctuated between 7.2 and 7.9 and ammonia levels varied from 1500 to 2200 ppm depending on loading rate. Ammonia levels were steady at around 2000 ppm in the effluent; well below inhibitory levels (22). As reported by Stafford et al. (4), digesters fed poultry manure have shown levels of up to 3150 ppm ammonia without inhibitory effects on digestion, even at a 10-d HRT.

As presented in Table 6, methane percent of biogas ranged from 54.5 to 56.0%, although volatile solids destruction from feed to effluent varied from 46.9 to 49.1%. These values are similar to reported data from poultry waste (layer) fed lab scale digesters at 10-d HRTs (4). The biogas generated referred to VS fed was 282.1 mL/g. This is well below reported values in a thermophilic reactor with 10-d HRT of 551 mL/g (4). The lower production of biogas may be attributed to the difference in the quality of the poultry litter (six flocks grow out and stored for 3 mo before feeding) and possibly to the disruption of microbial communities by the sheer force of the recirculation

Table 6
Chemical Changes During Anaerobic Digestion and Bioreactor Performance

Parameter	Minimum	Maximum	Mean[a] ± SD
Methane content (%)	54.5	56	55.3 ± 0.6
VS reduction (kg)	53.4	57.4	55.2 ± 1.7
VS reduction (%)	46.9	49.1	49.7 ± 0.9
Biogas/VS added (mL/g)	273.6	295.2	281.1 ± 10
COD reduction (mg/L)	9100	17,800	12,708 ± 3 04
COD reduction (%)	19.5	33.3	25.3 ± 5.8
BOD reduction (mg/L)	4400	6500	5156 ± 953
BOD reduction (%)	36.9	46.9	41.03 ± 4.2
Biogas/COD added (mL/g)	215.5	242.7	232 ± 11.8
Biogas/BOD added (mL/g)	800.2	1055.2	929.4 ± 104

[a]Mean of three replicates.

pump. Additionally, it is possible to see in Table 6 that COD and BOD had, respectively, 25.3% and 41% reduction during the process. The COD reduction value is considerably below to the 75% reported by Stafford et al. *(4)* for similar 10-d HRT. However, the performance observed in this work is similar to that reported by Güngor-Demirci and Demirer *(7)* for 100% broiler manure, having an initial COD of 53,500 mg/L, in batch process under mesophilic conditions (35°C). These authors reported a reduction of 37.9% in a period of 91 d. Considering the time used in the process, the thermophilic process reported here offers some advantages.

Conclusions

Results indicated that heating is the principal component of energy demand, representing 94% of the VPI. The energetic performance could be enhanced as the heating and insulation systems could be improved. Insulation, fouling, heat transfer area, combustion efficiency, operation conditions, and temperature control strategy are some of the aspects that may be considered. The vessel structure should be also reviewed; alternatives such as a jacketed tank or an internal heat exchanger may be included in the analysis. Additionally, an economical study may be used to define the best options to improve biodigester performance. The mechanical power used in this process is similar to the one used by stirred reactors. Pumping uses 56% of the total mechanical energy and plays the principal role in mixing. The slurry moved by pumping in 24 h represented around six times the fermentation mass. On the other hand, even when the power used for bubbling represents 30% of the total mechanical input, it seems to be useless on this pilot-scale reactor. The pumping process necessary to maintain temperature could possibly affect the consortia performance reflected in the reduced COD and BOD conversions and also in the low biogas yield. New strategies oriented to power consumption and mechanical stress reductions are needed.

In this particular case, the removal of the blower or replacement with a larger unit and the reduction of pumping could contribute to manage that issue. Future modifications to the operational conditions and process arrangement may enhance biodigester performance and therefore increase biogas yield. These issues will be discussed in additional communications.

Acknowledgments

The authors gratefully acknowledge the financial support from BIO-PLEX Project, EPSCoR, and Autonomous University of Chapingo. The study was funded by USDA Administrative Grant 2004-06200. We are grateful to Charles "Chip" Vincent, Chris Postalwait, Scot Shapero, and Ysabel Bombardiere for their contributions to the experiments.

References

1. Ceotto E. (2005), *Biores. Technol.* **96**, 191–196.
2. Ribaudo, M., Gollehon, N., Aillery, M., et al. (2003), *Manure Management for Water Quality* AER-824, USDA/ERS, Washington, DC, USA.
3. Wen, Z., Liao, W., and Chen, S. (2005), *Biores. Technol.* **96**, 491–499.
4. Stafford, D. A., Hawkes, D. L., and Horton, R. (1980), *Methane Production from Waste Organic Matter*, CRC Press, Boca Raton, Florida, USA.
5. Lettinga, G., Rebac, S., and Zeeman, G. (2001), *Trends Biotechnol.* **19**, 363–370.
6. Kim, M., Ahn, Y.-H., and Speece, R. E. (2002), *Wat. Res.* **36**, 4369–4385.
7. Güngör-Demirci, G. and Demirer, G. N. (2004), *Biores. Technol.* **93**, 109–117.
8. de Pinho, S. C., Fernandes, B. S., Rodrigues, J. A. D., Ratusznei, S. M., Foresti, E., and Zaiat, M. (2005), *Appl. Biochem. Biotechnol.* **120**, 109–120.
9. Himmelblau, D. M. (1997), *Principios Básicos y Cálculos en Ingeniería Química*, 6th ed., México, México City, DF.
10. Dukelow, S. G. (1991), *The Control of Boilers*, 2nd ed., USA.
11. AOAC (1990), *Official Methods of Analysis*. 15th ed., Arlington, Virginia, USA.
12. Hall, M. B., Hoover, W. H., Jennings, J. P., and Miller, T. K. (1999), *J. Sci. Food Agric.* **79**, 2079–2086.
13. APHA/AWWA/WEF (1998), *Standard Methods for the Examination of Water and Wastewater*, 20th ed., USA, Washington, DC.
14. Hach (2004), *Hach Water Analysis Handbook*, 4th ed., Colorado, USA.
15. Camacho, F., Sánchez, A., Cerón, M. C., García, F., Molina, E., and Chisti, Y. (2004), *Chem. Eng. Sci.* **59**, 4369–4376.
16. Gavrilescu, M. and Tudose, R. Z. (1997), *Chem. Eng. J.* **66**, 97–104.
17. Fu, C.-C., Lu, S. Y., Hsu, Y.-J., Chen, G.-C., Lin, Y.-R., and Wu, W.-T. (2004), *Chem. Eng. Sci.* **59**, 3021–3028.
18. Espinosa-Solares, T., Brito-De la Fuente, E., Tecante, A., and Tanguy, P. A. (2002), *Chem. Eng. Technol.* **25**, 723–727.
19. Pierkiel, A. and Lanting, J. (2004), *Proc. Anaerobic Digestion 2004* **2**, 851–855.
20. Espinosa-Solares, T., Brito-De la Fuente, E., Tecante, A., and Tanguy, P. A. (2001), *Chem. Eng. Technol.* **24**, 913–918.
21. Jin, B. and Lant, P. (2004), *Chem. Eng. Sci.* **59**, 2379–2388.
22. Sawayama, S., Tada, C., Tsukahara, K., and Yagishita T. (2004), *J. Biosci. Bioeng.* **97**, 65–70.

Ethyl Alcohol Production Optimization by Coupling Genetic Algorithm and Multilayer Perceptron Neural Network

Elmer Ccopa Rivera,* Aline C. da Costa, Maria Regina Wolf Maciel, and Rubens Maciel Filho

*DPQ/FEQ/UNICAMP, Campinas, SP, Brasil Cx. Postal 6066, 13081-970;
E-mail: elmer@feq.unicamp.br*

Abstract

In this present article, genetic algorithms and multilayer perceptron neural network (MLPNN) have been integrated in order to reduce the complexity of an optimization problem. A data-driven identification method based on MLPNN and optimal design of experiments is described in detail. The nonlinear model of an extractive ethanol process, represented by a MLPNN, is optimized using real-coded and binary-coded genetic algorithms to determine the optimal operational conditions. In order to check the validity of the computational modeling, the results were compared with the optimization of a deterministic model, whose kinetic parameters were experimentally determined as functions of the temperature.

Index Entries: Alcoholic fermentation process; artificial intelligence; design of experiments; modeling; penalty function.

Introduction

Demand for ethyl alcohol as a renewable fuel for automotive industries continues to be high as countries are pushing for greater energy self-sufficiency. As a large tropical country, Brazil has a high potential for the use of biomass (1). Sugarcane products are today the most economically important biomass source for energy cogeneration. One of the main cost contributive parameters for ethanol production from biomass is the cost of the raw material. Such cost can be reduced if the conversion efficiency of the raw material is maximized. This would also avoid the additional energy required to remove water from ethanol, which is one of the largest costs involved in the production of ethanol to be used as a fuel additive. The productivity, conversion and yield can be optimized both at the biochemical level and at the level of process operation (2). Some studies have used mathematical models based on fundamental mass balances and kinetic equations with experimentally determined parameters to investigate

---

*Author to whom all correspondence and reprint requests should be addressed.

the influence of operating variables on the conversion, productivity, and yield *(3,4)*.

Artificial intelligence (AI), such as artificial neural networks and genetic algorithms (GA), covers a wide range of techniques and tools that facilitate decision making. These methods have already been successfully applied in the optimization and control of bioprocesses for more than 20 yr *(5)*. Moreover, recent studies in chemical and biochemical research showed that these tools can often be favorably combined especially for modeling, optimization *(6,7)*, and control *(8)*.

This work focuses on the process operation aspects using model-based optimization of an extractive alcoholic fermentation process. Silva et al. *(9)* have shown that a scheme combining a fermentor with a vacuum flash vessel presents several positive features and better performance than a conventional industrial process *(10)*. The objective in optimizing this process is to maximize productivity, although maintaining a high conversion of substrates to ethanol in the fermentor. A comparison is made between the performances of two models when the process is optimized using a real-coded genetic algorithm and a binary-coded genetic algorithm. In real coded genetic algorithms, the decision variables are coded in real numbers, unlike the binary numbers used in the binary coded genetic algorithms. The first model used is a deterministic model, whose kinetics parameters were experimentally determined as functions of the temperature, the second is a nonlinear model represented by a multilayer perceptron neural network (MLPNN).

A methodology based on AI was used to determine the best conditions from the interactive effect of four variables: inlet substrate concentration, cells recycle rate, residence time, and flash recycle.

Methods

A Neural Network Mathematical Model for the Ethanol Production Process

The process to be optimized is shown in Fig. 1 *(9)*. The process consists of four interlinked units: the fermentor (ethanol production unit), the centrifuge (cell separation unit), the cell treatment unit, and the vacuum flash vessel (ethanol–water separation unit). The mathematical model is built up of five ordinary differential equations derived from mass and energy balances on the fermentor; algebraic equations for the mass balance on the centrifugue, cell treatment unit, and the purge, as well as the mass and energy balances on the flash tank. Details of the model can be found in Costa et al. *(11)*. The objective of the optimization is to maximize productivity and conversion, which are strongly influenced by the inlet substrate concentration, S_0; cell recycle rate, R; residence time, t_r; and flash recycle rate, r *(11)*.

Ethyl Alcohol Production Optimization

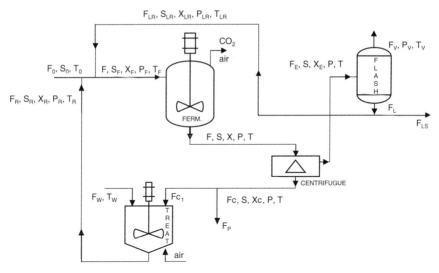

Fig. 1. Extractive alcoholic fermentation scheme.

Table 1
Coded Factor Levels and Equations to Determine the Real Values for the Input Variables Used in the Training of the Neural Networks

Coded factor level	S_0	t_r (h)	R	r
Level +2	$(+2) \cdot \Delta S_0(i) + 180$	$(+2) \cdot \Delta t_r(i) + 1.75$	$(+2) \cdot \Delta R(i) + 0.35$	$(+2) \cdot \Delta r(i) + 0.4$
Level +1	$(+1) \cdot \Delta S_0(i) + 180$	$(+1) \cdot \Delta t_r(i) + 1.75$	$(+1) \cdot \Delta R(i) + 0.35$	$(+1) \cdot \Delta r(i) + 0.4$
Central point (0)	180	1.75	0.35	0.4
Level -1	$(-1) \cdot \Delta S_0(i) + 180$	$(-1) \cdot \Delta t_r(i) + 1.75$	$(-1) \cdot \Delta R(i) + 0.35$	$(-1) \cdot \Delta r(i) + 0.4$
Level -2	$(-2) \cdot \Delta S_0(i) + 180$	$(-2) \cdot \Delta t_r(i) + 1.75$	$(-2) \cdot \Delta R(i) + 0.35$	$(-2) \cdot \Delta r(i) + 0.4$

In order to obtain simulation data for the training of the neural network, the inputs (S_0, R, t_r, and r) are distributed according to a design of experiment within the operational intervals (80, 280), (0.2, 0.5), (1.0, 2.5), and (0.2, 0.6), respectively. These intervals were defined based on prior knowledge of the process *(11)*.

A detailed description of the design of experiments theory can be found in *(12)*. In the present work a full factorial design 2^{4+} star configuration with a central point was used (Table 1). In order to have a large amount of training data, eight different values of ΔS_0, Δt_r, ΔR, and Δr were used, as shown in Table 2, so that eight factorial designs 2^{4+} star configuration were simulated. The data of productivity and conversion are shown in Fig. 2 as functions of S_0 and R.

Table 2
Values of ΔS_0, Δt_r, ΔR, and Δr to Simulate the Data Set, Used to Generate Table 1

i	$\Delta S_0(i)$ (kg/m^3)	$\Delta t_r(i)$ (h)	$\Delta R(i)$	$\Delta r(i)$
1	50	0.075	0.375	0.1
2	45	0.0675	0.3375	0.09
3	40	0.06	0.3	0.08
4	35	0.0525	0.2625	0.07
5	30	0.045	0.225	0.06
6	25	0.0375	0.1875	0.05
7	20	0.03	0.15	0.04
8	10	0.015	0.075	0.02

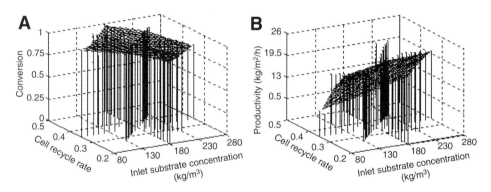

Fig. 2. Spatial distribution of the data set for identification (the partially collected data points would be located only on the mesh surface if only one full factorial 2^4+ star were done): **(A)** Conversion and **(B)** Productivity as functions of cells recycle rate and inlet substrate concentration.

Building the MLPNN Model

In the past years many works *(6,13)* have been proposed that use artificial neural networks as a substitute for deterministic models. The substitution of the deterministic model by an equivalent MLPNN at the optimization step presents the advantage of high speed processing.

One of the main current problems related to the alcoholic fermentation process is the lack of robustness of the fermentation in the presence of fluctuations in the quality of the raw material, which leads to changes in the kinetic behavior and influences yield, productivity, and conversion. These changes make the prediction of the process dynamic behavior with a single mathematical model difficult. In order to take the kinetic changes into account, the kinetic parameters of the model should be re-estimated. However, frequent re-estimation of these parameters in deterministic models is usually difficult, mainly owing to nonlinearities, great number of parameters, and interactions among them. The use of neural network

Ethyl Alcohol Production Optimization

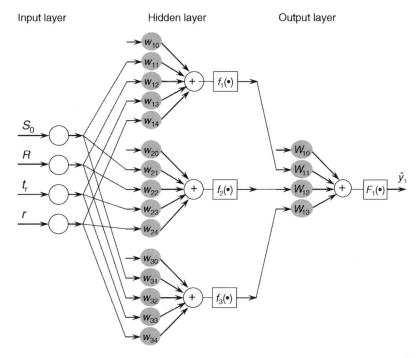

Fig. 3. Representation of a MLPNN with one hidden layer.

models is a good option in this case, because the changes in the kinetic behavior can be assessed by updating the network weights.

A MLPNN consists of three types of layers: an input layer, an output layer, and one or more hidden layers, as shown in Fig. 3 and described mathematically by Eq. 1. Each layer may have a different number of neurons, and even a different transfer function. The most used transfer functions are the sigmoidal, the hyperbolic tangent and the linear transfer functions. More details about neural networks can be found in the works of Chen et al. (6) and Chang and Hung (13). An appropriate architecture would help to achieve higher model accuracy.

$$\hat{y}_1 = g_1[\delta,\theta] = F_1\left[\sum_{j=1}^{3} W_{1j} f_j(\sum_{l=1}^{4} w_{jl} \delta_l + w_{j0}) + W_{10}\right] \quad (1)$$

In Eq. 1, θ specifies the parameters vector, which contains all the adjustable parameters of the network; i.e., the weights and biases $\{w_{j,l}, W_{1,j}\}$.

In this work, conversion and productivity were each one modeled with a MLPNN with four inputs (inlet substrate concentration, S_0, cell recycle rate, R, residence time, t_r, and flash recycle rate, r) and one hidden layer.

The appropriate number of nodes to include in the hidden layer was addressed with the cross-validation technique in order to avoid model over-fitting. This technique splits the sample data into a training sample and a validation sample. Then MLPNN with different numbers of hidden nodes are trained with the training sample, and their performance monitored with the validation sample in terms of the lowest normalized mean square error *(14)*, given by Eq. 2. In addition, the quality of the prediction of the neural models can be also characterized using the correlation coefficient *(15)*, given by Eq. 3.

The Normalized Mean Square Error (NMSE):

$$\text{NMSE} = \frac{1}{\sigma^2 np} \sum_{n=1}^{np} (d_p - x_p)^2 \qquad (2)$$

where σ^2 denoted the sample variance of the desired outputs in the test set, x_p and d_p are, respectively, the network and desired outputs and np is the number of patterns tested.

The Correlation Coefficient (COR):

$$\text{COR} = \left(1 - \frac{\text{SEE}}{\text{S}\tau\tau}\right) 100 \qquad (3)$$

where and $\text{SEE} = \sum_{p=1}^{np}(d_p - x_p)^2$ and $\text{S}\tau\tau = \sum_{p=1}^{np}(d_p - \bar{d}_p)^2$

In the present work, one half of the data sample was used for training and the other half was intended for validation. The parameters were adjusted using the Levenberg–Marquardt algorithm in Matlab's neural network toolbox.

After the neural networks for conversion and productivity were trained and validated, their parameters were fixed, and the neural network models operated in parallel (Fig. 4) to simulate a function, *f(conv, prod)*, that was used as the fitness criteria in the ensuing genetic algorithm optimization. In order to accomplish this task, the neural model was written in Fortran to be used by the optimization algorithm, also written in Fortran.

Optimization by Genetic Algorithms

GA is used in this work to optimize a given objective function *f* over a given search space. A population of individuals undergoes some artificial Darwinian evolution (genetic inheritance and Darwinian strife for survival) based on the fitness of each individual. The fitness of an individual is directly related to the value of the objective function of this individual. The manipulation is done by the genetic operators that work on the chromosomes, in which the parameters of possible solutions are encoded. In each generation of the GA, the new solutions replace the solutions in the population that are selected for deletion. The GA considered in the present article is based on the freeware versions written in Fortran of a real-coded genetic algorithm (RGA) developed by Yedder *(16)* and a binary-coded genetic

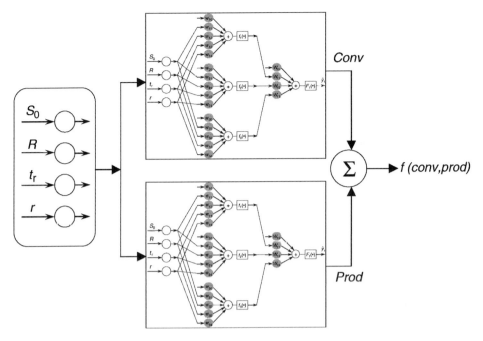

Fig. 4. Combination of the neural network models (hybrid model) for the optimization of the fermentation process.

algorithm (BGA) developed by Carroll (17), adapted to the specific needs of the process optimization. These algorithms are well suited to large-scale problems because they do not demand the computation of any Hessian matrix or its inverse, thus having relatively small memory and processing requirements. Also, it ensures the convergence of the optimization procedure (with a second-order rate).

The formulation of objective functions is one of the crucial steps in the application of optimization to a practical problem. In this work, the objective function was formulated as a nonlinear programming of the following type:

Optimize $f(\vec{x})$

subject to $g_i(\vec{x}) \leq 0, i = 1, \ldots, p,$

$$h_j(\vec{x}) = 0, j = p+1, \ldots, m, \quad (4)$$

$l_k \leq x_k \leq u_k, k = 1, \ldots, n,$

where x is a vector of n decision variables $(x_1, \ldots, x_n)$, $f(x)$ is the objective function, $g_i(\vec{x})$ is the ith inequality constraints, and $h_j(\vec{x})$ is the jth equality constraints. The l_k and u_k are specified lower and upper bounds on the variables with $l_k \leq u_k$.

A general way to deal with constraints, whatever the optimization method, is by penalty function method (18). This technique is the most common approach in the genetic algorithms community, but there are no general guidelines on designing. Some suggestions for genetic algorithms are given in *(19)*. The penalty method of nonlinear programming transforms a constrained problem into a sequence of unconstrained problem. The method used in this work, belong to the second category of constrained handling methods described by *(19)*. In this approach, for handling inequality constraints, the penalty function $P(\vec{x})$ is defined as the sum of the objective function $f(x)$ and a penalty term which depends on the constraints violation $g_i(x)$ and $h_j(x)$:

$$\left. \begin{array}{l} \text{Optimize } f(\vec{x}) \\ \text{Subject to } g_i(\vec{x}) \leq 0, i = 1, \ldots, p, \\ \quad h_j(\vec{x}) = 0, j = p+1, \ldots, m, \\ \quad l_k \leq x_k \leq u_k, k = 1, \ldots, n, \end{array} \right\} \text{Optimize} : P(f, g_i, h_j, C) \quad (5)$$

where $P(f, g_i, h_j, C)$ is a penalty function, and C is a positive penalty parameter. After the penalty function is formulated, it is minimized for a series of values of increasing C-values, which force the sequence to approach the optimum of the constrained problem.

The most significant advantages of the GA are that it avoids the initial guess selection problem and provides a systematic scanning of the whole population and several acceptable local solutions. Thus, it is important to properly define constraints on the possible parameter values. First, constraints limit the search space and thus speed up the optimization. Second, physically impossible values for the parameters are avoided, which improves the reliability of the approach. Additional constraints depend on the type of model that is used. In this work the parameters satisfy such conditions as: $80 < S_0 < 280$ kg/m^3, $0.2 < R < 0.5$, $1 < t_r < 2.5$ h and $0.2 < r < 0.6$. Costa et al. *(4)* already used these values successfully in a previous study.

Results

Choice of the Neural Network Architecture

The number of hidden nodes was varied from 2 to 8 and, using the cross-validation criterion, the network with three hidden nodes was found to present the lowest normalized mean square error (Eq. 2) for the validation sample. The results can be seen in Fig. 5.

Productivity and conversion were then modeled both with a MLPNN with 19 scalar parameter, which is calculated using Eq. 6.

$$\text{Number of parameters} = (n_i + 1)n_h + (n_h + 1)n_o \quad (6)$$

where $n_i = 4$, $n_h = 3$ and $n_o = 1$ are the number of neurons in the input, hidden, and output layers, respectively.

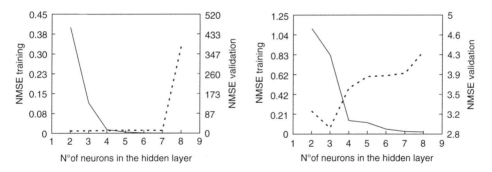

Fig. 5. Training (solid line) and validation (dashed line) error as functions of the number of neurons in the hidden layer: **(A)** Productivity and **(B)** conversion.

Table 3 shows the quality of the prediction for productivity and conversion for training and validation. As can be seen, in both cases by increasing the number of hidden neurons, the training accuracy improves, as indicated by the smaller NMSE and COR values approaching 1. However, using the cross-validation criterion, three neurons in the hidden layer appeared to be the optimal architecture to prevent overfitting.

Computation of an Optimal Solution Using Genetic Algorithms and Deterministic Model

The optimization was conducted with the deterministic steady state model of the process, which can be found in Costa et al. *(11)*. It consists of the steady state mass and energy balances for the fermentor and all the other process units (*see* Fig. 1). The equations that describe productivity and conversion can also be found in Costa et al. *(11)*.

The optimization problem is to maximize productivity. Thus, the objective function can be formulated as follows:

$$\text{Maximize } prod \tag{7}$$

subject to the equality constraints described by the steady state mass and energy balances for the fermentor and to the inequality constrains:

$$Conv > 0.99 \tag{8}$$

$$28 < T < 40°C \tag{9}$$

$$80 < S_0 < 280 \text{ kg/m}^3 \tag{10}$$

$$0.2 < R < 0.5 \tag{11}$$

$$1.0 < t_r < 2.5 \text{ h} \tag{12}$$

$$0.2 < r < 0.6 \tag{13}$$

Because the objective in Yedder's real-coded genetic algorithm driver *(16)* is to minimize the objective function, it is necessary to map the

Table 3
Normalized Mean Square Error (NMSE) and Correlation Coefficient (COR) for Productivity and Conversion

Neurons	NMSE				COR			
	Productivity (kg/m$^3 \cdot$h)		Conversion		Productivity (kg/m$^3 \cdot$h)		Conversion	
	Training	Validation	Training	Validation	Training	Validation	Training	Validation
2	0.4	10	1.1	3.2	99.06	97.59	98.72	95.99
3	0.1	11	0.8	2.9	99.33	97.11	99.13	92.53
4	0.01	11	0.1	3.6	99.96	96.76	99.68	92.44
5	0.004	13.2	0.1	3.9	99.97	96.48	99.87	80.31
6	0.001	12.5	0.05	3.9	99.99	96.18	99.88	91.98
7	0.0005	12	0.03	3.9	99.99	96.16	99.95	73.19
8	0.0003	380.1	0.02	4.3	99.99	16.26	99.97	82.82

formulated problem (Eqs. 7–13) into unconstrained objective function given by Eq. 14, following the penalty function method for handling inequality constraints.

The equality constraints were handled by converting them as inequality constraints as $g_i(\vec{x}) = \lambda - |h_j(\vec{x})| \geq 0$ for all j, where λ (a small positive value) is set to 10^{-3}, in order to allow some room for the search algorithm to work on.

$$\min P(\vec{x}, R) = f(\vec{x}) + R_1 |g_1(\vec{x})|^2 + R_2 \left(\sum_{j=1}^{5} (10^{-3} - |h_j(\vec{x})|)^2 \right) \quad (14)$$

where $f(\vec{x}) = 1/|prod|$, $g_1 = 0.99 - |conv|$ and $h_j(\vec{x})$ are the steady state mass and energy balances for the fermentor. The parameters ($\vec{x}$) satisfy such conditions as: 80 kg/m³ < S_0 < 280 kg/m³, 0.2 < R < 0.5, 1 h < t_r < 2.5 h, and 0.2 < r < 0.6.

The optimization variables are: concentration of viable cells, X_v; dead cells, X_d; substrate, S; product, P; and Temperature, T; as well as the variables used by Costa et al. *(11)* (S_0, R, t_r, and r).

With the best penalty parameters, $R_1 = 0.02$ and $R_2 = 1250$ for constraints, the calculated values were $S_0 = 216.29$ kg/m³, $R = 0.41$, $t_r = 1.77$ h, and $r = 0.48$. The optimal solution to this problem, using RGA, is: productivity = 13.1 kg/(m³ · h) and conversion = 0.99.

Carroll's binary-coded genetic algorithm *(17)* maximizes the objective function. In consequence, the problem (Eqs. 7–13) was formulated into the penalty function given by Eq. 15.

$$\max P(\vec{x}, R) = f(\vec{x}) + R_1 \left(|g_1(\vec{x})|^2 \right)^{-1} + R_2 \left(\sum_{j=1}^{5} (10^{-3} - |h_j(\vec{x})|)^2 \right)^{-1} \quad (15)$$

where $f(\vec{x}) = |prod|$, $g_1(\vec{x}) = 0.99 - |conv|$ and $h_j(\vec{x})$ are the steady state mass and energy balances for the fermentor. Here, the parameters ($\vec{x}$) also satisfy conditions given by Eqs. 9–13.

Using BGA the following values are obtained: $S_0 = 135.86$ kg/m³, $R = 0.28$, $t_r = 1.96$ h, and $r = 0.40$. The optimal solution to this problem with the best penalty parameter, $R_1 = 97$ and $R_2 = 1$ for constraints, are: productivity = 13.1 kg/(m³ · h) and conversion = 0.99.

Kalil et al. *(3)* optimized the industrial conventional process designed by Andrietta and Maugeri *(10)* and obtained productivity of 12 kg/(m³ · h), conversion of 0.99 and yield of 0.86. Although in the formulated problem only one inequality restriction was considered (conv > 0.99), the values obtained for yield: 0.87 and 0.9 for RGA and BGA, respectively, were better than the optimized conventional process of Kalil et al. *(3)*. The optimization using genetic algorithms led to productivity a little higher than when the extractive process was optimized using successive quadratic programming by Costa et al. *(4)*.

Computation of an Optimal Solution Using Genetic Algorithms and Neural Network Model

Because the objective of the optimization study was to maximize productivity, although maintaining a high conversion, the neural network models were structured (*see* Fig. 4) to explicitly estimate these output variables from the inlet variables of interest (S_0, R, t_r, and r). This structure models the mass and energy balances for the fermentor and all the other process units. After the solution of the optimization problem, the optimal values of S_0, R, t_r, and r were used in the deterministic model to determined if the MLPNN predictions for optimal productivity and conversion present deviations from the values calculated by the deterministic model.

The optimization problem is the function $f(prod, conv)$ transformed in the penalty function form given by Eqs. 16 and 17 for RGA and BGA, respectively:

$$\min P(\vec{x}, C) = f(\vec{x}) + C_1 |g_1(\vec{x})|^2 \tag{16}$$

where $f(\vec{x}) = 1 / |prod|$ and $g_1(\vec{x}) = 0.99 - conv |$.

In this problem, such as in the previous case, a constraint in conversion is included ($conv > 0.99$). With the best penalty parameter, $C_1 = 260$ for constraint, the calculated values were $S_0 = 131.93$ kg/m^3, $R = 0.28$, $t_r = 1.20$ h, and $r = 0.54$.

The optimal solution to this problem, using RGA, is: productivity = 13 kg/(m$^3 \cdot$ h) and conversion = 0.99. Using the values calculated for S_0, R, t_r, and r in the deterministic model, productivity obtained was of 13.2 kg/(m$^3 \cdot$ h) and conversion of 0.99.

$$\max P(\vec{x}, C) = f(\vec{x}) + C_1 \left(|g_1(\vec{x})|^2 \right)^{-1} \tag{17}$$

where $f(\vec{x}) = |prod|$ and $g_1(\vec{x}) = 0.99 - |conv|$.

Using BGA the following values are calculated: $S_0 = 161.4$ kg/m^3, $R = 0.44$, $t_r = 1.60$ h, and $r = 0.37$. The optimal solution to this problem with the best penalty parameter, $C_1 = 97$ for constraint, is: productivity = 13 kg/(m$^3 \cdot$ h) and conversion = 0.99. Using the values calculated for S_0, R, t_r, and r in the deterministic model it was found productivity of 13.0 kg/(m$^3 \cdot$ h) and conversion of 0.99.

It is possible to notice that, when the optimization problem is solved using genetic algorithms and the neural network models, the values of temperature and concentrations in the fermentor do not have to be considered as optimization variables and so, the number of the optimization variables is smaller than using deterministic model.

Table 4 shows the optimization results of productivity and conversion calculated by the RGA and BGA on the deterministic and MLPNN

Table 4
Optimization Variable Values Obtained Through RGA and BGA on the Deterministic and MLPNN Models

	Neural-RGA	Neural-BGA	Deterministic-RGA	Deterministic-BGA
S_0 (kg/m³)	131.93	161.4	216.29	135.86
R	0.28	0.44	0.41	0.28
t_r (h)	1.20	1.6	1.77	1.96
R	0.54	0.37	0.48	0.4
X_v (kg/m³)	–	–	36.66	24.26
X_d (kg/m³)	–	–	3.18	2.69
S (kg/m³)	–	–	2.39	2.06
P (kg/m³)	–	–	39.31	37.99
T (°C)	–	–	32.98	35.18
Prod (kg/m³·h)	13.2	13	13.1	13.1
Conv	0.99	0.99	0.99	0.99

models. It also shows the comparison among the optimization variables. The results of the optimization methods are expected to lead to different parameters values because they are numerical procedures with expected imprecision in finding the global optimum. Different mathematical models may lead to different predictions, however, a well-trained neural network is able to capture the process dynamic behavior. It is worthwhile mentioning that all the parameters values in Table 4 are within the accepted range for industrial process operation and this was taken into account in the definition of the constraints for the optimization problem.

Discussion

This work focused on modeling and optimization on the level of process operation of an extractive alcoholic fermentation. The computational intelligence techniques are sought to efficiently combine all available knowledge and to direct the development toward an improved process operation strategy. The genetic algorithms have shown a good capability to determine the best conditions of conversion and productivity using both the deterministic and the neural network models.

The NN model was developed using a data set generated from the deterministic model, which means the data are not noisy. However, this may not be considered a limitation, because the procedure may be used when experimental data are available. It can be observed that the values for productivity and conversion obtained using the MLPNN models are similar to that obtained using the deterministic model. MLPNN modeling is a powerful and flexible tool and its use in the alcoholic fermentation process is advantageous because updating the neural network parameters is a simpler procedure than re-estimating kinetic parameters of the deterministic models. Frequent re-estimation of kinetic parameters (or in the case of this work updating the neural network parameters) is necessary owing to changes in the dynamic behavior caused by fluctuations in the quality of the raw material, changes in microorganism metabolism, and variation of the dominant yeast strains present in the fermentation process, among other factors.

Normally the optimization methods based on GA require a large computer burden compared with the conventional optimization methods. Neural network models led to a mathematical representation of lower order than the deterministic approach, which intuitively makes the optimization procedure based on hybrid algorithms (NN-RGA and NN-BGA) to be significantly quicker. In fact, the neural network model requires only four optimization variables; whereas the deterministic model needs nine.

Experimental design is used to obtain data for the training of the neural network in order to guarantee that the region of interest is covered by the training data. This is a very important issue to be addressed as neural network models have no (or very limited) extrapolation properties. The penalty function optimization using RGA and BGA was done in terms of constraint violation to check the feasibility of the solution. In the case of infeasible solution, the one having smaller constraint violation is preferred. In real coded genetic algorithm, the decision variables are coded in real numbers unlike the binary numbers as in the case of binary coded genetic algorithms. The most important feature of the RGAs is their capacity to exploit local continuities, and the corresponding one of the BGA is their capacity to exploit the discrete similarities (20). According to Goldberg (21), BGA are less efficient when applied to multi-dimensional, high precision or continuous problems. The bit strings can become very long and the search space blows up. In RGA the chromosome/individual is simply a vector in which each real-valued element (gene) stands for an unknown parameter of the model.

Although the best result was the one obtained using the rigorous model, the values for productivity and conversion obtained using the MLPNN model are acceptable and can partly eliminate the difficulties of having to specify completely the structure of an alcoholic fermentation model. It is especially promising in situations in which rigorous model is excessively difficult computationally, time-consuming or costly.

Acknowledgment

The authors acknowledge FAPESP (process number 03/03630-3) for financial support.

Nomenclature

F	Feed stream flow rate (m³/h)
F_c	Cell suspension flow from centrifugue (m³/h)
F_{c1}	Cell suspension flow to treatment tank (m³/h)
F_E	Light phase flow rate to flash tank (m³/h)
F_L	Liquid outflow from the vacuum flash tank (m³/h)
F_{LR}	Liquid phase recycling flow rate (m³/h)
F_{RS}	Liquid phase flow to rectification column (m³/h)
F_O	Fresh medium flow rate (m³/h)
F_R	Cell recycling flow rate (m³/h)
F_V	Vapor outflow from the vacuum flash tank (m³/h)
F_W	Water flow rate (m³/h)
P	Product concentration into the fermentor (kg/m³)
P_F	Feed product concentration (kg/m³)
P_{LR}	Product concentration in the light phase from centrifuge (kg/m³)
P_R	Product concentration in the cells recycle (kg/m³)
P_V	Product concentration in the vapor phase from the flash tank (kg/m³)
$R = F_R/F$	Cells recycle rate
$r = F_{LR}/F_L$	Flash recycle rate
S	Substrate concentration into the fermentor (kg/m³)
S_F	Feed substrate concentration (kg/m³)
S_{LR}	Substrate concentration in the light phase from centrifugue (kg/m³)
S_O	Inlet substrate concentration (kg/m³)
S_R	Substrate concentration in the cells recycle (kg/m³)
T	Temperature into the fermentor (°C)
T_F	Feed temperature (°C)
T_{LR}	Light phase temperature (°C)
T_O	Inlet temperature of the fresh medium (°C)
T_R	Cells recycle temperature (°C)
T_V	Temperature of vapor from the flash tank (°C)
T_W	Water temperature (°C)
X_c	Biomass concentration in the heavy phase from centrifugue (kg/m³)
X_d	Dead biomass concentration into the fermentor (kg/m³)
X_E	Biomass concentration in the light phase flow rate to flash tank (kg/m³)
X_F	Feed biomass concentration (kg/m³)
X_{LR}	Biomass concentration in the light phase from centrifugue (kg/m³)
X_R	Cell recycling concentration (kg/m³)
$X_t = X_v + X_d$	Total biomass concentration into the fermentor (kg/m³)
X_r	Viable biomass concentration into the fermentor (kg/m³)

References

1. Goldemberg, J., Coelho, S. T., Nastari, P. N., and Lucon, O. (2004), *Biomass Bioenerg.* **26**, 301–304.
2. Parekh, S., Vinci, V. A., and Strobel, R. J. (2000), *Appl. Microbiol. Biol.* **54**, 287–301.
3. Kalil, S. J., Maugeri, F., and Rodrigues, M. I. (2000), *Process Biochem.* **35**, 539–550.
4. Costa, A. C. and Maciel Filho, R. (2004), *App. Biochem. Biotech.* **114/1–3**, 485–496.
5. Schügerl, K. (2001), *J. Biotechnol.* **85**, 149–173.
6. Chen, L., Nguang, S. K., Chen, X. D., and Li, X. M. (2004), *Biochem. Eng. J.* **22**, 51–61.
7. Nandi, S., Badhe, Y., Lonari, J., et al. (2004), *Chem. Eng. J.* **97**, 115–129.
8. Ahmad, A. L., Azid, I. A., Yusof, A. R., and Seetharamu, K. N. (2004), *Comput. Chem. Eng.* **28**, 755–766.
9. Silva, F. L. H., Rodrigues, M. I., and Maugeri, F. (1999), *J. Chem. Tech. Biotechnol.* **74**, 176–182.
10. Andrietta, S. R. and Maugeri, F. (1994), In *Adv. Bioprocess Engng.* 47–52.
11. Costa, A. C., Atala, D. I. P., Maciel Filho, R., and Maugeri Filho, F. (2001), *Process Biochem.* **37-2**, 125–137.
12. Montgomery, Douglas C. (1997), *Design and Analysis of Experiments*, 4th ed., John Wiley and Sons, New York.
13. Chang, J. and Hung, B. (2002), *Ind. Eng. Chem. Res.* **41**, 2716–2727.
14. Geva, A. B. (1998), *IEEE Trans. neural net.* **95**, 1471–1482.
15. Milton, J. S. and Arnold, J. C. (1990), *Introduction to probability and statistics*, McGraw Hill, New York.
16. Yedder, R. B. (2002), Ph.D. thesis, ENPC, Paris, France.
17. Carroll, D. L. (1998), FORTRAN Genetic Algorithm Driver, last accessed date 04-04-2005. University of Illinois, USA (carroll@cuaerospace.com).
18. Edgar, T. F. and Himmelblau, D. M. (1998), *Optimization of Chemical Processes*, 2nd ed., McGraw-Hill, New York.
19. Michalewicz, Z. and Schoenauer, M. (1996), *Evol. Comp.* **4**, 1–32.
20. Herrera, F., Lozano, M., and Verdegay, J. L. (1998), *A. I. Rev.* **12**, 265–319.
21. Goldberg, D. E. (1989), *Genetic algorithms in search, optimization and machine learning*. Addisson-Wesley Reading, Massachussets.

Lactic Acid Recovery From Cheese Whey Fermentation Broth Using Combined Ultrafiltration and Nanofiltration Membranes

YEBO LI* AND ABOLGHASEM SHAHBAZI

Bioenvironmental Engineering Program, Department of Natural Resources and Environmental Design, North Carolina A&T State University, 1601 East Market Street, Greensboro, NC 27411; E-mail: yli@ncat.edu

Abstract

The separation of lactic acid from lactose in the ultrafiltration permeate of cheese whey broth was studied using a cross-flow nanofiltration membrane unit. Experiments to test lactic acid recovery were conducted at three levels of pressure (1.4, 2.1, and 2.8 MPa), two levels of initial lactic acid concentration (18.6 and 27 g/L), and two types of nanofiltration membranes (DS-5DK and DS-5HL). Higher pressure caused significantly higher permeate flux and higher lactose and lactic acid retention ($p < 0.0001$). Higher initial lactic acid concentrations also caused significantly higher permeate flux, but significantly lower lactose and lactic acid retention ($p < 0.0001$). The two tested membranes demonstrated significant differences on the permeate flux and lactose and lactic acid retention. Membrane DS-5DK was found to retain 100% of lactose at an initial lactic acid concentration of 18.6 g/L for all the tested pressures, and had a retention level of 99.5% of lactose at initial lactic acid concentration of 27 g/L when the pressure reached 2.8 MPa. For all the tests when lactose retention reached 99–100%, as much as 64% of the lactic acid could be recovered in the permeate.

Index Entries: Cheese whey; fermentation; lactic acid; membrane; lactose; nanofiltration.

Introduction

Manufacturing of cheese produces large volumes of whey as a byproduct. The United States generates nearly 1.2×10^9 t of cheese whey/yr *(1)*. It is estimated that as much as 40–50% of the whey produced is disposed of as sewage or as fertilizer applied to agricultural lands with the rest being used primarily as animal feed. Cheese whey contains about 4.5–5% lactose, 0.6–0.8% soluble proteins, 0.4–0.5% w/v lipids, and varying concentrations of mineral salts *(2)*. Therefore, there is an interest to utilize lactose from cheese whey in the production of value-added products. Lactic acid is one such value-added product that is produced from processing cheese whey.

*Author to whom all correspondence and reprint requests should be addressed.

Lactic acid is a natural organic acid and has many applications in the pharmaceutical, food, and chemical industries. It is used as an acidulant and as a preservative, and also as a substrate in the production of biodegradable plastics and other organic acids (1,3).

Lactic acid can be produced by fermentation of sugar-containing substrates such as cheese whey using *Lactobacillus helveticus* (3,4), and *L. casei* (5,6). *L. helveticus* is a thermophilic and acidophilic bacterium that can grow under conditions inhibitory for most contaminant microorganisms (3). *Bifidobacterium longum* is a bacteria that can both convert lactose into lactic acid and also produce an anti-bacterial compound, which can boost the immune system in its host. *Bifidobacterium* spp. produces high yield of L-(+) lactic acid compared with D-(–) lactic acid (7).

Most previous studies examining lactic acid production have concentrated on increasing *B. longum* cell production by cell immobilization and optimized pH (8,9). Till date, there has been no report on using *B. longum* to produce lactic acid from cheese whey. By the *Bifidum* pathway, the fermentation of two moles of hexose results in 3 mol of acetate and 2 mol of lactate (7), but there has been no report on the metabolic pathway of *B. longum* to convert lactose to lactic acid.

The processes of lactic acid production include two key stages which are fermentation and product recovery. The biggest challenge in lactic acid production lies in the recovery and not in the fermentation step (10). A successful lactic acid recovery approach is that of continuous fermentation in a recycled reactor in which the cells, protein, and lactose are separated by a filtration unit and returned to the fermentor while the lactic acid is removed in the permeate.

Ultrafiltration can remove dissolved macromolecules with molecular weight cutoff (MWCO) between 1000 and 100,000 Da (11). An important hurdle in the application of membrane technology in whey processing is the decline in permeate flux during the operation. The permeate flux decline during ultrafiltration of cheese whey is attributed to concentration polarization and membrane fouling (12). Cells and proteins can be successfully separated from the cheese whey fermentation broth using ultrafiltration membrane with MWCO around 20,000 Da (13,14).

Nanofiltration is a pressure-driven membrane process with a MWCO situated between reverse osmosis and ultrafiltration. The nanofiltration membrane has already been used in the demineralization of salted, acid, and sweet cheese whey (15). The process could separate monovalent salts and organics in the molecular weight range 200–1000 Da (16). Nanofiltration membrane with MWCO around 400 Da was demonstrated to retain about 97% of lactose and 12–35% of lactate at pH 3.3 in a nanofiltration membrane reactor (17). Nanofiltration of cheese whey has been evaluated based on the permeate flux to improve the demineralization rate by Alkhatim et al. (18). In their research, they found that lower pH was observed to have higher permeability of sodium and potassium.

Jeantet et al. *(17)* also found out that decreasing pH resulted in decreased retention of lactic acid.

This study was the second step of a three-step membrane separation (ultrafiltration, nanofiltration, and reverse osmosis separation) process for lactic acid recovery. The protein and cells have been separated by ultrafiltration unit in the previous study *(14)*. The objectives of this study were twofold: (1) to evaluate membranes that could be used to retain lactose as concentrate while recover lactic acid in permeate; and (2) to study the effects of transmembrane pressure and initial lactic acid concentration of feed stream on the permeate flux, lactose retention, and lactic acid recovery using nanofiltration.

Materials and Methods

Cheese Whey Media

Cheese whey media was prepared by dissolving 50 g of deproteinized cheese whey powder (Davisco Foods International, Inc., Eden Prairie, MN) into a liter of deionized (DI) water and stirring for 5 min at ambient temperature. The composition of the deproteinized cheese whey powder was as follows: crude protein (total nitrogen × 6.38) 6.8%, crude fat 0.8%, lactose 78.6%, ash 9.4%, and moisture 4.4%. The solutions were autoclaved at 103°C for 10 min.

Microorganism and Culture Media

B. longum was obtained from the National Collection of Food Bacteria (NCFB 2259). Stock culture of this strain was maintained in 50% glycerol and Man Rogosa Sharpe (MRS) broth media at −80°C. Active cultures were propagated in 10 mL MRS broth at a temperature of 37°C for 18–24 h under anaerobic conditions. This was used as a preculture to initiate cell production of higher volume with a 1% inoculation into 100 mL fresh MRS broth, incubated at 37°C for 24 h.

Fermentation

Fermentation was conducted in a stirred 5-L bench top fermentor. The pH of the broth was maintained at 5.5 by neutralizing the acid with 5 N ammonium hydroxide during fermentation. The agitation speed of the fermentor was maintained at 150 rpm and the temperature was maintained at 37°C. Samples were withdrawn every 2 h during the first 8 h and every 12 h during the remaining fermentation process. Each fermentation test lasted for 48 h.

Ultrafiltration Separation

The fermentation broth was first filtrated through an ultrafiltration membrane unit. Two types of membranes (PES5 and PES20, Nadir Filtration

GmbH, Wiesbaden, Germany) with MWCO of 5000 and 20,000 Da were used in the ultrafiltration experiment. The membrane polymer consisted of hydrophilic polyethersulfone and polysulfone. The permeate of the ultrafiltration process was collected for this nanofiltration study. The average true protein concentration in the ultrafiltration permeate was less than 0.02% (14). As most of the lactose was converted after 48 h of fermentation, the lactose concentration of permeate was adjusted to 22 g/L by the following two methods. The first method involved adding 22 g of lactose in 1 L of permeate which produced a solution with lactose and lactic acid concentration of 22 and 27 g/L. The second method used the permeate obtained by ultrafiltration of the unfermented cheese whey to adjust the lactose concentration. The lactose and lactic acid concentrations obtained by the second method were about 22 and 18.6 g/L, respectively.

Nanofiltration

The nanofiltration system consisted of a recirculation pump, nanofiltration unit (SEPA CF II, Osmonics, Minneapolis, MN), and an online permeate weighting unit. The diagram of nanofiltration membrane system is shown in Fig. 1. The ultra filtered permeate was circulated from the fermentor to the membrane unit at a constant velocity via a positive pump (M03-S, Hydracell, Minneapolis, MN) which can run at pressure up to 8.4 MPa. Two Nano membranes (DS-5DK and DS-5HL, Osmonics, Minneapolis, MN) were used in the experiments. Both the membranes could retain 98% of $MgSO_4$ but had different levels of permeate flux. No MWCO information was provided by the manufacturer. The surface area of the membrane is 140 cm^2. The hold up volume of the membrane unit is 70 mL. The fermentor was used to maintain the ultra filtered permeate at constant temperature and pH. The concentrate was recycled to the fermentor via the pump while permeate was collected in a container placed on an electronic balance. The reading of the balance was continually recorded at 30-s intervals by a computer. The transmembrane pressure and cross-flow feed velocity were adjusted by a manual valve and pump controller. The pressure was measured by a standard pressure gauge. The cross-flow feed velocity was calculated based on the flow rate and section area of the membrane channel. The cross flow velocity used in the experiments was 0.5 m/s. The transmembrane pressure levels used were 1.4, 2.1, and 2.8 MPa. The pH of the ultra filtrated permeate was kept at 5.5 and the ultra filtrated permeate was agitated at 200 rpm in the fermentor. The nanofiltration process lasted for 1 h. Duplicate samples of initial feed stream, of permeate and of concentrate were collected at the end of nanofiltration for analysis to determine the lactose and lactic acid concentrations. An alkali–acid treatment method was applied to the membrane system in the following steps:

1. Fully open the recirculation and permeate valves.
2. Flush with tap water for 5 min.

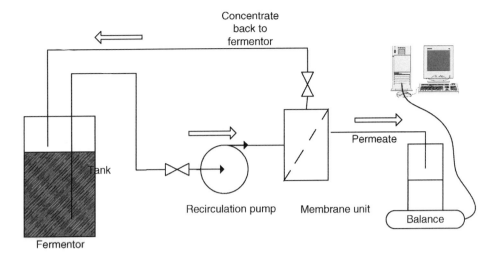

Fig. 1. Schematic diagram of the nanofiltration membrane separation system.

3. Circulate 2 L of 4% phosphoric acid for 10 min.
4. Rinse with tap water for 5 min.
5. Circulate 2 L of 0.1 N NaOH solution for 10 min.
6. Rinse with 10 L of DI water for 5 min.

Analyses

Lactose, lactic acid, and acetic acid were determined by high-performance liquid chromatography (Waters, Milford, MA) with a KC-811 ion exclusion column and a Waters 410 differential refractometer detector. The mobile phase was 0.1% H_3PO_4 solution at a flow-rate of 1 mL/min. The temperatures of the detector and of the column were maintained at 35°C and 60°C, respectively.

The lactic acid productivity was evaluated by, (a) lactic acid yield and (b) conversion ratio. The conversion ratio was expressed as follows:

$$\text{Conversion ratio (\%)} = \frac{\text{Initial lactose concentration - Residual lactose concentration}}{\text{Initial lactose concentration}} \times 100\%$$

The lactic acid yield was expressed as grams of lactic acid produced/g lactose used:

$$\text{Lactic acid yield (g/g)} = \frac{\text{Lactic acid produced}}{\text{Lactose used}}$$

The performance of membrane separation was evaluated by using three criteria: (a) permeate flux, (b) lactose retention, and (c) lactic acid recovery. The permeate flux was calculated by measuring the quantity of permeate collected during a certain time and dividing it by the effective membrane area for filtration.

$$\text{Permeate flux} = \frac{\text{Permeate volume}}{\text{Membrane area} \times \text{Time}} (L/m^2h)$$

The lactose retention (%) was defined as:

$$\text{Retention} = \left(1 - \frac{\text{Concentration of lactose in the permeate}}{\text{Concentration of lactose in the feed stream}}\right) \times 100$$

C_{L0} = concentration of component in the feed stream;
C_{LP} = concentration of component in the permeate.
The lactic acid recovery (%) was defined as:

$$\text{Recovery} = \frac{\text{Concentration of lactic acid in the permeate}}{\text{Concentration of lactic acid in the feed stream}} \times 100$$

Analysis of variance was performed using a statistical package from the SAS System (SAS Institute, Cary, NC).

Results and Discussion

Fermentation

The lactose, lactic acid, and acetic acid concentrations obtained during the 48 h of fermentation are shown in Table 1. The values in the table represent the averaged values of two runs. The results show that about 97.2% of the lactose was converted and that 0.73 g lactic acid was produced from 1 g of lactose used. The production of acetic acid was trivial in comparison to that of lactic acid. The production of acetic acid could be caused by the nonstrict anaerobic fermentation conditions and contamination of broth. The lactose conversion ratio and lactic acid yield are similar to results of other lactic acid producing bacteria such as *L. helveticus*. Tango and Ghaly (3) obtained a lactose conversion ratio of 92–95% and a lactic acid yield of 0.86 g lactic acid/g lactose when using immobilized *L. helveticus* with nutrient supplement at 36 h of fermentation. Shahbazi et al. (1) obtained a lactose conversion ratio of 79% and lactic acid yield of 0.84 g lactic acid/g lactose using immobilized *L. helveticus* in bioreactor. Most of the previous work targeted at obtaining high lactose conversion ratios and lactic acid yields have used immobilized cells and nutrient supplementation. In the current experiments, free cells of *B. longum* were grown with no nutrient supplementation, which would significantly reduce the cost of lactic acid production and be compatible with current fermentation facilities.

Nanofiltration

The testing results of permeate flux, lactose retention, and lactic acid recovery are shown in Table 2. The values in the table are the average

Table 1
The Lactose Conversion Ratio and Lactic Acid Yield of Cheese Whey
Fermentation Using Immobilized *B. longerm*

Time (h)	Lactose concentration (g/L)	Lactic acid concentration (g/L)	Acetic acid concentration (g/L)	Conversion ratio (%)	Yield (g lactic acid/ g lactose)
0	36.8	2.2	0.2		
2	34.1	3.9	0.6	7.3	0.63
4	31.5	5.7	0.6	14.3	0.67
6	28.4	7.2	0.6	22.8	0.6
8	27.2	8.2	0.7	26.1	0.62
12	23.6	11	0.6	35.9	0.66
24	14.3	18.6	0.6	61.2	0.73
36	6.5	24.3	0.6	82.3	0.73
48	1	28.4	0.6	97.2	0.73

Table 2
Permeate Flux, Lactose Retention, and Lactic Acid Recovery
During Nanofiltration

Membrane type	Initial lactic acid concentration (g/L)	Pressure (MPa)	Flux (L[/m²]h[/r])	Lactose retention (%)	Lactic acid retention (%)	Lactic acid recovery (%)
DS-5DK	18.6	1.4	27	100	45.7	54.4
		2.1	51.3	100	56.2	43.9
		2.8	80.6	100	63.4	36.6
	27	1.4	35.2	94.7	23.2	76.9
		2.1	64.8	96.8	30.8	69.3
		2.8	91.9	99.5	36.5	63.5
DS-5HL	18.6	1.4	56.5	82.2	33.7	66.4
		2.1	87.1	87.3	45.5	54.5
		2.8	126.2	90.7	49.6	50.4
	27	2.1	102.2	73.4	24.8	75.2

of two runs. Figure 2 shows the decreasing of permeate flux with time during nanofiltration at different transmembrane pressures. The permeate flux kept nearly constant when the transmembrane pressure was 1.4 MPa. Increased transmembrane pressure caused more decreasing of permeate flux during nanofiltration process.

Permeate Flux

Figure 3 shows that permeate flux increased with the increase of transmembrane pressure. When the transmembrane pressure increased

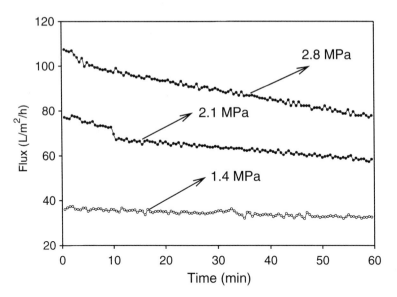

Fig. 2. Permeate flux profiles during nanofiltration (membrane: DS-5DK, initial lactic acid concentration: 27 g/L).

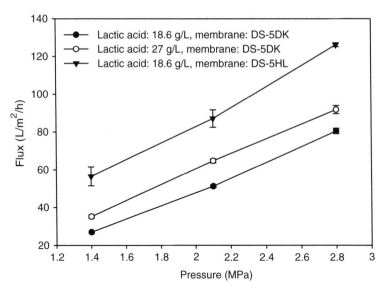

Fig. 3. Effects of pressure, membrane, and initial lactic acid concentration of feed stream on nanofiltration permeate flux.

from 1.4 to 2.8 MPa, the permeate flux increased 2–3 times for all of the tested membranes.

Figure 3 shows that higher permeate flux could be obtained at higher initial lactic acid concentration. When the initial lactic acid concentration increased from 18.6 to 27 g/L, the permeate flux increased about 30, 26, and 14% at pressure 1.4, 2.1, and 2.8 MPa, respectively, for membrane of DS-5DK

(Table 2). It can be concluded that accumulation of lactic acid during fermentation process could speed up the nanofiltration process.

Between the two tested membranes DS-5DK and DS-5HL, higher permeate flux was obtained with membrane of DS-5HL (Fig. 3). When the initial lactic acid concentration of 18.6 g/L, the permeate flux values obtained with membrane of DS-5HL were about 100, 70, and 57% higher than that of the membrane DS-5DL at pressures of 1.4, 2.1, and 2.8 MPa, respectively (Table 2). These results are in agreement with manufacturer provided permeate flux data (the permeate flux of membrane DS-5HL is 77% higher than that of the membrane DS-5DK at 140 KPa). Membranes with higher permeate flux have the potential to increase the capacity of the nanofiltration membrane unit.

The membrane, pressure, and initial lactic acid concentration showed significant ($p < 0.0001$) effects on the permeate flux. The interaction between these parameters were not significant ($p = 0.045$ and 0.11, respectively).

Retention of Lactose

Lactose retention increased with the increase of transmembrane pressure (Fig. 4). Figure 4 shows that lower lactose retention was obtained at higher initial lactic acid concentration. When the DS-5DK membrane was used, 100% retention of lactose was obtained at initial lactic acid concentration of 18.6 g/L for all tested transmembrane pressures. When the initial lactic acid concentration was increased to 27 g/L, lactose retention of 94.7, 96.8, and 99.5% were obtained at pressure levels of 1.4, 2.1, and 2.8 MPa, respectively (Table 2). This indicates, using this membrane, that at higher initial lactic acid concentrations, nearly 100% lactose retention can be obtained by increasing transmembrane pressure.

When the DS-5HL membrane was used to separate ultra filtered permeate with initial lactic acid concentration of 18.6 g/L, with the same pressure levels as for the DS-5DK membrane, lactose retention were 82.2, 87.3, and 90.7%, respectively. A lactose retention of 73.4% was obtained at initial lactic acid concentration of 27 g/L and transmembrane pressure of 2.1 MPa (Table 2). At most of the test conditions, the lactose retention of DS-5HL was lower than 91%, while the lactose retention for membrane of DS-5DK reached about 99–100%. These results indicate that, in comparison with the DS-5HL membrane, the DS-5DK membrane should be used for separating lactose from lactic acid in nanofiltration process. The transmembrane pressure, lactic acid concentration and membrane showed significant effects on the lactose retention ($p < 0.0001$, 0.0005, and 0.01, respectively).

Jeantet et al. *(17)* reported that lactose retention was kept constant at 97% ± 2% with transmembrane pressure for 400 Dalton Millipore nanofiltration membrane (spiral wound module). In our study, transmembrane pressure caused increase of lactose retention. At initial lactic acid concentration of 18.6 g/L, lactose retention was kept at 100% at all the studied transmembrane pressures.

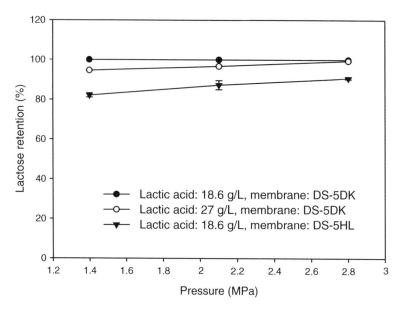

Fig. 4. Effects of pressure, membrane, and initial lactic acid concentration of feed stream on lactose retention during nanofiltration.

Recovery of Lactic Acid

It can be seen from Table 2 that increased retention of lactic acid corresponded positively with increased lactose retention. Lactic acid recovery in the permeate was used to evaluate the nanofiltration process in this study. Increases of transmembrane pressure were associated with lower levels of lactic acid recovery in the permeate (Fig. 5). Figure 5 shows higher lactic acid recovery was obtained at higher initial lactic acid concentration. At pressures of 1.4, 2.1, and 2.8 MPa, when the initial lactic acid concentration increased from 18.6 to 27 g/L, the lactic acid recovery increased from 54.4, 43.9, and 36.6% to 76.9, 69.3, and 63.5%, respectively (Table 2).

Between the two membranes tested, at the initial lactic acid concentration of 18.6 g/L, higher lactic acid recovery were obtained with membrane of DS-5HL. As the requirement of lactose retention around 99–100% needs to be met firstly, the higher lactic acid recovery of membrane DS-5HL could not change the previous conclusion on the membrane selection based on the lactose retention.

The membrane, pressure, and initial lactic acid concentration showed significant effects ($p < 0.0001$) on the lactic acid recovery. The interactions between these parameters were not significant ($p = 0.19$ and 0.18, respectively).

These results are in agreement with the work of Jeantet et al. (17), who observed that lower pH caused lower lactate retention. In our study, when the pH was kept constant at 5.5, we found out that increasing the initial lactic acid concentration reduces lactic acid retention, i.e., enables higher lactic acid recovery.

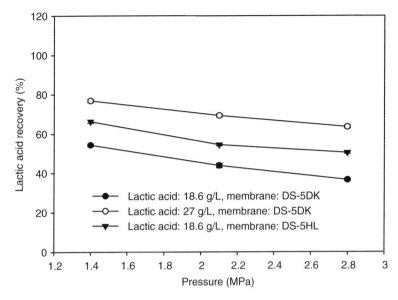

Fig. 5. Effects of pressure, membrane, and initial lactic acid concentration of feed stream on the recovery of lactic acid by nanofiltration.

Conclusions

The tested nanofiltration membrane system has successfully separated lactic acid from lactose in cheese whey fermentation broth. Nearly all the lactose (99–100%) was retained using a DS-5DK membrane at both of the tested initial lactic acid concentration of 18.6 and 27 g/L. Transmembrane pressure higher than 2.8 MPa needed to be applied when the initial lactic acid concentration reached 27 g/L to obtain nearly 100% of lactose retention. Among the tests when 99–100% of lactose was retained in the concentrate, the highest lactic acid recovery in the permeate reached 63.5%. Increasing transmembrane pressure caused higher permeate flux and higher lactose retention, but lower lactic acid recovery. Increased initial lactic acid concentration in the feed stream caused higher permeate flux and higher lactic acid recovery, but lower lactose retention. Further study needs to be conducted to determine the optimized pressure for each initial lactic acid concentration of feed stream to obtain near 100% lactose retention while keeping lactic acid recovery as high as possible.

References

1. Shahbazi, A., Mims, M. R., Li, Y., Shirley, V., Ibrahim, S. A., and Morris, A. (2005), *Appl. Biochem. Biotechnol.* **121–124,** 529–540.
2. Siso, M. I. G. (1996), *Bioresour. Technol.* **57,** 1–11.
3. Tango, M. S. A. and Ghaly, A. E. (2002), *Appl. Microbiol. Biotechnol.* **58,** 712–720.
4. Roy, D., Goulet, J., and LeDuy, A. (1986), *Appl. Microbiol. Biotechnol.* **24,** 206–213.
5. Bruno-Barcena, J. M., Ragout, A. L., Cordoba, P. R., and Sineriz, F. (1999), *Appl. Microbiol. Biotechnol.* **51,** 316–324.

6. Roukas, T. and Kotzekidou, P. (1998), *Enzyme Microb. Technol.* **22,** 199–204.
7. Gomes, A. M. P. and Malcata, F. X. (1999), *Trends Food Sci. Technol.* **10,** 139–157.
8. Doleyres, Y., Paquin, C., LeRoy, M., and Lacroix, C. (2002), *Appl. Microbiol. Biotechnol.* **60,** 168–173.
9. Song, S. H., Kim, T. B., Oh, H. I., and Oh, D. K. (2003), *World J. Microbiol. Biotechnol.* **19,** 721–731.
10. Atkinson, B. and Mavituna, F. (1991), In: *Biochemical Engineering and Biotechnology Handbook*, 2nd ed., Stockton Press, New York, pp. 1181, 1182.
11. Vigneswaran, S. and Kiat, W. Y. (1988), *Desalination* **70,** 299–316.
12. Caric, M. D., Milanovic, S. D., Krstic, D. M., and Tekic, M. N. (1999), *J. Membrane Sci.* **165,** 83–88.
13. Torang, A., Jonsson, A. S., and Zacchi, G. (1999), *Appl. Biochem. Biotechnol.* **76,** 143–157.
14. Li, Y., Shahbazi, A., and Kadzere, C. T. (2005), *J. Food Engn. 2006* (in press).
15. Van der Horst, H. C. (1995), *Int. Dairy Federation Special Issue* **9504,** 36–52.
16. Eriksson, P. (1988), *Environ. Prog.* 7, 58–62.
17. Jeantet, R., Maubois, J. L., and Boyaval, P. (1996), *Enzyme Microbial. Technol.* **19,** 614–619.
18. Alkhatim, H. S., Alcaina, M. I., Soriana, E., Iborra, M. I., Lora, J., and Arnal, J. (1998), *Desalination* **199,** 177–184.

Fermentation of Rice Straw/Chicken Manure to Carboxylic Acids Using a Mixed Culture of Marine Mesophilic Micoorganisms

Frank K. Agbogbo and Mark T. Holtzapple*

Department of Chemical Engineering, Texas A&M University, College Station, TX 77843-3122; E-mail: mth4500@chemail.tamu.edu

Abstract

Countercurrent fermentation of rice straw and chicken manure to carboxylic acids was performed using a mixed culture of marine mesophilic microorganisms. To increase the digestibility of the biomass, rice straw, and chicken manure were pretreated with 0.1 g $Ca(OH)_2$/g biomass. Fermentation was performed for 80% rice straw and 20% chicken manure at various volatile solid loading rates (VSLR) and liquid residence times (LRT). The highest acid productivity of 1.69 g/(L·d) occurred at a total acid concentration of 32.4 g/L. The highest conversion (0.69 g VS digested/g VS fed) and yield (0.29 g total acids/g VS fed) were at a total acid concentration of 25 g/L. A Continuum Particle Distribution Model of the process predicted the experimental total acid concentration and conversion results with an average error of 6.41% and 6.15%, respectively. Results show how total acid concentrations, conversions, and yields vary with VSLR and LRT in the MixAlco process.

Index Entries: Biomass; carboxylic acids; CPDM; digestion; fuels; mixed culture.

Introduction

Currently, rice straw is either burned in open fields or incorporated into the soil. Increasing environmental concerns and government legislation call for a decrease in the quantity of rice straw burned *(1)*. Incorporating rice straw into soil increases foliar disease, reduces crop yield, degrades soil conditions, and produces methane, a greenhouse gas *(2)*. Therefore, a low-cost technology to convert these wastes into useful fuels *(3)* and chemicals is valuable. Significant potential benefits result from fuel derived from cellulosic biomass, a renewable nonfood feedstock *(4)*.

The conversion of rice straw to ethanol has been studied using simultaneous saccharification and fermentation (SSF) *(5,6)* in which lignocellulose is simultaneously hydrolyzed to sugars and fermented to ethanol. Unfortunately, this process requires enzymes that contribute heavily to production costs.

*Author to whom all correspondence and reprint requests should be addressed.

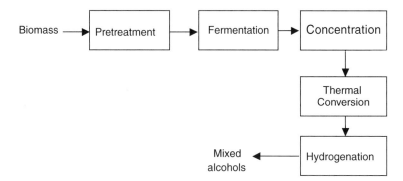

Fig. 1. MixAlco process.

Another option is the MixAlco process, which converts biodegradable materials into mixed alcohols *(7)* (Fig. 1). In the MixAlco process, to increase biomass digestibility, it is treated with lime, an inexpensive chemical. The biomass is then fermented with a mixed culture of microorganisms to produce carboxylic acids under nonsterile, anaerobic conditions. To maintain the pH, calcium carbonate is added as a buffer, which reacts with the acid to form carboxylate salts. These carboxylate salts can be concentrated, followed by "acid springing" to produce the corresponding organic acids *(8)*. Alternatively, fuels can be produced by converting carboxylate salts to ketones, and hydrogenating the ketones to mixed alcohol fuels. The MixAlco process has many benefits, such as no sterility requirement, adaptability to many feedstocks, and no enzyme addition.

To achieve high substrate conversions and product concentrations, countercurrent fermentation is used (Fig. 2). Fresh biomass is added to the fermentor containing the highest carboxylate concentration and fresh water is added to the fermentor with the most digested biomass. This countercurrent flow arrangement addresses two issues: (1) the recalcitrant portion of biomass remains after digestion of the easily converted fraction and (2) product carboxylate salts are extremely inhibitory *(9)*. The countercurrent fermentation allow fresh biomass (most reactive) to contact a high carboxylate acid concentration, leading to a higher product concentration. The most digested biomass (least reactive) contacts fresh water, which is less inhibitory allowing for high conversions.

Chang *(10)* showed that lime pretreatment removes all acetyl groups and a moderate amount of lignin. A lime loading of 0.1 g $Ca(OH)_2$/g dry biomass at 86–135°C for 1–3 h is optimum for pretreatment of biomass *(10)*. Chang *(10)* recommended 10 mL water/g dry biomass and Karr *(11)* recommended 5 g H_2O/g dry biomass. The pretreatment effect was independent of water loading *(11)* but there should be enough water to cover the biomass. Karr *(11)* showed that lime pretreatment increased the conversion of corn stover by about 90%.

Carboxylic acids are intermediates in the fermentation of biomass to methane. Zhang et al. *(12)* studied biogasification of rice straw to produce

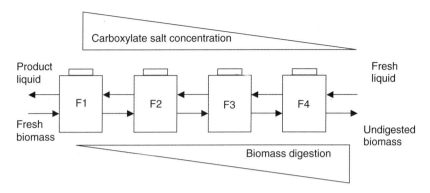

Fig. 2. Four-stage countercurrent fermentation.

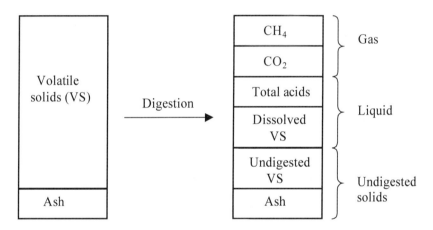

Fig. 3. The digestion of biomass.

biogas (CH$_4$ 50%), however, methane is a low-value product. In the MixAlco process, methane formation from carboxylic acids is inhibited. Methane analogs such as iodoform and bromoform are effective in inhibiting methanogenesis *(13)*, which eliminates a potential hydrogen sink and reducing power is used to produce higher carboxylic acids, such as propionate and butyrate *(9,13)*.

Biomass contains volatile solids (VS) (the major components are cellulose, hemicellulose, and lignin) and ash (Fig. 3). When the biomass is digested, volatile solids convert to gaseous and liquid products, plus solid residues. The gaseous products are principally methane and carbon dioxide, the liquid products are carboxylate salts, extracellular proteins, and energy storage polysaccharides *(9)*, and the solid residue contains ash and undigested VS. The following definitions were used:

$$\text{Volatile solids (VS)} = \text{Dry weight} - \text{Ash weight (g)} \quad (1)$$

$$\text{Conversion } (x) = \frac{\text{VS digested}}{\text{VS fed}} \text{ (g/g)} \quad (2)$$

$$\text{Yield } (y) = \frac{\text{Total carboxylic acids produced}}{\text{VS fed}} \text{ (g/g)} \quad (3)$$

Total acid productivity (p)
$$= \frac{\text{Total carboxylic acids produced}}{\text{Total liquid volume in all fermentors} \cdot \text{time}} \text{ [g/(L·d)]} \quad (4)$$

$$\text{Total acid selectivity} = \frac{\text{Total carboxylic acids produced}}{\text{VS digested}} \text{ (g/g)} \quad (5)$$

Liquid residence time (LRT)
$$= \frac{\text{Total liquid in all fermentors}}{\text{Flow rate of liquid out of the fermentor train}} \text{ (d)} \quad (6)$$

Volatile solids loading rate (VSLR)
$$= \frac{\text{VS fed to the system}}{\text{Total liquid in all fermentors} \cdot \text{time}} \text{ [g/(L·d)]} \quad (7)$$

Countercurrent fermentation requires an extended time to reach steady state (~4 mo). Because of the long residence time, it would be very time-consuming and cost-ineffective to explore a wide variety of operating conditions. To overcome this, Loescher developed the Continuum Particle Distribution Model (CPDM), a mathematical model that predicts the total acid concentration and substrate conversion using batch fermentation data (14). CPDM can save considerable time in determining the optimum operating conditions. The conversion penalty (15) was used in CPDM to account for the fact that biomass reactivity reduces at longer residence times because easy-to-digest portions react first.

Nutrients are required to enhance microorganism growth. Rice straw is rich in carbohydrates but low in nutrients. In contrast, chicken manure is rich in nutrients such as nitrogen, vitamins, and minerals. In this study, rice straw (80%) was used as a carbohydrate-rich source whereas chicken manure (20%) was used as a nutrient-rich source. Marine inoculum was used as a source of microorganisms. The volatile solid loading rates (VSLR) and liquid residence times (LRT) were varied to determine their effect on acid concentration, yield, conversion, and total acid productivity. CPDM was used to predict the acid concentrations and conversions, and the results were compared with experimental data.

Materials and Methods

Substrates

Rice straw (RS) was obtained from Lee Tarpley of the Texas A&M Agricultural Research and Extension Center. The substrate was milled in a

Thomas Wiley Laboratory mill and was passed through a 2-mm screen. Rice straw was pretreated with 0.1 g $Ca(OH)_2$/g dry biomass at 100°C for 1 h. The average moisture content of the pretreated RS was 0.028 g water/g raw RS, the average ash content was 0.274 g ash/g dry RS, and the VS was 0.726 g VS/g dry RS. Chicken manure was obtained from the Poultry Science Center, Texas A&M University, College Station, Texas. The manure was air dried and then pretreated with 0.1 g $Ca(OH)_2$/g dry biomass at 100°C for 1 h. The average moisture content of pretreated chicken manure was 0.052 g water/g raw chicken manure, the average ash content was 0.34 g ash/g dry chicken manure, and the volatile solids was 0.660 g VS/g dry chicken manure. Pretreated rice straw (80%) and pretreated chicken manure (20%) were used as substrates in all experiments.

Media and Nutrient

The liquid medium was deoxygenated water prepared by boiling distilled water under nitrogen purge for 5 min. After cooling the media to room temperature, 0.275 g/L sodium sulfide and 0.275 g/L cysteine hydrochloride were added under continuous nitrogen purge. Sodium sulfide and cysteine hydrochloride were added to further reduce the oxygen content of the media.

Dry nutrient mixture contained (g/100 g of mixture) K_2HPO_4 (16.3) KH_2PO_4 (16.3) $(NH_4)_2SO_4$ (16.3), NaCl (32.6), $MgSO_4 \cdot 7H_2O$ (6.8), $CaCl_2 \cdot 2H_2O$ (4.4), HEPES (0.86), hemin (0.71), nicotinamide (0.71), p-aminobenzoic acid (0.71), Ca-pantothenate (0.71), folic acid (0.35), pyridoxal (0.35), riboflavin (0.35), thiamine (0.34), cyanocobalamin (0.14), biotin (0.14), EDTA (0.35), $FeSO_4 \cdot 7H_2O$ (0.14), $MnCl_2$ (0.14), H_3BO_3 (0.021), $CoCl_2$ (0.014), $ZnSO_4 \cdot 7H_2O$ (0.007), $NaMoO_4$ (0.0021), $NiCl_2$ (0.0014), and $CuCl_2$ (0.0007) (8). The dry nutrients were added during the fermentation.

Inoculum

Marine inoculum was used from a previous fermentation of sugarcane bagasse/chicken manure (16). The original inoculum was previously collected from the sediments of three coastal swamps at Galveston, Texas. The sediment was collected from 0.5-m-deep holes and placed into bottles filled with deoxygenated media, consisting of 0.275 g/L sodium sulfide and 0.275 g/L cysteine hydrochloride.

Inhibitor

Iodoform (CHI_3) solution containing 20 g CHI_3/L ethanol was used as a methanogen inhibitor in this experiment. Owing to light and air sensitivity, the solution was kept in a tinted bottle and capped immediately after use.

pH Control

Calcium carbonate ($CaCO_3$) was added as a neutralizing agent to control the pH between 5.7 and 6.4.

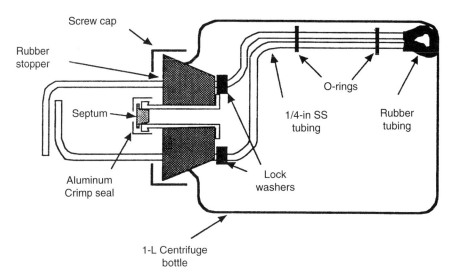

Fig. 4. Centrifuge bottle bioreactor *(8)*.

Fermentor

The fermentors (Fig. 4) were made from Beckman 1-L polypropylene centrifuge bottles (98 × 169 mm^2), Nalgene brand NNI 3120-1010. The bottles were closed with a size 11 rubber stopper with a hole drilled in the middle. A glass tube was inserted through the hole and capped with a rubber septum for gas sampling and release. The release of gas from the fermentors was necessary to prevent explosions because the fermentors could only withstand a pressure of 2 atm. The rubber septum was replaced when there was a visible hole owing to frequent gas venting. Two 0.25-in. stainless steel tubes with welded ends were also inserted into holes in the stopper. The tubes were used as stirrers to mix the components inside the fermentors. The fermentors were placed in a Wheaton Modular Cell Production Roller Apparatus (Model III) located in a 40°C incubator and were rotated horizontally at 1 rpm.

Experimental Procedure

The fermentations were performed at 40°C. Anaerobic conditions were maintained by purging with high-pressure nitrogen whenever fermentors were opened. Four fermentors were started as a batch fermentation with 80% pretreated rice straw and 20% pretreated chicken manure, calcium carbonate, urea, dry nutrients, and deoxygenated water. To establish the culture, a batch fermentation was maintained for 10 d, then the countercurrent fermentation was started. Every 2 d, liquid/solid were transferred (Fig. 2) and 2 g of CaCO$_3$ were added to each fermentor to neutralize the carboxylic acids.

A series of six countercurrent fermentation experiments were performed at various combinations of volatile solid loading rates (VSLR) and

Table 1
Operating Parameters for Rice Straw/Chicken Manure Countercurrent Fermentation With Marine Inoculum

Fermentation trains	A	B	C	D	E	F
VSLR (g VS/[L of liquid in all fermentors·d])	6.32	3.45	6.47	9.99	5.87	8.06
LRT (days)	23.4	23.7	24.0	24.1	19.2	26.3
VS fed at each transfer (g VS)	13.9	6.96	13.9	22.2	11.1	18.1
Liquid volume in all fermentors (L)	1.1	1.01	1.08	1.11	0.944	1.12

liquid residence time (LRT). The operating parameters for the fermentation trains are shown in Table 1. On each transfer, 0.2 g of dry nutrients, 40 µL of iodoform solution (20 g/L of iodoform dissolved in ethanol), and 0.1 g of nutrients (if pH < 6.0) were added. Deoxygenated water (0.1 L) was fed to F4 on each transfer. The single-centrifuge procedure, in which liquids are transferred in a single step, was used (9,17,18). After the system reached a steady state (±5 g/L average total acid concentration), fermentation data was collected for at least 10 transfers to determine acid productivity, carboxylic acid concentration, yield, selectivity, conversion, biotic CO_2 productivity, and CH_4 productivity.

Analytical Methods

The volume of gas produced during fermentation was measured using an inverted graduated glass cylinder apparatus (water displacement apparatus) that was filled with a solution of 300 g calcium chloride/L solution. A gas chromatograph (Agilent 6890 series Agilent Technologies, Palo Alto, CA) with thermal conductivity detector was used to determine the methane and CO_2 composition of the fermentation gas. Samples were taken directly from the fermentors using a 5-mL syringe. To calibrate the samples, a standard gas mixture of carbon dioxide (29.99 mol%), methane (10.06 mol%), and the balance nitrogen was used.

CO_2 produced during fermentation is the sum of biotic and abiotic CO_2. Abiotic CO_2 is produced by neutralizing the carboxylic acids with calcium carbonate and biotic CO_2 is produced from the fermentation. It is assumed that for every 2 mol of acid produced in the fermentor, 1 mol of abiotic CO_2 is produced. The biotic CO_2 produced directly from the fermentation was calculated by subtracting abiotic CO_2 from total CO_2. Only biotic CO_2 was used in the mass balance calculations.

During each transfer schedule, liquid from Fermentor 1 and solids from Fermentor 4 were collected. A liquid sample (~3 mL) was taken from Fermentor 1 and analyzed for carboxylic acid concentration. The remaining liquid collected from Fermentor 1 was analyzed for VS. Although every attempt was made to obtain solid-free liquid stream from Fermentor 1, the liquid contains some carryover solids. The solids collected from Fermentor 4 were analyzed for undigested volatile solids. Acid analysis was performed using an Agilent 6890 gas chromatograph with capillary column (J & W Scientific, model DB-FFAP, Agilent Technologies, CA). It was operated with a flame ionization detector and an Agilent 7683 Series Injector. The oven temperature in the gas chromatograph (GC) increased from 50°C to 200°C at 20°C/min and was held an additional 1 min at 200°C. Liquid samples were mixed with 1.162 g/L of internal standard (4-methyl-n-valeric acid) and acidified with 3 M phosphoric acid. Volatile solids in all samples were determined by first drying the samples at 105°C and then ashing at 550°C in an oven (18).

Mass balance closure on the entire system was calculated over the steady-state period.

The mass balance closure was calculated as:

$$\text{Closure} = \frac{\text{Mass out}}{\text{Mass in} + \text{Water of hydrolysis}} \text{ (g/g)} \quad (8)$$

$$\text{Closure} = \frac{\text{Undigested VS} + \text{Dissolved VS} + \text{Acids} + \text{Biotic } CO_2 + CH_4}{\text{VS in} + \text{Water of hydrolysis}} \text{ (g/g)} \quad (9)$$

Dissolved VS is the carryover solids in the liquid from Fermentor 1 and solubilized lignin and they were measured by drying at 105°C and then ashing at 550°C in an oven. Theoretically, the system should have a 100% closure; in practice, human errors in measurement and transfer processes caused some discrepancies. During cellulose hydrolysis, a mole of water is gained per mole of monomer resulting in mass increase. Ross (9) suggested that biomass could be represented as cellulose, with a monomer weight of 162 g/mol. The water of hydrolysis was calculated as:

$$\text{Water of hydrolysis} = \text{VS digested} \times \frac{18}{162} \text{ [g]} \quad (10)$$

Continuum Particle Distribution Modeling

CPDM was used to simulate data for the countercurrent fermentation using data collected from batch fermentations. Batch experiments at varying initial substrate concentrations (20, 40, 70, 100, and 100+ g dry substrate/L liquid) were used to obtain the data. The 100 and 100+ fermentors had the same initial substrate concentrations, but the 100+ fermentor contained a medium with a mixture of carboxylate salts (70 wt% calcium acetate, 20 wt% calcium propionate, and 10% calcium butyrate) at a concentration of 20 g carboxylic acids/L of liquid to better simulate the high acid concentration and low conversion existing in Fermentor 1. The inoculum for batch fermentations was taken from a steady-state countercurrent fermentation on the same substrates. Deoxygenated water was used for this fermentation and other components, such as urea (0.25 g/L), dry nutrient (0.5 g/L), and calcium carbonate (5 g/L), were added initially to the fermentors. To prevent methane production, iodoform (40 µL of iodoform solution) was added continuously. Liquid samples were taken daily from the five batches. Reaction rates at varying acid concentrations and biomass digestion were determined from the batch data.

The liquid samples were analyzed for carboxylic acid concentrations using the GC and the results were converted to acetic acid equivalent (α):

$$\alpha = \text{acetic (mol/L)} + 1.75 \times \text{propionic (mol/L)} + 2.5 \times \text{butyric (mol/L)} \\ + 3.25 \times \text{valeric (mol/L)} + 4 \times \text{caprioc (mol/L)} \\ + 4.75 \times \text{heptanoic (mol/L)} \; [\text{mol/L}] \quad (11)$$

On mass basis, the acetic acid equivalent can be expressed as:

$$A_e = 60.05 \; (\text{g/mol}) \times \alpha \; (\text{mol/L}) \; [\text{g/L}] \quad (12)$$

Acetic acid equivalents are based on the reducing power of the acids produced from the fermentation and allows the various acid products to be expressed on a common basis *(19)*. The acetic acid equivalents (A_e) from each of the five batch experiment was fit to the equation:

$$A_e = a + \frac{bt}{1+ct} \; [\text{g/L}] \quad (13)$$

where t is the time (d) of fermentation, and a, b, and c are constants fit by least squares analysis. The rate r was obtained from Eq. 14.

$$r = \frac{d(A_e)}{dt} = \frac{b}{(1+ct)^2} \; [\text{g/(L·d)}] \quad (14)$$

The specific rate, $\hat{r}$ (g A_e produced/(g VS·d)) was determined from Eq. 14 by dividing it by the initial substrate concentration, S_o (g VS/L), in each of the five fermentors:

$$\hat{r} = \frac{r}{S_o} \; [\text{g/(g·d)}] \quad (15)$$

The predicted rate, $\hat{r}_{pred}$, was obtained from Eq. 16, where the rate of acid production depends on volatile solids conversion (x) and product concentration (A_e):

$$\hat{r}_{pred} = \frac{e(1-x)^f}{1+g[\phi A_e]^h} \; [\text{g/(g·d)}] \quad (16)$$

where, $\hat{r}_{pred}$ is the g acetic acid equivalents produced/(g VS·d); x the dimensionless; ϕ the ratio of g total acid to g acetic acid equivalents; and A_e the g acetic acid equivalent produced.

Least square analysis was used to determine the empirical parameter constants e, f, g, and h for $\hat{r}_{pred}$ (Eq. 16) from the specific rate $\hat{r}$ (Eq. 15). A_e was converted back to carboxylic acid concentration by ϕ (the ratio of total grams of actual acids to grams of A_e). The $(1 - x)$ term in the numerator of Eq. 16, is the conversion penalty function by South and Lynd *(15)*. The conversion term $(1 - x)$ in the numerator (Eq. 16) shows that as the biomass is digested, the reaction rate decreases because the less reactive components remain. The acid concentration in the denominator (A_e) shows the inhibitory effect on the microorganisms when product concentration is high.

The conversion, x was calculated using

$$x(t) = \frac{A_e(t) - A_e(t=0)}{S_o \cdot \sigma} \quad (g/g) \tag{17}$$

where, σ is the selectivity (g A_e produced/g VS digested); and S_o is the initial substrate concentration (g VS/L).

The selectivity σ, for Eq. 17 was calculated from the selectivity s (g total acids produced/g VS digested) determined in the countercurrent experiment.

$$s = \phi\sigma \quad (g/g) \tag{18}$$

Equation 16 was used in a Mathematica program *(16)* to predict acetic acid equivalent concentration (A_e) and conversion (x) for the countercurrent fermentation at various VSLR and LRT. A_e was converted back to carboxylic acid concentration by multiplying by ϕ. Other system-specific parameters needed for the Mathematica program are selectivity, holdup (ratio of liquid–solid in wet solids), and moisture (ratio of liquid–solid in feed) (*see* Table 4).

Statistical Methods

The "solver" in Microsoft Excel was used to obtain the model fit parameters in Eqs. 13 and 16. The sum of the mean-square error between experimental and predicted values was minimized to obtain the model parameters.

Results and Discussion

Countercurrent Fermentation Using Marine Inocula

The results for the countercurrent fermentation (data collected for at least 10 transfers at steady state) are shown in Table 2. The highest acid productivity of 1.69 g/(L·d) occurred at a concentration of 32.4 g/L in Fermentation Train E (LRT = 19.2 d and VSLR = 5.87 g/[L·d]). This fermentation train had the shortest LRT; Fermentation Train F (LRT = 26.3 d and VSLR = 8.06 g/[L·d]) had the longest LRT. The highest selectivity of 0.57 g total acids/g VS digested was in Fermentation Train C (LRT = 24 d and VSLR = 6.47 g/[L·d]). The acid productivity, total acid concentration, conversion, and yield depend on the VSLR and LRT.

Table 2
Results for Rice Straw/Chicken Manure Countercurrent Fermentation With Marine Inoculum

Fermentation trains	A	B	C	D	E	F
Average pH in all fermentors	6.0 ± 0.18	6.4 ± 0.37	5.7 ± 0.06	5.7 ± 0.18	5.8 ± 0.16	5.8 ± 0.17
Total acid productivity (g/[L of liquid in all fermentors·d])	1.47	1.00	1.53	1.63	1.69	1.57
Total acid concentration (g/L)	35.1 ± 1.61	25.0 ± 2.80	36.7 ± 0.85	39.8 ± 4.30	32.4 ± 3.35	40.8 ± 3.37
Acetic acid (wt%)	42.8 ± 2.21	36.1 ± 3.18	48.6 ± 1.28	48.4 ± 4.36	44.0 ± 4.72	50.9 ± 2.56
Propionic acid (wt%)	8.55 ± 2.85	21.3 ± 2.78	3.64 ± 0.18	7.32 ± 4.1	5.46 ± 2.15	8.13 ± 2.8
Butyric acid (wt%)	24.8 ± 1.55	26.7 ± 3.46	27.2 ± 0.49	25.4 ± 2.55	29.9 ± 3.54	23.9 ± 3.41
Valeric acid (wt%)	8.44 ± 1.37	8.84 ± 0.99	4.45 ± 0.45	5.17 ± 1.55	5.59 ± 1.1	4.79 ± 0.8
Caproic acid (wt%)	13.4 ± 2.65	6.11 ± 0.81	15.6 ± 0.72	13.2 ± 2.48	14.5 ± 1.76	11.9 ± 1.65
Heptanoic acid (wt%)	2.01 ± 0.33	0.95 ± 0.43	0.56 ± 0.04	0.47 ± 0.14	0.55 ± 0.11	0.41 ± 0.05
VS digested (g VS/d)	3.54	2.41	2.91	4.13	3.39	3.77
Yield (g total acid/g VS fed)	0.23	0.29	0.24	0.16	0.29	0.19
Selectivity (g total acid/g VS digested)	0.46	0.42	0.57	0.44	0.47	0.47
Conversion (g VS digested/g VS fed)	0.51	0.69	0.42	0.38	0.61	0.42
Biotic CO_2 productivity (g CO_2/[L of liquid in all fermentors·d])	0.43	0.61	0.34	0.17	0.22	0.25
CH_4 productivity (g CH_4/[L of liquid in all fermentors·d])	0.003	0.043	0.015	0.001	0.001	0.001
Mass balance closure (g VS out/g VS in)	1.19	0.93	1.08	0.96	1.15	1.16

All errors are ±1 standard deviation.

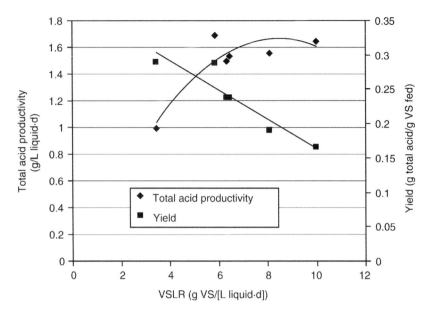

Fig. 5. Correlation of total acid productivity and yield with volatile solid loading rate.

The highest acid concentrations 40.8 g/L (Train F), and 39.8 g/L (Train D) were obtained at high VSLR of 8.06 g/(L·d) and 9.99 g/(L·d), for Trains F and D, respectively. This is because at high VSLR, the microorganisms have more substrates available for acid production. However, the conversion for Trains F and D were 0.42 g/g and 0.38 g/g, respectively, which are low. The low conversions resulted because the VSLR was so high that there was insufficient time to digest all of the solids entering the fermentors.

Fermentation Train B (LRT = 23.7 d and VSLR = 3.45 g/[L·d]), had the lowest total acid concentration of 25 g/L and the highest conversion (0.69 g VS digested/g VS fed) and yield (0.29 g total acids/g VS fed). At low VSLR, less substrate is available to the microorganisms and therefore the total acid concentration is low. However, the conversion is high because the microorganisms do not only digest the easily digestible fractions, but also the difficult portions of the biomass. The MixAlco process requires the selection of VSLR and LRT that is optimum for both high total acid concentration and conversion.

Correlation for Productivity, Selectivity, Conversion, and Yield

The correlations between VSLR and total acid productivity (p), and yield (y) are shown in Fig. 5. The data for the six fermentation trains were fit with a second order polynomial and linear regression and the following correlations were obtained:

$$p = -0.0255 \text{ VSLR}^2 + 0.4301 \text{ VSLR} - 0.1456 \qquad (19)$$

$$y = -0.0213 \text{ VSLR} + 0.3763 \qquad (20)$$

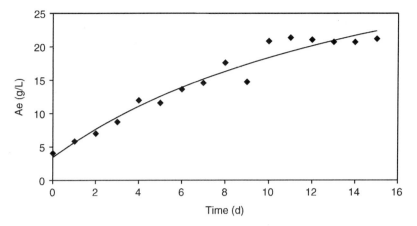

Fig. 6. Acetic acid equivalent of rice straw/chicken manure batch fermentor (70 g dry substrate/L liquid).

From Fig. 5, the acid productivity (p) increases from 1 g/(L·d) (VSLR = 3.45 g/[L·d]) to 1.6 g/(L·d) (VSLR = 6 g/[L·d]) with no apparent increase in acid productivity after VSLR of 6 g/(L·d). Selectivity (s) is essentially constant and does not depend on VSLR (Table 2). From Fig. 5, yield (y) decrease as VSLR increases.

At a high VSLR, the microorganisms digest more biomass leading to a high acid productivity; however, yield, and conversion are lower because only a small fraction of the total biomass fed is digested. On the other hand, at low VSLR, the microorganisms are limited by the amount of digestible biomass available. Therefore, the acid productivity is low but yield, and conversion are higher because the microorganisms consume both the reactive and recalcitrant biomass components.

CPDM Verification

Batch fermentations at different initial substrate concentrations were performed. Figure 6 shows the acetic acid equivalent profile in 70 g dry substrate/L liquid. The values for a, b, and c at the different initial substrate concentrations are shown in Table 3. The values of e, f, g, h, and other parameter constants used in the Mathematica program are shown in Table 4. The rate equation obtained for 80% rice straw/20% chicken manure fermentation with marine inocula is:

$$\hat{r}_{pred} = \frac{1.06(1-x)^{3.18}}{1+3.00[\phi A_e]^{0.917}} \tag{21}$$

Table 5 compares the experimental total carboxylic acid concentration and conversion to the CPDM predictions. The average error between the experimental and predicted total acid concentrations is 6.41%, and the average error between the experimental and predicted conversion is 6.15%. The average error in conversion is very similar to results from corn

Table 3
The Value of a, b, c in CPDM for Rice Straw/Chicken Manure Fermentation

Initial substrate concentration (g/L)	a (g/L liquid)	b (g/[L liquid·d])	c (/d)
20	3.44	1.33	0.061
40	2.49	1.83	0.067
70	3.47	2.34	0.057
100	4.51	2.04	0.023
100+	16.48	3.83	0.148

Table 4
Parameter Constant Values in CPDM for Rice Straw/Chicken Manure Fermentation

Parameter constant	Values
Hold up (g liquid/g VS wet cake)	4.4
Moisture (g liquid/g wet solid)	0.051
Selectivity (g Ae/g VS digested)	0.497
φ (g total acid/g Ae)	0.704
Liquid volume (L)	0.269
e (g Ae/g VS·d)	1.06
f (dimensionless)	3.18
g (L/g total acid)$^{1/h}$	3
h (dimensionless)	0.917

stover fermentation (20), when the selectivity was maintained as a constant. This shows that a constant selectivity in the fermentation improves the predictability of the CPDM model. The highest error in total acid total carboxylic acids is 16.2% and the highest error in conversion is 23.8%. Further improvements in the CPDM can be made by making replicate measurements of the CPDM batch data so as to obtain a more accurate rate equation.

Continuum Particle Distribution Modeling

Figure 7 shows the CPDM "map" at the average solid concentrations used in this study (129.5 g VS/L of liquid). At high VSLR and LRT, the total acid concentration is high but the conversion is low. This is because the microorganisms have more substrate and the liquids have a high residence time. The low conversions at high VSLR are because only the easily digestible portions of the biomass are consumed, leaving a high fraction of undigested substrate. At low VSLR and LRT, the microorganisms have less substrate to digest and the liquids have a short residence time. Conversion is higher at low VSLR because the microorganisms digest both the easily digestible and recalcitrant portions of the biomass. Both high product

Table 5
Comparison of Experimental and Predicted Carboxylic Acid Concentration and Substrate Conversion for Rice Straw/Chicken Manure Fermentation

Fermentation trains	A	B	C	D	E	F	Average (%)
Experimental carboxylic acid concentration (g/L)	35.1 ± 1.6	25 ± 2.8	36.7 ± 0.85	39.8 ± 4.3	32.4 ± 3.4	40.8 ± 3.4	
Predicted carboxylic acid concentration (g/L) (CPDM)	35.1	29.06	35.7	36.9	30.6	38.26	
Error[a] (%)	0	16.24	2.72	7.29	5.55	6.64	6.41
Experimental conversion	0.52	0.69	0.42	0.38	0.61	0.42	
Predicted conversion (CPDM)	0.52	0.73	0.52	0.38	0.58	0.43	
Error[a] (%)	0	5.8	23.8	0	4.92	2.38	6.15

All errors are ±1 standard deviation.
[a]Error = (Predicted − Experimental) × 100/Experimental.

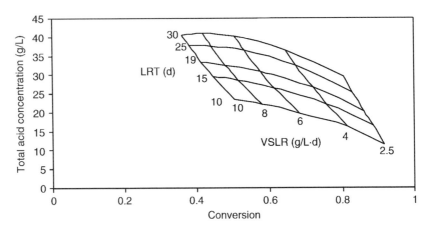

Fig. 7. The CPDM "map" for rice straw/chicken manure countercurrent fermentation (129.5 g VS/L of liquid).

concentrations and conversions are required in the application of this technology. The "map" (Fig. 7) predicts a total acid concentration of 40 g/L at LRT of 30 d, VSLR of 8 g/(L·d) and a conversion of 42%. At a VSLR of 2.5 g/(L(d) and LRT of 30 d, a total acid concentration of 30 g/L could be obtained at 80% conversion. From the map, VSLR of 2.5 g/(L·d) and LRT of 30 d will fit the requirement for both high product concentration and conversion. CPDM helped to determine this operating point, which would have been difficult without the use of the model. Further experimental work is underway to look at the effect of long residence times on total acid concentrations and conversion.

Conclusions

Fermentation results were obtained using a mixed culture of marine microorganisms for fermentation of rice straw (80%) and chicken manure (20%) at various VSLR and LRT. The highest acid concentrations were obtained at high VSLR but at the cost of low conversion. At low VSLR, the total acid concentration was low but the conversion was high. The fermentation had the highest acid productivity of 1.69 g/(L·d) at a total acid concentration of 32.4 g/L. The highest conversion (0.69 g VS digested/g VS fed) and yield (0.29 g total acids/g VS fed) occured at a total acid concentration of 25 g/L. Because it is impossible to experimentally explore all possible VSLR and LRT combinations to obtain both high product concentration and conversion, CPDM was applied. The CPDM predicted the experimental total acid concentration and conversion at an average error of 6.41% and 6.15% respectively. The model was used to predict product concentrations and conversions at other VSLR and LRT. The model predicts that, at a solids concentration of 129.5 g VS/L of liquid, a VSLR of 2.5 g/(L·d), and LRT of 30 d, a total acid concentration of 30 g/L can be obtained at 80% conversion.

Nomenclature

A_e	acetic acid equivalent concentration (g acetic acid equivalents/L)
a	parameter constant (g acetic acid equivalents/L)
b	parameter constant (g acetic acid equivalents/[L·d])
c	parameter constant (/d)
e	parameter constant (g acetic acid equivalent/[g VS·d])
f	parameter constant (dimensionless)
g	parameter constant (L/g total acid)$^{1/h}$
h	parameter constant (dimensionless)
LRT	liquid residence time (d)
p	total acid productivity (g total acid/[L·d])
r	reaction rate (g acetic acid equivalents/[L·d])
$\hat{r}$	specific rate (g acetic acid equivalents produced/[g VS·d])
$\hat{r}_{pred}$	predicted specific rate (g acetic acid equivalents produced/[g VS·d])
S_o	initial substrate concentration (g VS/L)
s	selectivity (g total acid produced/g VS digested)
t	time (d)
VSLR	volatile solid loading rate (g VS/[L·d])
x	conversion (g VS digested/g VS fed)
α	acetic acid equivalent concentration (mol acetic acid equivalents/L)
ϕ	ratio (g total acid/g acetic acid equivalents)
σ	selectivity (g acetic equivalents produced/g VS digested)

References

1. California Rice Commission's Library on Rice Straw Utilization, www.westbioenergy.org/ricestraw/one.html. Date accessed: April 13, 2004.
2. Chidthaisong, A. and Watanabe, I. (1997), *Soil Biol. Biochem.* **29,** 1173–1181.
3. Lynd, L. R., Weimer, P. L., van Zyl, W. H., and Pretorius, I. S. (2002), *Microbiol. Mol. Biol. Rev.* **66(3),** 506–577.
4. Lynd, L. R., Cushman, J. H., Nichols, R. J., and Wyman, C. E. (1991), *Science* **251,** 1318–1323.
5. Chadha, B. S., Kanwar, S. S., Saini, H. S., and Garcha, H. S. (1995), *Acta Microbiologica et Immunologica Hungarica*, **42(1),** 53–59.
6. Hoshino, K., Yamasaki, H., Chida, C., et al. (1997), *J. Chem. Eng. Japan* **30(1),** 30–37.
7. Holtzapple, M. T., Davison, R. R., Ross, M. K., et al. (1999), *Appl. Biochem. Biotechnol.* **77–79,** 609–631.
8. Chan, W. N. and Holtzapple, M. T. (2003), *Appl. Biochem. Biotechnol.* **111,** 93–112.
9. Ross, M. K. (1998), PhD dissertation, Texas A&M University, College Station, Texas.
10. Chang, V. S. and Holtzapple, M. T. (2000), *Appl. Biochem. Biotechnol.* **84-86,** 1–37.
11. Kaar, W. E. and Holtzapple, M. T. (2000), *Biomass Bioenergy* **18,** 189–199.
12. Zhang, R. and Zhang, Z. (1999). *Biores. Technol.* **68,** 235–245.
13. Bauchop, T. (1967), *J. Bacteriol.* **94,** 171–175.

14. Loescher, M. E. (1996), PhD dissertation, Texas A&M University, College Station, Texas.
15. South, C. R. and Lynd, L. R. (1994), *Appl. Biochem. Biotechnol.* **45, 46,** 467–481.
16. Thanakoses, P., Mostafa, N. A. A., and Holtzapple, M. T. (2003), *Appl. Biochem. Biotechnol.* **105,** 523–546.
17. Ross, M. K. and Holtzapple, M. T. (2001), *Appl. Biochem. Biotechnol.* **94,** 111–126.
18. Aiello-Mazzarri, C. (2002), PhD dissertation, Texas A&M University, College Station, Texas.
19. Datta, R. (1981), *Biotechnol. Bioeng.* **23,** 61–77.
20. Thanakoses, P., Black, A. S., and Holtzapple, M. T. (2003), *Biotechnol. Bioeng.* **83(2),** 191–200.

Construction of Recombinant *Bacillus subtilis* for Production of Polyhydroxyalkanoates

Yujie Wang,[1] Lifang Ruan,[1] Wai-Hung Lo,[1] Hong Chua,[2] and Hoi-Fu Yu*,[1]

[1]*State Key Laboratory of Chinese Medicine and Molecular Pharmacology, Shenzhen, Department of Applied Biology and Chemical Technology; and* [2]*Department of Civil and Structural Engineering, The Hong Kong Polytechnic University, Hong Kong, China, E-mail: bcpyu@polyu.edu.hk*

Abstract

Polyhydroxyalkanoates (PHAs) are polyesters of hydroxyalkanoates synthesized by numerous bacteria as intracellular carbon and energy storage compounds and accumulated as granules in the cytoplasm of cells. In this work, we constructed two recombinant plasmids, pBE2C1 and pBE2C1AB, containing one or two PHA synthse genes, respectively. The two plasmids were inserted into *Bacillus subtilis* DB104 to generate modified strains, *B. subtilis*/pBE2C1 and *B. subtilis*/pBE2C1AB. The two recombinants strains were subjected to fermentation and showed PHA accumulation, the first reported example of mcl-PHA production in *B. subtilis*. Gas Chromatography analysis identified the compound produced by *B. subtilis*/pBE2C1 to be a hydroxydecanoate-co-hydroxydodecanoate (HD-co-HDD) polymer whereas that produced by *B. subtilis*/pBE2C1AB was a hydroxybutyrate-co-hydroxydecanoate-co-hydroxydodecanoate (HB-HD-HDD) polymer.

Index Entries: *Bacillus subtilis*; cloning and expression; P (HB-co-mclHA); PHA synthase gene; polyhydroxyalkanoates (PHAs).

Introduction

Polyhydroxyalkanoates (PHAs) function as carbon and energy reserves in prokaryotic cells *(1)*. They are accumulated by a wide range of bacteria when a carbon resource is provided in excess and at least one essential growth nutrient is limited *(2,3)*. Because their physical characteristics are similar to those of petrochemical polymers such as polypropylene, PHAs have been studied intensively by academic and industry and are considered good candidates for biodegradable plastics and elastomers *(4)*. The synthesis of PHA requires the enzyme PHA synthase *(phaC)*, which uses β-hydroxyacyl-coenzyme as a substrate for polymerization. The production of such substrates can occur by a variety of pathways *(5)*, including the simplest using the enzymes β-ketothiolase (encoded by *phaA*), acetoacetyl-CoA

*Author to whom all correspondence and reprint requests should be addressed.

reductase (encoded by *phaB*), β-oxidation *(6)*, and a fatty acid *de novo* synthesis pathway *(7)*. Several genes have been cloned and expressed in *Escherichia coli*. However, the expression levels were much lower than that in the parent organisms and the expressed enzymes were accumulated as inclusion bodies inside the cells, which was a limiting factor in continuous-culture fermentation *(8–10)*. Because the method of transforming *Bacillus subtilis* with plasmid DNA *(11)* was discovered, *B. subtilis* has become an attractive alternative to *E. coli* as a host for the expression of cloned genes because it has several advantages over *E. coli*. Secretion of proteins may circumvent the formation of inactive inclusion bodies, which occurs during the overexpression of foreign genes in *E. coli*. Furthermore, *B. subtilis* is not a human pathogen and can be considered biologically safe. In this study, we report the expression of the PHA synthase gene(s) in *B. subtilis* DB104.

Methods

Strains and Plasmids

B. subtilis DB104 and *E. coli* HB101 were used as hosts. pBHR71 contains the *phaC1* gene from *Pseudomonas aeruginosa (12)*, pJM9131 contains the *phaAB* gene from *Ralstonia eutropha (13)*, pBE2 is an *E. coli–B. subtilis* shuttle vector.

Media and Culture

PHA fermentation medium was described by Ramsay et al. *(2)*. Nutrient broth, nutrient agar, and R2A agar were purchased from Sigma (Germany). The strain was first inoculated into 5-mL nutrient broth, and then 1% inoculum was used in fermentation. Fermentation was carried out in a 500-mL or 1-L shaking flask.

Construction of Recombinants

Plasmids were isolated using the Wizard Plus SV Minipreps DNA Purification System (Promega, Madison, WI). The isolated plasmids and digested DNA fragments were analyzed by electrophoresis in horizontal slab gels containing 0.7% (w/v) agarose, and a 1 kb DNA ladder (Promega, Madison, WI) was used as standard marker. DNA restriction fragments were isolated from the agarose gel using the QIAEX II Gel Extraction Kit (Qiagen, Valencia, CA). Restriction enzymes and T4 DNA ligase (Promega) were used according to the instructions provided by the supplier. A 1.8 kb Bam HI/Xba I restriction fragment containing *phaC1* gene obtained from plasmid pBHR71 was used as an insert, and plasmid pBE2 was employed as vector. A 2.5 kb *pst* I restriction fragment containing *phaAB* gene obtained from plasmid pJM9131 was used as an insert, and plasmid pBE2C1 was employed as vector. Ligation products were mixed with competent *B. subtilis* DB104 cells, and the resulting transformants were selected on nutrient agar plates containing 100 µg/mL of ampicillin. The positive transformants were proved by restriction digestion followed by agarose gel electrophoresis.

Gas Chromatography Analysis

Almost 1 mL esterification solution (3 mL 95–98% H_2SO_4, 0.29 g benzoate, and 97 mL methanol), 15 mg freeze-dried cells, and 1 mL chloroform were mixed and heated at 100°C for 4 h, 1 mL of double-distilled H_2O (ddH_2O) was added to the cooled mixture, which was then vortexed for phase separation. A 1 µL portion of the lower organic phase was subjected to gas chromatography (GC) analysis, which was performed on a Hewlett Packard 5890 Series II Gas Chromatograph, using a 6-ft Supelco (10% Carbowax 20 M with 80/100 in mesh size Chromosorb WAW) Packed Column. Nitrogen was used as the carrier gas at the flow rate of 20 mL/min. The analysis was started at 135°C and the temperature was kept stable for 10 min to determine both the content and composition of the polymer.

Fourier Transform Infrared Spectroscopy

Almost 2–5 mL of the cell culture was centrifuged at 2610g for 15 min. The cells were transferred onto an IR window (ZnSe Disc, Spectratech) and dried on it. A mirror was used to give the reflected infrared signal to the horizontally laid window. With a scan number of 32, resolution of 16 and autogain, spectra were recorded at wavenumbers (/cm) from 400 to 4000 using a Mangna-IR spectrometer 750 (Nicolet) *(14)*. The PHA peak was observed at wavelengths of 1726–1740/cm *(15)*.

Results

Construction of the Recombinant Strain

The recombinant plasmid pBE2C1 was obtained by using a 1.8 kb *Bam* HI/*Xba* I restriction fragment *phaC1* gene from *P. aeruginosa* obtained from plasmid pBHR71 as an insert, and plasmid pBE2 as vector. The recombinant plasmid pBE2C1AB was obtained by using a 2.5 kb *pst* I restriction fragment *phaAB* gene obtained from plasmid pJM9131 as an insert, and plasmid pBE2C1 as vector (Fig. 1). The constructed plasmids were confirmed by restriction digestions and agarose gel electrophoresis. The newly cloned plasmids were amplified in *E. coli*. The size of the vector pBE2 is around 6.2 kb and the inserts around 1.8 kb (*phaC1* gene) and 4.3 kb (*phaC1AB* gene). These two newly constructed plasmids were transformed into *B. subtilis*. The positive transformants were checked by restriction digestions followed by agarose gel electrophoresis (Fig. 2). These result showed that the foreign fragment had been successfully inserted into the plasmids.

Gene Expression in Recombinant Strain

The culture was incubated at 28°C and shaken at 280 rpm. The cells were subjected to Fourier transform infrared spectroscopy (FTIR) and appeared to have a PHA peak with wavelengths about 1726–1740/cm. The result suggested that the recombinant strains could produce PHA. After

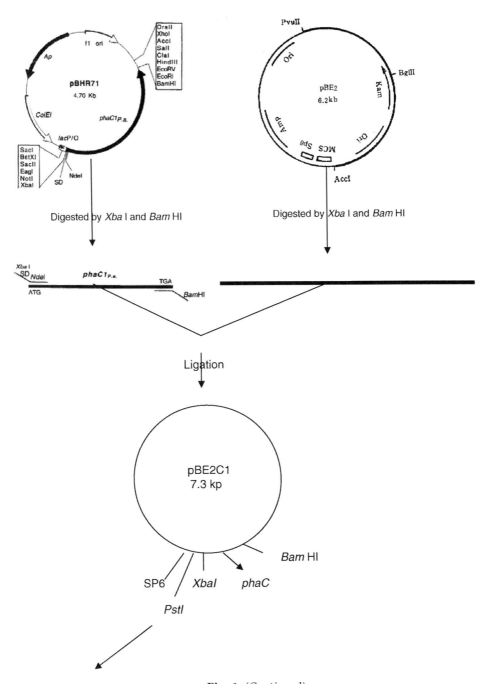

Fig. 1. (*Continued*)

fermentation, the cells were freeze-dried and the composition of the biopolymer was analyzed by GC; the result is showed in Fig. 3. The expression in the recombinant strain is controlled by the promoter SP6 of pBE2.

There was no PHA accumulation in *B. subtilis* DB104. However, *B. subtilis* DB104/pBE2C1 and *B. subtilis* DB104/pBE2C1AB, recombinants

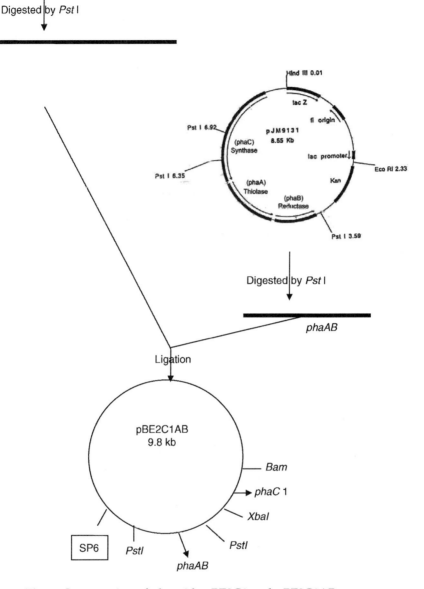

Fig. 1. Construction of plasmids pBE2C1 and pBE2C1AB.

containing *pha* genes, showed PHA accumulation—the first reported expression of *pha* genes from *P. aeruginosa* and *R. eutropha* into *B. subtilis*. As shown in Fig. 3, PHA production by the recombinant *B. subtilis* DB104/pBE2C1AB is higher than that by *B. subtilis* DB104/pBE2C1. GC analysis further identified the product synthesized by *B. subtilis* DB104/pBE2C1 to be a HD-HDD polymer whereas the product synthesised by *B. subtilis* DB104/pBE2C1AB was identified to be a HB-HD-HDD polymer (Table 1).

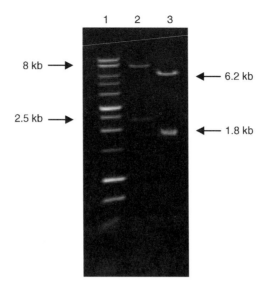

Fig. 2. Agarose gel electrophoresis. 1. DNA ladder 2. pBE2C1AB/*Pst* I 3. pBE2C1/ *Xba* I and *Bam* HI.

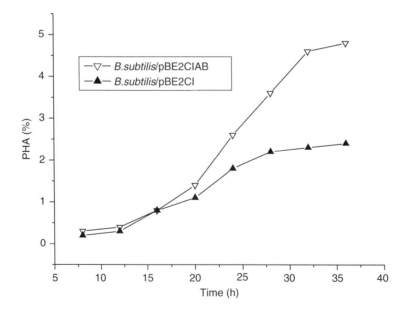

Fig. 3. Percentage the CDW of PHA-time curve of recombinants.

The *B. subtilis* DB104/pBE2C1 and *B. subtilis* DB104/pBE2C1AB carrying *pha* gene were subjected to fermentation and showed PHA accumulation, providing the first evidence that the *pha* gene from *P. aeruginosa* and *R. eutropha* can be expressed in *B. subtilis*.

Table 1
Monomer Composition of the PHA Accumulated
by Recombinant Strains

Strains	HB (%)	HD (%)	HDD (%)
B. subtilis DB104/pBE2C1	0	58	42
B. subtilis DB104/pBE2C1AB	16	62	22

Discussion

In this research, we used B. subtilis as a host because it offered advantages such as short generation time, absence of endotoxin, and secretion of amylases and proteinases that utilize food wastes for nutrients, reducing the cost of production of PHAs. Our work is the first to report the expression of the mcl-PHA synthase gene obtained from P. aeruginosa into B. subtilis. The recombinant strains successfully produced mcl-PHA. P. aeruginosa can synthesize mcl-PHA by β-fatty acid oxidation using fatty acid as substrate, and by the *de novo* fatty acid synthesis pathway, which uses simple molecules such as glucose as substrate, but production of the latter is lower. In this work, glucose was used as a substrate, so the production of the PHA is low. Some mcl fatty acids can be added in future experiments to improve the production of the PHA.

As the results indicate, the recombinant B. subtilis DB104/pBE2C1 and B. subtilis DB104/pBE2C1AB can synthesize monomer mcl-PHA, which agrees with the character of PHA synthase gene *phaC1*. The synthase (encoded by *phaC1*) is referred to as type II PHA synthase and prefers 3-hydroxyacyl CoA with chain lengths of 6–14 carbon atoms (mcl-HA) as substrates (16). Because the melting point temperature was low and the crystallization rate of mcl-PHA was slow, the recombinant can produce P (HB-co-mclHA). We are interested in the monomer composition of PHAs and monomer percentage. To follow up, we subcloned β-ketothiolase gene *(phbA)* and acetoacetyl-CoA reductase gene *(phbB)* from R. eutropha into plasmid pBE2C1 to obtain plasmid pBE2C1AB. The recombinant B. subtilis DB104/pBE2C1AB was found to produce PHB.

Recently, there had been growing interest in P (HB-co-mcl HA) because of its enhanced physical and chemical character. Our research results suggest that recombinant strain B. subtilis DB104/pBE2C1AB could produce P (HB-co-mcl HA) polymer.

Acknowledgments

The authors would like to express their sincere gratitude to The Hong Kong Polytechnic University and the research Grant Council of the Hong

Kong special administration Region, CHINA (Polyu5272/01M, Polyu5257/02M and Polyu5403/03M).

References

1. Anderson, A. J. and Dawes, E. A. (1990), *Microbiol. Rev.* **54,** 450–472.
2. Ramsay, B. A., Lomaliza, K., Chavarie, C., Dube, B., Bataille, P., and Ramsay, J. A. (1990), *Appl. Environ. Microbiol.* **56,** 2093–2098.
3. Steinbuchel, A. and Schlegel, H. F. (1991), *Mol. Microbiol.* **5,** 535–542.
4. Poirier, Y., Nawrath, C., and Somerville, C. (1995), *Bio. Technol.* **13,** 142–150.
5. Madison, L. L. and Huisman, F. W. (1999), *Microbiol. Mol. Biol. Rev.* **63,** 21–53.
6. Page, W. J. and Manchak, J. (1995), *Can. J. Microbiol.* **41 (Suppl 1)** 106–114.
7. Rehm, B. H. A., Kruger, N., and Steinbuchel, A. (1998), *J. Biol. Chem.* **273,** 24,044–24,051.
8. Bernier, R. J., Driguez, H., and Desrochers, M. (1983), *Gene* **26,** 59–65.
9. Panbangred, W., Kondo, T., Negoro, S., Shinmyo, A., and Okada, H. (1983), *Mol. Gen. Genet.* **192,** 335–341.
10. Shendye, A. and Rao, M. (1993), *Enzyme Microb. Technol.* **15,** 343–347.
11. Ehrlich, S. D. (1977), *Proc. Natl. Acad. Sci. USA* **70,** 1680–1682.
12. Langenbach, S., Rehm, B. H., and Steinbuchel, A. (1997), *FEMS Microbiol. Lett.* **150(2),** 303–309.
13. Valentin, H. E. and Steinbuchel, A. (1994), *Appl. Microbiol. Biotechnol.* **40,** 699–709.
14. Hong, K., Leung, Y. C., Kwok, S. Y., et al. (2000), *Appl. Biochem. Biotechnol.* **84–86,** 381–390.
15. Yoshiharu, D., Yasushi, K., Yoshiyuki, N., and Masao, K. (1989), *Appl. Environ. Microbiol.* **55,** 2932–2938.
16. Hein, S., Paletta, J., and Steinbüchel, A. (2002), *Appl. Microbiol. Biotechnol.* **58(2),** 229–236.

Microorganism Screening for Limonene Bioconversion and Correlation With RAPD Markers

GECIANE TONIAZZO,[1] LINDOMAR LERIN,[2] DÉBORA DE OLIVEIRA,*,[2] CLÁUDIO DARIVA,[2] ROGÉRIO L. CANSIAN,[2] FRANCINE FERREIRA PADILHA,[2] AND OCTÁVIO A. C. ANTUNES[1]

[1]*Departament of Biochemistry, Instituto de Química - UFRJ, CT, Bloco A, Lab 641 Rio de Janeiro - RJ, 21945 – 970 - Brazil; and* [2]*Departament of Food Engineering - URI - Campus de Erechim Av. Sete de Setembro, 1621 - Erechim - RS, 99700 – 000 – Brazil, E-mail: odebora@uricer.edu.br*

Abstract

The use of microorganisms for biotransformations of monoterpenes has stimulated the biotechnological market. Aiming at the highest efficiency in the process of strains screening, the application of molecular biology techniques have been proposed. Based on these aspects, the objective of this work was to select different strains able to convert limonene using fermentative process and random amplified polymorphic DNA (RAPD) markers. The results obtained in the fermentative screening, from 17 strains tested, pointed out that four microorganisms were able to convert limonene into oxygenated derivatives. The RAPD study showed a polymorphism of 96.02% and a similarity from 16.02 to 51.51%. Based on this it was possible to observe a high genetic diversity, even among strains of same species, concluding that the RAPD was not able to correlate the genetic characteristics of the microorganism with the results obtained from the biotransformation process.

Index Entries: Bioconversion; limonene; RAPD; screening.

Introduction

Terpenes and especially their oxygenated derivatives are extensively used in the flavor industry. Limonene (4-isopropenyl-1-methylcyclohexene), a monocyclic monoterpene, is a low-priced monoterpene and in most cases the major compound in essential oils of citrus fruits *(1–3)*. The production of flavors through a biotechnological route offers a number of advantages. One important attribute of microbial biocatalysis is the ability

---

*Author to whom all correspondence and reprint requests should be addressed.

to synthesize products that can be labeled as natural, if derived from natural substrates, and added to foods without being considered as additives. Another important attribute is to convert monoterpene precursors or intermediates into higher value products for flavors and fragrance industries *(1,2)*. Screening of microorganisms, plants, or animal cells represent a natural way to obtain new enzymes. The microorganisms, in this case, are of particular interest owing to the large diversity of metabolic processes and enzymes involved, and the unlimited number of microorganisms in nature that can be tested. They modify and degrade a variety of organic molecules and complexes, and so, it can be expected that some of them may be capable of catalyzing specific reactions *(4)*, such as biotransformation of limonene.

In this sense, the fermentative screen can be performed in order to confirm the strains that are able to convert limonene into oxygenated monoterpenes. As examples, Dhavalikar and Bhattacharyya *(5)* and Dhavalikar et al. *(6)* isolated strains of *Pseudomonas* able to grow on limonene as the sole carbon and energy source. Dehydrocarvone, carvone, carveol, 1,2-*cis*-diol-limonene among other compounds were accumulated during growth. Bowen *(7)* used the *Penicillium digitatum* strain, isolated from orange, and the strain was able to convert limonene into several compounds. Noma et al. *(8)* used the fungus *Aspergillus niger* to biotransform limonene, obtaining *trans*-1,2-diol-(+)-limonene as a major reaction product and, in smaller quantities, (+)-*cis*-carveol, (–)-carvone, and α-terpineol, among others. From these results, it can be observed that at least a few microorganisms are able to transform limonene into cis/trans-carveol and carvone *(5,7)*. In order to obtain a better efficiency in the screening process, techniques of molecular biology have been used *(9–11)*. In this sense, the random amplified polymorphic DNA (RAPD) is technically simple and automated. It requires small amounts of DNA, no previous information on the target genome sequence and the level of polymorphism obtained is usually high *(12)*. RAPD markers, which are based on the amplification of discrete DNA fragments in the genome by the use of oligonucleotide primers with random sequences, have been widely used to identify physiological strains of fungi *(9)*. The adoption of RAPD markers in detection, diagnostic, and determination of genetic diversity is related to the simplicity of the method, which has high sensitivity, low cost and is safe and fast. The RAPD technique has been proven to be useful for measuring and characterizing the genetic variability of microorganisms, plants, and animal cells *(13,14)*. The objective of this work was to use RAPD markers to analyze different strains reported in literature as able to biotransform limonene. Simultaneously, fermentations have been performed to bioconvert limonene, with the purpose of comparing the results with those obtained from the molecular biology study. Based on these aspects, this work screened 17 strains of microorganisms for their ability to bioconvert limonene.

Table 1
Microorganisms Tested in the Screening and Composition of Different Growth Media

Code	Microorganism	Culture medium	Time cultivation (h)
1	A. niger[a]	PD[e]	72
5	A. niger ATCC 16404[b]	PD[e]	72
12	A. niger ATCC 9642[b]	YM[f]	72
19	A. niger ATCC 1004[b]	PD[e]	72
17	A. oryzae ATCC 1003[b]	PD[f]	72
3	P. citrinum ATCC 28752[b]	Malt[g]	72
7	P. digitatum ATCC 26821[b]	PD[e]	72
10	P. notatum ATCC 9478[b]	PD[e]	72
15	P. brevicompactum[c]	PD[e]	72
29	P. camembertii (CT) ATCC 4845[b]	PD[e]	72
31	P. verrucosum[c]	PD[e]	72
33	P. simplicissimum[c]	PD[e]	72
35	P. duclauxii ATCC 9121[b]	PD[e]	72
13	Paecilomyces variotii ATCC 22319[b]	Czapek Dox[h]	120
21	P. putida[a]	LB[i]	24
25	P. aeruginosa ATCC 27853[d]	LB[i]	24
27	Candida spp. ATCC 34147	LB[i]	24

[a]Instituto de Microbiologia of the Universidade Federal do Rio de Janeiro (Rio de Janeiro Brazil).
[b]Fundação Instituto Oswaldo Cruz (Rio de Janeiro Brazil).
[c]These microorganisms were isolated from the reject of the industry of babassu for Freire et al. (22).
[d]Laboratório de Biotecnologia Vegetal of the Universidade Regional Integrada (Erechim, Brazil).
[e]PD (300 g infusion from potates and 20 g glucose).
[f]YM (3 g yeast extract, 3 g malt extract, 5 g peptone, and 10 g glucose).
[g]Malte (30 g malt extract, and 3 g peptone).
[h]Czapek Dox (0.5 g KCl, 1 g KH_2PO_4, 2 g $NaNO_3$, 30g sucrose, 0.1 g $FeSO_4 \cdot H_2O$, 0.5 g $MgSO_4 \cdot 7H_2O$).
[i]LB (10 g tryptone, 5 g yeast extract, and 10 g NaCl), all the media were prepared in 1000 mL of distilled water.

Materials and Methods

Cell Production

A total of 17 microorganisms were used in this study. The strains of microorganisms belonging to different genera were obtained from different institutions as shown in Table 1. All the strains were generously supplied by the respective institutions. Table 1 also presents the cultive medium and cultivation times used for each microorganism. Fungi were cultivated in potato dextrose (PD), yeast malt (YM), Malt, and Czapek Dox at 25°C for 72 h; except for Paecilomyces variotii that was cultivated for 120 h. Yeasts were cultivated in YM at 25°C for 24 h and bacteria were cultivated in Luria

Bertani (LB) at 28°C for 24 h. Stock cultures of organisms were maintained in medium slants and stored at 4°C after growth.

Screening Experiments

For biotransformation experiments the reactions were started 3 d after inoculation for fungi and 24 h after inoculation for bacteria and yeasts by adding 300 µL of substrate directly into culture flasks containing 30 mL of medium. These were then maintained for 6 d in an orbital shaker operating at 25°C and 150 rpm. The substrate was added as solutions in absolute ethanol (1:1 v/v); all experiments used 1% (v/v) of substrate in a background of 1% (v/v) of EtOH. All experiments were carried out in duplicate in parallel, blanks were run using control media, under the same conditions, without the microorganism present. The experiments were performed for each microorganism in closed stopped glass flasks in order to avoid substrate and product evaporation.

Extraction and Identification of Biotransformation Products

At the end of the experiments, the cells were removed by filtration for fungi and centrifugation for bacteria and yeasts. The product recovery was performed by liquid–liquid extraction with 3×12 mL of ethyl acetate (Et_2O). After extraction, the solution volume was brought up to 25 mL with the additional Et_2O. The final solution was then dried using anhydrous sodium sulphate. The reaction products were identified by gas chromatography (GC)/mass spectrometer (MS) (Shimadzu QP5050A), using a capillary DB-WAX column (30 m, 0.25 mm, 0.25 µm). The column temperature was programmed at 50°C for 3 min, then increased at 5°C/min up to 130°C and then further increased at 15°C/ min up to 210°C and then held at this temperature for 5 min. Helium was the carrier gas, and the injection and detector temperatures were set to 250°C 0.5 µL of the dried solution was injected into the GC/MS system. The apparatus operated with a flow rate of 1 mL/min in electronic impact mode of 70 eV and in split mode (split ratio 1:10). The identification of the compounds was accomplished by comparing the mass spectra with those from the Wiley library and by additional comparison of the GC retention time of standard compounds. The products were semi-quantified by the percent of area of each peak in relation to total peak area.

Genomic DNA Extraction

DNA extraction of bacteria and yeast was performed according to Sambrook et al. *(15)* with some modifications. Cultures of each strain were grown as previously presented (Table 1). Almost 1.5 mL of each culture was centrifuged (Eppendorf Centrifuge Model 5403) for 5 min at 21.467*g* to collect the bacterial or yeast cells, and the supernatant was discarded. Each pellet was mixed in 600 µL of buffer containing 25 m*M* Tris ([Hydroxymethyl] aminomethanol – Sigma)-HCl, pH 8.0, 10 m*M* ethylenediaminetetracetic acid (EDTA), (Disodium Salt, Dihydrate – Na_2 EDTA·$2H_2O$ – Gibco BRL) and then 50 mM glucose was added and the solution further mixed. Then, 50 µL of lysozyme (2 mg/mL, Sigma) was added, and the mixture was maintained in repose for 10 min. Afterwards, 66 µL of 10% (w/v) sodium dodecyl sulfate (SDS) and 3 µL of 2-mercaptoethanol was added to each tube, which was then briefly vortexed before being incubated at 65°C for 30 min. Thereafter, 190 µL of 3 M sodium acetate was added to each tube and mixed by inversion. This mixture was then incubated at 40°C for 30 min (in a refrigerator) and then centrifuged for 5 min at 21.467g. The supernatant was collected for use. The DNA was precipitated using one volume of ice-cold isopropanol mixed in by inversion, and then incubating at –20°C (in a freezer) for 10 min before centrifuging for 5 min at 21.467g. The supernatant was discarded. The recovered precipitate was air dried and then resuspended in 200 µL of TE buffer (10 mM Tris-HCl, pH 8.0, 1 mM EDTA).

The DNA extraction of fungi was based on the method reported by Roeder and Broda (16). Cultures of each strain were grown as previously presented (Table 1). Liquid nitrogen was added to 300 mg of mycelia in a mortar, and the cells were ground with a pestle. In brief, the powdered mycelium was transferred to an Eppendorf tube and 700 µL of buffer (3% [v/v] SDS, 50 mM EDTA, 50 mM Tris-HCl, pH 8.0, 1% [v/v] 2,-mercaptoethanol) were added. Each sample was incubated for 1 h at 65°C. Then, the tubes'contents were removed and cooled, extracted with an equal volume of chloroform-isoamyl alcohol (24:1), and centrifuged at 21.467g for 10 min. This procedure was repeated three times. Afterwards, 120 µL of 3 M sodium acetate pH 5.2 and 600 µL of ice-cold isopropanol were added and the solution was mixed by inversion, incubated at –20°C for 30 min and then centrifuged for 5 min at 21.467g. The suspension was resuspended at 500–700 µL of TE buffer, washed twice in 500 µL of 70% ethanol, air dried, and resuspended in 200 µL of TE buffer.

DNA Quantification

For DNA concentration 980 µL of Milli-Q sterile water and 20 µL of sample were added to an assay tube. DNA concentration was estimated by measuring the optical density at 260 nm, using a spectrophotometer (Agilent 8453), and this estimate was checked by performing electrophoresis on a 0.8% agarose gel (Gibco BRL) in TBE 1 × (0.89 M Tris, 0.89 M of H_3BO_3 and 0.08 M EDTA).

Amplification Reaction

For amplification procedure, the reaction reported by Williams et al. (10), was used with some modifications. RAPD reactions were conducted in total volumes of 25 µL. The reaction mixture contained buffer (50 mM Tris-HCl pH 9.0, 50 mM KCl) (Life Technologies, São Paulo, Brazil), dNTPs mix (200 mM of each nucleotide), 0.2 mM of primer, 3 mM $MgCl_2$, 0.25 mM

Table 2
Total and Polymorphic Numbers of Fragments Obtained for Each Primer
Used for the 17 Microorganisms

Primer	Sequence (5' to 3')	Total fragments	Polymorphic fragments
OPA-03	AGTCAGCCAC	17	17
OPA-04	AATCCGGCTG	30	30
OPA-13	CCACACCAGT	31	31
OPA-20	GTTGCGATCC	22	21
OPF-09	CCAAGCTTCC	12	6
OPH-18	GAATCGGCCA	20	20
OPH-19	CTGACCAGCC	27	27
OPY-03	ACAGCCTGGT	18	18
OPW-19	CAAAGCGCTC	24	23
TOTAL		201	193
Polymorphism (%)			96.02

Triton-X-100, 1.5 U of recombinant Taq DNA polymerase (Invitrogen), and 40 ng of DNA.

Random RAPD Primers

The kits OPA, OPF, OPH, OPW, and OPY from the Operon Technologies (Alameda, CA), were used with 20 primers of each one, in order to identify the primers that showed the better results, estimate the numbers, intensities, sizes of each band, reproducibility as well as to generate polymorphism, selecting nine primers (Table 2).

Amplification Procedure RAPD

The amplification was performed in a thermo cycles MJ Research Inc., Watertown, MA (model PTC100TM Programmable Thermal Controller) as follows: one initial cycle for 3 min at 92°C, followed by 40 cycles comprising denaturation (one min at 92°C); annealing (1 min at 35°C); and extension (two min at 72°C) and then a final extension for 3 min at 72°C.

Electrophoresis of the Amplifying Fragments

Amplification products were separated by electrophoresis in 1.4% agarose gels in buffer TBE 1X (0.89 M Tris, 0.89 M of H_3BO_3, and 0.08 M EDTA) on an electrophoresis horizontal cube using a constant voltage of 90 Volts. DNA Lambda was digested with *Eco*RI and *Hind* III (Gibco BRL- Life Technologies, São Paulo, Brazil) was included as a molecular size marker. Gels were visualized by staining with ethidium bromide (SIGMA Chemical CO, St Louis, MO) and banding patters were photographed over UV light. The gels were photographed using a GEL-PRO digital photographic system (Media Cybernetics, Silver Spring, MD).

Analysis of RAPD Products

To determine the genetic variability, the matrix formed by the presence or absence of bands was analyzed using NTSYS version 1.7 software (Numerical Taxonomy System for Multivariate Analysis). The dendrogram were constructed by the algoritm UPGMA (Unweighted Pair Group Method Using Arithmetic Averages) using the Jaccard coefficient of similarity *(17)*.

Results and Discussion

The strains used in this work, as presented in Table 1, were selected based on results reported in the literature related to bioconversion of monoterpenes.

Screening Experiments

The screening experiments confirmed that some strains were able to perform the desired biotransformation of limonene. Table 3 shows the subset of microorganisms that exhibited the ability for transformation and the end products they produced. Because the proposed reactions are slow, we believe the oxygen in the headspace of the closed flasks was sufficient to enable biotransformation. However, this was a screening effort and the evaluation of oxygen limitations was outside the scope of this work. Further investigations are being carried out to verify oxygen requirements and limitations in 2-L bioreactors equipped with oxygen monitoring and control.

The strains selected were *Aspergillus niger, Penicillium simplicissimum, A. niger* ATCC 9642 and *Pseudomonas putida*. The metabolites recovered after biotransformation of limonene were α-terpineol and cycle-hexanemethanol for *A. niger*, α-terpineol for *A. niger* ATCC 9642, *cis*-carveol for *P. simplicissinum*, dehydrocarveol, and perillyl alcohol for *P. putida*. Figure 1 presents the structures of these different metabolites. It is relevant to note that in spite of the low conversions obtained, *cis*-carveol, and dehydrocarveol are precursors to carvone production, an important and high value compound in the flavor industries *(18)*. Experiments are underway in our group to optimize the process conversion process, to improve carveol production.

RAPD Analysis

Based on the results obtained in the fermentative screening, the RAPD technique was used in order to test if genetic characteristics could be correlated with the microbial capacity of limonene bioconversion. These experiments used the nine primers from Operon Technologies listed in Table 2. The amplification reactions with the nine primers generated a total of 201 fragments, 193 (96.02%) of them being polymorphic (Table 2) with sizes ranging from 50 to 2200 bp (Fig. 2). The average number of fragments per primer was 22.33. Figure 2 shows the variability observed for different species analyzed using the primer OPA-04.

Table 3
Microorganisms Selected, End Products and Concentration of the Fermentative Screening

Microorganism	Products (% ratio of product in relation to all products and substrate present in the sample)				
	α-terpineol	Cycle-hexanemethanol	Cis-carveol	Dehydrocarveol	Perillyl alcohol
A. niger	0.68 ± 0.19	0.64 ± 0.27	–	–	–
A. niger ATCC 9642	0.13 ± 0.07	–	–	–	–
P. putida	–	–	–	0.07 ± 0.00	0.31 ± 0.01
P. simplicissimum	–	–	0.18 ± 0.05	–	–

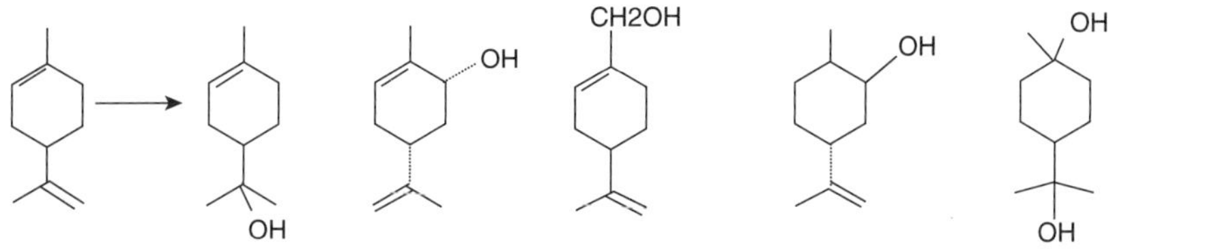

Microorganism Screening for Limonene Bioconversion

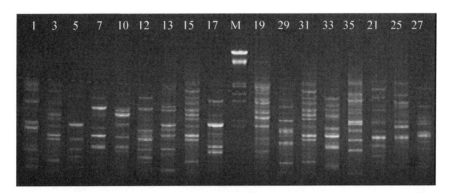

Fig. 1. Structures of metabolites recovered after biotransformation of limonene.

Fig. 2. RAPD patterns obtained with the OPA-04 primer in agarose gel 1.4%, where M is molecular size marker in base pairs (phase λ DNA digested wit *Eco*RI and *Hind*III, Gibco, BRL); and 1 represents *A. niger*, 3 *P. citrinum* ATCC 28752, 5 *A. niger* ATCC 16404, 7 *P. digitatum* ATCC 26821, 10 *P. notatum* ATCC 9478, 12 *A. niger* ATCC 9642, 13 *Paecilomyces variotii* ATCC 22319, 15 *P. brevicompactum*, 17 *A. oryzae* ATCC 1003, 19 *A. niger* ATCC 1004, 29 *P. camembertii* (CT) ATCC 4845, 31 *P. verrucosum*, 33 *P. simplicissimum*, 35 *P. duclauxi* ATCC 9121, 21 *P. putida*, *P. aeruginosa* ATCC 27853, and 27 *Candida* spp. ATCC 34147.

Similarity indices (Jaccard coefficient) among the studied microorganisms varied from 0.1527 to 0.6578 with a mean value of 0.2445 (Table 4). The similarity from 15.27% to 65.78% among the studied microorganisms can be considered low. These results indicate a high variability among the 17 strains of microorganisms tested. Low similarity between fungi and bacteria was expected, but it was not anticipated that fungi from the same species would have genotypic profiles exhibiting so much variability (Fig. 3).

The high variability among the strains prevented correlating RAPD analysis results with the capacity of limonene bioconversion. It was not possible to select the microorganisms tested by application of the RAPD technique. Different microorganisms, independent of genetic differences among them, are able to convert limonene, making it possible to suggest that several genes might be involved and probably no amplified fragment corresponds to the genes involved in desired bioconversion *(19)*. The high genetic difference among the studied microorganisms, even within single species, suggests that it is not possible to discard any microorganism. The biocatalytic a hydroxylation converts these molecules into different metabolites,

Table 4
Similarity of the Studied Microorganisms

Strain number	Mean	Minimum value	Maximum value	Standard deviation	Variation coefficient (%)
17	0.2445	0.1527	0.6578	0.0632	0.2584

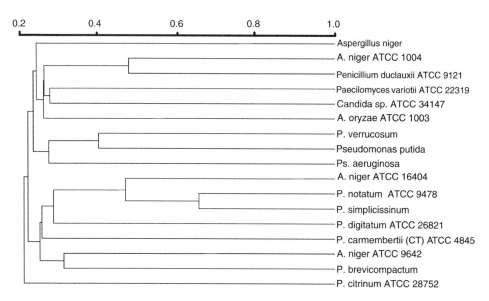

Fig. 3. Dendrogram generated by the UPGMA method from the RAPD data showing the relationship among the clusters of the strains *Aspergillus, Penicillium, Paecilomyces, Candida,* and *Pseudomonas*. Similarities were calculated using the Jaccard coefficient.

easier to be excreted by the cell. Thus, different microorganisms may present such metabolic routes as a way to survive *(20)*. Another interesting aspect of the enzymatic complex responsible for biotransformation is that their substrates induce the enzyme synthesis. This auto-induction can supply a mechanism of detoxification *(21)*.

Conclusions

The results obtained in the fermentative screening showed that, from 17 strains of microorganisms, four were able to bioconvert limonene into oxygenated monoterpenes. The metabolites recovered (and their relative abundance) were α-terpineol (0.68% ± 0.19) and cycle-hexanemethanol (0.64% ± 0.27) for *A. niger*, α-terpineol (0.13% ± 0.07) for *A. niger* ATCC 9642, dehydrocarveol (0.07% ± 0.00) and perillyl alcohol (0.31% ± 0.01) for *P. putida* and *cis*-carveol (0.18% ± 0.18) for *P. simplicissimum*.

The molecular biology study showed that it was not possible to establish a correlation between the genetic characteristics and the microbial capacity for limonene bioconversion. From a mean similarity of genomic profile of 24.45% it can be concluded that a high genetic variability exists among the strains. It was expected that there would be a low similarity between fungi and bacteria, but it was not expected that fungi within the same species would exhibit such dissimilar genotypic profiles.

Acknowledgment

The authors thank CAPES (PROCAD), FUJB, CNPq, FINEP, and FAPERJ for the financial support of this work and scholarships.

References

1. Adams, A., Demyttenaere, J. C. R., and Kimpe, N. (2003), *Food Chem.* **80**, 525–534.
2. Onken, J. and Berger, F. G. (1999), *J. Biotechnol.* **69**, 163–168.
3. Van der Werf, M. J., de Bont, J. A. M., and Leak, D. J. (1997), In: *Advances in Biochemical Engineering Biotechnology*, Scheper T., (ed.), Springer-Verlag, Berlin, vol. 55, pp. 147–176.
4. De Conti, R., Rodrigues, J. A. R. E., and Moran, P. J. S. (2001), *Quím. Nova* **24(5)**, 672–675.
5. Dhavalikar, J. C. R. and Bhattacharyya, P. K. (1966), *Indian J. Biochem.* **3**, 158–164.
6. Dhavalikar, J. C. R., Rangachari, P. N., and Bhattacharyya, P. K. (1966), *Indian J. Biochem.* **3**, 144–157.
7. Bowen, E. R. (1975), *Florida State Horticultural Society*. Florida, p. 304.
8. Noma, Y., Yamasaki, S., and Asakawa, Y. (1992), *Phytochemistry* **31(8)**, 2725–2727.
9. Guthrie, P. A. I., Magill C. W., Frederiksen, R. A., and Odvody, G. N. (1992), *Phytopathology* **82**, 832–835.
10. Williams, J. G. K., Kubelik, A. R., Livak, K. J., Rafalski, J. A., and Tingey, S. V. (1991), *Nucleic Acids Res.* **18**, 6531–6535.
11. Harry, M., Jusseaume, N., Gambier, B., and Garnier-Sillam, E. (2001), *Soil Biol. Biochem.* **33**, 417–427.
12. Williams, J. G. K., Hanafey, M. K., Rafalski, J. A., and Tingey, S. V. (1990), *Methods Enzimol.* **218**, 704–740.
13. Ikeh, E. I. (2003), *Afr. J. Clin. Exp. Microbiol.* **4(1)**, 48–52.
14. Fungaro, M. H. P. (2000), PCR na micologia. *Biotecnologia: Ciência e desenvolvimento*, **(3)**, 12–16.
15. Sambrook, J., Fritsch, E. F., and Maniatis, T. (1989), In: *Molecular cloning: a laboratory manual*, 2nd. ed., Cold Spring Harbor Laboratory, New York.
16. Roeder, V. and Broda, P. (1987), *Lett. Appl. Microbiol.* **1**, 17–20.
17. Sokal, R. R. and Michener, C. D. (1958), *Univ. Sci. Bull.* **38**, 1409–1438.
18. Tecelão, C. S. R., van Keulen, F., and da Fonseca, M. M. R. (2001), *J. Mol. Catal. B: Enzymatic.* **11**, 719–724.
19. van der Werf, M. J., Swarts, H., and de Bont, J. A. M. (1999), *Appl. Environ. Microbiol.* **65(5)**, 2092–2102.
20. Duetz, W. A., Bouwmeester, H., van Beilen, J. B., and Witholt, B. (2003), *Appl. Microbiol. Biotechnol.* **61**, 269–277.
21. Voet, D., Voet, J. G., and Pratt, C. W. (2000), *Fundamentos de Bioquímica*, Editora Artmed, Brasil.
22. Freire, D. M. G., Gomes, M. P., Bom, S. P. E., et al. (1997), *J. Braz. Soc. Microbiol.* **28**, 6–12.

Use of Different Adsorbents for Sorption and *Bacillus polymyxa* Protease Immobilization

Irem Kirkkopru, Cenk Alpaslan, Didem Omay,* and Yüksel Güvenilir

Istanbul Technical University, Department of Chemical Engineering, Maslak, Istanbul, Turkey; E-mail: omayd@itu.edu.tr

Abstract

Proteases constitute one of the most important groups of industrial enzymes, accounting for at least 25% of the total enzyme sales, with two-thirds of the proteases produced commercially being of microbial origin (1). Immobilized enzymes are currently the subject of considerable interest because of their advantages over soluble enzymes or alternative technologies, and the steadily increasing number of applications for immobilized enzymes. The general application of immobilized proteins and enzymes has played a central role in the expansion of biotechnology and synthesis-related industries. Proteases have been immobilized on natural and synthetic supports (2,3).

In the present work, a protease from *Bacillus polymyxa* was partially purified with 80% ammonium sulfate precipitation followed by dialysis and chromatography using a diethylaminoethyl (DEAE)-cellulose ion exchange column. Immobilizaiton was evaluated by using different adsorbents (chitin, chitosan, alginate, synthetic zeolite, and raw zeolite) and the storage stability and recycle of the immobilized protease determined. Immobilization yields were estimated to be 96% and 7.5%, by using alginate and chitosan, respectively, after 24 h. The yield of the immobilization was 17% for alginate at 16 h and the enzyme did not adsorb on the chitin, chitosan, synthetic zeolite, and raw zeolite.

Index Entries: Alginate; *Bacillus polymyxa*; chitin; immobilization; protease.

Introduction

Industrial proteases are used as additives in detergents, and in tanning and food manufacturing (baking, brewing, and dairy industries). The majority of theses enzymes are derived from microbial sources, with a few microbial strains being used for industrial applications. *Bacillus polymyxa* has been shown to produce starch-degrading enzymes that yield maltose in very high yield. In the course of investigations it was reported that this microorganism produced considerable quantities of an extracellular proteases in a basal starch–peptone medium (4).

Immobilization involves the coupling of the enzyme with a selected solid support via covalent and (or) noncovalent interactions. Several

---

*Author to whom all correspondence and reprint requests should be addressed.

noncovalent linkages can be specifically favored, including ionic and metal bindings as well as physical adsorption. Immobilization by adsorption results from weak forces between the protein and the solid surface, such as electrostatic and hydrophobic interactions or hydrogen bonds. Adsorption might proceed through conformational rearrangements, depending on the enzyme nature. As adsorptions constitute a mild immobilization process with respect to covalent linkage, it is expected to have less effect on enzymatic kinetic behavior. Despite its simplicity, immobilization by physical adsorption is significantly limited by the tendency of enzyme to desorb from the support (5–7).

The support material can effect the stability of the enzyme and the efficiency of enzyme immobilization, although it is difficult to predict in advance which support will be most suitable for a particular enzyme. The most important requirements for a support material are that it must be insoluble in water, have a high capacity to bind enzyme, be chemically inert and be mechanically stable. Owing to the often conflicting requirements of a good support, various materials have been used. The type of support can however be conveniently classified into one of three categories:

1. Hydrophilic biopolymers based on natural polysaccharides such as agarose, dextran, and cellulose.
2. Lipophilic synthetic organic polymers such as polyacrylamide, polystyrene, and nylon.
3. Inorganic materials such as controlled pore glass and iron oxide.

This work was undertaken to investigate production and immobilization of *B. polymyxa* protease using various carriers using physical adsorption.

Materials and Methods

The compounds obtained from Sigma Chemical Co., UK were bacto tryptone, bacto soytone, bacto agar, peptone, beef extract, ammonium sulfate, manganese chloride, calcium chloride, sodium carbonate, sodium bicarbonate, and diethylaminoethyl (DEAE)-cellulase. Manganese sulfate, acetic acid, tricholoro acetic acid, sodium acetate, boric acid, sodium hydroxide, casein, ammonium sulfate, hydrochloric acid, sodium chloride, ethyl alcohol, ethylenediamine-tetracetic (EDTA), tris, sodium dodecylsulfate, ammonium persulfate, glycerol, and tin sulfate were purchased from Merck Chemical Co., Germany. Acrylamide, *bis*-acrylamide, and glycine were obtained from BioRad Chemical Co., France. Chitin and chitosan were obtained from Fluka, BioChemica, Switzerland. Alginate was purchased from Riedel-de Haen, Germany. Synthetic zeolite and raw zeolite were synthesized.

Isolation, Screening, and Identification of Protease Producing Strains

Bacterial strains used in the study were isolated from industrial waste discharge samples using enrichment techniques, and screened for their ability to produce protease. Samples that were collected from different factories in Turkey were screened for microorganisms, especially for the genus *Bacillus*. A 200-µL aliquot of each sample was spread onto different nutrient agar plates containing: 5 g/L peptone; 3 g/L meat extract; 15 g/L agar (pH adjusted to 7.0) and cultured at 37°C for 24 h. After 24 h of incubation isolated colonies were transferred to new agar plates and were grown under same conditions *(8–10)*.

The Gram-stain was used to evaluate the purity and morphology of the isolates. Only Gram-positive spore forming rods were further evaluated. Specific identification of bacterial strains was done using API 50 CH test kits (Biomérieux, France). The API 50 CH test strips were inoculated and incubated according to the manufacturer's instructions. Using this method only one bacterial strain was identified as *B. polymyxa*.

Production of Enzyme

The production of enzyme was carried out by growth in 1-L Erlenmeyer flasks containing 200 mL nutrient broth (5 g peptone, 3 g meat extract, 10 mg $MnSO_4 \cdot H_2O$ for 1 L media) as liquid media. The isolates were cultivated for 20–22 h at 37°C with agitation of 200–250 rpm.

Partial Purification of Protease

After incubation, the culture broth was centrifuged at 4°C and 20,000g for 20 min. Ammonium sulfate was added slowly to the solution up to 80% saturation with gentle stirring and left for 30 min at 4°C *(11,12)*. The precipitate formed was collected by centrifugation at 25,000g for 30 min, dissolved in a minimum amount of 50 mM Tris-HCl buffer, pH 7.5, containing 5 mM $CaCl_2$, dialyzed against the same buffer overnight and then lyophilized. The precipitate that was obtained after ammonium sulfate saturation, dialysis, and centrifugation was loaded to a DEAE–cellulose ion-exchange chromatography column (2 × 60 cm^2) that had been pre-equilibrated with borax-NaOH buffer (0.01 mol/L and pH 9.3). The protein was eluted with the same buffer using a fraction size 10 mL at the rate of 1 mL/min. Fractions containing the majority of the protease activity were pooled for activity assay. The activity of protease enzyme at the end of each step was measured by spectrophotometric method *(13)*.

Protease Assay

Two methods were used to determine protease activity. With the first method activity was determined at 30°C using tubes containing 2.5-mL milk casein in water bath for 5 min. The reaction was carried out by adding 0.5 mL enzyme solution for 10 min. After 10 min, the reaction was stopped by adding 2.5 mL TCA solution, the solution was storage in water bath for 20 min sample *(14)*. With the second method activity was determined at

35°C using tubes containing 3 mL substrate solution incubated in a water bath at 35°C for 5 min. The reaction was carried out by adding 0.5 mL enzyme solution for and incubation 10 min. The reaction was stopped by the addition of 3.2 mL TCA solution, the solution was stored in water bath for 10 min. The rate of casein hydrolysis was determined using absorbance values 275 nm (12). One unit of activity was defined as the amount of enzyme required to liberate 1 µg of tyrosine in 1 min at 30°C (1 U/mL·1 µg tyrosine/min·mL). A tyrosine standard was prepared by dissolving different amounts of tyrosine in TCA solution.

Immobilization Method

B. polymyxa protease was immobilized by adsorption on alginate, chitin, chitosan, raw zeolite, and synthetic zeolite. For the immobilization, 100 mg of the carrier was incubated with 100 U B. polymyxa protease dissolved in 1 mL of 0.5 M Tris-HCl buffer (pH 7.2) at 4°C overnight under static conditions (in the case of chitosan 200 U of B. polymyxa protease was used). The unbound enzymes were removed by washing with Tris-HCl buffer (0.5 M, pH 7.2).

Results and Discussion

The specific activity and purification degrees of the partially purified protease are shown in Table 1. The column purification resulted in a 10-fold increase in specific activity compared with the culture supernatant, and a twofold increase compared with ammonium sulfate precipitation. Increases of the partial purified protease activity using ammonium precipitation (twofold increase) are consistent with published literature, which shows a purification (fold) range from 1.9 to 96 (15,16). A 10-fold increase using DEAE-chromatography column is also consistent with reported purification ranges from 7.8 to 555 (17,18).

In the present study a partially purified protease from B. polymyxa was immobilized on alginate, chitin, chitosan, raw zeolite, and synthetic zeolite by physical adsorption. The carriers were incubated with 100 U and 200 U protease dissolved in Tris-HCl buffer at +4°C for 16 h. The highest bound enzyme and immobilization yield was found with alginate. A low immobilization yield (17%) and bound enzyme (171.38 U/g solid) was detected with alginate. It was determined that protease enzyme was not adsorbed on the chitin, chitosan, synthetic zeolite, and raw zeolite (19,20).

Table 2 shows that an immobilization yield (5.8–96%) and bound enzyme (52.1–950 U/g solid) was achieved using raw zeolite, chitin, chitosan, and alginate. The enzymes physically bound to alginate showed the highest immobilization yield (96%) and the highest activity per gram carrier (950 U/g solid). The activity recovery of the immobilized protease increased with prolonged reaction times with the highest activity recovery obtained the immobilization was allowed to proceed for 24 h. Activity

Table 1
Protein Purification of Partial Purified Protease Enzyme From B. polymyxa

	Total protein (mg)	Total activity (U)	Specific activity (U/mg)	Fold
Crude homogenate	9797.6	155,467	15.9	1
40% saturation	8812.8	141,073	16	1
80% saturation	210	6369	30.3	2
DEAE-cellulose ion exchange chromatography	2.5	403.5	161.4	10

Table 2
Immobilization of B. polymyxa Protease (24 h)

Solid	Protease enzyme (U/g solid)	Bound enzyme activity (U/g solid)	Unbound enzyme activity (U/g solid)	Yield
Alginate	1000	950	10.97	96
Chitin	1000	65.36	74.2	7.5
Chitosan	2000	91.34	131.09	4.9
Raw zeolite	1000	52.1	107.52	5.8
Synthetic zeolite	1000	–	–	–

Table 3
Effect of Time on the Immobilization of B. polymyxa Protease

Solid	Protease enzyme (U/g solid)	Activity (U/g solid) 20 d after	Protected activity (%)
Alginate	950	611	64
Raw zeolite	52.1	32.8	62

recovery decreased if the reaction time was longer because of the increasing process deactivation. The immobilization yield was found as 7.5% and 5.8% for chitin and raw zeolite, respectively (21,22).

Immobilization by physical adsorption for 24 h showed considerable bound enzyme activity (good loading efficiency) and immobilization yield. This good loading efficiency for the immobilization might have been owing to the formation of stable cross linking between the carrier and the enzyme through a group (alginate).

Storage stability is a major concern in enzyme preservation. The storage stability of the B. polymyxa protease was examined for 20 d at +4°C as shown in Table 3. It was found that protease immobilized on alginate and synthetic zeolite maintained their activities 64 and 62%, respectively. After storage for 20 d, the immobilized protease still maintained about 64% of its initial activity, whereas the native enzyme only maintained about 40% of the initial activity.

Conclusions

Physical adsorption of enzymes onto solid supports is probably the simplest way of preparing immobilized enzymes. The method relies on nonspecific physical interaction between the enzyme protein and the surface of the matrix, brought about by mixing a concentrated solution of enzyme with the solid. A major advantage of adsorption as a general method of immobilization enzymes is that usually no reagents and only a minimum of activation steps are required. As a result, adsorption is cheap, easily carried out, and tends to be less disruptive to the protein than chemical means of attachment, the binding being mainly by hydrogen bonds, multiple salt linkages, and Van der Waal's forces.

In this study, a protease was partially purified from B. polymyxa using 80% $(NH_4)_2SO_4$ precipitation, dialysis and DEAE-ion exchange chromatography. Immobilization was achieved on alginate, chitin, chitosan, raw zeolite, and synthetic zeolite by noncovalent absorption. The maximum yield of absorption was 17% for 16 h and 96% for 24 h onto alginate. It was found that immobilized enzyme was stable to 20 d. Having storage stability the practical applications of enzyme is a potential advantage. The detailed study of immobilization on the same support with the different methods, such as the covalent attachment, and the usage of this immobilized protease, which has high storage stability in the batch, and continuous processes will be further investigated.

References

1. Moon, S. M. and Satish, J. (1993), *Biotech. Bioeng.* **41**, 43–54.
2. Sierecka, K. (1998), *Int. J. Biochem. Cell Biol.* **30**, 579–595.
3. Longo, A. M., Novella, I. S., Garcia, L. A., and Diaz, M. (1999), *J. Biosci. Bioeng.* **88**, 35–40.
4. Kokufuta, E. (1993), *Adv. Polym. Sci.* **110**, 157–177.
5. Liang, J. F., Li, Y. T., and Yang, C. (2000), *J. Pharm. Sci.* **89**, 979–990.
6. Cabral, M. S. (1993), In: *Thermostability of Enzymes*, Gupta M. N., ed., Springer-Verlag, Berlin, pp. 162–181.
7. Këstner, A. I. (1974), *Russian Chem. Rev.* **43**, 690–705.
8. Frost, G. M. and Moss, D. A. (1987), In: *Biotechnology*, Kennedy J., ed., New York.
9. Pazlarova, J. and Tsaplina, I. (1988), *Folia Microbial.* **33**, 267–272.
10. Jensen, D. E. (1972), *Biotechnol. Bioeng.* **14**, 647–662.
11. Bailey, J. E. and Ollis, D. F. (1987), In: *Biochemical Engineering Fundamentals*, McGraw-Hill, New York.
12. Yenigün, B. and Güvenilir, Y. (2003), *Appl. Biochem. Biotechnol.* **105-108**, 677–687.
13. Gerze, A., Omay, D., and Güvenilir, Y. (2005), *Appl. Biochem. Biotechnol.* **121(1–3)**, 335–346.
14. Orhan, E., Omay, D., and Güvenilir, Y. (2005), *Appl. Biochem. Biotechnol.* **121(1-3)**, 183–194.
15. Stoscheck, C. M. (1990), *Methods Enzymol.* **182**, 50–69.

16. McKevitt, A. S., Bajaksouzian, J. D., and Klinger, D. E. (1989), *Appl. Environ. Microbiol.* **57**, 771–778.
17. Sexton, M. M., Jones, A. L., and Chaowagul, W. (1994), *Can. J. Microbiol.* **40**, 903–910.
18. Rao, M. B., Tanksale, A. M., Ghatge, M. S., and Deshpande, V. V. (1998), *Microbiol. Mol. Biol. Rev.* **62**, 597–635.
19. Gao, J., Xu, J., Locascio, L. E., and Lee, C. S. (2001), *Anal. Chem.* **73**, 2648–2655.
20. Cooper, J. W., Chen, J., Li, Y., and Lee, C. S. (2003), *Anal. Chem.* **75**, 1067–1074.
21. Batra, R. and Gupta, M. N. (1994), *Biotech. Appl. Biochem.* **19**, 209–215.
22. Nisto, C., Emmenus, J., Gorton, L., and Ciucu, A. (1999), *Anal. Chim. Acta* **387(3)**, 309–326.

Simulation and Optimization of a Supercritical Extraction Process for Recovering Provitamin A

ELENISE BANNWART DE MORAES, MARIO EUSEBIO TORRES ALVAREZ, MARIA REGINA WOLF MACIEL,* AND RUBENS MACIEL FILHO

Separation Process Development Laboratory (LDPS), Chemical Engineering School, State University of Campinas (UNICAMP), CP 6066, Zip Code 13081-970, Campinas-SP, Brazil, E-mail: elenise@feq.unicamp.br and wolf@feq.unicamp.br

Abstract

In this work, a simulation procedure of a supercritical extraction process was developed through the use of the commercial simulator HYSYS™ (Hyprotech Ltd.), adapting the existing units to the operating conditions typical of the supercritical extraction process. The objective is to recover provitamin A (β-carotene) from palm oil (esterified) using carbon dioxide/ethanol as the supercritical mixed solvent. This example characterizes the problem for recovering high added value product from natural sources, as the palm oil, which is desired by the market. Owing to the fact that esterified palm oil is a complex mixture, made by several components, in order to characterize this system in the simulator, it was necessary to create hypothetical components using the UNIFAC (universal function-group activity coefficients model) group contribution, because they are not present in a conventional database and, then, their physical properties must be estimated and/or predicted before the simulation. The optimization was carried out in each simulation for each equipment, in terms of operating conditions (temperature and pressure), in order to obtain the maximum recovery of carotenes. According to the results, it was possible to concentrate carotenes through two cycles of supercritical extraction with high yield. Furthermore, ethyl esters (biodiesel) were also obtained, as a byproduct of the proposed process, which can also be used as an alternative fuel, with the important characteristic that it is renewable.

Index Entries: Biodiesel; carotenes; palm oil; supercritical extraction.

Introduction

The supercritical extraction process represents an alternative to conventional separation methods, especially in treating and processing natural compounds like, for example, vitamins *(1)* because of the favorable

---

*Author to whom all correspondence and reprint requests should be addressed.

properties of supercritical fluids, such as solvent recovery, simple separation, favorable thermal conditions and mass transfer properties, and solvent-free products.

CO_2 is the most commonly employed solvent, because it is relatively inert, cheap, nontoxic, recyclable and nonflammable and its critical properties (critical pressure = 7.37 MPa and critical temperature = 30.95°C) are easily reachable. The use of cosolvent, as the ethanol in the present work, is also possible, in order to improve the separation. However, the solvent choice depends on the system.

An extensive description of the properties of supercritical fluids, as well as an overview about SFE application is given in ref. 2.

Palm oil is the major edible oil in the south-east Asian region and has the second place in the world, after soy oil. The first step in palm oil processing is at the mill, in which the crude palm oil is extracted from the fruit. The oil extraction follows various steps: sterilization in large pressure vessels, stripping in rotating drum, extraction in a homogeneous oil mash, and purification in a continuous clarification tank. The oil is bright orange-red because of the high content of carotenoids (3).

Carotenoids are divided into two groups: carotenes and xanthophylls. Although carotenes are purely hydrocarbons, xanthophylls are oxygenated at the ends groups and as a result they are polar. The type and amount of pigment in vegetable oils depend fundamentally on the species, cultivars, state of ripeness, and agronomic conditions, and so on, and, in general, undergo a considerable variation during storage and preparation as edible oils (4).

Carotenoids, in particular β-carotene, are one of the major groups of natural pigments that find widespread utilization in the food industry. Addition of β-carotene to foods imparts a uniform color and some vitamin A activity. As many vegetables are sources of carotenoids, they offer the possibility of obtaining natural pigments, provided that a nontoxic extraction technology is used. β-Carotene is a hydrocarbon of 40 carbon atoms, a lipid-soluble, yellow to orange–red pigment, and highly unsaturated, as it contains 11 conjugated carbon–carbon double bonds (5). It is illustrated in Fig. 1.

In this work, an esterified palm oil was used as raw material, which means that the crude palm oil was esterified and, then, the glycerides were reduced to lighter components, as ethyl esters. Ethyl esters are also known as biodiesel, which has to be seen as an energy alternative with the advantage of being environment-friendly, biodegradable and renewable.

It is a simulation of the supercritical extraction process with CO_2/ethanol as supercritical solvent, using the HYSYS™ commercial simulator.

Methodology

It is known that the palm oil contains a high concentration of natural carotenoids, in the range of 500–3000 ppm, depending on the species of the palm fruit, which the oil is obtained.

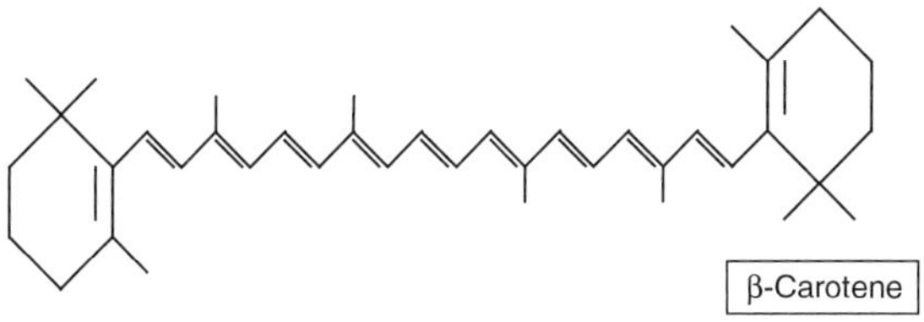

Fig. 1. Molecular structure of β-carotene.

The majority of the carotenoids in the palm oil are destroyed in the conventional refining process for producing clear oil. This represents a loss of the natural source of carotenoids. The importance of the carotenoids is well reported in the literature and some methods of extraction and recovery, from the palm oil, have been developed. This includes extraction by saponification, adsorption, and transesterification followed by molecular distillation and others. However, only transesterification processes and distillation have been fully developed in the commercial scale processes (6). Batistella et al. (7) has studied the recovery of biodiesel and carotenoids through the molecular distillation process.

Palm oil can be used as raw material for production of provitamin A (carotenes), but not in its natural form. It is necessary to carry out a transesterification reaction, which can reduce the glycerides in lighter components, as ethyl esters. So, after transesterification, the esterified palm oil is rich in ethyl esters and carotenes, whereas before, the crude palm oil was rich in triglyderides and carotenes.

Glycerides react with ethyl alcohol through a basic catalysis, producing ethyl esters. The chosen process of transesterification presents innumerable advantages in relation to other processes, as for example, the fermentative. This last one needs enzymes of difficult attainment and raised price, besides presenting high times of process. The chosen process presents high conversion rate, even for small reaction time. Reactions in reduced time, such as 15 min, present a conversion in ethyl esters of, approx 94%. The conversion of the triglycerides, a compound that has a larger molecular weight, reaches the levels of 98% in this same time, as it is presented ahead. The methodology for the preparation of the raw material is described in the following sections.

Neutralization of the Palm Oil

The crude palm oil presents, normally, 2–3% of free fatty acids (FFA). These acids represent an enormous problem in the transesterification reaction, because they deactivate the used catalyst. So, the FFA must be reduced from the palm oil to values of 0.3%. The adopted process is the controlled saponification (the neutralization of these acids) and the removal of the soaps by centrifugation. After this removal, the palm oil is washed, treated with Sorbamol™ (Süd Chemie AG) (which is an adsorbent that removes soap residues, phosphorous, and metallic particles of the oil), and it is dried by evaporation at reduced pressures. After the drying, the oil is filtered to remove the sorbamol used.

The yield was 98%, and the loss of carotenes was 3%. The neutralization of the palm oil, for this process, reduced the acidity in 0.2% of FFA, becoming possible to use the oil for the transesterification reaction, the next stage of preparation of the raw material.

Transesterification

This is the final step of preparation of the raw material and it has the purpose to transform the palm oil, which is rich in glycerides, to lighter components, as ethyl esters (biodiesel).

Because it is the last step, the oil must be without solid materials, volatiles and soap (elements present in the preparation processes), the conversion to ethyl esters must be high and the carotenoids loss must be minimum.

After neutralization, the oil was transesterified with ethanol (with 10% excess of ethanol solution), catalyzed by 0.4% (w/w) sodium ethoxide, at 60°C and 0.3 h, obtaining ethyl esters and glycerol. The ethyl esters were, then, separated from glycerol by decantation, washed with distillated water up to pH 6.0–8.0 and dried under reduced pressure by evaporation. The conversion was above 94% in ethyl esters. The obtained oil is denominated esterified palm oil.

Components Creation and Methodology for Simulation

Esterified palm oil is a complex mixture made by several complex components, and it is very hard to find their physical properties in a conventional simulator database. First, it was necessary to create in the simulator hypothetical components, which take into account the structure of the molecules, in order to be possible to count the groups and the number of groups present in each molecule according to the methodology presented in the software, using the UNIFAC group contribution (a tool of the commercial simulator HYSYS). So, they are automatically introduced in the simulator database and the physical properties can be predicted. After creating, the components can be used to simulate the process in the simulator HYSYS. Furthermore, it was necessary to adapt the existing units in the simulator to simulate the supercritical extraction process. In the HYSYS database, the thermodynamic package that was chosen was the EOS-Peng-Robinson to represent the vapor phase and an activity coefficient model (UNIQUAC [universal quasi-chemical model]) to represent the liquid phase. For this study, the components used are described *(8)*, in Table 1.

The unknown binary coefficients, a_{ij}, which are the nontemperature-dependent energy parameter between components i and j, of the UNIQUAC

Table 1
Esterified Palm Oil Composition Used
in the Simulation

Components	Mass (%)
Ethyl palmitate	42.3
Ethyl oleate	37.6
Ethyl stearate	3.76
Ethyl linoleate	10.34
β-carotene	0.06
α-tocopherol	0.01
Tripalmitin	0.69
Triolein	0.69
Dipalmitin	0.94
Diolein	0.93
Monopalmitin	1.34
Monoolein	1.34
Total	100

activity coefficient model were estimated through UNIFAC vapor-liquid equilibrium. The automatic UNIFAC generation of energy parameters in HYSYS is a very useful tool and it is available for all activity models. By default, HYSYS regresses only the a_{ij} parameter, whereas the b_{ij}, the temperature dependent energy parameter between components i and j, is set equal to 0.

To make the extraction, an absorber column was used as unit operation present in the simulator, called here as "extractor." The solvent was the supercritical CO_2/ethanol mixture and the feed was made by the components presented in Table 1 (esterified palm oil). The design and operating variables are described in Section 3.

Results and Discussions

A diagram for the supercritical extraction process was proposed for provitamin A recovering from the esterified palm oil, as it can be seen in Fig. 2. At this point, it is important to remember that ethyl esters (biodiesel) are also one of the components to be considered in the simulation. The mass flow rates of the main streams are presented in Table 2.

The first extractor has 10 stages and operates at pressure (P) = 30 MPa. The temperature of the feed is 100°C and the mass flow rate is 100 kg/h. The temperature of the solvent (CO_2/ethanol) is 100°C and the mass flow rate is 800 kg/h in the proportion 80/20 (640/160 kg/h).

Analyzing the first extractor, as can be seen in Table 2, the raffinate stream (bottom extractor) is composed by a phase that is rich in beta-carotene (0.06 kg/h), tocopherol (0.01 kg/h), ethyl esters (47.72 kg/h), mono- (2.52 kg/h), di- (1.87 kg/h), and triglycerides (1.38 kg/h). In the

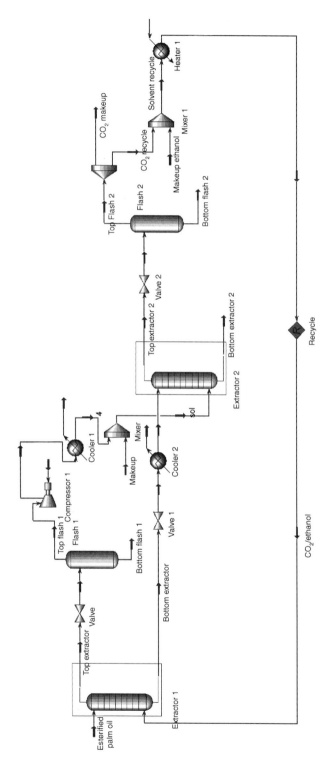

Fig. 2. Proposed process flow diagram for recovering provitamin A from esterified palm oil.

Table 2
Simulation Results of the Supercritical Extraction Process for Provitamin A Recovering From the Esterified Palm Oil

Components	Feed flow rate (kg/h)	Top extractor Mass flow rate (kg/h)	Bottom extractor Mass flow rate (kg/h)	Bottom flash 1 Mass flow rate (kg/h)	Bottom extractor 2 Mass flow rate (kg/h)	Bottom flash 2 Mass flow rate (kg/h)
Ethyl palmitate	42.3	25.89	16.41	25.88	0	16.41
Ethyl oleate	37.6	14.57	23.03	14.57	0	23.03
Ethyl stearate	3.76	1.26	2.5	1.26	0	2.49
Ethyl linoleate	10.34	4.56	5.78	4.56	0	5.78
β-carotene	0.06	0	0.06	0	0.06	0
α-tocopherol	0.01	0	0.01	0	0.06	0.01
Tripalmitin	0.69	0	0.69	0	0.69	0
Triolein	0.69	0	0.69	0	0.69	0
Dipalmitin	0.94	0	0.94	0	0.94	0
Diolein	0.93	0	0.93	0	0.93	0
Monopalmitin	1.34	0.11	1.23	0.11	0	1.23
Monoolein	1.34	0.05	1.29	0.05	0	1.29
Total (streams)	100	46.44	53.56	46.43	3.31	50.24
CO_2 (solvent)	640	635.58	4.42	6.46	0	5.66
Ethanol (solvent)	160	128.69	31.31	99.35	0.24	51.54
Total (solvent)	800	764.27	35.73	105.81	0.24	57.2
Total (streams + solvent)				152.24	3.55	107.44

extract stream (top extractor), there is a phase that is rich in the supercritical solvent CO_2/ethanol (635.58/128.69 kg/h), ethyl esters (46.28 kg/h) and monoglycerides (0.16 kg/h). As the carotenes are in the rafinatte stream of the first extractor (bottom extractor) with a great quantity of ethyl esters, it will be necessary another extractor in order to separate these components. So, this stream passes through a valve to reduce the pressure to 20 MPa and, later, it passes through Cooler 2 to reduce the temperature to 100°C and it will be used as the feed of extractor 2.

On the other hand, the top stream of extractor 1 (top extractor) passes through a valve to reduce the pressure to 6.5 MPa and goes to flash 1. At the top of the flash (top flash 1), there is the solvent CO_2/ethanol (629.13/29.34 kg/h) and at the bottom (bottom flash 1), there are ethyl esters (46.27 kg/h), monoglycerides (0.16 kg/h) and a bit of solvent (6.46/99.35 kg/h). This stream passes, then, through a compressor (compressor 1) to reach the pressure equal to 20 MPa, after through cooler 1 to reduce the temperature to 100°C and, later, goes to mixer for its future use in extractor 2. In this mixer, a make-up of solvent CO_2/ethanol (22.5/2.5 kg/h) is carried out.

The rafinatte stream of the first extractor (bottom extractor) feeds the extractor 2 and the stream of the top of the flash 1 (top flash 1) is used to offer the necessary quantity of solvent (651.63/31.84 kg/h) for carrying out the second extraction.

Extractor 2 has also 10 stages and operates at 20 MPa. At the top of extractor 2 (top extractor 2), there are ethyl esters (47.72 kg/h), tocopherol (0.01 kg/h), monoglycerides (2.52 kg/h), and the solvent CO_2/ethanol (656.04/62.91 kg/h), whereas that at the bottom of extractor 2 (bottom extractor 2), there are β-carotenes (0.06 kg/h), di- (1.87 kg/h) and triglycerides (1.38 kg/h). This stream (bottom extractor 2) is important because beta-carotenes can be concentrated.

The top stream of extractor 2 (top extractor 2) passes through a valve to reduce the pressure to 6.5 MPa and goes to flash 2. At the top of the flash (top flash 2), there is the solvent CO_2/ethanol (650.38/11.37 kg/h) and at the bottom there are ethyl esters (47.72 kg/h), tocopherols (0.01 kg/h), monoglycerides (2.52 kg/h), and a bit of solvent (5.66/51.54 kg/h). The stream top flash 2 is divided: one part goes to solvent recycle (640.06/11.19 kg/h) and the other one (10.32/0.18 kg/h) to a possible use in the make-up of solvent. This divisor must be used in order to reuse the solvent. This stream must be identical to the initial solvent stream, that is 640 kg/h of CO_2/160 kg/h of ethanol. So, this procedure avoids any problem with the recycle closing. As the part of the solvent recycle contains only a bit of ethanol, it is necessary to introduce a make-up of ethanol (148.9 kg/h), using a mixer (mixer 1). Then, it is already possible to have the initial proportion of 80/20 of CO_2/ethanol. Moreover, this stream (solvent recycle) passes through a heater to increase the temperature to 100°C and the pressure must be 30 MPa. Thus, the solvent stream is ready to be reused in extractor 1, closing the recycle, in order to reinitialize the process.

As can be verified in Table 2, in bottom extractor 2 stream, it was possible to concentrate β-carotenes. There are also di- and triglycerides, which come from the nonconversion fraction in the transesterification process, that remain with the carotenes, what is already expected because of similar characteristics among these components. The recovery and the purity in relation to the composition of the feed and the total mass flow of this stream can be calculated. As a result, carotenes can be recovered from esterified palm oil until 16,900 ppm that is, about 28 times in relation to the feed (600 ppm of carotenes). The recovery was about 100.00%.

The optimization was carried out in each simulation for each equipment, in terms of operational conditions (temperature and pressure), in order to obtain the maximum recovery of carotenes.

Because other interesting product was obtained, the ethyl esters or biodiesel, as a byproduct, it will be calculated the recovery and the purity in relation to the composition of ethyl esters in the feed (94 kg/h) and the total mass flow (259.68 kg/h) of the two streams, bottom flash 1 and bottom flash 2, that are rich in ethyl esters (93.98 kg/h). The recovery was of 99.98% and the purity of 36.19%, because of the presence of tocopherols (0.01 kg/h), monoglycerides (2.68 kg/h) and, also, solvent (12.12/150.89 kg/h). According to these results, it was possible to concentrate carotenes through two cycles of supercritical extraction. Biodiesel was also obtained, as a byproduct of the proposed process, which can be also used as an alternative fuel, with the important characteristic that it is renewable.

β-Carotene is a high added value product that may be obtained under specific operating conditions to have its properties preserved. This is especially true for temperature, so that it is imperative the use of process development taking into account such constraint. Supercritical extraction is a process, which fits well in such situation as shown for β-carotene in this work and for tocopherol and phytosterol recovery from deodorizer distillate of soy oil (9). A question commonly concerned to supercritical extraction process is that related to its cost. However, this is not a restriction for recovering valuable products with clean technology. In fact, for the case of vitamin E a rough product cost estimation leads to a value about half of the market price, and taking this into consideration as a basis, the value for the β-carotene is to be compatible and, in fact, very attractive for this case. It is worthwhile mentioning that the required quality is achieved without any addition of toxic chemical which is very much welcome. An interesting feature to be pointed out is the fact that in the developed process to obtain β-carotene is the production of the relatively valuable byproduct, to know, the biodiesel, which, in fact, leads to an additional gain in the process, as well as a significant decrease in environmental and energetic impacts, because a drastic reduction in the downstream treatment is achieved.

Conclusions

The results obtained in this work indicate that it is possible to concentrate carotenes approx up to 17,000 ppm, that is, about 28 times higher than the feed (600 ppm of carotenes). The recovery was about 100%. To accomplish this, two supercritical extractors in series were necessary what demonstrates the rigorous methodology used and its practical viability. Furthermore, the operating conditions were the typical ones encountered in supercritical process and the use of ethanol as cosolvent is very attractive, because it is a renewable source material. Moreover, with this study, it was observed that it was also possible to recover ethyl esters (biodiesel), with a recovery of 99.98% and a purity of 36.19%, considering the two streams that are rich in ethyl esters, what makes this process highly important, because the byproduct is a valuable material. This is a theoretical result, but extremely important in order to guide experimental works, both in terms of the rigorous calculations used and of the optimization conditions achieved for the design and operating variables.

Acknowledgments

The authors are grateful to FAPESP (Fundação de Amparo à Pesquisa do Estado de São Paulo) for the financial support of this project (99/04656-9) and CNPq (Conselho Nacional de Desenvolvimento Científico e Tecnológico).

References

1. Gast, K., Jungfer, M., Saure, C., and Brunner, G. (2005), *J. Supercrit. Fluids* **34**, 17–25.
2. Brunner, G. (1994), *Gas Extraction*, Steinkopff Verlag, Darmstadt.
3. Malaysian Palm Oil Promotion Council <http://www.mpopc.org.my> (accessed on: 22 April 2005).
4. Cert, A., Moreda, W., and Pérez-Camino, M. C. (2000), *J. Chromatogr. A.* **881**, 131–148.
5. Subra, P., Castellani, S., Ksibi, H., and Garrabos, Y. (1997), *Fluid Phase Equilib.* **131**, 269–286.
6. Ooi, C. K., Choo, Y. M., Yap, S. C., Barison, Y., and Ong, A. S. H. (1994), *J. AM. Oil Chem. Soc.* **71**, 423–426.
7. Batistella, C. B., Moraes, E. B., Maciel Filho, R., and Maciel, M. R. W. (2002), *Appl. Biochem. Biotechnol.* **98**, 1149–1159.
8. Batistella, C. B. and Maciel, M. R. W. (1998), *Comput. Chem. Eng.* **22(Suppl)**, S53–S60.
9. Mendes, M. F., Pessoa, F. L. P., Coelho, G. V., and Uller, A. M. C. (2005), *J. Supercrit. Fluids* **34**, 157–162.

Affinity Foam Fractionation of *Trichoderma* Cellulase

QIN ZHANG, CHI-MING LO, AND LU-KWANG JU*

Department of Chemical Engineering, The University of Akron, Akron, OH 44325-3906; E-mail: LukeJu@UAkron.edu

Abstract

Cellulase could not be selectively collected from fermentation broth by simple foam fractionation, because of the presence of other more surface-active compounds. A new approach of affinity foam fractionation was investigated for improvement. A hardwood hydrolysate (containing cellulose oligomers, substrates to cellulase) and two substrate analogs, i.e., carboxymethyl cellulose (CMC) and xylan hydrolysate, were added before the foaming process. The substrates and substrate analogs were indeed found to bind the cellulase selectively and form more hydrophobic complexes that partition more readily onto bubble surfaces. In this study, the effects of the type and concentration of substrate/analog as well as the presence of cells at different growth stages were examined. The foam fractionation properties evaluated included foaming speed, foam stability, foamate volume, and enrichment of filter paper unit (FPU) and individual cellulase components (i.e., endoglucanases, exoglucanases, and β-glucosidases). Depending on the broth and substrate/analog employed, the foamate FPU could be more than fourfold higher than the starting broth FPU. Addition of substrate/analog also deterred the enrichment of other extracellular proteins, resulting in the desired cellulase purification in the foamate. The value of E/P (enzyme activity-FPU/g/L of proteins) in the foamate reached as high as 18, from a lactose-based fermentation broth with original E/P of 5.6. Among cellulase components, exoglucanases were enriched the most and β-glucosidases the least. The study with CMC of different molecular weights (MW) and degrees of substitution (DS) indicated that the CMC with low DS and high MW performed better in cellulase foam fractionation.

Index Entries: Affinity foam fractionation; carboxymethyl cellulose; cellulose; cellulose hydrolysate; xylan hydrolysate.

Introduction

Foam fractionation is a simple, inexpensive, environment-friendly process for protein concentration, and purification. Proteins, which have both hydrophobic and hydrophilic moieties, are surface-active and easily collectable from aqueous solutions by foam (1,2). In addition, proteins of different surface hydrophobicity are expected to have different tendencies

---

*Author to whom all correspondence and reprint requests should be addressed.

for adsorption on bubble surface, which provides the opportunity for simple protein separation or fractionation. The simple foam fractionation has a significant potential of reducing the cost of protein recovery in the pharmaceutical and food industries (3). Simple foam fractionation, however, cannot be applied to the products that do not have the highest efficiency in partitioning onto bubble/foam surfaces, among all of the materials present in the product-bearing broth. Cellulase separation from its fermentation broth belongs to this category.

Cellulase is a group of enzymes that hydrolyzes cellulose to glucose. Economic, effective cellulase production is critical to utilization of the abundant cellulosic materials as renewable feedstock. Cellulase is typically considered to contain three component groups (4). Endoglucanases attack insoluble cellulose in the middle of the chain and produce oligosaccharides with more than six glucose units. Exoglucanases nick the end of oligosaccharides to generate cellobiose (dimer). β-Glucosidases cleave cellobiose to two glucose molecules.

Foaming occurred in the cellulase fermentation, particularly during the stationary phase. Because the increase in foaming intensity appeared to parallel the profile of cellulase production, cellulase had been assumed to be the cause of foaming. More recent studies, however, provided evidence against the above assumption (5). In particular, when applying foam fractionation to separate cellulase from the fermentation broth, Zhang et al. (6) found that cellulase was surface-active but not the strongest in the broth, and, therefore, could not be selectively foamed out. None of the individual cellulase components, i.e., endoglucanases, exoglucanases, and β-glucosidases, showed higher activities in the collected foamate than in the original broth (6).

Enzymes have selective binding affinity to their substrates, substrate analogs, and compounds containing moieties of the substrates or analogs. The affinity binding of these agents with the active sites of enzymes also helps to protect the enzymes from being deactivated during the foaming process. Protein deactivation in foaming resulted mainly from surface denaturation, while other potential factors such as shear stress and oxidation did not contribute significantly (7,8). To enhance foam fractionation of cellulase from the fermentation broth, the effects of adding the substrate (hardwood hydrolysate) and different substrate analogs, including various types of carboxymethyl cellulose (CMC) and a xylan hydrolysate (XH), were examined in this study. Cellulase was reported to hydrolyze both CMC and xylan (9–14), indicating the existence of certain binding affinity of cellulase to these substrate analogs.

Materials and Methods

Fermentation

Trichoderma reesei Rut C-30 (NRRL 11460) was obtained from US Department of Agriculture (Agricultural Research Service Patent Culture

Collection, Peoria, Illinois). The culture was maintained at 4°C on slants of Potato Dextrose Agar (Sigma; 39 g/L, as recommended), with regular subculturing every 3–4 wk. The inoculum for fermentation, which provided the broth for foam fractionation study, was prepared by transferring three loops of cells from an agar slant to a 250-mL shake flask containing 50 mL of Potato Dextrose medium (Sigma, St. Louis, MO). After 2 d of cultivation at room temperature, the broth was added to a 2-L flask containing 500 mL of a defined medium modified from that used by Mandels and Weber (15).

For cellulase production, the defined medium required not only C-substrates, as the source of material and energy for cell growth and maintenance, but also inducers, to activate the expression of all cellulase components. In this study, three medium systems were compared for their cellulase synthesis capacity and, more important to this study, for their foaming properties. The first medium included 5 g/L of glucose as C-substrate and 5 g/L of pure cellulose (Avicel, Sigma) as inducer and C-source (on hydrolysis by cellulase produced by cells). The second medium contained a hardwood hydrolysate, with 12–16 g/L of reducing sugars, preparation described elsewhere (16), as both C-substrate and inducer. The above two media were used in batch fermentation. The broths used in the foaming study were generally harvested on the fifth day when the cellulase activity (assayed in FPU) reached the highest level. The third medium was lactose based, i.e., with lactose serving as both C-substrate and inducer. The fermentation with this medium was conducted in a batch-then-continuous mode. The original medium had 10 g/L of lactose; the feed for continuous culture had 20 g/L of lactose. The culture was grown to late exponential-growth phase and then converted to the continuous culture, with the feed rate computer controlled according to a pH-based algorithm, details described elsewhere (17). The broth used in the foaming study was collected after about 1 wk into the continuous culture.

To avoid foaming, which would otherwise necessitate antifoam addition, surface aeration at the rate of 1 vvm was employed for oxygen supply into the broth that was magnetically stirred at the rate of 250 rpm. The fermentation was kept at room temperature (24 ± 1°C). Samples were taken daily. Cell-free media were used in most of the foaming experiments. In this case, the harvested fermentation broth was centrifuged at 8000 rpm (9300g) (Sorvall RC 5C Plus Superspeed Centrifuge, Sorvall, Newtown, CT) for 10 min to remove the biomass. The supernatant was collected for the subsequent affinity foaming study.

Affinity Foam Fractionation Study

The experiments were conducted in 250-mL volumetric cylinders. The volume of each liquid sample used was 40 mL. An air diffuser, placed at the bottom of the cylinder, was used to generate fine bubbles for the foaming study. The total volume before foaming, including both the liquid sample and the air stone, was 50 mL. The air bubbling rate was kept at 1 vvm (i.e., 40 mL/min) using a flow meter with a three-way valve. On bubbling, the foam rose and typically reached the maximal level within 5 min. The total foam volume was recorded. The bubbling was then stopped to allow the foam to collapse. The collapsing rate was also recorded as an indicator of the foam stability. The air bubbling was then resumed. While maintaining the foam at the highest level, the liquid broth remaining at the bottom (residue) was collected and its volume (V_r) measured. The foam in the foaming cylinder was next collapsed (if necessary, by blowing an air stream on the foam surface), and the cylinder wall and the air stone were rinsed with a known volume of deionized water (V_w). The diluted foamate was collected for analysis. The "actual" foamate volume (V_f) was obtained by subtracting V_r from the initial sample volume (40 mL). The dilution factor, $(1 + V_w/V_f)$, was used to adjust all the analysis concentrations obtained with the diluted foamate.

Analytical Methods

Reducing Sugar Concentration

The reducing sugar concentration was measured by the nonspecific dinitrosalicylic acid (DNS) method, based on the color formation of DNS reagent when heated in the presence of reducing sugars (18). The DNS reagent was prepared by dissolving 10 g of 3,5-dinitrosalicylic acid in 400 mL distilled water, adding 200 mL of 2 M NaOH, and then diluting the solution to a total volume of 1 L with distilled water.

Cell Concentration

Because of the presence of cellulose in some systems, cell dry-weight concentration could not always be measured directly. Intracellular protein concentration was measured instead (19), as described later. By centrifugation, the solids in broth samples were collected and washed twice with distilled water. The cells were then lysed in 3 mL of 0.2 N NaOH, at 100°C for 20 min. The protein concentration of the lysate was then measured by the standard Lowry method. The absorbance at 595 nm was measured with a UV/VIS spectrophotometer (Perkin-Elmer Lambda 3B). To establish the relationship between the intracellular protein concentration and cell dry-weight concentration, batch fermentation was made with glucose as the sole carbon source. Samples taken at different stages of the fermentation were analyzed for both cell dry-weight concentration and intracellular protein concentration. The relationship was established as:

Cell dry-weight concentration (g/L)
$$= \text{intracellular protein concentration (g/L)} \times 8 \ (\pm 0.5)$$

Cellulase Activity

The total activity of cellulase was measured by the standard filter paper assay method (15). Assays for the activity of individual enzyme components, i.e., endoglucanases, exoglucanases, and β-glucosidases are briefly described here.

Endoglucanases

A modified method of Berghem and Petterson *(20,21)* was used. A 1% CMC solution was prepared in 0.05 M sodium acetate buffer (pH 5.0). The CMC solution was incubated with 0.28 mL of the test enzyme solution at 50°C for 30 min. Almost 3 mL of 1% DNS reagent were added to terminate the reaction. The reducing sugar concentration produced from the enzymatic reaction was then measured and used to calculate the endoglucanase activity according to the following equation *(19,20)*.

Endoglucanase activity (U/mL) = reducing sugars released (mg) × 0.66

Exoglucanases

A modified method of Berghem and Petterson *(20,21)* was used. Almost 1 mL of the test enzyme solution was added to 1 mL of 2% Avicel suspension prepared in 0.05 M sodium acetate buffer (pH 5.0). After 30 min incubation at 40°C, 3 mL of 1% DNS reagent was added to end the reaction and the resultant reducing sugar concentration was measured. The exoglucanase activity was calculated according to the following equation *(22)*.

Exoglucanase activity (U/mL) = reducing sugars released (mg) × 0.185

β-Glucosidases

Three test tubes were used *(23)*. The test tube for cellobiose blank contained 1 mL each of 15 mM cellobiose solution, citrate buffer (pH 4.8), and water. A second test tube, for the sample blank, contained 1 mL sample and 2 mL water. The third tube, for the test sample, contained 1 mL each of the cellobiose solution, buffer, and the test sample. The test tubes were mixed, capped tightly, and incubated at 50°C for 30 min. Again, 3 mL of the DNS reagent were added and the resultant reducing sugar (glucose) concentration was measured by the DNS method. The absorbance of the sample, subtracted by those of the sample blank and the cellobiose blank, was used in determining the reducing sugar concentration. The β-glucosidase activity was determined according to the following equation *(22)*.

β-Glucosidase activity (U/mL) = glucose released (mg) × 0.0926

Results and Discussion

The presentation below is organized in three sections according to the different foaming agents added, i.e., cellulose (hardwood) hydrolysate (CH), CMC, and XH. The cross comparison between these foaming agents is included in the proper sections.

Table 1
Effects of Addition of Cellulose Hydrolysate and Cells of Different Culture Stages on Foaming Behaviors

Cells	Hydrolysate (%)	Foaming phase		Holding phase	
		Volume (mL)	Time (s)^a	Volume (mL)	Time (s)^b
None	0	250	153	250	393
	5	250	246	250	390
Growth phase	0	190	360	190	246
	5	250	311	250	310
Stationary phase	0	230	335	210	282
	5	180	351	160	351

^{a}Time required in the foaming phase, to reach the maximum foam volume reported in the previous column.
^{b}Time observed in the holding phase (with air bubbling being turned off), for the foam volume to change from the maximum volume to the volume reported in the previous column. In the first four systems, there were no foam volume changes observed during the time periods reported.

Affinity Foam Fractionation With Addition of Cellulose Hydrolysate

Many foaming experiments were carried out. Only some of them are reported here to present the general, reproducible phenomena observed and the effects of important factors identified. The results from an experiment displaying the typical effects of three factors on the foaming behaviors are summarized in Tables 1 and 2. The factors were (1) the presence or absence of cells in the foaming broth; (2) the different growth stages of the cells present; and (3) the hydrolysate addition. The cells used in cell-containing systems were pre-grown in a glucose-based medium (with 10 g/L glucose). For studying the effects of different growth stages, the cells were harvested either at the late exponential growth phase (the third day of batch cultivation) or at the stationary phase (the fifth day). The broth supernatant used as the basal cellulase-bearing medium in all of the systems was prepared by a fermentation using the hydrolysate-based medium. The broth was harvested on the fifth day, and centrifuged to remove the cells. The use of the same basal broth supernatant helped to ensure that different systems in the study differed only in the added cells and/or CH. The cell-containing systems were added with the same cell concentration, approx 3 g/L. The CH-containing systems were added with 5% CH shortly before the foaming study. The hydrolysate added was prepared to have the same medium composition (C-source omitted) so that the hydrolysate addition had minimal effects on other broth properties.

The observations on foaming speed and foam stability are summarized in Table 1. The hydrolysate-based broth foamed readily and the foam was quite stable, significantly more so than the lactose-based broth, as described in more detail later. The presence of cells slowed down the

Table 2
Effects of Addition of Cellulose Hydrolysate and Cells of Different Culture Stages on Foam Fractionation

Cells	Hydrolysate (%)	Cells (g/L)			Reducing sugars (g/L)			FPU			Extracellular proteins (g/L)		
		Broth	Foamate	ER	Broth	Foamate	ER	Broth	Foamate	ER	Broth	Foamate	ER
None	0	0	0	—	1.2	1.1	0.9	0.37	0.46	1.2	0.16	0.21	1.3
	5	0	0	—	1.6	1.2	0.8	0.34	0.7	2.1	0.14	0.1	0.7
Growth phase	0	2.3	1.9	0.8	1.2	0.9	0.7	0.37	0.45	1.2	0.14	0.1	0.8
	5	2	2.4	1.2	1.5	1.1	0.7	0.34	0.62	1.8	0.13	0.07	0.6
Stationary phase	0	2.2	2.1	1.0	1.1	0.8	0.8	0.37	0.42	1.1	0.15	0.16	1.1
	5	2.1	2.7	1.3	1.5	1.1	0.7	0.34	0.56	1.6	0.14	0.08	0.6

foaming speed and made the foam less stable. The addition of CH also slowed down the foaming speed in the cell-free systems but had minimal effect in the cell-containing systems.

The enrichment ratios (ER) for cells, reducing sugars, cellulase (FPU) activity, and extracellular proteins achieved are reported in Table 2. The ER for cells is defined as the ratio of cell concentration in the foamate to that in the original broth. ER for other parameters is similarly defined. The results in Table 2 indicated the following:

1. CH addition enriched cellulase in all systems. The ER was larger in the cell-free systems.
2. CH addition significantly decreased the partition of extracellular proteins that had no cellulase activity, resulting in much lower ER of proteins. Together with the increased enrichment of cellulase, the observation indicated a clear selectivity of CH toward cellulase. The complex formed between cellulase and the pertinent CH components (presumably the cellulose oligomers) out-competed the other proteins in partitioning onto the bubble/foam surface, consequently, causing the decrease in ER of proteins.
3. CH addition seemed to decrease the removal of reducing sugars, although the effect was significant only in the cell-free systems. ER of reducing sugars was smaller than 1 in all of the systems.
4. The presence of cells did not affect ER of FPU, proteins, and reducing sugars substantially, but the cells were removed by foaming. CH addition increased the extent of cell removal. Cells harvested at the two different growth stages behaved similar in the broth foaming.

The earlier observations were qualitatively reproducible in all of the subsequent foaming experiments conducted with the hardwood hydrolysate as the affinity foaming agent. Note that the hydrolysate had approx 12 g/L of reducing sugars. Thus, the 5% addition used in the above experiment corresponded to addition of approx 0.6 g/L of reducing sugars. The predominant majority of the reducing sugars in the hydrolysate were glucose (40%) and xylose (27%). Assuming that only the oligomers had the high affinity in binding cellulase and forming more hydrophobic complex for enhanced partition onto foam surface, the amount of "actual" foaming agents introduced was very low. On developing methods to produce CH with larger fractions of oligomers, the efficiency of affinity foam fractionation of cellulase may be greatly improved.

Further experiments were conducted in cell-free systems to evaluate the effects of increasing CH fractions, up to 75%, on the foam fractionation. The experiments were done with three combinations of the broth supernatant (from hydrolysate-based or glucose plus Avicel cellulose-based fermentation) and the type of CH as foaming agent (with or without autoclaving at 121°C for 15 min): System 1, nonautoclaved CH added to hydrolysate-based broth supernatant; System 2, autoclaved CH added

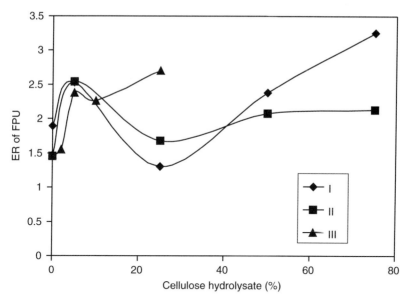

Fig. 1. ER of FPU at different percentages of CH for three cell-free systems: System I, hydrolysate-based broth + nonautoclaved CH; System II, hydrolysate-based broth + autoclaved CH; System III, glucose plus cellulose-based broth + autoclaved CH.

to hydrolysate-based broth supernatant; and System 3, autoclaved CH added to glucose plus cellulose-based broth supernatant. The ER of FPU achieved in the three systems, at different CH fractions, are summarized in Fig. 1. CH addition was found beneficial in all of the three combinations, giving maximal ER of 2.6–3.4. The cause for the dips at 25% CH in Systems 1 and 2 were unknown, but the trend was reproducible in both systems. Autoclaved and nonautoclaved CH behaved similar except at the very high fraction (75%).

Affinity foam fractionation by CH addition improved the purity, in addition to concentration/enrichment, of the cellulase in foamate. This is shown in Fig. 2, by the substantially higher values of E/P in foamate (enzyme-to-proteins, calculated by dividing FPU by the concentration of extracellular proteins) with increasing CH fractions for the three systems.

As described earlier, cellulase includes three groups of components: endoglucanases, exoglucanases, and β-glucosidases. It is important to evaluate the effect of CH addition on foam fractionation of individual groups of cellulase. The ER for cellulase components at different CH fractions is shown in Fig. 3 for System 2. (The profiles were essentially the same for System 1, but not measured for System 3.) Exoglucanases were the primary component enriched. Enrichment of the other two components, particularly endoglucanases, was also observed at the low CH fraction of 5%. The enrichment diminished at higher CH fractions, and for β-glucosidases, it even dropped below the level attained without CH addition. The poor enrichment in endoglucanases and/or β-glucosidases must

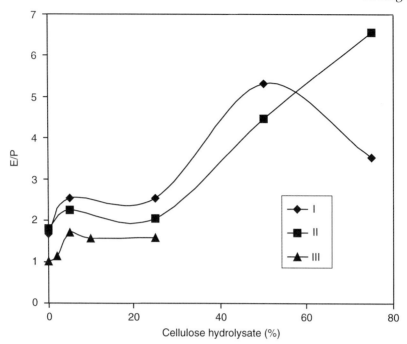

Fig. 2. Values of E/P at different percentages of cellulose hydrolysate for three cell-free systems (as described in Fig. 1).

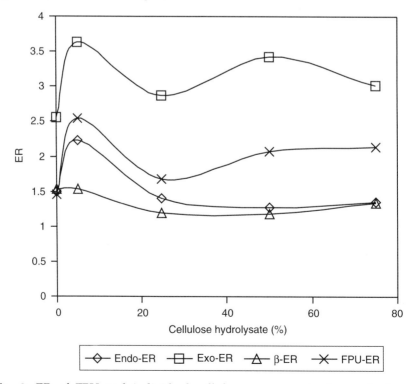

Fig. 3. ER of FPU and individual cellulase components, i.e., endoglucanases, exoglucanases, and β-glucosidases, at different percentages of cellulose hydrolysate for System II (as described in Fig. 1).

be responsible for the lower ER (2–2.5) of overall FPU than those (3–3.6) of exoglucanases.

The above observations are very important to the future improvement of affinity foam fractionation of cellulase. The different effects on cellulase components were probably associated with the different sizes of oligomers preferred by the different components as substrates. With the primary function of hydrolyzing cellobiose (dimer), β-glucosidases are expected to have higher affinity to smaller oligomers, which are very water soluble and tend not to partition onto the foam surface. To enhance the enrichment of β-glucosidases would require attaching the small oligomers to another hydrophobic entity so that the bound complexes would partition actively to the foam surface. On the other hand, endoglucanases function to cleave long cellulose chains. Presumably, they would have higher affinity to larger oligomers (more so than the exoglucanases, which can bind to shorter chains for their function of cleaving the chains at the end). The poor enrichment of endoglucanases observed in this work might be a result of the extremely low concentration of large oligomers present in CH, which was originally prepared to contain primarily glucose. Methods designed to obtain longer oligomers in the hydrolysate are desirable for optimizing the efficiency of the affinity foaming technology.

Affinity Foam Fractionation With Addition of CMC

CMC are modified, water-soluble, long-chain cellulose analogs. The potential use of CMC for affinity foam fractionation of cellulase was therefore examined. The experiment was carried out in cell-free supernatant of the broth collected from a hydrolysate-based fermentation. To obtain higher FPU (~0.7) than that from the earlier batch cultivation (~0.3–0.4 FPU), the fermentation was supplemented with a lactose-based continuous feed after reaching the stationary phase. Three systems were compared, the broth supernatant (control), the supernatant added with 5% (v/v) of a 5 g/L CMC solution, and the supernatant added with 5% of a hardwood CH (having 16 g/L of reducing sugars). The ER of FPU, extracellular proteins, and reducing sugars are shown in Fig. 4. The CMC solution was found to perform as well as, if not better than, CH in cellulase enrichment.

CMC is available commercially in several molecular weights (MW) and degrees of substitution (DS, in introduction of the carboxylic acid group). The CMC used in the above experiment belonged to the type "7L": "7" stands for a 70% DS (i.e., on average, 70% of the glucose units have an acid group attached), and "L" stands for low MW (~90,000). To study the effects of DS and MW on the affinity foaming performance, an experiment was conducted with five systems, each added with 5% (v/v) of a specific type of CMC (10 g/L solution), 7L, 7M, 7H, 9M8, and 12M8, where M and H refer to medium and high MW (~250,000 and 700,000, respectively), and

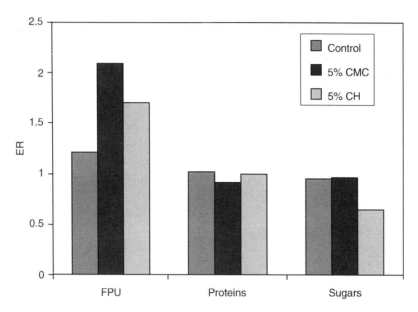

Fig. 4. Comparison of the effects of CMC and CH addition on foam fractionation, in terms of ER of FPU, extracellular proteins and reducing sugars. Control had no addition of CMC or CH.

9M8 and 12M8 refer to approx 90% and 120% DS (and the "8" indicates that the viscosity of a 2% solution is ~800 centipoises). The ER of FPU obtained is shown in Fig. 5. CMC with the lower DS (at 70%) performed significantly better than those of higher DS, presumably because the higher DS decreased the affinity between cellulase and the modified sugar chains. Increasing MW also had a positive effect on the cellulase enrichment.

Affinity Foam Fractionation With Addition of Xylan Hydrolysate

The results of an experiment comparing the effects of xylan hydrolysate (XH) and cellulose (hardwood) hydrolysate (CH) in affinity foam fractionation of cellulase are given in Table 3. The cell-free broth supernatant was collected from the lactose-based fermentation, as described in Materials and Methods. The lactose-based broth, despite its much higher FPU (~0.9), turned out to be not very foaming. The poor foaming correlated with its much lower concentration of extracellular proteins, confirming our earlier observation that cellulase did not cause active foaming compared to certain other proteins present in the broth, and the selective separation of cellulase by foam fractionation requires the affinity foaming developed in this work.

XH had a stronger foaming ability than CH, as indicated by the substantially larger foam volumes obtained with XH than with CH in Table 3. The two hydrolysates performed similar in enrichment of reducing sugars. Compared with CH, XH had slightly lower ER for both FPU and extracellular proteins. The FPU enrichment in the lactose-based broth supernatant

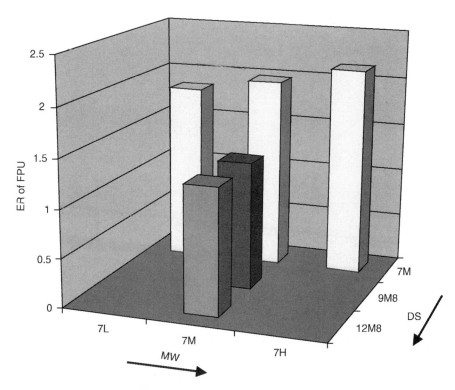

Fig. 5. Comparison of different types of CMC for enrichment ratios of FPU. DS refers to degree of substitution (at 70, 90, and 120%), and MW refers to molecular weight (L, low; M, medium; and H, high).

Table 3
Comparison of the Effects of XH and CH Addition on Foam Fractionation From Cell-Free, Lactose-Based Broth Supernatant, in Terms of Maximum Foam Volume, ER of FPU, Extracellular Proteins and Reducing Sugars, as well as Values of E/P (i.e., FPU/g/L of Proteins)

Hydrolysate (%)	Foam volume (mL)		ER of FPU		ER of proteins		ER of sugars		E/P	
	XH	CH	XH	CH	XH	CH	XH	CH	XH	CH
25	20	–	1.3	–	1	–	1		8.7	–
50	20	10	1.0	1.7	0.9	1.4	0.8	0.9	8.9	6.3
75	70	10	1.5	1.8	1	1.5	1.1	0.9	18.4	9.2

Lactose-based broth did not foam well. Addition of 25% CH did not allow meaningful collection of foamate; consequently, no measurements were available for that system.

was not very high, up to approx 1.8, as compared with that in the hydrolysate-based broth supernatant, up to 3.5–4.5. However, it should be noted that there was less room for enrichment and purification in the lactose-based supernatant because it was much richer and purer in cellulase

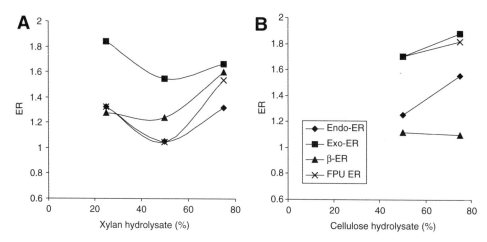

Fig. 6. Comparison of the effects of (A) XH addition and (B) (CH) addition on foam fractionation of FPU and individual cellulase components, i.e., endoglucanases, exoglucanases, and β-glucosidases, from cell-free, lactose-based broth supernatant.

(with E/P = 5.6) than the hydrolysate-based supernatant (with E/P = 1.1) to begin with. Although the E/P value for pure cellulase is yet to be determined, the value reached about 18 and 9 in the foamate produced with 75% XH and CH, respectively, from the lactose-based broth supernatant. Both were much higher than the E/P value (up to ~6.5) obtained with 75% CH from the hydrolysate-based broth supernatant (Fig. 2).

The effects of XH and CH, at different fractions, on foam fractionation of individual cellulase components are shown in Fig. 6. The CH-facilitated behaviors were similar in the hydrolysate-based broth supernatant (Fig. 2) and in the lactose-based broth supernatant (Fig. 6B): ER was highest for exoglucanases and lowest for β-glucosidase. XH also enriched exoglucanases the most from the lactose-based broth supernatant (Fig. 6A), but appeared to have least enrichment for endoglucanases. This last observation was rather surprising, because XH was prepared to have higher concentrations of oligomers than CH. Cellulase was nonetheless expected to have lower affinity to XH, than to CH, which might have played a role in the poorer enrichment of endoglucanases by XH.

The above results clearly demonstrated the beneficial enrichment and purification of cellulase, particularly exoglucanases, by affinity foam fractionation using CH, CMC, and XH. Conceivably, the fermentation broth can be fed into the foam fractionation columns for collecting the cellulase-enriched foamate in batch or continuous operation. The presence of cells did not seem to hamper the FPU enrichment and the cells would be present in the foamate collected. For applications stipulating cell-free formulations, a separate cell clarification step is required. Depending on the desired cellulase purity, the enriched foamate may or may not need further purification with other conventional methods. Nonetheless, the foamate volume would be significantly smaller than the original broth volume, allowing for more economical cell clarification and/or enzyme purification. Fermentation broths are, however, notoriously complex and the intertwining factors governing the interfacial phenomena occurring in foam fractionation are far from being clearly understood or quantitatively described. More studies and further development of affinity foaming agents will prove fruitful in realizing the significant potential of the affinity foam fractionation technology.

Acknowledgment

The project was supported by Initiative for Future Agriculture and Food Systems grant no. 2001-52104-11476 from USDA Cooperative State Research, Education, and Extension Service. Both hardwood hydrolysate and xylan hydrolysate were provided by Dr. Patrick Lee of Tennessee Valley Authority (Muscle Shoals, AL).

References

1. Montero, G. A., Kirschner, T. F., and Tanner, R. D. (1993), *Appl. Biochem. Biotechnol.* **39/40,** 467–475.
2. Varley, J. and Ball, S. K. In: *Separations for Biotechnology 3*, D. L. Pyle, ed., The Royal Society of Chemistry, London, pp. 525–530.
3. Wenzig, E., Lingg, S., Kerzel, P., Zeh, G., and Mersmann, A. (1993), *Chem. Eng. Technol.* **16,** 405–412.
4. Batt, C. A. (1991), In: Biomass. *In Biotechnology: The Science and the Business*, Moses, V., Cape, R. E., eds., Harwood Academic, New York, pp. 521–536.
5. Bailey, M. J., Askolin, S., Horhammer, N., et al. (2002), *Appl. Microbial. Biotechnol.* **58(6),** 721–727.
6. Zhang, Q., Lo, C. C., and Ju, L. K. (2005), *Bioresour. Technol.*, submitted.
7. Dubreil, L., Compoint, J. P., and Marion, D. (1997), *J. Agric. Food Chem.* **45,** 108–118.
8. Maa, Y. F. and Hsu, C. C. (1997), *Biotechnol. Bioeng.* **54(6),** 503–512.
9. Bravo Rodríguez, V., Páez Dueñas, P. M., Reyes Requena, A., Aoulad El-Hadj, M., and García López, A. I. (2001), *Can. J. Chem. Eng.* **79,** 289–295.
10. Sreenath, H. K. (1993), *Lebensm-Wisst u-Technol.* **26(3),** 224–228.
11. Bravo, V., Páez, M. P., Aould El-Hadj, M., Reyes, A., and García, A. I. (2001), *J. Chem. Technol. Biotechnol.* **77,** 15–20.
12. Biely, P., Vrsanska, M., and Claeyssens, M. (1991), *Eur. J. Biochem.* **200,** 157–163.
13. Vladimir, M. C., Olga, V. E., and Anatole, A. K. (1988), *Enzyme Microb. Technol.* **10,** 503–507.
14. Biely, P. and Markovic, O. (1988), *Biotech. Appl. Biochem.* **10(2),** 99–106.
15. Mandels, M., Andreotti, R., and Roche, C. (1976), *Biotechnol. Bioeng. Symp.* **6,** 21–33.
16. Lee, P. and Moore, M. Abstracts of papers, (2002), 223rd ACS National Meeting, Orlando, FL, United States.
17. Lo, C. -M., Zhang, Q., and Ju, L. K. (2005), submitted to 27th Symposium on Biotechnology for Fuels and Chemicals.
18. Miller, W. M., Blanch, H. W., and Wilke, C. R. (1988), *Biotechnol. Bioeng.* **32,** 947–965.
19. Ohnishi, S. T. and Barr, J. K. (1978), *Anal. Biochem.* **86,** 193–207.
20. Gunjikar, T. P., Sawant, S. B., and Joshi, J. B. (2001), *Biotechnol. Prog.* **17,** 1166–1168.
21. Berghem, L. E. R. and Petterson, L. G. (1973), *Eur. J. Biochem.* **37,** 21–30.
22. Afolabi, O. A. (1997), M.S. Thesis, The University of Akron, Akron, OH.
23. Klesov, A. A., Rabinoyich, M. L., Sinitsyn, A. P., Churulova, I. V., and Grigorash, S. Y. (1981), *Soy. J. Bioorg. Chem.* **6,** 662–667.

Molecular Distillation

A Powerful Technology for Obtaining Tocopherols From Soya Sludge

Elenise Bannwart de Moraes,* Patricia Fazzio Martins, César Benedito Batistella, Mario Eusebio Torres Alvarez, Rubens Maciel Filho, and Maria Regina Wolf Maciel*

Separation Process Development Laboratory (LDPS),
Faculty of Chemical Engineering, State University of Campinas,
(UNICAMP), CP 6066, ZIP CODE 13081-970, Campinas-SP,
Brazil; E-mail: elenise@feq.unicamp.br and wolf@feq.unicamp.br

Abstract

Molecular distillation was studied for the separation of tocopherols from soya sludge, both experimentally and by simulation, under different operating conditions, with good agreement. Evaporator temperatures varied from 100°C to 160°C and feed flow rates ranged from 0.1 to 0.8 kg/h. The process pressure was maintained at 10^{-6} bar, the feed temperature at 50°C, the condenser temperature at 60°C, and the stirring at 350 rpm. For each process condition, samples of both streams (distillate and residue) were collected and stored at –18°C before tocopherols analyses. Owing to the differences between molecular weights and vapor pressures of free fatty acids and tocopherols, tocopherols preferentially remained in the residue at evaporator temperatures of 100°C and 120°C, whereas for higher temperatures (140°C and 160°C) and lower feed flow rate, tocopherols tended to migrate to the distillate stream.

Index Entries: Free fatty acids; molecular distillation; soya sludge; tocopherols.

Introduction

From a market perspective, the interest in natural substances to be used in food products is higher than for synthetic ones. Natural substances of commercial interest are present in complex mixtures made up of a number of different molecules. This is the case of tocopherols obtained from soya sludge. Most of the substances that are present in soya sludge are molecules of high molecular weight and are thermally sensitive. These properties hinder the separation or purification of substances through traditional methods, because they are decomposed when subjected to high temperatures.

---

*Author to whom all correspondence and reprint requests should be addressed.

Molecular Distillation

An alternative separation/purification of such products is the use of molecular distillation, which is a special evaporation technique, it operates under low pressure and at relatively low temperatures. Furthermore, this process has advantages over other techniques that use toxic or flammable solvents (as the separating agent, avoiding toxicity, and environmental problems). In fact, this process shows potential in the separation, purification and/or concentration of natural products which are often complex and thermal sensitivity molecules, such as vitamins, because it can minimize losses by thermal decomposition. In lipid chemistry, it has been used for the purification of monoacylglycerols *(1)*, recovery of carotenoids from palm oil *(2)*, fractionation of polyunsatured fatty acids from fish oils *(3)*, recovery of squalene *(4)*, and recovery of tocopherols *(5)*, among others.

As previously described by Lutisan et al. *(6)*, in a falling film molecular distillator apparatus, the distilled liquid continuously passes down the heated evaporating cylinder, evaporates partially and the vapors then condense on the internally cooled condenser placed close to the evaporating cylinder (Fig. 1). A sufficiently low pressure around 10^{-3} mmHg, is maintained in the evaporator. Therefore, evaporated molecules can pass through the distillation gap to the condenser freely. The evaporation of the liquid on the evaporating cylinder is a key step in the molecular distillation *(6)*.

Soya sludge, also known as deodorizer distillate of soya oil (DDSO), is an important byproduct from the refining process of soybean oil, because of the content of interesting products. It contains tocopherols (3–12%), mainly the γ-isomer, triglycerides (25–50%), free fatty acids (FFA) (25–85%), phytosterols (7–8%), hydrocarbon and other unsaponifiables in trace amounts. Owing to the high content of FFA, separation of these compounds from deodorizer distillate is an important step to concentrate tocopherols to high purity. The composition of soya sludge depends on the source and process conditions employed for the refining process of the soybean oil. The recovery of tocopherols, phytosterols, and other components is important from a commercial point of view for making value-added products *(7)*. Vitamin E, which includes four isomers (α, β, γ, and δ), is a major natural antioxidants used for protection of fats and oils against atmospheric oxidation. Phytosterols are used as starting materials for the synthesis of steroids for pharmaceutical purposes.

So, the aim of this work is to incorporate in the DISMOL (molecular distillation process simulator) the characteristics of some molecules present in the DDSO to simulate the molecular distillation process and then to validate the results with experimental data, in order to find out the range of operating conditions to obtain tocopherols from soya sludge. Complex systems must be well characterized in terms of physical–chemical properties of the components (e.g., critical pressure, critical temperature, critical volume, and acentric factor) to represent the system that will feed the molecular distillator. These properties are necessary to calculate other properties, for example, mean free path, enthalpy of vaporization, mass diffusivity, vapour

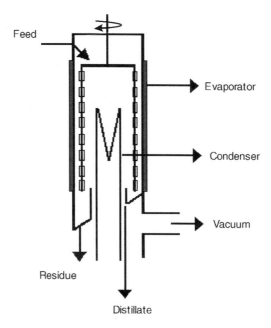

Fig. 1. Wiped Film Molecular Distillator.

pressure, liquid density, heat capacity, thermal conductivity, and viscosity of the system. Some properties are very difficult to find owing to the complexity of the involved components. Consequently, their determinations must be made through correlations and/or predictions, in order to be able to characterize the system and to simulate it. The simulation is very important to investigate the process viability as well as to establish the operating conditions, so that it is possible to improve the yield and purity of the final product. Therefore, in this work a comparison between simulated and experimental data is presented. The intention is to validate the simulation results for evaluating the tocopherols (vitamin E) recovery and, also, FFA elimination, using the DISMOL simulator. The equipment used in this study is the falling film molecular distillator. The problem may be stated as: for lower temperatures, FFA are eliminated from the residue stream along with a small amount of tocopherols; as the temperature increases, FFA concentrations in the distillate stream increase with an increasing amount of tocopherols as well. However, despite higher losses, operating at higher temperature has the advantage of minimizing the quantity of FFA in the desired product stream (tocopherols).

Methodology

Simulation

DDSO is a complex mixture owing to its large number of components. It includes thermal sensitive molecules such as tocopherols. Some physical properties can not be experimentally determined without the decomposition of these molecules. Consequently, it is very hard to find their physical properties and they must be estimated and/or predicted before the simulation. Firstly, it was necessary to create hypothetical components using the UNIFAC group contribution method (a tool of the Commercial Simulator HYSYS™, Hyprotech Ltd.), to estimate some physical properties, for example, critical pressure, critical temperature, critical volume, and acentric factor. These properties will be necessary to calculate other properties, for example, mean free path, enthalpy of vaporization, mass diffusivity, vapor pressure, liquid density, heat capacity, thermal conductivity, and viscosity of the system to be studied (DDSO) to insert in the DISMOL simulator (which allows simulation of the molecular distillation process). The DISMOL simulator requires, besides the component and mixture properties, equipment, process, and system characteristics which are simulation inputs. Evaporation rate, temperature and concentration profiles, residence time, stream compositions, and flow rates are the outputs from the simulation. All explanations of the equations used, of the solution methods and of the routine of solution to represent the molecular distillation process have been described (8) as well as calculations of these properties for the studied system (9). For molecular distillation there is no discussion about equilibrium, because this process is a nonequilibrium process.

In relation to the equipment, it is necessary to know its dimension, the feed flow rate, and its heating temperature. In this study, the equipment used was a falling film distillator and the DDSO composition, described in Table 1, was applied to the model. This results in concentrations and exit flow rates of the distilled and of the concentrated streams, the evaporation rate and the duration of distillation.

Experimental Procedure

Deodorizer distillate from industrial refining of soybean vegetable oil was obtained from Bunge Alimentos S. A. (São Paulo, Brazil). Solvents and reagents used for fatty acids analysis were of analytical grade. Hexane, isopropanol, acetonitrile, and water of high-performance liquid chromatography (HPLC) grade were used for the tocopherols and phytosterols analyses. Its composition is shown in Table 2. The DDSO composition was determined experimentally by the following methods.

Free Fatty Acid Analyses

The FFA were determined by titration with NaOH according to the Recommended Practice AOCS Ca 5a-40 using phenolphthalein as an indicator. The samples were dissolved in hot neutralized alcohol and the acids groups of FFA were neutralized with NaOH solution (1 N). The sample mass and the volume of alkali used in each determination were used to calculate the amount of free fatty acids. The percentage of FFA

Table 1
DDSO Composition Used in the Simulation

Components	Mass (%)
Palmitic acid	10.8
Stearic acid	2.4
Linoleic acid	26.2
Oleic acid	12.3
Lauric acid	1.9
Arachidic acid	4.2
Phytosterols	7.7
Tocopherols	8.9
Squalene	14.7
Glycerides	10.9
Total	100

Table 2
DDSO Composition Determined Experimentally (10)

Components	Mass (%)
Free fatty acids–oleic acid (FFA)	57.42
Tocopherols	8.97
Phytosterols	7.69
Triglyceride	8
Diglyceride	4.3
Monoglyceride	13.62
Total	100

in most types of fats and oils is calculated as oleic acid ($C_{18:1}$) using the formula:

$$\text{FFA (\%)} = \frac{\text{Alkali (mL)} \times N \times 28.2}{\text{Mass of sample (g)}}$$

where N corresponds to the normality of the sodium hydroxide solution.

Tocopherols Analysis

The composition of tocopherol was determined by normal phase HPLC (11) using a modular equipment composed by Waters 515 HPLC pump (Mildford, MA), equipped with a fluorescence detector (Waters model 2475 multi fluorescence). The separation was conducted in a microporasil column 125 Å, with particle size of 10 μ and 3.9 × 300 mm of dimension (Waters, Ireland). The mobile phase used was hexane:isopropanol (99:01). The feed flow rate of mobile phase was set at 1 mL/min. The data processing was done by the Millenium software 2010 Chromatography Manager Software

(Waters, Mildford, MA). The DDSO and samples of distillate and residue were dissolved in hexane (~1 mg/mL) and injected in the equipment. Each chromatografic run took about 10 min. This method determines α-, β-, γ-, and δ-tocopherol individually. The tocopherols detected in the chromatograms of DDSO were identified comparing the retention time of these compounds with the retention time of standards tocopherols.

Phytosterols Analyses

β-Sitosterol (24β-ethylcholesterol) standard with 98% of purity was purchased from Sigma (St. Louis, MO) to carry out phytosterols analyses by HPLC. This method determines the sitosterol content using UV detection with wavelength set at 206 nm. The mobile phase consisted of water: acetonitrile (78:22). Modular equipment consisted of HPLC (Waters, Milford, MA), equipped with a Waters 515 HPLC pump, Waters 2487 Dual Absorbance and an oven with a Waters temperature control module. Chromatographic separations were conducted in a Spherisorb S10 C8, 4.6 × 250 mm column (Waters, Ireland) under isocratic conditions, flow rate of 1 mL/min and temperature of 50°C. The samples were injected with no previous treatment into the 20 µL injection loop dissolved using ethanol before the analysis. Identification of compounds was achieved by comparing their retention times with the standards. For quantitative analysis, a calibration curve was used.

Triacylglycerols, Diacylglycerols, and Monoacylglycerols Analyses

Tri-, di-, and monoacylglycerols (Tg, DG, and MG, respectively) were determined according to literature procedure (12) using gel permeation chromatography (GPC). The samples were dissolved in tetrahydrofuran in the concentration of 1 mg/mL and injected into a modular equipment made up of an isocratic pump model Waters 515 HPLC (Mildford, MA), equipped with a differential refractometer detector model 2410 (Waters model, Mildford, MA) and an oven for columns. The samples were injected using a manual injector model Rheodyne 7725i with a 20-µL sample loop. The flow rate of the mobile phase was set at 1 mL/min. The separation was conducted using two GPC columns, Styragel HR1 and HR2 (Waters, Milford, MA), with dimensions of 7.8 × 300 mm, particle size of 5 µm and packed with styrenedivinylbenzene copolymer. The columns were connected in series using a U-shaped column joining tube. The acylglycerol content in the DDSO was based on the content of tocopherols, phytosterols, FFA, TG, DG, and MG, that are the main components present and their respective proportions were determined by GPC analysis.

Molecular Distillation

The distillation was performed using a laboratory wiped film molecular distillator model KDL 5, GmbH UIC (Alzenau, Germany) which is a variation from falling film molecular distillation with agitation. The major

Table 3
Molecular Weights and Vapor Pressures of FFA and Tocopherols (14)

Component	Molecular weight (g/gmol)	Vapor pressure at 200°C (mmHg)
FFA	180	4
Tocopherols	415	0.15

part of the equipment is made in glass. The heating of the evaporator was provided by a jacket circulating heated oil from an oil bath. The vacuum system included diffusion and mechanical pumps. The surface area of the evaporator is 0.048 m² and the surface area of the internal condenser was 0.065 m². The roller wiper speed inside the evaporator was fixed at 350 rpm.

The experiments were organized according to the following methodology: samples were melted to obtain a liquid and homogeneous mixture necessary to feed the equipment. The evaporator selected temperature was 100°C. Firstly, the evaporator temperature was fixed and, then, the feed flow rate was varied from 0.1 to 0.7 kg/h. For each process condition, samples of both streams (distillate and residue) were collected and submitted to tocopherols analyses. The process pressure was maintained at 10^{-6} bar, the feed temperature at 50°C, the condenser temperature at 60°C, and the stirring at 350 rpm. The collected samples of the distillate and residue streams were kept in a freezer at –18°C, for further analysis. The experiments were carried out in order to demonstrate that the FFA could be eliminated from the residue stream, facilitating the concentration of tocopherols through molecular distillation, manipulating operating conditions, such as evaporator temperature and feed flow rate, in order to remove from the residue stream a larger amount of FFA, and, increasing consequently, tocopherols concentration (13). It was a reasonable approach owing to the differences between molecular weights and vapor pressures of FFA and tocopherols (Table 3).

Owing to the values of molecular weights and vapor pressures, it was expected that FFA would be removed in the distillate stream and tocopherols preferentially concentrated in the residue stream. However, as discussed in the next section, depending on the level of temperature and feed flow rate, tocopherols may also be present in the distillate stream. Bearing this in mind, extensive evaluation was carried out to identify the temperature and feed flow rate conditions at which the tocopherol content would be maximized in the residue stream, while minimizing the amount of FFA.

Results and Discussion

The results of the analyses are presented in Figs. 2–5. It is observed that the simulated results agree with the experimental data. It is important to note that, although there are raw material differences as shown

Molecular Distillation

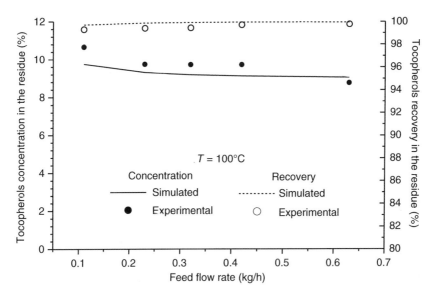

Fig. 2. Concentration of tocopherols in the residue (%) vs feed flow rate (kg/h) and recovery of tocopherols in the residue (%) vs feed flow rate (kg/h) at 100°C.

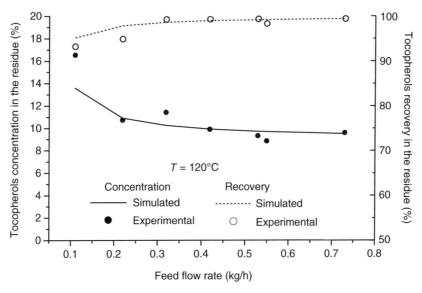

Fig. 3. Concentration of tocopherols in the residue (%) vs feed flow rate (kg/h) and recovery of tocopherols in the residue (%) vs feed flow rate (kg/h) at 120°C.

in Tables 1 and 2, a good agreement was reached between simulated concentration results and experimental data. Some of the deviations between experimental and simulated results occurred probably owing to (1) slight differences between real and simulated composition of the raw material and, (2) use of predicted properties of the components. Because the experimental curves follow a well defined behavior, although different for higher

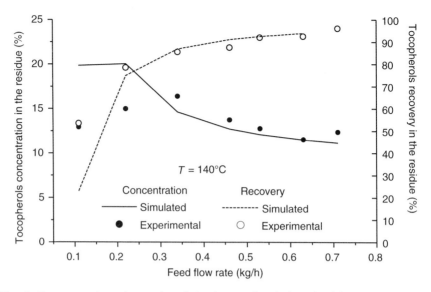

Fig. 4. Concentration of tocopherols in the residue (%) vs feed flow rate (kg/h) and recovery of tocopherols in the residue (%) vs feed flow rate (kg/h) at 140°C.

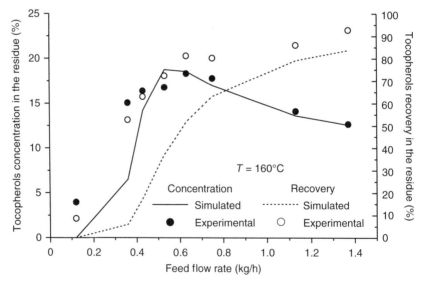

Fig. 5. Concentration of tocopherols in the residue (%) vs feed flow rate (kg/h) and recovery of tocopherols in the residue (%) vs feed flow rate (kg/h) at 160°C.

temperatures when compared with the lower ones (for higher operating temperatures, the tocopherols concentrations reach a maximum value), it can be said that there is no systematic experimental errors. So, probably, the deviations may be related to FFA and glyceride concentrations. Experimentally, FFA content was measured by titration and calculated as if FFA were made up of only oleic acid. The concentration of glycerides was estimated as a total mixture of mono-, di-, and triglycerides. On the other hand, the simulated raw material considered each FFA, for example, palmitic, stearic, oleic, linoleic, lauric, and arachidic acids, as individual species and glycerides as only one group of substances, including mono-, di-, and triglycerides. Even so, larger deviations were observed at lower feed flow rates and higher temperatures.

Good quality properties are necessary for simulation and these data are very difficult to find in the literature, owing to the complexity of the components. The composition of the raw material and the estimated properties of individual substances influence the mixture properties used in the molecular distillation simulation. As the agreement of the experimental data and the predictions of the model is quite good, it can be said that property prediction procedures are suitable to be used for process simulation. In fact, the procedure is robust enough to allow for process condition definitions and operating strategy discrimination before doing experiments. Moreover, through molecular distillation simulation, it is possible to obtain important information on the equipment capacity, on the fractionating of each component as well as on what is needed for future process development.

Figures 2 and 3 show that tocopherols remain in the residue at the lower temperatures investigated (100°C and 120°C). For higher temperatures (140°C and 160°C), (Figs. 4 and 5), the results show that tocopherols tend to migrate to the distillate stream, as can be seen by the presence of a maximum of tocopherol concentration in the residue stream.

In the raw material, according to Table 2, the FFA composition is 57.4%. If all FFA (the most volatile component) were removed from the residue stream by the molecular distillation, the concentration of all other components including tocopherols would be 2.3 times higher than the original composition (this means around 42% of the original mass). Taking this into consideration, and analyzing Figs. 2–5, they show that in the region of tocopherols maximum concentration, the values of such concentrations increase with temperature (higher elimination of FFA). For temperature of 160°C, a maximum of 19% of tocopherols was obtained, which represent a concentration factor of about 2.2 times over its original value (i.e., nearly maximum efficiency). Analyzing the results of tocopherols recovery, it can be observed that for lower temperature and higher feed flow rate, practically 100% recovery is obtained. For higher temperatures, the recovery is smaller owing to carryover of tocopherols into the distillate stream.

Conclusions

This work showed that the Molecular Distillation is a powerful process to recover tocopherols from soya sludge, enabling a concentration of about 2.2 times their original value, at a temperature of 160°C, in which the maximum concentration was achieved. The results indicated good agreement between simulated results using the DISMOL simulator and experimental results for tocopherols recovered from DDSO. This shows that the simulator can be used to evaluate new separations through molecular distillation, to define suitable operating conditions and to analyse different operating strategies, before carrying out experiments.

Acknowledgments

The authors are grateful to FAPESP (Fundação de Amparo à Pesquisa do Estado de São Paulo) for the financial support of this project (99/04656-9), CAPES (Coordenação de Aperfeiçoamento de Pessoal de Nível Superior) and CNPq (Conselho Nacional de Desenvolvimento Científico e Tecnológico).

References

1. Szelag, H. and Zwierzykowski, W. (1983), *Fette Seifen Anstrichmittel* **85,** 443–446.
2. Batistella, C. B. and Maciel, M. R.W. (1998), *Comput. Chem. Eng.* **22,** S53–S60.
3. Breivik, H., Haraldsson, G. G., and Kristinsson, B. (1997), *J. Am. Oil Chem. Soc.* **74,** 1425–1429.
4. Sun, H., Wiesenborn, D., Tostenson, K., Gillespie, J., and Rayas Duarte, P. (1997), *J. Am. Oil Chem. Soc.* **74,** 413–418.
5. Batistella, C. B., Moraes, E. B., Maciel Filho, R., and Maciel, M. R. W. (2002), *Appl. Biochem. Biotechnol.* **98,** 1187–1206.
6. Lutisan, J., Cvengros, J., and Micov, M. (2002), *Chem. Eng. J.* **85,** 225–234.
7. Nagesha, G. K., Manohar, B., and Udaya Sanar, K. (2004), *J. Supercritical Fluids* **32,** 137–145.
8. Batistella, C. B. and Maciel, M. R. W. (1996), *Comput. Chem. Eng.* **20(Suppl),** S19–S24.
9. Moraes, E. B., Batistella, C. B., Torres Alvarez, M. E., Maciel Filho, R., and Maciel, M. R. W. (2004), *Appl. Biochem. Biotechnol.* **113–116,** 689–711.
10. Martins, P. F. (2005), Master Thesis, Separation Process Development Laboratory (LDPS). Faculty of Chemical Engineering. State University of Campinas, (UNICAMP), Brazil.
11. AOCS, Ce 8-89, 1990, Official and Tentative Methods of the American Oil Chemist's Society, 4th ed., Champaign, IL.
12. Schoenfelder, W. (2003), *Eur. J. Lipid Sci. Techn.* **105,** 45.
13. Martins, P. F., Ito, V., Batistella, C. B., and Maciel, M. R. W. (2005), Separation and Purification Technology, vol 48/1, pp. 78–84.
14. Winters, R. L. (1986), In: *Proceedings World Conference on Emerging Technologies in the Fats and Oils Industry*, A. R. Baldwin, ed., American Oil Chemists' Society, Champaign, 184–188.

Application of Two-Stage Biofilter System for the Removal of Odorous Compounds

GWI-TAEK JEONG,[1,2] DON-HEE PARK,*[,2–6] GWANG-YEON LEE,[7] AND JIN-MYEONG CHA[8]

[1]Engineering Research Institute, [2]School of Biological Sciences and Technology, [3]Faculty of Applied Chemical Engineering, [4]Biotechnology Research Institute, [5]Research Institute for Catalysis, [6]Institute of Bioindustrial Technology, Chonnam National University, Gwangju, 500-757, Korea, E-mail: dhpark@chonnam.ac.kr; [7]Department of Ophthalmic Optics, Dong-A College, Jeonnam, 526-872; and [8]B & E Tech Co., Ltd., Jeonnam, 519-831, Korea

Abstract

Biofiltration is a biological process which is considered to be one of the more successful examples of biotechnological applications to environmental engineering, and is most commonly used in the removal of odoriferous compounds. In this study, we have attempted to assess the efficiency with which both single and complex odoriferous compounds could be removed, using one- or two-stage biofiltration systems. The tested single odor gases, limonene, α-pinene, and iso-butyl alcohol, were separately evaluated in the biofilters. Both limonene and α-pinene were removed by 90% or more EC (elimination capacity), 364 g/m^3/h and 321 g/m^3/h, respectively, at an input concentration of 50 ppm and a retention time of 30 s. The iso-butyl alcohol was maintained with an effective removal yield of more than 90% (EC 375 g/m^3/h) at an input concentration of 100 ppm. The complex gas removal scheme was applied with a 200 ppm inlet concentration of ethanol, 70 ppm of acetaldehyde, and 70 ppm of toluene with residence time of 45 s in a one- or two-stage biofiltration system. The removal yield of toluene was determined to be lower than that of the other gases in the one-stage biofilter. Otherwise, the complex gases were sufficiently eliminated by the two-stage biofiltration system.

Index Entries: Biofilm; biofilter; complex odor; two-stages; VOCs.

Introduction

Emissions of volatile organic compounds (VOCs) can be controlled using a variety of chemical, physical, or biological technologies, including incineration, adsorption, chemical scrubbing, bioscrubbing, and biofiltration (1). Within the two last decades, biological treatment protocols

---

*Author to whom all correspondence and reprint requests should be addressed.

including bioscrubbers, trickling beds, and biofilters have been employed successfully in the control of both VOCs and odors (2).

Biofiltration is a biological process which is considered to be one of the more successful examples of biotechnological techniques applied to environmental engineering, and has been most commonly employed in the removal of odoriferous compounds and VOCs. This biological process has been shown to be competitive in applications which involve the treatment of large volumes of air that contain low concentrations of odors. These processes tend to be associated with low operating costs, and are quite effective in the treatment of large volumes of moist air containing low concentrations of biodegradable compounds. One advantage of such biological processes is that they are typically of lower cost than other processes, including incineration, absorption, adsorption, and condensation. In general, biological processes tend to be both ecologically and economically desirable, particularly regarding off-gases at concentrations of up to 1–5 g/m^3 (1–3).

The biofiltration process utilizes microorganisms, which have been attached to porous support media, to breakdown both VOCs and odoriferous compounds. It has also been found to have potential applications in the control of emissions from industrial plants (3). A biofilter typically consists of a container filled with packing material and populated with microbes, through which the odor-containing air is passed, normally in an upward direction (4).

The contaminants are transferred from the air to the biofilm, in which they are subsequently biodegraded into carbon dioxide and water. Odor and VOCs are similarly transferred from the air into a biofilm (bio-active layer) which surrounds the organic or inorganic packing material in the biofilter. The odorous gases are then degraded into a variety of end products, or subsumed into the biomass. The end products appear to depend on the nature of the odors (1,2,4).

The packing material in the biofilter also performs as a carrier for the microbes, nutrients, and water. The packing material must possess a number of characteristics in order to ensure high deodorizing performance, such as, large surface area for gas contact, high levels of microbe immobilization, high water retention capacity, and easy removal of deodorization wastes. Also, the carrier is required to be extremely durable, with no clogging/blocking and a low pressure drop in the packed bed during its operation. Currently, a variety of carriers which fulfill these criteria are actually in use (2,5–7).

The air which is introduced into the biofilter may be prehumidified in order to maintain adequate moisture in the biofilm. Alternatively, or in addition, water may be sprinkled over the biofilm's surface and allowed to trickle downwards, counter to the flow of odorous air. This water must then contain nutrients required for the growth of the microbes (4).

Biofiltration has demonstrated, in many studies, an ability to remove alcohols, toluene, phenol, ketones, petroleum fuel vapors, and a variety of other VOCs (2,7,8). Most recently, simultaneous treatments of single or complex VOCs have been demonstrated using biofilters (2,9). Many industrial air emissions have been shown to contain VOC mixtures, which exhibit different physical and chemical characteristics which affect their biological treatment. In particular, the effectiveness of biofiltration in the treatment of VOC mixtures is highly dependent on the solubility of the compounds within the liquid layer of the biofilm. For example, methanol, a VOC which is both hydrophilic and readily biodegradable, could be expected to be removed effectively via biofiltration. However, α-pinene removal by biofiltration is more difficult owing to its extreme hydrophobicity, which results in low diffusivity through the biofilm (3). The operating conditions of a given biofilter, including temperature, pH, nutrient concentration, water contents, and relative humidity in the air, are the most relevant factors affecting the removal capacity of a biofilter (2,7–10).

The objective of this research, then, was to determine the effectiveness of biofilter packed with a reticulated polyurethane foam carrier to remove hydrophilic and hydrophobic components from VOC mixtures.

Materials and Methods

Microbes, Nutrient, and Packing Material

All of the biofilters were inoculated with a microbial consortium, primarily composed of a variety of microbes able to degrade ethanol, acetaldehyde, iso-butyl alcohol, limonene, α-pinene, and toluene, developed by enrichment of sludge isolated from compost at composting facilities in Seosan, Chungnam, Korea. The nutrient solution provided for the growth and maintenance of the microbes in the biofilter was a mineral medium consisting of basic mineral salt supplemented with: 1.5 g/L KH_2PO_4, 6 g/L Na_2HPO_4, 3 g/L $(NH_4)_2SO_4$, 0.05 g/L $MgSO_4$, and 0.01 g/L $CaCl_2$. In order to prepare the biofilter packing material, reticulated polyurethane foam (RPF; Yoowonurethane Co., Ltd., Korea) was cut to pieces of $1.9 \times 1.9 \times 1.9$ cm^3 in size, and these were used in all experiments. The characteristics of the packing material are summarized in Table 1.

Biofilter Experiments

The biofilter (Fig. 1) was constructed from a transparent acrylic tube with an inner diameter of 18 cm. This was then divided into three 80 cm sections, and upper two sections were filled to a volume of 3.1 L (packing ratio 30.5%), with equal amounts of the prepared filter-bed materials (Table 1).

In order to characterize the removal characteristics and yields at several different loading concentrations of limonene, α-pinene, and iso-butyl alcohol, gases containing each of these components were separately introduced to the biofilter. The biofiltration of single gases under continuous flow conditions was conducted at a variety of loading concentrations of limonene, α-pinene,

Table 1
Used Carrier and Characteristics

Bed material	Reticulated polyurethane foam (RPF)
Surface area (m²/g)	367.5
Absorptance (g·H$_2$O/g)	57
Porosity (%)	80
Bulk volume (L)	10.2 L/reactor
Bed volume (L)	3.1 L/reactor
Packing density (g/cm³)	0.0195
Packing ratio (%)	30.5

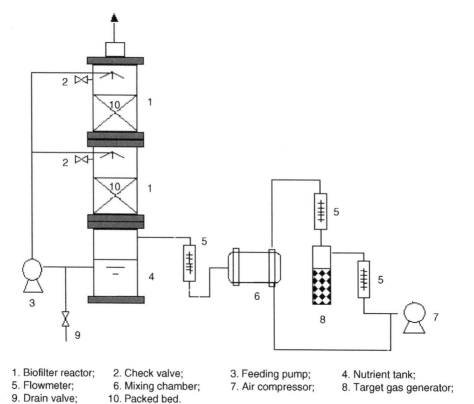

1. Biofilter reactor; 2. Check valve; 3. Feeding pump; 4. Nutrient tank;
5. Flowmeter; 6. Mixing chamber; 7. Air compressor; 8. Target gas generator;
9. Drain valve; 10. Packed bed.

Fig. 1. Scheme of two-stage biofilter for odor removal.

and iso-butyl alcohol, for a period of 1 mo. The gas samples were obtained from both the inlet and outlet streams, and also taken axially along the biofilter. The input gases, limonene, α-pinene, and iso-butyl alcohol in air, were prepared via the evaporation of liquid solutions with air flow, and the input concentrations were adjusted by controlling of flow rates of the prepared gases and air. In order to control the humidity of the input gas, air was passed though a humidifier before being mixed with the prepared gas, and then the mixed gas was introduced into the bottom portion of the biofilter. The biofilter was inoculated with a microbial consortium consisting of an enriched version of that reported previously (11). In order to feed the microorganisms, four times concentrated BSM nutrient solution was periodically fed into the biofilters via peristaltic pump. Water and nutrient were sprayed from the upper nozzles of each section of the biofilter in order to protect the biofilter media from drying. The operation temperature was set to 30°C, and pH levels were not controlled.

In order to determine the process characteristics and removal yields at a variety of loading concentrations of mixtures of ethanol, acetaldehyde, and toluene, the mixed gases were introduced to the bottom portion of the biofilter at a variety of loading concentrations, under continuous flow conditions for a period of 1 mo. In this study, we applied a two-stage biofilter and compared the results to those obtained using the one-stage biofilter. As for the single VOC gases, the introduced gas mixture of ethanol, acetaldehyde, and toluene was prepared via the evaporation of a liquid solution of this mixture by flowing air, and the input concentrations were adjusted by controlling of flow rates of the prepared gas mixture and air. Experiments concerning the removal of the gas mixtures were conducted as described earlier. The applied ranges of VOC concentrations in the ethanol, acetaldehyde, and toluene mixture were 50–200 ppm, 20–70 ppm, and 10–70 ppm, respectively.

Analytical Method

Both the VOCs and the odoriferous gases were collected at both the inlet and outlet of the biofilter, with a 10 mL Hamiltion gas tight syringe, and then dissolved in 15-mL sealed (septum) glass tubes containing 1 mL of methylene chloride. Samples were then extracted from the methylene chloride solution with a syringe, and 2 µL of each sample was injected onto a gas chromatograph. Different concentrations of VOCs and odoriferous gases were measured with a gas chromatograph (GC-14A, Shimadzu, Japan) which was equipped with a FID. We used a DB-WAX column (30 m × 0.53 mm × 1 mm, J&W Scientific, Folsom, CA) in this phase of the experiment. The GC oven temperature was set to 50°C, and detector temperature was 200°C. Helium was used as the carrier gas, and the column flow rate was set to 8 mL/min. The air flow rate was 140 mL/min, and the hydrogen flow rate was set at 40 mL/min. A portable VOC detector (Multi gas monitor PGM-50, RAE Systems Inc., CA) was also used.

Removal Yield and Elimination Capacity

Removal yield (RY) and elimination capacity (EC) were the variables used to determine the treatment capacity of the biofilter. Removal yield was expressed as the content (%) of odoriferous gas eliminated by the biofilter.

Elimination capacity was expressed as the amount of odoriferous gas removed in the bed volume/unit time.

$$RY\ [\%] = (C_{Gi} - C_{Go})/C_{Gi} \times 100$$

$$EC\ [g/m^3 \cdot hr] = (C_{Gi} - C_{Go}) \times Q/V_f$$

where Q is the gas flow rate (m^3/h), V_f is the volume of the filter bed (m^3) and C_{Gi} and C_{Go} are the inlet and outlet odoriferous gas concentration (ppm; g/m^3).

Results and Discussion

Removal of Single Odorous Compounds

A large quantity of ethanol, aldehydes, and various concentrations of aromatic compounds are emitted from recycling facilities (11). We performed the elimination of limonene, α-pinene, and iso-butyl alcohol separately, in a one-stage biofilter. For the initial adaptation of the microorganisms, we introduced one of the tested single odor gases into the biofilter under low concentration conditions, a maximum concentration of 20 ppm, using a 2 min residence time. After which the outlet concentration was maintained at 5 ppm, the input concentration was increased in a stepwise manner using a 30 s residence time.

The removal characteristics and removal yields of the tested odor gases, limonene, α-pinene, and iso-butyl alcohol, were separately observed in the biofilter (Figs. 2–4). More than 90% (EC 364 g/m^3/h) of the limonene was removed at a 50 ppm input concentration and a 30 s residence time, but we noted an 85% (EC 678 g/m^3/h) removal yield at a 100 ppm input concentration and a 30 s residence time (Fig. 2). Nearly 90% (EC 321 g/m^3/h) of the α-pinene was eliminated at a 50 ppm input concentration and a 30 s residence time, but the removal yield dropped to approx 82% (EC 631 g/m^3/h) at a 100 ppm input concentration and a 30 s residence time (Fig. 3). However, the biofiltration protocol in this experiment maintained an effective iso-butyl alcohol removal yield of above 90% (EC 375 g/m^3/h at 100 ppm), in an input concentration range between 20 and 150 ppm, after the adaptation period (Fig. 4). In general, the capacity of the biofiltration treatment for VOCs is highly dependent on the solubility of the odorous compounds within the liquid layer of the biofilm. α-Pinene is extremely hydrophobic (with a maximum water solubility of less than 5–10 mg/L), which appears to result in a low solubility in the biofilm of the biofilter (3).

Removal of Complex Odorous Compounds

The effectiveness with which a given biofiltration protocol can effect the removal of VOCs appears to be principally dependent on the solubility of the compounds in the biofilm of the biofilter. Hence, the hydrophilic

Application of Two-Stage Biofilter System

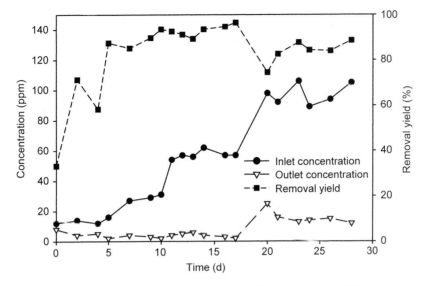

Fig. 2. Removal characteristics of limonene in biofilter.

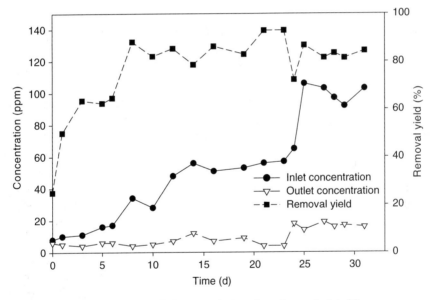

Fig. 3. Removal characteristics of α-pinene in biofilter.

and hydrophobic characteristics of a given pollutant is thought to significantly influence the capacity of a biofilter setup to remove it (3). Mohseni and Allen (3) reported that, when hydrophilic and hydrophobic materials were treated simultaneously, the microorganisms involved in the treatment of the hydrophilic materials exhibited overgrowth in the biofilter, and this created an obstacle to the treatment of the hydrophobic materials. Methanol, a VOC which is both hydrophilic and readily biodegradable, is expected to be easily removable via biofiltration. α-Pinene, on the other

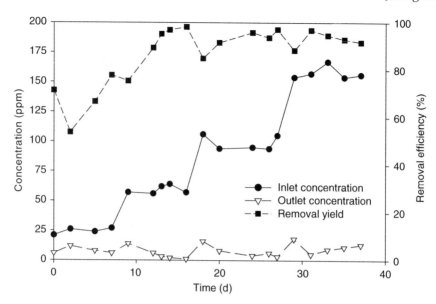

Fig. 4. Removal characteristics of iso-butyl alcohol in biofilter.

hand, is extremely hydrophobic, which may result in it being removed at a lower rate.

Therefore, in this work, we attempted the removal of hydrophilic compounds with a one-stage biofilter, and tried to remove hydrophobic compounds with a two-stage biofilter. We employed ethanol and acetaldehyde as hydrophilic compounds, and used toluene as the model hydrophobic compound. The mixture gas was then applied with the following inlet concentrations, ethanol 200 ppm, acetaldehyde 70 ppm, and toluene 70 ppm, with a residence time of 45 s. This part of the procedure was the same in the single- and double-stage biofiltration systems. Two identical 1-stage bench-scale biofilters (Reactor I and II) were operated in parallel, in order to characterize the influence of step loads on the system's efficacy for removing toluene, a hydrophobic VOC, and ethanol and acetaldehyde, both hydrophilic VOCs. The ethanol/acetaldehyde and toluene were introduced and stabilized in Reactors I and II, respectively. In order to stabilize the biofilter, 20 ppm of ethanol, and 10 ppm of acetaldehyde were introduced into the system (residence time of 1 min). The toluene input concentration was controlled at 10 ppm and stabilized (residence time of 2 min). After completion of the adaptation period, Reactors I and II were connected in order to simultaneously remove the three odoriferous gases in a combined two-stage biofilter. The ethanol, acetaldehyde, and toluene were then applied to the bottom portion of Reactor I, then passed through Reactor II.

The process characteristics and removal yields of the odoriferous gas mixture in the two-stage biofilter are shown in Fig. 5. The removal yields of the ethanol and acetaldehyde were maintained at over 95% throughout

Application of Two-Stage Biofilter System

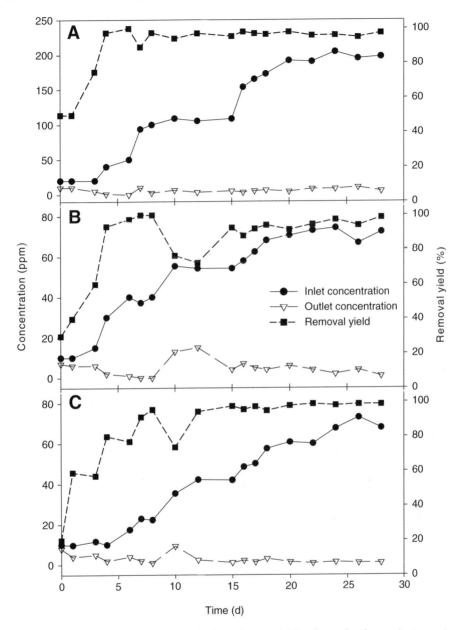

Fig. 5. Removal characteristics of ethanol, acetaldehyde and toluene in two-stage biofilter. **(A)** ethanol, **(B)** acetaldehyde, **(C)** toluene.

the 6 d of processing in reactor 1 (not shown), and after that period, the input concentration was increased in a stepwise manner using a 15 s residence time. When we had confirmed that the removal yields of the three odor gases were maintained at over 95% for 14 d, Reactor I and II were vertically connected, and then the mixed gas was supplied to Reactor II, after being passed through Reactor I. The input concentration was then increased in a stepwise manner using a 30 s residence time. After the two biofilters had been connected, ethanol and acetaldehyde had removal yields over 97% (EC 240 g/m^3/h) at 100 ppm and 92–98% (EC 165 g/m^3/h) at 70 ppm, respectively, (Fig. 5A,B). Also, the toluene removal yield was maintained at between 95 and 98% (EC 239 g/m^3/h) at 50 ppm (Fig. 5C).

In order to determine the process characteristics and removal yields in the one-stage biofilter, mixed odor gas (ethanol, acetaldehyde, and toluene) was applied directly to Reactor II, using initial input concentrations of 20, 10, and 10 ppm, respectively. During the adaptation period to stabilize the biofilter, the retention time was set at 3 min. After 8 d, the input concentrations and retention time (30 s) began to be controlled at different levels in order to determine the characteristics of the removal of the mixed odor gases. Figure 6 shows the performance characteristics and the removal yields of mixed odor gas obtained using the one-stage biofilter. After the completion of the adaptation period, the removal yield of ethanol was maintained at over 97% (EC 481 g/m^3/h at 100 ppm) (Fig. 6A). The 1-stage biofilter also removed approx 90% of the acetaldehyde (311 g/m^3/h at 70 ppm) and approx 70–80% of the toluene (EC 388 g/m^3/h at 50 ppm) (Fig. 6B,C).

When comparing Figs. 5 and 6, the removal profiles of ethanol and acetaldehyde exhibited similar patterns in the one- and two-stage biofilters. However, the toluene removal was generally poorer in the one-stage biofilter with the toluene removal yield determined to be approx 70–80% in the one-stage biofilter. However, the gas mixture was properly eliminated by the two-stage biofilter, with the toluene removal yield more than 95% using the two-stage biofiltration treatment.

Our results may be attributable to the fact that, in the continuous treatment of complex odors cause by hydrophilic and hydrophobic materials, the microorganisms which feed on hydrophilic materials exhibit overgrowth in one-stage biofilters, as reported previously (3). This can attenuate the efficacy inherent to the treatment of hydrophobic materials in such a system. Otherwise, we found that the hydrophilic and hydrophobic compounds could effectively be removed by a two-stage biofilter system.

Conclusions

The application of one- or two-stage biofilter systems was conducted in order to determine the efficiencies with which single or complex odoriferous compounds could be removed when reticulated polyurethane foam was used as a packing material for the biofilter. Firstly, we separately determined the removal characteristics and yields of the single odor gases, limonene, α-pinene, and iso-butyl alcohol, in a one-stage biofilter. More than 90% of the limonene and α-pinene were removed at a 50 ppm input concentration and a 30 s residence time, but removal yields dropped to approx 80% when the input concentration was increased to 100 ppm.

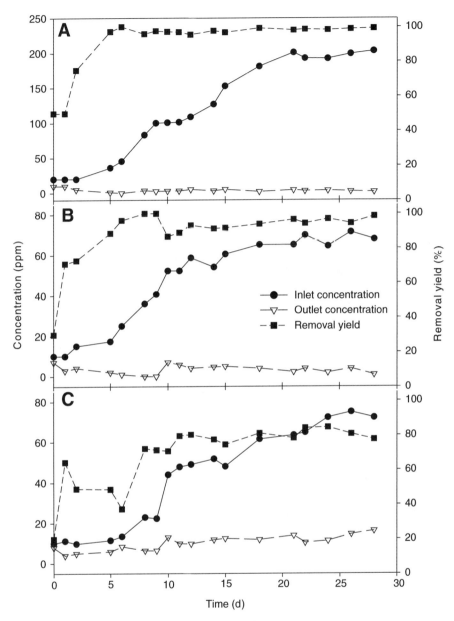

Fig. 6. Removal characteristics of ethanol, acetaldehyde and toluene in one-stage biofilter. **(A)** ethanol, **(B)** acetaldehyde, **(C)** toluene.

An effective iso-butyl alcohol removal yield above 90% was maintained at a 100 ppm input concentration. Secondly, we attempted to eliminate complex odoriferous gases with a two-stage filtration scheme using a residence time of 45 s and an ethanol inlet concentration of 200 ppm, an acetaldehyde inlet concentration of 70 ppm, and a toluene inlet concentration of 70 ppm, in both one- and two-stage biofiltration systems. Toluene removal yields were determined to be significantly lower than those of other gases in the one-stage biofilter, but not in the two-stage system. Mixture experiments showed that complex gases could be properly eliminated using a two-stage biofilter.

Acknowledgment

This study was financially supported by the Program for the Training of Graduate Students in Regional Innovation, which was conducted by the Ministry of Commerce, Industry, and Energy, of the Korean Government.

References

1. Ruokojärvi, A., Ruuskanen, J., and Martikainen, P. J. (1995), *J. Air Waste Manage. Assoc.* **51(1)**, 11–16.
2. Auria, R., Aycaguer, A. C., and Devinny, J. S. (1998), *J. Air Waste Manage. Assoc.* **48**, 65–70.
3. Mohseni, M. and Allen, D. G. (1999), *J. Air Waste Manage. Assoc.* **49(12)**, 1434–1441.
4. McNevin, D. and Barford, J. (2000), *Biochem. Eng. J.* **5(3)**, 231–201.
5. Shinabe, K., Oketani, S., Ochi, T., et al. (2000), *Biochem. Eng. J.* **5(3)**, 209–217.
6. Park, D. H., Liu, J. M., et al. (2002), *Biochem. Eng. J.* **11**, 167–173.
7. Morales, M., Perez, F., Auria, R., and Revah, S. (1994), in *Advances in Bioprocess Engineering* (Galindo, E. and Ramirez, O. T., eds.), Kluwer Academic, Dordrecht, The Netherlands, pp. 405–411.
8. Shareefdeen, Z., Baltzis, B., Oh, Y. -S., and Bartha, R. (1993), *Biotech. Bioeng.* **41**, 512–524.
9. Deshusses, M. A., Hamer, G., and Dunn, I. J. (1995), *Environ. Sci. Technol.* **29**, 1059–1068.
10. Williams, T. O. and Miller, F. C. (1992), *Biocycle Magazine* **33(10)**, 72–77.
11. Lee, G. Y. (2004), PhD Thesis, Chonnam National University, Gwangju, Korea.

Author Index

A
Agbogbo, F. K., 997
Akin, D. E., 104
Al-Dahhan, M. H., 887
Alpaslan, C., 1034
Alriksson, B., 599
Alvarez, M. E. T., 1041, 1066
Alves-Prado, H. F., 234, 289
Antunes, O. A. C., 1023
Arato, C., 55
Azargohar, R., 762

B
Ballesteros, I., 278, 496, 631, 922
Ballesteros, M., 278, 496, 631, 922
Barton, F. E., II, 104
Batistella, C. B., 405, 680, 716, 1066
Becker, J. G., 774
Bender, J. P., 951
Berlin, A., 55, 528
Berson, R. E., 612, 621
Berwanger, A. L. S., 942
Bohlin, C., 195
Bombardiere, J., 959
Borole, A. P., 887, 897
Bransby, D. I., 382
Buanafina, M. M. O., 416
Bura, R., 55, 528
Burapatana, V., 247

C
Cabañas, A., 496
Cairncross, R. A., 774
Calabria, G. M. M., 320
Cansian, R. L., 1023
Cara, C., 631
Carrier, D. J., 382
Carvalheiro, F., 461
Castellanos-Hernandez, N., 959
Castillo-Angeles, S., 959
Castro, A. M., 226
Castro, E., 631
Cha, J.-M., 1077
Chan, P.-L., 933
Chatfield, M., 959
Chen, G., 911
Chen, P., 563, 574
Chen, R., 586
Chen, S., 751, 844
Cherry, J. R., 179
Cholewa, O., 334
Christensen, B. H., 448
Christensen, H., 117
Chua, H., 1015
Chun, B.-W., 645
Ciapina, E. M. P., 880
Clausen, E. C., 382
Converti, A., 130

D
da Costa, A. C., 969
Da Silva, R., 234, 289
Dair, B., 645
Dalai, K., 762
Dalton, S. J., 416
Dam, B. P., 117
Damaso, M. C. T., 226
Damiano, V.B., 289
Dariva, C., 1023
Das Neves, L. C. M., 130
Dasari, R. K., 621
Davis, M. F., 427
Davison, B. H., 427
De Aguiar, P. F., 659
de Campos, N. P., 727
de França, F. P., 727
de Lima da Silva, N., 405
de Moraes, E. B., 1041, 1066
de Moraes, F. F., 864
de Oliveira, D., 951, 1023
De Souza, M. B., Jr., 256
Decker, S. R., 509
Delvigne, F., 392
Desbarats, A., 180
Di Luccio, M., 942, 951
Domaschko, M., 959
Domingues, N. M., 942
Drescher, S. R., 427
Duan, L., 382
Duarte, L. C., 461

E
Easter, M., 959
Easterly, J. L., 3
Eggeman, T., 361
Ekenseair, A. K., 382
Elander, R. T., 586
Erzinger, G. S., 787
Espinosa-Solares, 959
Etoc, A., 392

F
Ferreira, V., 226
Flynn, P. C., 1, 71, 88
Fraser, E. D. G., 22
Fregolente, L. V., 680
Freire, D. M. G., 880
Fu, R., 563

G

Galbe, M., 546
Gao, Y., 563
García-Aparicio, M. P., 278
Gáspár, M., 738
Gentille, T. R., 334
Gidh, A., 829
Gilkes, N., 55, 528
Gírio, F. M., 461
Gomes, E., 234, 289
Gonçalves, A. R., 320, 326
González, A., 278
Gray, K., 179
Gregg, D. J., 55
Gusakov, A., 528
Güvenilir, Y., 1034

H

Haltrich, D., 215
Hamilton, C. Y., 897
Han, K.-C., 870
Hanley, T. R., 612, 621
Haq, Z., 3
Hauck, B., 416
Henrich, E., 153
Hicks, K. B., 104
Himmel, M. E., 509
Himmelsbach, D. S., 104
Holland, J., 887
Holtzapple, M. T., 997
Hong, F., 303
Hon-Nami, K., 808
Hu, B., 751
Huang, C., 180

I

Ishida, N., 795
Ito, V. M., 716

J

Jeknavorian, A., 645
Jeong, G.-T., 165, 265, 436, 668, 705, 1077
Jönsson, L. J., 195, 303, 599
Jørgensen, H., 448
Jozala, A. F., 334
Ju, L.-K., 1051

K

Kálmán, G., 738
Karim, K., 887
Kearney, M., 349
Kilburn, D., 528
Kim, C. H., 486
Kim, H. J., 486
Kim, H.-O., 694
Kim, H.-S., 265
Kim, J-N., 694, 705
Kim, S. B., 486
Kim, S.-Y., 705
Kim, T. H., 586
Kirkkopru, I., 1034
Kitamoto, K., 795
Klasson, K. T., 887, 897
Kochergin, V., 349
Koegel, A., 153
Krogh, K. M., 117
Kumar, A., 71

L

Ladisch, M. R., 909
Langdon, T., 416
Larsen, J., 448
Lecomte, J. P., 392
Lee, D.-H., 870
Lee, G.-Y., 1077
Lee, H.-J., 265
Lee, Y. Y., 586
Lerin, L., 1023
Li, W., 911
Li, Yebo, 41, 985
Li, Yuhong, 563
Liao, W., 844
Lin, X., 563, 574
Linde, M., 546
Liu, Chengmei, 574
Liu, Chuanbin, 751, 844
Liu, S., 854
Liu, Yan, 751, 844
Liu, Yuhuan, 563, 574
Lo, C.-M., 1051
Lo, W.-H., 1015
Lundquist, K., 303

M

Mabee, W. E., 22, 55
Maciel Filho, R., 405, 680, 716, 969, 1041, 1066
Maciel, M. R. W., 405, 680, 716, 969, 1041, 1066
Macuch, P. J., 645
Mahmudi, H., 88
Manzanares, P., 496, 631, 922
Markov, A., 528
Martins, C. D. C., 659
Martins, P. F., 716, 1066
Matioli, G., 864
Maximenko, V., 528
Mazutti, M. A., 951
McFarlane, P. N., 22
McKeown, C. K., 897
McMillan, J. D., 509
Melo, W. C., 880
Meyer, A. S., 117
Miazaki, T. T., 130
Mikell, A., 829
Mirochnik, O., 55
Miyafuji, H., 476

Index

Miyamura, O., 130
Moraes, D. A., 130
Moriya, R. Y., 326
Morris, P., 416
Morrison, W. H., III, 104

N
Nagamori, E., 795
Nakata, T., 476
Nakkharat, P., 215
Negro, M. J., 278, 496, 922
Nghiem, N. P., 427
Nilvebrant, N.-O., 599
Nolasco, P. C., 226

O
O'Connor, R., 774
Oh, Y.-J., 870
Oh, Y.-T., 165
Ohnishi, T., 795
Okunev, O., 528
Oliva, J. M., 278, 496, 922
Oliveira, F. J. S., 727
Oliveira, L. R. M., 326
Olsson, L., 117
Omay, D., 1034

P
Padilha, F. F., 942, 1023
Pan, X., 55
Park, D.-H., 165, 265, 436, 668, 1077
Penna, T. C. V., 130, 334
Pereira, N., Jr., 226, 256, 880
Pessoa, A., Jr., 334
Porteneuve, C., 645
Portilho, M., 864
Prokop, A., 247
Pye, E. K., 55

R
Raffelt, K., 153
Ramaswamy, S., 774
Recseg, K., 738
Réczey, K., 738
Ridenour, W., 887
Rigsby, L. L., 104
Rivera, E. C., 969
Roth, R., 195
Ruan, L., 1015
Ruan, R., 563, 574
Ruiz, E., 631
Ryu, H.-W., 694, 705
Ryu, Y.-W., 870

S
Saddler, J. N., 22, 55, 528
Sáez, F., 922

Saitoh, S., 795
Saka, S., 476
Santa Anna, L. M. M., 880
Santos, A. S., 226, 880
Saville, B. A., 180
Scamparini, A. R. P., 864, 942
Selig, M. J., 509
Seo, J.-H., 870
Sérvulo, E. F. C., 659
Shahbazi, A., 41, 985
Silva, J. N. C., 226
Sinitsyn, A., 528
Sjöde, A., 599
Skomarovsky, A., 528
Soerensen, H. R., 117
Sokhansanj, S., 71
Solovieva, I., 528
Stafford, D. A., 959
Stahl, R., 153
Steinhardt, J., 153

T
Tadeu, J., 461
Takahashi, H., 795
Talreja, D., 829
Tanner, R. D., 247
Thomsen, A. B., 448
Thomsen, M. H., 448
Thornart, P., 392
Thygesen, A., 448
Tokuhiro, K., 795
Toniazzo, G., 1023
Treichel, H., 942, 951
Tsobanakis, P., 347
Tucker, M. P., 509
Tuskan, G. A., 427

V
van Zyl, W. H., 195
Vanzo, L. T., 942
Vásquez, M. P., 256
Vazquez, L., 727
Verser, D., 361
Viamajala, S., 509
Vinzant, T. B., 509, 829
Vitolo, M., 787

W
Wai, L., 933
Wang, Y., 1015
Ward, R., 289
Wee, Y.-J., 694, 705
Wei, Y., 303
Weirich, F., 153
Wiebe, D., 645
Williford, T. C., 829

Y
Yacyshyn, V., 180
Ying, M., 911
Young, J. S., 612
Yu, F., 563, 574
Yu, H.-F., 933, 1015
Yu, J.-H., 870
Yu, V., 933
Yun, J.-S., 694

Z
Zacchi, G., 546
Zanin, G. M., 864
Zhang, Q., 1051
Zhu, Y., 586

Subject Index

A
Acetoin, 854
Acetolactate synthase, 854
Acetosolv pulping, 326
Activated carbon, 612, 762
Aeration, 392
Affinity foam fractionation, 1051
AFM, 829
Agriculture, 22
Agricultural residues, 3
Agrobacterium sp., 864
Agroindustry waste, 942
Alcoholic fermentation process, 969
Alginate, 1034
Alkali, 599
Alkaline catalyst, 405
Alkalophilic bacteria, 289
als and *lysP* mutant, 854
Anaerobic digestion, 887
Animal manure, 887
Antifoam, 392
Aqueous ammonia, 586
Artificial chaperones, 247
Artificial intelligence, 969
Aspergillus niger, 195
Auger, 621

B
Bacillus licheniformis, 289
Bacillus polymyxa, 1034
Bacillus sphaericus, 659
Bacillus subtilis, 1015

B
Bacterial cellulose, 705
Barley straw, 278, 546
Beetles, 829
β-cyclodextrin, 247
β-glycosidase, 215
Bio-based polymers, 774
Biocatalysis, 265
Biocatalyst, 911
Biochar, 762
Biochemical oxidation, 645
Bioconversion, 265, 528, 1023
Biodegradability, 574
Biodegradable polymers, 774
Biodiesel, 165, 405, 911, 1041
Biodiesel fuel, 668
Bioenergy, 71
Bioethanol, 448, 727
Biofilm, 1077
Biofilter, 1077
Biofiltration, 130
Biogas, 887, 959
Bioinsecticide production, 659
Bioleaching, 897
Biomass, 621, 997
Biomass collection systems, 71
Biomass conversion, 509
Biomass pretreatment, 382
Biomass recalcitrance, 509
Biomass supply, 3
Biomass transportation, 88
Biomass transportation systems, 71
Biopolymers, 942
Bioreactor, 392
Biorefinery, 349
Biorefining, 22
Bioremediation, 130
Biosurfactant, 880
Biosurfactants production, 727
2,3-Butanodial, 854

C
Candida magnoliae, 870
Cane juice-based medium, 795
Carboxylic acids, 997
Carboxymethyl cellulose, 1051
Carotenes, 1041
Castor oil, 405
Cell reutilization, 226
Cell wall chemistry, 427
Cellulase, 104, 247, 528
Cellulose hydrolysate, 1051
Cellulose, 1051
Cement, 645
Centrifugal distillation, 716
CFD simulations, 621
Cheese whey, 985
Chelating agents, 320
Chemical activation, 762
Chitin, 1034
Chlamydomonas perigranulata, 808
Chromatography, 349
Cloning and expression, 1015
Complex odor, 1077
Concrete, 645
Constitutive vacuolar-targeted expression, 416
Corn cobs, 586
Corn fiber hydrolysis, 738
Corn fiber oil, 738
Corn steep liquor, 694
Corn stover, 3, 41, 563, 586
CPDM, 997
Crop residues, 41
Curdlan, 864
Cyclodextrin glycosyltransferase, 234

D

Debaryomyces hansenii, 461
Depolymerization, 829
Design of experiments, 969
Detoxification, 599, 612
Diffusion, 774
Digestibility, 416
Digestion, 997
Dilute acid pretreatment, 496, 509
Direct blue-I, 509
Distilled monoglycerides, 680

E

Edible oils, 933
EDTA, 334
Elicitor, 436
Ellaus guineensis fruit, 727
Endogenous starch fermentation, 808
Endoxylanase, 586
Energy crops, 382
Energy plantations, 22
Engine performance, 165
Entomopathogenic biomass, 659
Entrained-flow gasification, 153
Enzymatic digestibility, 486
Enzymatic hydrolysis, 278, 476, 546
Enzymatic oxidation, 320
Enzymatic pretreatment, 326
Erythritol, 870
Escherichia coli, 334
Esterase, 104
Esterification, 265
Ethanol, 41, 55, 117, 361, 476, 496, 546, 599, 631, 854, 922
Ethanolysis, 405
Ethylene carbonate, 563
Ethylene glycol, 563
Eupergit, 215
Exhaust emission, 165
Experimental design, 942

F

Factorial design, 680
Fatty acids, 933
Feedstock cost, 3
Fermentation, 180, 705, 751, 864, 922, 985
Ferroxidans, 897
Ferulic acid esterase, 416
Flash-pyrolysis, 153
Foam fractionation, 247
Forestry, 22
Fractal, 349
Free fatty acids, 1066
Fructophilic yeast, 870
Fructose utilization, 870
Fuel alcohol, 808
Fuels, 997

G

Gas recirculation, 887
Genetic variation, 427
Ginseng saponin, 436
Glucoamylase, 180
Gluconacetobacter, 705
Gluconic acid, 787
Glycerolysis, 680
Green power, 361

H

Hairy roots, 436
Hemicellulose, 528
Heterologous expression, 195, 226
Horseradish peroxidase, 303
HPLC, 829
H_2SO_4, 546
Hybrid poplar, 427
Hydrogen recovery, 808
Hydrolysis, 427, 486, 774
Hydrothermal, 448
Hyperoside, 382

I

IBSAL model, 71
Immobilization, 215, 1034
Immobilized enzyme, 265
Immobilized *R. oryzae* lipase, 911
Indirect ethanol process, 361
Industrial residues, 659
Inhibition, 278, 461
Inhibitor, 476, 599
Interaction effects, 461
Internode transport, 509
Inulinase, 951
Ion exchange, 349

J

Jerusalem artichoke, 922

K

Kinetics, 180
Kluyveromyces marxianus, 951

L

L. multiform, 416
Laccase, 195, 303
Lactase, 215
Lactic acid, 694, 844, 985
Lactic acid bacteria, 854
Lactobacillus sake, 334
Lactobacillus, 694
Lactococcus lactis, 334
Lactose, 694, 985
Lignin, 55, 104, 320, 427, 829
Lignin peroxidase, 303

Index

Lignocellulose, 448, 528, 599
Lignocellulosic biomass, 22
Lignocellulosic byproducts, 461
Lignocellulosics, 476
Limonene, 1023
Liquefaction, 563, 574
Liquid separations, 349
Livestock residual, 959
L-Lactic acid production, 795

M
Marine green microalga, 808
Mass spectra, 727
Medium-chain-length
    polyhydroxyalkanoates, 933
Membrane, 985
Mercury, 897
Methane, 959
Methoxy-substituted benzyl alcohols, 303
Microbial exopolysaccharides, 864
Microbial gums, 864
Mimosa, 382
Mixed culture, 751, 997
Mixing, 887
Modeling, 969
Molecular distillation, 680, 716, 1066
Multicriteria assessment, 71

N
Nanofiltration, 985
Natural products, 716
Newspaper, 486
Nickel, 436
NIR, 829
Nisin, 334, 751

O
Oil removal, 880
Olive tree wood, 631
Optical purity, 795
Optimization, 265, 705
Organic acids, 612
Organosolv, 55
Overliming, 612

P
P (HB-co-mclHA), 1015
Paenibacillus campinasensis, 234
Palm oil, 727, 1041
Pellet morphology, 844
Penalty function, 969
Persimmon vinegar, 705
PHA synthase gene, 1015
Phenolic acids, 104
Phenolic residues, 130
Phytosterol, 738
Pichia pastoris, 195, 226

Pilot plant, 448
PLA, 774
Policy reform, 22
Polyester, 574
Polyhydroxyalkanoates (PHAs), 1015
Polylactic acid, 774
Polyol, 563
Polyphenoloxidase, 320
Potassium hydroxide, 762
Pretreated corn stover hydrolyzate, 612
Pretreatment, 448, 486, 546, 586, 631
Pretreatment reactor, 621
PROMETHEE, 71
Protease, 1034
Protein denaturation, 247
Protein refolding, 247
Pseudomonas aeruginosa, 727, 933

Q
Quercitrin, 382

R
Rail transport, 88
Ranking, 71
RAPD, 1023
Rapeseed oil, 165, 668
Recombinant green fluorescent protein, 334
Response surface methodology analysis, 256
Rhamnolipids, 727
Rheology, 942
Rhizopus oryzae, 844
Rhodococcus, 880

S
Saccharification, 180
Saccharomyces cerevisiae, 795
Screening, 1023
Screw conveyor, 621
Selenium, 436
Short path distillation, 680
Silicone, 392
Slurry, 153
Softwoods, 55
Solid state, 951
Solid-state fermentation, 234
Solubility, 574
Sorbitan, 265
Sorbitol, 787
SO_2 steam explosion, 55
Soya sludge, 1066
Soybean oil, 680
Sphingomonas capsulata, 942
Spore/crystal toxins, 659
SSF, 448, 631
Starch free fibers, 117
Steam explosion, 278, 496
Straw, 88, 153

Strength, 574
Submerged fermentation, 234
Sugar extraction, 922
Sugarcane bagasse, 326
Sunflower stalks, 631
Supercritical extraction, 1041
Supercritical water, 476
Surfactant, 486
Synthesis gas, 153

T
Thermophilic anaerobic digestion, 959
Tocopherol, 716, 1066
Transesterification, 405, 668, 911
Transgalactosylation, 215
Transgenic grasses, 416
Transportation economics, 88
Truck transport, 88
Two-stage, 668, 1077

V
Value-added compounds, 382
Vinasse, 117
VOCs, 1077

W
Waste oils, 911
Wheat straw, 3, 41, 496
Whey, 694, 751

X
Xylan, 586
Xylanase, 278, 326, 528
Xylanase A, 226
Xylanase characterization, 289
Xylanase purification, 289
Xylan hydrolysate, 1051
Xylitol, 256, 461
Xylonic acid, 645
Xylooligosaccarides, 586
Xylose, 256
Xylose conversion, 117
Xylose utilization, 645

Z
Zea mays L., 104
Zymomonas mobilis, 787

Instructions for Authors and Reviewers

For Authors:

1. We prefer that manuscripts be submitted in MS Word, but other word processing formats are also acceptable. PDF *files cannot be accepted for the book production process.*
2. You must include an abstract and at least five (5) keywords.
3. Formatting will be done by the publishing company, just submit a plain, word processor manuscript.
4. Use line-numbering and double-spacing in your manuscript.
5. Artwork should be in .jpg, .tif, or .eps format. Do not use Freeware.
6. Figures should be saved as separate files and labeled "Figure 1,"etc. The figure legends should be listed in the manuscript.
7. You are permitted a *combined total* of ten (10) figures and tables. If you have more than this, see if any can be combined.
8. Only SI units will be acceptable in manuscripts.
9. Citations should follow *Applied Biochemistry and Biotechnology* format as in the following examples:
    1. Baker, A. "Feed Grains Database," Economic Research Service: US Department of Agriculture. http://www.ers.usda.gov/db/feedgrains/default.asp? ERSTab=3&freq=Monthly&Report=Commodity&data=PRICES (accessed March 11, 2005).
    2. Conner, A. H., Wood, B. F., Hill, C. G., and Harris, J. F. (1986), in *Cellulose: Structure, Modification, and Hydrolysis,* Young, R. A. and Rowell, R. M., eds., Wiley, New York, pp. 281–296.
    3. Doner, L. W., Chau, H. K., Fishman, M. L., and Hicks, K. B. (1998), *Cereal Chem.* **75,** 408–411.
10. You must submit your manuscript to the Editor's Desk at the *Biotechnology for Fuels and Chemicals Symposium* in Nashville, Tennessee, **April 30 – May 3, 2006.** Papers will not be accepted after the meeting, unless by special permission.
11. You are required to submit one (1) electronic copy and three (3) hard copies of your manuscript.
12. Failure to comply with these instructions may result in exclusion from the published proceedings.

For Reviewers:

Thank you for agreeing to review manuscripts for the *Biotechnology for Fuels and Chemicals Symposium* proceedings. To ensure rapid processing, please follow these guidelines:

1. All manuscripts must be reviewed and posted to the editors by **June 19, 2006**.
2. Each manuscript must be returned with a completed "**Manuscript Review Form**."
3. Please make your notes on the manuscript by reviewing electronically and using the "track changes" function.

We sincerely appreciate your prompt action and cooperation.